introduction to genetic principles

D0023484

David R. Hyde

University of Notre Dame

McGraw-Hill
Higher Education

Boston   Burr Ridge, IL   Dubuque, IA   New York   San Francisco   St. Louis
Bangkok   Bogotá   Caracas   Kuala Lumpur   Lisbon   London   Madrid   Mexico City
Milan   Montreal   New Delhi   Santiago   Seoul   Singapore   Sydney   Taipei   Toronto

INTRODUCTION TO GENETIC PRINCIPLES

Published by McGraw-Hill, a business unit of The McGraw-Hill Companies, Inc., 1221 Avenue of the Americas, New York, NY 10020. Copyright © 2009 by The McGraw-Hill Companies, Inc. All rights reserved. No part of this publication may be reproduced or distributed in any form or by any means, or stored in a database or retrieval system, without the prior written consent of The McGraw-Hill Companies, Inc., including, but not limited to, in any network or other electronic storage or transmission, or broadcast for distance learning.

Some ancillaries, including electronic and print components, may not be available to customers outside the United States.

This book is printed on acid-free paper.

2 3 4 5 6 7 8 9 0 VNH/VNH 0 9

ISBN 978–0–07–298760–7
MHID 0–07–298760–X

Publisher: *Janice Roerig-Blong*
Executive Editor: *Patrick E. Reidy*
Senior Developmental Editor: *Anne Winch*
Marketing Manager: *Barbara Owca*
Senior Project Manager: *Jayne Klein*
Senior Production Supervisor: *Sherry L. Kane*
Senior Media Project Manager: *Jodi Banowetz*
Cover/Interior Designer: *Laurie B. Janssen*
Lead Photo Research Coordinator: *Carrie Burger*
Photo Research: *Mary Reeg*
Supplement Producer: *Melissa Leick*
Compositor: *Precision Graphics*
Typeface: *10/12 Palatino*
Printer: *Von Hoffmann Graphics*
Cover Image: *Zebrafish, Jean Claude Revy - ISM/Phototake, Color enhanced scanning electron micrograph of a human chromosome in metaphase: Biophoto Associates/Photo Researchers, Inc., Computer artwork of the double helix strands of DNA, Alfred Pasieka/Photo Researchers, Inc., Confocal image of a BSC-1 monkey kidney epithelial cell during metaphase, Edward H. Hinchcliffe, Ph.D., Department of Biological Sciences, University of Notre Dame.*

The credits section for this book begins on page C-1 and is considered an extension of the copyright page.

Library of Congress Cataloging-in-Publication Data

Hyde, David (David Russell), 1958-
    Introduction to genetic principles / David Hyde. -- 1st ed.
        p. cm.
    Includes index.
    ISBN 978–0–07–298760–7 — ISBN 0–07–298760–X (hard copy : alk. paper) 1. Genetics. I. Title.
QH430.H93 2009
576.5--dc22

                                                                          2007038649

introduction *to* genetic principles

Multiple Choice Quiz
(See related pages)

Results Reporter

Out of 25 questions, you answered 3 correctly, for a final grade of 12%.

3 correct (12%)
4 incorrect (16%)
18 unanswered (72%)

Your Results:

The correct answer for each question is indicated by a 😊.

1 CORRECT   The nucleotide change _____ is an example of a transversion, while the nucleotide change _____ is an example of a transition.

- ○ A) A→G; C→G
- 😊 ● B) C→G; A→G
- ○ C) T→C; A→G
- ○ D) C→T; G→A
- ○ E) G→C; C→G

2 CORRECT   Which type of mutat on is most likely to revert?

- ○ A) deletion
- ○ B) translocation
- ○ C) inversion
- 😊 ● D) transposition
- ○ E) transition

3 CORRECT   Which type of mutation is least likely to revert?

- 😊 ● A) deletion
- ○ B) translocation
- ○ C) inversion
- ○ D) transposition
- ○ E) transition

4 INCORRECT   The hydrolysis of an -NH2 group from a base is called _____, causing _____:

- ○ A) deamination; transversions

Routing Information

Date: Tue Sep 09 09:13:24 CDT 2003
My name: [   ]

Email these results to:

	Email address:	Format:
Me:	[ ]	Text
My Instructor:	[ ]	Text
My TA:	[ ]	Text
Other:	[ ]	Text

[E-Mail The Results]

Test Yourself

Take a quiz at the Genetics Online Learning Center to gauge your mastery of chapter content. Each chapter quiz is specifically constructed to test your comprehension of key concepts. Immediate feedback on your responses explains why an answer is correct or incorrect. You can even e-mail your quiz results to your professor!

Interactive Activities

Fun and exciting learning experiences await you at the Genetics Online Learning Center! Each chapter offers interactive animation activities, vocabulary flashcards, and other engaging activities designed to reinforce learning.

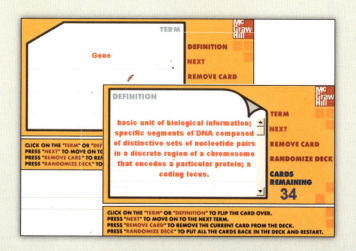

Slipped-strand Mispairing

View the animation below, then complete the quiz to test your knowledge of the concept.

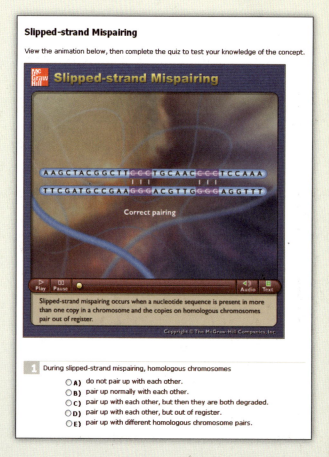

Slipped-strand mispairing occurs when a nucleotide sequence is present in more than one copy in a chromosome and the copies on homologous chromosomes pair out of register.

Copyright © The McGraw-Hill Companies, Inc.

1   During slipped-strand mispairing, homologous chromosomes

- ○ A) do not pair up with each other.
- ○ B) pair up normally with each other.
- ○ C) pair up with each other, but then they are both degraded.
- ○ D) pair up with each other, but out of register.
- ○ E) pair up with different homologous chromosome pairs.

about the author

Professor David R. Hyde is currently a Full Professor in the Department of Biological Sciences and the Rev. Howard J. Kenna C.S.C. Memorial Director of the Center for Zebrafish Research at the University of Notre Dame. He earned a B.S. degree in Biochemistry from Michigan State University in 1980 and a Ph.D. in Biochemistry from the Pennsylvania State University in 1985. His doctoral thesis research was with David C.-P. Tu and dealt with the characterization of the *E. coli* transposon, Tn21. He was a postdoctoral research fellow at the California Institute of Technology with Seymour Benzer from 1985 to 1988. His research at Caltech dealt with the characterization of gene expression patterns in the adult *Drosophila* head, particularly in the compound eye. He moved to the University of Notre Dame in 1988, where he began research projects dealing with the function of the *Dro-sophila* retina and genetic mechanisms behind retinal degeneration. His current research focuses on the development of the zebrafish eye and the mechanisms that underlie the regeneration of the retinal neurons. His research has been funded by the National Eye Institute of the National Institutes of Health (NIH), the Foundation Fighting Blindness, the Fight for Sight Foundation, and the Johnson and Johnson Focused Giving Program.

David Hyde has taught undergraduate genetics to Biology and Biochemistry majors since 1989. He has won several teaching awards including the Kaneb Faculty Teaching Award (twice) and was one of the 10 inaugural Kaneb Faculty Teaching Fellows in 2002. He regularly gives presentations to the faculty and graduate students on ways to improve the quality and efficiency of teaching. He teaches a course every summer for graduate students and postdoctoral fellows that covers how to develop an undergraduate lecture and laboratory course and methods to improve student learning and retention. His enthusiasm for teaching is appreciated by his students, who regularly invite him to participate in their graduation ceremony.

To my parents Barbara and Frank, for fostering my love of learning;

To my children Nathan, Brittany, and Jordan, who remain my greatest and proudest genetic experiments;

and

To my wife Laura, whose love and support remains an instrumental part of my life and career.

—David Hyde

brief contents

iv

contents

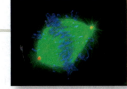

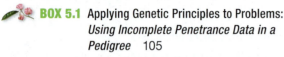
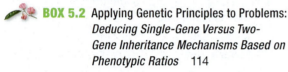

part two

Molecular Basis of Inheritance and Gene Expression

part four

Control of Gene Expression

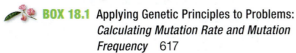
part five

Genetics and
Biological Processes

preface

When I first began teaching genetics in 1989, I developed a course that covered a wide array of topics in genetics, from Mendelian genetics and chromosome mechanics to regulating gene expression and recombinant DNA technology. As I thought about my goals for this course, I knew, as do many faculty teaching genetics, that I wanted the students to both learn and be able to apply their knowledge. I felt that students who really understood the concepts could use their knowledge to solve a variety of different problems and integrate the information to develop a deeper understanding of the concepts. My job as the instructor was to carefully cover the material in each chapter, stressing concepts and information that I thought were particularly important or interesting, and assigning end-of-chapter problems and old exam questions so the students could practice solving problems before the exam.

I found, however, that the students struggled on the assigned problems. I would then go over the material and problems in review sessions and during office hours to help them reach a clear understanding of the material and how to solve the problems before the exam. Though the exam had similar problems that were only slightly modified from the homework problems we reviewed, many of the students still failed to perform at the level that I expected.

Meeting with the students, I found they had a very difficult time in two general areas of my course:

1. They lacked the ability to extract and understand the relationship between the lecture material and the underlying major concepts.
2. The students had difficulty in applying the material learned in the lecture to the average problem. Because they continued to have trouble when the problem was slightly modified, it was clear that they lacked either the understanding of how to solve a problem or the training to recognize important details and concepts in a problem.

This problem was exacerbated when I asked them to combine information and concepts from more than one chapter. For example, asking the students to find an open reading frame in a specific DNA sequence could be solved by many of the students. However, if I gave the students an image of a DNA sequencing ladder and asked them to identify the open reading frame and determine which DNA strand corresponds to the transcriptional template, many of them were baffled.

I began to look at genetics books and thought about how they dealt with problem-solving skills and the integration of material. Most of the books concentrated on the historical, factual, and conceptual topics in genetics, with very little instruction about how to solve different types of problems in a chapter. At the end of each chapter, however, there were several problems that required the student to suddenly apply this information, without providing sufficient problem-solving instruction. I already knew from my own experience that this was demanding a lot from the students.

I also found many of the books provided short summaries throughout the chapter and longer end-of-chapter summaries. The short, interspersed summaries were too concise and often did not convey much information or put the key points into a context for the students to grasp. While the end-of-chapter summaries were more thorough, they usually just reiterated the material in the short, interspersed summaries rather than putting the information in a different context. While there were occasional end-of-chapter questions that required some knowledge from another chapter, they failed to significantly reinforce the previous topics or clearly demonstrate how it related to the material that was covered in the current chapter.

Keeping these shortcomings in mind, I set out to write a genetics book that was designed for both students and instructors. I wanted this book to help students overcome the two roadblocks that I saw many students had encountered. At the same time, I wanted a textbook that contained sufficient detail for students to supplement the material that the instructor presented during lectures. I also wanted to develop a book that would assist instructors in developing problem-solving skills in their students. My hope is that this book also provides the flexibility for faculty who cover the entire book in one semester and those, like myself, who prefer to concentrate on fewer topics in the semester.

Developing Problem Solving Skills

I believe that this genetics textbook demonstrates to the student how to solve problems, which is so critical for their success in genetics classes across the country. I committed significant space in this textbook to describe how data are analyzed while solving a problem, both through text discussions and through effective pedagogical features. Each of these different features provide alternative opportunities for students to test their comprehension of the material and their ability to apply their knowledge to solving problems.

A key feature of this text is "Applying Genetic Principles to Problems." Almost every chapter contains at least one of these boxes that leads the student step-by-step through analyzing experimental data and solving a problem. A good example is on pages 134 and 135 of the text where the student learns how to create a recombination map using data from a three-point testcross. Placing these clear instructions immediately after the relevant section in the text allows the student to directly compare the concept with the application. Furthermore, these boxes provide clear locations where the student can find assistance on how to solve particular types of problems.

The "It's Your Turn" feature, located at the end of key sections in each chapter, directs students to specific end-of-chapter problems that they can attempt based on the material that was just presented (see example, page 136). This provides students the opportunity to attempt specific problems immediately after reading about a particular principle or seeing an example of how to use a specific type of data set. This immediate reinforcement helps students gain confidence before they move on, provides an assessment of their understanding of sections within a chapter, and indicates the section of the chapter that should be reviewed if they are having difficulty with specific problems.

Every chapter also contains several Solved Problems, which provides another opportunity for the student to see how related problems can be solved. Solved Problems are followed by at least 20-25 problems in the Exercises and Problems section for students to practice their problem-solving skills. Solutions to even-numbered problems are provided for students on the text website at www.mhhe.com/hyde.

The end-of-chapter assessment materials also contain a series of questions called "Chapter Review Questions" that assesses the student's comprehension of the text through definitions, drawing relationships, and listing features of a process. The student's ability to see a new problem and know how to tackle it would be severely compromised without this basic understanding.

While using actual data in the problems adds a sense of reality to the student, this type of data is not often transparent so the student can see the concept at work. The use of actual data often requires much statistical analysis to determine how well it fits the model. Because I feel it is more important for the students to see, understand, and then apply the concepts while they are solving the problems, the problems in this book usually use fabricated data to better fit and demonstrate the principles being discussed. I believe that this makes the same instructional point in a more obvious way. As students become more comfortable with the data and its analysis, they will be able to move on to analyzing real scientific data.

Identifying and Integrating Key Concepts

Genetics is a broad discipline that is used in all areas of modern biology, from cell and developmental biology, through physiology to ecology and evolution. This textbook provides the student with several ways to see how each topic integrates within the larger landscape of genetics.

At key points throughout the chapter, "Restating the Concepts" distills the preceding section down to a few main points (example, page 189). Rather than trying to make these overly concise to fit within a restricted amount of space, I allowed sufficient space for these summaries to contain enough information to effectively transmit the key points.

Reviewers of early drafts of this manuscript agreed that the "Restating the Concepts" reviews eliminated the necessity of a standard end-of-chapter summary. I used that opportunity to conclude each chapter with the "Genetic Linkages" box, which describes how the different topics in that chapter are related to material covered in both previous and future chapters. Students then have the opportunity to be directed between chapters if a particular aspect interests them or is somewhat unclear. Providing this larger roadmap to the student gives him or her some freedom to explore genetics and gain a deeper appreciation of the field rather than focusing on, and getting lost in, the individual details.

Finally, at the end of almost every chapter is a Chapter Integration problem, such as the example on page 261. These problems require the student to draw on some information or concepts from previous chapters. This reinforces their existing knowledge from other chapters and integrates this information through a multi-part problem. Students in my course find these problems challenging, but successfully solving the problem gives them a clear understanding of the material and a renewed confidence in their ability to manipulate the data. Chapter Integration problems also provide an excellent assessment of how well students retain and integrate the material.

Visualizing Genetics Through Accurate and Instructional Art

While emphasizing the problem-solving and integrative nature of the textbook, we also placed great importance on the visual program. My students rely heavily on visual representations to sort out the many structures and processes in genetics, thus, we tried to provide a piece of art for every major concept, which could then complement the description in the text. This resulted in a much larger number of figures (15–30% greater) compared to the other genetics textbooks on the market. These visual aids are critically important for the current student to understand the concepts and to place that information into a larger context.

Great care was also placed on developing a consistent and informative art program for this textbook. You will find that not only is the color of the artwork striking, but that similar elements maintain the same color and shape from chapter to chapter. For example, generic DNA is always blue and mRNA is yellow. We used a hand structure for DNA polymerase and a bi-lobed oval for RNA polymerase. The ribosome is always a red triangular element. This continuity will reduce the amount of confusion for the student and allow them to be more comfortable when they see a familiar element within a new topic or concept.

Furthermore, many of the figures were designed to show the basic concept in a larger framework. See figure 17.25, page 609 for example; rather than simply showing the proteins that are in a signal transduction cascade, we placed the corresponding proteins in their correct cellular compartments so the students can see what restrictions are placed on the different proteins.

Modern Molecular Genetics With a Traditional Approach

There are two different views of how to present genetics. The first is a historical view that allows the students to see the progression of the field and how advances are built on the previous work of others. This is an important concept that students who want to stay in science need to understand. The second approach is a more modern approach, where DNA is presented as the genetic material at the start. This approach is based on the idea that the students have seen much of this material for many years prior to taking the genetics course and that this familiarity makes it easier for students to move into genetics.

The problem with the second approach is twofold. First, there are many different ways to proceed after introducing DNA as the genetic material. You can move into gene regulation or into a description of a mutation, or into classical genetics. This variety of downstream directions makes writing a textbook using this approach very difficult. More importantly, after trying to teach my class using the DNA first approach, I found that the students lose some of the appreciation of the subsequent genetic analyses because they know that the ultimate cause is the DNA sequence. They lack a desire to learn the importance of genetic crosses, they become very confused with the definitions of genetic terms, such as a null mutation, compared to the molecular definition of a null mutation. Possibly the most disturbing outcome of the DNA-first approach is the students try to simplify all the concepts down to a DNA sequence, which makes it difficult for them to understand how some forms of gene regulation, such as epigenetics, can occur without affecting the DNA sequence. Teaching a genetics course from a DNA-first approach is like trying to read a novel after first reading the last chapter. Once you know the outcome, you often second-guess the rationale behind every event as the story evolves. The students lose the beauty of seeing the story unfold and the excitement of discovering why different events happen and their consequences. Thus, based on my previous experiences, I wrote this book using the historical approach. However, the level of detail provided in each chapter would likely permit this book to be used in a DNA-first approach, if the chapters were not followed sequentially.

The Conflict Between the Breadth and the Depth of the Material

One of the most difficult decisions that I encountered in writing this book was how much breadth I wanted to expose the students to versus the level of depth that I wanted them to appreciate. In an ideal world, this book would convey the breadth of the topics within the field of genetics and provide sufficient detail to satisfy the most highly motivated student. Unfortunately, there are limits to the amount of material that can be placed in a genetics book at this level and the amount of material that can be covered in a one-semester genetics course.

I opted to fall somewhere in the middle of the two extremes. I admit that there were several aspects of genetics that were excluded from this book, such as the variety of different ways to generate the organismal body axes shortly after fertilization of the egg (chapter 21). Instead, I felt that it was more important to present one fairly well understood example in greater detail, hoping that the students would gain a deeper appreciation of both the details of the process and some of the experimental techniques used to study the concept. My intention is to provide enough breadth about genetics, without morphing into the areas that utilize genetics. For example, I wanted to demonstrate a variety of different ways that genetics is used to understand developmental biology (chapter 21), without writing a detailed chapter about developmental biology. It was important to give the students sufficient detail so they can become excited about genetics as a field and appreciate its application in other areas of biology.

To help generate student interest in genetics, I knew that the textbook had to be up-to-date. In some ways that is impossible to achieve in a published book, which is only updated every three years. However, it was important to me to keep the information and figures as current as possible. Thus, in our discussion of genome sequencing projects, I used the most recent available data from www.genomesonline.com and updated it as late in the publication process as possible.

I also updated the text and figures associated with several techniques. I made a sincere attempt to keep the discussion of some topics very current, including the generation and function of microRNAs, nonsense-mediated mRNA decay, the creation of a phenocopy through small molecule screens, and the use of TILLING to screen for mutants. For each of these topics there are multiple diagrams to illustrate these concepts and methodologies.

In the cases where the topics are more complex, I incorporated a figure that shows an example of the use of the concept or technique. For example, in the discussion of the small molecule screen, I used a figure that illustrates how different small molecules may affect two different transmembrane receptors that are linked through a common intracellular signaling pathway and how this could relate to the phenotype associated with a genetic mutation in one of the corresponding genes (Figure 20.37). This figure not only demonstrates the potential targets of a small molecule screen, but also how the resulting phenocopy may differ from a standard mutation. This is important for students who want to think that all processes can be reduced to a simple operation, which takes away much of the beauty of biology.

Similarly, I attempted to provide real examples in places to summarize complex topics. For example, after discussing a variety of mechanisms used to control the expression of a protein in eukaryotes, such as transcriptional regulation, translational regulation, and posttranslational regulation, I then discussed Prader-Willi syndrome and Angelman syndrome in humans (chapter 17), where many of these different processes converge on the complex topic of imprinting. I hope that these real examples, particularly as they relate to humans, will help to grab the interest of the students and motivate them to work through the processes in greater detail.

Acknowledgments

As with any large project, there are numerous people who must be thanked, for without their help or support, this textbook would still be languishing in one of the many piles in my office. First, I sincerely appreciate Robert H. Tamarin for providing me with *Principles of Genetics,* Seventh Edition as a starting point for this project. Even though this book represents a complete revision, with several sections produced from scratch, Tamarin's text was very instrumental and supportive by providing some initial text and a general roadmap for how to assemble a genetics textbook.

The Precision Graphics art group, especially Jen Gibas and Matt Mabry, did an outstanding job with the artwork. I attempted to show many of the pieces of art in a larger conceptual framework to allow the students an opportunity to place the information into their knowledge base, rather than simply having each figure represent a different, and separate, piece of information. The individuals at Precision Graphics took many of my rudimentary sketches and produced outstanding pieces of artwork that I am sure will greatly enhance the learning experience for the student.

Much of the credit for the artwork also needs to be shared with Jeremy N. Friedberg, of vive Technologies. Jeremy saw many of my early pieces of "art" and made fundamental suggestions for improving the conceptual development of the entire art program. In later stages, Jeremy was instrumental in ensuring the consistency of the art style, while maintaining the scientific accuracy. Jeremy has an excellent eye, not only for details, but also how the students interpreted different concepts. The result is artwork that will greatly clarify many of the more difficult concepts in introductory genetics.

Jody Larson, developmental copyeditor, poured over the first draft of each chapter and provided countless suggestions that shortened the chapters to a manageable length and improved the flow of the text. Linda Davoli also put in a fantastic effort to develop a consistent, clear, and accurate style sheet throughout the text.

Several individuals were invited to contribute preliminary draft chapter manuscript or end-of-chapter questions and problems. Still others provided constructive criticism of one or more chapters, which improved the accuracy of this textbook and helped to guide me as I attempted to balance the depth and breadth of the material covered. Their assistance was invaluable and I thank each of them for the time they took away from their already busy lives to help improve this project. In particular, Martin Tenniswood worked with me on chapter 22, Cancer Genetics, through the final draft. Martin is an accomplished cancer researcher, an outstanding teacher, and a valued colleague. I am sure that his contribution will make the complexities of cancer genetics more student friendly. Also, Johnny El-Rady was very helpful in developing the end-of-chapter pedagogy and providing a valuable accuracy check in the final stages of the book.

Preliminary Draft Revision Contributors

Kari Beth Krieger, *University of Wisconsin—Green Bay*

David Reed, *University of Mississippi*

Barbara Sears, *Michigan State University*

Martin Tenniswood, *University of Notre Dame*

Michael Windelspecht, *Appalachian State University*

End-of-Chapter Pedagogy

Aaron Cassill, *University of Texas—San Antonio*

Johnny El-Rady, *University of South Florida*

Kari Beth Krieger, *University of Wisconsin—Green Bay*

David Reed, *University of Mississippi*

Reviewers and Accuracy Checkers

Shivanthi Anandan, *Drexel University*

Mianm Ashraf, *Rust College*

John Belote, *Syracuse University*

Michael Benedik, *Texas A&M University*

Edward Berger, *Dartmouth College*

Andrew J. Bohonak, *San Diego State University*

Helen C. Boswell, *Southern Utah University*

Warren W. Burggren, *University of North Texas*

Diane Caporale, *University of Wisconsin-Stevens Point*

Joseph P. Caruso, *Florida Atlantic University*

J. Aaron Cassill, *University of Texas at San Antonio*

Helen Chamberlin, *The Ohio State University*

Joseph P. Chinnici, *Virginia Commonwealth University*

Francis Choy, *University of Victoria*

Alberto Civetta, *University of Winnipeg*

Craig E. Coleman, *Brigham Young University*

Wes Colgan III, *Pikes Peak Community College*

Brian G. Condie, *University of Georgia*

Gregory P. Copenhaver, *The University of North Carolina at Chapel Hill*

Jeff DeJong, *University of Texas at Dallas*

A. Elkharroubi, *Johns Hopkins University*

Johnny El-Rady, *University of South Florida*

Asim Esen, *Virginia Polytechnic Institute and State University*

Julia Frugoli, *Clemson University*

Gail E. Gasparich, *Towson University*

Allan W. Gibson, *Malaspina University College*

Jack R. Girton, *Iowa State University*

Elliott S. Goldstein, *Arizona State University*

Nels H. Granholm, *South Dakota State University*

Mike Grotewiel, *Virginia Commonwealth University*

Nancy Ann Guild, *University of Colorado, Boulder*

Rosalind C. Haselbeck, *University of San Diego*

Jutta B. Heller, *Loyola University Chicago*

Muriel B. Herrington, *Concordia University*

April Hill, *University of Richmond*

Nancy N. Huang, *Harvard University*

Colin Hughes, *Florida Atlantic University*

John Kasmer, *Northeastern Illinois University*

David Kass, *Eastern Michigan University*

Brian R. Kreiser, *University of Southern Mississippi*

Paula Lessem, *University of Richmond*

Mark Levinthal, *Purdue University*

Paul F. Lurquin, *Washington State University*

Philip M. Mathis, *Middle Tennessee State University*

Harry Nickla, *Creighton University*

John C. Osterman, *University of Nebraska-Lincoln*

James B. Parker, *Parker College of Chiropractic*

Bruce Patterson, *University of Arizona, Tucson*

James P. Prince, *California State University, Fresno*

Jennifer Regan, *University of Southern Mississippi*

James V. Robinson, *The University of Texas at Arlington*

Barbara B. Sears, *Michigan State University*

Ekaterina Sedia, *The Richard Stockton College of New Jersey*

Rebecca L. Seipelt, *Middle Tennessee State University*

Allan M. Showalter, *Ohio University*

Maureen Shuh, *Loyola University New Orleans*

Monica M. Skinner, *Oregon State University*

Traci L. Stevens, *Randolph-Macon College*

Harald Vaessin, *The Ohio State University*

Yunqiu Wang, *University of Miami*

Tamara L. Western, *McGill University*

D. S. Wofford, *University of Florida*

Andrew J. Wood, *Southern Illinois University-Carbondale*

Yang Yen, *South Dakota State University*

Jianzhi Zhang, *University of Michigan*

The publishing team at McGraw-Hill made this not only a project that could be completed in a reasonable length of time, but also an enjoyable experience. I would like to particularly thank Patrick Reidy, executive editor, for providing just the right amount of information to make this project sound appealing and very manageable at the start. His enthusiasm for producing a first-rate textbook for the students was contagious, as was his love for the Fighting Irish. Once the ball was rolling, he turned my everyday assignment, or oversight, to Anne Winch, senior developmental editor. Anne's dry sense of humor and continual "checking on my progress" kept me focused on my numerous goals and allowed this project to be completed in a relatively timely manner. Late in the writing process, we realized that the book was 50–75 pages longer than budgeted. Rather than removing or reducing the outstanding artwork or slashing large amounts of text, Anne and Patrick fought to keep the book intact. The result is a significantly better textbook, without the cost of the increased length being passed on to the students. Pat and Anne, you are two outstanding individuals who actually made writing this textbook an enjoyable experience.

When the text entered the production stage, Jayne Klein, senior project manager, expertly handled the difficult task of keeping manuscript, art proofs, and pages moving out to all of the core team members, freelancers, and vendors who work together to create the final text. I almost came to look forward to Jayne's regular overnight delivery of text and artwork sitting outside my office door when I returned from lunch. The efforts of Laurie Janssen, designer, resulted in the beautiful cover, interior design, and art program for this text. Carrie Burger, lead photo research coordinator, helped secure the many photos that are scattered throughout the book.

A project of this magnitude could not have been accomplished without the support of many other individuals. Tom Vihtelic has been an outstanding colleague in my research lab at Notre Dame. His intelligence and skill at directing lab personnel and research projects relieved me of many of the everyday responsibilities of running a research program and allowed me to make progress on this textbook. The remainder of my lab personnel did an outstanding job of advancing our research on the mechanisms of retinal regeneration in zebrafish, which allowed me the time to get this textbook completed, without our research program suffering. In some ways, I think that we increased our research productivity during this period, which might suggest that I should stay out of the lab's way more often. I have had the support of several dear friends, such as David and Barb Ponder, whose interest and support helped to reduce some of the pressure that I was under.

Finally, I am truly indebted to my wonderful family. My parents, Frank and Barbara Hyde, and mother-in-law, Fran Schaefer, have always been supportive of my career. Their regular questions about how the book was progressing revealed their interest, while making it clear that they did not want to know my difficulties in keeping the genomics and proteomics chapter to a reasonable length. My three children, Nathan, Brittany, and Jordan provided some important suggestions on the general design of the chapter features, which I think will truly benefit the students reading this book. Last, but by no means least, is my wonderful wife, Laura. Her support and love provided me with the strength to complete this book, especially at times when this project seemed endless.

Pedagogical Aids to Promote Learning

Each chapter is structured using the same set of pedagogical devices, which enables the student to develop a consistent learning strategy. These tools work together to provide a clear content hierarchy, break content into smaller, more accessible chunks, rephrase important concepts, and create connections between concepts from previous and upcoming chapters.

Chapter opening pages

Each chapter opener begins with a list of **Essential Goals** which provides students with specific outcomes they should be able to achieve after mastering the chapter content. A **Chapter Outline,** comprised of the numbered chapter headings, introduces the big picture view of the chapter organization.

Interim Summaries

These boxes don't just regurgitate the key points from the section the student has just read; they state the concept in a new way to aid comprehension.

Genetic Linkages

Each chapter ends with a discussion of how the different topics in that chapter are related to material covered in both previous and future chapters. This not only helps students see the "big picture," it also allows them to read ahead if a topic interests them or go back to material that needs review.

Mitosis and Meiosis

Essential Goals

1. Understand the key features of a chromosome and how chromosomes are classified.
2. Describe the key steps in the cell cycle and how the cycle is regulated.
3. Diagram and describe the relationship between mitosis and meiosis in both haploid and diploid cells.

Chapter Outline

3.1 Cell Structure and Function 42
3.2 Chromosomes 43
3.3 The Cell Cycle and Its Control Mechanisms 45
3.4 Mitosis 47
3.5 Meiosis 52
3.6 Comparison Between Mitosis and Meiosis 57
   Box 3.1 *The Relationship Between Meiosis and Mendel's Laws* 59
3.7 Meiosis in Mammals 60
3.8 Life Cycles 61
3.9 Chromosomal Theory of Inheritance 63
   Genetic Linkages 63
   Chapter Summary Questions 64
   Solved Problems 64
   Exercises and Problems 65

41

Restating the Concepts

▶ Meiosis is the specialized process that produces haploid gametes required for sexual reproduction from diploid cells.

▶ Meiosis involves two nuclear divisions. The first division requires pairing of homologous chromosomes and their separation (reductional division). The two haploid cells that result then enter the second nuclear division (equational division) to generate the gametes.

▶ The movement of the chromosomes through meiosis is consistent with the movement of alleles in Mendel's first and second laws.

▶ Meiosis produces genetic diversity through the independent assortment of chromosomes and through recombination, both of which occur during meiosis I.

---

genetic linkages

In this chapter, we discussed how eukaryotic cells divide and eukaryotic organisms propagate. Cell growth involves the cell cycle, in which the genomic DNA replicates prior to nuclear division. We describe the molecular details of DNA replication in chapter 9, which includes the enzymes and activities required for replicating linear DNA. One of the major features of mitosis is to produce daughter cells that are genetically identical; in chapter 9 you will see how DNA replication possesses the ability to minimize errors, and in chapter 18 you will learn the different mechanisms employed to repair mutations in the DNA.

We also described in this chapter the process of meiosis, which produces haploid gametes. The pairing of the homologous chromosomes and their separation during meiosis I is consistent with Mendel's first law of equal segregation, and the orientation and separation of one homologous pair is independent of another pair, which underlies Mendel's second law of independent assortment (chapter 2). The realization that Mendel's factors were associated with chromosomes was the basis for the chromosomal theory of inheritance. In chapter 4, we will discuss how this theory was proved using traits that are associated with the *Drosophila* X chromosome.

One of the key differences between mitosis and meiosis is that the homologous chromosomes pair during meiosis I, which leads to crossing over, or recombina-

tion, between nonsister chromatids. Synapsis ensures that homologs pair during meiosis I, and it also provides a powerful mechanism to localize or map genes and mutations on chromosomes. In chapter 6, we will discuss the mechanism of recombination and the application of recombination to mapping genes. Finally, in chapter 8 we will describe how altered chromosomal structures affect synapsis and recombination during meiosis.

As we discussed, homologous chromosomes pair during meiosis I. However, mammalian males contain two different sex chromosomes (X and Y) that pair during meiosis I. In chapter 4, we will describe how this occurs. You will also see how meiosis can be disrupted, such that the gametes that are produced fail to contain one chromosome from each homologous pair. In chapter 8, we will discuss how disrupting meiosis can lead to changes in the chromosome number in offspring.

Mitosis and meiosis are two fundamental processes that underlie the transmission of genetic material during cell division and from generation to generation. Therefore, a clear understanding of these processes will help to explain many of the genetic phenomena described throughout this book. You may want to refer back to this material as you read later chapters to help clarify how other processes are related to mitosis and meiosis.

The author collaborated with a team of scientific illustrators and expert consultants to create the visual program for *Introduction to Genetic Principles*. Focusing on consistency, accuracy, and pedagogical value, the team created an art program that is intimately connected with the text narrative. The resulting realistic, 3-D illustrations will stimulate student interest and help instructors teach difficult concepts.

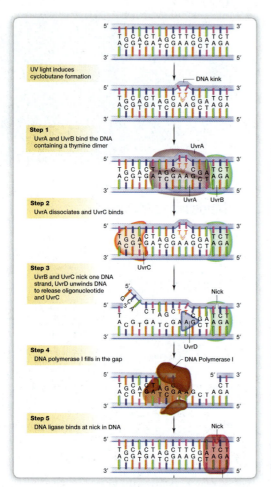

For complex processes, numbered text boxes lead the student step-by-step through the figure. This allows the student to easily follow between the description in the text and the parts of the figure. This also results in a relatively uncluttered figure, that remains very informative on its own.

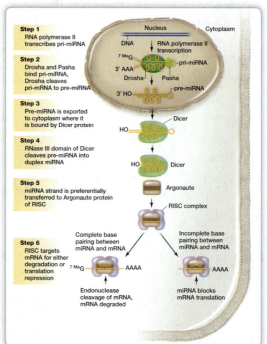

Consistent color coding means that students immediately recognize the biological structures used throughout the book. Their study time is spent learning concepts rather than orienting themselves to figure conventions.

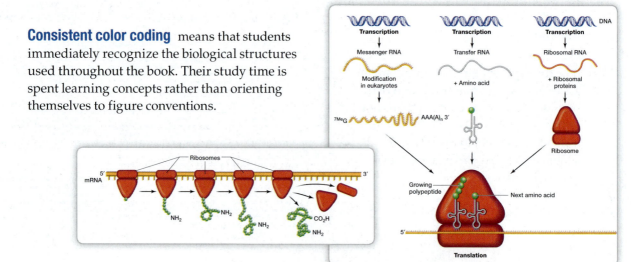

for Visual Learners

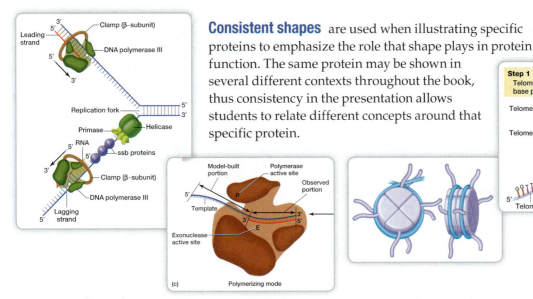

Consistent shapes are used when illustrating specific proteins to emphasize the role that shape plays in protein function. The same protein may be shown in several different contexts throughout the book, thus consistency in the presentation allows students to relate different concepts around that specific protein.

Biological processes are placed within their proper location in the cell to accurately convey to the student where they reside and potential obstacles that must be overcome for the process to occur.

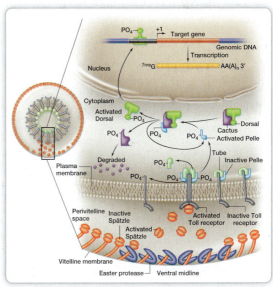

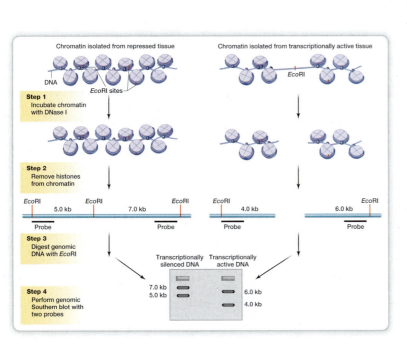

Experimental procedures are diagrammed in a sequential manner to highlight descriptions of complex or new methods that are described in the text. This will greatly facilitate students' understanding of concepts such as the identification of DNase I hypersensitive sites.

Tools for Building Problem-Solving Skills

One of the biggest stumbling blocks for introductory genetics students is how to solve problems. In addition to step-by-step explanations within the text narrative, *Introduction to Genetic Principles* employs several tools designed to help students master problem-solving. The author's approach highlights the basic genetic principles underlying the problem, focuses the reader's attention on key details, and teaches them to apply those strategies to related problems.

Applying Genetic Principles to Problems

boxes lead the reader through the process of analyzing data and solving problems.

It's Your Turn

directs the student to problems related to the material they just studied.

> #### It's Your Turn
>
> Problem 28 involves determining the frequency of generating a particular phenotype for a specific cross involving three linked genes. Calculate the frequency of each genotypic class in the female gametes, beginning with the double recombinant classes. Remember, interference = 1 − C (coefficient of coincidence) and *Drosophila* males do not exhibit recombination.

> ## 6.2 applying genetic principles to problems
>
> ### Calculating Recombination Frequencies from a Linkage Map
>
> In actual research, you may wish to calculate the frequency that a particular genotype will be produced from a three-point cross. This calculation is relatively straightforward if you have the linkage map of the three genes being analyzed.
>
> In the following example, you want to know the frequency of producing an $a\, b\, c^+$ phenotype from the cross of an $a^+\, b^+\, c/a\, b\, c^+$ male and female. You already know that the coefficient of coincidence is 0.4, and that the linkage map is:
>
> ```
>        5 mu              10 mu
> a------------b--------------------------c
> ```
>
> In this problem, you need to reverse the thinking you employed in box 6.1. The first thing you do is calculate the frequency of *observed* double recombinants. You know that the expected double recombinant frequency is defined as the frequency of a recombination event between a and b and the frequency of a recombination event between b and c (the product rule from chapter 2) which is calculated as:
>
> Percentage of expected double recombinants = $(0.05 \times 0.10)(100\%) = 0.5\%$
>
> You can then calculate the frequency of observed double recombinants by solving the equation as follows:
>
> $$C = \frac{\text{percentage of observed double recombinants}}{\text{percentage of expected double recombinants}}$$
>
> The value of C is given as 0.4, and the value of the expected double recombinants is 0.5%, so solving for the observed double recombinants:
>
> Percentage of observed double recombinants $= (0.4) \times (0.5\%) = 0.2\%$
>
> Based on the parental genotype of $a^+\, b^+\, c/a\, b\, c^+$, the double recombinant gametes will be $a^+\, b\, c$ and $a\, b^+\, c^+$. Because the total frequency of double recombinants is 0.2%, 0.1% of the gametes will be $a^+\, b\, c$ and 0.1% will be $a\, b^+\, c^+$. You can now calculate the observed frequency of each single recombinant class. Remember that the
>
> recombination distance between a and b represents the frequency of the single recombinants between a and b and the double recombinant class. Therefore, the frequency of single recombinants between a and b is:
>
> $$5\% - 0.2\% = 4.8\%$$
>
> of which 2.4% of the gametes will be $a^+\, b\, c^+$ and 2.4% will be $a\, b^+\, c$. Similarly, the frequency of single recombinants between b and c will be 10% − 0.2% = 9.8%, of which 4.9% will be $a^+\, b^+\, c^+$ and 4.9% of the gametes will be $a\, b\, c$.
>
> All you need to do now is calculate the frequency of the parental gametes, which will be the remainder of the gametes (100% − 9.8% − 4.8% − 0.2% = 85.2%). Thus, 42.6% of the gametes will be $a^+\, b^+\, c$, and 42.6% of the gametes will be $a\, b\, c^+$.
>
> You can now calculate the frequency of any desired genotype or phenotype that is produced from the cross of two individuals with the genotype of $a^+\, b^+\, c/a\, b\, c^+$. Because you are interested in the frequency of an $a\, b\, c^+$ phenotype, you need to calculate the potential genotypes that can produce this phenotype. In this problem, assume that recombination occurs in both the male and the female. The potential genotypes that must be considered and their frequency (based on the product rule) are
>
> $a\, b\, c^+/a\, b\, c^+$ (42.6%)(42.6%) = 18.15%
>
> $a\, b\, c^+/a\, b\, c$ (42.6%)(4.9%) = 2.09%
>
> The second genotype can be produced two ways: one in which the male contributes the first gamete, and one in which the female does. Therefore, the probability of producing the $a\, b\, c^+$ phenotype is
>
> $$18.15\% + 2.09\% + 2.09\% = 22.33\%$$
>
> If recombination failed to occur in the male (as it fails to occur in *Drosophila* males), then 50% of the male gametes would be $a\, b\, c^+$, and the female could contribute either the $a\, b\, c^+$ or the $a\, b\, c$ gamete. The result would be the same two genotypes that we listed earlier, but now generated only a single way each, at a frequency of
>
> $$(50\%)(42.6\%) + (50\%)(4.9\%) = 23.75\%.$$

Solved Problems

and plentiful end-of-chapter exercises and problems provide students with opportunities to practice what they've learned. Answers to even-numbered problems are found at www.mhhe.com/hyde.

Chapter Integration Problems

challenge students to utilize concepts and skills covered in previous chapters in a broader, multi-part problem. This also provides another learning experience for students as they realize the material from different chapters all come together to be a part of genetics.

> #### Chapter Integration Problem
>
> As a student in a genetics lab, you make a startling discovery: a *Drosophila* male and female that can do backflips! You name the gene *bf* and decide to investigate the mode of inheritance of this interesting trait. You and your labmates carry out the following two crosses:
>
> I: Backflipping male × wild-type female → 1/2 wild-type males : 1/2 wild-type females
>
> II: Backflipping female × wild-type male → 1/2 backflipping males : 1/2 wild-type females
>
> The backflipping female in cross II happens to have miniature wings, a known recessive mutation. You suspect that the *bf* gene is linked to gene *m*, so you perform another cross:
>
> III: Backflipping female with miniature wings   ×↓   Wild-type male for both traits
>
> The F_1 females are then testcrossed to produce the following F_2 offspring:
>
> | | |
> |---|---|
> | Backflipper, miniature wings | 251 |
> | Backflipper, normal wings | 130 |
> | Nonbackflipper, miniature wings | 125 |
> | Nonbackflipper, normal wings | 244 |
>
> Use this information to answer the following questions:
> a. Is the *bf* locus on an autosome, the X chromosome, or the Y chromosome? Explain.
> b. Is backflipping dominant or recessive? Explain.
> c. Are genes *bf* and *m* following Mendel's first law? Justify your answer.

Instructor Resources

McGraw-Hill is dedicated to providing high quality and effective supplements for instructors. We are proud to offer the following instructor supplements for Hyde: *Introduction to Genetic Principles*.

The **Online Learning Center** at **www.mhhe.com/hyde** offers a wealth of teaching and learning aids for instructors and students. Instructors will appreciate:

- Access to all of the illustrations, photographs, and tables from the text in convenient jpeg format and on PowerPoint slides for easy integration into existing lectures.
- PowerPoint lecture outlines that can be used as supplied or customized to fit individual course needs.
- An instructor's manual containing outlines, key terms, learning objectives, and instructional strategies.
- Access to **Presentation Center**, a searchable database of line art, photos, animations, and videos from multiple McGraw-Hill textbooks.
- A computerized test bank, **EZ Test Online**, that makes it easy to create, administer, score, and report quizzes and tests using questions written for Hyde: *Introduction to Genetic Principles*.

Presentation Center

Build instructional materials wherever, whenever, and however you want! Presentation Center is an online digital library containing assets such as photos, artwork, animations, PowerPoints, and other types of media that can be used to create customized lectures, visually enhanced tests and quizzes, compelling course websites, or attractive printed support materials.

Access to your book, access to all books! This ever-growing resource gives instructors the power to utilize assets specific to their adopted textbook as well as content from other McGraw-Hill books in the library. Presentation Center's dynamic search engine allows you to explore by discipline, course, textbook chapter, asset type, or keyword. Simply browse, select, and download the files you need to build engaging course materials. All assets are copyrighted by McGraw-Hill Higher Education but can be used by instructors for classroom purposes.

Course Delivery Systems

With help from our partners WebCT, Blackboard, Top-Class, eCollege, and other course management systems, professors can take complete control of their course content. Course cartridges containing Online Learning Center content, online testing, and powerful student tracking features are readily available for use within these platforms.

Computerized Test Bank Online

A comprehensive bank of test questions is provided within a computerized test bank powered by McGraw-Hill's flexible electronic testing program EZ Test Online (www.eztestonline.com). EZ Test Online allows you to create paper and online tests or quizzes in this easy to use program!

Test Creation

- Select questions from multiple McGraw-Hill test banks or author your own
- Author/edit questions online using the 14 different question type templates
- Create printed tests or deliver online to get instant scoring and feedback.
- Create question pools to offer multiple versions online—great for practice.
- Export your tests for use in WebCT, Blackboard, PageOut, and Apple's iQuiz
- Compatible with EZ Test Desktop tests you've already created
- Sharing tests with colleagues, adjuncts, TAs is easy

Online Test Management

- Set availability dates and time limits for your quiz or test
- Control how your test will be presented
- Assign points by question or question type with drop down menu

Student Resources

- Provide immediate feedback to students or delay until all finish the test
- Create practice tests online to enable student mastery
- Your roster can be uploaded to enable student self-registration

Online Scoring and Reporting

- Automated scoring for most of EZ Test's numerous question types
- Allows manual scoring for essay and other open response questions
- Manual re-scoring and feedback is also available
- EZ Test's grade book is designed to easily export to your grade book
- View basic statistical reports

Support and Help

- User's Guide and built-in page specific help
- Flash tutorials for getting started on the support site
- Support Website—www.mhhe.com/eztest
- Product specialist available at 1-800-331-5094
- Online Training: http://auth.mhhe.com/mpss/ workshops/

Electronic Books

McGraw-Hill and VitalSource have partnered to bring you innovative and inexpensive electronic textbooks. By purchasing E-books from McGraw-Hill & Vital-Source, students can save as much as 50% on selected titles delivered on the most advanced E-book platform available, VitalSource Bookshelf.

E-books from McGraw-Hill & VitalSource are smart, interactive, searchable, and portable. Vital-Source Bookshelf comes with a powerful suite of built-in tools that allow detailed searching, highlighting, note taking, and student-to-student or instructor-to-student note sharing. Contact your McGraw-Hill sales representative to discuss E-book packaging options.

Student Resources

We offer the following student learning tools that were designed to help maximize study time for success in genetics.

The **Online Learning Center** for *Introduction to Genetic Principles* is a great place to review chapter material and enhance your study routine.

Visit **www.mhhe.com/hyde** for access to the following online study tools:

- chapter quizzing
- animation quizzes
- interactive inheritance activities
- key term flashcards
- answers to even numbered questions and problems from the textbook

Student Study Guide/Solutions Manual

The study guide offers students a variety of tools to help focus your time and energy on important concepts:

1. **Chapter Goals**—defines what you should know or be able to do after studying the chapter content.
2. **Key Chapter Concepts**—provides an overview of the chapter, summarizes key points, and gives helpful hints on topics to consider.
3. **Key Terms**—lists the important terms in the order in which they occur.
4. **Understanding Key Concepts**—integrates the key terms into a fill-in-the-blank activity.
5. **Figure Analysis**—asks a series of questions about the key figures in the chapter to ensure you understand the most important concepts.
6. **General Questions**—are short answer and essay questions that test your comprehension.
7. **Multiple-Choice Questions**—provide a quick practice test to help you prepare for your upcoming test.
8. **Practice Problems**—include several problems as well as tips for solving them.
9. **Assessing Your Knowledge**—provides answers to all of the study guide questions.
10. **Solutions Manual**—contains answers and solutions for all questions and problems from the textbook.

Introduction to Genetics

Essential Goals

1. Know key events in the modern history of genetics

2. Gain an overview of the topics included in this book—the syllabus of genetics

3. Understand why certain organisms and techniques have been used preferentially in genetics research

4. See how genetics may be applied to human life

Chapter Outline

Photo: One characteristic of the iguana is its very colorful appearance.

enetics is the study of inheritance in all of its manifestations. It can range from the molecular level, such as the DNA sequence and the organization of the chromosome, to the generation of a phenotype, which is derived from the various genetic interactions and environmental conditions, to the changes of allele frequency in a population. This text is not intended to be a comprehensive treatment of genetics, but rather a demonstration of the breadth of the field and how genetics can impact all areas of modern biology. In this chapter, we examine the history of genetics to see how it has evolved to our current level of understanding and mastery. We then outline the areas to be covered in this book as a preview of our trek through the diverse field of genetics.

1.1 A Brief Overview of the Modern History of Genetics

We are fortunate to live at a time when incredible technological advances are commonplace. Computers are becoming ever smaller and faster, and they are now accessible to almost anyone. Space scientists are sending more sophisticated probes to more distant planets; medicine is finding more advanced methods to probe the tissues deep inside the human body; and gene therapy is currently being tested as a treatment for a number of different inherited human diseases.

All of these advancements, however, required previous studies, tests, and incremental modifications to reach their current status. An understanding of where the field started and a general idea of how it progressed allows us to fully appreciate these modern accomplishments. Before we delve into the specific areas of genetics and its many processes, therefore, we first take a very brief, encapsulated look at the history of genetics.

Early Human History

Genetics extends thousands of years back in history. Various forms of records clearly demonstrate that genetics was being practiced as early as 7000–5000 B.C.E., when the Mayans were cultivating maize, the Chinese were growing rice, and the Assyrians were caring for the date palm. These peoples' general ideas of crossing individual plants of a species to produce a hybrid plant with more desirable characteristics are still being used today.

The ancient Greeks also exhibited a passion for trying to understand the basis for life. The Hippocratic school of medicine (500–400 B.C.E.) believed that semen was composed of "humors" that are the hereditary traits of an individual. The humors were produced from the different tissues in an individual, and they represented the health status of the tissue. These humors were then combined to produce the newborn infant. This idea suggested that diseased tissue would produce an infant with birth defects, and that an infant could differ from the parent because of traits that the parent acquired during his or her life. Some individuals even believed that these "humors" were actually miniature organs present in the sperm and egg that would then produce the infant.

During the next 2000 years, general ideas about heredity did not change significantly. In the field of agriculture, investigators continued to produce hybrid plants and animals, and others helped clarify the process of reproduction, if not its specific details. For our purposes, we divide the recent history of genetics into four periods: before 1860, 1860–1900, 1900–1944, and 1944 to the present.

Before 1860: Discovery of the Cell and the Nucleus

Into the eighteenth century, the general beliefs of the Hippocratic school persisted, in which the gametes either contained the parts of the organism that would assemble into the individual or actually contained miniature individuals. This latter idea was the root of the **homunculus,** a miniature human carried in the head of a sperm (**fig. 1.1**).

This belief in preformation of offspring began to crumble in the 1600s, when William Harvey first proposed the ideas that became the **theory of epigenesis.** This theory states that *substances* in the gametes would produce the adult structure, rather than the growth of miniature adult structures that are already present in the gametes. The nature of the substances that directed

figure 1.1   A drawing of homunculus by Nicolas Hartsoeker in 1694. Some scientists incorrectly believed that a minature adult was located in the head of a sperm. The fully formed arms and legs are in front of the body in this drawing.

the development of the adult and how they performed this function remained unclear.

Before 1860, the major technical discovery that paved the way for our current understanding of genetics was the development of light microscopy. In 1665, Robert Hooke coined the term *cell* to describe the structures that he observed in cork under a light microscope. Between 1674 and 1683, Anton van Leeuwenhoek, who was a master lens maker, produced lenses that generated magnifications of several hundred power (**fig. 1.2**). Using these single lenses, van Leeuwenhoek discovered living organisms (protozoa and bacteria) in rainwater. More than a hundred years passed before compound microscopes could equal Leeuwenhoek's magnifications.

In 1830, Jan Purkinje first described the nucleus within a cell. In 1831, Scottish researcher Robert Brown coined the term *nucleus*. Brown, an influential botanist, also described Brownian motion of microscopic particles and discovered the distinction between gymnosperms and angiosperms. Between 1835 and 1839, Hugo von Mohl described mitosis in a cell. This era ended in 1858, when Rudolf Virchow, a prolific medical researcher, summed up the concept of the cell theory with his Latin aphorism *omnis cellula e cellula*: all cells come from preexisting cells. Thus, by 1858, biologists had an understanding of the continuity of cells and knew of the cell's nucleus.

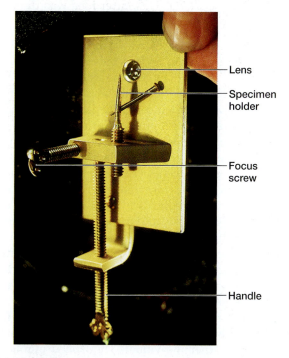

figure 1.2   One of Anton van Leeuwenhoek's microscopes, ca 1680. This single-lensed microscope magnifies up to 200×.

1860–1900: Mendelian Traits and Observation of Chromosomes

Gregor Mendel (**fig. 1.3**), an Austrian monk, performed his classical hybridization experiments with pea plants between 1856 and 1863. In this work, published in 1866, he described the statistical patterns of heritable phenotypes and proposed the theory that factors in the germ cells accounted for the underlying basis of inheritance. However, this publication went essentially unnoticed until it was rediscovered in 1900 by Hugo de Vries, Carl Correns, and Erich von Tschermak.

figure 1.3   The Augustinian monk, Gregor Mendel (1822–1884), the father of transmission genetics.

Although Mendel's work and its rediscovery were very important events, they were not the only important discoveries in this period. In 1875, German embryologist Oscar Hertwig described the fusion of sperm and egg to form the zygote. From 1879 to 1885, Walther Flemming, formerly a physician during the Franco-Prussian war, used the newly synthesized aniline dyes to view and describe chromosomes and the way they moved during mitosis. In 1888, Heinrich Waldeyer first used the term *chromosome*. Also in the 1880s, Theodor Boveri, as well as Karl Rabl and Édouard van Beneden, hypothesized that chromosomes are individual structures with continuity from one generation to the next despite their "disappearance" between cell divisions. In 1885, August Weismann stated that the inheritance of traits is based exclusively in the nucleus. Then, in 1887, Weismann predicted the occurrence of a reductional stage of cell division, which we now call meiosis. By 1890, Hertwig and Boveri had described the process of meiosis in detail.

1900–1944: The Chromosome Theory and Sex Linkage

From 1900 to 1944, modern genetics flourished with the development of the chromosomal theory, which stated that chromosomes are linear arrays of genes that contained the genetic information required by living organisms. In addition, geneticists made discoveries that provided the foundations of modern evolutionary and molecular genetics.

In 1902, Walter Sutton, a 25-year-old graduate student at Columbia University, hypothesized that the behavior of chromosomes during meiosis explained Mendel's rules of inheritance, thus leading to the discovery that genes are located on chromosomes. In the early 1900s, the eminent geneticist Thomas Hunt Morgan (**fig. 1.4**) introduced *Drosophila melanogaster* as a model genetic system, which played an important part of many genetic discoveries over the next century. Nettie Maria Stevens, a Bryn Mawr graduate student who studied with Morgan, investigated sex determination in mealworms, publishing their findings about the X and Y chromosomes in 1905. In 1911, Morgan demonstrated that the genes producing white eyes, yellow body, and miniature wings in *Drosophila* are located on the X chromosome.

Alfred Sturtevant, an undergraduate in Morgan's laboratory at Columbia University, used *Drosophila* to create the first genetic map in 1913, which demonstrated that genes existed in a linear order on chromosomes. Calvin Bridges, working with Morgan in 1914, described nondisjunction of *Drosophila* sex chromosomes to prove the chromosomal theory of inheritance. In 1927, Lewis Stadler and Hermann Muller showed that genes can be mutated artificially by X-rays.

As biologists increased their understanding of the underlying mechanisms of inheritance and generated hypotheses, they also created new terminology to describe the discoveries and predictions. The British biologist William Bateson first coined the terms F_1, F_2, *homozygote, heterozygote,* and *allelomorph* (which was later shortened to *allele*) in 1902, and created the term *genetics* in 1905. Danish botanist Wilhelm Johannsen introduced the terms *phenotype, genotype,* and *gene* in 1909.

Between 1930 and 1932, Ronald A. Fisher, Sewall Wright, and John B. S. Haldane developed the algebraic foundations for our understanding of the process of evolution. In 1943, Salvador Luria and Max Delbrück demonstrated that bacteria have genetic systems and phenotypes that could be studied. This allowed a variety of bacteria and their viruses to serve as models for studying fundamental genetic processes.

1944–Present: DNA, RNA, and Molecular Genetics

The period from 1944 to the present is the era of molecular genetics, beginning with the demonstration that DNA is the genetic material and culminating with our current explosion of knowledge based on recombinant DNA technology and the genomic sequences from several hundred different organisms.

Experiments by Oswald Avery and his colleagues at the Rockefeller Institute in 1944, and Alfred Hershey

figure 1.4   Thomas Hunt Morgan (1866–1945), the founder of *Drosophila* genetics and numerous genetic principles.

and Martha Chase at Cold Spring Harbor in 1952, showed conclusively that deoxyribonucleic acid—DNA—was the genetic material. You are probably already aware that James Watson and Francis Crick worked out the structure of DNA in 1953, based on this and other experimental data.

Between 1968 and 1973, Werner Arber, Hamilton Smith, and Daniel Nathans, along with their colleagues, discovered and described restriction endonucleases, the enzymes that opened up the ability to manipulate DNA through recombinant DNA technology. Arber's daughter Silvia, 10 years old when the three geneticists won the Nobel Prize in 1978, dubbed the restriction endonuclease "the servant with the scissors." In 1972, Paul Berg was the first to construct a recombinant DNA molecule containing parts of DNA from different species.

Since 1972, geneticists have cloned numerous genes, including a large number of human genes that correspond to inherited diseases. Cloning a gene involves isolating a specific DNA sequence from an organism and joining it to a **vector,** which is a piece of DNA that can be replicated in a cell. Gene cloning has been greatly simplified by the identification of restriction enzymes, polymerase chain reaction (PCR) amplification of DNA sequences, and the sequencing of the genomes from many different organisms. In 1995, *Haemophilus influenzae* was the first organism to have its complete genome sequenced. A **genome** corresponds to all the DNA, both the genes and the regions between the genes, in an organism. The human genome was sequenced and published in 2001. As of September of 2007, 634 different organisms have had their complete genome sequenced and published (www.genomesonline.org).

With the cloning and identification of genes, scientists can create **transgenic organisms,** organisms with functioning foreign genes. For example, genes have been introduced into farm animals to enable them to produce pharmaceuticals in their milk, which can then be harvested easily and inexpensively for human use. In 1997, the first transgenic cow, named Rosie, was created. Rosie expressed the human α-lactalbumin protein in her milk, which was more beneficial than normal cow's milk for babies and for the elderly with digestive problems.

In contrast to transgenic animals that express a foreign gene, **cloned animals** are derived from a somatic nucleus of another individual. The cloned animal is then genetically identical to the original animal. In 1996, the first mammal was cloned, a sheep named Dolly (**fig. 1.5**). Since then, pigs, mules, cattle, cats, and dogs

figure 1.5 The first cloned mammal, Dolly, a Finn Dorset sheep, was born on July 5, 1996. The surrogate mother (on right) is a Blackface ewe and could not conceive a white-faced lamb. This provides proof that Dolly was not genetically related to the surrogate mother. A variety of DNA tests demonstrated that Dolly was a clone of another animal.

are some of the animals that have been cloned. Cloned animals provide valuable resources for the study of specific diseases, the expression of pharmaceuticals, and the isolation of tissues and organs for human use.

Emerging Fields

New fields are rapidly emerging in the general area of genetics. **Genomics** is the analysis of the DNA content and gene organization in and between organisms. **Proteomics** is the study of all the proteins expressed by an individual or organism. Because of the availability of the entire genomic DNA sequence for a large number of organisms and the creation of computer programs to analyze very large sets of data, some scientists are beginning to examine global questions about genomes. Genomics and proteomics are discussed in detail in later chapters.

Information gained from these studies promises to lead to more individualized treatments for a wide range of diseases, from cancer to mental disorders. The application of **gene therapy,** the introduction of a foreign gene into the somatic cells of an individual to cure an inherited disease, is being tested on a variety of diseases. Although gene therapy has not completely cured any human disease to this point, success seems close at hand for many conditions.

The space in this chapter is much too brief to convey the details or the excitement surrounding the many discoveries of modern genetics. Throughout the rest of this text, you'll encounter more of the fascinating story that makes genetics such a vital and amazing field of study.

1.2 The Three General Areas of Genetics

Historically, geneticists have worked in three different areas: *classical, molecular,* and *population and evolutionary genetics.* Each has its own particular questions, terminology, tools, and organisms. We discuss all three areas briefly in the sections that follow.

Classical Genetics

In **classical genetics,** scientists are concerned with genes, mutations, and phenotypes. This study begins with the arrangement of genes on the chromosome and their transmission to the next generation through meiosis, the methods that generate mutations and identify mutants, and the various patterns of inheritance to produce a specific trait or phenotype.

Gregor Mendel published the basic rules of inheritance in 1866, which described his carefully controlled breeding experiments with the garden pea plant, *Pisum sativum.* At that time, the role of chromosomes and genes in inheritance and producing a trait was still 50 years in the future. Yet, Mendel correctly hypothesized that traits such as pod color were controlled by genetic elements that we now call *genes* (**fig. 1.6**). Alternative forms of a gene are called *alleles,* which can confer different characteristics to the organism. The *genotype* of an organism refers to the combination of alleles that it contains, and the *phenotype* is the characteristic it exhibits. All these terms and concepts are described in more detail in later chapters.

Mendel also predicted that adult organisms have two copies of each gene (*diploid* state); gametes receive just one of these copies (*haploid* state). In other words, one of the two parental copies segregates into any given gamete.

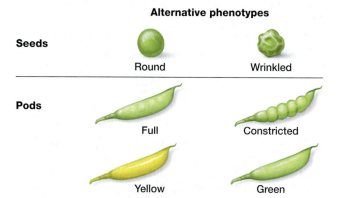

figure 1.6 Mendel worked with garden pea plants. He observed seven traits of the plant—each with two discrete forms—that affected attributes of the seed, the pod, and the stem. For example, all plants had either round or wrinkled seeds, full or constricted pods, or yellow or green pods.

Upon fertilization, the zygote gets one copy from each gamete, reconstituting the diploid number (**fig. 1.7**).

Mendel's work has been distilled into two rules, referred to as the *law of segregation* (two alleles separate randomly from each other during gamete formation) and the *law of independent assortment* (the alleles of different genes sort independently of each other into gametes). Scientists did not have a clue about how these rules could work until they observed the segregation of chromosomes in meiosis. At that time, in the year 1900, the science of genetics was born.

During much of the early part of the 20th century, geneticists discovered many genes by looking for organisms within a species that exhibited altered phenotypes (*mutants*). Crosses were made to determine the genetic control of mutant traits. From this research evolved chromosomal mapping, the ability to locate the relative positions of genes on chromosomes by crossing certain individuals with different genotypes.

From this work on mapping arose the **chromosomal theory of inheritance:** Chromosomes, which contain genes, are the carriers of the genetic material (**fig. 1.8**). A "beads on a string" model of gene arrangement, in which

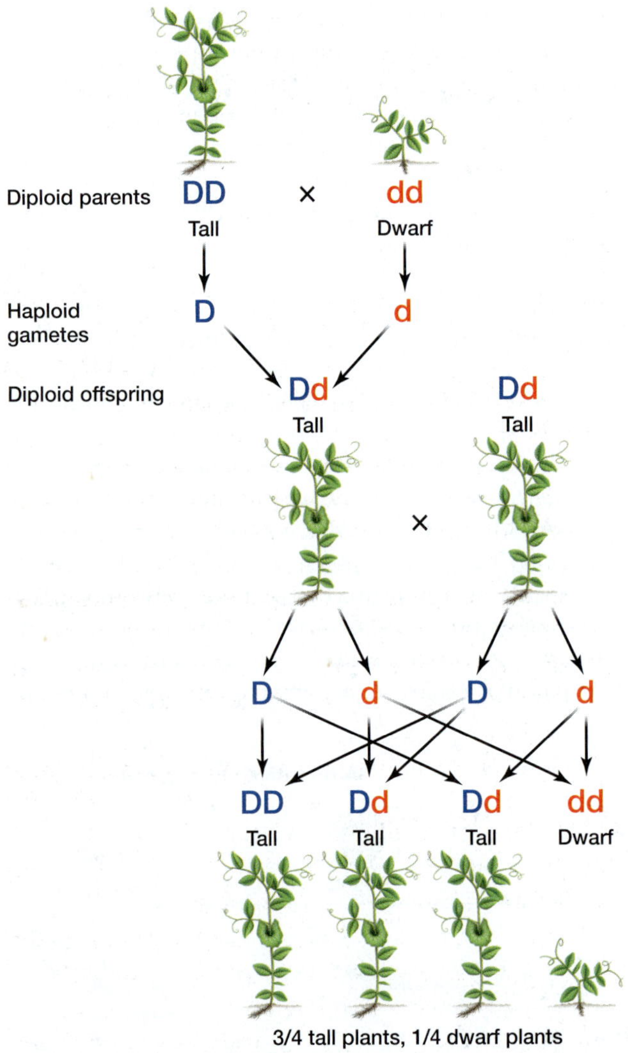

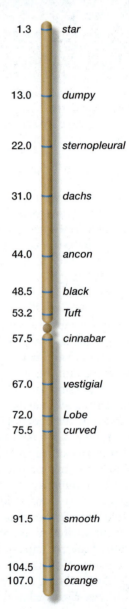

figure 1.7   Mendel crossed tall and dwarf pea plants and deduced which phenotype was dominant. Each parent passed one copy of every gene on to its progeny through a gamete. Fusion of the gametes resulted in the generation of a diploid individual. By crossing these two tall plants and observing the phenotypes of the progeny, Mendel deduced that the tall phenotype was dominant. The *d* allele, which confers the dwarf phenotype, is recessive because two copies of the *d* allele are required to observe the dwarf phenotype.

figure 1.8   Mutations, which correspond to genes, are located in linear order on chromosomes. Chromosome 2 of *Drosophila melanogaster*, the common fruit fly, shows the relative map position of several mutations to the left (in recombination map units), the location of each corresponding gene in blue, and the name of each mutation to the right. Dominant mutations are designated by the first letter of the mutation name being uppercase. The centromere is the constriction in the chromosome.

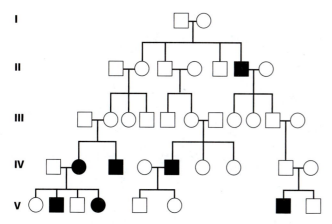

figure 1.9 A pedigree of an extended family exhibiting an autosomal recessive disease. Filled symbols are individuals who are affected with the recessive disease, and open symbols are unaffected individuals.

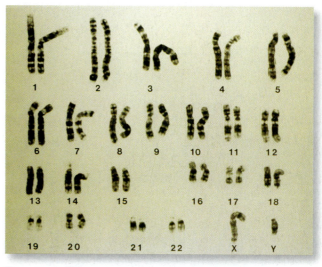

figure 1.10 A human karyotype. This karyotype is of a male, as indicated by the presence of the Y chromosome.

each gene was an indivisible unit located at a fixed position on the chromosome, was not modified to any great extent until the middle of the twentieth century, after Watson and Crick worked out the structure of DNA.

In general, the DNA sequence of genes encodes the amino acid sequence of proteins, including *enzymes,* which act as biological catalysts in biochemical pathways. In the early 1940s, George Beadle and Edward Tatum suggested that one gene encodes one enzyme. We now know that many enzymes are made up of more than one polypeptide, with each polypeptide often encoded by a different gene. Thus, a more accurate statement would be "one-gene-encodes-one-polypeptide."

The analysis of inheritance patterns for human genes is one current application of classical genetics. This involves the generation of a *pedigree* (**fig. 1.9**), which is a pictorial representation of related individuals and the phenotype that each individual exhibits. By examining the pedigree, it is possible to deduce the pattern of inheritance, such as either a dominant or recessive phenotype or whether the associated gene is located on either a sex chromosome or an autosome. By comparing two or more phenotypes (which are encoded by separate genes) in a pedigree, it is possible to calculate the linkage relationship or map distance between the genes. This pedigree and linkage analysis allows a geneticist to calculate the probability that an individual possesses a specific allele, or that a child will exhibit a specific phenotype.

An additional tool in classical genetics is the *karyotype,* which is the representation of all the chromosomes in an individual (**fig. 1.10**). Examining the karyotype reveals not only the sex of the individual, but if extra or fewer chromosomes are present and whether any chromosome has a significant abnormality or rearranged order. The study of chromosomes and their organization is called **cytogenetics.**

The major questions addressed in this book that pertain to classical genetics are

1. What are the common patterns of inheritance that are observed for one gene? (chapters 2, 4, and 5)
2. What are the common patterns of inheritance that are observed for two or more genes? (chapters 4, 5, and 6)
3. How do the movement of chromosomes during nuclear division (mitosis and meiosis) account for the patterns of inheritance? (chapter 3)
4. What factors determine an individual's gender, and what genes are linked to sex chromosomes? (chapter 4)
5. How does linkage of genes affect inheritance patterns and allow mapping of gene locations? (chapters 6, 13, and 15)
6. How do alterations in the structure and number of chromosomes occur, and how do they affect the patterns of inheritance? (chapter 8)
7. What are the mechanisims of nonnuclear inheritance in a eukaryotic cell? (chapter 19)
8. By what different methods are mutations generated, and how are the corresponding mutants analyzed? (chapters 18 and 20)
9. How is genetics used to study biological processes, such as development and cancer? (chapters 21 and 22)

Molecular Genetics

Molecular genetics is the study of the structure, replication, and expression of the genetic material and of the expressed protein. It also includes the methods involved in manipulating the expression and analysis

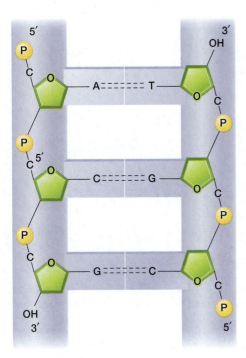

figure 1.11   A look at a DNA double helix, showing the sugar–phosphate units that form the molecule's "backbone" and the base pairs that make up the "rungs." We abbreviate a phosphate group as a "P" within a yellow circle; the pentagonal ring containing an oxygen atom is the sugar deoxyribose. Bases are either adenine, thymine, cytosine, or guanine (A, T, C, G). The 5' and 3' ends of the DNA strands are in an antiparallel orientation.

of the genetic material (recombinant DNA technology and genomics).

The genetic material for all cellular organisms is double-stranded DNA, a double-helical molecule shaped like a twisted ladder. The backbones of the helices are repeating units of sugars (deoxyribose) and phosphate groups. **Polarity** or directionality is an important aspect of nucleic acids and proteins. Each strand of linear DNA contains a 3' end, having a free hydroxyl (–OH) group on the 3' carbon of deoxyribose, and a 5' end, having a free phosphate ($-PO_4^-$) group on the 5' carbon of deoxyribose (**fig. 1.11**). The two strands in double-stranded DNA are **antiparallel;** the base at the 5' end of one strand pairs with a base at the 3' end of the other.

The two helices are held together by bases that extend from each sugar-phosphate backbone strand, similar to the rungs of the ladder (see fig. 1.11). Only four bases normally occur in DNA: adenine, thymine, guanine, and cytosine, abbreviated A, T, G, and C, respectively. The sequence of these bases on one strand dictates the sequence on the second strand through a relationship called *complementarity*. Adenine pairs with thymine, and guanine pairs with cytosine.

The complementary nature of the DNA base pairs provided clues about the process of DNA replication. The

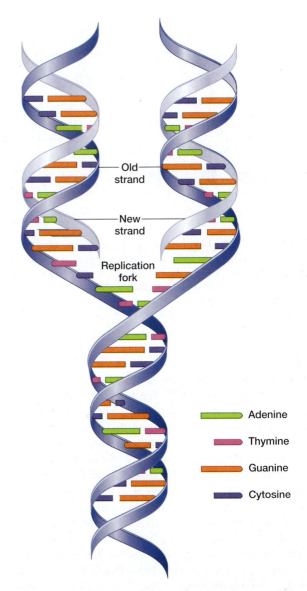

Adenine

Thymine

Guanine

Cytosine

figure 1.12   The two strands of the DNA double helix unwind during replication, and each strand then acts as a template for a new double helix. Because of the rules of complementarity, each new double helix is identical to the original, and the two new double helices are identical to each other. Thus, an AT base pair in the original DNA double helix replicates into two AT base pairs, one in each of the daughter double helices.

double helix "unzips," and each strand acts as a template for a new strand, resulting in two double helices exactly like the first (**fig. 1.12**). The enzyme that performs DNA replication is known as *DNA polymerase*, of which there are several different forms in the cell. Mutations could result from either an error in base-pairing during replication or some sort of damage to the DNA that was not repaired by the time of the next replication cycle.

Groups of three nucleotide bases form a *codon* that specifies one of the 20 naturally occurring amino acids

table 1.1 The Genetic Code Dictionary of RNA

Codon	Amino Acid	Codon	Amino Acid	Codon	Amino Acid	Codon	Amino Acid
UUU	Phe	UCU	Ser	UAU	Tyr	UGU	Cys
UUC	Phe	UCC	Ser	UAC	Tyr	UGC	Cys
UUA	Leu	UCA	Ser	UAA	**STOP**	UGA	**STOP**
UUG	Leu	UCG	Ser	UAG	**STOP**	UGG	Trp
CUU	Leu	CCU	Pro	CAU	His	CGU	Arg
CUC	Leu	CCC	Pro	CAC	His	CGC	Arg
CUA	Leu	CCA	Pro	CAA	Gln	CGA	Arg
CUG	Leu	CCG	Pro	CAG	Gln	CGG	Arg
AUU	Ile	ACU	Thr	AAU	Asn	AGU	Ser
AUC	Ile	ACC	Thr	AAC	Asn	AGC	Ser
AUA	Ile	ACA	Thr	AAA	Lys	AGA	Arg
AUG	**Met (START)**	ACG	Thr	AAG	Lys	AGG	Arg
GUU	Val	GCU	Ala	GAU	Asp	GGU	Gly
GUC	Val	GCC	Ala	GAC	Asp	GGC	Gly
GUA	Val	GCA	Ala	GAA	Glu	GGA	Gly
GUG	Val	GCG	Ala	GAG	Glu	GGG	Gly

Note: A codon, specifying one amino acid, is three bases long (read in RNA bases in which U replaced the T of DNA). There are 64 different codons, specifying 20 naturally occurring amino acids (abbreviated by three letters: e.g., Phe is phenylalanine—see fig. 11.3 for the names and structures of the amino acids). Also present is *stop* (UAA, UAG, UGA) and *start* (AUG) information for translation.

used in protein synthesis. The sequence of bases making up the codons are referred to as the genetic code (**table 1.1**).

Translation and Transcription

The information encoded in a gene is present within the sequence of bases on one strand of the DNA double helix. During gene expression, that information is *transcribed* from DNA into RNA, a different form of nucleic acid, which takes part in protein synthesis. RNA differs from DNA in several respects: it contains the sugar ribose in place of deoxyribose; it has the base uracil (U) rather than thymine (T); and it is usually single-stranded. RNA is transcribed from DNA by the enzyme *RNA polymerase,* using DNA–RNA rules of complementarity: A, T, G, and C in DNA pair with U, A, C, and G, respectively, in RNA (**fig. 1.13**). RNA also contains a polarity, with a 5' and 3' end. As described in later chapters, the polarities in both DNA and RNA are important for the movement of polymerases during DNA replication and RNA transcription.

There are several different types of RNA. **Messenger RNA** (mRNA) is the RNA that contains the information that will be *translated* into a protein's amino acid sequence. **Ribosomal RNA** (rRNA) forms a complex with several proteins to form the *ribosome,* which is where mRNA nucleotide sequences are translated into amino acid sequences (**fig. 1.14**). As the RNA moves

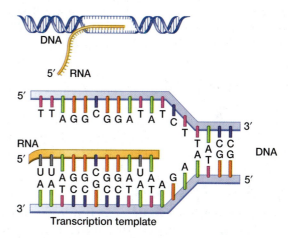

figure 1.13 The relationship between DNA and RNA during transcription. One strand of DNA serves as the template for the RNA that is being transcribed. The RNA "grows" in a 5' → 3' direction. The other DNA strand possesses the same sequence as the RNA, except for the presence of T in the DNA and U in the RNA.

along the ribosome, a three-nucleotide codon sequence in the mRNA base-pairs with a three-nucleotide anticodon sequence in the **transfer RNA** (tRNA). Each tRNA is attached to a specific amino acid, and the complementarity between the codon and anticodon sequences ensures that the correct amino acid enters the ribosome. A variety of small RNAs are also important in producing

figure 1.14 In translation, a ribosome attaches near the 5' end of the mRNA and reads the mRNA codon to begin translating the amino terminus (NH_2) of the protein. As the ribosome moves along the mRNA, amino acids attach to the carboxyl terminus (CO_2H)

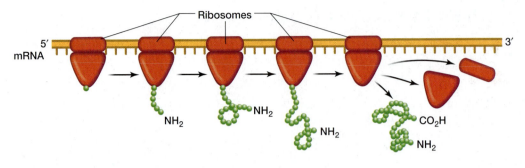

of the growing protein within the ribosome. As the first ribosome moves along, a second ribosome can attach at the 5' end of the mRNA, and so on, so that a mRNA strand may have many ribosomes attached at one time.

the mRNA and controlling the transcription of DNA and translation of mRNA.

Functional protein expression can be controlled at a variety of steps in the generation of the protein product. First, transcription can be regulated, such that a gene may be either transcribed or not transcribed. Second, the mRNA that is to be translated into protein can be regulated to permit or prevent translation. Finally, some proteins must be modified after translation to become active, which serves as an additional opportunity for regulation.

The **central dogma** describes the flow of genetic information; it states that DNA is transcribed to mRNA, which is then translated into protein (**fig. 1.15**). Although this principle served as the generally accepted process for many years, we now know that some viruses contain an RNA genome. Upon entering a cell, the RNA is converted into DNA that can then become inserted in the host's genome. Transcription of the viral genome then produces new viral genomes. Currently no known mechanism permits a protein sequence to direct the synthesis of a specific RNA sequence.

Recombinant Technology

Molecular genetics has vastly expanded in the last 30 years with the information learned using **recombinant DNA techniques.** This revolution began with the discovery of *restriction endonucleases*, enzymes that cut DNA at specific sequences to yield DNA fragments with reproducible base sequences at the ends. These ends allow DNA from different sources to be joined (**fig. 1.16**). Much of the recombinant DNA work utilizes *plasmids*, small, circular extrachromosomal DNA units found in some bacteria. Insertion of foreign DNA (cloning) into a plasmid allows the foreign genes to be "grown" inside bacteria in amounts that can be analyzed for sequence and function. For example, recombinant plasmids have been introduced into *E. coli* bacteria to produce human growth hormone, which can be harvested for pharmaceutical use.

The development of *complementary DNA* (*cDNA*), which is a reverse transcribed copy of mRNA, and

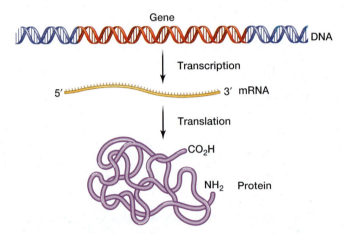

figure 1.15 The central dogma states that genetic information flows from DNA to RNA (by transcription) to protein (by translation).

the *polymerase chain reaction* (*PCR*), which amplifies a desired DNA sequence, have greatly increased our ability to clone and manipulate DNA sequences in vitro.

This technology has tremendous possibilities in medicine, agriculture, and industry. It has provided the opportunity to locate and study disease-causing genes, such as those for cystic fibrosis and muscular dystrophy, as well as suggesting potential treatments. Crop plants and farm animals, which people have always manipulated genetically through selective breeding, are being more directly modified for better productivity by improving growth and disease resistance. Industries that apply the concepts of genetic engineering, such as the pharmaceutical industry, are flourishing.

Cloning and Gene Therapy

Cloning has frequently been used in botany and agriculture to produce whole plants from a few meristem cells. Potatoes, for example, are rarely grown from seeds produced by flowers, but rather from "seed potatoes"— tubers or parts of tubers that are planted and grown.

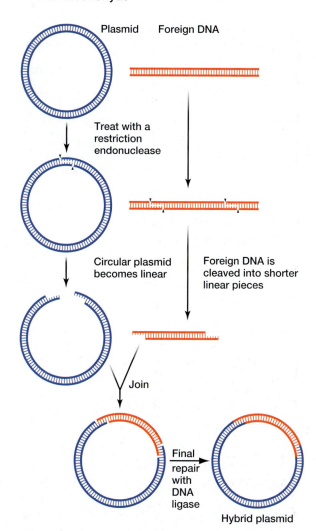

Plasmid   Foreign DNA

Treat with a
restriction
endonuclease

Circular plasmid
becomes linear

Foreign DNA is
cleaved into shorter
linear pieces

Join

Final
repair
with
DNA
ligase

Hybrid plasmid

figure 1.16   The basic features of cloning a gene. A small, circular plasmid, which can replicate in a bacterial cell, and a foreign piece of DNA are both cut with the same restriction endonuclease, which cuts the double-stranded DNA at a specific sequence to generate the same type of ends on the cut DNA. Cutting the circular plasmid once creates a linear DNA molecule. The foreign DNA and linear plasmid can be mixed together so the foreign DNA rejoins the plasmid into a circle using the enzyme DNA ligase. This hybrid or recombinant DNA molecule can be introduced into bacteria, where it will replicate.

The techniques of genetic engineering have now led to the ability to clone animals, which involves transplanting a nucleus from a somatic cell into an enucleated egg cell to produce a genetically identical individual. The first cloned mammal was Dolly, a Finn Dorset sheep, who was created in 1996 by Ian Wilmut at the Roslin Institute in Scotland (see fig. 1.5). Since Dolly, a variety of different organisms have been cloned.

In recent years, the cloning of genes and the sequencing of over 630 different genomes has shifted the genetic questions that are asked. Although it remains important and informative to study how the expression of a

single gene is controlled, scientists can now ask how the expression of all the genes in the cell changes during development or as a result of a disease or drug therapy. These large-scale questions have created the **"-omics"** fields of genomics and proteomics.

Genomics is the study of an organism's complete DNA or RNA sequence, the organization and expression of genes, and the relationship of these elements between different species. *Transcriptomics* is the study of all the transcribed genes and how their expression changes in response to different stimuli or diseases. This study carries the potential for designing drug therapies that are tailored for a specific disease in an individual. For example, many individuals develop colon cancer. Mutations in a number of different genes can produce colon cancer (discussed in chapter 22). Identification of specific mutations causing an individual's colon cancer might lead to a customized therapy that would be successful against that particular type of colon cancer, avoiding the "shotgun approach" of more conventional anticancer therapies.

Proteomics is the study of all the proteins in a cell or individual. It includes analyzing the expressed proteins and any changes in the amount of the protein or in modifications to it, such as phosphorylation. Proteomics can also be used to identify what proteins physically interact with other proteins, and which interactions are essential to the function of the protein.

The major questions we will address in this book that pertain to molecular genetics are

1. What is the structure of DNA and chromosomes and how is it replicated? (chapters 7 and 9)
2. What is the mechanism of RNA transcription and how is it regulated? (chapters 10, 16, and 17)
3. What is the mechanism of translation? (chapter 11)
4. What are the basic recombinant DNA techniques and how are they applied to modern problems? (chapters 12 and 13)
5. What techniques are used in global analysis of genomes, gene expression, and protein function? (chapter 14)
6. What basic techniques are used to study genes in prokaryotes? (chapters 15 and 16)
7. What are the mechanisms of mutation and how does the cell repair these errors? (chapter 18)
8. What are the genetic mechanisms underlying complex processes like development and cancer? (chapters 21 and 22)

Population and Evolutionary Genetics

Population and evolutionary genetics is the study of the mechanisms that result in changes in allele frequencies in *populations* over time and the consequences of those changes. Darwin's concept of evolution by natural selection finds a firm genetic footing in this area of the study of inheritance.

Charles Darwin described evolution as the result of natural selection. In the 1920s and 1930s, geneticists including Fisher, Wright, and Haldane provided algebraic models to describe evolutionary processes. The marriage of Darwinian theory and population genetics has been termed *neo-Darwinism.*

In 1908, British mathematician Godfrey Hardy and Wilhelm Weinberg, a German obstetrician with a life-long interest in genetics, independently discovered that a simple genetic equilibrium occurs in a population if the population is large, has random mating, and has negligible effects of mutation, migration, and natural selection. This observation has since been known as the *Hardy-Weinberg equilibrium.*

This equilibrium gives geneticists a baseline for comparing populations to see whether any evolutionary processes are occurring. We can formulate a statement to describe the equilibrium condition: If the assumptions are met, then the population is not experiencing changes in allelic frequencies, and these allelic frequencies will accurately predict the frequencies of *genotypes* (allelic combinations in individuals, e.g., *AA, Aa,* or *aa*) in the population.

Recently, several surprising results have been found in the area of evolutionary genetics. Electrophoresis, a method for separating DNA, proteins, and other molecules, and DNA sequencing have revealed that much more *polymorphism* (allelic variation) exists within natural populations than older mathematical models could account for. One interesting explanation for this variability is that much of the polymorphism is neutral. That is, natural selection, the guiding force of evolution, does not act differentially on many, if not most, of the genetic differences found so commonly in nature.

At first, this theory was quite controversial, attracting few followers. Now the majority of scientists accept this view to explain the abundance of molecular variation found in natural populations.

Another controversial theory concerns the rate of evolutionary change. Based on the fossil record, it was originally thought that most evolutionary change was gradual. However, recent evidence suggests that evolutionary change may actually occur in short, rapid bursts, followed by long periods of very little change. This theory is called *punctuated equilibrium.*

A final area of evolutionary biology that has generated much controversy is the theory of *sociobiology.* Sociobiologists suggest that social behavior is under genetic control and is acted on by natural selection, as is any morphological or physiological trait. This idea is controversial mainly as it applies to human beings; it calls altruism into question and suggests that to some extent we are genetically programmed to act in certain ways. This theory has been criticized because it appears to justify racism and sexism.

The major questions we will address in this book that pertain to population and evolutionary genetics are

1. How are allele frequencies maintained in a population, and what factors alter the prevalence of the alleles in a population? (chapter 23)
2. How do genetics and the environment influence the expression of quantitative traits? (chapter 24)
3. What factors influence evolution and how does it occur at the molecular level? (chapter 25)

1.3 Why Fruit Flies and Colon Bacteria?

As you read this book, you will see that certain organisms are used repeatedly in genetic experiments in the laboratory. If the goal of science is to uncover generalities about the living world, you may wonder why geneticists use these particular organisms. The answer is that certain organisms possess attributes, both practical and genetic, that make them desirable *model organisms* for research.

Some of the important features of a good model organism are that it

1. Has a short generation time to reach sexual maturity. This allows the researcher to quickly analyze multiple generations of specific crosses of individuals.
2. Produces a large number of offspring. Larger brood sizes have a greater chance of exhibiting the expected phenotypic and genotypic ratios. Furthermore, an increased number of offspring allows a greater chance of isolating rare individuals with the desired phenotype in different crosses.
3. Can be easily and inexpensively reared or grown in a small space in a laboratory. The ability to maintain a large number of individuals in a limited space at low cost allows the researcher to have larger populations of individuals for isolating and characterizing mutants.
4. Exhibits interesting features that correspond to a variety of organisms. Studying phenotypes that are present in other organisms makes the study of greater general importance. Additionally, if the phenotype can be related to humans in some way, such as the development of a nervous system or the control of gene expression, then the relevance of the study is significantly increased.
5. Has genomic DNA that has been largely or entirely sequenced. Knowing the genomic DNA sequence increases the opportunity to identify genes in both normal and mutant individuals. Knowledge about the genomic DNA sequence also increases the ease of performing a variety of recombinant DNA techniques.

1.1 applying genetic principles to problems

Lysenko, Government, and Failed Genetics

Science is a very human activity; people living within societies explore scientific ideas and combine their knowledge. The society in which a scientist lives can affect not only how that scientist perceives the world, but also what that scientist can do in his or her scholarly activities.

For example, the United States and other countries have decided that mapping and sequencing the entire human genome would be valuable (see chapter 14). Thus, granting agencies directed money to this purpose. Because much of scientific research is expensive, scientists often can study areas only for which funding is available. Thus, many scientists are working on the Human Genome Project. Examples also exist in which a societal decision has had negative consequences for both the scientific establishment and the society itself. One example is the Lysenko affair in the former Soviet Union during Stalin's and Krushchev's reigns.

The preeminent Soviet geneticist in the 1920s and 1930s was Nikolai Vavilov, who was interested in improving Soviet crop yields by growing and mating many varieties of crop species and selecting the best to be the breeding stock of the next generation. This approach, which had been the standard way of improving a plant crop or livestock breed (see chapter 24), was consistent with Mendel's ideas that the genotype of an individual controlled the phenotype. By mating varieties of crop species and selecting the best breeding stock, Vavilov and others were introducing genetic diversity into the plant and then selecting for the most productive genotype. This method, although successful, is a slow process that only gradually improves yields.

Trofim Lysenko was a poor Ukrainian biologist in the former Soviet Union researching the effects of temperature on plant development (fig. 1.A). His hope was to develop a method to more rapidly engineer crops with favorable characteristics to feed the starving masses in the Soviet Union. Lysenko suggested that crop yields could be improved quickly by treating the plant, which would make it acquire specific characteristics. For example, chilling the germinated seeds of summer crops would allow the subsequent plants

to be able to grow in a cold environment. Thus, Lysenko believed he could manipulate the organism to produce the desired phenotype, and this manipulation was the way for organisms to acquire new phenotypes.

Although Lysenko's ideas were doomed to fail because they did not follow the correct laws of genetic inheritance, they were greeted with much enthusiasm by the political elite. The enthusiasm was due not only to the fact that Lysenko promised immediate improvements in crop yields, but also that Lysenkoism was politically favored. That is, Lysenkoism fit in very well with communism; it promised that nature could be manipulated easily and immediately. If people could manipulate nature so easily, then communism could easily convert people to its doctrines.

Not only did Stalin favor Lysenkoism, but Lysenko himself was favored politically over Vavilov because Lysenko came from a peasant family, whereas Vavilov was from a wealthy family. (Communism was a revolution of the working class over the wealthy aristocracy.) Supported by Stalin, and then Krushchev, Lysenko gained inordinate power in his country. All visible genetic research in the former Soviet Union was forced to conform to Lysenko's views. People who disagreed with him were forced out of power; Vavilov was arrested in 1940 and died in prison in 1943. It was not until Nikita Krushchev lost power in 1964 that Lysenkoism fell out of favor. Within months, Lysenko's failed pseudoscience was repudiated, and Soviet genetic research got back on track.

For 30 years, Soviet geneticists had been forced into fruitless endeavors, forced out of genetics altogether, or punished for their heterodox views. Superb scientists died in prison as crop improvement programs failed, all because the Soviet dictators favored Lysenkoism and prevented the study and application of Mendelian genetics. The message of this affair is clear: Politicians can support research that agrees with their political agenda and can punish scientists whose findings disagree with this agenda, but politicians cannot change the truth of the laws of nature. These laws can be elucidated only through open inquiry and a free expression of ideas.

figure 1.A   Ukrainian scientist Trofim Denisovich Lysenko (1898–1976) shows branched wheat to collective farmers in the former Soviet Union.

The Ubiquitous Colon Bacterium (*Escherichia coli*)

In the middle of the twentieth century, when geneticists developed techniques to work with bacteria, the common colon bacterium, *Escherichia coli,* became a favorite organism of genetic researchers (**fig. 1.17**). Because *E. coli* has a generation time of only twenty minutes and only a small amount of genetic material, many research groups used it in their experiments. In addition, large numbers of bacteria can be tested on a single Petri dish, allowing the isolation of rare mutants relatively quickly. The sequence of the *E. coli* genome was completed in 1997, surprisingly after the completion of many larger and more complex eukaryotic genomes.

Mutations that disrupt the ability of the bacteria to synthesize a variety of compounds, such as amino acids, or to utilize specific sugars for energy were isolated. Much of our initial understanding of gene regulation developed from studying these mutants.

Still later, bacterial viruses, called *bacteriophages,* became very popular in genetics labs. These viruses are constructed of only a few different proteins and a very small amount of genetic material. Some can replicate 100-fold in an hour. *E. coli* and its bacteriophages became model organisms in understanding the gene and its relationship to the genome.

The Baker's Yeast (*Saccharomyces cerevisiae*)

Yeast commands a unique position among the model organisms (**fig. 1.18**). It is microscopic and easy to manipulate like bacteria, but it is a eukaryote. Yeast

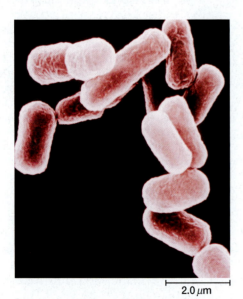

figure 1.17   Scanning electron micrograph of rod-shaped *Escherichia coli* bacteria.

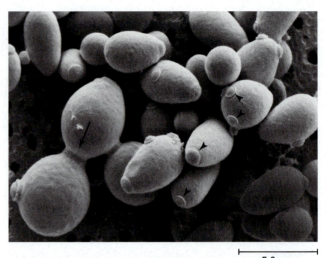

5.0 µm

figure 1.18   Scanning electron micrograph of the yeast, *Saccharomyces cerevisiae.* One yeast cell is actively undergoing cell division (arrow). Other yeast cells show the remnant of cell division, a bud scar (arrowheads).

is often found in a haploid state, which means that it contains one copy of each of its 16 different chromosomes. These two facts have allowed large numbers of baker's yeast, *Saccharomyces cerevisiae,* mutants to be isolated and analyzed.

Many of the unique features of linear eukaryotic chromosomes, such as the identification of the DNA sequences associated with the centromeres and telomeres, were heavily studied in yeast. Furthermore, yeast proved valuable in studying how centromeres function in chromosomal movement and the role of telomeres during replication of linear DNA. Yeast mutants were critical in revealing important details of the cell cycle.

Yeast has also proven to be a valuable component of most of the different genome sequencing projects. These projects require cloning very large pieces of genomic DNA for DNA sequencing and analysis. These cloned pieces are often too large to introduce and maintain in bacteria. Artificial yeast chromosomes can be built, however, which contain the cloned DNA from another organism. These artificial chromosomes can be maintained in yeast and sequenced to produce very long contiguous DNA sequences of genomic DNA from other organisms, including human DNA. The sequencing of the *S. cerevisiae* genome itself was completed in 1996.

The Fruit Fly (*Drosophila melanogaster*)

In the early stages of genetic research at the beginning of the twentieth century, techniques did not yet exist for doing genetic work with microorganisms or yeast. At that time, the organism of preference was the fruit fly, *Drosophila melanogaster* (**fig. 1.19**). Thomas Hunt

figure 1.19 Adult female fruit fly, *Drosophila melanogaster*. Mutations that disrupt this wild-type eye color, bristle type and number, and wing characteristics are easily visible.

Morgan pioneered the use of *Drosophila* in the early 1900s. It was not until 2000 that the sequence of the *Drosophila melanogaster* genome was completed.

Drosophila initially became a model organism because it has a relatively short generation time (about 10 days), it survives and breeds well in the lab with very little cost of upkeep, and it has many easily observable phenotypes. For example, researchers can easily see the results of mutations that affect eye color, bristle number and type, and wing characteristics such as shape or vein pattern in the fruit fly. *Drosophila* mutants are instrumental in studying a variety of multicellular eukaryotic processes such as development, neurobiology, and behavior.

One interesting feature of *Drosophila* is the chromosomes in the larval salivary gland are very large and stain in a highly reproducible manner to give a consistent pattern of bands. Because these bands are relatively easy to identify using a microscope, it was straightforward to correlate large chromosomal changes with specific regions on the chromosome. Thus, the early studies on the effects of mutagens (substances that alter the DNA) were pioneered in *Drosophila*.

Caenorhabditis elegans, the Developing Worm

In the 1960s, Sydney Brenner was looking for a new model organism that could easily be analyzed genetically and would permit the study of organ development. The worm *Caenorhabditis elegans* proved to be an ideal organism for a variety of reasons (**fig. 1.20**). First, it is transparent, which permitted the observation of internal cell movements and organ development under a microscope. Second, it exists as two genders, a male and a hermaphrodite, which contains both male and female sexual organs. Thus, the hermaphrodite can

be either mated to a male or to itself to generate the desired genotypic progeny. Third, its small size and rapid generation time provided it with similar advantages like *Drosophila*.

In addition, the *C. elegans* genome turned out to be approximately half the size of the *Drosophila* genome. Sequencing of its genome was completed in 1998.

Because *C. elegans* is transparent, several researchers were able to demonstrate that all of the 959 cells in the adult hermaphrodite and 1031 cells in the adult male were generated by reproducible cell divisions that started with the fertilized egg (see fig. 1.20). This finding demonstrated that cells were genetically programmed to divide and take on specific identities. Mutants that altered these identities were instrumental in revealing fundamental processes that are involved in transmitting signals between cells (cell–cell signaling). Programmed cell death (apoptosis) was also identified through observing the reproducible death of specific daughter cells after cell division during development.

Arabidopsis thaliana, a Flowering Plant

In addition to bacteria and animals, plants constitute a significant and important group of organisms. *Arabidopsis thaliana*, which is a member of the cabbage family of plants, gained popularity as a genetic model system in the 1980s (**fig. 1.21**). Its small size and 6-week life cycle make it suitable for genetic experiments in the laboratory. The sequencing of its genome was completed in 2000. The *Arabidopsis* genome is approximately 5% of the size of the maize genome and 1% of the wheat genome. This small genome size is one reason *Arabidopsis* has rapidly become one of the leading plant model systems in genetics.

Mutants have been isolated that affect a variety of different specific phenotypes in flower development (see fig. 1.21), such as missing or extra flower petals. Some of these flower mutants have revealed that the different parts of a flower are under a genetic control similar to that of the different parts of a fruit fly.

The Mouse (Mus musculus)

The mouse *Mus musculus* has become another model organism for a variety of reasons (**fig. 1.22**). The sequence of the mouse genome was completed in 2002, and it appears to have the same relative number of genes as the human genome. It is estimated that 99% of the human genes have a related mouse gene. This genetic similarity is a major feature for using the mouse as a model system.

Because the mouse is a vertebrate, and also a mammal, its anatomy is more similar to humans than

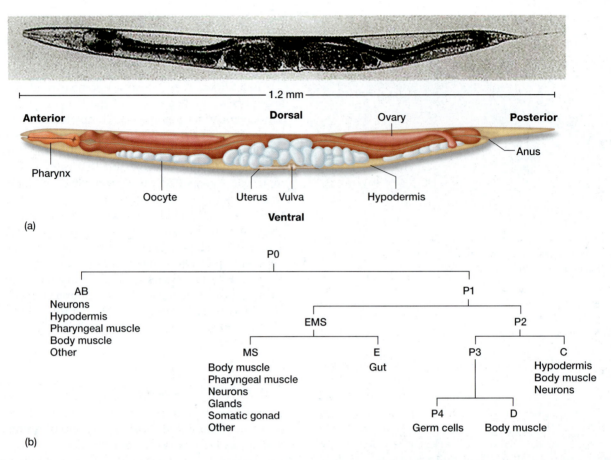

figure 1.20   Anatomy and the early cell lineage during development of *Caenorhabditis elegans.* A photograph of a *C. elegans* hermaphrodite is shown in (*a*). Below it is a schematic highlighting some of the anatomical features of the hermaphrodite. (*b*) A diagram showing the first four cell divisions of the *C. elegans* zygote. The daughter cells of each cell division are named (AB, P1, EMS, P2, MS, E, P3, C, P4, and D). Notice that the daughter cells, even after the first cell division, are already committed to becoming specific types of cells and excluded from becoming other types of cells. For example, only cells descended from the P1 cell, and not the AB cell, can become a germ cell. However, the P1 cell will produce a wide variety of different cell types that are derived from the MS, E, P4, D, and C daughter cells.

figure 1.21   The flowering weed, *Arabidopsis thaliana.* The wild-type flower (being held by a hand for scale) contains four white petals (*a*). Mutants that have an abnormal number of flower petals include *APETALA3* (*b*), which lacks petals, and *AGAMOUS* (*c*), which possesses extra petals. The additional petals are generated by other parts of the flower, such as the stamens and carpels, being transformed into flower petals.

figure 1.22   Three adult laboratory mice, *Mus musculus.* The mouse is a common research organism because it is a vertebrate like humans, and due to the variety of genetic manipulations that can be performed on it.

figure 1.23   An adult zebrafish, *Danio rerio.* Zebrafish are becoming a common research organism because of the large size of the brood and the genetic manipulations that can be employed. Furthermore, because the embryos are transparent and they develop externally, these vertebrates are rapidly becoming very popular to study questions that pertain to developmental biology, such as the genetic control of organ development.

bacteria, flies, or plants. These genetic and anatomical similarities result in many mouse mutants mimicking many inherited human diseases. For example, a number of mouse mutants have phenotypes similar to inherited human neurodegenerative diseases.

The major boost to mouse genetics was the ability to target specific mutations to any desired mouse gene, as either a *transgenic* or *knockout* mutation. These targeted mutations, which are analogous to the mutations that can be created in yeast, allowed scientists to create mouse mutations that correspond to mutations that cause human diseases. These mouse mutants may then serve as models for studying the mechanisms of the disease and to test a variety of drug therapies.

Zebrafish (*Danio rerio*)

In recent years, the zebrafish, *Danio rerio,* has become a new model organism (**fig. 1.23**). Large numbers of zebrafish, which is a vertebrate, can be maintained in a relatively small space in a laboratory. Zebrafish, unlike mice, produce large numbers of progeny (100–200 offspring per mating), which simplifies the identification of rare mutants. Because zebrafish embryos develop externally, they can be easily isolated and manipulated. Additionally, the embryos are transparent, which makes it possible to view the development of internal organs using a microscope. Thus, zebrafish are becoming a major model system for studying vertebrate development and identifying the mechanisms of organ differentiation.

Although it is currently impossible to generate targeted mutations in a zebrafish gene, as can be done in yeast or mice, it is relatively easy to block the expression of any desired protein in early zebrafish development. Furthermore, unfertilized zebrafish eggs can readily be manipulated to generate individuals that are diploid,

but contain only the maternal genome (via parthenogenesis). Although these individuals are not clones of the mother—that is, they are not genetically identical—they do permit the analysis of recessive mutations that are found in heterozygous females.

1.4 Application of Genetics to the Human Population

As you read the following chapters, you may come to realize that the field of genetics is more than just a class that you are taking. It has direct effects on your current life, and it is likely to become even more important in the future.

Treatment of Disease

The use of genetics to treat diseases will likely take two major paths in the future. First, clinical tests currently focus on introducing a wild-type ("normal") gene into the somatic cells of an individual who lacks this allele to restore the activity of the encoded protein. Although some trials have met with a degree of success, others have not; a few individuals have even died from complications of the treatment. Diseases that result from the loss of a single protein, such as cystic fibrosis, Duchenne's muscular dystrophy, Gaucher's disease, severe hemophilia, familial hypercholesterolemia, and some forms of cancer, have the greatest chance of being successfully treated by gene therapy.

One of the greatest gene therapy successes to date is the treatment of severe combined immunodeficiency

figure 1.24   A boy with severe-combined immuno-deficiency disease (SCID). Individuals with SCID lack an immune system and formerly had to live in a germ-free chamber, such as this plastic enclosure. David Vetter, shown in this picture, lived to the age of 12. One form of SCID is due to a mutation in a single gene that encodes the adenosine deaminase enzyme. Gene therapy introduces a wild-type adenosine deaminase gene into SCID patients to partially restore their immune system and allow them to live a relatively normal life.

("bubble-boy" disease, **fig. 1.24**), which results in the complete loss of the individual's immune system due to the absence of the functional enzyme adenosine deaminase (ADA). In 1990, Ashanti DeSilva was the first person with this rare condition to be treated by gene therapy. The wild-type *ADA* gene was introduced into some of her isolated T cells, which were then reintroduced in her body. Some of these T cells migrated to Ashanti's bone marrow, where they began dividing and producing ADA. Although the ADA levels are still below normal, they are sufficiently high to allow Ashanti to lead a fairly normal life.

Since that time, several setbacks have occurred in gene therapy trials, which have resulted in some deaths. In most cases, the deaths were due to the virus that was used to introduce the gene into the cells. In one case, the virus caused a massive inflammatory response in the patient. In a few other cases, the viral genome was inserted into the patient's genome at a location that caused a leukemia-like disorder. Currently, scientists are testing new viruses and alternative methods to introduce the wild-type copy of the gene into the patient's cells.

Because the gene is introduced into somatic cells, a treated individual still possesses the defective form of the gene in his or her germ cells and can pass it on to any offspring. Thus, the treatment must be reinitiated with every individual that suffers from the disease.

The second use of genetics involves complex diseases such as some forms of cancer, neurodegenerative diseases, and inherited forms of emotional and behavioral disorders. These diseases do not result from a single defective gene, but rather the clinical symptoms are more likely the consequences of the summation of several different mutations. Furthermore, the mutated genes may be different from one individual to the next. One promise of genomics and bioinformatics is the ability to determine the specific molecular defects associated with each individual suffering from a specific disease. This knowledge may allow the development of drug therapies that are tailored to the specific needs of an individual, rather than the current generic treatments that work well on some individuals and poorly on others.

Food Supply

We are already seeing the "fruits" of genetic research in the foods we eat. A variety of crops have been genetically modified to become herbicide-resistant, allowing farmers to kill the weeds that can compete with the food crop and increase the yield of the crop in more efficient ways. Crops are also being engineered to delay their spoiling, to increase their ability to thrive in more extreme climates, and to increase their nutrient content. Similar work is also being done on animals to increase milk and meat production.

Many individuals have expressed concern about these *genetically modified organisms* (*GMOs*) and the potential dangers they may pose for the environment and for human health. As a result, exhaustive testing must be done before exposing the public to these products to ensure their safety. In addition, studies have been undertaken to discover any effect farmed GMOs might have on the environment.

Genetic Counseling

A number of different tests are currently available to determine whether an unborn fetus will suffer from a specific inherited disease that exists in the families of the parents. For example, cystic fibrosis is an inherited disease that results in a thickened mucus in the lung that leads to life-threatening infections. Cystic fibrosis also can affect the pancreas, which blocks the efficient absorption of food. Individuals with cystic fibrosis will usually suffer with the symptoms until they die at approximately 30 years of age.

Individuals who have a history of cystic fibrosis in their family can have a simple blood test to determine if they possess the defective allele. If both parents are found to be carriers, a blood test can be performed on the fetus to determine if it is homozygous for the defective allele. A negative result is sure to be a relief; however, if the test reveals that the child will have the disease, then the couple may have to make an agonizing decision about whether to terminate the pregnancy or to allow the child to be born knowing that it will suffer from the symptoms of cystic fibrosis.

Science does not make the decision on what the couple should do; rather, science helps people to be better informed. Debates are ongoing about whether genetic testing should be done for diseases for which no available treatment exists. This discussion takes place in the context of ethics, philosophy, religious beliefs, and even public policy.

Concerns About Genetic Testing

Although we often think about genetic tests being performed on an unborn fetus to determine the presence or absence of a particular genotype, genetic testing can also be done on adults. For example, genetic tests are currently available to determine your predisposition to several different forms of cancer, Huntington disease, and early-onset familial Alzheimer disease to name a few.

Many of these tests only reveal a predisposition or likelihood for developing the disease. Thus, an individual who has a genetic test for colorectal cancer may test positive but never develop the cancer due to their diet or other genetic influences. However, a positive genetic test result for colonrectal or breast cancer may result in the individual being more vigilant in getting annual check-ups, which may lead to an early detection of the cancer and a greater chance of survival.

Recently, it has become possible for an individual to take a swab of their own cheek cells and send it directly to a company to have the test performed. This eliminates the need to go to a physician, where the test results may be entered in your medical record. Because many of these tests determine only a likelihood of developing the disease, it may not be desirable to have a positive test result in your medical record where it could be interpreted as an absolute forecast that you will develop the disease.

Although genetic testing can clearly provide many advantages for our future, many concerns still remain.

As an example, the results from genetic testing could be used by insurance companies to deny an individual either medical insurance or a life insurance policy. An insurance company might argue that an individual testing positive for a genetic defect will contract a disease or may be more susceptible to acquiring a disease, and therefore the company does not want to bear the risk of covering that person.

Similarly, a company might attempt to deny hiring a person because of the individual's genotype. The company may be concerned that combined with the job condition, the individual would experience an increased likelihood of developing the predisposed condition. The company, trying to reduce its potential liability, may make it a practice not to hire individuals that have a particular genotype.

Although the arguments of the insurance company and employer may appear to be valid on an economic level, we must ensure that no individual is discriminated against based on his or her genetic composition. In the latter nineteenth and early twentieth centuries, so-called "eugenics" laws in many states allowed for compulsory sterilization of individuals deemed "socially unfit," especially the mentally or emotionally disabled from poor classes. These laws grew out of an idea that not only the people, but also their genes could be eliminated from society. Today, such action would be considered a violation of human rights. We must be vigilant that such a misuse of information does not happen again.

This debate will likely continue for some time before a standard practice of using genetic information is decided on. It is reassuring that these concerns are being noticed and debated now, before genetic information is more accessible to the general public.

In the next chapter, we begin with the details of Mendel's work and his laws of segregation and independent assortment, two concepts that began the shift in thinking that led to our modern understanding of genetics.

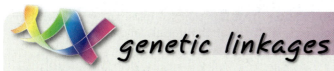

genetic linkages

The purpose of this chapter has been to provide a brief history of genetics and an overview of the book's chapters. This chapter differs from those that follow because it lacks some of the materials provided to help you with your studies: exercises and problems for review, as well as a worked-out example of the application of genetics to a problem. These features are presented chapter by chapter throughout the remainder of the book.

Many of the topics that were mentioned in this introductory chapter will be developed more fully in later chapters. Clearly, this chapter was intended to give you a taste of the breadth of genetics in this book. Thus, it is unnecessary and cumbersome to list every topic in this chapter and the chapter that they are "linked" to elsewhere in this book. However, the end of each of the three sections under "The Three General Areas of Genetics" does provide a list of topics and their associated chapters.

Chapter Summary Questions

1. Compare and contrast the concepts of preformation and epigenesis.

2. For each individual in column I, choose the best matching phrase in column II.

Column I	Column II
1. William Bateson	A. Provided algebraic models to understand population genetics
2. Francis Crick	
3. Rosie	
4. Trofim Lysenko	B. Coined the term *cell*
5. Robert Hooke	C. Introduced *Drosophila melanogaster* as a genetic model
6. Dolly	
7. Thomas Hunt Morgan	
8. August Weismann	D. Worked out the structure of DNA
9. Anton van Leeuwenhoek	E. First transgenic cow
10. Ronald Fisher	F. Discovered bacteria in rainwater
	G. Promoted flawed agricultural genetics
	H. Recognized the importance of the nucleus in heredity
	I. First cloned mammal
	J. Coined the term *genetics*

3. What are restriction endonucleases? How did their discovery contribute to modern genetics?

4. Define the following terms: **(a)** genomics, **(b)** transcriptomics, **(c)** proteomics.

5. Compare and contrast the three general areas of genetics.

6. List and briefly describe Gregor Mendel's two laws of inheritance.

7. What is a *pedigree* and why is it used in genetics?

8. What is a *karyotype* and why is it used in genetics?

9. Define the following terms: **(a)** allele, **(b)** haploid, **(c)** genotype, **(d)** phenotype.

10. Complete the following double-stranded DNA molecule by filling in the complementary bases and by indicating the polarity of the DNA strands with the appropriate terms.

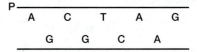

P——————————————
   A   C   T   A   G
        G   G   C   A
   ——————————————

11. List and briefly describe the three major types of cellular RNAs.

12. What is the term used to describe the flow of genetic information? Briefly discuss it.

13. Briefly describe the relationships between genes, DNA, and chromosomes.

14. Briefly describe two controversial concepts in evolutionary genetics.

15. Define the following terms: **(a)** cDNA, **(b)** PCR.

16. What are the characteristics of a model organism used in genetic research?

17. Complete the following table covering model organisms.

Genus and Species	Organism
Drosophila melanogaster	
Mus	Mouse
thaliana	
Escherichia	
	Baker's yeast
Danio rerio	
Caenorhabditis	Developing worm

18. Briefly describe two important ways that genetics affects our lives.

19. What is gene therapy? List some problems associated with the technique.

20. Briefly describe two concerns raised by genetic testing.

 Do you need additional review? Visit **www.mhhe.com/hyde** for practice tests, answers to end-of-chapter questions and problems, interactive exercises, and animation tutorials all designed to enhance your understanding of key genetic concepts.

Mendelian Genetics

Essential Goals

1. Understand Mendel's first law of equal segregation and manipulate data for the inheritance of a single trait.

2. Understand Mendel's second law of independent assortment and manipulate data for the inheritance of multiple traits.

3. Apply the concepts of dominance and recessiveness to interpret inheritance patterns.

4. Be able to apply the rules of probability to solve genetic questions.

5. Use the chi-square test to statistically analyze genetic data.

Chapter Outline

Photo: A family that demonstrates the visible similarities and differences between individuals.

Genetics is concerned with the transmission, expression, and evolution of genes, the molecules that control the function, development, and ultimate appearance of individuals. Gregor Mendel ushered in the field of modern genetics in the mid-1800s. Using the simple pea plant, he elucidated the basic rules that govern the transmission of genetic information. These rules hold true for everything from simple molds to human beings.

Although genetics is critical in the development and characteristics of an individual, the environment also plays a key role. For example, individuals raised in an environment that permits plenty of nutritious food will often grow larger and stronger than individuals raised on a severely restricted diet. Sometimes it is difficult to separate the importance of genetics from that of the environment in defining the traits of an organism.

Although genetics can be used to explain the underlying cause of many human traits, only a relatively small number of these traits exhibit a simple pattern of inheritance. However, most of these single-gene traits are associated with diseases. Some of these, such as sickle cell disease, result from recessive mutations; others, such as Huntington disease, from dominant mutations.

It is essential to understand the mode of inheritance of these diseases for two major reasons. First, this information allows us to calculate the probability of a disease being inherited in a future generation. This information can be very important to a couple who are contemplating starting a family and know that an inherited disease exists somewhere in their family tree. Second, the mode of inheritance will significantly affect our understanding of the underlying mechanism of the disease and potentially how to treat it through gene therapy.

Thus, as science begins to generate specific therapies for a variety of inherited diseases in the coming century, it will be essential to understand the underlying mode of inheritance. In this section of the book, we will look at the laws of transmission that govern genes and affect their passage from one generation to the next. These laws, which were discovered by Gregor Johann Mendel, allow us to infer the underlying mode of inheritance of the disease.

2.1   Mendel's Experiments

Gregor Mendel

Johann Mendel was born on July 22, 1822, in Heinzendorf, Austria (**fig. 2.1**). He was the second child of Anton and Rosine Mendel, who were farmers. Being a very bright student, Johann's parents wished for him to pursue higher education, but their finances were very limited. In 1843, Johann entered the Augustinian monastery of St. Thomas in Brno, in what is now the Czech Republic, where he continued his education. It was at this time that he took the name of Gregor. In August, 1847, he was ordained into the priesthood.

figure 2.1
Gregor Johann Mendel (1822–1884).

Mendel began teaching in a secondary school in 1849. He was fascinated with science and nature. In particular, he was interested in physics, theories of evolution, botany, and the natural sciences. In 1851, Mendel was sent to the University of Vienna to study physics and biology. Mendel considered himself to be an experimental physicist, which required analytical skills and a strong background in mathematics and statistics. This quantitative training would prove very beneficial when he began his studies in genetics. In 1854, he returned to Brno to continue his teaching and research.

It was after his return to Brno that Mendel began his classic experiments with the common garden pea plant, *Pisum sativum* (**fig. 2.2**). He also experimented with crossing mice to study if the inheritance of traits in animals differed from the process in plants. In 1866, he published his results from years of experimentally crossing pea plants, which laid the foundation for modern genetics. These experiments and his teaching of physics and natural science at the secondary school continued until 1868, when he was elected abbot of the monastery. This new responsibility demanded a significant amount of time and energy. Although he remained interested in genetics, he no longer could continue his teaching and scientific endeavors with his beloved pea plants.

Mendel died on January 6, 1884, in Brno of a kidney disorder and was buried in an Augustinian tomb in the central cemetery in Brno. A plaque near his grave reads, "Scientist and biologist, in charge of the Augustinian monastery in Old Brno. He discovered the laws of heredity in plants and animals. His knowledge provides a permanent scientific basis for recent progress in genetics."

Mendel's Experimental Design

One of the strengths of Mendel's experiments was his ability to specifically cross only the pea plants that possessed the desired traits. This was accomplished by two different methods. **Figure 2.3** shows a cross section of the pea flower that indicates the keel, in which the male and female parts develop. Because of the anatomy of the plant, pollination is easy to control.

Normally, **self-fertilization, or selfing,** occurs when pollen falls from the anther onto the stigma of the same flower before the bud opens. This *selfing* of individual plants was a useful technique that Mendel

figure 2.2   The garden pea plant, *Pisum sativum.* Notice the flower and pod colors (white and green, respectively), which are two of the seven traits that Mendel studied.

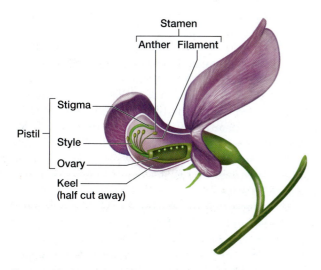

figure 2.3   Anatomy of a garden pea plant flower. The female part, the pistil, is composed of the stigma, its supporting style, and the ovary. The male part, the stamen, is composed of the pollen-producing anther and its supporting filament.

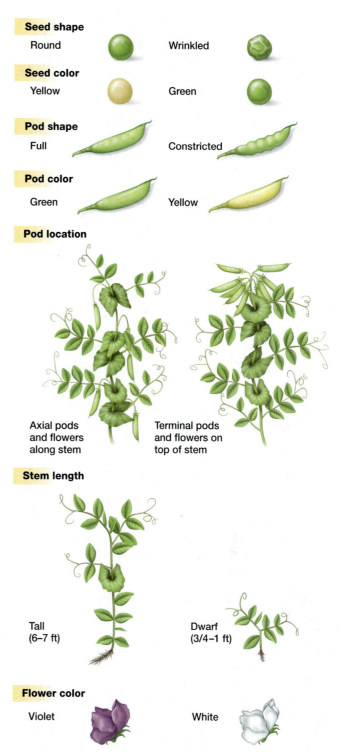

figure 2.4   Seven characteristics that Mendel observed in peas. Phenotypes in the left column are dominant, and the ones in the right column are recessive.

exploited in generating the purebred plants. The second method involves **cross-fertilization,** or **cross-breeding,** in which the pollen of one plant is used to fertilize a different plant. Mendel cross-fertilized the plants by opening the keel of a flower before the anthers matured and removed them to prevent self-fertilization. Mendel then collected pollen from the removed anther and placed it on the stigma of a second plant. In the more than 10 thousand plants Mendel examined, only a few were fertilized other than the way he had intended (either self- or cross-fertilized).

Another important factor in the success of Mendel's experiments was that he selected discrete, nonoverlapping characteristics and then observed the distribution of these characteristics over the next several generations. He chose to study the seven characteristics shown

in **figure 2.4**. Take as an example the characteristic of plant height (characteristic 6). Although height is often exhibited as a continuous range of values, Mendel used plants that displayed only two alternatives: tall or dwarf. By using nonoverlapping characteristics, he was able to unambiguously describe each plant.

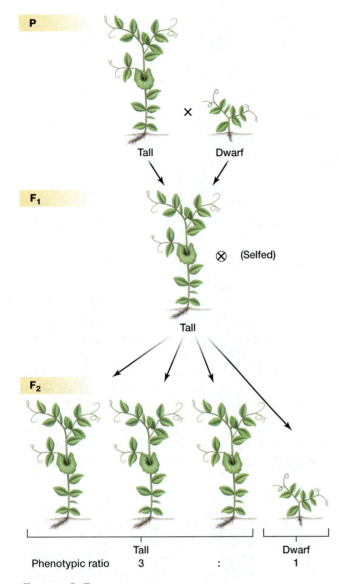

figure 2.5 First two offspring generations (F_1 and F_2) from the cross between a tall and dwarf plant. The recessive phenotype (dwarf) vanishes in the F_1 generation and reappears in 25% of the F_2 plants.

Before Mendel started his actual experiments, he grew the plants for 2 years. During this time, he identified plants that were homogeneous, or pure-breeding, for each of the particular characteristics he wanted to study. Knowing the plants were pure-breeding allowed Mendel to perform his crosses with confidence and yielded data that could be easily interpreted.

Let us look at one of Mendel's crosses, where he crossed tall and dwarf plants (**fig. 2.5**). In the **parental generation,** or P, pure-breeding dwarf plants pollinated pure-breeding tall plants. Offspring of this cross are referred to as the **first filial generation,** or F_1. Mendel also referred to these F_1 individuals as **hybrids** because the offspring were a mixture from parents

with different traits. We will refer to these offspring as **monohybrids** because they are hybrid for only one characteristic (height).

Because all the F_1 plants were tall, Mendel referred to tallness as the **dominant** trait. The alternative, dwarfness, he referred to as **recessive.** Notice that when crossing a pure-breeding dominant individual with a pure-breeding recessive individual, all the F_1 progeny express the dominant trait. Mendel wondered what happened to the dwarf trait in the F_1 generation.

When the F_1 offspring of figure 2.5 were self-fertilized to produce the **second filial generation,** or F_2, both tall and dwarf offspring occurred; the recessive (dwarf) characteristic reappeared. Among the F_2 offspring, Mendel observed 787 tall and 277 dwarf plants for a ratio of 2.84:1. Mendel recognized that this ratio approximated a 3:1 ratio, which suggested to him the mechanism of inheritance at work in pea plant height.

The mechanism that Mendel proposed involved units that were inherited to produce the trait. Mendel did not recognize that these units were **genes.** The term *gene* would first be used in 1909, by Danish botanist Wilhelm Johannsen, 43 years after Mendel published his results. Each trait studied by Mendel, such as plant height, was controlled by a gene. Mendel also proposed that each trait required two related, but different, determinants that we refer to as alleles. **Alleles** represent different forms of a gene. Although a single gene controls the height of pea plants, the gene exists in two forms or alleles: *D* (tall, dominant) and *d* (dwarf, recessive).

Phenotype refers to the observable characteristics of an organism. One allele (*D*) confers the dominant phenotype (tall), and the other allele (*d*) causes the recessive phenotype (dwarf). The combination of alleles that an organism possesses is called the **genotype.** In **figure 2.6,** the genotype of the pure-breeding parental tall plant is *DD;* that of the F_1 tall plant is *Dd.* Thus, two different genotypes can produce the dominant phenotype. Genotypes may be either **homozygous,** in which both alleles are the same (*DD* or *dd*), or **heterozygous,** in which the two alleles are different (*Dd*). William Bateson coined these last two terms in 1902.

Restating the Concepts

▶ Mendel proposed that an organism carries two forms of a genetic unit, which we now call the alleles of a gene.

▶ In homozygous individuals, both alleles are identical. In monohybrid or heterozygous individuals, the two alleles are different.

▶ A dominant trait is exhibited in the monohybrid individuals in the F_1 generation. A recessive trait is

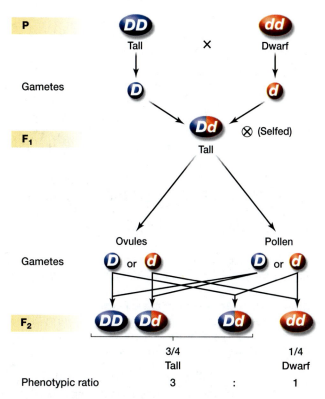

figure 2.6 Assigning genotypes to the cross in figure 2.5. When crossing homozygous tall and dwarf plants, each parent produces only one type of gamete, which fuse to produce only heterozygous (*Dd*) tall F_1 plants. Selfing the F_1 plants produces both tall (*DD* and *Dd* genotypes) and dwarf (*dd*) plants because both parents produce two different types of gametes that can combine in four different ways. Loss of the dwarf phenotype in the F_1 generation and its reappearance in the F_2 generation is consistent with a reccessive phenotype.

are inherited from generation to generation constitutes Mendel's first principle, the law of segregation.

Law of Segregation

The **law of segregation** states that during gamete formation, the two alleles separate (segregate) randomly, with each gamete having an equal probability of receiving either allele. **Fertilization** involves the fusion of two gametes, which reestablishes the two copies of the gene in the cell. From figure 2.6, we can see that Mendel's law of segregation explains several things:

1. The heterozygous F_1 progeny, which all have the dominant tall characteristic, get one allele from each parent.

2. The F_1 progeny are heterozygotes because they possess two different alleles.

3. The F_1 progeny possess the recessive allele (even though they are all phenotypically tall), which accounts for the reappearance of the dwarf phenotype in the F_2 generation.

4. The hybrid nature of the F_1 individuals accounts for the 3:1 ratio of tall-to-dwarf phenotype in the F_2 offspring.

You can see in figure 2.6 that the *DD* homozygote can produce only one type of gamete, which contains the dominant *D* allele, and the *dd* homozygote can similarly produce only gametes containing the recessive *d* allele. Thus, the F_1 individuals are uniformly heterozygous *Dd*. Each F_1 individual can produce two kinds of gametes in equal frequencies. These two types of gametes randomly fuse during fertilization to produce the F_2 generation. **Figure 2.7** shows three ways of picturing the gametes produced from the F_1 progeny and the various fusion arrangements to produce the F_2 individuals.

Testing the Law of Segregation

You can see in figure 2.5 that the F_2 generation would have a phenotypic ratio of 3:1, a standard Mendelian ratio for a monohybrid cross. But we would also expect a *genotypic* ratio of 1:2:1—that is, twice as many heterozygotes as either type of homozygote considered singly. (Do you see why?) The challenge is to demonstrate that this genotypic ratio exists in the F_2 offspring, when we can only observe phenotypes.

The simplest way to test the hypothesis is to self-fertilize the F_2 individuals to produce an F_3 generation, which Mendel did (**fig. 2.8**). The law of segregation predicts the frequencies of any phenotypic classes that would result.

The dwarf F_2 plants should be recessive homozygotes, and so, when selfed, they should produce only

absent in the monohybrid F_1 progeny, but reappears in the F_2 generation.

▶ The dominant to recessive trait ratio in the F_2 generation is 3:1 in a monohybrid cross.

▶ The genotype is the combination of alleles that an individual possesses. The phenotype is the characteristic the individual exhibits and depends on the genotype. One or two copies of the dominant allele produce the dominant phenotype, whereas two copies of the recessive allele yield the recessive phenotype.

2.2 Segregation

Although the genotype of an individual involves two alleles, only one of these alleles is passed on to the gamete, which is either the pollen or ovule in plants. The fusion of two gametes, or fertilization, forms a zygote that restores two alleles in the cell. The explanation of how alleles

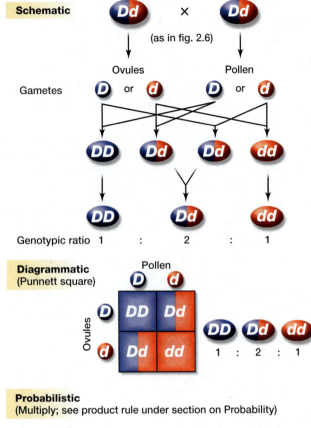

Schematic

(as in fig. 2.6)

Ovules   Pollen

Gametes

Genotypic ratio   1   :   2   :   1

Diagrammatic
(Punnett square)

Pollen

Ovules

1 : 2 : 1

Probabilistic
(Multiply; see product rule under section on Probability)

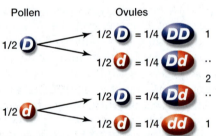

Pollen   Ovules

1/2 *D*   1/2 *D* = 1/4 *DD*   1
            1/2 *d* = 1/4 *Dd*   ··
                                   2
1/2 *d*   1/2 *D* = 1/4 *Dd*   ··
            1/2 *d* = 1/4 *dd*   1

figure 2.7   Methods of determining F$_2$ genotypic combinations in a self-fertilized monohybrid. All three approaches are based on each heterozygous parent producing 50% *D* gametes and 50% *d* gametes. With random fertilization, the F$_2$ progeny will be a 1:2:1 genotypic ratio. The Punnett square diagram is named after the geneticist Reginald C. Punnett.

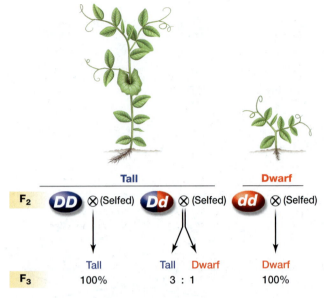

Tall   Dwarf

F$_2$ *DD* ⊗(Selfed)   *Dd* ⊗(Selfed)   *dd* ⊗(Selfed)

Tall   Tall   Dwarf   Dwarf
F$_3$ 100%   3 : 1   100%

figure 2.8   Mendel self-fertilized F$_2$ tall and dwarf plants. To determine the F$_2$ genotypes, Mendel selfed all the dwarf plants and found they produced only dwarf progeny, which is consistent with dwarf being a homozygous genotype. When he selfed the tall F$_2$ plants, 2/3 produced both tall and dwarf F$_3$ progeny in a 3:1 ratio, which is consistent with these tall F$_2$ plants being heterozygous. The remaining tall F$_2$ plants only produced tall F$_3$ progeny, suggesting these F$_2$ plants were homozygous.

d-bearing gametes and only dwarf offspring in the F$_3$ generation. The tall F$_2$ plants, however, should be a genotypically heterogeneous group: one-third should be homozygous *DD*, and two-thirds heterozygous *Dd*. The tall homozygotes, when selfed, should produce only tall F$_3$ offspring (genotypically *DD*). However, each F$_2$ heterozygote, when selfed, should produce tall and dwarf offspring in a ratio identical to that produced by the selfed F$_1$ plants: three tall to one dwarf.

Mendel found that all the dwarf (homozygous) F$_2$ plants bred true as predicted. Among the tall, 28%

(28/100) bred true (produced only tall offspring) and 72% (72/100) produced both tall and dwarf offspring, which was very close to the predicted one-third (33.3%) and two-thirds (66.7%), respectively. We thus can conclude that Mendel's progeny-testing experiment confirmed his hypothesis of segregation.

Another way to test the segregation law is to use a **testcross,** which crosses any organism with a recessive homozygote. Another type of cross, a **backcross,** refers to crossing progeny with a parent or an individual with the parental genotype. When the parent has the homozygous recessive genotype, a backcross is also a testcross.

Because the gametes of the recessive homozygote contain only recessive alleles, the alleles carried by the gametes of the other parent will determine the phenotypes of the offspring (**fig. 2.9**). This testcross can be used to distinguish the genotype of a phenotypically dominant individual. If the tested individual has a homozygous dominant genotype, then the testcross will only produce progeny with the dominant phenotype. In contrast, if the tested individual is heterozygous, the testcross will produce progeny of which 50% will be phenotypically dominant and 50% will be recessive (see fig. 2.9). A testcross of the tall F$_2$ plants in figure 2.6 would produce the results shown in **figure 2.10**. These results further confirm Mendel's law of segregation.

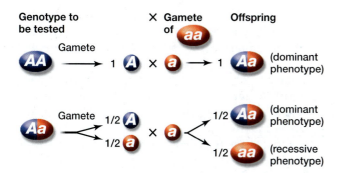

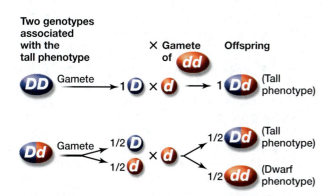

figure 2.9 Testcross. In a testcross, the genotype of an individual with the dominant phenotype is deduced by mating with a homozygous recessive individual. If all the offspring have the dominant phenotype, then the dominant parent is homozygous. If half the offspring are phenotypically dominant and half are recessive, then the dominant parent must be heterozygous.

figure 2.10 Testcrossing the phenotypically dominant F₂ individuals from figure 2.6. The phenotypically tall F₂ plants are of two genotypes. The homozygous tall F₂ plants will only produce tall progeny in the testcross. The heterozygous tall F₂ plants will produce an equal number of tall and dwarf progeny in the testcross.

Restating the Concepts

▶ The law of segregation states that each allele in a heterozygous individual has an equal probability of being present in the gamete.

▶ The F₂ progeny of a monohybrid cross will have a 1:2:1 genotypic ratio and a 3:1 phenotypic ratio.

▶ A testcross will reveal if an individual with the dominant phenotype is a dominant homozygote or a heterozygote.

2.3 Independent Assortment

Mendel also analyzed the inheritance pattern of two traits simultaneously. For example, he examined plants

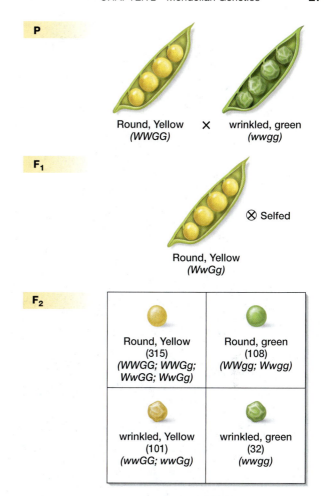

figure 2.11 Independent assortment in garden peas. In one of Mendel's experiments, he crossed a pure-breeding plant with round and yellow seeds with a pure-breeding plant with wrinkled and green seeds. The F₁ plants had round and yellow seeds, which corresponds to the two dominant phenotypes. When he selfed the F₁ plants, he isolated F₂ plants with four different phenotypes in the numbers shown.

that differed in both the form and color of their peas. He crossed homozygous plants that produced round, yellow seeds with plants that produced wrinkled, green seeds. Mendel's results appear in **figure 2.11**. The F₁ plants all had round, yellow seeds, which demonstrated that round was dominant to wrinkled and yellow was dominant to green.

When these F₁ plants were self-fertilized, they produced an F₂ generation of plants that had all four possible combinations of the two seed characteristics:

315 plants with round, yellow seeds
108 plants with round, green seeds
101 plants with wrinkled, yellow seeds
32 plants with wrinkled, green seeds.

Dividing the number of plants by 32 (the number in the smallest group) gives a 9.84 to 3.38 to 3.16 to 1.00

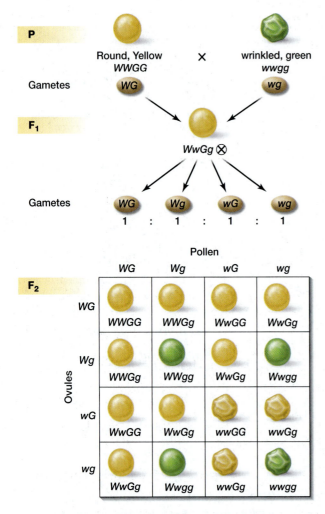

figure 2.12   Assigning genotypes to the cross in figure 2.11. The cross of the plant with yellow, round seeds with a plant possessing wrinkled, green seeds produces F_1 plants with round and yellow seeds. These double heterozygous F_1 plants can produce gametes with four different genotypes. Using a Punnett square to sort the F_2 genotypes, we observe four different phenotypes.

ratio, which is very close to a 9:3:3:1 ratio. This is the ratio we would expect if the genes governing these two traits behaved independently of each other, as the following discussion explains.

In figure 2.11, the letter W is assigned to the dominant allele, round, and w to the recessive allele, wrinkled; G and g are used for yellow and green color, respectively. In **figure 2.12**, we rediagrammed the cross in figure 2.11. The P generation plants in this cross produce only one type of gamete each, WG from the homozygous dominant parent and wg from the homozygous recessive parent. The resulting F_1 plants are heterozygous for both genes **(dihybrid).** Self-fertilizing the dihybrid ($WwGg$) produces the F_2 generation.

figure 2.13
Reginald C. Punnett (1875–1967).

One way to visualize the different gamete fusions that can occur to produce the F_2 generation is to use a **Punnett square** (invented by Reginald C. Punnett, **figure 2.13**). The Punnett square requires a critical assumption: The four types of gametes from each dihybrid parent will be produced in equal numbers, and hence every offspring category, or "box," in the square is equally likely.

Grouping the 16 different boxes (F_2 offspring) in the Punnett square by phenotype, we find 9/16 have round, yellow seeds; 3/16 have round, green seeds; 3/16 have wrinkled, yellow seeds; and 1/16 have wrinkled, green seeds. This is a graphical method of demonstrating the expected 9:3:3:1 F_2 ratio, which is the foundation for Mendel's law of independent assortment.

Law of Independent Assortment

This 9:3:3:1 ratio comes about because the alleles of two different genes are inherited independently. The F_1 plants produce four types of gametes (see fig. 2.12): WG, Wg, wG, and wg. These gametes occur in equal frequencies. Regardless of which allele for seed shape a gamete ends up with, it has an equal chance of getting either of the alleles for color—the two genes are segregating, or assorting, independently. This is the essence of Mendel's second law, the **law of independent assortment,** which states that alleles for one gene can segregate independently of alleles for other genes.

If Mendel's laws are functioning properly, we should observe Mendel's first law in action within the dihybrid cross that demonstrates the second law. Looking only at the seed shape phenotype in figure 2.11, self-fertilizing the heterozygous F_1 plants with round seeds (Ww) produced 315 + 108 plants with round seeds (WW or Ww), and 101 + 32 plants with wrinkled seeds (ww) in the F_2 generation. This phenotypic ratio is 423:133 or 3.18:1.00—very close to the expected 3:1 ratio. So the gene for seed shape is segregating normally. In a similar manner, looking only at the seed color phenotype, we see that the F_2 ratio of yellow to green seeds is 416:140, or 2.97:1.00—again, very close to a 3:1 ratio. Thus, the equal segregation of two genes (first law) allows their independent assortment (second law).

From the Punnett square in figure 2.12, you can see that because of dominance, each phenotypic class except the homozygous recessive one—wrinkled, green seeds—is actually composed of several genotypes. For example, the double dominant phenotypic class, with round, yellow seeds, represents four genotypes: $WWGG$, $WWGg$, $WwGG$, and $WwGg$. When we group all the genotypes by phenotype, we obtain the ratio shown in **figure 2.14**.

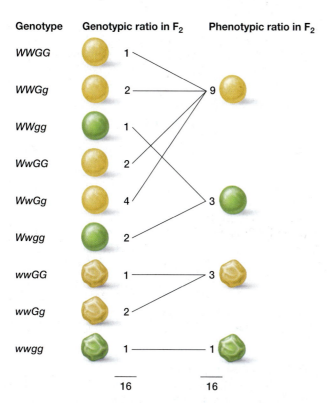

figure 2.14   The phenotypic and genotypic ratios of the offspring of dihybrid pea plants. The nine different genotypes observed in the F₂ generation from figure 2.12 are shown on the left, along with the relative number of each genotype. The genotypes can be grouped based on the resulting phenotypes, which produce the 9:3:3:1 phenotypic ratio on the right.

With complete dominance, then, a self-fertilized dihybrid gives a phenotypic ratio in the F₂ generation of 9:3:3:1 for four different phenotypes. This ratio results from a genotypic ratio of 1:2:1:2:4:2:1:2:1 for nine different genotypes. Can you see why? Look at the *WWGg* genotypic group: This genotype includes both the *WWGg* and *WWgG*—they are equivalent. Likewise, the *WwGg* genotype includes *WwGg*, *WwgG*, *wWGg*, and *wWgG*.

Testing the Law of Independent Assortment

A simple test of Mendel's law of independent assortment can be made by testcrossing the dihybrid plant. Remember when we testcrossed a monohybrid, we observed a 1:1 phenotypic ratio of dominant: recessive (see fig. 2.10). We would then predict that if we testcrossed a *WwGg* F₁ individual with a *wwgg* individual the progeny would include four phenotypes in a 1:1:1:1 ratio as shown in **figure 2.15**. Mendel's data verified this prediction. Thus, each allele for a given gene possesses an equal probability of being present in a gamete along with each allele of a second, independently assorting gene.

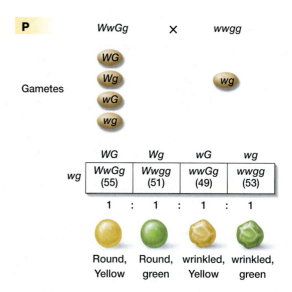

figure 2.15   Testcross of a dihybrid to prove Mendel's law of independent assortment. A dihybrid plant (*WwGg*) can produce four different gametes that are all equally likely based on independent assortment. Testcrossing this dihybrid parent would produce progeny with four different phenotypes that are equally likely. When Mendel performed this testcross, he observed the four different phenotypes in the numbers shown in the parentheses. He observed the expected 1:1:1:1 phenotypic ratio in the offspring.

Restating the Concepts

▶ Mendel's second law of independent assortment states that the alleles of one gene will segregate independently of the alleles at another gene.

▶ The F₂ progeny of a dihybrid cross will have a 1:2:1:2:4:2:1:2:1 genotypic ratio and a 9:3:3:1 phenotypic ratio.

2.4 Genetic Symbol Conventions

Throughout the last century, botanists, zoologists, and microbiologists have adopted different methods for naming alleles. Botanists and mammalian geneticists tend to prefer the uppercase–lowercase scheme to demonstrate dominant and recessive alleles, respectively. This nomenclature scheme has already been introduced for Mendel's alleles.

Microbiologists and geneticists who study *Drosophila melanogaster* (fruit fly) by contrast, have adopted schemes that relate to the **wild-type** phenotype, which is the characteristic that is most commonly found in nature. For *Drosophila*, the wild-type has red eyes and flat, oblong wings (**fig. 2.16**). Alternatives to the wild-type are referred to as **mutants.** Thus, red eyes are wild-type, and white eyes are mutant. Several different wing

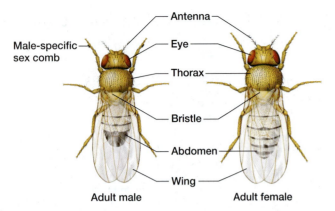

figure 2.16 Wild-type fruit fly, *Drosophila melanogaster*. Several of the morphological structures in male and female fruit flies are shown. The sex combs, which are only present on the male forelegs, are also shown.

table 2.1 Some Mutants of *Drosophila*

Mutant Designation	Description	Dominance Relationship to Wild-Type
abrupt (*ab*)	Shortened, longitudinal, median wing vein	Recessive
amber (*amb*)	Pale yellow body	Recessive
black (*b*)	Black body	Recessive
Bar (*B*)	Narrow, vertical eye	Dominant
dumpy (*dp*)	Reduced wings	Recessive
Hairless (*H*)	Various bristles absent	Dominant
white (*w*)	White eye	Recessive
white-apricot (*w^a*)	Apricot-colored eye (allele of white eye)	Recessive

mutants are shown in **figure 2.17**. *Drosophila* genes are usually named after the mutant phenotype, beginning with a capital letter if the mutation is dominant, and a lowercase letter if it is recessive. Figure 2.17 and **table 2.1** give some examples.

The wild-type allele often carries the symbol of the mutant with a plus sign ($^+$) added as a superscript. By definition, every mutant has a wild-type allele as an alternative. For example, w stands for the white-eye allele, a recessive mutation. The wild-type (red eyes) is thus assigned the symbol w^+. *Hairless* is a dominant allele with the symbol H. Its wild-type allele is denoted as H^+.

Sometimes geneticists use the + symbol alone for wild type, but only when there will be no confusion about its use. If we are discussing eye color only, then + is clearly the same as w^+: both mean red eyes. However, if we are discussing both eye color and bristle mor-

phology, the + alone could refer to either of the two phenotypes and should be avoided.

In some cases, the gene is designated by more than one letter. For example, the *Drosophila* mutant *wingless* is represented as wg and the wild type as +. By using more than one letter, we can easily differentiate between the white-eyed mutant (w) and the wingless mutant (wg). In some cases, the same letters are used to represent different genes. For example, the dominant *Eyeless* mutant is designated as Ey (notice that only the first letter is uppercase). A recessive *eyeless* mutant at a different gene is represented as ey. In both cases, the wild-type allele is shown as +. Thus, the combination of letters and the upper- or lowercase of the first letter are both important in properly designating the allele of a specific gene.

For some genes, multiple mutant alleles exist. In these cases, the mutant alleles are distinguished by either a superscript number (w^1) or a superscript letter(s) that may further define the mutant phenotypes. For example, two *white* alleles are designated w^a (*white-apricot*) and w^c (*white-crimson*) for their yellowish-orange and light reddish-orange pigmentation, respectively. We will discuss multiple alleles in more detail in chapter 5.

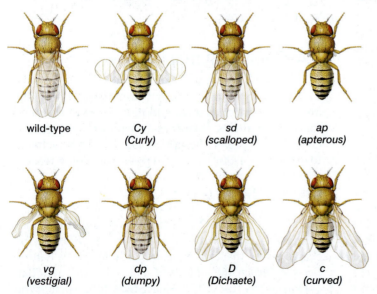

figure 2.17 Wing mutants of *Drosophila melanogaster* and their allelic designations. *Cy*, curly; *sd*, scalloped; *ap*, apterous; *vg*, vestigial; *dp*, dumpy; *D*, Dichaete; *c*, curved.

Restating the Concepts

▶ Uppercase letters denote dominant alleles and lowercase letters recessive alleles in botany and mammalian genetics.

▶ A plus sign denotes a wild-type allele and no plus sign a mutant allele in *Drosophila* genetics and microbiology. Multiple mutant alleles are shown with superscripted italic letters or numbers.

▶ The plus sign can be used alone for wild type only when no confusion will exist about its use. In this case, the uppercase mutant allele designates the mutant as dominant, and the lowercase mutant allele demonstrates that the wild-type allele is dominant.

2.5 Probability

Part of Gregor Mendel's success came from his ability to work with simple mathematics. He had the foresight to generate large numbers of progeny for his pea crosses, which he turned into ratios that allowed him to deduce the underlying mechanisms of inheritance. Taking numbers that did not exactly fit a ratio and rounding them off lay at the heart of Mendel's deductive powers. The underlying rules that allow us to make predictions for reasonable results are the rules of probability.

Scientists make predictions, perform experiments, and gather data to compare with their original predictions. Even if the bases for the predictions are correct, the data almost never exactly fit the predicted outcome. We live in a world permeated by random, or **stochastic,** events. A bright new penny when flipped in the air twice in a row will not always give one head and one tail. In fact, that penny, if flipped a hundred times, could conceivably give a hundred heads. In a stochastic world, we can guess how often a coin should land heads up, but we cannot know for certain what the next toss will bring. We can guess how often a pea should be yellow from a given cross, but we cannot know with certainty what the next pod will contain. **Probability theory** tells us what to expect from data. This section deals with some thoughts on probability to help us predict outcomes from experiments.

Types of Probabilities

The **probability** (P) that an event will occur is the number of times that the event is observed (a) divided by the total number of possible cases (n):

$$P = a/n$$

The probability can be determined either by observation (empirical) or by the nature of the event (theoretical). For example, investigators have observed that about 1 child in 10 thousand is born with phenylketonuria. Therefore, the probability that the next child born will have phenylketonuria is 1/10,000.

$$P = a/n = 1/10,000$$

The odds that a particular face of an ordinary six-sided die (singular of dice) being up (e.g., a 4) is one-sixth:

$$P = a/n = 1/6$$

The probability of drawing the seven of clubs from a deck of cards is

$$P = 1/52$$

The probability of drawing any spade from a deck of cards is

$$P = 13/52 = 1/4$$

The probability that an individual from a self-fertilized dihybrid cross will show both dominant phenotypes is

$$P = 9/16$$

From the probability formula, we can say that an event that is certain to occur has a probability of 1, and an event that is an impossibility has a probability of 0. If an event has the probability of P, all the other alternatives combined will have a probability, represented by Q, where $Q = 1 - P$. The probability of all the possible events must equal 1 or $P + Q = 1$.

Looking again at Mendel's dihybrid crosses, the probability P of exhibiting the dominant phenotype for both traits in the selfed F_2 generation is 9/16, and the probability Q of any other phenotype is 7/16. $P + Q = 16/16$, or 1.

The basic principle of probability can be stated as follows: If one event has c possible outcomes, and a second event has d possible outcomes, then there are cd possible outcomes of the two events. From this principle, we obtain two rules that concern us as geneticists: the *sum rule* and the *product rule*.

To understand these rules of probability requires a few definitions. **Mutually exclusive outcomes** are events in which the occurrence of one possibility excludes all other possibilities. For example, when throwing a die, only one face can land up. Thus, throwing a 4 precludes the possibility of any of the other faces.

Independent outcomes are events that do not influence one another. For example, if two dice are thrown, the face value of one die does not affect the face value of the other; they are thus independent of each other. With these definitions in mind, let us consider the two rules of probability that affect genetics.

Sum Rule   or (+)

When events are mutually exclusive, the **sum rule** is used: The probability that one of several mutually exclusive events will occur is the sum of the probabilities of the individual events. This rule is also known as the *either-or rule*. For example, what is the probability, when we throw a die, of its showing *either* a 4 *or* a 6? According to the sum rule,

$$P = 1/6 + 1/6 = 2/6 = 1/3$$

You should notice that the probability increases as the number of possible outcomes increases. This rule is not useful for traits that are expressed on a continuum, such as the range of heights in the adult human population.

Product Rule   and (×)

When the occurrence of one event is independent of the occurrence of other events, the **product rule** is used: The probability that two independent events will both occur is the product of their separate probabilities. This rule is also known as the *and rule*. For example, the probability of throwing a die two times and getting a 4 *and* then a 6, in that order, is

$$P = 1/6 \times 1/6 = 1/36$$

The probability will decrease as you increase the number of independent events being examined. This is in contrast to the sum rule.

Using Probabilities

You already observed these two rules in action. Remember the monohybrid cross of two tall pea plants (*Dd* × *Dd*)? You saw that the progeny were present in a 1:2:1 genotypic ratio (1 *DD*: 2 *Dd*: 1 *dd*). If we wanted to calculate the probability of the monohybrid cross producing a plant with the dominant tall phenotype, it would be the probability of being *either* homozygous dominant *or* heterozygous. This is the same as

$$P = 1/4 + 1/2 = 3/4$$

When we were discussing a dihybrid cross *WwGg* × *WwGg*, we could calculate the probability of producing a plant that has round *and* green seeds. Remember that round is a dominant phenotype (*W*), and green is a recessive (*g*) phenotype. Using the product rule and knowing that the probability of a round phenotype is 3/4 and the probability of a green phenotype is 1/4, the probability of generating a plant with round and green seeds is

$$P = 3/4 \times 1/4 = 3/16$$

It should be obvious that Mendel's law of independent assortment is based on the product rule. We can expand this analysis and apply it to more complex genotypes and questions. To do this, we will use the branch-line approach to calculate probabilities.

Branch-Line Approach to Calculate Probabilities

Mendel's first and second laws are the basis of inheritance, and the Punnett square provides a method of visualizing the production of gametes and their potential combinations to produce diploid offspring. When we discuss a dihybrid cross, the Punnett square contains 16 different diploid combinations that are produced from fusing gametes (see fig. 2.12). For a trihybrid, each gamete will have one of eight possible genotypes, which can combine to produce eight different phenotypes from 27 possible diploid genotypes (**table 2.2**). You would need a Punnett square with 64 cells (boxes) to analyze this trihybrid cross. Thus, the ability to analyze crosses

with a Punnett square becomes very cumbersome when we move beyond the dihybrid cross.

However, we can also apply the product and sum rules of probability to these more complicated genetic questions. One common method to analyze the more complicated questions is the **branch-line approach,** which is based on the product rule and Mendel's law of independent assortment. In the branch-line approach, we break the problem down to examine each gene or trait independently.

Let's see how the branched-line approach is applied in the following dihybrid cross:

$$AaBb \times AaBb$$

If we want to calculate the probability of each phenotypic class produced, we would independently calculate the probability of each trait and apply the *product rule*.

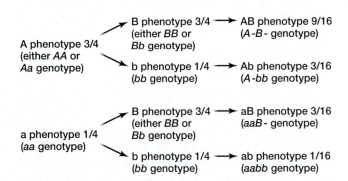

Thus, we can use the branch-line approach to generate the same 9:3:3:1 ratio from a dihybrid cross that would be produced if we used a Punnett square. This should help you realize that this approach is as valid as the Punnett square, but much simpler to calculate.

Let's see how the branched-line approach is applied to a more complex cross:

$$AaBbCc \times AabbCc$$

If we wanted to calculate the probability of each *phenotypic* class produced, we would independently calculate the probability of each trait and apply the product rule.

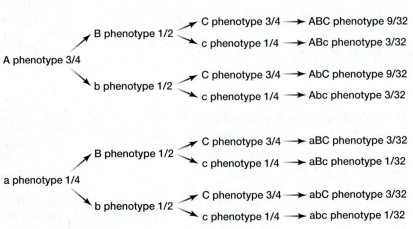

table 2.2 Multihybrid Self-Fertilization

	Monohybrid $n = 1$	Dihybrid $n = 2$	Trihybrid $n = 3$	General Rule
Number of F$_1$ gametic genotypes	2	4	8	2^n
Proportion of recessive homozygotes among the F$_2$ individuals	1/4	1/16	1/64	$1/(2^n)^2$
Number of different F$_2$ phenotypes, given complete dominance	2	4	8	2^n
Number of different genotypes (or phenotypes, if no dominance exists)	3	9	27	3^n

*n = number of genes segregating two alleles each.

2.1 applying genetic principles to problems

Calculating Probabilities from Complex Genotypes

We can use the branch-line approach and the probability rules to solve a number of different problems that involve either simple or complex genotypes. You should therefore become familiar with these approaches. Let's look at a few examples of the variety of problems that you might encounter and how these approaches can be used to solve them.

In the cross $AaBbCcdd \times AabbCcDd$, we already determined that the probability of producing an offspring that is heterozygous for all four genes would be 1/16, or 6.25%. We can also ask what the probability is that the same cross will produce an individual that is homozygous recessive for all four genes. We can break the genotype down into the four monohybrid crosses shown here:

$$Aa \times Aa = 1/4\ aa$$
$$Bb \times bb = 1/2\ bb$$
$$Cc \times Cc = 1/4\ cc$$
$$dd \times Dd = 1/2\ dd$$

We can then apply the product rule to determine that the probability of having an offspring with the *aabbccdd* genotype is $1/4 \times 1/2 \times 1/4 \times 1/2 = 1/64$, or 1.56%.

We can also calculate the probability that offspring will be phenotypically like one of the parents in the cross $AaBbCcdd \times AabbCcDd$. We can restate this as the probability of being either phenotypically like the first parent or the second parent, which means that we will use the sum rule. Because we are concerned about the phenotype, we remember that either the homozygous dominant (*AA*) or the heterozygous (*Aa*) genotype produces the dominant phenotype (A), and only the homozygous recessive genotype (*aa*) produces the recessive phenotype (a).

We first calculate the probability of producing a child that is phenotypically like the first parent, and then the probability of being phenotypically like the second parent:

Probability of producing ABCd phenotype (first parent) = $3/4 \times 1/2 \times 3/4 \times 1/2 = 9/64 = 14.06\%$

Probability of producing AbCD phenotype (second parent) = $3/4 \times 1/2 \times 3/4 \times 1/2 = 9/64 = 14.06\%$

Probability of producing a child like either parent = $9/64 + 9/64 = 18/64 = 9/32 = 28.12\%$

What is the probability of having offspring unlike either parent? The most straightforward way to address this question is to first calculate the probability of producing an offspring like either of the parents. You then deduce that everything other than one of the parents' phenotypes will be phenotypically unlike either parent.

Probability of producing a child that is phenotypically like either parent = $9/32 = 28.12\%$

Probability of producing a child that is phenotypically unlike either parent = $1 - 9/32 = 23/32 = 71.88\%$

Because of the law of independent assortment, the product rule underlies the principles of the Punnett square. Therefore, the Punnett square may be omitted in a complex cross, and the probability rules can be directly applied to calculate the desired solution.

First, we break this genotype into three independently assorting genes. Next, we determine that the cross $Aa \times Aa$ produces 3/4 of the progeny with the dominant A phenotype and 1/4 with the recessive a phenotype. The cross $Bb \times bb$ produces 1/2 of the progeny with the dominant B phenotype and 1/2 with the recessive b phenotype. Lastly, the cross $Cc \times Cc$ produces progeny that are 3/4 dominant C and 1/4 recessive c phenotypically.

You can see that the branched-line approach, which uses the product rule for independently assorting genes, allows for the rapid and direct calculation of all the possible phenotypes. A similar approach can be applied to calculate the possible genotypes. For example, we can calculate the probability of producing offspring that are heterozygous for each gene in the cross $AaBbCcdd \times AabbCcDd$.

We can break this genotype into the four independently assorting genes. In this case, the probability of $Aa \times Aa$ producing a heterozygote is 1/2. The probability of $Bb \times bb$ producing a heterozygote is 1/2. The probability of $Cc \times Cc$ yielding a heterozygote is 1/2, and the probability of $dd \times Dd$ producing a heterozygote is 1/2. Thus, the probability of producing the $AaBbCcDd$ heterozygote from the cross is $1/2 \times 1/2 \times 1/2 \times 1/2 = 1/16 = 6.25\%$.

In box 2.1, we see additional opportunities to apply the branch-line approach in genetic problems.

It's Your Turn

You have seen several examples of applying probabilities to complex genotypes. You are now ready to try calculating genotypic and phenotypic probabilities from crosses in problems 15, 16, and 22.

 Restating the Concepts

▶ The sum rule is used to examine questions that can be stated as *either* one outcome (phenotype) *or* another outcome. It is used when calculating the probability of two or more mutually exclusive events.

▶ The product rule is used to examine questions that can be stated as one outcome (phenotype) *and* another outcome. The product rule is used when calculating the probability of independent events and underlies Mendel's second law of independent assortment.

▶ The branch-line approach is based on the product rule and the law of independent assortment. A complicated genotype is broken into the individual genes and the probability of the genotype or phenotype for each gene (or trait) is calculated and then multiplied together to produce the entire genotype or phenotype.

2.6   Statistics

In one of Mendel's experiments, phenotypically tall F_1 heterozygous pea plants were self-fertilized. He found that 787 of the F_2 plants were tall and 277 were dwarf, which corresponded to a ratio of 2.84:1. Mendel interpreted this as a 3:1 ratio. From our brief discussion of probability, we expect some deviation from an exact 3:1 ratio (798:266), but how much of a deviation is acceptable? Would 780:284 (2.75:1) still support Mendel's

rule? Would 760:304 (2.5:1) support it? Would 709:355 (a 2:1 ratio)? Where do we draw the line? It is at this point that the discipline of statistics provides help.

We can never speak with certainty about stochastic events. For example, take Mendel's cross. Although a 3:1 ratio is expected on the basis of Mendel's hypothesis, chance could yield a 1:1 ratio (532:532), yet the mechanism could be the one that Mendel suggested. Conversely, Mendel could have gotten exactly a 3:1 ratio (798:266) in his F_2 generation, yet his hypothesis of segregation could have been wrong. Whenever we deal with probabilistic events, there is some chance that the data will lead us to support a bad hypothesis or reject a good one. Statistics quantify these chances and help us to analyze our data. We cannot say with certainty that a 2.84:1 ratio represents a 3:1 ratio; we can say, however, that we have a certain degree of confidence in the ratio. Statistics helps us ascertain these **confidence limits.**

Hypothesis Testing

Statistics is a branch of probability theory that helps the experimental geneticist in two major ways. First, statistics is helpful in summarizing data. Familiar terms such as *mean* and *standard deviation* are part of the body of descriptive statistics that takes large masses of data and reduces it to one or two meaningful values. We examine some of these terms and concepts further in the chapter on quantitative inheritance (chapter 24). The second way that statistics is valuable to geneticists is in the **testing of hypotheses:** determining whether to support or reject a hypothesis by comparing the data with the predicted results. This area is the most germane to our current discussion. For example, was the ratio of 787:277 really indicative of a 3:1 ratio (798:266)? Since we know that we cannot answer with an absolute yes, how can we decide to what level the data support the predicted 3:1 ratio?

To determine if the data are consistent with the hypothesis, we generate the **null hypothesis,** which assumes that the difference between the observed result and the expected result is due to chance. The statistical analysis that geneticists usually use to analyze breeding data is the chi-square test.

Chi-Square

When sample subjects are distributed among discrete categories such as tall and dwarf plants, geneticists frequently use the **chi-square distribution** to evaluate data. The formula for converting categorical experimental data to a chi-square value is

$$\chi^2 = \sum \frac{(O-E)^2}{E}$$

where χ is the Greek letter chi, O is the observed number for a category, E is the expected number for that category, and Σ means to sum the calculations for all categories.

The chi-square (χ^2) value for Mendel's data is calculated in the following manner. Mendel observed 787 tall plants and 277 dwarf plants for a total of 1064 F_2 progeny. The expected phenotypic ratio of 3:1 corresponds to 798 tall plants and 266 dwarf plants (**table 2.3**). Thus, the difference between the observed number of tall plants and the expected number of tall plants is –11, which corresponds to a $(O-E)^2$ equal to 121. The $(O-E)^2/E$ for the tall plants equals 0.15. Similarly, the $(O-E)^2$ for the dwarf plants is 121 and the $(O-E)^2/E$ for the dwarf plants equals 0.45. The chi-square (χ^2) value then is 0.60 (see table 2.3). If Mendel had originally expected a 1:1 ratio, the calculated chi-square value would be 244.45 (**table 2.4**).

However, these χ^2 values have little meaning in themselves; they are not probabilities. To properly use chi-square values, we need to correlate them to a **probability value (p)**. This is usually done with a chi-square table that contains the probability values that have already been calculated (**table 2.5**).

Reexamination of the chi-square formula and tables 2.3 and 2.4 reveals that each category (tall and dwarf plants) of data contributes to the total chi-square value, because chi-square is a summed value. We therefore expect the chi-square value to increase as the total number of categories increases. Even if the observed data fits relatively well against the expected data, the chi-square value will increase as the number of categories increases. Hence, we need some way of keeping track of categories. We can do this with degrees of freedom, which is basically a count of independent categories.

With Mendel's data, the total number of offspring is 1064, of which 787 had tall stems. Because the only other category is short stems, there must be 277 plants with short stems (1064 – 787). Thus, defining the size of one category conclusively revealed the size of the second group. For our purposes here, **degrees of freedom** equal the number of categories minus 1. Thus, with two phenotypic categories, there is only one degree of freedom.

table 2.3 Chi-Square Analysis of One of Mendel's Experiments, Assuming a 3:1 Ratio

	Tall Plants	Dwarf Plants	Total
Observed numbers (O)	787	277	1064
Expected ratio	3/4	1/4	
Expected numbers (E)	798	266	1064
$O-E$	–11	11	
$(O-E)^2$	121	121	
$(O-E)^2/E$	0.15	0.45	$0.60 = \chi^2$

table 2.4 Chi-Square Analysis of One of Mendel's Experiments, Assuming a 1:1 Ratio

	Tall Plants	Dwarf Plants	Total
Observed numbers (O)	787	277	1064
Expected ratio	1/2	1/2	
Expected numbers (E)	532	532	1064
$O-E$	255	–255	
$(O-E)^2$	65,025	65,025	
$(O-E)^2/E$	122.23	122.23	$244.45 = \chi^2$

Let us look at a second example. Assume we are studying the genotypes associated with the same cross. We know that there are three classes, homozygous dominant, homozygous recessive, and heterozygous. From the data, we know that 277 of the plants are homozygous recessive. With this information, we are unable to determine the size of the other two genotypic classes. However, if we knew that there are 275 homozygous dominant plants, then there must be 512 heterozygous plants (1064 – 277 – 275). Thus, there are two degrees of freedom in this genotypic problem. You should see that the degrees of freedom equals the total number of classes minus 1 because the size of the last class will be known if the number of individuals in all the other classes and the total number of individuals is known.

Table 2.5, the table of chi-square probabilities, is read as follows. Degrees of freedom appear in the left column. In our first problem dealing with the pea height phenotypes, we are interested in the first row, in which there is 1 degree of freedom. The numbers across the top of the table are the probabilities. We are interested in the next-to-the-last column, headed by the 0.05. This column tells us: *The probability is 0.05 of getting a chi-square value of 3.841 or larger by chance alone, given that the hypothesis is correct.* We can compare our calculated χ^2 values (0.60 and 244.45) with the **critical chi-square** (at $p = 0.05$, 1 degree of freedom) of 3.841. Because the chi-square value for the 3:1 ratio is 0.60 (see table 2.3), which is less than the critical value of 3.841, we cannot reject the hypothesis of a 3:1 ratio. Because the χ^2 for the 1:1 ratio (see table 2.4) is 244.45, which is greater than the critical value, we reject the hypothesis of a 1:1 ratio. Notice that once we did the chi-square test for the 3:1 ratio and failed to reject the hypothesis, no other statistical tests are needed: Mendel's data are consistent with a 3:1 ratio.

A word of warning when using the chi-square: If the expected number in any category is less than 5, the conclusions are not reliable. In that case, you can

table 2.5   Chi-Square Values

Degrees of Freedom	Probabilities						
	0.99	**0.95**	**0.80**	**0.50**	**0.20**	**0.05**	**0.01**
1	0.000	0.004	0.064	0.455	1.642	3.841	6.635
2	0.020	0.103	0.446	1.386	3.219	5.991	9.210
3	0.115	0.352	1.005	2.366	4.642	7.815	11.345
4	0.297	0.711	1.649	3.357	5.989	9.488	13.277
5	0.554	1.145	2.343	4.351	7.289	11.070	15.086
6	0.872	1.635	3.070	5.348	8.558	12.592	16.812
7	1.239	2.167	3.822	6.346	9.803	14.067	18.475
8	1.646	2.733	4.594	7.344	11.030	15.507	20.090
9	2.088	3.325	5.380	8.343	12.242	16.919	21.666
10	2.558	3.940	6.179	9.342	13.442	18.307	23.209
15	5.229	7.261	10.307	14.339	19.311	24.996	30.578
20	8.260	10.851	14.578	19.337	25.038	31.410	37.566
25	11.524	14.611	18.940	24.337	30.675	37.652	44.314
30	14.953	18.493	23.364	29.336	36.250	43.773	50.892

repeat the experiment to obtain a larger sample size, or you can combine categories. Note also that chi-square tests are always done on whole numbers, not on ratios or percentages.

It's Your Turn

Now that you had an opportunity to see how the chi-square test is applied to genetics data, you can apply those skills to problems 33 and 36.

Failing to Reject Hypotheses

When we test against the null hypothesis, we are assuming that there is no difference between the observed and the expected samples. If the null hypothesis is not rejected, then we say that the data are consistent with it, not that the hypothesis has been proved. (As previously discussed, it is always possible we are not rejecting a false hypothesis or are rejecting the true one.) If, however, the hypothesis is rejected, as we rejected a 1:1 ratio for Mendel's data, we fail to reject the alternative hypothesis: that there is a difference between the observed and the expected values. We may then retest the data against some other hypothesis. (We don't say "accept the hypothesis" but rather "fail to reject the hypothesis," because supportive numbers could arise for many reasons. Our failure to reject is tentative acceptance of a hypothesis. However, we are on stronger ground when we reject a hypothesis.)

The use of the 0.05 probability level as a cutoff for rejecting a hypothesis is a convention called the **level of significance.** When a hypothesis is rejected at that level ($p = 0.05$), statisticians say that the data depart *significantly* from the expected ratio. Another level of significance that is also often used is the 0.01 level, which corresponds to the data departing in a *highly significant* manner from the null hypothesis. Because the chi-square value at the 0.01 level is larger than the value at the 0.05 level, it is more difficult to reject a hypothesis at this level and hence more convincing when it is rejected.

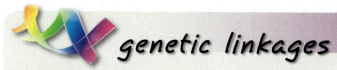
genetic linkages

In this chapter, we discussed Mendel's laws of equal segregation and independent assortment. These laws form the basis of the mode of inheritance for many eukaryotic genes. We will see in chapter 3 that the movement of chromosomes during meiosis provides the physical and cellular basis of Mendel's laws.

In chapter 5, we will see that some genes contain more than two alleles and that these alleles do not always act in either a completely dominant or recessive manner. We follow this in chapter 6 with a discussion of linkage or how genes that are located on the same chromosome are transmitted together to the same gamete through meiosis, rather than exhibiting independent assortment. Although chapters 5 and 6 may exhibit instances that do not appear to follow Mendelian patterns of inheritance,

we will see that Mendel's laws are still at work and that a higher level of complexity exists within genetics.

In our discussion of Mendel's pea plants and independent assortment, we examined how two or more genes are transmitted to affect independent traits in an organism. However, in chapter 5 a variety of phenomena reveal how two or more genes may interact to produce a common phenotype. In these cases, the standard phenotypic ratio of 9:3:3:1 (see fig. 2.14) is not observed, although the underlying law of independent assortment is not altered.

In summary, Mendel's choice of particular traits to study was fortunate because it enabled him to identify his laws. If he had selected traits that exhibit the phenomena that we will discuss in the next few chapters, the laws would not have been as obvious. However, these laws are still at work, even when they are not obvious.

Chapter Summary Questions

1. Distinguish between cross-fertilization and self-fertilization. Which method did Mendel use?

2. If you have a tall pea plant, how can you determine whether it is homozygous or heterozygous?

3. Is Mendel's second law universally true? Why or why not?

4. Complete the following table concerning three different genetic crosses.

Cross (P generation)	AABB × aabb	AABBCC × aabbcc	AABBCCDD × aabbccdd
No. of different F$_1$ gametes			
No. of different F$_2$ genotypes			
No. of different F$_2$ phenotypes			
Degrees of freedom in chi-square test			

5. If 2 black mice are crossed, 10 black and 3 white mice result.
   a. Which allele is dominant?
   b. Which allele is recessive?
   c. What are the genotypes of the parents?

6. Diagram the cross and determine which allele is dominant for each of the following *Drosophila* matings:
   a. Two red-eyed flies yield 110 red-eyed and 35 brown-eyed offspring.
   b. A dark-bodied fly and a tan-bodied fly yield 76 tan and 80 dark offspring.

7. In fruit flies, a new dominant trait, washed eye, was discovered. Describe different ways of naming the alleles of the *washed-eye* gene.

8. The following is a list of 10 genes in fruit flies, each with one of its alleles given. Are the alleles shown dominant or recessive? Are they mutant or wild-type? What is the alternative allele for each? Is the alternative allele dominant or recessive in each case?

Name of Gene	Allele
yellow	y^+
Hairy wing	Hw
Abruptex	Ax^+
Confluens	Co
raven	rv^+
downy	dow
Minute(2)e	$M(2)e^+$
Jammed	J
tufted	tuf^+
burgundy	bur

9. Define the following terms: a) chi-square test; b) null hypothesis; c) level of significance.

10. For each event in column I, choose the best matching phrase in column II.

Column I	Column II
1. Getting a heart or club, when picking one card	A. Independent outcomes
2. Having a boy, then a girl, in a family	B. Mutually exclusive outcomes
3. Showing a 2 or 3, when rolling a die	C. Nonmutually exclusive outcomes
4. Getting an ace or diamond, when picking one card	

Solved Problems

PROBLEM 1: In corn, rough sheath (*rs*) is recessive to smooth sheath (*Rs*), midrib absent (*mrl*) is recessive to midrib present (*Mrl*), and crinkled leaf (*cr*) is recessive to smooth leaf (*Cr*). What are the results of testcrossing a trihybrid?

Answer: The trihybrid has the genotype *Rsrs Mrlmrl Crcr.* This trihybrid parent produces eight different types of gametes in equal frequencies; *Rs Mrl Cr, Rs Mrl cr, Rs mrl Cr, rs Mrl Cr, Rs mrl cr, rs Mrl cr, rs mrl Cr,* and *rs mrl cr.* In a testcross, the other parent is homozygous recessive, *rsrs mrlmrl crcr,* and produces only one type of gamete (*rs mrl cr*). Thus, this cross produces zygotes with eight different genotypes (and phenotypes), one for each of the gamete types of the trihybrid parent: *Rsrs Mrlmrl Crcr* (smooth sheath, midrib present, smooth leaf); *Rsrs Mrlmrl crcr* (smooth sheath, midrib present, crinkled leaf); *Rsrs mrlmrl Crcr* (smooth sheath, midrib absent, smooth leaf); *rsrs Mrlmrl Crcr* (rough sheath, midrib present, smooth leaf); *Rsrs mrlmrl crcr* (smooth sheath, midrib absent, crinkled leaf); *rsrs Mrlmrl crcr* (rough sheath, midrib present, crinkled leaf); *rsrs mrlmrl Crcr* (rough sheath, midrib absent, smooth leaf); and *rsrs mrlmrl crcr* (rough sheath, midrib absent, crinkled leaf). Each genotype and phenotype should make up one-eighth of the total number of progeny.

PROBLEM 2: Mendel self-fertilized a dihybrid plant that had round, yellow peas. In the offspring (F_1) generation: What is the probability that a pea picked at random will be round? What is the probability that a pea picked at random will be round and yellow? What is the probability that five peas picked at random will be round and yellow?

Answer: The dihybrid plant would have a genotype of *WwGg.* The offspring peas will be 9/16 round and yellow (*W–G–*), 3/16 round and green (*W–gg*), 3/16 wrinkled and yellow (*wwG–*), and 1/16 wrinkled and green (*wwgg*). The probabil-

ity that a pea picked at random will be round is 9/16 (round and yellow) + 3/16 (round and green), which equals 3/4, or 0.75. The probability that a pea picked at random will be round and yellow is 9/16, or 0.563. The probability of picking five of these peas in a row is $(9/16)^5$, or 0.057.

PROBLEM 3: On a chicken farm, walnut-combed fowl were crossed with each other and produced the following offspring: walnut-combed, 87; rose-combed, 31; pea-combed, 30; and single-combed, 12. What might be your hypothesis about the control of comb shape in fowl? Do the data support that hypothesis?

Answer: The numbers 87, 31, 30, and 12 are very similar to 90, 30, 30, and 10, which corresponds to a 9:3:3:1 ratio. Thus, you should hypothesize that inheritance of comb type is by two genes, and that dominant alleles at both result in walnut combs, a dominant allele at one gene and homozygous recessive alleles at the other result in rose or pea combs, and homozygous recessive alleles at both genes result in a single comb. Crossing two dihybrids would then produce fowl with the four comb types in a 9:3:3:1 ratio of walnut-, rose-, pea-, and single-combed fowl, respectively.

To test this hypothesis, we must compare our observed numbers 87, 31, 30, and 10 (sum = 160), with the expected numbers for a 9:3:3:1 ratio, or 90, 30, 30, and 10 fowl.

This problem has 3 degrees of freedom because there are four categories of combs (4 – 1 = 3). The critical chi-square value with 3 degrees of freedom and a probability of 0.05 is 7.815 (see table 2.5). Because our calculated chi-square value (0.533) is less than this critical value, we cannot reject our hypothesis. In other words, our data are consistent with the hypothesis that the data represent a 9:3:3:1 phenotypic ratio that results from two genes, each possessing a dominant and recessive allele.

	Comb Type				
	Walnut	**Rose**	**Pea**	**Single**	**TOTAL**
Observed numbers (*O*)	87	31	30	12	160
Expected ratio	9/16	3/16	3/16	1/16	
Expected numbers (*E*)	90	30	30	10	160
O – *E*	–3	1	0	2	
$(O - E)^2$	9	1	0	4	
$(O - E)^2/E$	0.1	0.033	0	0.4	$0.533 = \chi^2$

Exercises and Problems

11. List all the different gametes produced by the following individuals, assuming all genes are on different chromosome pairs.
   a. *AAbbCCddEE*   **b.** *aaBBCcDDee*   **c.** *AABbCCDdEe*

12. Consider a gene with two alleles: *A* and *a*. List the different matings that
   a. could result in offspring that are heterozygous.
   b. only result in offspring that are homozygous.

13. In cats, long hair (*l*) is recessive to short hair (*L*). A short-haired female cat is crossed to a long-haired male cat. Surprisingly, only one kitten results from the mating. Can the genotype of the mother be determined if
    a. the kitten is long-haired?
    b. the kitten is short-haired?

14. Mendel crossed tall pea plants with dwarf ones. The F_1 plants were all tall. When these F_1 plants were selfed, the F_2 generation consisted of progeny in a 3:1 tall-to-dwarf ratio. Predict the genotypes and phenotypes and relative proportions of the F_3 generation produced when the F_2 generation was selfed.

15. Assume that Mendel looked simultaneously at four traits of his pea plants (and each trait exhibited dominance). If he crossed a homozygous dominant plant with a homozygous recessive plant, all the F_1 offspring would exhibit the dominant phenotype. If he selfed the F_1 plants, how many different types of gametes would be produced? How many different phenotypes would appear in the F_2 generation? How many different genotypes would appear? What proportion of the F_2 offspring would be of the fourfold recessive phenotype?

16. A geneticist crossed two corn plants, creating an F_1 deca-hybrid (10 segregating loci). She then self-fertilized this decahybrid. How many different kinds of gametes did the F_1 plant produce? What proportion of the F_2 offspring were homozygous for the recessive allele at all 10 genes? How many different genotypes and phenotypes were generated in the F_2 offspring? What would your answer be if the geneticist testcrossed the decahybrid instead?

17. To determine the genotypes of the offspring of a cross in which a corn trihybrid (*AaBbCc*) was selfed, a geneticist has three choices. He can take a sample of the progeny and (a) self-fertilize the individual plants, (b) testcross the plants, or (c) cross the individuals with a trihybrid (backcross). Explain which method is preferable?

18. Mendel self-fertilized dihybrid pea plants (*WwGg*) with round and yellow seeds and got a 9:3:3:1 ratio in the progeny. As a test of Mendel's hypothesis of independent assortment, predict the kinds and numbers of progeny produced when individual progeny are testcrossed.

19. Consider the following crosses in pea plants and determine the genotypes of the parents in each cross. Yellow and green refer to seed color; tall and dwarf refer to plant height.

	Progeny			
Cross	**Yellow, tall**	**Yellow, dwarf**	**Green, tall**	**Green, dwarf**
a. Yellow, tall × yellow, tall	89	31	33	10
b. Yellow, dwarf × yellow, dwarf	0	42	0	15
c. Yellow, dwarf × green, tall	21	20	24	22

20. A red-eyed, long-winged fly is mated with a brown-eyed, long-winged fly. The progeny are
    51 long, red   18 short, red
    53 long, brown   16 short, brown
    What are the genotypes of the parents?

21. True-breeding flies with long wings and dark bodies are mated with true-breeding flies with short wings and tan bodies. All the F_1 progeny have long wings and tan bodies. The F_1 progeny are allowed to mate and produce:
    44 long, tan   14 short, tan
    16 long, dark   6 short, dark
    What is the mode of inheritance?

22. In peas, dwarf (*d*) is recessive to tall (*D*), green (*g*) is recessive to yellow (*G*), and wrinkled (*w*) is recessive to round (*W*). From a cross of two triple heterozygotes, what is the chance of getting a plant that is
    a. tall, yellow, round?
    b. dwarf, green, wrinkled?
    c. dwarf, green, round?

23. Consider the following crosses in *Drosophila*. Based on the results, deduce which alleles are dominant and the genotypes of the parents. Orange and red are eye colors; crossveins occur on the wings.

	Progeny			
Cross	**Orange, cross-veins**	**Orange, cross-veinless**	**Red, cross-veins**	**Red, cross-veinless**
a. Orange, crossveins × orange, crossveins	83	26	0	0
b. Red, crossveins × red, crossveinless	20	18	65	63
c. Red, crossveins × red, crossveinless	0	0	74	81
d. Red, crossveins × red, crossveins	28	11	93	34

24. In *Drosophila*, a recessive gene, *ebony*, produces a dark body color when homozygous, and an independently assorting gene, *black*, has a similar effect. If homozygous *ebony* flies are crossed with homozygous *black* flies,
    a. what will be the phenotype of the F_1 flies?
    b. what phenotypic ratios would occur in the F_2 generation?
    c. what phenotypic ratios would you expect to find in the progeny of the backcrosses of F_1 × *ebony*? F_1 × *black*?

25. On average, about 1 child in every 10 thousand live births in the United States has phenylketonuria (PKU). What is the probability that
    a. the next child born in a Boston hospital will have PKU?
    b. after that child with PKU is born, the next child born will have PKU?
    c. two children born in a row will have PKU?

26. A plant with the genotype *AAbbccDDEE* is crossed with one that is *aaBBCCddee*. F_1 individuals are selfed. What is the chance of getting an F_2 plant whose genotype exactly matches the genotype of one of the parents?

27. Consider a family of four children. Calculate the probability of the following outcomes?
    a. Two girls followed by two boys
    b. Four girls
    c. Four children of the same sex
    d. Any combination other than four boys

28. Albinism and PKU are two unlinked recessive disorders in human beings. If two people, each heterozygous for both traits, produce a child, what is the chance that it will have
    a. albinism?
    b. only one of the two disorders (either albinism or PKU)?
    c. both disorders?

29. In human beings, the absence of molars is inherited as a dominant trait. If two heterozygotes have four children, what is the probability that
    a. all will have no molars?
    b. the first two will have molars and the second two will have no molars?

30. A city had 900 deaths during the year, and of these, 300 were from cancer and 200 from heart disease. What is the probability that the next death will be from
    a. cancer?   b. either cancer or heart disease?

31. In pea plants, the genes A, B, C, D and E assort independently. In the following cross, $AaBbCCDdee \times AaBbCcDdEe$, what is the probability of generating an individual who is
    a. phenotypically dominant for all five traits?
    b. phenotypically like either parent?
    c. phenotypically unlike either parent?
    d. genotypically heterozygous for all five genes?
    e. genotypically like the first parent?
    f. genotypically unlike either parent?

32. The ability to taste phenylthiocarbamide is dominant in human beings. If a heterozygous taster mates with a nontaster, what is the probability that of their five children, only one will be a taster?

33. The following data are from Mendel's original experiments. Suggest a hypothesis for each set of data and test this hypothesis with the chi-square test. Do you reach different conclusions with different levels of significance?
    a. Self-fertilization of round-seeded hybrids produced 5474 round seeds and 1850 wrinkled ones.
    b. One particular plant from a yielded 45 round seeds and 15 wrinkled ones.
    c. Of the 567 plants raised from F_2 round-seeded plants, 374, when self-fertilized, gave both round and wrinkled seeds in a 3:1 proportion, whereas 193 yielded only round seeds.
    d. A violet-flowered, long-stemmed plant was crossed with a white-flowered, short-stemmed plant, producing the following offspring:
       46 violet, long-stemmed plants
       40 white, long-stemmed plants
       37 violet, short-stemmed plants
       41 white, short-stemmed plants

34. Mendel self-fertilized pea plants with round and yellow peas. In the next generation he recovered the following numbers of peas:
    315 round and yellow peas
    108 round and green peas
    101 wrinkled and yellow peas
    32 wrinkled and green peas
    What is your hypothesis about the genetic control of the phenotype? Do the data support this hypothesis?

35. Two curly-winged flies, when mated, produce 61 curly and 35 straight-winged progeny. Use a chi-square test to determine whether these numbers fit a 3:1 ratio.

36. A short-winged, dark-bodied fly is crossed with a long-winged, tan-bodied fly. All the F_1 progeny are long-winged and tan-bodied. F_1 flies are crossed among themselves to yield 84 long-winged, tan-bodied flies; 27 long-winged dark-bodied flies; 35 short-winged, tan-bodied flies; and 14 short-winged, dark-bodied flies.
    a. What ratio do you expect in the progeny?
    b. Use the chi-square test to evaluate your hypothesis.

37. Two fair six-sided dice are rolled simultaneously. Calculate the probability of the following outcomes:
    a. Two 1's
    b. A 3 and a 5
    c. Two different numbers
    d. At least one 6
    e. No 6's

38. You are studying five unlinked genes in *Drosophila*. You set up the following cross:
    $AaBbccDdEe \times AabbCcDdEE$. Calculate the probability of producing an offspring who is
    a. genotypically heterozygous for all five genes.
    b. genotypically like one of the two parents.
    c. phenotypically like the second parent.
    d. phenotypically unlike either parent.
    e. recessive for all five traits.

39. In corn, the genotype A–C–R– is colored. Individuals homozygous for at least one recessive allele are colorless. Consider the following crosses involving colored plants, all with the same genotype. Based on the results, deduce the genotype of the colored plants.
    colored $\times$ $aaccRR \rightarrow$ 1/2 colored; 1/2 colorless
    colored $\times$ $aaCCrr \rightarrow$ 1/4 colored; 3/4 colorless
    colored $\times$ $AAccrr \rightarrow$ 1/2 colored; 1/2 colorless

40. A friend shows you three closed boxes, one of which contains a prize, and asks you to choose one. Your friend then opens one of the two remaining boxes, a box she knows is empty. At that point, she gives you the opportunity to change your choice to the last remaining box. Should you?

Do you need additional review? Visit **www.mhhe.com/hyde** for practice tests, answers to end-of-chapter questions and problems, interactive exercises, and animation tutorials all designed to enhance your understanding of key genetic concepts.

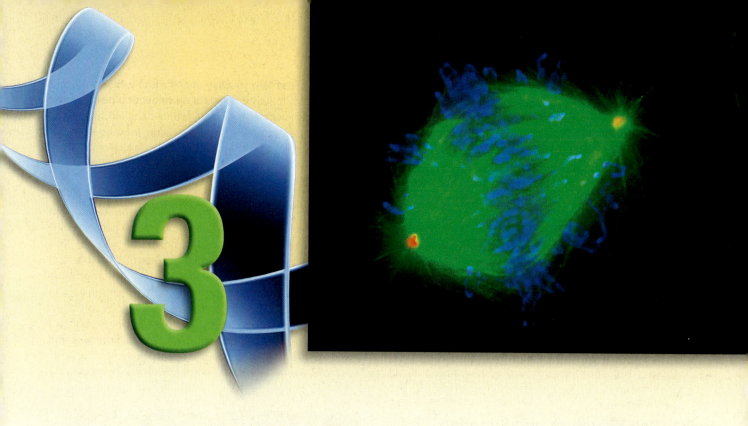

3

Mitosis and Meiosis

Essential Goals

1. Understand the key features of a chromosome and how chromosomes are classified.

2. Describe the key steps in the cell cycle and how the cycle is regulated.

3. Diagram and describe the relationship between mitosis and meiosis in both haploid and diploid cells.

4. Understand the relationship between Mendel's two laws and meiosis.

Chapter Outline

Photo: Confocal image of a BSC-1 monkey kidney epithelial cell during metaphase. The chromosomes (blue) are arranged on the metaphase plate and attached to the microtubule spindle fibers (green). The centrioles flouresce red, but appear yellow due to the close proximity of both microtubules (green) and the centrioles (red).

In this chapter, we turn our attention to critical processes in eukaryotic cell division, namely mitosis and meiosis. You are probably already familiar with these processes from other biology courses; however, the fundamental role of these two processes in understanding several important genetic concepts warrants their discussion early in this book. If you are already familiar with mitosis and meiosis, this chapter will give you a deeper appreciation as well as some new details.

Mitosis allows eukaryotic cells to reproduce their DNA with accuracy. When a cell divides mitotically, the resulting *daughter cells* each contain the same genetic material as the parental cell. Producing genetically identical daughter cells is required for the development of a multicellular organism.

When mitosis and other aspects of the cell cycle become disrupted, cancer can develop. Some cancers clearly possess a heritable component, often referred to as a *susceptibility,* and a random component. Cancer cells have altered gene expression, and they multiply uncontrollably, often invading other tissues. To fully appreciate the underlying mechanisms of cancer development, it is critical to understand several key topics of this chapter: cell cycle regulation, the process of mitosis, and **nondisjunction,** which is the failure of chromosomes to properly separate during cell division. (Cancer will be covered in depth in chapter 22).

Meiosis is different. This process produces cells, called *gametes,* that contain half of the genetic material that was present in the parental cell. Fusion of two gametes during fertilization produces an individual with a new combination of genetic information. One of the basic and critical events during meiosis is the pairing of homologous chromosomes, which allows crossing over and results in genetic diversity among the gametes.

To set the stage for the presentation of cell division, we begin this chapter with a brief review of cell structure and function, followed by the details of chromosome structure.

3.1 Cell Structure and Function

Modern biologists group organisms into three major categories: **prokaryotes,** which include bacteria and cyanobacteria, **archae,** and **eukaryotes,** which include all other organisms. Although a number of differences exist between prokaryotes and eukaryotes (**table 3.1**), the major difference is the presence of a nucleus in eukaryotes and the absence of one in prokaryotes. The nucleus (**fig. 3.1**) contains the primary genome of the cell, which is organized as linear, double-stranded DNA (deoxyribonucleic acid) that is complexed with protein (**nucleoprotein**). In most prokaryotes, the primary genetic material is a circle of double-stranded DNA with some associated proteins.

Eukaryotes contain several additional *organelles* ("small organs") that are functionally important for the survival of the cell (see fig. 3.1). For example, the **endoplasmic reticulum** is an organelle where lipid production and some protein translation occurs. The endoplasmic reticulum and the **Golgi apparatus** are the organelles where most proteins are modified to achieve their final functional structure.

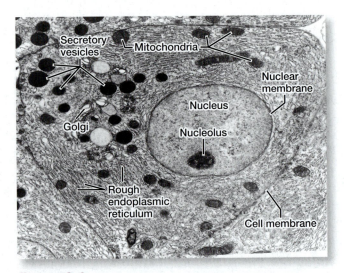

figure 3.1   Mouse lung cell magnified 4,270×.

Mitochondria are the structures where most of the cellular energy is produced in the form of adenosine triphosphate (ATP). Plants, algae, and some protozoans contain both mitochondria and chloroplasts. The **chloroplast** is the location of photosynthesis, where the

table 3.1   Differences Between Prokaryotic and Eukaryotic Cells

	Prokaryotic Cells	Eukaryotic Cells
Taxonomic groups	Bacteria, cyanobacteria	All plants, fungi, animals, protists
Size	Usually < 5 μm in greatest dimension	Usually > 5 μm in smallest dimension
Nucleus	No true nucleus, no nuclear membrane	Nuclear membrane
Genetic material	One circular molecule of DNA, little protein	Linear DNA molecules complexed with histones
Mitosis and meiosis	Absent	Present

energy of sunlight is used to convert carbon dioxide and water into sugar. The mitochondria and chloroplasts also possess their own genomic DNA. Unlike the nuclear genome, the mitochondrial and chloroplast genomes are circular and are not complexed with proteins.

Most animal and lower plant cells also contain a functional unit called the **centrosome** (see fig. 3.11). Unlike organelles, the centrosome is not bound by membranes to separate it from the surrounding cytosol. The centrosome consists of two cylindrical structures called **centrioles** (**fig. 3.2**). The centrosome serves as the organizing unit for **microtubules,** which are dynamic protein polymers composed of subunits that contain one molecule of alpha (α) tubulin and one molecule of beta (β) tubulin (**fig. 3.3**). Microtubules possess a polarity, such that their "minus," or fixed, end is associated with the centrosome and the "plus," or growing, end extends radially toward the cell's periphery. Microtubules extend and retract to provide shape and structure to a eukaryotic cell, and they form the network that internal components move along to their proper destination within the cell. The **spindle fibers** that attach to chromosomes during the early stages of mitosis and meiosis are also composed of microtubules. Thus, the centrosome is essential for the correct formation of spindle fibers and the proper movement of eukaryotic chromosomes during mitosis and meiosis. In some organisms, such as fungi, a different cell organelle, the **spindle pole body,** serves the function of the centrosome.

Restating the Concepts

▶ Eukaryotic cells contain many different organelles that perform specific functions. Of these organelles, nuclei, mitochondria, and chloroplasts contain genomes.

▶ The eukaryotic cell possesses some structures that are critical for mitosis and meiosis: centrosomes and microtubules/spindle fibers.

3.2 Chromosomes

Chromosomes were discovered by the botanist Karl von Nägeli in 1842. The term **chromosome,** which Heinrich Wilhelm Waldeyer coined in 1888, means "colored body." Von Nägeli discovered chromosomes after staining techniques were developed that made them visible. Linear eukaryotic chromosomes are composed of a complex of double-stranded DNA and protein, which is referred to collectively as **chromatin.** Chromatin can be found in either a loosely packed state (**euchromatin**) or a condensed and readily visible organization (**heterochromatin**). The functional differences between these different chromatin conformations are described in chapter 7.

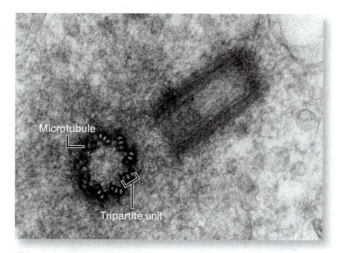

figure 3.2 A centriole is composed of two barrels at right angles to each other. Each barrel is composed of nine tripartite units and a central cartwheel. Each of the three parts of a tripartite unit is a microtubule. Magnification 111,800×.

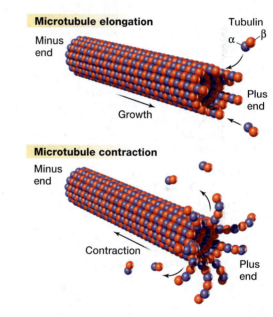

figure 3.3 Microtubules are hollow tubes made of α- and β-tubulin subunits. Microtubules grow by adding α and β tubulin dimers to the plus end, while contraction occurs by removing dimers from the plus end. The minus end is associated with the centrosome.

During the period between nuclear divisions, which is called **interphase,** the linear chromosomes are diffused throughout the nucleus and are usually not identifiable. Chromosomes are classified by the location of their spindle attachment points, which have distinct positions. The attachment point occurs at a constriction in the chromosome termed the **centromere,** which is composed of several specific DNA sequences (see chapter 7). The **kinetochore** is the proteinaceous structure on the surface of the centromere to which the microtubules attach.

Chromosomes can be classified according to whether the centromere is in the middle of the chromosome (**metacentric**), at the end of the chromosome (**telocentric**), near the end of the chromosome (**acrocentric**), or somewhere between the middle and near the end of the chromosome (**submetacentric; figs. 3.4** and **3.5**). For any particular chromosome, the position of the centromere is fixed. The location of the centromere often divides the chromosome into two parts that are referred to as the short arm (**p arm** for petite) and the long arm (**q arm;**

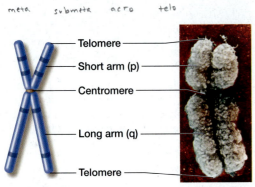

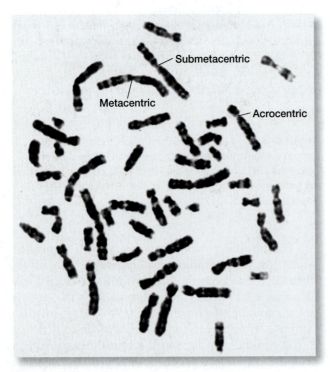

figure 3.4   Schematic of a submetacentric chromosome (a) and an electron micrograph of human chromosome 2 (b). The locations of the centromere, telomeres, short arm (p arm), and long arm (q arm) are shown. The human chromosome is best seen after DNA replication, but before the two identical double-stranded DNA molecules (sister chromatids) separate during mitosis.

figure 3.5   Metacentric, submetacentric, and acrocentric Giesma-stained chromosomes. In all three types, the centromere divides each chromosome into two arms.

see fig. 3.4). In various types of preparations, dark bands (**chromomeres**) are also visible (described in chapter 7).

Chromosome Complement

A particular species possesses a constant number of chromosomes. Most cells of eukaryotic organisms are **diploid; that is, they contain two sets of chromosomes.** In the diploid state, members of the same chromosome pair are referred to as **homologous chromosomes,** or **homologs;** the two chromosomes make up a homologous pair. One member of each pair comes from each parent.

Humans have 23 homologous chromosome pairs, which is often expressed as $2n = 46$. This expression indicates that humans are diploid ($2n$) and possess a total of 46 chromosomes. **Haploid** cells, which include some eukaryotic organisms and the reproductive cells (**gametes**), have only one set of chromosomes. The products of meiosis in humans, eggs and sperm, are haploid cells that contain one member of each homologous pair ($n = 23$).

Some species exist mostly in the haploid state or have long haploid intervals in their life cycle. For example, the pink bread mold *Neurospora crassa* has a chromosome number of seven ($n = 7$) in the haploid state. Its diploid number would then be fourteen ($2n = 14$). The diploid chromosome numbers of several species appear in **table 3.2**.

Karyotype

The total chromosomal complement of a cell can be photographed during mitosis and rearranged in pairs to make a picture called a **karyotype** (**fig. 3.6**). From the karyotype it is possible to see whether the chromosomes have any gross abnormalities and to identify the sex of the individual. As you can see from figure 3.6, all of the homologous pairs are made up of

table 3.2   Chromosome Number for Selected Species

Species	2n
Human being (*Homo sapiens*)	46
Garden pea (*Pisum sativum*)	14
Fruit fly (*Drosophila melanogaster*)	8
House mouse (*Mus musculus*)	40
Roundworm (*Ascaris* sp.)	2
Pigeon (*Columba livia*)	80
Boa constrictor (*Constrictor constrictor*)	36
Cricket (*Gryllus domesticus*)	22
Lily (*Lilium longiflorum*)	24
Indian fern (*Ophioglossum reticulatum*)	1260

Note: 2n is the diploid complement. The fern has the highest known diploid chromosome number.

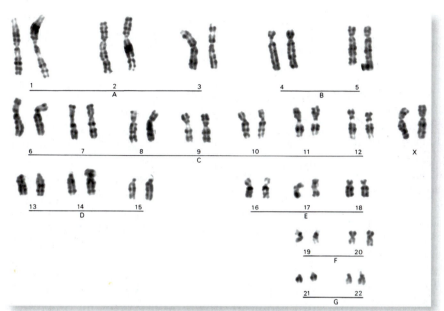

figure 3.6 Karyotype of a human female (two X chromosomes, no Y chromosome). A male would have one X and one Y chromosome. The chromosomes are grouped into categories (A–G, X, Y) based on their length and centromere position. Individual chromosomes are distinguished by their size, centromere location, and the pattern of dark and light G bands.

identical partners, and are thus referred to as **homomorphic chromosome pairs.** A potential exception is the sex chromosomes, which in some species are of unequal size and composition and are therefore called a **heteromorphic chromosome pair.** The X and Y chromosomes form the heteromorphic pair in humans. Thus, one sex is the **homogametic** sex (females in mammals) because all their gametes contain an X chromosome, and the other is referred to as the **heterogametic** sex (males in mammals) because their gametes contain either an X or a Y chromosome.

In eukaryotes, two processes partition the genetic material into offspring, or daughter, cells. In **mitosis,** the two daughter cells each receive an exact copy of the genetic material in the parental cell. The accompanying cellular process is called simple cell division. In **meiosis,** the genetic material must precisely halve, so that fertilization will restore the diploid complement. The accompanying cellular process is called gamete formation in animals and spore formation in plants. In both mitosis and meiosis, the chromosomes must properly separate from each other.

The division of the cytoplasm of the cell, **cytokinesis,** is less organized. In animals, a constriction of the cell membrane distributes the cytoplasm. In plants, the growth of a cell plate accomplishes the same purpose.

Restating the Concepts

▶ Chromosomes possess two structural units that are essential for mitosis and meiosis: the centromere and the kinetochore.

▶ Cells may either be haploid (possessing one set of unique chromosomes) or diploid (possessing two sets of unique chromosomes).

3.3 The Cell Cycle and Its Control Mechanisms

The continuity of life depends on cells growing, replicating their genetic material, and then dividing—a process called the **cell cycle** (fig. 3.7). Although cells usually divide when they have doubled in volume, the control of this process is very complex and precise. Not only do all the steps have to occur in sequence, but the cell must also "know" when to proceed and when to wait. Continuing at inappropriate moments— for example, before the DNA has replicated or when the chromosomes or spindle are damaged—could have catastrophic consequences to a cell or to a whole organism. Numerous points exist during the cycle to assess whether the next step should proceed.

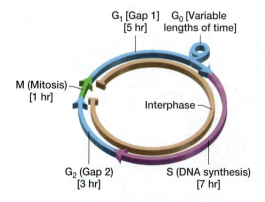

figure 3.7 Typical cell cycle for a human cell that is maintained in culture. These times can vary depending upon the cell type and culture conditions. The DNA content of the cell doubles during the S phase and is then reduced back to its original value by mitosis. G_0 is the point that some terminally differentiated cells, either temporarily or permanently, exit the cell cycle.

Cell Cycle

The cell cycle is composed of four principal stages (see fig. 3.7): gap 1 (G_1), DNA replication (S), gap 2 (G_2), and mitosis (M). The G_1, S, and G_2 phases are sometimes collectively referred to as interphase. DNA replication is required to synthesize a copy of the genomic DNA before mitosis occurs. This ensures that a complete set of the genomic DNA can be segregated into each of the two daughter cells. (The mechanism of DNA replication will be discussed in chapter 9.) The two gap phases are times when the cell is primarily devoted to growth, differentiation, and biological activity. This cycle can continue for the life of some cells. Other cells, such as neurons, do not continue to grow and divide after they completely differentiate. These cells leave the cell cycle and enter the G_0 phase (see fig. 3.7), where they remain metabolically active and viable. Some cells also enter the G_0 phase temporarily and then reenter the cell cycle. Normal cells must decide whether they should bypass the G_0 phase, enter and remain in G_0, or simply move through G_0, depending on their cell type. Occasionally, cells either fail to enter the G_0 phase or do not remain in the G_0 phase, which results in their continual **proliferation** (growth and division). This uncontrolled cell proliferation can lead to a cancerous state.

Initiation of Mitosis

Early research into the cell cycle involved fusing cells in different stages of the cycle to determine whether the cytoplasmic components of one cell would affect the behavior of the other. Results of these experiments led to the discovery of a protein complex called the **maturation-promoting factor (MPF)** because of its role in causing oocytes to mature. It is now also referred to as the **mitosis-promoting factor:** it initiates the mitotic phase of the cell cycle.

Further research has shown that MPF is made of two proteins: one that oscillates in quantity during the cell cycle and one whose quantity is constant. The oscillating component is referred to as **Cyclin B;** the constant gene product, CDC2, is encoded by the *cdc2* gene (**fig. 3.8**). CDC2 is a **kinase,** an enzyme that transfers a phosphate group from a molecule, such as ATP, to a specific amino acid of the protein it is acting on. This process of adding a phosphate group onto a protein is called **phosphorylation.** Because the CDC2 kinase is only functional when it is combined with cyclin, it is referred to as a **cyclin-dependent kinase (CDK).** Several of these kinase–cyclin combinations control stages of the cell cycle, with different cyclins acting at different points in the cell cycle. In general, cyclin-dependent kinases are regulated by phosphorylation and dephosphorylation, cyclin levels, and activation or deactivation of inhibitors.

Normally, CDC2 remains at high levels in the cell but does not initiate mitosis for two reasons. First, phosphate groups block its active site, the place on the enzyme that actually does the phosphorylating. Second, the enzyme can only function when it combines with a molecule of Cyclin B, the protein that oscillates during the cell cycle.

Cyclin B is at very low levels at the beginning of the G_1 phase. During G_1 phase, the number of Cyclin B molecules increases as the cell grows, combining with the constant number of CDC2 proteins until a critical quantity is reached. However, to activate the CDC2–Cyclin B complexes, another kinase must first phosphorylate two amino acids on CDC2. Another enzyme then **dephosphorylates** CDC2, by removing one of the two phosphate groups. At that point, the active MPF complex can initiate the changes that begin mitosis (see fig. 3.8). Presumably the cell is now ready for mitosis, having gone through G_1, S, and G_2 phases.

Once mitosis has been initiated, Cyclin B, along with other proteins that have served their purpose by this point in the cell cycle, breaks down with the help of a protein complex called the **anaphase-promoting complex (APC),** also called the **cyclosome.** The cyclosome attaches a **ubiquitin** molecule to the proteins as a signal that they are to be degraded. The removal of Cyclin B regenerates the inactive CDC2. As the cell enters G_1, virtually no functioning MPF complex remains (see fig. 3.8).

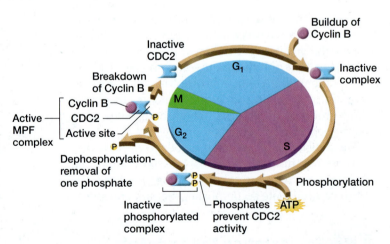

figure 3.8   The proteins CDC2 (CDK1) and cyclin B combine to form the maturation-promoting (MPF) factor. During mitosis, Cyclin B is broken down. During G_1 and S phases, Cyclin B builds up and combines with CDC2, which is then phosphorylated at the active site to render it inactive. Dephosphorylation, a process that begins to take place only after DNA replication is properly finished, produces an active MPF. This represents a single cell cycle checkpoint.

Cell Cycle Checkpoints

Some points in the cell cycle, such as the initiation of mitosis, can be delayed until all necessary conditions are in place. These **checkpoints** allow the cell to make sure that various events have been properly completed before the next phase begins. **Surveillance mechanisms** that involve dozens of proteins, many just discovered, oversee these checkpoints. In the cell cycle, three major checkpoints involve CDKs (see fig. 3.9); each has its own specific cyclin and is functional at only a single point in the cell cycle.

- The **G$_1$/S checkpoint** determines whether the cell has reached the proper size and determines if the DNA is damaged. For example, if the cell attempts to replicate damaged DNA, breaks will occur in the DNA or replication will be blocked. In either case, the daughter cells that are produced at the end of this cell cycle will likely not be viable. To prevent this problem, the cell cycle is delayed at this point until the necessary conditions have been met, such as the repair of the damaged DNA. After the problem has been corrected, the cell proceeds to the S phase.

   The CDK4-Cyclin D and CDC2-Cyclin E complexes phosphorylate the Retinoblastoma (Rb) protein, which releases the bound E2F transcription factor during G$_1$ phase. The freed E2F protein then allows transcription of genes that are required for DNA replication.

   The **S-phase-promoting factor (SPF)** is one of the proteins required at this checkpoint. The SPF, which is composed of Cyclin A bound to CDC2, enters the nucleus and prepares the DNA to begin replication.

- The **G$_2$/M checkpoint** evaluates whether DNA replication is completed and if any damaged DNA still needs to be repaired. Again, the cell cycle is arrested at this point until these criteria are met. MPF is required to progress through this checkpoint.

- The **M checkpoint** monitors whether spindle fibers are properly assembled and attached to the kinetochores. If either of these two events is not completed, the chromosomes cannot faithfully be separated into the daughter cells. This would result in daughter cells that are not genetically identical to the parental cell. Therefore, mitosis is delayed until these structures are properly assembled.

In addition to the cyclins and CDKs, the ubiquitin-dependent degradation of the cyclin proteins is an important activity in regulating the cell cycle. For example, the APC serves two important functions during mitosis. First, it degrades the Cyclin B protein of MPF. The second function of APC, which is to permit the separation of the sister chromatids at the start of anaphase, is described later in this chapter.

Some checkpoints don't involve CDKs. The **spindle attachment checkpoint** ensures that spindle fibers are attached to every kinetochore before the sister chromatids attempt to separate. This involves several proteins, including the **mitotic arrest-deficient protein 2 (MAD2),** binding to the kinetochore of each chromosome. As spindle fibers attach to each kinetochore, the MAD2 protein is released. Only after all the kinetochores are free of the MAD2 proteins, which confirms that every kinetochore is attached to spindle fibers, will the cell continue through mitosis.

The cell cycle routinely halts when genetic damage is present, giving the cell a chance to repair the damage before committing to cell division. If the damage is too extreme, the cell can enter a programmed cell death sequence (discussed in chapter 22). For example, if the G$_1$/S checkpoint detects DNA damage, the **p53 protein** targets the cell for regulated death (**apoptosis**). If the *p53* gene is defective, then the controlled death of the damaged cells would not take place, and the possible ensuing uncontrolled cell growth would result in cancer. In fact, a number of human cancers, including colon, breast, and lung cancers, have been shown to be associated with mutations in the *p53* gene. As a result, the genetic control of the cell cycle is one of the most active areas of current research. The 2001 Nobel Prize for physiology or medicine was awarded to Lee Hartwell, Tim Hunt, and Paul Nurse for their studies on the control of the cell cycle.

 Restating the Concepts

▶ The cell cycle describes the different stages in the life span of a cell. This includes DNA replication (S phase), nuclear division (mitosis, M phase), and two intervening phases (gap phases, G$_1$ and G$_2$).

▶ The cell cycle is regulated. Checkpoints monitor that the cell possesses the correct parameters, such as proper size, DNA content, and DNA integrity, before it moves to the next stage of the cycle. Misregulation of the cell cycle could lead to cancerous growth.

3.4 Mitosis

Consider the engineering problem that mitosis must solve. DNA replication during the S phase produces identical DNA molecules, called *sister chromatids*, that must separate so that each goes into a different daughter cell (**fig. 3.9**). Each chromatid is a double-stranded DNA molecule. The sister chromatids are initially held together at the centromeric region; each will be called a chromosome when they separate and become independent. Each of the two daughter cells then ends up with a chromosome complement identical to that of

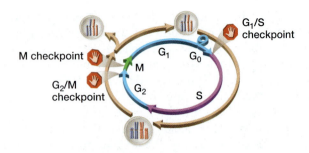

figure 3.9 The movement of a hypothetical diploid cell (2*n* = 4) through the cell cycle. The diploid cell contains one long pair of metacentric chromosomes and one short pair of telocentric chromosomes. During S phase, each double-stranded DNA molecule is duplicated to produce sister chromatids. The sister chromatids then separate during mitosis to produce daughter cells that contain equivalent DNA molecules as the parental cell. Also shown are three checkpoints that function to ensure the DNA and cell are prepared before progressing further in the cell cycle.

the parent cell. Mitosis is nature's elegant process for achieving that end—surely an engineering marvel.

Mitosis is a continuous process. However, for descriptive purposes, we can break it into four stages: *prophase, metaphase, anaphase,* and *telophase* (greek: *pro-*, before; *meta-*, mid; *ana-*, back; *telo-*, end). Replication (duplication) of the genetic material occurs during the S phase of the cell cycle (see fig. 3.7). The timing of the four stages varies from species to species, from organ to organ within a species, and even from cell to cell within a given cell type.

Prophase

Prophase, the first stage of mitosis, begins with the shortening and thickening of the chromosomes so that individual chromosomes become visible. Each chromosome is composed of two **sister chromatids,** which are identical double-stranded DNA molecules that are the products of DNA replication in S phase. The sister chromatids are held together by a complex called **cohesin,** made up of at least four different proteins. At this time also, the nuclear envelope (membrane) disintegrates and the nucleolus disappears (**fig. 3.10**). The **nucleolus**

is a darkly stained, diffuse region in the nucleus where **ribosomal RNA (rRNA)** is transcribed, and the initial stages of ribosome assembly take place.

The two centrioles, which are present in the centrosome, replicate during the S and G$_2$ phases of the cell cycle. During prophase, the centrosome divides and moves to opposite poles of the cell, around the nucleus (**fig. 3.11**). The newly divided centrosomes radiate microtubules, which begin at each centrosome and overlap in the middle of the cell. These microtubules are called *spindle fibers.* Microtubules also spread out from the centrosome in the opposite direction from the spindle itself, forming an **aster** (see fig. 3.11). A third form of tubulin, gamma (γ) tubulin, is also needed to begin the formation of a microtubule at the centrosome.

As the microtubules elongate and shrink at their plus end, some attach to a kinetochore on a sister chromatid; these fibers are called **kinetochore microtubules** (see fig. 3.11). At first, a spindle fiber attaches randomly to one of the kinetochores on a chromatid pair. As the microtubules continue to move, and as new microtubules attach and old microtubules break, each sister kinetochore eventually attaches to microtubules emanating from different poles (**fig. 3.12**), ensuring that sister chromatids move to opposite poles during anaphase. A second class of microtubules fail to attach to kinetochores and are called **polar microtubules.** The interactions between polar microtubules from opposite poles help to keep the two centrosomes apart during mitosis and assist in driving the separation of the chromosomes.

The number of microtubules that attach to each kinetochore differs in different species. It appears that 1 attaches to each kinetochore in yeast, 4 to 7 attach to each kinetochore in cells of a rat fetus, and 70 to 150 attach in the plant *Haemanthus katherinae* (see fig. 3.12).

Metaphase

During **metaphase,** the chromosomes move to the equator of the cell. With the attachment of the spindle fibers and the completion of the spindle itself, the chromosomes jockey into position in the equatorial plane of the spindle, called the **metaphase plate.** This happens as kinetochore microtubules exert opposing

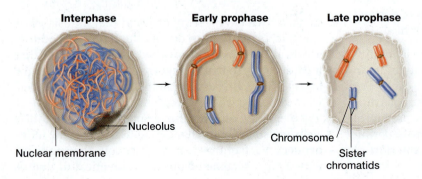

Interphase   **Early prophase**   **Late prophase**

Nucleolus

Nuclear membrane

Chromosome

Sister chromatids

figure 3.10 Nuclear events during interphase and prophase of mitosis. This cell, 2*n* = 4, consists of one pair of long and one pair of short metacentric chromosomes. By early prophase, the chromosomes are condensing and by late prophase, the nuclear membrane is broken down. Note that each chromosome consists of two chromatids when the cell enters mitosis. Maternal chromosomes are red; paternal chromosomes are blue.

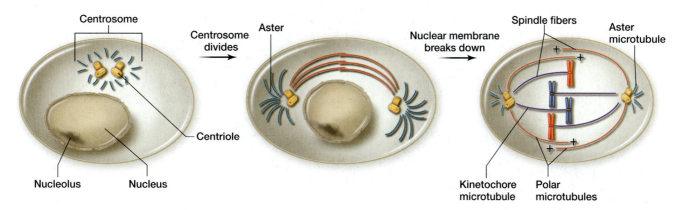

figure 3.11   In animal cells, the centrosome divides in prophase, and the separate halves move to opposite poles of the cell. This creates a spindle in the middle of the cell after the nuclear membrane breaks down. Each centromere becomes attached to a kinetochore microtubule. The polar microtubules, with their plus ends directed towards the center of the cell, overlap to provide tension for the subsequent chromosome movement.

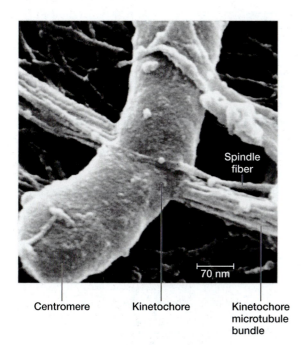

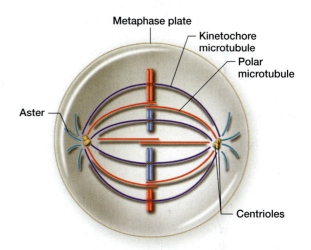

figure 3.13   During metaphase, the chromosomes align on the metaphase plate, with each centromere attached to a kinetochore microtubule from each centrosome. Maternal chromosomes are red; paternal chromosomes are blue.

figure 3.12   Scanning electron micrograph of the centromeric region of a metaphase chromosome from the plant *Haemanthus katherinae.* Spindle fiber bundles on either side of the centromere extend in opposite directions. A fiber not connected to the kinetochore is visible lying over the centromere. These fibers are 60 to 70 nm in diameter.

tension on the two sister kinetochores. Alignment of the chromosomes on this plate marks the end of metaphase (**fig. 3.13**).

Anaphase

Anaphase begins with the two sister chromatids separating and moving toward opposite poles on the spindle fibers. The physical separation of the sister chromatids and their movement to opposite poles are

two separate activities. Recall that chromatid separation represents a checkpoint in the cell cycle. A complex of proteins, which includes MAD2, is bound to a kinetochore until a spindle fiber attaches to the kinetochore. Only when all the kinetochores are attached to kinetochore microtubules from both centrosomes will anaphase proceed.

The next step involves the separation of the sister chromatids. After S phase, the sister chromatids are joined along their length by *cohesins.* To physically separate the sister chromatids, the cohesins must be destroyed.

The APC, which helps degrade Cyclin B, is also required for cohesin degradation. Binding of CDC20 to an inactive APC leads to APC activation (**fig. 3.14**). This active APC attaches ubiquitin to the securin protein,

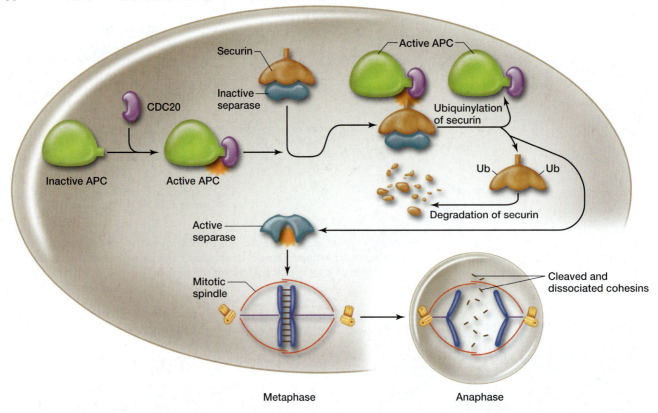

figure 3.14   Regulation of sister chromatid separation. The active APC–CDC20 complex adds ubiquitin to the securin protein, which leads to its degradation. The freed separase protease can then degrade the cohesin proteins that are located along the length of the sister chromatids. The loss of cohesin, which keeps the sister chromatids tightly associated, permits the sister chromatids to be separated.

which targets it for degradation. **Securin** is an inhibitory protein that binds and inactivates a protease called separase. Degradation of securin frees **separase,** which, when active, degrades cohesin and liberates the sister chromatids from each other. The separation of the sister chromatids during anaphase is the point at which they are referred to as daughter chromosomes.

The spindle fibers then separate the daughter chromosomes using two mechanisms (**fig. 3.15**). First, proteins within the kinetochore act as a microtubule motor, which hydrolyzes ATP to generate the energy to move along the microtubule toward the minus end (centrosome). Second, the chromosome is dragged toward the centrosome as the plus end of the microtubule is disassembled (**fig. 3.16**). As the chromosomes are pulled through the cytosol, metacentric chromosomes appear V-shaped (as in fig. 3.16), subtelocentrics appear J-shaped, and telocentrics appear rod-shaped.

Telophase

At the end of anaphase (**fig. 3.17**), the separated daughter chromosomes have been pulled to opposite poles of the cell. In telophase, the cell reverses the steps of prophase to return to the interphase state (**fig. 3.18**). The

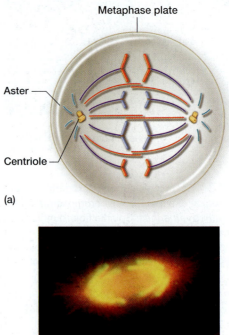

figure 3.15   (a) The mitotic spindle during anaphase. The kinetochore microtubules contract, which drag the chromosomes toward opposite poles. Maternal chromosomes are red; paternal chromosomes are blue. (b) Fluorescent microscope image of a cultured cell in anaphase. Microtubules are red; chromosomes (DNA) are stained yellow.

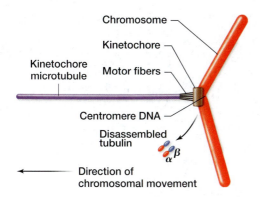

figure 3.16   The kinetochore acts as a microtubule motor, pulling the chromosome along the kinetochore microtubule towards one pole in the cell. One microtubule is shown, although many may be present within the cell.

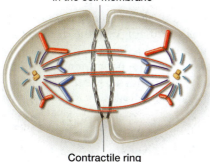

Late anaphase
Growing constriction point in the cell membrane

Contractile ring

figure 3.17   Late anaphase of mitosis. A constriction begins to form in the middle of the cell (in animals) as the contractile ring tightens. Maternal chromosomes are red; paternal chromosomes are blue.

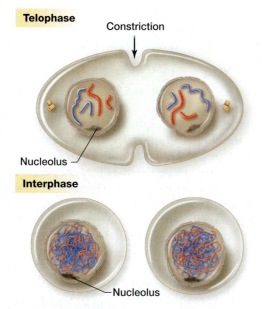

figure 3.18   Telophase and interphase of mitosis. Maternal chromosomes are red; paternal chromosomes are blue.

chromosomes uncoil and begin to direct protein synthesis. A nuclear envelope re-forms around each set of chromosomes. The nucleolus forms again around the **nucleolar organizers,** which are regions on the chromosomes where rRNA genes are located. Finally, cytokinesis, which is the division of the remaining cellular components, takes place.

In animals, cytokinesis is first apparent by a constriction between the two poles. A ring of actin around the cell seems to be associated with a constriction in the cell membrane and may assist in the continued tightening of the constriction until the two daughter cells separate. In plants, a cell plate grows in the approximate location of the metaphase plate. The completion of this plate results in the separation of the two daughter cells. After completing cytokinesis, the daughter cells enter the G_1 phase of the cell cycle. **Figure 3.19** summarizes mitosis as it occurs in onion root tip cells.

Significance of Mitosis

Mitosis and cytokinesis result in two daughter cells, each with genetic material identical to that of the parental cell. This exact distribution of the genetic material, in the form of chromosomes, to the daughter cells ensures the stability of the genetic material and the inheritance of traits from one cell generation to the next. Cells have evolved complex life functions; through mitosis they produce offspring cells with these same capabilities. With this stability ensured, single-celled organisms could thrive and multicellular organisms could evolve.

It's Your Turn

Using the information we discussed on the cell cycle and mitosis, you are ready to attempt a related question. Question 16 requires you to diagram the key events of mitosis and relate the effect of chromosome movements on the inheritance of genes and their alleles.

Restating the Concepts

▶ Mitosis is the process of nuclear division.

▶ Mitosis is a conservative process, through which two genetically identical daughter cells are generated with the same amount of DNA and the same number of chromosomes as the parental cell.

▶ Both haploid and diploid cells can undergo mitosis.

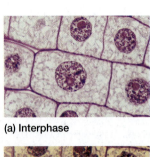

(a) Interphase

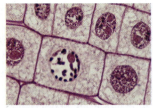

(b) Early prophase

(c) Late prophase

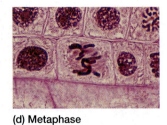

(d) Metaphase

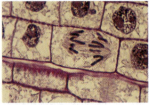

(e) Anaphase

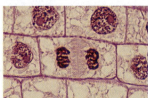

(f) Telophase

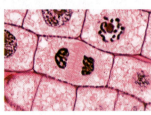

(g) Daughter cells

figure 3.19 Cells in interphase and in various stages of mitosis in the onion root tip. The average cell is about 50 μm long.

3.5 Meiosis

Reproduction in diploid organisms presents an interesting engineering problem. Haploid animal gametes or plant spores with half the amount of parental DNA must be formed from diploid cells, so that when gametes fuse, the diploid number and total amount of DNA is restored. In addition, the process of halving the diploid number must result in one copy of *every* chromosome being present in the haploid cells so that gamete fusion restores two copies of every chromosome. Simply separating chromosomes randomly into two equal-sized batches will not accomplish this goal.

From an engineering point of view, a method is needed to recognize the homologous chromosomes. Next, a process must move one of each pair of homologues into each daughter cell. Cells have solved these challenges by pairing the homologous chromosomes via the spindle apparatus. But there is one complication.

Cells entering meiosis have already replicated their chromosomes, just as in mitosis. Therefore, to reach the goal of a haploid number and half the amount of DNA, two nuclear divisions are necessary, without an intervening DNA replication step. As a result, meiosis is a two-division process that produces four haploid cells from each diploid parental cell. These two divisions are known as **meiosis I** and **meiosis II.**

Mitosis can occur in either haploid or diploid cells, but meiosis is restricted to diploid cells. In animals and plants, meiosis occurs only in certain kinds of cells, namely *germ cells*. Haploid organisms, such as fungi, can also undergo meiosis. However, two haploid cells must first fuse to form a diploid cell that then enters meiosis. We will describe these different life cycles after we discuss the general mechanism of meiosis.

Prophase I

Cytogeneticists have divided **prophase** of meiosis I into five stages to distinguish the different events that the chromosomes undergo. The five stages are named: **leptonema, zygonema, pachynema, diplonema,** and **diakinesis.** The names of these stages are derived from Greek roots as follows:

Term	Greek root and meaning	Adjective form
Leptonema	*lepto-*, thin	Leptotene stage
Zygonema	*zygo-*, yoke-shaped	Zygotene stage
Pachynema	*pachy-*, thick	Pachytene stage
Diplonema	*diplo-*, double	Diplotene stage
Diakinesis	*dia-*, across and *kinesis*, movement	(None)

Leptonema

As a cell enters the **leptotene stage** of prophase I, it behaves much like a cell entering prophase of mitosis, with the centrosomes duplicating and the spindles forming around the intact nucleus. As the chromosomes condense, they are visible as individual threads; sister chromatids are so close together that they are not distinct. The chromosomes are more spread out than they are in mitosis, with dark chromomeres interspersed.

The tips of the chromosomes are attached to the nuclear membrane in the leptotene stage (**fig. 3.20**). As the cell approaches the zygotene stage, the tips of the chromosomes move until most end up in a limited region near each other. This forms an arrangement called a **bouquet stage.** Presumably, this arrangement helps homologous chromosomes find each other and begin the pairing process without becoming entangled.

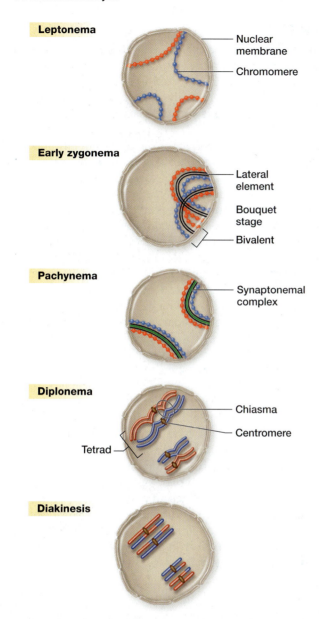

Zygonema

The pairing of homologous chromosomes marks the beginning of the **zygotene stage.** Initially, contact is generated between identical regions of homologous chromosomes, leading to a loose interaction along the entire length of the homologous chromosomes. A

proteinaceous complex, named the **lateral element,** is detected between the pairing homologs (see fig. 3.20). At this point, the paired homologous chromosomes are referred to as **bivalents,** because of the appearance of two chromosomes in the paired unit. The number of bivalents in a cell is equivalent to the haploid number (*n*) of the cell. The synapsis of all chromosomes marks the end of the zygotene stage.

Pachynema

The chromosomes continue to shorten and thicken, giving the **pachytene stage** its name. The loose interaction between homologous chromosomes becomes a close association through a process called **synapsis.** The lateral element becomes a more extensive protein network between the homologs, which is now called the **synaptonemal complex** (**fig. 3.21**). This complex appears to mediate synapsis. It is also likely that **crossing over** (also known as **recombination,** see chapter 6) takes place during this stage.

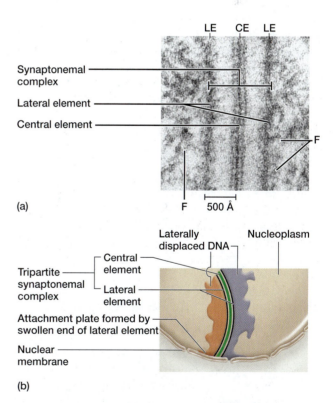

figure 3.20 Prophase I of meiosis. Prophase I is divided into five stages. During leptonema, the chromsomes begin condensing. During zygonema, the homologous chromosomes start aligning with the help of the proteins in the lateral element. Additional proteins form the synaptonemal complex in pachynema, which results in synapsis of the homologs. By the end of diplonema, the synaptonemal complex is gone and the tetrad of four chromatids is visible. Maternal chromosomes are red; paternal chromosomes are blue. Note that crossing over is evident at diplonema.

figure 3.21 The synaptonemal complex. (a) An electron micrograph shows the synaptonemal complex, which is composed of the proteinaceous central element (CE) that is flanked by the lateral elements (LE). The diffuse chromosome fibers (F) are present to the left and right of the synaptonemal complex. (b) Diagram of the synaptonemal complex (central element, green; lateral element, black) and the chromosome bivalent (paternal chromosome, blue; maternal chromosome, red).

When synapsis brings two chromatids into close proximity, enzymes can break both chromatid strands and reattach them differently (**fig. 3.22**). Although genes have a fixed position on a chromosome, crossing over may cause alleles that started out attached to a paternal centromere to end up attached to a maternal one. Before crossing over takes place, densely staining nodules are visible. These structures, called **recombination nodules** (**fig. 3.23a**), presumably correlate with crossing over and represent the enzymatic machinery present on the chromosomes.

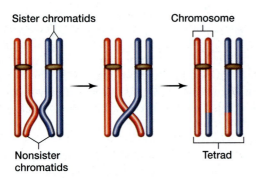

Sister chromatids   Chromosome

Nonsister chromatids   Tetrad

figure 3.22 Crossing over in a tetrad during prophase of meiosis I. The relationship between sister and nonsister chromatids are shown. Maternal chromosomes are red; paternal chromosomes are blue. Note the exchange of chromosome pieces after the process is completed.

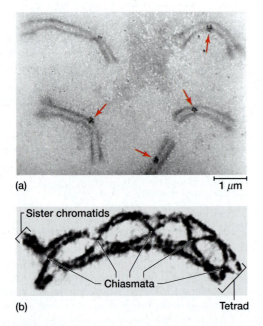

(a)   1 μm

Sister chromatids

Chiasmata

(b)   Tetrad

figure 3.23 (a) Recombination nodules (*arrowhead*) in spermatocytes of the pigeon, *Columba livia*. The darkly stained recombination nodules are marked by red arrows. (Bar = 1 μm.) (b) A tetrad from the grasshopper, *Chorthippus parallelus*, at diplonema with five chiasmata.

Diplonema

As the chromosomes shorten and thicken further and the homologs begin to separate in the **diplotene stage,** we see that each chromosome is composed of two chromatids. Now the paired chromosome figures are referred to as **tetrads** because each is made up of four chromatids (see fig. 3.22). The two chromatids in a single chromosome are sister chromatids, but chromatids from each of the homologous chromosomes are called **nonsister chromatids.**

About this time, the synaptonemal complex disintegrates in all but the areas of the **chiasmata** (singular: chiasma), the X-shaped configurations marking the places of crossing over (fig. 3.23b). The chiasmata are observed between nonsister chromatids, which serve as the last contacts between the homologous chromosomes as they separate. Virtually all tetrads exhibit at least one chiasma. In rare cases in which no crossing over occurs, the tetrads tend to fall apart and segregate randomly. Thus, crossing over provides two important functions for meiosis: first, it increases genetic diversity by moving alleles from one chromosome to its homolog to generate new combinations of alleles on a chromosome. Second, it solves an engineering problem by ensuring the recognition and separation of homologous chromosomes.

During the diplotene stage, chromosomes can again disperse and become active. This activity is especially obvious in amphibians and birds, which produce a great amount of cytoplasmic nutrient for the future zygote. Recondensation of the chromosomes takes place at the end of diplonema. This stage can be very long; in human females, it begins in the fetus and is not completed until an egg is shed during ovulation, sometimes more than 50 years later.

Diakinesis

As prophase I moves into diakinesis, the chromosomes become very condensed (see fig. 3.20). The homologous chromosomes continue to separate, with the chiasmata moving toward the ends of the tetrads, which is termed **terminalization.** The nucleolus vanishes and the nuclear membrane breaks down. The duplicated centrosomes are at opposite poles, and the spindle fibers from each centrosome attach to only one kinetochore in each tetrad. In this way, both sister kinetochores become attached to spindle microtubules coming from the same pole (**fig. 3.24**).

Metaphase I and Anaphase I

Metaphase I is marked by the movement of the tetrads to the metaphase plate of the cell. Once all of the tetrads are aligned, anaphase I begins. The chiasmata vanish as the homologous chromosomes are separated, moving toward opposite poles (**fig. 3.25**). Cohesin breaks down

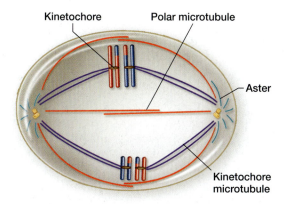

figure 3.24 Metaphase of meiosis I. Sister kinetochores (effectively single, merged kinetochores), which join sister chromatids, are attached to kinetochore microtubules from the same centrosome or pole. The tetrads are aligned on the metaphase plate. Maternal chromosomes are red; paternal chromosomes are blue.

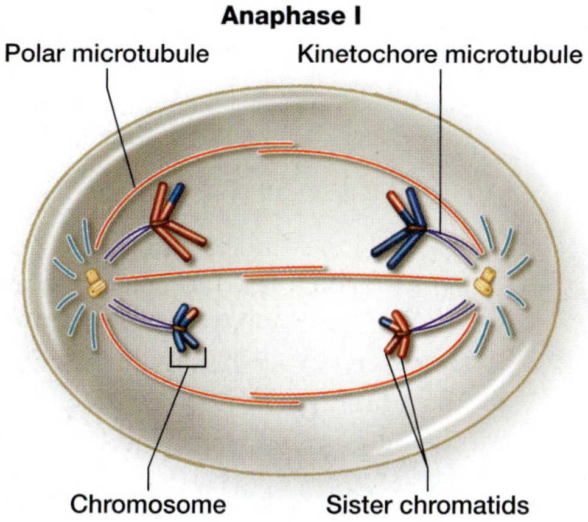

figure 3.25 Anaphase of meiosis I. Homologous chromosomes separate and move to opposite poles along the kinetochore microtubules. Maternal chromosomes are red; paternal chromosomes are blue.

everywhere but at the centromeres, which allows sister chromatids to be pulled to the same pole: One chromosome of each homologous pair moves to each pole. This meiotic division is therefore called a **reductional division** because it reduces the number of chromosomes ($2n$) by half in each daughter cell (n).

At this time, the two sister chromatids represent a single chromosome because they share a single centromere. As a result, for every tetrad there is now one chromosome in the form of a chromatid pair, known as a **dyad,** or **monovalent,** at each pole of the cell.

The initial objective of meiosis, separating homologs into different daughter cells, has been accomplished. Because each dyad consists of two sister chromatids, a second, mitosis-like division is required to reduce each chromosome to a single chromatid.

Telophase I and Prophase II

Depending on the organism, telophase I may or may not be greatly shortened in time. In some organisms, telophase I occurs and a nuclear membrane forms around the chromosomes as cytokinesis takes place. This abbreviated interphase is termed **interkinesis.** In these organisms, prophase II begins next, and meiosis II proceeds. In still other organisms, the late anaphase I chromosomes go almost directly into metaphase II, virtually skipping telophase I, interphase, and prophase II. In both cases, no chromosome duplication (DNA replication) occurs between meiosis I and meiosis II.

Meiosis II

Meiosis II is basically a mitotic division in which the chromatids of each chromosome are pulled to opposite poles. For each original cell entering meiosis I, four cells emerge at telophase II. Meiosis II is an **equational division;** although it reduces the amount of genetic material per cell by half through separation of the sister chromatids, it does not further reduce the chromosome number per cell (**fig. 3.26**). Sometimes it is simpler to concentrate on the behavior of centromeres during meiosis than on the chromosomes and chromatids. Meiosis I separates maternal from paternal centromeres, and meiosis II separates sister centromeres. **Figure 3.27** summarizes meiosis in corn (*Zea mays*).

In terms of chromosomes, meiosis begins with a diploid ($2n$) cell and produces four haploid (n) cells. In terms of DNA, the process is a bit more complex but has the same result. Let's say that the quantity of DNA in a gamete is C. A diploid cell before S phase has $2C$ DNA, and the same cell after S phase, but before mitosis, has $4C$ DNA. Mitosis reduces the quantity of DNA back to $2C$.

A cell entering meiosis also has $4C$ DNA. After the first meiotic division, each daughter cell has $2C$ DNA, and after the second meiotic division, each daughter cell has C DNA, the quantity appropriate for a gamete. When two gametes fuse, the original diploid state, $2n$, and the original quantity of DNA, $2C$, is reestablished.

Significance of Meiosis

Meiosis is significant for several reasons. First, it reduces the diploid number of chromosomes by half. During this reduction, half of the gametes get one homolog and the other half of the gametes receive the other homolog, which is consistent with Mendel's first law of equal segregation.

Second, the randomness of how the tetrads align on the metaphase plate and separate in meiosis I underlies Mendel's second law of independent assortment. Each gamete produced during meiosis has an equal probability as all the other gametes of receiving a particular

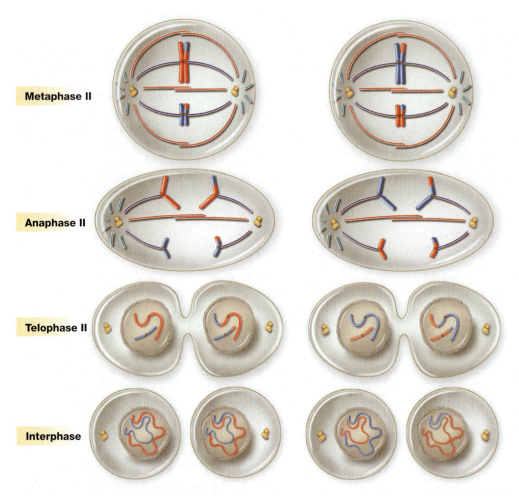

figure 3.26 Meiosis II. At the start of meiosis II, each cell is haploid (*n* = 2). Due to the separation of sister chromatids during meiosis II, the final cells at the end of meiosis II remain haploid (*n* = 2), compared to 2*n* = 4 at the start of meiosis I. Maternal chromosomes are red; paternal chromosomes are blue.

combination of chromosomes. A simple demonstration is a diploid cell with two sets of homologous chromosomes that generates four different types of gametes (**fig. 3.28a**). In anaphase I, the direction of homolog orientation is independent of the other chromosome pairs. Whereas one pole may get the maternal centromere from chromosomal pair 1, it could get either the maternal or the paternal centromere from chromosomal pair number 2, and so on (fig. 3.28a).

Third, meiosis produces genetic diversity. For example, independent assortment produces chromosome combinations that are not present in either parent. In humans, who have 23 chromosome pairs, 2^{23} or 8,388,608 different chromosome combinations are possible. This number of different combinations increases further when we consider recombination. Assuming 30 thousand genes in a human being, with two alleles each, $2^{30,000}$ different gametes could potentially arise by meiosis.

It is important to realize that both independent assortment and recombination occur simultaneously. The genetic diversity thus produced is significantly higher than by either process alone. If a model diploid cell with two sets of homologous chromosomes undergoes a single recombination event in only one of the two homologous pairs, a total of eight different types of gametes are possible (fig. 3.28b).

It's Your Turn

Applying your knowledge about chromosome pairing, crossing over, and movement during meiosis will help you to further understand meiosis. Try problems 17, 31, and 40 to test yourself.

Restating the Concepts

▶ Meiosis is the specialized process that produces haploid gametes required for sexual reproduction from diploid cells.

▶ Meiosis involves two nuclear divisions. The first division requires pairing of homologous chromosomes and their separation (reductional division). The two haploid cells that result then enter the second nuclear division (equational division) to generate the gametes.

▶ The movement of the chromosomes through meiosis is consistent with the movement of alleles in Mendel's first and second laws.

▶ Meiosis produces genetic diversity through the independent assortment of chromosomes and through recombination, both of which occur during meiosis I.

Four stages of Prophase I

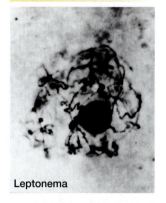

Leptonema

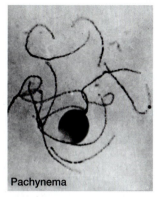

Pachynema

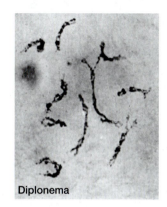

Diplonema

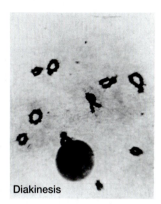

Diakinesis

Remaining stages of Meiosis I

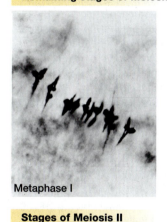

Metaphase I

Anaphase I

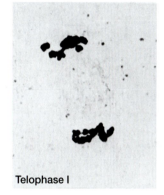

Telophase I

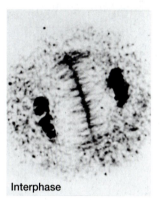

Interphase

Stages of Meiosis II

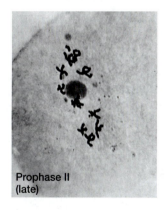

Prophase II
(early)

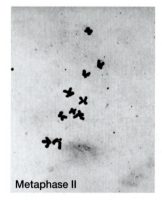

Prophase II
(late)

Metaphase II

Anaphase II

figure 3.27   Meiosis in corn (*Zea mays*).

3.6 Comparison Between Mitosis and Meiosis

It is worthwhile to note the similarities and differences between mitosis and meiosis to fully appreciate these two processes. The first significant difference is that both haploid and diploid cells can enter mitosis, but only diploid cells can proceed through meiosis. In **figure 3.29** we diagram side by side a diploid cell undergoing mitosis and one proceeding through meiosis to better illustrate the comparison. Prior to either mito-

sis or meiosis, DNA replication occurs to produce two DNA molecules (sister chromatids) from each chromosome. Although DNA replication occurs before meiosis I, it does not occur before the start of meiosis II.

In a broad sense, mitosis involves only a single round of cell division, but meiosis involves two (see fig. 3.29). Mechanistically, there are significant differences between mitosis and the first round of meiosis (meiosis I). The first obvious difference is the pairing of homologous chromosomes in prophase I of meiosis I and the absence of pairing in mitosis (see fig. 3.29). The pairing of homologous chromosomes underlies

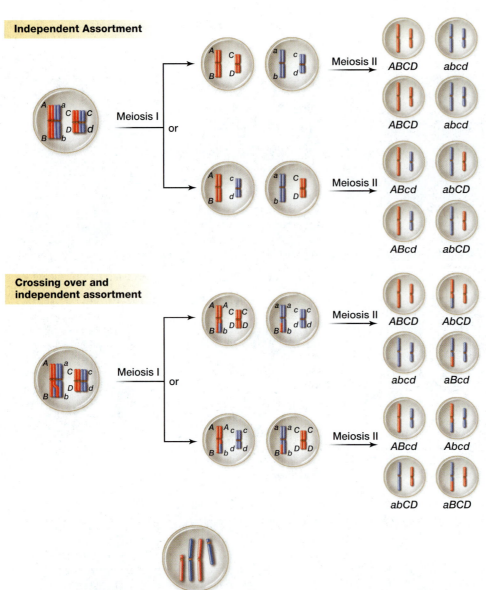

Independent Assortment

Crossing over and independent assortment

figure 3.28
Demonstration of how independent assortment and crossing over both lead to genetic diversity. Maternal (red) and paternal (blue) chromosomes separate independently in different tetrads. Independent assortment produces gametes with four possible genotypes (*ABCD, abcd, ABcd,* and *abCD*). Combining crossing over (between the centromere and the *B* gene) and independent assortment produces eight different genotypes. As the number of crossover events increases, the number of different genotypes in the gametes rises.

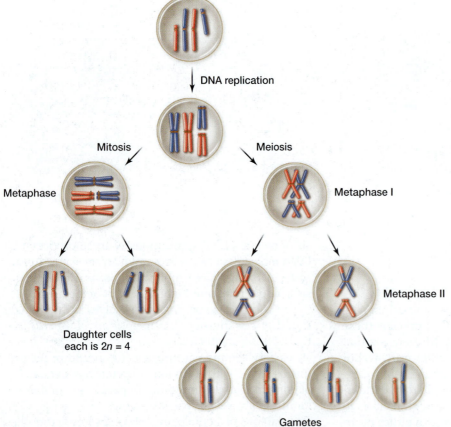

figure 3.29 Comparison of a diploid cell undergoing mitosis and meiosis. Each mitotic product is $2n = 4$, while each meiotic gamete is $n = 2$. Notice the genetic consistency in mitosis compared to the genetic diversity in the gametes due to crossing over. Maternal chromosomes are red; paternal chromosomes are blue.

3.1 applying genetic principles to problems

The Relationship Between Meiosis and Mendel's Laws

You learned in chapter 2 that Mendel's two laws (equal segregation and independent assortment) accounted for the inheritance of traits. We discussed in this chapter how chromosomes move during meiosis to produce gametes that fuse to produce progeny. We can now ask how chromosome movements account for Mendel's laws? Using a simple diploid cell that contains two nonhomologous (different) chromosomes, we can diagram the major features of meiosis that illustrate both of Mendel's laws.

We start by drawing a cell that contains long and short paternal chromosomes (blue) and long and short maternal chromosomes (red). Prior to meiosis, DNA replication takes place. This produces two identical sister chromatids that are attached at a single centromere (fig. 1). A gene on the long chromosome has two alleles, *L* and *l,* and a gene on the short chromosome also has two alleles, *S* and *s.*

As the cell enters metaphase I, the homologs synapse. The separation of the homologs during anaphase I accounts for Mendel's first law of equal segregation. Half of the gametes have the paternal *L* allele on the blue chromosome and half have the maternal *l* allele on the red chromosome.

As the homologs pair during meiosis I, the two nonhomologous chromosomes can align in two different orientations relative to each other. In the first orientation (see fig. 1, left column), both paternal chromosomes move toward the same pole and both maternal chromosomes migrate toward the other pole. This produces gametes that are genotypically *LS* and *ls*. In the second orientation (see fig. 1, right column), a paternal long chromosome and maternal short chromosome move toward one pole and the maternal long chromosome and paternal short chromosome migrate toward the other pole. This produces gametes that are *Ls* and *lS*. Thus, all four genotypes are equally likely, illustrating Mendel's second law of independent assortment.

The ability to draw the location and potential pairing of chromosomes during meiosis and mitosis and properly label the various features will help you to fully appreciate the processes of meiosis and mitosis. Further, diagramming these processes as we move into more complex phenomena will help you to understand more complicated situations.

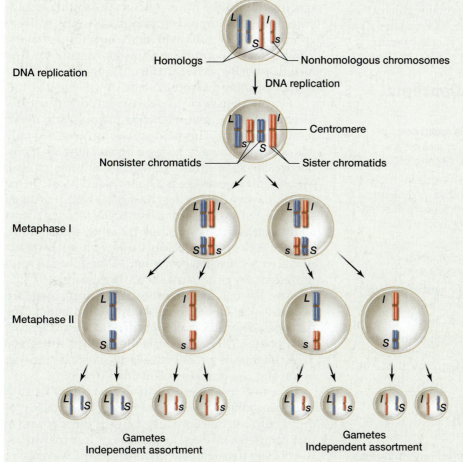

figure 1 Illustration of Mendel's two laws through meiosis of a diploid cell containing four chromosomes. The paternal chromosomes are blue, and the maternal chromosomes are red. The chromosomes can pair in two different orientations during metaphase I (left and right columns). In both orientations (columns), half the gametes receive a dominant *L* allele and half receive the recessive *l* allele (similarly for *S* and *s*). This demonstrates Mendel's first law of equal segregation. The arrangement of the short homologous chromosome pairing relative to the pairing of the long homologous pair (left and right columns) produces four genotypically different gametes that are all equally possible. This demonstrates Mendel's second law of independent assortment. The centromere, homologs, nonhomologous chromosomes, sister chromatids, and nonsister chromatids are labeled.

the reason why a cell must be diploid to enter meiosis. However, any cell that has an even number of chromosome sets (2*n*, 4*n*, 6*n*) can proceed through meiosis because every chromosome will have a homolog to pair with. Pairing of homologs during meiosis I makes crossing over between nonsister chromatids possible.

The second apparent difference occurs when the spindle fibers attach to the kinetochores. In mitosis, spindle fibers from opposite poles attach to sister chromatid kinetochores, whereas in meiosis I the spindle fibers attach to homologous chromosome kinetochores. The centromere splits in mitosis, and the sister chromatids migrate to opposite poles. However, the centromere does not divide in meiosis I, and *homologous chromosomes* migrate to opposite poles. In meiosis II, the centromere splits, and now the sister chromatids migrate to opposite poles. As you learned earlier, meiosis I is the stage at which the number of chromosomes is halved.

Mitosis is a genetically conservative process. There is no change in the amount of DNA, number of chromosomes, or allele combinations between the parental cell and the two daughter cells (see fig. 3.29). In contrast, meiosis is designed to generate genetic diversity. The four meiotic products have undergone crossing over to produce new allele arrangements (see fig. 3.29). If we include independent assortment into the meiotic model (not shown in fig. 3.29), the genetic diversity in the gametes would increase further. Combining the alleles from different individuals when haploid gametes form a diploid zygote further increases genetic diversity.

Restating the Concepts

▶ Meiosis occurs in specialized cells (germ cells) in plants and animals. The products of meiosis are usually four haploid cells per original germ cell.

▶ All eukaryotic organisms undergo meiosis, regardless of whether the organism exists primarily in haploid or diploid form. Thus, all eukaryotic organisms at some point in their life span pass through both a haploid and a diploid stage.

▶ Mitosis maintains genome integrity; meiosis introduces change in the next generation.

▶ Only diploid cells can enter meiosis; haploid or diploid cells may enter mitosis. Mitosis involves only a single nuclear division, but meiosis involves two.

▶ Homologous chromosomes pair during meiosis I to ensure their proper segregation and recombination. Homologs do not pair during mitosis.

3.7 Meiosis in Mammals

All mammals—and many other vertebrates and invertebrates as well—form gametes in a similar way. The process takes place in germ cells of the gonads of both sexes. Although the preceding discussion described a single mechanism to generate all gametes, male and female mammals exhibit some differences in how they produce gametes.

Spermatogenesis

In male mammals, each meiosis produces four equal-sized **sperm cells** in a process called **spermatogenesis** (**fig. 3.30**). A specific cell type in the testes, called the **spermatogonium,** undergoes mitosis to produce additional spermatogonia as well as **primary spermatocytes.**

When the male enters puberty, the primary spermatocytes enter meiosis. After the first meiotic division, these cells are known as **secondary spermatocytes;** after the second meiotic division, they are known as **spermatids.** The spermatids mature into **spermatozoa** (sperm cells) by **spermiogenesis.** The result is four sperm cells produced from each primary spermatocyte. Once meiosis begins in the male, he may produce several hundred million sperm cells per day for the remainder of his life.

Oogenesis

During embryonic development in female mammals, specific cells in the ovary, known as **oogonia,** proliferate through numerous mitotic divisions to form approximately a million **primary oocytes** per ovary. These primary oocytes are then able to enter the first meiotic division, but stop during a prolonged diplonema, called the **dictyotene** stage.

Once the female enters puberty, the primary oocytes resume meiosis, and ovulation takes place under hormonal control. This process usually occurs for only one oocyte per month during a human female's reproductive life span (from about 12 to 50 years of age). Meiosis only then proceeds in the ovulated oocyte. In the female, the two cells formed by meiosis I are of unequal size. One, termed the **secondary oocyte,** contains almost all the nutrient-rich cytoplasm; the other, a **polar body,** receives very little cytoplasm. Cells of unequal size are produced because the oocyte nucleus and meiotic spindle reside very close to the surface of this large cell.

The second meiotic division in the secondary oocyte yields another polar body and an **ovum.** The first polar body may or may not divide to form two new polar bodies. All polar bodies ultimately disintegrate. Thus, **oogenesis** produces cells of unequal size—an ovum and two or three polar bodies (**fig. 3.31**).

The male produces four functional sperm for each spermatogonium that enters meiosis, but the female produces only one functional egg for each primary oocyte that completes meiosis. In the case of fraternal twins, two oocytes have matured simultaneously. On average, a human female produces about 400 ova in her lifetime.

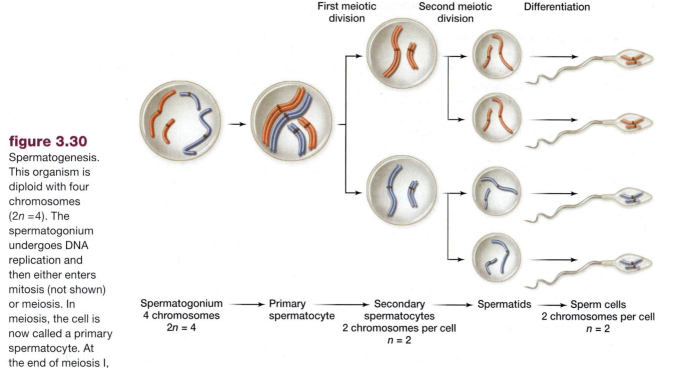

figure 3.30
Spermatogenesis. This organism is diploid with four chromosomes (2*n* =4). The spermatogonium undergoes DNA replication and then either enters mitosis (not shown) or meiosis. In meiosis, the cell is now called a primary spermatocyte. At the end of meiosis I,

Spermatogonium
4 chromosomes
2*n* = 4
→ Primary spermatocyte
→ Secondary spermatocytes
2 chromosomes per cell
n = 2
→ Spermatids
→ Sperm cells
2 chromosomes per cell
n = 2

the two secondary spermatocytes are both haploid (*n* = 2). Both of these cells proceed through meiosis II to yield a total of four spermatids that differentiate into male gametes, spermatozoa (sperm cells). Maternal chromosomes are red; paternal chromosomes are blue.

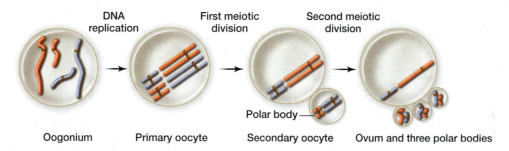

Oogonium   Primary oocyte   Secondary oocyte   Ovum and three polar bodies

Polar body

figure 3.31   Oogenesis. An oogonium in the developing ovary undergoes DNA replication and then either enters mitosis (not shown) or meiosis. The primary oocyte, which will enter meiosis, arrests during diplonema. At puberty, one primary oocyte per month completes meiosis. The result is one ovum (functional egg) and three polar bodies are produced from each primary oocyte. Maternal chromosomes are red; paternal chromosomes are blue.

Restating the Concepts

▶ Meiosis in mammals occurs in specialized cells located in the gonads.

▶ Spermatogenesis is the meiotic process in human males, which produces four sperm cells per primary spermatocyte.

▶ Oogenesis is the meiotic process in human females, which produces one ovum per primary oocyte. The remaining meiotic products, polar bodies, disintegrate.

▶ Spermatogenesis occurs continuously after the male reaches sexual maturity. Primary oocytes, which are produced during embryogenesis, are stalled during meiosis I. It is reinitiated, under hormonal control, in one primary oocyte per month after the female reaches sexual maturity.

3.8 Life Cycles

Mammals and many multicellular animals exist as diploid organisms, with a temporary haploid form as gametes that participate in sexual reproduction. However, exceptions to this basic model are numerous. For example, in the bees, wasps, and ants (Hymenoptera), males are haploid and produce gametes by mitosis; females are diploid and produce gametes through meiosis. Some fishes exist by **parthenogenesis,** in which the offspring come from unfertilized eggs that do not undergo meiosis. And, in some copepods, the sexual and parthenogenetic stages of their life cycles alternate.

The general pattern of the plant life cycle alternates between two distinct generations, each of which, depending on the species, may exist independently.

In lower plants, the haploid generation predominates, whereas in higher plants, the diploid generation is dominant. In flowering plants (**angiosperms**), the plant you see is the diploid **sporophyte** (**fig. 3.32**). It is referred to as a sporophyte because, through meiosis, it gives rise to spores. The spores germinate into the alternate generation, the haploid **gametophyte,** which produces gametes by mitosis. Fertilization then produces the next generation of diploid sporophytes.

In lower plants, the gametophyte has an independent existence; in angiosperms, this generation is radically reduced. For example, in corn (see fig. 3.32), an angiosperm, the mature corn plant is the sporophyte. The male flowers produce microspores by meiosis. After mitosis, three cells exist in each spore, a structure that we call a **pollen grain,** the male gametophyte. In female flowers, meiosis produces megaspores. Mitosis within a megaspore produces an embryo sac of seven cells with eight nuclei. This is the female gametophyte. A sperm cell fertilizes the egg cell. The two polar nuclei of the embryo sac are fertilized by a second sperm cell,

producing triploid (3*n*) nutritive endosperm tissue. The sporophyte grows from the diploid fertilized egg.

Many fungi and protista exist primarily as haploids in their mature form. These organisms proceed through mitosis to continually produce haploid cells (**fig. 3.33**). Some of these haploid cells (gametes) possess the ability to fuse with other haploid cells. Thus, in these haploid organisms, gametes are produced by mitosis. After gametes fuse, they generate a diploid cell that is able to undergo meiosis (see fig. 3.33). Meiosis then produces haploid cells that are called **sexual spores.** In some species, these spores are new unicellular adults; in other species the spores go through several rounds of mitosis to produce the multicellular haploid organism.

Why would a haploid organism want to go through meiosis when it could simply propagate through mitosis? This question brings us back to one of the major features of meiosis, which is to increase genetic diversity

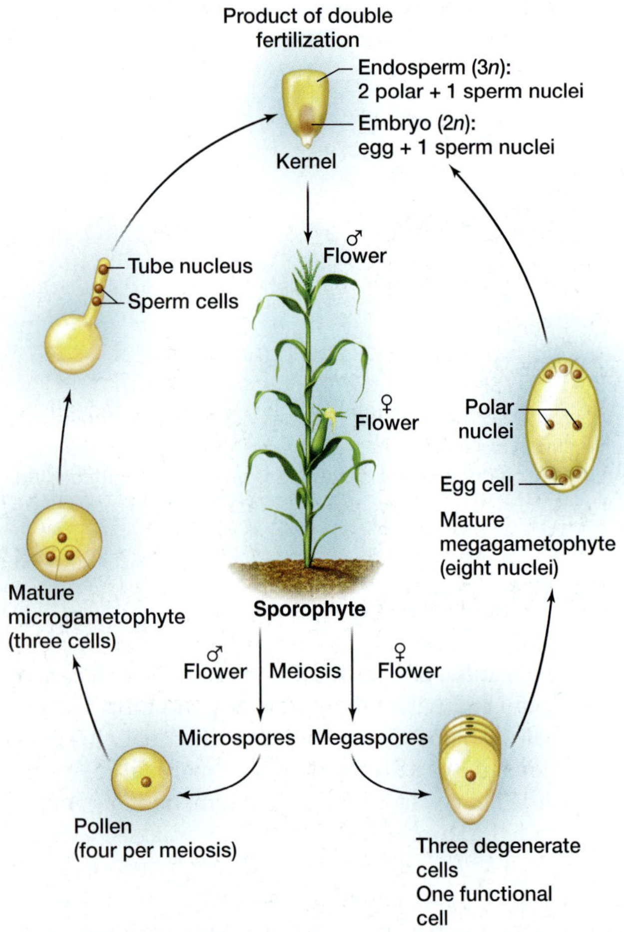

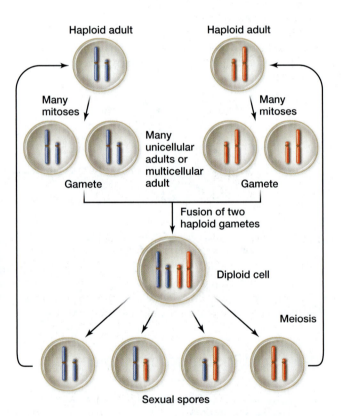

figure 3.32   Life cycle of the corn plant. Generation of a corn plant requires a double fertilization event. The embryo in the kernel is diploid and results from the fertilization of one sperm and one egg cell. The kernel's endosperm is the fusion of two polar nuclei and one sperm nucleus.

figure 3.33   A generalized diagram of the sexual life cycle of an organism in which the haploid generation is dominant. The sexual spores that are generated are either the mature haploid adult in unicellular organisms or go through several mitoses to produce the haploid multicellular adult. Haploid cells must fuse to generate a temporary diploid, which can then undergo meiosis. Notice that meiosis produces haploid cells that exhibit independent assortment of chromosomes and genotypically different haploid cells relative to the haploid parents. The meiotic products that result from recombination are not shown, but are also produced. The red and blue chromosomes represent the chromosomes that are inherited from one of the two parental lineages .

within a population. For a haploid organism, meiosis provides the opportunity to change combinations of alleles through both independent assortment of chromosomes and recombination between nonsister chromatids. This increased genetic diversity provides a significant advantage for the adaptation and survival of a species.

It's Your Turn

You should now try problems 30, 32, and 37 to test your understanding of the different types of life cycles.

 ## Restating the Concepts

▶ Diploid organisms undergo meiosis to produce gametes that fuse to reestablish the diploid life form.

▶ Haploid organisms can undergo meiosis. However, two haploid cells must fuse to generate a diploid cell that is then able to enter meiosis. The haploid meiotic products can then regenerate the organism through mitosis.

▶ One of the key purposes of meiosis for both haploid and diploid organisms is to increase genetic diversity through independent assortment and recombination.

3.9 Chromosomal Theory of Inheritance

Prior to the early twentieth century, biologists were not certain what portion of the cell carried and transmitted the genetic traits described by Mendel. In the early 1900s, Walter Sutton and Theodor Boveri independently published papers that described the behavior of chromosomes during meiosis. They noted that chromosomes moved in a manner that was consistent with Mendel's first and second laws. This idea, that chromosomes in meiosis behave in a manner that correlated with the movement of Mendel's genetic factors, became known as the **chromosomal theory of inheritance.** Sutton proposed that chromosomes were associated with the genetic unit factors described by Mendel. We now know that the genes (Mendel's unit factors) are located on the chromosomes.

The work required to confirm this theory depended on traits that are **sex-linked,** that is, their genes are carried on the chromosomes that determine the sex of an organism. The details of experiments to prove the chromosomal theory of inheritance are described in chapter 4.

In this chapter, we discussed how eukaryotic cells divide and eukaryotic organisms propagate. Cell growth involves the cell cycle, in which the genomic DNA replicates prior to nuclear division. We describe the molecular details of DNA replication in chapter 9, which includes the enzymes and activities required for replicating linear DNA. One of the major features of mitosis is to produce daughter cells that are genetically identical; in chapter 9 you will see how DNA replication possesses the ability to minimize errors, and in chapter 18 you will learn the different mechanisms employed to repair mutations in the DNA.

We also described in this chapter the process of meiosis, which produces haploid gametes. The pairing of the homologous chromosomes and their separation during meiosis I is consistent with Mendel's first law of equal segregation, and the orientation and separation of one homologous pair is independent of another pair, which underlies Mendel's second law of independent assortment (chapter 2). The realization that Mendel's factors were associated with chromosomes was the basis for the chromosomal theory of inheritance. In chapter 4, we will discuss how this theory was proved using traits that are associated with the *Drosophila* X chromosome.

One of the key differences between mitosis and meiosis is that the homologous chromosomes pair during meiosis I, which leads to crossing over, or recombina-

tion, between nonsister chromatids. Synapsis ensures that homologs pair during meiosis I, and it also provides a powerful mechanism to localize or map genes and mutations on chromosomes. In chapter 6, we will discuss the mechanism of recombination and the application of recombination to mapping genes. Finally, in chapter 8 we will describe how altered chromosomal structures affect synapsis and recombination during meiosis.

As we discussed, homologous chromosomes pair during meiosis I. However, mammalian males contain two different sex chromosomes (X and Y) that pair during meiosis I. In chapter 4, we will describe how this occurs. You will also see how meiosis can be disrupted, such that the gametes that are produced fail to contain one chromosome from each homologous pair. In chapter 8, we will discuss how disrupting meiosis can lead to changes in the chromosome number in offspring.

Mitosis and meiosis are two fundamental processes that underlie the transmission of genetic material during cell division and from generation to generation. Therefore, a clear understanding of these processes will help to explain many of the genetic phenomena described throughout this book. You may want to refer back to this material as you read later chapters to help clarify how other processes are related to mitosis and meiosis.

Chapter Summary Questions

1. For each term in column I, choose the best matching phrase in column II.

   Column I
   1. Spermatogonium
   2. S phase
   3. Metacentric
   4. Heterochromatin
   5. Spermatid
   6. Acrocentric
   7. G_0 phase
   8. Secondary spermatocyte
   9. Euchromatin
   10. Telocentric

   Column II
   a. Phase that has exited from the cell cycle
   b. Loosely packed DNA
   c. Chromosome with its centromere in the middle
   d. Product of meiosis I in males
   e. Chromosome with its centromere at one end
   f. Diploid male cell
   g. Tightly condensed DNA
   h. Cell cycle phase when DNA replication occurs
   i. Product of meiosis II in males
   j. Chromosome with its centromere near one end

2. What are the major differences between prokaryotes and eukaryotes?

3. Distinguish between a centromere and a kinetochore?

4. What is the difference between sister and nonsister chromatids?

5. What are homologous chromosomes? Are they identical? Explain.

6. What are the differences between a reductional and an equational division? What do these terms refer to?

7. Use diagrams to describe the life cycles of human beings and peas.

8. What keeps sister chromatids together after the S phase? What is the mechanism that triggers their separation in anaphase?

9. Name three major checkpoints in the cell cycle and briefly discuss their roles.

10. Meiosis ensures genetic diversity. What are the two mechanisms involved, and when do they occur during meiosis?

11. Indicate whether each of the following events occur in mitosis, meiosis I, or meiosis II.
    a. Chromosomes replicate prior to the initial stage.
    b. Sister chromatids attach to spindle fibers from opposite poles.
    c. Homologous chromosomes undergo synapsis.
    d. Diploid cells are produced.
    e. Chromosomes are maximally condensed.
    f. Centromeres divide.

12. Diploid cells contain an even number of chromosomes (refer to Table 3.2). Provide an explanation.

13. How is mitotic telophase different in animals and plants?

14. List three major differences between spermatogenesis and oogenesis in mammals.

Solved Problems

PROBLEM 1: What are the differences between chromosomes and chromatids?

Answer: In eukaryotes, a chromosome is a linear DNA molecule complexed with protein and, generally, with a centromere somewhere along its length. During the cell cycle, in the S phase, the DNA replicates and each chromosome is duplicated. The duplication is visible in the early stages of mitosis and meiosis when chromosomes condense. At this point, each duplicated chromosome is composed of two sister chromatids that are identical in sequence and joined by a centromere. The sister chromatids do not become chromosomes until the centromeres divide and the DNA molecules/chromosomes begin to be pulled to opposite centrosomes.

PROBLEM 2: A hypothetical organism has six chromosomes ($2n = 6$). How many different combinations of maternal and paternal chromosomes can appear in the gametes?

Answer: You could figure this empirically by listing all combinations. For example, let A, B, and C = maternal chromosomes and A', B', and C' = paternal chromosomes. Two combinations in the gametes could be A B C' and A' B' C;

obviously, several other combinations exist. It is easier to recall that 2^n = number of combinations, where n = the number of chromosome pairs. In this case, $n = 3$, so we expect $2^3 = 8$ different combinations.

PROBLEM 3: The arctic fox, *Alopex lagopus*, has a diploid complement of 50 small chromosomes. The red fox, *Vulpes vulpes*, has 38 larger chromosomes. Mating between an artic fox and a red fox produces a hybrid of these two species that is sterile. Cytological studies during meiosis in these hybrids reveal both paired and unpaired chromosomes.

a. How many chromosomes are found in the hybrid offspring?
b. Account for the sterility of the hybrids.
c. How can you explain the paired chromosomes?

Answer:

a. The arctic fox parent produces gametes with the haploid number of chromosomes for its species, or $50/2 = 25$. The red fox parent produces gametes with the haploid number of chromosomes for its species, or $38/2 = 19$. The union of these two gametes will produce a zygote containing 25 + 19 = 44 chromosomes. The zygote will develop into an adult hybrid with a chromosome number of 44.

b. The 44 chromosomes of the hybrid are derived from two different species. Even if all 19 of the red fox's chromosomes were identical to 19 of the arctic fox, the

remaining 6 artic fox chromosomes could not pair, and imbalanced gametes and zygotes would result.

c. There must be some homology between the large red fox chromosomes and the small arctic fox chromosomes. This homology allows some of the chromosomes to pair.

Exercises and Problems

15. In human beings, $2n = 46$. How many chromosomes would you find in a
   a. brain cell?
   b. red blood cell?
   c. polar body?
   d. sperm cell?
   e. secondary oocyte?

16. You are working with a species with a chromosome number of $2n = 6$, in which one pair of chromosomes is telocentric, one pair submetacentric, and one pair metacentric. The *A*, *B*, and *C* loci, each segregating a dominant and recessive allele (*A* and *a*, *B* and *b*, *C* and *c*), are each located on different chromosome pairs. Draw the labelled chromosomes through the different stages of mitosis.

17. Given the same information as in problem 16, diagram one of the possible meioses. How many different gametes can arise, absent crossing over? What variation in gamete genotype is introduced by a crossover between the *A* locus and its centromere?

18. Identify the stages of mitosis and meiosis that are associated with panels *a–f* in the following figure. Include the process, stage, whether the cell is haploid or diploid, and the chromosome number (e.g., meiosis I, prophase, $2n = 6$). Keep in mind that one picture could represent more than one process and stage.

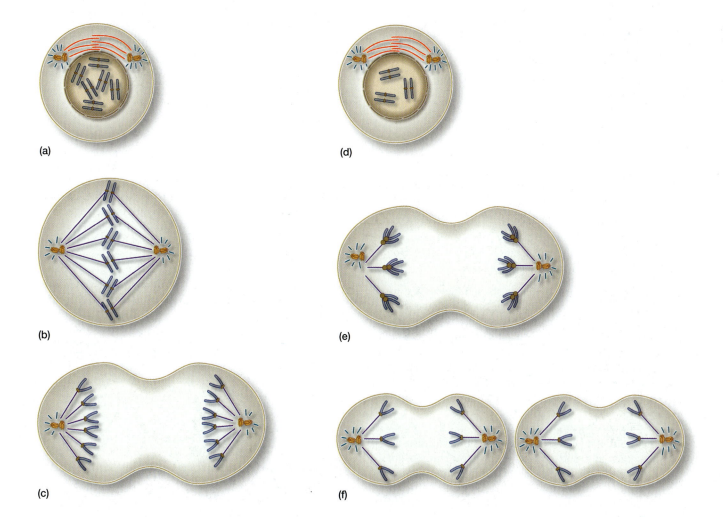

(a)

(b)

(c)

(d)

(e)

(f)

19. A mature human sperm cell has c amount of DNA. How much DNA (c, $2c$, $4c$, etc.) will a
    a. somatic cell have if it is in G_1?
    b. somatic cell have if it is in G_2?
    c. cell have at the end of meiosis I?

20. How many bivalents, tetrads, and dyads would you find during meiosis in human beings? in fruit flies? in the other species of table 3.2?

21. Can you devise a method of chromosome partitioning during gamete formation that would not involve synapsis—that is, can you reengineer meiosis without passing through a synapsis stage?

22. An organism has six pairs of chromosomes. In the absence of crossing over, how many different chromosomal combinations are possible in the gametes?

23. Wheat has $2n = 42$ and rye has $2n = 14$ chromosomes. Explain why a wheat–rye hybrid is usually sterile.

24. How do the quantity of genetic material and the number of chromosome sets change from stage to stage of spermatogenesis and oogenesis (see figs. 3.30 and 3.31)? (Consider the spermatogonium and the oogonium to be diploid, with the chromosome number arbitrarily set at two.)

25. In human beings, how many sperm cells will form from
    a. fifty primary spermatocytes?
    b. fifty secondary spermatocytes?
    c. fifty spermatids?

26. In human beings, how many eggs will form from
    a. fifty primary oocytes?
    b. fifty secondary oocytes?

27. In corn, the diploid number is 20. How many chromosomes would you find in a(n)
    a. sporophyte leaf cell?
    b. embryo cell?
    c. pollen grain?
    d. polar nucleus?

28. The plant *Arabidopsis thaliana* has five pairs of chromosomes: *AA*, *BB*, *CC*, *DD*, and *EE*. If this plant is self-fertilized, which of the following chromosome complements would be found in a root cell of the offspring?
    a. *A B C D E*
    b. *AA BB CC DD EE*
    c. *AAA BBB CCC DDD EEE*
    d. *AAAA BBBB CCCC DDDD EEEE*

29. Metacentric chromosomes are shaped like an X during metaphase of mitosis, but take on a V shape during anaphase. Telocentric chromosomes, on the other hand, have a V shape and an I shape in metaphase and anaphase, respectively. Explain.

30. In wheat, the haploid number is 21. How many chromosomes would you expect to find in
    a. the tube nucleus?
    b. a leaf cell?
    c. the endosperm?

31. A diploid plant consists of two pairs of telocentric chromosomes and one pair of metacentric chromosomes. After four generations of self-fertilization, what proportion of the progeny will have only telocentric chromosomes and what proportion will have the original parental composition? Explain.

32. If a plant of genotype *AA* pollinates one with a genotype *aa*, what is the genotype of the zygote and endosperm of the resulting seed?

33. The domestic cow, *Bos taurus,* has a diploid complement of 60 chromosomes. How many chromosomes and chromatids (if applicable) will be found in cells at the following stages of division
    a. G_1 of mitotic interphase
    b. metaphase of mitosis
    c. metaphase I of meiosis
    d. metaphase II of meiosis
    e. telophase II of meiosis

34. The house mouse, *Mus musculus,* has 20 pairs of chromosomes. When two mice mate, how many genetically different offspring are possible? Assume there is no crossing over.

35. Determine the diploid number of the organisms based on the following observations:
    a. 16 chromosomes per cell at anaphase of mitosis
    b. 38 centromeres per cell at anaphase of meiosis I
    c. 40 chromatids per cell at prophase of meiosis II

36. Drone (male) honeybees are haploid (arising from unfertilized eggs), and a queen (female) is diploid. Draw a testcross between a dihybrid queen and a drone. How many different kinds of sons and daughters might result from this cross?

37. If a dihybrid corn plant is self-fertilized, what genotypes of the triploid endosperm can result? If you know the endosperm genotype, can you determine the genotype of the embryo?

38. A hypothetical organism has two distinct chromosomes ($2n = 4$) and 50 known genes, each with two alleles. If an individual is heterozygous at all known loci, how many gametes can be produced if
    a. all genes behave independently?
    b. all genes are completely linked?

39. Can meiosis occur in a haploid cell? Can mitosis?

40. You are studying a diploid organism that contains a metacentric chromosome and a telocentric chromosome (a total of 0.3 pg of DNA in the diploid cell). Each of the circles in the following diagram represents the cell as the chromosome(s) arrange in the center of the cell during mitosis, meiosis I, or meiosis II.
    a. Diagram the chromosomes in each cell and draw a dashed line to represent the location/orientation of the cell division.
    b. Under each cell, state the number of chromosomes, chromatids, and picograms of DNA present in the diagrammed cell.

Mitosis	Meiosis I	Meiosis II

Chromosome number   _____   _____   _____
Chromatid number   _____   _____   _____
Picograms of DNA   _____   _____   _____

Chapter Integration Problem

Consider the following diploid cell, in which $2n = 8$. The blue chromosomes carrying genes A, B, C, and D are maternal in origin, whereas the red chromosomes carrying genes a, b, c, and d are paternally derived.

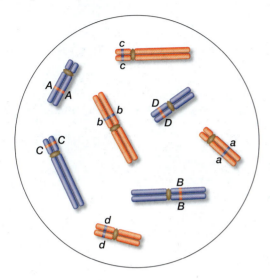

a. Use the behavior of these chromosomes/genes during cell division to explain Mendel's first and second laws.

b. Which gene is found on the p arm of an acrocentric chromosome? Which gene is found on the q arm of a metacentric chromosome?

c. What stage of mitosis or meiosis is this cell in? Explain.

d. If this cell undergoes mitosis, what will be the genotype of its daughter cells?

e. If this cell undergoes meiosis, how many tetrads will form during prophase I? In how many different ways can these tetrads align during metaphase I? Draw each of these possible tetrad alignments. How does this correspond to Mendel's laws?

f. What fraction of the gametes produced by this cell will have the genotype $ABCD$?

g. What fraction of the gametes produced by this cell will have two maternal chromosomes and two paternal chromosomes?

h. What fraction of the gametes produced by this cell will have some chromosomes of both maternal and paternal origin?

 Do you need additional review? Visit **www.mhhe.com/hyde** for practice tests, answers to end-of-chapter questions and problems, interactive exercises, and animation tutorials all designed to enhance your understanding of key genetic concepts.

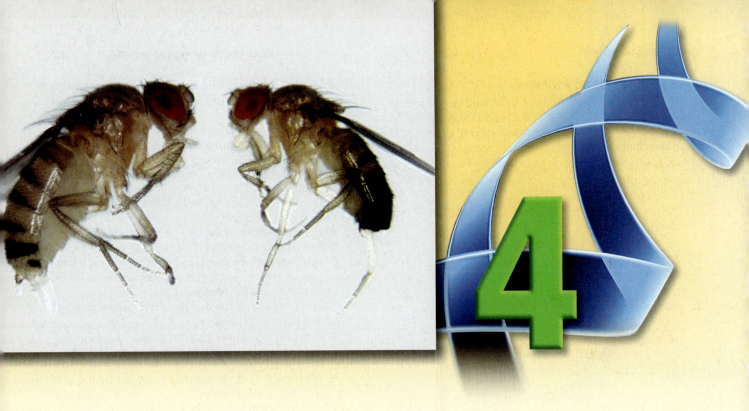

Sex Linkage and Pedigree Analysis

Essential Goals

1. Distinguish between traits that result from genes located on the sex chromosomes and those that are present on the autosomes and whose expression are influenced by gender.

2. Explain the mechanisms of sex determination in various organisms.

3. Describe the different mechanisms organisms use to achieve balanced expression of sex-linked genes.

4. Use pedigree analysis to deduce the mode of inheritance for species that produce small numbers of offspring.

5. Calculate the probability that a particular individual in a pedigree has a specific genotype by determining the pattern of inheritance and deducing the genotypes of some of the individuals in the pedigree.

Chapter Outline

Photo: A female (left) and male (right) fruit fly, *Drosophila melanogaster.*

*F*or eukaryotic organisms, sexual reproduction is one of the major mechanisms by which genetic diversity is introduced into an organism. In diploid organisms, sexual reproduction involves the fusion of two haploid gametes. The independent assortment of homologous chromosomes and the recombination between nonsister chromatids that occur while generating the gametes are significant sources of genetic diversity.

As you will see in this chapter, the topic of sex impinges on genetics in a variety of different ways. First is the question of how sex (gender) is determined in an organism. Several mechanisms exist in different organisms. In many cases, the two genders have differences in the number of sex chromosomes. Usually, one gender will have two copies of the same sex chromosome and the other has two different sex chromosomes, or only one sex chromosome.

Although this difference in sex chromosome number is a convenient way to distinguish between the sexes, it does produce a potential problem. Many important genes are present on the sex chromosome that must be expressed in both sexes. Does this suggest that one gender will have twice as many copies of these genes and potentially express twice the amount of encoded proteins as the other gender? This increased (or decreased) expression could be detrimental. As you will see, each organism has mechanisms to alter the expression of genes on the sex chromosome.

You know from previous discussions, genes can be present in two different alleles: one dominant and one recessive. In the Mendelian pattern of inheritance (chapter 2), each diploid individual has two copies of every gene. However, if a gene is present on a sex chromosome, one gender will have two copies of the gene and the other may only have one copy. These differences affect not only the interpretation of dominant and recessive alleles but also the pattern of inheritance we observe. Based on the pattern of inheritance of a trait in a family, it is possible to deduce not only if the trait is dominant or recessive but also if the corresponding gene is located on an autosome or sex chromosome.

4.1 Sex Linkage

The gender of an organism usually depends on a very complicated series of developmental changes under genetic and hormonal control. But often one or a few genes can determine which pathway of development an organism takes. Some of those control genes are located on the **sex chromosomes.** Unlike the homologous pairs of chromosomes that we discussed during meiosis (chapter 3), namely **autosomes,** the sex chromosomes often consist of two different chromosomes. **Sex linkage** refers to the genes that are located on the sex chromosomes and their inheritance.

Differences in the expression of some phenotypes between the sexes have been known for many centuries. For example, a common form of hemophilia (a failure of blood to clot) affects mostly men. Women often pass the disease on to their sons without showing any symptoms themselves. The general nature of the inheritance of this trait was known in biblical times. The Talmud—the Jewish book of laws and traditions—described this uncontrolled bleeding during circumcision. Certain males were exempt from circumcision if the uncontrolled bleeding occurred in certain relatives on the maternal side of the family. Without understanding the genetic details, the writers of the Talmud were nonetheless describing an X-linked recessive disease.

Sex-Linked Patterns of Inheritance

Two significant inheritance patterns are observed for X-linked recessive alleles. First, males will preferentially exhibit the recessive X-linked phenotype. This expression pattern of primarily affecting males is shown in **figure 4.1**.

Because females have two X chromosomes, they can produce phenotypically normal offspring that have either homozygous or heterozygous allelic combinations. Males, with only one copy of the X chromosome, are neither homozygous (two copies of the same allele) nor heterozygous (two different alleles). Instead, males are **hemizygous,** which describes genes that are present in only one copy, like the X-linked genes in males.

When only a single copy of a gene is present, a single recessive allele can determine the phenotype in a phenomenon called **pseudodominance.** Thus, a male fly with one recessive *w* allele is white-eyed, with the *w* allele acting like a dominant allele to confer the white-eyed phenotype. This demonstrates that expression of a recessive phenotype is not dependent upon the presence of two recessive alleles, but rather the absence of a dominant allele.

The second major inheritance pattern observed with X-linked genes and the corresponding phenotypes is a *crisscross pattern of inheritance.* **Figure 4.1d** demonstrates this pattern as the male parent passes his dominant trait (such as wild-type eyes, fig. 4.1d) to his female offspring, and the female parent passes her recessive phenotype (white eyes) to her male offspring. If we examine the reciprocal cross, we do not observe the crisscross pattern (fig. 4.1a). Thus, the crisscross pattern of phenotypic inheritance is not always observed for X-linked genes.

Because males do not transmit their X chromosome to their male offspring, a recessive X-linked phenotype such as color blindness in humans is never normally transmitted from father to son. However, the male will

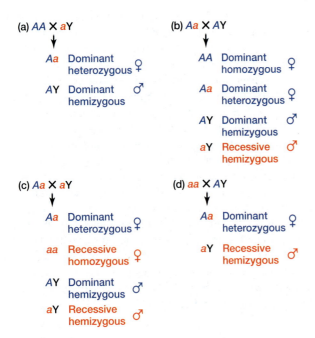

figure 4.1 Four different crosses involving an X-linked recessive allele. (a) If the mother is homozygous dominant, all the offspring will exhibit the dominant phenotype. (b) When the mother is heterozygous and the father has the dominant phenotype, the only offspring with the recessive phenotype will be half the male offspring. (c) When the mother is heterozygous and the father exhibits the recessive phenotype, half the sons and daughters will have the recessive phenotype. (d) When the female exhibits the recessive phenotype and the father the dominant phenotype, all the sons will have the recessive phenotype and all the daughters the dominant phenotype.

transmit his X chromosome to his daughter (fig. 4.1a). The daughter of an affected male will be heterozygous if she is phenotypically normal. We call this heterozygous individual a **carrier,** as she possesses the recessive allele. Thus, the transmission of the X chromosome from the male parent to only his female offspring to produce carriers, along with the pseudodominance of X-linked recessive alleles in males, combine to produce the crisscross pattern of inheritance.

Deviation from Mendelian Patterns

Notice that the two different crosses (see fig. 4.1a and 4.1d) produced different phenotypes in the F₁ progeny. These differences are not consistent with the Mendelian pattern of inheritance that we described in chapter 2. In figure 4.1a, the recessive white-eyed phenotype is lost in the F₁ generation as expected, but in the reciprocal cross it is present in half the progeny (fig. 4.1d).

This does not suggest that Mendel's laws are incorrect, but rather it demonstrates another level of complexity that Mendel did not observe in his pea crosses. In this case, the increased complexity is due to the two different sex chromosomes and the hemizygous

nature of the male. It is comforting that as we uncover the cause of the increased complexity, we see how Mendel's laws still apply, through the pairing of the X and Y chromosome during meiosis in the male. This observation demonstrates that Mendel's laws are actually more fundamental than we had expected.

In the following chapter, we will explore other examples of deviations from Mendel's patterns.

Y-Linked Traits

Because both the X and the Y are sex chromosomes, two different patterns of sex-linked inheritance are possible. The terms *sex-linked* and *X-linked* usually refer to loci found only on the X chromosome. The term **Y-linked** refers to loci found only on the Y chromosome, which controls **holandric traits** (traits found only in males).

There are 1,098 genes that are located on the human X chromosome, whereas 171 genes are known to be on the Y chromosome. The small number of loci on the Y chromosome makes the identification of Y-linked phenotypes difficult. In humans, the best example of a Y-linked trait is a form of retinitis pigmentosa, which results in night blindness that progresses into complete blindness. Retinitis pigmentosa has many different genetic causes, some of which are autosomal and others are sex-linked.

This particular form of retinitis pigmentosa was first described in a four-generation Chinese family. In this family, only males were affected. Furthermore, all of the sons, and none of the daughters, of an affected male were also affected. All of the daughters of an affected male also failed to produce an affected child. This transmission from father to all of his sons and no affected females is consistent with the trait being inherited in a Y-linked manner.

For many years, hairy ear rims was believed to be a Y-linked trait (**fig. 4.2**) because of the description that it was inherited only in fathers and their sons. NIH's database, the Online Mendelian Inheritance in Man (OMIM) describes recent data that strongly suggests that hairy ear rims result either from more than one locus on the Y chromosome, with one of the loci located in the pseudoautosomal region, or it is not Y-linked at all.

Sex-Limited and Sex-Influenced Traits

Aside from X-linked and Y-linked inheritance, two additional inheritance patterns show a bias in the phenotypic expression between the two sexes. However, the genes controlling these traits are located on the autosomes rather than either of the sex chromosomes.

Sex-limited traits are expressed in only one gender, although the genes are present in both. In women, breast development and milk production are sex-limited traits, as are facial hair distribution and

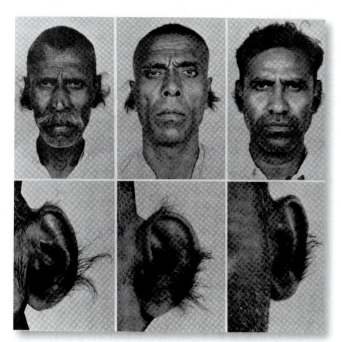

figure 4.2   Hairy ear rims. Hairy rims was thought to be a Y-linked trait because it is transmitted from father to son in three pedigrees. Further studies revealed that it is actually due to two or more genes that might be autosomal.

figure 4.3   Sex-limited trait. A sex-limited trait is the presence of horns on the bull elk (left) and their absence in the cow (female) elk.

sperm production in men. Nonhuman examples of sex-limited traits include plum age patterns in birds—in many species, the male is brightly colored—and horns found in male, but not female, elk (**fig. 4.3**).

Sex-influenced traits appear in both genders, but are recessive in one and dominant in the other. One potential example of a sex-influenced trait in humans is a heart condition called long QT syndrome (LQTS). This syndrome is characterized by a prolonged QT interval on the electrocardiograph, ventricular arrhythmias, potential seizure, and sudden death. LQTS appears to

affect more adult females than adult males, suggesting that a sex-related difference is present. However, the mutation associated with LQTS has been mapped to the voltage-gated potassium channel-1 gene (*KCNQ1* gene), which is located on chromosome 11.

The autosomal location of the gene suggests that males and females would be equally affected by LQTS. However, the predominance of this autosomal-linked phenotype in females confers the sex-influenced nature on LQTS. The reason that women preferentially exhibit LQTS remains unclear, although hormonal differences between the sexes is one possibility.

 Restating the Concepts

▶ Some genes are present on one or both of the sex chromosomes (X and Y chromosomes).

▶ Transmission of sex-linked genes results in differences of the phenotypic expression between the two sexes.

▶ X-linked recessive alleles are preferentially expressed in males due to pseudodominance, which aids in the crisscross pattern of inheritance. Affected males pass the recessive allele to their female offspring, which are called carriers.

▶ Autosomal genes that exhibit differences in their phenotypic expression between the sexes reflect either sex-limited or sex-influenced traits.

4.2 Proof of the Chromosomal Theory of Inheritance

We mentioned at the end of chapter 3 that in the early twentieth century, the *chromosomal theory of inheritance* proposed that the traits Mendel observed were found on chromosomes in the nucleus, which were passed on each time a cell divided. Fusion of haploid gametes at fertilization combined parental chromosomes and the traits they carried in a new individual.

The chromosomal theory focused significant effort on determining how chromosomes contained and passed on the genetic information within a species. Thomas Hunt Morgan and his student Calvin Bridges performed the major experiments in *Drosophila* that definitively demonstrated that genes were associated with chromosomes (**fig. 4.4**).

The Case of the White-Eyed Male

In 1910, Morgan described the pattern of **sex-linked inheritance** in *Drosophila*. He isolated a white-eyed mutant male in a culture of wild-type (red-eyed) flies (**fig. 4.5**). When this male was crossed with a wild-type female, all the F$_1$ progeny were wild-type (see fig. 4.5). The loss of the white-eyed phenotype in these F$_1$

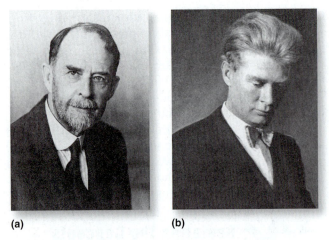

(a)   (b)

figure 4.4   Proof of the chromosomal theory of inheritance was shown by (a) Thomas Hunt Morgan (1866–1945) and one of his students, (b) Calvin B. Bridges (1889–1938). Using the X-linked *white* mutation in *Drosophila*, Morgan and Bridges correctly predicted the number of X and Y chromosomes in individual flies based on the sex and eye color of the individuals and their parents.

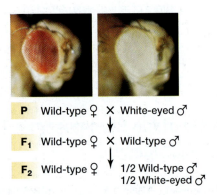

figure 4.5   The wild-type (red-eyed) and white-eyed phenotypes in *Drosophila*. Shown below the photos are the F$_1$ and F$_2$ phenotypes when a wild-type female is crossed with a white-eyed male. Notice that half of the F$_2$ males, but none of the F$_2$ females, exhibit the white-eyed phenotype.

progeny suggested that the white-eyed phenotype was recessive to the wild-type phenotype.

When these F$_1$ individuals were crossed with each other, both white-eyed and red-eyed progeny were produced (see fig. 4.5). However, the phenotypes were not equally distributed between both sexes. All the females and half the males were red-eyed, whereas the remaining half of the males were white-eyed.

Because the female possessed two X chromosomes and the male only one X chromosome, we can redraw the crosses to include the sex chromosomes (**fig. 4.6**). Furthermore, the X chromosome contains the *white* gene locus, but the Y chromosome does not. We denote the *white* allele as *w* and the wild-type allele as +.

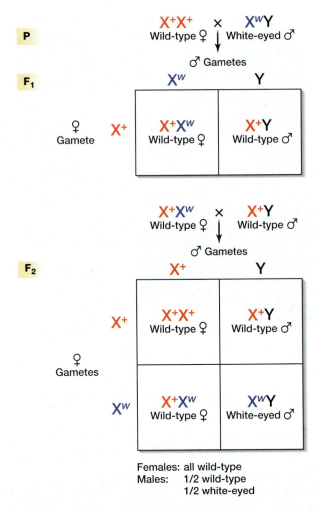

figure 4.6   Crosses of figure 4.5 redrawn to include the sex chromosomes. The dominant wild-type eye color allele on the X chromosome is represented as X$^+$ and the recessive white-eyed allele is represented as X^w.

Morgan then performed the **reciprocal cross,** which involved switching the phenotypes of the parents (**fig. 4.7**). In contrast to all the F$_1$ progeny having the dominant wild-type phenotype (see fig. 4.6), the reciprocal cross produced white-eyed F$_1$ males and red-eyed F$_1$ females (see fig. 4.7). This is a **crisscross pattern of inheritance,** in which the male passes his phenotype to his daughters and the female passes her phenotype to her sons. Crossing two F$_1$ individuals produced both white-eyed flies (half the males and half the females) and red-eyed flies (half the males and half the females).

The inheritance of the white-eyed phenotype from Morgan's crosses was consistent with the eye color phenotype being associated with the X chromosome. However, he was unable to provide conclusive proof of the chromosomal theory of inheritance. This proof was left to one of Morgan's students, Calvin Bridges.

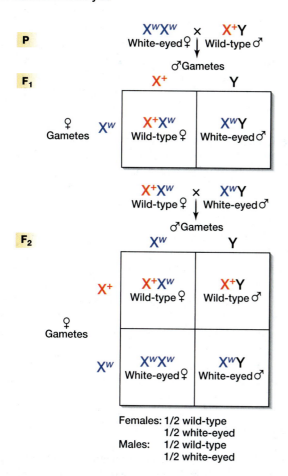

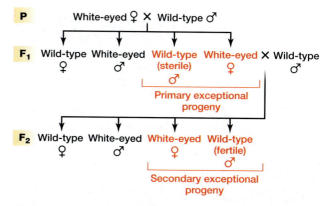

figure 4.7   Reciprocal cross to that shown in figure 4.6. A crisscross pattern of inheritance is seen between the parents and their F$_1$ progeny. Notice that half the F$_2$ females are wild-type and half have white eyes in this cross, but all the F$_2$ females in the reciprocal cross (fig. 4.6) are wild-type.

The Logic of Gender

Bridges repeated Morgan's cross of red-eyed males (X$^+$Y) and white-eyed females (X^{w}X^w) in very large numbers (**fig. 4.8**). The F$_1$ progeny consisted of white-eyed males and red-eyed females to observe the crisscross pattern of inheritance. However, 1 in approximately every 2000 males were red-eyed, and 1 in every 2000 females were white-eyed (see fig. 4.8). Bridges called these rare F$_1$ individuals **primary exceptional progeny.** Although the primary exceptional males were sterile, the primary exceptional females were fertile.

Bridges then crossed the white-eyed primary exceptional females with red-eyed males and again found the majority of the progeny were white-eyed males and red-eyed females (see fig. 4.8). Now, however, approximately 1 in 200 males were red-eyed and an equivalent ratio of females were white-eyed. Bridges called these unusual F$_2$ progeny **secondary exceptional progeny.** Unlike the primary exceptional males, the secondary exceptional males and females were both fertile.

figure 4.8   Pattern of inheritance of the white-eye trait in *Drosophila.* The recessive *white* allele is located on the X chromosome and exhibits the crisscross pattern of inheritance. However, at a low frequency, the crisscross pattern of inheritance is not observed in what are called the primary and secondary exceptional progeny. While the primary exceptional male is sterile, the primary exceptional female is fertile and crossed to wild-type males to produce the F$_2$ generation.

Bridges used the knowledge of *Drosophila* sex determination (see section 4.3) and the fact that the *white* gene was on the X chromosome to construct a model to explain this data. Bridges knew that *Drosophila* males only have one X chromosome, and that males lacking a Y chromosome were sterile. Therefore, the sterile primary exceptional males must have been X0. The X chromosome in these males must have been inherited from their male parent, as this parent possessed the only X chromosome with the wild-type eye color allele.

How did the male pass his X chromosome on to a male offspring? Normally, an X chromosome from the male parent would be passed to female offspring. Therefore, a male fly having a paternal X chromosome can occur only if the female parent produced an egg that lacks an X chromosome. The result is that the primary exceptional male receives the X chromosome from the male parent and no sex chromosome (0) from the female parent (see **fig. 4.9**). According to genic balance in *Drosophila,* the offspring's X:A ratio would be 0.5, and the fly would be male.

Then how does the female produce an egg that lacks an X chromosome? This occurs as a result of nondisjunction, in which the chromosomes failed to segregate during either meiosis I or meiosis II (see **fig. 4.10**). Thus, sperm that contained an X chromosome could fertilize an egg lacking an X chromosome to produce an X0 male (see fig. 4.9). Cytology of these primary exceptional males confirmed Bridges' hypothesis that they were X0.

The white-eyed primary exceptional females must be X^{w}X^w to account for their gender. The white-eyed phenotype suggested that both X chromosomes must contain the recessive *w* allele. Because the male parent

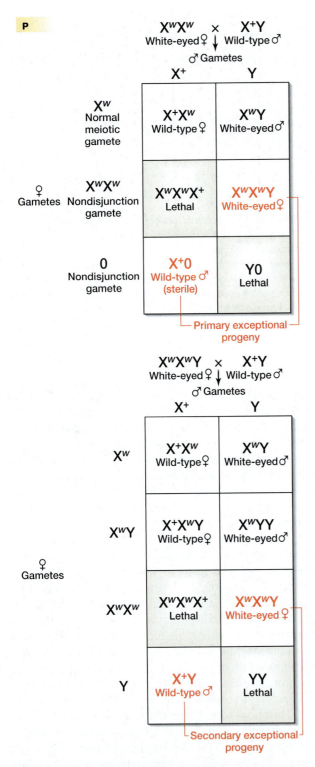

figure 4.9 Crosses in figure 4.8 redrawn to show the sex chromosomes. X^w represents the X chromosome containing the recessive *white* allele, and X^+ represents the X chromosome containing the dominant wild-type allele. The primary and secondary exceptional progeny genotypes are shown in red. Notice that the primary exceptional males are sterile because they lack the Y chromosomes, and the secondary exceptional males possess a Y chromosome (inherited from their mother) and are fertile.

contained an X chromosome with the wild-type allele (X^+), both X chromosomes in the primary exceptional females must have been inherited from their female parent (see fig. 4.9). Thus, the female parent must have produced an egg from nondisjunction that contained two X^w chromosomes (see fig. 4.10).

These $X^w X^w$ eggs could have been fertilized with sperm lacking a sex chromosome to produce the primary exceptional females; however, this result would require nondisjunction to also have occurred in the male parent. The likelihood of producing an individual from two gametes that both experienced nondisjunction would be significantly less than that of the single nondisjunction event that produced the primary exceptional males.

Because the primary exceptional males and females were found at the same frequency, they both must be produced by a nondisjunction event in only one parent. Thus, the primary exceptional females were generated from $X^w X^w$ eggs that were fertilized with a sperm containing the Y chromosome to produce $X^w X^w Y$ females (see fig. 4.9). Bridges knew that the Y chromosome would not change the gender of the XXY female because of genic balance, and his cytological analysis confirmed that the primary exceptional females were XXY.

Proof of the Theory

The ultimate proof of the chromosomal theory of inheritance came when Bridges examined the secondary exceptional progeny. The white-eyed primary exceptional females possessed an odd number of sex chromosomes. As these germ cells entered meiosis, only two of the three sex chromosomes could pair during meiosis I.

The preferred synapsis would occur between the two X chromosomes (**fig. 4.11** left), because they contain more sequences in common than exist between the X and Y pseudoautosomal regions. When the two X chromosomes paired, two types of gametes were produced: X^w and $X^w Y$ (see figs. 4.9 and 4.11). In the rarer cases when an X and Y synapsed in meiosis I (see fig. 4.11), the two types of gametes produced are $X^w X^w$ and Y (see figs. 4.9 and 4.11).

The Punnett square in figure 4.9 shows the various products resulting from these four primary exceptional female gametes and the two different wild-type male gametes. Again, secondary exceptional progeny are produced, namely the white-eyed female ($X^w X^w Y$) and wild-type male ($X^+ Y$). However, the secondary exceptional males would be fertile because they carry a Y chromosome (see fig. 4.9).

Bridges' predictions on what sex chromosomes would be present in flies expressing each phenotype was again confirmed by cytology. The ability to predict the correct combination of sex chromosomes that corresponded to a specific eye color provided compelling proof that the chromosomes were tightly associated with the genetic units that conferred a given phenotype.

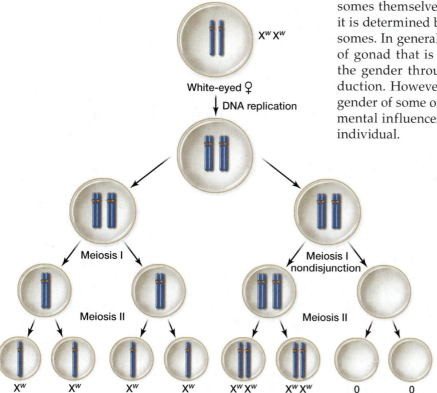

figure 4.10   Movement of the female sex chromosomes during meiosis in *Drosophila*. The left side depicts normal meiosis that produces four gametes that each contain an X^w chromosome. The right side shows the result of nondisjunction during meiosis I, which produces two gametes that each contain two X^w chromosomes and two gametes that each lack an X chromosome (0).

Restating the Concepts

▶ The chromosomal theory of inheritance states that the genetic traits (alleles) are tightly associated with the chromosomes.

▶ Nondisjunction is the improper segregation of chromosomes during meiosis or mitosis. The result is that the gametes have either an extra copy of a chromosome or lack one chromosome.

▶ Bridges used *Drosophila* mutants that experienced nondisjunction to predict the chromosomal content of different phenotypic classes of flies. This correct correlation between the observed phenotype and cytology proved the chromosomal theory of inheritance.

4.3 Sex Determination

Several different systems of sex determination occur in nature. These general systems include: XY, ZW, X0, haplo-diploid, and compound chromosomal mechanisms. It should be emphasized that the chromo-

somes themselves do not determine gender, instead it is determined by the genes located on the chromosomes. In general, the genotype determines the type of gonad that is generated, which then determines the gender through male or female hormonal production. However, as we will also discuss later, the gender of some organisms is determined by environmental influences, regardless of the genotype of the individual.

The XY System

You are probably already familiar with the XY system. Human females have 46 chromosomes arranged in 23 homologous pairs, including the two X chromosomes. Because females have two copies of the same sex chromosome, all of their gametes contain the same sex chromosome. For this reason, they are termed the **homogametic sex.** Males, with the same number

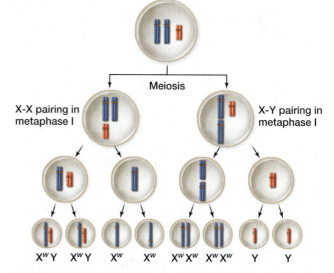

figure 4.11   The products of meiosis in the white-eyed primary exceptional female. The primary exceptional females from figure 4.9 contain two X chromosomes and one Y chromosome. During meiosis I, either the two X chromosomes can pair (preferred orientation, *left side*) or an X and Y chromosome can pair (*right side*). Four different gametes are generated, with X Y and X being the majority of the gametes (*left side*) and X X and Y being the two minor gametes.

of chromosomes, have 22 homologous pairs (autosomes) and one heterologous pair, the XY pair (**fig. 4.12**). Males are termed the **heterogametic sex** because they produce gametes with either an X or a Y sex chromosome.

The Y chromosome, which is known to encode approximately 171 genes (**fig. 4.13**), contains a single gene that determines if a developing mammal will be male. This Y chromosome gene, which is called either the **testis-determining factor (*TDF*)** or **sex-determining region Y (*SRY*)** gene, acts as a sex switch to initiate male development in mammals (see chapter 21). The role of the *SRY* gene in determining maleness has been conclusively shown in mice, where mice that contain two X chromosomes and the *SRY* gene inserted in an autosome develop as males. Additionally, mice containing one X chromosome and a Y chromosome, with the *SRY* gene deleted, will develop as females. Thus, it is not two X chromosomes that determine femaleness nor a Y chromosome that determines maleness, but rather the presence of the *SRY* gene that determines maleness, and the absence of *SRY* that determines femaleness.

Although most mammals use this XY system for gender determination, there are a few exceptions. Notably, the platypus utilizes a compound sex chromosome system that we will discuss later in this chapter.

Drosophila also utilizes the XY system. As with humans, the female flies are the homogametic sex, with six autosomes and two X chromosomes. Males, which are the heterogametic sex, contain six autosomes, one X chromosome and one Y chromosome. The *Drosophila* Y chromosome does not encode an *SRY* gene that determines maleness; however, the Y chromosome is known to carry at least six genes that are essential for male fertility. These fertility genes encode proteins needed during spermatogenesis. For example, *kl-5* encodes part of the dynein motor needed for sperm flagellar movement. We will discuss how *Drosophila* gender is determined in the next section.

The Y chromosome in both humans and flies contains two regions, one at either end of the chromosome, that are homologous to the X chromosome (see fig. 4.13). These regions are termed **pseudoautosomal,** and they permit the Y chromosome to pair with the X chromosome during meiosis in males to allow for their proper segregation.

Because both human and *Drosophila* females normally have two X chromosomes, and males have an X and a Y chromosome it seems impossible to know whether maleness is determined by the presence of a Y chromosome or by the absence of a second X chromosome. One way to resolve this problem would be to

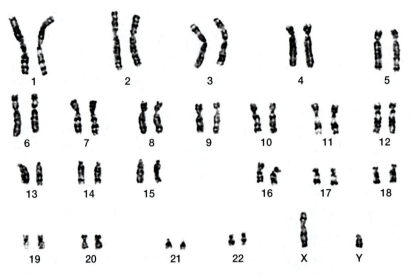

figure 4.12   Human male karyotype. Human chromosomes that are stained with Geisma reveal banding patterns that identify each chromosome. The presence of the Y chromosome determines that this is a male. A female would have a second X chromosome in place of the Y.

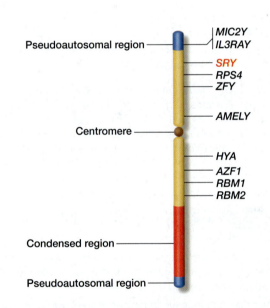

figure 4.13   The human Y chromosome. In addition to the genes shown on the right, the Y chromosome contains DNA sequences that are homologous to the X chromosome, pseudoautosomal regions, which allow synapsis between the Y and X chromosomes. The presence of the *SRY* gene (*Sex-determining region Y*) determines that the individual will be a male. The additional gene symbols shown include *MIC2Y*, T-cell adhesion antigen; *IL3RAY*, interleukin-3 receptor; *RPS4*, a ribosomal protein; *ZFY*, a zinc-finger protein; *AMELY*, amelogenin; *HYA*, histocompatibility Y antigen; *AZF1*, azoospermia factor 1 (mutants result in tailless sperm); and *RBM1*, *RBM2*, RNA binding proteins 1 and 2.

isolate individuals with unusual numbers of chromosomes. In chapter 8, we will examine the causes and outcomes of anomalous chromosome numbers. Here, we consider two facts from that chapter.

First, in rare instances, individuals can have extra sets of chromosomes, although they are not necessarily viable. These individuals are referred to as **polyploids** (*triploids* with 3*n*, *tetraploids* with 4*n*, etc.).

Second, also infrequently, individuals have more or fewer than the normal number of any one chromosome. These **aneuploids** usually result when a pair of chromosomes fails to separate properly during meiosis, an occurrence called **nondisjunction** (fig. 4.14). Thus, nondisjunction can produce gametes that either contain an extra copy of a chromosome or lack a chromosome. After fertilization, the gamete with an extra copy of a chromosome will produce an individual that has three copies, and the gamete lacking a copy of a chromosome will yield an individual that only has one copy.

The existence of polyploid and aneuploid individuals makes it possible to test whether the Y chromosome is male determining. For example, the gender of a human or a fruit fly that has the expected number of autosomes (44 in human beings, 6 in *Drosophila*), but only a single X without a Y would answer the question. If the Y is absolutely required in determining a male, then this X0 individual should be female. However, if the sex-determining mechanism is a result of the number of X chromosomes, this individual should be a male.

In humans, a single X chromosome (X0) results in a female. However, this female is not completely normal, a condition called **Turner syndrome.** Alternatively, humans with an extra sex chromosome, XXY, are male. Similar to the X0 condition, the XXY individuals are not completely normal, a condition called **Klinefelter syndrome.** (Details of both these conditions are described in chapter 8.) Because the sex of a Turner syndrome individual is female, it must be the presence of the Y chromosome and not a single X chromosome that determines maleness. Similarly, the sex of a Klinefelter syndrome individual reveals that a single Y chromosome determines maleness, regardless of the number of X chromosomes.

Although *Drosophila* also utilizes the XY system for sex determination, the Y chromosome does not determine maleness. Using polyploids and aneuploids, a different sex determination mechanism was revealed in the fruit fly (see following section).

You might wonder about an individual that does not contain an X chromosome. These Y0 individuals have not been identified in either mammals or *Drosophila*. This suggests that the X chromosome, unlike the Y chromosome, is essential for the viability of the organism.

Genic Balance in *Drosophila*

In 1921, geneticist Calvin Bridges, working with *Drosophila*, crossed a triploid female (3*n*) with a normal male (2*n*). He observed many combinations of autosomes and sex chromosomes in the offspring

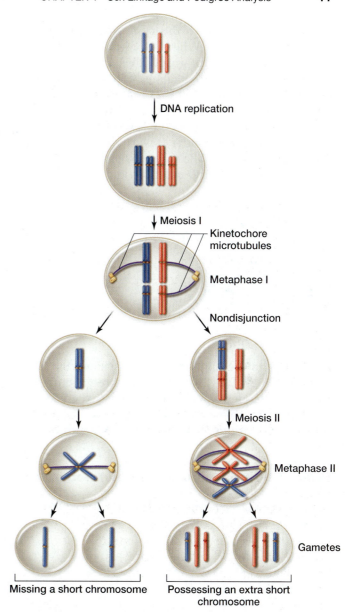

figure 4.14   Results of nondisjunction during meiosis I. Occasionally, the kinetochore microtubules fail to properly attach to the kinetochores or separate the chromosomes during either meiosis I or meiosis II. This may result in the homologous chromosomes (meiosis I) or sister chromatids (meiosis II) failing to separate, which results in the gametes either lacking a chromosome (acrocentric chromosome in the first two gametes) or possessing an extra chromosome (extra acrocentric chromosome in the last two gametes). Maternal chromosomes are red; paternal chromosomes are blue.

(**table 4.1**). From his results, Bridges suggested that sex in *Drosophila* is determined by the balance between (ratio of) autosomes that favor maleness and X chromosomes that favor femaleness. He calculated a ratio of X chromosomes to autosomal sets to see if this ratio would predict the sex of a fly. An **autosomal set** (A) in *Drosophila* consists of one chromosome from each autosomal pair, or three chromosomes. Thus, a diploid would have two autosomal sets.

Bridges' **genic balance theory** of sex determination was essentially correct. When the X:A ratio is 1.00, as in a normal female (XX), or greater than 1.00, the organism is a female. When this ratio is 0.50, as in a normal male (XY), or less than 0.50, the organism is a male. At 0.67, the organism is an **intersex,** which means that it possesses characteristics of both sexes. **Metamales** (X:A = 0.33) and **metafemales** (X:A = 1.50) are usually very weak and sterile. The metafemales usually do not even emerge from their pupal cases and die.

It's Your Turn

Now that we discussed how gender is determined in flies, work problems 14 and 17 to see if you understand this system.

The ZW System

Analogous to the XY chromosome system, the ZW system also utilizes two different chromosomes to determine gender. Males, which have two copies of the Z chromosome, are the homogametic sex (**table 4.2**). Females are the heterogametic sex, with one Z

chromosome and one W chromosome. The ZW system occurs in birds, some fishes, and moths.

Unlike the relatively small pseudoautosomal region that is shared between the X and Y chromosomes, the Z and W chromosomes share a relatively large pseudoautosomal region. In some species, the Z and W chromosomes are nearly indistinguishable. This suggests that the Z and W chromosomes evolved from a pair of homologous autosomes and are unrelated to the X and Y sex chromosomes in origin.

Regardless of their origin, the Z and W chromosomes evolved into the sex chromosomes in a variety of organisms. Rather than using a gene that is related to the mammalian Y chromosome's *SRY* gene to determine maleness, the ZW system appears to employ at least two different genes in gender determination.

The Z chromosome contains the *DMRT1* gene, which appears to be important in determining the male gender. Although the *DMRT1* gene seems to be unrelated to the mammalian *SRY* gene, it may function in an analogous manner to control the formation of the male testis. However, this mechanism is more complicated than the *SRY* determination process in the XY system because both males and females contain a Z chromosome, and therefore, a *DMRT1* gene. The two Z chromosomes in the male ostensibly produce more

table 4.1   Data Supporting Bridges' Theory of Sex Determination by Genic Balance in *Drosophila*

Number of X Chromosomes	Number of Autosomal Sets (A)	Total Number of Chromosomes	X:A Ratio	Sex
3	2	9	1.50	Metafemale
4	3	13	1.33	Female
4	4	16	1.00	Female
3	3	12	1.00	Female
2	2	8	1.00	Female
1	1	4	1.00	Female
2	3	11	0.67	Intersex
1	2	7	0.50	Male
1	3	10	0.33	Metamale

table 4.2   Summary of Different Gender Determination Mechanisms

Gender Determination Mechanism	Male	Female	Organisms Using this Mechanism
XY	XY, presence of Y chromosome	XX, absence of Y chromosome	Most mammals
ZW	ZZ	ZW	Birds, some fish
X0	X0, X:A ratio = 0.5	XX, X:A ratio = 1.0	Many insects
Haplo-diploid	Haploid	Diploid	Bees
Compound chromosome	$X_1 Y_1 X_2 Y_2 X_3 Y_3 X_4 Y_4 X_5 Y_5$	$X_1 X_1 X_2 X_2 X_3 X_3 X_4 X_4 X_5 X_5$	Duck-billed platypus

DMRT1 protein than the single copy in the female. Thus, maleness may be controlled by the amount of DMRT1 protein compared with the presence or absence of the SRY protein in the XY system. The *DMRT1* gene, which is present in human chromosome 9, must be present in two copies for normal testis differentiation in males.

The W chromosome also contains a gene, *ASW* or *Wpkci,* which is thought to be important in determining the female gender. In addition to activating the expression of genes that must be expressed in females, the Wpkci protein may also help prevent the expression of genes that must be expressed in males.

The X0 (X-zero) System

In the X0 system, which is observed in a variety of insects, only a single type of sex chromosome is designated the X chromosome (see table 4.2). This X chromosome should not be confused with the X chromosome in the XY system. In the X0 system, the female usually has two X chromosomes (XX) and the male a single X chromosome (X0). Whereas the X0 male is sterile in the XY system due to the absence of the Y chromosome, the X0 male in the X0 system is fertile. The number of X chromosomes appears to be all that is required for determining the gender of the individual, which is analogous to the number of Z chromosomes determining gender.

The nematode *Caenorhabditis elegans*, which is a popular genetic model, also uses the X0 system. As described earlier, an individual with a single X chromosome (X0) will be a male. However, a *C. elegans* individual with two X chromosomes (XX) will be a **hermaphrodite,** which is an individual with both sex organs (ovaries and testis). There are no females. In this case, the hermaphrodite can either self-fertilize or mate with a male to produce offspring. It is hypothesized that the X chromosome:autosome ratio, similar to that described for *Drosophila,* determines the gender of the individual. An X:A ratio of 1.0 would produce a hermaphrodite, and a ratio of 0.5 would yield a male.

The Haplo-Diploid System

The most common example of the haplo-diploid system is found in bees. In this system, there is no sex chromosome. Instead, males (drones) are haploid, and females (workers and the queen) are diploid (see table 4.2). During meiosis, the queen produces haploid eggs. If those eggs are not fertilized, they will develop into males. If they are fertilized, they will become either workers or a queen.

The Compound Chromosomal System

In the compound chromosome case, several X and Y chromosomes combine to determine sex. Compound chromosomal systems tend to be complex, with the

precise mechanism of gender determination unclear. For example, the nematode *Ascaris incurva* has eight X chromosomes and one Y. The species has 26 autosomes. Males have 35 chromosomes (26A + 8X + Y), and females have 42 chromosomes (26A + 16X). During meiosis, the X chromosomes unite end to end and behave as one unit. Thus, the male produces gametes that contain either eight X chromosomes or a single Y chromosome. Females produce gametes that contain eight X chromosomes.

The duck-billed platypus (*Ornithorhynchus anatinus*) is a monotreme, an egg-laying mammal. Unlike other mammals that utilize the XY system, the platypus employs a compound chromosome system. Only in the last 2 years has the sex chromosome composition been determined in the platypus.

The five different X chromosomes in the platypus (X_1, X_2, X_3, X_4, X_5) differ in their size and gene composition. It also has five different Y chromosomes (Y_1, Y_2, Y_3, Y_4, Y_5). The male contains all five X and Y chromosomes and the female contains 10 X chromosomes, which are composed of two copies of each X chromosome (see table 4.2). Cells that are undergoing meiosis in a male arrange their X and Y chromosomes in a long chain in a specific order ($X_1 Y_1 X_2 Y_2 X_3 Y_3 X_4 Y_4 X_5 Y_5$). It is thought that this chromosome chain is held together by homologous sequences that are shared between specific pairs of X and Y chromosomes.

Although the number and arrangement of the sex chromosomes have been determined in the platypus, the genetic basis of sex determination remains unknown. Unlike other mammals, the *SRY* gene is not present on any of the five Y chromosomes. However, the X_5 chromosome does possess a *DMRT1* gene, which is present in one copy in the male platypus and two copies in the female. This is surprising because the *DMRT1* gene, which determines maleness in the ZW system, is present in two copies in the male and one copy in the female. Furthermore, two copies of the *DMRT1* gene are required for testis development in other male mammals. The actual genetic basis of gender determination remains unknown in the platypus, but the identification of the sex chromosome composition will greatly speed the discovery of the critical gene(s).

Environmental Control of Sex Determination

Sex chromosomes are not the only determinants of an organism's gender. Environmental sex determination (ESD) is an alternative mechanism that controls the gender of individuals. One mechanism of ESD depends on the substrate where some marine worms and gastropods develop. The environmental temperature during egg development also plays a major role in gender determination in turtles, lizards, crocodiles, and alligators.

The effect of temperature on gender determination varies from species to species (**fig. 4.15**). At low temperatures, some organisms develop only as males, but other organisms develop only as females. At the high temperature, development is directed into the other gender. It is the temperatures that lie between the two extremes that allow both males and females to develop. However, some species develop as only females at both the low and high temperatures (see fig. 4.15). At the intermediate temperatures, the percentage of individuals that develop as males increases and may even reach 100%.

Although ESD is controlled by environmental conditions and not the presence of specific sex chromosomes, it is clear that the environment must act on a metabolic or physiological process in the developing embryo. The processes controlled by the environment are unclear, but they likely involve steroid biosynthesis. Two potential target enzymes are reductase and aromatase, which control the conversion of testosterone into either dihydrotestosterone or estradiol, respectively.

Consistent with a potential role in ESD, levels of aromatase were increased in eggs that were incubated at female-producing temperatures, but not at male-producing temperatures. To test the potential role of aromatase and reductase, aromatase inhibitors were injected into eggs that were incubated at female-producing temperatures and primarily males were produced. Injection of reductase inhibitors into eggs that were incubated at male-producing temperatures yielded mostly females.

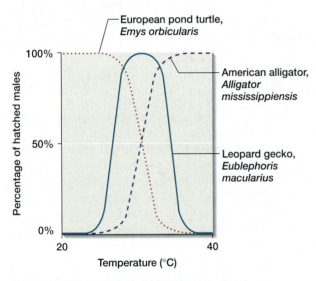

figure 4.15   Effect of temperature on gender determination. Eggs of several different species were incubated at the designated temperature, and the percentage of males that hatched from the eggs were determined. The European pond turtle, *Emys orbicularis*, produces only males at the lower temperature and only females at higher temperatures (*dotted line*). In contrast, the American alligator, *Alligator mississippiensis*, produces only males at the higher temperature and females at the lower temperatures (*dashed line*). A third class of reptiles, the Leopard gecko (*Eublepharis macularis*), only produces males at the intermediate temperatures and only females at both the low and high extreme temperatures (*solid line*).

It also appears that temperature is affecting the expression of the aromatase and reductase genes either directly or indirectly. At female-producing temperatures, the expression of aromatase is increased and reductase remains low. At male-producing temperatures, reductase is expressed at higher levels and aromatase remains low. Thus, ESD is mediated through the regulation of gene expression, as it is in other mechanisms of sex determination.

Restating the Concepts

▶ Gender is determined by either environmental conditions or the presence of specific genes that are located on sex chromosomes. The systems that utilize sex chromosomes include the XY, ZW, XO, and compound chromosomal sets.

▶ In both mammals and *Drosophila*, XX corresponds to a female and XY corresponds to a male. However, the sex determination mechanism is different between mammals and fruit flies. In most mammals, the presence of the Y chromosome determines that the individual will be a male, whereas in *Drosophila* the X-chromosome:autosome ratio determines the gender of the individual, with the Y chromosome only being required for male fertility.

▶ The ZW system is similar to the XY system of sex determination. However, the male (ZZ) is the homogametic sex in the ZW system and the female (ZW) is the heterogametic sex. The ZW system does not utilize the *SRY* gene to determine maleness, but rather uses the level of expression of the *DMRT1* gene, which is located on the Z chromosome.

4.4 Gene Balance

One of the major challenges that must be resolved after the sex of an organism is determined early in development is how to maintain gene balance. **Gene balance** refers to the need to keep the same amount of gene expression between different individuals. The concern about gene balance in this discussion is that the two genders often differ in the number of one or more chromosomes, as described in the first section of this chapter. If genes are transcribed at a similar rate in both genders, one would produce twice as much mRNA, and possibly twice the amount of protein product, as the other.

We call the genes that are located on the sex chromosome **sex-linked genes,** or **X-linked genes** if the X chromosome is involved. Balancing gene expression of sex-linked genes in different genders is termed **dosage compensation.** The cells within the organism must either increase the expression of the genes on the sex chromosome in the heterogametic sex or reduce the expression of those genes in the homogametic sex.

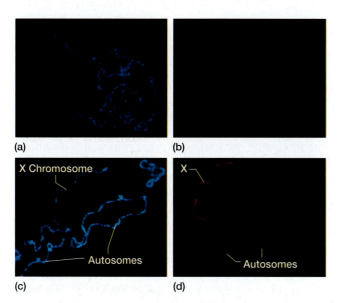

(a)   (b)

X Chromosome

Autosomes   X

                                 Autosomes

(c)   (d)

figure 4.16   Localization of the male-specific lethal-2 (MSL-2) protein. Stained chromosomes from a female cell (a) are not bound with the antibody that detects the MSL-2 protein (b). Stained chromosomes from a male cell (c) are bound by an antibody that detects MSL-2 (d, shown in red). Notice that only the X chromosome, but not the autosomes, are bound by MSL-2 in the male cell.

Increasing X-Linked Gene Expression in Male *Drosophila*

In fruit flies, the X-linked genes in the male are hyper-transcribed relative to the female. By increasing the transcription of the X-linked genes approximately two-fold in males, the males and females express roughly equivalent levels of the gene products.

This increased transcription in males requires four proteins that are encoded by the *maleless* and *male-specific lethal-1, -2,* and *-3* genes. Mutations in any of these four genes result in the male dying early in development because it cannot express the X-linked genes at a high enough level for survival.

The proteins encoded by these four genes form a complex called the dosage compensation complex. This complex also requires at least two different RNAs that are encoded by the *rox1* and *rox2* genes. The dosage compensation complex can be detected bound to the X chromosome in males, but not in females (**fig. 4.16**).

X Chromosome Inactivation in Female Mammals

In mammals, dosage compensation is achieved when the X chromosome undergoes **X chromosome inactivation.** In this process, all of the X chromosomes in a cell, except one, are modified so that they are transcriptionally inactive. The result is that normal females (XX), males (XY), and Klinefelter males (XXY) all exhibit approximately the same level of expression for

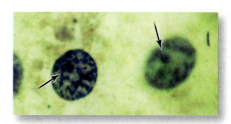

figure 4.17
Barr body. In mammals, every normal female cell contains a single Barr body (arrow).

X-linked genes. The inactive X chromosomes, called **Barr bodies** (**fig. 4.17**), are named after Murray Barr who first identified them. Both a normal female and a Klinefelter male have one Barr body in each nucleus.

It should be noted that only part of the X chromosome is inactivated. Approximately nine genes in the pseudoautosomal regions of the X chromosome and several other genes scattered along the X chromosome remain active in the Barr body. All the pseudoautosomal genes are present on both the X and Y chromosome. Thus, the female has two copies of each of these genes (one on each X chromosome) and the male also has two copies of each of these genes (one on the X and the other on the Y chromosome). By not inactivating these particular genes on the Barr body, both the normal female and male transcribe two copies of each of these genes.

What determines which X chromosome will be inactivated? In marsupials, the paternal X chromosome is specifically inactivated. In all other mammals, the selection of the X chromosome that will be inactivated appears to be random. In humans, the selection point occurs when the embryo is composed of approximately 500–1000 cells. Once the inactivation decision is made in one cell, all of the cells descended from that cell exhibit the same inactive X chromosome (**fig. 4.18a**).

This random inactivation leads to **mosaic** females that possess different **patches,** or regions, that have an active paternal X chromosome and other patches that have an active maternal X chromosome. These random patches can be seen in the Calico cat (see fig. 4.18b), where the orange and black patches are due to the inactivation of different X chromosomes during early development.

The mechanism by which X inactivation takes place is described in chapter 21.

Restating the Concepts

▶ Gene balance refers to the maintenance of equivalent gene expression for all individuals. This is a problem for genes on the sex chromosomes when one gender has more copies of the chromosome than the other gender.

▶ *Drosophila* maintains gene balance by increasing expression of X-linked genes in the male.

▶ Mammals maintain gene balance by inactivating all but one X chromosome in every cell. These inactive X chromosomes are called Barr bodies.

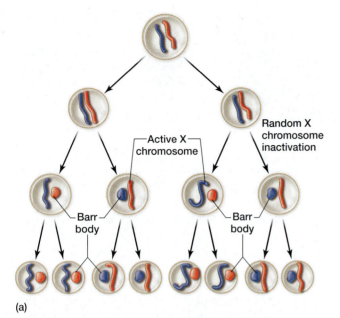

(a)

figure 4.18

X chromosome inactivation in most mammals is random. (a) Depending on the species, the inactivation occurs when the embryo contains between 64 and 1000 cells. Each cell randomly selects the X chromosome that will be inactivated. Some cells will inactivate the maternal chromosome (*red*) and some the paternal chromosome

(b)

(*blue*). (b) The random inactivation produces patches that may express different alleles that are present on the different active X chromosomes. These patches lead to the random patches of orange and black fur on the Calico cat.

4.5 Pedigree Analysis

Inheritance patterns in many organisms are relatively easy to determine when large numbers of progeny are generated. Mendel's observation of a 3:1 phenotypic ratio in the F_2 generation led him to suggest the law of equal segregation. If Mendel's sample sizes had been smaller, he might not have seen this ratio. Furthermore, testcrosses to confirm the second law of independent assortment required crossing specific genotypes (dihybrid and doubly homozygous recessive plants). But in many cases, most notably in humans, crossing and testcrossing is not an option or is otherwise not feasible. In these cases, geneticists rely on analysis of a trait within many generations of a family to deduce the type of inheritance at work.

Human Genetics

Think of the difficulties Mendel would have faced had he decided to work with human beings instead of pea plants. Human geneticists face the same problems today. Humans take a long time to reach reproductive maturity, and the number of offspring that result from two parents are not large—rarely over single-digit numbers in the lifetime of the mother. And of course, it is unethical to "arrange" matings between individuals with desired genotypes to suit the geneticist's wishes. Under these circumstances, it is unlikely that Mendel would have observed his standard ratios in human families.

To determine the inheritance pattern of many human traits, geneticists often have little more to go on than a family history that many times does not include critical mating combinations. This family history can be diagrammed as a **pedigree,** which represents all the individuals in the family and their phenotypes. Frequently, uncertainties and ambiguities plague human genetic analysis. In chapter 5, we will discuss particular phenomena (*penetrance* and *expressivity*) that can further increase the difficulty in elucidating human inheritance patterns. In the following sections, we assume for the sake of discussion that the simplest explanations are correct.

One assumption often made in pedigree analysis is that the trait being examined is rare. This simplifies the pedigree because the presence of the mutant allele will also be rare. In these cases, it is assumed that someone marrying into the pedigree lacks the rare allele, unless evidence exists to the contrary.

The NIH's OMIM database contains information on a large number of human genes and genetic diseases. As of 2007, OMIM contains information on 11,823 human genes and 6,195 human traits and diseases, including autosomal dominant, autosomal recessive, and sex-linked genes. Pedigrees are useful in deducing both the mode of inheritance and the genotypes of individuals. Genotype information is important in determining the probability that a couple will have a child that suffers from a particular trait.

To review, humans have 23 pairs of chromosomes. One pair is called the sex chromosomes because they determine the gender and secondary sexual characteristics of the individual. The rest are autosomes. In the following sections, we will differentiate the dominant from the recessive patterns of inheritance as well as distinguishing patterns associated with the sex chromosomes from those connected to the autosomes.

Family Trees

One way to examine a pattern of inheritance is to draw a pedigree. **Figure 4.19** defines some of the symbols used in constructing a pedigree. Circles represent females, and squares represent males. Filled symbols represent individuals who exhibit the trait under study. These individuals are said to be **affected** when the trait being studied is an inherited disease. The open symbols represent those who do not exhibit the trait (unaffected).

A horizontal line connecting two individuals (one male, one female) is called a *marriage* or *mating line*, and a double horizontal line represents a **consanguineous** mating, which is a mating between related individuals. Offspring (children) are attached to a mating line by a vertical line. All the brothers and sisters (**siblings** or **sibs**) from the same parents are connected by a horizontal line above their symbols.

Figure 4.20, which is a pedigree for the trait polydactyly (having more than 10 fingers or toes), shows other conventions. Siblings are numbered below their symbols according to birth order, and generations are numbered on the left in Roman numerals. When the sex of a child is unknown, the symbol is diamond-shaped (for example, the children of III-1 and III-2 in fig. 4.20). A number within a symbol represents the number of siblings not separately listed.

Individuals IV-7 and IV-8 in figure 4.20 are fraternal (dizygotic or nonidentical) twins: they originate from the same point. Individuals III-5 and III-6 are identical (monozygotic) female twins: they originate

from the same short vertical line and their lines are connected by a horizontal line.

When other symbols occur in a pedigree, they are usually defined in the legend. Individual V-5 in figure 4.20 is identified as the **proband** or **propositus** (female, **proposita**) by the arrow pointing to his symbol. The proband is the individual that is first identified with the trait, usually by a physician or clinical investigator, and the pedigree is usually constructed by assembling the ancestors of that individual.

On the basis of the information in a pedigree, geneticists attempt to determine the mode of inheritance of a trait. Two types of questions might be answered through pedigree analysis. First, do patterns occur within the pedigree that are consistent with a particular mode of inheritance? Second, are patterns present that are not consistent with a particular mode of inheritance? Often, it is impossible to determine the mode of inheritance of a trait with certainty.

Autosomal Dominant Inheritance

Looking again at the pedigree in figure 4.20, several points become apparent. First, polydactyly, (shown in **fig. 4.21**), occurs in every generation. Every affected child has an affected parent—no generations are skipped. This pattern suggests dominant inheritance.

Second, the trait occurs about equally in both sexes; there are eight affected males and six affected females in the pedigree. This equal ratio of affected genders indicates autosomal rather than sex-linked inheritance. Because of this, we would categorize polydactyly as an autosomal dominant trait.

Note that individual IV-13 (see fig. 4.20), a male, passed on the trait to two of his three sons. This inheritance from father to son would rule out X-linked dominant inheritance, because a male gives his X chromosome to all of his daughters and none of his sons. Consistency in many such pedigrees has confirmed that an autosomal dominant gene causes polydactyly.

Autosomal Recessive Inheritance

Figure 4.22 shows a pedigree with a different pattern of inheritance for hypotrichosis (lack of hair growth). Unlike the autosomal dominant pedigree, affected individuals are not found in each generation of this pedigree. Also, the affected daughters, identical triplets, come from unaffected parents. They represent, in fact, the first appearance of the trait in three generations.

A telling point for the pattern of inheritance in hypotrichosis is that the parents of the triplets are first cousins, that is, they represent a consanguineous mating. Consanguineous matings often produce offspring that have

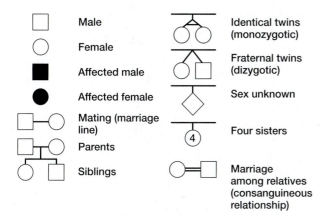

figure 4.19   Symbols used in a pedigree.

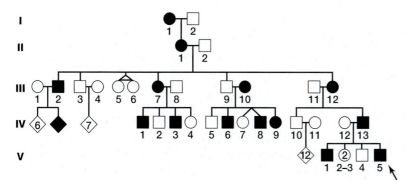

figure 4.20   Part of a pedigree showing the inheritance of polydactyly, a dominant autosomal trait, in a family. Notice that there are affected individuals in every generation, which is an indication of a dominant trait. The affected individuals are represented as filled symbols.

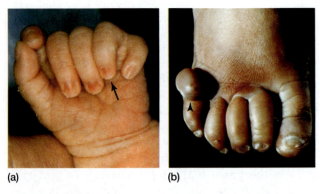

(a)   **(b)**

figure 4.21   Polydactyly phenotype. Manifestations of polydactyly range in severity from an extra finger (arrow) or toe to a portion of one (arrowhead) or more extra digits.

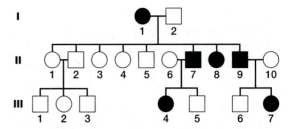

figure 4.23   Part of a pedigree of the inheritance of vitamin-D-resistant rickets, a sex-linked dominant trait, in a family. Affected individuals have low blood-phosphorus levels. Although the sample is too small for certainty, dominance is indicated because every generation was affected, and sex linkage is suggested by the distribution of affected individuals.

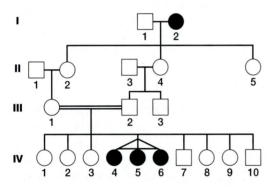

figure 4.22   Autosomal recessive pedigree. This pedigree shows the inheritance of hypotrichosis (the inability of hair to grow), an autosomal recessive trait, in a family. Notice that two unaffected individuals (III-1 and III-2) have some affected children, which suggests a recessive trait. The affected individuals are represented by filled symbols.

rare recessive, and often deleterious, traits. The reason this occurs is that through common ancestry, a rare allele that is heterozygous in a single common ancestor can be passed on to both sides of the pedigree and produce a homozygous child. In this case, the parents share a common set of grandparents. Because of the danger of combining deleterious alleles, marriage between first cousins is prohibited in 30 states.

Although the appearance of a trait in a consanguineous pedigree is often good evidence for an autosomal recessive mode of inheritance, all modes of inheritance appear in consanguineous pedigrees, and recessive inheritance is not confined to consanguineous pedigrees. It is possible that two unrelated individuals may carry a recessive allele for the same gene. Although not shown in this pedigree, all the children of two affected individuals (homozygous recessive) must also be affected to demonstrate an autosomal recessive trait.

Sex-Linked Dominant Inheritance

Figure 4.23 is a pedigree with yet a different pattern of inheritance. The phenotype under consideration is the

distribution of low blood phosphorus levels, which is one characteristic of vitamin-D-resistant rickets. This pedigree suggests a dominant mode of inheritance because the trait does not skip generations.

Unlike the autosomal dominant pedigree in figure 4.20, this pedigree shows affected males passing the trait on to all their daughters, but not to their sons. This pattern is consistent with the transmission of the X chromosome, which fathers pass on to all of their daughters and none of their sons. Therefore, this likely represents a sex-linked dominant trait.

Although this pedigree is consistent with a sex-linked dominant mode of inheritance, it does not rule out autosomal dominant inheritance. An instance of transmission of the trait from father to son, however, would eliminate the X-linked mode of inheritance as a possibility.

In figure 4.23, a slight possibility exists that the trait is recessive. This could be true if individuals I-2, II-6, and II-10 were all heterozygotes. But because this is a rare trait, the possibility that all three of these individuals are heterozygous is small. For example, if 1 person in 50 (0.02) is a heterozygote, then the probability of three heterozygotes mating within the same pedigree is $(0.02)^3$, or 8 in 1 million. The rareness of this event further supports the hypothesis of dominant inheritance.

Sex-Linked Recessive Inheritance

Figure 4.24 is the pedigree of Queen Victoria of England and the expression of the blood clotting disorder hemophilia. Through her children, hemophilia was passed on to many of the royal houses of Europe.

Several interesting aspects of this pedigree help to confirm the method of inheritance. First, the hemophilia trait skips generations, which suggests a recessive mode of inheritance. Although Alexis (1904–18) was a hemophiliac, his parents, grandparents, great-grandparents, and great-great-grandparents in this pedigree were unaffected. This pattern of affected individuals having unaffected parents occurs in several other places in the pedigree. From this and other pedigrees, and from the

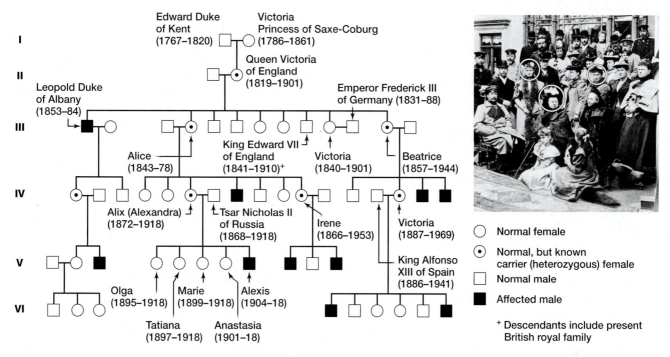

figure 4.24   Part of a pedigree showing the inheritance of hemophilia in the ancestors of Queen Victoria of England. In the photograph, three carriers—Queen Victoria (center), Princess Irene of Prussia (right), and Princess Alix (Alexandra) of Hesse (left)—are indicated by circles.

biochemical nature of the defect, scientists confirmed that hemophilia is a recessive trait.

Second, all the affected individuals are males, strongly suggesting sex linkage. Because males are hemizygous for the X chromosome, more males than females should have the sex-linked recessive phenotype because of pseudodominance.

Third, the trait is never passed from father to son in this pedigree. This result would defy the route of inheritance of an affected X chromosome. We can conclude from the pedigree, therefore, that hemophilia is an X-linked recessive trait.

It is important to realize that many human diseases exhibit similar symptoms, but are due to mutations in different genes. This suggests that different pedigrees for the same clinical disease (described on the basis of symptoms) may show different patterns of inheritance. For example, several different inherited forms of hemophilia are known, each of which is deficient in one of the steps in the pathway that forms fibrinogen, the blood clot protein. Two of these forms, "classic" hemophilia A and hemophilia B, are sex-linked. Other hemophilias are autosomal.

We can make several predictions about the pedigree in figure 4.24, if it indeed reveals an X-linked recessive trait. First, because all males get their X chromosomes from their mothers, affected males should be the offspring of carrier (heterozygous) females. A female must be a carrier if her father had the disease, as the father passes his X chromosome to all of his daughters. However, a female has a 50% chance of being a

carrier if her brother has the disease and her father is unaffected. In that case, her mother is a carrier.

It is also possible for a woman to be homozygous for this condition—that is, her father would be affected, and her mother would be a carrier. Because both the father and mother must have the recessive allele to have an affected daughter, X-linked recessive traits are less common in females than males; where only the mother must be a carrier. Hemophilia in a woman, however, could be fatal with the start of menses, and childbirth would be extremely dangerous. Such a situation is not found in the pedigree in figure 4.24.

Summary of the Different Patterns of Inheritance

The following lists summarize the major points of the four patterns of inheritance that we discussed. Keep in mind that a pedigree showing all of the features associated with one pattern does not necessarily eliminate other patterns. The other patterns may simply be less likely, as you saw with an autosomal recessive inheritance mode in figure 4.24. The only way to conclusively eliminate a pattern of inheritance is to identify a part of the pedigree that is inconsistent with that pattern.

Autosomal Dominant Inheritance

1. Trait should not skip generations (unless the trait lacks full penetrance, see Chapter 5).

2. When an affected person mates with an unaffected person, approximately 50% of their offspring should be affected (indicating also that the affected individual is heterozygous).

3. The trait should appear in almost equal numbers between males and females.

Autosomal Recessive Inheritance

1. Trait often skips generations.

2. An almost equal number of males and females will be affected.

3. Traits are often found in the offspring of consanguineous matings.

4. If both parents are affected, all children should be affected.

5. In most cases when an unaffected individual mates with an affected individual, all the children will be unaffected. When at least one child is affected (indicating that the unaffected parent is heterozygous), approximately half the children should be affected.

6. Most affected individuals have unaffected parents who are carriers.

Sex-Linked Dominant Inheritance

1. The trait does not skip generations.

2. Affected males must come from affected mothers.

3. Approximately half the children, both sons and daughters, of an affected heterozygous female are affected.

4. Affected females may have either affected mothers or fathers.

5. All the daughters, but none of the sons, of an affected father are affected (if the mother is unaffected).

Sex-Linked Recessive Inheritance

1. Most affected individuals are male.

2. All the daughters of an affected male will be carriers and all of the sons will be unaffected.

3. Affected females have affected fathers and affected or carrier mothers.

4. All the sons of affected females should be affected.

5. Approximately half the sons of carrier (heterozygous) females should be affected.

It's Your Turn

You now know how to read a pedigree and deduce the different patterns of inheritance. Try problems 31 and 36 to use this information to correctly deduce the patterns of inheritance in the various pedigrees.

Deducing Genotypes from Pedigrees

Once a pedigree is generated and the mode of inheritance for the observed trait is determined, we can then calculate the probability of a particular genotype for any individual in the pedigree. We can also calculate the probability that an offspring produced from the pedigree will possess a specific genotype or phenotype. These calculations require the sum and product rules we first discussed in chapter 2.

Let's look at the pedigree in **figure 4.25**. This pedigree represents a recessive trait because two individuals (III-2 and III-5) are both affected but have unaffected parents. Because both of these affected individuals are females and the majority of the affected individuals are not males, this is likely an autosomal recessive trait.

For this autosomal recessive trait, we can assign genotypes to several individuals: I-1 would be homozygous dominant (*RR*) and I-2 would be homozygous recessive (*rr*). Individuals II-2 and II-3 must therefore be heterozygotes based on their parents' genotypes. Individuals II-1, II-5, and II-6 must be heterozygotes to produce the affected daughters III-2 and III-5. If this trait is rare, then individual II-4 would likely be homozygous dominant (*RR*).

The genotypes of these individuals can be conclusively deduced, but two individuals have ambiguous genotypes. There is a 50% chance that the female III-3 is a heterozygote, based on her parents' genotypes (*Rr* × *RR*). Individual III-4 is phenotypically normal and the son of two heterozygotes, which means that his genotype could be either *RR* or *Rr*.

We know that a child from a monohybrid cross (which is II-5 × II-6 in this pedigree) has a 25% chance of being homozygous dominant, 50% of being heterozygous, and 25% chance of being homozygous recessive. We know that individual III-4 is not homozygous recessive because he does not express the recessive trait. Of the possible genotypes that exhibit the dominant phenotype, being heterozygous is twice as likely as being homozygous. Thus, there is a 66.7% chance that

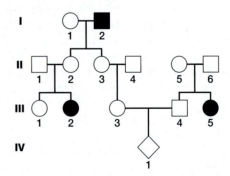

figure 4.25 The probability of a child exhibiting a recessive trait can be calculated from an analysis of a pedigree. This pedigree shows the inheritance of a recessive trait in a family. The probability of the child (IV-1) exhibiting the recessive trait will be based on the probability that the child's parents (III-3 and III-4) are both carriers (heterozygotes).

individual III-4 is heterozygous and a 33.3% chance he is homozygous dominant.

We can also calculate the probability that individual IV-1 will exhibit the recessive trait. We deduced that the probability of individual III-3 being heterozygous is 50% and individual III-4 being heterozygous is 66.7%. We also know that in a monohybrid cross between two heterozygotes, the probability of the offspring being homozygous recessive is 25%. Using the product rule,

$$P = 50\% \times 66.7\% \times 25\% = 8.3\%$$

When studying *rare* traits in a pedigree, as mentioned earlier, it is assumed that someone marrying into the pedigree lacks the rare allele, unless evidence exists to the contrary. Figure 4.25 shows an example of this. Individual II-4 is assumed to be homozygous dominant,

because his offspring show no evidence of possessing the rare recessive phenotype. However, individual II-1 cannot be homozygous dominant, even though he is also marrying into the family, because individual II-1 has an affected daughter. Both parents (II-1 and II-2) must be heterozygotes for an offspring to be affected.

It's Your Turn

Using a pedigree to calculate the probability of having a child affected with a particular disease is an important skill that must be practiced. Try problems 25 and 40 to see how comfortable you are in applying the techniques we just described to this type of problem.

4.1 applying genetic principles to problems

Calculating Genotypic Probabilities from Pedigrees

Knowing how to deduce the mode of inheritance from a pedigree and determining the probabilities that individuals have a given genotype are important aspects of genetics. This type of problem can be clearly approached in the way described here.

Suppose that you are presented with a pedigree for the Smith and Jones families, in which a very rare inherited form of anemia, a blood disorder, occurs. You are asked to determine the mode of inheritance and the probability that John Smith and Mary Jones will have a child with anemia. You are initially given this family information:

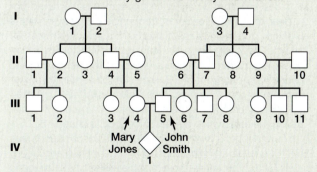

You question the available family members about the individuals that show the anemia trait. From this information, you can modify the pedigree to reveal the three affected individuals:

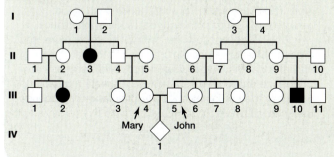

From this modified pedigree you can determine the mode of inheritance. You notice that all three affected individuals have unaffected parents. This situation is inconsistent with a dominant trait. Because the two affected females have unaffected fathers, it is not an X-linked trait. Therefore, you can deduce that this form of anemia is an autosomal recessive trait.

You can now assign the unambiguous genotypes to individuals. The three affected individuals must be homozygous recessive (*aa*). The unaffected parents of these three children must therefore be heterozygous (*Aa*). You also know that for individuals I-3 and I-4, one must be heterozygous and the other homozygous dominant to produce a heterozygous female (II-9, who has an affected child). Finally, because this is a rare trait, you can assume individuals II-5 and II-6 (marrying into the family) are homozygous dominant because there is no data indicating that their children inherited the recessive allele from them.

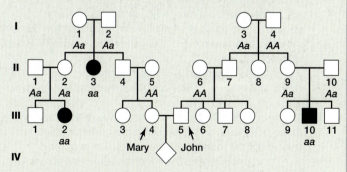

The remaining individuals in the pedigree do not suffer from anemia and are either homozygous dominant (*AA*) or heterozygous carriers (*Aa*).

You are now in a position to begin calculating the probabilities that some of the remaining individuals in the

(continued)

pedigree possess the recessive allele. Because you are interested in the probability that Mary Jones (III-4) and John Smith (III-5) will have a child that suffers from anemia, we need to calculate the probability that Mary and John are carriers. The probability that Mary's unaffected father, whose parents are both carriers, is himself a carrier is 2/3, or 66.7%. Remember, because he is unaffected, he must be either *AA* or *Aa* and there is a 66.7% probability that an unaffected individual is heterozygous if both of the parents are heterozygotes. If Mary's father is a carrier, there is a 50% chance that she is also a carrier. Thus, the probability that Mary is a carrier is 66.7% × 50% = 33.3%.

Because John's father is the child of a union between a carrier and a homozygous dominant individual, there is a 50% chance that John's father is a carrier. If John's father is a carrier, there is a 50% chance that John is a

carrier. Thus, the probability that John is a carrier is 50% × 50% = 25%.

You can now determine the probability that Mary and John will have a child that suffers from anemia. You already know that in a monohybrid cross, two heterozygotes will produce a homozygous recessive individual 25% of the time. If Mary is a carrier (33.3%) and John is a carrier (25%), the probability that they will have an affected child is:

$$33.3\% \times 25\% \times 25\% = 2.1\% \qquad \text{or about 1 in 50}$$

To recap, you first identify the affected individuals in a pedigree and determine the mode of inheritance. Next, you assign all the known genotypes. You then determine the probability that the relevant undetermined individuals have the mutant allele. Finally, you calculate the probability that the interested individual will exhibit the trait.

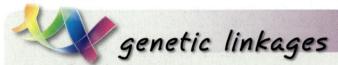

genetic linkages

In this chapter, we discussed the proof of the chromosome theory of inheritance (chapter 3), which is based on the movement of chromosomes during meiosis and the inheritance of traits. We will reexamine how aneuploid individuals are produced through abnormal segregation of chromosomes and the consequences of abnormal chromosome numbers as the cell proceeds through meiosis (chapter 8). We will also discuss the importance of chromosome loss during mitosis in cancer development in chapter 22.

In this chapter, we discussed the general mechanisms of sex determination in a variety of organisms. We will revisit the topic of sex determination in *Drosophila* in chapter 21, because much is known about the genes and corresponding proteins that are involved in this process. You will see that some of the genes involved in *Drosophila* sex determination are alternatively spliced, such that two different mRNAs and proteins are produced from a single gene (chapter 10).

We also described sex determination in mammals. In both the *Drosophila* and mammalian systems, we used individuals that had an abnormal number of chromosomes to reveal the underlying mechanism. We will discuss how these polyploids and aneuploids are generated and how they affect the individual in more detail in chapter 8.

We also discussed the importance of gene balance in regard to the need to maintain the same level of expression for genes that are present on the sex chromosomes in both sexes. The problem of gene balance was solved by X chromosome inactivation in mammals. Determining which X chromosomes will be inactivated is usually a ran-

dom event. However, in marsupials, the paternal X chromosome is preferentially inactivated. This sex-specific inactivation of genes is called epigenetics and will be discussed in more detail in chapter 17. The mechanism of random X chromosome inactivation in humans will be examined in chapter 21.

You saw that a variety of inheritance patterns do not appear to follow Mendelian frequencies. However, this is primarily due to the transmission of the sex chromosomes and pseudodominance in hemizygous individuals. Upon closer examination, Mendel's laws and the genotypic and phenotypic ratios (chapter 2) were seen to be still at work for genes on the sex chromosomes. This was further supported by understanding how the different sex chromosomes pair during meiosis, which we first described in chapter 3.

In chapter 5, we will describe additional phenomena that affect the standard Mendelian phenotypic and genotypic ratios. You will see how multiple alleles of a single gene and how different genes can interact to alter the final phenotype. You will also be exposed to the concept of incomplete penetrance, which was briefly mentioned in this chapter, and how this can change the phenotype of a specific genotype. The consequence of this phenomenon is that some pedigrees will suggest an incorrect mode of inheritance.

Finally, we applied the probability rules (sum and product) that you learned in chapter 2 to the calculation of an individual with a specific genotype. We discussed how to analyze a pedigree as a series of simple monohybrid crosses to determine the mode of inheritance and deduce the genotype of each individual.

Chapter Summary Questions

1. For each term in column I, choose the best matching phrase in column II.

   Column I
   1. Homogametic
   2. X0
   3. Polyploidy
   4. Sex-limited traits
   5. ZW
   6. Sex-linked traits
   7. Aneuploidy
   8. ZZ
   9. Sex-influenced traits
   10. Heterogametic

   Column II
   A. Rooster
   B. Having more or fewer than the normal number of any one chromosome
   C. Traits expressed in only one sex that are controlled by genes on the autosomes
   D. Human female
   E. Hen
   F. Traits expressed more often in one sex that are controlled by genes on the sex chromosomes
   G. Having extra sets of chromosomes
   H. Human male
   I. Male nematode worm
   J. Traits expressed more often in one sex that are controlled by genes on the autosomes

2. What is the difference between an X and a Z chromosome?

3. What is a sex switch? What gene serves as sex switch in mammals?

4. In humans, Duchenne muscular dystrophy is inherited as an X-linked recessive trait. Is an affected male homozygous or heterozygous? Explain.

5. What is the pseudoautosomal region? What is the mode of inheritance of genes located in this region?

6. Compare and contrast the mechanisms of sex determination in *Drosophila melanogaster* and *Caenorhabditis elegans*.

7. Calico cats have large patches of colored fur. What does this indicate about the age of onset of X chromosome inactivation (is it early or late)? Tortoiseshell cats have very small color patches. Explain the difference between the two phenotypes.

8. What is ESD? Describe the roles of the two enzymes that mediate it.

9. What family history of hemophilia would indicate to you that a newborn male baby should be exempted from circumcision?

10. List the characteristics of Y-linked inheritance.

11. What is crisscross inheritance? What genes exhibit this pattern of inheritance?

Solved Problems

PROBLEM 1: A female fruit fly with a yellow body is discovered in a wild-type culture. The female is crossed with a wild-type male. In the F_1 generation, all the males are yellow-bodied and all the females are wild-type. When these flies are crossed among themselves, the F_2 progeny are both yellow-bodied and wild-type, equally split among males and females (see fig. 1). Explain the genetic control of this trait.

Answer: Since the results in the F_1 generation differ between the two sexes, we suspect that a sex-linked locus is responsible for the control of body color. If we assume that it is a recessive trait, then the female parent must have been a recessive homozygote, and the male must have been a wild-type hemizygote. If we assign the wild-type allele as X^+, the yellow-body allele as X^y, and the Y chromosome as Y, then **figure 1**, showing the crosses into the F_2 generation, is consistent with the data. Thus, a recessive X-linked gene controls yellow body color in fruit flies.

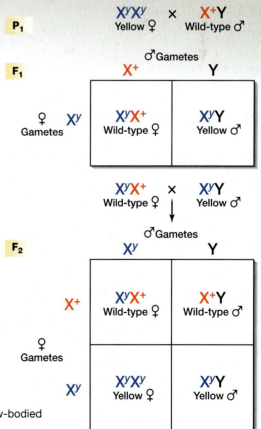

figure 1   Cross between yellow-bodied and wild-type fruit flies.

PROBLEM 2: The affected individuals in the pedigree in **figure 2** are chronic alcoholics. What can you say about the inheritance of this trait?

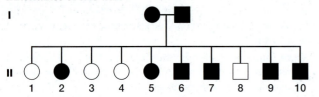

figure 2   A pedigree for alcoholism.

Answer: We can rule out either sex-linked or autosomal recessive inheritance because both parents had the trait, yet they produced some unaffected children. Nor can the mode of inheritance be by a sex-linked dominant gene because an affected male would have only affected daughters, since each daughter gets a copy of his single X chromosome. We are thus left with autosomal dominance as the mode of inheritance. If that is the case, then both parents must be heterozygotes; otherwise, all the children would be affected. If both parents are heterozygotes, we expect a 3:1 ratio of affected to unaffected offspring (a cross of $Aa \times Aa$ produces offspring of $A-$:aa in a 3:1 ratio); here, the ratio is 6:4.

PROBLEM 3: A female fly with orange eyes is crossed with a male fly with short wings. The F_1 females have wild-type (red) eyes and long wings; the F_1 males have orange eyes and long wings. The F_1 flies are crossed to yield:

47 long wings, red eyes

45 long wings, orange eyes

17 short wings, red eyes

14 short wings, orange eyes

with no differences between the sexes. What is the genetic basis of each trait?

Answer: In the F_1 flies, we see a difference in eye color between the sexes, indicating some type of sex linkage. Since

the females are wild-type, wild-type is probably dominant to orange. We can thus diagram the cross for eye color as:

P: (female) X^oX^o × X^+Y (male)
         orange   red
                          ↓
F_1: X^+X^o × X^oY
              red   orange
                          ↓
F_2: X^+X^o X^oX^o X^+Y X^oY
            red   orange   red   orange

We would thus expect to see equal numbers of red-eyed and orange-eyed males and females, which is what we observe. Now look at long versus short wings. If we disregard eye color, wing length seems to be under autosomal control with short wings being recessive. Thus, the parents are homozygotes (ss and s^+s^+), the F_1 offspring are heterozygotes (s^+s), and the F_2 progeny have a phenotypic ratio of 3:1, wild-type (long) to short wings.

We can combine this information to produce the complete genotype of each individual in the cross.

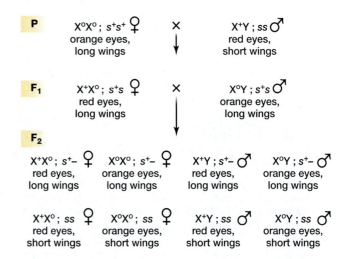

Exercises and Problems

12. Do honeybee males produce sperm via mitosis or meiosis? Explain using diagrams.
13. How many Barr bodies would you see in the nuclei of persons with the following sex chromosomes?
    a. X0
    b. XX
    c. XY
    d. XXY
    e. XXX
    f. XXXXX
    g. XX/XY mosaic
    What would the sex of each of these persons be?
14. Assuming they are viable, determine the sex of each of the following *Drosophila* flies.

	Chromosome Composition	
	Sex chromosomes	Autosomes
a.	X	Diploid
b.	XX	Diploid
c.	XY	Diploid
d.	XXX	Diploid
e.	XXY	Diploid
f.	X	Triploid
g.	XX	Triploid
h.	XY	Triploid
i.	XXX	Triploid
j.	XXY	Triploid
k.	XX	Tetraploid
l.	XXX	Tetraploid
m.	XXYY	Tetraploid

15. Two *Drosophila* flies, each with a normal complement of chromosomes, produce an offspring with a normal number of autosomes and an XYY sex chromosome composition.
    a. Is this fly male or female?
    b. Diagram how this could have been produced?

16. A grandson does not inherit a sex chromosome from which of his four grandparents? Does the answer change in the case of a granddaughter?

17. The autosomal recessive *transformer* (*tra*) mutation converts chromosomal females into sterile males. A female *Drosophila* heterozygous for the *transformer* allele is mated with a normal male homozygous for *transformer*. What is the sex ratio of their offspring? What is the sex ratio of their offspring's offspring?

18. The autosomal recessive *doublesex* (*dsx*) gene converts males and females into developmental intersexes. Two fruit flies, both heterozygous for the allele, are mated. What are the sexes of their offspring?

19. In *Drosophila*, the recessive X-linked *lozenge* allele (*lz*) produces narrow eyes. Diagram to the F_2 generation a cross of a lozenge male and a homozygous normal female. Diagram also the reciprocal cross.

20. In chickens, barred plumage (white bars on black feathers) is under the control of a dominant gene *B*, located on the Z chromosome. The recessive *b* allele produces nonbarred (plain) plumage.
    a. What F_1 and F_2 phenotypic ratios are expected in the offspring of a barred hen and a nonbarred rooster?
    b. Two barred chickens produce a nonbarred offspring. Is it male or female? Explain.

21. In *Drosophila*, cut wings are controlled by a recessive X-linked allele (*ct*), and fuzzy body is controlled by a recessive autosomal allele (*fy*). When a fuzzy female is mated with a cut male, all the members of the F_1 generation are wild-type. What are the proportions of F_2 phenotypes, by sex?

22. Consider the following crosses in canaries:

Parents	Progeny
a. pink-eyed female × pink-eyed male	all pink-eyed
b. pink-eyed female × black-eyed male	all black-eyed
c. black-eyed female × pink-eyed male	all females pink-eyed, all males black-eyed

Explain these results by determining which allele is dominant and how eye color is inherited.

23. Consider the following crosses involving yellow and gray true-breeding *Drosophila*:

Cross	F_1	F_2
gray female × yellow male	all males gray, all females gray	97 gray females, 42 yellow males, 48 gray males
yellow female × gray male	all females gray, all males yellow	?

a. Is color controlled by an autosomal or an X-linked gene?
b. Which allele, gray or yellow, is dominant?
c. Assume 100 F_2 offspring are produced in the second cross. What kinds and what numbers of progeny do you expect? List males and females separately.

24. Sex linkage was originally detected in 1906 in moths with a ZW sex-determining mechanism. In the currant moth, the recessive *pale color* allele (*p*) is located on the Z chromosome. Diagram reciprocal crosses to the F_2 generation involving pale color and true-breeding wild-type moths.

25. A man with brown teeth mates with a woman with normal white teeth. They have four daughters, all with brown teeth, and three sons, all with white teeth. The sons all mate with women with white teeth, and all their children, three daughters and two sons, have white teeth. One of the daughters (A) mates with a man with white teeth (B), and they have two brown-toothed daughters, one white-toothed daughter, one brown-toothed son, and one white-toothed son.
    a. Explain these observations.
    b. Based on your answer to part (*a*) what is the chance that the next child of the A–B couple will have brown teeth?

26. In human beings, red-green color blindness is inherited as an X-linked recessive trait. A woman with normal vision whose father was color-blind marries a man with normal vision whose father was also color-blind. This couple has a color-blind daughter with a normal complement of chromosomes. Is infidelity suspected? Explain.

27. In *Drosophila*, white eye is an X-linked recessive trait, and ebony body is an autosomal recessive trait. A homozygous white-eyed female is crossed with a homozygous ebony male.
    a. What phenotypic ratio do you expect in the F_1 generation?
    b. What phenotypic ratio do you expect in the F_2 generation?
    c. Suppose the initial cross was reversed: ebony female × white-eyed male. What phenotypic ratio would you expect in the F_2 generation?

28. In *Drosophila*, abnormal eyes can result from mutations in many different genes. A true-breeding wild-type male is mated with three different females, each with abnormal eyes. The results of these crosses are as follows:

	Females	Males
male × abnormal-1 →	all normal	all normal
male × abnormal-2 →	1/2 normal, 1/2 abnormal	1/2 normal, 1/2 abnormal
male × abnormal-3 →	all abnormal	all abnormal

Explain the results by determining the mode of inheritance for each abnormal trait.

29. Based on the following *Drosophila* crosses, explain the genetic basis for each trait and determine the genotypes of all individuals:

P: white-eyed, dark-bodied female × red-eyed, tan-bodied male

F₁: females are all red-eyed, tan-bodied; males are all white-eyed, tan-bodied

F₂: 27 red-eyed, tan-bodied

24 white-eyed, tan-bodied

9 red-eyed, dark-bodied

7 white-eyed, dark-bodied

(No differences between males and females in the F₂ generation.)

30. In *Drosophila*, vermillion (orange-red) eye color is determined by an X-linked recessive gene, *v*. Wild-type (brick-red) eye color is determined by the dominant v^+ allele.

   a. A true-breeding vermillion female is crossed to a wild-type male. What is the sex ratio and phenotypes of the F₁ offspring?

   b. Mating two F₁ siblings produced an F₂ generation containing:

   65 vermillion males
   60 wild-type males
   62 vermillion females
   58 wild-type females

   Use a chi-square test to determine whether these numbers are consistent with the expected results for X-linked genes.

31. What are the possible modes of inheritance in pedigrees *a–c* in **figure 3**? What modes of inheritance are not possible for a given pedigree?

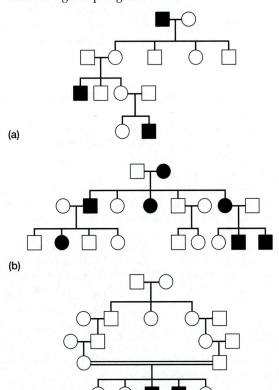

(a)

(b)

(c)

figure 3   Three pedigrees showing different modes of inheritance.

32. Construct pedigrees for traits that could *not* be
   a. autosomal recessive.   **c.** sex-linked recessive.
   b. autosomal dominant.   **d.** sex-linked dominant.

33. In chicken, the pattern of hen feathering (*H*) is dominant to cock-feathering (*h*). Hen feathering is expressed in both sexes, but cock feathering is sex-limited to the male. Diagram to the F₂ generation a cross between a cock-feathered rooster and a homozygous hen-feathered hen.

34. Determine the possible modes of inheritance for each trait in pedigrees *a–c* in **figure 4**.

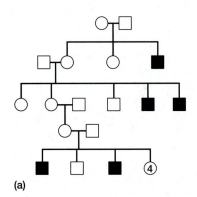

(a)

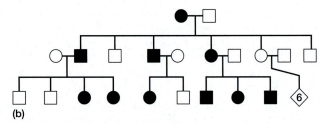

(b)

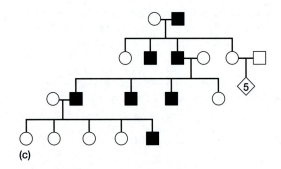

(c)

figure 4   Varying modes of inheritance.

35. Consider the following four genes in *Drosophila*: Gene *A* is on the X chromosome, and genes *B*, *C*, and *D* are located on three different autosomes. Each of the four genes segregates two alleles (*A/a*, *B/b*, *C/c*, and *D/d*) that exhibit complete dominance/recessiveness. A true-breeding female who is dominant for all four traits is crossed to a male that is recessive for all four traits.

   a. How many types of gametes can be produced by an F₁ female?

   b. What proportion of the F₂ generation is expected to be dominant for all four traits?

   c. What proportion of the F₂ generation is expected to be heterozygous at all four loci?

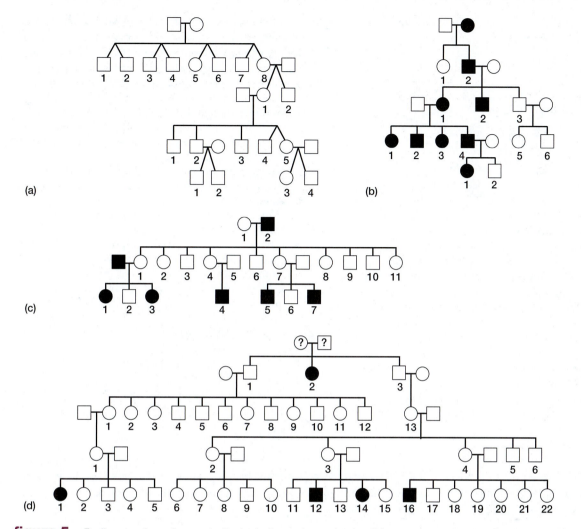

(a)

(b)

(c)

(d)

figure 5   Pedigrees of rare human traits, including twin production (a).

36. Pedigrees *a–d* in **figure 5** show the inheritance of rare human traits, including twin production. Determine which modes of inheritance are most probable, possible, or impossible.

37. Homer, who is affected with a particular disease, is married to Marge, a phenotypically normal woman. They have 10 children: 6 normal boys and 4 girls with the disease. Characterize each of the following modes of inheritance as: impossible, unlikely, probable. Justify your answers.

   a. Autosomal dominant   **d.** X-linked dominant
   b. Autosomal recessive   **e.** X-linked recessive
   c. Y-linked

38. The following pedigree shows the inheritance of pattern baldness, a potential sex-influenced trait that is dominant in males and recessive in females.

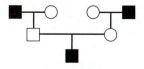

Using the symbols *B/b*, assign genotypes to the extent that is possible.

39. Which mode of inheritance is consistent with the following pedigree? Explain.

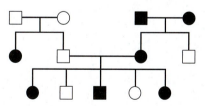

40. In humans, hemophilia A is an X-linked recessive trait characterized by an abnormality in blood clotting. Albinism is an autosomal recessive trait characterized by the absence of pigment in the skin. A man and woman, both with normal skin pigmentation, have two daughters: one is albino and the other hemophilic. The maternal grandfather had normal blood clotting. The couple plan to have another child. What is the probability that their next child will be:

   a. a son that is normal for both skin pigmentation and blood clotting?
   b. a daughter that is albino and hemophilic?
   c. a child that is a carrier for both traits?

41. What is the most likely mode of inheritance for the following pedigree? Why?

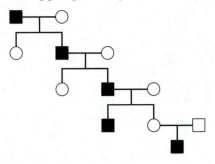

Chapter Integration Problem

A hypothetical genetic disease, Nanunanu syndrome, is caused by a deficiency of the enzyme nazrat, which is involved in the formation of skin. Affected individuals have skin that is strikingly hyperextensible, fragile, and has a tendency to bruise quite easily. Mork OrKool and Mindy McCool, who are phenotypically normal, are planning to start a family. They visit a genetic counselor and provide him with the following information:

1. Mork's parents are normal, and so are his three older sisters.
2. Mork's father has an older sister who has the disease. Their parents are normal.
3. Mindy's parents are normal, but she has a younger brother who has Nanunanu syndrome.
4. Mindy's father has two younger sisters, who are fraternal twins and normal. Their parents are also normal.
   a. Construct the pedigree that encompasses three generations of the OrKools and McCools.
   b. Which of the following modes of inheritance can be excluded: autosomal dominant, autosomal recessive, Y-linked, X-linked dominant, and X-linked recessive? Assume complete penetrance.
   c. For each mode that you excluded, indicate the specific individual (e.g., II-1, III-2, etc.) that led you to that decision.
   d. Using the symbols N/n, assign genotypes to the extent that is possible.
   e. What is the probability that Mork and Mindy's first child will have Nanunanu syndrome? Assume Mork's mother is homozygous.
   f. If Mork and Mindy have two children, what is the probability that neither of them will be affected by the disease?

Suppose Mork and Mindy's first child actually develops the disease.
   g. What is the probability that their next child will also have Nanunanu syndrome?
   h. What is the probability that *at least* one of their next three children will be normal?

Do you need additional review? Visit **www.mhhe.com/hyde** for practice tests, answers to end-of-chapter questions and problems, interactive exercises, and animation tutorials all designed to enhance your understanding of key genetic concepts.

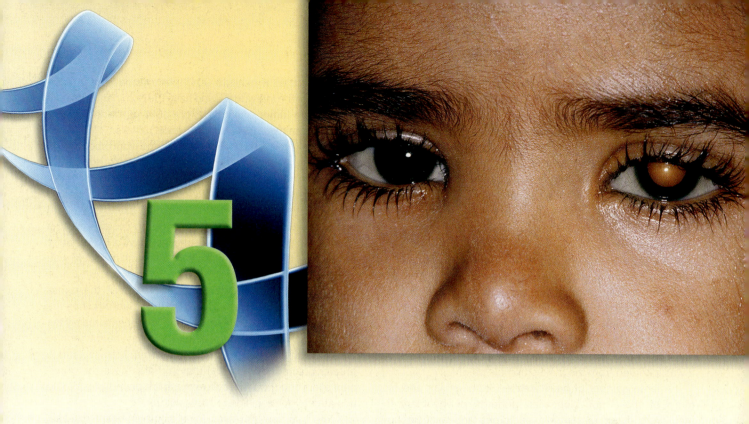

Modifications to Mendelian Patterns of Inheritance

Essential Goals

1. Compare the various types of dominance and how they affect the phenotype.

2. Understand how penetrance and expressivity can affect the expression of an allele.

3. Understand how epistasis and suppression of nonallelic genes affect the phenotype.

4. Using phenotypic ratios, deduce the underlying genetic interactions.

5. How do epigenetic phenomena affect the observed pattern of inheritance?

Chapter Outline

Photo: A child with a tumor in the left eye due to retinoblastoma. The unusual light reflection from the back of the eye results from the tumor.

*I*n chapter 2, we discussed Mendel's laws of inheritance and defined the principle of dominant and recessive traits. Together, these properties generated very clear phenotypic and genotypic ratios in monohybrid crosses. In dihybrid crosses, the phenotypic and genotypic ratios were derived by applying the product rule to independently acting genes.

In this chapter, you will see that a gene may have a large number of alleles, and they may not function in a simple dominant–recessive manner. The result is an increase in the number of phenotypes that a single gene can produce in a diploid organism.

In contrast, different genes can interact to produce a common phenotype, causing the number of phenotypes observed in a dihybrid cross to be reduced. In both these cases, however, the Mendelian principles discussed in chapter 2 still underlie these results.

Additionally, in some instances, termed **epigenetic inheritance,** the expression of an allele depends on whether it was inherited from the male or female parent.

A large number of health-related traits exhibit one or more of these non-Mendelian patterns of inheritance. Identifying these patterns in pedigrees and associating them with specific human diseases are essential components of accurate genetic counseling.

As an example, retinoblastoma is an inherited human disease that occurs in children primarily younger than 2 years of age and very rarely in children older than age 5. The disease causes the growth of eye tumors, which, if untreated, may proceed up the optic nerve and reach the brain. It occurs in 1 in every 15,000–25,000 births.

Retinoblastoma is inherited as an autosomal dominant disease. In approximately 90% of the families that exhibit retinoblastoma, it exhibits 90% penetrance. This means that 90% of the individuals who have the dominant allele will express the tumor phenotype, but 10% of the individuals with the dominant allele fail to develop tumors.

Imagine for a moment that you and your spouse are considering the possibility of having children. You know that your father and brother both suffered from retinoblastoma, but you appear to be perfectly normal. Thus, you assume that either you do not have the dominant allele, or you are one of the 10% of the individuals who do not express the phenotype. With the knowledge that you have a 10% chance of passing the dominant allele to your child, would you decide to have children?

In 10% of the affected families, however, the penetrance of the disease is very low. An individual who has the dominant allele has a high probability of not expressing the tumor phenotype. In this case, knowing that your father and brother both suffered from retinoblastoma, there is a much greater chance (30%–40%) that you also have the dominant allele and have failed to express the phenotype. Would this affect your decision to have children?

You can see that even for a single inherited disease, great variation can exist in how it is inherited and manifested in offspring. With incorrect knowledge about the mode of inheritance, parents may suffer a large amount of guilt that they were the cause of their child's suffering. Genetic counseling must therefore be guided by the most up-to-date information on how alleles affect a trait's expression.

In this chapter, we will consider in detail the types of modifications that have been observed in Mendel's basic inheritance patterns and what they imply about the actions and interactions of alleles and genes.

5.1 Variations on Dominance

A dominant allele was defined in chapter 2 as one that determined the phenotype of a heterozygous individual, and the recessive allele would not affect the phenotype. Another way to think about this is that in a diploid organism, a dominant allele requires only one copy to determine the phenotype, but a recessive allele requires two copies (except where only one copy of the gene is present, such as on the X chromosome in male mammals).

Crossing two heterozygotes when dominance is complete always produces a 3:1 dominant to recessive phenotypic ratio. In several instances, however, this simple dominant–recessive relationship does not appear to function. We will first consider two of these phenomena—incomplete dominance and codominance—in which the heterozygous genotype does not exhibit a simple dominant phenotype. A little later on, we will discuss the situation in which a single allele can produce both a dominant and recessive phenotype.

Incomplete Dominance

When we examine four-o'clock plants (*Mirabilis jalapa*), we find two different types of homozygous flower color. The first has red flower petals, and the other has white. If a pure-breeding red-flowered plant is crossed with a pure-breeding white one, the heterozygous offspring have pink petals. Thus, based on the original definition of dominance, namely, the phenotype observed in heterozygotes, neither red nor white is dominant. Rather, an intermediate phenotype is observed.

If these pink-flowered F_1 plants are crossed, the F_2 plants appear in a ratio of 1:2:1, having red, pink, or white flower petals, respectively (**fig. 5.1**). This 1:2:1 ratio is the same genotypic ratio that Mendel described in a monohybrid F_1 cross.

In **incomplete dominance,** the phenotype of the heterozygote falls between those of the two homozygotes. A 1:2:1 phenotypic ratio in a monohybrid cross, where 1/2 of the progeny have a phenotype intermediate to the phenotypes of the two parents, is an indication of incomplete dominance.

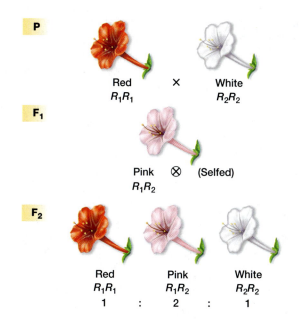

figure 5.1 Flower color inheritance in the four-o'clock (*Mirabilis*) plant: an example of incomplete dominance. Incomplete dominance results when the heterozygote (R_1R_2) has a phenotype (pink) that is between the two homozygous phenotypes (red and white).

In the case of four-o'clock plants, one allele (R_1) specifies red pigment color, and another (R_2) specifies no color. Because neither allele is dominant over the other, we do not use an uppercase/lowercase designation for the alleles. Rather, both alleles are designated with uppercase letters, and the subscript numbers reveal the true identity of the allele.

Flowers in heterozygotes (R_1R_2) have about half the red pigment of the flowers in red homozygotes (R_1R_1) because they have only one copy of the allele that produces color.

Codominance

Another genetic interaction that fails to exhibit complete dominance is the human ABO blood type system. If an individual who is homozygous for blood type A has children with an individual who is homozygous for blood type B, the offspring have blood type AB. Like incomplete dominance, the phenotype of the offspring is neither blood type A nor type B. But unlike incomplete dominance, the phenotype of the offspring is not intermediate between blood type A or B. Rather, the phenotype includes both A and B. (See the section entitled "Multiple Alleles," page 98, for more information about blood types.)

Cases in which the heterozygote expresses both phenotypes simultaneously are referred to as **codominance.** As with incomplete dominance, the offspring of two AB heterozygotes will have both genotypic

and phenotypic ratios of 1:2:1 (type A: type AB: type B). The alleles for blood type are shown as I^A and I^B, because neither is dominant to the other.

Levels of Dominance

If we reexamine the pink four-o'clock flowers described earlier at the subcellular level, we find that the color results from organelles within the flower cells called **plastids.** Red flower cells contain red plastids, and white flower cells contain white ones. Pink flower cells contain both red and white plastids. Thus, when we describe the phenotype of the pink flowers at the level of whole-flower color, it represents incomplete dominance because of the intermediate color between red and white. But if we choose to describe the pink phenotype at the level of the cell, it would be classified as codominance because of the presence of both red and white plastids. It is therefore very important to clarify the level of phenotype description when differentiating between these two processes.

As technology has improved, researchers have found more and more cases in which the heterozygous phenotype results from codominance. For example, in Tay–Sachs disease, homozygous recessive children usually die before the age of 3 after suffering severe nervous system degeneration. In contrast, both homozygous normal and heterozygous individuals appear to be phenotypically normal. Thus, at the level of the individual's viability, the wild-type Tay–Sachs gene exhibits complete dominance over the recessive lethal allele (**fig. 5.2a**). As biologists discovered how this and other diseases work, however, they made the detection of the heterozygous individuals possible.

As with many genetic diseases, the culprit in Tay–Sachs is a defective enzyme. This enzyme, **hexosaminidase-A,** is needed for proper lipid metabolism. Modern techniques allow technicians to assay the blood for this enzyme. Afflicted homozygotes have no enzyme activity; heterozygotes have about half the normal level; and, of course, homozygous normal individuals have the full level of activity (fig. 5.2b). At the level of enzyme activity, this disease exhibits incomplete dominance: Heterozygous individuals exhibit enzyme activity that is intermediate between those of the two homozygotes (100% and 0% enzyme activity).

Scientists have identified several different mutations in the hexosaminidase-A gene that result in the Tay–Sachs phenotype. One of these mutations produces a smaller protein that can be identified by gel electrophoresis. Afflicted individuals possess only the smaller mutant protein; heterozygotes have both the smaller and larger protein; and, of course, homozygous normal individuals have only the larger wild-type protein (fig. 5.2c). Thus, at the protein level, Tay–Sachs disease exhibits codominance: Heterozygous individuals

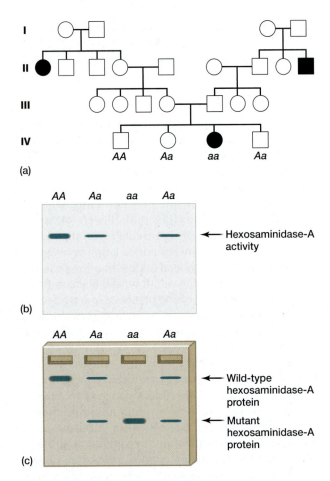

(a)

(b)

← Hexosaminidase-A
activity

(c)

← Wild-type
hexosaminidase-A
protein

← Mutant
hexosaminidase-A
protein

figure 5.2   Analysis of Tay–Sachs in humans.
(a) A pedigree reveals that the Tay–Sachs disease is
due to a recessive allele. (b) The four individuals in the
youngest generation (IV) of the pedigree are analyzed for
hexosaminidase-A activity. The homozygous dominant
individual exhibits complete activity, the two heterozygous
siblings possess half the activity (thinner band compared
to the *AA* individual), and the affected individual exhibits
no activity. (c) The same four individuals in the pedigree
are analyzed by gel electrophoresis for hexosaminidase-A
protein. The homozygous dominant individual possesses only
the larger wild-type protein, the two heterozygous siblings
possess both the larger and smaller protein, and the affected
individual exhibits only the smaller mutant protein.

possess both the smaller mutant protein and the larger
wild-type protein.

   In human genetics, the identification of heterozygotes has a profound impact on the ability to evaluate
potential inheritance in the subsequent generation.
As you already know for a monohybrid cross, two
heterozygotes have a 25% chance that any child they
bear will have the recessive phenotype. By being able
to distinguish heterozygous individuals, scientists
can more accurately calculate the probability that prospective parents will have children with the recessive
phenotype.

Restating the Concepts

▶ In addition to complete dominance, alleles may
exhibit incomplete dominance or codominance.

▶ Incomplete dominance produces a heterozygous
phenotype that is intermediate between the two
homozygous phenotypes.

▶ Codominance produces a heterozygous phenotype
that expresses both of the homozygous phenotypes.

▶ The level at which the phenotype is being described
is critical for determining the type of dominance.

▶ Crossing two individuals who are heterozygous for
incomplete dominant or codominant alleles produces
progeny in a 1:2:1 phenotypic ratio, which matches
the genotypic ratio.

5.2   Multiple Alleles

Our discussion of genetic interactions up to this
point has dealt with genes that only have two alleles.
However, a given gene can have more than two alleles.
Although any particular diploid individual can have
only two, many alleles of a given gene may exist in a
population. In fact, multiple alleles are the rule rather
than the exception. The following three examples help
explain how these alleles interact.

The ABO Blood Group

A classic example of multiple human alleles is in the
ABO blood group, which Karl Landsteiner discovered
in the early 1900's. The presence of four blood type
phenotypes, A, B, AB, and O, are the best known of all
the red cell antigen systems, primarily because of the
importance of these phenotypes in blood transfusions.

   Antigens are substances that induce an immune
response. The ABO system is unusual because antibodies
can be present in an individual without prior exposure to
the antigen. People with a particular AB antigen on their
erythrocytes (red blood cells) have the antibody against
the other antigen in their serum, as shown in **table 5.1**.

   Adverse reactions to blood transfusions primarily
occur because the antibodies in the recipient's serum
react with the antigens on the donor's red blood cells.
Thus, type A persons cannot donate blood to type B
persons because the anti-A antibodies in the serum of
type B persons react with the A antigen on the donor
red blood cells and cause the cells to clump. Type O
blood can be donated to people with any blood type,
because it lacks both antigens. In contrast, type AB
individuals can receive blood transfusions from any
of the four blood types, because these individuals lack
the antibodies against either the A or B antigens.

The four blood type phenotypes in the ABO system are produced by three alleles, I^A, I^B, and i (see table 5.1). The I^A and I^B alleles are responsible for the production of the A and B antigens. The I^A and I^B alleles code for glycosyl transferase enzymes; each allele causes a different modification to the terminal sugars of a mucopolysaccharide (termed an H structure), which is found on the surface of red blood cells (**fig. 5.3**). They are codominant because both modifications (antigens) are present in a heterozygote. In fact, whichever enzyme reaches an H structure first will modify it. Once modified, the H structure will not respond to the other enzyme. Therefore, both A and B antigens are produced in the heterozygote in roughly equal proportions.

The third allele, i, causes no change to the H structure because i produces a nonfunctioning enzyme. The i allele is recessive to both the I^A or I^B alleles, because either of the glycosyl transferase enzymes encoded by the I^A or I^B will modify the H product, masking the fact that the i allele is present. Thus, we observe both codominance (I^A and I^B) and complete dominance (I^A and I^B completely dominant to i) between the three alleles of this single gene.

Eye Color in *Drosophila*

In chapter 4, we described the *Drosophila white* mutation that was identified by Thomas Hunt Morgan (w^1). As you remember, the white-eyed mutant phenotype was completely recessive to the wild-type (brick red color) phenotype. However, since the first *white* allele was identified in 1910, over 1000 additional *white* alleles have been identified. The most severe *white* allele (w^1) produces an eye that completely lacks color. The *white-apricot* allele (w^a) produces a yellowish-orange eye color, the *white-buff* allele (w^{bf}) produces a buff eye color, the *white-eosin* allele (w^e) produces a yellowish-pink eye color, the *white-speck* allele (w^{sp}) produces a yellowish-brown eye color, and the *white reddish* allele (w^r) produces a reddish eye color that is nearly wild-type. Thus, all of the alleles for the *white* gene form an **allelic series,** which produce a range of phenotypes from the most severe (w^1) to nearly normal (w^r).

In this allelic series, you can detect a variety of genetic interactions. The wild-type *white* allele, w^+, is completely dominant over all the other *white* alleles because any heterozygous fly that contains w^+ and any mutant *white* allele will exhibit the wild-type eye color. The majority of the mutant *white* alleles are incompletely dominant with each other because a heterozygous fly will have an intermediate eye color relative to the two homozygous mutant flies. For example, the $w^e w^a$ fly has an eye color that lies between $w^e w^e$ and $w^a w^a$ flies.

There are some notable exceptions to this incomplete dominance. The $w^a w^{sp}$ flies have a darker brown color than either $w^a w^a$ or $w^{sp} w^{sp}$ flies. Thus, an allelic series may exhibit some complex genetic interactions that cannot be defined as either complete dominance, codominance, or incomplete dominance.

table 5.1   ABO Blood Types with Immunity Reactions

Blood Type Corresponding to Antigens on Red Blood Cells	Antibodies in Serum	Genotype	Reaction of Red Blood Cells to Anti-A Antibodies	Reaction of Red Blood Cells to Anti-B Antibodies
O	Anti-A and anti-B	ii	−	−
A	Anti-B	$I^A I^A$ or $I^A i$	+	−
B	Anti-A	$I^B I^B$ or $I^B i$	−	+
AB	None	$I^A I^B$	+	+

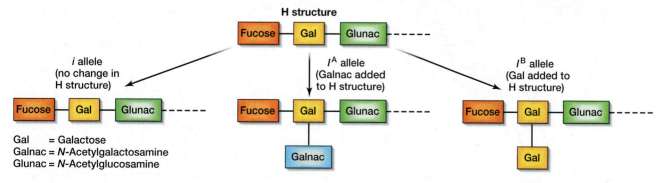

H structure

Gal   = Galactose
Galnac = *N*-Acetylgalactosamine
Glunac = *N*-Acetylglucosamine

figure 5.3   Function of the I^A, I^B, and i alleles of the ABO gene. The gene products of the I^A and I^B alleles of the ABO gene affect the terminal sugars of a mucopolysaccharide (H structure) found on red blood cells. The gene products of the I^A and I^B alleles are the enzymes alpha-3-N-acetyl-D-galactosaminyltransferase and alpha-3-D-galactosyltransferase, respectively.

Fur Color in the Mouse

Mouse fur color is controlled by several genes, one of which is named *agouti*. Researchers have identified over 25 different agouti alleles, most that exhibit either a unique phenotype or different genetic interaction with other agouti alleles. We will discuss only four of the agouti alleles here, to demonstrate the potentially complex genetic interactions between different alleles.

The wild-type *agouti* allele (A) produces a dark gray fur color, which results from the deposition of both yellow and black pigment in the hairs (**fig. 5.4**). The white-bellied agouti allele, A^w, has dark gray fur on the back and white or cream-colored fur on the belly. The nonagouti allele, a, is recessive to all the agouti alleles. The aa mouse will be black on both the back and belly (fig. 5.4). A mouse that is homozygous for the black and tan allele, a^t, will have a black back with a yellow belly.

The phenotypes of the heterozygous mice reveal that the A^w allele is dominant to the other three agouti alleles (**table 5.2**). The a allele is recessive to the other three alleles. However, a complex interaction occurs between the wild-type A allele and the a^t allele. The dark gray back reveals that the A allele is dominant to a^t on the back, but the yellow belly shows that a^t is dominant to A on this part of the body (fig. 5.4).

Thus, an allelic series may exhibit a hierarchy of dominance such as A^W being dominant to A, and both being dominant to a. Alleles within the series may also exhibit complex dominant interactions. For example, A is dominant to a^t on the back, but a^t is dominant to A on the belly. We will revisit the *agouti* gene in our discussion on recessive lethal mutations.

table 5.2   Mouse Fur Color

Genotype	Phenotype
AA	Dark gray (agouti)
A^wA^w	Dark gray back and white or cream-colored belly
a^ta^t	Black back with cream-colored belly
aa	Solid black
A^w-	Dark gray back and white or cream-colored belly
Aa^t	Dark gray back with cream-colored belly
Aa	Dark gray
a^ta	Black back with cream-colored belly

It's Your Turn

Most genes have many mutant alleles that exhibit complex interactions, such as exhibiting dominant and recessive phenotypes. Try working problems 18, 20, and 44 that deal with an allelic series.

Restating the Concepts

▶ Most genes have more than two alleles. Within the allelic series for a gene, any particular allele may exhibit complete dominance, incomplete dominance, or codominance with any other allele.

5.3 Testing Allelism

As you saw in the preceding discussion, a gene may have a large number of alleles that may produce different phenotypes. Additionally, mutations in different genes may cause similar or even identical phenotypes. To accurately study the mode of inheritance for a particular phenotype, it is essential to know which mutations are different versions of the same gene, which are called **alleles**, and what mutations are in different genes.

The **complementation test** is used to determine whether two recessive mutations are alleles. An individual homozygous for one recessive mutation is crossed with an individual homozygous for a different recessive mutation. The phenotypes of the heterozygous

figure 5.4   Phenotypes of different allelic combinations at the mouse *agouti* gene. The phenotype associated with the AA, A^YA, and a^ta^t (left to right) genotypes are shown. The a^ta^t mouse has a black back compared to the yellow coat of the A^YA mouse. The wild-type mouse (AA) has a brown or agouti coat.

progeny are then examined. If the phenotype is normal, then the mutations occur in different genes and are said to complement each other and are nonallelic. If the phenotype is mutant, the mutations fail to complement and are allelic.

Complementation

An example of complementation is shown in **figure 5.5a**. Two flies are both homozygous for a recessive wingless mutation. All of the F_1 offspring have wings (wild-type phenotype). The production of a wild-type phenotype reveals that the mutations complement, and therefore they occur in different genes. This result is diagrammed in figure 5.5a.

Notice that the male contributes a wild-type allele for gene 1 and a recessive mutant allele for gene 2, while the female contributes a recessive mutant allele for gene 1 and a dominant wild-type allele for gene 2. Thus, the F_1 progeny are heterozygous for both gene 1 and gene 2. These flies exhibit the dominant wild-type phenotype because they have a dominant wild-type allele for both genes. Notice also that the mutations are arranged in the F_1 progeny such that each homolog contains one mutation. This arrangement of the two recessive mutations on different chromosomes is called the *trans* configuration.

An example of noncomplementation is shown in figure 5.5b. In this case, two flies are both homozygous for a recessive wingless mutation. All the F_1 progeny from this cross are wingless. Because the progeny fail to exhibit the wild-type phenotype, these two mutations fail to complement and are alleles.

A second example of noncomplementation is shown in figure 5.5c. In this case, two flies are both homozygous for a recessive mutation. Notice that the phenotypes are not identical; the male is wingless, and the female has small, wrinkled wings. The F_1 progeny from this cross have very small, wrinkled wings. Because the progeny fail to exhibit the wild-type phenotype, these two mutations fail to complement and must be alleles.

Notice that the two alleles do not have to produce the same phenotype, and the progeny do not have to exhibit one of the allele phenotypes (remember that the alleles may be codominant or incompletely dominant). The critical point is whether the progeny exhibit the wild-type or a mutant phenotype. The diagram of this cross reveals that each parent contributes one mutant recessive allele for a single gene. Again, the mutations are in a *trans* configuration.

The *Cis–Trans* Test

This complementation test is also called the ***cis–trans* test.** This test was devised by Edward B. Lewis, who was awarded a Nobel Prize in physiology or medicine in 1995 for his analysis of mutations that affect

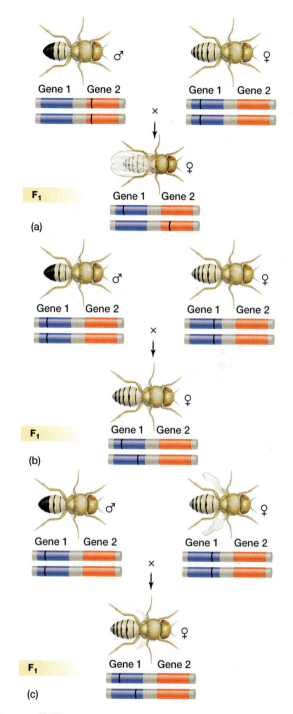

figure 5.5   The complementation test determines if two mutations are alleles. (a) Crossing two flies that are homozygous for a wingless mutation produces wild-type flies. The two mutations must be in different genes and *complement* each other in the F_1 progeny to produce wild-type phenotype. (b) Crossing two flies that are homozygous for a wingless mutation produces wingless flies. The two mutations must be alleles in the same gene. (c) Crossing a homozygous wingless male and a small-winged female produces small-winged flies. Because the F_1 progeny are not phenotypically wild-type, the two mutations must be alleles.

Drosophila development. He developed the *cis–trans* test to examine a large number of mutations and group them into complementation groups, or genes. **Figure 5.6** shows the potential arrangement of mutations in both the *cis* and *trans* configurations.

The ***cis* configuration** places both mutations on the same chromosome. Notice that regardless of whether the mutations are allelic or nonallelic (in different genes), the *cis* configuration always produces a wild-type phenotype because the chromosome lacking the mutations provides the wild-type alleles. The *cis* configuration then serves as a **control,** which confirms that the cross can produce a wild-type phenotype.

The ***trans* configuration** places one mutation on each of the two chromosomes. The *trans* configuration reveals whether the mutations are allelic or not. If the mutations are alleles, they fail to complement and produce a mutant phenotype (fig. 5.6, top right). Mutations that are in different genes (nonallelic) will complement in the *trans* configuration and exhibit a wild-type phenotype (fig. 5.6, bottom right).

This *cis–trans* test is important because it identifies both mutations in separate genes and independent alleles of a single gene. Beyond this, the test also helps to clarify gene-controlled processes. For example, researchers may attempt to mutate every gene in a particular pathway, an approach termed **saturation.** The mutations are then analyzed using the *cis–trans* test to reveal how many genes control the components of the pathway. This knowledge may be useful for treating genetic diseases as well as increasing our understanding of life processes.

Restating the Concepts

▶ The complementation test (*cis–trans* test) is used to determine if different mutations are alleles.

▶ Mutations that are allelic will produce a mutant phenotype when they are in the *trans* configuration.

▶ Mutations that are nonallelic (in different genes) will produce a wild-type phenotype when they are in the *trans* configuration.

5.4 Lethal Alleles

In the above examples, we described how incomplete dominance and codominance could produce phenotypic frequencies in a monohybrid cross that deviated from the expected 3:1. Another class of mutations that can produce a variation of the 3:1 or 1:2:1 ratios are **lethal alleles,** which cause death of some genotypes and shift the phenotypic frequencies. Lethal alleles can exhibit the standard Mendelian dominant–recessive relationship that we described in chapter 2; however, if one or more of the

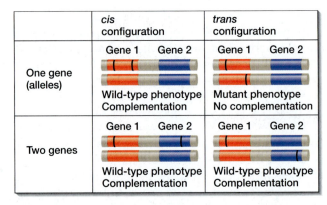

figure 5.6 The possible arrangement of mutations in the *cis–trans* test. When the mutations (vertical black bars) are in the *cis* configuration, the wild-type phenotype is always produced. Only when the mutations are in the *trans* configuration is it possible to determine if the mutations are alleles, based on their failure to complement and retain the mutant phenotype.

genotypic classes fail to survive, the standard Mendelian ratio will be skewed due to this missing class.

Recessive Lethal Mutations

Earlier we discussed four alleles of the *agouti* gene, which are involved in determining fur color in mice. We saw that the agouti phenotype, which is dark brownish-gray fur on the back and belly, is due to a dominant allele (*A*). A fifth allele, *yellow* (A^Y), produces a mouse with yellow or cream-colored fur on both the back and belly (fig. 5.4).

Two crosses are important in examining the dominant–recessive relationship of *yellow* with the wild-type *agouti* allele. First, mating homozygous agouti mice and yellow mice always produces 1/2 yellow mice and 1/2 agouti mice (**fig. 5.7a**). This 1:1 relationship is the expected ratio in a monohybrid testcross. Because the agouti mice are known to be homozygous, the yellow mice must be heterozygous, with the yellow phenotype being dominant to agouti. The second cross is between two yellow mice, which produces 2/3 yellow progeny and 1/3 agouti (fig. 5.7b). This ratio was unexpected; the phenotypic ratio produced from a monohybrid cross should have been 3:1 yellow:agouti.

This 2:1 yellow:agouti ratio does not correspond to either the 3:1 or 1:2:1 Mendelian monohybrid ratios. The dilemma was resolved when pregnant yellow female mice were examined and 25% of the developing embryos were dead! A Punnett square reveals that if the homozygous *yellow* allele is lethal ($A^Y A^Y$), 2/3 of the surviving mice will be yellow and 1/3 will be agouti (fig. 5.7c).

The wild-type allele is recessive for coat color (*A*). The mutant allele is dominant for the yellow coat color (A^Y). When we describe viability as a phenotype, however, we need two copies of the mutant A^Y allele to

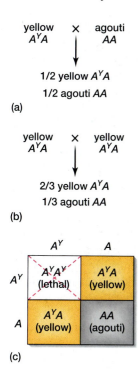

(a)

(b)

(c)

figure 5.7 Coat color in mice reveals a recessive lethal mutation. A yellow mouse and an agouti mouse are shown in figure 5.4. (a) Crossing a yellow mouse with a homozygous agouti mouse produces 50% yellow mice and 50% agouti mice in the F_1 generation. This 1:1 phenotypic ratio in the progeny is consistent with testcrossing a heterozygous mouse (yellow). (b) Crossing two yellow mice always produces 2/3 yellow mice and 1/3 agouti mice in the F_1 generation. (c) A Punnett square diagrams the resulting genotypes from crossing two yellow mice. If the homozygous $A^Y A^Y$ phenotype is lethal, the remaining viable mice will be 2/3 yellow ($A^Y A$) and 1/3 agouti (AA).

observe the mutant lethal phenotype. Requiring two copies of an allele to observe the associated phenotype is the definition of a recessive allele. So in this case, the A^Y mutant allele produces both a dominant phenotype (coat color) *and* a recessive phenotype (lethality). **Pleiotropy** is the term used when a single mutation causes multiple phenotypes. Pleiotropy is not uncommon.

The A^Y allele is described as a recessive lethal mutation, but we do not need to change the symbol to lowercase. Describing the lethality as a recessive mutation is sufficient. Geneticists adopt a symbol system usually based on the first phenotype that was observed. In this case, it was the yellow coat color. If the lethal phenotype had been identified first, it would have been assigned the recessive mutant phenotype and likely a lowercase letter ($a^Y a^Y$). This notation would also be confusing, because the mutant yellow phenotype would still be dominant (Aa^Y), even though it was denoted lowercase. Therefore, it is important to become comfortable with the use of the terms dominant and recessive, and not to rely solely on whether the mutant allele is uppercase or lowercase.

Dominant Lethal Mutations

While a recessive lethal mutation must be present in two copies for death, a **dominant lethal** requires only a single mutant allele. It may not be obvious how a dominant lethal mutation can exist, since the presence of either one or two copies of this allele would result in the death of the organism and the inability to pass it on to progeny.

Two possible mechanisms can permit the inheritance of dominant lethal mutations. One possibility is that the lethality may not be expressed until after

the individual has reached sexual maturity. This is the mechanism by which Huntington disease is inherited in humans. The neurological phenotype that occurs prior to death is not usually expressed until after age 40, which is often after children have been born. Because Huntington disease is a fairly rare disease, individuals who develop it are often heterozygotes. Thus, a person whose parent dies from Huntington disease has a 50% chance of having inherited the dominant mutant allele. Death from the disease will eventually result, barring other causes.

Another mechanism that allows dominant lethal mutations to be inherited is incomplete penetrance of the lethal phenotype. A description of **incomplete penetrance,** which is the failure to express the phenotype associated with a particular genotype, is described in the next major section of this chapter.

Deleterious Mutations

Deleterious alleles reduce the viability of the organism, without always causing death. Whereas lethal mutations affect the Mendelian ratio in a predictable manner, deleterious mutations will alter the ratio by differing degrees that are specific for a given mutation. For example, a particular mutation may have no phenotypic effect in *Drosophila* when the flies are raised at 25°C (standard temperature). However, increasing the temperature by 3°C may affect the mutation such that the flies have a reduced metabolism. This lower metabolism may cause random deaths in the population. Because the lethality is based on both the genotype and the environment, a consistent frequency of dying individuals will not be found. Although it is fairly straightforward to use the phenotypic ratio to identify either dominant or recessive lethality, it is very difficult to determine that a deleterious mutation is present in a population because of the variability of the phenotypic expression.

 Restating the Concepts

▶ Lethal mutations, both dominant and recessive, can skew the observed phenotypic ratio due to the death of particular genotypes.

▶ Dominant mutations can exist in a population when they are either not lethal until after sexual maturity or they exhibit a reduced penetrance.

5.5 Penetrance and Expressivity

Researchers have encountered situations in which a particular genotype fails to exhibit the expected phenotype. In some cases, this failure is due to one gene affecting the expression of a second gene. We will discuss those examples in the next section of this chapter. In other cases, a single allele may or may not express the expected phenotype, or a range of phenotypes may exist for a single allele.

Incomplete Penetrance

In **incomplete penetrance** (or reduced penetrance), some percentage of the individuals with a specific genotype fail to express the expected phenotype. In **figure 5.8**, all the flies in the top row are genotypically identical and should be expressing the mutant white eye phenotype; on the second row, only 6 of the 10 homozygous mutant flies actually express the white eye phenotype. We describe this phenotype, therefore, as being 60% penetrant (6 out of 10 express the expected phenotype). Knowing the level of penetrance for an allele allows more accurate predictions about the possible genotype of an individual in a pedigree or more precisely calculating the probability of having affected offspring in a cross.

All of the previously discussed genes are 100% penetrant, as are most genotypes. For example, no known cases exist of individuals homozygous for albinism who do not actually lack pigment.

An example of reduced penetrance can be found in the case of vitamin-D-resistant rickets (a bone disease). This disease is caused by a sex-linked dominant allele and is distinguished from normal vitamin D deficiency by a failure to respond to low levels of vitamin D. Affected individuals do respond, however, to very high levels of vitamin D, and the disease is thus treatable. In some family trees, affected children are born to unaffected parents. This outcome would violate the rules of dominant inheritance; one of the parents must have had the dominant allele, yet did not express it.

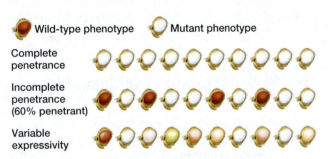

figure 5.8 Examples of incomplete penetrance and variable expressivity. All of the individuals have the same mutant genotype, white eyes in *Drosophila.* In incomplete penetrance only some of the individuals exhibit the phenotype that matches their mutant genotype (40% have the wild-type dark red phenotype even though they are genotypically mutant). In variable expressivity, individuals will show a range of phenotypes from the most extreme mutant, that matches the genotype (white eyes) to nearly a wild-type phenotype (red eyes). Again, all of these individuals have the same white-eyed mutant genotype.

The fact that one parent actually possesses the allele is demonstrated by the occurrence of low levels of phosphorus in the blood, a pleiotropic effect of the same allele. The low-phosphorus aspect of the phenotype is always fully penetrant.

For a single gene, therefore, one phenotype may be 100% penetrant and another may exhibit reduced penetrance. You can see how these differences could confuse predicting the mode of inheritance in a pedigree, especially when multiple phenotypes are associated with the same mutation.

Vitamin-D-resistant rickets also illustrates a case in which a phenotype that is not genetically determined mimics a phenotype that is. This **phenocopy** effect is the result of dietary deficiency or environmental trauma. A dietary deficiency of vitamin D, for example, produces a form of rickets that is virtually indistinguishable from genetically caused rickets.

It's Your Turn

As we just discussed, incomplete penetrance cannot only affect our deduced mechanism of inheritance in a pedigree, but also the calculated probability of having affected offspring. Try working problem 27 to determine your level of understanding of incomplete penetrance.

Variable Expressivity

In **variable expressivity,** the range of phenotypes associated with a genotype is increased. In figure 5.8, each of the flies in the bottom row are homozygous recessive for the white eye mutation. Unlike incomplete penetrance, in which some of these flies express the wild-type phenotype, all of the flies showing variable expressivity express a mutant phenotype. The range of mutant phenotypes varies, however, from a light shade of red to an extreme of white. These phenotypes are associated with a single mutant allele.

A range of phenotypes that are associated with an allelic series (that is, a series of different alleles) is not variable expressivity because each allele has a single phenotype that is associated with the mutant genotype.

Many developmental traits are not only incompletely penetrant, but they also show variable expressivity, from very mild to very extreme, when they do penetrate. In chapter 4, we described polydactyly, an autosomal dominant trait that results in extra fingers and/or toes. Polydactyly (**fig. 5.9**) shows both incomplete penetrance and variable expressivity. The most extreme manifestation of the trait is an extra digit on each hand and one or two extra toes on each foot. But

5.1 applying genetic principles to problems

Using Incomplete Penetrance Data in a Pedigree

In Chapter 4, we discussed some of the rules for deducing the mode of inheritance, such as dominant versus recessive and autosomal versus sex linked. For example, dominant alleles fail to skip generations, but recessive alleles often do.

Look at the following pedigree (fig. 5.A) and determine whether the phenotype is autosomal dominant or autosomal recessive. Based on what we already discussed, you likely deduced that it was autosomal recessive because the phenotype skipped two different generations. However, our discussion about incomplete penetrance in this chapter raises another possibility, namely that the phenotype is autosomal dominant with incomplete penetrance.

If we assume that this allele is responsible for a rare *recessive* disease, what is the probability that Bob and Mary will have an affected child?

We first determine the probability that Mary is heterozygous. Mary's probability of being heterozygous is 100% because her father exhibited the recessive disease and he was therefore homozygous recessive.

We next determine the probability that Bob is heterozygous, which is:

$$(2/3 \times 1/2 \times 1/2) = 1/6$$

The probability that Bob's grandfather is heterozygous is 2/3. Because Bob's grandfather's sister is affected, both of his great-grandparents must have been heterozygous, which means that if Bob's grandfather is unaffected, there is a 2/3 probability that he is heterozygous. Can you see why? Only three genotypic classes will have normal phenotypes, and of those, two will be heterozygotes.

The probability that Mary and Bob will have an affected child is then:

1 (Mary) × 1/6 (Bob) × 1/4 (chance of an affected child from monohybrid cross) = 1/24, or 4.17%

But what if this disease is a rare *dominant* disease with incomplete penetrance? In this case, we need to know the level of penetrance. Because we do not know all the genotypes, we cannot determine it from this pedigree; the information would need to be provided from other sources such as biochemical analyses of large family studies. Assume for our purposes that we know that the phenotype is 40% penetrant. Because Mary's father expresses the dominant phenotype, he must be heterozygous (he is not likely to be homozygous because this is a rare disease, and because both of his parents would have to have had the dominant allele). We can then determine the probability that Mary is heterozygous, which is:

(1/2 × 60%) = 30% probability that Mary is heterozygous and *phenotypically normal*

50% probability that Mary is homozygous recessive and *phenotypically normal*

30%/(30% + 50%) = 37.5% that phenotypically normal Mary is genotypically heterozygous

The probability that Bob is heterozygous is:

37.5% × 37.5% × 37.5% = 5.3%

Mary and Bob could have an affected child if either Mary or Bob or both are heterozygous. If Mary is heterozygous and Bob is not, the probability that they will have an affected child is:

37.5% × (1/2 × 40%) = 7.5%

(We use 40% in the calculation because we want to know the probability that the child will have the dominant allele *and* express the dominant phenotype.)

If Bob is heterozygous and Mary is not, the probability that they will have an affected child is:

5.3% × (1/2 × 40%) = 1.1%

If both Mary and Bob are heterozygous, the probability that they will have an affected child is:

37.5% × 5.3% × (3/4 × 40%) = 0.6%

The probability of Mary and Bob having an affected child could be due to any of the three scenarios or:

7.5% + 1.1% + 0.6% = 9.2%

Notice that if the disease is dominant with 40% penetrance, the chance of Bob and Mary having an affected child is more than double the chance if the disease is recessive and completely penetrant (9.2% versus 4.17%).

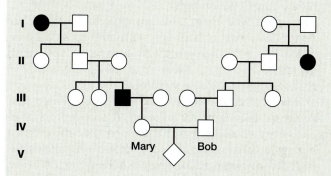

figure 5.A   A human pedigree for a disease that is either autosomal recessive or autosomal dominant with incomplete penetrance.

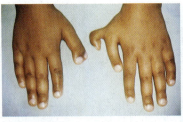

(a)

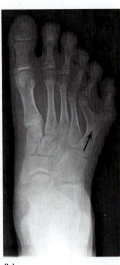

(b)

figure 5.9 Polydactyly is an example of variable expressivity. (a) The number of fingers on the hand (or toes on a foot) will vary from five to seven, and the number can differ on the two hands of a single individual. (b) This X-ray shows the duplicated little toe is only a partial duplication, as the two toes share a common bone (arrow).

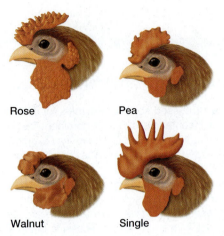

Rose

Pea

Walnut

Single

figure 5.10 Four types of combs in fowl.

some individuals have only extra toes, some have extra fingers, and some have only a portion of an extra digit, such as the partial extra thumbs (fig 5.9a) or the partial extra little toe (the two smallest toes join into a single bone, fig. 5.9b).

Another example is cleft palate. When the genotype penetrates, the severity of the impairment varies considerably, from a very mild external cleft to a very severe clefting of the hard and soft palates.

Incomplete penetrance and variable expressivity are not unique to human traits, but are characteristic of developmental traits in many organisms. Experiments have shown that reduced penetrance and variable expressivity are affected by both the genetic background and environmental factors.

Restating the Concepts

▶ Incomplete penetrance occurs when the mutant genotype produces some individuals with the wild-type phenotype.

▶ Variable expressivity is seen when the mutant genotype produces a range of phenotypes from extreme mutant to almost wild type.

▶ Some alleles can have both dominant and recessive phenotypes or can exhibit both reduced penetrance and variable expressivity.

5.6 Genotypic Interactions

As you recall from chapter 2, two independent phenotypes, such as plant height and seed shape in garden peas, can be tracked in dihybrid crosses. Crossing het-

erozygotes for both traits, Mendel found the phenotypic ratio for the two traits to be 9:3:3:1, where 9 out of 16 plants would have both dominant phenotypes, and 1 out of 16 both recessive phenotypes. Three out of 16 plants would exhibit the dominant phenotype for one trait and the recessive phenotype for the other trait, whereas the other 3 out of 16 plants would exhibit the recessive phenotype for the first trait and the dominant phenotype for the second.

Often, several genes can contribute to a single phenotype, just as they can affect different phenotypes. The results of crosses reveal the nature of these genotypic interactions by their effects on the phenotype.

An example occurs in the combs of chickens (*Gallus gallus*; fig. 5.10). If we cross a rose-combed hen with a pea-combed rooster (or vice versa), all the F$_1$ offspring are walnut-combed. If we cross the hens and roosters of this heterozygous F$_1$ group, the F$_2$ generation contains walnut-, rose-, pea-, and single-combed fowl in a ratio of 9:3:3:1. Based on what you know, can you deduce the genotypes of each of the four phenotypic classes in the F$_2$ population? An immediate indication that two genes are involved is the 9:3:3:1 phenotypic ratio of the F$_2$ generation.

Figure 5.11 shows the analysis of this cross. When dominant alleles of both genes are present in an individual (*R– P–*), the walnut comb appears. (The dash indicates any second allele; so, *R– P–* could be *RRPP*, *RrPP*, *RRPp*, or *RrPp*.) A dominant *rose* allele gene (*R–*) and homozygous recessive *pea* alleles (*pp*) gives a rose comb. A dominant *pea* allele (*P–*) and homozygous recessive *rose* alleles (*rr*) gives pea-combed fowl. When both genes are homozygous for the recessive alleles, the fowl are single-combed. Thus, a 9:3:3:1 F$_2$ ratio arises from crossing dihybrid individuals, whether the two genes affect different phenotypes or, as in this case, a single phenotype.

In the remainder of this section, we will discuss instances in which two genes affect a single trait to produce only two or three different phenotypes. Keep in

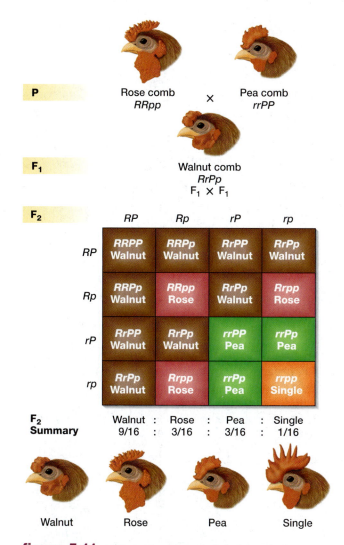

figure 5.11 Dihybrid cross reveals independent assortment in the determination of comb type in fowl. The genetic interaction between two genes produces the four different comb phenotypes in chickens. The 9:3:3:1 phenotypic ratio is consistent with a dihybrid cross.

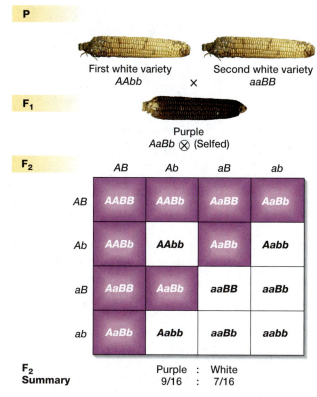

figure 5.12 Dihybrid cross of *Zea mays* reveals complementary gene action in the production of kernel color. Crossing two varieties of *Zea mays* with white kernels produces a dihybrid F_1 with purple-colored kernels. The 9:7 phenotypic ratio in the F_2 progeny results from the *A- B-* genotype producing purple kernels and the *A- bb*, *aa B-*, and *aa bb* genotypes yielding white kernels.

mind that regardless of the number of different phenotypes produced, a dihybrid cross will produce progeny in a 9:3:3:1 phenotypic ratio or some variation of it.

Complementary Gene Action

In corn (or maize, *Zea mays*), several different field varieties produce white kernels. In certain crosses, two white varieties will result in an F_1 generation with all purple kernels. If plants grown from these purple kernels are selfed, the F_2 individuals have both purple and white kernels in a ratio of 9:7. How can we explain this?

We must be dealing with the offspring of dihybrids, with each gene segregating two alleles, because the ratio is in sixteenths (9 + 7). Furthermore, we can see that the F_2 9:7 ratio is a variation of the 9:3:3:1 ratio. The three smaller genotypic classes produce a single

phenotype that makes up 7/16 of the F_2 offspring. **Figure 5.12** outlines the cross. You can see from this figure that the purple color appears only when at least one dominant allele for both genes is present. When one or both genes have only recessive alleles, the kernels will be white.

The color of corn kernels illustrates the concept of **complementary gene action:** A dominant allele for two different genes must both be present to produce a specific phenotype (purple). Without the dominant allele at either or both genes, a single phenotype is produced (white).

How can we explain this type of genetic interaction? The answer requires knowing that most genes encode proteins or enzymes. It is usually (but not always) the protein that performs a particular function in the cell, and that ultimately confers a phenotype. We can think of a wild-type allele as encoding a normal protein or enzyme and mutant alleles as encoding forms of the enzymes that lack the necessary function.

Assume that genes encode enzymes that synthesize a specific pigment that determines the color of a corn kernel. Because the genetic analysis suggests that

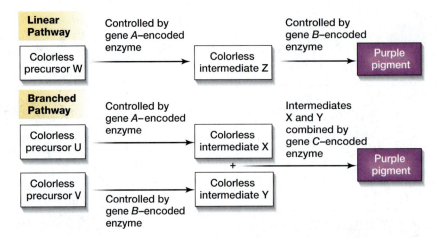

figure 5.13   Two possible biochemical pathways that account for complementary gene action.

two genes are required to produce the 9:7 phenotypic ratio, we assume that each encodes a different enzyme. We can then place these enzymes in a hypothetical biochemical pathway that ultimately produces the purple color in wild-type kernels. This can be drawn as the linear pathway shown in **figure 5.13**.

In this pathway, the wild-type *A* allele encodes an enzyme that converts a colorless precursor to a colorless intermediate, and the wild-type *B* allele encodes an enzyme that converts the colorless intermediate into a purple pigment. An *A–B–* genotype should therefore produce purple kernels.

Let's look at the other three general genotypic classes. The *A–bb* genotype would be able to convert the precursor to the intermediate, but it could not convert the intermediate into the purple pigment because it lacks the wild-type *B* allele. The *A–bb* genotype thus produces a colorless or white kernel. In contrast, the *aaB–* genotype could not convert the precursor to the colorless intermediate because it lacks the wild-type *A* allele. Because the intermediate is not made, the purple pigment cannot be synthesized, and the phenotype is also white. The *aabb* genotype also cannot convert the precursor to the intermediate, so the phenotype is white.

(Notice that the absence of the *A* allele prevents either the *B* or *b* allele from ever having an effect on pigmentation. This action is a special case of epistasis, which is described in detail shortly.)

An alternative branched pathway is also shown in figure 5.13. In this case, the enzymes encoded by genes *A* and *B* each produce an intermediate from different precursors. These colorless intermediates must be combined in another biochemical reaction, regulated by an enzyme encoded by a third gene, to produce the purple pigment. It is not possible to differentiate between these two pathways based on the information we have. Thus, complementary gene action implies that a biochemical pathway may be affected, and that two gene products are both required to generate the phenotype, without necessarily providing details.

Duplicate Gene Action

The shepherd's purse plant, *Capsella bursa-pastoris*, has fruit that is either "heart-shaped" or "narrow." When pure-breeding plants for each trait are crossed, all the F₁ progeny have heart-shaped fruit (**fig. 5.14**). The appearance of only heart-shaped fruit in the F₁ progeny and the loss of the narrow-fruit phenotype is consistent with heart-shaped being dominant and narrow fruit being recessive.

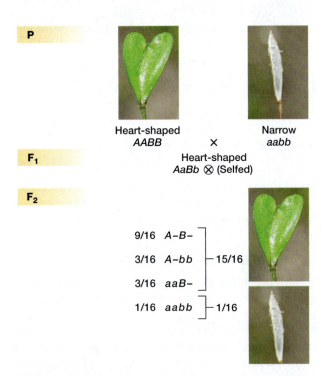

figure 5.14   *Capsella bursa-pastoris* produces fruit that is either "heart-shaped" or "narrow" through a duplicate gene action mechanism. The 15:1 phenotypic ratio in the F₂ progeny results from the *A-B-*, *A-bb*, and *aa B-* genotypes all producing the heart-shaped phenotype and the *aa bb* genotype yielding the narrow fruit phenotype.

But when we self the F_1 progeny, the F_2 progeny fail to exhibit one of the standard monohybrid ratios (3:1, 2:1, or 1:2:1). Rather, the F_2 exhibit a 15:1 ratio of heart-shaped to narrow (see fig. 5.14). You can recognize that this ratio is a modification of the 9:3:3:1 dihybrid cross ratio, suggesting that two genes are involved in determining the phenotype of fruit shape. The one narrow-fruit class must represent the 1/16 that are homozygous recessive for both genes (*aabb*). The 15/16 that are heart-shaped must correspond to the following three genotypic classes: *A–B–*, *A–bb*, and *aaB–*. In this case, a dominant allele for either gene is sufficient to produce the heart-shaped phenotype, which suggests that the two genes have the same function. **Duplicate gene action** describes the phenomenon of two genes being involved in producing a phenotype, with both genes appearing to be equivalent.

As we did with complementary gene action, we can describe the mechanism of duplicate gene action with a biochemical pathway. Because the dominant allele for either gene is sufficient to produce the heart-shaped phenotype, we can envision that the enzymes encoded by the dominant *A* and dominant *B* alleles are equivalent. Thus, we would draw the biochemical pathway as shown in **figure 5.15**.

In this mechanism, one gene (either *A* or *B*) likely was duplicated in the evolution of the species, so that two different genes encode identical enzymes. One of the genes was then able to mutate and through evolution develop a slightly different function.

These gene duplications and the evolution of related, but different, functions are sometimes present as **gene families**. In humans, the β-globin gene family is composed of five genes that are expressed at different times in development. Each gene encodes a different form of the β-globin protein found in red blood cells, which is essential in binding oxygen. Because a fetus and an adult are exposed to different relative percentages of oxygen, forms of the fetal and adult β-globin allow the red blood cells of an individual to efficiently bind oxygen in the womb, and later in the external environment.

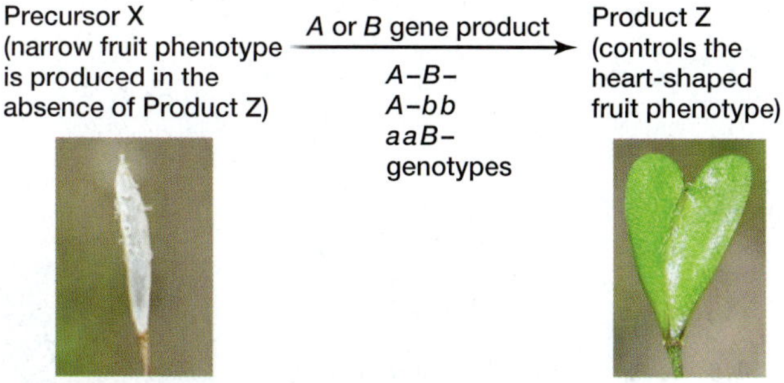

Epistasis

In **epistasis,** the phenotype produced by one gene (the **epistatic gene**) masks the phenotype produced by a second gene (the **hypostatic gene**). Dominance, in contrast, involves the interaction between different alleles at a *single gene*. Epistasis and dominance can exhibit analogous effects, but these effects proceed from different causes.

For example, the recessive *apterous* (wingless) gene in fruit flies is epistatic to any gene that controls wing characteristics. In other words, when the recessive *apterous* allele is homozygous, it masks the presence of the curly wing gene, because, obviously, without wings, it is impossible to tell if the wing should be curly or straight!

It is also important to realize that two genes that affect the same phenotype do not always exhibit epistasis. For example, the previous discussion on the genetic control of comb type in domestic chickens does not involve epistasis. No allelic combinations at one locus mask genotypes at another: The 9:3:3:1 ratio is not an indication of epistasis. A clear indication of epistasis would be the production of a *modified 9:3:3:1 ratio* in which *only three phenotypes* are observed.

Finally, it is possible that an epistatic allele may be either dominant or recessive, and that the mutant phenotype that is hypostatic may also be either dominant or recessive. In these situations, we can observe several different modified Mendelian ratios.

Recessive Epistasis

Crossing a pure-breeding black mouse with a pure-breeding albino mouse (pure white because all pigment is lacking) produces only agouti offspring (the typical dark gray mouse color). When the F_1 agouti mice are crossed with each other, agouti, black, and albino offspring appear in the F_2 generation in a ratio of 9:3:4 (**fig. 5.16**). What are the genotypes in this cross?

You probably recognize by now that the F_2 ratio of 9:3:4 is a variant of the 9:3:3:1 ratio. The 9/16 agouti phenotypic class contains a dominant allele for both genes (*A–C–*), the 3/16 black phenotypic class is dominant for one gene and homozygous recessive for the other (*aaC–*), and the 4/16 albino phenotypic class is composed of the other 3/16 class (*A–cc*) and the 1/16 class (*aacc*).

Based on this assumption, any genotype that includes *cc* will be albino and will mask the phenotype associated with the *A* gene (agouti or black). As long as at least one dominant *C* allele is present, the *A* gene can express itself, with agouti (*A*) being dominant to black (*a*). The homozygous recessive *c* allele produces the albino phenotype, regardless of the alleles that are present at the *A* gene. Thus, the *c* allele is epistatic to the *A* gene (and the *A* gene is hypostatic to the *c* allele). This

| Precursor X (narrow fruit phenotype is produced in the absence of Product Z) | $\xrightarrow{\text{A or B gene product}}$ *A–B–* *A–bb* *aaB–* genotypes | Product Z (controls the heart-shaped fruit phenotype) |

figure 5.15 A possible biochemical pathway that is consistent with duplicate gene action. Either the dominant *A* allele or the dominant *B* allele is required to produce Product Z that generates the heart-shaped fruit phenotype. In the absence of both dominant *A* and *B* (*aa bb* genotype), Precursor X cannot be converted to Product Z, which results in the narrow fruit phenotype.

particular phenomenon is termed **recessive epistasis** because the recessive phenotype at the C gene (albinism) masks the phenotype at the A gene.

Dominant Epistasis

Another example of epistasis is demonstrated by the color of summer squash, the fruits of which can be white, yellow, or green. In a cross between pure-breeding white and pure-breeding green squash plants, the F_1 progeny all produce white squash (**fig. 5.17**).

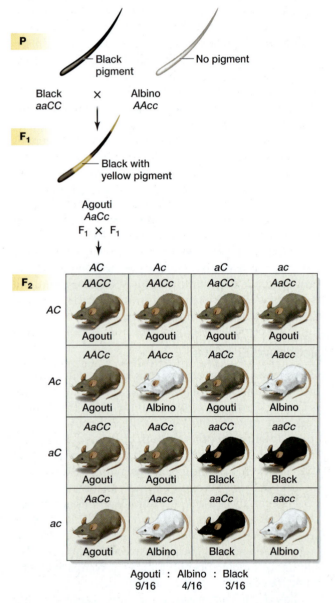

Agouti : Albino : Black
9/16   4/16   3/16

figure 5.16 The coat color of mice can be black, albino, or agouti. The albino phenotype results from the absence of pigment in the hair, while the agouti phenotype is due to the presence of two different pigments in a striped pattern in the hair. Crossing a pure-breeding black-haired mouse with a pure-breeding *albino* mouse produces agouti mice. Crossing two agouti dihybrid mice produces a 9:4:3 ratio of agouti to albino to black mice due to recessive epistasis.

Crossing the F_1 progeny gives an F_2 generation with a 12:3:1 ratio of white to yellow to green squash.

It should be apparent that this ratio is a modified 9:3:3:1, with the 12 representing the $A-B-$ and $A-bb$ genotypes, the 3 representing the $aaB-$ genotypes, and the 1 corresponding to the $aabb$ genotype. This outcome is an example of **dominant epistasis,** with the dominant A allele masking the phenotype that would be generated from the B gene. Thus, depending on whether the epistatic allele is dominant or recessive, you will observe either a 9:3:4 or a 12:3:1 ratio.

Our previous example of complementary gene action (white and purple corn kernel color) is a specialized case of epistasis, where the absence of the dominant A allele prevents the expression of the B gene's phenotype. It just happens that the epistatic aa phenotype and the recessive bb phenotype are the same, yielding white pigmentation. Thus, the only time you would observe purple pigmentation is when both a dominant A allele and a dominant B allele are present ($A-B-$). You see only two phenotypes, rather than the three phenotypes that you would normally observe in epistasis.

The Mechanism of Epistasis

In the previous example of epistasis in mouse coat color, researchers have determined the physiological mechanism. The pigment melanin is present in both the black and agouti phenotypes. The agouti phenotype exhibits a modified black hair in which yellow stripes (from a

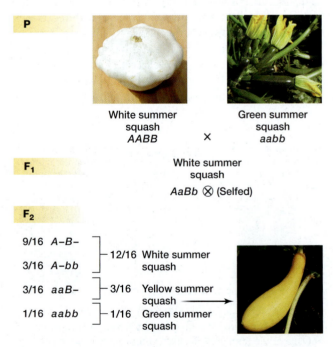

figure 5.17 Dihybrid cross of white summer squash produces a 12:3:1 F_2 ratio of white to yellow to green squash through dominant epistasis.

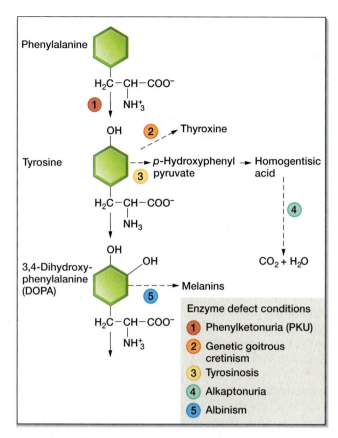

figure 5.18   The biochemical pathway for the production of melanin from phenylalanine. Each solid arrow represents a biochemical step that is catalyzed by an enzyme, while each dashed arrow represents several steps that are catalyzed by different enzymes. The circled numbers represent points in the pathway that are associated with specific genetic diseases due to mutations in the genes encoding the enzymes at those biochemical steps.

related pigment, phaeomelanin) have been added to the black phenotype (see fig. 5.16). Thus, when melanin is present, agouti is dominant to black. Without melanin, the mice fail to produce any pigment, and we observe the albino phenotype, regardless of the effect of the agouti gene because both agouti and black require melanin. Albinism is the result of one of several defects in the enzymatic pathway for the synthesis of melanin (**fig. 5.18**).

Knowing that epistatic modifications of the 9:3:3:1 ratio come about through gene interactions at the biochemical level, we can look for biochemical explanations for the 9:3:4 ratio. Using the mouse coat color as an example, the epistatic gene functions early in the biochemical pathway, which suggests that the recessive c allele produces the albino phenotype by blocking the biochemical pathway shown in **figure 5.19**. In this case, the pigmentation pathway cannot proceed beyond the albino phenotype.

The 4/16 ratio of albino progeny suggests that c is a recessive epistatic mutation. A dominant epistatic mutation would be represented by the sum of both the genotypic groups that have a dominant allele for one gene that exhibits the same phenotype (9 + 3 = 12). In the case of both the wild-type dominant A and C alleles, the pigmentation pathway is functional and produces the agouti phenotype. If the recessive a allele is homozygous, then the production of yellow pigment is blocked and the agouti phenotype cannot be generated. However, the black phenotype will still be produced if the dominant C allele is present.

Notice again that the biochemical pathway can be logically generated if we assume that the epistatic gene functions earlier than the hypostatic gene in the pathway. This represents one of the strengths of this type of analysis, which is the ability to order the genes,

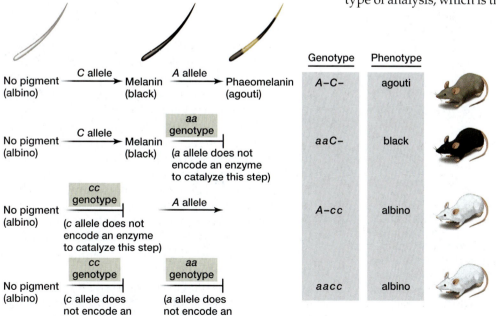

figure 5.19   A possible biochemical pathway that is consistent with recessive epistasis. To produce the wild-type agouti phenotype, the dominant C allele is required to deposit melanin in the hair and the dominant A allele is required to deposit phaeomelanin. In the absence of the dominant A allele ($aa\ C$-), melanin is deposited and phaeomelanin is not, which produces the black hair. In the absence of the dominant C allele (A- cc or $aa\ cc$), neither melanin nor phaeomelanin are deposited in the hair and the albino phenotype is observed.

Genotype	Phenotype
A–C–	agouti
aaC–	black
A–cc	albino
$aacc$	albino

or better, the corresponding encoded proteins in a biochemical pathway. This allows a genetic analysis to complement a biochemical analysis.

Suppression

In *Drosophila*, a particular recessive autosomal mutation called *vestigial* (*vg*) exhibits either a very small wing or no wing (**fig. 5.20**). A certain phenotypically wild-type fly was identified, which, when crossed with a *vestigial* fly, produced all vestigial F_1 progeny. Furthermore, when the F_1 progeny were crossed, the F_2 progeny were found in a 13:3 phenotypic ratio of normal to vestigial (see fig. 5.20). Examining large numbers of progeny confirmed that this was actually a 13:3 ratio and not 3:1.

This 13:3 ratio is also a modified 9:3:3:1 ratio, which suggests that two genes are interacting in this phenotype. The wild-type *vestigial* allele (*vg*$^+$) is dominant to the recessive mutant allele (*vg*). We will designate the second gene that interacts with vestigial as *su(vg)*$^+$ for the dominant wild-type allele and *su(vg)* for the recessive mutant allele. Based on the 13:3 ratio, the wild-type phenotype results from the 9/16 class, one of the 3/16 classes and the 1/16 class. We can therefore deduce that the wild-type phenotype is due to *vg*$^+$– *su(vg)*$^+$–, *vg*$^+$– *su(vg)su(vg)*, and *vgvg su(vg)su(vg)*. The mutant *vestigial* phenotype would then be due to the *vgvg su(vg)*$^+$– genotype.

The *vgvg su(vg)su(vg)* genotypic class is homozygous for the recessive *vestigial* mutation, yet it exhibits the wild-type phenotype. Geneticists refer to this phenomenon as *suppression*.

Suppression occurs when one gene pushes the mutant phenotype of a second gene toward the wild-type phenotype. In this case, the *su(vg)* recessive mutation pushes the vestigial phenotype towards wild type. The *vg*$^+$– *su(vg)*$^+$– and *vg*$^+$– *su(vg)su(vg)* genotypes are both wild-type for *vestigial* and exhibit the wild-type phenotype. The *vgvg su(vg)su(vg)* genotype is recessive for the *vestigial* mutation, but two copies of the recessive suppressor mutation produces the wild-type phenotype. Thus, the *su(vg)* allele is termed a **recessive suppressor,** and the *su(vg)* designation reveals that this is a *recessive suppressor of vestigial*. The *vgvg su(vg)*$^+$– genotype is *vestigial* because the dominant *su(vg)*$^+$ allele fails to suppress the *vestigial* phenotype.

There are several things to note about suppressors. First, suppressors may have no observable phenotype other than the ability to suppress another gene's mutant phenotype. As a result, the dihybrids produce only two phenotypes, unlike the three phenotypes usually observed in epistasis. Second, suppressors only make the second gene's mutant (*vg*) phenotype more like wild-type. They usually have no effect on the second gene's wild-type allele (*vg*$^+$). Third, suppressors can be either dominant or recessive, and they may suppress either a dominant or a recessive mutant allele at a second gene. This results in several different phenotypic ratios. And finally, suppressors, like any alleles may exhibit *both* dominant and recessive phenotypes. For example, a recessive suppressor may exhibit a dominant phenotype, but the suppression requires two copies of the mutant (suppressor) allele.

The Mechanisms of Suppression

Suppressors can exert their effect through a variety of different mechanisms that depend on how the mutant phenotype is generated. Some suppressors specifically affect only a single allele or gene, but other suppressors can affect several different genes. We will discuss suppression mechanisms throughout this book, but is it worthwhile to mention a few here.

One mechanism of suppression is to increase or decrease the expression of a gene. As expected, this mechanism often has the ability to suppress a number of different mutant genes.

A second mechanism is to affect the translation of an mRNA. This often involves specifically suppressing mutations that introduce a translational stop codon (*nonsense mutations*). In this case, the suppressors are often mutations in tRNA genes and would nonspecifically affect any nonsense mutation.

A third class of suppressor would be specific for a given mutation, or more likely for a specific allele. Imagine two different proteins that must interact in

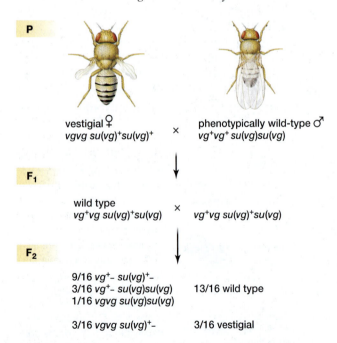

P

vestigial ♀
vgvg su(vg)$^+$*su(vg)*$^+$ ×   phenotypically wild-type ♂
vg$^+$*vg*$^+$ *su(vg)su(vg)*

F₁

wild type
vg$^+$*vg su(vg)*$^+$*su(vg)*   ×   *vg*$^+$*vg su(vg)*$^+$*su(vg)*

F₂

9/16 *vg*$^+$– *su(vg)*$^+$–
3/16 *vg*$^+$– *su(vg)su(vg)*   13/16 wild type
1/16 *vgvg su(vg)su(vg)*

3/16 *vgvg su(vg)*$^+$–   3/16 vestigial

figure 5.20   The recessive *suppressor of vestigial* allele (*su(vg)*) can suppress the vestigial small wing phenotype and produces a 13:3 ratio in a dihybrid cross. This 13:3 phenotypic ratio for a recessive suppressor is in contrast to the 9:4:3 phenotypic ratio observed in a dihybrid cross exhibiting recessive epistasis.

the wild-type cell to convert a substrate to a product (**fig. 5.21**). The shape of both of these proteins will dictate their ability to interact. If one of the proteins changes its shape (conformation) due to a mutation in the corresponding gene (m^-), it could not interact with its partner. This change would prevent the conversion of the substrate to the product and would result in a mutant phenotype. If a mutation occurred in a second gene (su^-), the interaction between the two proteins could be restored and the wild-type phenotype would be observed. In this example, the suppressor mutation (su^-) identified a gene that encodes a protein that specifically interacted with the first protein. Thus, suppressors are a powerful method for identifying components in a complex of proteins.

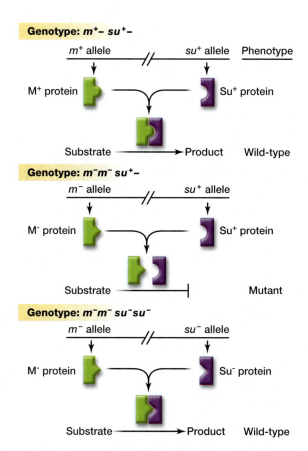

figure 5.21   Two genes (*m* and *su*) encode different proteins that interact to convert a substrate to a product and produce the wild-type phenotype. Mutation of the *m* gene (m^-) prevents the two proteins from interacting and the substrate cannot be produced. The resulting $m^-\ m^-\ su^+$ - genotype yields a mutant phenotype. A new mutation, this time in the *su* gene (su^-), changes the S protein so that it now interacts with the mutant M protein and produces the product from the substrate. The resulting $m^-\ m^-\ su^-\ su^-$ genotype produces a wild-type phenotype even though the individual is homozygous for the recessive mutant m^- allele. The su^- mutation is a suppressor, as it restored the wild-type phenotype to the m^- mutant.

It's Your Turn

Epistasis and suppression are important genetic interactions in producing a phenotype. Both yield modified 9:3:3:1 ratios, although epistasis often has three phenotypes and suppression two. Try problems 22, 26, and 29, which involve epistasis and suppression so that you become more comfortable at recognizing these processes and manipulating their associated data.

Restating the Concepts

▶ Several different mechanisms can alter the standard dihybrid 9:3:3:1 phenotypic ratio, including duplicate gene action, complementary gene action, epistasis, and suppression.

▶ Examining the modified Mendelian phenotypic ratio and deducing the corresponding genotypes allows the underlying genetic principle to be elucidated.

▶ These mechanisms suggest particular defects in enzyme–protein interactions and help to define how the proteins encoded by the genes function in the cell or organism.

5.7 Epigenetics

Epigenetics is the heritable modification of gene function without a change in the DNA sequence. There are many different examples of epigenetics; one of which is X-chromosome inactivation that generates Barr bodies in female mammals (see chapter 4). One of the two female X chromosomes is randomly inactivated during embryonic development. Because this process is random and occurs after the fertilized egg has already begun dividing, some of the cells in the female will have the paternal X chromosome inactivated, and other cells will have the maternal X chromosome inactivated.

If a female is heterozygous for an X-linked gene, some of the cells will express the dominant allele and others the recessive allele (due to the inactivation of the X chromosome containing the dominant allele). Thus, the female contains both dominant and recessive phenotypes for this gene in nonoverlapping cells or tissues (**fig. 5.22**).

A **mosaic** individual is composed of two or more genetically distinct tissues or cell types. Because the phenotype may only be expressed in a single tissue, a female may exhibit either the dominant or recessive phenotype, depending on which X chromosome was inactivated in that tissue. If heterozygous, the individual may express only the recessive phenotype, which would

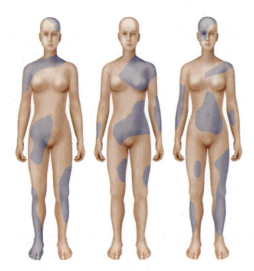

figure 5.22   Mosaicism in human females. The X-linked condition anhidrotic ectodermal dysplasia results in the absence of sweat glands. Because of the randomness of X chromosome inactivation, women heterozygous for this condition will be mosaic for the presence of sweat glands (shaded areas correspond to the location of sweat glands).

alter the expected Mendelian phenotypic ratio. However, the germ cells remain heterozygous, and half of the eggs will contain the dominant allele and half the recessive allele; no permanent change occurs in the female's DNA. Because many of the mechanisms of epigenetics involve changes in the expression of a gene, we will explore this phenomenon in more detail in chapter 17.

Restating the Concepts

▶ Epigenetics refers to heritable changes in gene function that do not alter the gene's sequence. An example is X-chromosome inactivation, where once a cell inactivates one of the X chromosomes, all subsequent daughter cells continue to inactivate the same X chromosome.

▶ In female mammals, the random inactivation of one X chromosome can result in mosaicism if the individual is heterozygous for the X-linked genes. In the heterozygote, some cells will express the dominant phenotype, and the remaining cells will have the recessive phenotype.

5.2 applying genetic principles to problems

Deducing Single-Gene Versus Two-Gene Inheritance Mechanisms Based on Phenotypic Ratios

When a phenotype is being studied, an obvious question is whether the trait is the result of a single gene or the interactions of multiple genes. Making the correct deduction is critical in uncovering the pattern of inheritance, in making accurate predictions of the transmission to children during genetic counseling, and ultimately, in potentially developing a genetic therapy for the disease. What particular clues allow us to deduce that a single gene is involved in the phenotype being studied?

Usually, the most informative cross will be a cross of two heterozygotes. If a single gene is involved, the genotypes of the progeny will be 1/4 homozygous dominant, 1/2 heterozygous, and 1/4 homozygous recessive. We discussed situations where this genotypic ratio will produce phenotypic ratios of 3:1 (complete dominance), 1:2:1 (codominance and incomplete dominance), and 2:1 (recessive lethal). If incomplete penetrance or variable expressivity is acting, then any of these three phenotypic ratios will be altered.

If two genes are involved, the four phenotypic classes would occur as follows: 9 (*A–B–*), 3 (*A–bb*), 3 (*aaB–*), and 1 (*aabb*). Interactions between two genes will regroup these phenotypic classes into either two classes (suppressors) or three classes (epistasis).

If the ratio observed in a problem is very close to one of the modified monohybrid or dihybrid ratios, then the genetic phenomenon associated with that ratio is most likely being observed. However, sometimes the observed ratio lies between a monohybrid and a dihybrid ratio. This usually occurs when trying to discern between a 1:2:1 ratio (codominance or incomplete dominance) and a 9:3:4 ratio (recessive epistasis) or between a 3:1 ratio (complete dominance) and a 13:3 ratio (recessive suppressor acting on a recessive mutation). In these cases, it is possible to apply the chi-square test to possibly eliminate one of the options.

Let's look at a problem that requires us to deduce the correct genetic phenomenon. A pure-breeding red-eyed fly (wild type) is crossed to a pure-breeding orange-eyed fly (mutant). All the F_1 progeny are wild type. Crossing two F_1 individuals produced 779 wild-type and 201 mutant flies. This result could be due to wild type being completely dominant to orange eyes, or to a recessive suppressor acting on the recessive orange eye mutation (fig. 5.B).

Recall from chapter 2 that the formula for converting experimental data to a chi-square value is

$$\chi^2 = \sum \frac{(O - E)^2}{E}$$

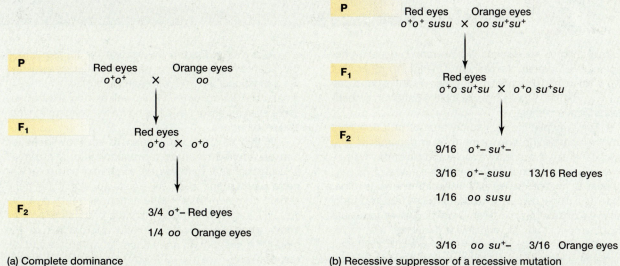

figure 5.B   Models to account for a F$_1$ cross that produces 779 wild-type and 201 mutant flies. These numbers could be generated by two different genetic models. (a) The genotypes that would account for complete dominance of a single gene to explain the observed data. This model predicts 735 wild-type and 245 orange-eyed flies in the F$_2$ progeny. (b) The genotypes that would account for a recessive suppressor (*su*) acting on a recessive mutation (*o*) to explain the observed data. This model predicts 796 wild-type and 184 orange-eyed flies in the F$_2$ progeny. A statistical analysis, such as the chi-square test, is required to determine what model best fits the data.

where χ is the Greek letter chi, O is the observed number for a category, E is the expected number for that category, and Σ is the summation of the calculations for all categories.

For the complete dominance model, we would expect a 3:1 ratio or 735 wild-type and 245 mutant flies from 980 progeny. We can generate the following table to show the chi-square calculations:

Phenotype	Observed	Expected	$(O-E)^2$	$(O-E)^2/E$
Wild type	779	735	1936	2.634
Orange	201	245	1936	7.902
				$\chi^2 = 10.536$

Because there are only two phenotypic classes, we have only one degree of freedom. Reading across the chi square table (p. 36), we find that the value 10.536 corresponds to a probability value less than 1%. This means that devia-tions this large or larger for complete dominance would occur less than 1% of the time. Because this value is less than 5%, we can reject the complete dominance model.

For the recessive suppressor acting on a recessive mutation model, we would expect a 13:3 ratio or 796 wild-type and 184 mutant flies from 980 progeny. We can generate the following table to show the chi-square calculations:

Phenotype	Observed	Expected	$(O-E)^2$	$(O-E)^2/E$
Wild type	779	796	289	0.363
Orange	201	184	289	1.571
				$\chi^2 = 1.934$

This yields a probability value of slightly less than 20%. Because this probability is greater than 5%, we are unable to reject the model of a recessive suppressor acting on a recessive mutation and assume that the observed data are consistent with the suppressor model.

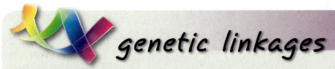

genetic linkages

In this chapter, we discussed how various genetic phenomena will modify the standard Mendelian ratios that were described in chapter 2. These modifications can have a profound effect on deducing the patterns of inheritance from a pedigree.

For example, incomplete penetrance of a dominant trait will cause the phenotype to skip generations and suggest that it is a recessive allele, and pleiotropy with either incomplete penetrance or reduced expressivity may appear as multiple genes with distinct phenotypes rather than as a single gene. These complexities require careful analysis of several (and often large) pedigrees to ensure that the correct genetic principle is deduced.

The concept of different alleles exhibiting different patterns of inheritance will become more obvious through future discussions on transcription (chapter 10), protein structure (chapter 11), and the genetic description of mutations (chapter 20). These three chapters will help you realize that a variety of mutations can affect the transcription and translation of a gene, as well as the structure and function of the encoded protein. Based on the mechanism by which the mutation causes the mutant phenotype, you may be able to make certain predictions about how the mutant allele will behave.

As we discussed in the section on suppression in this chapter, some proteins physically interact to generate a functional unit. We will examine protein structure in greater detail in chapter 11, as well as how tRNAs may serve as suppressors by allowing translation through mutant stop codons.

In this chapter, you began to see how genetics can be applied to study the relationship of enzymes in a biochemical pathway by observing particular genetic phenomena such as epistasis, duplicate gene action, complementary gene action, and suppression. You will learn about the application of these principles again when we discuss genetic approaches to the study of development and cancer (chapters 21 and 22, respectively).

We discussed epigenetics as a broad class of genetic phenomena that affect the stable and heritable expression of a gene, without changing the DNA sequence. This includes somatic expression, such as X-chromosome inactivation in female mammals. Many of the underlying mechanisms are known (to some extent) and involve specific changes in transcription. Therefore, we will discuss these mechanisms in chapter 17.

Chapter Summary Questions

1. For each term in Column I, choose the best matching phrase in Column II.

   Column I
   1. Mosaicism
   2. Suppression
   3. Epistatic gene
   4. Epigenesis
   5. Phenocopy
   6. Recessive epistasis
   7. Duplicate gene action
   8. Dominant epistasis
   9. Hypostatic gene
   10. Complementary gene action

   Column II
   A. Gene that masks the phenotype of another
   B. 12:3:1 ratio in offspring
   C. Mimics a genetically determined phenotype
   D. 9:3:4 ratio in offspring
   E. 13:3 ratio in offspring
   F. 9:7 ratio in offspring
   G. Individual composed of two different genotypes, producing two different phenotypes
   H. 15:1 ratio in offspring
   I. Gene that is masked by the effects of another
   J. Differential gene expression based on parent of origin

2. How do codominance and incomplete dominance differ?

3. Compare and contrast dominance and epistasis?

4. What is the difference between incomplete penetrance and variable expressivity?

5. What is pleiotropy? How can you determine that pleiotropic effects, such as those seen in the mouse dominant fur color allele (A^Y), are not due to different genes?

6. Explain how Tay–Sachs disease can be both a recessive and an incomplete dominant trait.

7. In blood transfusions, which blood type of the ABO system is called the "universal donor" and which one is the "universal recipient"? Explain.

8. How does the biochemical pathway in figure 5.3 explain how alleles I^A and I^B are codominant, yet both dominant to allele i?

9. What does gene interaction mean? Give an example.

10. What is the complementation test? Why is it used?

Solved Problems

PROBLEM 1: Summer squash comes in three shapes: disk, spherical, and elongate. In one experiment, researchers crossed two squash plants with disk-shaped fruits. The first 160 seeds planted from this cross produced plants with fruit shapes as follows: 89 disk, 61 sphere, and 10 elongate. What is the mode of inheritance of fruit shape in summer squash?

Answer: The numbers are very close to a ratio of 90:60:10, or 9:6:1, an epistatic variant of the 9:3:3:1, with the two 3/16ths categories combined. If this is the case, then the parent plants with disk-shaped fruits were dihybrids (*AaBb*). Among the offspring, 9/16ths had disk-shaped fruit, indicating that it takes at least one dominant allele of each gene to produce disk-shaped fruits (*A–B–: AABB, AaBB, AABb, or AaBb*). The 1/16th category of plants with elongate fruits indicates that this fruit shape occurs in homozygous recessive plants (*aabb*). The plants with spherical fruit would be plants with a dominant allele of one gene but homozygous recessive at the other gene (*AAbb, Aabb, aaBB, or aaBb*). In summary, then, two genes combine to control fruit shape in summer squash.

PROBLEM 2: Human blood groups are characterized by the presence of antigens (glycoproteins) on the surface of red blood cells. The ABO blood group is controlled by three alleles of a single gene on chromosome 9: I^A and I^B are codominant, and both are dominant to *i*. The MN blood group is controlled by a gene on chromosome 4 that has two codominant alleles, L^M and L^N. Rh is the most complex of the blood group types, involving at least 45 different antigens. The most clinically important antigen, D or Rh_O, is encoded by the gene *RhD* which is found on chromosome 1. Rh-positive individuals have either one or two *RhD* genes, whereas the Rh-negative phenotype is caused by the absence of the *RhD* gene.

Four babies born on the same day were accidentally mixed up in a maternity ward. Given the following blood types, can you match each baby to its correct set of parents.

BLOOD TYPES

BABY	COUPLE
a. O, MN, Rh+	1. O, MN, Rh– x B, N, Rh–
b. AB, MN, Rh+	2. A, M, Rh+ x B, M, Rh+
c. O, M, Rh–	3. A, M, Rh+ x A, N, Rh–
d. B, N, Rh–	4. A MN, Rh– x AB, M, Rh+

Answer: The three blood type genes are on three different chromosomes. Therefore, they are independently assorting and the three-factor crosses can be considered as three separate crosses. Probably the best way to solve this problem is by a systematic approach: take each couple's blood type and figure out whether they will be able to produce the child in question. Let's start with couple **1**: they can have children that are B or O (ABO system), MN or N (MN system), and Rh-negative (Rh system). They could not have produced child **a** because it is Rh-positive. They could not have produced child **b** because it is AB and Rh-positive. They could not have

produced child **c** because it is M. Therefore, child **d** belongs to couple **1**. We can double check our answer by making sure that all the blood types of child **d** can be accounted for by the blood type of its parents. Let's move on to couple **2**: they can have children that are A, B, AB, or O (ABO system), M (MN system), and Rh-positive or Rh-negative (Rh system). They could not have produced **a** or **b** because both children are MN. Therefore, child **c** belongs to couple **2**. Using the same approach, children **a** and **b** can be matched to couples **3** and **4,** respectively. Therefore, the successful relationship is

> Couple **1** and child **d;** Couple **2** and child **c**
> Couple **3** and child **a;** Couple **4** and child **b**

PROBLEM 3: In the four-o'clock plant, *Mirabilis jalapa*, two independently assorting genes *R* and *T* control flower color and plant height, respectively. Each of the two genes segregates two incompletely dominant alleles that provide the following phenotypes:

R_1R_1: red flower	T_1T_1: tall plant
R_1R_2: pink flower	T_1T_2: medium height plant
R_2R_2: white flower	T_2T_2: dwarf plant

a. Give the proportions of genotypes and phenotypes produced if a dihybrid plant is self-fertilized.

b. A plant of unknown genotype and phenotype is crossed with a white, tall plant, producing 152 pink, tall; 149 pink, medium height; 150 white, tall; and 154 white, medium-height offspring. What are the plant's genotype and phenotype?

Answer:

a. $R_1R_2\ T_1T_2$ × $R_1R_2\ T_1T_2$
    (pink, medium)   (pink, medium)

This problem can be worked out using the branched-line approach. To calculate the probability of each class produced, we can independently calculate the probability of each trait and apply the *product rule*.

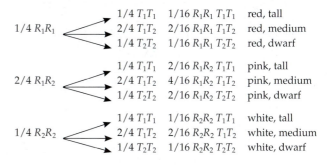

b. Because the two genes are independently assorting, we can consider each trait separately. First, examine the color phenotype. The offspring are in an approximate ratio of 1 pink: 1 white (301:304). The white parent must be R_2R_2. To produce offspring in a 1:1 ratio, the unknown parent must be heterozygous (R_1R_2) and therefore pink. Now, examine the shape phenotype. The offspring are in an approximate ratio of 1 tall: 1

medium (302:303). The tall parent is T_1T_1. To produce offspring in a 1:1 ratio, the unknown parent must be heterozygous (T_1T_2) and therefore have medium height.

Thus, the genotype of the unknown parent must be $R_1R_2\ T_1T_2$ with a corresponding phenotype of pink flowers and medium height.

Exercises and Problems

11. A plant with red flowers is crossed with a plant with white flowers. All the progeny have pink flowers. When two plants with pink flowers are crossed, the progeny are 11 red, 23 pink, and 12 white. What is the mode of inheritance of flower color?

12. Two short-eared pigs are mated. In the progeny, three have no ears, seven have short ears, and four have long ears. Explain these results by diagramming the cross.

13. In a paternity dispute, a type AB woman claimed that one of four men, each with different blood types, was the father of her type A child. Which of the following could be the blood type of the father of the child on the basis of the evidence given?
    a. Type A   c. Type O
    b. Type B   d. Type AB

14. Consider a mating between a blood type A individual and a blood type B individual. List *all* possible crosses and the expected phenotypic and genotypic ratios in the offspring?

15. Rather than clearly demonstrating paternity, blood types are usually used to exclude someone from paternity. List all circumstances under which the phenotypes of the ABO system can be used to exclude paternity?

16. Among the genes having the greatest number of alleles are those involved in self-incompatibility in plants. In some cases, hundreds of alleles exist for a single gene. What types of constraints might exist to set a limit on the number of alleles a gene can have?

17. A premed student, Steve, plans to marry the daughter of the dean of nursing. The dean is sterile, and his daughter was conceived by artificial insemination. Having served as an anonymous sperm donor, Steve's father is concerned that Steve and his fiancée may be half brother and sister. Given the following information, deduce whether Steve and his fiancée could be related. (The MN and Ss systems are two independent, codominant blood type systems.)

	Blood Type
Dean's wife	A, MN, Ss
Her daughter	O, M, S
Steve's father	A, MN, Ss
Steve	O, N, s
Steve's mother	B, N, s

18. A particular variety of corn has a gene for kernel color and a gene for height with the following phenotypes:
    CC, Cc: purple kernels T_1T_1: tall stem
    cc: white kernels T_1T_2: medium height stem
                                T_2T_2: dwarf stem
    Give the proportions of genotypes and phenotypes produced if a dihybrid plant is selfed.

19. In the shepherd's purse plant, the seed capsule comes in two forms: triangular and rounded. If two dihybrids are crossed, the resulting ratio of capsules is 15:1 in favor of triangular seed capsules. Diagram a biochemical pathway that might explain this ratio.

20. In humans, a hypothetical trait is under the control of a single gene, A. Four alleles have been found: A_1 which is dominant to all others; A_2 which is dominant to A_3 and A_4, which are codominant. The dominance hierarchy can be expressed as: $A_1 > A_2 > A_3 = A_4$.
    a. If gene A is on an autosome, how many different genotypes and phenotypes are possible?
    b. If gene A is on the X chromosome, how many different genotypes and phenotypes would now be possible?

21. When studying an inherited phenomenon, a geneticist discovers a phenotypic ratio of 9:6:1 among offspring of a given mating. Give a simple genetic explanation for this result. How would you test this hypothesis?

22. In a variety of onions, three bulb colors segregate: red, yellow, and white. A plant with a red bulb is crossed to a plant with a white bulb, and all the offspring have red bulbs. When these are selfed, the following plants are obtained:
    Red-bulbed   119
    Yellow-bulbed   32
    White-bulbed   9

    What is the mode of inheritance of bulb color, and how do you account for the ratio?

23. Two agouti mice are crossed, and over a period of a year they produce 48 offspring with the following phenotypes:
    28 agouti mice
     7 black mice
    13 albino mice

    What is your hypothesis about the genetic control of coat color in these mice? Do the data support that hypothesis?

24. The color and shape of radishes are each under the control of a single gene segregating two incompletely dominant alleles. Color may be red (C^RC^R), white (C^WC^W) or purple (C^RC^W), shape may be long (S^LS^L), spherical (S^SS^S) or ovoid (S^LS^S). If two dihybrid radish plants are crossed, what phenotypic classes and proportions are expected in the offspring?

25. You notice a rooster with a pea comb and a hen with a rose comb in your chicken coop. Outline how you would determine the nature of the genetic control of comb type. How would you proceed if both your rooster and hen had rose combs?

26. You are working with the exotic organism *Phobia laboris* and are interested in obtaining mutants that work hard. Normal phobes are lazy. Perseverance finally pays off, and you successfully isolate a true-breeding line of hard workers. You begin a detailed genetic analysis of this trait. To date you have obtained the following results:

P: Hard worker × Lazy
↓
F₁: All lazy of both sexes
F₁ female × Hard worker male
↓
3/4 Hard workers : 1/4 Lazy of both sexes

From these results, predict the expected phenotypic ratio from crossing two F₁ lazy individuals.

27. Polydactyly (extra fingers or toes) is a rare autosomal dominant trait that is 80% penetrant. What is the chance that a child born to an affected father and normal mother will have the disorder? Assume the paternal grandfather is normal.

28. In screech owls, crosses between red and silver individuals sometimes yield all red; sometimes 1/2 red : 1/2 silver; and sometimes 1/2 red : 1/4 white : 1/4 silver offspring. Crosses between two red owls yield either all red, 3/4 red : 1/4 silver, or 3/4 red : 1/4 white offspring. What is the mode of inheritance?

29. Labrador retriever dogs may be black, brown, or yellow. The inheritance of coat color involves two independently assorting genes, B and E. Gene B controls the production of the pigment melanin. The allele B (black) is completely dominant to b (brown). Gene E controls the deposition of pigment in the hair. The dominant allele E allows normal deposition of the pigment, whereas the recessive e allows only a small amount of pigment deposition, producing a yellow coat. What is the expected F₂ phenotypic ratio if the parental cross was *BBEE* × *bbee*? What type of genetic interaction is observed in this problem?

30. In the mythical beast *Godzilla bigfootus*, coat color is under the control of a single gene C, with two codominant alleles, C^B and C^Y. Heterozygotes are green, and homozygotes are either blue (C^BC^B) or yellow (C^YC^Y). An independently assorting gene I, segregates two alleles: I (color) is completely dominant to i (color inhibition). What phenotypic ratio is expected from the mating of two individuals that are both $C^BC^Y Ii$?

31. The following ratios are obtained from the dihybrid cross *AaBb* × *AaBb*. What phenotypic ratios would be expected if one of the dihybrid parents is testcrossed?
   a. 9:3:4
   b. 9:7
   c. 12:3:1
   d. 13:3
   e. 1:2:1:2:4:2:1:2:1

32. A particular species of *Drosophila* has four strains with differing eye color: wild-type, orange-1, orange-2, and pink. The following matings of true-breeding individuals were performed.

Cross	F₁
wild-type × orange-1	all wild-type
wild-type × orange-2	all wild-type
orange-1 × orange-2	all wild-type
orange-2 × pink	all orange-2
F₁ (orange-1 × orange-2) × pink	1/4 orange-2 : 1/4 pink : 1/4 orange-1 : 1/4 wild-type

What F₂ ratio would you expect if the F₁ siblings from orange-1 × orange-2 were crossed?

33. In chicken, the creeper phenotype is characterized by shortened legs and wings. Affected chickens cannot walk normally and have to creep along. Creeper chickens are heterozygous for the creeper allele, C. In the homozygous state, C disrupts embryological development and leads to death of the chicken before hatching.
   a. What are the expected genotypic and phenotypic ratios from the cross of two creeper chickens?
   b. Is the creeper allele dominant or recessive for leg shape?
   c. Is the creeper allele dominant or recessive for lethality?

34. In mallard ducks, *Anas platyrhynchos*, plumage is under the control of a single gene with three alleles: M, which codes for wild-type mallard plumage; M^R, which codes for restricted plumage (a modified mallard pattern with white on the wing fronts at maturity); and M^D, which codes for dusky plumage (darker and plainer than the mallard). All three types of plumage breed true. Deduce the dominance hierarchy of the three alleles based on the following crosses:

Cross	Offspring
a. Dusky × mallard	1/2 dusky : 1/2 mallard
b. Dusky × restricted	1/2 dusky : 1/2 restricted
c. Mallard × mallard	3/4 mallard : 1/4 dusky
d. Mallard × restricted	1/2 restricted : 1/4 mallard : 1/4 dusky

35. A certain species of sheep may have either gray or black coats. Based on the following crosses, determine the inheritance of coat color.

Cross	Offspring
a. Black × black	all black
b. Black × gray	1/2 black : 1/2 gray
c. Gray × gray	2/3 gray : 1/3 black

36. Consider the cross *AaBbCc* × *AaBbCc*, in which the three genes are independently assorting. What is the expected phenotypic ratio in the offspring if allele B is homozygous lethal?

37. In a hypothetical plant species, flower color is determined by the following biochemical pathway:

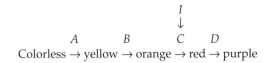

$$\begin{array}{cccc} & & I & \\ & & \downarrow & \\ A & B & C & D \\ \text{Colorless} \rightarrow \text{yellow} \rightarrow \text{orange} \rightarrow \text{red} \rightarrow \text{purple} \end{array}$$

Genes *A*, *B*, *C*, *D*, and *I* are independently assorting. The dominant alleles, *A*, *B*, *C*, and *D*, encode active enzymes that catalyze the corresponding reactions, and the recessive alleles, *a*, *b*, *c*, and *d* are nonfunctional. The dominant *I* allele totally inhibits the action of gene *C*, whereas *i* has no effect.
  a. Determine the F_1 and F_2 phenotypes from a cross between *AAbbCCDDII* and *AABBCCDDii* plants.
  b. What proportion of the offspring of an *AaBBCCCDdIi* × *AaBBCcDdIi* cross is expected to have colored flowers?

38. In a certain species of animals, two independently assorting genes, *A* and *B*, interact in a complementary manner to produce horns. Animals that are homozygous recessive for one or both genes (*aaB–*, *A–bb*, or *aabb*) lack horns.
  a. What genotypic and phenotypic ratios would be expected from this mating: *AaBb* × *AaBB*?
  b. Can two hornless animals produce offspring with horns? If so how?

39. The extremely rare Bombay phenotype in humans (first discovered in Bombay, India) provides yet another example of epistasis. The I^A and I^B alleles of the ABO blood group system encode glycosyl transferase enzymes that modify the terminal sugars of a mucopolysaccharide, called the H substance. H is produced by an independently assorting gene. Individuals with the genotype *HH* or *Hh* have a complete H structure. However, individuals with the *hh* genotype produce an incomplete H substance that the glycosyl transferase enzymes cannot use as a substrate. The *hh* condition thus masks the expression of the I^A and I^B alleles, and therefore *hh* individuals express the O phenotype, regardless of the alleles at the ABO locus. If many couples of the genotype $I^A I^B Hh \times I^A I^B Hh$ have children, what is the expected phenotypic ratio in their offspring?

40. In a hypothetical animal, joy and movement are under the control of the independently assorting genes *J* and *M*, respectively. Individuals can laugh ($J_1 J_1$), frown ($J_2 J_2$), or smile ($J_1 J_2$). They can also be runners ($M_1 M_1$), walkers ($M_2 M_2$), or joggers ($M_1 M_2$). A mating of two animals produces the following offspring: 10 laughing runners, 11 laughing walkers, 20 laughing joggers, 9 smiling runners, 10 smiling walkers, and 19 smiling joggers. What are the phenotypes of the two parents?

41. Consider four independently assorting genes, *A*, *B*, *C*, and *D*. Alleles *A* and *B* are completely dominant to *a* and *b*, respectively, whereas C_1 and D_1 are codominant to C_2 and D_2, respectively. How many different phenotypes are expected from the self-cross of $AaBbC_1C_2D_1D_2$?

42. *A*, *B*, and *C* are independently assorting genes controlling the production of black pigment in a rodent species. Alleles of these genes are indicated as *a*, *b*, and *c*, respectively. Assume that *A*, *B*, and *C* act in this pathway:

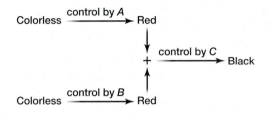

A black *AABBCC* individual is crossed with a colorless *aabbcc* to give black F_1 individuals. The F_1 individuals are mated with each other to give F_2 progeny.
  a. What proportion of the F_2 progeny is colorless?
  b. What proportion of the F_2 progeny is red?

43. Johnny, who has blood type AB, is married to Rosalie with blood type B. Rosalie's father has blood type O. Johnny and Rosalie plan to have two children. What is the probability that
  a. both children will have blood type B?
  b. the first child will be a boy of blood type AB and the second a girl of blood type A?
  c. *at least* one child will have blood type A?

44. Coat color in rabbits is under the control of a single gene with four alleles: *C* (full gray color); c^{ch} (chinchilla, light gray); c^h (Himalayan, white with black extremities); and *c* (albino, absence of pigmentation). The dominance hierarchy is represented as: $C > c^{ch} > c^h > c$. A series of matings involving six different rabbits produced the following offspring:

Cross	Offspring
a. Gray-1 × gray-2	3/4 gray : 1/4 chinchilla
b. Gray-2 × gray-3	3/4 gray : 1/4 chinchilla
c. Gray-1 × chinchilla-1	1/2 gray : 1/4 chinchilla : 1/4 himalayan
d. Chinchilla-2 × himalayan	1/2 chinchilla : 1/4 himalayan : 1/4 albino

Deduce the genotype of each rabbit, to the extent that it is possible.

45. A white-eyed male fly is mated with a pink-eyed female. All the F_1 offspring have wild-type red eyes. F_1 individuals are mated among themselves to yield:

Females		**Males**	
red-eyed	450	red-eyed	231
pink-eyed	155	white-eyed	301
		pink-eyed	70

Provide a genetic explanation for the results.

Chapter Integration Problem

In rats, several independently assorting autosomal genes affect coat color. Gene *A* controls the distribution of yellow pigment in hair, and gene *B* causes black pigmentation. The two genes interact as follows: *A–B–* (gray), *A–bb* (yellow), *aaB–* (black), and *aabb* (cream). These genotypes are only expressed in the presence of the dominant allele of a third gene, *C*; rats with genotype *cc* are albino.

a. Deduce the genotype of each albino mouse, to the extent that is possible, in the following table. Explain your answers.

True-breeding parents	F$_1$	F$_2$ offspring
Gray × albino-1	All gray	3/4 gray : 1/4 albino
Gray × albino-2	All gray	9/16 gray : 3/16 yellow : 4/16 albino
Gray × albino-3	All gray	9/16 gray : 3/16 black : 4/16 albino

b. Deduce the genotype and phenotype of each parent in the following table. Explain your answers.

Parents	Numbers of offspring
Cross 1	135 gray, 83 albino, 47 yellow, 44 black, 16 cream
Cross 2	103 albino, 74 black, 25 cream
Cross 3	179 gray, 62 yellow, 57 black, 18 cream

c. A gray-colored rat is mated with one that is yellow. The offspring include an albino rat and a cream-colored rat. Diagram this cross. Be sure to include the Punnett square and the phenotypic ratio in the offspring.

Assume the presence of a hypothetical gene *D* (dotted), which controls the polka-dotting trait. Polka-dotted rats have regularly spaced orange spots on their skin. The *D* gene segregates two alleles (*D, d*) and assorts independently from *A*, *B*, and *C*. The *D* allele is dominant in males and recessive in females. Rats with genotype *dd* have skin with no dots.

Consider the following cross:
AabbCcDd (male) × *aaBbCcDd* (female).

d. Is the *dotted* gene obeying Mendel's second law with respect to any X-linked gene? Explain.
e. How many different genotypes are possible for a secondary spermatocyte? List them all.
f. What is the probability of each of the following offspring: (1) An albino rat? (2) A yellow-colored, polka-dotted male rat? (3) A gray rat with no dots? (4) A polka-dotted rat?

Do you need additional review? Visit **www.mhhe.com/hyde** for practice tests, answers to end-of-chapter questions and problems, interactive exercises, and animation tutorials all designed to enhance your understanding of key genetic concepts.

Linkage and Mapping in Eukaryotes

Essential Goals

1. Understand how genes that are linked on a chromosome affect their transmission and the genotypic and phenotypic ratios in the progeny.

2. Analyze and manipulate data from two- and three-point crosses to generate linkage maps.

3. Calculate genotypic frequencies from a linkage map.

4. Understand the underlying mechanism of recombination and how tetrad analysis can be used to explain this mechanism.

5. Understand the difference between meiotic and mitotic recombination and the outcomes of both events for an organism.

6. Utilize additional methods to map eukaryotic genes.

Chapter Outline

Photo: Five chiasmata are apparent for this synapsing homologous chromosome pair in a grasshopper testis.

*I*n the preceding chapter, we described situations in which the standard Mendelian ratios in offspring were not observed. These cases included variable expressivity, incomplete penetrance, and codominance for a single gene, as well as epistasis and suppression between two different genes.

But investigators found other cases in which the law of independent assortment did not appear to hold. For example, in *Drosophila*, two characteristics that would be expected to produce offspring in a 1:1:1:1 phenotypic ratio instead produced unusual ratios like 1:250:250:1. Clearly, neither independent assortment nor the genetic interactions described in the previous chapter could explain these results; something odd was happening.

Although Mendel worked with observable traits and knew nothing of genes or chromosomes, researchers in the early 1900s were aware of chromosomes. They theorized that chromosomes encoded the traits and transmitted this information from one generation to the next—the *chromosomal theory of inheritance*. In chapter 3, we reviewed the process of meiosis that produces gametes and noted that chromosomes have been observed to cross over. Crossing over presumably leads to new combinations of genetic material, an effect termed **recombination.**

If two traits are carried on a single chromosome, then according to the chromosomal theory of inheritance, they would not be expected to sort independently, but rather would be inherited together. This theory would explain the 250:250 part of the ratio previously described. But what about the few odd ones on each end of the ratio? Geneticists deduced that these groups might result from crossing over between nonsister chromatids during meiosis, and therefore they represented the products of recombination.

Researchers have used the frequency of recombination between mutations to determine the relative location of genes on a chromosome that cause particular phenotypes. These **recombination maps** have been used to identify the genes that cause a large number of different inherited human diseases. For example, the figure on the first page of this chapter reveals the relative location of a few of the over 300 disease genes that are located on the human X chromosome. Recombination, thus, provides a powerful approach to identify genes associated with mutant phenotypes, which is a critical prerequisite in learning the cause of the phenotype and determining a potential therapy.

In this chapter we explore linkage and crossing over, and how frequencies of crossover are used to create recombination maps of chromosomes.

6.1 Linkage, Crossing Over, and Recombination: An Overview

After Sutton suggested the chromosomal theory of inheritance in 1902, evidence accumulated that genes were located on chromosomes. For example, Morgan showed by an analysis of inheritance patterns that the white-eye locus in *Drosophila* is located on the X chromosome (chapter 4). Because any organism has many more genes than it has chromosomes, it is obvious that each chromosome must contain many genes or loci. And, because chromosomes in eukaryotes are linear, it follows that genes are arranged in a linear fashion on chromosomes, like beads on a string. This linear relationship of genes is called **linkage.**

Alfred H. Sturtevant (**fig. 6.1**), working in Morgan's laboratory, first demonstrated linkage in 1913 using mutations that were known to exist on the *Drosophila* X chromosome. Sturtevant's mapping technique was based on the critical recombination event that occurs between nonsister chromatids during meiosis I. The structure at which nonsister chromatids crossover was termed the **chiasma** (plural, *chiasmata*), from the Greek word meaning "to cross."

An Example of Recombination Results

In *Drosophila*, the recessive *band* mutation (*bn*) causes a dark transverse band on the thorax, and the *detached* mutation (*det*) causes the crossveins of the wings to be either detached from the longitudinal veins or absent (**fig. 6.2**). A banded fly was crossed with a detached fly to produce wild-type, dihybrid F₁ offspring. F₁ females were then testcrossed to banded, detached males (**fig. 6.3**). If the loci assorted independently, following Mendel's second law, we would expect a 1:1:1:1 phenotypic ratio. However, of the first 1000 offspring examined, experimenters recorded a ratio of 2:483:512:3.

Two main points emerge from the data in figure 6.3. First, no simple ratio is apparent. If we divide by 2, we get a ratio of 1:241:256:1.5. Although the first and last categories seem about equal, as do the middle two, no simple numerical relation seems to exist between the middle and end categories.

Second, the two largest phenotypic categories in the F₂ generation have the same phenotypes as the original parents in the cross (fig. 6.3). That is, banded flies and detached flies were the original parents as

figure 6.1   Alfred H. Sturtevant (1891–1970), a student of Thomas Hunt Morgan, generated the first linkage map in 1913. The mapping technique devised by Sturtevant is the standard for determining recombination distances.

figure 6.2   The *detached* mutant in *Drosophila* lacks the CV2 crossvein in the wing (right panel, arrow). The vein and crossvein designations are shown in the wild-type wing.

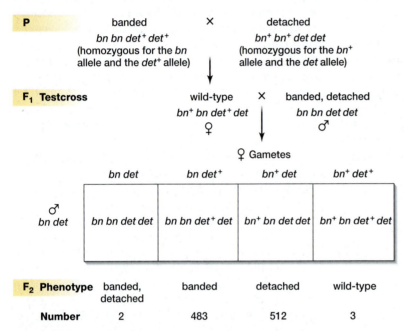

figure 6.3   Results from recombination between two *Drosophila* genes. Crossing homozygous *banded* flies with homozygous *detached* flies produces dihybrid, wild-type F₁ progeny. Testcrossing these dihybrid females produces F₂ progeny with four different phenotypes that are not in the standard 1:1:1:1 phenotypic ratio.

well as the great majority of the testcross offspring. We call these phenotypic categories in the F₂ progeny the **nonrecombinants,** or **parentals,** because they possess the same phenotypes as the individuals in the P generation. The testcross offspring with the lowest frequencies (banded, detached, and wild-type flies) possess a combination of the parental phenotypes. These two categories are referred to as **recombinants,** or **nonparentals.**

Because the F₂ progeny fail to exhibit independent assortment ratios, the *banded* and *detached* loci are likely located near each other on the same chromosome (or linkage group). This would suggest that the allele arrangements in the parental flies usually remain together (*bn det⁺* and *bn⁺ det*), and only at a low frequency would they rearrange to produce *bn det* and *bn⁺ det⁺*.

Analysis of Results

We can analyze this cross by drawing the loci as points on a chromosome (**fig. 6.4**). This representation shows that 99.5% of the testcross offspring, the nonrecombinants, come about through the simple linkage of the two loci, *detached* and *banded*. The remaining 0.5%, the recombinants, must have arisen through a physical exchange between the two loci on nonsister chromatids of two homologs during meiosis I (**fig. 6.5**).

The crossover event is viewed as a breakage and reunion of two chromatids lying adjacent to each other during prophase I of meiosis. We will discuss the cytological proof for this in the following sections.

It is important to understand what these crosses reveal. First, they inform us that the *banded* and *detached* genes are located on the same chromosome. Second, they do not reveal the location of the centromere relative to the two genes. Thus, the centromeres are not included in figure 6.5. Third, these crosses do not identify which chromosome contains the loci.

To summarize, the testcross in figure 6.4 shows that 99.5% of the gametes produced by the dihybrid female are nonrecombinant, whereas only 0.5% are recombinant. This very small frequency of recombinant offspring indicates that the two loci lie very close to each other on their particular chromosome.

The Physical Basis of Crossing Over

In chapter 3, you learned about the various events that occur during meiosis. In particular, DNA replication occurs prior to the start of meiosis to produce sister chromatids that are attached by a single centromere, homologous chromosomes pair during prophase I, and chiasmata are observed between nonsister chromatids of a homologous pair. Let's examine how this might relate to the mechanics of recombination.

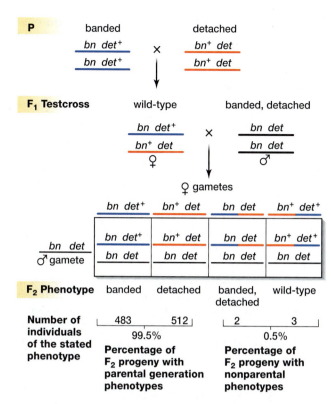

figure 6.4　Chromosomal arrangement of the two loci in the crosses of figure 6.3. A line represents the chromosomes on which these alleles are situated. Testcrossing the F_1 females produces F_2 progeny with four different phenotypes in the numbers shown.

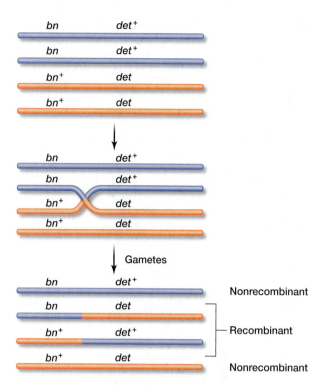

figure 6.5　Crossover between the *bn* and *det* loci in the tetrad stage of meiosis I in a dihybrid female. A chiasma forms between two nonsister chromatids, ultimately resulting in two nonrecombinant gametes and two recombinant gametes.

Linkage Versus Independent Assortment

For a pair of genes, we can imagine three different arrangements of the genes. First, the two genes may be located on different chromosomes, in which case they will exhibit independent assortment (**fig. 6.6, top**). The resulting gametes will consist of four different, but equally likely, genotypes. Because the two genes are located on different chromosomes, this represents one extreme where the two genes exhibit *no linkage*.

Second, the two genes may be located very close to each other on the same chromosome (fig. 6.6, middle). In this case, the two alleles that are present on the same chromosome will always be inherited together. This scenario produces only two types of gametes. Notice that each gamete contains only one parental chromosome. This scenario represents the other extreme, in which the two genes exhibit **complete linkage,** or the inability of recombination to separate the two genes.

The third possibility occurs when the two genes are located on the same chromosome, but recombination occurs between them (fig. 6.6, bottom). When a crossover occurs between nonsister chromatids, the resulting gametes consist of four different genotypes like independent assortment. However, the genotypes are not equally likely. Two of the genotypes (*A B* and *a b*) correspond to the parental chromosomes and will be present greater than 50% of the time. The recombinant genotypes (*A b* and *a B*) will be present less than 50% of the time.

You should see that the farther two genes are separated on a chromosome, the greater the probability that a crossover event will occur between them. This results in an increased frequency of recombinant gametes as the physical distance between two genes increases. Later in this chapter, you will see that the frequency of recombination for two genes that are present on the same chromosome will lie between the two extremes (0% and 50%).

Physical Proof of Recombination

If crossing over between nonsister chromatids is the mechanism by which recombination occurs, the recombination between genes should be accompanied by cytological exchange of physical parts of homologous chromosomes. But such an exchange can be demonstrated only if it is possible to distinguish between two homologous chromosomes. In 1931, Harriet Creighton and Barbara McClintock demonstrated this phenomenon using maize (corn); Curt Stern accomplished a similar result in the same year using *Drosophila*.

Two independently assorting genes

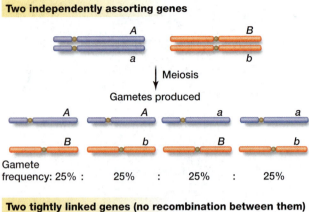

Two tightly linked genes (no recombination between them)

Two linked genes (recombination between the genes)

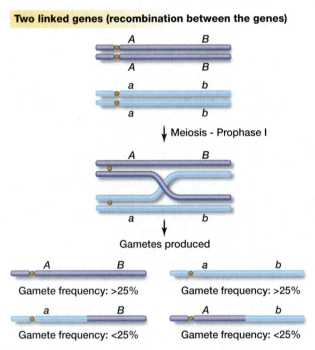

figure 6.6   Production of gametes from three possible gene arrangements. Two genes are unlinked and exhibit independent assortment (top), which produces four different gametes in equal amounts. Two genes are tightly linked and recombination rarely occurs between them (middle). Only two different gametes are usually produced, which correspond to the parental chromosome arrangement. Two genes are linked and recombination occurs between them (bottom). Four different gametes will be produced. Because recombination will not always occur between the two genes, the majority of the gametes will have the parental allele arrangements and the minority will have the recombinant arrangements.

Creighton and McClintock worked with chromosome 9 in maize (*Zea mays*). In one strain, they found a chromosome 9 with abnormal ends. One end had a knob, and the other had an extra piece of chromosome, called a translocation (**fig. 6.7**). This knobbed chromosome was clearly structurally different from its normal homolog. It also carried the dominant *colored* (*C*) allele and the recessive *waxy texture* (*wx*) allele.

After mapping studies showed that *C* was very close to the knob and *wx* was close to the translocation, Creighton and McClintock made the cross shown in figure 6.7. The dihybrid plant (*c Wx/C wx*) with heteromorphic (different-shaped) chromosomes was crossed with the normal homomorphic plant (only normal chromosomes) that had the genotype of *c wx/c wx* (colorless and nonwaxy phenotype).

If a physical crossover occurred during meiosis I between *C* and *wx* in the dihybrid plant, then the cytology of the recombinant chromosomes should also be apparent under the microscope. The *C* allele and the knob would become linked to the *Wx* allele, and the recessive *c* allele would become associated with the *Wx* allele and the translocation. Four types of gametes would result (see fig. 6.7).

Of 28 offspring examined, all were consistent with the predictions of the Punnett square in figure 6.7. Most of the colorless, nonwaxy individuals had the predicted normal (parental) chromosome (class 1). All of the colored, waxy individuals (class 4) had a parental knobbed and translocation chromosome.

The colorless, waxy individuals (class 2) were knobless and had the translocation.

Lastly, all of the colored, nonwaxy individuals (class 3) had the recombinant knobbed chromosome lacking the translocation. Thus, all four phenotypic classes were detected, and they possessed the predicted cytology.

One unexpected class was also detected. Two colorless, nonwaxy individuals were identified that were knobless and possessed the translocation. Although class 1 is colorless and nonwaxy, it lacks the translocation. Class 2 is colorless and possesses the translocation, but is waxy instead of nonwaxy. Can you predict how colorless and nonwaxy with the translocation were generated?

These individuals are recombinants because the *Wx* allele is joined to the translocation. However, instead of the recombination event occurring between the *c* and *Wx* genes as in classes 2 and 3, it occurred between the *Wx* gene and the translocation (**fig. 6.8**). The identification of this unexpected individual demonstrated that the recombination event can occur anywhere along the chromosome and is not restricted to the regions between genes. The reciprocal recombinant chromosome was not detected due to the small sample size of the experiment.

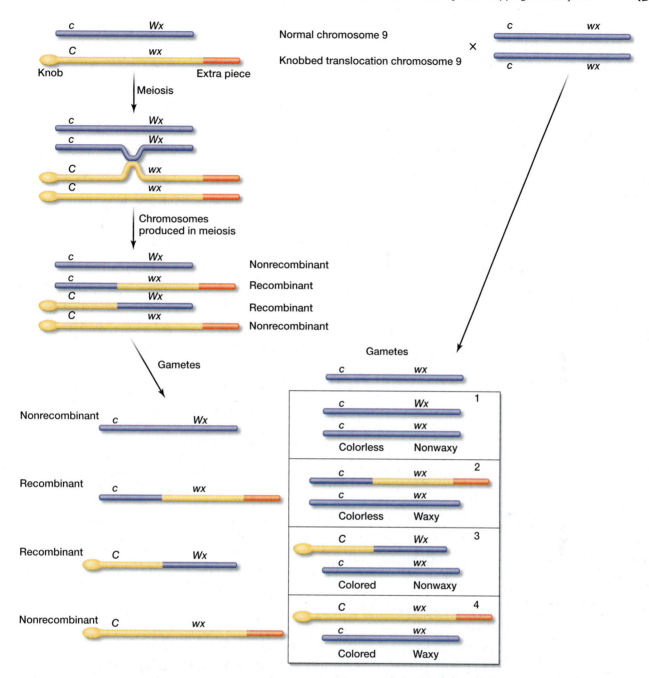

figure 6.7   Creighton and McClintock's experiment in maize demonstrated that genetic crossover correlates with cytological crossing over. Using chromosomes containing morphological differences, a knob and an extra piece of DNA, it was possible to correlate the presence of a specific allele and a physical exchange of chromosomal material. The Punnett square shows the potential chromosomes and phenotypes in the progeny of the cross shown.

The identification of plants that exhibited the recombinant morphological phenotypes with the expected cytological alterations was sufficient proof that recombination resulted in a physical exchange between homologous chromosomes. Creighton and McClintock concluded: "Pairing chromosomes, heteromorphic in two regions, have been shown to exchange parts at the same time they exchange genes assigned to these regions."

In the following sections, you will see how results of crosses between linked genes are used to generate **linkage maps.** Using Sturtevant's mapping technique, it is possible to determine the order of genes or DNA sequences along the length of a chromosome and the relative distances between them. Such a map can be used to calculate the odds of an individual in a pedigree exhibiting a specific genotype. We will address this point in more detail later in the chapter.

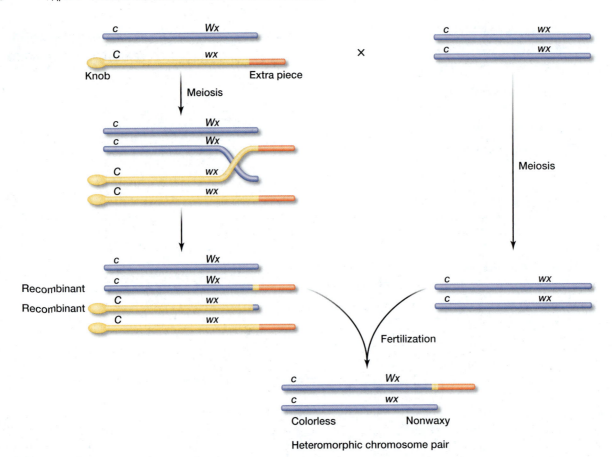

figure 6.8   An individual with an unexpected cytology in Creighton and McClintock's experiment revealed that recombination can occur outside of the region between the *C* and *Wx* genes. Recombination occurs between the *wx* allele and the translocation (red) at the end of the chromosome, which yields a nonwaxy individual that possesses the extra piece of DNA. The colorless and nonwaxy phenotype is consistent with no recombination event between the two genes.

 Restating the Concepts

▶ Linkage may be indicated when observed offspring ratios deviate markedly from ratios expected on the basis of independent assortment.

▶ The largest categories of gametes when traits are linked have the same allele arrangements as the parents and are termed *parentals* or *nonrecombinants*. The frequency of recombinant gametes will vary from just slightly larger than 0% to 50%, depending on the distance between the two genes.

▶ Creighton and McClintock first demonstrated in maize that recombination between linked traits corresponds to a physical exchange of chromosomes.

6.2 Diploid Mapping

Looking again at the testcross in figure 6.4, we noted that the two loci lie very close to each other on their particular chromosome. Sturtevant realized this, and he was the first to deduce that the frequency of generating recombinants corresponded to the rela-

tive physical distance between two genes. He used the percentage of recombinant testcross progeny, which is equivalent to the percentages of recombinant gametes, as a relative distance between loci on a chromosome.

The Definition of Map Units

In recombination analysis, 1% recombinant offspring is referred to as one **map unit** (mu, or one **centimorgan, cM,** in honor of geneticist Thomas Hunt Morgan). A map unit is a relative distance that reveals the order of and relative separation between loci on a chromosome. In the case of *bn* and *det*, the two loci are 0.5 mu apart.

We know the relationship between map units and DNA basepairs is highly variable and depends on the species, sex, and chromosome region. For example, in human beings, 1.0 mu can vary between 100,000 and 10,000,000 basepairs, whereas in the fission yeast, *S. pombe,* 1.0 mu is only about 6000 base pairs.

An important point needs to be mentioned. The measured recombination frequency between *any* pair of genes will be between 0.0 and 50.0 mu. The maximal recombination frequency of 50% (50 mu) represents the frequency for two genes that exhibit independent

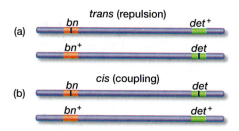

figure 6.9   *Trans* (repulsion) and *cis* (coupling) arrangements of dihybrid chromosomes. (a) In the *trans* arrangement, the mutant alleles are on different chromosomes. (b) In the *cis* arrangement, wild-type alleles are all on one chromosome, and all the mutant alleles are on the homologous chromosome.

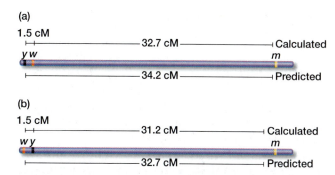

figure 6.10   Two possible recombination maps can be generated with the information of the *yellow* to *white* distance and the *white* to *miniature* distance. (a) If *white* lies between *yellow* and *miniature*, the distance between the latter two is 34.2 cM. (b) In contrast, if *yellow* lies between *white* and *miniature*, the distance between *yellow* and *miniature* is 31.2 cM.

assortment. Thus, two genes that exhibit 50% recombination may be located either on different chromosomes, or very far apart on the same chromosome. In fact, of the seven genes that Mendel randomly selected for his studies, two were located on one chromosome and three on another chromosome. Because of the large distance between the genes on the same chromosome, however, they exhibited independent assortment. It is interesting to speculate on what Mendel would have concluded had he selected one or two pairs of genes that were linked and that exhibited less than 50% recombination.

In figure 6.4, the *bn* and *det* alleles in the dihybrid F_1 female are in the **trans-configuration** (**fig. 6.9**); the two mutations are on separate homologs, as are the two wild-type alleles. If one chromosome carries both mutations and the homologous chromosome carries both wild-type alleles (see fig. 6.9), this is called the **cis-configuration**. (**Repulsion** and **coupling** have the same meanings as *trans* and *cis,* respectively.) In chapter 5, you learned how the *cis-* and *trans-* configurations were used to distinguish whether mutations were alleles or in different genes.

Two-Point Testcrosses

Sturtevant determined the recombination distance between several pairs of genes present on the *Drosophila* X chromosome. A cross involving two loci is referred to as a **two-point cross**. A **two-point testcross** involves crossing a dihybrid parent to a homozygous recessive individual. Using a number of two-point testcrosses, Sturtevant determined the recombination distance between several different pairs of genes on the X chromosome. Sturtevant's next challenge was to devise a method to combine all of these pair-wise distances into a single X chromosome map that revealed the relative positions of all the genes.

As one example, Sturtevant determined that the recombination distance between the *white* (*w*) and *yellow* (*y*) genes was 1.5 mu, while the *white* and *miniature* (*m*) genes were 32.7 mu apart. Based on this information, he devised two different maps that included all

three genes (**fig. 6.10**). The two maps differed by the distance between the *yellow* and *miniature* genes. One map predicted the distance to be 34.2 mu and the other map predicted 31.2 mu.

How was the distance from *yellow* to *miniature* calculated in the first map? The frequency of a crossover event between *yellow* and *miniature* will equal the frequency of a crossover event between either *yellow* and *white* or *white* and *miniature*. Notice that this is analogous to the sum rule (chapter 2). Thus, we add the two smaller distances to produce the larger one. In the second map, we can predict the *yellow* to *miniature* distance by subtracting the *white* to *yellow* distance from the *white* to *miniature* distance. The ability to add two or more smaller distances to generate a larger one is critical in generating the linkage maps that cover entire chromosomes.

Sturtevant measured the *yellow* to *miniature* recombination distance, which corresponded to 35.4 mu. Although this value did not perfectly match either of the two maps, it was more consistent with the first map. Subsequent crosses and analyses confirmed that this was the correct gene order. Sturtevant completed his X chromosome map by performing the additional two-point testcrosses to confirm the gene order.

Restating the Concepts

▶ The recombination frequency is the percentage of recombinant progeny in a population and represents a relative measurement of how close two genes are to each other.

▶ The recombination frequency can vary from 0.0 (two genes that are very close together) to 50% (for two genes that are very far apart on the same chromosome or physically unlinked on different chromosomes).

▶ A recombination map can be generated by comparing the recombination distances between all the possible pairwise gene combinations.

Three-Point Testcrosses

An alternative to examining pairwise combinations of genes is to look at three loci simultaneously. This has two distinct advantages over the two-point cross. First, we can determine the relative order of all three genes on the chromosome. Second, we can analyze the effects of multiple crossovers on map distances, which cannot be detected in a two-point cross.

Analysis of three loci, each segregating two alleles, is referred to as a **three-point cross.** In this example, we will examine wing morphology, body color, and eye color in *Drosophila*. Black body (*b*), purple eyes (*pr*), and curved wings (*c*) are all recessive mutations. Because the most efficient way to study linkage is through the testcross of a multihybrid, we will study these three loci by means of the crosses shown in **figure 6.11**.

One point in this figure should be clarified. Because the organisms are diploid, they have two alleles at each locus or gene. Geneticists use various means to represent this situation. For example, the recessive homozygote can be pictured as

1. *bb prpr cc*   or
2. *b/b pr/pr c/c*   or
3. *b pr c/b pr c*

A slash (also called a rule line) is used to separate alleles on homologous chromosomes. Thus, form (1) is used tentatively, when we do not know the linkage arrangement of the loci; form (2) is used to indicate that the three loci are on different chromosomes, and (3) indicates that all three loci are on the same chromosome.

When testcrossing the trihybrid organism (fig. 6.11), independent assortment would produce eight types of gametes with equal frequencies, and thus the eight phenotypic classes would each make up one-eighth of the offspring. However, if there is *complete linkage,* so the loci are so close together on the same chromosome that virtually no crossing over takes place, we would expect the trihybrid to produce only two gamete types in equal frequency, the *b pr c* genotype and the *b⁺ pr⁺ c⁺* genotype (namely, the parental genotypes). Finally, if crossing over occurs between linked loci, it would produce eight phenotypic classes in various proportions that would depend on the distances between loci. The actual data, which appear in **table 6.1**, support the model of crossing over between linked loci.

The data in the table are arranged in reciprocal classes. Two classes are **reciprocal** if between them they contain each mutant phenotype just once. Wild-type

($b^+ pr^+ c^+$) and black, purple, curved ($b\ pr\ c$) classes are thus reciprocal, as are the purple, curved ($b^+ pr\ c$) and the black ($b\ pr^+ c^+$) classes. Individuals within reciprocal classes occur in roughly equal numbers: 5701 and 5617; 388 and 367; 1412 and 1383; and 60 and 72.

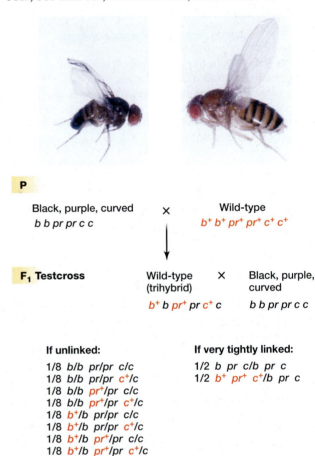

figure 6.11   Possible results in the testcross progeny of the *b pr c* trihybrid. The dominant wild-type alleles are in red.

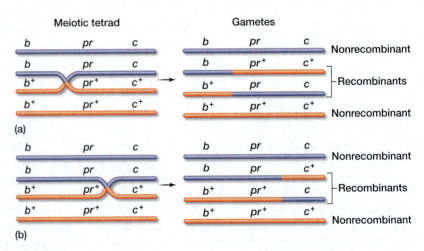

figure 6.12   Results of two different single crossovers in the *b-pr-c* region. (a) The diagram of a single crossover between the *b* and *pr* genes and the four different gametes produced. (b) A diagram of a single crossover between the *pr* and *c* loci and the resulting gametes.

table 6.1   Results of Testcrossing Female *Drosophila* Heterozygous for Black Body Color, Purple Eye Color, and Curved Wings ($b^+b\ pr^+pr\ c^+c \times bb\ prpr\ cc$)

Phenotype	Genotype	Number	Alleles from Trihybrid Female	Number of Recombinants Between		
				b and *pr*	*pr* and *c*	*b* and *c*
Wild-type	$b^+\ pr^+\ c^+/b\ pr\ c$	5,701	$b^+\ pr^+\ c^+$			
Black, purple, curved	$b\ pr\ c/b\ pr\ c$	5,617	$b\ pr\ c$			
Purple, curved	$b^+\ pr\ c/b\ pr\ c$	388	$b^+\ pr\ c$	388		388
Black	$b\ pr^+\ c^+/b\ pr\ c$	367	$b\ pr^+\ c^+$	367		367
Curved	$b^+\ pr^+\ c/b\ pr\ c$	1,412	$b^+\ pr^+\ c$		1,412	1,412
Black, purple	$b\ pr\ c^+/b\ pr\ c$	1,383	$b\ pr\ c^+$		1,383	1,383
Purple	$b^+\ pr\ c^+/b\ pr\ c$	60	$b^+\ pr\ c^+$	60	60	
Black, curved	$b\ pr^+\ c/b\ pr\ c$	72	$b\ pr^+\ c$	72	72	
Total		15,000		887	2,927	3,550
Percent				5.9	19.5	23.7

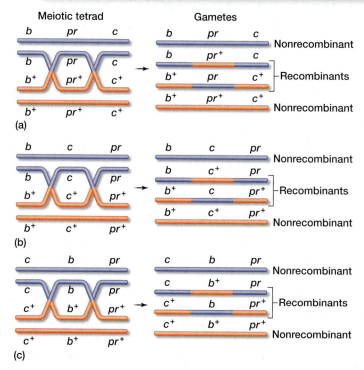

(a)

(b)

(c)

figure 6.13   The different types of gametes produced from double crossovers involving three different orders of the *b pr c* genes. The three gene orders vary based on whether the *pr* (a), *c* (b), or *b* (c) gene is in the middle. In each gene order, the double crossover recombinant class, which will be the rarest class, will indicate the correct gene order.

(Reprinted from Bridges, Calvin B. "Salivary Chromosome Maps: With a Key to the Banding of the Chromosomes of Drosophila Melanogaster," *The Journal of Heredity*, February 1935, © 1935, by permission of Oxford University Press.)

Wild-type and black, purple, curved are the two phenotypes that compose the parental (nonrecombinant) classes. The purple, curved class of 388 is grouped with the black class of 367. These two would be the products of a single crossover between the *b* and the

pr loci, if we assume that the three loci are linked and that the gene order is *b pr c* (**fig. 6.12a**). The next two classes, containing 1412 and 1383 flies, would result from a single crossover between *pr* and *c* (fig. 6.12b), and the last set, 60 and 72, would result from a **double crossover,** one between *b* and *pr* and the other between *pr* and *c* (**fig. 6.13a**). The numbers in columns 5 and 6 of table 6.1 are based on these recombinant events.

In the final column of table 6.1, recombination between *b* and *c* is scored. Only those recombinant classes that have a new arrangement of *b* and *c* alleles, as compared with the parentals, are counted. This last column shows us what a *b×c* two-point cross would have revealed, had we been unaware of the *pr* locus in the middle.

Gene Order

We assumed that the *pr* gene is the middle gene of the three; however, we can use the data in table 6.1 to determine that the actual gene order is *b pr c*.

Of the four pairs of reciprocal phenotypic classes in table 6.1, one pair has the highest frequency (5701 and 5617) and one pair has the lowest (60 and 72). The pair with highest frequency is always the parental group and reveals the linkage arrangement of alleles in the parental chromosomes. In this case, all three recessive alleles are linked on one parental chromosome, and all three dominant alleles are linked on the other parental chromosome. (This linkage arrangement will likely appear in every cross, but will not always be the parental arrangement.)

The pair with the lowest frequency is always the double recombinant group. The double recombinants represent the class that contains one crossover between the first pair of genes, and one crossover between the second (fig 6.12). The double recombinant class always corresponds to the smallest pair because it represents the frequency of two independent crossovers (which uses the product rule—the frequency of the first crossover times the frequency of the second). In the double recombinant, the middle locus changes position relative to the outside loci in the parental arrangement.

We can visualize this situation by drawing the parental chromosomes in all three possible arrangements, with the middle gene changed in each variant (see fig. 6.13). The allele arrangement on the parental chromosomes corresponds to the largest pair of phenotypes, and in this case, it corresponds to $b^+ pr^+ c^+$ and $b\ pr\ c$. We draw the homologous chromosomes with all three dominant alleles on one chromosome, and all three recessive alleles on the homolog. We then vary which gene is the middle one (the order of the first and last genes makes no difference).

Although two recombination events in all three gene orders correspond to one of the phenotypic pairs in table 6.1, only the first gene order ($b\ pr\ c$) allows for the two crossovers to produce the smallest class. Thus, the pr gene must be located between the b and c genes. Note that the double recombinant (smallest) class produces a pair of phenotypes in which only the middle gene has switched positions relative to the flanking genes.

Map Distances

The percent row in table 6.1 reveals that 5.9% (887/15,000) of the offspring in the *Drosophila* trihybrid testcross resulted from recombination between b and pr, 19.5% between pr and c, and 23.7% between b and c. These numbers allow us to form a tentative map of the loci (**fig. 6.14**). A discrepancy, however, is evident. The distance between b and c can be calculated in two ways. By adding the two distances, b–pr and pr–c, we get 5.9 + 19.5 = 25.4 map units; yet by directly counting the recombinants—the last column of table 6.1—we get a distance of only 23.7 map units. What causes this discrepancy of 1.7 map units?

Returning to the last column of table 6.1, you can observe that the double crossovers (60 and 72) are not included in the calculated distance from b to c, yet each double crossover event actually represents two crossovers between the b and c genes. The reason they are not counted in this column is that when only the b and c genes are studied, you are unable to observe any recombination between them. Do you see why? The first one of the two crossovers causes a recombination between the two end loci, whereas the second one returns these outer loci to their original configuration (see fig. 6.13).

If we took the 3550 recombinants between b and c and added in twice the total of the double recombinants, namely 264, we would get a total of 3814, which represents the 25.4 mu (3814/15,000) that we calculated from the sum of the b–pr and pr–c distances. We add twice the total number because each double recombination event is composed of two single recombination events. Thus, the correct map would show the distance between the b and c genes to be 25.4 mu.

This illustrates an important point about recombination mapping. The farther two loci are apart on a chromosome, the more double crossovers will occur between them. Double crossovers tend to reduce the observed recombination distance, as in our example, so that distantly linked loci usually appear closer than they really are. The most accurate map distances are those established using closely linked loci. In other words, summed short distances are more accurate than directly measured larger distances.

The results of the previous experiment show that we can obtain at least two map distances between any two loci: *measured* and *actual*. **Measured map distance** between two loci is the value obtained from either a two-point or three-point cross. **Actual map distance** is an idealized, more accurate value obtained from summing short distances between many intervening loci that accounts for multiple crossover events.

When the measured map distance is plotted against actual map distance, a **mapping function** curve is obtained (**fig. 6.15**). Notice that the observed recombination distance is smaller than the actual recombination distance. This is largely due to the failure to detect double or multiple crossover events.

This graph is of both practical and theoretical value. Pragmatically, it allows conversion of a measured map distance into a more accurate map distance. Theoretically, it shows that the measured map distance never exceeds 50 mu in any one cross. Multiple crossovers (double and triple) reduce the apparent distance between two loci to a maximum of 50 mu, the same value that independent assortment produces. For this reason, it is impossible for the recombination distance measured between any two genes to be greater than 50 mu.

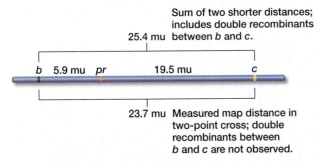

Sum of two shorter distances; includes double recombinants between b and c.

25.4 mu

b 5.9 mu pr 19.5 mu c

23.7 mu   Measured map distance in two-point cross; double recombinants between b and c are not observed.

figure 6.14 Tentative map of the *black, purple,* and *curved* chromosome in *Drosophila*. The top map distance (25.4 mu) is calculated by summing the distances between adjacent pairs of genes and includes the double recombinants. The lower map distance (23.7 mu) is based on the two-point cross data between b and c, which lacks the frequency of double recombinants.

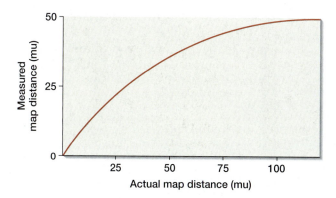

figure 6.15 A mapping function demonstrates how the measured recombination distance underestimates the actual recombination distance.

Coefficient of Coincidence and Interference

The next question in analyzing these three-point cross data is, are the crossovers occurring independently of one another? That is, does the observed number of double recombinants equal the expected number?

In our example, the observed frequency of double recombinants was (60 + 72)(100%)/15,000, or 0.88%. The expected frequency of double recombinants is based on the independent occurrence of crossing over in the two regions measured. That is, 5.9% of the time there is a crossover in the *b–pr* region, and 19.5% of the time there is a crossover in the *pr–c* region. Applying the product rule (chapter 2), the expected frequency of double crossovers is 0.059 × 0.195 = 0.0115. This means that 1.15% of the gametes are expected to be double recombinants (1.15% of 15,000 = 172.5).

In this example, the observed number of double recombinants is lower than expected (132 observed, 172.5 expected). This implies a **positive interference,** in which the occurrence of one crossover reduces the chance of the second. We can express this as a **coefficient of coincidence (C),** which is defined as:

$$C = \frac{\text{frequency of observed double crossovers}}{\text{frequency of expected double crossovers}}$$

The coefficient of coincidence for our example is 132/172.5 = 0.77. In other words, only 77% of the expected double crossovers occurred. Sometimes we express this reduced quantity of double crossovers as the **degree of interference,** which is defined as

$$\text{Interference} = 1 - C$$

In the example, the interference is 23%.

It is also possible to have **negative interference,** in which more double recombinants are observed than expected. In this situation, the occurrence of one crossover seems to enhance the probability that a second crossover will occur in adjacent regions.

Geneticists have mapped the chromosomes of many eukaryotic organisms from three-point crosses of this type—those of *Drosophila* have probably been the most extensively studied. Part of the *Drosophila* chromosomal map is presented in **figure 6.16.** Locate the loci we mapped so far to verify the map distances.

These *Drosophila* linkage maps illustrate an important feature about recombination frequency and distance that we already mentioned: Genes on a single chromosome can be greater than 50 mu apart. For example, the *dumpy wings* and *orange eyes* genes on chromosome 2 are located 94 mu apart (107.0 − 13.0 = 94.0). If you directly measured the recombination frequency between the *dumpy wings* and *orange eyes* genes, you would obtain 50% recombination. Thus, two genes that are located far apart on the same chromosome can exhibit independent assortment due to the multiple crossover events between the genes. The *dumpy wings* and *orange eyes* genes were shown to be linked by determining several smaller recombination distances that covered the entire region between the two genes.

In summary, two or more loci are linked if offspring do not fall into simple Mendelian ratios. Map distances are the percentage of recombinant gametes, which are usually detected in a testcross. When analyzing three loci, determine the parental (nonrecombinant) and double recombinant groups first. Then establish the locus in the middle, and recast the data in the correct gene order. The most accurate map distances are those obtained by summing shorter distances. A coefficient of coincidence is determined as the ratio of the number of observed double recombinants relative to the expected number.

Restating the Concepts

▶ The ability to detect double crossovers in a three-point cross allows it to yield a more accurate map than a two-point cross.

▶ The parental class (the largest reciprocal class) reveals the arrangement of the alleles on both of the parental chromosomes.

▶ In a three-point cross, comparing the parental class and the double recombinant class (the smallest reciprocal class) identifies the gene that is in the middle.

▶ The recombination distance between two genes is the percentage of crossover events (both single and double) in all the progeny.

▶ The coefficient of coincidence and interference provides information on how one crossover event affects a second crossover between the same homologs.

▶ The sum of two shorter recombination distances is more accurate than the calculation of the recombination distance between the two farthest genes because, as the recombination distance increases, the difference between the observed and actual recombination distances becomes greater.

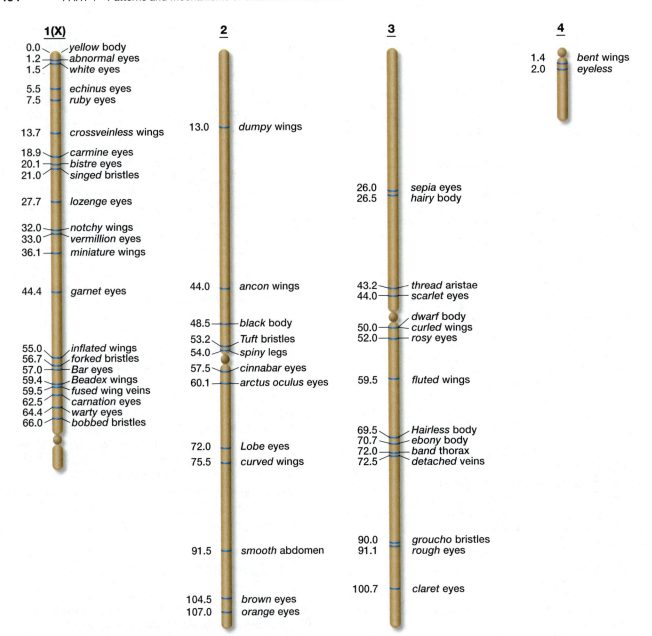

figure 6.16   Partial map of the *Drosophila melanogaster* chromosomes. The location of several mutations are shown along with their map position in mu. The centromere is marked by a filled circle.
(From C. Bridges, "Salivary Chromosome Maps," *Journal of Heredity*, 26:60–64, 1935. Reprinted with permission of Oxford University Press.)

6.1 applying genetic principles to problems

Generating a Recombination Map from Three-Point Testcross Data

In this problem, you will work through a three-point cross in which neither the middle gene nor the *cis–trans* relationship of the alleles in the trihybrid F₁ parent is known.

The *Drosophila*, *hairy* (*h*) mutation causes extra bristles on the body, *thread* (*th*) causes a thread-shaped arista (antenna tip), and *rosy* (*ry*) causes the eyes to be reddish brown. All three traits are recessive.

Trihybrid females were testcrossed; the phenotypes from 1000 offspring are listed in the following table. You can use this data to determine the parental genotypes, the gene order, the map distances, and the coefficient of coincidence. The table presents the data in no particular order, just as a scientist might have recorded them. Phenotypes are tabulated and, from these, the genotypes can be reconstructed.

Notice that the data can be regrouped into a large set (359 and 351) with reciprocal phenotypes ($h^+ ry^+ th$ and $h\ ry\ th^+$), a small set (4 and 6) with reciprocal phenotypes ($h\ ry\ th$ and $h^+ ry^+ th^+$), and large and small intermediate sets (98 and 92, 47 and 43).

box table 1

Phenotype	Genotype (order unknown)	Number
Thread	$h^+ ry^+ th/h\ ry\ th$	359
Rosy, thread	$h^+ ry\ th/h\ ry\ th$	47
Hairy, rosy, thread	$h\ ry\ th/h\ ry\ th$	4
Hairy, thread	$h\ ry^+ th/h\ ry\ th$	98
Rosy	$h^+ ry\ th^+/h\ ry\ th$	92
Hairy, rosy	$h\ ry\ th^+ /h\ ry\ th$	351
Wild-type	$h^+ ry^+ th^+/h\ ry\ th$	6
Hairy	$h\ ry^+ th^+/h\ ry\ th$	43

From the data presented, are the three loci linked? Because the four sets of data are not present in a 1:1:1:1:1:1:1:1 ratio the three genes are not independently assorting. But are only two of the three genes linked, with the third gene assorting independently?

If one of the three genes was assorting independent of the other two, you would find this unlinked gene produced four phenotypic classes present in a 1:1:1:1 ratio with either of the two linked genes (box table 2). Table 2 reveals that for each pair of genes, the 1:1:1:1 ratio was not observed. This result demonstrates that **all three genes are linked** (see box table 1).

Now that you determined that all three genes are linked, you can deduce the allelic arrangement in the trihybrid parent. The reciprocal group of offspring with

the greatest number represents the nonrecombinant, or parental, class. In this example, the $h\ ry\ th^+/h\ ry\ th$ and $h^+ ry^+ th/h\ ry\ th$ genotypes represent the parental genotypes. Because these classes correspond to the nonrecombinant gametes from the parents, the parents' genotypes must therefore have been $h\ ry\ th^+/h^+ ry^+ th$ and $h\ ry\ th/h\ ry\ th$.

Once you determined the allelic arrangement in the parents, it is now possible to determine the order of the three linked genes. The first step is to identify the double recombinant classes, which correspond to the reciprocal group of offspring with the fewest number. In this example, the $h\ ry\ th/h\ ry\ th$ and $h^+ ry^+ th^+/h\ ry\ th$ genotypes correspond to the double recombinants.

You then compare the parental chromosomes ($h\ ry\ th^+$ and $h^+ ry^+ th$) with the double recombinant chromosomes ($h\ ry\ th$ and $h^+ ry^+ th^+$). In this comparison, the alleles of the middle gene in the double recombinant chromosome will switch positions relative to the flanking alleles in the parental chromosome. You can see that the th alleles switched chromosomes when you compare $h\ ry\ th$ with $h\ ry\ th^+$ and $h^+ ry^+ th$ with $h^+ ry^+ th^+$. Thus, the *thread* gene is located between the *hairy* and *rosy* genes. You can now rewrite the genotypes of the parents as $h\ th^+ ry/h^+ th\ ry^+$ and $h\ th\ ry/h\ th\ ry$.

Having determined the gene order and the parental genotypes, you can calculate the recombination distances in a straightforward manner. Remember that recombination distance is the frequency of recombinant progeny in the cross. The distance from *hairy* to *thread* is therefore:

$$\frac{(100\%)(98 + 92 + 4 + 6)}{1000} = 20\%, \text{ or } 20\ \text{mu}$$

The distance from *thread* to *rosy* is:

$$\frac{(100\%)(47 + 43 + 4 + 6)}{1000} = 10\%, \text{ or } 10\ \text{mu}$$

Because the double recombinant class has a crossover between the first pair of genes and between the second pair of genes, they must be included in both calculations. The distance from *hairy* to *rosy* will simply be the sum of the two distances, or 30 map units.

Once you calculated the map distances, you can determine the coefficient of coincidence. The expected double recombination frequency is the product of the two shorter recombination frequencies (20% × 10% = 2%). The observed double recombination frequency is calculated from the numbers in the previous table:

$$\text{observed double recombination frequency} = \frac{(100\%)(4 + 6)}{1000} = 1\%$$

The coefficient of coincidence is therefore

$$\frac{1\%}{2\%} = 0.5\ C$$

This result reveals that only 50% of the expected double recombination events actually were observed in the previous cross. What is the implication of this finding?

box table 2

Phenotype		hairy and thread		hairy and rosy		rosy and thread	
Thread	359						
Rosy, thread	47			$ry\ th$	47	$h^+ ry$	47
Hairy, rosy, thread	4	$h\ th$	4	$ry\ th$	4		
Hairy, thread	98	$h\ th$	98			$h\ ry^+$	98
Rosy	92	$h^+ th^+$	92			$h^+ ry$	92
Hairy, rosy	351						
Wild-type	6	$h^+ th^+$	6	$ry^+ th^+$	6		
Hairy	43			$ry^+ th^+$	43	$h\ ry^+$	43
Total	1000		200		100		280

It's Your Turn

Using recombination frequencies to generate a linkage map is one of the underlying tools throughout this book. It is critical that you have a solid understanding of this concept. Problem 22 is a straight forward three-point testcross. Problem 57, however, is a variant with an additional component. Remember that recombination does not occur in *Drosophila* males, so a $+ + +/a\,b\,c$ male can only produce two types of gamete. Will it be possible to detect the recombinant chromosomes from the female if they are fertilized with sperm containing a $+ + +$ chromosome? If not, how should we deal with these individuals?

6.2 applying genetic principles to problems

Calculating Recombination Frequencies from a Linkage Map

In actual research, you may wish to calculate the frequency that a particular genotype will be produced from a three-point cross. This calculation is relatively straightforward if you have the linkage map of the three genes being analyzed.

In the following example, you want to know the frequency of producing an $a\,b\,c^+$ phenotype from the cross of an $a^+\,b^+\,c/a\,b\,c^+$ male and female. You already know that the coefficient of coincidence is 0.4, and that the linkage map is:

$$a\text{-------------}\overset{\text{5 mu}}{}\text{-------------}b\text{-------------}\overset{\text{10 mu}}{}\text{-------------}c$$

In this problem, you need to reverse the thinking you employed in box 6.1. The first thing you do is calculate the frequency of *observed* double recombinants. You know that the expected double recombinant frequency is defined as the frequency of a recombination event between a and b and the frequency of a recombination event between b and c (the product rule from chapter 2) which is calculated as:

Percentage of expected double recombinants =
$(0.05 \times 0.10)(100\%) = 0.5\%$

You can then calculate the frequency of observed double recombinants by solving the equation as follows:

$$C = \frac{\text{percentage of observed double recombinants}}{\text{percentage of expected double recombinants}}$$

The value of C is given as 0.4, and the value of the expected double recombinants is 0.5%, so solving for the observed double recombinants:

$$\text{Percentage of observed double recombinants} = (0.4) \times (0.5\%) = 0.2\%$$

Based on the parental genotype of $a^+\,b^+\,c/a\,b\,c^+$, the double recombinant gametes will be $a^+\,b\,c$ and $a\,b^+\,c^+$. Because the total frequency of double recombinants is 0.2%, 0.1% of the gametes will be $a^+\,b\,c$ and 0.1% will be $a\,b^+\,c^+$.

You can now calculate the observed frequency of each single recombinant class. Remember that the recombination distance between a and b represents the frequency of the single recombinants between a and b and the double recombinant class. Therefore, the frequency of single recombinants between a and b is:

$$5\% - 0.2\% = 4.8\%$$

of which 2.4% of the gametes will be $a^+\,b\,c^+$ and 2.4% will be $a\,b^+\,c$. Similarly, the frequency of single recombinants between b and c will be $10\% - 0.2\% = 9.8\%$, of which 4.9% will be $a^+\,b^+\,c^+$ and 4.9% of the gametes will be $a\,b\,c$.

All you need to do now is calculate the frequency of the parental gametes, which will be the remainder of the gametes ($100\% - 9.8\% - 4.8\% - 0.2\% = 85.2\%$). Thus, 42.6% of the gametes will be $a^+\,b^+\,c$, and 42.6% of the gametes will be $a\,b\,c^+$.

You can now calculate the frequency of any desired genotype or phenotype that is produced from the cross of two individuals with the genotype of $a^+\,b^+\,c/a\,b\,c^+$. Because you are interested in the frequency of an $a\,b\,c^+$ phenotype, you need to calculate the potential genotypes that can produce this phenotype. In this problem, assume that recombination occurs in both the male and the female. The potential genotypes that must be considered and their frequency (based on the product rule) are

$$a\,b\,c^+/a\,b\,c^+ \quad (42.6\%)(42.6\%) = 18.15\%$$

$$a\,b\,c^+/a\,b\,c \quad (42.6\%)(4.9\%) = 2.09\%$$

The second genotype can be produced two ways: one in which the male contributes the first gamete, and one in which the female does. Therefore, the probability of producing the $a\,b\,c^+$ phenotype is

$$18.15\% + 2.09\% + 2.09\% = 22.33\%$$

If recombination failed to occur in the male (as it fails to occur in *Drosophila* males), then 50% of the male gametes would be $a\,b\,c^+$, and the female could contribute either the $a\,b\,c^+$ or the $a\,b\,c$ gamete. The result would be the same two genotypes that we listed earlier, but now generated only a single way each, at a frequency of

$$(50\%)(42.6\%) + (50\%)(4.9\%) = 23.75\%.$$

It's Your Turn

Problem 28 involves determining the frequency of generating a particular phenotype for a specific cross involving three linked genes. Calculate the frequency of each genotypic class in the female gametes, beginning with the double recombinant classes. Remember, interference = 1 – C (coefficient of coincidence) and *Drosophila* males do not exhibit recombination.

6.3 Mapping DNA Sequences

Up to this point, we examined recombination between genes. One powerful approach in linkage mapping is mapping DNA sequences. Various DNA sequences can be mapped, but before we delve too deeply into this, consider the simple linkage map in **figure 6.17a**. This map reveals that 10% of the gametes will exhibit crossing-over between two genes. Based on this map, we can expect the following results:

P   *Ab/aB* × *ab/ab*
F$_1$ *Ab*   P   45%
       aB   P   45%
       AB   R   5%
       ab   R   5%

It is more accurate to draw a linkage map that represents not only the physical distance between genes, but also the physical size of the genes. The map can be redrawn as shown in figure 6.17b.

Alleles of a gene are DNA sequences that differ by as little as a single nucleotide. These DNA differences result in variations in the function of the encoded protein, which produce different phenotypes. When mapping the locus of a gene by recombination, we are actually mapping the location of the allelic differences in the DNA sequence.

For example, a recombination event between the *A* and *B* genes in the *trans*-configuration (**figure 6.18a**) produces *ab* and *AB* gametes. But the crossover does not necessarily occur between the genes. If a crossover occurs *within* the *A* gene, but between the *a* allelic site and the *B* gene, then the *ab* and *AB* recombinant gametes would still be produced (figure 6.18b). The reason is that this single site determines whether the gene is the *A* or *a* allele. So, technically speaking, the calculated map

distance is the frequency of a crossover between the *A* allelic difference and the *B* allelic difference.

The essential elements of all linkage mapping experiments are (1) at least one of the individuals must be heterozygous for each gene or locus being analyzed, and (2) each allele must produce a different phenotype. In the previous examples in this chapter, the phenotype was a **morphological characteristic,** such as size, color, and shape. However, some phenotypes are not

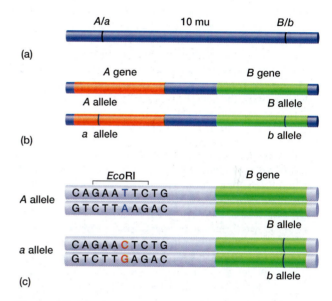

figure 6.17 Representation of alleles as (a) genes, (b) sites within a gene, and (c) DNA sequences. Recombination may occur within a gene as well as between genes.

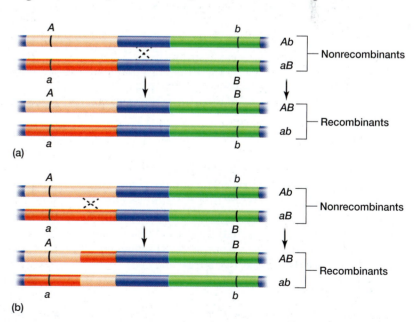

figure 6.18 Crossing over between genes in the *trans* configuration (a) produces recombinant allele arrangements (*AB* and *ab*). If the recombination event occurs within the first gene, but still between the mutant sites (b), the same recombinant allele arrangements are produced.

6.3 applying genetic principles to problems

Using Linked DNA Markers to Calculate Genotypic Probabilities

In genetic counseling, it is often not possible to directly test if an individual possesses a recessive allele that is associated with an inherited disease. However, the individual may be tested for the presence of a particular DNA marker that is located a known distance from the disease gene. It is then possible to calculate the probability that the individual possesses the disease allele.

Let's assume that there is an RFLP that is located 4 mu from a gene that causes a rare recessive disease that results in death within the first year. The wild-type allele is designated *D* and the disease allele is represented as *d*. We can draw the linkage map as follows:

$$RFLP\text{-----------}\overset{4 \text{ mu}}{\text{-----------}}d$$

The RFLP results from the presence or absence of an *Eco*RI restriction enzyme site. The presence of the *Eco*RI site results in the production of 5-kb and 3-kb restriction fragments that can be visualized on a gel. The absence of the *Eco*RI site results in a 8-kb fragment. We can designate the two RFLP alleles as *RFLP5.3* and *RFLP8*.

Let's assume that Mary is expecting a child. One of Mary's four sisters died from this disease. Mary is concerned about passing the recessive allele onto her child. What is the probability that Mary is a carrier?

The first step in solving this problem is to draw the pedigree. Based on the information in the problem, you should be able to draw the following pedigree:

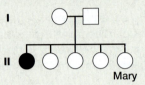

box figure 6.1   Pedigree of Mary's family. Mary's youngest sister died from the recessive disease.

You can now assign genotypes to the individuals. Because Mary's sister died from the disease she must have been *d/d*, and Mary's parents both must be heterozygous, *D/d*. Mary and each of her three remaining sisters must have the dominant *D* allele, either as a homozygote or a heterozygote. We can now add this information to the pedigree.

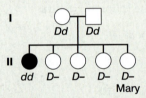

box figure 6.2   Pedigree of Mary's family. The deduced genotype for each individual is shown.

as obvious, such as the ABO blood group. Consider the DNA sequence that relates to the difference between the *A* and *a* alleles, shown in figure 6.17c. What if the two alleles differed by only a single nucleotide? The *A* allele may contain the sequence:

CA<u>GAATTC</u>TG

The underlined sequence, as it turns out, is recognized by the restriction enzyme *Eco*RI. This enzyme will cleave the DNA at this point in a reaction termed a restriction digest reaction. We will discuss in chapter 12 how researchers could distinguish the phenotype that results from the presence of this *Eco*RI sequence. The sequence of the *a* allele, in contrast, may be

CA<u>GAAC</u>TCTG

Notice that one phenotypic difference between the two alleles is in this DNA nucleotide sequence, with the *a* allele having a **C** (in bold) instead of a T. In this case, the *Eco*RI enzyme cannot cleave the altered DNA sequence. It is also possible to identify the absence of this *Eco*RI sequence.

Restriction fragment length polymorphism (RFLP) is a nucleotide change that results in the elimination or creation of a restriction enzyme site, as seen for the *A* and *a* alleles. In chapter 13, we will revisit the topic of recombination and the current technologies used to analyze this and other genetic elements, such as restriction enzymes and the separation of different DNA fragment sizes.

At this point, it is sufficient to realize that the RFLP represents two different alleles of a DNA sequence. These alleles can be mapped by recombination in the same manner as the mutations that we already described. It is possible to map the recombination distance between either two RFLPs or a gene that exhibits a morphological phenotype and an RFLP.

One advantage of the RFLP is that it does not have to be a part of a gene that produces an obvious phenotype. Because RFLPs do not have to be associated with genes, there may be many more RFLPs than there are genes in the genome. This large number of RFLPs should yield at least one RFLP near every gene, which will produce a small and accurate recombination distance between the RFLP and the gene.

You can now calculate the probability that Mary inherited the recessive *d* allele. Mary cannot be *d/d* because she is alive. Thus, there is a 66.7% chance that Mary is heterozygous. If this is confusing, see the discussion on recessive lethality in chapter 5.

If you had some information about the RFLP in the pedigree, the problem changes significantly. Assume that DNA testing revealed Mary's dead sister was homozygous for the *RFLP5.3* allele, each of Mary's sisters are either homozygous for the *RFLP8* allele or are heterozygous, and Mary is heterozygous for the RFLP alleles. Because the only individual who is homozygous for the *RFLP5.3* allele is dead, we assume that the *d* allele is linked to the *RFLP5.3* allele in Mary's family and the genotypes of both parents are *D RFLP8/d RFLP5.3*.

You can calculate the probability of each genotype that is heterozygous for the *D* allele. If the two loci are 4 mu apart, then 96% of the gametes will have a parental chromosome (*D RFLP8* or *d RFLP5.3*) and 4% of the gametes will have a recombinant chromosome (*D RFLP5.3* or *d RFLP8*). The four possible genotypes and the probability of each is as follows:

D RFLP8/d RFLP5.3 = 2 × (48% × 48%) = 46.08%

D RFLP5.3/d RFLP5.3 = 2 × (48% × 2%) = 1.92%

D RFLP8/d RFLP8 = 2 × (48% × 2%) = 1.92%

D RFLP5.3/d RFLP8 = 2 × (2% × 2%) = 0.08%

You multiply each of these genotypic probabilities by 2 because either parent can contribute the first chromosome in each genotype.

Summing the four probabilities gives

46.08% + 1.92% + 1.92% + 0.08% = 50%

which corresponds to the probability of two heterozygotes producing a heterozygote!

Because there is a 66.7% chance that Mary is a heterozygote, these genotypes must also total 66.7%. Remember, if Mary was homozygous for the recessive *d* allele, she would be dead. You can make the necessary conversion by determining the percentage each genotype is of the total amount of D/d heterozygotes. This produces the following probabilities:

D RFLP8/d RFLP5.3 = 66.7% × (46.08%/50%) = 61.47%

D RFLP5.3/d RFLP5.3 = 66.7% × (1.92%/50%) = 2.56%

D RFLP8/d RFLP8 = 66.7% × (1.92%/50%) = 2.56%

D RFLP5.3/d RFLP8 = 66.7% × (0.08%/50%) = 0.11%

If DNA testing revealed that Mary was heterozygous for the RFLP, she has two potential genotypes that will also be *D/d*. The first, *D RFLP8/d RFLP5.3*, results from each parent contributing a nonrecombinant chromosome, and the second, *D RFLP5.3/d RFLP8*, results from each parent contributing a recombinant chromosome. Using the sum rule, there is a 61.58% chance that if Mary is heterozygous for the RFLP, she will also be heterozygous for the *D* gene.

Notice that if the DNA test revealed Mary was homozygous for the *RFLP8* allele, there would be a 2.56% chance that she would also have the recessive *d* allele. Thus, the information on the RFLP alleles that Mary possesses greatly affects the calculated probability that she also possesses specific alleles at a linked locus.

It's Your Turn

When using an RFLP or any DNA sequence in a recombination mapping problem, treat it like a gene with two alleles. The absence of a morphological phenotype does not affect the principles underlying recombination mapping. You are now ready to attempt problem 58.

 Restating the Concepts

▶ Recombination mapping always requires that at least one of the parents is heterozygous for each gene being studied.

▶ Recombination distance is always the number of recombinant offspring divided by the total number of offspring.

▶ Mutations can be mapped by recombination. Because mutations are variations of DNA sequences, any DNA sequence, including RFLPs, can also be mapped if the individual is heterozygous for the sequence.

6.4 Haploid Mapping (Tetrad Analysis)

For *Drosophila* and other diploid eukaryotes, recombination mapping involves sperm and eggs, each of which carries only one chromatid of a meiotic tetrad, fusing to generate a diploid zygote. Zygotes, therefore, result from the random uniting of chromatids. Because we are unable to directly observe the phenotype of the meiotic products, we need to perform informative crosses, such as a testcross, to deduce the genotype of one of the gametes.

Fungi of the class *Ascomycetes* retain all four haploid products of a single meiosis in a sac called an **ascus.** These organisms provide a unique opportunity to directly examine the four products of a single meiosis, which is called a **tetrad.** Furthermore, these meiotic products, which are called **ascospores,** are not required to fuse to generate the individual expressing the phenotype that will be examined; the haploid meiotic products grow through several rounds of cell division (mitosis) to produce the mature haploid individual.

Having all four products of a single meiosis contained in one ascus allowed geneticists to determine such basics as the reciprocity of crossing over and that DNA replication occurs before crossing over. We will look at two fungi, the common baker's yeast, *Saccharomyces cerevisiae*, and pink bread mold, *Neurospora crassa*.

Phenotypes of Fungi

At this point, you might wonder what phenotypes fungi express. In general, microorganisms have phenotypes that fall into three broad categories: colony morphology, drug resistance, and nutritional requirements.

Many microorganisms can be cultured in petri plates or test tubes that contain a supporting medium such as agar, to which additional substances can be added (**fig. 6.19**). Wild-type *Neurospora* generally grows in a filamentous form, whereas yeast tends to form colonies. Mutations exist that change the growth morphology. In *Neurospora*, *fluffy* (*fl*), *tuft* (*tu*), *dirty* (*dir*), and *colonial* (*col4*) are all mutants of the basic filamentous growth form. In yeast, the *ade* gene causes the colonies to be red on specific medium.

In terms of drug resistance, wild-type *Neurospora* is sensitive to the sulfa drug sulfonamide, whereas one of its mutants (*Sfo*) actually requires sulfonamide to survive and grow. Yeast shows similar sensitivities to antifungal agents.

Nutritional requirement phenotypes provide great insight not only into genetic analysis but also into the biochemical pathways of metabolism (see chapter 11). Wild-type *Neurospora* (called **prototrophs**) can grow on a medium containing only sugar, a nitrogen source, some organic acids and salts, and the vitamin biotin. This is referred to as **minimal medium.** However, several different *Neurospora* mutants (called **auxotrophs**) cannot grow on minimal medium unless some additional, essential nutrient is added. For example, one mutant strain will only grow on minimal medium if the amino acid arginine is added (**fig. 6.20**). From this we can infer that the wild-type has a normal, functional enzyme that is required to synthesize arginine. The arginine-requiring mutant has an allele that

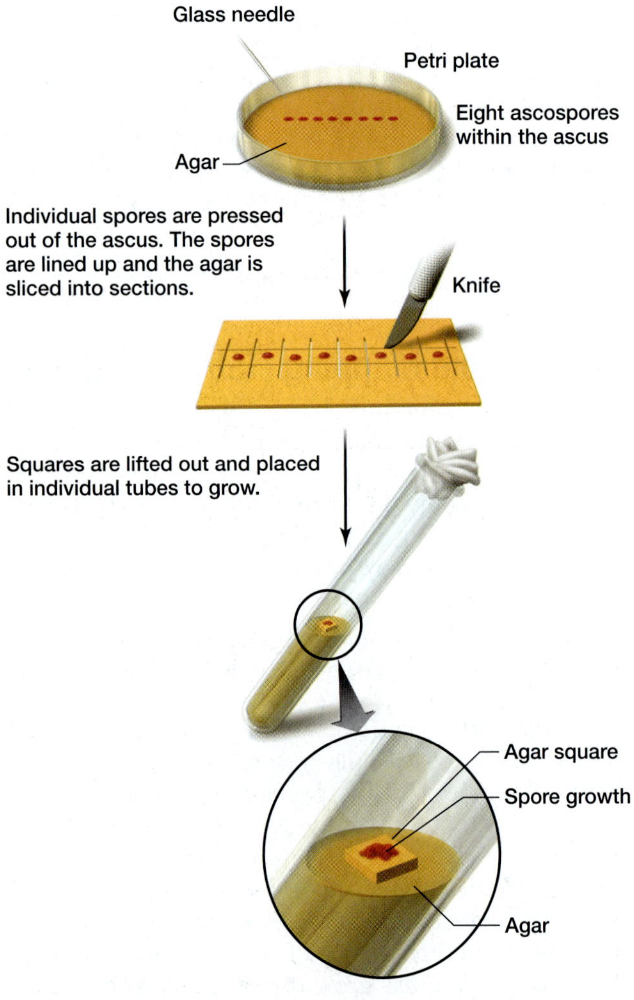

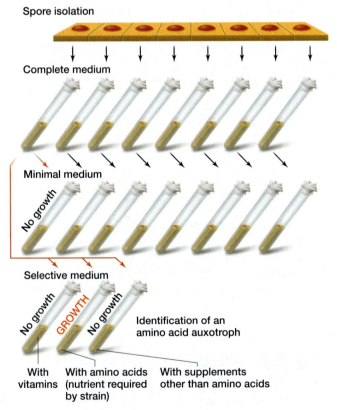

figure 6.20 Isolation of *Neurospora* nutritional mutants. The absence of growth on minimal medium (second row) indicates the presence of a nutritional mutant or auxotroph. The auxotroph is then picked from the corresponding complete medium tube and then placed on minimal medium with specific nutrients (third row). Once the correct supplement is identified to support growth of the auxotroph on minimal medium, nutrient-requiring strains can be cultured on supplemented medium for future use.

figure 6.19 *Neurospora* spores can be isolated from an ascus and cultured to reveal the phenotype of each spore.

lacks this functional enzyme, so arginine cannot be produced. If the synthetic pathway is long, then many different loci may have alleles that cause the strain to require arginine (**fig. 6.21**). This is the case, and the different loci are named arg_1, arg_2, and so on. Yeast and *Neurospora* contain numerous biosynthetic pathways, and mutants exhibit many different nutritional requirements. Mutants can be induced experimentally by radiation or by chemicals and other treatments.

These, then, are the tools we use to analyze and map the chromosomes of microorganisms, including yeast and *Neurospora*.

Unordered Spores (Yeast)

Baker's, or budding, yeast, *Saccharomyces cerevisiae*, exists in either a haploid or diploid state, depending on the point in its life cycle (**fig. 6.22**). The haploid form is usually produced under nutritional stress (starvation). When better conditions return, haploid cells of the two sexes, called **a** and **α-mating types,** fuse to form the diploid. The haploid state is regenerated from a diploid cell through meiosis under starvation conditions.

Let's consider a mapping problem. When an *a b* spore (or gamete) fuses with an $a^+ b^+$ spore (or gamete), the resulting diploid cell can then undergo meiosis. The four meiotic products from a single diploid

cell (tetrad) are maintained together in an ascus. The meiotic products from one ascus can be isolated and grown as haploid colonies, which are then observed for the phenotypes the two loci control.

Only three types of asci are observed, which are grouped by the genotype of the haploid cells (**table 6.2**). The class 1 ascus has two types of spores, which are identical to the parental haploid genotypes and is referred to as a **parental ditype (PD).** The second ascus class also contains spores with only two genotypes, but they are recombinants. This ascus type is referred to as a **nonparental ditype (NPD).** The third ascus class has spores with all four possible genotypes and is referred to as a **tetratype (TT).**

All three ascus types can be generated regardless of whether the two loci are linked. If the loci are linked (**fig. 6.23**), parental ditypes come from the lack of a

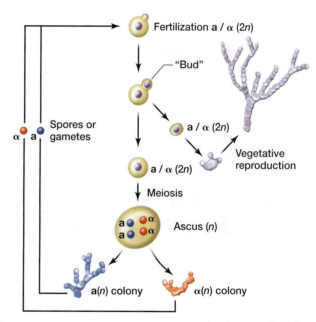

figure 6.22   The life cycle of yeast. Mature cells are mating types **a** or α; *n* is the haploid stage; 2*n* is diploid.

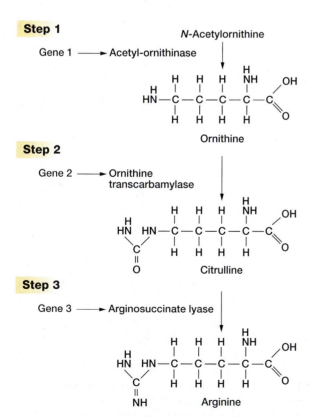

figure 6.21   Simplified biosynthetic pathway of arginine in *Neurospora*. The identity of the enzymes that catalyze each biochemical step and the intermediate molecules are shown.

table 6.2   The Three Ascus Types in Yeast Resulting from Meiosis in a Dihybrid, $a^+ b^+/a\ b$

1	2	3
Parental ditype (PD)	**Nonparental ditype (NPD)**	**Tetratype (TT)**
a b	*a b$^+$*	*a b*
a b	*a b$^+$*	*a b$^+$*
a$^+$ b$^+$	*a$^+$ b*	*a$^+$ b*
a$^+$ b$^+$	*a$^+$ b*	*a$^+$ b$^+$*
75	5	20

crossover, whereas nonparental ditypes come about from double crossovers involving all four chromatids. Thus, we expect parental ditypes to be *significantly more numerous* than nonparental ditypes for linked loci. If the loci are not linked, however, then both parental and nonparental ditypes are produced through independent assortment—and they should occur in equal frequencies. We can therefore determine whether the loci are linked by comparing the numbers of parental ditypes and nonparental ditypes.

In table 6.2, the parental ditypes greatly outnumber the nonparental ditypes; the two loci are, therefore, linked. The next question is, What is the map distance between the loci?

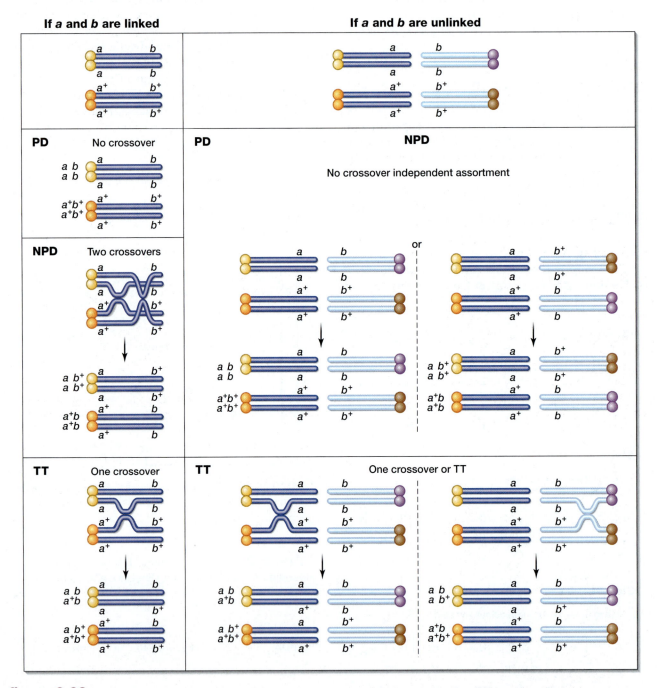

figure 6.23 Formation of parental ditype (PD), nonparental ditype (NPD), and tetratype (TT) asci in a dihybrid yeast by recombination or independent assortment. For linked genes (left column), the parental ditype (PD) is produced from no crossovers, the nonparental ditype (NPD) is produced from two crossovers, and the tetratype (TT) is produced from one crossover. If the two genes are unlinked (right column), both the PPD and NPD are produced from no crossovers through independent assortment. The TT is also produced by a single crossover between the centromere and one of the two genes. Filled circles represent centromeres.

Figure 6.23 shows that in a nonparental ditype, all four chromatids are recombinant. Tetratypes are primarily generated by one crossover (fig. 6.23), in which only half the tetratype chromatids are recombinant. Remembering that 1% recombinant offspring equals 1 map unit, we can use the following formula:

Map units =

$$\frac{[(1/2)(\text{number of TT asci}) + \text{number of NPD asci}]}{\text{total number of asci}} \times 100\%$$

For the data of table 6.2, then:

$$\text{Map units} = \frac{[(1/2)20 + 5]}{100} \times 100\% = \frac{[10 + 5]}{100} \times 100\% = 15 \text{ mu}$$

Therefore, the a and b loci are 15 mu apart.

Ordered Spores (Neurospora)

The *Neurospora* life cycle is shown in **figure 6.24**. Fertilization takes place within an immature fruiting body after a spore or filament of one mating type contacts a special filament extending from the fruiting body of the opposite mating type (mating types are referred to as *A* and *a*). Unlike yeast, *Neurospora* does not have a diploid phase in its life cycle. Rather, the zygote undergoes meiosis immediately after the diploid nucleus forms.

Because the *Neurospora* ascus is narrow, the meiotic spindle is forced to lie along the cell's long axis. The two nuclei that are produced from the first meiotic division are also oriented along the long axis of the ascus for the second meiotic division. The result is that the spores are ordered in the ascus according to the positions of their centromeres (fig. 6.25). That is, at the end of meiosis in *Neurospora*, the four ascospores are in the order *A A a a* or *a a A A*.

It is simpler to discuss separation of centromeres rather than chromosomes or chromatids because of the complications associated with crossing over. The *A* centromere is always an *A* centromere, but the chromosome attached to that centromere may be partly from the *A* parent and partly from the *a* parent due to crossing over.

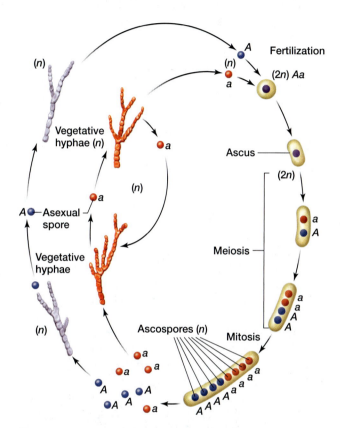

figure 6.24   Life cycle of *Neurospora*. *A* and *a* represent the two different mating types; *n* is a haploid stage; 2*n* is diploid.

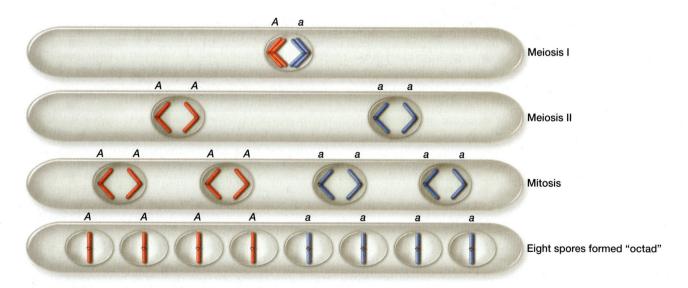

figure 6.25   Meiosis in *Neurospora*. Although *Neurospora* has seven pairs of chromosomes at meiosis, only one pair is shown. *A* and *a*, the two mating types, represent the two centromeres of one homologous chromosome pair.

Before the ascospores mature in *Neurospora,* a single round of mitosis takes place in each nucleus so that four *pairs,* rather than just four spores, are formed. The eight haploid cells that are produced from a single diploid cell in *Neurospora* are called an **octad.** In the absence of mutation or gene conversion, which will be discussed later in this chapter, pairs are always identical (see fig. 6.25). As you will see in a moment, because of the ordered spores, we can map loci in *Neurospora* in relation to their centromeres.

First- and Second-Division Segregation

At the end of the single round of mitosis following meiosis, a 4:4 segregation of centromeres has taken place in the ascus. Two kinds of patterns appear among the loci on these chromosomes, depending on whether there was a crossover between the locus and its centromere (**fig. 6.26**).

If no crossover takes place between the locus and the centromere, the allelic pattern is the same as the centromeric pattern. This result is referred to as **first-division segregation (FDS),** because the alleles (*a* and a^+) separate from each other at the end of meiosis I. But if a crossover occurs between the locus and the centromere, patterns of different types emerge (2:4:2 or 2:2:2:2). These patterns are referred to as **second-division segregation (SDS),** because the alleles do not completely separate from each other until the end of meiosis II.

Because the spores are ordered, the centromeres always follow an FDS pattern. For this reason, we should be able to map the distance from any locus to the centromere of the chromosome on which it occurs. Under the simplest circumstances (see fig. 6.26), every SDS octad has four recombinant and four nonrecombinant chromatids (spores). Thus, half of the chromatids (spores) in an SDS ascus are recombinant. Because 1% recombinant chromatids equal 1 map unit,

$$\text{Map units} = \frac{(1/2) \times (\text{number of SDS asci})}{\text{total number of asci}} \times 100\%$$

An example using this calculation appears in **table 6.3**.

Gene Order

In a dihybrid cross, we can determine the distance from each locus to its centromere and the link-

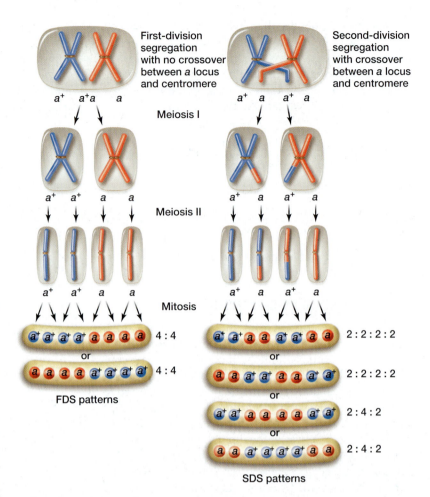

First-division segregation with no crossover between *a* locus and centromere

Second-division segregation with crossover between *a* locus and centromere

Meiosis I

Meiosis II

Mitosis

4:4

or

4:4

FDS patterns

2:2:2:2

or

2:2:2:2

or

2:4:2

or

2:4:2

SDS patterns

figure 6.26 The six possible *Neurospora* ascospore patterns with respect to one locus. With no crossing over (left column) the first-division segregation (FDS) patterns are produced. If crossing over occurs between the gene and centromere (right column) second-division segregation (SDS) patterns are produced.

age arrangement of the loci relative to each other. In **table 6.4**, the meiotic products are shown for a cross of two haploids, $a\ b \times a^+\ b^+$. We can establish whether the two loci are linked to each other—and therefore to the same centromere—by comparing the number of parental ditypes (classes 1 and 5) with nonparental ditypes (classes 2 and 6) in table 6.4. If the two loci are unlinked, then the total number of parental ditypes and nonparental ditypes should be very similar. Because the total number of parental ditypes (879) greatly outnumbers the nonparental ditypes (3), we can be sure the two loci are linked.

To determine the distance of each locus to the centromere, we calculate one-half the percentage of SDS patterns for each locus. For the *a* locus, classes 4, 5, 6, and 7 are SDS patterns. For the *b* locus, classes 3, 5, 6, and 7 are. Therefore, the distances to the centromere, in map units, for each locus are

table 6.3   Genetic Patterns Following Meiosis in an $a^+ a$ Heterozygous *Neurospora* (Ten Asci Examined)

Spore Number	Ascus Number									
	1	2	3	4	5	6	7	8	9	10
1	a	a	a^+	a	a	a^+	a	a^+	a^+	a^+
2	a	a	a^+	a	a	a^+	a	a^+	a^+	a^+
3	a	a	a^+	a^+	a^+	a^+	a	a	a	a^+
4	a	a	a^+	a^+	a^+	a^+	a	a	a	a^+
5	a^+	a^+	a	a^+	a	a	a^+	a	a^+	a
6	a^+	a^+	a	a^+	a	a	a^+	a	a^+	a
7	a^+	a^+	a	a	a^+	a	a^+	a^+	a	a
8	a^+	a^+	a	a	a^+	a	a^+	a^+	a	a
	FDS	FDS	FDS	SDS	SDS	FDS	FDS	SDS	SDS	FDS
Number	10	11	8	9	9	9	12	10	12	10

Note: Map distance (*a* locus to centromere)   = (1/2)% SDS
                                      = (1/2) 40%
                                      = 20 map units
FDS: first-division segregation; SDS: second-division segregation

table 6.4   The Seven Unique Classes of Asci Resulting from Meiosis in a Dihybrid *Neurospora*, $a\ b/a^+\ b^+$

Spore Number	Ascus Number						
	1	2	3	4	5	6	7
1	$a\ b$	$a\ b^+$	$a\ b$	$a\ b$	ab	$a\ b^+$	$a\ b$
2	$a\ b$	$a\ b^+$	$a\ b$	$a\ b$	ab	$a\ b^+$	$a\ b$
3	$a\ b$	$a\ b^+$	$a\ b^+$	$a^+\ b$	a^+b^+	a^+b	$a^+\ b^+$
4	$a\ b$	$a\ b^+$	$a\ b^+$	$a^+\ b$	a^+b^+	$a^+\ b$	$a^+\ b^+$
5	$a^+\ b^+$	$a^+\ b$	$a^+\ b^+$	$a^+\ b^+$	$a^+\ b^+$	$a^+\ b$	a^+b
6	$a^+\ b^+$	$a^+\ b$	$a^+\ b^+$	$a^+\ b^+$	a^+b^+	$a^+\ b$	$a^+\ b$
7	$a^+\ b^+$	$a^+\ b$	$a^+\ b$	$a\ b^+$	$a\ b$	$a\ b^+$	$a\ b^+$
8	$a^+\ b^+$	$a^+\ b$	$a^+\ b$	$a\ b^+$	$a\ b$	$a\ b^+$	$a\ b^+$
Number of each ascus type	729	2	101	9	150	1	8
SDS for *a* locus?	No	No	No	Yes	Yes	Yes	Yes
SDS for *b* locus?	No	No	Yes	No	Yes	Yes	Yes
Ascus segregational type	PD	NPD	TT	TT	PD	NPD	TT

NPD: nonparental ditype; PD: parental ditype; SDS: second-division segregation; TT: tetratype

a to centromere distance =

$$\frac{(1/2)(9 + 150 + 1 + 8)}{1000} \times 100\% = 8.4 \text{ mu}$$

b to centromere distance =

$$\frac{(1/2)(101 + 150 + 1 + 8)}{1000} \times 100\% = 13.0 \text{ mu}$$

Unfortunately, these two distances do not provide a unique determination of gene order. In **figure 6.27**, we see that two alternatives are possible: One alternative has a map distance between the loci of 21.4 mu; the other has a distance of 4.6 mu. How do we determine which of these is correct?

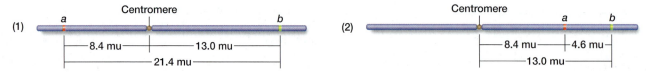

figure 6.27 Two possible gene maps of the *a* and *b* loci with respect to the centromere. In both maps, the distance from the centromere to gene *a* and from the centromere to gene *b* remains constant. However, the two maps differ in the distance from gene *a* to gene *b*.

The simplest way is to calculate the *a*–*b* distance using the equation we discussed with unordered spores. That is, the map distance is

Map units

$$= \frac{(1/2)(\text{number of TT asci}) + \text{number of NPD asci}}{\text{total number of asci}} \times 100\%$$

$$= \frac{(1/2)(118) + 3}{1000} \times 100\%$$

$$= 6.2 \text{ mu}$$

Because 6.2 mu is much closer to the *a*–*b* distance expected if both loci are on the same side of the centromere, the second alternative in figure 6.27 is most likely correct.

Occasionally, the two maps may be very similar and differ only by whether both genes are on the same side of the centromere or the centromere lies between the two genes. In this case, you may use a qualitative assessment of the gene and centromere order. If two genes are on the same side of the centromere, then an SDS pattern for the gene closer to the centromere will usually also be an SDS pattern for the farther gene (because a crossover between the centromere and the closer gene is also a crossover between the centromere and the farther gene).

You first determine which gene is closer to the centromere based on the SDS patterns. For the data in table 6.4, this corresponds to the *a* gene (8.4 mu versus 13.0 mu). Because 159 of the 168 SDS asci for the *a* gene are also SDS for the *b* gene, there is a **coincidence of SDS patterns,** which implies that both genes are on the same side of the centromere. This is also consistent with the second map shown in figure 6.27.

It's Your Turn

When generating a linkage map from tetrad data, first determine if you are working with an ordered or unordered tetrad to ensure that you are calculating appropriate distances. Then determine if the genes are linked (PD>>NPD) before calculating distances. You are now ready to try problems 35 and 42.

Restating the Concepts

▶ In all haploids, two genes are linked if the number of PD asci are significantly greater than the NPD asci.

▶ In all haploids, it is possible to determine the recombination distance between two loci by adding half the percentage of tetratype asci and the percentage of NPD asci.

▶ In an ordered ascus (*Neurospora*), the position of the haploid gametes in the ascus is dependent on the location of the centromeres during meiosis. The distance from the centromere to a locus is half the frequency of SDS patterns. The distance from centromere to a locus cannot be calculated in an unordered ascus (yeast).

6.5 The Mechanism Of Recombination

We described earlier how Creighton and McClintock used cytologically distinguishable homologous chromosomes in corn to demonstrate that recombination involves physical exchange of chromosome regions (see fig. 6.7). However, elucidating the actual mechanism of recombination required the analysis of asci in fungi. We will first explore the unusual asci types that were critical in this analysis and then describe how these data yielded a model for recombination.

Unusual Asci: Evidence for a Recombination Mechanism

As discussed earlier, the linear arrangement of the octad in the *Neurospora* ascus corresponds to the position of their chromosomes' centromeres. Two segregation patterns can occur, namely, the FDS pattern of 4:4, and the SDS pattern, which appears as either 2:4:2 or 2:2:2:2.

In *Neurospora*, we also can detect unusual segregation patterns at a very low frequency. These unusual ratios include 3:1:1:3, 5:3, and 6:2. First, we will discuss how the 3:1:1:3 ratio is generated, which will help reveal the recombination mechanism. We will then apply this knowledge to the latter two ratios in a process called *gene conversion*.

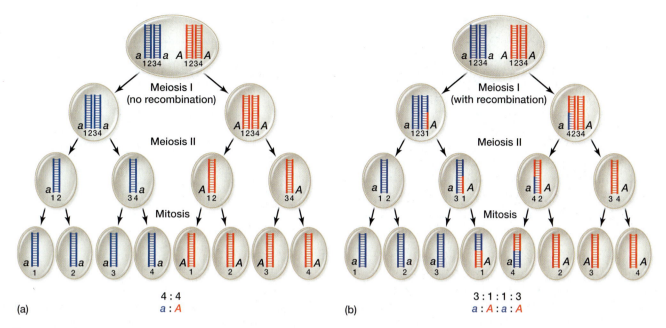

4 : 4
a : *A*
(a)

3 : 1 : 1 : 3
a : *A* : *a* : *A*
(b)

figure 6.28 The segregation of DNA strands during meiosis, followed by mitosis, in generating a *Neurospora* ascus. (a) In the absence of recombination, the octad contains a 4:4 ratio of *a* to *A*. (b) If recombination between nonsister chromatids occurs, only a single strand of the double-stranded DNA chromatid actually crosses over to produce a heteroduplex DNA molecule (labeled 3 1 and 4 2 in the middle two meiotic products) to generate the 3:1:1:3 phenotypic ratio in the octad.

The first unusual ratio (3:1:1:3) contains an equal number of both alleles, just as do the standard ratios (4:4 or 2:4:2). The first and fourth meiotic products proceed through mitosis, and each generates two daughter cells that contain the same alleles (**fig. 6.28b**). However, the second and third meiotic products each produce two mitotic daughter cells that contain different alleles.

Remember the haploid meiotic products undergo one round of DNA replication to produce two DNA molecules that will segregate away from each other during mitosis. The only way a haploid ascospore in the tetrad can produce two different daughter spores in the octad is if the double-stranded DNA molecule in the meiotic product contains two different sequences (see fig. 6.28). **Heteroduplex DNA** is a double-stranded DNA molecule in which the two strands are not completely complementary. This implies that crossing over between homologs during meiosis I actually involves only a single strand of the double-stranded DNA molecule to produce heteroduplex DNA. The previous diagrams (such as fig. 6.26) that show a breakage and rejoining of the double-stranded DNA on both nonsister chromatids was not entirely correct because this could not produce the heteroduplex DNA molecule required to yield the 3:1:1:3 ratio.

How are the 5:3 and 6:2 ratios produced? This is surprising because it results in an increased number of one allele (from four copies in the octad to either five or six) and less of the other allele. This is in contrast to the 3:1:1:3 ratio, which retains four copies of each allele.

Starting with the heteroduplex DNA, the cell can proceed through meiosis and mitosis to produce the 3:1:1:3 ratio as described previously. Alternatively, the cell can recognize the heteroduplex region as an abnormality and repair the mismatches in the DNA sequence. The cell's repair system recognizes one strand as the "correct" strand and "repairs" the other strand. Depending upon which strand is selected as the "correct" strand, a CA mismatched base pair can be converted to either a CG or a TA base pair (**fig. 6.29**). If the original base pair had been TA, then repair of the CA heteroduplex to either CG or TA would affect the allele frequency in either the tetrad or octad.

Gene conversion is the alteration of progeny ratios that indicate one allele has been converted to another (**fig. 6.30**). This phenomenon is observed in up to 10% of yeast asci. The mismatched AC is changed to an AT or a GC base pair; the mismatched GT base pair is changed to AT or GC. The result of the repair, as shown at the bottom of figure 6.30, can be gene conversion in which an expected ratio of 2:2 in the tetrad is converted to a 3:1 ratio or a 1:3 ratio of meiotic products. After one round of DNA replication and mitosis, the octad would have either a 6:2 or 2:6 ratio.

In some cases, one of the heteroduplexes is repaired and the other is not. This again produces a situation in which a meiotic product will generate two daughter cells with different genotypes. This scenario yields the 5:3 ratio.

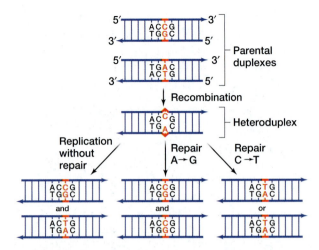

figure 6.29 Fate of heteroduplex DNA. Recombination produces heteroduplex DNA with mismatched bases. Replication without repair yields two different daughter molecules each with a different parental sequence (left). Repair converts the mismatched base pair to one or the other normal base pair. This produces two double-stranded DNA molecules with either the first parental sequence (center) or the second parental sequence (right).

Holliday Structure and Resolution

These unusual ratios reveal that only a single strand of the double-stranded DNA molecule is involved in generating a heteroduplex during recombination.

Figure 6.31 shows a single strand invading the non-sister chromatid and generating heteroduplex regions on both homologs. As you can see, the heteroduplex region is flanked by homoduplex regions on both homologs. This **Holliday structure** was first proposed by Robin Holliday in 1964 to account for the products observed in recombination. It is possible to rotate the lower DNA molecule 180° to convert the crossover portion of the diagram to a more completely base-paired structure. Both of these structures are equivalent, but the rotated structure will better visualize the next event.

To complete recombination, the nonsister chromatids must be separated. This is done by cutting two strands on the Holliday structure and rejoining the cleaved ends to produce two separate chromatids in a process called **resolution**. It is possible to cut the Holliday structure two ways, either horizontally (at the h sites) or vertically (at the v sites).

Cutting horizontally and rejoining the ends produces *AB* and *ab* chromatids (see fig. 6.31). Because these are the parental allele arrangements, recombination does not appear to have occurred, except for the presence of a heteroduplex region in both chromatids. Once again, we see a situation in which the frequency of crossing over between markers will be underestimated. Vertical resolution of the Holliday structure produces the *Ab* and *aB* recombinant chromatids that flank the heteroduplex region (see fig. 6.31).

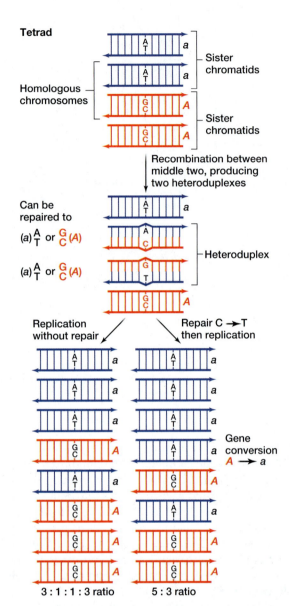

figure 6.30 Recombination and repair can cause gene conversion. During recombination, heteroduplex DNA is formed. If these mismatches are not repaired before DNA replication (left side), the resulting ascus possesses a 3:1:1:3 ratio. If the mismatch is randomly repaired before DNA replication (right side), the resulting ascus has a 5:3 ratio (right side). Because there is an increase in one of the alleles (5 *a* rather than 4), gene conversion took place. Since the repair enzymes randomly convert the mismatch to a complementary base pair, an AC base pair can be converted to either an AT (*a* allele) or a GC (*A* allele) base pair. Thus, the repair will convert a 3:1:1:3 ratio into either 5:3 (if one mismatch is repaired), 6:2 ratio (if both mismatches are repaired to the same base pair), or a 4:4 ratio (if each mismatch is repaired into a different base pair).

This mechanism of single DNA strands exchanging between nonsister chromatids to generate a Holliday structure in recombination is largely based on the rare 3:1:1:3 ratio. This ratio is rare because it can only be observed when the gene being analyzed is located in

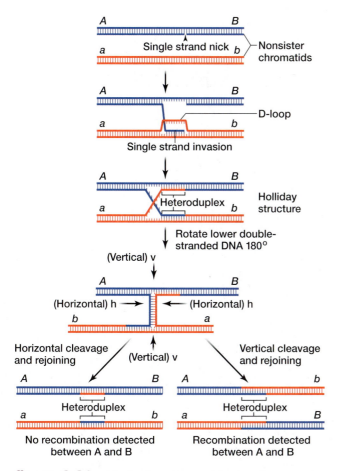

figure 6.31 Mechanism of recombination. One strand of a DNA molecule is nicked, which permits a single strand to invade the nonsister chromatid. Strand invasion produces a single-stranded D-loop that can be nicked. Ligation between nonsister chromatid strands produces two heteroduplexes, which is the Holliday structure. The Holliday structure can be rotated 180° and resolved by either horizontal (h) or vertical (v) cleavage. Horizontal cleavage and ligation produces two chromatids that contain heteroduplex regions that are flanked by the parental allele arrangements. Vertical cleavage and ligation yields two chromatids that contain heteroduplex regions that are flanked by recombined allele arrangements.

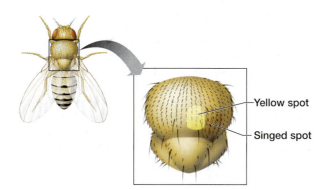

figure 6.32 Mitotic recombination produces twin spots on the thorax of a female *Drosophila*. A *singed* (*sn*) and *yellow* (*y*) dihybrid fly in the *trans* configuration can undergo recombination during mitosis to produce a patch with the yellow phenotype adjacent to a patch with the singed bristle phenotype.

the heteroduplex region. Because all the gametes of a single meiosis cannot be analyzed in a diploid organism, we are left hypothesizing that this recombination mechanism occurs in all eukaryotes.

Restating the Concepts

▶ Unusual spore ratios in ordered asci reveal that recombination involves a single-strand invasion (or exchange) between nonsister chromatids that produces a heteroduplex region.

▶ A Holliday structure interlocks the two nonsister chromatids that recombined. The Holliday structure can be cut and rejoined in one of two orientations, one that results in the parental arrangement of flanking alleles and the other produces the recombinant arrangement.

▶ Horizontal resolution of the Holliday structure is another reason why observed recombination distances underestimate the actual recombination distance.

6.6 Additional Recombination Events

We discussed crossing over as a process that normally occurs between nonsister chromatids during meiosis. However, it also occurs in mitotic cells at a very low rate and between sister chromatids. These additional recombination processes have proven useful in mapping genes in some organisms and in the evolution of the genome.

Somatic (Mitotic) Crossing Over

In the fungus *Aspergillus nidulans*, mitotic crossing over occurs about once in every hundred cell divisions. Mitotic recombination apparently occurs when two homologous chromatids come to lie next to each other and breakage and reunion follow, most likely as a consequence of DNA repair (see chapter 18) or irradiation.

Mitotic recombination was discovered in 1936 by Curt Stern, who noticed the occurrence of *twin spots* in *Drosophila* that were dihybrid for the *yellow* allele (*y*) for body color and the *singed* allele (*sn*) for bristle morphology (**fig. 6.32**). A twin spot could be explained by mitotic crossing over between the *sn* locus and its centromere (**fig. 6.33**). A crossover between the *sn* and *y* genes would produce only a yellow spot. (Verify this for yourself.)

These two phenotypes were found in the expected relative frequencies. That is, given that the gene locations are drawn to scale in figure 6.33 (with the centromere to *sn* distance being greater than the *sn* to *y* distance), we would expect twin spots to be more frequent than yellow spots.

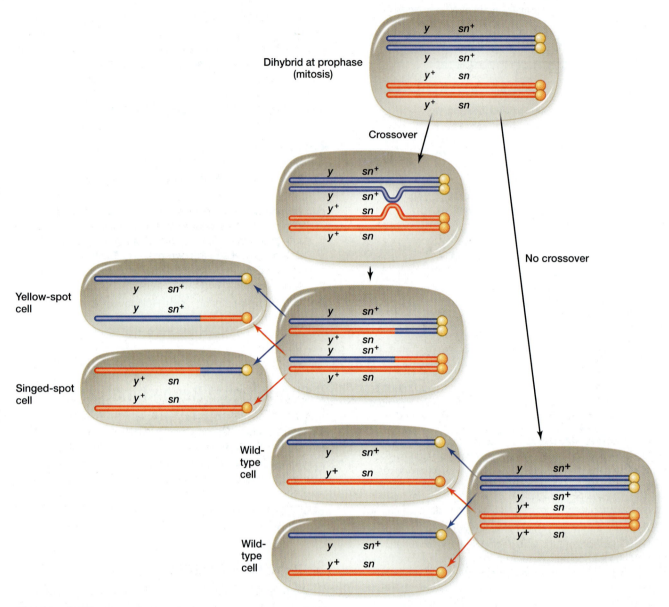

figure 6.33 Formation of twin spots by somatic crossing over. Homologous chromatids undergo crossing over (left side) between the centromere and closest gene (*sn*). The colored arrows show the direction in which the centromeres separate. In the absence of recombination (right side), mitosis yields cells with the same genotype as the parental cell.

Mitotic crossing over has been used in fungal genetics as a method for determining linkage relations. Although gene orders are consistent between mitotic and meiotic mapping, relative distances are usually not. This is not totally unexpected because neither meiotic nor mitotic crossing over is uniform along a chromosome.

Sister Chromatid Exchange

Several types of experiments have demonstrated that crossing over can occur between sister chromatids, which is called **sister chromatid exchange (SCE).** Because the recombination is between sister chromatids, which are genetically identical, new allelic combinations are not usually produced.

How do we know that SCEs occur? If a cell is grown in the presence of bromodeoxyuridine, it will be incorporated into replicating DNA in place of thymidine (DNA replication is described in chapter 9). After two rounds of DNA replication, one sister chromatid will have bromodeoxyuridine in both DNA strands and the other will have it in only one strand (**fig. 6.34**). Staining these chromosomes causes the chromatid with bromodeoxyuridine in both strands to fluoresce less (appear darker) than the sister chromatid containing bromodeoxyuridine in only one strand.

Examining the fluorescence of these mitotic chromosomes reveals that individual sister chromatids contain both bright and dim fluorescing segments (**fig. 6.35**). Because the fluorescence is complementary

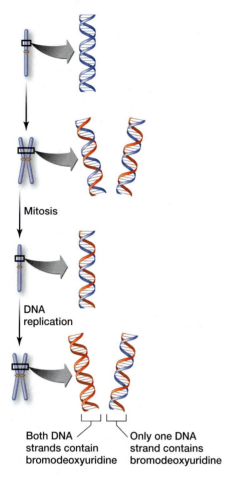

Both DNA strands contain bromodeoxyuridine   Only one DNA strand contains bromodeoxyuridine

figure 6.34   DNA is labeled by incorporating bromodeoxyuridine during DNA replication. After one round of DNA replication, each chromatid contains bromodeoxyuridine (in place of some thymines) in only the newly synthesized strand (red). After two rounds of DNA replication, one sister chromatid contains bromodeoxyuridine in both strands, and the sister chromatid contains bromodeoxyuridine in only one strand.

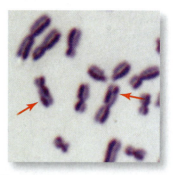

figure 6.35   Visualization of sister chromatid exchanges (SCEs) by fluorescing chromatids. The dark chromatids contain bromodeoxyuridine in only one strand of the DNA, and the lighter chromatids contain bromodeoxyuridine in both DNA strands (see fig. 6.34). Notice that sister chromatids contain complementary fluorescing segments: where one chromatid is dim, the sister chromatid is dark. The SCEs are found at the points where a chromatid changes from light to dark (marked with red arrows).

Restating the Concepts

▶ Somatic, or mitotic, crossing over occurs rarely between homologous chromosomes. It can produce a mosaic phenotypic pattern (single or twin spots) in a multicellular organism.

▶ Sister chromatids can also recombine in somatic cells. This appears to be an important process to repair damaged DNA and the generation of gene families. A variety of abnormalities result from the increased frequency of SCEs in Bloom syndrome.

6.7 Human Chromosomal Maps

Generation of a complete human genetic map will help identify the defective genes for inherited diseases and improve prenatal diagnosis of genetic diseases. A large number of human genetic diseases are known, as are the location and identity of most of the human genes (through the Human Genome Project). However, identifying the defective gene that is associated with a genetic disease requires significantly more knowledge than simply the DNA sequence of the genome. The disease defect (mutation) must first be localized to a specific chromosome and then a small region of the chromosome before we can attempt to identify the defective gene.

Let's consider retinitis pigmentosa, which is a group of inherited human diseases that initially cause night blindness and may progress to complete blindness at later ages. Retinitis pigmentosa can be either a dominant or a recessive disease, and the 24 different loci that cause this form of blindness have been mapped to the X chromosome and 11 different autosomes.

between sister chromatids, crossing over must have occurred between them.

Two potential functions have been associated with SCEs. First, the process of SCE appears to be important in repairing DNA damage. This is suggested by the increased frequency of SCEs by compounds that damage DNA, such as irradiation or some mutagens. Second, SCEs are thought to play an important role in the evolution of the genome through the creation of gene families. We will discuss gene families in chapter 7.

The genetic disease **Bloom syndrome** results in an increased amount of SCE in human somatic cells. This rare recessive syndrome results in prenatal and postnatal growth retardation, an increased sensitivity of the skin to sunlight, a predisposition to tumors, and behavioral abnormalities. The human chromosomes in somatic cells exhibit increased instability and a number of breaks. This suggests that abnormalities in the frequency of SCEs can be deleterious to an organism.

Retinitis pigmentosa 17 (*RP17*) is an autosomal dominant disease. The *RP17* locus was mapped to chromosome 17 in 1995. In 1999, the *RP17* locus was mapped to a 1.0-cM region using two different South African pedigrees. Mapping DNA markers further localized the *RP17* gene to a 6-megabase (Mb) (6000-kilobase, [kb]) region. The human genome sequence defined nine genes within this region.

To identify the *RP17* gene, the nine genes were sequenced in several affected individuals and their unaffected siblings. This revealed that a mutation in a carbonic anhydrase gene (*CA4*) caused *RP17*. Studying the mutant and wild-type carbonic anhydrase enzymes revealed potential mechanisms for the blindness in *RP17* individuals. With this information, potential therapies, including gene therapy, are being developed to combat this disease.

Without the linkage map, it would have been very difficult to identify what gene in the human genome, or even on chromosome 17, corresponded to *RP17*. Using the mapping data, researchers narrowed their search to only nine potential candidate genes out of the approximately 1500 that are likely located on chromosome 17. Linkage maps therefore provide an important companion to the large body of DNA sequence data being generated to determine the location of genes.

We can map human genes, mutations, and DNA sequences to chromosomes in the same manner as described earlier in this chapter. However, the inability to make specific crosses coupled with the relatively small number of offspring make the analysis of this data more difficult in humans.

Despite these problems, significant progress has been made, especially in assigning genes to the X chromosome. As the pedigree analysis in chapter 4 showed, X chromosomal traits have unique patterns of inheritance that easily identify loci on the X chromosome. Additionally, there are fewer restrictions on the parental genotypes to mapping X-linked genes. Currently over 300 genetic disease loci are known to be on the X chromosome, which is thought to contain 1098 genes.

X Linkage

After determining that a human gene is X-linked, the next problem is to determine the position of the locus on the X chromosome. Sometimes investigators do this with the proper pedigrees, if crossing over can be ascertained.

An example of this "grandfather method" appears in **figure 6.36**. In this example, a grandfather has one of the X-linked traits in question (here, color blindness). We then find that he has a grandson who is deficient in glucose-6-phosphate dehydrogenase (G6PD). From this we can infer that the boy's mother was dihybrid for the two alleles (colorblindness and G6PD deficiency) in the *trans*-configuration. The reasoning is that she must have inherited the only X chromosome that her father had,

which contained the color blindness mutation. Because the grandfather was not G6PD-deficient, the mother must have the G6PD-deficient allele on the X chromosome that she inherited from her mother.

The two grandsons on the left in figure 6.36 are nonrecombinant, because each has only one of the two mutations. However, the two grandsons on the right must both be recombinants, since they are both phenotypically different from their grandfather (one has both mutations and the other has neither). Theoretically, we can determine the map distance between these two mutations by simply totaling the number of recombinant grandsons and dividing by the total number of grandsons.

The methodology would be the same if the grandfather is both color-blind and G6PD deficient. The mother would then be dihybrid in the *cis*-configuration, and the recombinant sons would have only one of the two mutations. The point is that the grandfather's phenotype gives us information that allows us to infer that the mother was dihybrid, as well as telling us the *cis-trans*-arrangement of her alleles. We can then score her sons as either recombinant or nonrecombinant.

Because we only examine the grandsons, the genotype of their father is irrelevant. (Do you see why?) We can identify a larger number of informative pedigrees because we only need to identify those in which the mother is heterozygous for the two mutations.

To circumvent the small number of children in human families, we can total the number of recombinant grandsons from all the informative pedigrees we can find and divide by the total number of grandsons in these pedigrees to generate the map distance between two mutations.

Autosomal Linkage

As you have just seen, it is relatively easy to map the human X chromosome because we only need the mother to be heterozygous for the two mutations, and the genotype of the father is irrelevant. Recombination mapping on the autosomes, by contrast, is significantly more difficult.

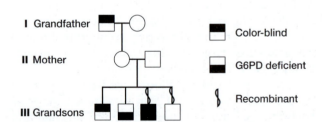

figure 6.36   Determining the frequency of recombination for X-linked genes. The grandfather expresses an X-linked recessive phenotype (color blindness) and the grandmother is heterozygous for a different X-linked trait (G-6-PD deficiency). The mother is heterozygous for both traits. The recombinant sons either express both traits or neither trait.

Pedigrees can provide us with sufficient information to determine whether two loci are linked. In **figure 6.37**, the nail–patella syndrome includes, among other things, abnormal nail growth coupled with the absence or underdevelopment of kneecaps. It is a dominant trait. The male in generation II is dihybrid. His AB blood type is derived from inheriting the I^A allele from his father and the I^B allele from his mother. Because he demonstrates the nail–patella syndrome, he must have inherited that dominant allele from his father. He must also be heterozygous for nail–patella syndrome because his mother lacked the dominant allele (and therefore must be homozygous recessive).

We then observe that all four siblings with nail–patella syndrome have type A blood, which is the same genotypic arrangement as their grandfather. Furthermore, three of the siblings with type B blood do not demonstrate the nail–patella syndrome. Thus it appears that the two genes are linked, with the I^A allele of the ABO blood type system associated with the nail-patella allele (*NPS1*) and the I^B allele with the normal nail-patella allele (*nps1*). Thus only one child in eight (III-5) is recombinant.

These data suggest that the map distance between the two genes is 12.5 mu (100% × 1/8). The difficulty arises when trying to use this information to assign either or both of these linked genes to a chromosome. Although pedigrees provide clues that suggest particular loci are located on the X chromosome, no analogous clues predict which of the 22 autosomal chromosomes are associated with a particular locus.

We now turn our attention to the localization of loci to particular human autosomes. The first locus definitely established to be on a specific autosome was the Duffy blood group on chromosome 1. This locus was ascertained in 1968 from a family that had a morphologically odd, or "uncoiled," chromosome 1. Inheritance of the Duffy blood group followed the pattern of inheritance of the "uncoiled" chromosome.

Real strides have been made since then. Two techniques, chromosomal banding and somatic-cell hybridization, have been crucial to this autosomal mapping effort.

Chromosomal Banding

Techniques were developed around 1970 that use certain histochemical stains to generate reproducible banding patterns on the chromosomes. For example, Giemsa staining results in the appearance of bands called **G-bands.** Using different dyes and staining techniques, it is possible to produce a variety of staining patterns that can be used to differentiate each chromosome. More detail on these techniques is presented in chapter 7.

By examining the relative size of the chromosome and the pattern of G-bands, it is possible to identify the chromosome and particular regions on that chromosome. The ability to reproducibly identify each human chromosome allows scientists to create a *karyotype,* which is a pictorial representation of each chromosome in an individual (**fig. 6.38**).

Some individuals possess chromosomes that contain specific abnormalities, similar to the knob and extra piece of DNA on chromosome 9 in maize (see figure 6.7). Occasionally, specific human genetic diseases are inherited in a pattern that correlates with the transmission of a chromosome with an abnormal staining pattern. In these cases, the genetic locus is usually located on the abnormal chromosome.

Somatic-Cell Hybridization

The ability to distinguish each human chromosome is required to perform **somatic-cell hybridization.** In this technique, human and mouse (or hamster) cells are fused in culture to form hybrid cells (**fig. 6.39**). The fusion is usually mediated either chemically with

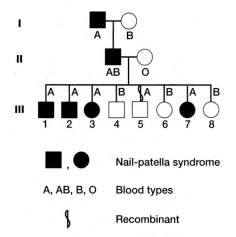

■ , ●   Nail-patella syndrome

A, AB, B, O   Blood types

∫   Recombinant

figure 6.37   Pedigree showing the inheritance of nail-patella syndrome and ABO loci.

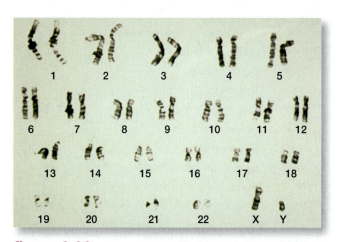

figure 6.38   Karyotype of a normal human male. Notice the presence of two copies of every autosome, and a single X chromosome and a single Y chromosome.

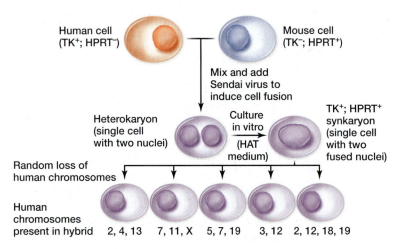

figure 6.39   Generation of somatic cell hybrids between a human cell and a mouse cell using Sendai virus. The hybrid cells are selected on a special medium that isolates only the fused cells. The fused cells randomly lose human chromosomes so that each hybrid contains a different combination of human chromosomes.

polyethylene glycol or with a virus such as Sendai virus, which can fuse to more than one cell simultaneously.

When two cells fuse, their nuclei remain separate, forming a **heterokaryon,** a cell with nuclei from different sources (see fig. 6.39). When the nuclei ultimately fuse, a hybrid cell is formed. The hybrid cell tends to lose human chromosomes preferentially through succeeding generations. Upon stabilization, the resulting cell contains one or more random human chromosomes in addition to the original mouse or hamster chromosomal complement. Chromosomal staining techniques identify the human chromosomes in the hybrid cell.

After successful cell hybrids are formed, two tests are used to map human genes. A **synteny test** determines whether two loci are in the same linkage group, which is the case if the

table 6.5   Definition of Selected Loci of the Human Chromosome Map (figure 6.40)

Locus	Protein Product	Chromosome	Locus	Protein Product	Chromosome
ABO	ABO blood group	9	IGH	Immunoglobulin heavy-chain gene family	14
AG	α-Globin gene family	16	IGK	Immunoglobulin κ-chain gene family	2
ALB	Albumin	4	INS	Insulin	11
AMY1	Amylase, salivary	1	LDHA	Lactate dehydrogenase A	11
AMY2	Amylase, pancreatic	1	MDI	Manic depressive illness	6
BCS	Breast cancer susceptibility	16	MHC	Major histocompatibility complex	6
C2	Complement component-2	6	MN	MN blood group	4
CAT	Catalase	11	MYB	Avian myeloblastosis virus oncogene	6
CBD	Color blindness, deutan	X			
CBP	Color blindness, protan	X	NHCP1	Nonhistone chromosomal protein-1	7
CML	Chronic myeloid leukemia	22	NPS1	Nail-patella syndrome	9
DMD	Duchenne muscular dystrophy	X	PEPA	Peptidase A	18
FES	Feline sarcoma virus oncogene	15	PVS	Polio virus sensitivity	19
FY	Duffy blood group	1	Rh	Rhesus blood group	1
GLB1	β-galactosidase-1	3	RN5S	5S RNA gene(s)	1
H1	Histone-1	7	RNTMI	Initiator methionine tRNA	6
HBB	Hemoglobin β chain	11	RWS	Ragweed sensitivity	6
HEMA	Classic hemophilia	X	S1	Surface antigen 1	11
HEXA	Hexosaminidase A	15	SIS	Simian sarcoma virus oncogene	22
HLA	Human leukocyte antigens	6	STA	Stature	Y
HP	Haptoglobin	16	TF	Transferrin	3
HYA	Y histocompatibility antigen, locus A	Y	XG	Xg blood group	X
IDDM	Insulin-dependent diabetes mellitus	6	XRS	X-ray sensitivity	13
IFF	Interferon, fibroblast	9			

phenotypes produced by the two loci are either always together or always absent in various hybrid cell lines. An **assignment test** determines which chromosome contains a particular locus by the concordant appearance of the phenotype whenever that chromosome is present in a cell line, or the lack of the phenotype when the chromosome is absent. The first autosomal synteny test, performed in 1970, demonstrated that the *B* locus of lactate dehydrogenase (LDH_B) was linked to the *B* locus of peptidase (PEP_B). Later, these loci were shown to reside on chromosome 12.

The current human map, compiled by Victor McKusick at Johns Hopkins University, contains over 9000 loci that were assigned either to a specific autosome or to a sex (X and Y) chromosome. A sample of the genes and disease loci that have been mapped is shown in **table 6.5** and **figure 6.40**. At present, geneticists studying human chromosomes are hampered not by a lack of techniques, but by a lack of marker loci. When a new locus is discovered, it is now relatively easy to assign it to its proper chromosome.

Restating the Concepts

▶ Human genes that are located on the X chromosome are relatively straightforward to map by recombination. If the mother is heterozygous for the alleles to be mapped, then regardless of the father's genotype, the phenotypes of all the sons can be used to deduce their genotype and identify the recombinant progeny.

▶ Alleles that are present on the autosomes must also be assigned to a specific chromosome, in addition to determining map distances. Chromosomal banding and somatic cell hybridization are two useful techniques for determining the chromosomal assignment.

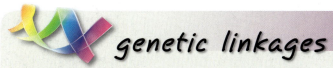

genetic linkages

In this chapter, we discussed the determination of genetic linkage maps that are based on the frequency of recombination between genes, DNA sequences, or both. Although linkage appears to complicate the law of independent assortment (chapter 2), it seems reasonable that genes that are physically linked on a single chromosome would not exhibit independent assortment.

Recombination mapping requires that at least one of the individuals in any mapping cross be heterozygous for each gene or DNA sequence being mapped. You learned that recombination distances are calculated as the percentage of recombinants in the offspring, which is stated in either map units or centimorgans (1.0% recombinant offspring = 1.0 mu = 1.0 cM). We also discussed how calculated recombination distance underestimates the actual recombination frequency because of multiple crossover events and horizontal resolution of the Holliday structure. The observed recombination distance can be determined more accurately by performing a three-point cross, in which the middle gene is used to detect double crossover events.

We described the mechanism of recombination, which builds on the events that occur during meiosis I (chapter 3). In particular, the synapsis of homologous chromosomes, the formation of the synaptonemal complex, and the physical exchange of material between nonsister chromatids are meiotic events associated with recombination. Rather than proceeding from a double-strand DNA break and exchange with a nonsister chromatid, recombination actually involves the invasion of one strand of the double-stranded DNA molecule to generate a heteroduplex and the Holliday structure. Depending on how the Holliday structure is resolved, markers that flank the heteroduplex region will exhibit either the parental or recombinant allele arrangement.

In chapter 13, we will discuss additional approaches to mapping genes. The mapping of genes and DNA sequences will be explored again in chapter 14, when we discuss the field of genomics. You will then see how mapping genes and DNA sequences is critical to determine the final genomic map and to locate specific genes associated with mutations. In chapter 15, you will learn how bacterial genes are mapped. Although the production of recombinant progeny will differ from the methods described in this chapter, the basic principle of recombination distance equaling the percentage of recombinant progeny will remain the same.

Chapter 7 will expand your knowledge of DNA sequences. Some of these DNA sequences are associated with genes, and others are located in the vast regions between genes. In both cases, the identification and mapping of these sequences facilitates generation of genomic maps and cloning of specific genes.

Chapter 7 will also continue the discussion of karyotype that we started in this chapter. Further details about the organization of chromosome structure and staining will be presented. These elements are important in unambiguously assigning the chromosome identities that were presented in this chapter.

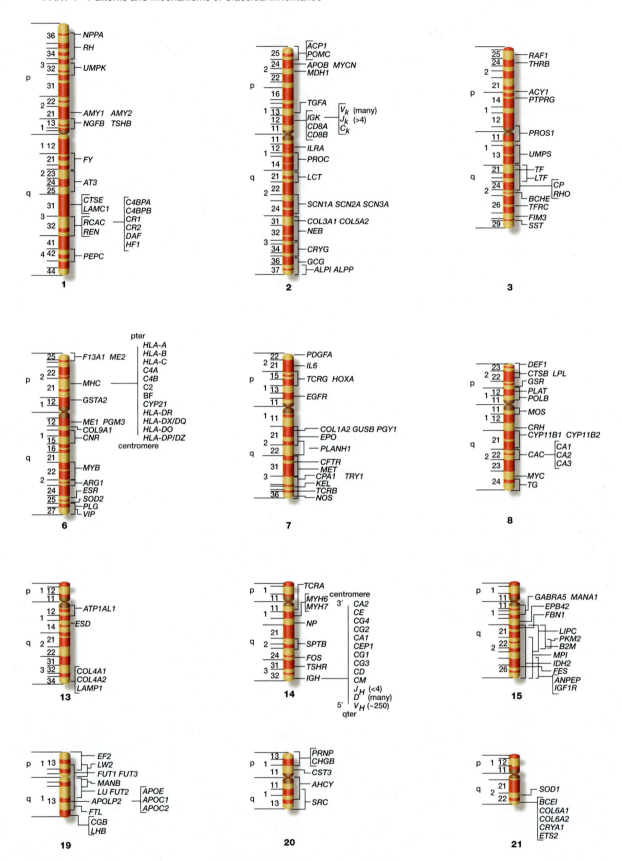

figure 6.40 Human G-banded chromosomes with their accompanying assigned loci. The p and q refer to the short and long arms of the chromosomes, respectively. A key to the loci is given in McKusick (1994).

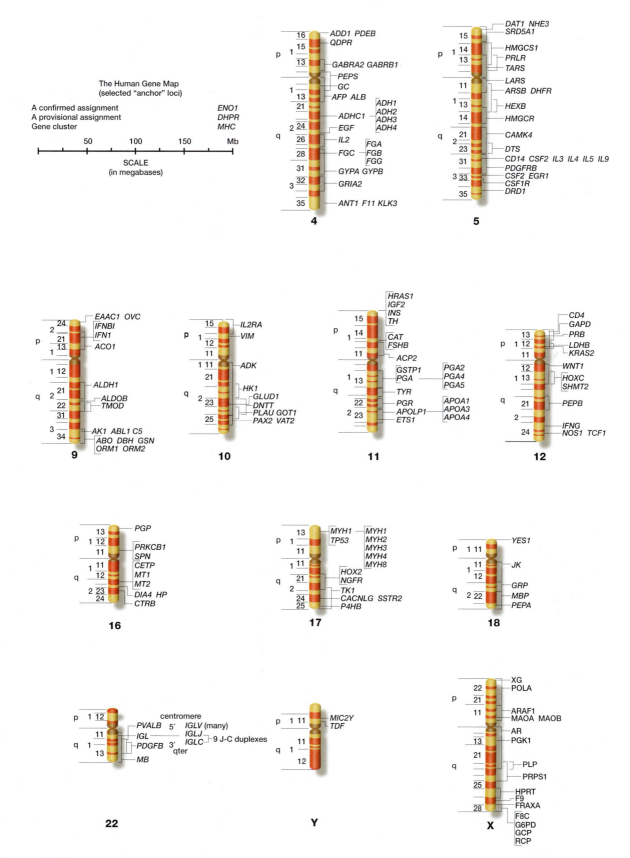

figure 6.40   Continued.

Chapter Summary Questions

1. For each term in column I, choose the best matching phrase in column II.

Column I
1. Synteny test
2. Ascus
3. Locus
4. Nonparental ditype
5. First-division segregation
6. Centimorgan
7. Tetratype
8. Second-division segregation
9. Parental ditype
10. Assignment test

Column II
A. Determines which chromosome a gene is on
B. Chromosomal position of a gene
C. Contains spores of four different genotypes
D. Indicates recombination between a gene and its centromere
E. Contains spores of only two nonrecombinant types
F. Determines whether two genes are on the same chromosome
G. Spore-holding sac found in certain fungi
H. Contains spores of only two recombinant types
I. One map unit
J. Indicates lack of recombination between a gene and its centromere

2. Linkage appears to violate one of Mendel's laws. Which one and why? If linkage violates one of Mendel's laws, then under what circumstances does this law prevail?

3. Consider an organism with the genotype $AaBb$, in which genes A and B are linked. Draw the chromosomal composition of this organism if the genes are
   a. in coupling.
   b. in repulsion.

4. What is a crossover? When does it occur? What is the physical evidence for it?

5. What is the maximum recombination frequency between two linked genes? Explain.

6. In three-point testcrosses, why are the double-crossover types expected less frequently than either of the single-crossover types?

7. Explain how the offspring of a three-point testcross can be used to determine the middle locus of three linked genes.

8. What is the difference between the measured map distance and actual map distance?

9. Compare and contrast positive and negative interference.

10. Define RFLPs. Discuss how they are useful in gene mapping.

11. Do three-point crosses in fruit flies capture all the multiple crossovers in a region?

Solved Problems

PROBLEM 1: A homozygous *claret* (*ca*, claret eye color), *curled* (*cu*, upcurved wings), *fluted* (*fl*, creased wings) fruit fly is crossed with a pure-breeding wild-type fly. The F_1 females are testcrossed with the following results:

fluted	4
claret	173
curled	26
fluted, claret	24
fluted, curled	167
claret, curled	6
fluted, claret, curled	298
wild-type	302

a. Are the loci linked?
b. If so, give the gene order, map distances, and coefficient of coincidence.

Answer: The pattern of numbers among the eight phenotypic classes is the pattern we are used to seeing for three linked loci. We can tell from the two groups with the largest numbers (the nonrecombinants—fluted, claret, curled and wild-type) that the alleles are in the coupling (*cis*) arrangement. The two groups with the smallest numbers (fluted and

claret, curled) represent the double recombinants. If we compare the parental classes with the double-crossover classes, we see that the fluted locus is in the center. For example, comparing claret, curled a double-crossover, with the fluted, claret, curled parental; clearly, fluted changes position relative to the flanking genes. Thus, the trihybrid female parent had the following arrangement of alleles:

$$\frac{ca \quad fl \quad cu}{ca^+ \, fl^+ \, cu^+}$$

A crossover in the *ca–fl* region produces offspring that are either claret or fluted, curled. A crossover in the *fl–cu* region produces offspring that are either fluted, claret or curled. The map distances are given by the recombination frequency, which is the average number of crossovers multiplied by 100%. A count of the crossovers in each region, including the double crossovers, provides the following map distances

$$ca - fl = \frac{173 + 167 + 6 + 4}{1000} \times 100 = 35 \text{ mu}$$

$$fl - cu = \frac{26 + 24 + 6 + 4}{1000} \times 100 = 6 \text{ mu}$$

The coefficient of coincidence is calculated by

$$\frac{\text{frequency of observed double crossovers}}{\text{frequency of expected double crossovers}}$$

The observed frequency of double crossovers is $(4 + 6)/1000 = 0.01$. The expected frequency of double crossovers is the product of the two single crossovers, or $0.35 \times 0.06 = 0.021$. Therefore, the coefficient of coincidence is $0.01/0.021 = 0.48$.

Interference, which corresponds to 1-(coefficient of coincidence), must then be $1 - 0.48 = 0.52$.

PROBLEM 2: The *ad5* locus in *Neurospora* encodes an enzyme in the synthesis of the DNA base adenine. A wild-type strain ($ad5^+$) is crossed with an adenine-requiring strain, $ad5^-$. The diploid undergoes meiosis, and 100 asci are scored for their segregation patterns with the following results:

Ascus Type

1	2	3	4	5	6
$ad5^+$	$ad5^-$	$ad5^+$	$ad5^-$	$ad5^-$	$ad5^+$
$ad5^+$	$ad5^-$	$ad5^+$	$ad5^-$	$ad5^-$	$ad5^+$
$ad5^+$	$ad5^-$	$ad5^+$	$ad5^-$	$ad5^+$	$ad5^-$
$ad5^+$	$ad5^-$	$ad5^-$	$ad5^+$	$ad5^+$	$ad5^-$
$ad5^-$	$ad5^+$	$ad5^-$	$ad5^+$	$ad5^-$	$ad5^+$
$ad5^-$	$ad5^+$	$ad5^-$	$ad5^+$	$ad5^-$	$ad5^+$
$ad5^-$	$ad5^+$	$ad5^+$	$ad5^-$	$ad5^+$	$ad5^-$
$ad5^-$	$ad5^+$	$ad5^+$	$ad5^-$	$ad5^+$	$ad5^-$
40	46	5	3	4	2

What can you say about the linkage arrangements at this locus?

Answer: You can see that 14 asci $(5 + 3 + 4 + 2)$ are of the second-division segregation type (SDS) and 86 are of the first-division segregation type (FDS). To map the distance of the *ad5* locus to its centromere, we divide the percentage of SDS types by 2: $14/100 = 14\%$; divided by 2 is 7%. Thus, the *ad5* locus is 7 mu from its centromere.

PROBLEM 3: In yeast, the *his5* locus encodes an enzyme required for the synthesis of the amino acid histidine, and the *lys11* locus encodes an enzyme needed in the synthesis of the amino acid lysine. A haploid wild-type strain ($his5^+ \ lys11^+$) is crossed with the double mutant ($his5^- lys11^-$). The diploid is allowed to undergo meiosis, and 100 asci are scored with the following results:

$his5^+ \ lys11^+$	$his5^+ \ lys11^+$	$his5^+ \ lys11^-$
$his5^+ \ lys11^+$	$his5^- \ lys11^-$	$his5^+ \ lys11^-$
$his5^- \ lys11^-$	$his5^- \ lys11^+$	$his5^- \ lys11^+$
$his5^- \ lys11^-$	$his5^+ \ lys11^-$	$his5^- \ lys11^+$
62	30	8

What is the linkage arrangement of these loci?

Answer: Of the 100 asci analyzed, 62 were parental ditypes (PD), 30 were tetratypes (TT), and 8 were nonparental ditypes (NPD). To map the distance between the two loci:

$$\frac{1/2(\text{total number of TT}) + (\text{total number of NPD})}{\text{total number of asci}} \times 100\%$$

$$\frac{1/2(30) + 8}{100} \times 100\% = 23\% = 23 \text{ mu}$$

Because yeast does not produce an ordered ascus, we are unable to map the centromere-to-gene distances.

Exercises and Problems

12. A double heterozygote for two linked genes is test-crossed. In the offspring, the two parental classes appear in a frequency of 40% each, and the two recombinant classes appear in a frequency of 10% each. What is the distance in centimorgans between these two genes?

13. Consider an individual with genotype *AaBb* who has inherited the dominant alleles from his mother and the recessive alleles from his father. Determine the types of gametes produced and their proportions in each of the following cases:
    a. The two loci are unlinked.
    b. The two loci are completely linked.
    c. The two loci are incompletely linked and 10 mu apart.
    d. The two loci are on opposite ends of the same chromosome and 70 mu apart.

14. Use the following two-factor recombination frequencies to construct a linkage map of these loci.

$i-e = 27$	$e-k = 19$	$a-g = 10$
$i-k = 8$	$g-e = 6$	$l-n = 12$
$k-g = 13$	$a-l = 16$	$i-a = 11$
$n-e = 20$	$a-e = 16$	

15. In *Drosophila*, curled wings (*cu*) and ebony body (*e*) are recessive traits that are produced by two genes that are 20 mu apart on chromosome 3. Two flies heterozygous for both traits ($cu^+cu \ e^+e$) are crossed. Determine the proportion of flies with curled wings and ebony bodies expected in the offspring in each of the following cases:
    a. Both parents have the *cis* configuration.
    b. Both parents have the *trans* configuration.
    c. The male has the *cis* configuration, and the female has the *trans* configuration.

16. A homozygous *groucho* fly (*gro*, bristles clumped above the eyes) is crossed with a homozygous *rough* fly (*ro*, eye abnormality). The F_1 females are testcrossed, producing these offspring:

groucho	518
rough	471
groucho, rough	6
wild-type	5
	1000

   a. What is the linkage arrangement of these loci?
   b. What offspring would result if the F_1 dihybrids were crossed among themselves instead of being testcrossed?

17. In the house mouse, the autosomal phenotypes Trembling and Rex (short hair) are dominant mutations. Heterozygous Trembling, Rex females were crossed with normal, long-haired males and yielded the following offspring:

Trembling, Rex	42
Trembling, long-haired	105
normal, Rex	109
normal, long-haired	44

   a. Are the two genes linked? How do you know?
   b. In the heterozygous females, were Trembling and Rex in *cis* or *trans* arrangement? Explain.
   c. Calculate the percent recombination between the two genes.

18. A female fruit fly with abnormal eyes (*abe*) of a brown color (*bis*, bistre) is crossed with a wild-type male. Her sons have abnormal, brown eyes; her daughters are wild type. When these F_1 flies are crossed among themselves, the following offspring are produced:

	Sons	**Daughters**
abnormal, brown	219	197
abnormal	43	45
brown	37	35
wild-type	201	223

   What is the linkage arrangement of these loci?

19. In *Drosophila*, the loci *inflated* (*if*, small, inflated wings) and *warty* (*wa*, abnormal eyes) are 10 mu apart on the X chromosome. Construct a data set that would allow you to determine this linkage arrangement. What differences would be involved if the loci were located on an autosome?

20. Consider the two X-linked genes, *z* and *w*, in *Drosophila*. The two loci are 30 mu apart, and each segregates as either a dominant wild-type (z^+, w^+) or a recessive mutant allele (*z*, *w*). For each of the following crosses, predict the offspring's phenotypic frequencies by sex.
   a. Heterozygous female in *cis* configuration × hemizygous *zw* male.
   b. Heterozygous female in *cis* configuration × hemizygous z^+w male.
   c. Heterozygous female in *trans* configuration × hemizygous *zw* male.
   d. Heterozygous female in *trans* configuration × hemizygous z^+w male.

21. A geneticist crossed female fruit flies that were heterozygous at three electrophoretic loci, each with fast and slow alleles, with males homozygous for the slow alleles. The three loci were *got1* (glutamate oxaloacetate transaminase-1), *amy* (α-amylase), and *sdh* (succinate dehydrogenase). The first 1000 offspring isolated had the following genotypes:

Class 1	$got^s\ got^s\ amy^s\ amy^s\ sdh^s\ sdh^s$	441
Class 2	$got^f\ got^s\ amy^f\ amy^s\ sdh^f\ sdh^s$	421
Class 3	$got^f\ got^s\ amy^s\ amy^s\ sdh^s\ sdh^s$	11
Class 4	$got^s\ got^s\ amy^f\ amy^s\ sdh^f\ sdh^s$	14
Class 5	$got^f\ got^s\ amy^f\ amy^s\ sdh^s\ sdh^s$	58
Class 6	$got^s\ got^s\ amy^s\ amy^s\ sdh^f\ sdh^s$	53
Class 7	$got^f\ got^s\ amy^s\ amy^s\ sdh^f\ sdh^s$	1
Class 8	$got^s\ got^s\ amy^f\ amy^s\ sdh^s\ sdh^s$	1

   What are the linkage arrangements of these three loci, including map units? If the three loci are linked, what is the coefficient of coincidence?

22. The following three recessive markers are known in lab mice: *h*, hotfoot; *o*, obese; and *wa*, waved. A trihybrid of unknown origin is testcrossed, producing the following offspring:

hotfoot, obese, waved	357
hotfoot, obese	74
waved	66
obese	79
wild-type	343
hotfoot, waved	61
obese, waved	11
hotfoot	9
	1000

   a. If the genes are linked, determine the relative order and the map distances between them.
   b. What was the *cis–trans* allele arrangement in the trihybrid parent?
   c. Is there any crossover interference? If so, how much?

23. The following three recessive genes are found in corn: *bt1*, brittle endosperm; *gl17*, glossy leaf; *rgd1*, ragged seedling. A trihybrid of unknown origin is testcrossed, producing the following offspring:

brittle, glossy, ragged	236
brittle, glossy	241
ragged	219
glossy	23
wild-type	224
brittle, ragged	17
glossy, ragged	21
brittle	19
	1000

   a. If the genes are linked, determine their relative order and map distances.
   b. Reconstruct the chromosomes of the trihybrid.
   c. Is there any crossover interference? If so, how much?

24. In *Drosophila,* the loci *ancon* (*an*, legs and wings short), *spiny legs* (*sple*, irregular leg hairs), and *arctus oculus* (*at*, small narrow eyes) have the following linkage arrangement on chromosome 3:

an----------10.0 mu----------*sple*--------6.1 mu--------*at*

   a. Devise a data set that would yield these map units assuming no interference.

   b. Devise a data set for the same map if interference = 0.4.

25. *Ancon* (*an*) and *spiny legs* (*sple*), from problem 24, are 10 mu apart on chromosome 3. The *notchy* locus (*ny*, wing tips nicked) is on the X chromosome (chromosome 1).

   a. Create a data set that would result if you were making crosses to determine the linkage arrangement of these three loci.

   b. How would you know that the *notchy* locus is on the X chromosome?

26. In *Drosophila,* kidney-shaped eye (*k*), cardinal eye (*cd*), and ebony body (*e*) are three recessive phenotypes. If homozygous kidney, cardinal females are crossed with homozygous ebony males, the F_1 offspring are all wild-type. If heterozygous F_1 females are mated with kidney, cardinal, ebony males, the following 2000 progeny appear:

880	kidney, cardinal
887	ebony
64	kidney, ebony
67	cardinal
49	kidney
46	ebony, cardinal
3	kidney, ebony, cardinal
4	wild-type

   a. Determine the allele arrangement and gene order in the F_1 females.

   b. Derive a map of the three genes.

27. In corn, a trihybrid Tunicate (*T*), Glossy (*G*), Liguled (*L*) plant was crossed with a nontunicate, nonglossy, liguleless plant, producing the following offspring:

Tunicate, liguleless, Glossy	58
Tunicate, liguleless, nonglossy	15
Tunicate, Liguled, Glossy	55
Tunicate, Liguled, nonglossy	13
nontunicate, Liguled, Glossy	16
nontunicate, Liguled, nonglossy	53
nontunicate, liguleless, Glossy	14
nontunicate, liguleless, nonglossy	59

   a. Determine which genes are linked.

   b. Determine the genotype of the heterozygote; be sure to indicate which alleles are on which chromosome.

   c. Calculate the map distances between the linked genes.

28. Following is a partial map of the third chromosome in *Drosophila.*

mu
19.2 *javelin* bristles (*jv*)
43.2 *thread* arista (*th*)
66.2 *Delta* veins (*Dl*)
70.7 *ebony* body (*e*)

   a. If flies heterozygous in the *cis* arrangement for *javelin* and *ebony* are mated among themselves, what phenotypic ratio do you expect in the progeny?

   b. A true-breeding *thread, ebony* fly is crossed with a true-breeding *Delta* fly. An F_1 female is testcrossed to a *thread, ebony* male. Predict the expected progeny and their frequencies for this cross. Assume no interference.

   c. Repeat **b,** but assume interference equals 0.4.

29. Suppose that you determined the order of three genes to be *a, c, b,* and that by doing two-point crosses you determined the map distances as *a–c* = 10 and *c–b* = 5. If interference is –1.5, and the three-point cross is *ACB/acb* × *acb/acb*, what is the observed frequency of double crossovers?

30. Consider three X-linked recessive *Drosophila* genes, *a, b,* and *c*. A triple heterozygote female was crossed to a male that is mutant for all three traits. The first 1000 offspring isolated had the following phenotypes:

a	*b*	+	462
a	+	*c*	27
a	+	+	16
+	+	*c*	458
+	*b*	+	23
+	*b*	*c*	14

   a. What phenotypes are missing in the offspring and why?

   b. What is the order of these genes on the X chromosome?

   c. Construct a linkage map that includes the distance between the genes.

   d. Calculate interference.

31. You carry out a four-point testcross $p^+q^+r^+s^+/pqrs$ × *pqrs/pqrs* in *Drosophila.* If all four loci are incompletely linked, how many recombinant classes will be expected in the offspring? Describe these classes in terms of number and location of crossovers.

32. Given the following cross in *Neurospora: ab* × a^+b^+, construct results showing that crossing over occurs in two of the four chromatids of a tetrad at meiosis. What would the results be if crossing over occurred during interphase before DNA replication? If each crossover event involved three or four chromatids?

33. A strain of yeast requiring both tyrosine (*tyr⁻*) and arginine (*arg⁻*) is crossed to the wild type. After meiosis, the following 10 asci are dissected. Classify each ascus as to segregational type (PD, NPD, TT). What is the linkage relationship between these two loci?

1	*arg⁻ tyr⁻*	*arg⁺ tyr⁺*	*arg⁺ tyr⁺*	*arg⁻ tyr⁻*
2	*arg⁺ tyr⁺*	*arg⁺ tyr⁺*	*arg⁻ tyr⁻*	*arg⁻ tyr⁻*
3	*arg⁻ tyr⁺*	*arg⁻ tyr⁺*	*arg⁺ tyr⁻*	*arg⁺ tyr⁻*
4	*arg⁻ tyr⁻*	*arg⁻ tyr⁻*	*arg⁺ tyr⁺*	*arg⁺ tyr⁺*
5	*arg⁻ tyr⁻*	*arg⁻ tyr⁺*	*arg⁺ tyr⁻*	*arg⁺ tyr⁺*
6	*arg⁺ tyr⁺*	*arg⁺ tyr⁺*	*arg⁻ tyr⁻*	*arg⁻ tyr⁻*
7	*arg⁻ tyr⁻*	*arg⁺ tyr⁺*	*arg⁻ tyr⁺*	*arg⁺ tyr⁻*
8	*arg⁺ tyr⁺*	*arg⁺ tyr⁺*	*arg⁻ tyr⁻*	*arg⁻ tyr⁻*
9	*arg⁺ tyr⁺*	*arg⁻ tyr⁻*	*arg⁻ tyr⁻*	*arg⁺ tyr⁺*
10	*arg⁻ tyr⁻*	*arg⁺ tyr⁺*	*arg⁺ tyr⁺*	*arg⁻ tyr⁻*

34. A certain haploid strain of yeast was deficient for the synthesis of the amino acids tryptophan (trp^-) and methionine (met^-). It was crossed to the wild-type and meiosis occurred. One dozen asci were analyzed for their tryptophan and methionine requirements. The following results, with the inevitable lost spores, were obtained:

1	$trp^- met^-$	?	?	$trp^- met^-$
2	?	$trp^- met^-$	$trp^+ met^+$	$trp^+ met^+$
3	$trp^- met^+$	$trp^- met^-$	$trp^+ met^-$	$trp^+ met^+$
4	$trp^- met^-$	$trp^- met^+$	$trp^+ met^+$	?	$trp^+ met^-$
5	$trp^- met^+$	?	?	$trp^+ met^-$
6	$trp^+ met^+$	$trp^+ met^+$	$trp^- met^-$	$trp^- met^-$
7	$trp^+ met^+$	$trp^+ met^-$	?	$trp^- met^-$
8	$trp^+ met^+$	$trp^- met^-$	?	$trp^+ met^+$
9	$trp^- met^-$	$trp^+ met^-$	$trp^- met^+$	$trp^+ met^+$
10	$trp^- met^-$	$trp^+ met^+$	$trp^- met^-$	$trp^+ met^+$
11	$trp^+ met^+$	$trp^+ met^+$	?	?
12	?	$trp^+ met^-$	?	$trp^- met^+$

a. Classify each ascus as to segregational type (noting that some asci may not be classifiable).
b. Are the genes linked?
c. If so, how far apart are they?

35. In *Neurospora*, a haploid strain requiring arginine (arg^-) is crossed with the wild type (arg^+). Meiosis occurs, and 10 asci are dissected with the following results. Map the *arg* locus.

1	arg^+	arg^+	arg^-	arg^-	arg^+	arg^+	arg^-	arg^-
2	arg^-	arg^-	arg^+	arg^+	arg^-	arg^-	arg^+	arg^+
3	arg^+	arg^+	arg^+	arg^+	arg^-	arg^-	arg^-	arg^-
4	arg^+	arg^+	arg^+	arg^+	arg^-	arg^-	arg^-	arg^-
5	arg^-	arg^-	arg^-	arg^-	arg^+	arg^+	arg^+	arg^+
6	arg^+	arg^+	arg^-	arg^-	arg^-	arg^-	arg^+	arg^+
7	arg^-	arg^-	arg^+	arg^+	arg^+	arg^+	arg^-	arg^-
8	arg^+	arg^+	arg^+	arg^+	arg^-	arg^-	arg^-	arg^-
9	arg^-	arg^-	arg^+	arg^+	arg^+	arg^+	arg^-	arg^-
10	arg^-	arg^-	arg^-	arg^-	arg^+	arg^+	arg^+	arg^+

36. A haploid strain of *Neurospora* with fuzzy colony morphology (f) was crossed with the wild type (f^+). Twelve asci were scored. The following results, with the inevitable lost spores, were obtained:

1	?	f	f	?	?	f^+	f^+	f^+
2	f	f	f^+	f^+	f^+	f^+	f	f
3	f	?	?	?	f^+	?	?	?
4	f^+	?	?	?	f	f	f	f
5	f	f	?	?	?	f^+	?	f^+
6	?	f	f	?	?	?	?	?
7	f^+	f^+	f	f	f	f	f^+	f^+
8	f	f	f	?	?	f^+	f^+	f^+
9	f^+	?	?	?	?	f	f	?
10	f	f	f^+	f^+	f	f	f^+	f^+
11	f	f	f	f	f^+	f^+	f^+	f^+
12	f	f	?	?	?	?	f^+	f^+

a. Classify each ascus as to segregational type and note which asci cannot be classified.
b. Map the chromosome containing the *f* locus with all the relevant measurements.

37. In yeast, the *a* and *b* loci are 12 mu apart. Construct a data set to demonstrate this.

38. In *Neurospora*, the *a* locus is 12 mu from its centromere. Construct a data set to show this.

39. An $ab \times a^+b^+$ cross was performed in *Neurospora*. Meiosis occurred, and 1000 asci were dissected. Using the classes of table 6.4, the following data resulted:

Class 1	700
Class 2	0
Class 3	190
Class 4	90
Class 5	5
Class 6	5
Class 7	10

What is the linkage arrangement of these loci?

40. Given the following linkage arrangement in *Neurospora*, construct a data set similar to that in table 6.4 that is consistent with this map (*cm* is centromere).

```
        15 mu                    15 mu
a--------------------- cm ---------------------b
```

41. Determine crossover events that led to each of the seven classes in table 6.4.

42. In *Neurospora*, a cross is made between ab^+ and a^+b individuals. The following 100 ordered tetrads are obtained:

Spores	I	II	III	IV	V	VI	VII	VIII
1, 2	a^+b	a^+b	a^+b	a^+b^+	a^+b^+	a^+b	a^+b	ab^+
3, 4	a^+b	a^+b^+	a^+b^+	a^+b	a^+b	ab^+	ab^+	a^+b
5, 6	ab^+	ab	ab^+	ab	ab^+	a^+b	ab^+	a^+b
7, 8	ab^+	ab^+	ab	ab^+	ab	ab^+	a^+b	ab^+
	85	2	3	2	3	3	1	1

a. Are genes *a* and *b* linked? How do you know?
b. Calculate the gene-to-centromere distances for *a* and *b*.

43. *Neurospora* has four genes—*a*, *b*, *c*, and *d*—that control four different phenotypes. Your job is to map these genes by performing pairwise crosses. You obtain the following ordered tetrads:

$$ab^+ \times a^+b \qquad\qquad bc^+ \times b^+c$$

Spores	I	II	III		Spores	I	II	III
1, 2	ab^+	ab	ab^+		1, 2	bc^+	b^+c^+	b^+c
3, 4	ab^+	ab	a^+b^+		3, 4	bc^+	b^+c^+	b^+c^+
5, 6	a^+b	a^+b^+	a^+b		5, 6	b^+c	bc	bc
7, 8	a^+b	a^+b^+	ab		7, 8	b^+c	bc	bc^+
	45	43	12			70	4	26

$$cd^+ \times c^+d$$

Spores	I	II	III	IV	V	VI	VII
1, 2	cd^+	cd	cd	cd	cd^+	cd	cd^+
3, 4	cd^+	cd	cd^+	c^+d	c^+d	c^+d^+	c^+d
5, 6	c^+d	c^+d^+	c^+d^+	c^+d^+	c^+d	c^+d^+	c^+d^+
7, 8	c^+d	c^+d^+	c^+d	cd^+	cd^+	cd	cd
	42	2	30	15	5	1	5

a. Calculate the gene-to-centromere distances.
b. Which genes are linked? Explain.
c. Derive a complete map for all four genes.

44. You isolated a new fungus and obtained a strain that requires arginine (*arg*⁻) and a second strain that requires adenine (*ad*⁻). You cross these two strains and collect 400 random spores that you plate on minimal medium. If 25 spores grow, what is the distance between these two genes?

45. Three distinct genes, *pab, pk,* and *ad,* were scored in a cross of *Neurospora*. From the cross *pab pk⁺ ad⁺ × pab⁺ pk ad,* the following ordered tetrads were recovered:

Spores	I	II	III	IV	V	VI	VII	VIII
1, 2	*pab pk⁺ ad⁺*	*pab pk⁺ ad⁺*	*pab pk⁺ ad⁺*	*pab pk⁺ ad⁺*	*pab pk⁺ ad⁺*	*pab pk⁺ ad⁺*	*pab pk⁺ ad*	*pab pk⁺ ad*
3, 4	*pab pk⁺ ad⁺*	*pab⁺ pk ad*	*pab pk ad*	*pab pk⁺ ad*	*pab⁺ pk ad*	*pab⁺ pk ad*	*pab⁺ pk ad*	*pab⁺ pk ad⁺*
5, 6	*pab⁺ pk ad*	*pab pk⁺ ad⁺*	*pab⁺ pk⁺ ad⁺*	*pab pk ad⁺*	*pab pk ad*	*pab pk⁺ ad*	*pab pk ad⁺*	*pab pk⁺ ad⁺*
7, 8	*pab⁺ pk ad*	*pab⁺ pk ad*	*pab⁺ pk ad*	*pab⁺ pk ad*	*pab⁺ pk⁺ ad⁺*	*pab⁺ pk ad⁺*	*pab pk⁺ ad⁺*	*pab ⁺pk ad*
	34	35	9	7	2	2	1	3

Based on the data, construct a map of the three genes. Be sure to indicate the relative positions of the centromeres.

46. The Duffy blood group with alleles FY^a and FY^b was localized to chromosome 1 in human beings when an "uncoiled" chromosome was associated with it. Construct a pedigree that would verify this.

47. What pattern of scores would you expect to get, using the hybrid clones in the following eoc table 6.1 below, for a locus on human chromosome 6? 14? X?

48. A man with X-linked color blindness and X-linked Fabry disease (α-galactosidase-A deficiency) mates with a normal woman and has a normal daughter. This daughter then mates with a normal man and produces 10 sons (as well as 8 normal daughters). Of the sons, five are normal, three are like their grandfather, one is only color-blind, and one has Fabry disease. From these data, what can you say about the relationship of these two X-linked loci?

49. In humans, the ABO system (I^A, I^B, i alleles) is linked to the aldolase-B locus (*ALDOB*), a gene that functions in the liver. Aldolase-B deficiency, which is recessive, results in fructose intolerance. A man with blood type AB and fructose-intolerant, whose father is type B and normal and whose mother is type AB, marries a woman with blood type O and fructose-intolerant and they have 10 children. Five are type A and normal, three are fructose-intolerant and type B, and two are type A and fructose intolerant. Determine the map distance between the two genes.

eoc table 6.1   Human Chromosomes Present in Several Different Human–Mouse Hybrid Cell Lines

Hybrid Cell Line Designation	1	2	3	4	5	6	7	8	9	10	11	12	13	14	15	16	17	18	19	20	21	22	X
WILI	−	−	−	−	−	−	−	+	−	−	−	−	−	+	−	−	+	−	−	−	+	−	+
WIL6	−	+	−	+	+	+	+	+	−	+	+	−	−	+	−	−	+	−	+	+	+	−	+
SIR3	+	+	+	+	+	+	+	−	+	+	+	+	+	−	−	+	+	+	+	+	+	+	+
SIR8	+	+	+	+	+	−	+	+	+	+	+	+	+	+	+	+	+	−	−	+	+	+	+
REW15	+	+	+	+	+	+	+	+	−	+	−	+	+	+	+	−	+	+	+	+	+	+	+
DUA1CsAzF	−	−	−	−	−	−	+	−	−	−	−	−	−	−	−	−	−	−	−	−	−	−	−
TSL1	−	−	+	+	−	−	−	−	+	+	−	+	+	−	+	+	−	+	−	−	−	−	−
TSL2	−	+	+	−	+	+	−	−	−	+	−	+	−	−	−	+	+	−	+	+	−	+	+
TSL2CsBF	−	−	−	−	+	−	−	−	−	−	−	−	−	−	−	−	−	−	−	−	−	−	−
XTR3BsAgE	+	−	+	−	+	+	+	+	+	+	−	−	+	+	−	−	+	+	+	−	+	−	+
XTR22	−	+	+	+	+	+	−	+	−	+	+	−	−	+	−	+	−	+	+	+	+	+	+
XER11	+	−	+	+	−	+	+	+	−	+	+	+	+	−	+	+	+	+	+	+	+	+	+
REX12	−	−	+	−	−	−	+	−	−	−	+	−	+	−	−	−	−	−	−	−	+	+	+
JWR22H	+	+	−	+	−	+	−	−	−	+	+	+	−	+	+	−	+	+	−	+	+	−	−

50. The results of an analysis of five human–mouse hybrids for five enzymes are given in the following tables along with the human chromosomal content of each clone (+ = enzyme or chromosome present; − = absent). Deduce which chromosome carries which gene.

| | Clone | | | | |
Human Enzyme	A	B	C	D	E
glutathione reductase	+	+	−	−	−
malate dehydrogenase	−	+	−	−	−
adenosine deaminase	−	+	−	+	+
glactokinase	−	+	+	−	−
hexosaminidase	+	−	−	+	−

Human Chromosome

	1	2	3	4	5	6	7	8	9	10	11	12	13	14	15	16	17	18	19	20	21	22
Clone A	−	−	−	−	+	+	+	+	−	+	−	−	−	−	+	+	−	−	−	−	+	+
Clone B	+	+	−	+	−	−	−	+	−	−	+	−	−	+	−	−	+	−	−	+	−	−
Clone C	−	−	−	+	−	−	+	−	−	+	−	+	+	+	+	−	+	−	+	−	−	+
Clone D	+	−	+	−	+	−	−	−	−	+	−	−	−	+	+	−	−	+	+	+	+	−
Clone E	−	−	−	+	−	−	−	−	+	+	+	+	−	+	−	+	−	+	−	+	+	+

51. Hemophilia A and color blindness are two X-linked recessive traits whose loci are 10 cM apart. Mickey, a phenotypically normal man is married to Minnie, a phenotypically normal woman with a hemophiliac father and a color-blind mother. What is the probability that Mickey and Minnie's first child will have
  a. hemophilia but not color-blindness?
  b. both hemophilia and color-blindness?
  c. neither hemophilia nor color-blindness?

52. You selected three mouse–human hybrid clones and analyzed them for the presence of human chromosomes. You then analyze each clone for the presence or absence of particular human enzymes (+ = presence of human chromosome or enzyme activity). Based on the following results indicate the probable chromosomal location for each enzyme.

Human Chromosomes

Clone	3	7	9	11	15	18	20
X	−	+	−	+	+	−	+
Y	+	+	−	+	−	+	−
Z	−	+	+	−	−	+	+

Enzyme

Clone	A	B	C	D	E
X	+	+	−	−	+
Y	+	−	+	+	+
Z	−	−	+	−	+

53. Three mouse–human cell lines were scored for the presence (+) or absence (−) of human chromosomes, with the following results:

Human Chromosomes

Clone	1	2	3	4	5	14	15	18
A	+	+	+	+	−	−	−	−
B	+	+	−	−	+	+	−	−
C	+	−	+	−	+	−	+	−

If a particular gene is located on chromosome 3, which clones should be positive for the enzyme from that gene?

54. Assume 4% of all tetrads have a single crossover between two loci.
  a. What is the map distance between these loci if these are fruit flies?
  b. What is the proportion of second-division segregants if these are *Neurospora*?
  c. What is the proportion of nonparental ditypes if these are yeast?

55. In the fictional organism, *Nogard dabgib*, fire production (*f*) is recessive to inability to produce fire, green color (*g*) is recessive to yellow, and presence of horns (*h*) is recessive to absence of horns. The three loci are linked on an autosome as shown in the following map:

f------- 5 mu -------*g*-------------- 20 mu ---------------*h*

Matings in this organism, although difficult to carry out, always produce 1000 offspring. Ness is a triple heterozygote female whose father was a true-breeding yellow, fire-producer with horns. Ness is mated to Puff who is homozygous recessive for all three traits. Assuming no interference, determine the expected *number* of offspring that are
  a. fire-producing, green, with horns?
  b. fire-producing, green, without horns?
  c. fire-producing, yellow, without horns?
  d. fire-producing, yellow, with horns?

56. The following three recessive markers are known in *Drosophila*: *y*, yellow body; *rb*, ruby eyes; and *m*, miniature wings. A trihybrid female of unknown origin is crossed to a male, producing the following offspring:

wild-type females	1000
yellow, ruby, miniature males	3
yellow, ruby males	31
miniature males	36
yellow males	135
wild-type males	5
ruby, miniature males	143
yellow, miniature males	321
ruby males	326
	2000

   a. Determine the allele arrangement in both parents and describe their genotypes accurately.
   b. Determine the correct linkage map using these data.
   c. Is there any crossover interference? If so, how much?

57. You are studying three recessive mutations in *Drosophila* (*a*, *b*, and *c*). You cross a heterozygous female with a phenotypically wild-type male and the following 2000 progeny appear:

+ + +	1321
a b c	318
a b +	3
+ + *c*	4
a + *c*	131
+ *b* +	123
a + +	47
+ *b c*	53

   a. What are the parental genotypes?
   b. Derive a map of the linkage arrangement and the recombination distances.
   c. Calcutate the interference value from these data.

58. David and Barb are expecting their first child. They met in a genetics class during college and dream that their child will become a genetics professor. However, both David and Barb have sisters that suffer from geneticitis, a rare recessive condition that results in the individual being unable to understand genetics. The *geneticitis* gene is located 7 cM from a RFLP, which corresponds to either two DNA fragments or one fragment (*RFLP2* and *RFLP1*, respectively). Both affected siblings are homozygous for *RFLP1*.

   a. If Barb knows that she is heterozygous for the RFLP, what is the probability that her child will suffer from geneticitis?
   b. If Dave then gets tested and learns that he is homozygous for RFLP2, what is the new probability that their child will be affected?

Chapter Integration Problem

As a student in a genetics lab, you make a startling discovery: a *Drosophila* male and female that can do backflips! You name the gene *bf* and decide to investigate the mode of inheritance of this interesting trait. You and your labmates carry out the following two crosses:

   I: Backflipping male × wild-type female → 1/2 wild-type males : 1/2 wild-type females

   II: Backflipping female × wild-type male → 1/2 backflipping males : 1/2 wild-type females

The backflipping female in cross II happens to have miniature wings, a known recessive mutation. You suspect that the *bf* gene is linked to gene *m*, so you perform another cross:

III. Backflipping female with miniature wings	× ↓	Wild-type male for both traits

The F_1 females are then testcrossed to produce the following F_2 offspring:

Backflipper, miniature wings	251
Backflipper, normal wings	130
Nonbackflipper, miniature wings	125
Nonbackflipper, normal wings	244

Use this information to answer the following questions:

   a. Is the *bf* locus on an autosome, the X chromosome, or the Y chromosome? Explain.
   b. Is backflipping dominant or recessive? Explain.
   c. Are genes *bf* and *m* following Mendel's first law? Justify your answer.
   d. Do genes *bf* and *m* exhibit epistasis? Why or why not?
   e. Perform a test that can evaluate the likelihood that genes *bf* and *m* are independently assorting.
   f. If genes *bf* and *m* are linked, calculate the distance between them.
   g. What specific genetic mechanism, and in which F_1 parent, gave rise to the backflipper, normal-winged flies and the non-backflipper, miniature-winged flies in cross III. What gametes can that F_1 parent produce? Explain with the help of a diagram.
   h. A mating is carried out between the following cross III flies: F_1 females × F_2 backflipping, normal-winged males. What proportion of the overall offspring is expected to be wild-type for *at least* one of the two traits?

 Do you need additional review? Visit **www.mhhe.com/hyde** for practice tests, answers to end-of-chapter questions and problems, interactive exercises, and animation tutorials all designed to enhance your understanding of key genetic concepts.

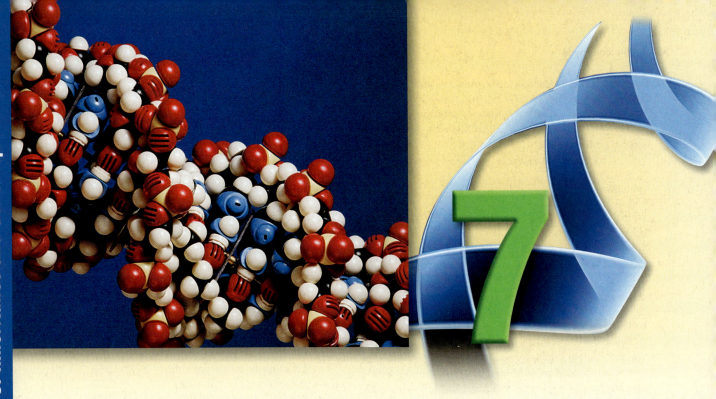

DNA Structure and Chromosome Organization

Essential Goals

1. Learn the key requirements for genetic material and describe the significant experiments that established the identity of the genetic material.

2. Understand the structure of DNA and its organization in chromatin.

3. Be able to distinguish the roles of the different functional units in a chromosome.

Chapter Outline

Photo: A space-filling model of DNA to show how the various atoms are exposed in both the sugar-phosphate backbone and in the major and minor grooves.

The genetic material in any organism must possess certain properties. It must contain information for cellular structure and function, be able to accurately reproduce itself, be mutable, and be located in the chromosomes, which had been found to carry traits from one generation to the next (chapter 4). Chromosomes are composed of DNA, some RNA, and many different proteins—which made it possible that the genetic material was either a nucleic acid or protein.

In 1953, James Watson and Francis Crick published a two-page paper in the journal *Nature* entitled "Molecular Structure of Nucleic Acids: A Structure for Deoxyribose Nucleic Acid." It began as follows: "We wish to suggest a structure for the salt of deoxyribose nucleic acid (D.N.A.). This structure has novel features which are of considerable biological interest." This paper, which first put forth the correct model of DNA structure, is a milestone in the modern era of molecular genetics, compared by some to the work of Mendel and Darwin. Watson, Crick, and X-ray crystallographer Maurice Wilkins won Nobel Prizes for this work; Rosalind Franklin, also an X-ray crystallographer, was acknowledged posthumously to have played a major role in the discovery of the structure of DNA.

Although the basic DNA structure was elucidated over 50 years ago, the packaging of DNA into **chromatin,** which represents the complex of DNA and proteins found in eukaryotic chromosomes, remains an area of active research and affects our changing views of gene regulation, development, and genetic diseases.

We already discussed one important aspect of chromatin organization, condensation of the X chromosome in female mammals to produce a Barr body (chapter 4). Most of the genes on the inactivated X chromosome are not expressed. This balances the level of X-linked gene expression in males and females, even though female mammals have twice the number of X chromosomes as males.

Although X chromosome inactivation blocks the expression of most X-linked genes, it is also possible to alter the chromatin organization of only a region of either a sex chromosome or autosome. This changes the expression of a single or a few genes. This is an important aspect of gene regulation in eukaryotes. However, if this regulation is disrupted, disease phenotypes can be produced.

Rhabdoid tumors are a highly malignant class of tumors that usually occur in children younger than 2 years of age. They affect soft tissues, such as the liver and kidney, and the central nervous system. One cause of these tumors is a mutation in the *SNF5* gene, which encodes a protein component of a chromatin reorganization complex. It is unclear how abnormal regulation of chromatin organization leads to these early-onset tumors. This makes the study of chromatin structure an important aspect of cancer research.

In this chapter, we will discuss the basic structure of nucleotides into DNA and RNA, the organization of bacterial and eukaryotic genomic DNA into chromosomes, and the functional units that are found in chromosomes. We will describe the nucleosome as the basic building block of eukaryotic chromatin organization; however, it will be important to remember that chromosomes exist in a fluid state that allows for genes to be either transcriptionally active or inactive. The concepts introduced here will be useful in the following chapters, in which we discuss DNA replication, transcription, and translation of proteins.

7.1 Required Properties of a Genetic Material

We will begin with a look at the data that proved DNA or RNA is the genetic material. Later sections will describe the chemistry of nucleic acids and their forms in both bacterial and eukaryotic cells. We concentrate on the molecular structure of DNA because, generally, structure reveals function: molecules have shapes that define how they work.

Geneticists knew that whatever the biochemical composition of the genetic material was, it had to meet certain criteria: (1) It must contain information that controls the synthesis of enzymes and other proteins; (2) it must be able to self-replicate with high fidelity, yet allow a low level of mutation; and finally, (3) it must be located in the chromosomes, which had been found to carry traits from one generation to the next (chapter 4).

Property 1: Control of Protein Expression and Function

The growth, development, and functioning of a cell are controlled by the proteins within it, primarily its enzymes. We can therefore say that the proteins within a cell control most of the cell's phenotype. Different cells will contain different proteins, which in turn confer different functions on those cells. For example, a mammalian red blood cell primarily produces hemoglobin to bind oxygen, while a rod photoreceptor cell in the eye primarily synthesizes rhodopsin to detect photons of light.

Most metabolic processes occur in pathways, with an enzyme facilitating each step in the pathway. As an example, the metabolic pathway for the conversion of threonine into isoleucine (two amino acids) appears in **figure 7.1**. The enzyme threonine dehydratase converts threonine into α-ketobutyric acid. The next enzyme in the pathway, acetolactate

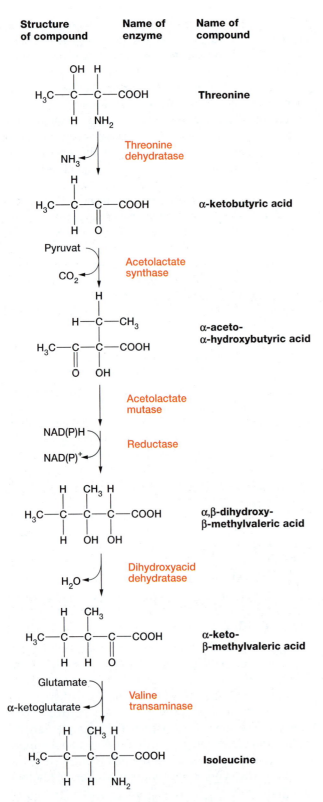

Structure of compound	Name of enzyme	Name of compound

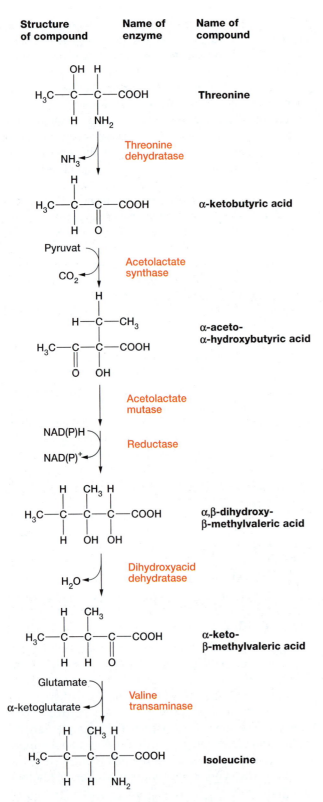

figure 7.1   Metabolic pathway to produce the amino acid isoleucine from threonine. The name of the enzymes that catalyze each biochemical step are in red.

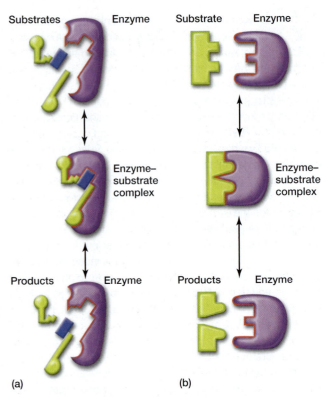

figure 7.2   Two models for the action of an enzyme. (a) The lock-and-key model suggests that the active site of an enzyme (purple) recognizes two substrates. Binding of the substrates by the active site catalyzes the transfer of a molecule between the substrates. The active site is outlined in red. The molecular group that is transferred between the two substrates is blue. (b) The induced-fit model suggests that binding of the substrate causes the active site to change shape, which provides the energy to generate the product.

The sequence of amino acids determines the three-dimensional structure of an enzyme, which is essential for its function. (The structure of proteins is covered in more detail in chapter 11.) The substrate or substrates interact with a part of the enzyme called the **active site** (**fig. 7.2a**). The shape of the active site allows only specific substrates to enter or fit into it. This view of the way an enzyme interacts with its substrates is called the *lock-and-key model* of enzyme functioning. When the substrates are in their proper position in the active site of the enzyme, the particular reaction that the enzyme catalyzes takes place.

Alternatively, the *induced-fit model* proposes that the shape of the enzyme's active site *changes upon binding* of the specific substrate (fig. 7.2b). The altered conformation of the active site is now able to catalyze the necessary reaction. Some enzymes appear to function through the lock-and-key model and others the induced-fit model.

Not all of the cell's proteins function as enzymes. Some are structural proteins, such as keratin, the main component of hair. Other proteins are regulatory—they control the rate at which other enzymes work. Still others are involved in different functions; albumins, for example, help regulate the osmotic pressure of blood.

synthase, converts α-ketobutyric acid into α-aceto-α-hydroxybutyric acid—and so it continues through this five-step path, ending in isoleucine. The end product of each step becomes the substrate for the next.

The genetic material must therefore be capable of producing proteins with a wide variety of amino acid sequences and structures to accomplish an organism's vast number of biochemical processes.

Property 2: Accurate Replication with Mutability

The genetic material must be capable of precisely directing its own replication so that every daughter cell receives an exact copy. In their 1953 paper, Watson and Crick hinted at a replication process based on the structure of DNA. Each strand of the double-stranded DNA molecule could serve as a template for a new strand using the proposed complementary base-pairing. The **fidelity** or accuracy of the replication process is very great, with only about one error in a billion bases. The details pertaining to the mechanism of DNA replication will be discussed in chapter 9.

In contrast to this need to accurately preserve the DNA sequence from generation to generation, the genetic material must be able to mutate or change. Mutation in the genetic material is required over a period of time to allow organisms to evolve. Thus, accurate replication is required to maintain the existence of an organism, while mutability provides a means to introduce change.

Property 3: Location in Chromosomes

Since the early twentieth century, geneticists were fairly certain that the genetic material was located in chromosomes within the nuclei of eukaryotic cells. This idea was based on the correlation between the way chromosomes behaved during cellular division and Mendel's laws of inheritance that described the behavior of genes. The inference was that genetic material in eukaryotes must be part of chromosomes.

However, it was known that chromosomes were composed of both nucleic acids and proteins. For a long time, proteins were considered the most probable genetic material because they possessed the necessary molecular complexity to encode the genetic information. Proteins were composed of 20 naturally occurring amino acids that could be combined in an almost unlimited variety, creating thousands upon thousands of different proteins. In contrast, DNA was composed of only four bases, which appeared to provide only a limited amount of variation. Nevertheless, many researchers speculated that DNA, and not proteins, constituted the genetic material.

Restating the Concepts

▶ Based on scientific observations, the genetic material in any organism must control the expression of proteins, be able to self-replicate with high fidelity, be altered through mutations, and be located in the chromosomes.

7.2 Evidence for DNA as the Genetic Material

The first proof that the genetic material is DNA came in 1944 from Oswald Avery and his colleagues, but their experiments followed other important research findings. By 1953, when Watson and Crick solved the structure of DNA, it was already clear that DNA is the genetic material in most organisms. The classical experiments that revealed nucleic acids as the genetic material are excellent examples of how scientists build on previous findings to ultimately reach an unambiguous explanation.

Transformation

In 1928, Frederick Griffith reported that heat-killing one type of *Streptococcus pneumoniae* could "transform" a different type of *S. pneumoniae,* causing the latter to exhibit traits of the first strain.

The S strain of *S. pneumoniae* produces smooth colonies when grown on media in a petri plate because the cells have outer capsules composed of polysaccharide (**fig. 7.3**). This strain also causes a fatal bacterial infection (pneumonia) in mice.

In contrast, the R strain lacks polysaccharide capsules and produced rough colonies on petri plates (fig. 7.3). The R strain does not have a pathological effect on mice because bacteria of this strain are engulfed by white blood cells, whereas the polysaccharide coat protected the virulent S strain from the white blood cells.

Griffith found that neither heat-killed S strain nor live R strain bacteria, by themselves, killed the mice. However, if he injected a mixture of live R strain and heat-killed S strain bacteria into mice, the mice developed pneumonia identical to that caused by living S strain cells (**fig. 7.4**). Furthermore, live S strain bacteria were recovered from the dead mice. Thus, something in the heat-killed S strain transformed the R strain into live S strain cells. The process by which this trait

Streptococcus pneumoniae

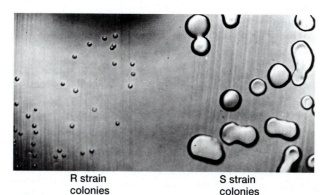

R strain   S strain
colonies   colonies

figure 7.3   Petri plate with rough (R strain) and smooth (S strain) colonies of *Streptococcus pneumoniae.*
(Reproduced from the *Journal of Experimental Medicine,* 1944, 79: 137–158. Copyright 1944 Rockefeller University Press.)

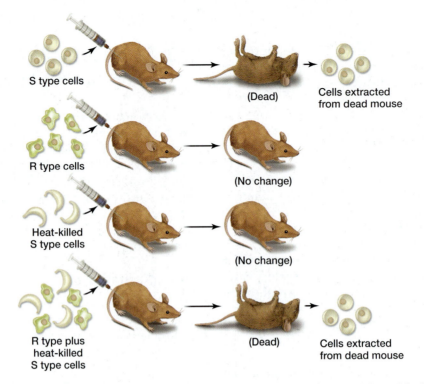

figure 7.4 Griffith's experiment with *Streptococcus*. The live S strain bacteria killed infected mice, while the live R strain bacteria failed to kill infected mice. Similarly, heat-killed S strain bacteria did not kill infected mice. However, mixing heat-killed S strain with live R strain bacteria resulted in death of the mice. Furthermore, live S strain bacteria were recovered from the dead mice.

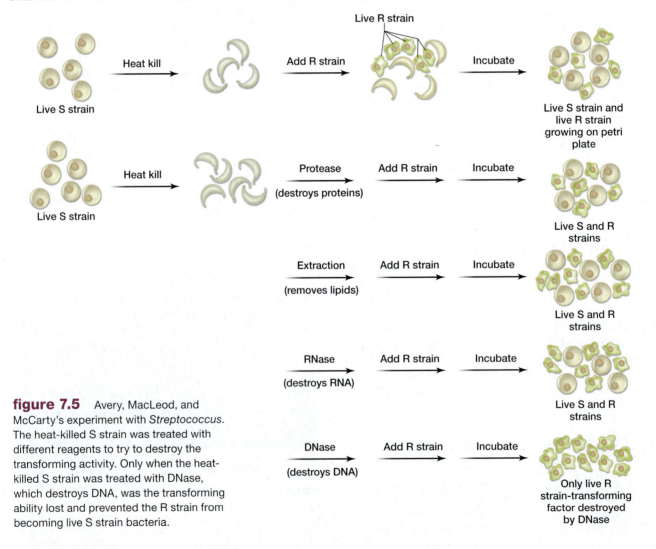

figure 7.5 Avery, MacLeod, and McCarty's experiment with *Streptococcus*. The heat-killed S strain was treated with different reagents to try to destroy the transforming activity. Only when the heat-killed S strain was treated with DNase, which destroys DNA, was the transforming ability lost and prevented the R strain from becoming live S strain bacteria.

moved from one bacterial strain to another was termed **transformation.**

In 1944, Oswald Avery, Colin MacLeod, and Maclyn McCarty expanded on this experiment and identified the transforming substance. Avery and his colleagues did their work **in vitro** (literally, in glass), using the morphology of the bacterial colonies on petri plates, rather than production of pneumonia in mice, as evidence of transformation. The heat-killed S strain was treated with a variety of additional methods to destroy different types of molecules before mixing with the live R strain cells. They demonstrated that destruction of proteins, carbohydrates, and lipids still resulted in the R strain being transformed into the S strain (**fig. 7.5**). In contrast, destruction of DNA blocked transformation of the R strain into the S strain. This provided the first experimental evidence that DNA was the genetic material: only loss of DNA prevented the transformation of the nonvirulent R strain of *S. pneumoniae* into the virulent S strain.

Phage Labeling

Valuable information about the nature of the genetic material has also come from viruses. Of particular importance are studies of bacterial viruses—the bacteriophage, or phage for short. Because phage consist only of nucleic acid surrounded by protein, they lend themselves nicely to the determination of whether the protein or the nucleic acid is the genetic material (**fig. 7.6a**).

In 1952, Alfred Hershey and Martha Chase published results supporting DNA's role in the generation of phage. These experiments were based on the understanding that all nucleic acids, but not bacterial proteins, contained phosphorus. In contrast, most proteins contain sulfur (in the amino acids cysteine and methionine), whereas nucleic acids do not.

Hershey and Chase used radioactive isotopes of sulfur (^{35}S) and phosphorus (^{32}P) to differentially label the viral proteins and nucleic acids during the infection process. They labeled the T2 bacteriophage by infecting

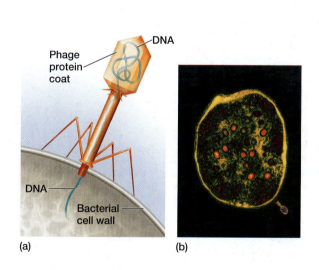

(a)

(b)

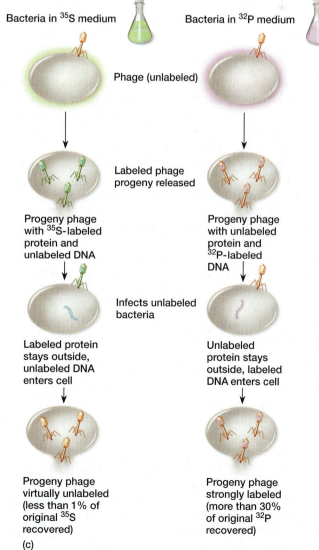

(c)

figure 7.6 The Hershey and Chase experiments using ^{35}S- and ^{32}P-labeled T2 bacteriophage. (a) A drawing of the T2 bacteriophage showing the general location of the DNA and protein coat. (b) A colorized electron micrograph shows T2 phage attached to the lower right portion of an *E. coli* cell wall. (c) Bacteria were grown in either the presence of ^{35}S or ^{32}P. The bacteria were infected with T2, which then incorporated radioactive proteins (^{35}S) or radioactive DNA (^{32}P) in newly synthesized bacteriophage. The labeled T2 were used to infect bacteria grown in nonradioactive media. The protein label (^{35}S) did not enter the bacteria during infection (left); the nucleic acid label (^{32}P) entered the bacteria during infection and was present in the newly synthesized bacteriophage (right).

Escherichia coli that were being grown in culture medium containing either ^{35}S or ^{32}P (fig. 7.6c). Hershey and Chase then mixed either the ^{35}S- or ^{32}P-labeled phage particles with unlabeled bacteria. They removed the phage after they attached to the bacterial wall and identified the phage material injected into the cell. It was known that the phage replicated inside the bacterial cell, which implied that the phage's genetic material must be transferred into the bacterial cell.

When ^{32}P-labeled phage were mixed with unlabeled *E. coli* cells, Hershey and Chase found that the ^{32}P label entered the bacterial cells, and that the next generation of phage released from the infected cells carried a significant amount of the ^{32}P label (fig. 7.6c). When ^{35}S-labeled phage were mixed with unlabeled *E. coli*, the researchers found that the ^{35}S label stayed outside the bacteria for the most part (fig. 7.6c). Hershey and Chase thus demonstrated that the outer protein coat of a phage does not enter the bacterial cell that it infects, whereas the phage's DNA does enter the cell. Because only the phage DNA is found inside the bacterial cell, it must be responsible for the production of the new phage during the infection process. This result demonstrated that the phage's DNA, not the protein, must be the genetic material.

RNA as Genetic Material

In some viruses, RNA (ribonucleic acid) serves as the genetic material. The tobacco mosaic virus that infects tobacco plants consists only of RNA and a protein coat. The single, long RNA molecule is packaged within a rodlike structure formed by over 2 thousand copies of a single protein. No DNA is present in tobacco mosaic virus particles (**fig. 7.7a**).

In 1955, Heinz Fraenkel-Conrat and Robley Williams showed that a virus can be separated, in vitro, into its component parts and reconstituted as a viable virus. This finding led Fraenkel-Conrat and Bea Singer to reconstitute tobacco mosaic virus with parts from different strains (fig. 7.7b). For example, they combined the RNA from the common tobacco mosaic virus with the protein from the masked (M) strain of tobacco mosaic virus. They then made the reciprocal combination of common-type protein and M-type RNA. In both cases, the tobacco mosaic virus produced after infection of tobacco plants corresponded to the RNA type, not the protein. Therefore, the nucleic acid (RNA in this case) was the genetic material.

Subsequently, scientists rubbed purified tobacco mosaic virus RNA into plant leaves. Normal infection and a new generation of typical, protein-coated tobacco mosaic virus resulted, confirming RNA as the genetic material for this virus.

From these and other landmark experiments, geneticists have concluded that DNA (and in some cases RNA) is the genetic material.

(a)

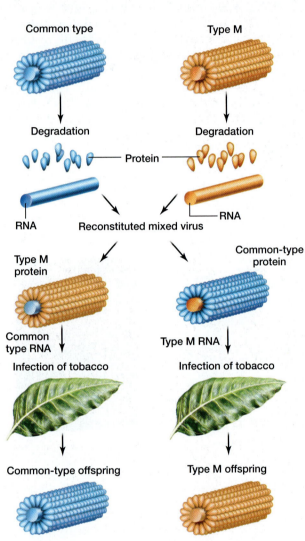

(b)

figure 7.7 (a) Electron micrograph of tobacco mosaic virus. Magnification 37,428. (b) Reconstitution experiment of Fraenkel-Conrat and Singer. The protein coat and RNA from two different viral strains were mixed and used to infect tobacco leaves. The nucleic acid (RNA), not the protein component of the virus, controls inheritance and the viral strain produced after infection.

Restating the Concepts

▶ Experiments that transformed nonvirulent strains of bacteria into virulent strains provided evidence that DNA is the genetic material.

▶ Experiments with bacteriophage showed that DNA, not protein, is the agent that leads to production of more virus particles inside a host cell.

▶ DNA is the most common genetic material, but in some cases, such as in certain viruses, RNA may play this role.

7.3 Chemistry of Nucleic Acids

Having identified the genetic material as DNA (or RNA in some viruses), we will now examine the chemical structure of nucleic acids. Their structure will tell us a good deal about how they function.

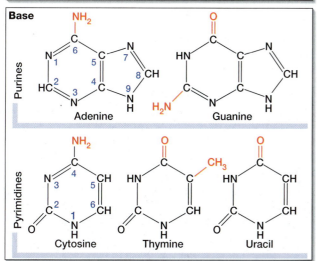

figure 7.8 Components of nucleic acids: phosphate, sugars, and bases. Primes are used to number the ring positions in the sugars to differentiate them from the ring positions in the bases. Ribose and deoxyribose differ in the hydroxyl or hydrogen at the 2' C position.

table 7.1 Components of Nucleic Acids

Base	Phosphate	Sugar	Purines	Pyrimidines
DNA	Present	Deoxyribose	Guanine	Cytosine
			Adenine	Thymine
RNA	Present	Ribose	Guanine	Cytosine
			Adenine	Uracil

The Basic Structure of Nucleic Acids

Nucleic acids are made by joining nucleotides in a repetitive way into long, chainlike polymers. **Nucleotides** are made of three components: (1) a phosphate, (2) a sugar, and (3) a nitrogenous base (**table 7.1** and **fig. 7.8**). A **nucleoside,** by contrast, is a sugar-base compound that lacks phosphate (**fig. 7.9**).

When incorporated into a nucleic acid, a nucleotide contains one of each of these three components. But, when free as a monomer in the cell, a nucleotide usually exists in a triphosphate form. The energy held in the two extra phosphate bonds is used, among other purposes, to synthesize the polymer. Adenosine triphosphate (ATP), the major energy form in the cell, is a nucleotide triphosphate. Similarly, the nucleotide guanosine triphosphate (GTP) is a major signaling molecule within the cell.

The sugars in DNA and RNA differ at a single position. In DNA, the sugar is deoxyribose, which contains a hydrogen at the 2' carbon. The sugar in RNA is ribose, which contains a hydroxyl at the 2' position. (The carbons of the sugars are numbered 1' to 5'. The

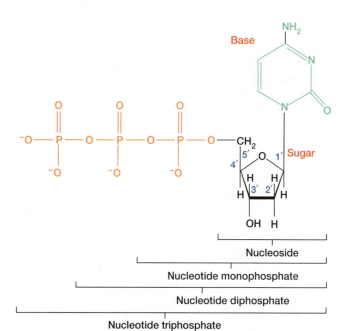

figure 7.9 The structure of a nucleoside and the three nucleotides.

7.1 applying genetic principles to problems

Analysis of the Genetic Material in Prions

Without exception, the genetic material of all living organisms is either DNA or RNA, with RNA being the genetic material in only a few viruses. Because virtually all transmissible diseases are of bacterial or viral origin, you could conclude that transmissible diseases result from organisms with DNA or RNA as their genetic material. This can be demonstrated using the classical experiments already described in this chapter. Let's look at one set of interesting infectious diseases and examine how the transmissible material was identified.

A group of four human diseases and six animal diseases exhibit slow-onset neurological symptoms. The human diseases are kuru, Creutzfeldt-Jakob disease, Gerstmann–Sträussler–Scheinker syndrome, and a recently discovered fatal familial insomnia. The animal diseases are scrapie (sheep and goats), four encephalopathies (bovine, feline, ungulate, and mink), and chronic wasting disease (deer and elk). All of these diseases are extremely slow to develop, with the first symptoms being a loss of coordination. In humans, there is minimal or no dementia associated with the disease. Ultimately the individual is unable to stand or eat, which results in a comatose-like state before dying. There is no known cure for any of these diseases.

In England, a recent epidemic of bovine spongiform encephalopathy ("mad cow" disease) peaked in 1992–1993, affecting over 160,000 cattle. At least 14 cases of a variant of Creutzfeldt–Jakob disease in people in England and France were attributed to eating affected beef, creating a panic in England. A smaller outbreak occurred in Canada in recent years and several infected cattle were identified in the United States in 2003 and 2004.

Clearly, these diseases have grave implications on health of the human population, the world's food supply, and the commerce associated with agriculture. Identifying the infectious agent would be the first step toward generating either a cure or an immunization against the disease.

Stanley Prusiner began studying how these diseases were transmitted and the resulting molecular processes in infected individuals in the 1970s. In 1982, he published a paper that summarized a variety of experiments that examined the possible transmission material in the infectious particles. Using an assay in which hamsters developed the symptoms of scrapie in a relatively short time of 200 days after infection (compared with the years for the symptoms to be exhibited in infected sheep or goats), he essentially repeated the Avery, MacLeod, and McCarty experiment. The treatments Prusiner described are summarized in table 7A.

If DNA is the material responsible for transmission, which of the treatments shown in the table would inactivate the infectivity of the scrapie particle? Because DNA degradation would prevent scrapie from replicating in the infected cell, DNase and UV irradiation would inactivate the scrapie infectivity.

If RNA is responsible for transmission, which treatments would inactivate the infectivity? In this case, RNase and UV irradiation would prevent infectivity. Any of the remaining treatments would likely not affect the infectivity.

If protein is the material that is responsible for transmission, which treatments would inactivate infectivity? In this case, SDS, urea, phenol, or proteinase K would all cause the loss of infectivity. If DNA and RNA had no role in the infectivity, then the DNase, RNase, and UV irradiation would have no effect.

When the experiments were completed, results supporting the last scenario were observed (table 7B). Although the experiments demonstrated the essential role of only protein in scrapie infectivity, many scientists

table 7A   Treatments Used to Explore the Transmissible Material of Scrapie

Treatment	Target and Effect
DNase	DNA degraded
RNase	RNA degraded
UV irradiation	DNA and RNA degraded
SDS	Protein denatured
Urea	Protein denatured
Phenol	Protein denatured
Proteinase K	Protein degraded

SDS: sodium dodecylsulfate; UV: ultraviolet.

table 7B   Effects of Various Treatments on Blocking Scrapie Infectivity

Treatment	Target and Effect	Effect on Infectivity
DNase	DNA degraded	None
RNase	RNA degraded	None
UV irradiation	DNA and RNA degraded	None
SDS	Protein denatured	Infectivity blocked
Urea	Protein denatured	Infectivity blocked
Phenol	Protein denatured	Infectivity blocked
Proteinase K	Protein degraded	Infectivity blocked

argued that this was impossible, and that either DNA or RNA, which was well protected by a protein coat, was the real agent of transmission.

Prusiner named the infective particle a **prion** (for *proteinaceous infectious particle*) because it behaved unlike any known virus or bacterium. He received the 1997 Nobel Prize in Physiology or Medicine for his work on prions.

We now know that the prion is a mutant protein referred to as PrP^{Sc}. The normal protein, PrP^{C} is a glycoprotein found on the membrane of brain neurons and some other tissues. PrP^{C} is encoded by a gene found in most eukaryotes.

The PrP^{Sc} protein, when it is ingested through infected tissues, somehow moves to the brain and causes the PrP^{C} protein in the recipient individual to assume an abnormal form, which is identical to PrP^{Sc}. This altered form of PrP^{C} causes additional PrP^{C} molecules to adopt the abnormal PrP^{Sc} form, which produces the phenotype. The altered PrP^{C}, which is now PrP^{Sc}, is capable of infecting another individual through ingestion.

Thus, these diseases are not propagated through the replication of genetic material and creation of additional viruses, but rather through the modification of a host-encoded protein by an infecting, altered form of the same protein. The only need for genetic material (DNA) is to encode the production of the wild-type host protein (PrP^{C}), which serves as a substrate for the PrP^{Sc} protein to make additional PrP^{Sc} protein.

You might then assume that mutations in the PrP^{C} gene would encode an abnormal form of the PrP protein that could mimic the neurological disease symptoms without requiring infection by the PrP^{Sc} protein. A fatal familial insomnia has been identified that also corresponds to a mutation in the PrP^{C} gene. The symptoms include attention and memory deficits, severe insomnia, and abnormal motor system control. The disease results in death, with a mean duration of the symptoms being 18 months. The identification of a mutation in the PrP^{C} gene that produces similar symptoms further supports the idea that an altered form of the PrP protein, like PrP^{Sc}, is the cause of the prion diseases.

With a change away from using animal matter in cattle feed and a culling of cattle herds, the epidemic of bovine spongiform encephalopathy has ended. However, new human cases may show up in the future owing to the long incubation period of this prion disease. Meanwhile, research continues in an effort to identify the function of the PrP^{C} protein in normal individuals and to find a way to stop the cascade effect.

---

primes are used to avoid confusion with the numbering system of the bases; see fig. 7.8.)

DNA and RNA both have four bases, two **purines** and two **pyrimidines,** in their nucleotide chains. Both molecules have the purines **adenine** and **guanine** and the pyrimidine **cytosine.** DNA contains the pyrimidine **thymine,** while RNA has the pyrimidine **uracil.** To summarize, DNA and RNA differ only in the 2' position of the sugar and a single base, thymine in DNA and uracil in RNA. These two structural differences account for the functional differences between DNA and RNA.

A nucleotide is formed in the cell when a base attaches to the 1' carbon of the sugar and a phosphate attaches to the 5' carbon of the same sugar (**fig. 7.10**); the name of the nucleotide is derived from the base (**table 7.2**). Nucleotides are linked together (polymerized) by the formation of a bond between the phosphate at the 5' carbon of one nucleotide and the hydroxyl (OH) group at the 3' carbon of an adjacent molecule—a linkage termed a **phosphodiester bond** (**fig. 7.11**).

Although the identities of the individual nucleotides that are polymerized to form a strand of DNA or RNA were known, the actual structure of the functional DNA polymer was not determined until 1953. The general feeling was that the biologically active structure of DNA had to be more complex than a single string of nucleotides linked together by phosphodiester bonds. Scientists throughout the world were testing several different models for the structure of DNA, with variations ranging from the number of strands in the molecule to the location of the sugar–phosphate backbone (i.e., either interior or exterior). Watson and Crick built potential models to determine that DNA existed as a two-stranded structure with the sugar–phosphate backbone on the exterior. This model proved to be correct for most DNA molecules in that it was the most consistent with the available data, namely the chemical nature of the components of DNA, Chargaff's ratios for base composition, and X-ray diffraction data.

Chargaff's Ratios

Until Erwin Chargaff's work, scientists had labored under the erroneous **tetranucleotide hypothesis.** This hypothesis proposed that DNA was made up of equal quantities of the four bases; therefore, a subunit of DNA consisted of one copy of each base.

Chargaff carefully analyzed the base composition of DNA in various species (**table 7.3**). He found that although the relative amount of a given nucleotide differs between species, the amount of adenine equaled that of thymine, and the amount of guanine equaled that of cytosine. That is, in the DNA of all the organisms he studied, a 1:1 correspondence existed between the purine and pyrimidine bases. This relationship is known as **Chargaff's rule.**

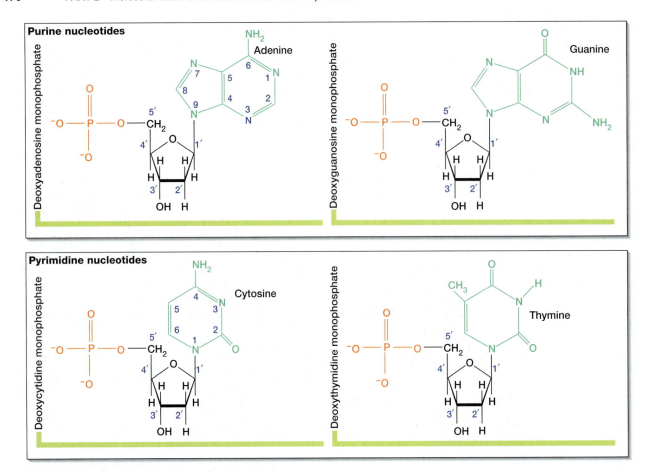

figure 7.10   Structure of the four deoxyribose nucleotides.

table 7.2   Nucleotide Nomenclature

Base	Nucleotide (nucleoside monophosphate)	Monophosphate		Diphosphate		Triphosphate	
		Ribose	Deoxyribose	Ribose	Deoxyribose	Ribose	Deoxyribose
Guanine	Guanosine monophosphate	GMP		GDP		GTP	
	Deoxyguanosine monophosphate		dGMP		dGDP		dGTP
Adenine	Adenosine monophosphate	AMP		ADP		ATP	
	Deoxyadenosine monophosphate		dAMP		dADP		dATP
Cytosine	Cytidine monophosphate	CMP		CDP		CTP	
	Deoxycytidine monophosphate		dCMP		dCDP		dCTP
Thymine	Deoxythymidine monophosphate		dTMP		dTDP		dTTP
Uracil	Uridine monophosphate	UMP		UDP		UTP	

5′-PO$_4$ end

Phosphodiester bond

H$_2$C   5′   Base

Nucleotide component

3′-OH end

figure 7.11 A polymer of nucleotides. Nucleotides in a DNA strand are connected by phosphodiester bonds between the 3' C of the sugar and the phosphate that is attached to the 5' C of the adjacent sugar.

table 7.3 Percentage Base Composition of Some DNAs

Species	Adenine	Thymine	Guanine	Cytosine
Human being (liver)	30.3	30.3	19.5	19.9
Mycobacterium tuberculosis	15.1	14.6	34.9	35.4
Sea urchin	32.8	32.1	17.7	18.4

The sum of the amount of adenine and thymine (A + T), however, did not equal the sum of the amount of cytosine and guanine (C + G). Depending on the species examined, Chargaff found that the percentage of (A + T) in DNA varied from 30% to over 60%. Chargaff's observations disproved the tetranucleotide hypothesis; the four bases of DNA did not occur in a 1:1:1:1 ratio. Chargaff's results provided important insight to Watson and Crick in the development of their model.

DNA X-Ray Diffraction Studies

While Watson and Crick were building models to elucidate the structure of DNA, Rosalind Franklin, working in Maurice Wilkins' laboratory (**fig. 7.12**), was collecting critical data that revealed the larger structure of DNA. The data was based on **X-ray diffraction.** In this technique, DNA molecules are first isolated in a crystal, which arranges the individual DNA molecules in an orderly way. When a beam of X-rays hits the crystal, the beam scatters in an orderly fashion, and the diffraction pattern can be recorded on photographic film or computer-controlled devices. The pattern that is produced reveals important structural features of the DNA in the crystal.

The cross in the center of the photograph in **figure 7.13** indicates that the molecule is a helix; the dark areas at the top and bottom come from the bases, stacked perpendicular to the main axis of the molecule. This image can also be used to determine the distance between adjacent bases and the number of bases in a helical unit. The image also revealed that the helix was composed of two parallel strands that remained an equal distance apart from each other. This image of the DNA molecule and the corresponding calculations greatly contributed to Watson and Crick's understanding of the DNA structure.

(a)   (b)

figure 7.12 Maurice H. F. Wilkins (1916–2004) and Rosalind E. Franklin (1920–1958).

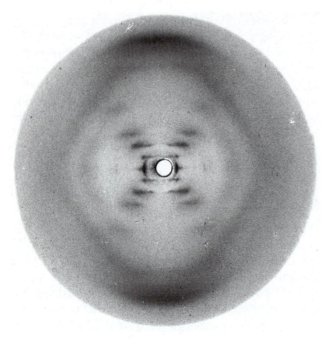

figure 7.13   Diffraction pattern of a beam of X-rays that passed through a DNA crystal. This image, produced by Rosalind Franklin, revealed that DNA was a double helix with a fixed diameter. This data was very helpful to Watson and Crick as they worked to deduce the structure of DNA.

(Reprinted by permission from Macmillan Publishers Ltd. From R. E. Franklin and R. Goslin, "Molecular configuration in sodium thymocucleate," *Nature* 171: 740–41, 1953.)

figure 7.14   A photograph of James D. Watson (1928–), left, and Francis H. C. Crick (1916–2004), right, was taken with their DNA model at the Cavendish Laboratories in 1953.

The Watson-Crick Model

With the data available, Watson and Crick began constructing molecular models (**fig. 7.14**). They found that a possible structure for DNA was one in which two helices coiled around one another (a **double helix**), with the sugar–phosphate backbones on the outside and the bases on the inside. This structure would fit the dimensions X-ray diffraction had established for DNA if the bases from the two strands were opposite each other and formed "rungs" in a helical "ladder" (**fig. 7.15**). The diameter of the helix could only be kept constant at about 20 Å (10 angstrom units = 1 nm) if one purine and one pyrimidine base made up each rung. Two purines per rung would be too big, and two pyrimidines would be too small.

After further experimentation with models, Watson and Crick found that the hydrogen bonding necessary to form the rungs of their helical ladder could occur readily between certain base pairs, namely the pairs that Chargaff found in equal frequencies. Thermodynamically stable hydrogen bonding occurs between thymine and adenine, and between cytosine and guanine (**fig. 7.16**). Two hydrogen bonds connect adenine and thymine and three connect cytosine and guanine. The relationship is one of **complementarity:** for any given base, there is one and only one other base that can optimally hydrogen bond with it.

Another point about DNA structure relates to the **polarity** that exists in each strand. That is, one end of a linear DNA strand has a 5' phosphate and the other end has a 3' hydroxyl group. Watson and Crick found that hydrogen bonding would occur if the polarity of the two strands ran in opposite directions; that is, if the two strands were **antiparallel** (**fig. 7.17**). Notice that the antiparallel nature requires knowledge of the opposite orientation of the two strands, which could not be predicted from X-ray diffraction data.

Alternative Forms of DNA

Although Watson and Crick's model of DNA has proved to be correct, research has subsequently shown that DNA can be either double-stranded or single-stranded, as in some DNA viruses. We also now know that double-stranded DNA exists in three different major forms; A-DNA, B-DNA, and Z-DNA.

The DNA form based on Wilkins' and Franklin's X-ray diffraction data is called **B-DNA.** It is a right-handed helix, that is, it turns in a clockwise manner when viewed down its axis. (Curl your fingers in your right hand and point your right thumb up. A right-handed helix will turn in the direction of your fingers as it moves upward.) In this form, the bases are stacked almost exactly perpendicular to the main axis, with about 10 base pairs (bp) per turn (34 nm per turn; see

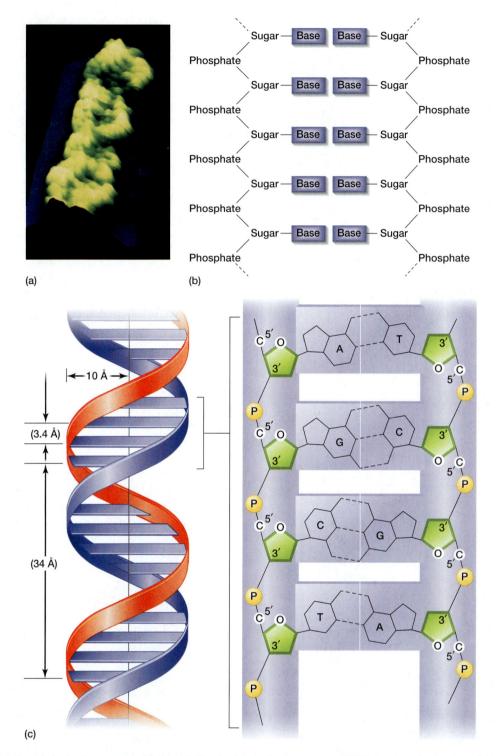

figure 7.15   Double helical structure of DNA. (a) The double-helical structure of DNA is magnified twenty-five million times by scanning tunneling microscopy. The yellow represents the sugar-phosphate backbone. (b) The organization of the basic components of DNA. (c) A drawing of DNA showing the double helix on the left and the polarity of the sugar–phosphate backbone and base pairing on the right. ([a] © John D. Baldeschweiler.)

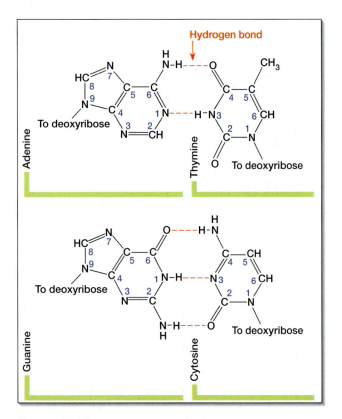

figure 7.16   Hydrogen bonding between the nitrogenous bases in DNA. The atom positions in the four bases are shown.

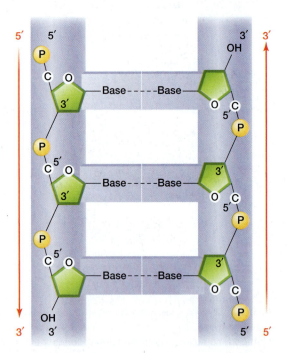

figure 7.17   Polarity of the DNA strands. Polarity is established by the 3' and 5' carbons of the deoxyribose sugar. For example, moving down the left strand, the polarity is 5' → 3' (read as five-prime to three-prime), while moving up the right strand, the polarity is 5' → 3'. This opposite polarity on the two strands is the antiparallel organization of double-stranded DNA.

fig. 7.15c). B-DNA represents the major form of DNA that is present in a cell.

If the water content increases to about 75%, DNA takes on another form, **A-DNA.** The A-DNA is also a right-handed helix. In contrast to B-DNA, the bases of A-DNA are tilted relative to the axis, and there are more base pairs per turn (approximately 11.3 bp per turn). It is thought that double-stranded RNA and hybrid molecules, in which one strand is DNA and the other is RNA, exist in the A form.

In 1979, Alexander Rich and his colleagues at MIT discovered a left-handed helix that they called **Z-DNA** because its backbone formed a zigzag structure (**fig. 7.18**). The structure of Z-DNA was determined in very small DNA molecules composed of repeating G–C sequences on one strand with the complementary C–G sequences on the other (alternating purines and pyrimidines). Z-DNA looks like B-DNA with each base rotated 180 degrees, resulting in a zigzag, left-handed structure. Z-DNA is slightly more compact than either B-DNA or A-DNA, with approximately 12 bp per turn.

Originally, it was thought that Z-DNA would not prove of interest to biologists because it required very high salt concentrations to become stable. However, it was found that Z-DNA can be stabilized under physiologically normal conditions if methyl groups are added

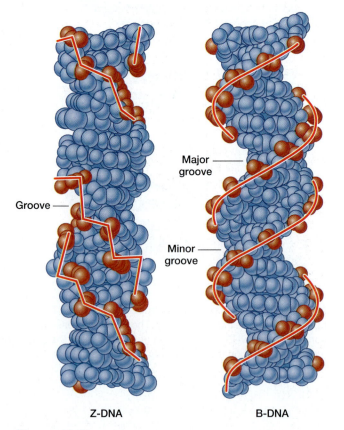

figure 7.18   Z-form (left) and B-form (right) DNA. The lines connect the phosphate groups (red).

(Reprinted, with permission, from the *Annual Review of Biochemistry*, Volume 53 © 1984 by Annual Reviews. www. annualreviews.org.)

to the cytosines. Thus, a single DNA molecule may have one region that contains B-DNA and an adjacent region containing Z-DNA, if the Z-DNA region contains alternating C–G sequences and the cytosines are methylated. Z-DNA may be involved in regulating gene expression in eukaryotes, because cytosine methylation is one important mechanism in gene regulation. We return to this topic in chapter 17.

Four additional forms of right-handed helical DNA have been observed in the laboratory. *C-DNA* occurs under very dehydrated conditions and contains only 9.3 bp per helical turn. *D-DNA* and *E-DNA* are found in synthetic DNA molecules that lack guanines and possess only 7 or 8 bp per helical turn. *P-DNA* was discovered when DNA was stretched so that only 2.6 bp are present per helical turn. It remains to be demonstrated that these latter four forms of DNA exist under physiological conditions and what function, if any, they possess.

 Restating the Concepts

▶ Both DNA and RNA are composed of nucleotides, each of which contains a nitrogenous base, a sugar, and a phosphate group.

▶ DNA contains the sugar deoxyribose and the bases adenine, thymine, guanine, and cytosine. RNA contains the sugar ribose, and uracil is substituted for thymine.

▶ DNA is usually a double helix, with two antiparallel strands that are held together by hydrogen bonding between the bases.

▶ The structure of DNA was determined from X-ray diffraction data, Chargaff's rule, and model building by Watson and Crick.

▶ DNA exists in three different major forms, with the right-handed B-DNA being the most prevalent in the cell and the unique, left-handed Z-DNA potentially being important in gene regulation.

7.4 The Eukaryotic Cell

Eukaryotes, bacteria, and archaea represent the three superkingdoms of organisms. From a genetics standpoint, the archaea have not been as extensively studied as either bacteria or eukaryotes. Although genetic research with prokaryotic organisms has provided a great deal of information, the prokaryotic system is different from that of eukaryotes in several ways. The following comparisons, using *E. coli* as a general model for prokaryotes, demonstrate the greater complexity that is observed in eukaryotes:

1. *E. coli* exists as a simple, single cell. Although some prokaryotes do aggregate, sporulate, and show a few other limited forms of differentiation, they are primarily one-celled organisms. And, although some eukaryotes are single-celled (e.g., yeast), the essence of eukaryotes is differentiation. In human beings, a zygote gives rise to every other cell type in the body in a relatively predictable manner.

2. An *E. coli* cell is small (0.5–5.0 μm in length for bacteria). Eukaryotic cells are generally larger than prokaryotes (10–50 μm in length for animal tissue cells).

3. An *E. coli* cell has very little internal structure. Eukaryotes have a number of internal organelles and an extensive lipid membrane system, including the nuclear envelope itself.

4. The single circular *E. coli* chromosome contains approximately 4.7×10^6 bp of DNA. The haploid human genome, with 23 (females) or 24 (males) linear chromosomes, contains nearly a thousand times more DNA.

5. Most *E. coli* genes are grouped as *operons* that are regulated through a common mechanism; almost all eukaryotic genes are separate and each possesses its own regulatory elements.

6. The chromosomal DNA of *E. coli* is not highly complexed with proteins, although some histonelike proteins are found in the cell. Eukaryotic DNA, by contrast, exists in the form of nucleoprotein, a DNA–histone protein complex.

Because of eukaryotes' greater complexity, we next take a look at the structure and characteristics of the eukaryote chromosome. In later chapters, we cover mechanisms of regulating gene expression and patterns of eukaryotic development. Bacterial genetics is described in its own chapter, chapter 15.

7.5 The Eukaryotic Chromosome

The eukaryotic chromosome is composed of both proteins and DNA. When DNA was shown to be the genetic material, the role of chromosomal proteins became of secondary importance as geneticists raced to identify genes and how their expression was regulated. We now know that the chromosomal proteins play an integral role in both the structure of the chromosome and the expression of the genes in the DNA.

In the following sections we will discuss several of the structural components of eukaryotic chromosomes. Because structure will help to reveal function, we will begin to describe the functional relevance of the different chromosomal structures. However, the details of the various functions will be presented in later chapters.

One DNA Molecule per Chromosome

Evidence that each eukaryotic chromosome contains one double helix of DNA comes from several sources. The best data are provided by radioactive-labeling studies, first done by J. Herbert Taylor, Philip Woods, and Walter Hughes in 1957. In these experiments, eukaryotic cells were grown in culture and allowed to undergo one round of DNA replication in the presence

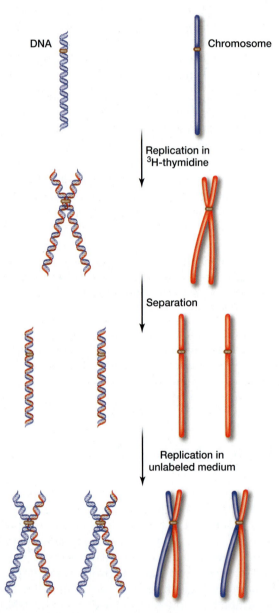

figure 7.19   Radioactive labeling of a eukaryotic chromosome following semiconservative replication (see chapter 9). Replication occurs first in the presence of ^{3}H-thymidine and then in its absence. Red represents the presence of ^{3}H-thymidine. After the second round of replication, one chromatid of each chromosome is labeled, whereas the other is not, confirming that there is only one DNA molecule per chromatid.

of tritiated (^{3}H-) thymidine (dTTP). The radioactive thymidine would be incorporated into the newly replicated daughter strands of DNA.

As we discussed in chapter 3, a single chromosome goes through the S phase of the cell cycle to produce two sister chromatids that are joined at the centromere. If each of the sister chromatids corresponds to a double-stranded DNA molecule, we would expect each chromatid (or each chromosome, after the completion of mitosis) to consist of one unlabeled DNA template strand and one labeled strand of newly synthesized bases (**fig. 7.19**).

A second round of DNA replication, in the absence of ^{3}H-thymidine, should produce chromosomes in which one chromatid would have unlabeled DNA and one would have labeled DNA (see fig. 7.19). **Figure 7.20** shows the chromosomes after this second replication in a nonlabeled medium. As expected, one chromatid of every pair is labeled and one is not. If a chromosome contained more than one double-stranded DNA molecule, we would not always find one labeled chromosome and one unlabeled chromosome after two rounds of replication. (The three arrows in figure 7.20 show sites of sister chromatid exchange that was described in chapter 6.)

In another experiment, Ruth Kavenoff, Lynn Klotz, and Bruno Zimm demonstrated that *Drosophila* chromosomes contained pieces of DNA that corresponded to the size predicted from their DNA content. They isolated DNA from nuclei and measured the size of the largest DNA molecules using the *viscoelastic* property of DNA, which is the rate at which stretched molecules relax. From other sources, primarily UV absorbance studies, it was estimated that the largest *Drosophila* chromosome had about 43×10^9 daltons of DNA. Results from the viscoelastic measurements indicated the presence of DNA molecules of between 38 and 44×10^9 daltons.

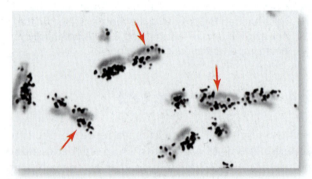

figure 7.20   Metaphase hamster chromosomes after one round of replication in the presence of ^{3}H-thymidine followed by one round in nonradioactive medium. The ^{3}H-thymidine corresponds to the silver grains (black dots) over a single chromatid on each chromosome, which verified that the eukaryotic chromosome contains a single DNA molecule. Where the label apparently switches from one chromatid to the other represents sites of sister chromatid exchanges (arrows).

Chromatin Organization

Based on these and other experiments, we know that each eukaryotic chromosome consists of a single, relatively long piece of duplex DNA that encompasses both arms of the chromosome. Because the average eukaryotic cell contains many chromosomes, this can amount to a large amount of DNA per cell. For example, the DNA in a single human cell would be nearly 2 m long if stretched from end to end. The problem is how to package this DNA inside the nucleus, which is only 2–4 μm (10^{-6} m) in diameter.

Other mechanical challenges exist; for example, in this packaged state, the DNA must remain accessible to RNA polymerase so that transcription of genes can occur. Furthermore, the chromosomes must condense to a greater extent during mitosis and meiosis, so that they can be more easily managed during their distribution to each daughter cell. This packaging of chromosomal DNA occurs at different levels, which allows for a high level of organization.

Nucleosome Structure

The first level of chromatin packing is the formation of nucleosomes. **Nucleosomes** consist of DNA wrapped around a core of basic proteins. The nucleosomes can be observed by placing interphase nuclei in a hypotonic liquid, such as water, which releases the higher order packaging of the chromosomes. Using an electron microscope, the DNA can be seen wrapped around the protein core (**fig. 7.21**), which is composed of **histone** proteins (**table 7.4**). The histones, a group of arginine- and lysine-rich basic proteins, are especially well suited to bind to the negatively charged DNA (**table 7.5**).

The structure and composition of nucleosomes have been well studied. When chromatin is treated with DNase, large amounts of DNA are degraded, leaving only small particles composed of DNA and proteins. The DNA in these small DNA–protein complexes consists of fragments of approximately 210 bp or multiples of 210 bp (**fig. 7.22**). This DNA is protected from DNase digestion due to its tight association with the proteins.

Treatment of the 210-bp DNA–protein complex with micrococcal nuclease removes additional DNA. At this point, the protected DNA is 165 bp in length. Harsh treatment of these smaller DNA–protein complexes releases the basic H1 histone protein, and the DNA is digested down to a final size of approximately 145 bp (see fig. 7.22).

When the quantities of the various histones in the final core were measured, it was found that the protein part of the core consisted of two molecules of each of four different histones: H2A, H2B, H3, and H4. These four histone proteins possess nearly identical amino acid sequences and are functionally equivalent in all eukaryotes.

These data allow us to build a model for the organization of the nucleosome. Approximately 145 bp of genomic DNA is intimately associated with the octomer of four histone proteins. We know that the DNA wraps approximately 1.75 times around this histone core (**fig. 7.23**). The DNA located between the nucleosomes, which amounts to approximately 55–75 bp depending on the species, is termed the linker DNA (see fig. 7.23).

table 7.4 The Constituency of Calf Thymus Chromatin

Constituent	Relative Weight*
DNA	100
Histone proteins	114
Nonhistone proteins	33
RNA	1

* Weight relative to 100 units of DNA.

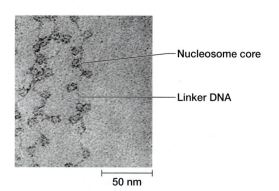

figure 7.21 Electron micrograph of chromatin fibers. Photo shows nucleosome core structures (spheres) connected by strands of linker DNA.

table 7.5 Composition of Histones

Class	Amino Acids	Number of Amino Acids	Percentage of Basic Amino Acids
H1	Very lysine-rich	213	30
H2A	Lysine-, arginine-rich	129	23
H2B	Moderately lysine-rich	125	24
H3	Arginine-rich	135	24
H4	Arginine-, glycine-rich	102	27

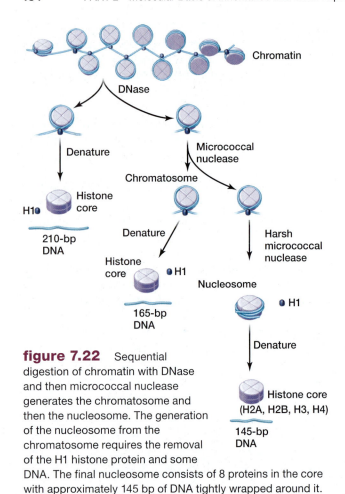

figure 7.22 Sequential digestion of chromatin with DNase and then micrococcal nuclease generates the chromatosome and then the nucleosome. The generation of the nucleosome from the chromatosome requires the removal of the H1 histone protein and some DNA. The final nucleosome consists of 8 proteins in the core with approximately 145 bp of DNA tightly wrapped around it.

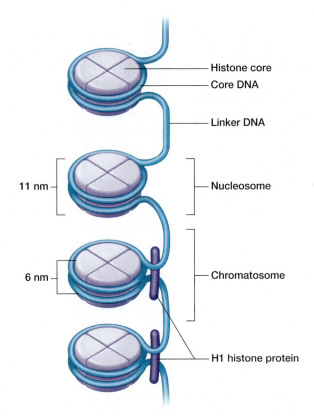

figure 7.23 The eukaryotic chromosome is associated with histone proteins to form nucleosomes. The DNA wraps 1.75 times around the protein core and a length of DNA (linker DNA) connects nucleosomes. The chromatosome contains a nucleosome, linker DNA, and H1 histone protein.

The H1 histone interacts with the nucleosome and some of the linker DNA as it enters and exits the nucleosome, which implies that it may serve to cross-link nucleosomes (see fig. 7.23). The term **chromatosome** has been suggested for the core nucleosome plus the H1 protein, a unit that includes approximately 165 bp of DNA (**fig. 7.24**). Nucleosomes, then, are a first-order packaging of DNA; they reduce its length approximately seven-fold and undoubtedly make the coiling and contraction required during mitosis and meiosis more efficient (**fig. 7.25**).

Fluid Nature of the Nucleosome

Although nucleosomes serve as a general, first-order packing mechanism in eukaryotic DNA, they are not static structures. The binding of the histone core by the DNA is a major factor in controlling gene expression. The histones and nucleosome can be altered to regulate transcription in three major ways.

First, the histone proteins can be exchanged for slightly different forms of the protein. For example, the Barr body in female mammals replaces histone H2A with histone mH2A to assist in the inactivation of the X chromosome. In yeast, histone H2A can be replaced with H2A.F/Z. Variant histone proteins are found in all eukaryotes and participate in regulating chromatin conformation and gene expression.

Second, some histone proteins can be modified—in particular, phosphate groups, methyl groups, and acetyl groups added and removed from specific amino acids in the H3 and H4 histone proteins. These modifications can lead to enhanced or repressed transcription. Finally, histone cores can be removed from the DNA or moved to a different location on the DNA.

A change in the location or composition of the histone core may make the DNA more or less accessible to proteins that are essential for transcription, such as RNA polymerase, or for transcription factors that regulate transcription initiation. Because not all genes are actively transcribed in every eukaryotic cell, the location of nucleosomes changes during development or varies in different tissues, to account for the activation or repression of gene transcription. Details of the processes are discussed in chapter 17.

Higher-Order Structure of Chromatin

Because the nucleosome has a width of only 11 nm, and each sister chromatid of a metaphase chromosome has an approximate diameter of about 700 nm (**fig. 7.26**), several additional levels of chromatin compaction must lead

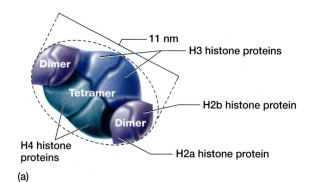

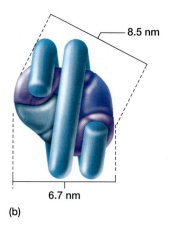

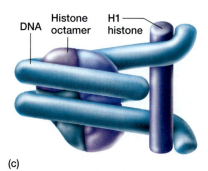

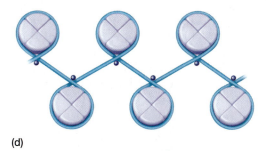

(d)

figure 7.24   Nucleosome structure. (a) Schematic comparison of the eight histones comprising the nucleosome in salt solution. A dimer consists of one H2A and one H2B histone molecule; a tetramer consists of two H3 and two H4 histones. (b) DNA fits in surface grooves on the more compacted histone structure found in physiological conditions. (c) The structure in panel b is rotated 90° to show the presumed position of the H1 histone, encompassing 165 base pairs of DNA in the chromatosome. (d) One potential model of how the H1 histone packages the nucleosomes into a higher order structure.

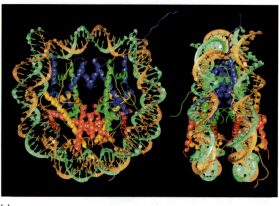

(a)

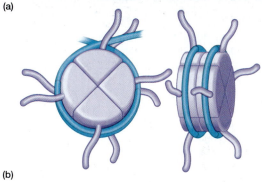

(b)

figure 7.25   Nucleosome structure. (a) The nucleosome core particle is shown at 2.8 Å resolution. Shown are 145 base pairs of DNA (brown and turquoise) wrapped around the eight histone proteins (purple: H3; green: H4; yellow: H2A; and red: H2B). Note the protein tails (N-terminal ends) of the histone polypeptides extending out of the nucleosome. On the left is a view of the nucleosome face, and on the right is the side view. (b) Below each ribbon diagram is the corresponding diagram of the nucleosome core from figure 7.24.

(Reprinted by permission from Macmillan Publishers Ltd. From Karolin Luger, et al., "Crystal structure of the nucleosome core particle at 2.8Å resolution" *Nature* 389: 251–260, 1997.)

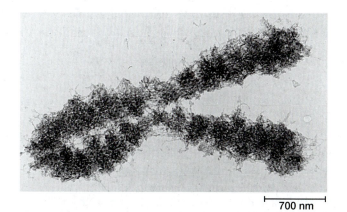

700 nm

figure 7.26   Chinese hamster chromosome. Note the fibers making up the chromosome; they are approximately 240 nm in diameter.

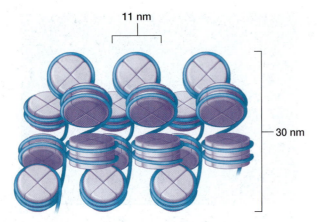

figure 7.27   Solenoid model for the formation of the 30 nm chromatin fiber. Nucleosomal DNA wraps in a helical fashion, forming a hollow core. Although histone H1 is not shown, it is known to be on the inside of the solenoid.

to the metaphase chromosome. The next level of chromatin packaging involves the 11 nm nucleosomal DNA to spontaneously form a **30-nm solenoid fiber,** a hollow helix resulting from the coiling of the nucleosomal DNA (**fig. 7.27**), with approximately six nucleosomes per turn.

The helix of the solenoid appears to shift between right-handed and left-handed orientations, but the purpose of the different orientations remains unclear. Unlike in the uncoiled nucleosomal DNA, which contains linker DNA separating adjacent nucleosomes, the nucleosomes likely make physical contact in the 30-nm solenoid. The formation of the solenoid results in an additional six- to seven-fold compaction of the DNA.

The 30-nm solenoid is then packaged into a **radial loop–scaffold organization.** In this structure, specific sequences of genomic DNA, called **scaffold-associated regions (SARs),** bind to a scaffold of nonhistone

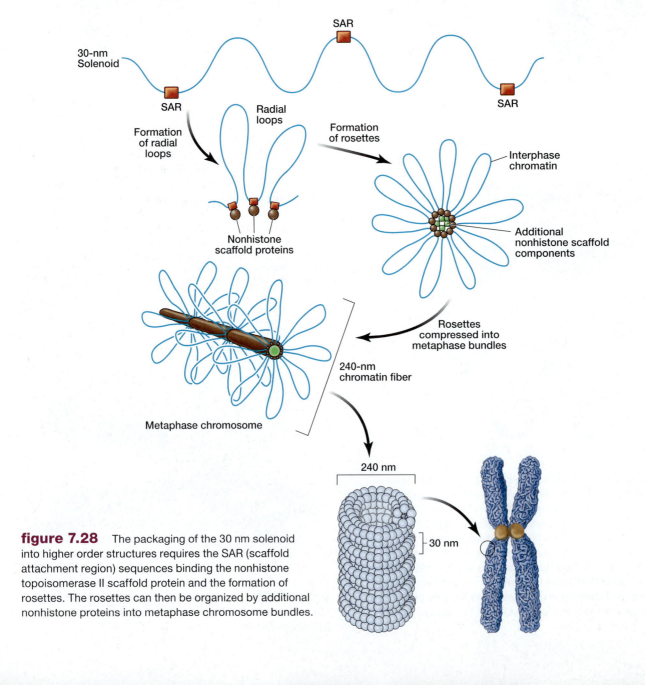

figure 7.28   The packaging of the 30 nm solenoid into higher order structures requires the SAR (scaffold attachment region) sequences binding the nonhistone topoisomerase II scaffold protein and the formation of rosettes. The rosettes can then be organized by additional nonhistone proteins into metaphase chromosome bundles.

proteins that include topoisomerase II. These radial loops of approximately 50–100 kb of genomic DNA are then gathered together by additional nonhistone proteins to form a rosette (**fig. 7.28**). As the cell enters prophase, the formation of loops and rosettes may increase to generate the most compact chromatin structure, which is observed in the metaphase chromosome.

The evidence for this radial loop–scaffold model comes from the removal of histone proteins from a chromosome. When this occurs, the DNA billows out and leaves the proteinaceous scaffold (**fig. 7.29**). The ultimate result of the chromatin packing is a 10,000-fold compaction of the naked genomic DNA.

Polyteny, Puffs, and Balbiani Rings

The larval salivary glands, as well as some other tissues of *Drosophila* and other diptera, contain giant banded chromosomes (see **fig. 7.30a**) that result from the replication of the chromosomes and the synapsis of homologs without cell division in a process called **endomitosis.** These chromosomes, termed **polytene chromosomes,** consist of more than a thousand copies of the same chromatid that are precisely aligned and produce alternating dark bands and lighter interband

regions when stained. The dark bands are referred to as *chromomeres.*

Polytene chromosomes also reveal diffuse areas called **chromosome puffs** (fig. 7.28b), which are also referred to as **Balbiani rings.** These puffs were originally identified in the midge *Chironomus* whose polytene chromosomes were discovered by Edouard-Gérard Balbiani in 1881. Currently, the term applies to all puffs, or at least the larger ones, in all species with polytene chromosomes.

The structure of the polytene chromosome can be explained by the diagram in **figure 7.31**. Dark bands

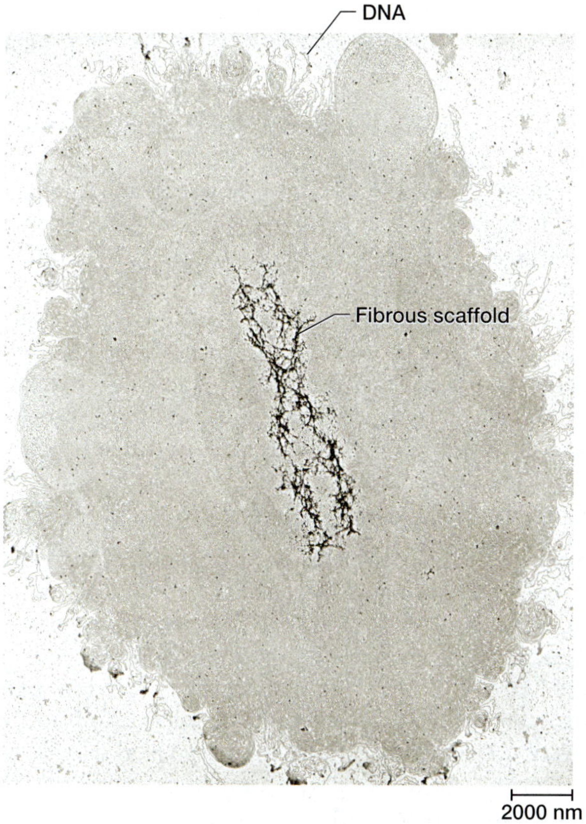

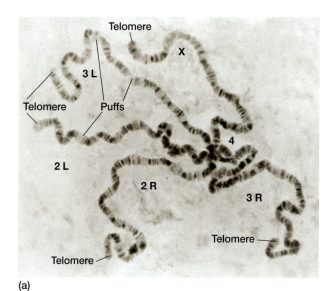

(a)

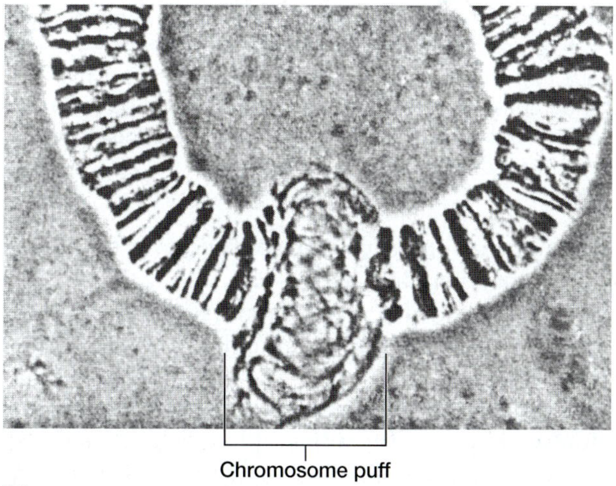

(b)

figure 7.30 (a) The *Drosophila* salivary gland contains polytene chromosomes, which are composed of more than a thousand copies of each chromosome. The precise alignment of all the chromosomes in the polytene chromosome allows dye staining to produce a high resolution banding pattern. (b) A chromosome puff on the left arm of chromosome 3 of the midge *Chironomus pallidivittatus.*

2000 nm

figure 7.29 Electron micrograph of a chromosome's scaffold and DNA. When the histones are removed from the eukaryotic chromosome, a fibrous scaffold remains. The DNA loops out from this scaffold.

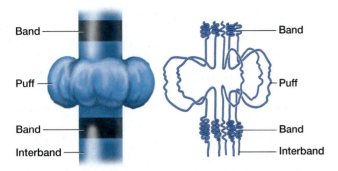

figure 7.31 Schematic of how bands and a puff could be produced in a polytene chromosome. Five of the several thousand precisely aligned chromatids in a polytene chromosome are diagrammed on the right.

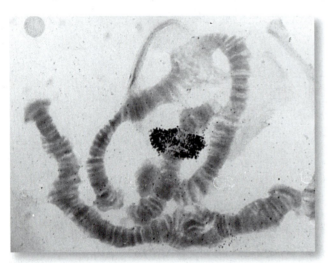

figure 7.32 Hybridization at a *Chironomus tentans* polytene salivary gland chromosome puff. The chromosomes are denatured and then hybridized with a radioactively labeled probe that produces the silver grains (black dots) on the photographic emulsion that lies over the chromosome. Thus, the black dots show the location of the corresponding gene in the chromosomal puff.

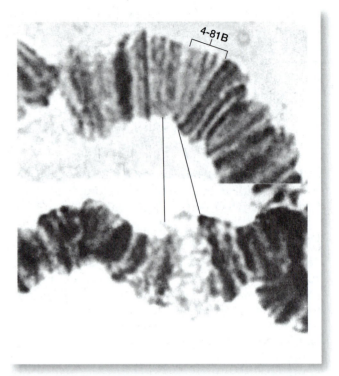

figure 7.33 Puff 4-81B in the *Drosophila hydei* salivary gland is induced by heat shock (37° C for one-half hour). The chromosome at the top lacks the puff, while the lower chromosome reveals the heat shock-induced puff that represents increased transcription of the genes in this region.

(chromomeres) are due to tight coiling of the 30-nm fiber; light interband regions are due to looser coiling. The figure also shows how chromosome puffs would come about as fibers unfold in regions of active transcription.

Staining with reagents specific for RNA, such as toluidine blue, or autoradiography with tritiated (^{3}H) uridine, have been used to demonstrate that puffs correspond to sites of large amounts of active transcription. Although active transcription may be occurring in neighboring regions of the polytene chromosome, it is at a much lower level relative to the puff regions. The mRNA isolated from cells with puffs has also been shown to hybridize primarily to the puffed regions of the chromosomes (**fig. 7.32**). Modern recombinant DNA techniques have also shown that many puffs probably represent the transcription of only one gene, although there are exceptions.

Puffs generally fall into four categories. *Stage-specific puffs* appear during a certain stage of development, such as molting in insects. *Tissue-specific puffs* are active in one tissue but not another. *Constituitive puffs* are present and actively transcribed almost all the time in a specific tissue. And *environmentally induced puffs* appear after some environmental change, such as heat shock (**fig. 7.33**). In all these cases, the appearance of puffs is associated with an increase in transcription.

Lampbrush Chromosomes

Lampbrush chromosomes, which occur in amphibian oocytes, are so named because their looped-out configuration has the appearance of a brush for cleaning kerosene lamps, now a relatively uncommon household item (**fig. 7.34**). The loops of the lampbrush chromosomes are covered by an RNA matrix and are the sites of active transcription. Presumably, the loops are unwindings of the single chromosome, similar to the unwindings in the polytene chromosome shown in figure 7.31. Thus, under certain circumstances, such as in polytene chromosomal puffs and in lampbrush chromosomes, active transcription can be seen in the light microscope.

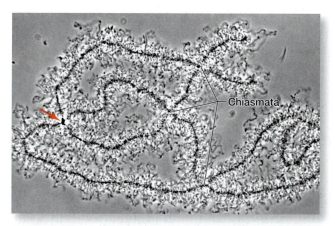

figure 7.34   Lampbrush chromosome of the newt, *Notophthalmus viridescens*. Centromere is at the left (arrow); the two long homologs are held together by three chiasmata (labeled).

 Restating the Concepts

▶ Each eukaryotic chromosome is composed of a single DNA molecule that must be properly packaged to generate a compact structure that can fit in the nucleus.

▶ The first order of chromatin organization is the nucleosome, which is a core of four different histone proteins with DNA wrapped around the outside. The nucleosome is not a static structure; the interaction between the histones and DNA can change to help regulate gene expression.

▶ The second order of chromatin organization is the formation of the 30-nm solenoid, which then forms radial loops on a nonhistone protein scaffold. The formation of the radial loop–scaffold results in a 10,000-fold compaction of the genomic DNA.

▶ Polytene chromosomes are thousands of sister chromatids that synapse in a cell that has not undergone cell division. The formation of polytene chromosomes allows for the visualization of puffs, which are sites where the chromatin becomes less compacted and high levels of gene transcription occur. The location of puffs can change, which highlights the dynamic nature of chromatin organization.

▶ Lampbrush chromosomes are another physical proof that chromatin organization is controlled to help regulate transcription.

Chromosomal Banding

Chromosomes can be identified and characterized based on the banding patterns that are generated using several different staining techniques (see fig. 3.34). This cytogenetic map, which generates a series of visible bands along each chromosome, is useful in

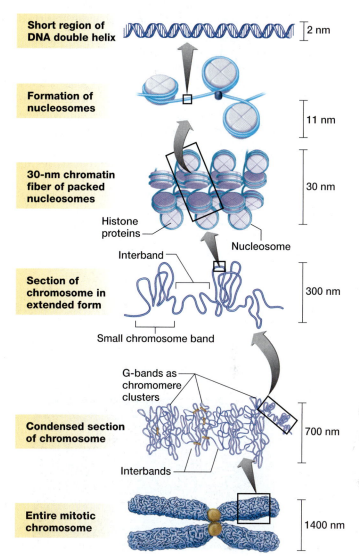

figure 7.35   Model of eukaryotic (mammalian) chromosomal banding. G-bands are chromomere clusters, which result from the contraction of smaller chromomeres. These, in turn, result from looping of the 30 nm fiber.

the localization of genes. Of possibly greater importance is the fact that these staining techniques have provided some insight into the structure of the chromosome. The techniques for staining the C, G, and R chromosomal bands will serve as an illustration.

G bands are obtained with **Giemsa stain,** a complex of stains that are specific for the phosphate groups of DNA. Treatment of fixed chromatin with trypsin or hot salts brings out the G bands. Giemsa stain enhances banding that is already visible in mitotic chromosomes, which is caused by the arrangement of chromomeres. Under careful observation, the major G bands consist of many smaller chromomeres, which may be produced by the mechanism of chromosomal folding shown in **figure 7.35**.

C bands are Giemsa-stained bands that are observed after the chromosomes are treated with sodium hydroxide. The C in C bands is for "centromere," because these bands represent highly condensed DNA surrounding the centromeres (**fig. 7.36**). This DNA is also usually rich in **satellite DNA,** which has a base content different from the majority of the cell's DNA. As you will see, this satellite DNA consists of numerous repetitions of a short sequence.

R bands are visible with a technique that stains the regions between G bands. The chromosomes are fixed, stained with Giemsa, and then viewed with a phase-contrast microscope. Because the dark–light pattern is the opposite of the G band pattern, these bands are called *reverse (R) bands.*

The consequence of these different staining procedures is that three types of chromatin were identified; euchromatin, constitutive heterochromatin, and facultative heterochromatin (**table 7.6**). Presumably, the only chromatin involved in transcription is **euchromatin. Constitutive heterochromatin,** which includes satellite DNA surrounding the centromere, is transcriptionally inactive during all stages of development and in all cells. **Facultative heterochromatin,** which is found throughout the chromosome, has the capacity to shift between a heterochromatin and euchromatin state.

Genes are found in both the euchromatin and the facultative heterochromatin. The facultative heterochromatin usually shifts to euchromatin when transcription of the gene located in that region is required. Because actively transcribed genes are primarily found in the euchromatin, the expression of some genes is regulated by shifting between a euchromatin and a heterochromatin state. It should be apparent to you that the eukaryotic chromosome is a complex and dynamic structure.

Centromeres and Telomeres

Two regions of the eukaryotic chromosome—the centromere and the telomeres—have specific functions. The **centromere** is involved in chromosomal movement during mitosis and meiosis, whereas the **telomeres** terminate the linear chromosomes.

Centromeres

As we pointed out in chapter 3, the terms *centromere* and *kinetochore,* although occasionally used interchangeably, are distinct. The centromere is the visible constriction in the chromosome that contains specific DNA sequences, whereas the kinetochore is the proteinaceous interface between the centromere and the spindle microtubules.

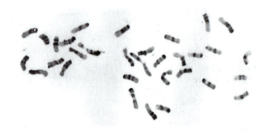

(a)

(b)

figure 7.36   (a) C banding of chromosomes from a cell in the bone marrow of the house mouse, *Mus musculus.* (b) Fluorescent *in situ* hybridization of human chromosomes. The location of the yellow fluorescence indicates where the satellite DNA probe hybridized to the centromeres of human chromosomes.

table 7.6   The Three Major Types of Chromatin in Eukaryotic Chromosomes

	Euchromatin	Centromeric Constitutive Heterochromatin	Facultative Heterochromatin
Relation to bands	In R bands	In C bands	In G bands
Location	Chromosome arms	Usually centromeric	Chromosome arms
Condition during interphase	Usually dispersed	Condensed	Usually condensed
Genetic activity	Usually active	Inactive	Inactive in the heterochromatin state and active in the euchromatin state
Relation to chromomeres	Interchromomeric	Centromeric chromomere	Chromomeres

The eukaryotic kinetochore contains proteins and some RNA. Microscopically, it is a trilaminar structure, attached to chromatin at the inner layer and to microtubules at the outer layer (**fig. 7.37**).

Centromeres appear to serve two important functions in the cell. First, they are the sites at which the kinetochores associate. Second, they are essential in keeping the sister chromatids together during mitosis and meiosis. Although this mechanism is not clearly understood, researchers have identified proteins that bind specifically to the centromere until anaphase II of meiosis. This observation suggests that these proteins are involved in keeping the sister chromatids together, and that they do this by binding to the centromere.

Most of our knowledge of the genetics of centromeres has come from work in yeast (*Saccharomyces cerevisiae*). It is possible to create artificial yeast chromosomes (*Yeast Artificial Chromosome, YAC*) that are linear molecules that faithfully replicate and properly segregate during mitosis. YACs require a centromere, telomeres, and an origin of replication to be functional. These three elements were cloned from yeast chromosomes and then reduced in size to identify the minimal functional unit of each element. After sequencing the centromeres of 15 of the 16 yeast chromosomes, investigators concluded that the yeast centromere is about 250 bp long with three consensus regions (**fig. 7.38**), in which two regions flank a central region of approximately 80 to 85 bp that is extremely AT-rich. Recent data indicate that this centromere region may contain a single, modified nucleosome. The 250-bp long yeast centromere is about 20 nm, the same as the diameter of a microtubule, indicating that only one microtubule

attaches to each centromere during mitosis or meiosis in a yeast cell. This region is called a **point centromere** because a single microtubule attaches to the centromere (fig. 7.38).

Higher eukaryotes have larger centromeric regions that attach more microtubules. These regions are referred to as **regional centromeres** (see figs. 7.37 and 3.12). We know much less about regional centromeres than we do about point centromeres.

Centromeres also serve a practical use to cytogeneticists. As we discussed in chapter 3, the location of the centromere along the length of the chromosome can be used to categorize chromosomes within a species and even help identify a specific chromosome. We described how chromosomes are classified as metacentric, submetacentric, acrocentric, and telocentric depending on the location of the centromere from the center of the chromosome (metacentric) to near the end (telocentric).

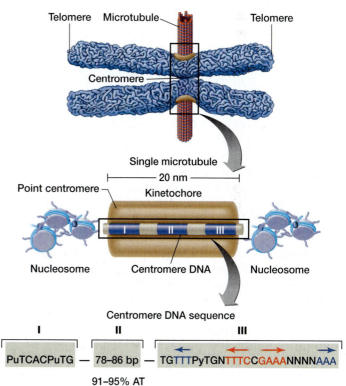

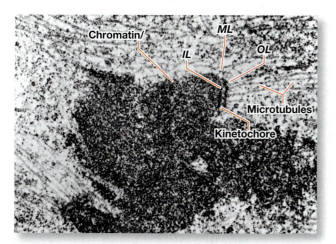

figure 7.37 The kinetochore of a metaphase chromosome of the rat kangaroo. IL, ML, and OL refer to inner, middle, and outer layers, respectively, of the kinetochore. Note the microtubules attached to the kinetochore and the large mass of dark-staining chromatin making up most of the figure. Magnification 30,800×.

figure 7.38 Organization of a yeast centromere. The yeast centromere is attached to the kinetochore. The width of one kinetochore suggests that it attaches to a single microtubule. The yeast centromere is approximately 250 bp and composed of three regions. Region I is a highly conserved sequence, Region II is 78–86 bp that consists of almost entirely A-T base pairs, and Region III is another conserved sequence that contains two different inverted repeats. Pu represents any purine, Py represents any pyrimidine, and N represents any base. The arrows appear over inverted repeat sequences.

Telomeres

Because eukaryotic chromosomes are linear, each has two ends that are referred to as **telomeres** (see fig. 7.30a). The telomeres not only mark the ends of the linear chromosome, but also have several specific functions. Telomeres must prevent the chromosomal ends from being degraded by exonucleases and allow chromosomal ends to be properly replicated. This is critical in preventing the cell from losing genomic DNA as it goes through many rounds of DNA replication and mitosis.

Most telomeres isolated so far are a sequence of five to eight bases that is repeated hundreds or thousands of times at each end of every chromosome. In human beings, the telomere sequence is 5'-TTAGGG-3', repeated 300–5000 times at the end of each chromosome. This highly conserved sequence is found at the telomere in all vertebrates studied as well as in unicellular trypanosomes. Similar sequences are found in various other eukaryotes (**table 7.7**).

We will discuss in chapter 9 how the telomeres are replicated. At this point, you only need to know that after replication, there is a short single-stranded region at the 3' end of the linear DNA molecule, which is called a **3' overhang** (**fig. 7.39**). This 3' overhang, of usually less than 50 bases, is always the G-rich strand of telomeric DNA. This single-stranded DNA is very susceptible to nucleases, which would remove the single-stranded sequence and permanently remove an increasing amount of DNA from the telomeres. Three different structures have been identified that protect the 3' overhang of telomeres from degradation.

First, the guanine-rich DNA in the telomere repeat can form complex structures. Biochemists have discovered that four guanines can form a planar **G-tetraplex,** with the four bases hydrogen bonded to each other (**fig. 7.40a**). One possible G-tetraplex structure at the end of a chromosome is shown in figure 7.40b. You should notice that multiple G-tetraplex complexes may exist on a telomere if there are sufficient guanines in the 3' overhang DNA.

table 7.7 Telomere Sequences in Eukaryotes; The G-Rich Strand of the Double Helix Is Shown

Organism	Telomere Repeat (shown 5' to 3')
Human beings, other mammals, birds, reptiles	TTAGGG
Trypanosomes	TTAGGG
Holotrichous ciliates (*Tetrahymena*)	TTGGGG
Hypotrichous ciliates (*Stylonychia*)	TTTTGGGG
Yeast	TG, TGG, and TGGG
Plants	TTTAGGG

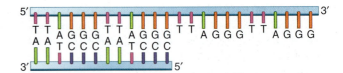

figure 7.39 The telomere region of a linear chromosome contains a 3' single-stranded DNA overhang.

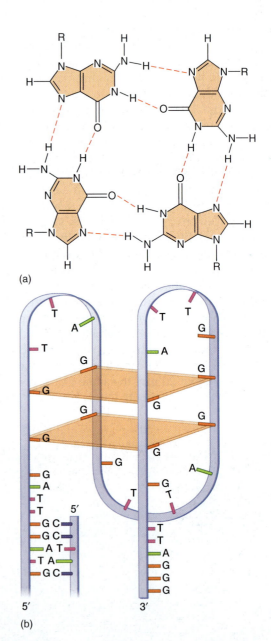

figure 7.40 Formation of a G-tetraplex in the telomere. (a) A diagram showing the hydrogen bonding (dashed red lines) between four guanines in a single plane of the G-tetraplex. The deoxyribose (R) attached to each guanine base (tan fill) is shown. (b) The structure of the G-tetraplex that can form within the single-stranded region of the telomere is shown. The repeating telomere sequence (TTAGGG) forms the G-tetraplex, which is a four-stranded structure, through hydrogen bonding between four guanines in a single plane. This structure shows two planes of guanine hydrogen bonding.

(a)

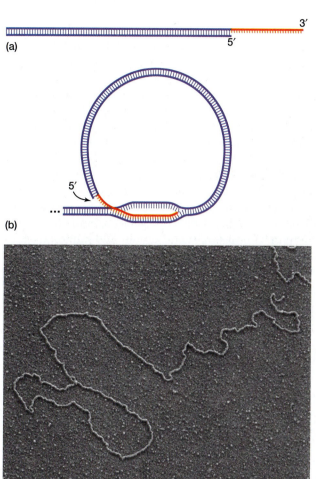

(a)

(b)

(c)

figure 7.42 The t-loop at the end of the mammalian telomere. (a) A diagram of how the t-loop is formed by the interdigitation of the 3' end of the telomere into the double helix. (b) Electron micrograph of a t-loop from a mouse liver cell. The loop is about 10,000 bases around.

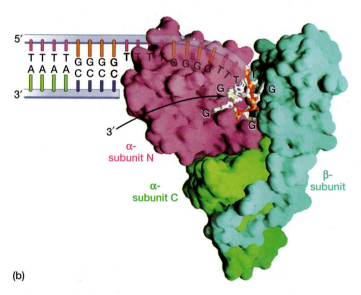

(b)

figure 7.41 The telomere end-binding protein (TEBP) was first identified in the ciliate *Oxytricha nova* (a). The TEBP is composed of two protein subunits (b). The α and β subunits form a deep cleft in which lies the 3' end of the telomere. The folding of the protein into its final form around the DNA may only occur after the DNA has bound, explaining how the DNA could be recognized and placed into such a deep cleft.

A second structure involves proteins that bind to the 3' ends of telomeres. In the ciliate *Oxytricha nova*, a protein called the telomere end-binding protein (TEBP) attaches to the single-stranded 3' ends of telomeres and protects them (**fig. 7.41**).

Finally, a novel structure called the **t-loop** has been discovered at mammalian telomeres. This loop forms under the direction of a protein called TRF2 (*telomere repeat-binding factor*), which causes the 3' overhang to loop around and interdigitate into the double helix, forming the loop (**fig. 7.42**). In all three structures, the single-stranded 3' overhang either base-pairs with another DNA sequence or is bound by proteins to hide the single-stranded end from nucleases.

Restating the Concepts

▶ Various staining techniques have revealed that the chromosome is composed of different types of DNA. Although heterochromatin is transcriptionally inactive, the euchromatin contains the vast majority of the actively transcribed genes.

▶ Centromeres, which are a physically distinct chromosomal compaction, are composed of a repetitive DNA sequence that binds specific proteins to hold the sister chromatids together and to generate the kinetochore.

▶ Telomeres are a repetitive DNA sequence that is located at the end of linear chromosomes.

▶ The telomere contains a single-stranded 3' overhang region. To protect this sequence from being degraded by exonucleases, the single-stranded region either forms a G-tetraplex, or a t-loop, or is bound by specific proteins.

The C-Value Paradox

When you examine the genome size of different species, you would expect that simpler organisms would have smaller genomes than complex organisms, largely because simpler organisms would require fewer genes. Often this is the case—but when two different eukaryotes are examined, the size of the genome does not always correspond to the complexity of the organism. For example, human beings have 3.3 billion base pairs in the haploid genome, whereas an amoeba has more than 200 billion base pairs. Why does an amoeba have almost 70 times more DNA than the haploid human genome, which must encode genes to generate a functionally complex nervous system and numerous specialized internal organs?

Even closely related organisms can possess large differences in the size of their genomes. The average bony fish has over 300 billion base pairs of DNA in its haploid genome, but the Japanese puffer fish has less than half a billion base pairs. If the basic bony fish can be created with less than half a billion base pairs, why does the average bony fish have over 600 times more DNA? What is this excess DNA doing?

These questions define the **C-value paradox,** in which *C* refers to the quantity of DNA in a cell. To explain the C value paradox, researchers examined the repetitiveness of DNA, and more recently, probed and sequenced DNA to understand its properties.

One of the surprises that came from genomic analysis was that some of the genomic DNA corresponds to highly repetitive sequences. Some of this DNA represents **transposable genetic elements,** which are DNA sequences that can move to different locations in the genome. In prokaryotes, some transposable elements have evolved to contain genes that encode resistance to different antibiotics.

In addition to the standard transposable elements, eukaryotes also possess **retrotransposons,** which are transposable elements that move by way of an RNA intermediate. That is, the retrotransposon DNA is transcribed into RNA and then *reverse transcribed,* which is an enzymatic process that converts single-stranded RNA into double-stranded DNA, before it is inserted into the genome. These elements can make up 50% of the eukaryotic genome, existing in hundreds of thousands of copies.

Retrotransposons generally fall into two categories: LINES and SINES. **Long interspersed elements (LINES),** are up to 7 thousand base pairs long and contain genes for reverse transcription, RNA binding, and endonuclease activity. **Short interspersed elements (SINES)** are generally derivatives of transfer RNAs that were converted to DNA by reverse transcription, and then reinserted into the host's genome. They rely on the reverse transcriptase encoded by LINES or retroviruses. In chapter 18, we will describe the variety of transposable elements that are present in both bacteria and eukaryotes.

At this point, we can see some potential explanations for the C-value paradox. Much eukaryotic DNA is so-called **junk DNA,** not having a purpose but apparently doing no harm. In some cases, 97% of the host genome is composed of junk DNA. Recent work seems to indicate that gross differences in DNA content between higher organisms may be due to the differing abilities of various species to rid themselves of this repetitive DNA rather than the complexity of the organism.

It is also possible that this repetitive DNA may perform some function within the genome and our inability to recognize that function results in the label. In future years, we may find that junk DNA actually serves a valuable service to either the genome or the organism.

Restating the Concepts

▶ The C-value paradox refers to the unexpected variation in the quantity of DNA (C value) in different eukaryotic organisms. Sometimes complex organisms have less DNA than simpler ones.

▶ Junk DNA refers to repetitive sequences of DNA that have no apparent value in an organism's genome. In particular, retrotransposons can exist in tens of thousands of copies in a eukaryotic genome.

▶ Two categories of retrotransposon elements are under study: long interspersed elements (LINEs) and short interspersed elements (SINEs).

genetic linkages

In previous chapters, we dealt with a variety of quantitative analyses, such as deducing genetic principles from the phenotypic ratios (chapters 2, 4, and 5) and linkage maps (chapter 6). Although this chapter does not have the same level of quantitative analysis, its material is fundamental to a large number of topics in genetics.

We reviewed the experiments that revealed DNA corresponded to the genetic material. These experiments have been used since as standard tests to confirm the identity of genetic material in a variety of organisms. In particular, the Avery, MacLeod, and McCarty experimental approach has been used to confirm that the genetic material of some viruses is RNA, and that prions lack either DNA or RNA as genetic material. In this way, you can see that classical experiments continue to have a bearing in modern genetics.

We discussed the general structure of DNA and its bases. The complementary nature and polarity of the double-stranded DNA molecule is essential for understanding the mechanism of DNA replication (chapter 9), DNA sequencing (chapter 12), and how some spontaneous mutations are generated and how those mutations are detected (chapter 18). We touched very briefly on the topic of DNA replication in this chapter and will explore the mechanistic details in chapter 9.

Although we discussed the structure of DNA in this chapter, you will learn more about its similarities and differences with RNA in chapter 10. Understanding the differences between DNA and RNA will reveal why only certain types of processes, such as splicing out introns in eukaryotic mRNA (chapter 10), can occur with RNA and not DNA.

You also were introduced to the structure of chromosomes, including chromatin packaging, chromosome banding, centromeres, and telomeres. The formation of nucleosomes and higher-order packing were explained, a topic we will revisit when we discuss regulation of eukaryotic transcription (chapter 17). We also touched on the nucleotide sequences that are particular to the centromere. We finished our discussion with information about transposable elements, which will be revisited in chapter 18.

In chapter 8, we will consider the effects of changes in the structure of chromosomes as well as the effect of changes in chromosome number. One of the best-known human examples of a change in chromosome number is trisomy 21, in which three copies of chromosome 21 exist. The result is the phenotype known as Down syndrome.

Chapter Summary Questions

1. For each term in column I, choose the best matching phrase in column II.

   Column I
   1. Griffith
   2. Purine
   3. Histone
   4. Nucleotide
   5. Solenoid
   6. Prion
   7. Nucleoside
   8. Hershey and Chase
   9. Chromatosome
   10. Pyrimidine

   Column II
   A. Nucleosome + H1 protein
   B. Nitrogenous base with two rings
   C. Performed the transformation experiment
   D. Base + sugar
   E. Base + sugar + phosphate
   F. Basic protein that binds DNA
   G. Contains six nucleosomes per turn
   H. Nitrogenous base with one ring
   I. Infectious protein
   J. Bacteriophage experiment

2. What evidence led to the idea that DNA was the genetic material?

3. How does DNA fulfill the requirements of a genetic material?

4. For a double-stranded DNA molecule, the sequence of one strand is 5' CATTAGACCGGTAGAC 3'. What is the sequence of the complementary strand? Label the 5' and 3' ends.

5. Describe the various ways, both structurally and in components, that DNA and RNA differ.

6. Nucleic acids, proteins, carbohydrates, and fatty acids could have been mentioned as potential genetic material. What other molecular moieties (units) in the cell could possibly have functioned as the genetic material?

7. Roughly sketch the shape of B and Z-DNA, remembering that B-DNA is a right-handed helix and Z-DNA is a left-handed helix.

8. Summarize the major differences between eukaryotes and prokaryotes, including the structures of their DNAs.

9. Why is higher order chromosomal structure expected in eukaryotes but not in prokaryotes?

10. Summarize the evidence that the eukaryotic chromosome contains a single DNA molecule.

11. What are the major protein components of the eukaryotic chromosome? What are their functions?

12. What evidence is used to determine the length of DNA associated with a nucleosome?

13. What is the protein composition of a nucleosome? What function does histone H1 have?

14. What are the relationships among the 11-nm, 30-nm, and 240-nm fibers of the eukaryotic chromosome?

15. Draw a mitotic chromosome during metaphase. Diagram the various kinds of bands that can be brought out by various staining techniques. What information is known about the DNA content of these bands?

16. Diagram a 30-nm fiber model of the chromosome that explains the existence of G bands.

17. Define the following terms: (a) centromere; (b) kinetochore; (c) telomere; (d) polytene chromosome.

18. Under what circumstances does a chromosomal puff occur? What does it signify?

19. Describe three ways in which cells protect their telomeres.

20. What is the C-value paradox, and how is it explained?

Solved Problems

PROBLEM 1: What can be concluded about the nucleic acids in the following table?

Nucleic Acid Molecule	%A	%T	%G	%C	%U
a	28	28	22	22	0
b	31	0	31	17	21
c	15	15	35	35	0

Answer: We must first look to see if U or T is present, for this will indicate whether the molecule is RNA or DNA, respectively. Molecule b is RNA; a and c are DNA. Now we look at base composition. In double-stranded molecules, A pairs evenly with T (or U) and G pairs with C. This relationship holds for molecules a and c, so they are double-stranded; molecule b is single-stranded. Finally, the higher the amount of G-C, the more thermodynamically stable the molecule; so molecule c is more stable than molecule a.

PROBLEM 2: The single *E. coli* chromosome contains approximately 4.7×10^6 bp of DNA. If you linearize the circular genome and the DNA is in the B form, how long would the DNA molecule be?

Answer: The B form of DNA contains 10 bp per helical turn, which corresponds to 34 Å, or 3.4 nm. Thus, each base pair represents 0.34 nm. The DNA molecule would then correspond to

$$4.7 \times 10^6 \text{ bp} \times 0.34 \text{ nm/bp} = 1.6 \times 10^6 \text{ nm}$$

Because 1.0 nm represents 1×10^{-9} m, the *E. coli* genome would be 1.6×10^{-3} m, or 1.6 mm long.

PROBLEM 3: Human chromosome 21 contains 3.3×10^7 bp. How many H1 and H3 histone proteins are normally associated with this chromosome?

Answer: The nucleosome contains an average of 210 bp, and there are two H3 histone proteins in each nucleosome and one H1 protein associated with each nucleosome. In 3.3×10^7 bp, there would be $3.3 \times 10^7/210 = 1.6 \times 10^5$ nucleosomes. Thus, human chromosome 21 would have 1.6×10^5 H1 proteins and 3.2×10^5 H3 proteins.

Exercises and Problems

21. Deduce whether each of the nucleic acid molecules in the following table is DNA or RNA and single-stranded or double-stranded.

Nucleic Acid Molecule	%A	%G	%T	%C	%U
a	33	17	33	17	0
b	33	33	17	17	0
c	26	24	0	24	26
d	21	40	21	18	0
e	15	40	0	30	15
f	30	20	15	20	15

22. A double-stranded DNA molecule contains 28% guanosine (G).
    a. What is the complete base composition of this molecule?
    b. Answer the same question, but assume the molecule is double-stranded RNA.

23. Why were the ^{32}P and ^{35}S radioactive isotopes used in the Hershey–Chase experiment? Could ^{3}H and ^{14}C have worked? Explain.

24. If the tetranucleotide hypothesis were correct regarding the simplicity of DNA structure, under what circumstances could DNA be the genetic material?

25. A double-stranded DNA molecule contains 60 A's and 105 G's. Calculate the total number of bases in this molecule.

26. The melting temperature, T_m, of double-stranded DNA refers to the temperature at which 50% of the DNA has denatured into single strands. The following are melting temperatures for five DNA molecules: 73°C, 69°C, 84°C, 78°C, 82°C. Arrange these DNAs in increasing order of percentage of G–C pairs.

27. Which of the following base compositions is possible in double-stranded DNA?
    a. Only T.   b. Only A and G.   c. Only G and C.

28. We normally think that single-stranded nucleic acids should not melt, but many, in fact, do have a T_m. How can you explain this apparent mystery? Can you think of other ways to express these relationships?

29. Which of the following is true of all double-stranded DNA samples?
    a. $(A + T)/(C + G) = 1$
    b. $(A + C)/(G + T) = 1$
    c. $(GA)/(CT) = 1$
    d. $G/C = 1$
    e. $A/T = 1$
    f. $C/T = 1$

30. How could mutations involving telomeres lead to cancer?

31. Assuming that the four bases are found in equal proportions, how many possible sequences are there for a five-base stretch of single-stranded DNA?

32. One strand of a double-stranded DNA molecule contains 30% A, 20% C, 10% G, and 40% T. What are the percentages of the four bases in the *entire* molecule?

33. What results would you get in the experiment shown in figures 7.19 and 7.20 if the eukaryotic chromosome did not contain a single DNA molecule, but instead had several DNA molecules?

34. The genome of a particular organism consists of 3×10^8 bp of B-DNA.
    a. How many phosphorus atoms are found in the double helix?
    b. What is the length of the DNA in microns?

35. Kavenoff and colleagues determined the size of DNA in *Drosophila* chromosomes in two ways: (1) Spectrophotometric measurements were made on the largest intact chromosome. These measurements were then used to calculate the amount of DNA in each chromosome. (2) Nuclei were gently lysed and the chromosomes were isolated. The lengths of the longest DNA molecules were measured, and those lengths were used to determine the amount of DNA in each molecule. What results for each method would you expect if
    a. the chromosomes contain one DNA molecule?
    b. the chromosomes contain more than one DNA molecule?

36. When chromatin is partially digested with an endonuclease, the proteins removed, and the DNA separated in a sizing gel, DNA fragments in multiples of 200 bp are found. Provide an explanation for this observation.

37. The yeast *Cryptococcus neoformans* has a genome of 2.1×10^7 bp of B-DNA. Assume that all nuclear DNA is associated with nucleosomes and that the average length of linker DNA is 65 bp. What is
    a. the number of nucleosomes in this genome?
    b. the total number of *core* histone molecules?

38. In a single-stranded DNA molecule, the amount of G is twice the amount of A, the amount of T is three times the amount of C, and the ratio of pyrimidines to purines is 1.5:1. What is the base composition of the DNA?

39. If chromatin is digested with an endonuclease to produce 200-bp fragments, and these fragments are then used for transcription experiments, very little RNA is made. Provide an explanation for this observation.

40. A double-stranded DNA molecule consists of 100,000 bp. Calculate the number of complete turns in this molecule if the DNA were in the
    a. B-form.   b. Z-form.

41. Can nucleosomes contain the DNA for one gene? Explain.

42. A particular DNA molecule has an (A+G)/(C + T) ratio of 1.0. Can you determine for sure if this molecule is double-stranded or single-stranded? Explain your answer.

43. Would you expect archaeal species to have nucleosomes? Why or why not?

44. DNA and RNA differ in two major ways: DNA has deoxyribose sugar, whereas RNA has ribose, and DNA has thymine, whereas RNA has uracil. Why might those differences exist other than accidents of evolution?

45. The genome of *Mycobacterium tuberculosis* consists of a single circular chromosome that has a diameter of about 478 μm in the relaxed (unsupercoiled) state. Calculate the number of base pairs in this bacterial chromosome.

46. How could comparative DNA studies aid us in understanding the roles of the different kinds of DNA present in the eukaryotic chromosome?

47. A double-stranded DNA molecule contains 20% adenine. Determine the *number* of cytosine bases if the DNA molecule is
    a. 1000 bp long.   b. 1 cm long.

Chapter 7 Integration Problem

The animal *Fauna hypotheticus* has a diploid chromosome number of 28. It has a nuclear genome consisting of 2×10^9 bp of B-form DNA, with a G + C content of 60%. Sex in this organism is determined by the XY chromosomal system.

a. Determine the number of chromosomes and the number of DNA molecules found in a cell: (i) during G1 of interphase; (ii) at the start of prophase of mitosis; (iii) at the end of anaphase of mitosis; (iv) after cytokinesis of mitosis; (v) at the start of prophase I of meiosis; (vi) at the start of prophase II of meiosis; and (vii) after cytokinesis of meiosis II?

b. Determine the total number of base pairs of DNA and the number of adenines found in the nuclear DNA of a cell at: (i) the start of mitosis; (ii) the end of mitosis; (iii) the start of meiosis I; and (iv) the end of meiosis II?

c. Which is heavier: (i) the nucleus of a secondary oocyte or the nucleus of a first polar body; and (ii) the nucleus of a secondary oocyte or the nucleus of a secondary spermatocyte? Explain your answers.

d. Assuming that nucleotides are arranged at random, how many times is the sequence 5' TAGAC 3' expected to occur in this genome?

e. If one chromosome has a strand of DNA with an (A + G)/(C + T) ratio of 0.5, what is the same ratio in the complementary strand?

f. The average molecular weight of a DNA base pair is 660 Daltons (*Note:* 1 Dalton [Da] = mass of a single hydrogen atom or 1.67×10^{-24} g). The body of the animal contains approximately 0.33 g of nuclear DNA. How far, in kilometers, would this DNA stretch if placed end to end?

Do you need additional review? Visit **www.mhhe.com/hyde** for practice tests, answers to end-of-chapter questions and problems, interactive exercises, and animation tutorials all designed to enhance your understanding of key genetic concepts.

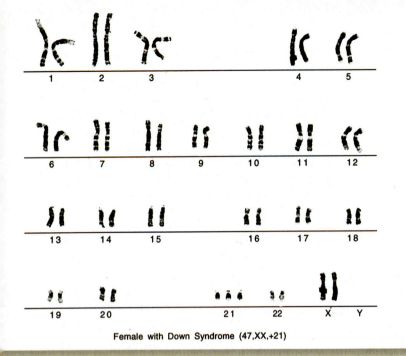

Female with Down Syndrome (47,XX,+21)

Changes in Chromosome Structure and Number

Essential Goals

1. Understand the cytogenetic and inheritance features associated with inversions, deletions, duplications, and translocations.

2. Describe the genetic consequences of each of the four chromosomal rearrangements delineated in goal number 1.

3. Define the differences between aneuploidy and euploidy.

4. Understand how aneuploidy and euploidy affect mitosis and meiosis.

5. Compare how aneuploidy and euploidy affect the viability of the organism.

Chapter Outline

Photo: Karyotype of a female with trisomy 21, Down syndrome.

198

*I*n this chapter, we will discuss changes in the structure and number of chromosomes, a field termed **cytogenetics.** Although some of these changes are visible with a light microscope using stained chromosomes (chapter 7), many of the changes are too small to be visible. In particular, analysis of human chromosomes reveals only very large changes because of the lack of resolution of stained bands.

This chapter presents you with information on how these alterations occur and what their consequences are to the organism. As you might expect, these changes have profound implications for the organism's survival and phenotype.

You are probably already familiar with trisomy 21, a condition in which an extra chromosome 21 occurs and results in Down syndrome. Similarly, a number of human syndromes have been described that result from an extra X or Y chromosome, as well as a missing sex chromosome. However, only a few syndromes exist in which an individual exhibits an extra or missing autosome. This suggests that the number of autosomes must be very tightly regulated. In contrast, structural rearrangements within and among chromosomes can result in effects ranging from lethal syndromes to essentially normal phenotypes.

As one example, over 90% of individuals who suffer from chronic myelogenous leukemia (CML) have a particular translocation of genomic DNA between chromosome 9 and chromosome 22. This translocation has been called the Philadelphia chromosome (named after the city where it was discovered), and its identification now constitutes a major diagnostic criterion for CML. Similarly, over 80% of the patients with follicular lymphoma possess a translocation between chromosomes 14 and 18. This appears to prolong the lifespan of B lymphocytes, which increases the likelihood that these cells may become cancerous.

It is important to also realize that not all chromosomal rearrangements are deleterious to the organism. The normal deletion of DNA during B-cell maturation forces every B cell to commit to the production of only a single immunoglobulin chain. This process produces hundreds of millions of different antibodies from only a few hundred coding sequences.

8.1 Variation in Chromosome Structure: An Overview

Chromosomes may break due to ionizing radiation, physical stress, or chemical compounds. Every chromatid break produces two ends. These ends have been described as "sticky," which means that enzymatic processes in the cell tend to rejoin broken ends rather than attaching a broken end to an undamaged end of another chromosome. If broken ends are not brought together, they will remain broken. However, this will result in the loss of some of the genetic material, which will likely affect the phenotype of the cell or organism.

If broken chromatid ends are brought together, they may rejoin in several different ways. First, the two broken ends of a single chromatid can reunite. Second, the broken end of one chromatid can fuse with the broken end of another chromatid, resulting in an exchange of chromosomal material and a new combination of alleles. Multiple breaks can lead to a variety of alternative recombinations.

Types of Breaks

The types of breaks and reunions discussed in this chapter can be summarized as follows:

I. Noncentromeric breaks
  A. Single breaks
    1. Restitution
    2. Deletion
    3. Dicentric bridge
  B. Two breaks (same chromosome)
    1. Deletion
    2. Inversion
  C. Two breaks (nonhomologous chromosomes)
    1. Reciprocal translocation
    2. Dicentric bridge
II. Centromeric breaks
  A. Fission
  B. Fusion

Single Breaks

If a single chromosome breaks, the broken ends may rejoin in a process called **restitution.** This type of break has no consequences because the original chromosome is regenerated. But when the ends do not rejoin, the result is an **acentric fragment,** without a centromere, and a **centric fragment,** with a centromere (**fig. 8.1**). The acentric fragment, however, will fail to faithfully segregate properly. It will subsequently be excluded from the nuclei formed, and eventually it degrades.

In contrast, the centric fragment migrates normally during mitosis or meiosis because it has a centromere. However, it lacks a telomere, which will prevent it from faithfully replicating the broken end in subsequent rounds of DNA replication. This will lead to the continual loss of larger amounts of DNA from the deleted end during each round of DNA replication.

A single break can have yet another effect. Occasionally, the two centric fragments of a single chromosome may join, forming a two-centromere, or **dicentric chromosome** and leaving the two acentric fragments to join or, alternatively, remain as two fragments (see fig. 8.1).

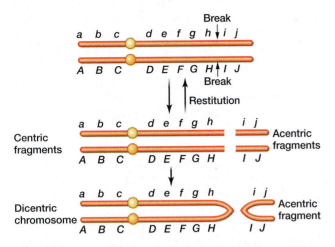

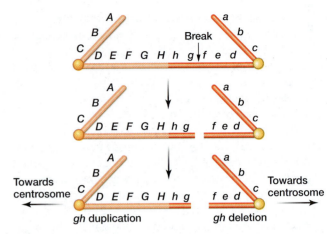

figure 8.1 The consequences of a chromosomal break in homologous chromosomes. A chromosomal break in both homologs produces two centric fragments (containing a centromere) and acentric fragments (lacking centromeres). The broken ends may rejoin to regenerate two homologous chromosomes (restitution) or the two acentric fragments may rejoin and the two centric fragments may rejoin to form a dicentric chromosome (containing two centromeres).

figure 8.2 The consequences of a dicentric chromosome. The dicentric chromosome that was generated in figure 8.1 is shown here. During mitosis or meiosis, the two centromeres are pulled to the opposite centrosomes, which results in a random break in the chromosome between the centromeres. The result is one chromosome has a deletion (*g* and *h*) and the other chromosome has a duplication (*g* and *h*). This is in addition to the deletions they both underwent in figure 8.1 (loss of *i* and *j* or *I* and *J*).

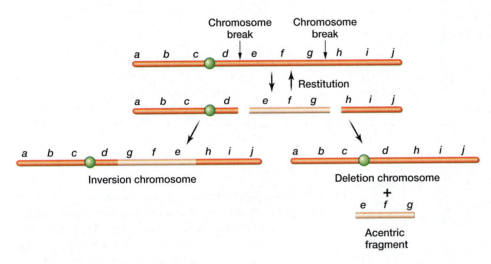

figure 8.3 Three possible consequences of a double break (*top arrows*) in the same chromosome. The internal fragment (*e f g*) may rejoin between the two end pieces in the opposite orientation to generate an inversion chromosome (left). Alternatively, the two end fragments could join to form a deletion chromosome and the internal fragment remains an acentric fragment (right). A third possibility is restitution of the original chromosome from the three fragments.

The acentric fragments are lost, as mentioned before. Because the centromeres are on sister chromatids, the dicentric fragment is pulled to opposite ends of a mitotic cell, forming a bridge; or, if meiosis is occurring, the dicentric fragment is pulled apart during the second meiotic division. The ultimate fate of this bridge is breakage as the spindle fibers pull the centromeres to opposite poles (or possibly exclusion from a new nucleus if the bridge is not broken).

The dicentric chromosome does not necessarily break in the middle, however, and the subsequent off-center break exacerbates the genetic imbalance: duplications occur on one strand, whereas more deletions occur on the other (**fig. 8.2**). In addition, the "sticky" ends produced on both fragments increase the likelihood of repeating this **breakage–fusion–bridge cycle** in each generation. The great imbalances resulting from the duplications and deletions usually cause the cell to die within several generations.

Two Breaks in the Same Chromosome

Figure 8.3 shows three possible results when two breaks occur in the same chromosome. One alternative is a reunion that omits an acentric fragment, which is then lost. The resulting chromosome then has a deletion or loss of material (*e-f-g* in fig. 8.3). Two breaks in the same chromosome can also lead to an **inversion,** in which the middle section is reattached but in the inverted configuration (see fig. 8.3). The third possibility is restitution of the original chromosome.

Two Breaks in Nonhomologous Chromosomes

While the above rearrangements involve two breaks in the same chromosome, it is also possible to produce two breaks simultaneously in two nonhomologous chromosomes. Reunion can then take place in various ways. The most interesting case occurs when the ends of two nonhomologous chromosomes are switched with each other in a **reciprocal translocation** (**fig. 8.4**). The organism in which this has happened, a reciprocal translocation heterozygote, retains all the genetic material. The outcome of a reciprocal translocation, like that of an inversion, is a new linkage arrangement in the translocated chromosomes.

Centromeric Breaks

An interesting variant of the simple reciprocal translocation occurs when two acrocentric chromosomes join at or very near their centromeres. The process, which essentially is the fusion of the long chromosomal arm from two nonhomologous chromosomes, is called a **Robertsonian translocation** or **fusion.** The short chromosomal arms from the two original chromosomes are lost because they both lack a centromere. A Robertsonian translocation produces a decreased number of chromosomes because the long arms of two chromosomes are fused into one.

In humans, Robertsonian translocations can occur between any two of the acrocentric chromosomes 13, 14, 15, 21, and 22. Down syndrome occurs when the Robertsonian translocation contains chromosome 21.

Notation for Chromosome Abnormalities

In the standard nomenclature system, a normal human chromosome complement is 46,XX for a female and 46,XY for a male. The total chromosome number appears first, then the description of the sex chromosomes, and, finally, a description of autosomes if an autosomal anomaly is evident. Thus, the number of any extra chromosomes would be added to 46, and the number of missing chromosomes would be subtracted.

For example, a male with an extra X chromosome would be 47,XXY. A female with a single X chromosome would be 45,X. Because all the autosomes are numbered, we describe their changes by referring to their addition (+) or deletion (−). For example, a female with trisomy 21 would be 47,XX,+21.

The shorter arm of a chromosome is designated p, and the longer arm as q. When a change in part of the chromosome occurs, a plus sign (+) after the arm indicates an increase in the length of that arm, whereas a minus sign (−) indicates a decrease in its length.

For example, a female with a translocation (t) that transfers part of the short arm of chromosome 9 to the short arm of chromosome 18 would be designated as

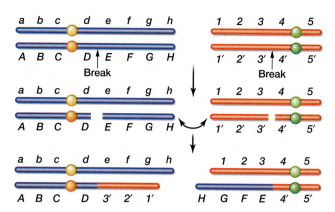

figure 8.4　Generation of a reciprocal translocation. A single break in two nonhomologous chromosomes (between D and E, and between 3' and 4') to generate two acentric fragments. The acentric fragments exchange places and rejoin with a nonhomologous chromosome. The result is a wild-type blue chromosome, a wild-type red chromosome, and the reciprocal translocated chromosome pair.

46,XX, t(9p−;18p+). The semicolon indicates that both chromosomes kept their centromeres.

You will see this notation in descriptions of common human anomalies later in this chapter.

Restating the Concepts

▶ Chromosome breakage can be caused by a number of chemical and environmental factors. Sometimes a broken chromosome is simply repaired, a process called restitution.

▶ Breakage in noncentromeric regions include a single break, two breaks in the same chromosome, and two breaks in nonhomologous chromosomes. Rejoining broken chromosomes can produce an acentric fragment that lacks a centromere, a dicentric chromosome that contains two centromeres, a deletion, translocation, or inversion.

▶ Breakage in the centromeric region may result in a Robertsonian translocation, which is a fusion of the centromeres from both chromosome arms.

▶ Geneticists use a three-element notational convention to describe chromosome abnormalities. This notation reveals the total number of chromosomes, the make-up of the sex chromosomes, and any changes in the autosomes.

8.2 Deletions

When a break produces an acentric fragment, it may be lost from subsequent cell divisions. The remaining chromosome will continue through cell divisions and become part of the gametes. Upon fertilization, the resulting zygote will have one normal chromosome and one chromosome missing material, termed a **deletion** or deficiency.

An individual that has a deletion chromosome can be cytologically identified if the individual also possesses a normal homolog. During meiosis, a bulge will be present in the tetrad if the deleted section is large enough (**fig. 8.5a**). This synapsed structure is often called a **deletion loop.**

By staining the chromosomes, it is possible to determine the precise bands that are present in the loop, which corresponds to the DNA that is missing in the homolog. This type of analysis has permitted the characterization of the deletions associated with particular human genetic anomalies, such as cri du chat syndrome (see later section).

This loop can also appear in the paired, polytene giant salivary gland chromosomes of *Drosophila* (fig. 8.5b). Note that the presence of a deletion loop actually signifies that one chromosome possesses less DNA than its homolog. In this example, the loop region corresponds to the wild-type chromosome that contains DNA missing in the deletion chromosome.

A diagnostic feature of a deletion is its inability to revert to wild type. The loss of a portion of a chromosome can not be regenerated.

Pseudodominance

Deletions exhibit a phenomenon called pseudodominance. **Pseudodominance** is the expression of the recessive phenotype when only a single recessive allele is present. We previously described pseudodominance in chapter 4, when we discussed the expression of recessive alleles on the X chromosome in males. Pseudodominance is also observed if the region absent in a deletion chromosome corresponds to the location of recessive alleles on the homologous full-length chromosome (**fig. 8.6**). In contrast, if a recessive mutation corresponds to a region that is *not* absent on the deletion chromosome, the dominant phenotype will be observed (assuming that the deletion chromosome contains the dominant allele; see fig. 8.6). Thus, it is possible to map recessive mutations to chromosomal regions using deletions by determining if the recessive allele on the intact chromosome results in the recessive phenotype or not.

This approach provides a powerful method to localize recessive mutations on chromosomes. If the location and size of the deletion is known from the stained chromosome banding patterns, then the location of the recessive alleles can also be localized.

For example, the high resolution banding pattern observed in *Drosophila* polytene chromosomes allows for a precise determination of the cytogenetic bands missing in specific deletion chromosomes. Combining this knowledge with the pseudodominance results, it is possible to localize a recessive mutation to a single polytene band (**fig. 8.7**).

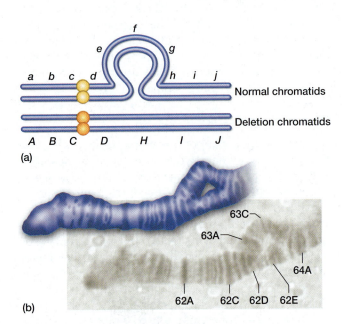

(a)

(b)

figure 8.5   Formation of a deletion loop during meiosis. (a) A bulge, also called a deletion loop, can occur in a wild-type chromosome if it pairs during meiosis I with a deletion chromosome. The deletion loop is formed to try to maximize synopsis between the homologous chromosomes. (b) In the *Drosophila* polytene chromosomes, the deletion loop in the wild-type homolog reveals the bands that are missing in the deletion chromosome.

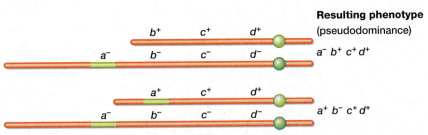

figure 8.6   Pseudodominance in a deletion heterozygote. Two deletion heterozygotes are shown, where the individual has a deletion chromosome (top of each pair) and a wild-type chromosome (bottom of each pair). The deletion chromosome contains all dominant alleles and the intact chromosome has all recessive alleles. The top example has a deletion of the *a* gene. Pseudodominance occurs when the cell expresses the recessive *a⁻* mutant phenotype because the deletion chromosome lacks an *a* gene. In the lower example, the deletion chromosome possesses the *a⁺* allele and the dominant wild-type *a⁺* phenotype is produced. However, the lower deletion chromosome now lacks the *b+* allele, which results in the expression of the recessive *b⁻* phenotype due to pseudodominance of the *b⁻* allele on the intact chromosome.

It's Your Turn

The basic concept of deletion mapping a recessive mutation involves the heterozygote expressing, either the recessive mutant phenotype if the wild-type copy of the gene is removed by the deletion or the dominant wild-type phenotype if the wild-type allele is located outside the deleted region. You should now try deletion mapping problems 30 and 52.

Genetic Imbalance

A second possible effect of a deletion is that, depending on the length of the deleted segment and the specific loci lost, the heterozygote may exhibit a genetic imbalance. **Genetic imbalance** is the consequence of having two copies of some genes and only a single copy (or three copies) of other genes, which results in an unnatural ratio of gene expression. The reduction of some genes by half (due to the loss of one copy in the deletion chromosome) leads to an insufficient amount of expression from the remaining gene.

Haploinsufficiency occurs when a deletion results in a lethal phenotype because the expression of a single wild-type allele cannot support the normal phenotype. In other cases, the genetic imbalance may be viable, but the resulting heterozygote may express an altered phenotype. An example of this viable genetic imbalance is the deletion of part of chromosome 5 in humans that produces cri du chat syndrome.

Cri du Chat Syndrome

The syndrome known as cri du chat (from the French, meaning cry of the cat) is so named because of the cat-like cry that about half the affected infants

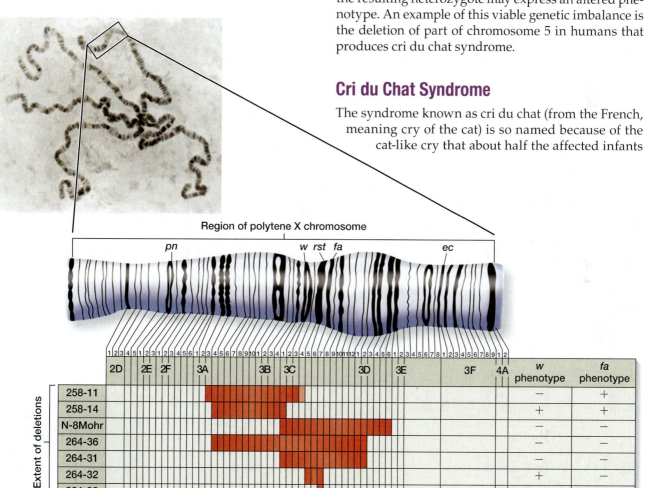

figure 8.7   Deletion mapping is demonstrated in this figure for two *Drosophila* X-linked mutations, *white* (*w*) and *facet* (*fa*). An image of the *Drosophila* polytene X chromosome is shown at the top. The boxed region is blown up below, showing the cytological location of five genes on the X chromosome. Shown below the X chromosome diagram is a schematic of seven deletions in this region. The deleted region for each of these X chromosomes is highlighted in red. Thus, deletion 258-11 is missing DNA from bands 3A3 to 3C4-5. The right of the table shows if the deletion heterozygote exhibits either the wild-type (+) or mutant (−) phenotype when placed *in trans* to either the *w* or *fa* mutation. By comparing the regions deleted and whether it yields the recessive mutant *w* phenotype, the location of the *white* gene can be mapped to 3C2-4. Using additional deletions, the location of the *white* gene was shown to be in band 3C2. Similarly, the *fa* gene was mapped to band 3C7 (deletion 264-33).

make. Microcephaly (an abnormally small head), congenital heart disease, and severe mental retardation are also common symptoms. This disorder arises from a deletion in the short arm of chromosome 5; the karyotype for this syndrome is 46,XX or XY,5p– (**fig. 8.8**).

Most other deletions studied (4p–, 13q–, 18p–, 18q–) also result in microcephaly and severe mental retardation. The rarity of viable heterozygotes possessing a large deletion is consistent with the fact that viable monosomics (having a single chromosome of a pair) are rare. An individual heterozygous for a deletion is, in effect, monosomic for the deleted region of the chromosome. Evidently, either monosomy or heterozygosity for large chromosome deletions is generally lethal in humans. This lethality is due to a combination of pseudodominance of recessive lethal or deleterious mutations, genetic imbalance, and haploinsufficiency.

Restating the Concepts

▶ Deletion chromosomes lack some genomic DNA.

▶ Deletions are characterized by the formation of a deletion loop during synapsis with a wild-type homolog during meiosis I and pseudodominance of recessive mutations.

▶ Pseudodominance is a powerful method to map recessive mutations and to correlate their location with a cytogenetic map.

▶ In humans, large deletions are generally lethal.

8.3 Inversions

As shown in figure 8.3, an **inversion** results from two breaks in a single chromosome and the subsequent reorientation of the central fragment before rejoining the broken ends. As a result, there is no net loss or gain in genomic DNA sequences. However, the chromosomal breaks can occur either between or within genes. If the break occurs between genes, the inverted chromosome usually does not create any new alleles (**fig. 8.9**).

If the breaks occur within a gene, however, two outcomes are possible. First, the break could disrupt the gene and produce a nonfunctional allele, which is usually a recessive mutation (unless it is haploinsufficient, see previous discussion; fig. 8.9). Alternatively, it could move a *promoter sequence*, which controls DNA transcription, near a different gene to express that wild-type gene in an abnormal location or time (see fig. 8.9). In either outcome, if the inversion produces a change in the chromosomal banding pattern, then it is possible to correlate the location of the mutation with the defect in the cytogenetic banding pattern, and in this way to map the location of the defective gene relative to the cytogenetic map.

There are two different types of inversions. A **pericentric inversion** contains the centromere within the inverted section (**fig. 8.10**, left). A **paracentric inversion** has the centromere outside of the inverted region (see fig. 8.10, right). When synapsis occurs in an inversion heterozygote (an individual with an inversion chromosome and a wild-type homolog), an inversion loop often forms to accommodate the point-for-point

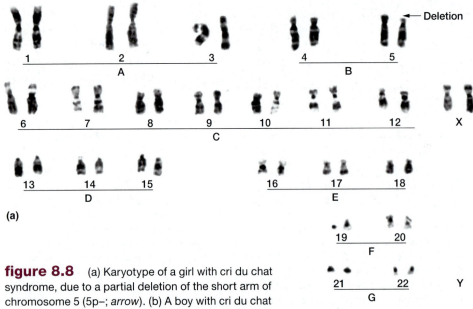

figure 8.8 (a) Karyotype of a girl with cri du chat syndrome, due to a partial deletion of the short arm of chromosome 5 (5p–; *arrow*). (b) A boy with cri du chat syndrome shows the wide-set and slightly downward-slanting eyes, small head (microcephaly), small jaw, and low-set ears.

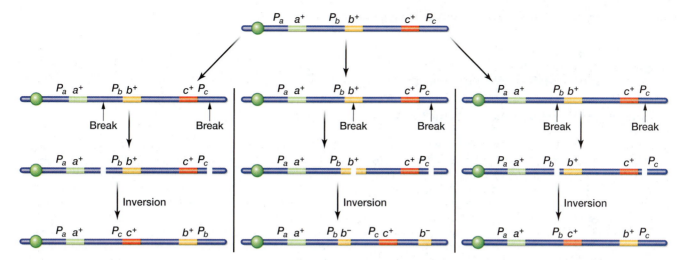

figure 8.9 Three different types of inversions relative to three genes. The breaks may occur between genes, which will not affect their expression (left). One of the breaks may occur within a gene, which will disrupt its expression and generate a mutant allele (b^-, center). The break may also occur between a promoter and its gene, which places the promoter upstream of a different gene (right). In this example, the b promoter (P_b) now controls the expression of the c^+ gene and the c promoter (P_c) controls the expression of the b^+ gene. This will likely produce b^- and c^- mutant alleles due to their altered expression patterns.

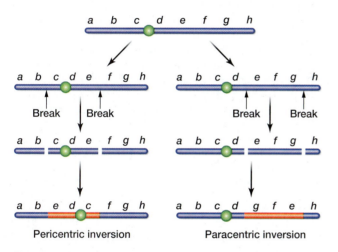

figure 8.10 A double chromosome break can produce either a pericentric inversion if the centromere is located between the two breaks or a paracentric inversion if the centromere is located outside of the two breaks. The inverted region is marked in red.

pairing along the chromosomes during meiosis—or, in *Drosophila* salivary glands, during endomitosis (**figs. 8.11** and **8.12**). A pericentric inversion will have the centromere located within the inversion loop, and a paracentric inversion will have the centromere located outside of the inversion loop.

Suppression of Recombination

One hallmark of an inversion is the suppression of recombination in an inversion heterozygote by two different mechanisms. The first is a real suppression of recombination or the inability to completely synapse during meiosis. This results from the difficulty of complete synapsis between the two homologs in the regions at the base of the inversion loop (fig. 8.11). In

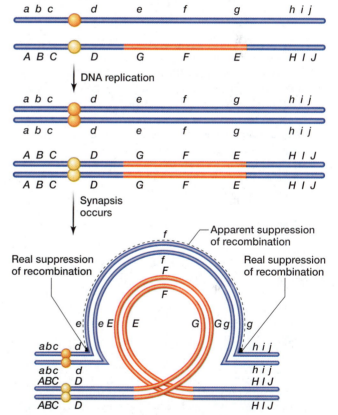

figure 8.11 Synapsis of an inversion heterozygote, which possesses one wild-type chromosome (blue) and one paracentric inversion chromosome, produces an inversion loop. During synapsis, the inverted region (red) is incorporated into the inversion loop to maximize synapsis along the length of both chromosomes. A real suppression of recombination occurs near the inversion break points (at the base of the loop) due to the difficulty in the chromosomes synapsing in this region. An apparent suppression of recombination occurs within the inversion loop because recombination within the loop produces chromatids with deletions and duplications that will not yield viable progeny.

205

8.1 applying genetic principles to problems

Deletion Mapping of Recessive Alleles

Assume you isolated two recessive mutations and you would like to map their location in the genome. You could map their relative locations using recombination if one of the individuals to be mated was heterozygous for each gene being mapped and a sufficient number of progeny can be generated.

Alternatively, you can employ deletion mapping of recessive mutations.

Let's assume that you generated a collection of deletion stocks that are shown below.

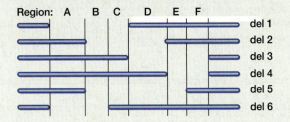

These six deletions define the limits of six different regions (A–F) that are either present or absent in each deletion chromosome.

Assume you have two recessive mutations *a* and *b* that you want to map using these deletions. If you cross *a* homozygotes and *b* homozygotes with each of the deletion stocks, you generate the following data (with + signifying the dominant wild-type phenotype and − signifying the recessive mutant phenotype):

Deletion	Mutation	
	a	*b*
del 1	−	−
del 2	−	−
del 3	+	−
del 4	+	+
del 5	−	−
del 6	−	+

The *a* gene must be located in the genomic DNA that is present in del 3 and del 4, which corresponds to regions A, B, or C. Additionally, the *a* gene must be absent in del 1, del 2, del 5, and del 6. The only common region that is absent in all four of these deletions is region B. Because region B is also present in del 3 and del 4, the *a* mutation must be located in region B.

In this example, comparing the deletions that exhibit pseudodominance was sufficient to map the location of the recessive mutation. The deletions lacking pseudodominance only confirmed the location. However, in some cases it will be necessary to combine both groups to identify the minimal region.

Mapping the *b* mutation is such an example that requires using all the available information. You can see that both del 4 and del 6 contain the dominant wild-type *b* gene, based on the dominant wild-type *b* phenotype. Regions C and D are present in both del 4 and del 6. We can also compare the common regions that are absent in del 1, del 2, del 3, and del 5 (which all exhibit the recessive *b* mutant phenotype). However, no common region is missing in all four of these deletions. In fact, del 1 ends where del 3 begins.

Looking at the common regions C and D in del 4 and del 6, if the *b* gene is in region C, then del 3 would not exhibit the recessive phenotype. Similarly, if gene *b* is in region D, then del 1 would not exhibit the recessive phenotype. Thus, the simplest explanation is that mutation *b* must be located in both regions C and D, such that the vertical line between these two regions is within gene *b*.

It is important to remember that some mutations have both dominant and recessive phenotypes (see chapter 5). For these mutations, we can deletion map the recessive phenotype. For example, the *Drosophila Cy* mutation exhibits a dominant curly wing phenotype and a recessive lethal phenotype. We can deletion map the recessive *Cy* lethal phenotype, but not the dominant curly wing phenotype. If a *Cy/+* stock is crossed with each of the previously diagrammed six deletion stocks, the following progeny would be generated:

Deletion	Phenotype
	Curly wings
del 1	−
del 2	+
del 3	+
del 4	+
del 5	+
del 6	−

Because the curly phenotype is dominant, it should be present in half the progeny. However, no curly-winged progeny were identified when the *Cy/+* stock was crossed to either the del 1 or the del 6 stocks. This suggests that any *Cy/del 1* or *Cy/del 6* flies died because of pseudodominance of the recessive *Cy* lethal phenotype. This suggests that the lethal mutation lies in either region A or B. Because region A, and not B, is present in the other four deletion stocks, region A presumably contains the wild-type *Cy* allele (+). These *Cy/+* individuals have the dominant curly-winged phenotype and survive because they have only one copy of *Cy*+. Therefore, you observe and map the pseudodominance of the recessive *Cy* mutant phenotype.

other words, as you move closer to the actual inversion breakpoints, the crossover frequency approaches zero.

The second effect is an apparent suppression of recombination, which occurs from crossovers within the inversion loop (see fig. 8.11). In reality, recombination can occur at a fairly normal frequency within the inversion loop relative to the same region in a normal individual. However, the gametes produced from recombination within the inversion loop are usually unable to produce viable offspring.

Mechanisms of Suppression

Let's look at the mechanism of the apparent suppression in an inversion heterozygote with a paracentric inversion. **Figure 8.13** shows a crossover within the inversion loop. The two nonsister chromatids that are not involved in the crossover will end up in normal gametes (carrying either the wild-type or inverted chromosome). The products of the crossover, rather

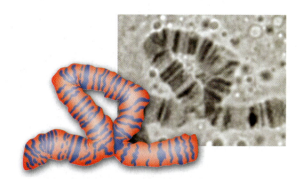

figure 8.12   A *Drosophila* paracentric inversion heterozygote will possess an inversion loop in the salivary gland chromosomes.

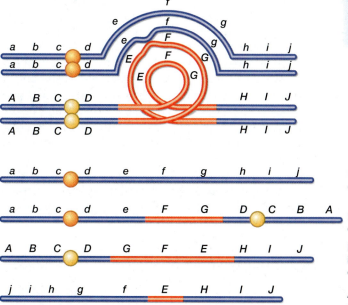

Nonrecombinant chromosome

Dicentric chromosome

Nonrecombinant inversion chromosome

Acentric chromosome

than being a simple recombination of alleles, are a dicentric and an acentric chromatid.

The acentric chromatid is not incorporated into a gamete nucleus and this recombinant will be lost. The dicentric chromatid begins a breakage–fusion–bridge cycle, as the two centromeres are pulled to opposite centrosomes during meiosis I. Ultimately, the dicentric chromosome randomly breaks between the two centromeres and each chromatid, containing deletions, produces a genetically imbalanced gamete. Thus, the gametes derived from the recombinant chromatids are unable to produce viable offspring.

A pericentric inversion also suppresses crossovers, but for slightly different reasons (**fig. 8.14**). All four chromatid products from a single crossover within the loop will have centromeres and are therefore incorporated into the nuclei of gametes. However, the two recombinant chromatids are not balanced—they both have duplications and deficiencies. In the figure, one chromatid has a duplication for *a-b-c-d* and is deficient for *h-i-j*, whereas the other is the reciprocal—being deficient for *a-b-c-d* and duplicated for *h-i-j*. These duplication–deletion gametes tend to form nonviable zygotes.

Thus, both paracentric and pericentric inversion heterozygotes will not exhibit recombinant gametes if the crossover occurs within the inversion loop. It should be mentioned that recombination occurs at normal frequencies outside, but not at the base, of the inversion loop. For this reason, recombination is only suppressed very near to or within the inversion loop in an inversion heterozygote.

It's Your Turn

Using what we discussed about the effect of an inversion heterozygote on recombination frequencies, you should be able to map the inversion in problem 33.

figure 8.13   Consequences of a crossover in the paracentric inversion heterozygote loop. This recombination event yields two chromatids that contain either the wild-type sequence or the inversion sequence. Two recombinant chromatids are also produced. The first is a dicentric chromatid and the second is an acentric chromatid. These recombinant chromatids produce gametes that are genetically imbalanced and unable to yield viable progeny. Thus, the recombinant chromatids will not be observed in the progeny and recombination appears to be blocked.

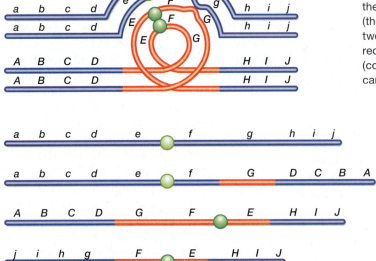

figure 8.14   Consequences of a crossover in the loop region of a pericentric inversion heterozygote. This recombination event yields the two nonrecombinant parental chromosomes (the wild-type and inversion chromosomes) and two recombinant chromosomes. Because both recombinant chromosomes are not balanced (containing both deletions and duplications) they cannot yield viable progeny.

Parental wild-type chromosome

Imbalanced recombinant (duplication and deletion) chromosome

Parental inversion chromosome

Imbalanced recombinant (duplication and deletion) chromosome

Inversions Cause Reduced Fertility

Because half the chromatids produced during recombination within the inversion loop are not balanced and cannot produce viable zygotes, inversions cause reduced fertility in individuals. But since recombination *outside* of the inversion loop is not affected in an inversion heterozygote, greater than half of all gametes produced are balanced and will produce viable zygotes. The percentage of unbalanced gametes will be proportional to the size of the inversion and whether multiple inversions occur on a single chromosome.

Balancer Chromosomes

In *Drosophila*, several **balancer chromosomes** have been generated containing multiple inversions along the majority of their length. A balancer chromosome suppresses recombination along the entire length of a given chromosome. This is a useful feature when characterizing recessive mutations or trying to create specific genotypes because the arrangement of alleles along a specific chromosome will not change between generations. However, the flies that carry these balancer chromosomes produce low numbers of offspring due to their *reduced fertility*, which approaches 50% when the inversions cover nearly the entire chromosome.

Properties of Inversions

An inversion has several interesting properties. First is the reduced fertility and suppression of recombinant offspring that results from recombination within the inversion loop of an inversion heterozygote.

Second, individuals who are homozygous for an inversion show novel linkage relationships when their chromosomes are mapped by recombination. Because both homologs contain the same inversion, an inversion loop is not generated, and recombination distances can be accurately mapped along the entire length of the inversion chromosome. Yet, you can see this novel linkage relationship in figure 8.11 where the distance from point *e* to *h* is very different in the wild-type chromosome (top) and the inversion chromosome (bottom).

Third, the inversion chromosome may produce a **position effect,** which is a change in the expression of a gene due to an altered linkage arrangement. Position effects are either stable, as in the *Bar* eye mutant of *Drosophila* (to be discussed later), or variegated, as with *Drosophila* eye color. For example, a heterozygous female fly ($X^w X^+$) will have phenotypically wild-type red eyes. If the *white*$^+$ locus is moved from the euchromatin region near the end of the X chromosome to the heterochromatin through an inversion (**fig. 8.15**), the fly shows **variegation**—patches of the eye are white.

The variegation is due to the *white*$^+$ locus on the inversion chromosome not being expressed in parts of the eye, so the $X^w X^+$ genotype produces either a red or white eye phenotype. Presumably the heterochromatin encroaches on the *white*$^+$ allele in some cells to "turn off" the expression of the *white*$^+$ locus. In the heterozygote, if the wild-type allele is turned off, then the cell will express the normally recessive white eye phenotype. Depending on what happens in each eye cell, patches of red and white eye color result.

Evolutionary Consequences of Inversions

Inversions have several evolutionary ramifications. Alleles found together within an inversion loop tend to be inherited together because of the suppressed

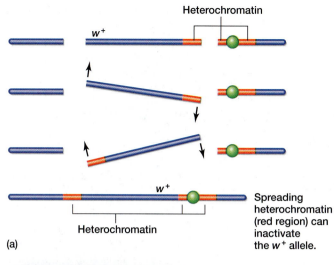

Heterochromatin

w^+

Heterochromatin

Spreading heterochromatin (red region) can inactivate the w^+ allele.

(a)

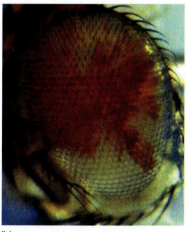

(b)

figure 8.15   (a) An inversion in the X chromosome of *Drosophila* produces a variegated eye color in a female if the inversion brings the *white*⁺ (*w*⁺) allele from the euchromatin to within or adjacent to the heterochromatin. (b) In a X⁺Xʷ female, the X⁺ chromosome contains an inversion of the *w*⁺ allele to the heterochromatin and the homologous X chromosome contains the *w* allele. In this heterozygote, the eye exhibits a variegated phenotype of both red and white patches due to the heterochromatin affecting the expression of the *w*⁺ allele on the inversion chromosome.

recombination within the inversion. If several loci within an inversion affect a single trait, the alleles are referred to as a **supergene.** Until careful genetic analysis is done, the loci in a supergene could be mistaken for a single locus; they affect the same trait and are inherited apparently as a single unit. Examples include shell color and pattern in land snails and mimicry in butterflies. Supergenes can be beneficial when they involve favorable gene combinations. But at the same time, their inversion structure prevents the formation of new allele combinations and the corresponding phenotypes. Supergenes, therefore, have evolutionary advantages and disadvantages. Chapter 25 discusses these evolutionary topics in more detail.

Sometimes the generation of inversions produces a record of the evolutionary history of a species. As species evolve, inversions can occur on preexisting inversions, leading to very complex arrangements of loci. We can readily study these patterns in Diptera by noting the changed patterns of bands in salivary gland chromosomes. Since certain genomic arrangements can only be produced from a specific sequence of inversions, it is possible to determine the order in which different species evolved.

Restating the Concepts

▸ Inversions result in the rearrangement of genetic material, with no net loss or gain. Pericentric inversions have the centromere located in the inverted region, and paracentric inversions have their centromere outside the inverted region.

▸ Inversion heterozygotes are characterized by the formation of an inversion loop during synapsis of meiosis I, suppression of recombination, and reduced fertility.

▸ Inversions suppress recombination by an apparent mechanism (due to the recombinant chromosomes producing gametes that will not generate genetically balanced and viable zygotes) and a real mechanism (physical constraint on complete homolog synapsis near the inversion breakpoints).

8.4 Translocations

As mentioned earlier, two breaks that occur simultaneously in nonhomologous chromosomes can result in a number of reunions, the most interesting of which is the *reciprocal translocation* (see fig. 8.4). Here, parts of two chromosomes have been switched without loss of genetic material. Pairing during synapsis is still possible.

Segregation After Translocation

During synapsis in meiosis, a point-for-point pairing in the reciprocal translocation heterozygote can be accomplished by the formation of a cross-shaped figure (**fig. 8.16**). Such a structure is diagnostic of a reciprocal translocation. In this arrangement, a single crossover in a reciprocal translocation heterozygote does not produce chromatids that are further unbalanced, as is the case for an inversion heterozygote.

Because two homologous pairs of chromosomes are involved in a reciprocal translocation, the independent segregation of the centromeres of the two tetrads has dramatic consequences on the gametes produced. Geneticists have observed two common segregation patterns and one that occurs less often (**fig. 8.17**). The first, called **alternate segregation,** occurs when the

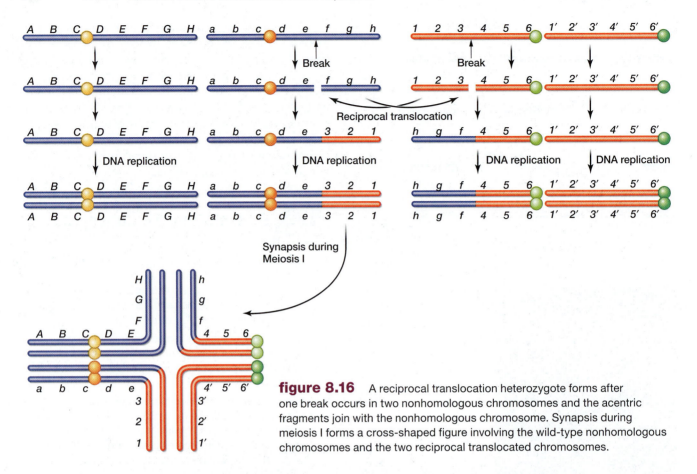

figure 8.16 A reciprocal translocation heterozygote forms after one break occurs in two nonhomologous chromosomes and the acentric fragments join with the nonhomologous chromosome. Synapsis during meiosis I forms a cross-shaped figure involving the wild-type nonhomologous chromosomes and the two reciprocal translocated chromosomes.

first centromere segregates with the fourth centromere, leaving the second and third centromeres to go to the opposite pole. This corresponds to the segregation of homologous centromeres, with each gamete receiving one yellow and one green centromere. The result is balanced gametes, one with normal chromosomes and the other with the reciprocal translocation.

Equally likely is the **adjacent-1** segregation pattern, in which the first and third centromeres segregate together in the opposite direction from the second and fourth centromeres (again, segregation of homologous centromeres). Here, all the resulting gametes are unbalanced, carrying duplications and deficiencies that will usually lead to nonviable offspring.

The third type of segregation, termed **adjacent-2** segregation (see fig. 8.17), involves the homologous centromeres moving to the same pole (first with second, third with fourth). This outcome can occur when the cross-shaped double tetrad opens into a circle in late prophase I. In the German cockroach, adjacent-2 patterns were observed in 10–25% of meiotic events, depending on which chromosomes are involved.

Alternate and adjacent-1 segregation patterns are preferred (relative to the adjacent-2 pattern) and are nearly equally likely because they both involve segregation of homologous centromeres. This suggests that alternate and adjacent-1 segregation will both occur

nearly 50% of the time. Because 50% of the gametes are unbalanced and cannot produce viable zygotes, we describe this phenomenon as **semisterility.**

Notice that semisterility occurs at a higher frequency than the reduced fertility described for inversion heterozygotes, which allows us a diagnostic method for differentiating between these two types of chromosomal rearrangements.

Pseudolinkage

Recombination in a reciprocal translocation exhibits an effect termed **pseudolinkage.** This phenomenon results from the absence of crossing over between loci located near the translocation breakpoints. For example, **figure 8.18** shows the cross-shaped structure produced during synapsis in meiosis I of a reciprocal translocation heterozygote. The wild-type alleles are located on the wild-type chromosomes, and the mutant alleles are on the translocation chromosomes.

Based on independent assortment, we would expect gametes containing four different allelic arrangements of the *f* and *3* loci: *f 3*, *F 3'*, *F 3*, and *f 3'*. However, alternate segregation of the centromeres will produce balanced gametes with only the *F 3'* and *f 3* allelic combinations. If recombination occurs, as shown between the translocation breakpoint and the *4* locus (see fig. 8.18), then

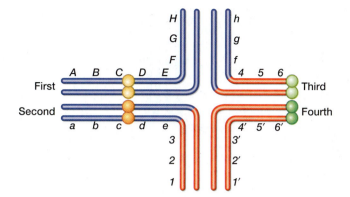

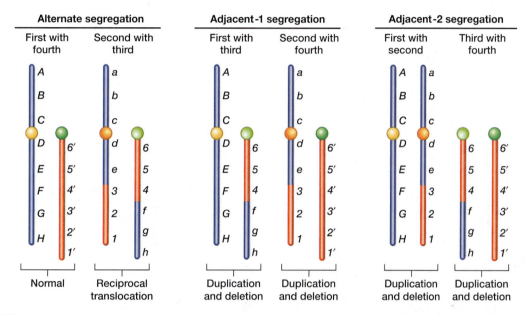

figure 8.17   Three possible results of chromosome segregation during meiosis in a reciprocal translocation heterozygote. Alternate segregation produces balanced gametes, while adjacent segregation produces unbalanced gametes (containing duplications and deletions) that cannot yield viable progeny.

alternate segregation also produces unbalanced gametes. If adjacent-1 segregation occurs after the recombination event, then balanced gametes can be produced. Even in this case, only the *f 3* and *F 3'* allelic combinations are observed.

The consequence of pseudolinkage is that two loci that are located near the reciprocal translocation breakpoint and on nonhomologous chromosomes will appear to be tightly linked. Pseudolinkage can be used to assist in the mapping of the translocation breakpoint.

8.5 Centromeric Breaks

Breaks at or near the centromeres of two nonhomologous chromosomes can result in a specific type of translocation, the Robertsonian translocation, in which two long arms of acrocentric chromosomes are joined

together (**fig. 8.19**). The two short arms may also fuse, and if they contain a centromere they will be maintained. Otherwise, the fused acentric fragment is lost.

Often, closely related species undergo Robertsonian translocations between different acrocentric chromosomes and end up with markedly different chromosome numbers without any significant difference in the quantity of their genetic material. Therefore, cytologists may count the number of chromosomal arms rather than the number of chromosomes to get a more accurate picture of the relatedness between different species. The number of arms is referred to as the **fundamental number,** or **NF** (from the French, *nombre fondamentale*).

In a similar fashion, **centromeric fission** occurs when a chromosome breaks right at the centromere, and each arm contains sufficient centromeric sequences to function as an independent chromosome. Thus, a centromeric fission increases the chromosome number without changing the fundamental number.

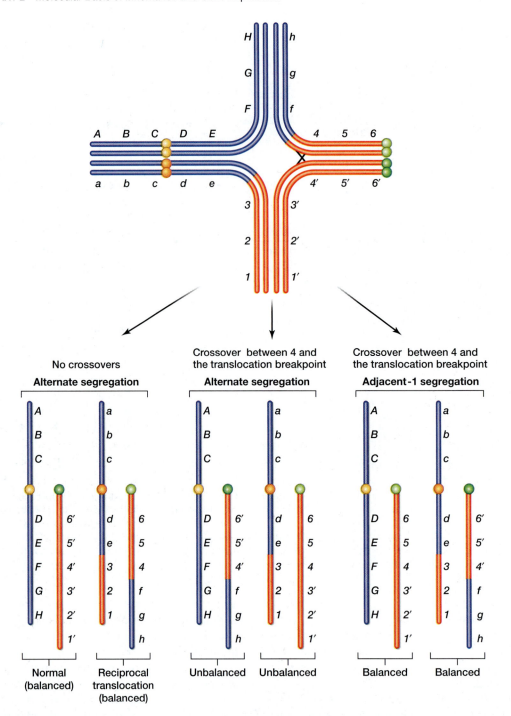

figure 8.18 Pseudolinkage in reciprocal translocations occurs when alleles on nonhomologous chromosomes near the translocation breakpoints appear to be tightly linked. For example, balanced gametes contain either the *F 3'* combination or the *f 3* in the case of alternate segregation and no crossovers or the crossover and adjacent-1 segregation. However, the combinations of *F 3* and *f 3'* that are produced by recombination and alternate segregation produce unbalanced gametes that cannot yield viable progeny. Thus, only the parental genotypes (*F 3'* and *f 3*) are observed.

Restating the Concepts

▶ Reciprocal translocation heterozygotes exhibit pseudolinkage and semisterility, which results from the segregation of chromosomes during meiosis I and the production of 50% of the gametes containing unbalanced chromosomes.

▶ Centromeric breaks in acrocentric chromosomes can result in a Robertsonian translocation that is the fusion of the long arms of nonhomologous chromosomes.

▶ The NF, or fundamental number, refers to the number of chromosomal arms and is sometimes preferred to the chromosome number when evaluating species relationships.

8.6 Duplications

Duplications of chromosomal segments can occur by the breakage–fusion–bridge cycle or by crossovers within an inversion loop. Unequal crossing over can also produce duplications in small adjacent regions of a chromosome. We illustrate this with a particularly interesting example, the *Bar* eye phenotype in *Drosophila* (**fig. 8.20**).

Bar Eye in *Drosophila*

The wild-type fruit fly has about 800 facets or ommatidia in each compound eye (see fig. 8.20). The dominant *Bar* (*B*) heterozygote has approximately 350 facets, and the *Bar* homozygote about 70. Another allele called *Doublebar* (*BB*: sometimes referred to as *Ultrabar*, B^U), decreases the facet number of the eye to about 45 when heterozygous and to about 25 when homozygous. Around 1920, researchers showed that roughly

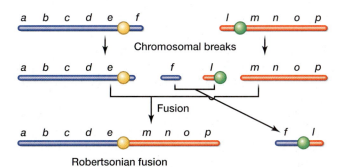

figure 8.19 Breaks in or near the centromeres of two acrocentric chromosomes and the subsequent fusion of the two long arms produces a Robertsonian translocation. The two small arms may fuse and be maintained in the cell if it contains a centromere as diagrammed here. However, if both small fragments lack a centromere, their fusion will produce an acentric fragment that will be lost.

one in 1600 progeny from homozygous *Bar* females is *Doublebar*. This frequency is much higher than we would expect from a mutational event.

Alfred Sturtevant found that every *Doublebar* fly experienced a crossover adjacent to the *Bar* locus. He suggested that the creation of *Doublebar* from *Bar* was due to **unequal crossing over** rather than a simple mutation of *Bar* to *Doublebar* (**fig. 8.21**). If the homologous chromosomes do not line up exactly during synapsis, then unequal crossing over between two *Bar* mutations produces the *Doublebar* and a wild-type chromosome. This demonstrates that duplications can revert to wild type, unlike a deletion chromosome, by loss of chromosomal DNA.

Later, an analysis of the polytene chromosome banding pattern confirmed Sturtevant's hypothesis. It was found that *Bar* is a duplication of the 16A region of the X chromosome (**fig. 8.22**), whereas *Doublebar* is a triplication of the segment.

A *position effect* also occurs in the *Bar* system. Unlike the position effect that the heterochromatin exerts on the *white⁺* allele in an inversion, the *Bar* position effect results from the arrangement of the *Bar* sequences relative to each other. A *Bar* homozygote (*B/B*) and a *Doublebar*/wild-type heterozygote (*BB/B⁺*) both have four copies of the 16A region (see fig. 8.21). It would therefore be reasonable to expect that both genotypes would produce the same phenotype. However, the *Bar* homozygote has about 70 facets in each eye, whereas the *Doublebar* heterozygote only has about 45. Thus, not only the amount of genetic material, but also its configuration (that is, position), determines the extent of the phenotype. *Bar* eye was the first position effect discovered.

Fragile-X Syndrome in Humans

The most common cause of inherited mental retardation is **fragile-X syndrome.** This condition occurs in

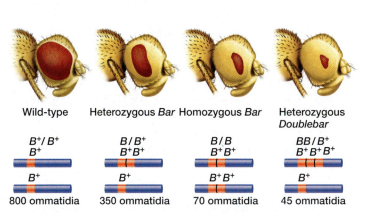

figure 8.20 *Bar* mutation in *Drosophila* females. The wild-type female fly (*B⁺/B⁺*) contains one copy of the *B⁺* gene on each X chromosome. Increasing the number of *B⁺* genes (3 in *B/B⁺*, 4 in *B/B* and *BB/B⁺*) results in eyes with fewer ommatidia and narrower width.

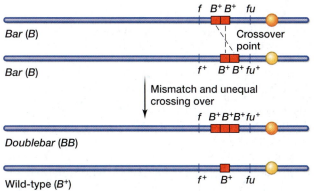

figure 8.21 Unequal crossing over in a *Bar*-eyed homozygous *Drosophila* female as a result of improper pairing. A *Doublebar* chromosome and wild-type chromosome is produced by a crossover between two flanking loci, forked (*f*) and fused (*fu*).

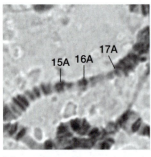
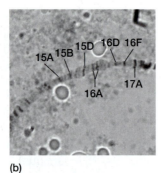

(a)

(b)

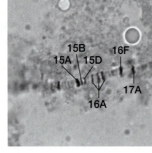

(c)

figure 8.22 *Bar* region of the X chromosome of *Drosophila*. The 16A band of the X chromosome contains the wild-type *Drosophila* *Bar* gene (a). Duplication of the 16A band (b) causes the *Bar* mutant phenotype. Three copies of the 16A region (c) corresponds to the *Doublebar* mutation.

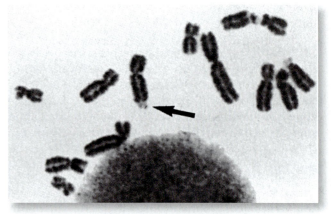

figure 8.23 Human metaphase chromosomes with the fragile-X site indicated by an arrow.
(Reprinted by permission from Macmillan Publishers Ltd. From Ian Craig, "Methylation and the Fragile X," *Nature* 349: 742, 1991.)

about 1 in every 1500 males (accounting for 4–8% of all males with mental retardation) and about 1 in every 2000–5000 females. It exhibits variable expressivity with the symptoms, including mental retardation, altered speech patterns, and physical attributes (such as protruding ears, a long jaw, and a long narrow face).

The condition is called fragile-X syndrome because it is related to a region at the tip of the X chromosome that appears to break more frequently than other chromosomal regions. However, the break is not required for the syndrome to occur. The fragile-X chromosome is usually identified by the lack of chromatin condensation at the site; in fact, under the microscope, it appears that the tip of the chromosome is being held in place by a thread (**fig. 8.23**). The gene responsible for the syndrome is called *FMR-1*, for fragile-X mental retardation-1.

Fragile-X syndrome has a highly unusual inheritance pattern, with the chance of inheriting the disease increasing through generations. This situation is unlike incomplete penetrance, in which the probability of expressing the phenotype remains constant from generation to generation. It is also unlike age-related diseases, in which the probability of expressing the phenotype increases with age.

Approximately 20% of males with the fragile-X chromosome do not have symptoms (termed **transmitting males**). All of their daughters, who inherited the fragile-X chromosome from their fathers, exhibit the symptoms about 50% of the time. The children of these females (grandchildren of the male with the fragile-X chromosome) show an increased probability of expressing the symptoms. As generations proceed, the percentage of affected sons of carrier mothers increases.

Basically, the *FMR-1* gene encodes a protein that regulates a fundamental process in translation suppression. The *FMR-1* gene normally has between 6 and 50 copies of a three-nucleotide repeat, CCG. Chromosomes with the fragile-X site appearance have between 230 and 2000 copies of the repeat. Repeat numbers above 230 inactivate the gene and thus cause the syndrome in men, who have only one copy of the X chromosome.

The number of repeats is very unstable; when carrier women transmit the chromosome, the number of repeats usually goes up. This mechanism of repeat expansion is due to unequal crossing over (**fig. 8.24**), which is analogous to the mechanism by which *Doublebar* is produced from a homozygous *Bar* *Drosophila* female (see fig. 8.21). This expansion of the trinucleotide repeat on the X chromosome occurs in females because the process requires unequal recombination between homologous chromosomes. Because males contain a single X chromosome, it is not possible for unequal crossing over to occur. However, the number of CCG repeats can increase in males (and females) through abnormal DNA replication.

This unusual form of inheritance, with unstable trinucleotide repeats in a gene, seems to be the mechanism in several other diseases as well, including myotonic muscular dystrophy and Huntington disease (described in chapter 5).

Restating the Concepts

▶ Duplications can occur by the breakage–fusion–bridge cycle or by crossovers within an inversion loop. They can also result from unequal crossing over in small regions of homologous chromosomes.

▶ Expansion of a trinucleotide repeat, through unequal recombination, results in the fragile-X syndrome and several other inherited human diseases.

www.mhhe.com/hyde
CHAPTER 8   Changes in Chromosome Structure and Number
215

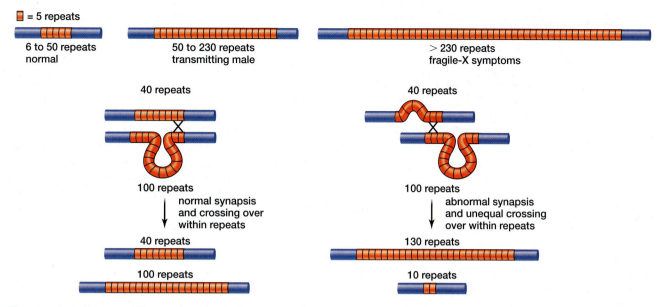

figure 8.24   The number of the CCG triplet repeats in the *FMR-1* gene determines if the individual is normal, is a transmitting male, or exhibits the fragile-X symptoms. The number of triplet repeats increases through unequal crossing over in a female. The two X chromosomes in the female may either synapse normally (left), which will not change the repeat number, or may synapse abnormally (right), in which case unequal crossing over can occur.

table 8.1   Partial List of Terms to Describe Aneuploidy, Using *Drosophila* as an Example (Eight Chromosomes: X, X, 2, 2, 3, 3, 4, 4)

Type	Formula	Number of Chromosomes	Example
Normal	$2n$	8	X, X, 2, 2, 3, 3, 4, 4
Monosomic	$2n - 1$	7	X, X, 2, 2, 3, 4, 4
Nullisomic	$2n - 2$	6	X, X, 2, 2, 4, 4
Double monosomic	$2n - 1 - 1$	6	X, X, 2, 3, 4, 4
Trisomic	$2n + 1$	9	X, X, 2, 2, 3, 3, 3, 4, 4
Tetrasomic	$2n + 2$	10	X, X, 2, 2, 3, 3, 3, 3, 4, 4
Double trisomic	$2n + 1 + 1$	10	X, X, 2, 2, 2, 3, 3, 3, 4, 4

8.7 Variation in Chromosome Number: An Overview

Anomalies of chromosome number occur as either aneuploidy or euploidy. **Aneuploidy** involves changes in chromosome *number* by additions or deletions of less than a whole set (usually only a single chromosome). **Euploidy** involves changes in the number of whole *sets* of chromosomes.

Aneuploidy results either from nondisjunction during meiosis (described in chapter 4) or by chromosomal lagging. In chromosomal lagging, one chromosome moves more slowly than the others during anaphase and is thereby excluded when the nucleus

reforms. A diploid cell missing a single chromosome is **monosomic,** whereas a cell missing both copies of a given chromosome is **nullisomic.** A cell missing two nonhomologous chromosomes is a double monosomic. Similar terminology exists for extra chromosomes resulting from nondisjunction. For example, a diploid cell with an extra chromosome is **trisomic.** A list of the terminology used in describing different states of aneuploidy appears in **table 8.1**.

Euploid organisms have varying numbers of complete haploid chromosomal sets. You are already familiar with haploids (*n*) and diploids (2*n*). Organisms with higher numbers of sets, such as **triploids** (3*n*) and **tetraploids** (4*n*), are called **polyploids.**

Restating the Concepts

▶ Variations in the number of single chromosomes is termed aneuploidy. Aneuploid conditions are named on the basis of number of chromosomes, for example, nullisomic, monosomic, trisomic, tetrasomic.

▶ Variations in the number of sets of chromosomes is termed euploidy. Euploid conditions are named according to the numbers of sets, such as, haploid, diploid, triploid, tetraploid.

8.8 Aneuploidy

Four different examples of nondisjunction are illustrated in **figure 8.25** using the sex chromosomes in XY organisms such as humans or fruit flies. Nondisjunction may occur in either the male or female during either the first or second meiotic divisions.

Figure 8.26 shows the types of zygotes that can result when gametes (both normal and nondisjunction) fuse. The resulting zygotes either have a normal or abnormal complement of sex chromosomes. The names and kinds of these sex chromosome imbalances in humans are detailed later in this chapter.

Calvin Bridges first showed the occurrence of nondisjunction in *Drosophila* in 1914 with crosses involving the *white* eye locus, which was a key proof of the chromosomal theory of inheritance (see chapter 4). This proof was based on Bridges' ability to correctly deduce the sex chromosome content in an individual based on the eye color phenotype of the individual and the parents. This deduction suggests further that if an individual is aneuploid and the genotypes of the parents are known, it is occasionally possible to determine which parent underwent nondisjunction and at which stage of meiosis it occurred.

Approximately 50% of the spontaneous abortions (miscarriages) among women in the United States involve fetuses with some chromosomal abnormality; about half of these are autosomal trisomics. About 1 in 160 live human births has some sort of chromosomal anomaly; most are balanced translocations, autosomal trisomics, or sex-chromosomal aneuploids. In the rest of this section, we describe viable human aneuploids who survive long enough after birth to have a named syndrome.

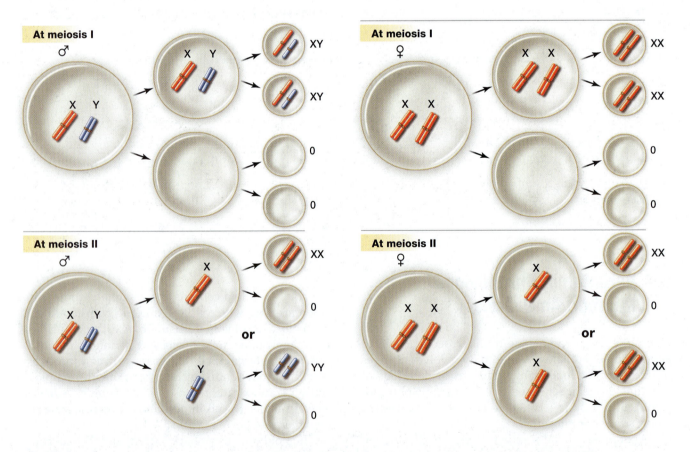

figure 8.25   Nondisjunction of the sex chromosomes in *Drosophila* or humans. The two diagrams on the left show a male (XY) cell going through nondisjunction in either meisis I (top) or meiosis II (bottom). The two diagrams on the right show the corresponding events in a female (XX) cell. "0" refers to the lack of sex chromosomes.

Trisomy 21 (Down Syndrome)

Down syndrome (47,XX or XY,+21) affects about 1 in 800 live births. Approximately 95% of all children with Down syndrome have three individual copies of chromosome 21. Most affected individuals are mildly to moderately mentally retarded, have congenital heart defects, and a high (1/100) risk of acute leukemia. They are usually short, have a broad, short skull, and possess hyperflexible joints. **Figure 8.27** shows the karyotype of a Down syndrome male and a girl with this syndrome.

The physician John Langdon Down first described this syndrome in 1866. Down syndrome was the first human syndrome shown to be associated with a chromosomal disorder that was published by the physician Jérôme Lejeune in 1959.

An interesting aspect of this syndrome is its increased incidence among children born to older mothers (**fig. 8.28**), a fact known more than 25 years before the discovery of its genetic cause. Because the future ova are maintained in prophase I of meiosis (diplonema) since before the mother's birth, all ova are the same age as the female. Presumably, as the mother grows older, the similarly older ova are more likely to be susceptible to nondisjunction of chromosome 21. This type of Down syndrome is also called **sporadic Down syndrome** because it randomly occurs in families that lack any previous history of Down syndrome.

Recently, molecular genetic techniques identified the origins of the three chromosome 21 copies in a large sample of individuals with Down syndrome. As expected, 95% of the extra copies of chromosome 21 were of maternal origin. Of these cases, approximately 75% resulted from errors in meiosis I and 25% were errors in meiosis II. Less than 1% of the Down syndrome children resulted from nondisjunction in meiosis II in the father. The remaining 5% of Down syndrome individuals that contain three free copies of chromosome 21 were produced by mitotic nondisjunction in the zygote or embryo. In these cases, the extra copy of chromosome 21 originated equally from either parent and there was no increased incidence with older parents.

Familial Down Syndrome

About 5% of individuals with Down syndrome do not have trisomy 21, but rather have a translocation of the long arm of chromosome 21 to usually chromosome 14 or 15. This type of Down syndrome is termed **familial**, based on the pattern of inheritance from generation to generation. **Figure 8.29** diagrams a typical Robertsonian translocation that is involved in familial Down syndrome having karyotype 14q;21q during synapsis of meiosis I.

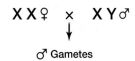

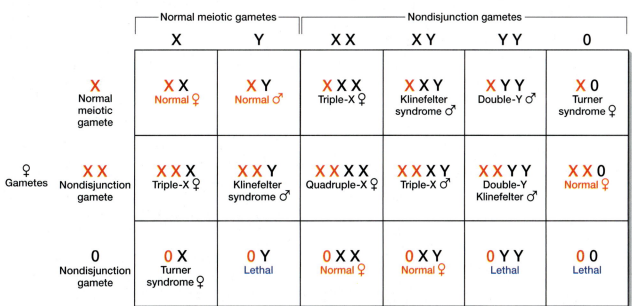

figure 8.26 Genotypes resulting from gametes that underwent nondisjunction of sex crhomosomes. The possible normal and nondisjunction gametes of the X and Y chromosomes are shown. The potential genotypes and resulting genders of the progeny are shown. Human syndromes corresponding to abnormal X and Y chromosome numbers are listed in black, normal gender genotypes are shown in red, and lethal genotypes are in blue. The maternal contribution to the genotype (X or 0) is in red. The 0 represents no sex (X or Y) chromosomes.

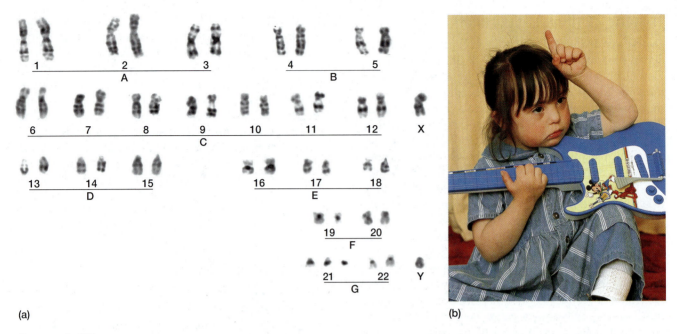

(a)

(b)

figure 8.27   (a) Karyotype of a male with Down syndrome is characterized by an extra chromosome 21 (total of three chromosome 21's). (b) This Down syndrome girl (47, +21) shows the characteristic short stature, broad and short skull and hands, and upward slanting eyes.

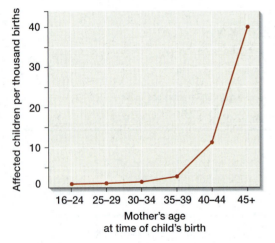

Mother's age at time of child's birth	Frequency of child with Down syndrome	Number of affected children per thousand births
16–24	1/1700	0.58
25–29	1/1100	0.91
30–34	1/770	1.30
35–39	1/250	4
40–44	1/80	12.5
45+	1/25	40

figure 8.28   Increased risk of trisomy 21 attributed to the age of the mother. As the age of the mother at the time of the child's birth increases, there is a correlative increase in the frequency of trisomy 21.

As discussed previously in this chapter, a reciprocal translocation can undergo primarily alternate segregation or adjacent-1 segregation, with adjacent-2 segregation occurring at a low frequency. In figure 8.29, you see all the gametes that can be produced from these three segregation patterns, and the zygotes produced when these gametes are fertilized with a normal gamete. Three of the combinations produce nonviable zygotes due to genetic imbalance. Of the remaining three possibilities, one produces a normal individual, a second produces a child with Down syndrome, and the third is a phenotypically normal child with the Robertsonian translocation.

Because the phenotypically normal individual with the Robertsonian translocation will produce gametes that undergo the same three segregation patterns, familial Down syndrome can skip one or more generations. Familial Down syndrome also does not exhibit the age-dependent expression pattern observed with sporadic Down syndrome. Although each of the three viable zygotic combinations is equally likely, Down syndrome children occur only 15% of the time because the Down syndrome fetuses often spontaneously abort due to inviability.

It is worth mentioning that aside from trisomy and translocation, Down syndrome can also come about through mosaicism (described later in this section) or through a centromeric event. In extremely rare cases, Down syndrome is caused by an abnormal chromosome 21 that has, rather than a short and long arm, two identical long arms attached to the centromere. This type of chromosome, called an **isochromosome,** presumably occurs by an odd centromeric fission

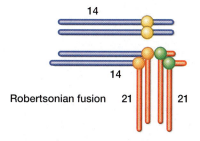

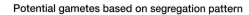

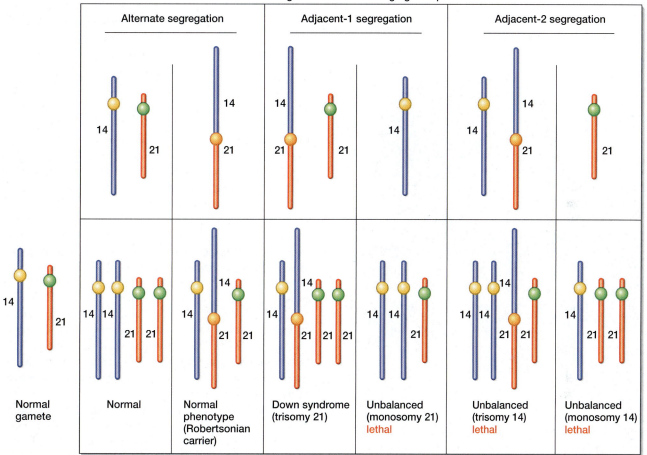

figure 8.29   Synapsis of a Robertsonian translocation (14q and 21q) with a normal chromosome 14 and 21 produces 6 different types of gametes. Fertilization of these gametes with normal gametes (containing one chromosome 14 and one chromosome 21) produces 3 types of viable individuals, one of which has Down syndrome. The other two genotypes are phenotypically normal, with one genotype being diploid for chromosome 14 and 21 and the second genotype containing the Robertsonian translocation.

(**fig. 8.30**). Approximately half of the isochromosomes that result in Down syndrome are of paternal origin. Hence, a person with a normal chromosome 21 and an isochromosome 21 has three copies of the long arm of the chromosome and exhibits Down syndrome.

Trisomy 18 (Edwards Syndrome)

Edwards syndrome affects 1 in 10 thousand live births (**fig. 8.31**). The karyotype for this syndrome is 47, XX or XY,+18. Most affected fetuses fail to survive to birth.

Those that do survive are usually female, with 80–90% mortality by 2 years of age.

An affected infant is usually born with major heart defects and a displaced liver. In addition, the infant usually has a small nose and mouth, a receding lower jaw, abnormal ears, and a lack of distal flexion creases on the fingers. The distal joints have limited motion, and the fingers display a characteristic posturing in which the little and index fingers overlap the middle two. The syndrome is usually accompanied by severe mental retardation.

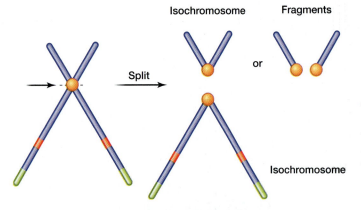

Isochromosome   Fragments

Split   or

Isochromosome

figure 8.30   If the centromere of chromosome 21 breaks perpendicular to the normal division axis, it can form an isochromosome consisting of both long arms and either an isochromosome consisting of both short arms or two separate short chromosomal fragments. This unusual centromeric split can occur during anaphase of either mitosis or meiosis II.

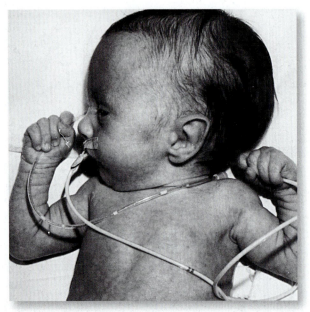

figure 8.31   Child with trisomy 18. This infant (47, +18) with Edwards syndrome exhibits the characteristic low-set ears, small jaw, and persistant clenched hands. Most infants with Edwards syndrome usually have a low birth weight, mental deficiency, congenital heart and kidney problems, and usually do not live longer than a few months of age.

Trisomy 13 (Patau Syndrome) and Other Trisomic Disorders

Patau syndrome, karyotype 47,XX or XY,+13, affects 1 in 20 thousand live births. Again, a very high percentage of fetuses with trisomy 13 fail to survive until birth. Individuals who are born with trisomy 13 often have small or missing eyes, although in rare cases they exhibit a single fused eye structure in the center of the face. Additional features include cleft palate, cleft lip, congenital heart defects, polydactyly, and severe mental retardation. Mortality is very high in the first year of life.

Other autosomal trisomics are known but are extremely rare. These include trisomy 8 (47,XX or XY,+8) and cat's eye syndrome, a trisomy of an unidentified, small acrocentric chromosome (47,XX or XY,[+acrocentric]).

XO (Turner Syndrome)

Turner syndrome occurs in about 1 in 2000 live female births, with approximately 99% of Turner syndrome fetuses failing to survive until birth. The karyotype designation is 45,X (**fig. 8.32**). Individuals with Turner syndrome usually exhibit normal intelligence but underdeveloped ovaries, infertility, abnormal jaws, webbed necks, and shieldlike chests. They are often short in stature. The severe consequences of missing one entire chromosome allow only three different human chromosomes to be monosomic (X, 21, or 22) and still yield a viable individual.

XYY Karyotype

Approximately 1 in 1000 live male births is an individual with a 47,XYY karyotype (**fig. 8.33**). We avoid the term *syndrome* here because XYY men have no clearly defined series of attributes, other than often being taller than normal.

Some controversy has surrounded this karyotype because it was once reported to occur in abundance in a group of mentally subnormal and overly violent males in a prison hospital. Seven XYY males were found among 197 inmates, whereas only 1 in about 2000 control men were XYY. This study has subsequently been expanded and corroborated. Although it is now fairly well established that the incidence of XYY males in prison is about 20 times higher than in society at large, the statistic is somewhat misleading: The overwhelming majority of XYY men seem to lead normal lives. At most, about 4% of XYY men end up in penal or mental institutions, where they make up about 2% of the total institution population.

There is some indication that XYY men may have some speech and reading problems, which could lead to lower test scores and lowered intelligence test results. Criminal tendencies could therefore be attributed to lower intelligence, rather than to the extra Y chromosome. But this idea is highly speculative.

A research project at Harvard University on XYY males, under the direction of Stanley Walzer (a psychiatrist) and Park Gerald (a geneticist), involved screening

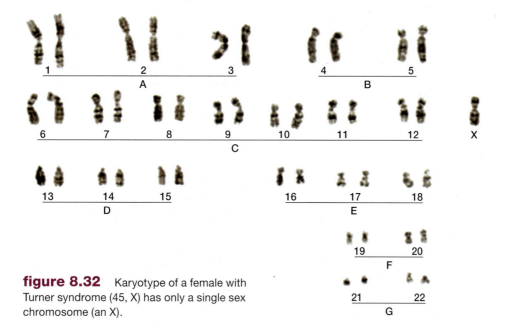

figure 8.32 Karyotype of a female with Turner syndrome (45, X) has only a single sex chromosome (an X).

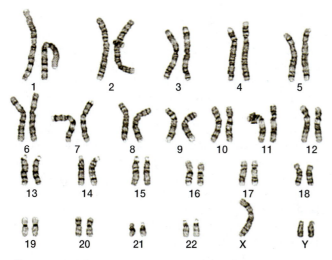

figure 8.33 A male with the XYY karyotype possesses two Y chromosomes rather than the standard single Y chromosome.

all newborn boys at the Boston Hospital for Women (now Brigham and Women's Hospital) and then following the development of those with chromosomal anomalies.

This study was mainly criticized on the necessity of informing parents that their sons had an XYY karyotype that might be associated with behavioral problems. Opponents of this work claimed that telling the parents could trigger a self-fulfilling prophecy; that is, parents who heard that their son was not normal and *might* cause trouble could behave toward their son in a manner that would increase the probability that he *would* cause trouble. The opponents claimed that the risks of this research outweighed the benefits. The project was terminated in 1975 primarily because of the harassment Walzer faced.

An unusual event occurs with the extra Y chromosome. At approximately age 35, one of the Y chromosomes often degenerates. This means that older XYY males will undergo normal meiosis with a single X and single Y chromosome and will conceive children with a very low probability of passing an extra Y chromosome onto their sons.

XXY (Klinefelter Syndrome)

The incidence of Klinefelter syndrome (47,XXY) is about 1 in 1000 live births. Tall stature, underdeveloped testes, and infertility are common symptoms. Diagnosis is usually by buccal (cheek tissue) smear to ascertain the presence of a Barr body, which in a male would indicate an XXY karyotype.

Some problems with behavior and speech development are associated with this syndrome. Although some Klinefelter men fail to learn that they have the condition until they find that they are infertile, many men with Klinefelter syndrome are fertile and will never know they have the condition. If discovered early, testosterone injections during adolescence can help to induce the secondary sexual characteristics (pubic and facial hair) that are often reduced in Klinefelter individuals.

Triple-X Female and Other Aneuploid Disorders of Sex Chromosomes

A triple-X female (47,XXX) appears in about 1 in 1000 female live births (**fig. 8.34**). Fertility can be normal, but these individuals are usually mildly mentally retarded. Delayed growth, as well as congenital malformations, are also sometimes present. In contrast, a triplo-X genotype in *Drosophila* is usually lethal.

This difference demonstrates the effectiveness of X-chromosome inactivation in mammals. X inactivation, by which all X chromosomes except one are set

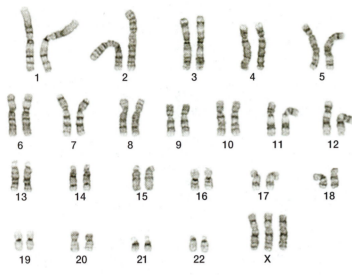

figure 8.34 Karyotype of a triple-X female (47,XXX) possessing three X chromosomes rather than the standard two.

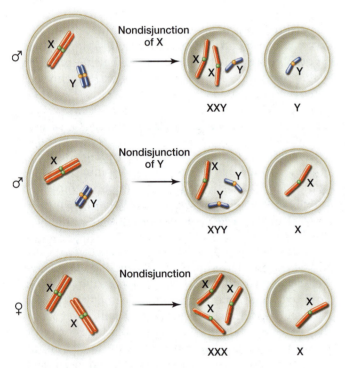

figure 8.35 Mitotic nondisjunction of the sex chromosomes. Mitotic nondisjunction can occur by the centromere splitting during mitosis without the sister chromosomes segregating to different daughter cells. The heterogametic male cell can undergo nondisjunction of either the X or Y chromosome to produce a total of four different genotypes. The homogametic female cell can undergo nondisjunction to produce two different genotypes.

aside as Barr bodies, maintains approximately the same level of expression of X-linked genes in both males and females. By contrast, in the *Drosophila* system, no X chromosomes are inactivated; male flies increase their expression of X-linked genes to approach the level

observed in females with two X chromosomes. Thus, extra X chromosomes in flies lead to increased expression of X-linked genes and lethal genetic imbalance.

Other sex-chromosomal aneuploids, including 48,XXXX, 49,XXXXX, and 49,XXXXY, are extremely rare. All seem to be characterized by mental retardation and growth deficiencies.

Mosaicism

Under most circumstances, an individual is derived from a fertilized egg, and all the cells within the individual have the same chromosome number. On rare occasions, however, an individual is made up of cells that contain one or more different chromosome numbers. These individuals are referred to as **mosaics** or **chimeras,** depending on the sources of the different cells.

One mechanism that can produce mosaic individuals is nondisjunction, which is again demonstrated with the sex chromosomes in **figure 8.35.**

Another mechanism to produce a mosaic individual is chromosome lagging (**fig. 8.36**), in which the X chromosome is lost in one of the dividing somatic cells and results in an XX cell lineage and an X0 cell lineage. In *Drosophila*, if this chromosomal lagging occurs early in development, an organism that is part male (X0) and part female (XX) develops. **Figure 8.37** shows a fruit fly in which chromosomal lagging has occurred at the one-cell stage, causing the fly to be half male and half female. In this case, the female must have been heterozygous for both *white* and *miniature* in the *cis* configuration (*w m*/+ +) to produce the male half with both the *white* and *miniature* recessive mutant phenotypes. A mosaic of this type, involving male and female phenotypes, has a special name—**gynandromorph.** (A **hermaphrodite** is an individual, not necessarily mosaic, with both male and female reproductive organs.)

Many sex-chromosomal mosaics are also known in humans, including XX/X, XY/X, XX/XY, and XXX/X.

About 2% of individuals with Down syndrome are mosaic, having cells with both two and three copies of chromosome 21. Some evidence suggests that the original zygotes were trisomic, but then a daughter cell lost one of the copies of chromosome 21. The severity of the symptoms in these individuals relates to the percentage of trisomic cells they possess. Mosaicism increases with maternal age, just as trisomy in general does.

Restating the Concepts

▶ Aneuploidy is the loss or addition of one or more individual chromosomes. In humans, most aneuploid fetuses spontaneously abort.

▶ For the human autosomes, aneuploidy is not common due to genetic imbalance. However,

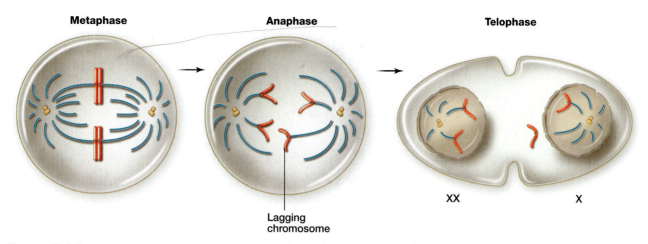

Metaphase — **Anaphase** — **Telophase**

Lagging chromosome

XX — X

figure 8.36 Mitotic nondisjunction due to chromosome lagging. The lagging X chromosome is apparent during mitotic anaphase and fails to be within the reassembling nucleus during telophase, which results in one daughter cell being monosomy X. Because gender determination in *Drosophila* is based on the X:autosome ratio, this daughter cell will express a "male" phenotype in the female fly.

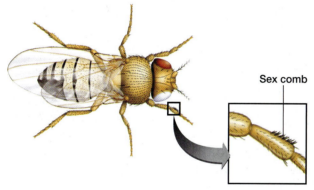

Sex comb

figure 8.37 *Drosophila* gynandromorph. The *left* side of this fly is wild-type XX female; the *right* side is X0 male, hemizygous for the *white* eye and *miniature wing* mutations. The right front leg contains the sex comb that is found on the front legs of male flies.

trisomy is more common than monosomy, possibly due to haploinsufficiency or the pseudodominance of recessive deleterious and lethal mutations in monosomies.

▶ For the human sex chromosomes, both monosomy and trisomy for the X chromosome and disomy for the Y chromosome can produce viable individuals. The severity of the symptoms is largely a reflection of the presence of essential genes that are either over- or underexpressed on the sex chromosomes.

▶ Mosaicism occurs when an individual possesses two different genotypes. Mosaicism can result from either nondisjunction or chromosome lagging early in development.

8.9 Changes in Euploidy

Three kinds of problems plague polyploids. First, the potential exists for a general chromosomal imbalance in the organism because of the extra genetic material

in each cell. As these polyploid cells progress through the cell cycle, they must properly segregate the extra chromosomes to maintain the genetic balance. As you saw with aneuploids, only a few human chromosomes can exist in either a monosomic or trisomic state. For example, a triploid human fetus has about a one in a million chance of surviving to birth, at which time death usually occurs from problems in all organ systems. Second, organisms with a chromosomal sex-determining mechanism may have development disrupted by polyploidy. And third, meiosis produces unbalanced gametes in many polyploids, which results in semisterility for the individual. If a polyploid has an odd number of chromosome sets, such as triploid ($3n$), two of the three homologs will tend to pair at prophase I of meiosis, producing a bivalent and leaving an unpaired chromosome, or univalent (**fig. 8.38**). The bivalent separates normally, but the third chromosome goes independently to one of the poles. This separation results in a 50% chance of aneuploidy in each of the n-different chromosomes, rapidly decreasing the probability of a balanced gamete as n increases. Therefore, as n increases, so does the likelihood of sterility. An alternative to the bivalent–univalent type of synapsis is the formation of trivalents, which have similar problems (fig. 8.38, red chromosomes).

Even-numbered polyploids, such as tetraploids ($4n$), can fare better during meiosis. If the centromeres segregate two by two in each of the n meiotic figures, balanced gametes can result. Often, however, the multiple copies of the chromosomes form complex arrangements during synapsis, including monovalents, bivalents, trivalents, and quadrivalents, tending to result in aneuploid gametes and sterility.

Some groups of organisms, primarily plants, have many polyploid members. An estimated 30–80% of all flowering plant species (angiosperms) are polyploids, as are 95% of ferns. (Polyploidy is apparently rare in

224

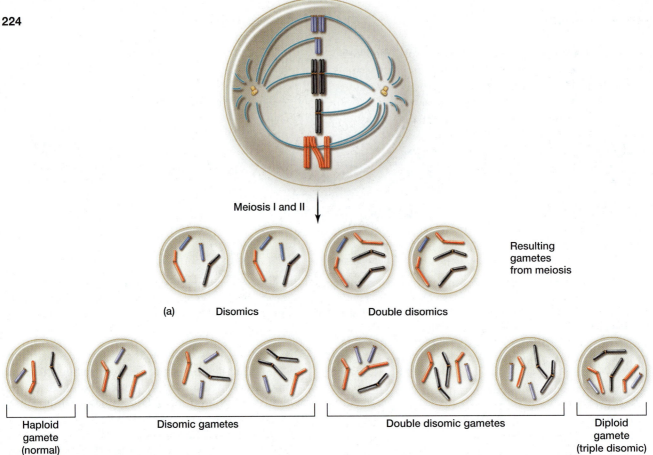

Meiosis I and II

Resulting
gametes
from meiosis

(a)   Disomics   Double disomics

Haploid
gamete
(normal)

Disomic gametes

Double disomic gametes

Diploid
gamete
(triple disomic)

(b)

figure 8.38   Meiosis in a triploid ($3n = 9$). (a) Shown is one possible pairing arrangement of the three chromosomes in this triploid and the resulting gametes. (b) The probability of a "normal" gamete is $(1/2)^n$ where n equals the haploid chromosome number. Here, $n = 3$ and $(1/2)^3 = 1/8$. The eight different gametes produced from the above triploid are shown.

gymnosperms and fungi.) For example, the genus of wheat, *Triticum*, has members with 14, 28, and 42 chromosomes. Because the basic *Triticum* chromosome number is $n = 7$, these forms are $2n$, $4n$, and $6n$ species, respectively. Chrysanthemums have species of 18, 36, 54, 72, and 90 chromosomes. With a basic number of $n = 9$, these species represent a $2n$, $4n$, $6n$, $8n$, and $10n$ series. In both these examples, the even-numbered polyploids are viable and fertile, but the odd-numbered polyploids are not.

Autopolyploidy

Polyploidy can come about in two different ways. In **autopolyploidy,** all of the chromosomes come from the same species. In **allopolyploidy,** the chromosomes come from hybridizing two different species (fig. 8.39).

Autopolyploidy occurs in several different ways. One way is by fusing of nonreduced gametes. For example, if a diploid gamete fertilizes a normal haploid gamete, the result is a triploid. Similarly, if two diploid gametes fuse, the result is a tetraploid.

The equivalent of a nonreduced gamete comes about in meiosis if the parent cell is polyploid to begin with.

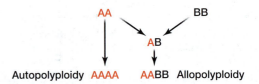

Autopolyploidy AAAA   AABB Allopolyploidy

figure 8.39   Autopolyploidy and allopolyploidy. If A and B are the haploid genomes of two diploid species A and B, respectively, then autopolyploidy produces a species with an AAAA karyotype, and allopolyploidy (with chromosome doubling) produces a species with AABB karyotype. If A represents seven chromosomes, then an AA diploid has 14 chromosomes and an AAAA tetraploid has 28 chromosomes. If B represents five chromosomes, then a BB diploid has 10 chromosomes and an AABB allotetraploid has 24 chromosomes.

example, if one branch of a diploid plant is tetraploid, its flowers produce diploid gametes. These gametes are not the result of a failure to reduce chromosome numbers meiotically, but rather the successful meiotic reduction by half of a tetraploid flower to a diploid gamete ($4n$ to $2n$).

The tetraploid tissue of the plant in the preceding example can originate by the **somatic doubling** of diploid tissues. Somatic doubling can occur spontaneously or can be induced by disrupting the normal sequence of a nuclear division. For example, the drug colchicine induces somatic doubling by inhibiting microtubule formation. This prevents the formation of a spindle and stops the chromosomes from moving apart during either mitosis or meiosis. The result is a cell with double the chromosome number. Other chemicals, temperature shock, and physical shock can produce the same effect.

Allopolyploidy

Allopolyploidy comes about by cross-fertilization between two species. The resulting offspring have the sum of the reduced (gametic) chromosome number of each parent species.

If each chromosome set is distinctly different, the new organisms have difficulty in meiosis because no two chromosomes are sufficiently homologous to pair. Then every chromosome forms a univalent (unpaired) figure, and they separate independently during meiosis, producing aneuploid gametes. However, if an organism (most commonly, a plant) can survive by vegetative growth until somatic doubling takes place in gamete precursor cells ($2n \rightarrow 4n$), or alternatively, if a zygote is formed by fusion of two unreduced gametes ($2n + 2n$), the resulting offspring will be fully fertile because each chromosome has a pairing partner at meiosis.

In 1928, the Russian geneticist G. D. Karpechenko was working with the radish (*Raphanus sativus*, $2n = 18$) and cabbage (*Brassica oleracea*, $2n = 18$). When these two plants are crossed, an F_1 results with $n + n = 18$ (9 + 9). This plant, which is an allodiploid, has characteristics intermediate between the two parental species (**fig. 8.40**). If somatic doubling then takes place, the chromosome number is doubled to 36, and the plant becomes an allotetraploid ($4n$). Because each chromosome has a homolog, it is also referred to as an **amphidiploid.**

If we did not know its past history, we might simply classify this plant as a diploid with $n = 18$. In this case, however, the new amphidiploid cannot successfully be crossed with either parent, because the offspring are sterile triploids. It is, therefore, a new species and has been named *Raphanobrassica.*

For agricultural purposes, however, the experiment was not a success because it did not combine the best features of the cabbage and radish.

Polyploidy in Plants and Animals

Although polyploids in the animal kingdom are known (in some species of lizards, fish, invertebrates, and a tetraploid mammal, the red viscacha rat), polyploidy as a successful evolutionary strategy is primarily a plant phenomenon. Several characteristics explain this observation.

To begin with, many more animals than plants have chromosomal sex-determining mechanisms.

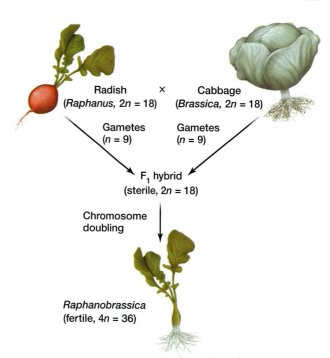

figure 8.40 Karpechenko crossed the radish and cabbage in hopes to generate a fertile amphidiploid with the root of the radish and the leaves like cabbage, which would make the entire plant edible. However, after doubling the chromosome number in the sterile F_1 hybrid, the fertile *Raphanobrassica* tetraploid had roots like a cabbage and leaves like a radish. While the fertile tetraploid was produced, it failed to exhibit the most desirable characteristics from the two diploid plants.

Polyploidy severely disrupts these mechanisms. For example, Bridges discovered a tetraploid female fruit fly, but it has not been possible to produce a tetraploid male. The tetraploid female's progeny were triploids and intersexes, which rendered them infertile.

Second, plants can generally avoid the meiotic problems of polyploidy longer than most animals. Some plants can exist vegetatively, allowing more time for the rare somatic doubling event to occur that will produce an amphidiploid. In contrast, animal life spans are more precisely defined, allowing less time for a somatic doubling.

Third, many plants depend on wind or insect pollinators to fertilize them and thus have more of an opportunity for hybridization. Many animals have relatively elaborate courting rituals that tend to restrict hybridization.

Polyploidy has been used in agriculture to produce "seedless" as well as "jumbo" varieties of crops. Seedless watermelon, for example, is a triploid. Its seeds rarely develop, and those that do are mostly sterile. It is produced by growing seeds from a cross between a tetraploid variety and a diploid variety. Jumbo Macintosh apples are another example of a tetraploid.

Restating the Concepts

▶ Changes in euploidy occur when the number of complete chromosome sets is either increased or decreased. Haploid and diploid organisms are very common in nature. More than two genome sets is generically referred to as polyploidy.

▶ An odd number of genome sets, such as triploid, is uncommon because of significant problems during synapsis of meiosis I. Triploidy usually produces aneuploid gametes that will produce nonviable zygotes. Thus, an odd number of genome sets usually results in sterility.

▶ An even number of genome sets, such as a tetraploid, is more common because each chromosome may synapse with only one homolog during meiosis I.

▶ Plants are more commonly polyploid than animals. Plants do not utilize the same sex-determining mechanisms as animals, which reduces the possibility of genetic imbalance, and many plants can exist in a vegetative state that allows for the rare somatic doubling of genome sets to occur.

 ## genetic linkages

In this chapter, we discussed a variety of situations in which the chromosome structure or number was altered relative to wild type.

We described four different changes to the chromosome structure: deletion, duplication, inversion, and translocation. Each has characteristic structures when pairing with wild-type chromosomes during meiosis I (chapter 3). In addition, each has diagnostic features when studied in genetic crosses. For example, you saw how deletions exhibit pseudodominance and the advantage of this feature in mapping genes. In chapter 15, we will describe how Seymour Benzer used deletions to map a large number of *rII* mutations in a bacteriophage. This technique reveals another important function associated with deletions, which is the inability for recombination to occur within the DNA that corresponds to the deleted region.

Duplications play an important role in evolution and the generation of genes with related, but different functions. We discussed an example of this in chapter 5 when the effect of duplicate gene action was explored in relation to the Mendelian ratios observed in a dihybrid cross. We also saw the effects of duplications when we discussed gene families in chapter 5.

Deletions are not the only chromosomal mutations important in mapping mutations and the corresponding genes. We will describe in chapter 20 how chromosomal breakpoints (primarily inversions and translocations) play an important role in mapping mutations on the cytogenetic map and ultimately in cloning the corresponding gene. The roles of inversions and translocations in some forms of cancer are also discussed in chapter 22.

We also described how changes in chromosome number can affect an organism, both the number of complete chromosome sets and individual chromosomes. You learned that a change in the number of chromosome sets is one mechanism by which new species are created. Speciation and other mechanisms in this process are further described in chapter 25.

We also discussed how individual chromosomes can change in number. Nondisunction is one mechanism leading to aneuploidy. You first encountered nondisjunction in the discussion of Bridges' proof of the chromosome theory of inheritance using the exceptional progeny of X-linked recessive mutations (chapter 4). You will also see in chapter 22 how some forms of cancer are associated with changes in the chromosome number of the malignant cells.

Chapter Summary Questions

1. For each term in column I, choose the best matching phrase in column II.

Column I	Column II
1. Mosaicism	A. Segregation of adjacent nonhomologous centromeres in a translocation heterozygote
2. Adjacent-1 segregation	B. Changes in entire sets of chromosomes
3. Euploidy	C. Most common cause of inherited mental retardation
4. Fragile-X syndrome	D. Presence of patches with different phenotypes
5. Pseudodominance	E. Segregation of adjacent homologous centromeres in a translocation heterozygote
6. Variegation	F. A chromosomal deletion causing mental retardation
7. Cri du chat syndrome	G. Expression of the recessive phenotype in the presence of only one recessive allele
8. Adjacent-2 segregation	H. Changes in subsets of chromosome sets
9. Patau syndrome	I. Trisomy 13
10. Aneuploidy	J. Presence of two or more different cell lines

2. Compare and contrast paracentric and pericentric inversions and how they each affect recombination when heterozygous with a wild-type chromosome.

3. Compare and contrast autopolyploids and allopolyploids.

4. Why is polyploidy more common in plants than in animals?

5. What are the consequences of monosomy in humans?

6. Provide a concise definition for the following terms: acentric chromosome; dicentric chromosome; and isochromosome. Which of these types of chromosomes would be expected to undergo normal segregation during cell division?

7. What kind of chromosomal structure is observed during meiosis I of an individual that is heterozygous for a reciprocal translocation? Describe the various ways that this structure can segregate during meiosis I and the consequences of each on the gametes that are ultimately produced?

8. Answer question 7 for an individual that is homozygous for a reciprocal translocation.

9. Can a deletion result in the formation of a variegation position effect? If so, how?

10. Does crossover suppression occur in an inversion homozygote? Explain.

11. Which rearrangements of chromosomal structure cause semisterility and why?

12. What are the consequences of single crossovers during tetrad formation in a reciprocal translocation heterozygote?

13. Give the gametic complement, in terms of acentrics, dicentrics, duplications, and deficiencies, when a three-strand double crossover occurs within a paracentric inversion loop.

14. Is a tetraploid more likely to show irregularities in meiosis or mitosis? Explain. What about these processes in a triploid?

15. Do autopolyploids or allopolyploids experience more difficulties during meiosis and why? Do amphidiploids have more or less trouble than auto- or allopolyploids and why?

16. Compare and contrast sporadic Down syndrome and familial Down syndrome.

17. Which would you expect to have more severe consequences in humans: aneuploidy for sex chromosomes or for autosomes? Explain your answer.

Solved Problems

PROBLEM 1: What are the consequences of an inversion?

Answer: In an inversion homozygote, the consequences are a change in linkage arrangements, including new gene orders and map distances, and the possibility of position effects if a locus is placed into or near heterochromatin. In an inversion heterozygote, crossover suppression causes semisterility because zygotes that carry genetic imbalances are lost. Inversion heterozygotes can be seen as meiotic loop structures or loops formed in endomitotic chromosomes such as those found in the salivary glands of fruit flies. In an evolutionary sense, inversions result in supergenes, locking together allelic combinations.

PROBLEM 2: Ebony body (*e*) in fruit flies is an autosomal recessive trait. A true-breeding ebony female (*ee*) is mated with a true-breeding wild-type male that has been irradiated. Among the wild-type progeny is a single ebony male. Explain this observation.

Answer: The cross is $ee \times e^+e^+$, and all F_1 progeny should be e^+e (wild type). The use of irradiation alerts us to the possibility of chromosomal breaks, as well as simple mutations. What type of chromosomal aberration would allow a recessive trait to appear unexpectedly? A deletion, which creates pseudodominance when there is no second allele, is a good possibility. The male in question could have gotten the *ebony* allele from its mother and no homologous allele from its father. Alternatively, the wild-type allele from the father could have mutated to an ebony allele.

PROBLEM 3: In *Drosophila melanogaster*, eyeless (*ey*) is a recessive trait in which the eyes are tiny or absent. The *ey* locus is found on chromosome 4. Monosomy and trisomy for this very small chromosome are compatible with survival and fertility. A trisomic wild-type male with the genotype *ey⁺ ey ey* is crossed to an *ey⁺ ey* female. Assuming random segregation of chromosomes into gametes, determine the chromosome constitution and phenotypic ratios in the offspring.

Answer: It appears that only one dominant *ey⁺* allele is required for the formation of wild-type eyes because flies with the *ey⁺ ey ey* genotype have normal eyes.

Let's start by giving a different designation to each of the two homologs in the male that are carrying the same allele: *ey*-1 and *ey*-2. Therefore, the genotype of the male is now: *ey⁺ ey*-1 *ey*-2. Three different types of segregation can occur in this trisomic individual during gamete formation in meiosis: *ey⁺ ey*-1 to one pole and *ey*-2 to the other pole; *ey⁺ ey*-2 to one pole and *ey*-1 to the other; and finally, *ey*-1 *ey*-2 to one pole and *ey⁺* to the other.

Thus, the trisomic parent has two different ways to generate an *ey⁺ ey* gamete: either *ey⁺ ey*-1 or *ey⁺ ey*-2. If we now ignore the different designations for the *ey* recessive allele, then there is a 2/6 probability of producing an *ey⁺ ey* gamete. Similarly, there is a 2/6 probability of producing an *ey* gamete. Finally, there is a 1/6 probability of producing the *ey ey* gamete and 1/6 probability of generating an *ey⁺* gamete. The disomic parent on the other hand, can only produce two types of gametes: 1/2 *ey⁺* and 1/2 *ey*.

The various possible fertilizations can be obtained by using the forked-line method for gametes.

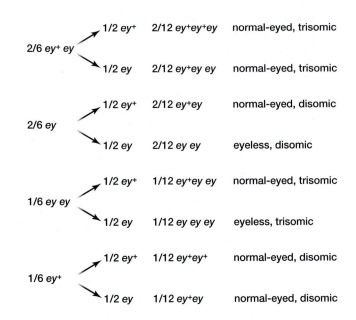

Therefore, the phenotypic ratio in the offspring is

5/12 normal-eyed, trisomic : 4/12 normal-eyed, disomic : 2/12 eyeless, disomic : 1/12 eyeless, trisomic.

Exercises and Problems

18. In humans, how many chromosomes would be found in the following conditions?
    a. triploidy   **b.** trisomy
    c. double monosomy   **d.** nullisomy

19. The house mouse, *Mus musculus,* has a diploid chromosome number of 40. How many chromosomes would be found in the following?
    a. monosomic cell   **b.** trisomic cell
    c. nullisomic cell   **d.** double tetrasomic cell
    e. triploid cell   **f.** 4*n* + 2 cell
    g. disomic gamete   **h.** nullisomic gamete

20. In studying a new sample of fruit flies, a geneticist noted phenotypic variegation, semisterility, and the nonlinkage of previously linked genes. What probably caused this, and what cytological evidence would strengthen your hypothesis?

21. In a second sample of flies, the geneticist found a position effect and semisterility. The linkage groups were correct, but the order was changed and crossing over was suppressed. What probably caused this, and what cytological evidence would strengthen your hypothesis?

22. In humans, the chromosome complements of normal males and females are designated 46, XY, and 46, XX, respectively. Even though some of the conditions are not compatible with survival, give the notations for the chromosomal abnormalities in the following individuals?
    a. Male with trisomy 11
    b. Female with monosomies for chromosomes 5 and 7

    c. Male with a duplication in the long arm of chromosome 20
    d. Female with a translocation of part of the short arm of chromosome 10 onto the long arm of chromosome 15
    e. An individual with one Y chromosome, one chromosome 3, three chromosomes 18, and two copies of all other autosomes

23. Consider a chromosome whose normal sequence of segments and centromere (heavy dot) location is shown here

    Determine the type (or types) of chromosomal aberrations necessary to produce each of the following chromosomes.

    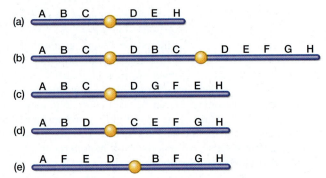

24. In *Drosophila melanogaster*, eyeless (*ey*) is a recessive trait in which the eyes are tiny or absent. The *ey* locus is found

on chromosome 4. Monosomy, trisomy, and tetrasomy for this very small chromosome are compatible with survival and fertility. Assuming random segregation of chromosomes into gametes, determine the chromosome constitution and phenotypic ratios in the offspring of the following crosses:

a. $ey^+\ ey^+\ ey$ male × $ey^+\ ey^+\ ey$ female
b. $ey^+\ ey^+\ ey$ male × $ey^+\ ey$ female

25. Diagram the results of alternate segregation for a three-strand double crossover between a centromere and the cross center in a reciprocal translocation heterozygote.

26. Consider two nonhomologous chromosomes whose normal sequences of segments and centromere (heavy dot) locations are shown here

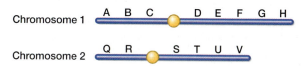

Determine the type (or types) of chromosomal aberrations necessary to produce the following chromosomes.

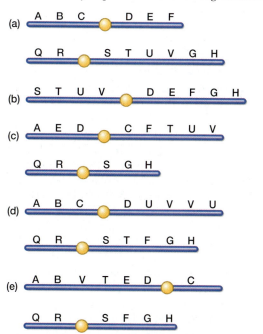

27. The haploid chromosome complement of *Drosophila melanogaster* consists of four chromosomes: one very small acrocentric, one large acrocentric, and two large metacentric chromosomes. The haploid chromosome complement of another species, *Drosophila virilis*, consists of six chromosomes that are all acrocentric. If the fundamental number (NF or *nombre fondamentale*) is constant, suggest a mechanism to explain the difference in chromosome number between the two species.

28. A heterozygous plant *ABCDE/abcde* is testcrossed with an *abcde/abcde* plant. Only the following progeny appear.

ABCDE/abcde   *abcde/abcde*
Abcde/abcde   *aBCDE/abcde*
ABCDe/abcde   *abcdE/abcde*

What is unusual about the results? How can you explain them?

29. White eye color in *Drosophila* is an X-linked recessive trait. A wild-type male is irradiated and mated with a white-eyed female. Among the progeny is a white-eyed female.
   a. Why is this result unexpected, and how could you explain it?
   b. What type of progeny would you expect if this white-eyed female is crossed with a normal, non-irradiated male?

30. You are trying to locate an enzyme-producing gene in *Drosophila*, which you know is located on the third chromosome. You have five strains with deletions for different regions of the third chromosome:

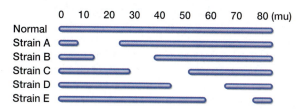

You cross each strain with wild-type flies and measure the amount of enzyme in the F_1 progeny. The results appear as follows. In what region is the gene located?

Strain Crossed	Percentage of Wild-Type Enzyme Produced in F_1 Progeny
A	100
B	45
C	54
D	98
E	101

31. Consider the following table, which shows the number of viable progeny produced by a plant under standard conditions. Provide an explanation for the results.

P:	Strain A × Strain A	Strain B × Strain B	Strain A × Strain B
F_1:	765	750	775
F_2:	712	783	416

32. The map position for three X-linked recessive genes in *Drosophila* (*v*, vermilion eyes; *m*, miniature wings; and *s*, sable body) is:

v	*m*	*s*
33.0	36.1	43.0

A wild-type male is irradiated and mated to a vermilion, miniature, sable female. Among the progeny is a single vermilion-eyed, long-winged, tan-bodied female. The following shows the progeny when this female is mated with a *v m s* hemizygous male.

Females	Males
87 vermilion, miniature, sable	89 vermilion, miniature, sable
93 vermillion	1 vermillion

Explain these results by drawing a genetic map.

33. In *Drosophila*, recessive genes clot (*ct*) and *black* body (*b*) are located at 16.5 and 48.5 mu, respectively, on the second chromosome. In one cross, wild-type females that are *ct⁺ b⁺/ct b* are mated with *ct b/ct b* males. They produce these progeny:

    wild-type   1,250
    clot, black 1,200
    black   30
    clot   20

    What is unusual about the results? How can you explain them?

34. You have four strains of *Drosophila* (1–4) that were isolated from different geographic regions. You compare the banding patterns of the second chromosome and obtain these results (each letter corresponds to a band):

    (1) m n r q p o s t u v   (2) m n o p q r s t u v
    (3) m n r q t s u p o v   (4) m n r q t s o p u v

    If strain (3) is presumed to be the ancestral strain, in what order did the other strains arise?

35. The broad bean, *Vicia faba*, has a diploid chromosome number of 12.
    a. How many different trisomics are possible?
    b. How many different double trisomics are possible?

36. In *Drosophila*, the recessive *white* eyes gene is located near the tip of the X chromosome. A wild-type male is irradiated and mated with a white-eyed female. Among the progeny is one red-eyed male. How can you explain the red-eyed male, and how could you test your hypothesis?

37. If a diploid species of $2n = 16$ hybridizes with one of $2n = 12$, and the resulting hybrid doubles its chromosome number to produce an allotetraploid (amphidiploid), how many chromosomes will it have?

38. A linkage group is a group of genes whose loci are on the same chromosome. In an organism, there are as many linkage groups as there are chromosomes in the basic set (i.e., the monoploid number). Consider a polyploid plant with 64 chromosomes. How many linkage groups are present if the plant is an
    a. autotetraploid?   b. allotetraploid?

39. The modern garden strawberry, *Fragaria ananassa*, is an autooctoploid. Consider a gene on chromosome 1 that segregates two alleles: *A* and *a*.
    a. How many copies of this gene does an individual plant have?
    b. How many different genotypes are possible? List them all.

40. Assume that nondisjunction of the sex chromosomes occurs in a human female in only one of the two daughter cells during meiosis II, and that all types of eggs are functional. If the egg is fertilized by a normal sperm, indicate all possible chromosomal complements for the resulting zygote?

41. Assume that nondisjunction of the sex chromosomes occurs in a human male during meiosis I (with meiosis II normal), and that all types of sperm are functional. If the sperm fertilizes a normal egg, indicate all possible chromosomal complements for the resulting zygote?

42. Turner syndrome mosaics (45, X/46, XX) and Klinefelter mosaics (47, XXY/46, XY) have variable phenotypes, ranging from the classical phenotype of the syndromes to almost normal appearance.
    a. How does each of these mosaics arise?
    b. Explain the phenotypic variances.

43. How might an XO/XYY human mosaic arise?

44. Plant species P has $2n = 18$, and species U has $2n = 14$. A fertile hybrid is found. How many chromosomes does it have?

45. Assume all Down syndrome individuals are fertile and that chromosome segregation in them occurs at random, with each gamete getting at least one chromosome 21. What is the expected phenotypic ratio in the offspring of the following matings?
    a. Down syndrome male × normal female
    b. Down syndrome male × Down syndrome female

46. Color blindness is a recessive X-linked trait in humans. A woman with normal vision whose father was color-blind mates with a man with normal vision. They have a color-blind daughter with Turner syndrome. In which parent did nondisjunction occur?

47. A color-blind man mates with a woman with normal vision whose father was color blind. They have a color-blind son with Klinefelter syndrome. In which parent did nondisjunction occur?

48. In humans, faulty tooth enamel is an X-linked dominant trait. A couple with abnormal teeth have a son with Klinefelter syndrome and normal teeth.
    a. In which parent did nondisjunction occur?
    b. Did nondisjunction occur in meiosis I or meiosis II?

49. Describe a genetic event that can produce an XYY man.

50. Chromosomal analysis of a spontaneously aborted fetus revealed that the fetus was 92,XXYY. Propose an explanation to account for this unusual karyotype.

51. Various species in the grass genus *Bromus* have chromosome numbers of 14, 28, 42, 56, 70, 84, 98, and 112. What can you tell about the genetic relationships among these species and how they might have arisen?

52. Consider seven recessive genes, *a, e, d, j, l, r,* and *y*. The seven loci are closely linked in the centromeric region of an autosome but their order is unknown. Six deletions are examined in an individual that is heterozygous for all seven genes. All six deletions resulted in pseudodominance of at least two genes, and two deletions spanned the region of the centromere.

Deletion	Centromere Included in Deletion	Genes Exhibiting Pseudodominance
del 1	No	*e* and *j*
del 2	Yes	*a, l,* and *r*
del 3	No	*a* and *d*
del 4	Yes	*e, l,* and *r*
del 5	No	*d* and *y*
del 6	No	*e* and *l*

Determine the gene order and the location of the centromere.

53. An autotetraploid with the genotype *AAaa* is self-fertilized. Assume that no crossing over of any kind occurs, that random segregation of chromosomes always produces gametes with two alleles, and that *A* is completely dominant regardless of the number of *a* alleles present.
    a. What types of gametes can this plant produce and in what ratios?
    b. Determine the genotypic and phenotypic ratios in the offspring?

54. There was a humorous television commercial in which someone accidentally discovered the desirability of combining chocolate and peanut butter. Could this combination be achieved by crossing peanut and cocoa plants?

Chapter Integration Problem

In *Drosophila melanogaster,* the gene *N* (Notch) is located near the *w* (*white* eyes) locus on the X chromosome. Female flies that are heterozygous for *Notch* have wing margins that fail to develop properly and appear to have small indentations or notches in them.

A female with normal wings and red eyes is heavily irradiated with X-rays and then mated to a normal-winged and red-eyed male. The offspring consisted of:

240 normal-winged, red-eyed females
 20 notch-winged, red-eyed females
119 normal-winged, red-eyed males
121 normal-winged, white-eyed males

a. What was the genotype of the P generation female before X-ray irradiation? Explain your answer.
b. Did any chromosomal aberration occur because of X-ray irradiation? If so, what is it and which of the two chromosomes did it involve? Justify your reasoning.
c. Provide a reason for the absence of Notched males.
d. Diagram this cross.
e. Predict the phenotypic classes and their proportions in the offspring of the following crosses:
  (i) White-eyed, Notch females × Red-eyed males with normal wings.
  (ii) White-eyed, Notch females × White-eyed males with normal wings.
f. Is it possible to design a cross that will yield females that are homozygous for *Notch?* Why or why not?
g. The Notch phenotype may be the result of a mutation in the *Notch* gene itself, or a deletion that also spans the *white* gene locus. Design a cross that can distinguish between these two causes.

 Do you need additional review? Visit **www.mhhe.com/hyde** for practice tests, answers to end-of-chapter questions and problems, interactive exercises, and animation tutorials all designed to enhance your understanding of key genetic concepts.

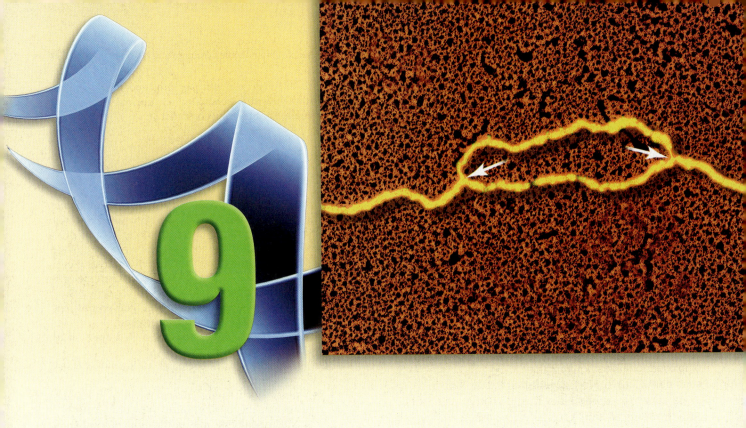

DNA Replication

Essential Goals

1. Understand and summarize the key experiments that revealed the mechanism of DNA replication in both prokaryotes and eukaryotes and what the results from those experiments would have looked like if semiconservative replication with bidirectional replication forks did not occur.

2. Diagram a DNA replication fork showing the leading and lagging strand, the location of RNA primers, Okazaki fragments, DNA ligase, helicase, single-stranded binding proteins, DNA polymerase, and the 5' and 3' ends of all the DNA ends.

3. Describe and diagram the major features associated with replicating a circular DNA molecule and a linear DNA molecule.

4. Understand how DNA replication resolves the issues of the leading and lagging strands.

5. Name the three different *E. coli* DNA polymerases and describe their roles during DNA replication and the activities that they possess.

Photo: A false-colored DNA molecule with a replication bubble containing two replication forks (arrows).

Chapter Outline

ear the conclusion of their landmark paper describing the structure of DNA, James Watson and Francis Crick stated, "It has not escaped our notice that the specific pairing we have postulated immediately suggests a possible copying mechanism for the genetic material." This brief understatement initiated a series of studies to determine the details of DNA replication. Although many of these details have been elucidated over the last 50 years, further questions about DNA replication remain unanswered, and the application of this information to modern medicine continues.

Human immunodeficiency virus (HIV) is a retrovirus that possesses an RNA genome. After infecting a cell, the RNA is converted into DNA by an RNA-dependent DNA polymerase (also known as **reverse transcriptase,** because it converts RNA to DNA). This DNA can insert into the host cell's genome and serve as a template for transcription of HIV-encoded genes and additional RNA genomes.

To combat HIV infections, a variety of drugs have been employed, one class of which are the nucleotide analogs, such as **AZT** (3'-azido-2', 3'-dideoxythymidine). This analog is converted in cells into the "AZT-triphosphate" form. As a thymidine analog, it can be incorporated into replicating DNA. However, when it is incorporated at the 3' end of the replicating DNA strand, it prevents the addition of the next nucleotide because it lacks the 3' hydroxyl group. Thus, it terminates DNA replication. Because the HIV-encoded reverse transcriptase is more efficient at utilizing AZT-triphosphate than DNA polymerase, the replication of the HIV DNA is preferentially blocked. However, the incorporation of AZT into the replicating host genomic DNA results in the toxic side effects.

Many viruses (including HIV) exhibit a high mutation rate, which can render them resistant to drugs in a short period of time and before the viral infection is eradicated. For this reason, combinations of nucleotide analogs are usually administered to reduce the probability that the virus will become resistant to all of them. Other nucleotide analogs that are currently being used to treat HIV include ddC (dideoxycytidine) and ddI (dideoxyinosine).

Another approach to treat the HIV infection is to determine the structure of the viral reverse transcriptase and identify molecules that may specifically bind to this enzyme and not to the host DNA polymerase. These compounds should inhibit the virus's synthesis of DNA, without significantly affecting the replication of the host's genomic DNA. These nonnucleotide reverse transcriptase inhibitors, such as calanolide A and capravirine, are being added to the arsenal of drugs to combat HIV.

In addition to viral diseases, a number of other disease processes involve abnormal DNA replication, including spontaneous mutations and cancers. Further understanding of these processes should help to reveal how replication is disrupted in particular situations and could potentially lead to new therapies for certain types of cancers. The future of medicine may lie in being able to design very specific drugs or treatments for distinct diseases or defects. This specificity should result in reduced side effects and suffering as a result of treatment, along with increased effectiveness.

9.1   The Mechanism of DNA Synthesis

Watson and Crick hinted that replication of the double helix could take place as the DNA strands separate, so that each strand would serve as a **template** for a newly synthesized strand (**fig. 9.1**). For example, suppose one strand had a small section with the nucleotide sequence AATC. Following the rules of base pairing (chapter 7), its complementary strand would have the matching sequence TTAG. When this double helix is unwound during replication, the AATC section would be used to produce a new TTAG section as the complementary nucleotides are added. Likewise, the original TTAG section would act as template for a new AATC section. The result, repeated for all the base pairs in both strands, would be two identical double helices.

Three models were hypothesized to account for the synthesis of daughter DNA molecules. In all three models, each strand of the parental DNA molecule can serve as the template for a new strand. The three models differ in how the daughter molecules are produced.

In **semiconservative replication,** each strand of the parental molecule would serve as a template for the synthesis of a new strand. The resulting daughter

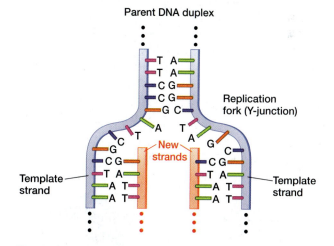

figure 9.1   Base complementarity provides a mechanism for accurate DNA replication. The parent DNA duplex opens, and each parental strand becomes a template for a new double-stranded DNA duplex.

DNA molecules consist of one intact template strand (that is, half the original helix) and a newly replicated complementary strand.

In **conservative replication,** one daughter molecule consists of the original parental DNA (or the parental molecule is conserved), and the other daughter is composed of totally new DNA. This model requires one strand in the parental molecule to serve as the template for a new strand. After synthesizing this new strand, the original parental molecule reassociates and the newly synthesized strand serves as a template for a complementary strand. Alternatively, each strand of the parental molecule serves as a template for a new strand, and then the two parental strands reassociate, as do the two newly synthesized strands.

In **dispersive replication,** some parts of the original double helix are conserved, and some parts are not. Each daughter molecule consists of part template and part newly synthesized DNA, without either being restricted to a single strand.

In reality, the dispersive category is the all-inclusive "other" category, including any possibility other than conservative or semiconservative replication.

To differentiate among these three models, one of the classic experiments in molecular biology was performed by Matthew Meselson and Franklin Stahl, who reported their findings in 1958 (**fig. 9.2**).

The Meselson–Stahl Experiment: Proof of Semiconservative Replication

Some science historians and philosophers consider the Meselson–Stahl experiment to be the most elegant scientific experiment ever designed. Its purpose was to distinguish the way in which new nucleotides are added during DNA replication.

Meselson and Stahl grew *E. coli* in a medium containing a heavy isotope of nitrogen, ^{15}N. (The normal form of nitrogen is ^{14}N.) The ^{15}N would then be incorporated into the base of the four dNTPs. After growing for several generations on the ^{15}N medium, these four "heavy" bases were incorporated into the DNA which resulted in this DNA possessing a greater than normal density.

Detecting Density Differences

The researchers determined the density of the double-stranded DNA using a technique known as **density-gradient centrifugation.** In this technique, a cesium chloride (CsCl) solution is spun in an ultracentrifuge at a high speed for several hours. As the centrifugal force drives the CsCl to the bottom of the tube, diffusion acts to move the CsCl up the tube. This establishes the equilibrium gradient in the tube with the CsCl concentration increasing from top to bottom (**fig. 9.3**).

If DNA (or any other substance) is added to a gradient tube, it concentrates and forms a band in the tube at the point where its density matches that of the CsCl. If several types of DNA having different densities are added, they form several bands. The bands are detectable under ultraviolet light at a wavelength of 260 nm, which nucleic acids absorb strongly.

Control preparations of DNA established band locations for DNA containing only ^{14}N and for DNA containing only ^{15}N.

Interpreting the Findings

Meselson and Stahl transferred the bacteria with heavy (^{15}N) DNA to a medium containing only ^{14}N. When the new DNA, replicated in the ^{14}N medium, was loaded onto the density gradient, it formed a band that was intermediate in density between light (^{14}N) and heavy (^{15}N) DNA (**fig. 9.4**).

If conservative replication had occurred, two bands would have appeared after this first round of replication: the original ^{15}N DNA band, and a new ^{14}N band (see fig. 9.4). The presence of an intermediate band meant that conservative replication was not the process at work. But both the semiconservative and dispersive models predicted that an intermediate density would appear after one round of replication.

To differentiate betweeen these two models, Meselson and Stahl also allowed some of the *E. coli* with heavy DNA to replicate for a second round in the

figure 9.2   Matthew Meselson (1930–)(a), and Franklin W. Stahl (1929–)(b), worked together to demonstrate the semiconservative mechanism of DNA replicaton in *E. coli.*

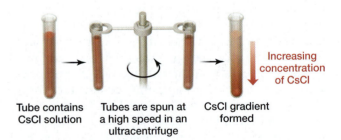

Tube contains CsCl solution

Tubes are spun at a high speed in an ultracentrifuge

CsCl gradient formed

Increasing concentration of CsCl

figure 9.3   Density-gradient ultracentrifugation. A cesium chloride (CsCl) solution is placed in a tube and spun at a very high rate in an ultracentrifuge. The CsCl generates a density gradient in the tube, with the lowest density at the top and the highest density at the bottom. Molecules, such as DNA, that are centrifuged with the CsCl will move to a point in the density gradient that is equivalent to their own density.

^{14}N medium. In this case, the semiconservative model predicted that 50% of the DNA molecules would contain two light strands, and the other 50% would continue to be intermediate in density (one light strand and one heavy strand). The dispersive model predicted that again only an intermediate band would appear, but now this intermediate density band would lie closer to the light DNA band than to the heavy DNA band.

Meselson and Stahl identified two bands that corresponded to the intermediate and light forms (see fig. 9.4). This result demonstrated that the semiconservative model was correct.

Autoradiographic Demonstration of DNA Replication

Although the Meselson–Stahl experiment established the semiconservative mechanism of DNA replication in *E. coli*, two questions remained. First, did eukaryotes also replicate their DNA in a semiconservative fashion, or was this process unique to prokaryotes? Second, what did the replicating DNA actually look

like? The Meselson–Stahl experiment failed to provide a visualization of the replication intermediate.

To answer both of these questions, researchers turned to autoradiography. **Autoradiography** incorporates radioactive compounds into a molecule, in this case radioactive bases into DNA (analogous to the ^{15}N-labeled bases incorporated into Meselson and Stahl's *E. coli* DNA). A photographic emulsion is then placed over the radioactive DNA molecules that are on a microscope slide. The radioactive particles produce visible silver grains on the emulsion, much like a photon of light does to make a photograph on film. Using a microscope, the silver grains in the developed emulsion can then be visualized over the chromosomes and the grains counted to provide an estimate of the quantity of radioactive material present.

Labeling of Eukaryotic Chromatids

In 1958, J. Herbert Taylor used autoradiography to visualize the products of DNA replication in bean root-tip cells. In this experiment, tritium (^{3}H)-labeled dTTP was added to the medium containing the cells. The cells were allowed to go through one round of mitosis

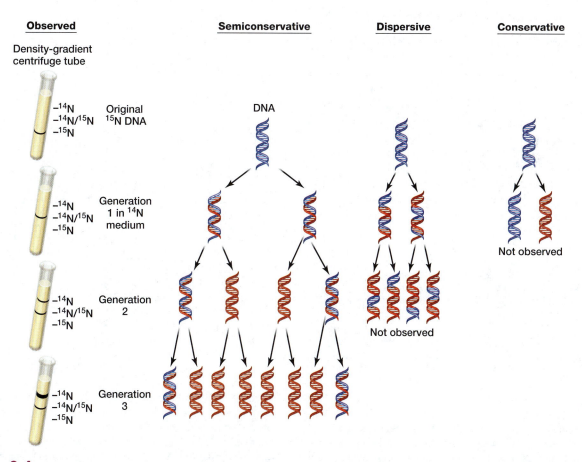

figure 9.4   The Meselson and Stahl experiment to determine the mode of DNA replication. The observed DNA bands in the centrifuge tube are shown on the left. The location and relative amount of the DNA bands in the gradient (left) are consistent with a semiconservative DNA replication mechanism of ^{15}N DNA (blue) replicating in a ^{14}N medium (red). The dispersive DNA replication mechanism is only consistent with the density gradient after one generation. The conservative replication mechanism is not consistent with any of the gradient patterns.

and then were transferred to a medium that lacked the tritium-labeled dTTP (**fig. 9.5**). Based on a semiconservative replication model, one strand of the DNA molecule should be ^{3}H-labeled, but the other should not.

Colchicine was added to the medium to inhibit the formation of spindle fibers, so that the sister chromatids, each of which would contain a daughter DNA molecule, would not separate but instead remained joined at the centromere (see fig. 9.5). In this way, the products of replication (sister chromatids) could easily be identified and autoradiography performed.

On the basis of semiconservative replication, we would expect each double helix to contain one radioactive strand, formed during mitosis in the presence of ^{3}H-labeled dTTP, and one nonradioactive strand—the original template (see fig. 9.5). The following mitosis in the unlabeled medium would then be expected to produce one daughter DNA containing one radioactive strand, and another daughter DNA with no radioactivity. The results of the autoradiography were consistent with semiconservative replication: one sister chromatid (containing one daughter DNA) was ^{3}H-labeled, and the other was not.

The Structure of DNA During Replication

In 1963, John Cairns used autoradiography to visualize the semiconservative mechanism of replication in a circular DNA molecule. Cairns grew *E. coli* bacteria in a medium containing ^{3}H-labeled dTTP. Cairns then carefully extracted the DNA from the bacteria and placed it on photographic emulsion to detect the radioactive decay (**fig. 9.6**).

Interpretation of this autoradiograph revealed several points. The first, which was already known at the time, is that the *E. coli* DNA is circular. Second, DNA is replicated while maintaining the integrity of the circle. Because the circle does not break during DNA

replication, an intermediate called the **theta (θ) structure** forms. (The theta structure is named after the Greek letter θ, which is similar to the topology of the replicating molecule.) Third, replication initiates at a single point in the circular DNA molecule and proceeds from one or two **Y-junctions** or **replication forks** in the circle.

Thus, the DNA is unwound at a given point, and replication proceeds at a replication fork, in a semiconservative manner, in one or both directions (see fig. 9.1). Cairns' experiment was the first visualization of a replication fork in any cell type.

Although the θ-structure is consistent with semiconservative replication, it does not make clear whether one or both of the Y-junctions functions as a replication fork. **Figure 9.7** diagrams the way in which the two replication forks move along the circle to ultimately form two new circles (bidirectional replication). However, only a single replication fork will also generate a theta structure that will produce two circular DNA molecules.

Autoradiography can be used to differentiate between one and two functional replication forks. If replication initiates and then the radioactive nucleotide is added, only the DNA at the functional replication forks will be radioactively labeled. If DNA replication is bidirectional, then both Y-junctions will be radioac-

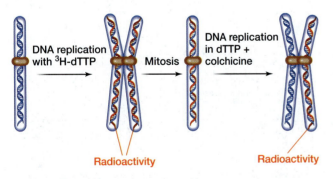

figure 9.5   Taylor demonstrated that linear eukaryotic chromosomes use a semiconservative replication mechanism. Bean root-tip cells were grown in a medium containing tritiated dTTP for one round of cell division and then transferred to medium lacking tritium. Colchicine was added to the medium to inhibit separation of sister chromatids during metaphase. The DNA strand containing the ^{3}H-labeled dTTP is in red and DNA strands lacking the tritiated dTTP are in blue.

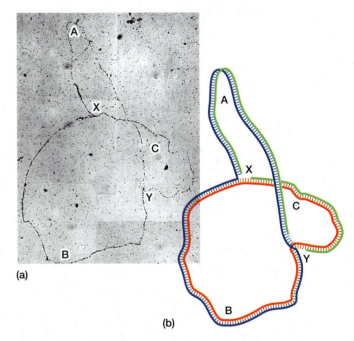

figure 9.6   (a) Autoradiograph of *E. coli* DNA during replication. (b) Loops A, B, and C are created by the existence of two replication forks, X and Y, in the circular DNA. Loop A contains an unlabeled DNA template strand (blue) and a newly synthesized strand (green). Loop C contains an H^3-labeled template DNA strand (red) and a newly sythesized ^{3}H-labeled strand (green), while loop B contains the two template strands, with the red strand being labeled with ^{3}H in the previous round of replication. Both replication forks are converging on each other in loop B. Chromosome is about 1,300 μm.

tively labeled, and the region between the junctions will be unlabeled (**fig. 9.8**). If replication is unidirectional, then only one actively replicating Y-junction will be radioactively labeled (see fig. 9.8). Cairns found growth to be bidirectional, proceeding from both Y-junctions. Autoradiographic and genetic analyses have subsequently verified this finding.

Sites of Initiation in Eukaryotes

Similar experiments were also performed with eukaryotes. The linear DNA chromosomes also exhibited

bidirectional DNA replication. However, another interesting feature was found with the larger eukaryotic chromosomes that had not been previously observed with the circular *E. coli* genome. In 1966, Cairns visualized with autoradiography that a single mammalian chromosome has multiple initiation sites of DNA replication. Each eukaryotic chromosome is composed of many replicating units, or **replicons**—stretches of DNA with a single origin of replication. In comparison, the *E. coli* chromosome is composed of only one replicon.

In eukaryotes, each replicon forms a "bubble" in the DNA during replication, with DNA replication proceeding bidirectionally at each Y-junction in the bubble (**fig. 9.9**). The basic process of DNA replication is therefore the same in both prokaryotes and eukaryotes, except that each eukaryotic DNA molecule consists of multiple replicons relative to the single replicon in the prokaryotic genome.

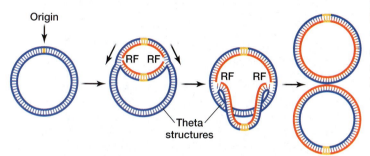

figure 9.7   Observable stages in the bidirectional replication of a circular DNA. The two replication forks (RF) move in opposite directions around the DNA molecule from a single replication origin (yellow). The template DNA strand is blue and the newly synthesized DNA strand is red. The intermediate figures are called theta structures.

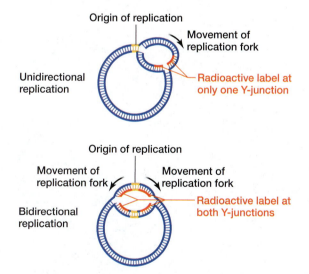

figure 9.8   Radioactive labeling distinguishes unidirectional from bidirectional DNA replication. In these hypothetical experiments, DNA replication was allowed to begin, and then a radioactive label (^{3}H-labeled dTTP) was added. After a short period of time, the process was stopped and the autoradiographs prepared. In unidirectional replication, the tritiated label (red) is found at only one of the Y-junctions, which represents a single active replicaton fork. In bidirectional replication, the tritiated label (red) appears at both Y-junctions, which reveals two active replication forks moving in opposite directions.

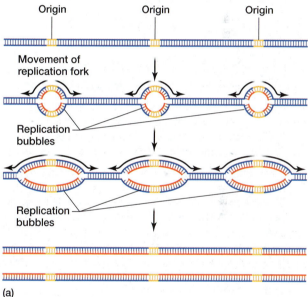

(a)

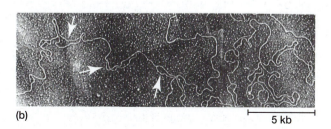

(b)

figure 9.9   Replication bubbles in eukaryotic chromosomes. (a) Replication bubbles form in eukaryotic DNA at multiple origins of DNA replication (yellow). The template DNA strand (blue) and newly synthesized DNA strand (red) are shown as the bidirectional DNA replication increases the bubbles size until they join and generate two double-stranded DNA molecules. (b) Electron micrograph of replicating *Drosophila* DNA showing several replication bubbles (marked by arrows) in the same linear chromosome.

9.1 applying genetic principles to problems

The Application of Reasoning to Solve the Structure of DNA

You just learned how Meselson and Stahl designed experiments to distinguish among three different models of DNA replication in prokaryotes. Similarly, J. Herbert Taylor designed an autoradiography experiment to confirm the semiconservative model of DNA replication in eukaryotes. Finally, John Cairns used autoradiography to confirm that replication proceeded bidirectionally from an origin of replication.

In each case, there were two or more potential explanations or models for how the process could occur. The results from critical experiments revealed which model was correct and also yielded additional models for a subsequent process or question.

James Watson and Francis Crick solved the problem of DNA structure by applying logic to the data and other information available to them. Their 1953 paper from the journal *Nature* is reprinted here so that you can see how they went about it. The principles of scientific reasoning they demonstrated remain the bedrock of any scientific analysis.

First, let's review the components of the research. The key pieces of data that Watson and Crick had at their disposal were:

1. Chargaff's data on the ratio of bases in DNA
2. A basic understanding of chemical bonds
3. Rosalind Franklin's X-ray diffraction patterns

These data were discussed in chapter 7, if you need to review it.

Watson and Crick built scale models of the four different deoxyribonucleotides. They then put these structures together like pieces of a puzzle. In each attempt, they had to determine whether the model violated or was consistent with the three major pieces of data. These attempts continued until they had generated a model consistent with all of the data.

Model building, which can encompass the construction of physical structures like that of Watson and Crick, to abstract mathematical equations that describe changes in a phenotype in a population, have remained an integral and important part of biology. The creation of a model should not only support the available data, but also generate hypotheses that can be tested through additional experimentation.

Watson and Crick describe two previous models for the structure of DNA as part of their paper, explaining why these models did not work. Try sketching these two DNA structures. Using your diagram, can you see instances where the models fail to support the data?

Molecular Structure of Nucleic Acids: A Structure for Deoxyribose Nucleic Acid

We wish to suggest a structure for the salt of deoxyribose nucleic acid (D.N.A.). This structure has novel features which are of considerable biological interest.

A structure for nucleic acid has already been proposed by Pauling and Corey.[1] They kindly made their manuscript available to us in advance of publication. Their model consists of three intertwined chains, with the phosphates near the fibre axis, and the bases on the outside. In our opinion, this structure is unsatisfactory for two reasons: (1) We believe that the material which gives the X-ray diagrams is the salt, not the free acid. Without the acidic hydrogen atoms it is not clear what forces would hold the structure together, especially as the negatively charged phosphates near the axis will repel each other. (2) Some of the van der Waals distances appear to be too small.

Another three-chain structure has also been suggested by Fraser (in the press). In his model the phosphates are on the outside and the bases on the inside, linked together by hydrogen bonds. This structure as described is rather ill-defined, and for this reason we shall not comment on it.

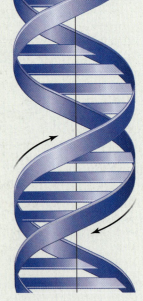

figure 9A This figure is purely diagrammatic. The two *ribbons* symbolize the two phosphate–sugar chains, and the *horizontal* rods represent the pairs of bases holding the chains together. The *vertical line* marks the fiber axis.

We wish to put forward a radically different structure for the salt of deoxyribose nucleic acid. This structure has two helical chains each coiled round the same axis (see diagram [figure 9A]). We have made the usual chemical assumptions, namely, that each chain consists of phosphate diester groups joining β-D-deoxyribofuranose residues with 3', 5' linkages. The two chains (but not their bases) are related by a dyad perpendicular to the fibre axis. Both chains follow right-handed helices, but owing to the dyad the sequences of the atoms in the two chains run in opposite directions. Each chain loosely resembles Furberg's[2] model No. 1; that is, the bases are on the inside of the helix and the phosphates on the outside. The configuration of the sugar and the atoms near it is close to Furberg's 'standard configuration,' the sugar being roughly perpendicular to the attached base. There is a residue on each chain every 3.4 Å in the z-direction. We have assumed an angle of 36° between adjacent residues in the same chain, so that the structure repeats after 10 residues on each chain, that is, after 34 Å. The distance of a phosphorus atom from the fibre axis is 10 Å. As the phosphates are on the outside, cations have easy access to them.

The structure is an open one, and its water content is rather high. At lower water contents we would expect the bases to tilt so that the structure could become more compact.

The novel feature of the structure is the manner in which the two chains are held together by the purine and pyrimidine bases. The planes of the bases are perpendicular to the fibre axis. They are joined together in pairs, a single base from one chain being hydrogen-bonded to a single base from the other chain, so that the two lie side by side with identical z-coordinates. One of the pair must be a purine and the other a pyrimidine for bonding to occur. The hydrogen bonds are made as follows: purine position 1 to pyrimidine position 1; purine position 6 to pyrimidine position 6.

If it is assumed that the bases only occur in the structure in the most plausible tautomeric forms (that is, with the keto rather than the enol configurations), it is found that only specific pairs of bases can bond together. These pairs are: adenine (purine) with thymine (pyrimidine), and guanine (purine) with cytosine (pyrimidine).

In other words, if an adenine forms one member of a pair, on either chain, then on these assumptions the other member must be thymine; similarly for guanine and cytosine. The sequence of bases on a single chain does not appear to be restricted in any way. However, if only specific pairs of bases can be formed, it follows that if the sequence of bases on one chain is given, then the sequence on the other chain is automatically determined.

It has been found experimentally[3,4] that the ratio of the amounts of adenine to thymine, and the ratio of guanine to cytosine, are always very close to unity for deoxyribose nucleic acid. It is probably impossible to build this structure with a ribose sugar in place of the deoxyribose, as the extra oxygen atom would make too close a van der Waals contact.

The previously published X-ray data[5,6] on deoxyribose nucleic acid are insufficient for a rigorous test of our structure. So far as we can tell, it is roughly compatible with the experimental data, but it must be regarded as unproved until it has been checked against more exact results. Some of these are given in the following communications. We were not aware of the details of the results presented there when we devised our structure, which rests mainly though not entirely on published experimental data and stereo-chemical arguments.

It has not escaped our notice that the specific pairing we have postulated immediately suggests a possible copying mechanism for the genetic material.

Full details of the structure, including the conditions assumed in building it, together with a set of coordinates for the atoms, will be published elsewhere.

We are much indebted to Dr. Jerry Donohue for constant advice and criticism, especially on interatomic distances. We have also been stimulated by a knowledge of the general nature of the unpublished experimental results and ideas of Dr. M. H. F. Wilkins, Dr. R. E. Franklin and their coworkers at King's College, London. One of us (J. D. W.) has been aided by a fellowship from the National Foundation for Infantile Paralysis.

J. D. Watson

F. H. C. Crick

Medical Research Council Unit for the Study of the Molecular Structure of Biological Systems, Cavendish Laboratory, Cambridge. April 2.

1. Pauling, L., and Corey, R. B., *Nature,* 171, 346 (1953); *Proc. U.S. Nat. Acad. Sci.,* 39, 84 (1953).
2. Furberg, S., *Acta Chem. Scand.,* 6, 634 (1952).
3. Chargaff, E., for references see Zamenhof, S., Brawerman, G., and Chargaff, E., *Biochim. et Biophys. Acta,* 9, 402 (1952).
4. Wyatt, G. R., *J. Gen. Physiol.,* 36, 201 (1952).
5. Astbury, W. T., *Symp. Soc. Exp. Biol. 1, Nucleic Acid* 66 (Camb. Univ. Press, 1947).
6. Wilkins, M. H. F., and Randall, J. T., *Biochim. et Biophys. Acta,* 10, 192 (1953).

It's Your Turn

Now that we have discussed the various experiments used to define the general mechanisms of DNA replication, try problems 24 and 32 to determine how well you understand these experiments and their possible results.

Restating the Concepts

▶ The Meselson–Stahl experiment revealed that DNA replication was a semiconservative process, by which each strand of the double-stranded molecule serves as a template for a new strand in both of the daughter DNA molecules.

▶ Autoradiography allowed researchers to visualize DNA replication and revealed that replication was bidirectional.

▶ A replicon is the DNA replicated from a single origin of replication. The *E. coli* genome is a single replicon; each eukaryotic linear chromosome contains multiple replicons.

9.2 The Process of Strand Synthesis

Like virtually all metabolic processes, DNA replication is under the control of enzymes. The details we will describe in the following sections come from physical, chemical, and biochemical studies of enzymes and nucleic acids and from the analysis of mutations that influence replication. Recently, recombinant DNA technology and nucleotide sequencing have allowed us to determine the nucleotide sequences of many of the key regions in DNA and RNA. We will look first at replication in *E. coli*.

DNA Polymerases and the Linkage of Nucleotides

Three major enzymes polymerize nucleotides into a growing strand of DNA in *E. coli*: *DNA polymerase I, II,* and *III*.

DNA polymerase I, discovered by Arthur Kornberg, who subsequently won the 1959 Nobel Prize in Physiology or Medicine for this work, is primarily utilized for filling in small DNA segments during replication and repair processes. **DNA polymerase II** can serve as an alternative polymerase to replicate damaged DNA. **DNA polymerase III** is the polymerase that normally replicates the majority of the DNA. The numbering system for DNA polymerases is based on

the sequence in which they were discovered and not on the order in which they act on the DNA.

In the simplest model of DNA replication, the replication process would move along both template strands in the direction of the fork, adding nucleotides to both daughter strands according to the rules of complementarity. But DNA's antiparallel nature creates a problem. As replication moves in the direction of the replication fork, it moves along one template strand in a 5' → 3' direction and in a 3' → 5' direction along the other template strand. These directions refer to the numbering of carbon atoms in the sugar molecule (described in chapter 7).

In **figure 9.10**, going from the bottom of the figure to the top, the template strand on the left is a 3' → 5' strand, and the template strand on the right is a 5' → 3' strand. Because of the antiparallel nature of the double-stranded DNA, one newly synthesized strand would be generated in a 5' → 3' direction (the new strand on the left) and the other in a 3' → 5' direction (the new strand on the right).

All the known polymerase enzymes, however, add nucleotides in only the 5' → 3' direction. That is, the polymerase catalyzes a *phosphodiester bond* between the first 5'–PO_4 group of a new nucleotide and the 3'–OH carbon of the last nucleotide in the newly synthesized strand (see fig. 9.10). Cleaving the two phosphates from the deoxyribonucleotide triphosphate that is being added to the daughter strand produces a large amount of energy that drives this reaction. The polymerases cannot create the same phosphodiester bond with the 5' phosphate of a nucleotide already in the DNA and the 3' end of a new nucleotide. Therefore our simple model needs some revision, since the growing strand on the right cannot add bases in the direction of the replication fork.

Replication in Two Directions

Autoradiographic evidence leads us to believe that replication occurs simultaneously on both DNA strands, and that both strands are growing in the direction of the replication fork. However, the activity of DNA polymerase allows only one of the two growing strands to add bases in the direction of the replication fork. Replication in this direction is called **continuous replication,** and it is only possible on the template strand that is oriented 3' → 5' toward the replication fork (left-strand template in fig. 9.10).

Continuous replication requires a DNA template and a **primer,** which is a nucleic acid sequence that has a free 3'–OH where the next base can be added (in this case, the strand that is being synthesized, as shown on the right in **fig. 9.11**). As we will see shortly, the primer can be either a DNA or RNA strand that is base-paired with the DNA template.

Discontinuous replication takes place on the complementary strand, where the template is oriented

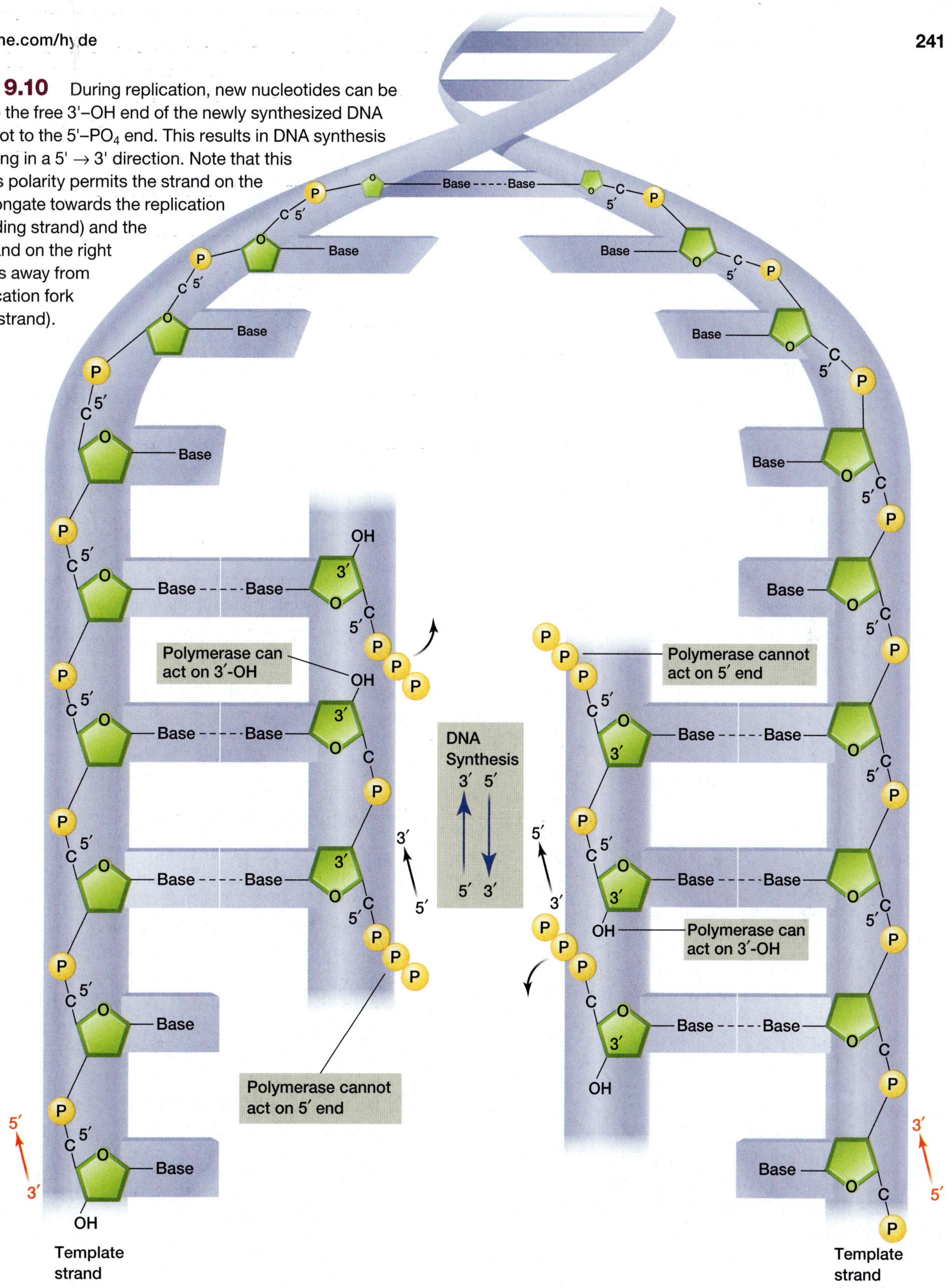

figure 9.10 During replication, new nucleotides can be added to the free 3'–OH end of the newly synthesized DNA strand, not to the 5'–PO$_4$ end. This results in DNA synthesis proceeding in a 5' → 3' direction. Note that this synthesis polarity permits the strand on the left to elongate towards the replication fork (leading strand) and the DNA strand on the right elongates away from the replication fork (lagging strand).

Polymerase can act on 3'-OH

Polymerase cannot act on 5' end

DNA Synthesis

Polymerase cannot act on 5' end

Polymerase can act on 3'-OH

Template strand

Template strand

in a 5' → 3' direction toward the replication fork (as shown on the right in fig. 9.10). To allow for the 5' → 3' polymerase activity, this growing strand must replicate away from the replication fork. As it does, the replication occurs in short segments, hence the term *discontinuous replication*. These short newly synthesized DNA segments, called **Okazaki fragments** after

Reiji Okazaki, who first identified them, average about 1500 nucleotides in prokaryotes and 150 nucleotides in eukaryotes.

The strand synthesized continuously toward the replication fork is referred to as the **leading strand** (top of **fig. 9.12**), and the strand synthesized discontinuously is referred to as the **lagging strand** (bottom of fig. 9.12).

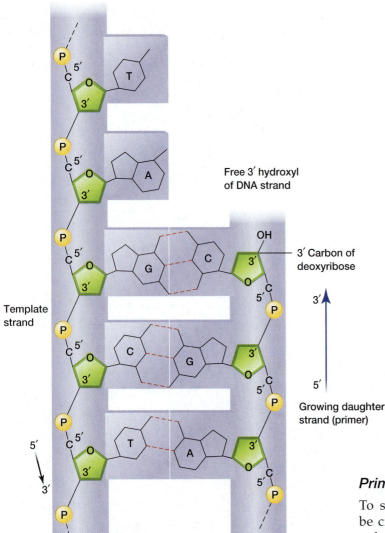

Figure 9.11   Primer configuration for DNA replication. A free 3'–OH group must be available on the daughter strand to act as a primer for DNA replication off of the single-stranded template.

Once initiated, continuous DNA replication can proceed indefinitely. DNA polymerase III, on the leading-strand template has high **processivity**, meaning that once it attaches to the template, it doesn't release until a very long complementary strand is synthesized. Discontinuous replication, however, is not such a simple story.

Details of the Discontinuous Synthesis Process

Discontinuous DNA synthesis requires the repetition of four steps: primer synthesis, elongation, primer removal with gap filling, and ligation.

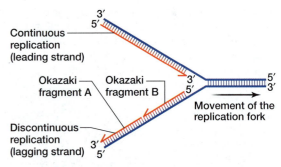

figure 9.12   Discontinuous model of DNA replication. DNA replication proceeds in a 5' → 3' direction on the newly synthesized strand (red). One strand (leading strand) exhibits continuous replication as it proceeds in the same direction as the migrating replication fork. The other strand (lagging strand) exhibits discontinuous replication because its 5' → 3' synthesis proceeds away from the replication fork. The Okazaki fragments on the lagging strand allow the sequential initiation of DNA replication near the fork and the 5' → 3' synthesis away from the fork (A), followed by a new fragment (B) initiating again near the fork (which has moved since the start of Okazaki fragment A) and being synthesized away from the fork.

Primer Synthesis by Primase

To synthesize Okazaki fragments, a primer must be created *de novo*, that is, anew. None of the DNA polymerases can create that primer. Instead, **primase,** a special RNA polymerase coded for by the *dnaG* gene, creates an RNA primer of 10 to 12 nucleotides (**fig. 9.13**). As with DNA polymerases, primase synthesizes this short strand in a 5' → 3' direction. This RNA primer, which has a free 3'–OH group, serves as the site of Okazaki fragment initiation (**fig. 9.14**).

Elongation and the Action of 3' → 5' Exonuclease

If the 3' base on the RNA primer is properly base-paired with the template, DNA polymerase III can add deoxyribonucleotides to synthesize the DNA portion of the Okazaki fragment (see fig. 9.14). DNA polymerase III will continue to extend (elongate) this Okazaki fragment until it reaches the RNA primer of the previously synthesized Okazaki fragment. At that point, it stops and releases from the DNA.

All three *E. coli* DNA polymerases not only can add new nucleotides to a growing strand in the 5' → 3' direction, but can also remove nucleotides in the opposite 3' → 5' direction. This property is referred to as 3' → 5' *exonuclease activity.*

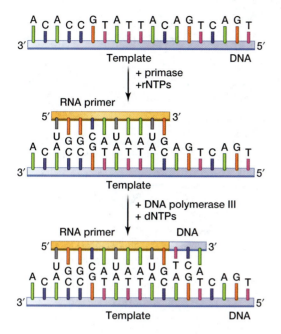

figure 9.13   Generation of an RNA primer. An RNA polymerase (primase) adds ribonucleotides (gold color) in a 5' → 3' direction into an RNA primer that utilizes the DNA template strand. After the RNA primer reaches 10-12 nucleotides, DNA polymerase can then begin to add deoxyribonucleotides onto the RNA primer in a 5' → 3' direction.

Enzymes that degrade nucleic acids are **nucleases.** They are classified as **exonucleases** if they remove nucleotides from the end of a nucleotide strand or as **endonucleases** if they can break the sugar–phosphate backbone in the middle of a nucleotide strand.

At first glance, exonuclease activity seems like an extremely curious property for a polymerase to have. However, if the DNA polymerase adds a wrong base to the growing strand, the result would be a mutation. To minimize this type of mutation, DNA polymerase possesses a mechanism to ensure that the complementarity is correct. If the base at the 3' end of the growing strand is not properly paired with the template, then the 5' → 3' polymerase activity is blocked (**fig. 9.15**). This allows the 3' → 5' exonuclease activity to start, which results in the 3' base on the newly synthesized strand being removed (fig. 9.15). The 5' → 3' polymerase activity can then restart and reattempt to insert the proper base in the growing strand.

Because the 3' → 5' exonuclease activity removes bases that are incorrectly paired, it is referred to as the **proofreading** function of DNA polymerase. As you will see in the following discussion, a different exonuclease activity can remove the RNA primers of Okazaki fragments. (Other correction and repair mechanisms are discussed later on, in chapter 18.)

The *E. coli* DNA polymerases can add approximately 400 nucleotides per second to the growing daughter strand. The proofreading activity ensures that an incorrect base pair is present only once in every 10^9 nucleotides.

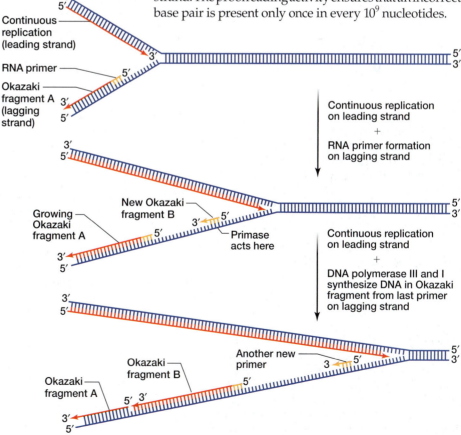

figure 9.14   Primer formation and elongation create Okazaki fragments during discontinuous DNA replication. The leading strand exhibits continuous replication as the DNA strand is synthesized 5' to 3' in the same direction as the moving replication fork. The lagging strand exhibits discontinuous replication through the synthesis of an RNA primer near the replication fork, DNA synthesis 5' to 3' away from the fork, and the synthesis of a new RNA primer close to the new location of the replication fork. This produces short discontinuous DNA strands (Okazaki fragments) on the lagging strand.

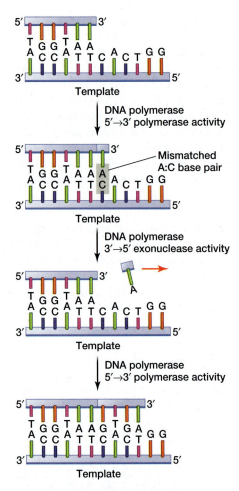

figure 9.15 DNA polymerase occasionally incorporates the wrong base into the growing strand, such as the adenosine instead of a guanosine at the 3' end. Because of the incorrect A:C base pair (highlighted) at the 3' end of the growing strand, the 5' → 3' polymerase activity is blocked. This permits the 3' → 5' exonuclease activity of DNA polymerase to remove the mismatched base on the synthesized strand. This leaves a properly base-paired 3' base (A), which allows the 5' → 3' polymerase activity to begin adding bases to the growing 3' end.

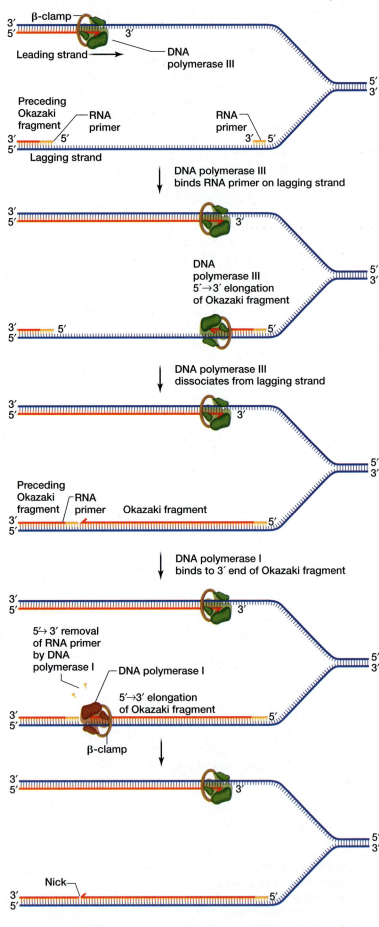

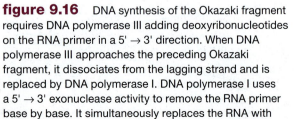

figure 9.16 DNA synthesis of the Okazaki fragment requires DNA polymerase III adding deoxyribonucleotides on the RNA primer in a 5' → 3' direction. When DNA polymerase III approaches the preceding Okazaki fragment, it dissociates from the lagging strand and is replaced by DNA polymerase I. DNA polymerase I uses a 5' → 3' exonuclease activity to remove the RNA primer base by base. It simultaneously replaces the RNA with deoxyribonucleotides using the 5' → 3' polymerase activity. A final nick in the DNA backbone remains (arrow).

Primer Removal and Gap Filling: Polymerase and 5′ → 3′ Exonuclease Activities

DNA polymerase I takes over on the lagging strand when DNA polymerase III extends an Okazaki fragment to the previous RNA primer. To generate a contiguous DNA strand, DNA polymerase I utilizes both the 5′ → 3′ polymerase and 5′ → 3′ exonuclease activities. This exonuclease activity is different from the 3′ → 5′ exonuclease that takes place during elongation. Although all three DNA polymerases possess 5′ → 3′ polymerase and 3′ → 5′ exonuclease activities, only DNA polymerase I has the 5′ → 3′ exonuclease activity.

When DNA polymerase III approaches the RNA primer in the preceding Okazaki fragment, it dissociates from the lagging strand. DNA polymerase I attaches to the lagging strand and completes the Okazaki fragment by removing the previous RNA primer with the 5′ → 3′ exonuclease activity and replacing it with DNA nucleotides using the 5′ → 3′ polymerase activity (**fig. 9.16**). When DNA polymerase I has removed the RNA primer and filled in the gap, the two Okazaki fragments are almost continuous, with only a single phosphodiester bond missing between two adjacent nucleotides.

Why is RNA used to prime DNA synthesis when using a DNA primer directly would avoid the exonuclease and resynthesis activity seen in figure 9.16? Probably, using RNA primers lowers the frequency of errors in the DNA after replication. Specifically, priming is an inherently error-prone process because nucleotides are initially added without a stable primer configuration. To reduce this high percentage of errors in the primer from being incorporated into the final DNA sequence, an RNA primer is used that can later be easily recognized and replaced by DNA polymerase I.

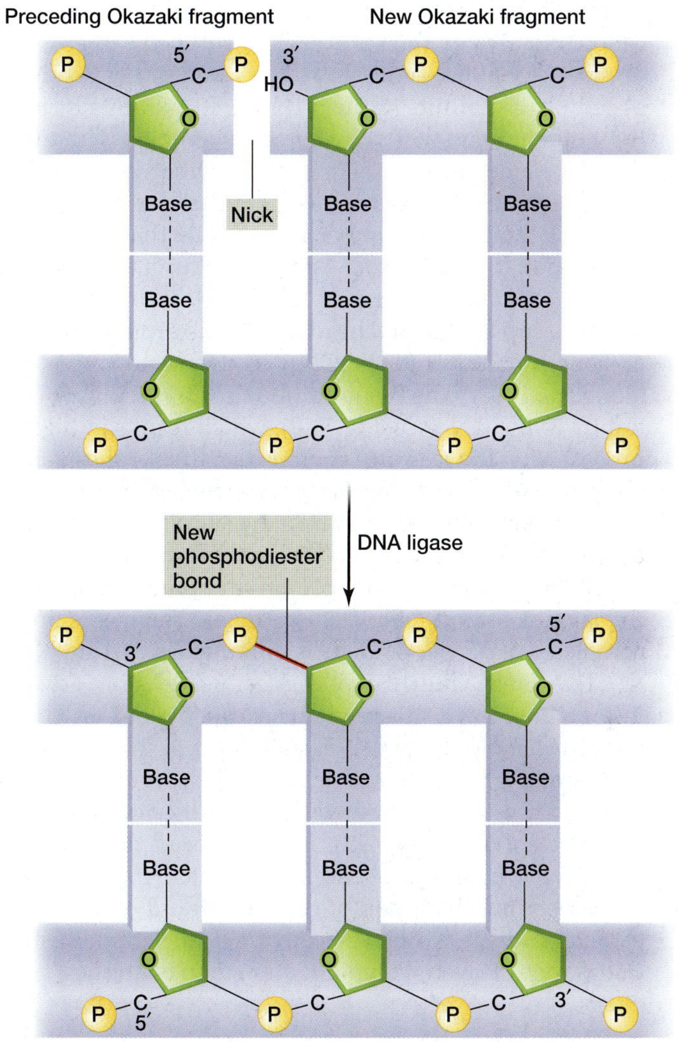

Ligation by DNA Ligase

Between the two Okazaki fragments, a "nick" remains where the final phosphodiester bond is needed (see fig. 9.16). DNA polymerase I cannot catalyze the final bond to join these fragments because, as you learned earlier, it can only generate a phosphodiester bond between a 5′ *tri*phosphate and a 3′ hydroxyl (see fig. 9.10). **DNA ligase** seals the nick (**fig. 9.17**). In *E.coli*, DNA ligase uses energy from nicotine adenine dinucleotide (NAD), but in other organisms these ligases may use ATP. As you will learn in chapter 18, DNA ligase also has an essential role in DNA repair.

figure 9.17 After DNA polymerase I removes the RNA primer in a 5′ → 3′ direction on the preceding Okazaki fragment and simultaneuosly extends the DNA strand in a 5′ → 3′ direction on the new Okazaki fragment, a final nick remains. DNA polymerase cannot seal this nick. DNA ligase seals the nick by generating the missing phosphodiester bond.

Restating the Concepts

▶ *E. coli* possesses three main DNA polymerases that have different functions in the replication of DNA. DNA polymerase III is the major replicating enzyme that synthesizes the leading strand (continuous replication) and the Okazaki fragments (discontinuous replication). DNA polymerase I primarily removes the RNA primers on Okazaki fragments and synthesizes their replacement by DNA. DNA polymerase II is primarily involved in repair mechanisms.

▶ Every DNA polymerase possesses a 5′ → 3′ polymerase and a 3′ → 5′ exonuclease activity (proofreading), whereas DNA polymerase I also has a 5′ → 3′ exonuclease activity to remove the RNA primers on the lagging and leading strands.

▶ Continuous DNA replication proceeds from beginning to end on the leading strand, moving 5′ → 3′.

▶ Discontinuous DNA replication on the lagging strand initially requires an RNA polymerase (primase) to generate an RNA primer to which the DNA polymerase can add bases. DNA ligase seals the final phosphodiester bond to connect Okazaki fragments.

9.3 The Point of Origin of DNA Replication

Each replicon must have a region where DNA replication initiates. As you learned earlier, the *E. coli* chromosome is a single replicon, while each eukaryotic chromosome consists of multiple replicons.

For DNA replication to begin, several steps must occur. First, the appropriate initiation proteins must recognize the specific origin site. Then the double-stranded sequence at the initiation site must be separated and stabilized. And, finally, a replication fork must be initiated in both directions, involving continuous and discontinuous DNA replication.

Events at the Origin Site

In *E. coli,* the origin of replication is referred to as the genetic locus *oriC;* which occurs at a specific location in the circular genome (**fig. 9.18**). The *oriC* is a specific DNA sequence that is about 245 bp long and is recognized by **initiator proteins** (**fig. 9.19**). These proteins, the product of the *dnaA* locus, bind to specific sequences in *oriC* (consensus sequence of TTATNCANA) and **denature,** or separate, the double-stranded DNA into a single-stranded region (see fig. 9.19).

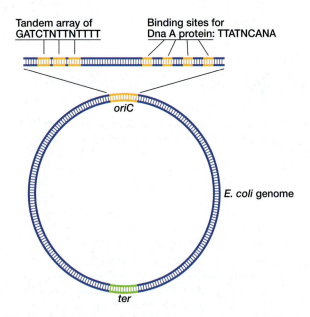

figure 9.18 Organization of the *oriC* locus in the *E. coli* genome. The *oriC* locus is composed of two different repeat sequences, which are shown in an expanded view at the top. DNA replication begins at *oriC* and proceeds bidirectionally around the circular genome and ends at the *terminus* (*ter*) locus, which is approximately equal distance for both replication forks.

DNA **helicase,** the product of the *dnaB* gene, binds to the DnaA-induced single-stranded region. DNA helicase unwinds DNA at the replication fork (see fig. 9.19). Helicase is also responsible for recruiting (binding) the rest of the proteins that form the replication initiation complex. The first of the proteins in this complex is primase, which synthesizes the RNA primers. Together, helicase and primase constitute a **primosome,** which is attached to the lagging-strand template (see fig. 9.19).

The primosome initially synthesizes a complementary RNA on the lagging strand template at one Y-junction, which will serve as the primer for the leading strand synthesis at the other Y-junction in the replication bubble (see fig. 9.19). This will be the only primer required for leading strand synthesis for each replicon. As the primosome moves along the lagging-strand template, it begins to synthesize RNA primers that DNA polymerase III uses to initiate lagging-strand replication (Okazaki fragment production; see fig. 9.19). The replication fork activity then proceeds as outlined earlier.

DNA polymerase III is a very large protein composed of 10 subunits (**table 9.1**). The 10 subunits form the **holoenzyme,** which is a protein complex with all the associated subunits. Three of the subunits—alpha (α), epsilon (ε), and theta (θ)—form the **core enzyme,** which is the fewest number of subunits that possess the enzymatic activities. Another subunit, the beta (β)-subunit, is a "processivity clamp." As a dimer (two identical copies attached head to tail), the protein forms a "doughnut" around the DNA so it can move freely along the DNA. The β-subunit holds the core enzyme tightly to the DNA (see fig. 9.19), which results in high processivity: the leading strand is usually synthesized entirely without the enzyme leaving the template.

The remaining polymerase subunits are involved in controlling processivity and replisome formation. They allow the polymerase to move off and on the DNA of the lagging-strand template as Okazaki fragments are completed (a process known as **polymerase cycling**).

Tom Steitz and his colleagues, using X-ray crystallography, have given us an excellent look at the structure of a polymerase. The enzyme is shaped like a cupped right hand with enzymatic activity taking place at two sites, separated by a distance of about 2 to 3 nucleotides (**fig. 9.20**). It is proposed that when the polymerization site senses a mismatch, the DNA is moved so that the 3' end of the new strand enters the exonuclease site, where the incorrect nucleotide residue is then cleaved. Polymerization then continues.

Events at the Replication Fork

To summarize the replication process, DNA polymerase III moves along the leading-strand template to synthesize the daughter leading strand by continuous DNA replication in a 5' → 3' direction. A second DNA

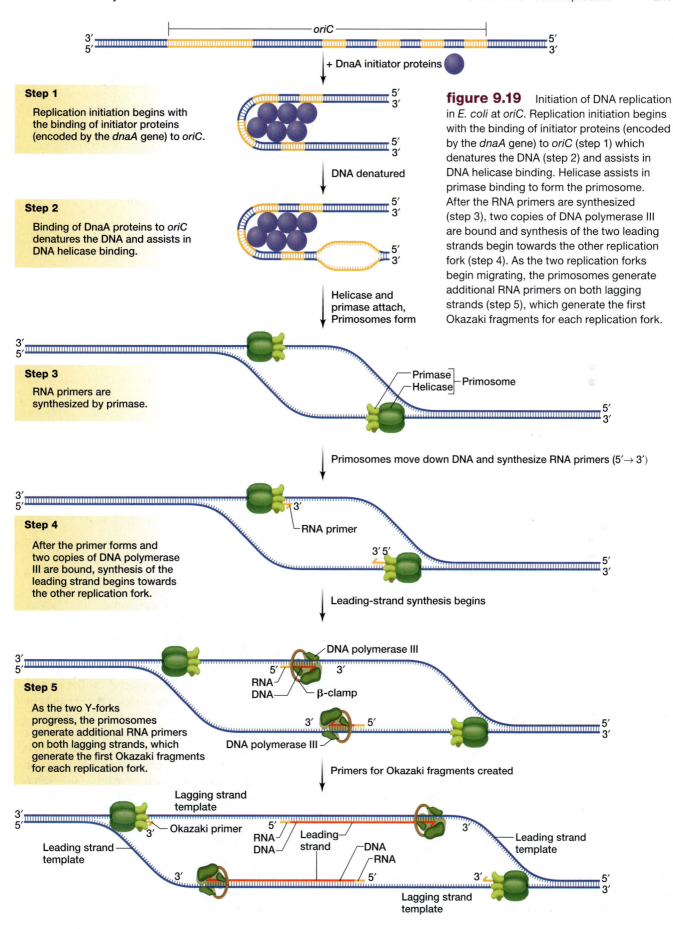

Step 1

Replication initiation begins with the binding of initiator proteins (encoded by the *dnaA* gene) to *oriC*.

Step 2

Binding of DnaA proteins to *oriC* denatures the DNA and assists in DNA helicase binding.

Step 3

RNA primers are synthesized by primase.

Step 4

After the primer forms and two copies of DNA polymerase III are bound, synthesis of the leading strand begins towards the other replication fork.

Step 5

As the two Y-forks progress, the primosomes generate additional RNA primers on both lagging strands, which generate the first Okazaki fragments for each replication fork.

figure 9.19   Initiation of DNA replication in *E. coli* at *oriC*. Replication initiation begins with the binding of initiator proteins (encoded by the *dnaA* gene) to *oriC* (step 1) which denatures the DNA (step 2) and assists in DNA helicase binding. Helicase assists in primase binding to form the primosome. After the RNA primers are synthesized (step 3), two copies of DNA polymerase III are bound and synthesis of the two leading strands begin towards the other replication fork (step 4). As the two replication forks begin migrating, the primosomes generate additional RNA primers on both lagging strands (step 5), which generate the first Okazaki fragments for each replication fork.

table 9.1   Summary of the Enzymes Involved in DNA Replication in *E. coli*

Enzyme or Protein	Genetic Locus	Function
DNA polymerase I	*polA*	Gap filling and primer removal
DNA polymerase II	*polB*	Replicating damaged templates
DNA polymerase III		
α-subunit	*dnaE*	Polymerization core; 5' → 3' polymerase
ε-subunit	*dnaQ*	Polymerization core; 3' → 5' exonuclease
θ-subunit	*holE*	Polymerization core
β-subunit	*dnaN*	Processivity clamp (as a dimer)
τ-subunit	*dnaX*	Preinitiation complex; dimerization of core
γ-subunit	*dnaX*	Preinitiation complex; loads clamp
δ-subunit	*holA*	Processivity core
δ'-subunit	*holB*	Processivity core
χ-subunit	*holC*	Processivity core
ψ-subunit	*holD*	Processivity core
Helicase	*dnaB*	Primosome; unwinds DNA
Primase	*dnaG*	Primosome; creates leading strand and Okazaki fragment primers
Initiator protein	*dnaA*	Binds at origin of replication
DNA ligase	*lig*	Closes Okazaki fragments
Ssb protein	*ssb*	Binds single-stranded DNA
DNA topoisomerase I	*topA*	Relaxes supercoiled DNA
DNA topoisomerase type II (DNA Gyrase)		
α-subunit	*gyrA*	Relaxes supercoiled DNA; ATPase
β-subunit	*gyrB*	Relaxes supercoiled DNA
Topoisomerase IV	*parE*	Decatenates DNA circles
Termination protein	*tus*	Binds at termination sites

polymerase III also synthesizes the daughter lagging strand (Okazaki fragments) in a 5' → 3' direction, but must move away from the replication fork to perform this synthesis. While replication proceeds, **single-stranded binding proteins** (ssb proteins) bind to the single-stranded DNA that is generated by helicase and prevents the strands from reannealing until the new strand is synthesized (**fig. 9.21**). Finally, DNA polymerase I acts on the Okazaki fragments to remove the RNA primers, and DNA ligase seals the phosphodiester nick (see fig. 9.16).

A question remained, however, regarding how the lagging strand can be replicated away from the replication fork while the overall replication still appears to move in the direction of the replication fork (see the earlier fig. 9.8). Bruce Alberts proposed a model to explain the coordination of leading-strand and lagging-strand replication. According to this model, two DNA polymerase III molecules, one replicating the leading strand and the other the lagging strand, are attached to each other and work in concert with the primosome at the replication fork (**fig. 9.22**).

According to this model, the two copies of DNA polymerase III combine with a helicase and a primase to form a single unit termed a **replisome,** which moves along the DNA. The leading-strand template is immediately fed to one DNA polymerase that replicates the leading strand (fig. 9.22*a*). In contrast, the lagging-strand template is not acted on by the other polymerase until an RNA primer has been placed on the strand, meaning that a long (1500-base) single strand has been opened up (fig. 9.22*a*).

As the replisome moves in the direction of the advancing replication fork, the Okazaki fragment is synthesized by pulling the single-stranded loop region of the lagging strand template through the DNA polymerase III (fig. 9.21*a*). At about the time that the Okazaki fragment is completed, a new RNA primer has been created (fig. 9.22*b*). The Okazaki fragment is released (fig. 9.22*c*), and a new Okazaki fragment is begun (*polymerase cycling*), starting with the latest primer (fig. 9.22*d*). This takes the replisome back to the same configuration as in fig. 9.22*a*, but one Okazaki fragment farther along.

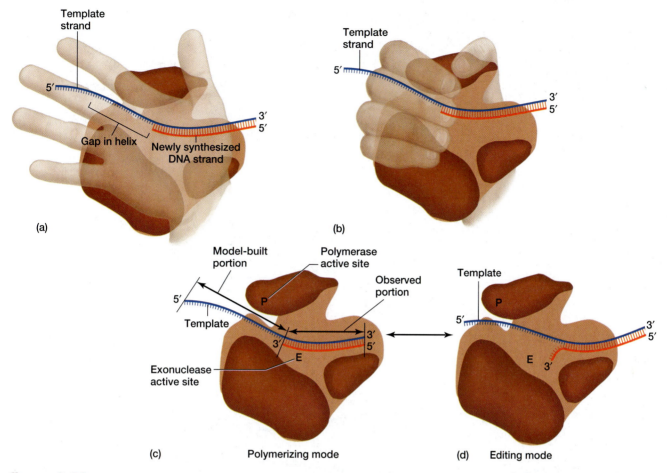

figure 9.20 Structure and activities of DNA polymerase. DNA polymerase is shaped like a hand grasping the DNA with the "thumb" pointing in the 5' → 3' direction of the newly synthesized DNA strand (red). The relative locations of the DNA polymerase (P) and exonuclease (E) active sites in *E. coli* DNA polymerase I are shown. On the lower left, 5' → 3' polymerization is occurring. On the lower right, the 3' end of the nascent strand has been backed up into the 3' → 5' exonuclease (proofreading) site, presumably when a mismatch was detected.

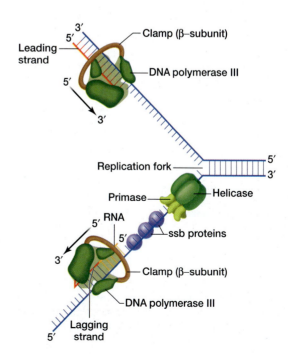

figure 9.21 Schematic drawing of DNA replication at a replication fork. Two copies of DNA polymerase III, single-stranded binding (ssb) proteins, and a primosome (helicase and primase) are present. One molecule of DNA polymerase III synthesizes the leading strand and a separate molecule of DNA polymerase III synthesizes the lagging strand. A single primosome moves along the lagging strand template to separate the two template strands and synthesize the RNA primers for the Okazaki fragments.

Step 1

Primase synthesizes lagging strand RNA primer and DNA polymerase III begins synthesizing Okazaki fragment B in 5′→ 3′ direction.

Step 2

DNA polymerase III synthesizes Okazaki fragment B toward Okazaki fragment A. Primase synthesizes RNA primer for Okazaki fragment C.

Step 3

DNA polymerase III releases lagging strand when Okazaki fragment B approaches Okazaki fragment A. DNA polymerase I binds 3′ end of Okazaki fragment B.

Step 4

DNA polymerase I extends Okazaki fragment B in a 5′→ 3′ direction and removes Okazaki fragment A RNA primer in a 5′→ 3′direction leaving only a nick between Okazaki fragments A and B.

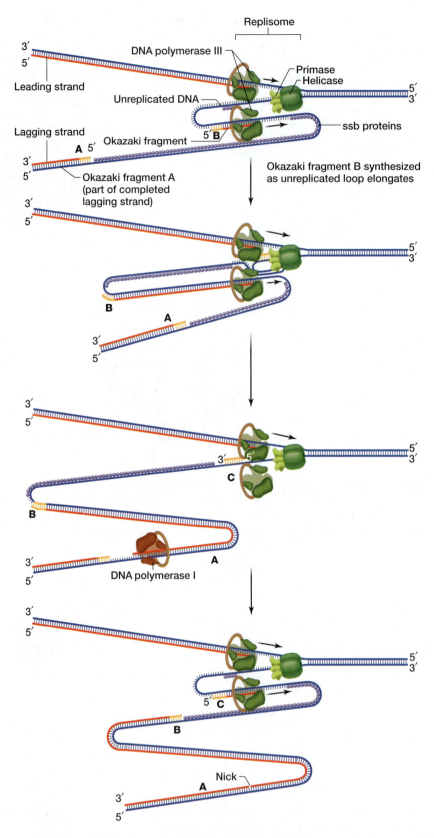

figure 9.22   The replisome, which consists of two DNA polymerase III holoenzymes and a primosome (helicase and primase complex), coordinates replication at the Y-junction. "Polymerase cycling," is shown in steps 2–4, in which the DNA polymerase III on the lagging-strand template releases a completed Okazaki fragment and then begins synthesizing the next one. DNA polymerase I binds the newly released Okazaki fragment, fills the gap, and removes the adjacent RNA primer.

Figure 9.23 gives us a closer look at the details of the replication fork at the moment of polymerase cycling. Primase, which is not highly processive, must be in contact with a single-stranded binding protein to stay attached to the DNA when forming a primer. At the appropriate moment, after the primer is formed, a clamp loader protein contacts the single-stranded binding protein, dislodging the primase. The clamp loader also loads a sliding β-clamp, which then recruits the DNA polymerase III to generate the lagging strand. The polymerase then synthesizes the Okazaki fragment in the 5' → 3' direction away from the replication fork. The primase can later attach at a new point on the lagging-strand template nearer the replication fork to create the next primer.

Supercoiling

The simplicity and elegance of the DNA molecule masks an inevitable problem: coiling. Because the DNA molecule is made from two strands that wrap about each other, certain operations, such as DNA replication and its termination, face topological difficulties.

Up to this point, we have diagrammed the circular *E. coli* chromosome in its "relaxed" state (for example, figs. 9.6 and 9.7). But certain enzymes in the cell cause DNA to become **supercoiled,** that is the circular molecule is coiled about itself. Positively supercoiled DNA is overcoiled, and negatively supercoiled DNA is undercoiled. Positive supercoiling comes about in two ways: either the DNA takes too many turns in a given length, or the molecule wraps around itself (**fig. 9.24**). You can imagine what this situation is like if you think of a coiled telephone cord that gets too many twists in it and coils back on itself.

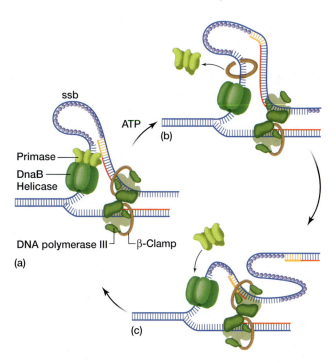

figure 9.23   A close-up view of the replication fork during polymerase cycling. The two DNA polymerase III complexes are held together by γ-subunits. Also pictured are the sliding β-clamp, primase, helicase, and ssb proteins. In part (a), the primase has just finished creating a primer. (b) The clamp loader dislodges the primase and recruits a β-clamp. (c) The DNA polymerase III on the lagging strand is cycled to the clamp to begin the next Okazaki segment.

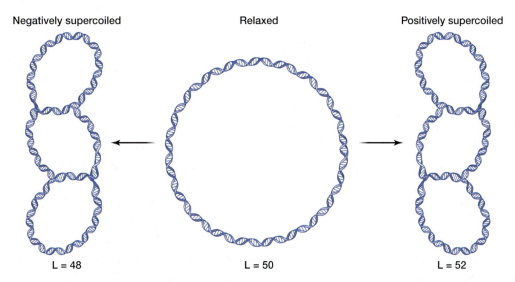

figure 9.24   Positive and negative supercoils. Enzymes called topoisomerases can take relaxed DNA (center) and add negative (left-handed) or positive (right-handed) supercoils. L is the linkage number.

Positive supercoiling occurs when the circular duplex winds about itself in the same direction as the helix twists (right-handed). This action increases the number of turns of one helix around the other (the **linkage number, L**). Negative supercoiling comes about when the duplex winds about itself in the opposite direction as the helix twists (left-handed). This action decreases the linkage number. The three forms of DNA in figure 9.24 all have the same sequence, yet they differ in linkage number. Accordingly, they are referred to as topological isomers (**topoisomers**). The enzymes that create or alleviate these states are called **topoisomerases.**

Topoisomerases affect supercoiling by one of two methods. Type I topoisomerases break one strand of a double helix and, while binding the broken ends, pass the other strand through the break. The break is then sealed and the linkage number is changed by 1 (**fig. 9.25**). Type II topoisomerases (for example, **DNA gyrase** in *E. coli*) do the same sort of thing, only instead of breaking one strand of a double helix, they break both strands and pass another double helix through the temporary gap. Because two strands pass through the break, a type II topoisomerase changes the linkage number in multiples of two. Four topoisomerases are active in *E. coli*, with somewhat confusing nomenclature: topoisomerases I and III are type I; topoisomerases II and IV are type II.

As DNA replication proceeds, positive supercoiling builds up ahead of the replication fork. As the supercoiling increases, it becomes more difficult for helicase to separate the two DNA strands. This positive supercoiling is analogous to separating two strands in a rope with both ends fixed. This positive supercoiling is eliminated by DNA gyrase, which either creates negative supercoiling ahead of the fork in preparation for replication or alleviates the positive supercoiling that helicase generates from strand separation.

It's Your Turn

Understanding the various proteins, their site of action, and their function are important to comprehending the overall process of DNA replication. Try problems 21 and 27 to determine your level of understanding this material.

Restating the Concepts

▶ A replicon is the region of genomic DNA that is replicated from a single origin of replication. The primosome (composed of DNA helicase and primase) binds to the origin of replication to denature the double-stranded DNA at the replication fork and synthesize the RNA primers.

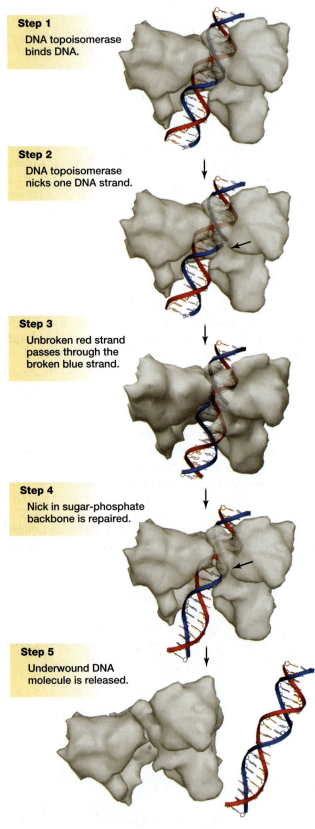

Step 1
DNA topoisomerase binds DNA.

Step 2
DNA topoisomerase nicks one DNA strand.

Step 3
Unbroken red strand passes through the broken blue strand.

Step 4
Nick in sugar-phosphate backbone is repaired.

Step 5
Underwound DNA molecule is released.

figure 9.25   Topoisomerase I can reduce DNA coiling by breaking one strand (blue) of the double helix and passing the other strand (red) through it. This passage of a single strand of DNA through the break results in the change of the linkage number by one.

▶ The replisome is composed of the primosome and two molecules of DNA polymerase III, with each polymerase synthesizing one of the two DNA strands.

▶ Because the replisome moves in the direction of the replication fork, the lagging-strand template must loop through the DNA polymerase III to allow synthesis of the Okazaki fragments in a 5' → 3' direction away from the replication fork while staying associated with the replisome and moving in the direction of the replication fork.

▶ Topoisomerases regulate the supercoiling of the DNA. A type II topoisomerase, DNA gyrase, relieves the positive supercoiling created in front of the replication fork.

9.4 Termination of Replication

Terminating the replication of a circular chromosome presents no major topological problems. At the end of the θ-structure replication (see figure 9.7), both replication forks have proceeded around the molecule. The region of replication termination on the *E. coli* chromosome, the terminus region (*ter*), is opposite from *oriC* on the circular chromosome (fig. 9.18).

The *ter* region contains six sites that are each approximately 20 bp in length. Three of the *ter* sites arrest the replication fork from the left, and three arrest the one from the right. Replication is terminated when the *ter* sites are bound by a termination protein, which is encoded by the *tus* gene (*tus* stands for *t*erminus *u*tilization *s*ubstance).

Completion of DNA replication results in a catenated structure, in which the two circular daughter DNA molecules are interlinked (**fig. 9.26**). A type II topoisomerase, topoisomerase IV, then releases the two circles. DNA polymerase I repairs any gaps in the two circles and ligase seals any nicks.

One interesting aspect of the termination of *E. coli* DNA replication is that some cells are viable even if the whole terminator region is deleted. Although loss of the *ter* region results in most of the cells dying or exhibiting severe growth problems, *E. coli* can successfully terminate DNA replication even without formal termination sites.

Restating the Concepts

▶ In *E. coli,* termination of replication takes place at the terminus region, 180 degrees from the *oriC* origin site. Six *ter* sequences arrest replication, three for each of the two replication forks.

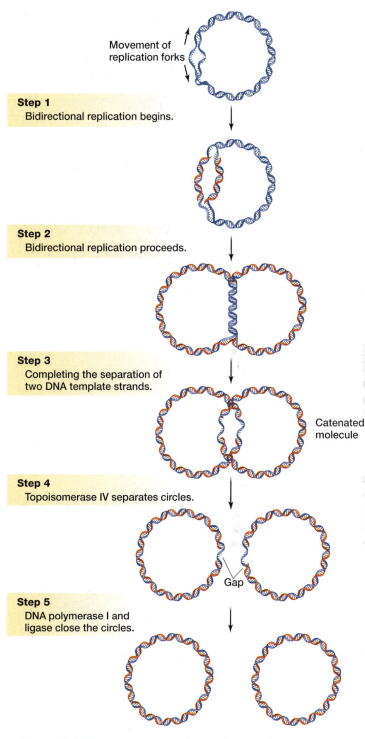

Movement of replication forks

Step 1
Bidirectional replication begins.

Step 2
Bidirectional replication proceeds.

Step 3
Completing the separation of two DNA template strands.

Catenated molecule

Step 4
Topoisomerase IV separates circles.

Gap

Step 5
DNA polymerase I and ligase close the circles.

figure 9.26 Completing the replication of circular DNA. As the two template strands unwind, the bidirectional replication of the daughter strands (red) proceeds. Eventually, the two template strands (blue) are double-stranded with each other for only a short segment. Separating this final region produces the interlocking catenated circles with single-stranded gaps corresponding to the final region to be replicated. Topoisomerase IV separates the catenated circles and DNA polymerase I completes DNA synthesis in the gaps in each circle. DNA ligase seals the final nicks.

9.5 Other Replication Structures

The *E. coli* model of DNA replication presented earlier relies on the intermediate θ-structure (see fig. 9.7). Two other modes of replication occur in circular chromosomes: rolling-circle and D-loop.

The Rolling-Circle Model

In the **rolling-circle** mode of replication, a break in one phosphodiester bond, or "nick," is made in one strand of the double-stranded circular DNA, resulting in replication of a circle and a tail (**fig. 9.27**). The nick creates a 3'–OH, which can then serve as a primer sequence for leading strand replication using the intact circular strand as the template (see fig. 9.27). The tail serves as a template for discontinuous replication. To complete replication of the tail, DNA polymerase must remove the RNA primers and fill the gaps; DNA ligase then must seal the nicks. This process is identical to the lagging-strand replication discussed earlier.

This form of replication occurs in the *E. coli* F plasmid, or Hfr chromosome, during bacterial conjugation (see chapter 15). The F$^+$ or Hfr donor cell retains the circular daughter plasmid, while passing the linear single-stranded tail into the F$^-$ recipient cell, where replication of the tail takes place. Several phages also use this method, filling their heads (protein coats) with linear DNA replicated from a circular parent molecule.

The D-Loop Model

Chloroplasts and mitochondria in eukaryotic cells have their own circular DNA molecules (see chapter 19), which appear to replicate by a slightly different mechanism. The origin of replication is at a different point on each of the two parental template strands. Replication begins on one strand, displacing the other while forming a displacement loop or **D-loop** structure (**fig. 9.28**). Replication continues until the process passes the origin of replication on the other strand. Replication then initiates on the second strand, in the opposite direction. Normal Y-junction replication, as described earlier, also occurs in mitochondrial DNA under some growth conditions.

Restating the Concepts

▶ Two specialized DNA replication mechanisms, rolling-circle and D-loop replication, are found in replicating circular molecules. Although they differ in the mechanism of replication initiation, they both involve a displacement of a DNA strand to permit the replication of the second strand.

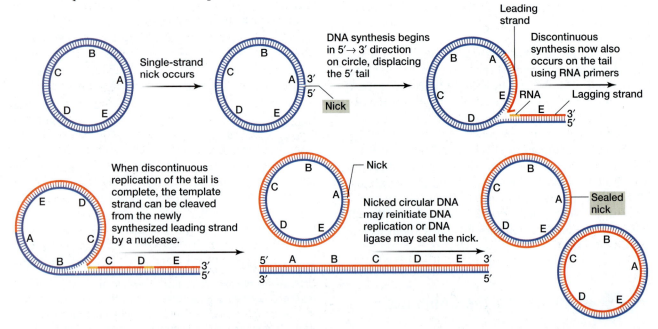

figure 9.27   Rolling-circle model of DNA replication. The letters A-E provide landmarks on the circular DNA. Nicking one template strand provides a 3'–OH for leading strand synthesis of the circular template. The nicked 5' end serves as the end of discontinuous or lagging-strand synthesis of a linear DNA molecule. After completing lagging strand synthesis, the template strand is cleaved from the synthesized leading strand to produce a linear and nicked circular molecules. The linear molecule is circularized by DNA ligase. The nicked circular molecule may immediately reinitiate DNA replication or DNA ligase may seal the nick. Template DNA is blue, synthesized DNA is red, and RNA primers are yellow.

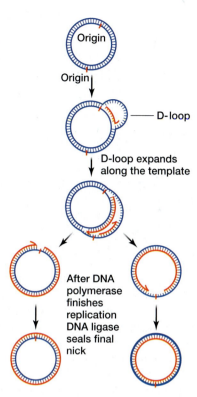

figure 9.28 D-loops form during mitochondrial and chloroplast DNA replication because the origins of replication are at different places on the two strands of the double helix. This results in unidirectional leading-strand synthesis from both origins.

9.6 Eukaryotic DNA Replication

As mentioned earlier, linear eukaryotic chromosomes usually have multiple origins of replication, resulting in figures referred to as "bubbles" (see fig. 9.9). Multiple origins allow eukaryotes to replicate their larger quantities of DNA in a relatively short time, even though eukaryotic DNA replication is considerably slowed by the presence of histone proteins associated with the DNA to form chromatin (see chapter 7). For example, the *E. coli* replication fork moves through about 25,000 bp/min, whereas the eukaryotic replication fork moves through only about 2000 bp/min. The number of replicons in eukaryotes varies from about 500 in yeast to as many as 60,000 in a diploid mammalian cell.

Eukaryotic Polymerases and Other Enzymes

Eukaryotes have evolved at least nine DNA polymerases, named DNA polymerase α, β, γ, δ, ε, ξ, η, θ, and ι. DNA polymerase delta (δ) seems to be the major replicating enzyme in eukaryotes, forming replisomes as DNA polymerase III does in *E. coli*.

In eukaryotes, the DNA polymerase α-primase complex generates the Okazaki fragment primers, first synthesizing an RNA primer and then a short length of DNA nucleotides. Polymerase epsilon (ε) may be involved in repair or in normal DNA replication. DNA polymerase gamma (γ) appears to replicate mitochondrial DNA. The remaining polymerases are probably involved in DNA repair, with polymerase β being the major repair polymerase, as polymerase II is in *E. coli*. Several of the polymerases most likely both replicate and repair DNA.

Eukaryotes also have a clamp-loader complex, called replication factor C, and a six-unit clamp called the proliferating cell nuclear antigen (PCNA). In eukaryotes, the completion (maturation) of Okazaki fragments involves RNase enzymes to remove the RNA primers rather than the 5' → 3' exonuclease activity of DNA polymerase. The resulting gap is repaired by a DNA polymerase, and a DNA ligase seals the final nick.

Helicases, topoisomerases, and single-stranded binding proteins play roles similar to those they play in prokaryotes. The completion of the replication of linear eukaryotic chromosomes involves the formation of specialized structures at the tips of the chromosomes called *telomeres*, which we will discuss at the end of this chapter. Thus, all of the enzymatic processes are generally the same in prokaryotes and eukaryotes.

Replication Origins in Eukaryotes

In budding yeast, a single-celled eukaryote that is often used as a model organism, DNA replication initiates at sites called **autonomously replicating sequences (ARS).** Each ARS consists of a specific 11-bp sequence plus two or three additional short DNA sequences encompassing 100–200 bp. Six proteins form a complex that binds to this sequence, referred to as the **origin recognition complex (ORC).** These proteins seem to be bound to the ORC all the time, and thus, additional proteins are needed to initiate DNA replication.

Some of these additional proteins are cyclin-dependent kinases, proteins involved in the control of the cell cycle (chapter 3). This makes sense because in eukaryotes, DNA replication can take place only once during the cell cycle, namely during the S phase. The initiation of DNA replication must therefore be tightly controlled to avoid multiple starts of replication of all replicons, except in the rare instances of gene amplification.

Telomeres and Termination of Replication

In a linear DNA molecule, the leading strand can undergo continuous replication to the end of the linear chromosome. However, the lagging strand is replicated away from the end with RNA primers that are then degraded, which leaves a short gap on the daughter strand (**fig. 9.29**).

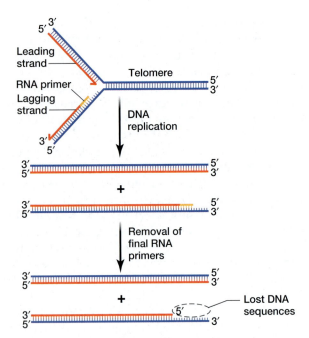

figure 9.29 Loss of DNA seqences at the end of linear chromosomes. As the replication fork moves towards the end of the chromosome (telomere), the leading strand (red) replicates to the 5' end of its template strand. This preserves all the sequence from the template strand. However, the discontinuous replication of the lagging strand (red) requires an RNA primer (yellow) before the Okazaki fragment is synthesized. This results in an RNA primer being located at some point along the template DNA. After DNA replication is completed, the RNA primer may be degraded. This results in the loss of the 5' end of the newly synthesized strand. When this strand serves as the template in the next round of DNA replication, these sequences cannot be synthesized.

As you learned in chapter 7, telomeres end each eukaryotic chromosome, with a repeating G-rich DNA sequence that forms a single-stranded 3' overhang. Thus, the normal replication process of a linear DNA molecule leaves an incomplete terminus, which would increase in length after every round of DNA replication. Additionally, this single-stranded DNA is very susceptible to nucleases, which would degrade the single-stranded sequence and permanently remove an increasing amount of DNA from the telomeres.

We also discussed three different mechanisms to protect this single-stranded telomeric sequence from being degraded. These include forming a planar *G-tetraplex*, the binding of proteins to the single-stranded 3' ends, or generating a *t-loop*. In all three mechanisms, the 3' single-stranded region either base-pairs with another DNA sequence or is bound by proteins to hide the 3' end from nucleases.

An additional mechanism used to protect against the loss of the ends of linear chromosomes is the rep-

lication of the telomere. An enzyme called telomerase adds copies of the telomeric repeat sequence *de novo*, without the need for a DNA template. Telomerase contains a segment of about 160 bases of RNA that has a region complementary to the G-rich repeat of telomeric DNA. Telomerase thus performs as a reverse transcriptase, using this RNA region as a template for DNA nucleotide polymerization.

Elizabeth Blackburn and her colleagues proposed in 1989 that the first step in telomere extension is hybridization of the RNA component of telomerase with the single-stranded 3' end of the genomic telomere DNA (**fig. 9.30a**). Then, with the telomerase RNA as a template, the 3' end of the telomere is extended (fig. 9.30*b*). A translocation step takes place where the RNA in the telomerase dissociates from the DNA and base-pairs with the newly synthesized DNA sequence. This returns the telomerase and its RNA to the same configuration as the beginning of the process, except the RNA is now annealed to the new telomeric repeat that was added onto the single-stranded 3' end of the genomic DNA telomere (fig. 9.30*c*).

This process is repeated several times until several telomere repeats have been added onto the 3' end of the DNA (**fig. 9.31**). The number of repeats in the telomere is not completely random, as you will see later. At this point, primase can generate an RNA primer that is complementary to the telomeric sequence (fig. 9.31). The selection of which telomere repeats will serve as the template for the RNA primer appears to be somewhat random. Once this RNA primer is synthesized, DNA polymerase and DNA ligase can synthesize the complementary DNA strand to the telomere in a manner similar to the generation of Okazaki fragments (see fig. 9.31).

The randomness of selecting the site where the RNA primer will anneal and the ultimate removal of this primer results in some loss of the newly added telomere sequences. However, the goal of this process is not to increase the length of the telomere, but rather to preserve its existence. Thus, the size of the telomere may fluctuate from one round of DNA replication to the next, but the addition of telomere repeats and the subsequent loss of some repeats ensures that some will remain for the next round of DNA replication, which will preserve the genomic DNA that is adjacent to the telomere.

How do cells keep track of the number of their telomeric repeats? Proteins have been isolated that bind to telomeres (Rap1 in *Saccharomyces cerevisiae*, TRF1 in human beings). By mutating these proteins or the telomeric sequences, scientists have changed the equilibrium number of telomeric repeats. This led to the current model that the cell counts the number of these proteins bound to the telomeres, not the number of telomeres directly, to know whether telomeres repeat sequences should be added.

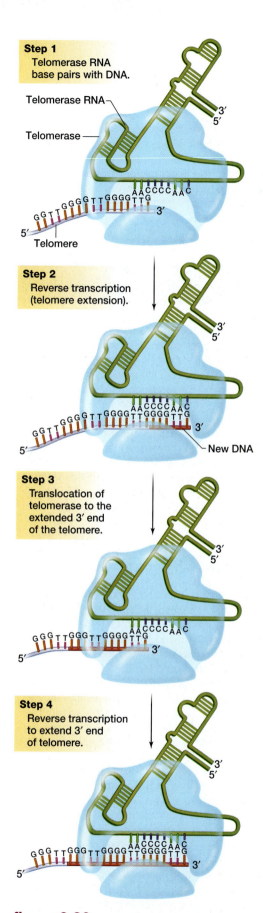

Step 1
Telomerase RNA base pairs with DNA.

Telomerase RNA

Telomerase

Telomere

Step 2
Reverse transcription (telomere extension).

New DNA

Step 3
Translocation of telomerase to the extended 3′ end of the telomere.

Step 4
Reverse transcription to extend 3′ end of telomere.

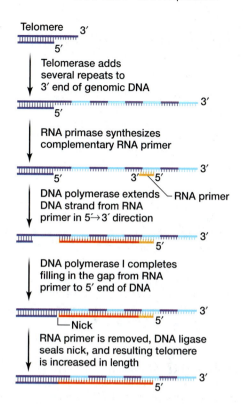

Telomere   3′
5′

Telomerase adds several repeats to 3′ end of genomic DNA

3′
5′

RNA primase synthesizes complementary RNA primer

3′
5′   3′ 5′

DNA polymerase extends DNA strand from RNA primer in 5′→3′ direction

RNA primer

3′
5′

DNA polymerase I completes filling in the gap from RNA primer to 5′ end of DNA

3′
Nick   5′

RNA primer is removed, DNA ligase seals nick, and resulting telomere is increased in length

3′
5′

figure 9.31   Telomere extension at the 3′ end of the linear chromosome. Telomerase adds several repeat sequence units to the 3′ end of the single-stranded DNA. RNA primase synthesizes a complementary RNA sequence to the single-stranded DNA. DNA polymerase extends a DNA strand from the RNA primer in a 5′ → 3′ direction. DNA ligase seals the nick. The RNA primer is lost. The number of telomeric repeat units is increased from the start of this figure.

figure 9.30   Telomerase extends the single-stranded 3′ end of telomeres. The RNA (green) component of telomerase base pairs with the single-stranded 3′ end of the telomere. The telomerase enzyme then synthesizes DNA in a 5′ → 3′ direction (red) using the RNA as a template. The telomerase RNA can then base pair at the new 3′ end of the single-stranded telomere and continue to add repeat unit sequences (GGGGTT) onto the single-stranded 3′ end of the DNA.

Restating the Concepts

▶ Eukaryotic chromosomes contain multiple origins of replication per linear chromosome. The multiple origins are required because eukaryotic DNA is larger than prokaryotic DNA, and replication is slower in eukaryotes, largely due to the difficulty of replicating DNA that is present in nucleosomes.

▶ The ends of linear chromosomes, telomeres, are replicated by telomerase, which is a reverse transcriptase enzyme that has an RNA molecule associated. The RNA is complementary to the telomeric repeat sequence and serves as the template for the reverse transcriptase enzyme.

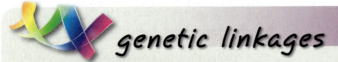

genetic linkages

The process of replication is associated with a variety of topics that are present throughout genetics. In chapter 3 we discussed the role of DNA replication prior to mitosis and meiosis. We established that the product of DNA replication, prior to the start of either mitosis or meiosis, was two daughter chromatids that were attached at the centromere. When we described Taylor's experiments on DNA replication in this chapter (see fig. 9.5), we showed the relationship between the DNA template strand and the sister chromatids, when colchicine disrupted the microtubules during mitosis. Furthermore, you learned that some of the proteins associated with the eukaryotic origin replication complex were cyclin-dependent kinases, which you also learned about in chapter 3 as a mechanism of cell cycle regulation. DNA replication's role as a cell cycle checkpoint has thus been underscored.

During recombination (chapter 6), a Holliday structure is generated that requires DNA polymerase to synthesize one of the DNA strands in the heteroduplex region. Thus, DNA synthesis is critical prior to and in the early stages of meiosis I.

In this chapter, we discussed the role of many enzymes that take part in replication, beginning with primase, which generates an RNA primer through which DNA polymerase III initiates replication. Because DNA polymerase adds bases in a 5' → 3' direction, it should not be surprising that primase also adds bases in a 5' → 3' direction. And, because primase is an RNA polymerase, you will probably not be surprised to learn in chapter 10 that all RNA polymerases synthesize RNAs in a 5' → 3' direction. Thus, all nucleic acid synthesis, both DNA and RNA, occurs with the same polarity.

We described the ability of DNA polymerases to remove replication errors with their 3' → 5' exonuclease activity. But in some cases a mutation is introduced after the polymerase has finished replicating a genomic region. In these instances, several different mechanisms can repair the damaged sequence. In all of these mechanisms, a wild-type DNA polymerase is a key component for adding one or more nucleotides after the removal of the mutations (chapter 18). We will also discuss in chapter 18 how the cell recognizes what strand of the newly replicated double-stranded DNA

corresponds to the template. This is important because a cell will assume that the template strand is correct and will try to correct a mutation in the newly synthesized strand. Altering the mutation in the proper strand will correct the mutation, whereas changing the sequence in the wrong strand will permanently incorporate the mutation in the DNA.

You will also see in later chapters how an understanding of DNA replication can be exploited in molecular biology techniques. For example, you now know that DNA synthesis in the 5' → 3' direction requires a 3'–OH at the end of the strand to which a new base will be added. You also learned that 2', 3'-dideoxyribonucleotides are inhibitors of DNA replication because they lack the 3'–OH. This knowledge has been exploited in DNA sequencing technology, in which the incorporation of a specific dideoxyribonucleotide by DNA polymerase reveals the complementary base, and the size of the terminated strand signifies the length from a given point to the incorporated dideoxyribonucleotide (chapter 12).

The ability to amplify DNA sequences in vitro, using the polymerase chain reaction (PCR), requires a DNA polymerase and two primers (chapter 12). From our discussion in this chapter, you should realize that the polymerase will add nucleotides onto the 3' end of the primer to synthesize a new DNA strand in a 5' → 3' direction.

Many of the enzymes required for DNA replication have become standard tools in recombinant DNA technology. As you will see in chapters 12 through 14, understanding the functions and activities of these proteins will allow you to better appreciate the various ways that DNA sequences can be manipulated in vitro. These proteins have permitted the potential of recombinant DNA technology to be maximized.

As we briefly mentioned in this chapter, the rolling-circle model of DNA replication is utilized by the *E. coli* F plasmid. This mechanism of replication, which requires the generation of a nick at a specific location in the F plasmid to initiate DNA synthesis, underscores the biology of not only the transfer of the F plasmid sequences in bacterial conjugation but also is critical for the mapping of genes by interrupted conjugation (chapter 15). Thus, DNA replication underlies many of the biological processes and in vitro manipulations associated with genetics.

Chapter Summary Questions

1. For each term in column I, choose the best matching phrase in column II.

   Column I
   1. Replicon
   2. Postive supercoiling
   3. DNA pol II
   4. DNA pol III
   5. Replication fork
   6. *oriC*
   7. Negative supercoilng
   8. DNA pol β
   9. DNA pol γ
   10. ARS

   Column II
   A. Major bacterial repair polymerase
   B. Major eukaryotic repair polymerase
   C. Y-junction where leading and lagging strand synthesis occurs
   D. Undercoiling
   E. Overcoiling
   F. Origin of replication in yeast
   G. Structure containing a single origin of replication
   H. Origin of replication in bacteria
   I. Major bacterial DNA polymerase
   J. Mitochondrial DNA polymerase

2. Diagram the results that Meselson and Stahl would have obtained after two rounds of
   a. conservative DNA replication.
   b. dispersive DNA replication.

3. What type of photo would Cairns have obtained if DNA replication were
   a. conservative.
   b. dispersive.

4. Why does DNA replication require the enzyme primase?

5. What is the significance of the 3' → 5' exonuclease activity during bacterial DNA replication? In which DNA polymerase(s) is it found?

6. What is the significance of the 5' → 3' exonuclease activity during bacterial DNA replication? In which DNA polymerase(s) is it found?

7. Indicate the order of participation of the following proteins during DNA replication in *E. coli*: (a) primase, (b) DNA polymerase I, (c) DnaA proteins, (d)DNA polymerase III, and (e) helicase.

8. List three major similarities and three major differences between the mechanisms of replication in eukaryotes and bacteria.

9. Compare and contrast the activities of DNA helicase and DNA topoisomerase IV.

10. What is a primosome in *E. coli*? A replisome? What enzymes make up each? What is the relationship between these structures?

11. What are the differences between continuous and discontinuous DNA replication? Why do both exist?

12. Describe the synthesis of an Okazaki fragment.

13. Eukaryotes require telomerase, but prokaryotes do not. Why is that so?

14. DNA polymerase III joins two different groups during DNA replication, and so does DNA ligase. Describe the chemical differences between the substrates of these two enzymes.

15. SSB proteins play a key role in DNA replication.
    a. Describe that role.
    b. Would you expect these proteins to be used more frequently during continuous or discontinuous replication? Explain your answer.

Solved Problems

PROBLEM 1: What enzymes are involved in DNA replication in *E. coli*?

Answer: The replisome consists of a primosome (a primase and a helicase) and two polymerase III holoenzymes, which forms at a replication fork on DNA. One polymerase acts processively, synthesizing the leading strand, while the other generates Okazaki fragments initiated by primers created by the primase. DNA polymerase I completes the Okazaki fragments, eliminating the RNA primer of the previous Okazaki fragment and replacing it with DNA. Finally, DNA ligase connects the fragments. Also involved in the process are single-stranded binding proteins and topoisomerases that relieve the DNA's supercoiling. Initiation involves initiation proteins at *oriC*, and termination requires termination proteins bound to the termination sites and a topoisomerase.

PROBLEM 2: As was discussed in this chapter, heavy ($^{15}N^{15}N$) DNA, light ($^{14}N^{14}N$) DNA, and hybrid ($^{15}N^{14}N$) DNA can be separated by density-gradient ultracentrifuga- tion. A solution containing equal amounts of heavy and light DNA is denatured at a high temperature. The solution is then slowly cooled, thereby allowing the single strands to hybridize. What is the expected proportion of heavy, hybrid, and light DNA in the renatured double-stranded molecules?

Answer: The formation of a particular type of duplex DNA is directly proportional to the availability of its single-stranded components. Heavy ($^{15}N^{15}N$) DNA will be formed when two ^{15}N strands hybridize to each other, and light ($^{14}N^{14}N$) DNA forms with the hybridization of two ^{14}N strands. Hybrid ($^{15}N^{14}N$) DNA on the other hand can be formed in one of two ways: (i) strand 1 can be ^{14}N-labeled while strand 2 is ^{15}N-labeled; or (ii) strand 1 is ^{15}N-labeled and strand 2 is ^{14}N-labeled.

Therefore, this problem can be solved using the product rule of probability.

The original solution contained equal amounts of heavy and light DNA. So, there are 1/2 heavy single strands: 1/2 light single strands.

$0.5\,^{15}N \times 0.5\,^{15}N = 0.25\,^{15}N^{15}N$
$0.5\,^{14}N \times 0.5\,^{14}N = 0.25\,^{14}N^{14}N$
$0.5\,^{15}N \times 0.5\,^{14}N = 0.25\,^{15}N^{14}N$
$0.5\,^{14}N \times 0.5\,^{15}N = 0.25\,^{14}N^{15}N$

Consequently, the proportions are $1/4\,^{15}N^{15}N : 1/2\,^{15}N^{14}N : 1/4\,^{14}N^{14}N$.

PROBLEM 3: A line of cultured human cells has an S-phase that lasts 5 h and a total of 2.04 m of DNA (B-form) per nucleus. The rate of DNA synthesis is 2500 bp/min at each replication fork. What is the minimum number of origins of replication required in the nucleus of each cell?

Answer: First, let's determine the number of base pairs in each cell. Remember that the B-form of DNA contains 1 bp per 0.34 nm (10 bp/turn), and that $1\,m = 1 \times 10^9$ nm. Therefore, 2.04 m of DNA contains

$$\frac{2.04 \times 10^9\,nm}{0.34\,nm/bp} = 6 \times 10^9\,bp$$

Each replication fork can synthesize 2500 bp/min × 60 min/h × 5 h (length of S-phase) = 750,000 bp. With two replications forks per origin, there are 2 × 750,000 = 1.5×10^6 bp per origin of replication.

Therefore, to replicate the entire genome, the minimum number of origins of replication is

$$\frac{6 \times 10^9\,bp}{1.5 \times 10^6\,bp} = 4 \times 10^3, \text{ or 4000 replication origins!}$$

Exercises and Problems

16. Consider the following three DNA molecules:

    I. 5' GATACGTAGCGCGCAGCTGT 3'
        3' CTATGCATCGCGCGTCGACA 5'
    II. 5' GATACTTATACACAAGCTAT 3'
        3' CTATGAATATGTGTTCGATA 5'
    III. 5' GACACGGAGCCACCCGCTGT 3'
         3' CTGTGCCTCGGTGGGCGACA 5'

    Which sequence is most likely to serve as an origin of replication and why?

17. The circular plasmid ColE1 of *Escherichia coli* replicates unidirectionally from a single origin. Would ColE1 be expected to utilize Okazaki fragments during DNA replication? Specify why or why not?

18. Following is a section of a single strand of DNA. Supply a strand, by the rules of complementarity, that would turn this into a double helix. What RNA bases would primase use if this segment initiated an Okazaki fragment? In which direction would replication proceed?

    5' ATTCTTGGCATTCGC 3'

19. Given the following problems in bacterial DNA replication, identify the defective enzyme or enzymes?
    a. Mismatched base pairs are found in the newly synthesized DNA.
    b. RNA bases are found in the newly synthesized DNA.
    c. Replication on the leading strand is not initiated.
    d. Replication forks are not formed.

20. Consider a mixture containing heavy ($^{15}N^{15}N$) and light ($^{14}N^{14}N$) DNA in a 4:1 ratio, respectively. The mixture is denatured at high temperature, and then slowly cooled, thereby allowing the single strands to hybridize. What is the expected proportion of heavy, hybrid ($^{15}N^{14}N$) and light DNA in the renatured double-stranded molecules?

21. Consider the following diagram of a DNA replication fork.

    a. Is the replication fork moving right-to-left or left-to-right?
    b. Is continuous synthesis expected to occur along the upper or lower template strand? Explain your answer.
    c. Okazaki fragments are expected to form along which template strand? Explain your answer.

22. It takes 30 min to copy a certain bacterial chromosome via theta replication. How long will it take if DNA synthesis were to occur via the rolling-circle replication mechanism? Assume no pauses and ignore discontinuous replication of the single-stranded tail.

23. Describe the mechanism by which topoisomerase II (gyrase) might work.

24. Suppose Meselson and Stahl had allowed *E. coli* to replicate for several generations in the ^{14}N medium. If the DNA was subjected to density-gradient centrifugation, what proportion of heavy, hybrid, and light double-stranded DNA would be observed after
    a. three generations?
    b. six generations?
    c. eight generations?

25. Answer question 24 again, assuming DNA replication is conservative.

26. The chromosome of the bacterium *Micrococcus hypotheticus* consists of 1.8 μm of B-form DNA. The chromosome, which has a single bidirectional origin of replication, is copied in 60 min under optimum conditions.
    a. What distance does each replication fork cover in 1 min?
    b. What is the speed of DNA synthesis, in base pairs per minute, at each replication fork?

27. Assuming DNA replication occurs continuously on *both* strands, which of the following will not be required and why?
    a. Primase   d. DNA pol I
    b. Helicase   e. Ligase
    c. Okazaki fragments

28. Describe the enzymology of the origin, continuation, and termination of DNA replication in *E. coli*.

29. Under what circumstances would you expect to see a DNA θ structure? D-loop? Rolling-circle? Bubbles? What function does each structure serve?

30. Progeria is a human disorder that causes affected individuals to age prematurely; a 9-year-old often resembles a 60- to 70-year-old individual in appearance and physiology. Suppose you extract DNA from a progeric patient and find mostly small DNA fragments rather than the expected long DNA molecules. What enzyme(s) might be defective in patients with progeria?

31. The chromosome of *E. coli* consists of 4.64×10^6 bp. The rate of DNA synthesis is 1000 bp/sec at each replication fork.
    a. Assuming no pauses in DNA synthesis, how long does it take for *E. coli* to replicate its entire chromosome?
    b. Under optimum conditions, *E. coli* has a generation time of 20 min. Suggest a mechanism to account for the discrepancy between the length of time for cell division and that of DNA replication.

32. The largest chromosome in a particular eukaryote contains 4×10^7 bp. The rate of DNA synthesis is 4000 bp/min at each replication fork.
    a. How long does it take to replicate the entire chromosome from a single origin of replication located exactly in the middle of the chromosome. Assume no pauses.
    b. Research reveals that it takes only 8 min for an actively growing cell to replicate this chromosome. What is the minimum number of replicons present on this chromosome?

33. Consider the following diagrams representing three different DNA molecules.

$$5' \text{━━━━━━} 3'$$

$$\begin{aligned} 3' &\text{━━} 5' \\ 5' &\text{━━━━} 3' \end{aligned}$$

$$\begin{aligned} 5' &\text{━━} 3' \\ 3' &\text{━━━━} 5' \end{aligned}$$

Assuming that DNA polymerase is the only enzyme present and that there is no additional DNA, state whether each of these molecules can serve as a substrate for DNA synthesis. Give reasons as to why or why not?

34. Can you think of any other mechanisms besides topoisomerase activity that could release supercoiling in replicating DNA?

35. In developing sea urchins, just after fertilization, the cells divide every 30–40 min. In the adult, the cells divide once every 10–15 h. The amount of DNA per cell is the same in each case, but the DNA obviously replicates much faster in developing cells. Propose an explanation to account for the difference in replication time.

36. Mutants are used to study various aspects of the phenotype and genotype. How can we study genes that are critically important in the functioning of an organism? For example, how do we study mutations in the gene for DNA polymerase III in *E. coli*, when changes in this

gene are usually lethal? Remember, to study the genes in bacteria, we need the bacteria to grow and form colonies in order to be scored for their phenotypes.

37. Propose a mechanism by which a single strand of DNA can make multiple copies of itself.

Chapter Integration Problem

A diploid eukaryotic organism has $2n = 10$ chromosomes. A cell line from this organism is grown in culture. The G_1, S, G_2, and M phases of the cell cycle last for 12 h, 8 h, 3 h, and 1 h, respectively.

a. Suppose a cell undergoes mitosis and then meiosis. Complete the following graph by tracing the amount of DNA over time in such a cell. Label the following stages: G_1, S, G_2, M, interphase, mitosis, meiosis I, and meiosis II.

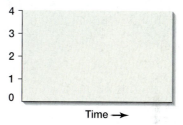

b. How many telomeres are found in a cell at: (i) prophase of mitosis; and (ii) prophase II of meiosis?
c. How many strands of DNA are found in a bivalent during metaphase I of meiosis?
d. Consider a submetacentric chromosome, containing three origins of replication: *O1* in the p arm, and *O2* and *O3* in the q arm. The order in which replication begins at the three origins is *O1–O2–O3*. Draw this chromosome as it would appear during S phase, knowing that all three origins of replication are activated. Show the old and the newly synthesized DNA strands.
e. If the cell contains one wild-type homolog of this chromosome and one homolog that contains a 10-kb deletion between the short arm telomere and *O1*, which homolog will replicate to the short arm telomere first and why? If the deletion was between the centromere and *O1*, which homolog would replicate to the short arm telomere first and why? If the deletion was between *O2* and *O3*, which homolog would replicate to the short arm telomere first and why? If the deletion was between *O3* and the telomere, which homolog would replicate to the short arm telomere first and why?

Tritium (^{3}H)-labeled dTTP is added to the culture medium and left for 10 min. The label is then removed and the cells are transferred to a culture medium containing nonradioactive dTTP.

f. What percentage of cells is expected to be labeled with tritium?
g. How long will it take for an ^{3}H-labeled acrocentric metaphase chromosome to appear?
h. Draw such a chromosome. Use different colors to represent labeled and nonlabeled strands.

Do you need additional review? Visit **www.mhhe.com/hyde** for practice tests, answers to end-of-chapter questions and problems, interactive exercises, and animation tutorials all designed to enhance your understanding of key genetic concepts.

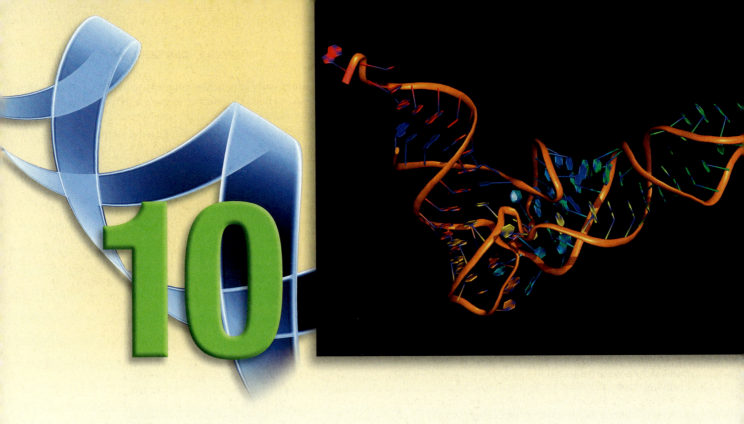

Gene Expression: Transcription

Essential Goals

1. Be able to name and describe the different types of RNAs produced in prokaryotes and eukaryotes and what functions they have in the cell.

2. Understand and compare the sequences and mechanisms that regulate transcription in prokaryotes and eukaryotes.

3. Diagram the posttranscriptional processing events that occur in eukaryotes and describe the functions of these events.

4. Be able to discuss microRNAs and how they can regulate gene expression.

Chapter Outline

Photo: A computer model of the serine transfer RNA structure.

*I*n this chapter, we continue our study of genetics at the molecular level. In 1958, Francis Crick originally described the flow of genetic information as the **central dogma:** DNA transfers information to RNA, which then directly controls protein synthesis (**fig. 10.1**). DNA also controls its own replication. **Transcription** is the process of synthesizing RNA from a DNA template using the rules of complementarity—the DNA information is rewritten, but in the same nucleotide language. RNA controls the synthesis of proteins in a process called **translation** because the information in the language of nucleotides is *translated* into information in the language of amino acids.

Years later, the central dogma needed to be modified when work with RNA viruses demonstrated that RNA could be converted into DNA using the enzyme reverse transcriptase. This represented the first of the major surprises in what had appeared to be a very straightforward genetic system.

For many years, geneticists thought that only three types of RNAs were transcribed from DNA: *messenger RNA*

(mRNA), transfer RNA (tRNA), and *ribosomal RNA (rRNA).* Investigators felt that these represented all the different RNA classes in the cell. However, a class of *small nuclear RNAs (snRNAs)* were identified that proved to be critical for proper splicing of eukaryotic mRNAs. Approximately 10 years ago, another class of small RNAs was identified that have been named *microRNAs (miRNAs).* The role of these RNAs in the cell was initially unclear.

The novel miRNAs, which are processed to be approximately 22 nucleotides in length, have revealed unique mechanisms of regulating gene expression. These miRNAs bind to complementary sequences in an mRNA and either induce cleavage of the mRNA or block translation by forming a double-stranded region. They are also able to block transcription of a gene by changing the conformation of the chromatin. (We will discuss these mechanisms in more detail later in this chapter.) These miRNAs regulate the expression of a variety of genes in fungi, plants, and animals, including humans.

A further indication of the importance of these small RNAs is the recent location of an miRNA locus in a chromosomal region that maps with neurological diseases, an X-linked mental retardation and an early-onset parkinsonism, Waisman syndrome. This raises the possibility that disrupting a 22 nucleotide miRNA may be sufficient to produce a clinically diagnosed neurological disease in humans.

Therefore, even though models for the transfer of genetic information (the central dogma) and the regulation of that information transfer were developed many years ago, we must remember that these models are based only on the current knowledge at that time. New discoveries in well-studied processes, such as composition and types of RNAs and transcription, may reveal additional novel and fundamental features.

Because much of the early work on DNA transcription was done in prokaryotic organisms, this chapter begins with a review of that process and the role of mRNA. Next, ribosomes and rRNA are discussed, followed by information on tRNA. We then turn to eukaryotic transcription. The chapter concludes with information on miRNAs and a summary of recent developments.

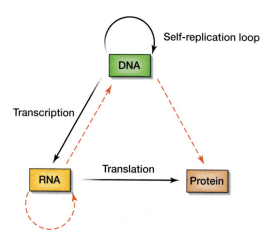

figure 10.1   Crick's original central dogma that depicted the flow of genetic information was one-way from DNA to RNA to protein. Dashed *red* lines indicate the information transfers that were confirmed after 1958, when Crick proposed the central dogma.

10.1   Types of RNA

In the protein synthesis process, five different kinds of RNA serve critical functions. The first type is **messenger RNA (mRNA),** which carries the DNA sequence information and will be directly translated into an amino acid sequence at the **ribosomes.** The second type is **transfer RNA (tRNA),** which brings the amino acids to the ribosomes, where it base-pairs with the complementary sequence in the mRNA. The third type of RNA is a structural and functional part

of the ribosome, called **ribosomal RNA (rRNA).** The general relationships of the roles of these three classes of RNAs are diagrammed in **figure 10.2**.

Although these three types of RNAs are found in all organisms, the final two types are found only in eukaryotic cells. The **small nuclear RNAs (snRNAs)** are essential for the proper production of mRNA by directing proteins to the exon–intron junctions where they remove the introns from the primary transcript to generate the mRNA. The final type of RNA consists of **microRNAs (miRNAs),** which play an important role in regulating both transcription of RNAs and translation of proteins.

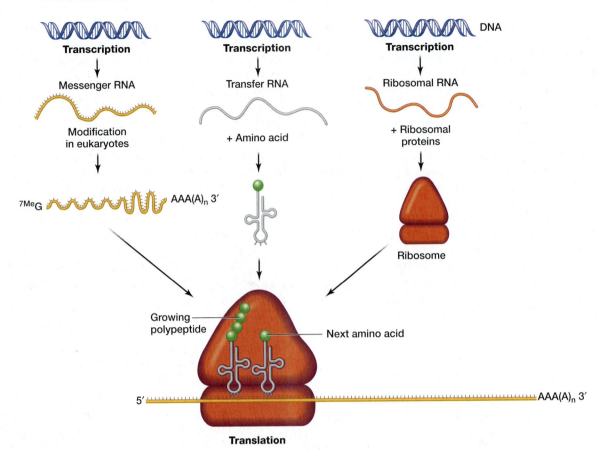

figure 10.2   Relationship among three of the different RNAs—messenger, transfer, and ribosomal—during protein synthesis. These three types of RNAs are found together at the ribosome during translation, with the ribosomal RNA being a component of the ribosome.

10.2 Bacterial Transcription: Preliminaries

We know that DNA does not take part directly in protein synthesis because, in eukaryotes, translation occurs in the cytoplasm, whereas DNA remains in the nucleus. Geneticists suspected for a long time that the genetic intermediate in prokaryotes and eukaryotes was RNA because the cytoplasmic RNA concentration increases with increasing protein synthesis. Proof of an RNA intermediate came when it was shown that mRNA directs protein synthesis and that the nucleotide sequences of mRNAs are complementary to nuclear DNAs.

DNA–RNA Complementarity

Investigators needed proof that gene-sized RNAs (not tRNAs or rRNAs) found in the cytoplasm were complementary to the DNA in the nucleus. Experiments yielded at least two lines of evidence supporting this hypothesis. First, it was shown that the RNAs produced by various organisms have base ratios very similar to the base ratios in the same organism's DNA (**table 10.1**).

table 10.1   Correspondence of G-C Content Between DNA and RNA of the Same Species

	RNA % G + C	DNA % G + C
E. coli	52	51
T2 phage	35	35
Calf thymus gland	40	43

The second line of evidence came from experiments by several researchers using **DNA–RNA hybridization.** This technique denatures DNA by heating, which causes the two strands of the double helix to separate (**fig. 10.3**). When the solution cools, a certain proportion of the DNA strands reanneal—that is, complementary strands "find" each other and re-form double helices. When RNA isolated from the cytoplasm was added to the denatured DNA solution and the solution was cooled slowly, some of the RNA molecules formed double helices with one strand of DNA. This would occur if the RNA fragments were complementary to a section of the DNA (see fig. 10.3). The existence of extensive complementarity

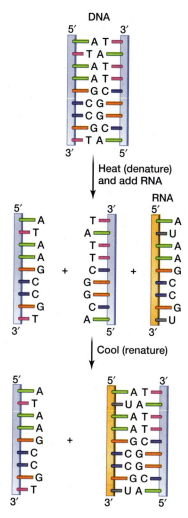

DNA

figure 10.3   DNA-RNA hybridization occurs between DNA and complementary RNA. The two strands in a DNA–RNA hybrid are held together by hydrogen bonds between the bases, with U in RNA paring with A in DNA. Thus, the two strands in an RNA–DNA hybrid have complementary sequence of bases.

table 10.2   Base Composition in RNA (percentage)

	Adenine	Uracil	Guanine	Cytosine
E. coli	24	22	32	22
Euglena	26	19	31	24
Poliovirus	30	25	25	20

then directs protein synthesis, two significant questions became apparent. First, is the RNA single- or double-stranded? Second, is it synthesized (transcribed) from one or both strands of the parental DNA?

How Many Strands Does RNA Have?

For the most part, cellular RNA does not exist as a double helix. It can form double-helical sections when complementary parts come into apposition (for example, see figs. 10.11 and 10.17), but its general form is not a double helix. The simplest, and most convincing, evidence for this is that complementary bases do not occur in corresponding equal proportions in RNA, as they do in DNA (Chargaff's ratios). That is, in RNA, uracil does not usually occur in the same quantity as adenine, nor does cytosine occur in the same quantity as guanine (**table 10.2**).

Is RNA Transcribed from One DNA Strand or Both?

The answer to the second question is that the strands of any given segment of the DNA double helix are not both used to synthesize RNA, although rare exceptions do occur. Similarly, a specific RNA is not synthesized from either DNA strand.

Suppose that a sequence of nucleotides on one strand of a DNA duplex specifies a sequence of amino acids for a protein, and the complementary nucleotide sequence also specifies the amino acid sequence for another functional protein. Because most enzymes are 300–500 amino acids long, the virtual impossibility of this entire region properly encoding two proteins with complementary nucleotide sequences should be obvious. For this to occur, the amino acid sequence in one protein, encoded by the DNA, would dictate the complementary nucleotide sequence, and therefore the amino acid sequence of the other protein. It was, therefore, assumed that for any particular gene only one DNA strand is transcribed, while the complementary strand is not transcribed in that region. Considerable evidence now supports this assumption.

The most impressive evidence came from work done with bacteriophage SP8, which infects *Bacillus subtilis*. This phage has an interesting property—a great disparity exists in the purine–pyrimidine ratio between the two strands of its DNA. The disparity is significant enough that the two strands can be

between DNA and RNA is a persuasive indication that DNA acts as a template for complementary RNA.

In another experiment, DNA–RNA hybridization showed that infection of bacteria by bacteriophage led to the production of bacteriophage-specific mRNAs. RNAs were extracted from *Escherichia coli* before and after bacteriophage T2 infection and were then tested to see whether they annealed with either the T2 phage DNA or the *E. coli* DNA. The RNA in the *E. coli* cell before infection was found to anneal with only the *E. coli* DNA. However, a large amount of the RNA in the *E. coli* cell after bacteriophage T2 infection annealed to the T2 phage DNA and not the *E. coli* DNA. From this it is apparent that when the phage infects the *E. coli* cell, it starts to manufacture RNA complementary to its own DNA and stops the *E. coli* DNA from serving as a template.

While this supported the hypothesis that RNA is transcribed (synthesized) from a DNA template and separated using density-gradient ultracentrifugation (see the Meselson and Stahl experiment in chapter 9). The SP8 genomic DNA was denatured, and the two strands were separated by density-gradient ultracentrifugation. DNA–RNA hybridization could then be carried out on each individual DNA strand with the RNA produced after the virus infects the bacterium.

Julius Marmur and his colleagues found that the RNA only hybridized to the heavier of the two SP8 DNA strands. Thus, only the heavy strand acted as a template for the production of RNA during the infection process. The DNA strand that is complementary to the RNA molecule is called the **noncoding** or **template strand.** The **coding** or **nontemplate strand** is the DNA strand that is not complementary to the RNA molecule. In the SP8 DNA molecule, the heavy strand is the noncoding strand and the light strand is the coding strand.

The idea that only one strand of DNA serves as a transcription template for RNA has also been verified for several other small phages. When larger viruses and cells are examined, however, we find that either of the strands may be transcribed, but only one strand is primarily used as a template in any one region (**fig. 10.4**). This was clearly shown with the T4 bacteriophage of *E. coli*, in which certain RNAs hybridize with one DNA strand, and other RNAs hybridize with the other.

Bacterial RNA Polymerase

In prokaryotes, the enzyme **RNA polymerase** controls transcription of RNA. We previously discussed RNA polymerase during DNA replication. Primase, which is an RNA polymerase, transcribes a short RNA primer using the DNA as the template. The RNA primer is extended in a 5' → 3' direction by DNA polymerase to synthesize the new DNA strand (chapter 9). Using DNA as a template, RNA polymerase adds ribonucleotide triphosphates (ribonucleotides) in a 5' → 3' direction, which is the same polarity that DNA polymerase adds deoxyribonucleotides.

Ribonucleotides differ from deoxyribonucleotides in two ways. First, ribonucleotides use the sugar ribose, whereas deoxyribonucleotides use the sugar 2'-deoxyribose, which differs from ribose by the absence of the hydroxyl group on the 2' carbon (**fig. 10.5**). Second, ribonucleotides have the base uracil instead of thymine, which is used in deoxyribonucleotides (see fig. 10.5). Thus, the RNA uracil will base-pair with adenine through two hydrogen bonds (analogous to the two hydrogen bonds between thymine–adenine base pairs).

The complete *E. coli* RNA polymerase enzyme—the holoenzyme—is composed of a core enzyme and a **sigma (σ) factor.** The core enzyme is composed of four subunits: two copies of an α-subunit, and one copy of a β and β'; this core is the component of the holoenzyme that actually carries out the RNA synthesis. The σ-factor is involved in recognizing transcription start signals on the DNA. Following the initiation of transcription, the σ-factor disassociates from the core enzyme.

Logically, transcription should not be a continuous process like DNA replication. If transcription were not regulated, then all the cells of a higher organism would be identical, and a bacterial cell would transcribe all of its genes all of the time and synthesize all of its proteins.

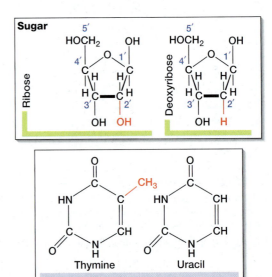

figure 10.5   RNA differs from DNA in two ways. RNA uses ribose as a sugar and DNA uses 2'-deoxyribose (shown in red in top panel). RNA possesses uracil in place of thymine as one of the four bases. While uracil lacks the methyl group that is found on thymine (in red in lower panel), uracil still forms two hydrogen bonds when it pairs with adenine as does thymine.

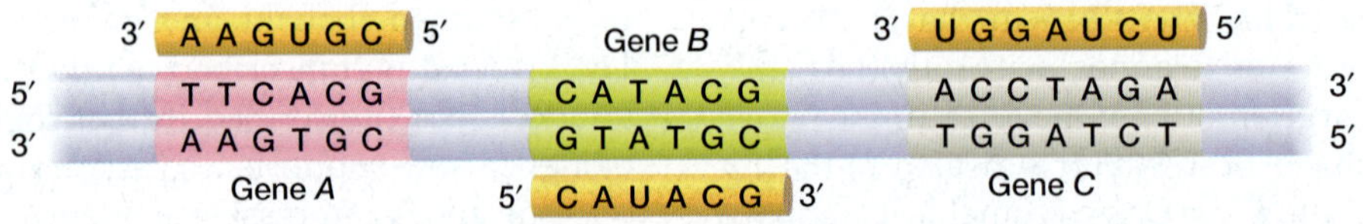

figure 10.4   Relationship of gene, DNA sequence, and RNA sequence. In a given region of genomic DNA (blue), three genes are located and a portion of each gene's sequence is shown. RNA, shown in gold, is complementary to only one strand of the DNA for each gene. However, different DNA strands can be complementary to the RNA for different genes. The RNA possesses a 5' → 3' polarity that is antiparallel to the strand with the complementary sequence.

Because some enzymes depend on substrates not present all of the time, and some reactions in a cell occur less frequently than others, the cell—be it a bacterium or a human liver cell—needs to regulate its protein synthesis.

One of the most efficient ways for a cell to exert the necessary control over protein synthesis is to regulate the amount of RNA that is transcribed from each gene. Transcription of regions outside of genes or of genes coding for unneeded enzymes is wasteful. Therefore, RNA polymerase should be selective. It should transcribe only those DNA segments (genes or small groups of genes) that code for products required by the cell at a particular time.

The mechanisms of transcriptional control can be examined in two ways. First, we need to understand how the beginnings and ends of transcriptional units (a single gene or a series of adjacent genes) are marked. Second, we want to elucidate how the cell can selectively turn transcription either on or off at different times and in different cells. The latter issues, which are the keys to transcriptional efficiency and to eukaryotic growth and development, are covered in later chapters.

RNA polymerase must be able to recognize both the beginnings and the ends of genes (or gene groups) on the DNA double helix in order to properly initiate and terminate transcription. It must also be able to recognize the correct DNA strand to avoid transcribing the DNA strand that is not informational. RNA polymerase accomplishes those tasks by recognizing certain start and stop signals in DNA, called initiation and termination sequences, respectively.

Restating the Concepts

▶ There are several different forms of RNAs. mRNA is translated to produce a protein, tRNA binds the amino acids and transports them to the ribosome for translation, and rRNA is a structural component of the ribosome. Two additional types of small RNAs are important in splicing eukaryotic mRNAs (snRNAs) and regulating gene expression (miRNAs).

▶ RNA differs from DNA in the presence of ribose as the sugar, the presence of uracil instead of thymine, and in usually being single-stranded.

▶ A gene is transcribed from only a single strand of the double-stranded DNA. The DNA strand that is complementary to the RNA sequence is the noncoding or template strand. However, in a DNA molecule, both strands may serve as template strands for different genes.

▶ RNA polymerase holoenzyme binds to the DNA through the σ-subunit. After transcription initiation, the σ-factor is released and the RNA polymerase core enzyme (two α-subunits, one β-subunit, and one β'-subunit) carries out RNA elongation.

10.3 Prokaryotic Transcription: The Process

The DNA region that RNA polymerase associates with immediately before beginning transcription is known as the **promoter.** The promoter is an important part of a gene for expression in both prokaryotes and eukaryotes. Promoters contain the information for transcription initiation, and they are the major sites where gene expression is controlled. As you will see, the promoters for prokaryotic and eukaryotic genes are very different, with the eukaryotic promoters possessing significantly more complexity to account for the larger variety of cells and tissues in higher organisms.

The bacterial RNA polymerase holoenzyme possesses high affinity for DNA sequences in the promoter region. This specific recognition of the promoter DNA sequence is due to the σ-factor; without the σ-factor, the core enzyme of RNA polymerase has been shown to bind randomly along the DNA.

Once transcription is initiated, the mRNA strand is elongated much the same way as a DNA strand is elongated during replication. Termination of transcription occurs when the RNA polymerase enzyme recognizes a DNA region known as a **terminator sequence.**

Promoter Sequences and Transcription Initiation

The bacterial RNA polymerase molecule binds a promoter region consisting of about 60 bp of DNA. This size was determined by allowing the RNA polymerase holoenzyme to bind DNA and then digesting the complex with nucleases, in a technique known as **footprinting** (fig. 10.6). The polymerase "protects," or prevents degradation of, the DNA region it binds. The undigested DNA is then isolated and its size determined by gel electrophoresis.

Further information about the nature of the promoter sequences was gained through recombinant DNA technology and nucleotide sequencing techniques (described in chapter 12). Sequencing many promoters from different genes revealed that they contain common sequences.

Bacterial Promoter Consensus Sequences

If several nucleotide sequences align with one another, and each has *exactly* the same series of nucleotides in a given segment, we say that the sequence in that segment comprises an invariant, or **conserved sequence.** If, however, some variation appears in the sequence, but certain nucleotides occur at a *high frequency* (significantly greater than by chance), we refer to those nucleotides as making up a **consensus sequence.** Surrounding a point in

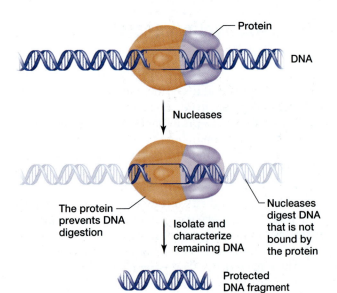

figure 10.6 DNA footprinting technique. DNA that is bound by a protein (e.g., RNA polymerase) will be protected from nuclease degradation. After the flanking DNA has been degraded, the bound protein can be released from the DNA and the protected DNA sequence can be determined and characterized.

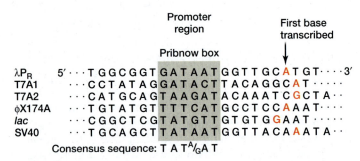

Consensus sequence: T A T^A/$_G$A T

figure 10.7 Nucleotide sequences of the nontemplate DNA strand for the bacterial –10 promoter region (Pribnow box) and the first base transcribed from several different genes (red). Lambda (λ), T7, and φ×174 are bacteriophages. *lac* is an *E. coli* operon, and SV40 is an animal virus. Only the SV40 promoter has the actual 5' TATAAT 3' consensus sequence. Including additional sequenced promoters not shown here, no base is found 100% of the time at the same position (invariant). The consensus sequence for the Pribnow box is aligned underneath.

bacterial promoters, about 10 nucleotides before the first transcribed base, is a consensus sequence 5' TATAAT 3'. This sequence is known as a **Pribnow box** after one of its discoverers, David Pribnow (**fig. 10.7**).

The nucleotides in the Pribnow box are mostly adenines and thymines, so the region is primarily held together by only two hydrogen bonds per base pair. Because local DNA denaturation must occur prior to the start of transcription so that complementary pairing of ribonucleotides with the DNA template strand can occur,

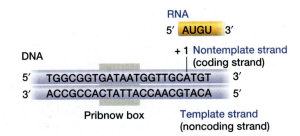

figure 10.8 The relationship between the DNA strands and RNA. The sequence of the template (noncoding) strand of DNA is complementary to both the coding strand of DNA and the transcribed RNA (shown above). The first base of the RNA molecule and the corresponding position in the DNA is designated as +1. The sequences are from the promoter of the λP$_R$ region (see fig. 10.7).

fewer hydrogen bonds in the DNA molecule make this denaturation easier energetically. When the polymerase is bound at the promoter region (see fig. 10.7), it is in position to begin polymerization 6–8 nt down from the Pribnow box.

The sequences shown in figure 10.7 are those of the coding strand of DNA. It is a general convention to show the coding strand because it possesses the same sequence, substituting U for T in the DNA, as the mRNA. Thus, the sequence of the coding strand can be used to directly determine the translated amino acid sequence (see chapter 11). Both the DNA coding strand and the mRNA are complementary to the template strand (also referred to as the **noncoding strand;** **fig. 10.8**). Another convention used in describing the DNA strands is the sense and antisense strands.

Another important convention used in labeling the DNA is to indicate the first transcribed base by the number **+1** and to use positive numbers to count farther along the DNA in the **downstream** direction of transcription. So, if transcription is proceeding to the right, then the direction to the right of +1 is downstream and to the left is called **upstream,** with bases indicated by negative numbers (**fig. 10.9**). Notice that there is not a base labeled as 0, as the +1 base is preceded by –1. Under this convention, the Pribnow box is often referred to as the **–10 sequence.**

Figure 10.9 shows a second consensus sequence, 5' TTGACA 3', which is present in many *E. coli* promoters. It is centered near position –35 and consequently is referred to as the **–35 sequence.** Mutations were generated in the –10 and –35 sequences to determine how they affected transcription initiation. Mutations in either the –10 or –35 sequences usually decreased the amount of transcription and increased the frequency of transcription initiation at positions other than +1. The σ-factor, which is required for correct transcription initiation, recognizes and binds both the –35 and the –10 sequences. Thus, the more a –10 or –35

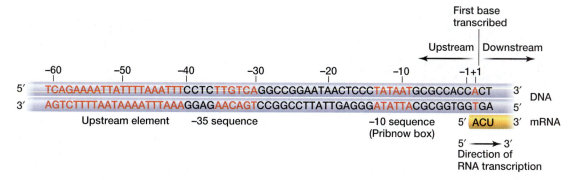

figure 10.9   The promoter of the *Escherichia coli* rRNA gene, *rrnB*. The complete promoter is comprised of the –10 (Pribnow box) and –35 sequences and an upstream element. The first base transcribed (the transcriptional start site) is noted as +1 on the nontemplate (upper) strand, as well as upstream and downstream. Notice that the RNA sequence (yellow) and polarity are complementary and antiparallel to the DNA template (lower) strand.

sequence differs from the consensus sequence, the less efficient σ-factor and RNA polymerase binds and the less frequently that particular promoter initiates transcription. The σ-factor was also found to be sensitive to the spacing between these sequences, with the most efficient distance being 17 bp.

Farther upstream from the –35 sequence is a third sequence that is present in bacterial promoters, which are very strongly expressed, such as the rRNA genes (see fig. 10.9). This **upstream element** is about 20 bp long, is centered at –50, and is rich in A and T bases. Mutational studies have shown that when this element is added at the proper location to promoters that do not normally have it, the rate of transcription of the modified gene greatly increases.

Other recognition sites have been found in prokaryotes, both upstream and downstream, at which various regulatory proteins attach. These proteins can enhance or inhibit transcription by direct contact with the polymerase (specifically the α- and σ-subunits). Chapter 16 contains a discussion of these proteins.

Variable Action of σ-Factors

Because the holoenzyme recognizes consensus sequences in a promoter, rather than binding to a conserved sequence, it is not surprising that some promoters are bound more efficiently than others or that different σ-factors exist within a cell.

In *E. coli,* the major σ-factor is a protein of 70,000 D (daltons), referred to as σ^{70}. (One dalton is an atomic mass of 1.0000, approximately equal to the mass of a hydrogen atom.) *E. coli* also possesses about five less common σ-factors that provide the cell with a mechanism to transcribe different genes under different circumstances. For example, at higher temperatures, the *E. coli* cell expresses a group of new proteins, referred to as **heat shock proteins,** which helps to protect the cell against the elevated temperatures. These various

heat shock proteins are all expressed simultaneously because their corresponding genes contain promoters that are recognized by the same σ-factor, one with a molecular weight of 32,000 D (σ^{32}). The heat shock gene promoters are actively transcribed only after the temperature increases because σ^{32} is only produced by the cell after heat shock. Both prokaryotic and eukaryotic heat shock proteins and other systems of transcriptional control are described in chapters 16 and 17.

From mutational studies of promoters and the proteins in the RNA polymerase holoenzyme, we now have a picture of a holoenzyme that binds to a DNA promoter because the σ-factor recognizes and binds the –10 and –35 DNA sequences, and the α-proteins recognize the upstream element when it is present (**fig. 10.10a**). Additionally, the α- and σ-subunits recognize proteins bound to various other upstream elements when these elements are present. This initiation complex is initially referred to as a **closed promoter complex** because the DNA has not denatured in the promoter region (see fig. 10.10a). After the holoenzyme induces the DNA to denature in a small region, the RNA polymerase and DNA form an **open promoter complex** (fig. 10.10b). The denatured DNA can then base-pair with ribonucleotides and permit the initiation of transcription (fig. 10.10c). After the transcription of 5–10 bases, the σ-factor is released and transcription proceeds (fig. 10.10d).

Transcription Elongation

Transcription, like DNA replication, always proceeds in the 5' → 3' direction. That is, a single ribonucleotide is added anew to the 3'-OH free end of the RNA (see fig. 10.10d). Due to the antiparallel nature of double-stranded nucleic acids, transcription proceeds along the DNA template in a 3' → 5' direction.

Unlike DNA polymerase, however, prokaryotic RNA polymerase does not possess a significant 3' → 5'

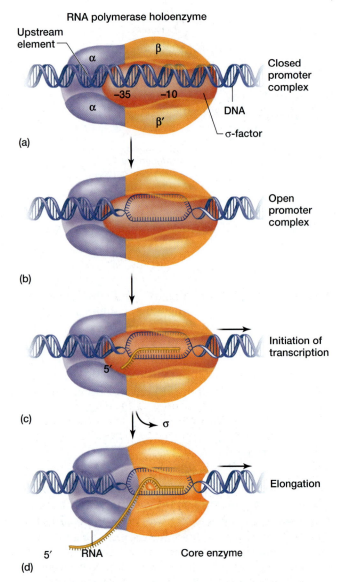

RNA polymerase holoenzyme

Upstream element

α

β

−35   −10

α

β′

Closed promoter complex

DNA

σ-factor

(a)

Open promoter complex

(b)

Initiation of transcription

5′

(c)

σ

Elongation

5′   RNA   Core enzyme

(d)

figure 10.10   Bacterial transcription initiation involves the σ-subunit (red) of the RNA polymerase recognizing and attaching to the −10 and −35 sequences in the promoter, which produces the closed promoter complex (a). The relative position of the upstream element (fig. 10.9) is shown. Local denaturation of the DNA forms the open promoter complex (b), which permits transcription to begin (c). After transcription has begun, the σ-factor leaves the holoenzyme, and transcription elongation continues with the RNA polymerase core enzyme (d).

exonuclease activity to verify the complementarity of the new bases added to the growing RNA strand. This deficiency is not serious; because many mRNAs are short-lived, and because many copies are made from actively transcribed genes, an occasional mistake probably does not produce permanent or overwhelming damage. If a particular RNA is generated that contains a mutation, a new RNA will be made soon. Evolutionarily speaking, it seems more important to

make RNA quickly than to proofread each RNA made. In contrast, if an error is made in DNA replication, the daughter cell that receives the mutant copy and all the subsequent daughter cells will possess the mutation.

To illustrate this difference, imagine that a mutation occurs in a gene during DNA replication once every 100 cell divisions. If this mutation results in the expression of a nonfunctional protein that causes a mutant phenotype, then this mutant cell and all the subsequent daughter cells will have the mutant phenotype. In contrast, if 1% of the RNAs transcribed from this gene in a wild-type cell contain mutations, then the cell will still possess 99% of the wild-type RNAs and will exhibit the wild-type phenotype. Furthermore, the subsequent daughter cells will still have the possibility of expressing only wild-type RNAs from this gene because mutations in the RNA are not inherited. This example shows that transcription can tolerate a higher error rate than replication.

Terminator Sequences and Transcription Termination

Transcription continues as RNA polymerase adds nucleotides to the growing RNA strand in a 5′ → 3′ direction according to the rules of complementarity (see fig. 10.10d). The polymerase moves down the DNA until it reaches a stop signal, or terminator sequence.

Rho-Dependent and Rho-Independent Terminators

Two types of terminator sequences exist in *E. coli*, namely rho-dependent and rho-independent, which differ in their requirement for the **rho (ρ) protein**. **Rho-dependent terminators** utilize both the ρ-protein and a specific RNA structure. In the absence of the ρ-protein, RNA polymerase continues to transcribe past the terminator sequence in a process known as read-through. **Rho-independent terminators** require a specific RNA structure, but not the ρ-protein.

Both types of terminators require the same type of RNA structure, namely a particular sequence and its inverted form, separated by another short sequence. This **inverted-repeat sequence** is shown in **figure 10.11**. In this example, the sequence 5′ AAAGGCTCC 3′ is present in both the coding strand and the template strand. Notice that the polarity dictates the orientation of the sequence within each strand. A 4-nt sequence (UUUU) separates the inverted repeats. Inverted repeats can form a **stem-loop structure** by pairing complementary bases within the transcribed mRNA.

Both rho-dependent and rho-independent terminators have a stem-loop structure in the RNA just before the last base transcribed. Rho-independent terminators also have a sequence of uracil-containing nucleo-

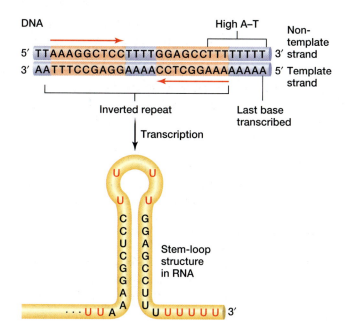

figure 10.11   An inverted-repeat base sequence characterizes terminator regions of DNA transcription. The inverted repeat sequences in the DNA are highlighted in red and marked with arrows to show their orientation. Stem-loop structures can occur in the transcribed RNA because of basepairing between the complementary sequences. The 3' poly-U tail indicates a rho-independent terminator.

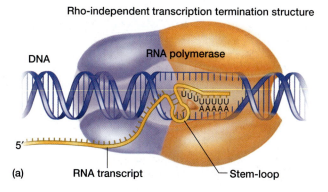

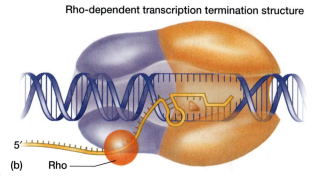

figure 10.12   Rho-independent (*top*) and rho-dependent (*bottom*) termination of transcription are preceded by a pause of the RNA polymerase at a stem-loop terminator structure in the nascent RNA. In rho-independent termination (a), the RNA is only associated with the DNA template strand by A-U basepairs when the polymerase pauses. This relatively weak base-pairing is insufficient to retain the RNA, which then dissociates from the DNA template. In rho-dependent termination (b), the rho protein binds RNA and moves along the RNA in a 5' → 3' direction. When the RNA polymerase pauses, rho reaches and contacts the polymerase, which causes the RNA to dissociate from the DNA template. Notice that there is no sequence requirement in the RNA after the stem-loop structure in rho-dependent transcription termination.

tides in the RNA after the inverted repeat, whereas rho-dependent terminators do not (see fig. 10.11).

Although the exact sequence of events at the terminator is not fully known, it appears that the RNA stem-loop structure forms and causes the RNA polymerase to pause just after completing it. This pause may then allow termination under two different circumstances.

In rho-independent termination, the pause may occur just after the sequence of uracils is transcribed (**fig. 10.12**). Uracil–adenine base pairs have two hydrogen bonds and are thus less stable thermodynamically than guanine–cytosine base pairs. Perhaps during the pause, the uracil–adenine base pairs spontaneously denature, releasing the transcribed RNA and the RNA polymerase, which terminates transcription.

Rho-dependent terminators, by contrast, do not have the uracil sequence after the stem-loop structure. Here, termination depends on the presence and action of the rho protein (ρ), which binds to a particular sequence in the newly forming RNA. In an ATP-dependent process, ρ travels along the RNA in a 5' → 3' direction at a speed comparable to the transcription process itself (see fig. 10.12). Possibly, when RNA polymerase pauses at the stem-loop structure, ρ catches up to the polymerase and denatures the DNA–RNA hybrid, which releases the transcribed RNA and RNA polymerase from the DNA template. Rho can do this because it has DNA–RNA helicase (unwinding) properties.

The process of transcription termination is probably more complex than described. Significant interactions may require other proteins and particular sequences surrounding the termination sequence to be present. This is an area of active research.

A Summary of Bacterial Transcription

Figure 10.13 shows an overview of bacterial transcription. The information of a gene, coded in the sequence of nucleotides in the DNA, has been transcribed into a complementary sequence of nucleotides in the RNA. This RNA transcript contains a sequence complementary to the template strand of the gene's DNA and thus either acts as a messenger (mRNA) from the gene to the cell's protein-synthesizing complex or produces a functional RNA (tRNA or rRNA) that is not translated.

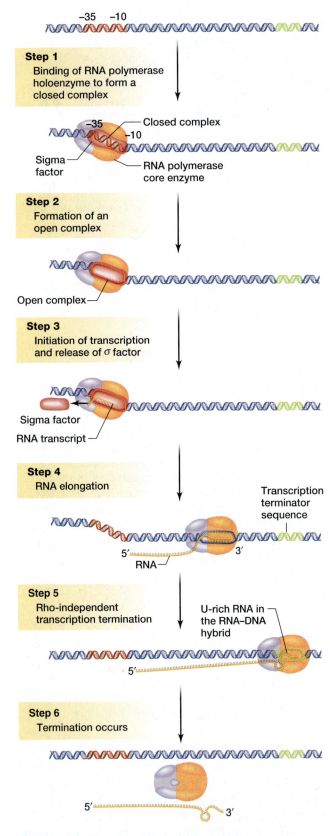

figure 10.13 Overview of bacterial transcription initiation, elongation, and termination. This summarizes the steps that were shown in previous figures.

An mRNA transcript contains nucleotide sequences that are to be translated into amino acids—coding segments—as well as noncoding segments before and after. The translatable segment, or **open reading frame (ORF)**, almost always begins with a three-base sequence, AUG, which is known as a translation initiator codon (**fig. 10.14**). The ORF ends with one of the three-base sequences UAA, UAG, or UGA, known as translation termination, or nonsense, codons. (We will discuss these signals in chapter 11.)

The portion of the mRNA transcript that begins at the start of transcription and goes to the translation initiator codon (AUG) is referred to as a **leader**, or **5' untranslated region (5' UTR)**. The length of mRNA from the nonsense codon (UAA, UAG, or UGA) to the last nucleotide transcribed is the **3' untranslated region (3' UTR)**. These sequences play a role in ensuring the structural stability of the mRNA and regulate translation at the ribosome.

Figure 10.14 diagrams the relationship between a bacterial gene and its mRNA transcript. The promoter sequences (–10 and –35) are located upstream of where transcription will begin (+1) and are *not* present on the mRNA. The mRNA contains a 5' UTR sequence that is located upstream of the translation initiator codon (AUG). The ORF is the region of the mRNA that will be translated and is followed by a 3' UTR sequence. Notice that the sequence of the coding strand of the DNA is analogous to the mRNA sequence; both contain a recognizable translation initiator codon (AUG in the mRNA and ATG in the DNA) and a translation termination codon (UAA in the mRNA and TAA in the DNA). Unlike the promoter sequence, the transcription termination signal is present in both the DNA and the 3' end of the mRNA.

In this simplified drawing of a bacterial gene (see fig. 10.14), the transcript represents only one gene and has a single ORF that extends from the AUG codon to the UAA codon. Many prokaryotic transcripts, however, contain the information for several genes. An **operon** is several genes that encode proteins that are involved in a related process, which are expressed on a single mRNA that is under the transcriptional control of a single promoter. This is a common mechanism to express several genes in a coordinated manner in a bacterial cell. Now we turn our attention to the other two types of bacterial RNAs: ribosomal and transfer RNAs.

It's Your Turn

It is important to understand the relationship between sequences in the DNA and the transcribed mRNA. Now that we have discussed the important components for correct transcription, see if you comprehend their relative positions in the gene and mRNA in problem 25.

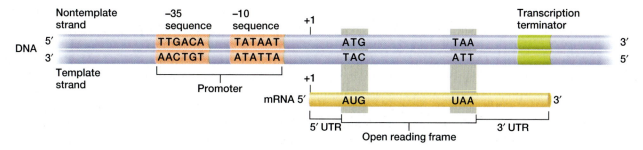

figure 10.14   Relationship between a bacterial mRNA and its DNA. Shown are the promoter (−10 and −35 sequences) and the transcription terminator region on the DNA. Notice that the mRNA lacks the sequence corresponding to the promoter, but possesses the sequence corresponding to the transcription terminator. The translation initiation (AUG) and nonsense (UAA) codons in the mRNA for protein synthesis are shown. Translation of the mRNA, beginning with the AUG and ending with the UAA, occurs at the ribosome. The region of the mRNA that will be translated is termed the open reading frame and is flanked on the mRNA by two regions that are not translated, the 5' untranslated region (UTR) and the 3' UTR.

 ### Restating the Concepts

▶ Transcription initiation requires specific DNA sequences (promoter) to bind the RNA polymerase holoenzyme. The consensus sequences in the promoter include a −10 (TATAAT), −35 (TTGACA) and occasionally an A–T-rich upstream element (the latter is only found in promoters of genes that are expressed at very high levels).

▶ Transcription elongation proceeds in a 5' → 3' direction on the RNA.

▶ Transcription terminates in either a rho-dependent or rho-independent mechanism. Both mechanisms require a stem-loop structure at the 3' end of the RNA. The rho-independent mechanism also requires a string of uracil residues immediately following the stem loop. The rho-dependent mechanism does not require the uracils, but does require the ρ-protein binding to the RNA to terminate transcription.

10.4 Ribosomes and Ribosomal RNA

Ribosomes are complexes in the cell where protein synthesis occurs. They are composed of proteins and RNA (ribosomal RNA or rRNA). In a rapidly growing *E. coli* cell where many proteins must be synthesized, ribosomes can make up as much as 25% of the mass of the cell.

Measuring rRNA Components: The Svedberg Unit

Ribosomes and other small molecules and particles are measured in units that describe their rate of sedimentation during density-gradient centrifugation in sucrose. This technique, developed in the 1920s by

physical chemist Theodor Svedberg, gives information on the size and shape of the particle due to the speed of sedimentation. This unit of sedimentation is named the **Svedberg unit, S.**

In sucrose density-gradient centrifugation, the gradient is formed by physically layering decreasingly concentrated sucrose solutions on top of each other in a tube. In the related cesium chloride density-gradient centrifugation described in chapter 9, the gradient develops as a result of centrifugation, and the cesium chloride gradient tube is spun until it reaches equilibrium. The sucrose method tends to be more rapid because the gradient is formed prior to centrifugation. The molecules or particles being studied, or a mixture, are then added to the sucrose density gradient, and the tube is spun in a centrifuge for a period of time. As in cesium chloride density centrifugation, the densest molecules (with the highest S values) will be at the bottom of the gradient tube.

Ribosomal Structure and Production

Ribosomes in all organisms are made of two subunits of unequal size. In *E. coli,* the sedimentation value is 50S for the large subunit and 30S for the smaller one. When together, the subunits have a sedimentation value of about 70S. Because the Svedberg value corresponds to density, simply adding the S values for the two subunits will not accurately determine the density of the protein complex. Eukaryotic ribosomes vary from 55S to 66S in animals, and 70S to 80S in fungi and higher plants. Most of our discussion will be confined to the well-studied ribosomes of *E. coli.*

Each ribosomal subunit contains one or two rRNA molecules and a fixed number of proteins. The 30S subunit of *E. coli* has 21 proteins and a 16S molecule of rRNA, and the 50S subunit has 34 proteins and two rRNAs: one 23S rRNA and one 5S rRNA (**fig. 10.15**). Details of ribosomal structure have been studied by isolating and purifying all the proteins, by experimenting with the proper sequence of assembly of subunits, and through immunological techniques that have clarified the positions of many proteins in completed ribosomal subunits. It is believed that the rRNA is essential in catalyzing the bond between adjacent peptides during protein translation.

In *E. coli*, all three rRNA genes are transcribed as a single long RNA molecule that is then cleaved and modified to form the final three rRNAs (16S, 23S, and 5S). The RNA that contains the three rRNA molecules also contains four tRNAs (**fig. 10.16**). This RNA is transcribed from five to ten different locations in the *E. coli* chromosome.

The occurrence of the three rRNA segments on the same piece of transcribed RNA ensures a final ratio of 1:1:1, the ratio needed for ribosomal construction. This transcription of three different rRNAs as a single RNA is analogous to an operon. It allows for the three different rRNA genes to be regulated in the same manner to ensure that the proper ratio of the gene products (rRNAs) is produced. However, it is *not* an operon because the single RNA produced from the three genes is not translated. Three of the four rRNAs are also initially transcribed as a long, single RNA that

is then cut to produce the three correctly sized rRNAs, as discussed later in this chapter.

The rRNAs are functional molecules, they are required for the ribosome to translate a protein, they are not translated. Because the rRNAs are not translated, they lack many of the key components that are present in an mRNA, such as an ORF and translation initiation and termination codons. Without an ORF, they also cannot possess either a 5' UTR or a 3' UTR.

Restating the Concepts

▶ Translation takes place in the ribosomes, which are composed of two subunits, with each subunit containing proteins and rRNAs. The rRNAs are functional RNAs and are not translated into a protein, rather they are required for the correct translation of proteins from mRNAs.

▶ Svedberg units are a measure of the size of large proteins and other particles based on their sedimentation rates in a sucrose density gradient. Ribosome components can be distinguished by their Svedberg values.

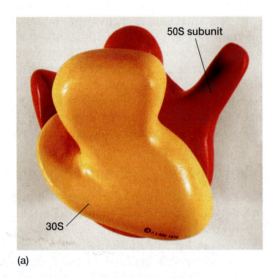

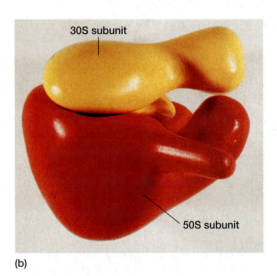

(a)   (b)

figure 10.15   The *E. coli* ribosome. (a) and (b) show models of the 70S ribosome of *E. coli*, revealing the relationship of the small 30S (*yellow*) and large 50S (*red*) subunits at the time of translation. The 30S ribosomal subunit is composed of twenty-one proteins and one 16S piece of rRNA. The 50S subunit is composed of thirty-four proteins and two pieces of rRNA, 23S and 5S.

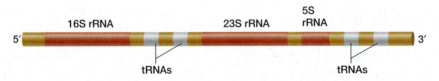

figure 10.16   The *E. coli* transcript that contains the three rRNA segments (16S, 23S, and 5S; red) and also four tRNAs (gray). Notice that some spacer RNA (gold) separates the tRNA and rRNA sequences.

10.5 Transfer RNA

During protein synthesis (see fig. 10.2), an mRNA molecule, carrying the information transcribed from the gene (DNA), is bound to a ribosome. The transfer RNAs (tRNAs), which are attached to specific individual amino acids, interact with the mRNA in the ribosome. The genetic code in the mRNA is read in groups of three adjacent nucleotides, called **codons** (**fig. 10.17a**). The nucleotides of the mRNA codon can pair with the complementary three bases—the **anticodon**—on a tRNA (see fig. 10.17a). Each different tRNA carries a specific amino acid. In this way, the tRNA recognizes the specificity of the genetic code.

The correct amino acid is attached to its proper tRNA by one of a group of enzymes called **aminoacyl-**

tRNA synthetases. One specific aminoacyl synthetase exists for every amino acid, but each synthetase may recognize more than one tRNA because there are more tRNAs (and codons) than there are amino acids. (The genetic code is described in detail in the following chapter, which focuses on translation.)

Robert W. Holley and his colleagues were the first to discover the nucleotide sequence of a tRNA. In 1965, they published the structure of the alanine tRNA in yeast (see fig. 10.17a), which ultimately led to Holley receiving a Nobel Prize in Physiology or Medicine in 1968. The average tRNA is about 80 nt long. Like rRNAs, the tRNAs are functional molecules and are not translated, which means that they also lack an ORF, translation initiation and termination codons, and 5' and 3' UTRs.

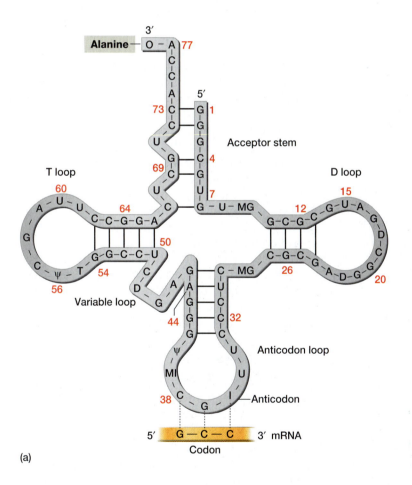

(a)

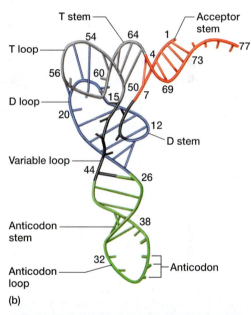

(b)

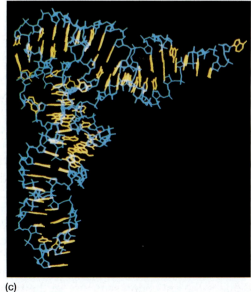

(c)

figure 10.17   Structure of yeast alanine tRNA. (a) A diagram showing the standard cloverleaf structure of the tRNA. The acceptor stem contains the 5' and 3' ends of the tRNA. The tRNA anticodon is shown paired with its complementary mRNA codon. Note the modified bases in the loops that include: ψ, Pseudouridine; I, Inosine; D, Dihydrouridine; T, Ribothymidine; MG, Methylguanosine; MI, Methylinosine. (b) A diagram showing coiling of the sugar-phosphate backbone. (c) A molecular model with bases in *yellow* and backbone in *blue.* The latter two parts of the figure (b and c) are in the same orientation.

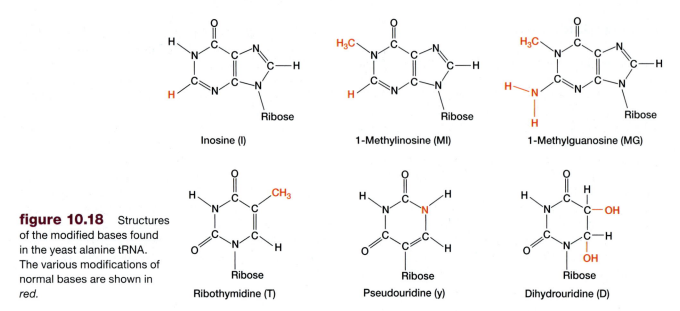

figure 10.18 Structures of the modified bases found in the yeast alanine tRNA. The various modifications of normal bases are shown in *red*.

Inosine (I)

1-Methylinosine (MI)

1-Methylguanosine (MG)

Ribothymidine (T)

Pseudouridine (ψ)

Dihydrouridine (D)

Structural Similarities of All Transfer RNAs

Transfer RNAs have several unusual properties. One obvious feature is that the tRNA contains unusual bases (fig. 10.17a). These bases include **dihydrouridine (D), inosine (I), methylguanosine (MG), methylinosine (MI), pseudouridine (ψ),** and **ribothymidine (T)** (**figure 10.18**). Presumably, these unusual bases disrupt normal base pairing and are in part responsible for the loops the unpaired bases form.

A second similarity is that all the different tRNAs have the same general shape. When the tRNAs are purified from cells, the heterogeneous mixture of all the tRNAs can form very regular crystals. This regularity of the shape of tRNAs makes sense. During translation, two tRNAs attach next to each other in a ribosome, and a peptide bond forms between the amino acids that are attached to each tRNA. Any two tRNAs must therefore have the same general dimensions as well as similar structures so that they can be recognized and positioned correctly in the ribosome.

This regularity in shape is due to the presence of a double-stranded acceptor stem and three stem-loop structures (T-loop, anticodon-loop, and D-loop) to form a cloverleaf shape (see fig. 10.17a). However, the helical twisting that results from the pairing of complementary sequences in the four stem regions produces a three-dimensional shape that ensures all tRNAs have a similar structure in the cell (fig. 10.17, panels *b* and *c*).

The aminoacyl-tRNA synthetases attach the specific amino acid to the 3'-hydroxyl of the tRNA, which is composed of the sequence 5' CCA 3'. Because each aminoacyl-tRNA synthetase attaches a specific amino acid to a tRNA, it would appear that this enzyme recognizes the anticodon sequence in the center loop of the tRNA. Because some synthetases attach the same amino acid to more than one tRNA, they must recognize more than only the anticodon sequence, which could be different in the various tRNAs. In reality, each aminoacyl-tRNA synthetase recognizes several points all over the tRNA to ensure that it is attaching the correct amino acid to the proper tRNA. The three-dimensional structure of the tRNA helps to bring the different stem loops together, which would simplify the synthetase interacting with these various structures before attaching the amino acid.

The first loop on the 3' side (the T-loop or T-ψ-C-loop because it contains the T-ψ-C-G sequence) is believed to be involved in making the tRNA recognizable to the ribosome. The ribosome must hold each tRNA in the proper orientation to check the complementarity of the anticodon of the tRNA with the codon of the mRNA.

Transfer RNA Genes

When a tRNA is originally transcribed from DNA, it is about 50% longer than its final length (70–90 nucleotides). In fact, some transcripts contain two copies of the same tRNA, or sometimes several different tRNA genes are part of the same transcript (see fig. 10.16). The original tRNA transcript also lacks the unusual bases that are found in the mature tRNA.

The initial transcript is processed down to the final size of a tRNA through the action of nucleases that remove trailing and leading pieces of RNA. An example of this sequential trimming of the RNA to produce the functional *E.coli* tRNA that binds tyrosine (tRNA[Tyr]) is shown in **figure 10.19**. In eukaryotes, a CCA trinucleotide sequence is added at the 3' end by a nucleotidyl transferase enzyme. Then the tRNA is further modified, frequently by the addition of methyl groups to the bases already in the RNA (see fig. 10.19).

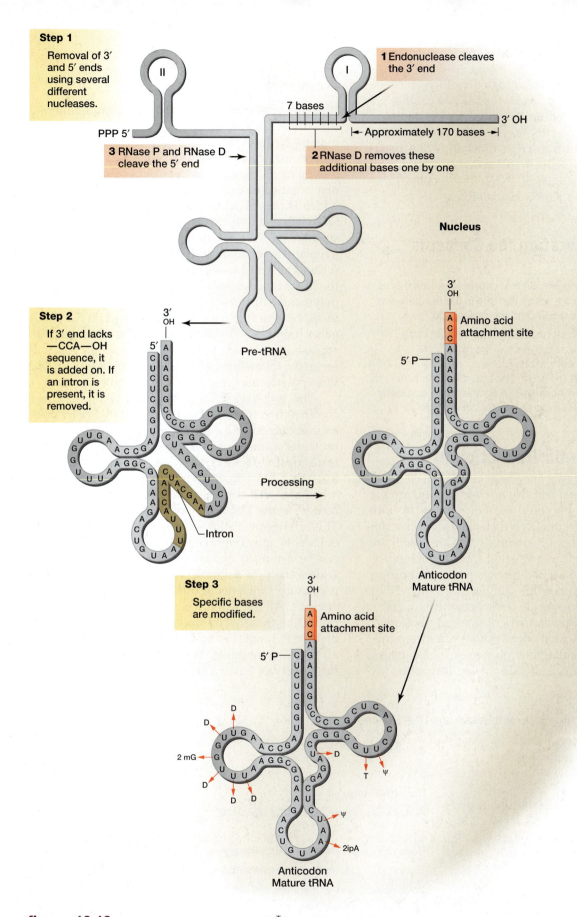

figure 10.19 Processing of the *E. coli* tRNA^Tyr. The initial RNA transcript is cleaved by an endonuclease at the 3' end (1), followed by the removal of seven additional bases at the 3' end by RNase D (2). RNase P and RNase D then cleave away the 5' end of the transcript (3). Bases are added to the 3' end if the 5' CCA 3' sequence is absent and any introns in the pre-tRNA are removed. Several bases are then modified to produce pseudouridine (ψ), 2-isopentyladenosine (2ipA), 2–o–methylguanosine (2mG), and 4-thiouridine (4tU) in the final mature tRNA.

Earlier we considered a rough definition of a gene as a length of DNA that codes for one protein. But we have just encountered an inconsistency—genes code for both tRNAs and rRNAs, yet neither is translated into a protein. Their transcripts function as final products without ever being translated. We should now think of a gene as *a segment of DNA that encodes a functional product,* either a protein or an RNA.

Restating the Concepts

▶ The tRNA is the molecule that delivers the amino acid to the ribosome for translation and is the link between the codon in the mRNA and the amino acid.

▶ The tRNA molecules undergo extensive processing, trimming of the 5' and 3' ends of the initial RNA transcript and modification of several bases, to generate the final molecule. This final tRNA interacts with a specific aminoacyl-tRNA synthetase to covalently attach the proper amino acid to the correct tRNA.

10.6 Eukaryotic Transcription

Many of the general features of transcription are the same in both prokaryotes and eukaryotes: In both cases, promoters in the DNA dictate where transcription will begin; ribonucleotides (including uracil) are used to make the transcript; and transcription of the RNA moves in a 5' → 3' direction. Significant differences exist, however, in the number of different RNA polymerases, the complexity of the promoter sequences, the relationship between transcription and translation, and the modifications made to mRNA after transcription. In addition, eukaryotic ribosomes are assembled in the nucleolus, which is located inside the nucleus.

Promoters

Whereas prokaryotes have a single RNA polymerase that transcribes all the genes, eukaryotes have three major RNA polymerases, each of which transcribes a particular class of nuclear genes. Eukaryotic RNA polymerase I transcribes three genes from the nucleolar organizer DNA (the nucleolus) as a single RNA, which is processed to yield the three largest rRNAs (28S, 18S, and 5.8S rRNAs). RNA polymerase II transcribes mRNAs, or the RNAs that will serve as templates for translation. RNA polymerase II also transcribes the genes encoding most of the snRNAs and is the likely polymerase to transcribe the miRNAs. RNA polymerase III transcribes small genes, primarily the 5S rRNA gene, the U6 snRNA gene, and all the tRNA genes (**table 10.3**). In addition, mitochondria and chloroplasts have other RNA polymerases. As you will see in the following discussion, each of the eukaryotic RNA polymerases responds to a different promoter.

The bacterial RNA polymerase recognizes and contacts the promoter, but the eukaryotic RNA polymerases have only a minimal effect on recognizing the promoter sequence in the DNA. Instead, a variety of **transcription factors,** which are proteins that bind to different DNA sequences within the promoter to form a complex that regulates RNA polymerase binding, recognize the DNA promoter for RNA polymerase. This arrangement allows another level of complexity in eukaryotic transcription, in that an array of sequences can be arranged in various combinations to form a wide variety of promoters that exhibit very precise regulatory functions at different times and in different cells. All the while, a single RNA polymerase can transcribe these genes because the RNA polymerase binds the complex of transcription factors rather than all the different promoter elements. We briefly discuss the variety of eukaryotic promoters in the following

table 10.3   Prokaryotic and Eukaryotic RNA Polymerases

Enzyme	Function
Prokaryotic	
RNA polymerase	Transcribes all genes
Primase	RNA primer synthesis during DNA replication
Eukaryotic	
RNA polymerase I	Transcribes large rRNA genes (nucleolar organizer genes)
RNA polymerase II	Transcribes protein-coding genes (mRNAs) and most snRNA and miRNA genes
RNA polymerase III	Transcribes 5S rRNA, U6 snRNA, and all the tRNA genes
Primase	RNA primer synthesis during DNA replication
Mitochondrial RNA polymerase	Transcribes genes in mitochondrial DNA
Chloroplast RNA polymerase	Transcribes genes in chloroplast DNA

section, but we will reexamine this issue in chapter 17 in the context of eukaryotic gene expression.

RNA Polymerase I and RNA Polymerase III Promoters

The simplest eukaryotic promoter is the RNA polymerase I promoter. This class of promoter is characterized by two sequence components: the *core promoter* and the *upstream promoter element* (**fig. 10.20**). The **core**

promoter is a G–C-rich sequence, except for a small A–T-rich region at the transcription start point (surrounding +1). This element is sufficient for RNA polymerase I transcription initiation; however, the presence of the **upstream promoter element** (another G–C-rich region) increases transcription efficiency.

RNA polymerase I transcription from this promoter requires several transcription factors. The **upstream binding factor** (**UBF**) binds to the upstream promoter element (see fig. 10.20). Binding of UBF assists

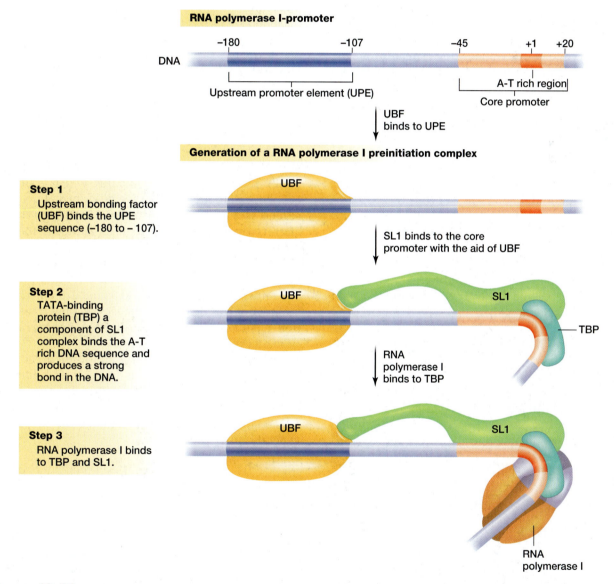

figure 10.20 The organization and binding proteins of the eukaryotic RNA polymerase I promoter. The promoter contains two sequence components, the core promoter (−45 to +20) and the upstream promoter element (−180 to −107). Both sequences are primarily composed of G-C base pairs. However, an A-T rich region surrounding the transcription initiation site (+1) is present in the core promoter. The generation of a transcription preinitiation complex for RNA polymerase I first requires an upstream binding factor (UBF) binding to the upstream promoter element. The UBF assists the SL1 protein complex to bind to the core promoter. TATA-binding protein (TBP), which is one of the components of SL1, binds the A-T rich sequence and produces a strong bend in the DNA. The RNA polymerase I binds the TBP, which positions the RNA polymerase to begin transcription at the proper location.

in another complex of transcription factors binding to the core promoter. This core promoter complex is composed of four different proteins, one of which is the **TATA-binding protein** (**TBP**). The TBP does not bind the core promoter. However, it is recognized and bound by the RNA polymerase I to generate the transcription **preinitiation complex** (**PIC**). Thus, the TBP positions the RNA polymerase such that transcription begins at the correct location. The formation of the transcription preinitiation complex appears to be required before efficient transcription of a gene can begin.

In contrast to the RNA polymerase I promoters, there are three different types of RNA polymerase III promoters (**fig. 10.21**). The first two types are rather unusual because they require sequences that are located between 55 and 80 nucleotides downstream from the transcription initiation point. In the first type of promoter, **transcription factor IIIA** (**TFIIIA**) binds to box A and helps **transcription factor IIIC** (**TFIIIC**) bind to box C. In the second type, TFIIIC binds to both box A and box C.

Both of these promoter types serve the same common purpose of binding TFIIIC, which in turn permits **transcription factor IIIB** (**TFIIIB**) to bind near the transcription initiation site. TFIIIB is composed of three proteins, one of which is TBP; in this case, TBP functions to bind RNA polymerase III at the correct transcription initiation site. Experiments have demonstrated that TFIIIA and TFIIIC do not have a direct effect on RNA polymerase III binding.

The RNA polymerase III type 3 promoter, which occurs less frequently than the type 1 and type 2 internal RNA polymerase III promoters, possesses all of the essential DNA sequences upstream of the transcription initiation site (see fig. 10.21). The **TATA box,** which has the consensus sequence of 5' TATAAA 3', is centered at approximately –25. The TATA sequence, which binds TFIIIB, is sufficient by itself for transcription initiation, but binding of accessory transcription factors to the upstream sequences will enhance TFIIIB

binding to the TATA region. The binding of proteins to the upstream sequences and to the TATA box generates a transcription preinitiation complex, which permits RNA polymerase III to bind at the proper location and initiate transcription.

RNA Polymerase II Promoters

There are two types of RNA polymerase II core promoters (**fig. 10.22**). The first type, which is termed a **TATA-less promoter,** contains two elements: One DNA sequence is termed the **initiator region** (**InR**), located at –3 to +5. The InR is composed of the +1 nucleotide, which is usually an adenine that is flanked by pyrimidines (cytosines and thymines). The second sequence is termed the **downstream promoter element** (**DPE**), which is located at approximately +28 to +34 and consists of the consensus sequence 5' AGACGTG 3'. The TATA-less promoters are usually associated with genes that are transcribed by RNA polymerase II in all cells throughout development.

The second type of RNA polymerase II promoter also contains at least two DNA sequences (see fig. 10.22). The first sequence is the InR, which is similar, or identical, to the InR that we described above. The second is a TATA box, also named the *Hogness box* after its discoverer, David Hogness. Although the consensus sequence of the TATA box is similar to the –10 region in bacterial promoters, it is located approximately 25 bp upstream of the transcription initiation site rather than centered at –10 as in bacteria. The remainder of this discussion focuses on how transcription from this second type of RNA polymerase II promoter is controlled.

RNA polymerase II is a large multiprotein complex, which consists of 12 subunits in yeast and humans. RNA polymerase II, like RNA polymerases I and III, cannot locate promoters or attach to DNA in a stable fashion by itself; instead, it must interact with several general transcription factors.

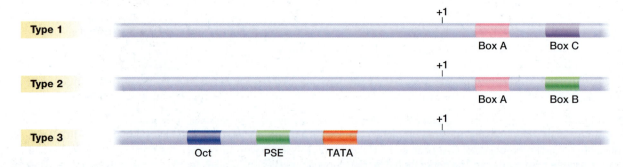

figure 10.21   The eukaryotic RNA polymerase III promoter. There are three types of RNA polymerase III promoters. The first and second both contain two promoter elements that are located downstream of the transcription initiation point (+1). Both of these promoters contain a Box A sequence, and either a Box C or Box B sequence. The third contains three sequence elements (Oct, PSE, and TATA) that are all located upstream of the transcription initiation point (+1).

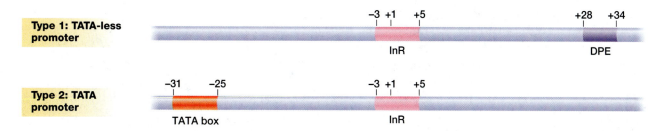

figure 10.22   Two types of the eukaryotic RNA polymerase II promoter. The first, a TATA-less promoter, contains a downstream promoter element (DPE) rather than an upstream TATA box. The second, a TATA core promoter, possesses a TATA box located approximately 25 bases upstream of the transcription initiation point. Both promoter types contain a conserved sequence (InR) that is centered on the transcription start site (+1).

As with RNA polymerase I and III, TBP plays a critical role in generating the transcription preinitiation complex for RNA polymerase II. Unlike our previous discussion, TBP directly binds to the DNA sequence in the RNA polymerase II TATA box. The bound TBP then recruits additional proteins, which are called **TBP-associated factors** (**TAFs**), to generate a protein complex that is called **transcription factor IID** (**TFIID**). One interesting aspect of TBP is that it binds the minor groove of the DNA, which causes a significant bending and opening of the DNA (**fig. 10.23**). This bending may be an important signal for other binding proteins.

One of the proteins in the TFIID complex, TAF$_{II}$250, is a histone acetyltransferase. This enzyme adds an acetyl group on the amino terminal lysine residues that are found on histones H3 and H4. As we will discuss in chapter 17, acetylation of histones is an important mechanism for altering the chromatin organization and activating transcription. It is possible that binding of TFIID leads to the acetylation of particular histones and transcription of a gene. Acetylation of these histones may then facilitate the assembly of subsequent transcription preinitiation complexes at the same gene.

Once TFIID binds to the TATA box, a cascade of other transcription factors bind to the DNA in a specific order to generate the transcription preinitiation complex (**fig. 10.24**). Binding of TFIIA either coincides or immediately follows TFIID binding. TFIIB is the next to bind and forms the DAB (TFIID, TFIIA, TFIIB) preinitiation complex. RNA polymerase II, which must be bound by TFIIF, recognizes TFIIB and binds to the DAB complex. TFIIE and TFIIH are then able to bind and form the transcription preinitiation complex. TFIIH then phosphorylates several amino acids in the carboxyl terminus of RNA polymerase II.

To activate transcription at a high level, other proteins are needed. These additional proteins are **activators,** which are specific transcription factors that bind to DNA sequences called **enhancers** (**fig. 10.25**). Enhancers are often hundreds or thousands of base pairs either upstream or downstream from the promoter. These specific transcriptional activators have

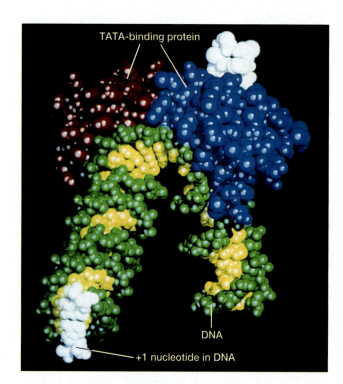

figure 10.23   Molecular space-filling model of a yeast TATA-binding protein attached to a TATA box on the DNA. The DNA sugar-phosphate backbone is *green* and the bases are *yellow.* The protein has two-fold symmetry (*red* and *blue*). Note the bending of the DNA through 80 degrees, which also opens up the minor groove of the DNA. The *upper white* atoms are the N-terminus of the TATA-binding protein; the *lower white* atoms are the first base pair (+1) at which transcription begins.

one or more protein domains (regions) that recognize DNA enhancer sequences, proteins associated with the RNA polymerase (general transcription factors), and other transcription factors. For specific transcription factors to attach to both enhancers and the polymerase machinery, which may be thousands of base pairs apart, the DNA must bend to allow them to come into the range of the polymerase.

Similar to activators and enhancers, **repressors** are proteins that can bind to silencers. **Silencers** are DNA

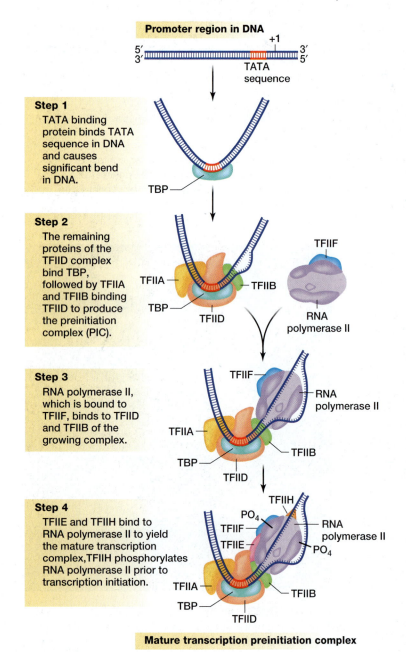

Step 1
TATA binding protein binds TATA sequence in DNA and causes significant bend in DNA.

Step 2
The remaining proteins of the TFIID complex bind TBP, followed by TFIIA and TFIIB binding TFIID to produce the preinitiation complex (PIC).

Step 3
RNA polymerase II, which is bound to TFIIF, binds to TFIID and TFIIB of the growing complex.

Step 4
TFIIE and TFIIH bind to RNA polymerase II to yield the mature transcription complex, TFIIH phosphorylates RNA polymerase II prior to transcription initiation.

Mature transcription preinitiation complex

figure 10.24   Generation of the mature eukaryotic RNA polymerase II transcription complex. The TBP protein, which is one of the subunits of TFIID, binds the TATA box (*red*). The TFIID complex recruits the TFIIA and TFIIB complexes. RNA polymerase II, which is associated with TFIIF, binds the growing complex. This complex recruits TFIIE and TFIIH, which phosphorylates RNA polymerase II to generate the mature transcription preinitiation complex.

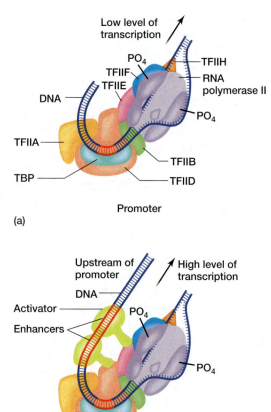

(a)

(b)

figure 10.25   The role of enhancers in eukaryotic transcription. (a) The formation of the mature transcription preinitiation complex allows for low levels of transcription of a gene. (b) Upstream of many promoters are enhancer sequences (red) that can bind activator proteins (lime green), which are DNA binding proteins that can further increase the level of transcription. One part of the green activator binds the enhancer sequences and another part (the transcriptional activation domain) binds proteins in the transcription preinitiation complex. This binding results in altered conformation of the genomic DNA that can be imagined as a loop.

sequences that are often located far upstream of the promoters and are involved in repressing transcription.

The RNA polymerase II complex contains proteins that act as mediators between activators and the polymerase holoenzyme. This complex coordination of the initiation of transcription in eukaryotes has been termed **combinatorial control;** the huge initiation complex may contain 85 or more different polypeptides. We will discuss these and other aspects of eukaryotic transcriptional regulation in chapter 17.

At this point, the mature transcription preinitiation complex is assembled and ready to begin transcription. However, only very short RNAs of 10 to 12 nucleotides in length, called **abortive RNAs,** are initially synthesized (**fig. 10.26**). To successfully transcribe a gene, one of these short RNAs must stably base pair with the DNA template and TFIIH utilizes a second activity, that of a DNA helicase to increase the size of the transcription bubble by unwinding more of the downstream DNA. This releases the transcription complex, containing RNA polymerase II and TFIIF, which uses the short base-paired RNA as a primer and progressively transcribes through the gene. **Table 10.4** summarizes the postulated roles of the general transcription factors.

It's Your Turn

We have finished discussing the critical elements of bacterial and various eukaryotic promoters. The relative locations of these sequence elements are essential for proper transcription of the corresponding genes. Problem 5 examines your understanding of the relative locations of these sequence elements.

Transcription Elongation

Once the RNA polymerase has cleared the promoter, most of the transcription initiation factors either disassociate from the RNA polymerase and fall off the DNA or remain at the promoter, which the TBP likely does. The RNA molecule then grows in a 5' → 3' direction.

As discussed in chapter 7, eukaryotic DNA is complexed with histone proteins to produce chromatin, which can interfere with transcription elongation by reducing the ability of the double-stranded DNA to unwind. It is not surprising, therefore, that part of the RNA polymerase II complex is made up of proteins that can disrupt the histones bound to the DNA.

Another recently identified component of the RNA polymerase II transcription elongation complex is the TFIIS protein. TFIIS performs two important functions in transcription. First, it increases the rate of transcription elongation, by reducing the frequency and length of time that RNA polymerase II pauses. Second, it

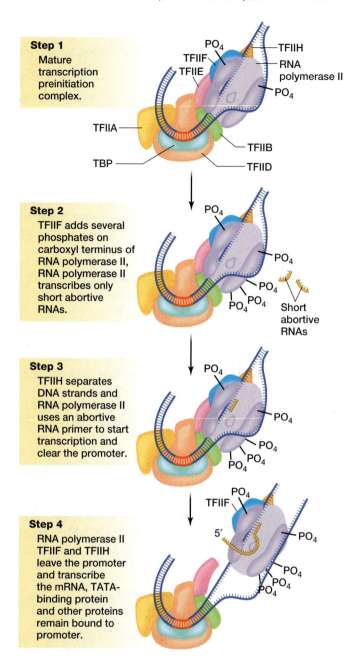

Step 1 Mature transcription preinitiation complex.

Step 2 TFIIF adds several phosphates on carboxyl terminus of RNA polymerase II, RNA polymerase II transcribes only short abortive RNAs.

Step 3 TFIIH separates DNA strands and RNA polymerase II uses an abortive RNA primer to start transcription and clear the promoter.

Step 4 RNA polymerase II TFIIF and TFIIH leave the promoter and transcribe the mRNA, TATA-binding protein and other proteins remain bound to promoter.

figure 10.26   The initiation of RNA polymerase II transcription in eukaryotes. TFIIF adds several phosphates on the carboxyl terminus of RNA polymerase II, which transcribes only short abortive RNAs (gold) that are 10 to 12 nucleotides long. The TFIIH DNA helicase activity hydrolyzes ATP to further unwind the double-stranded DNA downstream of the transcription start site and one of the abortive RNAs serves as the primer for transcription. The transcription elongation complex, RNA polymerase II, TFIIH, and TFIIF, are now cleared from the promoter and begin transcribing the entire gene. The remainder of the transcription factors either remain associated with the promoter, such as the TATA binding protein (TBP), or disassociate from the DNA.

table 10.4   Putative Roles of the General Transcription Factors of RNA Polymerase II

General Transcription Factor	Function
TFIID, TBP	Recognizes TATA box
TFIID, TAFs	Recognizes initiator element and regulatory proteins
TFIIA	Stabilizes TFIID
TFIIB	Aids in start-site selection by RNA polymerase II
TFIIE	Controls TFIIH functions; enhances promoter melting
TFIIF	Destabilizes nonspecific interactions of RNA polymerase II and DNA
TFIIH	Melts promoter with helicase activity; activates RNA polymerase II with kinase activity

Source: Data from R. G. Roeder. 1996. The role of general initiation factors in transcription by RNA polymerase II. *Trends in Biochemical Sciences,* 21:327–335.

stimulates a proofreading activity during transcription. It is thought that when RNA polymerase II incorporates an incorrect ribonucleotide, the RNA polymerase II complex backtracks and extrudes a short region of RNA containing the mismatched base (**fig. 10.27**). Unlike pausing, the RNA polymerase cannot reinitiate transcription past this extruded RNA. It is thought that TFIIS stimulates an intrinsic RNase activity in RNA polymerase II that cleaves this short single-stranded RNA. At this point, the RNA polymerase can reinitiate transcription elongation.

Transcription elongation proceeds till the RNA polymerase reaches the end of the gene. Although RNA polymerases I and III seem to have termination signals similar to rho-independent promoters in prokaryotes, termination of transcription of RNA polymerase II genes is more complex, coupled with further processing of the mRNA.

Differences Between Eukaryotic and Prokaryotic Transcription

We described a variety of differences between eukaryotic and prokaryotic transcription. Two additional significant differences are found in the coupling of transcription and translation that is possible in prokaryotes, and in the extensive **posttranscriptional modifications** that occur in eukaryotic mRNA.

In *E. coli,* translation of the newly transcribed mRNA into a protein can take place before transcription is complete (**fig. 10.28**). The prokaryotic mRNA is synthesized in the 5' → 3' direction, and translation begins near the 5' end. As soon as the 5' end of the mRNA is available, a ribosome can attach and move along the mRNA in the 5' → 3' direction, lengthening the growing polypeptide as it moves. When the first ribosome moves far enough away from the 5' end of the transcript, a second ribosome can attach and begin translation. These processes are repetitive, as electron micrographs clearly show (fig. 10.28b).

In eukaryotes, however, mRNA is synthesized in the nucleus, but protein synthesis takes place in the cytoplasm. This regional division of labor, which is not present in *E. coli* and other bacteria and archaea that lack a nucleus, is one reason that transcription and translation are not coupled in eukaryotes.

Before a eukaryotic mRNA leaves the nucleus, it is highly modified by processes that generally do not occur in bacteria. These modifications are important in helping to transport the mRNA from the nucleus and stabilizing the mRNA in the cytoplasm. Furthermore, these modifications are essential for enhancing the initiation of translation and translating the correct protein from an RNA. These eukaryotic mRNA modifications can also increase the complexity of proteins that are produced from a gene, as is discussed later on.

Restating the Concepts

▸ Whereas prokaryotes have one RNA polymerase, eukaryotes have three RNA polymerases (I, II, and III) that transcribe different classes of nuclear genes.

▸ Each RNA polymerase has a different promoter, with RNA polymerases II and III transcribing from two and three different classes of promoters, respectively.

▸ Whereas bacterial RNA polymerase binds the DNA at the promoter, eukaryotic RNA polymerases do not recognize the promoter. Instead, a variety of transcription factors bind the promoters, including the TATA-binding protein, which binds all three RNA polymerases and correctly positions them to begin transcription at the correct site.

▸ In prokaryotes, transcription and translation are coupled (occur simultaneously), whereas eukaryotic transcription occurs in the nucleus and translation occurs in the cytoplasm (transcription and translation are uncoupled). Before a eukaryotic mRNA can be translated, it must be processed and then transported out of the nucleus.

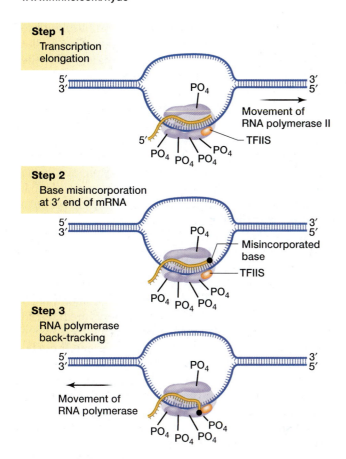

Step 1
Transcription elongation

Movement of RNA polymerase II

TFIIS

Step 2
Base misincorporation at 3′ end of mRNA

Misincorporated base

TFIIS

Step 3
RNA polymerase back-tracking

Movement of RNA polymerase

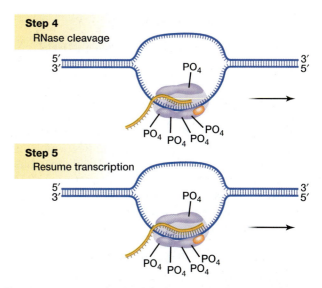

Step 4
RNase cleavage

Step 5
Resume transcription

figure 10.27   Proofreading by the RNA polymerase II complex. As RNA polymerase II transcription elongation proceeds, an incorrect base (black filled circle) may be inserted into the 3′ end of the growing mRNA. This causes the RNA polymerase II to pause and then back up along the DNA template in a 5′ → 3′ direction, which extrudes the 3′ end of the mRNA out of the RNA polymerase II active site. The RNase activity of the RNA polymerase, which is stimulated by the TFIIS subunit, cleaves several nucleotides off the 3′ end of the RNA, including the incorrect base. Removal of this short RNA segment permits the RNA polymerase II molecule to continue synthesizing the RNA in a 5′ → 3′ direction.

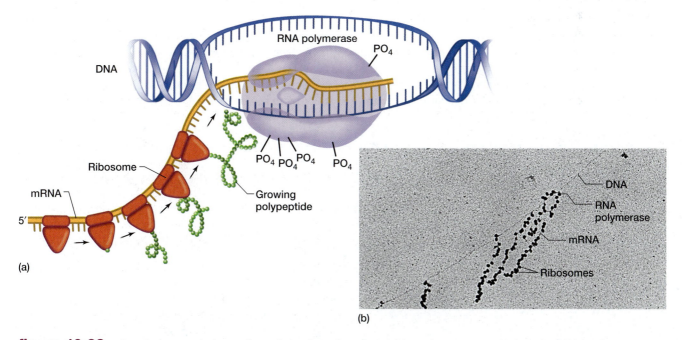

(a)

(b)

figure 10.28   Coupled transcription and translation in prokaryotes. (a) In prokaryotes, translation of mRNA by ribosomes begins before transcription is complete. Ribosomes attach to the 5′ end of the growing mRNA strand and translate along the elongating RNA. When the first ribosome moves from the 5′ end, a second ribosome can attach, and so on. (b) Electron micrograph of events diagrammed in (a). The growing polypeptides cannot be seen in this preparation. Magnification 44,000×.

10.7 Posttranscriptional Modifications in Eukaryotes

Whereas most prokaryotic transcripts contain information from several genes, virtually all transcripts from higher eukaryotes contain the information from just one gene. The major exception is the presence of several tRNAs or rRNAs in a single RNA molecule, which is processed (RNA is cleaved and some bases are modified) to produce the final functional RNAs. A different form of processing is required to produce eukaryotic mRNAs. Transcription of the eukaryotic gene results in a **primary transcript** that must be modified before it reaches the mature state. Primary transcripts that are synthesized by RNA polymerase II usually undergo three different modifications before being transported into the cytoplasm: modifications to the 5' end and the 3' end and removal of intervening sequences. We refer to all of these changes as posttranscriptional modifications.

The Nucleolus in Eukaryotes

As we discussed in chapter 3, the nucleolus is a structure within the nucleus that contains the nucleolar organizer region in the chromosomes. The nucleolar organizer is the location of the rRNA genes, and its chromosomal location differs between species. For example, the fruit fly, *Drosophila melanogaster,* has the nucleolar organizer on the sex (X and Y) chromosomes, whereas they are located on chromosomes 13, 14, 15, 21, and 22 in humans. It is in the nucleolus that the ribosomal proteins and rRNAs are assembled into the functional ribosomes. Although the ribosomal proteins are synthesized in the cytoplasm and migrate to the nucleolus, the rRNA genes are transcribed and processed in the nucleolus before ribosomal assembly.

Eukaryotes have four different rRNAs in the ribosome, compared with three in prokaryotes. The smaller ribosomal subunit (usually 40S, compared with the *E. coli* 30S) has an 18S rRNA molecule, and the larger subunit (usually 50S relative to the *E. coli* 40S) has 5S, 5.8S, and 28S rRNAs. The three larger rRNAs are transcribed as part of the same precursor RNA, which corresponds to 45S (**fig. 10.29**). Analogous to the prokaryotes, transcribing all three eukaryotic rRNAs on a single transcript ensures that they are expressed in the proper ratio. After this precursor RNA is transcribed, it is processed by trimming the 5' and 3' ends to produce the three mature rRNAs (see fig. 10.29).

Like tRNAs, some of the ribonucleotides in the rRNAs are also modified: some uridines are converted to pseudouridines, and some ribose sugars are methylated. These conversions also take place in the nucleolus, orchestrated by particles composed of small RNA segments and protein. The RNA segments are referred to as **small nucleolar RNAs** (**snoRNAs**) and, when combined with protein, are referred to as **small nucleolar ribonucleoprotein particles** (**snoRNPs**). Each different snoRNP has a snoRNA that is complementary to the region surrounding the nucleotide to be modified. Thus, sites for modification are chosen based on complementarity to a snoRNA, which then somehow directs the modification to take place.

Because a significant amount of protein translation must occur in a cell, large quantities of ribosomes and rRNAs are necessary. This quantity is generated by having multiple copies of the rRNA genes. For example, the fruit fly, *Drosophila melanogaster,* has about 130 copies of the repeat unit that contains the three larger rRNA genes. In the nucleolar organizer, an untranscribed region of spacer DNA separates each repeat of the large rRNA gene (**fig. 10.30a**). Using electron microscopy, the polarity of transcription is evident from the short RNA at one end of the transcribing

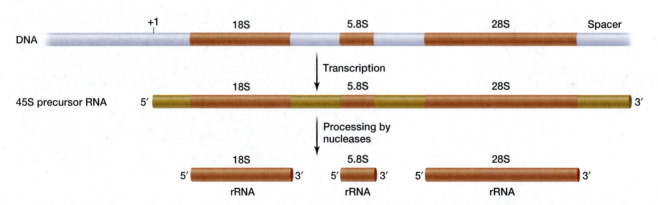

figure 10.29 Expression of the large eukaryotic rRNAs. The eukaryotic large ribosomal RNA genes (28S, 18S, and 5.8S) are transcribed as a single precursor RNA from a single promoter. This 45S precursor RNA is processed by nucleases to produce the three mature rRNAs.

segment and the long RNA at the other end, with a uniform gradation between (fig. 10.30b). Notice that many RNA polymerases are transcribing each region at the same time. The regions between the transcribed DNA segments are the spacer DNA regions.

Processing of RNA Polymerase II Transcripts

RNA polymerase II initially transcribes a **pre-mRNA,** or primary transcript, that must undergo several modifications before it is transported out of the nucleus as an mRNA that can be translated. These modifications, which include altering the 5' and 3' ends of the mRNA and removing the introns, are important for efficient export of the mRNA out of the nucleus, for maintaining the stability of the mRNA, and for efficient translation initiation.

5' Cap

At the 5' end of polymerase II transcripts, 7-methylguanosine is added in the "wrong" direction, 5' → 5' by the enzyme guanylyl transferase (**fig. 10.31**). This **5' cap** is linked to the +1 nucleotide of the RNA through three phosphate groups, rather than the single phosphate in the remainder of the RNA molecule. This process, termed *capping* of the RNA, occurs on the primary transcript shortly after transcription initiation. Thus, the mature mRNA does not possess a free 5' end. Nevertheless, we still call the capped end of the mRNA the 5' end as it corresponds to the start of the mRNA.

This 5' cap serves four purposes. First, it protects the RNA from a 5' → 3' degradation, which increases the lifetime of the mRNA. Second, the 5' cap is essential for the proper removal of the first intron in the pre-mRNA. Third, it is important for the transport of

figure 10.30　Transcription of the large eukaryotic rRNA genes. (a) Diagram of RNA polymerase I transcribing the large 45S rRNA gene cluster (fig. 10.29). The polarity of transcription is represented by the length of the RNA that extends from the DNA template, with the shorter RNAs closer to the transcription start site. The copies of the large rRNA gene are separated by spacer DNA. (b)Transcription of the large rRNA genes in the newt, *Triturus.* The polarity of transcription (progressing from small to large transcripts), as well as the spacer DNA (*thin lines* between transcribing areas), is clearly visible. Magnification 18,000×.

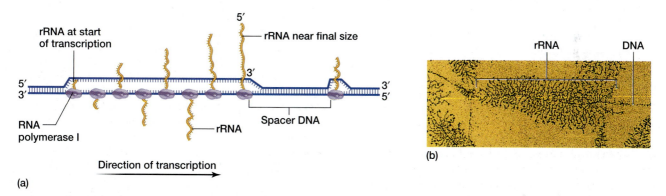

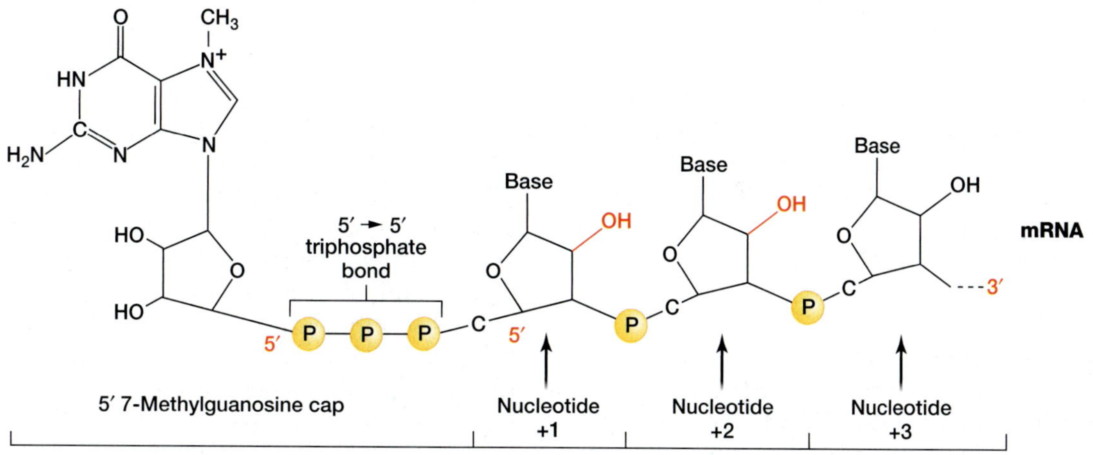

figure 10.31　The 5' cap on mRNA. A 7-methyl guanosine is added through a 5' → 5' triphosphate linkage to the 5' end of the +1 nucleotide on a eukaryotic mRNA. This mature mRNA does not have a free 5' end. The first nucleotide transcribed by RNA polymerase II (designated +1) is actually the second nucleotide in the mature mRNA.

10.1 applying genetic principles to problems

Determining Which Eukaryotic RNA Polymerase Transcribes a Gene

The three eukaryotic RNA polymerases each transcribe specific sets of genes. However, when microRNAs were identified, they did not fall into one of the previous categories of RNAs that were transcribed by an RNA polymerase. How can we determine if a new class of RNA is transcribed by one of the previously characterized RNA polymerases or by a yet to be identified RNA polymerase? Two types of experiments can be performed that will reveal the identity of this RNA polymerase.

The first type of experiment examines a specific property of all three RNA polymerases—their sensitivity to α-amanitin. **α-Amanitin** is a compound that is expressed in several poisonous mushrooms of the *Amanita* genus. Each of the three RNA polymerases exhibits a different sensitivity to α-amanitin. At very low α-amanitin concentrations (0.02 μg/mL), RNA polymerase II activity is inhibited 50% and RNA polymerases I and III activities are unaffected. At higher concentrations (20 μg/mL), RNA polymerase II activity is completely inactivated, RNA polymerase III activity is 50% inactivated, and RNA polymerase I activity is unaffected. At very high α-amanitin concentrations (200 μg/mL), RNA polymerase II and III activities are completely inactivated and RNA polymerase I activity is unaffected.

Cells that express the gene of interest are isolated from the organism and grown in culture. Different concentrations of α-amanitin are added to the cells and grown for several hours. The RNA is isolated from the cells and the RNA of interest is detected by either PCR amplification or Northern hybridization (see chapter 12).

Assume that we are studying gene X, which is expressed in mouse epithelial cells. We can isolate and grow mouse epithelial cells in several different culture dishes that contain 0, 0.02, 20, and 200 μg/mL of α-amanitin, respectively. We isolate RNA from cells in each of the treatments and determine if gene X RNA is present. If we find it only in the culture lacking α-amanitin, what is the likely RNA polymerase that transcribes gene X?

Because RNA polymerase I acitivty is not affected by any of these α-amanitin concentrations, RNA polymerase I would transcribe gene X in all four culture conditions. RNA polymerase III is completely inactivated at 200 μg/mL α-amanitin and 50% inactivated at 20 μg/mL. We would expect RNA polymerase III to transcribe gene X RNA at the 0 and 0.02 μg/mL α-amanitin concentrations, at a reduced amount at 20 μg/mL, and unable to transcribe gene X in the presence of 200 μg/mL α-amanitin. If RNA polymerase II transcribed gene X, then the absence of α-amanitin would permit gene X transcription, 0.02 μg/mL α-amanitin would permit reduced levels of transcription, and the other two concentrations would completely prevent gene X transcription. Because this is closest to the pattern that we observed, we would deduce that gene X is transcribed by RNA polymerase II.

The second type of analysis is to perform in vitro transcription of the gene using the purified RNA polymerases. In this type of experiment, the gene would only be transcribed when the correct RNA polymerase is added to the reaction containing the genomic DNA, four ribonucleotides, and the mixture of transcription factors.

In the case of the miRNAs, an entirely different approach was used. Characterization of the miRNA precursor (pri-miRNA) revealed that it is both capped at the 5' end and polyadenylated at the 3' end. Because only RNA polymerase II transcripts contain these post-transcriptional modifications, it was assumed that the miRNA genes are transcribed by RNA polymerase II. By using one or more of these approaches, it is possible to identify the RNA polymerase that transcribes any gene, even without any knowledge of the promoter sequences.

---

the mRNA out of the nucleus. Fourth, the ribosome recognizes and binds to the cap. Without the 5' cap, ribosome binding and translation efficiency would be significantly reduced.

3' Poly(A) Tail

At the 3' end of polymerase II transcripts, a **poly(A) tail** is added. In mammals, the primary transcript is cleaved 15 to 30 nucleotides after the 5' AAUAAA 3' polyadenylation consensus sequence by an endonuclease (**fig. 10.32**). Although, plants also utilize the AAUAAA consensus sequence, they exhibit greater variability in this sequence than mammals. In contrast, yeast rarely use the AAUAAA sequence, opting instead for a very diverse collection of polyadenylation signal sequences. The cleavage reaction of the RNA requires a complex of several proteins, which includes RNA polymerase II and poly(A) polymerase. After cleavage of the RNA, the enzyme **poly(A) polymerase** adds from 15 to over 250 adenines to the new 3' end of the RNA. The poly(A)-binding protein binds to the growing poly(A) tail to stimulate the polyadenylation reaction and to ensure that the proper number of adenines are added to the 3' tail.

The poly(A) tail is required for the proper removal of the last intron in the pre-mRNA. The 3' tail also enhances the ability of the mRNA to be translated and improves the stability of the mRNA in the cytoplasm.

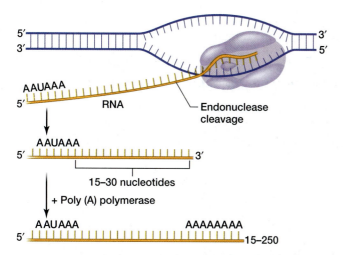

figure 10.32   A poly(A) tail is added to the 3' end of most RNA polymerase II transcripts. The primary transcript is cleaved by an endonuclease 15 to 30 nucleotides after the AAUAAA consensus sequence. Poly(A) polymerase then adds 15 to 250 adenines to the 3' end of the cleaved RNA.

However, the poly(A) tail is a dynamic sequence in the cytoplasm, as it continually increases and decreases in length. Cytoplasmic RNases are constantly removing adenines from the 3' tail, and poly(A) polymerase is continually trying to replace the lost adenines. However, the poly(A) tail is eventually lost, which targets the mRNA for rapid degradation. As you will see in chapter 12, this poly(A) tail can also be exploited to purify mRNA for recombinant DNA techniques.

Introns

In eukaryotes, some segments of DNA are transcribed into RNA but never appear in the final mature RNA species. This loss of RNA sequences occurs during the generation of mRNAs, rRNAs, and tRNAs. Introns are also found in the rRNA and tRNA genes of both bacteria and archaea.

These **intervening sequences,** or **introns,** are transcribed RNA sequences that are removed in the nucleus before the mature RNA is transported into the cytoplasm (see **fig. 10.33**). Phillip Sharp and Richard Roberts independently discovered introns in 1977, which earned them the 1993 Nobel Prize in Physiology or Medicine (**fig. 10.34**). The segments of the gene between the introns, which are transcribed and present in the mature RNA molecule are termed **exons.**

The nucleotide sequence and corresponding amino acid sequence for the β-globin gene with introns appears in figure 10.33. The results of intron removal are clear when the mRNA, which lacks its introns, is hybridized with the original gene (**fig. 10.35**). The DNA forms double-stranded structures with the exons in

the mRNA. The introns in DNA have nothing to pair with in the mRNA, so they form single-stranded loops. While we usually think of introns being removed to generate mRNA, they also occur in eukaryotic tRNA and rRNA genes.

For introns to be removed, the ends of the exons must be brought together and joined in a process called *splicing*. At least two types of splicing occur, although they are related: self-splicing and protein-mediated splicing.

Self-Splicing

In 1982, Thomas Cech and his colleagues discovered self-splicing by RNA. Working with an intron in the 35S rRNA precursor in the ciliated protozoan, *Tetrahymena,* they found that the intron was removed in vitro with no proteins present. However, a guanine-containing nucleotide (GMP, GDP, or GTP) was required to be present for the self-splicing activity. **Figure 10.36** diagrams how self-splicing occurs. In this case, the intron acts as an enzyme. We call an RNA with enzymatic properties a **ribozyme.**

During self-splicing, the U–A bond at the 5' end of the exon–intron junction is broken. The freed A at the 5' end of the intron creates a phosphodiester bond with the free guanine nucleotide (GTP). The U that is now free at the 3' end of the exon displaces the G at the 3' side of the intron, which connects the 5' exon with the 3' exon and releases the linear intron (see fig. 10.36).

Because all bonds are reversible transfers (transesterifications) rather than new bonds, no external energy source is required. Self-splicing introns of this type are called **group I introns.** An extensive secondary structure of RNA stem loops must form for intron removal.

Although the first enzymatic activity of the ribozyme is its own removal, the secondary structure of the released intron has the ability to catalyze additional reactions. The *Tetrahymena* ribozyme catalyzes transesterification and the hydrolytic cleavage of RNA molecules into two parts. Ribozymes can also perform other functions, including peptide bond formation. Currently, at least seven different classes of ribozymes are known, based on their enzymatic properties. These ribozymes can be modified in the laboratory for specific purposes, such as the cleavage of specific RNA sequences, which has the potential to be a clinical treatment of genetic diseases.

Self-splicing has also been found in genes in the yeast mitochondria. These introns are referred to as **group II introns** because they use a mechanism of splicing that does not require an external nucleotide. Instead, cleavage of the phosphodiester bond at the 5' end of the intron allows the first nucleotide in the intron (G) to be available to form a phosphodiester bond with an adenosine (A) in the intron, which results in the formation

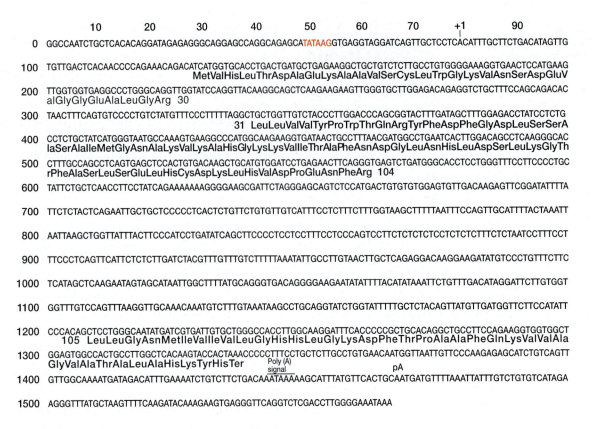

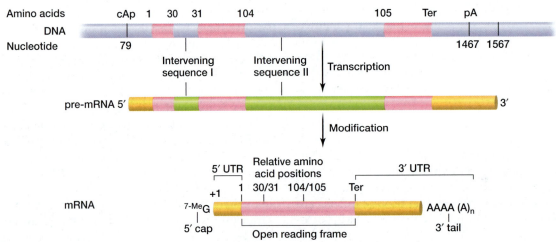

figure 10.33 Nucleotide sequence of the mouse β-globin major gene. The coding DNA strand is shown; the TATA sequence is in red; +1 (position 79) indicates the transcription initiation site; the poly (A) signal sequence is labelled and marked with a bar; pA indicates the start of the poly-A tail (position 1467); numbers inside the sequence are adjacent amino acid positions; Ter is the translation termination codon (position 1334). The three-letter abbreviations (e.g., Met, Val, His) refer to amino acids (see chapter 11). Below the sequence is a schematic of the gene, the pre-mRNA and the mRNA, showing the positions of the introns (green) and exons and the corresponding positions of the +1 nucleotide, the beginning and end of the open reading frame, and the start of the poly(A) tail (pA). Notice the relationship of the 5' and 3' UTRs and open reading frame relative to the pre-mRNA and the genomic DNA organization.

figure 10.34   Phillip A. Sharp (1944–) and Richard J. Roberts (1943–) independently discovered the existence of introns in DNA.

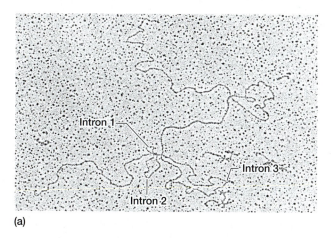
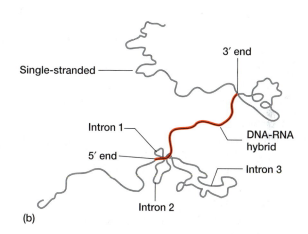

(a)   (b)

figure 10.35   The adenovirus mRNA hybridized with its DNA. Three introns are visible as single-stranded DNA loops that lack complementary sequences in the mRNA. (a) Electron micrograph, (b) explanatory diagram.

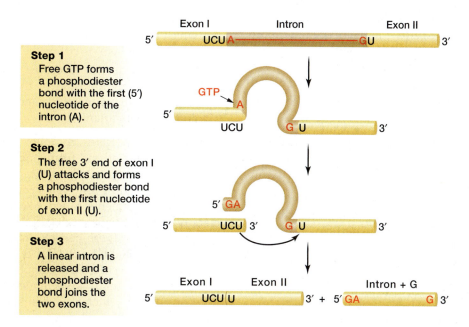

figure 10.36   Self-splicing of an rRNA precursor in *Tetrahymena*. This group I intron requires an external GTP molecule to form a phosphodiester bond with the 5' adenosine of the intron. The free 3' end of exon I (U) then attacks the 5' nucleotide of exon II (U) to release a linear intron and generate a phosphodiester bond between the two exons.

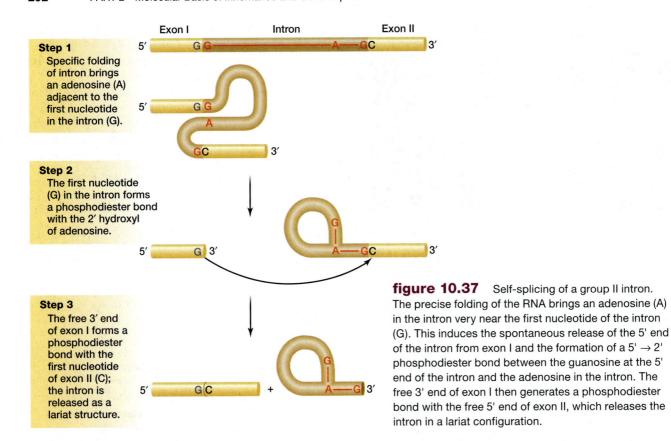

Step 1
Specific folding of intron brings an adenosine (A) adjacent to the first nucleotide in the intron (G).

Step 2
The first nucleotide (G) in the intron forms a phosphodiester bond with the 2' hydroxyl of adenosine.

Step 3
The free 3' end of exon I forms a phosphodiester bond with the first nucleotide of exon II (C); the intron is released as a lariat structure.

figure 10.37 Self-splicing of a group II intron. The precise folding of the RNA brings an adenosine (A) in the intron very near the first nucleotide of the intron (G). This induces the spontaneous release of the 5' end of the intron from exon I and the formation of a 5' → 2' phosphodiester bond between the guanosine at the 5' end of the intron and the adenosine in the intron. The free 3' end of exon I then generates a phosphodiester bond with the free 5' end of exon II, which releases the intron in a lariat configuration.

of a lariat structure (**fig. 10.37**). This lariat requires the ribose of the adenosine to make three phosphodiester bonds (**fig. 10.38**). The 5' and 3' carbons of adenosine's ribose are already involved in 5' → 3' phosphodiester bonds with the adjacent ribonucleotides. The third bond is between the 5' carbon of the guanosine at the beginning of the intron and the 2' carbon in the adenosine. The presence of the 2'-hydroxyl in the ribose sugar permits the formation of this 5' → 2' phosphodiester bond. Producing the free 3' end of the exon allows it to form a phosphodiester bond with the 5' carbon of the first nucleotide in the next exon. This joins the two exons and releases the lariat-shaped intron.

Protein-Mediated Splicing (the Spliceosome)

Eukaryotic nuclear pre-mRNAs also have their introns removed by way of a lariat structure, just as in group II introns. However, the removal of these nuclear introns requires highly conserved sequences in the pre-mRNA and several different RNA–protein particles.

Figure 10.39 shows the three consensus sequences in the nuclear mRNA's primary transcript that identify the majority of introns. We can summarize the sequences as follows:

- First is the splice donor, which includes the 3' end of the upstream exon and the 5' side of the intron. The splice donor consensus sequence contains a highly conserved 5' GU 3' sequence at the 5' end of the intron.

The consensus 5' AG 3' at the 3' end of the exon and the consensus 5' AAGU 3' following the conserved GU help to further define the proper splice site.

- Second is the splice acceptor, which includes the 3' end of the intron and the 5' end of the following exon. The splice acceptor consensus sequence contains an invariant 5' AG 3' at the 3' end of the intron, which is preceded by a pyrimidine-rich tract and a cytosine that helps to define the splice recipient site.

- The third sequence is located entirely within the intron. The rightmost A of the 5' UACUAAC 3' sequence will become the branch point of the lariat and is also invariant.

The protein–RNA complex required for the removal of the introns from nuclear pre-mRNA is called a **spliceosome**. The spliceosome is a 40–60S complex, depending on the organism, that consists of several components called **small nuclear ribonucleoproteins,** abbreviated as **snRNPs** and pronounced "snurps." Five of these particles take part in splicing and are designated U1, U2, U4, U5, and U6.

Each of these snRNPs is composed of one or more proteins and a small RNA molecule (*small nuclear RNA* or *snRNA*), which ranges in size from 100 to 215 bases. The snRNPs and their associated snRNAs are located in 20 to 40 small regions in the nucleus called *speckles* because of their appearance in the fluorescent microscope.

The sequence of each of the snRNAs has been determined, and they contain regions that are complementary to sequences in the exons, the intron, or other snRNAs (**table 10.5**). A variety of experiments have demonstrated that the base pairing between the snRNAs and the pre-mRNA transcript are essential for proper splicing. The following sequence of events is deduced from a variety of molecular analyses and is summarized in **figure 10.40**.

1. The U1 snRNP binds at the 5' site of the intron and the U2 auxiliary factor (U2AF) binds to the pyrimidine-rich tract and the 3' splice junction.
2. The U1 snRNP and U2AF help the U2 snRNP bind to the branch site sequence.
3. The U4, U5, and U6 snRNPs form a single particle that binds to the RNA with U5 snRNP, displacing the U1 snRNP at the 5' junction site and U6 snRNP binding to the U2AF.
4. The U4 snRNP releases, which causes the U5 snRNP to bind the exon at the 3' junction site and allows the U6 snRNP to bind to both the 5' splice junction and the U2 snRNP.
5. The U6-U2–snRNP complex catalyzes the transesterification reaction that generates the lariat structure.
6. The U5 snRNP binds the two exons together, allowing the splice to be completed as the lariat is removed.

Notice that the splicing reaction is not initiated until the U2 and U6 snRNPs bring the two junctions close together and catalyze the transesterification.

Over 98% of introns undergo this splicing process. The remainder of the introns are identified by GC—intron—AG (approximately 1% of the introns) or AU—intron—AC (less than 0.1% of the introns) conserved sequences at the ends of the intron. In these rarer classes, different snRNPs are required because the complementary RNA sequences are slightly different.

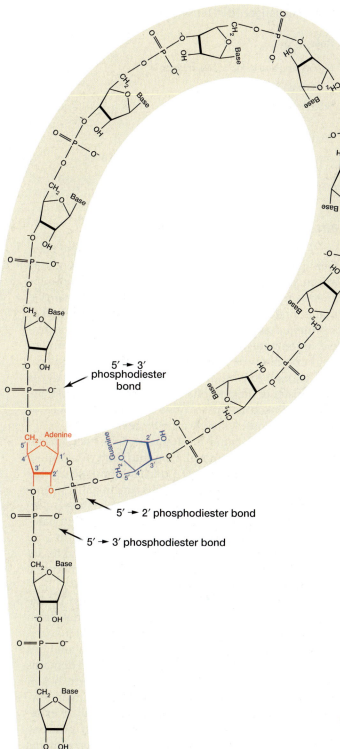

figure 10.38   Structure of nucleotides in the lariat intron. The lariat branch point adenosine within the intron (see fig. 10.37) is shown in red. To form the lariat, this adenosine is linked by three phosphodiester bonds: the two normal 5' → 3' bonds and the lariat forming 5' → 2' phosphodiester bond with the first nucleotide of the intron, guanosine (in blue).

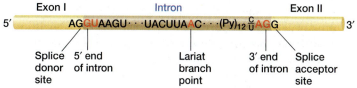

figure 10.39   The three consensus sequences that are essential for intron splicing are the 5' and 3' splice junction sites and the branch point. Letters in red (GU, A, AG) represent invariant bases in all introns, while Py represents either pyrimidine. The last A (in red) of the UACUAAC sequence is the lariat branch point that will form a 5'→ 2' phosphodiester bond with the 5' end (G) of the intron in the lariat.

table 10.5   The Five Small Nuclear RNAs Involved in Nuclear Messenger RNA Intron Removal

snRNA	Partial Sequence	Complementarity	Role
U1	3'-UCCAUUCAUA	5' end of intron	Recognizes and binds 5' site of intron
U2	3'-AUGAUGU	Branch point of intron	Binds branch point of intron
U4	3'-UUGGUCGU . . . AAGGGCACGUAUUCCUU	U6	Binds to (inactivates) U6
U5	3'-CAUUUUCCG	Exon 1 and exon 2	Binds to both exons
U6	3'-CGACUAGU . . . ACA	U2, 5' site	Displaces U1 and binds 5' site and U2 at branch point

Source: Reprinted with permission, from the *Annual Review of Genetics*, Volume 28 © 1994 by Annual Reviews www.AnnualReviews.org.

In these cases, U11 and U12 snRNPs replace the U1 and U2 snRNPs, respectively. Because the splicing mechanism described earlier begins with the binding of U1 snRNP to the 5' splice junction, the less common introns require binding of U11 snRNP to initiate splicing.

Besides producing the final contiguous mRNA sequence, splicing is also important in mRNA transport. Only spliced mRNA is transported from the nucleus to the cytoplasm where it is translated; unspliced or partially spliced RNAs do not efficiently reach the cytoplasm and are not translated. This requirement allows a cell to reduce the production of nonfunctional proteins, which would waste a significant amount of cellular energy.

Although we separated RNA polymerase II transcription and pre-mRNA processing into different sections, they are interrelated. It appears that capping and splicing occur during transcription elongation. Some data also suggest that the enzymes involved in pre-mRNA processing associate with the largest RNA polymerase II subunit. The relationship between these different processes is further observed by the requirement of the 5' cap to efficiently splice out of the first intron and the need for polyadenylation before the final intron can be spliced out. By making these processes interdependent and having them occur in association with transcription elongation, the cell increases the likelihood that the only mRNAs that reach the cytoplasm to be translated are ones that have been completely transcribed and properly processed.

One other mode of protein-mediated intron removal is known. Nuclear tRNAs have introns that are not self-splicing but are removed by an endonuclease; the exons are subsequently joined by a ligase. Archaea seem to have this type of intron.

It's Your Turn

Processing the primary transcript is essential for correct expression of the eukaryotic mRNA. Test your understanding of the different processing events in problem 14.

Intron Function and Evolution

Since the discovery of introns, geneticists have been trying to figure out why they exist. Several views have arisen. Walter Gilbert suggested in 1977 that introns separate exons (coding regions) into functional domains—that is, one or more adjacent exons encode a portion of a protein that possesses a specific function. For example, an enzyme that phosphorylates a specific protein might possess one region that binds the substrate protein, another that binds ATP, and a third that catalyzes the transfer of the phosphate from ATP to the substrate. The gene encoding this enzyme therefore might have at least three exons, with one exon for each activity.

By recombinational mechanisms, or by excluding an exon during intron removal, **exon shuffling** would allow the rapid evolution of new proteins whose structures would be conglomerates of various functional domains. By changing a single exon, the processes of natural selection could generate a new enzyme that phosphorylates a different substrate protein. Gilbert calculated in 1990 that all proteins in eukaryotes can be accounted for by as few as 1000–7000 exons; all proteins may therefore be conglomerates of this primordial number.

Gilbert's exon-shuffling view is supported by the increasing amount of genomic DNA sequence data from a variety of organisms and the protein structural analysis of the encoded proteins. Combining both analyses reveals a pattern of exons coding for functional domains of a protein. For example, the second of three exons of the globin gene binds heme. Similarly, the human low-density lipoprotein (LDL) receptor is a mosaic of exon-encoded modules shared with several other proteins. Gilbert's exon-shuffling hypothesis, however, remains controversial.

A second question arises about when introns first appeared in evolution. In 1978, W. Ford Doolittle proposed the **introns-early view**, which suggests that introns were present in the earliest ancestors of modern eukaryotes. If this is true, then different organisms that have related genes should possess introns in the same location. Genomic DNA sequencing has revealed many examples where flies, mice, and humans all

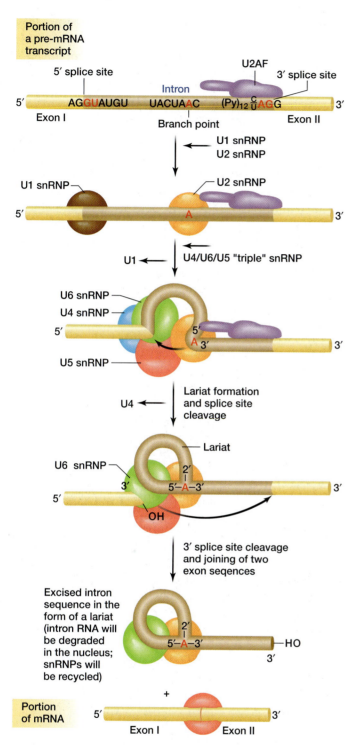

figure 10.40 Sequence of steps involving U1, U2, U4, U5, and U6 snRNPs in the removal of introns in RNA polymerase II RNAs. First, U1 binds to the 5' end of the intron and U2AF binds to the 3' end. The U2 snRNA basepairs with the branch point consensus sequence. The U4/U5/U6 snRNP complex causes the release of U1 and brings the two exons together. The release of U4 activates the remaining snRNPs to cleave the 5' end of the exon and catalyze the formation of the 5' → 2' phosphodiester bond and generate the intron lariat. The U5 snRNP then drives the release of the intron lariat and the formation of the phosphodiester bond between the two exons.

contain introns at the same location in related genes. This view is also consistent with the opinion that the original genetic material was RNA. In this "RNA world," introns arose as part of the genetic apparatus; they were the first enzymes (ribozymes).

Why then would prokaryotes, which have introns in tRNA and rRNA genes, lack introns in the mRNA genes and yeast contain only limited numbers of mRNA genes with introns? It is possible that prokaryotes and lower eukaryotes were under strong selective pressure to minimize the size of their genomes to reduce the energy required for DNA replication and transcription.

Others failed to believe this argument and proposed an **introns-late view.** In this model, introns were not introduced until after prokaryotes and eukaryotes split in evolution. Their introduction in this relatively later eukaryotic ancestor would still be in agreement with the conserved location of introns in related genes between very different organisms.

Why would introns be introduced later in evolution? Some suggest that introns give the organism the ability to adapt quickly to new environments by an exon-shuffling type of mechanism. Alternatively, introns may actually be invading "selfish DNA," DNA that can move from place to place in the genome without necessarily providing any advantage to the host organism. In some cases, important transcriptional regulatory sequences, enhancers and silencers, are present within some introns.

Evidence for the introns-early hypothesis includes the discovery of several introns in bacteriophage genes and the tRNA and rRNA genes in archaea. An intron was even identified in a tRNA gene in seven species of cyanobacteria (bacteria formerly called blue-green algae). This intron was suspected to exist because it occurred in the equivalent chloroplast gene; the chloroplast evolved from an invading cyanobacterium. These findings have been used to support both the introns-early and introns-late view. The introns-early supporters say this evidence confirms that introns arose before the eukaryotes–prokaryotes split. Introns-late supporters say they expect to see some introns in prokaryotes because of the mobility these bits of genetic material have.

Both the introns-early and the introns-late views may be correct. It is possible that introns arose early, were lost by the prokaryotes (which prioritized small genomes and rapid, efficient DNA replication), and later evolved to produce exon shuffling in eukaryotes.

Restating the Concepts

▶ The three large eukaryotic rRNAs are initially transcribed as a single RNA that is processed by trimming the 5' and 3' ends to release each rRNA and then some of the bases are modified, similar to tRNAs.

▶ The 5' end of eukaryotic pre-mRNA is modified by adding a 7-methylguanosine in a 5' → 5' triphosphate linkage with the +1 nucleotide of the RNA transcript. This 5' cap is important for ribosome binding and efficient translation, to protect the mRNA from degradation, to correctly splice the first intron, and to transport the mRNA out of the nucleus.

▶ The 3' end of the pre-mRNA transcript in eukaryotes is cleaved 15 to 30 nucleotides after the consensus 5' AAUAAA 3' sequence, and poly(A) polymerase adds approximately 15 to over 250 adenosines to this new 3' end. This poly(A) tail is involved in protecting the mRNA from degradation and correctly splicing the final intron.

▶ Splicing is the processing event that removes RNA sequences from the primary transcript to produce the final mRNA. There are two types of self-splicing introns. Group I introns utilize a free GTP to bond with the free 5' end of the intron. Group II introns do not require a free GTP, as the free 5' end of the intron forms a 5' → 2' phosphodiester bond with an adenosine in the intron.

▶ Protein-mediated splicing involves small RNAs (snRNAs), which are complementary to consensus sequences that flank and are within the intron, and snRNPs to form the spliceosome. Splicing produces a free lariat-shaped intron, due to a 5' → 2' bond between the first guanosine of the intron and an internal adenosine. Splicing is also an important event in the transport of the RNA from the nucleus to the cytoplasm, where it can be translated.

▶ It remains unclear what the function and evolution of introns are. Although some data support the exon-shuffling hypothesis, which states that introns separate functional domains in what will become the translated protein, arguments have been made about whether this was initially present in the first cells or appeared after the evolutionary split between prokaryotes and eukaryotes.

Variations in Splicing

The preceding discussion covered the general description of splicing in eukaryotes, but two variations of this model should be discussed. The first involves alternative splicing, which is the production of multiple mRNAs from a single gene. The different mRNAs are produced when certain splice junctions are either used or ignored. The second involves *trans*-splicing, which occurs when the exons on two different primary transcripts are spliced together to generate a hybrid mRNA.

Alternative Splicing

In some instances, a single eukaryotic gene will produce multiple mRNAs that possess some sequences that are identical and others that are completely different. These variant mRNAs are produced when the pre-mRNA transcript undergoes splicing, with some RNAs using one splice junction and other RNAs ignoring that splice junction in favor of another.

An example of alternative splicing is the rat α-tropomyosin gene, which encodes a protein that is expressed in both smooth muscles (such as the small intestine) and striated muscles (such as skeletal muscle), as well as some nonmuscle cells. In both types of muscle, α-tropomyosin is involved in regulating muscle contraction. But smooth muscles exhibit a slower and steadier contraction than do striated muscles, implying that α-tropomyosin is regulated differently in the two systems.

We know that α-tropomyosin regulates contractions by interacting with different proteins in the two muscle types. This difference may occur as a result of having different α-tropomyosin genes (that encode different protein forms) or having a single gene in which a single part of the encoded protein is changed.

The rat α-tropomyosin gene and its two alternatively spliced mRNAs are shown in **figure 10.41**. The primary transcript contains 14 exons, whereas each of the processed mRNAs contains either 9 or 10 exons. Although both mRNAs contain exons 1, 4, 5, 6, 8, 9, and 10, the α-tropomyosin in striated muscle also contains exons 2 and 14. In contrast, the α-tropomyosin in smooth muscle contains exons 3, 11, and 12 in place

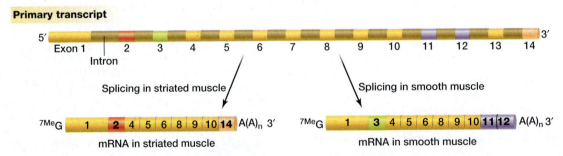

figure 10.41   Alternative splicing of the eukaryotic α-tropomyosin primary transcript. A single primary transcript, which contains 14 exons (top), may be alternatively spliced in different muscle types. In striated muscle, exons 3, 7, 11, 12, and 13 are removed along with the introns (lower left). In contrast, the primary transcript is spliced in smooth muscle to remove exons 2, 7, 13, and 14 along with the introns (lower right). Notice that the two mRNAs contain identical exons (gold) and exons that are found in the mRNA in specific muscle types (red, green, orange, and purple). The different exons in the two mRNAs produce different forms of α-tropomyosin protein in each muscle type.

of exons 2 and 14. Furthermore, nonmuscle forms of α-tropomyosin possess exon 7 in place of exon 8. These mRNAs can be translated to generate different forms of the protein that possess both common regions (such as the ability to bind actin) and unique regions (such as the ability to bind different regulatory proteins).

Through alternative splicing, a large number of different proteins can be produced from a single gene. In fact, it has been estimated that over half of the human genes exhibit some level of alternative splicing, with many of these genes encoding three or more alternatively spliced mRNAs. It is possible that the unexpectedly low number of genes in the human genome (estimated near 25,000) may be compensated by the relatively high percentage of genes that are alternatively spliced and yield different proteins.

How can some splice junctions be used in generating one mRNA and not be used in generating another? Because the snRNPs bind to the splice junctions to per-

form splicing, the accessibility of these RNA sequences to the snRNAs affect their use during splicing. Just as transcription factors activate and suppress transcription, splicing factors enhance or repress the availability of RNA sequences for splicing. **Figure 10.42** demonstrates both how splicing factors can suppress a splice junction site to remove the incorporation of an exon in the final mRNA and how enhancers can incorporate an exon in the final mRNA.

In some cases, alternative splicing can be used to affect the transcription of the gene. Assume that we have a gene that needs to be expressed for a short time in the developing nervous system and then again in the functional retina. One possibility would be to have two genes that are each expressed in one of those patterns. A single gene that possesses a complex promoter to regulate the two expression patterns is a second possibility. A third option would be two different promoters that are upstream of different first exons could be

Splicing repressors

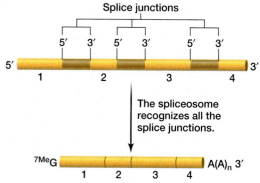

Splicing enhancers

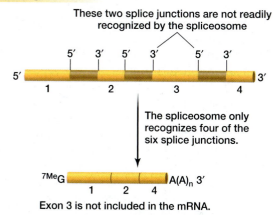

figure 10.42   Mechanisms of splice repressors and enhancers in regulating alternative splicing. In one model (top), all three introns are removed and all four exons are present in the mRNA. However, a splice repressor binding to the 3' end of the intron (upper right) will prevent binding of U2AF. Blocking this splice junction forces the spliceosome to splice exon I to exon 3 in the mRNA. In splice enhancers (bottom), the splice junctions that flank exon 3 are not normally recognized by U1 and U2AF, which results in exon 2 splicing to exon 4. However, binding of splice enhancers to the RNA flanking exon 3 (lower right) makes these splice junctions more recognizable by the spliceosome and yields an mRNA that contains all four exons.

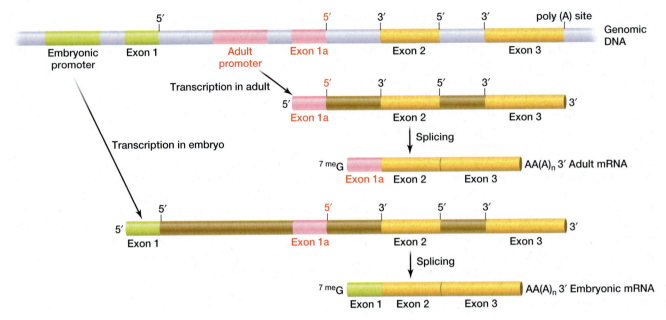

figure 10.43   Alternative splicing may be utilized in a gene that contains more than one promoter. This gene contains two separate promoters, one that is transcriptionally active in the embryo (green) and one that is recognized in the adult (pink). In the adult tissue, the primary transcript contains three exons, 1a, 2, and 3. The introns (blue) between these exons are removed by splicing to yield the mRNA diagrammed. In the embryo, a much longer primary transcript is produced that contains four exons 1, 1a, 2, and 3. During splicing, the 5' end of exon 1 splices to the 3' end of exon 2, because exon 1a lacks a 3' splice acceptor site. The result is an embryonic mRNA that contains three exons 1, 2, and 3.

alternatively spliced into the same downstream exons (**fig. 10.43**). This alternative splicing of the first exon is not uncommon in eukaryotic genes that exhibit complex expression patterns of the same protein.

The presence of different splicing enhancers or suppressors in different cell types can produce a dramatic effect on the number of alternatively spliced mRNAs from a single gene. We will discuss alternative splicing in the context of *Drosophila* sex determination in chapter 21.

It's Your Turn

Alternative splicing is an important mechanism by which multiple mRNAs and proteins can be generated from a single gene. Problem 40 examines your understanding of alternative splicing.

Trans-Splicing

All of the splicing events that have been discussed up to now are *cis*-acting, in which two exons that are present on the same primary transcript are spliced to produce the mRNA. In some cases, however, the final mRNA is composed of exons that were originally present on two different primary transcripts. The process of ***trans*-splicing** joins exons from different genes and primary transcripts into a new mRNA.

In a variety of free-living (*C. elegans*) and parasitic worms, *trans*-splicing occurs to attach a common 5' end onto the mRNA produced from several RNAs transcribed from different genes. The first RNA, called the **spliced leader** (**SL**), contains the equivalent of the first exon and the 5' half of an intron (**fig. 10.44**). The second RNA contains the 3' half of an intron followed by exons and introns.

During splicing, the U1 RNA base-pairs with the SL RNA, and U2AF attaches to the second RNA. Because the leader lacks a 3' splice acceptor site and the second RNA lacks a 5' splice donor site, neither is able to assemble a complete spliceosome on its own. However, if the two molecules join together, a splicesome can be generated.

When *trans*-splicing occurs, the released intron is actually the two half introns that are joined by the 5' → 2' phosphodiester linkage at the conserved adenosine (see fig. 10.44). Because this is two different introns, the product is a Y rather than the previously described lariat. The identification of this Y intermediate provided the proof of the mechanism of *trans*-splicing. Although this is not a common splicing event in eukaryotes, *trans*-splicing has been described for several human genes recently.

RNA Editing

In the last few years, several examples have arisen in which a DNA sequence did not completely predict the

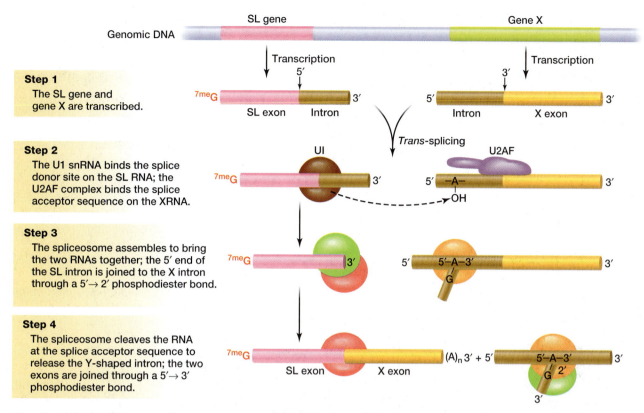

figure 10.44   Mechanism of *trans*-splicing. An RNA is transcribed from the SL (spliced leader) gene, which is capped (^{7me}G) at the 5' end. This RNA contains a 5' splice donor site, but lacks a 3' splice acceptor site. Gene X is transcribed to produce an RNA that begins with an intron sequence containing the branch point adenosine (A) and the 3' splice acceptor site, but no upstream 5' splice donor site. The U1 snRNP anneals with the 5' splice donor and U2AF anneals with the 3' splice acceptor. A spliceosome complex is produced that brings these two different RNAs together. Completing the *trans*-splicing reaction results in the exons from two different RNAs now joined by a phosphodiester bond and the released intron is Y-shaped, rather than lariat-shaped. This Y-shaped free intron has been identified in splicing reactions, which supports this model of *trans*-splicing.

corresponding sequence of the encoded protein. In several cases, the only way that the altered amino acid sequence could have occurred was by the insertion or deletion of nucleotides in the mRNA before translation. This insertion or deletion consists almost exclusively of uridines. The process is termed **RNA editing.**

Guide RNA (gRNA) Editing

RNA editing was particularly evident in the mitochondrial proteins of a group of parasites, the trypanosomes (some of which cause African sleeping sickness); in one case, more than 50% of the nucleotides in the mRNA were additional uridines not complementary to the DNA sequence. Uridines were also deleted from the original sequence.

Researchers showed in 1990 that these parasites encoded a new class of RNA called **guide RNA (gRNA)**, which directs the process of mRNA editing.

The gRNA base pairs with the mRNA to be edited; however, the sequence of the gRNA is complementary to the *final* mRNA, the one with additional bases. Because these extra bases have not yet been added, a bulge

occurs in the gRNA where the complementary sequence to be added occurs (**fig. 10.45**). The mRNA is then cleaved opposite the bulge by an editing endonuclease. A uridine (UTP) is brought into the mRNA as a complement to the adenine (A) in the gRNA using the enzyme terminal-U-transferase. An RNA ligase then closes the nick in the mRNA, which now has a uridine added.

An exciting outcome of this research, aside from learning about a novel mechanism of mRNA processing, is the possibility of clinical therapies. Any time a specialized pathway occurs in a parasite that is not found in its host, it is possible to use that pathway to attack the parasite. Thus, this research might lead to new ways of combating parasitic trypanosome infections.

Nucleotide Substitution

RNA editing also occurs in other species by a different mechanism. For example, in the mammalian *apolipoprotein-B* (*apoB*) gene, two forms of the protein are produced. In the liver, the ApoB protein is 4563 amino acids long, and in the intestine, the ApoB protein is 2153 amino acids. When the *apoB* mRNAs were iso-

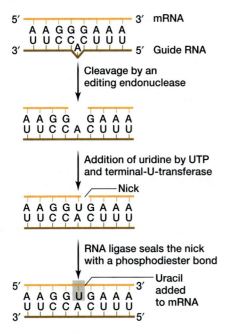

figure 10.45   RNA editing. The guide RNA (brown) is a complementary to the mRNA sequence (gold), except for an extra adenosine. When the guide RNA basepairs with the mRNA, an endonuclease nicks the phosphodiester bond in the mRNA across from the extra adenosine. Terminal–U-transferase inserts a uridine in the gap and RNA ligase seals the phosphodiester bond. After the cycle shown, a uridine-containing nucleotide has been added to the mRNA.

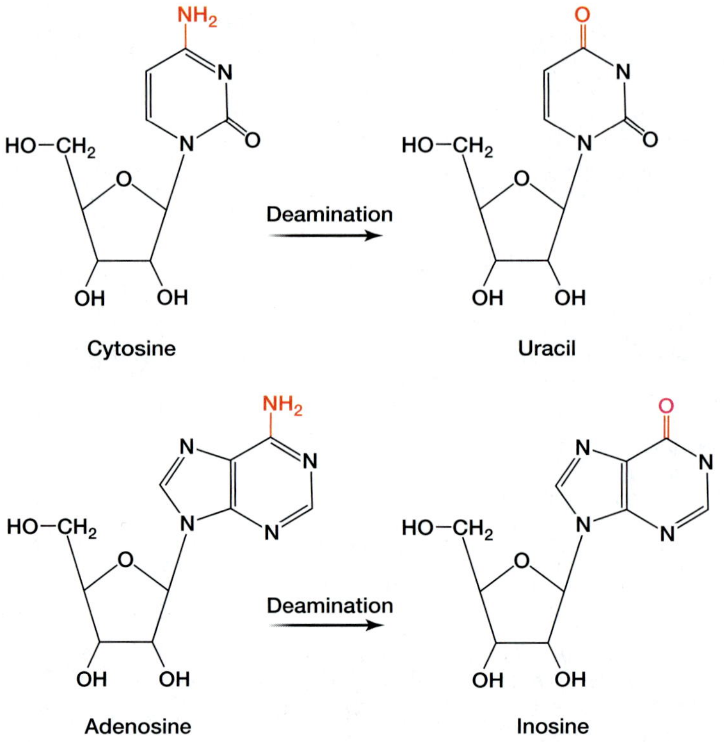

figure 10.46   RNA editing can occur through deamination of specific bases. Deamination of a base will result in the change of a single codon in the mRNA. Cytosine can be deaminated to uracil and adenosine can be deaminated to inosine.

lated from both tissues, it was found that there was only a single nucleotide difference.

In the liver mRNA, nucleotide 6666 is a cytosine, which is part of a 5' CAA 3' codon that encodes the amino acid glutamine. By contrast, nucleotide 6666 in the intestinal mRNA is a uracil, which is the 5' UAA 3' translation termination codon. The amino acid sequences are identical, except that the intestinal form of apoB terminates early.

The mechanism that produces the single nucleotide change from a cytosine to a uracil in apoB is deamination of the cytosine in the mRNA (**fig. 10.46**). This reaction is catalyzed by a cytidine deaminase. Although this deaminase is nonspecific and should act on any cytidine, it is in fact highly specific in its action. The specificity of the reaction is conferred by accessory proteins that possess an RNA-binding activity. These accessory proteins recognize a specific mRNA sequence and bind the deaminase to the mRNA, correctly positioning the deaminase to act on the correct base.

RNA editing is also observed in a class of glutamate receptors in the rat brain. The ribonucleotide change results in a single glutamine amino acid being changed to arginine. This single amino acid change results in a physiologically altered response by the brain neuron when glutamate binds the glutamate receptor. In this case, an adenosine deaminase results in the deamination of adenosine to inosine (see figure 10.46). The specificity of the deaminase is again regulated by accessory proteins that bind the mRNA. These RNA-editing mechanisms allow for an increased variety of proteins that are encoded in a single gene.

Restating the Concepts

▶ Alternative splicing is the process by which a gene produces a single primary transcript that is spliced such that not all of the exons are present in each mRNA. This can produce different forms of the encoded proteins that may have slightly different activities. Alternative splicing can also be used to regulate the expression of a single gene from multiple promoters. This provides a mechanism to increase the genetic diversity with a limited number of genes.

▶ *Trans*-splicing is the process by which two exons that are present in different primary transcripts (from different genes) are spliced together to generate a single novel mRNA. The splicesome produces a Y-shaped intron product from *trans*-splicing rather than the lariat-shaped intron product.

▶ RNA editing is the process by which the mRNA sequence is specifically altered to encode a protein with a different amino acid sequence and different activity. RNA editing occurs through either gRNAs inserting specific uridines or through deamination of specific cytidines or adenosines.

10.8 MicroRNA

In the last few years, a new class of RNA has been identified in eukaryotes. The *microRNAs (miRNAs)* have been implicated in a variety of regulatory mechanisms within the cell.

The miRNAs were first discovered in the early 1990s when Victor Ambros and his colleagues determined that the *lin-4* gene in *C. elegans,* which is required for development, encoded only a 22-nucleotide RNA rather than a protein. It was shown that the *lin-4* RNA was complementary to the 3' UTR of the *lin-14* mRNA, and that translation of the lin-14 protein was blocked when the *lin-4* RNA base paired with the *lin-14* mRNA. Other examples of miRNAs regulating gene expression have been identified in a variety of other eukaryotes.

The miRNAs appear to be RNA polymerase II transcripts that originate as RNAs ranging from several hundred to thousands of bases in length, and that form a stem-loop structure (**fig. 10.47**). This initial transcript is called a pri-miRNA (primary microRNA). The pri-miRNAs are processed in the nucleus to pre-miRNAs by nuclease digestion of the RNA that flanks the stem loop. After being transported to the cytoplasm, the

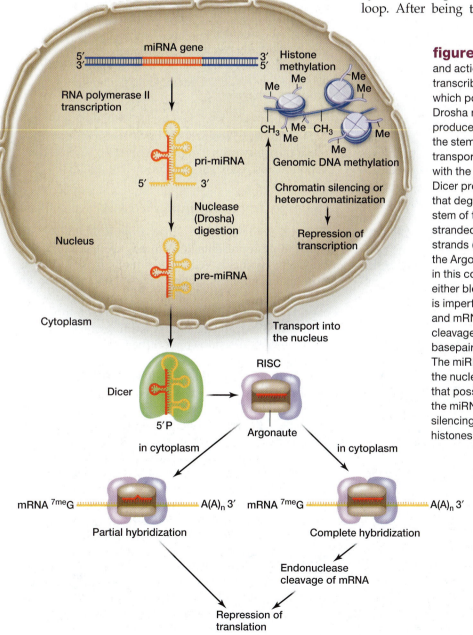

figure 10.47   Model for the formation and action of miRNAs. RNA polymerase II transcribes a gene that yields a pri-miRNA, which possesses a stem-loop structure. The Drosha nuclease trims the 5' and 3' ends to produce the pre-miRNA, which is primarily the stem-loop structure. The pre-miRNA is transported out of the nucleus and associates with the Dicer protein in the cytoplasm. The Dicer protein possesses a nuclease activity that degrades the loop region and some of the stem of the pre-miRNA to yield a short double-stranded RNA. One of the 22 nucleotide strands (miRNA) is preferentially transferred to the Argonaute protein of the RISC. If the miRNA in this complex basepairs with an mRNA it will either block translation of the mRNA (if there is imperfect annealing between the miRNA and mRNA) or result in the endonuclease cleavage of the mRNA (if there is complete basepairing between the miRNA and mRNA). The miRNA-Argonaute complex can also enter the nucleus and repress transcription of a gene that possesses a complementary sequence to the miRNA (termed gene silencing). This gene silencing likely involves methylation of the H3 histones and DNA.

Dicer protein degrades the remaining single-stranded regions of the pre-miRNA and one strand of the double-stranded stem to generate the mature 22-nt miRNA (see fig. 10.47). The miRNA then forms a **miRNP** (**microribonucleoprotein**) complex with the protein **Argonaute,** which is a component of the **RNA-induced silencing complex (RISC).** We will discuss the generation of miRNAs in greater detail in chapter 17.

This complex can perform three different functions (see fig. 10.47). First, the miRNA can base-pair with a specific mRNA, and if some of the bases do not precisely pair, then translation of the mRNA is repressed. Second, if the miRNA completely base pairs with the mRNA, then an endonuclease activity generates a single cut in the mRNA phosphodiester backbone (within the double-stranded region), which truncates the mRNA and also results in the suppression of translation. Third, and most dramatic, is gene silencing. **Gene silencing** involves the miRNA/Dicer/Argonaute complex entering the nucleus and stimulating the methylation of the H3 histone and genomic DNA, resulting in transcriptional repression. It is thought that the miRNA targets the methylation to a specific region in the chromatin by extensive base pairing with the genomic DNA.

The miRNA genes have been identified in both plants and animals. The number of potential miRNA genes varies between organisms, with nearly 100 in *Arabidopsis thaliana,* between 200 and 225 in *C. elegans,* between 100 and 125 in *Drosophila,* and over 300 in humans. Although many of these genes have only been identified based on computer analysis of genomic DNA sequences, the expression of several hundred have been verified. In this relatively new field, significant effort in the coming years will be devoted to identifying the mRNA targets of the miRNAs and how widely involved miRNAs are in different developmental and physiological processes.

Restating the Concepts

▶ MicroRNA, discovered in the 1990s, involves the generation of short single-stranded RNAs that are complementary to specific mRNAs. These miRNAs, in complex with the Dicer and Argonaute proteins, can block translation of the mRNA or repress transcription of a gene by methylating the H3 histone and genomic DNA in a process called gene silencing.

10.9 Update: The Flow of Genetic Information

The original description of the central dogma included three information transfers that were presumed to occur even though they had not been observed (see fig.

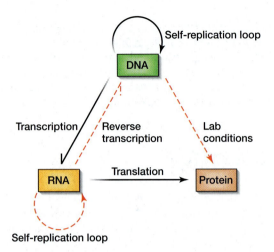

figure 10.48 An updated version of Crick's central dogma, showing all known paths of genetic information transfer. Paths confirmed since Crick proposed the original central dogma appear as dashed *red* lines (reverse transcription, RNA self-replication, and direct DNA translation). Direct DNA translation is observed only under specific laboratory conditions: the process apparently does not occur naturally. There is no known information flow beginning with protein.

10.1). Since then, researchers have documented two of these transfers occurring in organisms: reverse transcription and RNA self-replication (**fig. 10.48**).

Reverse Transcription

First, the return dashed arrow from RNA to DNA in figure 10.48 represents RNA that is used as a template for DNA synthesis. All RNA tumor viruses, such as human immunodeficiency virus (HIV), encode an RNA-dependent DNA polymerase (**reverse transcriptase**). This polymerase uses a viral RNA template to synthesize a complementary strand of DNA. This enzyme is required to convert the tumor virus's RNA genome into DNA.

This enzyme synthesizes a DNA–RNA double helix, which then is enzymatically converted into a DNA–DNA double helix that can integrate into the host chromosome. After integration, the DNA is transcribed into copies of the viral RNA, which are both translated and packaged into new viral particles that are released from the cell to repeat the infection process. In chapter 12, we will discuss the use of reverse transcriptase in generating DNA copies of mRNA (termed cDNA), which is an integral step in recombinant DNA technology.

RNA Self-Replication

The second modification to the original central dogma is the identification that RNA can serve as a template for its own replication. This process has been observed in a small class of both bacterial and eukaryotic **RNA phages,** such as R17, f2, MS2, and Qβ. MS2 contains about 3500 nucleotides and codes for only three proteins: a coat protein, an attachment protein (responsible for attachment of the phage to and subsequent penetration of the host cell), and a subunit of the enzyme **RNA replicase.** The RNA replicase subunit combines with three of the cell's proteins to form RNA replicase, which replicates the single-stranded phage RNA.

Because the phage-encoded RNA replicase enzyme must be synthesized before the phage can replicate its own RNA, the phage RNA must first act as an mRNA when it infects the cell. Thus, protein synthesis is taking place without a preceding transcription process. After the proteins are translated, the viral RNA is replicated as a complementary RNA sequence. In another major departure from the central dogma, this complementary RNA serves as the template for transcription for more messenger RNA. Some of this mRNA is translated and other is packaged in phage particles for the infection of additional bacterial cells (where it is again translated on entering the cell).

Restating the Concepts

▶ The central dogma originally stated that DNA could be replicated or transcribed into RNA. Some RNAs, such as mRNA, could be translated, but could not be replicated. Thus, the flow of genetic information was one-way, from DNA to RNA to protein. Reverse transcription, which occurs with some RNA viruses, reverses a portion of the flow so that RNA is converted to DNA (using a reverse transcriptase enzyme), which is then transcribed into RNA.

▶ A second exception to the original central dogma is observed with some bacterial RNA phages, in which the RNA can replicate itself without going through a DNA intermediate. This process also necessitates that RNA is transcribed from an RNA template.

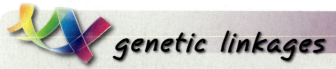

In this chapter, we discussed the variety of different RNAs that are present in both prokaryotes and eukaryotes and how the RNA is produced through transcription. Although eukaryotes are significantly more complex than prokaryotes in the process of transcription, the basic mechanisms are similar. We see this throughout genetics, where the simple systems (bacteria and phage) provide a good general understanding of the process, and analysis of higher eukaryotes provides the details and specific differences.

Although we discussed general mechanism of transcription initiation in both prokaryotes and eukaryotes, we only briefly described the various mechanisms utilized to regulate transcription. In chapter 16, we will discuss operons, which are clusters of prokaryotic genes that are transcribed on a single RNA, and how their transcription is regulated. This involves several DNA-binding proteins that can either activate or repress transcription. In chapter 17, we will discuss transcription regulation in eukaryotes, including various DNA sequences and DNA-binding proteins. We will also discuss an important issue in eukaryotic transcription, the conformation of the chromatin and how it can be modified to regulate transcription. You will then have a better understanding of the similarities and differences in the transcriptional regulatory mechanisms in prokaryotes and eukaryotes. Additionally, we will discuss how a single gene can utilize multiple promoters and alternative splicing to produce complex transcription patterns in an organism. You will then have a greater appreciation of the complexity that underlies how transcription is controlled in higher organisms.

In chapter 11, we will discuss the next step in gene expression, which is the translation of the mRNA into a protein. You will see the consequences of the various transcriptional events on translation. For example, what is the role of the 5' cap in eukaryotic translation initiation and how prokaryotic mRNA achieves the same goal without the 5' cap.

The importance of alternative splicing will be examined in our discussion of sex determination in *Drosophila* in chapter 21. You will see that two alternatively spliced mRNAs can encode functionally different proteins that will yield the two sexes. In that chapter we will also discuss how regulating eukaryotic RNA polymerase II gene transcription is involved in developing different body parts.

As with much of genetics, scientists attempt to apply what is learned in biological processes to study other problems. In our discussion of recombinant DNA techniques in chapter 12, you will see how the 3' poly(A) tail has simplified the isolation of mRNA and its conversion to DNA for a variety of studies. We will also discuss in chapter 20 how the identification of miRNA led to a rapid and informative approach for geneticists to study the role of most genes in either the cell or organism. This application of short interfering RNAs (siRNAs) has had an enormous influence on modern genetic analysis.

This knowledge of RNA structure, function, and transcription has propelled molecular genetics and medicine to points that were unimaginable 20 years ago. Increasing our understanding of the complexity of transcriptional regulation will similarly allow us to reduce birth defects, combat genetic and infectious diseases, and lead to other as yet undreamed of benefits in coming decades.

Chapter Summary Questions

1. For each term in column I, choose the best matching phrase in column II.

Column I	Column II
1. mRNA	A. Template strand
2. miRNA	B. RNA pol I
3. snRNA	C. Found in bacterial promoters
4. snoRNA	D. Involved in modification of rRNA
5. 28S rRNA	E. RNA pol II
6. tRNA	F. Involved in regulation of gene expression
7. Coding strand of DNA	G. Nontemplate strand
8. Noncoding strand of DNA	H. Found in eukaryotic promoters
9. Pribnow box	I. RNA pol III
10. Hogness box	J. Involved in splicing of introns

2. What is the central dogma of molecular biology?

3. Diagram the relationships of the three types of RNA at a ribosome. Which relationships make use of complementary base pairing?

4. Transcription and translation are coupled in prokaryotes but not in eukaryotes. Explain.

5. Draw and label the promoter sequence elements of the following enzymes:
   a. RNA polymerase of *E. coli*
   b. RNA polymerase II of humans

6. Distinguish between the holoenzyme and core enzyme of RNA polymerase in bacteria.

7. In what ways does the transcriptional process differ in eukaryotes and prokaryotes?

8. What is a stem-loop structure? An inverted repeat? A tandem repeat? Draw a section of a DNA double helix with an inverted repeat of 7 bp. Write an RNA sequence that has a stem-loop structure and then an inverted repeat closer to the 3' end? How do these two structures differ from each other?

9. What is DNA footprinting? How did it help define promoter sequences?

10. Compare and contrast the two main types of transcription terminators in bacteria.

11. Assume that bacterial RNA polymerase does not proofread. Do you expect high or low levels of error in transcription compared with DNA replication? Why is it more important for DNA polymerase than RNA polymerase to proofread?

12. What is unusual about the promoters of genes transcribed by RNA pol III?

13. List all the similarities and differences between a promoter and an enhancer?

14. Outline four ways in which the mature eukaryotic mRNA may be different from its primary transcript?

15. What are the differences between group I and group II introns?

16. Describe the function of the following: **(a)** spliceosome, **(b)** guide RNA, and **(c)** Argonaute protein?

17. How is *trans*-splicing different from alternative splicing?

Solved Problems

PROBLEM 1: What would be the order of sequence elements on a prokaryotic mRNA that contains more than one gene?

Answer: The transcript would have unmodified 5' and 3' untranslated regions (UTRs). Reading the sequence of nucleotides on the RNA from the 5'-to-3' end, you would come across a translation initiation codon (AUG), followed by an open reading frame (ORF) and then a translation termination codon (UAA, UAG, or UGA). The ORF would be those codons that will be translated into the protein. Then there would be a spacer region of nucleotides, followed by another translation initiation codon, ORF, and a translation termination codon. This sequence of initiation codon, codons to be translated, a termination codon, and spacer RNA would be repeated for as many genes as are present in the mRNA. At the 3' end of the mRNA, a transcription termination signal would be present as either a stem-loop structure (rho-dependent transcription termination signal) or a stem-loop structure that is followed by several continuous uracils (rho-independent transcription termination signal).

PROBLEM 2: Can one nucleotide be a conserved sequence?

Answer: Conserved sequences are invariant sequences of DNA or RNA recognizable to either a protein or a complementary sequence of DNA or RNA. However, in group II introns, an adenine is needed near the 3' end of the intron for lariat formation. Thus, this single nucleotide, given its relative position in the intron and possible surrounding bases, is a conserved sequence of one.

PROBLEM 3: Why might *E. coli* not have a nucleolus?

Answer: The nucleolus is the site of ribosomal construction in eukaryotes. It is centered at the nucleolus organizer, the tandemly repeated genes coding for the three larger rRNAs. *E. coli* has only 5–10 copies of the rRNA gene cluster, whereas eukaryotes have an order of magnitude or more copies. Thus, the simplest reason that a nucleolus is not visible in *E. coli* is because it has too few copies of the gene around which a nucleolus forms.

PROBLEM 4: If the following sequence of bases represents the start of a gene on double-stranded DNA, what is the sequence of the transcribed RNA, what is its polarity, and what is the polarity of the DNA?

> GCTACGGATTGCTG
> CGATGCCTAACGAC

Answer: Begin by writing the complementary strand to each DNA strand: C G A U G C C U A A C G A C for the top, and G C U A C G G A U U G C U G for the bottom. Now look for the start codon, AUG. It is present only in the RNA made from the top strand, so the top strand must have served as the template for transcription. The polarity of the start codon is always 5'-A U G-3'. Because transcription occurs 5' → 3', and since nucleic acids are antiparallel, the left end of the top DNA strand is the 3' end. Once we have determined the polarity at one end of the DNA molecule, all the other ends can be determined based on the 5' → 3' polarity and antiparallel nature of the DNA. Thus, the top left end is 3', the top right end must be 5', the lower right end must be 3', and the lower left end is 5'.

Exercises and Problems

18. What is a consensus sequence? Propose one for the following DNA molecules:

> GAATTAG
> GATCTGA
> CTTCTGG
> AATCAAG
> GCTTTGG
> GACATAG
> GATCTAC

19. Determine the sequence of both strands of the DNA from which the following RNA was transcribed. Indicate the 5' and 3' ends of the DNA, which was the template strand, and the direction of transcription.

5'-CCAUCAUGACAGACCUUGCUAACGC-3'

20. What would the effect be on transcription if a bacterial cell had no sigma factors? No rho protein?

21. Why do you think most promoter regions are A–T-rich?

22. How could DNA–DNA or DNA–RNA hybridization be used as a tool to construct a phylogenetic (evolutionary) tree of organisms?

23. The following DNA fragment was isolated from the beginning of a gene. Determine which strand is transcribed, indicate the polarity of the two DNA strands, and then give the sequence of bases in the resulting mRNA and its polarity.

> CCCTACGCCTTTCAGGTT
> GGGATGCGGAAAGTCCAA

24. A bacterial mRNA molecule is isolated and determined to have the following base composition: 18% A, 27% C, 33% G, and 22% U. What is the base composition of the corresponding segment in
    a. the template strand of DNA?
    b. the nontemplate strand of DNA?
    c. the double-stranded DNA?

25. The following sequence of bases in a DNA molecule served as the template for transcribing an RNA:

CCAGGTATAATGCTCCAG TATGGCATGGTACTTCCGG

If the T (*underlined*) is the first base transcribed, determine the sequence and polarity of bases in the RNA, and identify the Pribnow box and the initiator codon.

26. The following DNA fragment represents the beginning of a gene. Determine which strand is transcribed, and indicate polarity of both strands in the DNA.

ATGATTTACATCTACATTTACATT
TACTAAATGTAGATGTAAATGTAA

27. Lysozyme, which is found in a variety of animal secretions, is an enzyme that breaks down bacterial cell walls. The primary sequence of human lysozyme consists of 129 amino acids.
    a. Draw and label the mature mRNA for the lysozyme protein.
    b. Calculate the minimum number of nucleotides in this mRNA.

28. You isolated a mutant that makes a temperature-sensitive rho molecule; rho functions normally at 30°C, but not at 40°C. If you grow this strain at both temperatures for a short time and isolate the newly synthesized RNAs, what relative size RNA do you expect to find in each case?

29. Suppose you repeat the experiment in problem 28 and find the same size RNAs are made at both temperatures. Provide two possible explanations for this unexpected finding.

30. What product would DNA–RNA hybridization produce in a gene with five introns? No introns? Draw these hybrid molecules.

31. Would introns be more or less likely than exons to accumulate mutations through evolutionary time?

32. Enhancers can often exert their effect from a distance; some enhancers are located tens of thousands of bases upstream from the promoter. Propose an explanation to account for this.

33. Given the following tRNA molecule. Determine the corresponding codon, 5' → 3' polarity of this tRNA and the corresponding codon, and the strand of DNA that this tRNA is transcribed from.

34. RNA–DNA hybrids are formed by using mRNA for a given gene that is expressed in the pituitary and the adrenal glands. The DNA used in each case is the full-length gene. Based on the figure, provide an explanation for the different hybrid molecules. DNA is the blue line; RNA is the yellow line.

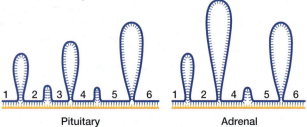

35. In the following drawing of a eukaryotic gene, solid red lines represent the exons and blue lines correspond to the introns. Draw the RNA–DNA hybrid that would result if cytoplasmic mRNA is hybridized to nuclear DNA.

36. How do prions (see chapter 7) relate to the central dogma of figure 10.48?

37. What are the upper limits to the size of a gene in eukaryotes?

38. Consider a double-stranded DNA molecule with an $(A + T)/(G + C)$ ratio of 1.4 and which is transcribed along its entire length.
    a. What is the $(A + G)/(T + C)$ ratio in the DNA molecule?
    b. If *both* strands of DNA are transcribed, what will the $(A + U)/(G + C)$ ratio be in the mRNA?
    c. If only one of the two DNA strands is transcribed, what will the $(A + U)/(G + C)$ ratio be in the mRNA?
    d. Assume the mRNA $(A + G)/(U + C)$ ratio is approximately 1.2. Is this mRNA transcribed from a single DNA strand or from both DNA strands? Explain your answer.

39. The following schematic representation of transcription in *E. coli* contains a number of mistakes.

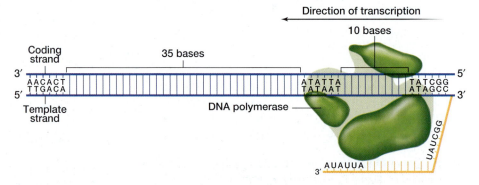

    a. List as many mistakes as you can find, and briefly explain why they are so.
    b. Redraw the diagram without the mistakes.

40. You are studying a genetic disease in humans called nerditis, which results in afflicted individuals constantly reading and studying. You cloned the *nerditis* gene and determined the sizes and locations of the exons from the DNA sequence (shown below). In studying a particular family, you found that the unaffected individuals have two different *nerditis* mRNAs (3000 and 3500 nucleotides long) and the affected individuals only have the 3000 nucleotide mRNA.

	Exon 1	Exon 2	Exon 3	Exon 4
Size in nucleotides:	1000	500	1000	800

If the polyadenylation site is 50 nucleotides before the end of exon 4 and both mRNAs possess a 250 nucleotide poly (A) tail, how are the two mRNAs in the normal individual produced? Describe, as accurately as possible, two different mutations that would produce the affected individuals in this family.

Chapter Integration Problem

In the animal *Fauna hypotheticus* (first encountered in Chapter 7's integration problem), the *large feet* trait has an autosomal recessive mode of inheritance. The *laf* gene, found on chromosome 7, segregates two alleles: laf^+, small feet, and laf^-, large feet. The mRNA transcript of the wild-type gene is given below, in the 5' → 3' orientation:

```
  1   AUCCAGUAAC   UUGAUACUGA   ACGAAACAGA   CGUGGCCGAA   CCGUACACCU
 51   ACCGACUGCC   UUCACGUUAC   CGCGAUUAAC   GAAUGAAUUA   UGAUGAUUUC
101   UUUCGGGUAA   GUAAACCAUG   UACUAACAAU   UAUCGCUCUA   UCCAGGGAUU
151   CUUCGUGGAU   AUGUUAAUCA   CCUGGUGGGU   ACGUCAGAUA   UUGACUCGUA
201   UAUGUACCUA   CUAACACGAC   ACCAGCUUGU   CAGUAUGACG   ACUUUUUCCG
251   CUAACUAUCU   UCACUGCAUU   AUUUGUGAAA   UAAACUUCGU   AAUGGGUAGA
301   AAAAAAAAAA   AAAAAAAAAA   AAAAAAAAAA   AAAAAAAAAA   AAAAAAAAAA
```

a. How many nanometers long is the double-stranded DNA that codes for this mRNA transcript?

b. Predict the sequence of the DNA coding strand of the laf^+ gene.

c. Outline the differences between replication and transcription of the laf^+ gene.

d. Does the promoter for the laf^+ gene contain a TATA box? Why or why not?

e. What RNA polymerase transcribes the *laf* gene? Justify your answer.

f. The wild-type Laf protein actually comes in two sizes: Laf^L (large) and Laf^S (small). How many amino acids does each form contain? Justify your answer.

g. The *laf* DNA from four mutants was isolated and compared with the wild-type laf^+ sequence. In all cases, the cause of the mutation was a single-base substitution in the DNA. Predict the likely location of each mutation, based on the following results:

Mutant 1 = No mRNA transcript.

Mutant 2 = Normal length mRNA transcript, but no protein.

Mutant 3 = Nonfunctional protein, Laf^{XL}, which has 33 amino acids.

Mutant 4 = Nonfunctional protein, Laf^{XS}, which has 7 amino acids.

h. Heterozygous ($laf^+ laf^-$) individuals have small feet, whereas homozygous recessive ($laf^- laf^-$) individuals have large feet. Provide a molecular explanation.

Do you need additional review? Visit **www.mhhe.com/hyde** for practice tests, answers to end-of-chapter questions and problems, interactive exercises, and animation tutorials all designed to enhance your understanding of key genetic concepts.

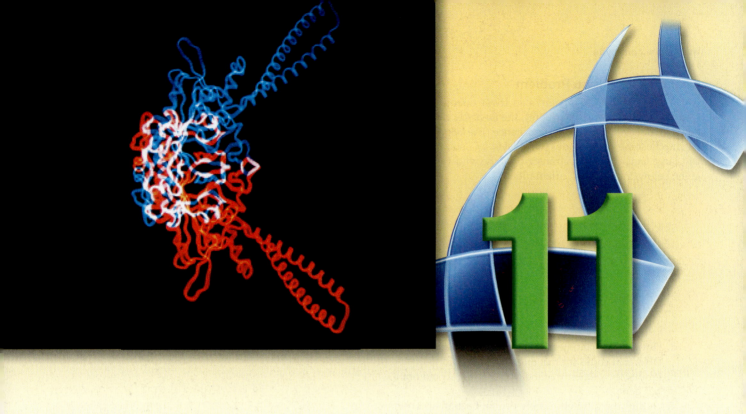

Gene Expression: Translation

Essential Goals

1. Understand the one-gene/one-enzyme hypothesis and how it was elucidated.

2. Describe the general structure of amino acids and how they confer physical properties to a protein.

3. Distinguish between the levels of protein structure.

4. Describe the differences and similarities in translation initiation in bacteria and eukaryotes.

5. List and describe the events that occur during protein translation.

6. Give examples of posttranslational modifications.

7. Understand the experimental steps that led to solving the genetic code.

Chapter Outline

Photo: Computer generated model of the enzyme serine tRNA synthetase, the enzyme that charges tRNAs with the amino acid serine.

ranslation involves the synthesis of an amino acid chain based on an RNA sequence. However, translation alone is insufficient to generate a functional polypeptide. The protein must be folded and possibly post-translationally modified, much as eukaryotic mRNA is post-transcriptionally processed. The failure of a protein to produce the proper shape can have dramatic effects on a cell. In fact, several human diseases are the result of improperly folded proteins.

Some **trinucleotide repeat diseases** in humans result from protein misfolding. These diseases include myotonic dystrophy, fragile X syndrome, and Huntington disease. In this group of diseases, a trinucleotide repeat, which often corresponds to a repeating codon in the open reading frame of the mRNA transcript, is present a variable number of times. When the repeat number is low, the individual is normal. However, the number of repeats can increase from generation to generation, and when the repeat number increases above a certain level (that varies depending on the disease), the neurological symptoms become apparent.

Because the repeated triplet corresponds to a codon, the increased number of repeats results in an increased number of the corresponding amino acid in the protein. This larger number of a specific amino acid at a particular location in the protein results in the protein either misfolding or failing to properly interact with another protein.

Additional human diseases also result from protein misfolding, but do not involve the expansion of triplet repeats. Examples of this class of diseases include Alzheimer disease and Creutzfeldt–Jakob disease (CJD). Alzheimer disease can be due to a misfolded protein called β-**amyloid,** which forms *plaques* in the learning and memory regions of the brain. Creutzfeldt–Jakob disease (CJD) is caused by a proteinaceous infectious particle (**prion**) disrupting the normal folding of a brain protein, which ultimately disrupts neuronal and brain function.

As you read about translation, keep in mind that the translation of the nucleotide sequence into the amino acid sequence of a protein is only part of the process, although a very important one. The resulting protein must be properly folded and, in the case of eukaryotes, may have to be modified several different ways to produce a functional protein. Thus, after all the care to properly transmit the genetic information (replication) and express it as mRNA (transcription), the production of a functional protein still relies on the correct translation of mRNA and the subsequent modification of the product.

11.1 The One-Gene/One-Enzyme Hypothesis

For the most part, dominant alleles produce functional enzymes, whereas recessive alleles do not. In earlier chapters we reviewed a number of characteristics, such as snapdragon flower color and round versus wrinkled peas, that demonstrate dominant and recessive phenotypes. Heterozygous organisms often appear normal because one allele of a gene pair is capable of producing a sufficient quantity of functional enzyme to produce the wild-type phenotype. From a wealth of evidence, geneticists began to suspect that each enzyme was coded for by a single gene.

Biochemical Genetics

The study of the relationship between genes and enzymes is generally called **biochemical genetics** because it involves the genetic control of biochemical pathways. Archibald Garrod, a British physician, pointed out this general concept in his work *Inborn Errors of Metabolism,* published in 1909. Garrod described several human conditions, such as albinism and alkaptonuria, which occur in individuals who are homozygous for recessive alleles. For example, humans normally degrade homogentisic acid (alkapton) into maleylacetoacetic acid. Persons with the disease alkaptonuria are homozygous for a nonfunctional form of an enzyme essential to this process: homogentisic acid oxidase, found in the liver. Absence of this enzyme blocks the degradation reaction so that homogentisic acid builds up. This acid darkens on oxidation. Thus, affected persons can be identified by the black color of their urine after its exposure to air. Eventually, alkaptonuria causes problems in the joints and a darkening of cartilage that is visible in the ears and the eye sclera.

Figure 11.1 shows part of the pathway in phenylalanine and tyrosine degradation in humans. You can also see that tyrosine can be converted into either melanin or thyroxine. At several points in these pathways, a metabolic defect causes an identifiable phenotypic condition, one of which is alkaptonuria. In all these cases, a defect in a single gene results in a defective enzyme in homozygous individuals.

The Work of Beadle and Tatum

The concept that genes control the production of enzymes was greatly advanced by George Beadle and Edward Tatum (**fig. 11.2**), who eventually shared the 1958 Nobel Prize in Physiology or Medicine for their work. In the early 1940s, they united the fields of biochemistry and genetics by using strains of the pink bread mold *Neurospora crassa* that had specific nutritional requirements. In 1941, Beadle and Tatum were the first scientists to isolate mutants with nutritional requirements that defined steps in a biochemical pathway, the biosynthesis of niacin (vitamin B$_3$) in *Neurospora.*

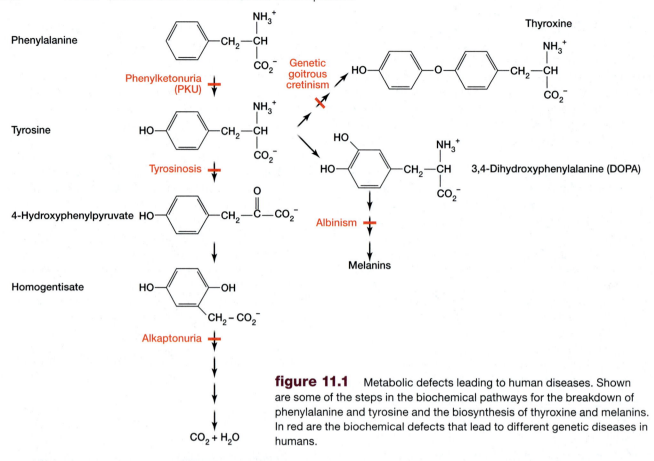

figure 11.1 Metabolic defects leading to human diseases. Shown are some of the steps in the biochemical pathways for the breakdown of phenylalanine and tyrosine and the biosynthesis of thyroxine and melanins. In red are the biochemical defects that lead to different genetic diseases in humans.

(a)   **(b)**

figure 11.2 George W. Beadle (1903–1989) and Edward L. Tatum (1909–1975), who isolated *Neurospora* mutants to define the biochemical steps in niacin biosynthesis.

The Niacin Biosynthetic Pathway

Normally, *Neurospora* synthesizes the niacin it requires for viability through a biochemical pathway (**fig. 11.3**). Beadle and Tatum isolated *Neurospora* mutants that could not grow unless niacin was provided in the culture medium. They deduced that these mutants had enzyme deficiencies in the synthesizing pathway that ends with niacin.

Beadle and Tatum had a general idea, based on the structure of niacin, as to what intermediates would occur in the niacin biosynthetic pathway. They made educated guesses about which substances they might add to the culture medium to enable the mutants to grow. Mutant *B* (**table 11.1**), for example, could grow if given either niacin or 3-hydroxyanthranilic acid. Mutant *B* could not grow if given only kynurenine. Therefore, Beadle and Tatum knew that the *B* mutation affected

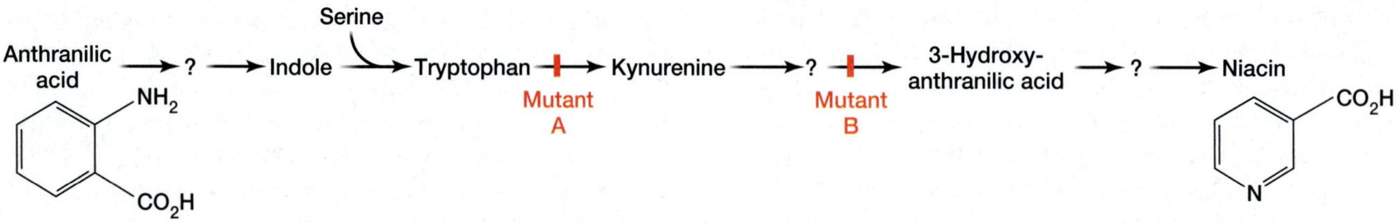

figure 11.3 The biochemical pathway defined by Beadle and Tatum for the biosynthesis of niacin from anthranilic acid. The structures of anthranilic acid and niacin are shown. Several of the predicted intermediates are shown, while others were unknown (shown as a question mark). The biochemical defects associated with two *Neurospora* mutants (mutant A and mutant B) are shown in red.

table 11.1   Viability of Neurospora Niacin Mutants.

Strain	Additive to Media			
	Tryptophan	**Kynurenine**	**3-Hydroxyanthranilic acid**	**Niacin**
Wild type	+	+	+	+
Mutant *A*	–	+	+	+
Mutant *B*	–	–	+	+

Plus sign (+) signifies growth, and minus sign (–) indicates no growth on supplemented media.

the pathway between kynurenine and 3-hydroxyan-thranilic acid. Furthermore, this confirmed their deduction that kynurenine was an earlier intermediate in the pathway than 3-hydroxyanthranilic acid.

Similarly, mutant *A* could grow if given 3-hydroxy-anthranilic acid or kynurenine instead of niacin. Therefore, these two products must be in the pathway *after* the step interrupted in mutant *A*.

Conversely, because neither of these mutant organisms could grow when given only tryptophan, Beadle and Tatum knew that tryptophan occurred earlier in the pathway than the steps with the deficient enzymes.

By this type of analysis, and by also observing the accumulation of substances when a pathway was blocked, Beadle and Tatum discovered the steps in several biochemical pathways of *Neurospora*. Many biochemical pathways are similar in a huge range of organisms, and thus, Beadle and Tatum's work was of general importance.

Beadle and Tatum concluded from their studies that one gene controls the production of one enzyme. This **one-gene/one-enzyme hypothesis** is an oversimplification, as you will see shortly. As a rule of thumb, however, the hypothesis is valid, and it has served to direct attention to the functional relationship between genes and their corresponding enzymes in biochemical pathways.

Pleiotropic Effects

Although a change in a single enzyme usually disrupts a single biochemical pathway, it frequently has more than one effect on phenotype. Multiple effects are referred to as pleiotropy (see chapter 5). An example can be seen in figure 11.1. The mutation that causes tyrosinosis will prevent the degradation of phenylalanine and tyrosine. This will cause an increase in the amount of tyrosine and phenylalanine in the organism as well as a reduced amount of the subsequent products, such as 4-hydroxyphenylpy-ruvate and homogentisate. Individuals suffering from tyrosinosis may experience ulcers in the corneas of their eyes, lesions on the skin, and mental retardation.

These symptoms, which are seemingly unrelated, can be reduced in severity if the individual's diet is low in both phenylalanine and tyrosine. This supports the hypothesis that tyrosinosis results from the inability to properly degrade tyrosine and phenylalanine. The pleiot-

ropy results from a single mutation, leading to increased levels of phenylalanine and tyrosine in different tissues of the body and from the way those tissues respond to the increased amount of these two amino acids.

It's Your Turn

The ability to use mutants to determine the order of intermediates and enzyme activities in a biochemical pathway is one of the strengths of genetics. Use the reasoning approach that Beadle and Tatum used to assemble the biochemical pathway from the mutants in problem 15.

 Restating the Concepts

▶ Each gene encodes an enzyme (or protein), which usually carries out the functions in the cell that determines the phenotype. This principle is the one-gene/one-enzyme hypothesis.

▶ Enzymes catalyze steps in a biochemical pathway to generate a product molecule from precursors.

▶ Pleiotropy is a phenomenon in which a single mutation may affect multiple aspects of phenotype, either directly or indirectly.

11.2 Protein Structure

As you have learned, nucleic acids have a particular structure with some versatility; they can be single-stranded or double-stranded, linear or circular. The linear forms of nucleic acids also possess a polarity, with one end lacking a phosphodiester bond on the 5' carbon of the sugar (either ribose or deoxyribose) and the other end lacking the bond on the 3' carbon.

Proteins also possess structure and polarity, although differently from the structure and polarity of nucleic acids. These differences result from the composition of proteins.

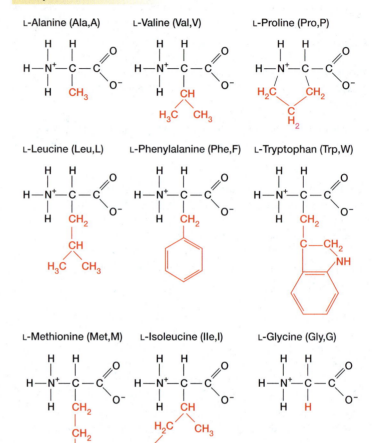

figure 11.4   The twenty common amino acids found in proteins. The structure and three- and one-letter abbreviation for each amino acid is given. At physiological pH, the structure of each amino acid, which usually exist as ions, is shown. The amino acids are divided into four groups, acidic, basic, nonpolar, or polar, based on the properties of their different R groups, which are shown in red.

The Structure and Properties of Amino Acids

Proteins are composed of subunits called **amino acids,** which have the generalized structure seen in **figure 11.4**. The 20 standard amino acids each possess a different side group, termed an R group, that gives the amino acid its identity. Every amino acid has both an amino (NH_2) terminus (end) and a carboxyl (CO_2H) terminus. At physiological pH (pH 6.8), an amino acid has an electrical charge.

Each amino acid is capable of taking on a charge because the amino end can donate an electron and the carboxyl end can accept an electron. Thus, each amino acid is a *dipolar ion* (or *zwitterion*) and has properties of both acids and bases. In addition, the R group may also have an electrical charge at physiological pH.

The amino acids can be divided into four groups based on the identity of the R group: (1) acidic R groups, (2) basic R groups, (3) nonpolar (hydrophobic) R groups, and (4) polar but uncharged R groups (see fig. 11.4).

A peptide bond is the covalent bond that is found between any two amino acids in a protein (**fig. 11.5**). Notice that these peptidyl bonds involve the carboxyl end of one amino acid and the amino end of another. The result is that the peptide chain still possesses a single free carboxyl end (*C terminus*) and an amino end (*N terminus*). The amino acids that are components of a polypeptide are often referred to as *residues*.

Levels of Protein Structure

As discussed previously, the active site is a part of an enzyme that provides both the specific structure to control the binding of the substrate and the amino acids to catalyze the enzymatic reaction (chapter 7). Four different levels of protein structure account for the overall organization and function of a polypeptide, but it is important to realize that the lower levels of protein structure have an effect on the higher levels.

The **primary structure** of a protein corresponds to the amino acid sequence of the polypeptide. The primary structure of a protein can often be inferred from the gene or mRNA nucleotide sequence, as discussed in more detail shortly. Included in the primary structure is the formation of disulfide bridges between cysteine residues (**fig. 11.6**).

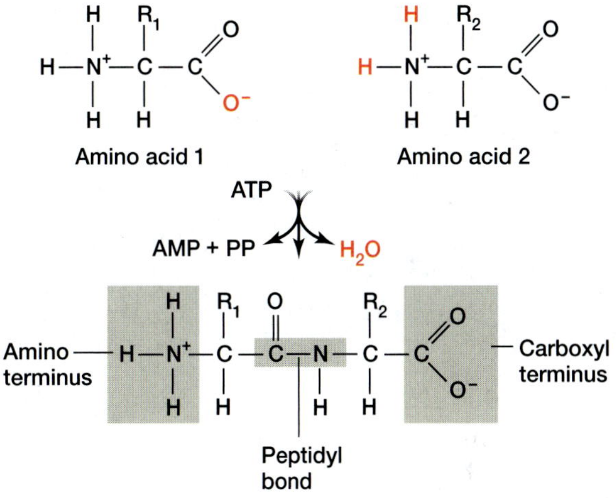

figure 11.5   Formation of a peptide bond between two amino acids. To generate a peptide, a bond is produced between the carboxyl group of one amino acid and the amino group of the other. Formation of this peptidyl bond requires ATP hydrolysis for energy. After forming the peptidyl bond, an amino (N) terminus and carboxyl (C) terminus remain on the peptide.

(a)

1									
Met	Val	His	Leu	Thr	Pro	Glu	Glu	Lys	Ser

					11				
					Ala	Val	Thr	Ala	Leu

					21				
Trp	Gly	Lys	Val	Asn	Val	Asp	Glu	Val	Gly

31									
Arg	Leu	Leu	Val	Val	Gly	Glu	Ala	Leu	Gly

					41				
					Tyr	Pro	Trp	Thr	Gln

Arg Phe Phe Glu Ser

					51				
Phe	Gly	Asp	Leu	Ser	Thr	Pro	Asp	Ala	Val

Met Gly Asn Pro Lys

61									
Val	Lys	Ala	His	Gly	Lys	Lys	Val	Leu	Gly

					71				
					Ala	Phe	Ser	Asp	Gly

					81				
Leu	Ala	His	Leu	Asp	Asn	Leu	Lys	Gly	Thr

Phe Ala Thr Leu Ser

91									
Glu	Leu	His	Cys	Asp	Lys	Leu	His	Val	Asp

					101				
					Pro	Glu	Asn	Phe	Arg

					111				
Leu	Leu	Gly	Asn	Val	Leu	Val	Cys	Val	Leu

Ala His His Phe Gly

121									
Lys	Glu	Phe	Thr	Pro	Pro	Val	Gln	Ala	Ala

					131				
					Tyr	Gln	Lys	Val	Val

					141				
Ala	Gly	Val	Ala	Asn	Ala	Leu	Ala	His	Lys

Tyr His

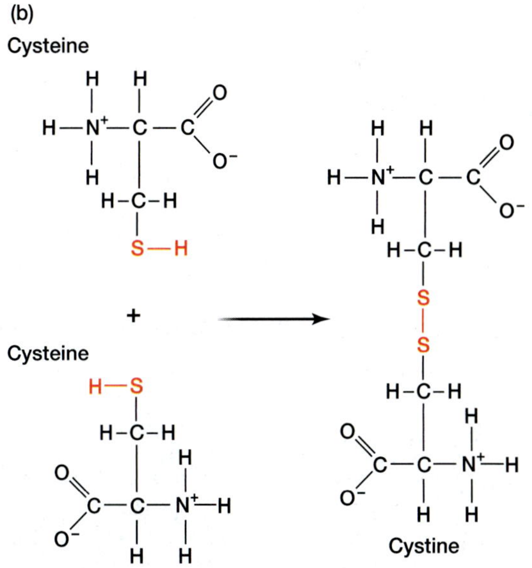

(b)

figure 11.6   Primary protein structure. (a) The primary structure of a protein is the amino acid sequence. Shown is the amino acid sequence of the human β-globin protein in the three letter code. (b) A second piece of the primary structure is the formation of a disulfide bond between two cysteines in the same polypeptide.

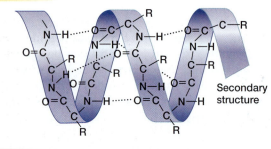

α–Helix

Secondary structure

β–Sheet

figure 11.7 The secondary protein structure is a repeating amino acid configuration within a protein. The two most common secondary structures in polypeptides are α-helices (top) and β-sheets (bottom). Both of these structures are stabilized by hydrogen bonding (dotted lines) between the oxygen atom and the amine hydrogen.

The amino acids that are nearby in the primary structure can interact to form the **secondary structure.** Two common secondary structure motifs are α-helices and β-sheets (**fig. 11.7**). Although the R groups are not usually involved in these interactions, different amino acids are more likely to form one structure or another. For example, alanine, glutamic acid, and methionine are more likely to be found in an α-helix, whereas isoleucine, tyrosine, and valine are more likely to be found in a β-sheet. These secondary structures are often stabilized by hydrogen bonding between amino acids (see fig. 11.7).

Further folding of the polypeptide brings the secondary structures into a precise three-dimensional configuration, which creates the **tertiary structure** of the protein (**fig. 11.8**). This tertiary structure creates the active site region in many proteins.

Some proteins contain only a single polypeptide and therefore, possess only three levels of structure. However, many proteins are composed of several subunits that together make up the **quaternary structure** of the protein (see fig. 11.8). The interactions between the different proteins or subunits are essential in the enzymatic function of the molecule. For example, you learned in the preceding chapter that the *E. coli* RNA polymerase is composed of several subunits.

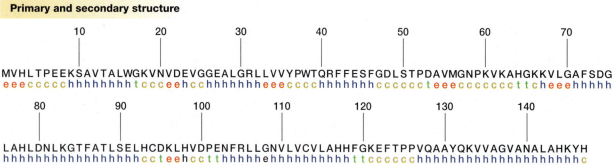

Primary and secondary structure

```
        10        20        30        40        50        60        70
        |         |         |         |         |         |         |
MVHLTPEEKSAVTALWGKVNVDEVGGEALGRLLVVYPWTQRFFESFGDLSTPDAVMGNPKVKAHGKKVLGAFSDG
eeeccccchhhhhhhhtccceehcchhhhhhhheeecccchhhhhhhhcccccteeecccccccttcheeehhhhh

        80        90        100       110       120       130       140
        |         |         |         |         |         |         |
LAHLDNLKGTFATLSELHCDKLHVDPENFRLLGNVLVCVLAHHFGKEFTPPVQAAYQKVVAGVANALAHKYH
hhhhhhhhhhhhhhhhhccteehcctthhhhhehhhhhhhhhhhhttcccccchhhhhhhhhhhhhhhhhhhhc
```

Tertiary structure   **Quaternary structure**

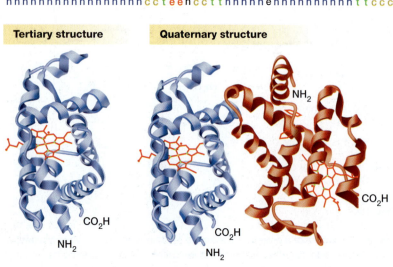

figure 11.8 Different levels of the human beta globin protein structure are shown. The secondary structure corresponds to common protein motifs that can be predicted from the amino acid sequence (primary structure, black letters). The predicted motifs in the human beta globin are alpha helices (h), extended strands (e), turn (t), and random coils (c). The tertiary structure, which represents the three-dimensional structure of the human beta globin protein is shown. The amino (NH$_2$) and carboxyl (CO$_2$H) termini are labeled. The heme structure, which binds oxygen, is shown in red. The quarternary structure, or the subunit organization, for human hemoglobin is shown, which contains one beta-globin protein (blue color) and one alpha-globin protein (red protein).

314

Restating the Concepts

▶ Twenty different amino acids are the components of every protein. Each amino acid has a carboxyl end and an amino end, which confers on each protein a free amino terminus and a carboxyl terminus. Each amino acid also has a side group (R group), which confers biochemical properties on the amino acid and on the protein.

▶ Each protein has either three or four levels of structure ranging from the amino acid sequence (primary), the common protein domains (secondary), the three-dimensional structure (tertiary), and the subunit composition and interactions (quaternary). Each level of protein structure is dependent on the lower levels. Proteins that contain a single polypeptide lack a quaternary structure.

11.3 The Colinearity of mRNA and Protein Sequences

Geneticist Charles Yanofsky, working in the 1960s, offered several research conclusions that affected our understanding of the correlation between nucleic acid sequence and amino acid sequence. One of the most important of these findings is the **colinearity** of gene and protein, which means that a linear relationship exists between the nucleotide sequence of a gene and the amino acid sequence of the encoded protein.

Yanofsky isolated a large number of mutations in the *trpA* gene of *E. coli.* The *trpA* gene encodes the A subunit of the tryptophan synthetase enzyme, which is essential for the synthesis of tryptophan. Thus, a *trpA⁻* mutant can grow only on minimal media if it is supplemented with tryptophan.

Yanofsky used recombination mapping to determine the relative order of the mutations in the *trpA* gene (**fig. 11.9**). Some of the *trpA⁻* mutations were nonsense mutations, resulting in early translation termination of the protein (discussed later in this chapter). Yanofsky found that the relative position of the mutation in the *trpA* gene corresponded to the relative position of the termination of the mutant proteins (see fig. 11.9). In other words, the closer the mutation was to the beginning of the gene, the shorter the mutant protein. This finding indicates that the nucleotide sequence of the gene and the amino acid sequence of the protein are colinear.

Some of the *trpA⁻* mutations were missense mutations, which resulted in one amino acid in the wild-type protein sequence becoming a different amino acid. In these cases, the wild-type and mutant proteins were nearly identical in molecular weight. Yanofsky, therefore, could not rely on the molecular weight of the mutant protein to predict the location of the mutation, as he had with the nonsense mutations. Rather, he had to determine the exact amino acid that was changed in the mutant protein.

In each of the missense cases he examined, he again found colinearity between the gene mutation and the altered amino acid (see fig. 11.9). In addition, both the missense and nonsense mutations could be placed on the same gene map, which was again colinear with the amino acid sequence. Therefore, a protein must be encoded in a gene in a linear relationship.

Although colinearity appears to be a logical prediction, the actual demonstration by Yanofsky was critical for subsequent studies on translation. Yanofsky also determined that each amino acid was encoded by more than two nucleotides, and that each altered nucleotide affected no more than a single amino acid. We will address both of these points later in this chapter.

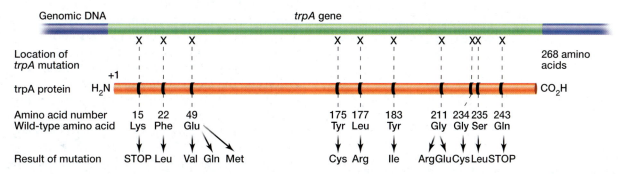

figure 11.9   Colinear relationship between the nucleotide sequence of a gene and the amino acid sequence of the encoded protein. Charles Yanofsky generated several mutations in the *E. coli trpA* gene and determined the location of the defects in the corresponding mutant proteins (either an early termination of the protein – STOP, or a change in the amino acid sequence). Colinearity indicates that a one-to-one correspondence exists between the nucleotide sequence of the gene and the amino acid sequence of the protein. Notice that some mutations could change the wild-type amino acid into either two or three different amino acids.

Restating the Concepts

▶ A colinear relationship exists between the nucleotide sequence of a gene and the amino acid sequence of the corresponding protein.

▶ An amino acid is coded by more than two nucleotides, and each nucleotide change in a gene affects no more than a single amino acid.

11.4 Translation: Preliminaries

Before proceeding to the details of translation, a snapshot of the process may help to keep the details in perspective (**fig. 11.10**). The ribosome is the site of protein synthesis. The genetic information from the gene is encoded in a messenger RNA (mRNA), which is associated with the ribosome. Each group of three nucleotides—a codon—in the mRNA specifies either an amino acid or a translation stop signal.

Amino acids are carried to the ribosome attached to transfer RNAs (tRNAs). Each tRNA has a 3 nucleotide anticodon, which is located at the end opposite the amino acid attachment site. This anticodon in the tRNA is complementary to a 3 nucleotide codon in the mRNA.

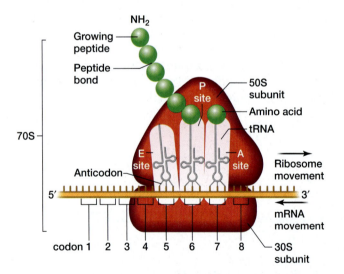

figure 11.10   An overview of bacterial translation. The ribosome (70S) is composed of a large subunit (50S) and a small subunit (30S). The ribosome contains three sites, the E site, P site, and A site. The E site is where the uncharged tRNA exits the ribosome. The P sites contains the tRNA bound to the growing peptide chain, and the A site contains the amino acyl-charged tRNA. The growing peptide chain is synthesized from the amino terminus to the carboxyl terminus as the ribosome moves along the mRNA in a 5' → 3' direction. The tRNA contains an anticodon sequence that can basepair with a complementary sequence in the mRNA (codon).

The tRNA in the E site leaves the ribosome. The peptide in the P site (codon 6) is cleaved from the tRNA and forms a peptide bond with the amino acid in the A site (codon 7). The ribosome shifts to the right (moves 5' → 3' along the mRNA), such that the tRNA in the P site, which now lacks an amino acid, shifts into the E site, and the tRNA in the A site, which now is attached to the growing peptide chain, moves into the P site. This brings codon 8 into the A site. A new tRNA with an attached amino acid can now enter the A site and basepair with codon 8. This cycle is repeated, with the polypeptide lengthening by one amino acid each time.

We will now take a look at the tRNAs and attachment of amino acids to them, and then we will review the components of ribosomes. The focus is on the bacterial system; details about eukaryotes are noted as appropriate.

Attachment of an Amino Acid to Transfer RNA

The function of the tRNA is to ensure that each codon in an mRNA chain is properly interpreted as a specific amino acid. Transfer RNA serves this function through its structure.

A tRNA molecule has an anticodon at one end (the middle loop) and an amino acid attachment site at the other end, with a double-stranded stem, as described in chapter 10. The "correct" amino acid, that is, the one corresponding to the anticodon, is attached to the tRNA by enzymes known as **aminoacyl-tRNA synthetases.** For example, arginyl-tRNA synthetase attaches an arginine to a tRNA having an anticodon that is complementary to the arginine codon. A **charged tRNA** is a tRNA with the correct amino acid attached. The tRNAs for each amino acid are designated by the convention $tRNA^{Leu}$ (for leucine), $tRNA^{His}$ (for histidine), and so on.

Aminoacyl-tRNA Synthetases

Aminoacyl-tRNA synthetases are a heterogeneous group of enzymes. In *E. coli*, they vary from monomeric proteins (one subunit) to tetrameric proteins, made up of two copies each of two different subunits. The aminoacyl-tRNA synthetases bind both the anticodon and the acceptor stem of the tRNA.

There are two classes of synthetases, which differ in the direction in which they bind the tRNA. To appreciate the synthetase–tRNA interaction, we must first twist the familiar cloverleaf structure of a tRNA (**fig. 11.11a**) to better represent the three-dimensional structure of the tRNA (fig. 11.11c). In this representation (fig. 11.11b and c), the D loop (*blue*) is below the plane of the page, the variable arm (*black*) is above the plane of the page, and the acceptor stem and anticodon loop are both in the plane of the page. Class I synthetases approach the tRNA from the D loop side or from behind the page, whereas Class II synthetases approach the tRNA from the variable arm side or from above the page.

An aminoacyl-tRNA synthetase joins an amino acid to a tRNA in a two-stage reaction. This critical reaction is based on the synthetase performing two important recognition events, the recognition of the amino acid and the recognition of specific sequences embedded in the structure of the tRNA. In the first

stage, the amino acid is activated with ATP, creating an amino acid linked to AMP (adenosine monophosphate) through a high-energy bond between the carboxyl end of the amino acid and the phosphate (**fig. 11.12**). High-energy bonds, which liberate a lot of free energy when hydrolyzed, are denoted as "~." In the second stage of the reaction, the amino acid is attached with a high-energy bond to either the 2' or 3' carbon of the ribose sugar at the 3' end of the tRNA (see fig. 11.12).

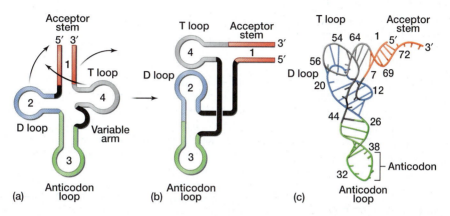

figure 11.11   Three different representations of the tRNA structure. (a) The cloverleaf tRNA structure clearly shows the three different loops and the acceptor stem in two dimensions. The arrows show how the cloverleaf structure can be twisted to form a two dimensional structure that is more similar to the actual tRNA structure (b). (c) A representation of the three dimensional structure of a tRNA showing the base pairing in the double-stranded regions. The colors correspond to the T loop (gray), D loop (blue), anticodon loop (green), and acceptor stem (red). The variable arm is shown in black.

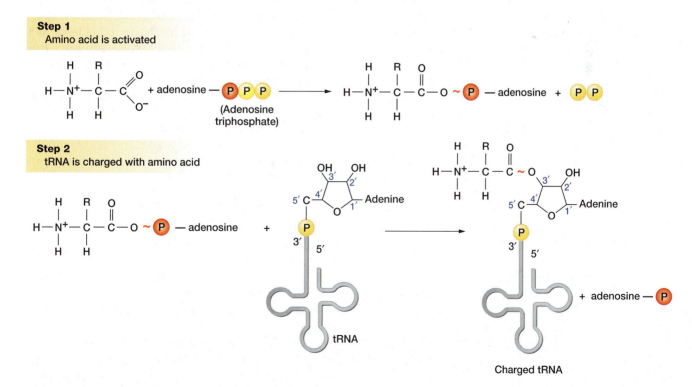

figure 11.12   The charging of a specific amino acid to its tRNA by an aminoacyl synthetase. High-energy bonds are indicated by ~. In the first step, an amino acid is attached to AMP with a high-energy bond. In the second step, the high-energy bond is transferred to the 3' hydroxyl on the tRNA, which results in the tRNA being referred to as "charged."

Proofreading and Editing by Aminoacyl-tRNA Synthetases

It is important that the correct amino acid is charged to the proper tRNA, otherwise the effort devoted to minimizing errors in DNA replication and transcription will be lost, as any amino acid could be inserted in the position of a particular mRNA codon. To increase the accuracy of the charging reaction, the aminoacyl-tRNA synthetases possess a mechanism to reduce errors.

For example, isoleucyl-tRNA synthetase incorrectly activates a valine only once for every 150 isoleucines. It is also possible to approximate the frequency of charging the wrong amino acid (valine) on a tRNA^{Ile} based on the misincorporation of valine for isoleucine in a protein. It has been determined that valine is misincorporated for isoleucine once in every 3000 positions in the ovalbumin and globin proteins. There must be a fairly effective error-reducing mechanism, as it differentiates between the correct (isoleucine) and incorrect (valine) amino acids that differ by only a single methyl group.

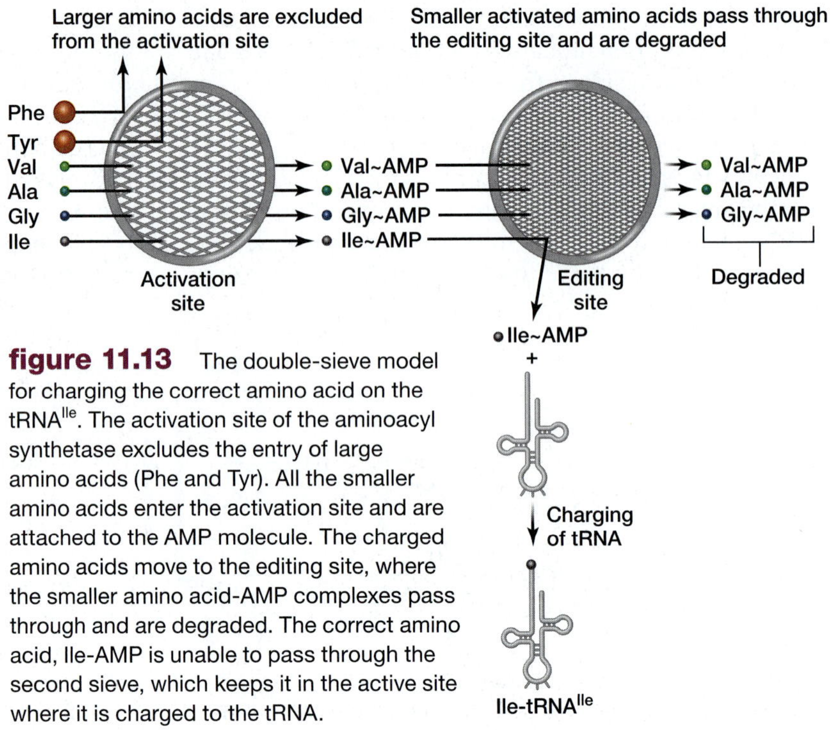

In 1977 Alan Fersht proposed a *double-sieve model* for aminoacyl-tRNA synthetases to account for minimizing charging errors. In this model, amino acids must enter the **activation site,** which is the region of the aminoacyl-tRNA synthetase where the amino acids are activated by ATP. Amino acids that are too large cannot enter the activation site (**fig. 11.13**). Amino acids that are activated then approach a second sieve, where the ones that are too small enter the **editing site,** where they are degraded. Only the activated amino acids that are the correct size will fail to enter the editing site and remain in the active site to be charged to the proper tRNA.

This double-sieve mechanism can account for the percentage of incorrect amino acids that are activated (1 in 150 for isoleucyl-tRNA synthetase) and the lower percentage of incorrectly charged tRNAs (1 in 3000). This also provides an efficient mechanism for removing incorrectly activated amino acids by a single enzyme.

The Number of Components

Bacteria have 20 aminoacyl-tRNA synthetases, one for each amino acid. Each enzyme recognizes a specific amino acid, as well as all the tRNAs that code for that amino acid. In eukaryotes, separate sets of 20 cytoplasmic and 20 mitochondrial synthetases are encoded by genes in the nucleus.

Several organisms have fewer than 20 aminoacyl-tRNA synthetases. For example, some archaea have no cysteinyl-tRNA synthetase. However, the prolyl-tRNA synthetase activates the tRNAs for both cysteine and

figure 11.13 The double-sieve model for charging the correct amino acid on the tRNA^{Ile}. The activation site of the aminoacyl synthetase excludes the entry of large amino acids (Phe and Tyr). All the smaller amino acids enter the activation site and are attached to the AMP molecule. The charged amino acids move to the editing site, where the smaller amino acid-AMP complexes pass through and are degraded. The correct amino acid, Ile-AMP is unable to pass through the second sieve, which keeps it in the active site where it is charged to the tRNA.

proline with their appropriate amino acids. Similarly, some bacteria have no glutaminyl-tRNA synthetase; the glutaminyl tRNA is charged with glutamic acid, rather than glutamine. An amido transferase enzyme then converts the glutamic acid to glutamine (see fig. 11.4).

The genetic code contains 64 possible codons (four nucleotide bases in groups of three = $4 \times 4 \times 4 = 64$). As we will discuss later in this chapter, three of these codons are used to terminate translation. Thus, 61 tRNAs are needed for the 61 different nonterminator codons. About 50 tRNAs are known in *E. coli*. The number 50 can be explained by the wobble phenomenon, which occurs in the third position of the codon. We examine this phenomenon in a later section on the genetic code.

Recognition of the Aminoacyl-tRNA During Protein Synthesis

Although amino acids enter the ribosome attached to tRNAs, it was theoretically possible that the ribosome recognized the amino acid itself, rather than the tRNA, during translation. A simple experiment was done to determine which part of the aminoacyl-tRNA was recognized by the ribosome.

In 1962, François Chapeville, Seymour Benzer, and their colleagues isolated charged Cys-tRNA^{Cys}. They chemically converted the charged cysteine to alanine by using the catalyst Raney nickel, which removes the SH group of cysteine (**fig. 11.14**). When this alanine-tRNA^{Cys} was used in protein synthesis, alanine was incorporated where cysteine should have been. This

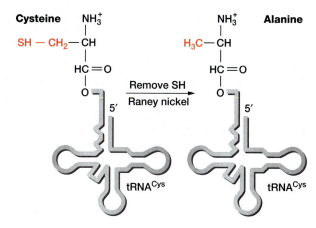

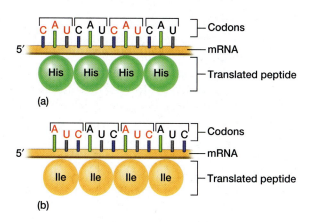

figure 11.14 Cysteine-tRNA^{Cys} treated with Raney nickel becomes alanine-tRNA^{Cys} by the removal of the SH group of cysteine. During protein synthesis, alanine will be incorporated in place of cysteine in proteins, indicating that the specificity of amino acid incorporation into proteins resides with the tRNA and not with the amino acid.

figure 11.15 The effect a frameshift mutation has on the translated protein. (a) In the wild-type reading frame of the mRNA, these codons are read as CAU repeats, coding for histidine. (b) Loss of the first cytosine (C) in the mRNA results in a shift in the mRNA reading frame, which causes the codons to be read as AUC repeats that code for isoleucine.

result demonstrated that the tRNA, not the amino acid, was recognized during protein synthesis.

The synthetase puts a specific amino acid on a specific tRNA; then, during protein synthesis, the anticodon on the tRNA—not the amino acid itself—determines which amino acid is incorporated.

 Restating the Concepts

▶ Aminoacyl-tRNA synthetases are the enzymes that attach the correct amino acid to the 3' end of the tRNA. There is usually a single synthetase for each amino acid.

▶ Charging, or attaching the amino acid to the tRNA, is a two-step process. First, the amino acid is activated by forming a high-energy bond with ATP. Second, the activated amino acid is coupled to the correct tRNA. A double-sieve model reduces errors in charging by limiting the size of amino acids that can be activated by the aminoacyl-tRNA synthetase, followed by degrading the activated amino acids that are smaller than the properly activated amino acid.

▶ The anticodon on the charged tRNA, and not the amino acid, determines which charged tRNA will be incorporated during translation. This implies that the anticodon on the tRNA, and not the amino acid, is recognized during the translation process.

11.5 Translation: The Process

Translation can be divided into three stages: initiation, elongation, and termination. **Initiation** involves the formation of the ribosome, recognition of the first codon (3 nucleotides) to be translated, and the posi-

tioning of the initiation-charged tRNA in the P site of the ribosome. **Elongation** is the repetitive process of adding amino acids via charged tRNAs through the ribosome's A site to a growing peptide chain (see fig. 11.10). **Termination** involves the recognition of the termination codon, stopping the growth of the polypeptide, and release of the polypeptide from the ribosome. In prokaryotes, once a tRNA has released an amino acid, it leaves the ribosome via the exit (E) site.

Of these three steps, it is extremely important that the translation process start precisely. If the reading of the mRNA begins one base too early or too late, the reading frame is shifted so that an entirely different set of codons is read (**fig. 11.15**). The protein produced, if any, will probably bear no structural or functional resemblance to the protein encoded by the gene.

The Role of N-Formyl Methionine

The synthesis of every protein in *E. coli* begins with the modified amino acid **N-formyl methionine** (**fig. 11.16**). However, none of the completely translated proteins in *E. coli* contains *N*-formyl methionine. Many of these proteins do not even have methionine as their first amino acid. Therefore, this initial amino acid must either be modified or removed before the final protein is achieved. In eukaryotes the initial amino acid, also methionine, does not have an *N*-formyl group.

If *N*-formyl methionine is not present in the mature proteins in bacteria, then why is it the first amino acid inserted during translation of every *E. coli* protein?

Methionine is encoded by the codon of 5'-AUG-3', which is known as the **initiation codon**. In bacteria, two tRNAs have the same anticodon sequence (3'-UAC-5'), but they differ at several nucleotide positions throughout the tRNA (**fig. 11.17**). One of these tRNAs (tRNA$_f^{Met}$), which is used only during translation initiation, is charged with methionine and then chemically modified to *N*-formyl methionine (fMet). The mismatched 5' cytosine on the tRNA$_m^{Met}$ and the three G–C pairs in the anticodon loop prevent the charged methionine from being formylated and from being used in translation initiation (see fig. 11.17). The properly base-paired C–G on the tRNA$_f^{Met}$ (the other form) and the absence of the three G–C pairs in the anticodon loop allows the charged methionine to be formylated and used only in translation initiation. The tRNA$_m^{Met}$ is used only to insert methionine in the peptide chain after the first position.

Because both tRNAs recognize the UAC codon, the bacterial cell must possess a system to ensure that the tRNA$_f^{Met}$ is only used during initiation and tRNA$_m^{Met}$ is only used during elongation. As discussed in the following section, the initiator tRNA$_f^{Met}$ must bind to the P site of the ribosome, and the internal tRNA$_m^{Met}$ must bind to the A site. The three G–C base pairs in the anticodon stem of tRNA$_m^{Met}$ (fig. 11.17b) are required for the charged Met-tRNA$_m^{Met}$ to enter the A site, which allows it to be used for translation elongation. However, the absence of the three G–C base pairs in the tRNA$_f^{Met}$ (fig. 11.17a) prevents this charged tRNA from entering the A site, and therefore, cannot be used in translation elongation.

Because the initiation methionine is not formylated in eukaryotes, the eukaryotic initiator tRNA is designated tRNA$_i^{Met}$; a separate internal methionine tRNA, termed tRNA$_m^{Met}$, is present in eukaryotes, as in bacteria.

figure 11.16   The structures of the amino acids methionine and *N*-formyl methionine. The *N*-formyl methionine, which contains an aldehyde in place of an amine hydrogen, is used as the first codon during bacterial translation initiation, while methionine is used for internal positions.

figure 11.17   The two tRNAs for methionine in *E. coli*. (a) The initiator tRNA, tRNA$_f^{Met}$. (b) The interior tRNA, tRNA$_m^{Met}$. Notice that the 5' base, a G, is basepaired in tRNA$_f^{Met}$, while the 5' end, a C, is not in tRNA$_m^{Met}$. Also, note the three consecutive G-C basepairs in the anticodon stem in tRNA$_m^{Met}$ that are absent in the tRNA$_f^{Met}$. The key of the modified bases in the tRNA are shown in red.

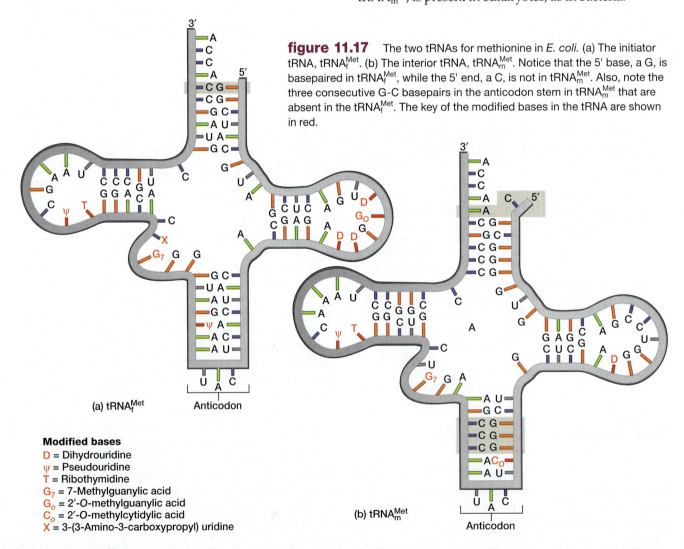

Modified bases

D = Dihydrouridine
ψ = Pseudouridine
T = Ribothymidine
G$_7$ = 7-Methylguanylic acid
G$_o$ = 2'-O-methylguanylic acid
C$_o$ = 2'-O-methylcytidylic acid
X = 3-(3-Amino-3-carboxypropyl) uridine

(a) tRNA$_f^{Met}$ Anticodon

(b) tRNA$_m^{Met}$ Anticodon

Translation Initiation

The subunits of the bacterial ribosome (30S and 50S) usually dissociate from each other when not involved in translation. To begin translation, an **initiation complex** forms, consisting of the following components in bacteria:

- the 30S subunit of the ribosome,
- an mRNA,
- the charged *N*-formyl methionine tRNA (fMet-tRNA$_f^{Met}$), and
- three **initiation factors (IF1, IF2, IF3).**

Initiation factors (as well as elongation and termination factors) are proteins loosely associated with the ribosome, which are involved in bringing different components together in the ribosome.

The components that form the initiation complex interact in a series of steps, summarized in **figure 11.18**:

1. First, IF1 and IF3 bind to the 30S ribosomal subunit (see fig. 11.18, step 1). IF1 binds to what will become the A site of the ribosome, which prevents the fMet-tRNA$_f^{Met}$ from binding to that site and forces it to bind to the P site. IF1 also helps the other two initiation factors bind to the 30S ribosomal subunit and stabilizes

the 30S initiation complex. IF3 binding permits the 30S subunit to bind the mRNA (see fig. 11.18, step 1).

2. Second, a complex forms between IF2, the charged fMet-tRNA$_f^{Met}$, and GTP (guanosine triphosphate; fig. 11.18, step 2). Because IF2 only binds the charged fMet-tRNA$_f^{Met}$, this initiator tRNA can only be involved in translation initiation.

3. Finally, these two complexes are brought together to form the initiation complex (fig. 11.18, step 3).

The hydrolysis of GTP to GDP + P$_i$ (inorganic phosphate) produces conformational changes that allow the 50S ribosomal subunit to join the initiation complex, which forms the complete ribosome. After GTP hydrolysis, the initiation factors and GDP are released.

Frequently, the hydrolysis of a ribonucleotide triphosphate (such as ATP or GTP) in a cell occurs to release the high energy in the phosphate bonds for use in a metabolic process. In translation, however, the hydrolysis apparently changes the shape of the guanosine so that it and the initiation factors can be released from the newly formed 70S particle. Thus, hydrolysis of GTP during translation is required to induce a conformational change rather than covalent bond formation.

The bacterial ribosome apparently recognizes the mRNA through complementarity of a region at the 3'

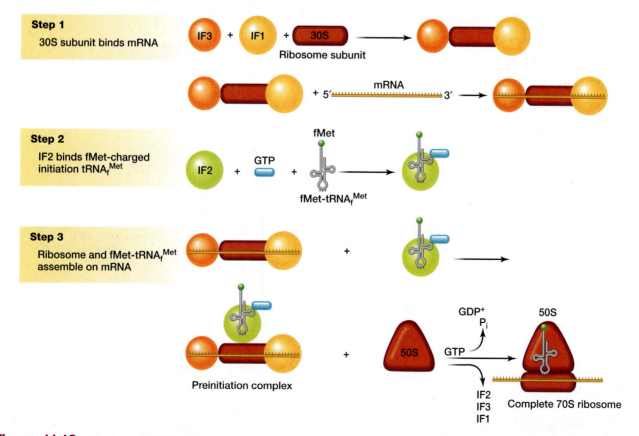

figure 11.18   The bacterial 70S ribosome forms in a three-step process. In the first step, the 30S ribosomal subunit combines with IF1 and IF3 to form a complex that binds the mRNA. Second, the initiator tRNA is bound by IF2. In the final step, the components from steps 1 and 2 combine to form the preinitiation complex, followed by the binding of the 50S subunit to form the 70S ribosome.

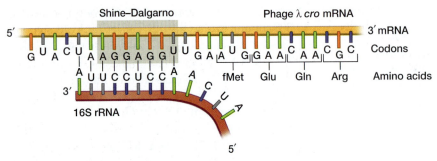

figure 11.19 The Shine-Dalgarno sequence is required for proper bacterial translation initiation. The Shine-Dalgarno sequence (AGGAGGU) is on the bacterial mRNA just upstream from the initiation codon AUG, which is complementary to a sequence (UCCUCCA) on the 3' end of the 16S ribosomal RNA. Basepairing between these sequences positions the small ribosomal subunit in the correct position to begin translation on the correct codon.

end of the 16S ribosomal RNA, which is a component of the 30S subunit, and a region slightly upstream from the initiation codon (AUG) on the mRNA. This consensus sequence, the **Shine–Dalgarno sequence** (5'-AGGAGGU-3', **fig. 11.19**), is named after the people who first identified it. Although a good deal of homology exists between bacterial and eukaryotic small ribosomal RNAs, the Shine–Dalgarno region is absent in eukaryotic mRNA. The binding of the 16S rRNA to the Shine–Dalgarno sequence positions the 30S ribosomal subunit at the proper location on the mRNA to begin translation at the initiation codon.

Differences in Eukaryotes

Translation initiation in eukaryotes is generally similar to *E. coli*, but more complex. The eukaryotic initiation factor abbreviations are preceded by an "e" to denote that they are eukaryotic (eIF1, eIF2, etc.). At least 11 different initiation factors are involved, including a specific 5' Cap-binding protein (**eIF4F**) and a poly(A)-binding protein (**PABP**).

Initiation Codon Recognition   The mechanism for recognizing the initiation codon on eukaryotic mRNA differs from the mechanism in bacteria. Initially, the translation preinitiation complex forms at the 5' cap of the mRNA (chapter 10). This complex includes the mRNA and eIF4F, which possesses both a 5' Cap-binding activity and an RNA helicase to remove any mRNA secondary structure (**fig. 11.20**). This complex binds a 43S protein complex that is composed of the small 40S ribosomal subunit, eIF3, and the eIF2 ternary complex to generate a 48S complex. The presence of a poly(A) tail on the mRNA allows the poly(A)-binding protein to bind to the 3' tail and simultaneously stabilize the growing protein complex at the 5' cap. As in bacteria, the eIF2 ternary complex includes eIF2, GTP, and the initiator-charged tRNA, Met–tRNA$_i^{Met}$.

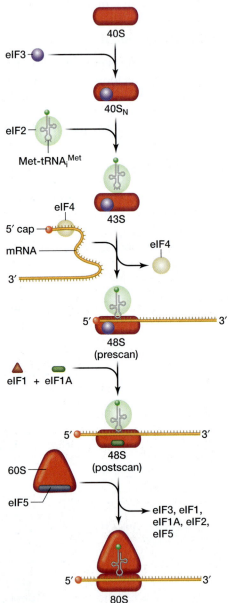

figure 11.20 Translation initiation in eukaryotes. The 40S ribosomal small subunit binds eIF3, which then binds the initiator Met- tRNA$_i^{Met}$ and eIF2 to form the 43S complex. This complex, in association with eIF4, binds the 5' cap on the mRNA. Binding of the eIF1 and eIF1A proteins allows the small ribosomal complex to scan from the 5' cap to the translation initiation codon. Binding of eIF5 allows the 60S large ribosomal complex to associate and form the translation initiation complex.

Binding of eIF1 and eIF1A stimulates the 48S complex to move along the mRNA until it recognizes the initiation codon (see fig. 11.20). This model is referred to as the **scanning hypothesis.** When the complex reaches the initiation codon, the eIF2-GTP hydrolyzes its bound GTP to GDP, which releases eIF3 and eIF2-GDP, which allows the large ribosomal subunit to associate with the small subunit and generate the functional ribosome. The eIF-2B protein catalyzes the exchange of GDP for GTP on eIF2-GDP, which regenerates the eIF2-GTP to perform another round of translation preinitiation complex formation.

Note that the eIF2 and eIF-2B are different protein complexes, with the former being required to bring the Met-tRNA$_i^{Met}$ to the small ribosomal subunit, and the latter being required to reconstitute the active form of eIF2 (eIF2-GTP) from the inactive form (eIF2-GDP).

In several known cases, a process called **shunting** occurs in eukaryotes, in which the first AUG does not serve as the initiation codon (**fig. 11.21**). During

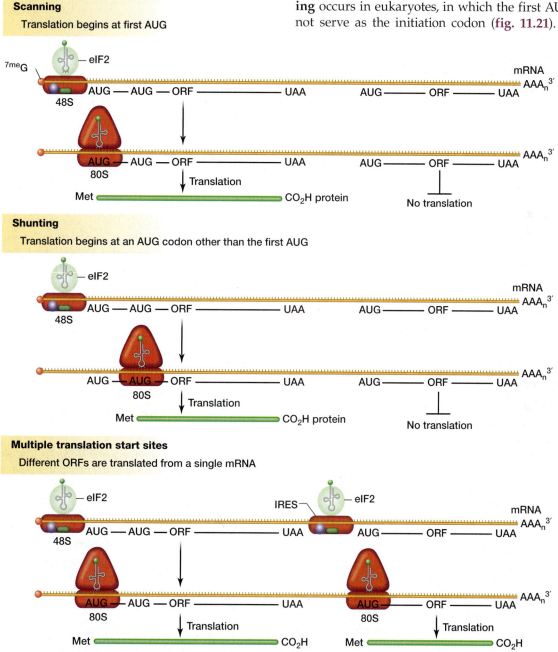

figure 11.21   Three different mechanisms used to identify the translation initiation codon in eukaryotes. Shown are the eukaryotic 5' cap, the translation initiation AUG codon, translation termination codon UAA, and the 3' poly(A) tail. In the scanning model, the small ribosomal subunit moves along the mRNA to the first AUG encountered. Only the first open reading frame (ORF) will be translated because the ribosome can only bind to the 5' cap and begins translating at the first AUG. In shunting, the first AUG is not recognized as the translation initiation codon, due to the presence of secondary structure in the mRNA or the absence of a Kozak sequence. The ribosome continues on to a subsequent AUG that is properly recognized and translation begins. Again, downstream open reading frames will not be translated. Finally, an internal ribosome entry site (IRES) can be used for ribosome assembly, which permits multiple open reading frames to be translated from the same mRNA.

shunting, the small ribosomal subunit bypasses AUG codons in a region of the mRNA upstream of the initiation codon, called the *leader,* or **5' untranslated region (5' UTR),** in favor of an AUG farther down the mRNA.

Two mechanisms appear to be at work in selecting the correct initiation codon. First, the correct AUG codon is usually found within a consensus sequence of PuNN*AUG*G, where Pu is a purine (A or G) and N is any ribonucleotide. This sequence is called the **Kozak sequence,** named after Marilyn Kozak who first identified it. Experiments have demonstrated that the purine and G residues can affect translation initiation frequency by more than tenfold. The second mechanism in selecting the correct initiation codon is the secondary structure in the mRNA upstream from the initiation AUG codon. In some cases, very small genes, called **open reading frames (ORFs),** are present in this region of the mRNA and play some role in shunting.

Initiation at More Than One Site   Under most circumstances, the eukaryotic preinitiation complex assembles at the 5' cap, moves to the translation initiation codon, translates the ORF, and then dissociates from the mRNA (see following discussion). In this situation, a second ORF on the mRNA cannot be translated because the translation preinitiation complex must reassemble at the 5' cap, where it again will translate the first ORF (see fig. 11.21).

Under certain circumstances, the translation preinitiation complex can assemble on the mRNA, between two different ORFs (see fig. 11.21). In this case, the preinitiation complex assembles on a sequence that is called an **internal ribosome entry site (IRES).** These sequences, which are several hundred nucleotides long, were discovered in the poliovirus RNA and in several cellular mRNAs. Although scanning accounts for the initiation of most eukaryotic mRNAs at their 5' ends, some initiation can take place internally in mRNAs that have internal ribosome entry sites.

Aminoacyl (A) and Peptidyl (P) Sites in the Ribosome

As mentioned earlier, when the initiator tRNA joins the 30S subunit of the bacterial ribosome with its mRNA attached, three potential sites are present in the ribosome. These sites, or cavities, in the ribosome, are referred to as the **A (aminoacyl) site,** the **P (peptidyl) site,** and the **E (exit) site** in bacteria (**fig. 11.22**). Here, we concentrate on the A and P sites, each of which contains a tRNA just before forming a peptide bond.

The P site contains the tRNA with the growing peptide chain (peptidyl-tRNA), and the A site contains a new tRNA with its single amino acid (aminoacyl-tRNA, see fig. 11.22). The E site helps eject the tRNAs that have lost their charged amino acid after a peptide

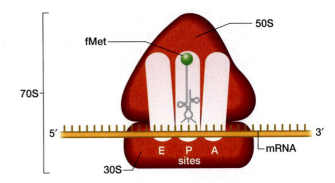

figure 11.22   Translation initiation in bacteria occurs when the initiation fMet-tRNA$_f^{Met}$ binds in the P site of the 70S ribosome. The A site is available for the second charged tRNA to enter and basepair with the codon on the mRNA. The mRNA runs through the bottom of the sites.

bond forms. When the complete 70S ribosome has formed, the initiation fMet-tRNA$_f^{Met}$ is located in the P site (see fig. 11.22) for two reasons. First, IF1 is already bound to and blocking the A site. Second, the three G–C base pairs in the anticodon stem (see fig. 11.17) prevent it from entering the A site, which leaves the P site as the only accessible site for a charged tRNA.

The binding of fMet-tRNA$_f^{Met}$ to the P site of the ribosome during translation initiation is the only time that a charged tRNA enters the ribosome's P site. The association of tRNA and ribosome is aided by the two unpaired cytosines in the 5'-CCA-3' terminus of all tRNAs and a guanine in the 23S ribosomal RNA.

Restating the Concepts

▶ Translation initiation requires *N*-formyl-Met-tRNA$_f^{Met}$ in bacteria. Structural features cause this charged tRNA to be located within the P site of the assembling ribosome. In contrast, bacterial Met-tRNA$_m^{Met}$ is charged with an unmodified methionine and only enters the A site of the ribosome and, therefore, is only used in translation elongation.

▶ Bacterial translation initiation requires three initiation factors. IF1 and IF3 are required for generating the preinitiation complex with the 30S ribosomal subunit. IF1 is bound to what will become the A site, which helps to prevent a charged tRNA from entering the A site until the initiator tRNA is present in the P site. IF2 forms a complex with the fMet-tRNA$_f^{Met}$ and delivers it to the P site of the ribosome. Because IF2 specifically binds fMet-tRNA$_f^{Met}$ and not the Met-tRNA$_m^{Met}$, bacterial translation begins with the *N*-formyl-Met-tRNA$_f^{Met}$.

▶ Eukaryotic ribosome assembly begins on the 5' cap of the mRNA. The scanning mechanism allows the small ribosomal subunit to move along the mRNA to the first AUG, where translation begins. Shunting allows the first AUG to be skipped, based on the presence of either secondary structure in the mRNA or the Kozak consensus sequence. Only the presence of an internal ribosome entry site between ORFs will allow multiple ORFs to be translated on a eukaryotic mRNA.

▶ Unlike bacteria, eukaryotic translation begins with a Met-tRNA$_i^{Met}$ and does not use an *N*-formyl-methionine.

Translation Elongation

The next event in translation, after the initiator tRNA is present in the P site, is to position the second tRNA, which is specified by the codon at the A site, so that it forms hydrogen bonds between its anticodon and the second codon on the mRNA. This represents the first step in translation elongation.

Positioning a Second Transfer RNA

The introduction of a charged tRNA into the A site of the ribosome requires the correct charged tRNA, another GTP, and two proteins called **elongation factors** (**EF-Ts** and **EF-Tu**). EF-Tu, bound to GTP, is

required to position a charged tRNA into the A site of the ribosome (**fig. 11.23**). After the charged tRNA is positioned, the GTP is hydrolyzed and the EF-Tu/GDP complex is released from the ribosome. EF-Ts displaces the GDP on EF-Tu to generate the EF-Tu/EF-Ts complex. Then a new GTP displaces EF-Ts, which regenerates the EF-Tu/GTP complex that can bind another tRNA.

Here again, GTP hydrolysis results in a changed conformation so that the EF-Tu/GDP complex can depart the ribosome after the tRNA is placed in the A site (**fig. 11.24**). **Figure 11.25** shows the ribosome at the end of this step. EF-Tu does not bind to fMet-tRNA$_f^{Met}$, which is another reason why this formylated methionine cannot be inserted into a growing peptide chain.

It takes several milliseconds for the GTP to be hydrolyzed, and another few milliseconds for the EF-Tu/GDP to actually leave the ribosome. During those two brief intervals of time, the base-pairing between the codon (mRNA) and anticodon (tRNA) is scrutinized. If the correct tRNA is in place, a peptide bond forms as described in the next section. If not, the charged tRNA is released, and a new cycle of EF-Tu/GTP-mediated testing of tRNA begins.

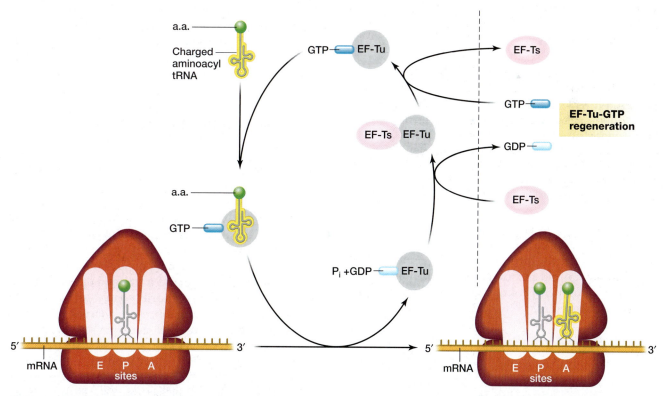

figure 11.23   The EF-Ts/EF-Tu cycle. EF-Ts and EF-Tu are required for an amino acyl-charged tRNA to basepair with the mRNA in the A site of the ribosome. At *lower right*, the EF-Tu is attached to a GDP. The GDP is displaced by EF-Ts, which in turn is displaced by GTP. A charged tRNA attaches to the EF-Tu/GTP complex and is brought to the ribosome. If the codon-anticodon fit is correct, the tRNA basepairs with the mRNA in the A site. GTP is hydrolyzed to GDP + Pi, allowing the EF-Tu/GDP complex to release. Since EF-Tu has a strong affinity for GDP, the role of EF-Ts is to displace the GDP, and later to be replaced by GTP.

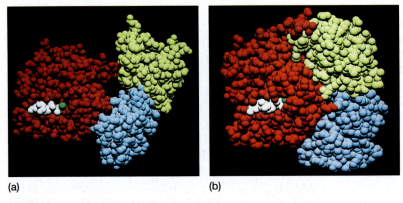

(a)   (b)

figure 11.24 Space-filling model of EF-Tu bound with (a) GDP and (b) GTP, showing the change in the EF-Tu protein's structure. The *yellow, blue,* and *red* structures are domains of the EF-Tu protein. The GTP and GDP molecules are in *white,* with a magnesium ion, Mg^{2+}, in *green.* When EF-Tu is bound with GDP (a), there is a visible hole in the EF-Tu molecule (arrow). The hole disappears when GTP is bound. The aminoacyl-tRNA is believed to bind between the *red* and *yellow* domains.

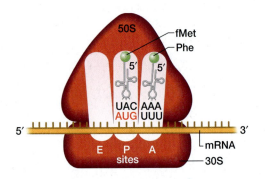

figure 11.25 A ribosome with two tRNAs attached. The P site in the ribosome contains the initiation fMet- tRNA$_f^{Met}$. Only at translation initiation will a charged tRNA with only a single amino acid attached be located in the P site. In this example, the second codon (UUU) is for the amino acid phenylalanine. The Phe-tRNA^{Phe} is located in the A site, with the two amino acids next to each other.

The error rate of this process is only about 1 in 10,000 amino acids incorporated into protein. The speed of amino acid incorporation is about 15 amino acids per second in prokaryotes and about 2–5 per second in eukaryotes.

Peptide Bond Formation

The two amino acids on the two tRNAs in the A and P sites are now in position to form a peptide bond between them; both amino acids are juxtaposed to an enzymatic center, **peptidyl transferase,** in the 50S subunit. This enzymatic center, an integral part of the 50S subunit, resides in a ribosomal RNA. The peptidyl transferase activity involves a bond transfer from the carboxyl end of *N*-formyl methionine to the amino end of the second amino acid (phenylalanine in fig. 11.25).

Every subsequent peptide bond formation is identical, with the carboxyl end of the growing peptide chain in the P site forming a peptide bond with the amino end of the amino acid in the A site. Cleavage of the high-energy ester bond between the tRNA in the P site and its amino acid provides the energy to drive the peptide bond formation (**fig. 11.26**). Immediately after forming the peptide bond, the tRNA with the dipeptide is in the A site, and a depleted tRNA is in the P site.

Translocation

The next stage in elongation is translocation of the ribosome in relation to the tRNAs and the mRNA. Elongation factor EF-G catalyzes the translocation process of converting the ribosome from the *pretranslocational state* to the *posttranslocational state,* which physically moves the mRNA and its associated tRNAs (**fig. 11.27**).

This movement is accomplished by the hydrolysis of a GTP to GDP after EF-G enters the ribosome at the A site. After the first posttranslocational state is reached, the depleted tRNA in the E site is ejected, leaving the ribosome ready to accept a new charged tRNA in the A site. A computer-generated diagram of a ribosome with all three tRNA sites occupied is shown in figure 11.27b.

When translocation is complete, the situation is repeated as diagrammed in figure 11.22, except that instead of fMet-tRNA$_f^{Met}$, the P site contains the second tRNA (in this case, tRNA^{Phe}) with a dipeptide attached to it. The process of elongation is then repeated, with a third tRNA coming into the A site. The process repeats from here to the end (**fig. 11.28**), synthesizing a peptide starting from the amino (N terminal) end and proceeding to the carboxyl (C terminal) end.

During the repetitive aspect of protein synthesis, two GTPs are hydrolyzed per peptide bond: one GTP in the release of EF-Tu from the A site, and one GTP in the translocational process of the ribosome after the peptide bond has formed. In addition, every charged tRNA has had an amino acid attached at the expense of the hydrolysis of an ATP to AMP + PP$_i$.

In eukaryotes, three elongation factors perform the same tasks that EF-Tu, EF-Ts, and EF-G perform in bacteria. The factor eEF1α replaces EF-Tu; eEF1βγ replaces EF-Ts; and eEF2 replaces EF-G. Additionally, the eukaryotic ribosome lacks the E site. Thus, after the peptide is cleaved from the tRNA in the P site, this free tRNA is released directly from the ribosome.

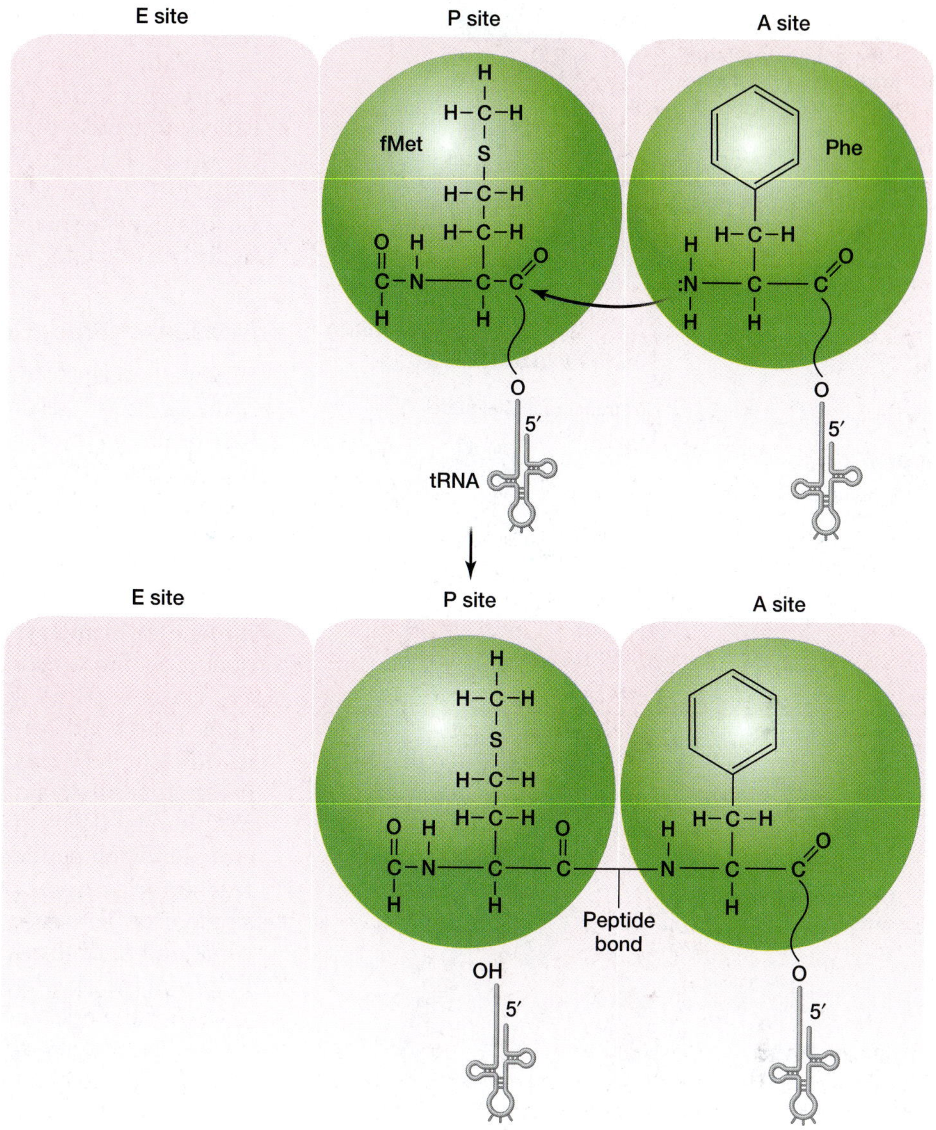

E site   P site   A site

E site   P site   A site

figure 11.26   This figure shows the formation of the peptide bond in the ribosome between the translation initiation N-formyl methionine (fMet) and phenylalanine (Phe). The high-energy bond attaching the carboxyl end of the first amino acid (fMet) in the P site to its tRNA is transferred to the amino end of the second amino acid (Phe) in the A site. The first tRNA (still in the P site) is now uncharged, whereas the second tRNA (in the A site) has a dipeptide attached.

Restating the Concepts

▶ Translation elongation requires the initiator tRNA or a tRNA with a growing peptide chain in the P site, EF-Tu/GTP binding, and delivery of a charged tRNA to the A site. Hydrolysis of the GTP releases the EF-Tu/GDP, which is regenerated to EF-Tu/GTP by EF-Ts.

▶ Elongation involves a peptidyl transferase activity in the ribosome catalyzing the cleavage of the high-energy ester bond in the P site (between the amino acid and the tRNA), which allows the free carboxyl end of this amino acid to form a peptide bond with the free amino end of the charged amino acyl-tRNA in the A site. EF-G binds the A site and hydrolyzes GTP

to drive the translocation of the ribosome along the mRNA. This causes the tRNA in the P site to move to the E site, where it exits the ribosome, and the tRNA and peptide in the A site to move to the P site. The empty A site is now ready to accept another EF-Tu/GTP-charged tRNA.

▶ In eukaryotes, the three elongation factors have different names, but perform the same functions. Because the eukaryotic ribosome lacks the E site, the tRNA in the P site immediately exits the ribosome after its peptide has been transferred to the next charged amino acid in the A site.

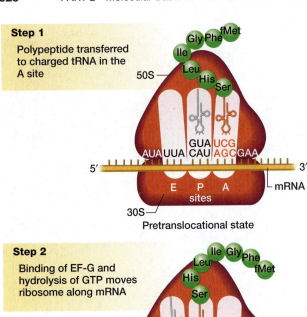

Step 1

Polypeptide transferred to charged tRNA in the A site

Pretranslocational state

Step 2

Binding of EF-G and hydrolysis of GTP moves ribosome along mRNA

First posttranslocational state

Step 3

Uncharged tRNA exits from E site

Final posttranslocational state

(a)

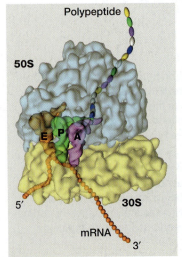

Polypeptide

50S

E   P   A

50S

30S

5′

mRNA

3′

(b)

Translation Termination

Termination of protein synthesis in both bacteria and eukaryotes occurs when one of three **stop codons,** also called **nonsense codons,** appears in the A site of the ribosome. These codons are 5′-UAG-3′, 5′-UAA-3′, and 5′-UGA-3′. In bacteria, three proteins called **release factors (RF)** are involved in termination, and a GTP is hydrolyzed to GDP + P_i.

Nonsense (Stop) Codons

When a nonsense codon enters the A site on the ribosome, a class I release factor (RF1 or RF2) binds it. RF1 binds the stop codons UAA and UAG, and RF2 recognizes UAA and UGA (**fig. 11.29**). Both release factors bind these specific codons because they have tripeptides that mimic anticodons and recognize the stop codons: proline–alanine–threonine in RF1, and serine–proline–phenylalanine in RF2. In this **molecular mimicry,** a protein mimics the shape of a nucleic acid to function properly.

The bound RF1 or RF2 then promotes the hydrolysis of the ester bond between the terminal amino acid and its tRNA in the P site. This hydrolysis appears to utilize the same mechanism as elongation, except the absence of a charged tRNA in the A site forces the free carboxyl end of the polypeptide to use H_2O as the acceptor. This reaction releases the polypeptide from the ribosome. A single class II release factor, RF3, then binds the ribosome and hydrolyzes its GTP to release either RF1 or RF2 from the ribosome.

After the release factors act, with the hydrolysis of a GTP and the release of either RF1 or RF2, the ribosome has completed its task of translating mRNA into a polypeptide. A **ribosome recycling factor (RRF)** enters the A site of the ribosome, which then, with the assistance of EF-G, translocates the RRF to the P site. With RRF bound to the large ribosomal subunit and the entry of IF3 to bind to the small ribosomal subunit, the ribosome dissociates. Binding of IF3 to the small subunit nearly regenerates the complex that is necessary to reinitiate translation, with only IF1 and mRNA binding necessary to complete the complex (see fig. 11.18, step 1). **Table 11.2** compares bacterial and eukaryotic translation.

figure 11.27   Ribosome translocation along the mRNA. (a) Binding of EF-G and GTP hydrolysis drives translocation of the ribosome along the mRNA and converts it from a pretranslocational state (P and A sites occupied) to a posttranslocational state (E and P sites occupied). The uncharged tRNA in the E site is then ejected, leaving the ribosome ready to accept the next charged tRNA in the A site. (b) A model of the *E. coli* 70S ribosome with tRNAs in the P and E sites. The structure was determined by electron microscopy of rapidly frozen samples. The positions of the mRNA and growing polypeptide chain are shown.

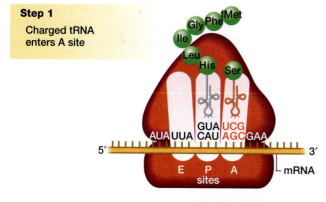

Step 1

Charged tRNA enters A site

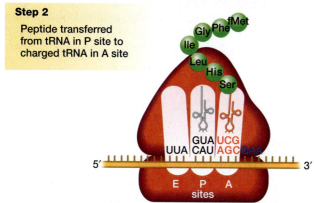

Step 2

Peptide transferred from tRNA in P site to charged tRNA in A site

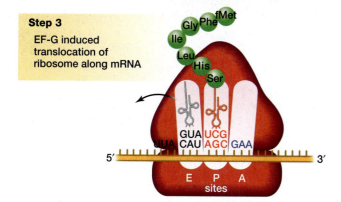

Step 3

EF-G induced translocation of ribosome along mRNA

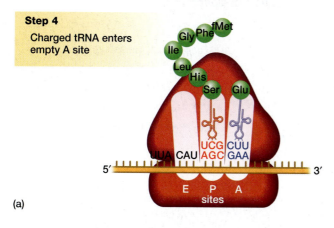

Step 4

Charged tRNA enters empty A site

(a)

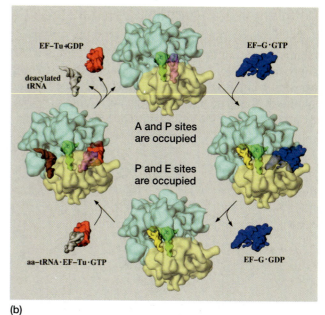

(b)

figure 11.28 Cycle of peptide bond formation and translocation on the ribosome. (a) After the peptide bond is produced (fig. 11.26), EF-G binds and hydrolyzes GTP to drive ribosome translocation one codon along the mRNA. The tRNA with the peptide is in the P site, and the A site is again open. The uncharged tRNA in the E site exits the ribosome. In this example, the next tRNA that moves into the A site is charged with glutamic acid. (b) Three-dimensional model of the translocation process minus the mRNA and amino acids. The tRNA in the A site is *pink,* the tRNA in the P site is *green,* and the tRNA in the E site is *yellow,* then *brown* when ready to leave. Going clockwise from the top, in which the A and P sites are occupied: EF-G translocates the ribosome after peptide bond formation and then evacuates the A site. EF-Tu brings a new charged tRNA to the A site while the E site is emptied.

The Rate and Cost of Translation

The average speed of protein synthesis is about 15 peptide bonds per second in prokaryotes (and 2–5 amino acids per second in eukaryotes). Discounting the time for initiation and termination, an average *E. coli* protein of 300 amino acids is synthesized in about 20 sec. An equivalent eukaryotic protein takes between 1 and 2.5 min.

The energy cost is at least four high-energy phosphate bonds per peptide bond formed: two from an ATP during tRNA charging, and two from GTP hydrolysis during tRNA binding at the A site and translocation—or about 1200 high-energy bonds per protein. This cost is very high; about 90% of the energy production of an *E. coli* cell goes into protein synthesis. This high energy cost is presumably the price a living system has to pay for the speed and accuracy of its protein synthesis.

Coupling of Transcription and Translation

In prokaryotes, which lack a nuclear envelope, translation begins before transcription is completed. **Figure 11.30** shows a length of an *E. coli* chromosome that is being transcribed by RNA polymerase. The mRNA, still being synthesized, can be seen extending away from the DNA. Attached to the mRNA are about a dozen ribosomes.

Because translation begins at the 5' end of the mRNA, an initiation complex can form and translation can begin shortly after transcription begins. As translation proceeds along the mRNA, its 5' end will again become exposed, and a new initiation complex can form. The occurrence of several ribosomes translating the same mRNA is referred to as a **polyribosome,** or simply a **polysome** (**fig. 11.31**).

In prokaryotes, most mRNAs contain the information for several genes. These mRNAs are said to be **polycistronic** (**fig. 11.32**). (*Cistron* is another term for gene; see chapter 16). Each gene on the mRNA is translated independently: Each has its own Shine–Dalgarno sequence for ribosome recognition and a translation initiation codon (AUG) for fMet (see fig. 11.32). The ribosome that completes the translation of the first gene may or may not

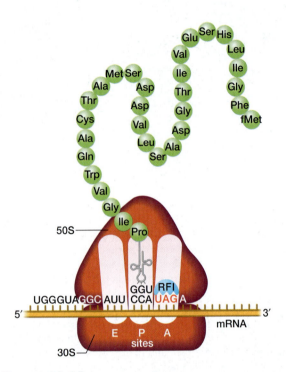

figure 11.29 Chain termination in the bacterial ribosome. One of two release factors (RF1 or RF2) recognizes a nonsense codon in the A site. In this case, RF1 binds the UAG in the A site. When the peptide bond to the tRNA in the P site is broken, the peptide and mRNA are released, and the ribosome disassociates into the 30S and 50S subunits.

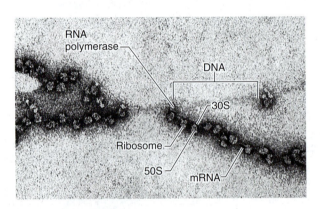

figure 11.30 A bacterial polysome (i.e., multiple ribosomes on the same strand of mRNA). Each ribosome is approximately 25 nm across. Also visible in this illustration are a region of the DNA molecule and the RNA polymerase that is transcribing the mRNA being translated. Thus, transcription and translation are occurring simultaneously.

table 11.2 Some Comparisons Between Bacterial and Eukaryotic Translation

	Bacteria	Eukaryotes
Initiation codon	AUG, occasionally GUG, UUG	AUG, occasionally GUG, CUG
Initiation amino acid	*N*-formyl methionine	Methionine
Initiation tRNA	$tRNA_f^{Met}$	$tRNA_i^{Met}$
Interior methionine tRNA	$tRNA_m^{Met}$	$tRNA_m^{Met}$
Initiation factors	IF1, IF2, IF3	At least 11 different eIF proteins
Elongation factor	EF-Tu	$eEF1\alpha$
Elongation factor	EF-Ts	$eEF1\beta\gamma$
Translocation factor	EF-G	eEF2
Release factors	RF1, RF2, RF3, RRF	eRF1, eRF3
Site of ribosome assembly	At initiation codon with Shine–Dalgarno sequence	At 5' cap and then scanning to first AUG or shunting to a downstream AUG; IRES after the first open reading frame
tRNA-binding sites in ribosome	A site, P site, E site	A site, P site

continue to the second gene after dissociation. The translation of any gene follows all the steps we have outlined.

In eukaryotes, however, almost all mRNAs contain the information for only one gene (**monocistronic**). Because most ribosomal small subunits recognize and bind to the eIFs that bind the 5' cap, and each eukaryotic mRNA has only one cap, usually only one polypeptide can be translated for any given mRNA. Exceptions occur when the mRNAs contain internal ribosome entry sites as described earlier.

Although it is certainly not the rule, the translated peptide can be modified or cleaved into smaller functional peptides. For example, in mice, a single mRNA codes for a protein that is later cleaved into epidermal growth factor and at least seven other related peptides. In addition, the same sequence can, in some cases, give rise to alternative proteins through alternative start codons, termination read-through, or alternative splicing.

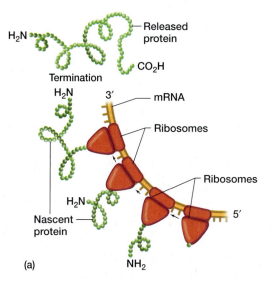

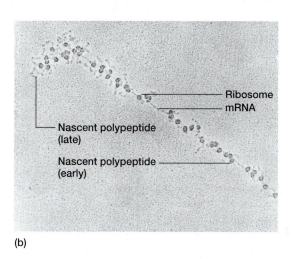

figure 11.31   (a) Schematic of protein synthesis at a bacterial polysome. Messenger RNA is translated by multiple ribosomes simultaneously while the nascent polypeptides exit from a tunnel in the 50S subunit. The ribosomes located at the 5' end of the mRNA contain shorter peptides, while ribosomes located closer to the 3' end of the mRNA are associated with longer peptides. This is consistent with translation occurring along the mRNA in a 5' → 3' direction. (b) A mRNA from the midge, *Chironomus tentans,* showing attached ribosomes and nascent polypeptides emerging from the ribosomes. Note the 5' end of the mRNA at the *right* (small peptides). Magnification 165,000×.

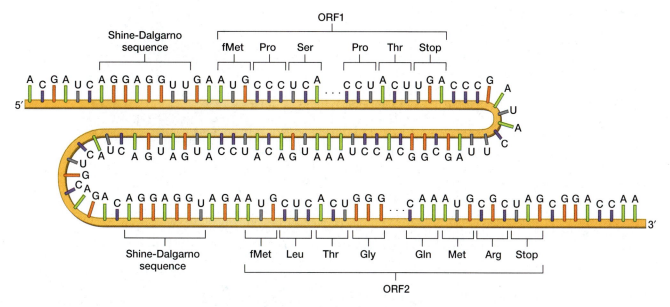

figure 11.32   Organization of a bacterial polycistronic mRNA. This mRNA contains two different open reading frames (ORF) that correspond to two different genes. Each open reading frame is preceded by a Shine-Dalgarno sequence to bind the small ribosomal subunit and permit translation of that specific open reading frame. Each open reading frame also contains a translation initiation codon (AUG) and termination codon (UGA).

It's Your Turn

Now that we have discussed the various molecules and steps that are involved in translation, put this knowledge to work and try problem 17.

Restating the Concepts

▶ Three different codons are recognized for translation termination. Release factors bind to the termination codon, and hydrolysis of the ester bond in the P site releases the polypeptide from the ribosome. A ribosome recycling factor and EF-G are responsible for dissociating the two ribosomal subunits after completing translation.

▶ Prokaryotes possess polycistronic mRNAs that are simultaneously transcribed and translated. Eukaryotes primarily have monocistronic mRNAs because the ribosome can only bind to an internal site on the mRNA and translate a second gene on an mRNA when an internal ribosome entry site (IRES) is present. Transcription and translation are not coupled in eukaryotes because the mRNA is transcribed and processed in the nucleus and then transported to the cytosol to be translated.

The Signal Hypothesis

In chapter 10, we briefly discussed the shape and composition of the ribosomal subunits. All of the protein and RNA components of ribosomes have been isolated. Assembly pathways are known. We know approximately where the mRNA, initiation factors, and EF-Tu are located on the 30S subunit during translation (see fig. 11.27). We also know where peptidyl transferase activity and EF-G reside on the 50S subunit, which has a cleft leading into a tunnel that passes through the structure. At present, it seems that the nascent peptide passes through this tunnel, emerging close to a membrane-binding site (**fig. 11.33**). The tunnel can hold a peptide of about 40 amino acids in length. Note that although every ribosome has a membrane-binding site, not all active ribosomes are bound to membranes.

In eukaryotes, ribosomes are either free in the cytoplasm or associated with the endoplasmic reticulum, depending on the type of protein being synthesized. Membrane-bound ribosomes, indistinguishable from free ribosomes, synthesize proteins that either pass through the membrane or enter into the endoplasmic reticulum where they may be transported outside the cell membrane. The **signal hypothesis** of Günter Blobel, a 1999 Nobel laureate in Physiology or Medicine, describes a mechanism by which the ribosome attaches to the endoplasmic reticulum in eukaryotes.

In eukaryotes, the signal for membrane insertion (the **signal peptide**) is encoded in approximately the first 20 amino acids of membrane-bound proteins. The signal peptide does not have a consensus sequence like a transcriptional promoter or the Kozak sequence for translation. Rather, the signal peptide usually contains a positively charged (basic) amino acid, commonly lysine or arginine, near the N-terminal end, followed by approximately 10–15 hydrophobic (nonpolar) amino acids, commonly alanine, isoleucine, leucine, phenylalanine, and valine. The signal peptide sequence for the bovine prolactin protein is shown in **table 11.3**.

When a membrane-bound protein is translated in eukaryotes, the signal peptide first becomes accessible outside of the ribosome. A ribonucleoprotein particle

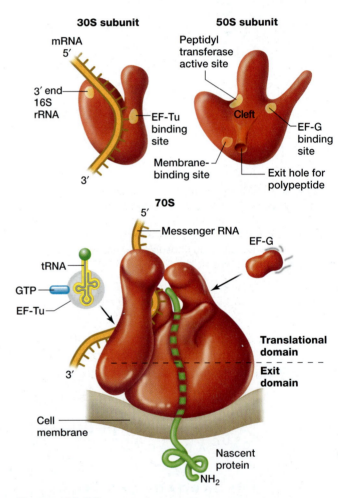

figure 11.33 Functional sites on the bacterial ribosome. The ribosome is synthesizing a protein involved in membrane passage. Note the position of the mRNA, the 3' end of the 16S rRNA that binds the Shine-Dalgarno sequence, and the EF-Tu binding site on the 30S subunit. The location of the peptidyl transferase domain, EF-G binding site, the polypeptide exit hole, and membrane-binding site are shown on the 50S subunit. The 50S ribosomal subunit also contains a membrane binding site that allows the ribosome to bind the cell membrane. As the protein is translated it passes through the ribosome and through the membrane.

called the **signal recognition particle (SRP),** which consists of six different proteins and a 7S RNA about 300 nucleotides long, binds the signal peptide. The complex of signal recognition particle, ribosome, and signal peptide then moves to the endoplasmic reticulum mem-

table 11.3 The Signal Peptide of the Bovine Prolactin Protein*

NH_2 – Met Asp Ser Lys Gly Ser Ser Gln Lys Gly Ser Arg Leu Leu Leu Leu Leu Val Val Ser Asn Leu Leu Leu Cys Gln Gly Val Val Ser I Thr Pro Val . . . Asn Asn Cys – COOH

Source: Reprinted, with permission, from: Sasavage, N.L., J. H. Nilson, S. Horowitz, et al. 1982. Nucleotide sequence of bovine prolactin messenger RNA. Evidence for sequence polymorphism. *Journal of Biological Chemistry* 25 (January): 678-81.

* The vertical line separates the signal peptide from the rest of the protein, which consists of 199 residues.

brane, where the SRP binds to the **docking protein** or signal recognition particle receptor (**fig. 11.34**).

While this complex is moving from the cytosol to the endoplasmic reticulum membrane, protein synthesis is halted. Several proteins help anchor the ribosome in direct contact with the membrane. Protein synthesis then resumes, with the signal peptide and the remaining protein passing through a **translocation channel (translocon)** into the interior (lumen) of the endoplasmic reticulum. Once through the membrane, the signal peptide is cleaved from the protein by an enzyme called *signal peptidase.* Thus, the mature protein lacks the first 15–25 amino acids that were translated.

A striking verification of this hypothesis came about through recombinant DNA techniques (chapter 12). DNA encoding a signal sequence was placed in front of the α-globin gene, whose protein product is normally translated in the cytosol and not transported through a membrane. When this fusion mRNA was translated, the ribosome became membrane-bound, and the protein passed through the membrane into the endoplasmic reticulum. Furthermore, the signal peptide that was originally on the α-globin protein

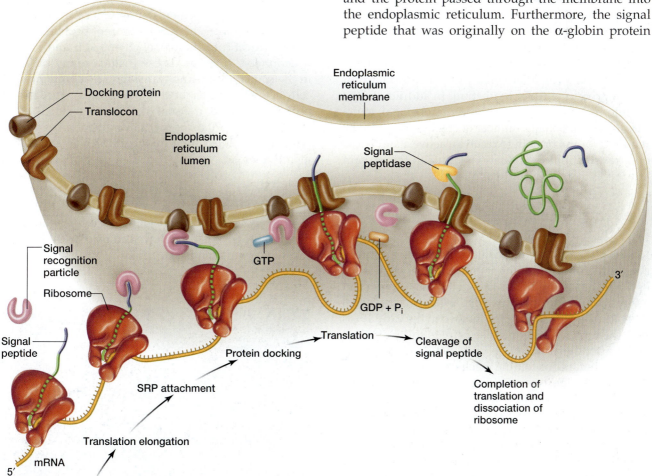

figure 11.34 The signal hypothesis. A signal recognition particle (SRP) binds a newly translated signal peptide and then binds a docking protein on the endoplasmic reticulum (ER) membrane. This brings the ribosome near a translocon and the large ribosomal subunit binds the translocon. With the addition of GTP, the signal recognition particle releases the signal peptide; hydrolysis of the GTP to GDP + P_i causes the signal recognition particle to leave the docking protein. Peptide synthesis then resumes, with the newly synthesized peptide passing through the translocon and membrane into the ER lumen. A signal peptidase in the lumen cleaves the signal peptide. When translation is completed, the ribosome dissociates and drops free of the translocon.

was cleaved from the translated protein to restore the original α-globin protein sequence.

Restating the Concepts

▶ Some bacterial and eukaryotic proteins must be either inserted into the membrane or secreted from the cell. These proteins possess a hydrophobic signal peptide at their amino terminus that binds the signal recognition particle. This bound ribosome moves to the endoplasmic reticulum, where it attaches to the docking protein. Translation resumes, with the protein passing through a translocon to the inside of the endoplasmic reticulum, where the signal peptide is removed by the enzyme signal peptidase.

11.6 Posttranslational Changes to Polypeptide Chains

We know that the three-dimensional structure of a protein is often essential for the function of the protein in the cell. Yet there are thousands of potential structures a protein can attain, many of which would produce a protein without a function. In addition to the protein needing a specific three-dimensional structure to be functional, a number of changes may also be required for the protein to reach an active state. These posttranslational modifications occur to the protein after translation is usually completed.

The Protein-Folding Problem

How does a protein "know" what is its proper structure? The final structure could result from an inherent coding of the protein, it could be a spontaneous optimal shape, or it might be that another protein can assist it to fold properly. Christian Anfinsen, Nobel laureate in Chemistry (1972), showed in the early 1970s that the enzyme ribonuclease refolds to its original shape after denaturation in vitro, which is consistent with a spontaneous folding mechanism.

Evidence indicates that many other proteins only form their final active shape in vivo with the help of proteins called **chaperones.** Chaperones bind to a protein in the early stages of folding to either prevent unproductive folding or to allow denatured proteins to refold correctly. Like human chaperones, they prevent or undo "incorrect interactions," according to John Ellis. Molecular chaperones allow proteins to fold into a thermodynamically stable and functional configuration. Each cycle of refolding requires ATP energy.

A well-studied class of chaperones is known as the *chaperonins*, or Hsp60 proteins, because they are heat shock proteins about 60 kilodaltons (kD) [60,000 daltons (D)] in size. They are present in both bacteria and eukaryotes. One of the best studied chaperonins is the GroE protein of *E. coli*, which is composed of two components: GroEL and GroES. GroEL (Hsp60) is made up of two disks, each composed of seven copies of a polypeptide. GroES (Hsp10) is a smaller component composed of seven copies of a small subunit.

GroEL forms a barrel in which protein folding takes place (**fig. 11.35**). The barrel is shaped so that entering proteins of a certain size make contact at interior points in either the upper or lower ring of GroEL (upper ring shown in **fig. 11.36**). The attachment of GroES, the cap, causes the ring to open outward at the top, stretching or denaturing the protein inside. This stretching utilizes the energy from the hydrolysis of ATP molecules located inside the rings. When GroES dissociates, the protein can fold into a new, more functional, configuration. If it doesn't, the cycle is repeated.

There are several classes of molecular chaperones, proteins of different sizes and shapes that recognize different groups of proteins or protein conformations. GroEL recognizes about 300 different proteins, small enough to fit into the barrel (20–60 kD) and having hydrophobic surfaces. These include many proteins in the transcription and translation machinery of the cell. Hsp90, another heat shock protein, recognizes proteins involved in signal transduction, discussed in chapter 21. Hsp70 recognizes hydrophobic regions in polypeptide side chains, many of which extend across membranes.

Posttranslational Modifications

A large variety of posttranslational modifications may occur to proteins, although we can place them into the following four general categories.

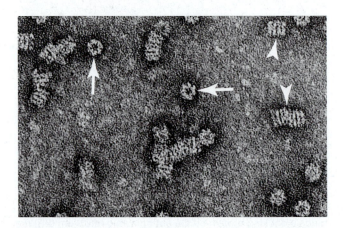

figure 11.35   Electron micrograph of the *E. coli* chaperone protein, GroEL. Note the hollow, barrel shape of the protein (arrow) and the cylindrical side view (arrowhead).

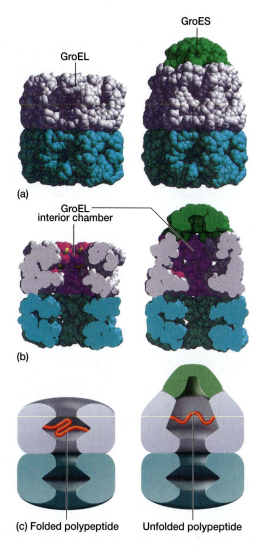

(a)

(b)

(c) Folded polypeptide   Unfolded polypeptide

figure 11.36   The change in structure of GroEL with GroES attached explains how the chaperonin can unfold a partially folded polypeptide to allow it to refold in a different way. (a) A space-filling model of GroEL is shown without (left) and with (right) GroES attached. GroEL's rings are blue and purple, and GroES is *green*. (b) The same structures are seen in a cutaway view. Binding of GroES causes an increase size of the GroEL interior chamber. (c) This diagram shows how the attachment of GroES causes the top part of the upper GroEL ring to pull apart an improperly folded polypeptide (red). Removal of GroES allows the protein to fold in a different shape.

1. First is the cleavage of the amino terminus. As already discussed, proteins that are either located in a membrane or secreted from the eukaryotic cell have the amino terminal signal sequence cleaved. This cleavage results in an irreversible loss of amino acid sequence.

2. Second, phosphate groups may be added to specific amino acids by the action of kinases. We have discussed several instances in which the addition of one or more phosphate groups is required to either activate or inactivate a protein.

3. Third, sugar or carbohydrate groups may be added to certain amino acids in a protein. In figure 11.8, you can see the quaternary structure of hemoglobin, which has several sugar groups attached. Regions of proteins that contain sugar groups are located either inside of vesicles or on the exterior of the eukaryotic cell. In the latter case, they may play a role in cell recognition.

4. Fourth, certain lipids can also be attached to some amino acids of a protein, which results in the protein being attached to the membrane without being actually embedded in the membrane. Thus, these lipoproteins do not need an amino terminal signal peptide and do not need to be translated into the endoplasmic reticulum.

Two important features are associated with posttranslational modifications. First, they are more common in eukaryotes than in bacteria. However, the kinase enzyme that adds the phosphate group often recognizes only specific proteins. Thus, the expression of a recombinant eukaryotic protein in a prokaryotic cell may not be phosphorylated because the necessary kinase is not present in the prokaryotic cell. The inability to properly modify a protein often will render the enzyme inactive (see chapter 13). Second, it is often not obvious what posttranslational modifications occur to a protein from only the primary protein structure. Thus, the cell must possess additional mechanisms to know specifically how to modify a protein to generate its functional state.

Restating the Concepts

▶ Some proteins will spontaneously fold to reach their functional tertiary structure. Other proteins require the assistance of chaperones to help them achieve their final shape.

▶ Several different types of changes occur to proteins after translation is completed. These posttranslational modifications usually occur only in eukaryotes and are essential to their final function.

11.7 The Genetic Code

Once researchers realized that the genetic code was in the nucleic acids and that mRNA was the intermediate between the DNA and the protein, they naturally wanted to learn how nucleotides could encode amino acids. Although they correctly assumed that the genetic code consisted of simple nucleotide sequences that specified particular amino acids, many questions remained. Is the code overlapping? Do nucleotides occur between code words (punctuation)? How many letters make up a code word (codon)? Logic, along with genetic experiments, supplied some of the answers, but only with the rapidly improving techniques of biochemistry did geneticists eventually decode the genetic language.

The Triplet Nature of the Code

Several lines of evidence suggested that the nature of the code was triplet (three bases in mRNA specifying one amino acid). The most basic evidence was the knowledge that there were only four different bases and 20 different amino acids. If codons contained only one base, then they would only be able to specify four amino acids. A two-base codon would have 4 × 4, or 16, different codons. Both of these possibilities were insufficient to account for all the amino acids.

A triplet code allows for 4 × 4 × 4, or 64, codons. Although this number is greater than the number of different amino acids, a triplet codon was the smallest codon that could account for 20 different amino acids.

Evidence for the Triplet Code

This logic can be used to infer that a codon is 3 nucleotides, but real proof is ultimately necessary. Francis Crick and his colleagues provided this proof using the mutagen proflavin. Proflavin, which is an acridine dye, serves as a mutagen by either adding or deleting one or a few nucleotides from the DNA (as is discussed in chapter 18).

The goal of Crick's experiment was to inactivate the rapid lysis gene (*rIIB*) of the bacteriophage T4. The *rII* gene controls the plaque morphology of this bacteriophage when growing on a plate of *E. coli* cells. Rapid-lysis mutants produce large plaques; the wild-type form of the gene, *rII*⁺, results in normal small plaques.

Figure 11.37 shows the consequences of adding or deleting a nucleotide from an ORF in a gene. From the point of either the addition or deletion onward, all the codons are **frameshifted,** which means the nucleotides are read in different groups of three.

If a deletion is combined with the addition of the same number of nucleotides, the double mutant gene contains the frameshift only in the region between the two mutations (see fig. 11.37). In a double mutant, then, the function of the gene's product may be restored if this frameshifted region is small or does not encode crucial amino acids. This type of double mutation failed to provide any insight into the size of the codon, because the second mutation restored the reading frame by simply acting in the opposite direction as the first mutation (first a deletion of one nucleotide and then an insertion of one nucleotide).

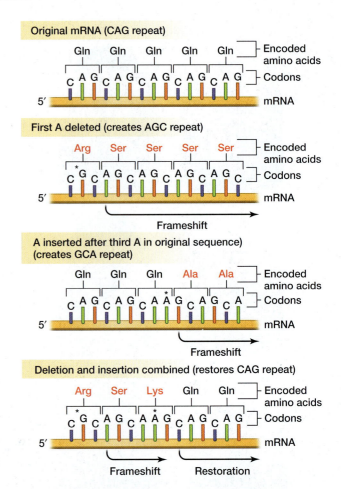

figure 11.37 The effect of a frameshift mutation on protein translation. The original mRNA has a repeating CAG codon that translates as polyglutamine. A single-base deletion of the first A shifts the reading frame to a repeating AGC that encodes polyserine. Insertion of an A after the third repeat in the original sequence shifts the reading frame to a repeating GCA (encoding polyalanine). However, if the deletion and insertion were both introduced into the same DNA molecule, then the reading frame will only be shifted between the two mutations, as the second mutation (insertion) restores the original CAG repeat. *Asterisks* (*) indicate points of deletion or insertion.

If two deletion events or two insertion events are combined, the reading frame is often not restored and the resulting protein remains nonfunctional. However, Crick and his colleagues found that if the two deletion events or two insertion events resulted in a net loss or addition of a multiple of 3 nucleotides, the *rII*⁺ function could be restored. This finding demonstrated that the genetic code was a triplet, because a gene function that is disrupted by the insertion of one nucleotide and restored by the insertion of two additional nucleotides could only occur in a triplet code (**fig. 11.38**). Similarly, a gene function that is disrupted by the deletion of a single nucleotide could be restored by the deletion of two additional nucleotides in a triplet code, but not in a doublet or quartet code.

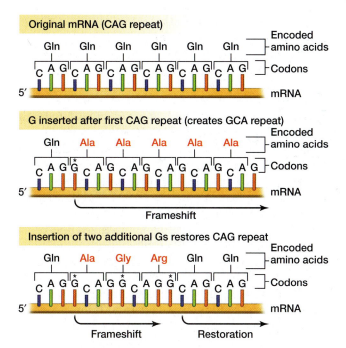

figure 11.38   The use of insertions to prove the codon is a triplet. The mRNA shown here normally is translated as a CAG repeat that encodes polyglutamine. A single-base insertion (G) after the first CAG repeat shifts the reading frame to a series of GCA repeats that encodes polyalanine. The insertion of two additional G's after the first G insertion restores the reading frame to a series of CAG repeats. The region between the first and third insertion remains in an altered reading frame. *Asterisks* (*) indicate insertions.

Overlap and Punctuation in the Code

Questions still remained: Was the code overlapping? Did it have punctuation? Several logical arguments favored a no-punctuation, nonoverlapping model (**fig. 11.39**).

In an **overlapping code,** in which each nucleotide would be used in either two or three different codons, two restrictions would be operative. First, a change in one base (a mutation) would affect more than one codon and thus affect more than one amino acid. However, studies of mutant amino acid sequences almost always showed that only one amino acid was changed, which argued against codon overlap. (An exception is a frameshift mutation, in which all the amino acids are changed after a given point.)

Second, an overlapping code would place certain restrictions on which amino acids could occur next to each other in proteins. For example, the amino acid encoded by UUU could never be adjacent to the amino acid encoded by AAA; the amino acids corresponding to UUA or UAA, depending on the number of bases overlapped, would always have to be inserted between the amino acids encoded by UUU and AAA. Overlap, then, seemed to be ruled out because every amino acid appears next to every other amino acid in one protein or another.

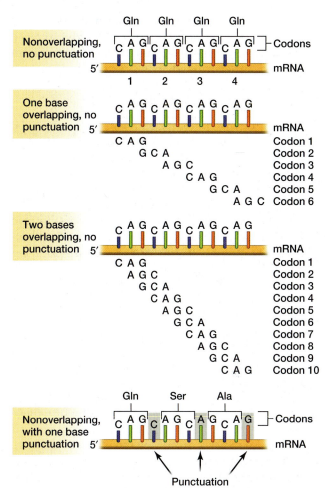

figure 11.39   The genetic code is read as a nonoverlapping code with no punctuation (*top*). Before that was proven, it was suggested that the code could overlap by one or two bases (*middle*) or have noncoded bases (punctuation) between codons (*bottom*). In an overlapping code, some nucleotides will be used in two or three adjacent codons. With punctuation, some bases within the open reading frame are not a part of a codon.

Punctuation between codons was also ruled out. *Punctuation,* in which one or more nucleotides would separate each codon (see fig. 11.39), would require the ORF in the mRNA to be at least 33% longer than what would be required to encode only the amino acids. The mRNA in the tobacco necrosis satellite virus has just about enough codons to specify its coat protein with no room left for punctuating bases between each codon.

Breaking the Code

Once geneticists deduced that the genetic code is in nonoverlapping triplets, they turned their attention to determining the amino acids encoded by each of the 64 codons. The work was done in two stages. First, Marshall Nirenberg and his colleagues made long artificial mRNAs and determined which amino acids these mRNAs incorporated into protein. Second, specific triplet RNA sequences were synthesized, and the charged amino acid/tRNA that bound to each triplet sequence was determined.

Synthetic Messenger RNAs

To determine the genetic code, it was necessary to be able to synthesize mRNAs with specific sequences. This became possible in 1955, when Marianne Grunberg-Manago and Severo Ochoa discovered the enzyme polynucleotide phosphorylase. **Polynucleotide phosphorylase** randomly joins any diphosphate nucleotides (NDPs) that are available into long-chain, single-stranded polynucleotides. Unlike a polymerase, polynucleotide phosphorylase does not need either a primer or a template on which to act.

In 1961, Marshall Nirenberg and Heinrich Matthaei added artificially formed RNA polynucleotides of known composition to an *E. coli* cell-free ribosomal system and looked for the incorporation of amino acids into proteins (**fig. 11.40**). This ribosomal system is called a **cell-free system** because cells are disrupted and the cellular contents are separated to produce a mixture of the cytoplasmic cellular components, without the nucleic acids and membranes. Although these systems contain virtually all the components needed for protein synthesis except for the mRNAs, they are relatively short-lived (several hours) and are fairly inefficient in translation. However, the *E. coli* cell-free system will translate, although inefficiently, RNAs that lack translation initiation signals and are not normally translated in vivo. This cell-free system could therefore translate artificial mRNAs that lacked a Shine-Dalgarno sequence for ribosomal binding.

Nirenberg and Matthaei first used the enzyme polynucleotide phosphorylase to synthesize poly-U RNA. This synthetic RNA was added to 20 different tubes that each contained the *E. coli* cell-free translation system and 1 of the 20 amino acids that was radioactively labeled (**fig. 11.41**). Of the 20 tubes, only 1 produced a radioactively labeled protein; the tube containing labeled phenylalanine. When the radioactively labeled protein was studied, it was found to consist of repeating phenylalanine residues. The first codon established, therefore, was UUU for phenylalanine. Nirenberg and Matthaei then synthesized the other three homopolymeric RNAs (poly-A, poly-C, and poly-G) and determined that AAA encoded lysine, CCC encoded proline, and GGG encoded glycine.

Unfortunately, this approach revealed only 4 of the possible 64 codons. To elucidate additional codons, the investigators synthesized RNAs from mixtures of the various diphosphate nucleotides in known proportions. This would produce multiple RNAs, each with its own random nucleotide sequence. However, the frequency of each potential codon in the total RNA population can be calculated. For example, if UDP and GDP were mixed in a 3:1 ratio prior to adding the polynucleotide phosphorylase, the frequency of each potential triplet codon would depend on the abundance of each nucleoside diphosphate. For example, 5'-UUU-3' would account for 42.2% ($3/4 \times 3/4 \times 3/4$) of all the possible codons. Similarly, the frequency of 5'-UGU-3' would account for 14.1% ($3/4 \times 1/4 \times 3/4$) of all the codons. Using this logic, the frequency of 5'-UUG-3' and 5'-GUU-3' would be the same as UGU. Thus, the frequency for all the possible codons in this experiment can be calculated (**table 11.4**).

The in vitro synthesized random RNA sequences were added to the *E. coli* cell-free system, and the translated proteins were isolated. Because the RNAs were a mixture of different sequences, it was not possible to determine the precise relationship between a specific RNA sequence and the corresponding amino acid sequence. In contrast, the frequency of each RNA codon could be calculated, and this could be correlated with the frequency of each amino acid in all the translated proteins. This was done by simply hydrolyzing the total mixture of translated proteins and determining the amino acid content (see table 11.4).

Several points can be inferred from their data. First, there are more codons than there are amino acids. This is consistent with the prediction that a triplet code will have more codons than amino acids. Thus, some codons either do not code for an amino acid or multiple codons code for the same amino acid (redundancy). Second, the frequency of the UUU codon was very similar to the frequency of phenylalanine. In contrast, the

figure 11.40   Heinrich Matthaei (1929-), and Marshall W. Nirenberg (1927–), who played significant roles in determining the genetic code.

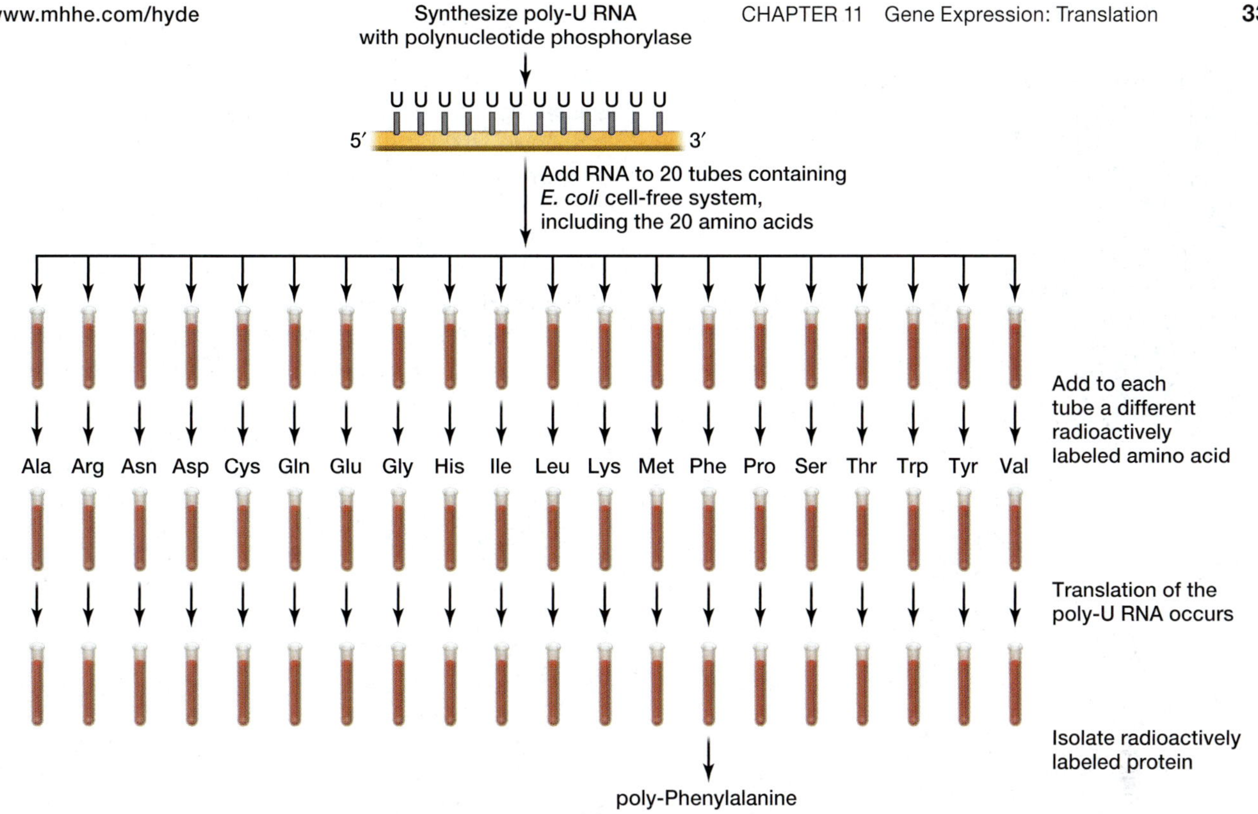

figure 11.41   Nirenberg and Matthaei's experiment to determine that UUU encodes phenylalanine. A poly-U RNA was synthesized in vitro and added to an *E. coli* cell-free extract in 20 tubes that each contained a different radioactively-labeled amino acid. The radioactive amino acid was charged to the correct tRNA in the cell-free extract. Translation of the same protein, poly-phenylalanine, occurred in each of the 20 tubes. However, the only tube containing a radioactively labeled protein was the tube that had radioactive phenylalanine.

frequency of GGG (1.4%), which was previously shown to encode glycine, was much less than the frequency of glycine in the amino acid mixture (5.1%). This supported the idea that at least one other codon in the table also encodes glycine. Third, it is possible to identify a subset of codons that correspond to a particular amino acid. For example, cysteine, leucine, and valine are all encoded by one G and two U's and would all be present approximately 14% of the time. Three amino acids were present approximately 15% of the time, leucine, valine, and cysteine. It is likely that these three codons encode these amino acids, but you cannot determine from this experiment what codon corresponds to what amino acid. Assigning absolute amino acids to each codon required an extra step in sophistication—that is, the synthesis of known trinucleotides.

Synthetic Codons

Once trinucleotides of known composition could be synthesized, Marshall Nirenberg and Philip Leder developed a "binding assay" (**fig. 11.42**). In 1964, they found that isolated *E. coli* ribosomes, in the presence of high-molarity magnesium chloride, could bind trinucleotides as though they were mRNAs and also bind the charged tRNA that carried the anticodon complementary to the trinucleotide.

table 11.4   Relationship Between Randomly Assembling Uracil- and Guanine-Containing Ribose Diphosphate Nucleotides with a Ratio of 3U:1G and Amino Acid Composition of Translated Proteins

Codon	Predicted Frequency of Occurrence
UUU	$(3/4 \times 3/4 \times 3/4) = 0.42 = 42.2\%$
UUG	$(3/4 \times 3/4 \times 1/4) = 0.14 = 14.1\%$
UGU	$(3/4 \times 1/4 \times 3/4) = 0.14 = 14.1\%$
GUU	$(1/4 \times 3/4 \times 3/4) = 0.14 = 14.1\%$
UGG	$(3/4 \times 1/4 \times 1/4) = 0.05 = 4.7\%$
GUG	$(1/4 \times 3/4 \times 1/4) = 0.05 = 4.7\%$
GGU	$(1/4 \times 1/4 \times 3/4) = 0.05 = 4.7\%$
GGG	$(1/4 \times 1/4 \times 1/4) = 0.01 = 1.4\%$

Amino Acid	Actual Frequency of Occurrence (%)
Phenylalanine	42.7
Leucine	15.8
Valine	15.4
Cysteine	15.0
Tryptophan	6.0
Glycine	5.1

figure 11.42 Phillip Leder (1934–) played a critical role in determining a large number of codon-amino acid identities.

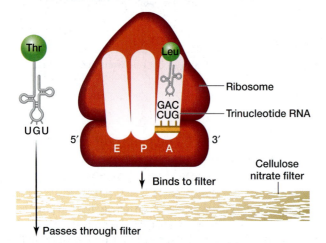

figure 11.43 The filter binding assay revealed the amino acid associated with a given trinucleotide codon. Trinucleotide RNAs were synthesized in vitro and mixed with ribosomes and different radioactively-labeled amino acids charged to tRNAs. The tRNAs with noncomplementary codons to the bound trinucleotide RNA passed through the membrane (*left*). Charged tRNAs with an anticodon complementary to the trinucleotide RNA bound the RNA and ribosome and become trapped by the filter. When the tRNA is charged with a radioactive amino acid, the radioactivity is trapped on the filter.

Growing *E. coli* in the presence of a single radioactive amino acid resulted in that amino acid being covalently attached to its correct tRNA. Next, a given synthetic trinucleotide of a known sequence was mixed with ribosomes and aminoacyl-tRNAs, including one that contained a radioactively labeled amino acid. The reaction mixture was passed over a filter that bound only the ribosome and anything tightly associated with it, including the trinucleotide and the base-paired aminoacyl-tRNA. If the radioactivity passed through the filter, it meant that the radioactive amino acid was not associated with the ribosome. The experiment was then repeated with another labeled amino acid.

When radioactivity appeared on the filter, the investigators knew that the amino acid was encoded by the selected trinucleotide codon. In other words, the radioactive amino acid was attached to a tRNA whose anticodon was complementary to the trinucleotide codon and was therefore bound at the ribosome.

Figure 11.43 shows an example. In the figure, the trinucleotide codon is 5'-CUG-3'. The tRNA with the anticodon 5'-CAG-3' is charged with leucine. The mixture is passed through a filter. If leucine is radioactively labeled, the radioactivity is trapped by the filter. If threonine or any other amino acid except leucine is radioactive, the radioactivity passes through the filter.

In a short period of time, all of the codons were deciphered. **Table 11.5** shows the correspondence between the triplet codons and the amino acids.

Restating the Concepts

▶ The genetic code is a nonoverlapping triplet code, which lacks punctuation.

▶ The genetic code was elucidated by Nirenberg and several colleagues using two major experiments. They first used polynucleotide phosphorylase to synthesize RNAs that were translated in a cell-free system into peptides. Second, they used filters to bind ribosomes that contained specific trinucleotide RNAs that were basepaired to tRNAs that were charged with radioactively labeled amino acids.

table 11.5   The Genetic Code

First Position (5' End)	Second Position				Third Position (3' End)
	U	**C**	**A**	**G**	
U	Phe	Ser	Tyr	Cys	U
	Phe	Ser	Tyr	Cys	C
	Leu	Ser	*stop*	*stop*	A
	Leu	Ser	*stop*	Trp	G
C	Leu	Pro	His	Arg	U
	Leu	Pro	His	Arg	C
	Leu	Pro	Gln	Arg	A
	Leu	Pro	Gln	Arg	G
A	Ile	Thr	Asn	Ser	U
	Ile	Thr	Asn	Ser	C
	Ile	Thr	Lys	Arg	A
	Met (*start*)	Thr	Lys	Arg	G
G	Val	Ala	Asp	Gly	U
	Val	Ala	Asp	Gly	C
	Val	Ala	Glu	Gly	A
	Val	Ala	Glu	Gly	G

The Wobble Hypothesis

If there are 64 codons and only 20 amino acids, what do the extra 44 codons do? As described earlier, there are 3 stop or nonsense codons (UAA, UGA, and UAG), which still leaves 41 additional codons.

Looking at table 11.5, you can see that a given amino acid may have more than one codon. The term used to describe this is **degeneracy,** and the genetic code is said to be a **degenerate code.** (In this usage, the term degenerate means "redundant.")

Eight of the 16 boxes in table 11.5 contain just one amino acid per box. (A box is determined by the first and second positions, for example, the UUN box, in which N is any of the four bases.) These eight groups of codons are termed **unmixed families** of codons. In each unmixed family, four codons beginning with the same two bases specify a single amino acid. For example, the four codons in the GUN family all encode valine.

Mixed families of codons, by contrast, encode two amino acids or stop signals and one or two amino acids. Six of the mixed-family boxes are split in half so that the codons are differentiated by the presence of a purine or a pyrimidine in the third base. For example, CAU and CAC both code for histidine; in both, the third base, U (uracil) or C (cytosine), is a pyrimidine. Only two of the eight mixed families of codons (UGN and AUN) are split differently.

We should mention, to avoid confusion, that both mRNA and tRNA bases are usually numbered from the 5' side. The first base of the codon is complementary to the third base of the anticodon (**fig. 11.44**); the third base of the codon is therefore of lesser importance, and it is complementary to the first base of the anticodon.

The lesser importance of the third position in the genetic code ties in with two facts about tRNAs. First, although there would seem to be a need for 61 tRNAs—of the 64 possible codons, 3 correspond to stop codons—there are actually only about 50 different tRNAs in an *E. coli* cell. Second, a rare base such as inosine can appear in the anticodon, usually in the position that is complementary to the third position of the codon. These two facts suggested that some tRNAs could base-pair with multiple codons and that rare bases may be involved.

Because the first base of the anticodon (5') is not as constrained as the other two positions, a given base at that position may be able to pair with any of several bases in the third position of the codon. Crick characterized this ability as **wobble** (**fig. 11.45**). **Table 11.6** shows the possible pairings that produce a tRNA system compatible with the known code. For example, an isoleucine tRNA with the anticodon 5'-IAU-3' is compatible with the three codons for that amino acid (see table 11.5): 5'-AUU-3', 5'-AUC-3', and 5'-AUA-3'. That is, inosine in the first (5') position of the anticodon can recognize U, C, or A in the third (3') position of the codon, and thus one tRNA can basepair with all three isoleucine codons.

Universality of the Genetic Code

Until 1979, scientists thought that the genetic code was universal—that is identical for all species. The universality of the code was demonstrated, for example, by mixing the ribosomes and β-globin mRNA from rabbit reticulocytes with the aminoacyl-tRNAs and other translational components of *E. coli*. The result was the translation of rabbit β-globin protein.

In 1979 and 1980, however, researchers noted discrepancies when sequencing mitochondrial genes for structural proteins (see chapter 19). Two kinds of deviations from universality were found in the way mitochondrial tRNAs read the code. First, fewer tRNAs were needed to read the code, and second, in several instances the mitochondrial and cellular systems interpreted a codon differently.

Yeast Mitochondria

According to Crick's wobble rules, 32 tRNAs (including 1 for initiation) were needed to basepair with all 61 nonterminating codons (see table 11.6). Unmixed families

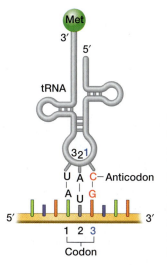

figure 11.44 Pairing of the mRNA codon and tRNA anticodon during translation. The codon and anticodon base positions are numbered from the 5' end and base-pair in an antiparallel manner. The 3' position in the codon (5' in the anticodon) is the wobble base (shown in red).

table 11.6 Pairing Combinations at the Third Codon Position

Number-one Base in tRNA (5' End)	Number-three Base in mRNA (3' End)
G	U or C
C	G
A	U
U	A or G
I	A, U, or C

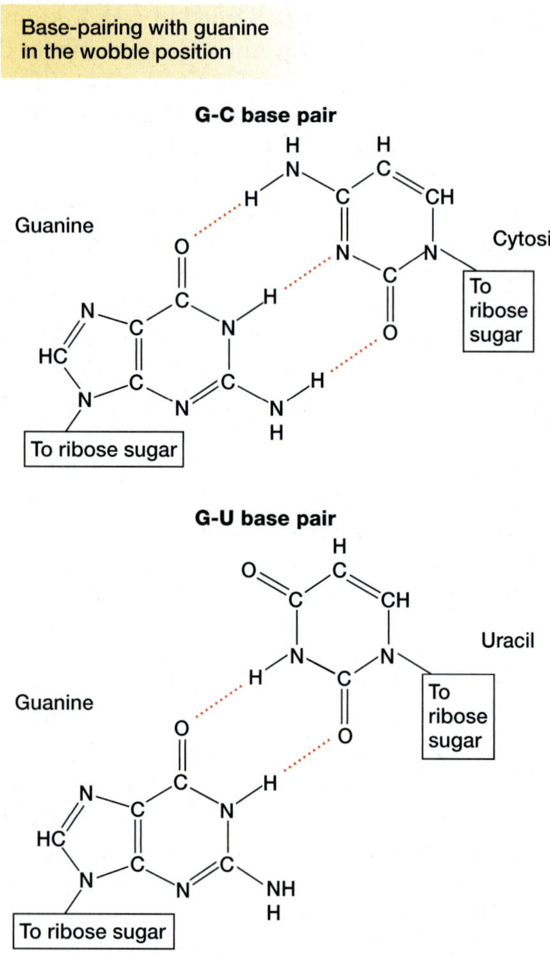

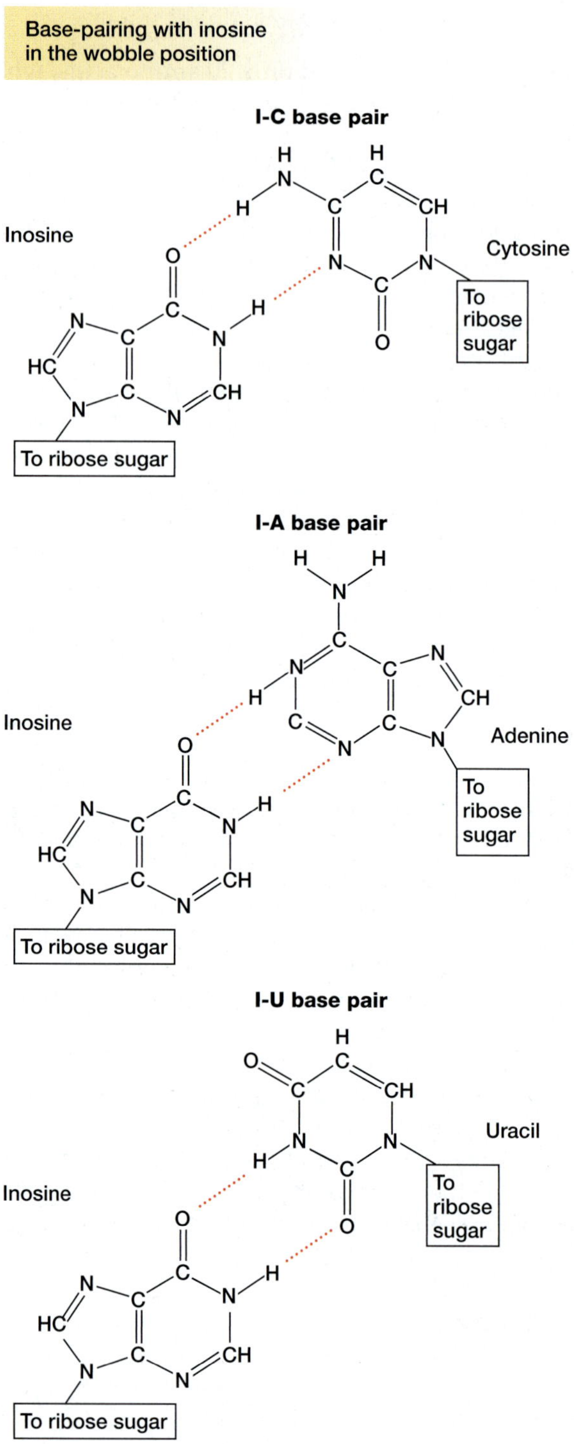

Base-pairing with guanine in the wobble position

G-C base pair

Guanine   Cytosine

G-U base pair

Guanine   Uracil

Base-pairing with inosine in the wobble position

I-C base pair

Inosine   Cytosine

I-A base pair

Inosine   Adenine

I-U base pair

Inosine   Uracil

figure 11.45   Base-pairing possibilities for guanine and inosine in the third (3') position of a codon. In the wobble position, guanine can form base pairs with either cytosine or uracil (left side). Inosine, in the wobble position, can pair with cytosine, adenine, or uracil (right side).

require two tRNAs, and mixed families require one, two, or three tRNAs, depending on the family. In contrast, the yeast mitochondria apparently needed only 24 tRNAs. The reduction from 32 to 24 tRNAs is accomplished primarily by having only one tRNA recognize each unmixed family (compare **table 11.7** with table 11.5). Because mitochondrial tRNAs for unmixed codon families have a U in the first (wobble) position of the anticodon, the structure of the mitochondrial tRNAs must allow the U to basepair with U, C, A, or G in the third position of the codon. Presumably, evolutionary pressure minimized the number of tRNA genes in the mitochondrial DNA to maintain its small size. Reduction from 32 to 24 tRNAs is a 25% savings.

It was also found that that the yeast mitochondria reads the CUN codon family as threonine rather than as leucine (see tables 11.5 and 11.7) and the UGA codon as tryptophan rather than as a terminator. Several groups of organisms, however, have differences in reading the CUN family. Human and *Neurospora* mitochondria appear to read the CUN codons as leucine, just as isolated cellular systems do. Of the groups

table 11.7 The tRNA Anticodon Dictionary of Yeast Mitochondria*

First Position (5' End)	Second Position				Third Position (3' End)
	U	**C**	**A**	**G**	
U	Phe (AAG)	Ser (AGU)	Tyr (AUG)	Cys (ACG)	U or C
	Leu (AAU)		*stop*	Trp (ACU)	A or G
C	Thr (GAU)	Pro (GGU)	His (GUG)	Arg (GCA)	U or C
			Gln (GUU)		A or G
A	Ile (UAG)	Thr (UGU)	Asn (UUG)	Ser (UCG)	U or C
	Met (UAC)		Lys (UUU)	Arg (UCU)	A or G
G	Val (CAU)	Ala (CGU)	Asp (CUG)	Gly (CCU)	U or C
			Glu (CUU)		A or G

Source: Data from Bonitz, S.G., R. Berlani, G. Coruzzi, et al. 1980. Codon recognition rules in yeast mitochondria, *Proceedings of the National Academy of Sciences USA* 77: (June) 3167–70.
* Anticodons (3' → 5') are given within parentheses. (The ACU Trp anticodon is predicted.)

table 11.8 Common and Alternative Meanings of Codons

Codon	General Meaning	Alternative Meaning
CUN	Leu	Thr in yeast mitochondria
AUA	Ile	Met in mitochondria of yeast, *Drosophila*, and vertebrates
UGA	*Stop*	Trp in mycoplasmas and mitochondria other than higher plants
AGA/AGG	Arg	*Stop* in mitochondria of yeast and vertebrates
		Ser in mitochondria of *Drosophila*
CGG	Arg	Trp in mitochondria of higher plants
UAA/UAG	*Stop*	Gln in ciliated protozoa
UAG	*Stop*	Ala or Leu in mitochondria of some higher plants

so far analyzed, only yeast reads the CUN family as threonine. Similarly, human and *Drosophila* mitochondria read AGA and AGG as stop signals rather than as arginine (**table 11.8**).

Other Variations in the Code

In 1985, investigators found that *Paramecium* species read the UAA and UAG stop codons as glutamine within the cell. In addition, a prokaryote (*Mycoplasma capricolum*) reads UGA as tryptophan. We can thus conclude that the genetic code seems to have universal tendencies among prokaryotes, eukaryotes, and viruses. Mitochondria, however, read the code slightly differently: Different wobble rules apply, and mito-

chondria and cells read at least one terminator and one unmixed family of codons differently. Also, the mitochondrial discrepancies are not universal among all types of mitochondria. Further work, involving the sequencing of more mitochondrial DNAs, should elucidate the pattern of discrepancies.

A second type of variation in codon reading occurs: **site-specific variation,** in which the interpretation of a codon depends on its location. In bacteria, GUG and UUG can occasionally serve as translation initiation codons, which means that they are recognized by $tRNA_f^{Met}$; however, they are not recognized by $tRNA_m^{Met}$ (that is, GUG and UUG are not misread internally in mRNAs). This interpretation requires the GUG or UUG codons to be properly located relative to the

11.1 applying genetic principles to problems

Reverse Translation from Amino Acid Sequence to Deduced Nucleotide Sequence

The nearly universal nature of the genetic code allows investigators to take the nucleotide sequence of an open reading frame and deduce the amino acid sequence of the encoded protein. As you can imagine, this is a very powerful application. The molecular weight of the encoded protein (prior to any posttranslational modification) and its potential function based on sequence similarities to other proteins can be determined.

An equally important technique is **reverse translation,** which involves using the amino acid sequence of a protein to deduce the corresponding nucleotide sequence. This process is more complicated, largely because of the redundancy in the genetic code. For example, if we know that a codon is 5'-CUA-3', then it must encode a leucine amino acid (see table 11.5). But if we observe the amino acid threonine in a protein, we do not know if the corresponding codon is ACU, ACC, ACA, or ACG.

Why would we be interested in reverse translation? As you will see in chapter 12, many molecular biology techniques require synthesizing a short DNA sequence that corresponds to a specific gene. Sometimes the nucleic acid sequence is unknown, but the corresponding protein sequence has been determined. If we could use this protein sequence to deduce the corresponding nucleic acid sequence, we could then use a machine to synthesize a DNA molecule with that sequence.

Let's look at an example of reverse translation of a short peptide sequence:

NH₂ – Cys – Thr – Tyr – Met – Arg – Pro – Leu – Val – His – Gly – Ile – COOH

Some amino acids have two potential codons, some have three, some have four, and some have six! Fortunately, this sequence also contains a methionine, which is encoded by a single amino acid (AUG). Placing all the potential codons under the corresponding amino acid produces the diagram in table 11A.

How many different nucleotide sequences can encode this peptide? Using the product rule from chapter 2, there are two codons for cysteine, four for serine, two for tyrosine, and so on, which means that there are

2 × 4 × 2 × 1 × 4 × 6 × 4 × 4 × 2 × 4 × 3, or 147,456 possible DNA sequences for this peptide. DNA synthesizing machines can make these sequences individually or simultaneously.

Assume you want to make a DNA sequence of 15 nucleotides that corresponds to a portion of the previous sequence with the lowest redundancy. Just a quick glance tells you to include the Cys, Tyr, and Met because they possess only two, two, and one possible codons. In contrast, including Arg would involve six different codons. The lowest redundant region would then be the first five amino acids (Cys–Thr–Tyr–Met–Pro), which possess only 64 possible nucleotide sequences.

Reverse translation can also reveal the nucleotide sequence when both a wild-type and a mutant protein sequence are available. Let's use the preceding peptide sequence as the wild-type sequence and a mutation produces this amino acid sequence:

Mutation 1

NH₂ – Cys – Thr – Tyr – Met – Pro – Arg – His – Val – His – Gly – Ile – COOH

The mutation changed only the seventh amino acid from leucine to histidine. Because there is only a single amino acid difference, the mutation cannot be either an insertion or deletion, which would result in a frameshift. Thus, the mutation must be a single nucleotide change.

Histidine is encoded by only two codons: CAU and CAC (see table 11.5). Leucine is encoded by six codons (CUN, AGU, and AGC). You could eliminate the last two leucine codons, AGU and AGC, as possibilities, because the first two nucleotides in these codons differ from the first two in histidine and we just determined that we are changing a single nucleotide.

Both the wild-type and mutant codons have C in the 5' position, and both could have either U or C in the last (3') position. The middle position is the one that differs between the two amino acids, so you can deduce that the mutation is a change of U for A in the second position of the codon. This results in the wild-type leucine at this

table 11A Deduced nucleotide sequence for a known peptide.

NH₂ –	Cys –	Thr –	Tyr –	Met –	Pro –	Arg –	Leu –	Val –	His –	Gly –	Ile – COOH
	UGU	ACU	UAU	AUG	CCU	CGU	CUU	GUU	CAU	GGU	AUU
	UGC	ACC	UAC		CCC	CGC	CUC	GUC	CAC	GGC	AUC
		ACA			CCA	CGA	CUA	GUA		GGA	AUA
		ACG			CCG	CGG	CUG	GUG		GGG	
						AGA					
						AGG					

position being either CUU or CUC. Thus, you used the mutation to eliminate four of the six possible codons for leucine in this wild-type sequence.

This analysis can also be used when a mutation affects more than one codon. Assume a second mutation was isolated with the peptide and corresponding nucleotide sequences shown in table 11B.

Both the wild-type and mutant sequences are identical for the first five amino acids, and differ at every amino acid after that. This suggests that the mutation causes a frameshift, which results from either an insertion or a deletion, changing the reading frame and all the amino acids after the point of the nucleotide mutation.

Comparing the wild-type and mutant amino acid sequences will reveal the point where the insertion or deletion occurred. Because the first difference between the wild-type and mutant amino acid sequences is an arginine to alanine, the mutation likely is within that codon.

Arginine is encoded by CGN, AGA, or AGG, and alanine is encoded by GCN. We cannot add or delete a nucleotide from either AGA or AGG to produce GCN. However, we can convert CGN to GCN by either inserting a G before the arginine codon to produce GCG or deleting the first nucleotide in the arginine codon (assuming the N in GCN is a G).

Let's consider the insertion model first. The wild-type sequence CGN CUN GUN CA(U/C) GGN AU(U/C/A) would become, with the addition of the G, **G**CG NCU NGU NCA (U/C)GG NAU (U/C/A). In this case, the GCG codes for alanine as expected. The NCU could code for serine, proline, threonine, or alanine, depending on whether the N represented a U, C, A, or G. Because serine follows alanine in the mutant sequence, UCU could be the second codon. The third codon NGU could code for cysteine, arginine, serine, or glycine. However, the mutant sequence has a phenylalanine following the second serine. Thus, the insertion model does not correctly predict this codon.

Consider the deletion model now. Deleting the first C in CGN CUN GUN CA(U/C) GGN AU(U/C/A) would yield GNC UNG UNC A(U/C)G GNA U(U/C/A). The GNC would encode alanine if the N is a C. The next amino acid, serine, could be encoded by UNG if N is a C. Phenylalanine could be encoded by UNC if the N is a U, and methionine is encoded by AUG. The valine could be encoded by GUA, and serine is encoded by UCN. Replacing the N's in the wild-type sequence with the deduced nucleotides, you could rewrite the wild-type sequence as shown in table 11C.

By deleting the underlined C in the wild-type sequence (table 11C), we generate the mutant nucleotide and amino acid sequence as shown in table 11D.

Notice that whatever nucleotide follows the wild-type isoleucine codon, it can be incorporated into the third position of the mutant serine codon.

This analysis accurately revealed the correct nucleotide sequence after the mutation site in both the wild-type and mutant sequences.

table 11B Mutation 2 Deduced nucleotide sequence for a second mutation.

NH$_2$ –	Cys –	Thr –	Tyr –	Met –	Pro –	Ala –	Ser –	Phe –	Met –	Val –	Ser –	COOH
	UGU	ACU	UAU	AUG	CCU	GCU	UCU	UUU	AUG	GUU	AUU	
	UGC	ACC	UAC		CCC	GCC	UCC	UUC		GUC	AUC	
		ACA			CCA	GCA	UCA			GUA	AUA	
		ACG			CCG	GCG	UCG			GUG		
							AGU					
							AGC					

table 11C Deduced wild-type sequence from deletion mutant.

NH$_2$ –	Cys –	Thr –	Tyr –	Met –	Pro –	Arg –	Leu –	Val –	His –	Gly –	Ile –	COOH
	UGU	ACN	UAU	AUG	CCN	<u>C</u>GC	CUC	GUU	CAU	GGU	AUC	
	UGC		UAC									

table 11D Deduced deletion mutant sequence.

NH$_2$ –	Cys –	Thr –	Tyr –	Met –	Pro –	Ala –	Ser –	Phe –	Met –	Val –	Ser –	COOH
	UGU	ACN	UAU	AUG	CCN	GCC	UCG	UUC	AUG	GUA	UCN	
	UGC		UAC									

Shine–Dalgarno sequence, so that they are correctly positioned during ribosome assembly.

In some cases, the UGA and UAG termination codons (but not UAA) are occasionally misinterpreted as codons for amino acids. That is, termination fails to occur at the normal place and results in a longer-than-usual protein. In some cases, these read-through proteins are vital—the organism depends on their existence. For example, in the phage Qβ, the coat-protein gene is read through about 2% of the time. Without this small number of read-through proteins, the phage coat cannot be constructed properly.

Evolution of the Genetic Code

Is the genetic code highly evolved, or just a "frozen accident?" In other words, is there a relationship between the codons and the amino acids they code for, or is the code just one of many random possibilities? Recent computer simulations of random codes indicate that the current genetic code is far outside the range of random possibility in its ability to protect the organism from mutation. This suggests that the genetic code is not a frozen accident, but rather is highly evolved.

For example, in the unmixed codon family 5'-CUN-3', any mutation in the third position produces another codon for leucine. This effect of a mutation in the third position of the codon not altering the encoded amino acid occurs in all eight unmixed codon families. In the eight mixed codon families, approximately one-third of the mutations in the third position of the codon will not alter the encoded amino acid. Thus, greater than half of the total mutations possible in the third position of the codon will not alter the encoded amino acid.

Patterns can also be found in the genetic code in which the mutation of one codon to another results in an amino acid that possesses similar properties. In these types of mutations, a high probability exists that the mutation will produce a functional protein. For example, all the codons with U as the middle base encode hydrophobic amino acids (phenylalanine, leucine, isoleucine, methionine, and valine). Mutation in either the first or third positions for any of these codons still encodes a hydrophobic amino acid. Both of the two negatively charged amino acids, aspartic acid and glutamic acid, have codons that start with GA. All of the aromatic amino acids—phenylalanine, tyrosine, and tryptophan (see fig. 11.4)—have codons that begin with uracil. Such patterns minimize the negative effects of mutation.

Restating the Concepts

▶ The wobble hypothesis states that the first base in the tRNA anticodon can basepair with one or more different bases in the last position of the triplet codon. This allows for redundancy, in which multiple codons encode a single amino acid.

▶ The genetic code is largely universal. The same codons are used to encode the same amino acids in nearly all known living organisms.

▶ Some exceptions have been found to the universality of the genetic code. These exceptions are usually restricted to a small number of codons in a particular organism or organelle within a eukaryotic cell.

▶ The triplet genetic code appears to be highly evolved to minimize the effect of mutation. For example, mutations in the third position of the codon are likely not to change the encoded amino acid. Also, functionally related amino acids are encoded by similar codons, such that a mutation in the codon will likely result in a different, but functionally related amino acid being encoded.

genetic linkages

Chapter 10 described mRNA as the RNA template for translation, rRNA as an integral component of the ribosome, and tRNA as the carrier of amino acids. This chapter expanded on those topics. For example, you learned about sequences in the mRNA that are essential or that enhance translation, including the AUG translation initiation codon, and a termination codon at the end of the ORF. Bacterial mRNA also requires a Shine–Dalgarno sequence immediately upstream of the translation initiation codon, which basepairs with the 16S rRNA to properly position the 30S ribosomal subunit on the mRNA.

Bacterial translation initiation permits multiple ORFs to be present on the same mRNA. The mRNA containing these multiple ORFs, called polycistronic mRNA, is a common feature of bacterial operons, which are discussed in greater detail in chapter 16.

Efficient translation initiation in eukaryotes requires the 5' cap structure and the 3' poly(A) tail on the mRNA (chapter 10). They both assist in the efficient assembly of the small ribosomal subunit on the mRNA, which then translocates along the mRNA until the translation initiation codon is reached. This process of the small ribosomal subunit assembling on the 5' end of the mRNA and its movement along the mRNA to the first ORF is the underlying reason that polycistronic mRNAs are not usually found in eukaryotes.

Our discussion on translation included aspects of translation initiation, elongation, and termination. We also introduced the subject of translation regulation, through the presence of specific sequences in the mRNA; the Shine–Dalgarno sequence in bacteria; and the 5' cap, poly(A) tail, and Kozak sequences in bacteria. Later in this book, we will further address mechanisms of translation regulation in bacteria (chapters 15 and 16) and eukaryotes (chapters 17 and 21).

The basic processes of transcription and translation are essential to the study of molecular genetics. We discussed the critical experiments that revealed the genetic code and the concept of reverse translation (box 11.1). In the following chapter, you will see how the material on reverse translation applies to the creation of nucleotide primers for the polymerase chain reaction and the generation of probes for the techniques called Southern and Northern hybridization. Understanding the basic components of transcription promoters and translation initiation will simplify the discussion on expressing foreign genes described in chapter 13.

Chapter Summary Questions

1. For each term in column I, choose the best matching phrase in column II.

   Column I
   1. A site
   2. Monocistronic
   3. Peptidyl transferase
   4. Methionine
   5. Signal peptidase
   6. P site
   7. *N*-formyl methionine
   8. Polynucleotide phosphorylase
   9. Polycistronic
   10. E site

   Column II
   A. First amino acid translated in eukaryotic proteins
   B. Bacterial mRNA
   C. Charged tRNAs usually enter through it
   D. Involved during protein synthesis
   E. The growing amino acid chain is found in it
   F. Involved in posttranslational modification
   G. Eukaryotic mRNA
   H. Uncharged tRNAs leave from it
   I. Involved in artificial mRNA synthesis
   J. First amino acid in bacterial proteins

2. What is the one-gene/one-enzyme hypothesis? Who proposed it in the 1940s? Does it still hold true today?

3. Describe the different levels of protein structure and how they relate to one another.

4. Summarize in equation form the process of aminoacyl-tRNA formation.

5. How does translation initiation occur in eukaryotes?

6. Compare and contrast the two tRNAs that bacteria use to carry methionine.

7. Describe the roles that EF-Tu, EF-Ts, and EF-G play in bacterial translation. Name their eukaryotic equivalents.

8. Which site(s) of the bacterial ribosome (A, P, and/or E) could contain the following:
   a. tRNA carrying only *N*-formyl methionine
   b. tRNA carrying any single amino acid
   c. tRNA carrying a chain of amino acids
   d. uncharged tRNA
   e. release factor

9. Outline the events that occur during termination of protein synthesis in bacteria.

10. The genetic code is degenerate, nonpunctuated, and almost universal. Explain what these terms mean.

11. List the four main types of posttranslational modifications that proteins can undergo.

12. Define the following terms: **(a)** polysome; **(b)** chaperone; and **(c)** wobble.

13. What is a signal peptide? What role does it play in eukaryotes? What is its fate?

14. What are the two main restrictions for an overlapping genetic code?

Solved Problems

PROBLEM 1: What is the energy requirement of protein biosynthesis?

Answer: The cost of adding one amino acid to a growing polypeptide is four high-energy bonds: two from an ATP during the charging of the tRNA, and two from the hydrolysis of GTPs during tRNA binding to the A site of the ribosome and during translocation. Thus, for an average protein of 300 amino acids, there is a cost of 1200 high-energy bonds.

PROBLEM 2: What amino acids could replace methionine if a one-base substitution occurred?

Answer: The codon for methionine (internal as well as initiation) is AUG. If the A is replaced, we would get UUG (Leu), CUG (Leu), or GUG (Val); if the U is replaced, we would get AAG (Lys), ACG (Thr), or AGG (Arg); and if the G is replaced, we would get AUA (Ile), AUU (Ile), or AUC (Ile). Hence, a one-base change in the codon for methionine could result in any of six different amino acids.

PROBLEM 3: The anticodons of three different tRNA molecules are shown as follows. For each of these tRNAs, determine the codon(s) they can bind to and the amino acid they carry.

   a. 5' AUG 3'   **b.** 5' UAG 3'   **c.** 5' IAU 3'

Answer: Remember that the binding of the anticodon and codon has to be in a complementary and antiparallel fashion. You also have to take into account any wobble base pairing. (Refer to table 11.6.)

a. The A in the 5' position of the anticodon (5'-AUG-3') has no complementary base other than U in the 3' position of the mRNA. Therefore, the codon that this tRNA can bind to is 3'-UAC-5'. To figure out the amino acid, the codon has to be considered in the conventional 5'-to-3' manner. According to table 11.5, the codon 5'-CAU-3' specifies the amino acid histidine.

b. The U in the 5' position of the anticodon (5'-UAG-3') can pair with either an A or a G (wobble base pairing) in the 3' position of the mRNA codon. Therefore, there are two codons that this tRNA can bind to: 3'-AUC-5' or 3'-GUC-5'. According to table 11.5, these two codons specify the amino acid leucine.

c. The inosine (I) in the 5' position of the anticodon (5'-IAU-3') can pair with three different bases at the 3' position of the mRNA codon: A, U or C (wobble base pairing). Therefore, there are three codons that this tRNA can bind to: 3'-AUA-5', 3'-UUA-5', and 3'-CUA-5'. According to table 11.5, these three codons specify the amino acid isoleucine.

PROBLEM 4: Artificial mRNA is synthesized from a reaction mixture containing only U and C ribonucleotides, in a ratio of 2:1, respectively. If this mRNA is translated in vitro in a cell-free system, name the amino acids and the proportions in which they would be incorporated into protein.

Answer: The ribonucleotides U and C can combine to form eight different codons: UUU, UUC, UCU, UCC, CUU, CUC, CCU, and CCC. The frequency of each of these codons is determined by the abundance of the U and C ribonucleotides. The probability of obtaining a U at any position in a codon is 2/3, and that of C is 1/3. Using the product rule we can determine the frequencies of the codons. Then we can use table 11.5 to determine the amino specificities.

UUU = (2/3)(2/3)(2/3) = 8/27 Phenylalanine
UUC = (2/3)(2/3)(1/3) = 4/27 Phenylalanine
UCU = (2/3)(1/3)(2/3) = 4/27 Serine
UCC = (2/3)(1/3)(1/3) = 2/27 Serine
CUU = (1/3)(2/3)(2/3) = 4/27 Leucine
CUC = (1/3)(2/3)(1/3) = 2/27 Leucine
CCU = (1/3)(1/3)(2/3) = 2/27 Proline
CCC = (1/3)(1/3)(1/3) = 1/27 Proline

Therefore, there are 12/27 (or 44.45%) phenylalanine, 6/27 (or 22.22%) serine; 6/27 leucine (or 22.22%), and 3/27 (or 11.11%) proline.

Exercises and Problems

15. You are studying the biosynthesis of Compound X in a deep sea microorganism. You isolated five mutants that cannot synthesize Compound X. Each mutant is tested for growth on minimal media containg a different compound. The results of that test are found in the following table.

Additive to Media

Strain	Compound A	Compound B	Compound C	Compound D	Compound E	Compound X
Wild type	+	+	+	+	+	+
Mutant 1	−	−	−	−	−	+
Mutant 2	−	−	+	−	−	+
Mutant 3	−	+	+	−	−	+
Mutant 4	−	+	+	+	−	+
Mutant 5	−	+	+	+	+	+

Using these data, define the biosynthetic pathway and show what step is defective in each mutant.

16. Given the following 3' end of an open reading frame (ORF), which will be transcribed and then translated into a pentapeptide, provide the base sequence for its mRNA. Give the anticodons on the tRNAs by making use of wobble rules. What amino acids are incorporated? Draw the actual structure of the pentapeptide.

```
3'-TACAATGGCCCTTTTATC-5'
5'-ATGTTACCGGGAAAATAG-3'
```

17. Draw the details of a moment in time at the ribosome during the translation of the mRNA produced in problem 16. Include in the diagram the ribosomal sites, the tRNAs, and the various nonribosomal proteins involved.

18. What are the similarities and differences among the three nonsense codons? Using the wobble rules, what could be their theoretical anticodons?

19. How many different DNA sequences can encode the pentapeptide Ile-Trp-Gly-Leu-Tyr?

20. Describe an experiment that demonstrates that the tRNA, and not its amino acid, is recognized at the ribosome during translation.

21. How many single-base deletions are required to restore the ORF of a messenger RNA? Give an example.

22. Compare and contrast ribosomes and spliceosomes?

23. Part of a DNA strand to be transcribed has the following sequence:
    a. What is the sequence of RNA transcribed from this part of the strand?
    b. What sequence of amino acids does the RNA produce?

    3'-TACTAACTTACGCTCGCCTCA-5'

24. A strain of *Escherichia coli* produces colicin E3, an endoribonuclease that cleaves 16S ribosomal RNA (rRNA) at the forty-ninth phosphodiester bond from the 3' end. What effect would you expect colicin E3 to have on bacterial protein synthesis? Be specific.

25. A *nonsense mutation* changes a codon for an amino acid into a translation termination codon. Give an example. What are its consequences?

26. The reverse situation to problem 25 is a mutation from a nonsense codon to a codon for an amino acid. Give an example. What are its consequences?

27. What are the consequences when an internal methionine codon recognizes a bacterial initiator tRNA?

28. What is the minimum number of tRNA molecules necessary to bind the six codons specifying arginine? List the anticodons in the 3'-to-5' orientation.

29. A peptide, 15 amino acids long, is digested by two methods, and each segment is sequenced. The 15 amino acids are denoted by the letters A through O, with F as the N-terminal amino acid. If the segments are as follows, what is the sequence of the original peptide?

    Method 1: CABHLN; FGKI; OEDJM
    Method 2: KICAB; JM; FG; HLNOED

30. What would the genetic code dictionary (see table 11.5) look like if wobble occurred in the second position rather than the third (that is, if an unmixed family of codons were of the form GNU)?

31. In experiments using repeating polymers, $(GCGC)_n$ incorporates alanine and arginine into polypeptides, and $(CGGCGG)_n$ incorporates arginine, glycine, and alanine. What codon can probably be assigned to glycine?

32. Artificial mRNA is synthesized from various reaction mixtures and then subjected to in vitro translation in a cell-free system. For each of the following mixtures, determine the amino acids and the proportions in which they would be incorporated into protein.
    a. 5A : 1C
    b. 3A : 1U : 1G

33. If poly-G is used as an mRNA in an incorporation experiment, glycine is incorporated into a polypeptide. If poly-C is used, proline is incorporated. However, if both the poly-G and poly-C are used in the same reaction, no amino acids are incorporated into protein. Why?

34. A protein has leucine at a particular position. If the codon for leucine is CUC, how many different amino acids might appear as the result of a single-base substitution?

35. How many ORFs are possible in a double-stranded DNA molecule?

36. The sixth amino acid in the β-chain of normal human hemoglobin is glutamate. Two different mutations of this codon substitute valine and lysine. What is the likely codon for glutamate?

37. Polymers of $(GUA)_n$ result in the incorporation of only two different amino acids rather than three, as for most other three-base polymers. Why?

38. A particular protein is composed of four identical polypeptides. The total molecular weight of the quaternary protein is 200,000 D. Assume the average molecular weight of an amino acid is 125 D. Determine the number of bases of exon DNA that encode this protein.

39. A normal protein has the following C-terminal amino acid sequence: Ser-Thr-Lys-Leu-COOH. A mutant is isolated with the following sequence: Ser-Thr-Lys-Leu-Leu-Phe-Arg-COOH. What probably happened to produce the mutant protein? Be specific.

40. The space probe HEA1 returns to earth following a trip to the planet Tantoun. Analysis of samples from this planet reveals that proteins are composed of 16 different amino acids, and DNA contains only two bases: rosine (R) and ziadine (Z).
    a. What is the minimum codon size in this genetic code?
    b. How many of the codons in this minimum genetic code will contain at least two Z's?

41. A normal protein has histidine in a given position. Four mutants are isolated and determined to have tyrosine, glutamine, proline, or leucine in place of histidine. What are the possible codon assignments, and what codon is probably used for histidine?

42. A segment of a normal protein and three different mutants appears as follows:
    normal   _____Gly-Ala-Ser-His-Cys-Leu-Phe_____
    mutant 1   _____Gly-Ala-Ser-His
    mutant 2   _____Gly-Ala-Ser-Leu-Cys-Leu-Phe_____
    mutant 3   _____Gly-Val-Ala-Ile-Ala-Ser_____

    What is the probable sequence of bases in the normal mRNA?

43. Complete the following table covering the first five amino acids in a eukaryotic protein. Columns I and II need to be filled with: 5', 3', C or N. Assume no wobble pairing.

I							C	T	II	
T			G							DNA double helix
		A	C					G		Transcribed mRNA
				C	U	A				tRNA anticodons
Met					Trp					Encoded amino acids

Chapter Integration Problem

Gene expression in the hypothetical bacterium *Bim nailliuqra* follows the same basic bacterial transcriptional and translational mechanisms described in chapters 10 and 11, with the following exceptions:

1. Transcription initiation always occurs at a purine.

2. Translation initiation can occur at an AUG, GUG, CUG or UUG, with all four codons specifying the amino acid N-formyl methionine at the start of an ORF.

3. UGA is not a stop codon; Rather, it encodes the twenty-first amino acid, jertzine (jer).

The genetic code of this bacterium, with the three-letter and one-letter amino acid abbreviations is shown below.

Second position of codon

		U	C	A	G	
First position of codon	U	Phe, F Phe, F Leu, L Leu, L	Ser, S Ser, S Ser, S Ser, S	Tyr, Y Tyr, Y *Stop* *Stop*	Cys, C Cys, C Jer, J Trp, W	U C A G
	C	Leu, L Leu, L Leu, L Leu, L	Pro, P Pro, P Pro, P Pro, P	His, H His, H Gln, Q Gln, Q	Arg, R Arg, R Arg, R Arg, R	U C A G
	A	Ile, I Ile, I Ile, I Met, M	Thr, T Thr, T Thr, T Thr, T	Asn, N Asn, N Lys, K Lys, K	Ser, S Ser, S Arg, R Arg, R	U C A G
	G	Val, V Val, V Val, V Val, V	Ala, A Ala, A Ala, A Ala, A	Asp, D Asp, D Glu, E Glu, E	Gly, G Gly, G Gly, G Gly, G	U C A G

(Third position of codon)

A cloned gene from *Bim* is given the name *fun* (function *un*known). The following DNA sequence, shown in the 5' → 3' orientation, was obtained from the *nontemplate* strand of the gene:

```
  1   CAATATGGAC   GGCTAACGAC   GAATACGCAG   ATCCGGTAAC   GGCTCCAGAC
 51   CTAAGTTGGC   AGGCCAGTAG   CAGGACTGAC   CATATAATGG   ACTTACCTGG
101   TATGCAGGTT   TATCGCGGCT   AACGGCAGTG   TAGGAGGCCG   GTGGAGAACA
151   TCAACGATGC   CAGAAAATAA   GAGACATCCG   GGGAAACATG   AATAGGAGGC
201   TACTGGATAA   CACACATAAC   GTAAGGGGAG   AATTAATGCC   CCGGGATGAA
251   GCTATTCATC   CTTTTTTTCG   CCGATGTTGA   CATGCACTAC   GTTGCAATGT
```

a. If codons in *Bim* were five bases long instead of three, how many different codons will the genetic code contain?

b. Determine the number of different primary structures possible for a *Bim* polypeptide that is 50 amino acids long.

c. What is the type of transcription termination for the *fun* gene? Explain with the help of a specific diagram.

d. Identify the promoter elements for the *fun* gene and predict the sequence and proper polarity of the *fun* mRNA. Discuss your reasoning.

e. What is the amino acid sequence of the Fun protein? Use the single-letter abbreviations for the amino acids.

Six different mutants are found to produce a Fun protein whose primary structure is three times longer than that of the wild-type protein. Sequence analyses revealed interesting results: All six mutants differ in only one or two nucleotides from each other, and all produce mutant Fun proteins with identical amino acid sequences. The differences in DNA sequence between the mutant and wild-type *fun* genes are shown as follows. (*Key:* R = purine and Y = pyrimidine)

Mutant 1 = A single R-to-Y base substitution.
Mutant 2 = A Y insertion and an R deletion.
Mutant 3 = A Y insertion, an R deletion, and a single R-to-Y base substitution.
Mutant 4 = A Y insertion, an R deletion, and a single R-to-Y base substitution.
Mutant 5 = A Y insertion, an R deletion, and a single R-to-R base substitution.
Mutant 6 = A Y insertion, an R deletion, and a single R-to-R base substitution.

f. What is the cause of this larger Fun protein in each of the six mutants?

g. Predict the different base(s) between each of the six mutants and the wild-type *fun* gene. Explain how you got to your answers.

h. What is the amino acid sequence of this larger mutant Fun protein? Use the single-letter abbreviations for the amino acids.

Do you need additional review? Visit **www.mhhe.com/hyde** for practice tests, answers to end-of-chapter questions and problems, interactive exercises, and animation tutorials all designed to enhance your understanding of key genetic concepts.

Recombinant DNA Technology

Essential Goals

1. Understand the basic features of restriction enzymes and how they are used in DNA cloning.

2. Compare the features of the various types of vectors and how they affect DNA cloning.

3. Describe nucleic acid hybridization techniques and be able to properly interpret the data.

4. Be able to prepare a restriction map using double-digest, partial-digest, and end-labeling data.

5. Diagram how the polymerase chain reaction works.

6. Understand the methods used to sequence DNA and be able to correctly read the DNA sequence from a sequencing gel.

Chapter Outline

Photo: Artificially colored transmission electron micrograph of plasmids from the bacterium *Escherichia coli*. These plasmids are used in genetic engineering.

Since the mid-1970s, the field of molecular genetics has undergone explosive growth, noticeable not only to geneticists, but also to medical practitioners and researchers, agronomists, animal scientists, venture capitalists, and the public in general. Medical practitioners and researchers have new treatments for diseases available. Agronomists see the possibility of greatly improved crop yields, and animal scientists have gained the possibility of greatly improving food production from domesticated animals. Geneticists and molecular biologists are gaining major new insights into understanding gene expression and its control.

The molecular genetics' revolution exploded when it became possible to isolate a specific DNA sequence and then to stably express it in a different cell or organism (**fig. 12.1**). The ability to clone and rapidly sequence DNA made the analysis of genes significantly easier. By cloning both wild-type and mutant genes, it became possible to take the wealth of genetic data that had been accumulating for decades and analyze a molecular defect at the DNA level. The ability to purify genes by cloning and the polymerase chain reaction procedure also allowed scientists to determine more precisely when and where a gene is transcribed in a multicellular organism. This overall technology is variously referred to as **gene cloning**, **recombinant DNA technology**, or **genetic engineering**.

In the coming chapters, we will discuss the techniques of gene therapy, genomics, and bioinformatics. However, those approaches are grounded in the fundamental molecular genetics' techniques presented in this chapter. It is important to fully understand these techniques to appreciate the power of molecular genetics and to see where this field is going. And, it is equally important to understand that molecular genetics is firmly based in the classical genetics that began this book. Rather than looking at whole-organism phenotypes, molecular genetics allows us to examine the phenotype at the DNA sequence level. By combining classical and molecular genetics, the true power of genetics can be brought to bear on a wide range of fields, from ecology and evolution to medicine.

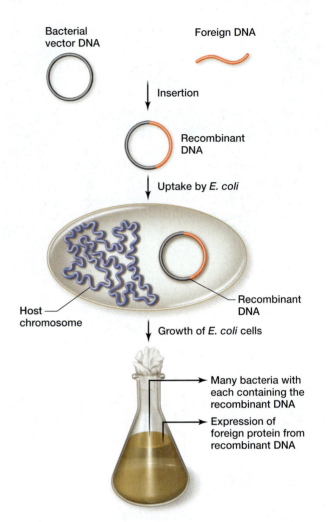

figure 12.1   Overview of recombinant DNA techniques. A recombinant DNA molecule is created that contains an insert of foreign DNA in a bacterial DNA vector. The recombinant DNA vector is then introduced into a bacterial cell, such as *E. coli*. Replication of the bacterial cell results in many cells that each contain the recombinant DNA and, if the gene is expressed, quantities of the gene product. All DNA shown is double-stranded.

12.1 Cloning DNA

The variety of recombinant DNA techniques are built on several fundamental processes, such as cloning DNA sequences and nucleic acid base pairing (hybridization). The three general steps involved in cloning DNA are:

1. Cleaving DNA into segments, which is accomplished by restriction endonucleases
2. Rejoining DNA segments, termed DNA ligation, and
3. Adding a replication origin, so that the newly made DNA can be replicated in a cell.

In this section, we discuss the discovery and details of methods that accomplish these processes.

Restriction Endonucleases

In 1978, the Nobel Prize in Physiology and Medicine was awarded to Werner Arber, Hamilton Smith, and Daniel Nathans for their pioneering work in the study of **restriction endonucleases.** These are enzymes that bacteria use to destroy foreign DNA, presumably the DNA of invading bacteriophages. The enzymes recognize specific nucleotide sequences (**restriction sites** or **restriction sequences**) that are usually 4–8 bp long and then cleave (digest) the DNA at or near those sites. Additionally, these enzymes have allowed us to easily clone DNA, which made it possible to generate transgenic animals and plants, complete the Human Genome Project, and begin gene therapy trials for several human diseases.

Types of Restriction Endonucleases

Three types of restriction endonucleases are known. Their groupings are based on the types of sequences they recognize, the nature of the cleavage made in the DNA, and the enzyme structure. Type I and III restriction endonucleases cleave DNA at sites other than the recognition sites. This causes random, unpredictable cleavage patterns, which does not make them useful for gene cloning. In contrast, type II endonucleases recognize and cleave specific sequences, which produces predictable cleavage patterns and makes them useful for cloning.

The type II endonucleases recognize **palindromic sequences,** which are sequences that read the same from either direction. (The name Hannah and the numerical sequence 1238321 are palindromes.) For example, the type II restriction endonuclease **BamHI** (pronounced bam-h-one) recognizes this palindromic DNA sequence:

$$5'\text{-GGA} \mid \text{TCC-}3'$$
$$3'\text{-CCT} \mid \text{AGG-}5'$$

Notice that both strands possess the same sequence, 5'-GGATCC-3'. You should also notice that the sequence possesses two-fold symmetry. Reading in a 5' → 3' direction from the center (vertical line), both the top and bottom strands contain the TCC sequence. **Figure 12.2** shows some palindromic sequences that type II restriction endonucleases recognize; well over 200 type II enzymes are known.

Restriction endonucleases are named after the bacteria from which they were isolated: *BamHI* from *Bacillus amyloliquefaciens* strain H; *EcoRI* (echo-r-one) from *Escherichia coli* strain RY13; *HindII* (hin-dee-two) from *Haemophilus influenzae* strain Rd; and *BglI* (bay-gul-one) from *Bacillus globigii.* Note that the I and II in the names does not correspond to the type of endonuclease; all of these are type II endonucleases. From here on, we will refer to type II restriction endonucleases simply as restriction enzymes.

Cleavage Site Frequency

The frequency of a specific restriction enzyme site depends on two factors. First, the general GC content of the DNA will affect the frequency. For example, the restriction enzyme *SmaI* cuts DNA at the sequence 5'-CCCGGG-3'. If the GC content is high in the genome, the *SmaI* restriction enzyme sites will be more frequent relative to a genome that possesses a low GC content.

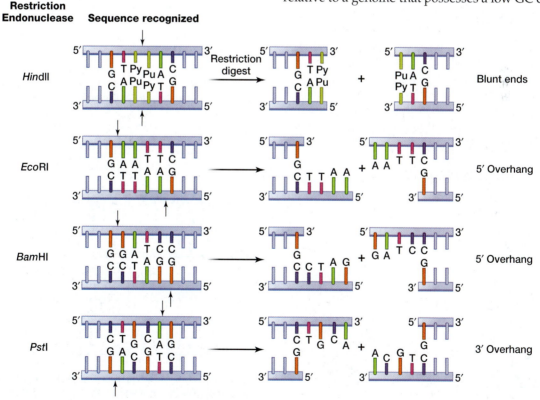

figure 12.2   Sequences cleaved by various type II restriction endonucleases. Py is either pyrimidine (cytosine or thymine) and Pu is either purine (adenine or guanine). The vertical arrows denote sites where the endonucleases always cleave the phosphodiester bond in the DNA to create either staggered cuts (with 3' or 5' overhangs) or blunt ends. This precision and reproducibility of enzyme digestion of the DNA made these enzymes useful for further research.

Second, the frequency of a site is the product of the probabilities of the presence of each correct base in the restriction enzyme site. For example, the *Eco*RI restriction enzyme cuts at the site 5'-GAATTC-3'. The frequency of this site equals the probability of having a G at the first position (1/4), an A at the second site (1/4), an A at the third position (1/4), and so on. The calculated frequency of an *Eco*RI site is therefore $1/4 \times 1/4 \times 1/4 \times 1/4 \times 1/4 \times 1/4$, or 1 in 4096.

Calculating frequencies provides a convenient way to predict the average length of DNA fragments after digestion with a specific restriction enzyme. In the preceding example, the average length of a DNA fragment produced from an *Eco*RI digestion is 4096 bp. If short restriction fragments are desired, it is best to use a restriction enzyme that recognizes a 4-bp sequence, which is found once in 256 bp. Longer fragments can be generated by using a restriction enzyme that recognizes a 6- or 8-bp sequence.

Protective Methylation of Host DNA

The natural function of restriction enzymes is to protect the host bacterial cell against the introduction of foreign DNA, usually by viruses. Cutting the viral DNA into pieces renders its replication impossible. However, these restriction sites will be present in both the foreign genome and the host genome.

To prevent the restriction enzyme from cutting the host DNA, a methylase enzyme specifically modifies the restriction sequences in the host DNA (**fig. 12.3**). The methylated sequence cannot be cut by the restriction enzyme, but the unmethylated sequence (foreign DNA) is cut. During host DNA replication, the new double helices are hemi-methylated; that is, the old strand is methylated but the new one is not. After completing replication, the new strand is quickly methylated (**fig. 12.4**).

Blunt and Staggered Cuts

Restriction enzymes cut DNA to produce two different types of ends: blunt or staggered. *Hin*dII, which cuts in the middle of its 6-bp recognition sequence, generates a blunt end; in contrast, *Bam*HI, which cleaves its 6-bp restriction site off center, produces a staggered cut (see fig. 12.2).

Additionally, two types of staggered cuts are possible. In the case of *Bam*HI-cleaved DNA, a single-stranded sequence is located at the 5' end of the DNA (termed a **5' overhang**). In contrast, the *Pst*I enzyme (see fig. 12.2) cleaves DNA to produce a single-stranded sequence that contains the 3' end of the DNA (termed a **3' overhang**). Notice that two DNA fragments, digested with the same restriction enzyme and possessing staggered ends, can reanneal and form hydrogen bonds between the complementary bases in the single-stranded regions (**fig. 12.5**). The ability to reanneal these *sticky ends,* first demonstrated by Stanley Cohen, Herbert Boyer, and colleagues in 1971, opened up the field of gene cloning.

DNA Ligation

Ends of DNA fragments that are able to anneal through complementary single-stranded sequences, as well as two blunt-ended DNA fragments, are called **compatible ends.** To clone a gene, these DNA fragments must be joined in a reproducible manner.

The most common method of joining two DNA molecules is ligation. In chapter 9 we discussed DNA ligase, the enzyme that seals the sugar–phosphate (phosphodiester) bond between Okazaki fragments. In the case of staggered ends, if two compatible ends anneal, then DNA ligase seals the phosphodiester bond

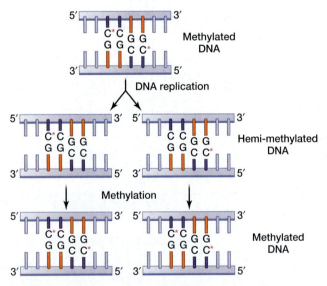

figure 12.4   DNA is methylated at the *Hpa*II restriction site. Red asterisks indicate methyl groups on specific cytosines. After DNA replication, the DNA is hemi-methylated; the cytosine on the template strand is methylated, but the cytosine on the new strand lacks the methyl group. Hemi-methylated DNA is then fully methylated by cellular enzymes before the next round of DNA replication.

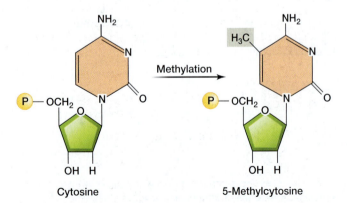

figure 12.3   A methylase enzyme adds a methyl group to cytosine, converting it to 5-methylcytosine.

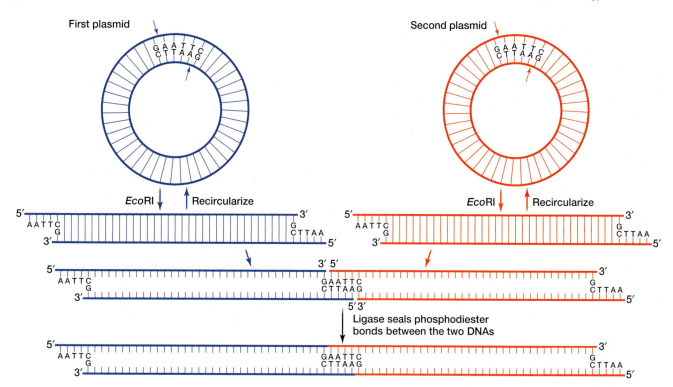

figure 12.5   Circular plasmid DNA with a palindrome recognized by *Eco*RI. After the DNA is cleaved by the *Eco*RI endonuclease, two short single-stranded ends are generated that can join by complementary basepairing to unite the two linear DNA molecules. DNA ligase is added to generate the phosphodiester bonds to seal the two nicks. The *Eco*RI sequences at the ends of this linear hybrid molecule can also anneal and be ligated to produce a hybrid circular molecule.

between the two DNA molecules (see fig. 12.5). In the case of blunt ends, DNA ligase will also ligate the molecules together. Because DNA ligase is nonspecific about which blunt ends it joins, many different, unwanted products result from its action. Notice that in all DNA ligations, the 3' end of one DNA strand is linked to the 5' end of another DNA strand.

A variation of blunt-end ligation uses **linkers**— short, artificially synthesized pieces of DNA containing a restriction endonuclease recognition site. When these linkers are attached to blunt-ended DNA and then treated with the appropriate restriction enzyme, staggered ends are created. In **figure 12.6**, the linkers are 10-bp segments of DNA with an *Eco*RI site in the middle. They are attached to the DNA to be cloned with T4 DNA ligase. Subsequent treatment with *Eco*RI will result in DNA with *Eco*RI sticky ends.

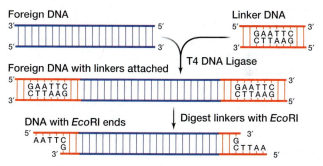

figure 12.6   Creating restriction sites at ends of linear DNA with linkers. Linkers are small segments of DNA with an internal restriction site. Linkers can be added to blunt-ended DNA by T4 DNA ligase. In this example, the linker DNA contains an *Eco*RI restriction site. Restriction digestion of this ligated DNA with *Eco*RI creates DNA with ends that are compatible with any DNA cut by *Eco*RI.

It's Your Turn

Restriction enzymes are a fundamental tool in molecular genetics. Try problem 18 to check your understanding of the different types of ends produced by restriction digests. If you wish to test your understanding of a palindromic DNA sequence, then problem 17 is for you.

Restating the Concepts

▶   Type II restriction enzymes are bacterial enzymes that recognize and cut within specific palindromic DNA sequences.

▶   Restriction digestion creates either blunt- or staggered DNA ends.

▶   DNA ligase can join two different DNA molecules, provided they each possess compatible ends.

Bacterial Vectors and Replication Origins

Although restriction enzymes provide a mechanism to isolate DNA fragments that can be joined together with DNA ligase, these DNA molecules cannot be faithfully maintained in a cell. These DNA fragments must be joined with an origin of replication so that they can replicate in a cell independently of the cellular genome.

Vectors are DNA molecules that can be maintained in a cell and can have foreign DNA cloned into them. Thus, vectors can serve as carriers for foreign DNA in a cell. A variety of different bacterial vectors exist, all which can replicate in *E. coli*, which is the major bacterial species used in molecular genetics. These vectors include *plasmids*, the *λ phage, cosmids, bacterial artificial chromosomes*, and *shuttle vectors*. Three of the major differences between these various types of vectors are: (1) the size of foreign DNA that can be cloned in them, (2) the process by which these vectors are introduced into bacterial cells, and (3) the mechanism by which the vectors replicate in the bacterial cell. In chapter 13, we will expand this discussion to include a variety of eukaryotic vectors.

Bacterial cloning vectors share two important features that simplify the isolation of a recombinant DNA molecule. First, vectors must have a mechanism to replicate autonomously in the cell, which means that they can replicate on their own without inserting into the bacterial genome. This simplifies the isolation of the cloning vector and cloned DNA from the bacterial genomic DNA. Second, vectors must have at least one selectable marker. This marker, which can be a gene encoding resistance to an antibiotic or the β-galactosidase (*lacZ*) gene, allows the identification of bacterial cells that contain the vector. In some cases, it can also be used to identify cells that contain a vector with a segment of DNA cloned within it. Other features either increase the ease of using the vector, such as a multiple cloning site, or are specific to the class of vectors, such as multiple origins of replication in shuttle vectors.

Plasmids

The simplest of all the vectors is a plasmid. **Plasmids** are circular DNA molecules that possess an origin of replication (**fig. 12.7**). The origin of replication (usually designated as *ori*) allows the plasmid to replicate autonomously in the cell and does not have to insert into the bacterial genome. Because the plasmid replicates autonomously, the number of copies of a given plasmid in a bacterial cell (*copy number*) can vary. Some plasmids, such as pSC101 (plasmids are designated with a lower case *p* in the name), have a low copy number and will be present in only four to seven copies in a cell; other plasmids, such as pUC13, have a high copy number and may be present in 30 or more copies per cell. The copy num-

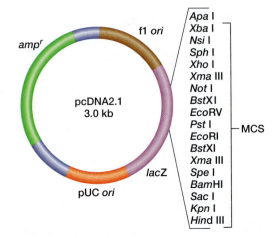

figure 12.7 The pcDNA2.1 plasmid contains several genetic elements that are common among most plasmids. This plasmid is a small circular DNA molecule that includes a bacterial plasmid origin of replication (pUC *ori*), a single-stranded bacteriophage origin of replication (*f1 ori*), a multiple cloning site region (MCS) that contains several unique restriction enzyme sites, and a selectable marker (ampicillin resistance, *amp*r). Cells containing this plasmid can be selected by resistance to ampicillin (*amp*r) and for the presence of a foreign piece of DNA cloned into the plasmid (generating *lacZ*−).

ber of a plasmid is controlled by the plasmid's origin of replication.

In addition to an origin of replication, most plasmids also contain two additional features. The first is a **multiple cloning site region** or MCS (see fig. 12.7). Several different restriction sites are located in this region. What makes this region particularly useful for cloning is that each of these restriction sites is present only once in the plasmid. Cutting a circular plasmid with any of these restriction enzymes results in the formation of a single linear DNA molecule (**fig. 12.8**).

The second feature is that plasmids contain selectable markers; these allow identification of bacterial cells that contain a plasmid and distinguishes the plasmids that contain a cloned foreign DNA. When the plasmid is digested with a restriction enzyme and ligated to foreign DNA, most plasmids will ligate into a circle either with or without an insert (**fig. 12.9**). This mixture of plasmids will be introduced into bacterial cells using either an electrical current (*electroporation*) or a chemical treatment (*transformation*). Some bacterial cells take in a plasmid that either contains an insert (**recombinant plasmid**) or lacks the insert, but other cells will not take in any plasmid (see fig. 12.9). The selectable markers allow us to distinguish among these three different types of bacteria.

The first selectable marker is often an antibiotic resistance gene, such as *amp*r, which makes the bacterial cell resistant to ampicillin (a derivative of penicillin).

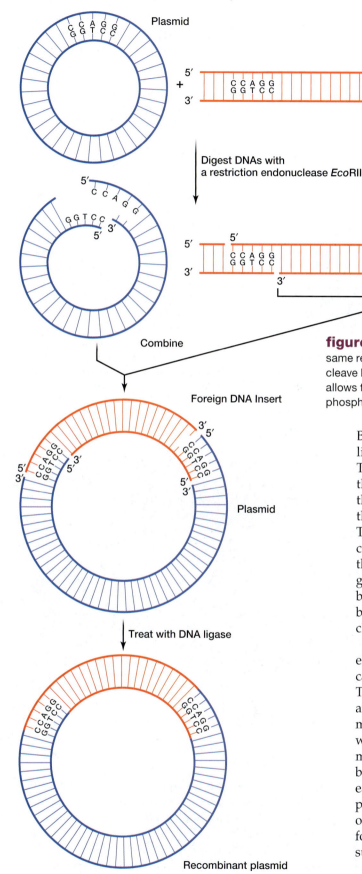

figure 12.8   Formation of a recombinant plasmid. The same restriction endonuclease, in this case *Eco*RII, is used to cleave both host and foreign DNA. Combining the digested DNAs allows the single-stranded ends to base pair. The two nicks in the phosphodiester backbone are sealed by DNA ligase.

Bacterial cells growing on media that contains ampicillin must possess a plasmid containing the *amp*^{r} gene. The second selectable marker is *lacZ*, which encodes the β-galactosidase protein. The MCS is located within the *lacZ* gene, so cloning foreign DNA into one of these restriction enzyme sites disrupts the *lacZ* gene. The β-galactosidase enzyme can cleave the artificial compound X-gal, which is colorless, into a molecule that forms a blue precipitate. Thus, bacterial cells that grow on a solid medium that contains X-gal will be blue (*lacZ*$^+$) if the plasmid lacks foreign DNA or will be white (*lacZ*$^-$) if the plasmid contains foreign DNA cloned in the MCS region.

Note that in the process of inserting a piece of foreign DNA into a plasmid, the restriction site is duplicated, with one copy at either end of the insert (fig. 12.9). This property makes it easy to remove the cloned insert at some future time, by cutting the recombinant plasmid with the same restriction enzyme. This cleavage will produce two linear molecules, the original plasmid and the cloned foreign DNA insert. Furthermore, because the plasmid is a free episome, it is relatively easy to isolate either pure wild-type or recombinant plasmid DNA from bacterial cells. This combination of easily isolating plasmid DNA and removing the foreign insert from the plasmid simplifies a number of subsequent molecular biology techniques.

Because plasmids must circularize during ligation, the maximal size of DNA that can efficiently form a

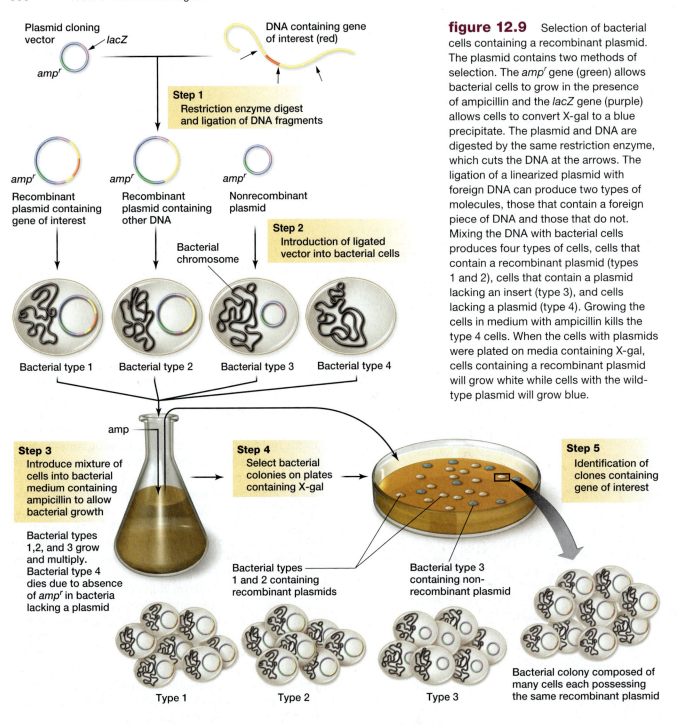

figure 12.9 Selection of bacterial cells containing a recombinant plasmid. The plasmid contains two methods of selection. The *amp^r* gene (green) allows bacterial cells to grow in the presence of ampicillin and the *lacZ* gene (purple) allows cells to convert X-gal to a blue precipitate. The plasmid and DNA are digested by the same restriction enzyme, which cuts the DNA at the arrows. The ligation of a linearized plasmid with foreign DNA can produce two types of molecules, those that contain a foreign piece of DNA and those that do not. Mixing the DNA with bacterial cells produces four types of cells, cells that contain a recombinant plasmid (types 1 and 2), cells that contain a plasmid lacking an insert (type 3), and cells lacking a plasmid (type 4). Growing the cells in medium with ampicillin kills the type 4 cells. When the cells with plasmids were plated on media containing X-gal, cells containing a recombinant plasmid will grow white while cells with the wild-type plasmid will grow blue.

circle affects the size of the foreign DNA that can be cloned into a plasmid. Usually, the maximum size of a circular recombinant DNA molecule is approximately 15 kb (kilobase pairs), although larger molecules can be generated at a lower efficiency. The size of the plasmid, which is usually 3–5 kb, then affects the size of the foreign DNA that can be cloned. Generally, plasmids are used to clone foreign inserts of 100 bp to 10 kb (**table 12.1**), which includes cDNAs (discussed later in this chapter) and genomic DNA fragments.

Lambda Phage

Lambda (λ) is a **bacteriophage,** a virus that infects bacterial cells. The λ genome is a linear DNA molecule that is approximately 48.5 kb in length. Outside of the *E. coli* cell, the λ DNA is present within a protein coat (**fig. 12.10**). The λ genome encodes the proteins that form the coat. When λ infects a bacterial cell, the protein coat attaches to the cell wall and injects the λ genomic DNA inside the bacterial cell.

table 12.1   Common Vectors Used in *E. coli*

Vector	DNA Form	Typical Insert Size (kb)	Common Uses
Plasmid	Circular	0.1–10	Cloning cDNA or DNA fragments
Lambda	Linear	2–25	Cloning cDNA or Genomic libraries
Cosmid	Circular	30–45	Genomic libraries
Bacteriophage P1	Linear	60–90	Genomic libraries
BAC	Circular	100–1,000s	Genomic libraries

On entering the cell, the λ DNA can do one of two things (also see chapter 16). It may integrate into the *E. coli* genome and reside in a silent state, until it generates more infectious particles. Alternatively, it may replicate immediately to form more DNA molecules, which are then transcribed and translated to make coat proteins. A single λ DNA molecule then enters each protein coat to make new infectious particles. The bacterial cell lyses, and the λ particles are then free to infect other bacterial cells.

The ends of the linear λ DNA contain short single-stranded sequences (12 nucleotides in length) that are complementary. These sequences, called *cos* **sites,** can base-pair and allow λ to form a circle in the *E. coli* cell (**fig. 12.11**). This circular form of λ is the molecule that integrates into the bacterial genome. The *cos* sites are also used to pack the λ DNA into the protein coat in a process termed *head filling.* Multiple λ genomes base-pair end to end through the *cos* sites to form long linear molecules called *concatamers* that are then loaded into the protein coat.

The ends of λ are essential for its use as a cloning vector. λ can be cut with a restriction enzyme to generate three fragments: the left arm, the right arm, and the internal fragment (see fig. 12.10). Foreign DNA, which was digested with a restriction enzyme that produced the compatible type of ends, can be cloned between the left and right arms to create a recombinant molecule of approximately 49 kb. The size of the DNA insert, between 2 and 25 kb, depends on the size of the internal fragment that is removed (see table 12.1). As the recombinant λ molecules are produced, they are ligated into concatamers (**fig. 12.12**). The proteins that form the coat are added directly to the ligated DNA, which permits the recombinant λ to be *packaged* into the protein coat and used to infect *E. coli* cells.

During λ packaging, the DNA between two *cos* sites in the linear concatamer is introduced into the λ particle's head. This protein head must not be overfilled or underfilled with DNA (see fig. 12.12). The ideal length of the total λ DNA molecule is 49 kb; significantly larger or smaller DNA molecules will alter the outer protein coat, causing it to either overexpand or collapse, both which render the particle noninfectious.

On one hand, λ is a vector that permits cloning larger DNA fragments than plasmids. The efficiency of λ infection is greater than either electroporation

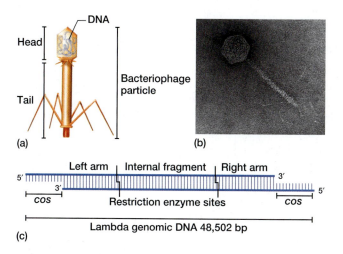

figure 12.10   Bacteriophage λ is a common vector in *E. coli*. (a) The bacteriophage particle is composed of a protein coat (head and tail) with the λ genomic DNA packaged in the head. (b) An electron micrograph of a λ particle. (c) The λ genome is 48,502 bp in length. It contains two complementary 12 nt single-stranded sequences (*cos* sites) at the ends of the linear molecule. The molecule can be divided by restriction enzymes into a left arm, right arm, and internal fragment.

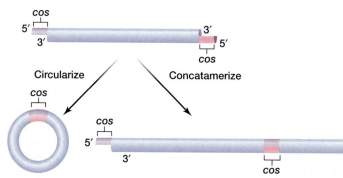

figure 12.11   The 12 nucleotide single-stranded *cos* sequences at the ends of λ are complementary so they can base-pair within a single molecule to form a circle (lower left) or they can base-pair between different molecules and form a linear concatamer (lower right).

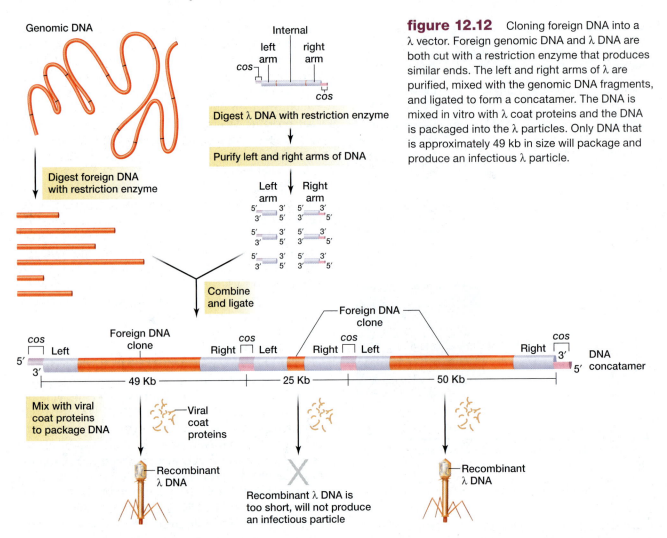

figure 12.12   Cloning foreign DNA into a λ vector. Foreign genomic DNA and λ DNA are both cut with a restriction enzyme that produces similar ends. The left and right arms of λ are purified, mixed with the genomic DNA fragments, and ligated to form a concatamer. The DNA is mixed in vitro with λ coat proteins and the DNA is packaged into the λ particles. Only DNA that is approximately 49 kb in size will package and produce an infectious λ particle.

or transformation of *E. coli,* which results in a greater number of bacteria that contain a recombinant λ molecule relative to a recombinant plasmid. Its inability to replicate autonomously without killing the cell, however, makes it harder to utilize as a cloning vector. The selection for the presence of λ is the killing of the bacterial cells, which creates a plaque (clear area of lysed bacterial cells) on an agar plate.

Cosmids

As a compromise between the ability to clone large DNA fragments (λ) and the ease of working with a plasmid, another cloning vector was created. A **cosmid** is a hybrid that contains the λ *cos* sites in a plasmid. Cosmids contain an origin of replication (so they can replicate like a plasmid), a selectable marker, and a single *cos* site (**fig. 12.13**). Large foreign DNA fragments are ligated to the linearized cosmid. The concentration of DNA fragments is adjusted to produce the recombinant linear concatamer.

The linear concatamer is mixed with the proteins that form the λ coat, and the recombinant DNA is

packaged into the coats (see fig. 12.13). (As mentioned, the DNA between *cos* sites must be approximately 49 kb.) When the recombinant phage infects a bacterial cell, the cosmid DNA is injected inside the *E. coli,* and the *cos* sites base-pair. *E. coli* DNA ligase seals the phosphodiester nicks. The resulting circular molecule, which contains an origin of replication and a selectable marker, now behaves as a plasmid and is unable to generate infectious particles (due to the absence of most of the λ genome).

Bacterial Artificial Chromosomes

As the genome projects of various organisms, including humans, began to move forward, the desire to isolate fragments that were greater than 1000 kb required the creation of a new class of vectors. **Bacterial artificial chromosomes (BAC)** were developed to meet this demand (**fig. 12.14**).

A BAC vector contains the origin of replication and the partitioning element sequences from the **F factor,** a naturally occurring plasmid in *E. coli* that is thousands of kilobase pairs in length (discussed in chapter 15).

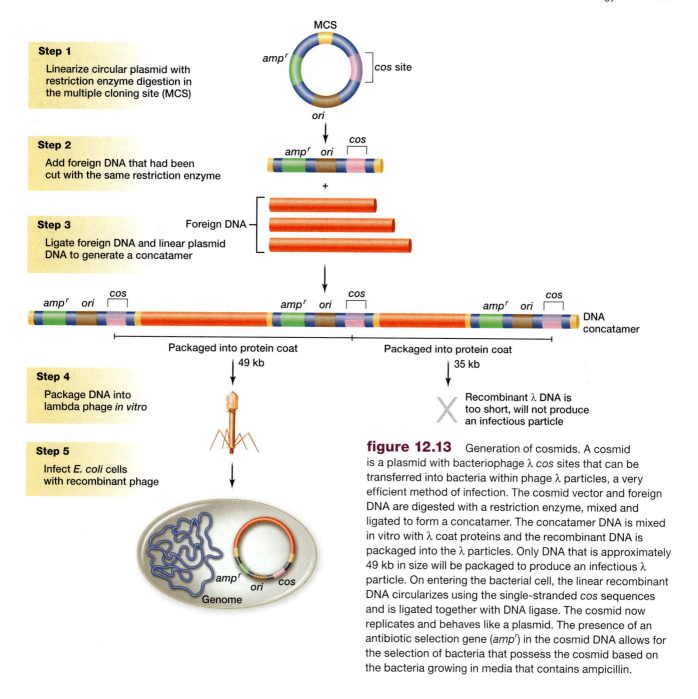

Step 1

Linearize circular plasmid with restriction enzyme digestion in the multiple cloning site (MCS)

Step 2

Add foreign DNA that had been cut with the same restriction enzyme

Step 3

Ligate foreign DNA and linear plasmid DNA to generate a concatamer

Step 4

Package DNA into lambda phage *in vitro*

Step 5

Infect *E. coli* cells with recombinant phage

Recombinant λ DNA is too short, will not produce an infectious particle

figure 12.13 Generation of cosmids. A cosmid is a plasmid with bacteriophage λ *cos* sites that can be transferred into bacteria within phage λ particles, a very efficient method of infection. The cosmid vector and foreign DNA are digested with a restriction enzyme, mixed and ligated to form a concatamer. The concatamer DNA is mixed in vitro with λ coat proteins and the recombinant DNA is packaged into the λ particles. Only DNA that is approximately 49 kb in size will be packaged to produce an infectious λ particle. On entering the bacterial cell, the linear recombinant DNA circularizes using the single-stranded *cos* sequences and is ligated together with DNA ligase. The cosmid now replicates and behaves like a plasmid. The presence of an antibiotic selection gene (*amp^r*) in the cosmid DNA allows for the selection of bacteria that possess the cosmid based on the bacteria growing in media that contains ampicillin.

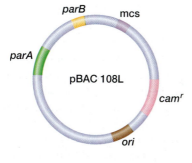

figure 12.14
Organization of a bacterial artificial chromosome (BAC). The BAC vector contains an origin of replication (*ori*) and two sequences (*parA* and *parB*) that are required to partition the BAC molecules into both daughter cells after DNA replication. A multiple cloning site (mcs) is present to help with cloning the foreign DNA. A gene conferring resistance to the antibiotic chloramphenicol (*cam^r*) is present to select bacterial cells that contain the BAC vector.

The BAC is present in a single copy in a bacterial cell and replicates prior to cell division, which allows the vector to enter each daughter cell. The partitioning element sequences (*parA* and *parB*) ensure that a single BAC moves into each daughter cell. While plasmids and cosmids lack *par* sequences, they are present at many copies in each cell, which allows at least one copy of the vector to randomly enter each daughter cell.

A gene that confers resistance to an antibiotic is included to allow selection of *E. coli* cells that contain the BAC. A multiple cloning site is added to provide several unique restriction enzyme sites. These sites simplify the cloning of the foreign DNA insert, which may be 100 to several thousand kilobase pairs in length.

Shuttle Vectors

The **shuttle vector** is simply a circular plasmid that contains two origins of replication. One origin usually allows the vector to replicate in *E. coli*. The second origin allows the vector to replicate in a different organism, such as yeast.

The advantage of this type of vector is that foreign DNA (such as human genomic DNA) can be cloned and manipulated in *E. coli*, and then the recombinant vector can be isolated and immediately introduced into a eukaryotic organism such as yeast for a different set of analyses, such as the expression and post-translational modification of proteins (see chapter 11). Shuttle vectors will be discussed in greater detail in chapter 13.

 Restating the Concepts

▶ Plasmids are small circular DNA molecules that replicate autonomously in the cell. They usually contain a multiple cloning site region and selectable markers and may be used for cloning DNA inserts of several hundred to ten thousand base pairs (10 kb) in length.

▶ Lambda bacteriophage vectors permit the cloning of approximately 2–25 kb of insert DNA.

▶ Cosmids represent a hybrid between a plasmid and λ vector and can accommodate approximately 30–45 kb of insert DNA. They reside as circular molecules in the *E. coli* cell.

▶ Bacterial artificial chromosomes (BACs) are used to clone very large DNA inserts, up to several thousand kilobase pairs in length.

▶ Shuttle vectors are plasmids that contain two origins of replication that allow the DNA to replicate autonomously in two different organisms.

12.2 DNA Libraries

With the use of restriction enzymes, DNA ligase, and vectors, researchers have the methods available to clone fragments of foreign DNA. We now shift our attention to the source of the foreign DNA that will be cloned. Only double-stranded DNA can be used in cloning.

There are two major sources of DNA: The first is genomic DNA from chromosomes. The second is DNA that is derived from mRNA through reverse transcription.

Creating a Genomic Library

Two basic types of cloning projects are undertaken in research. The first is cloning designed to isolate a specific sequence, which involves the techniques described earlier. The second type of project is cloning all the possible sequences that are present in an organism—for example all the human genomic DNA sequences. This latter collection can be maintained for the isolation of any specific sequence at some later time. Because of its similarity to a library of books, a collection of recombinant genome sequences for a particular organism is termed a **genomic library.**

Choice of Restriction Enzymes

In generating a genomic library, it is important to try to clone genomic DNA inserts that are as large as possible, because this will reduce the total number of clones required to account for all the sequences. Isolating large genomic DNA fragments is often achieved by cutting the genomic DNA with a restriction enzyme that recognizes an 8-bp restriction sequence, such as either *Not*I or *Spe*I. The frequency of either of these sites will be $(1/4)^8$, or an average of once every 65,536 bp. This length is clearly too large to clone into a λ or cosmid vector, but it would be appropriate for a vector designed from another bacteriophage, P1 (see table 12.1).

Although this strategy increases the chances of getting large genes in a single fragment, some small fragments will also be produced. Unfortunately, these small fragments cannot be cloned into many of the vectors because the final size of the recombinant molecule will be too small (**fig. 12.15**).

The partial digestion approach was designed to isolate all the DNA sequences on fragments that are approximately the same size. In a restriction digestion, millions to billions of identical DNA molecules are being cleaved. For a **partial digestion,** either the amount of restriction enzyme or the reaction time is limited, so that not every restriction site in every DNA molecule is cut. The partial digestion is usually performed with a restriction enzyme that recognizes a fairly common sequence, such as *Sau*3AI, which cleaves at 5'-GATC-3'. Using a restriction enzyme that recognizes a sequence every 256 bp on average ensures that every sequence will be on a fragment that is the correct size to clone. As you can see in figure 12.15, the complete digestion can produce fragments that are either too small or too large to clone into a bacteriophage vector. The genomic library produced from such a digestion will be missing these sequences. In contrast, every sequence in the genomic DNA is contained in one or more fragments produced from the partial digestion. Whereas some fragments are also either too small or too large to clone in a bacteriophage vector by the partial digestion approach, all of these sequences are also present in DNA fragments of the proper size. The chance of missing particular sequences in the partial digestion approach is therefore dramatically reduced.

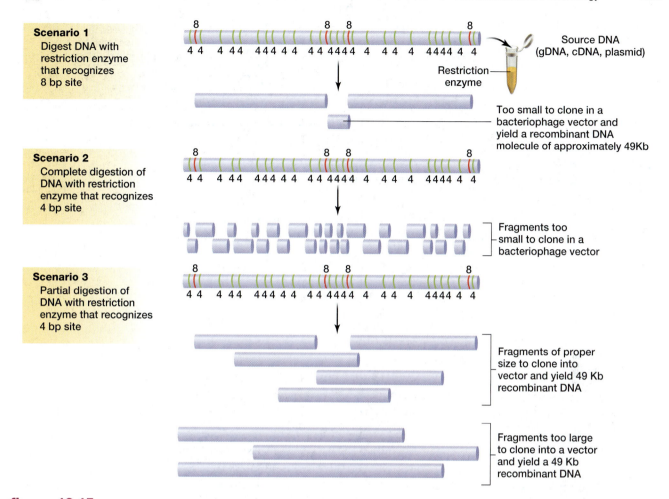

Scenario 1
Digest DNA with restriction enzyme that recognizes 8 bp site

Scenario 2
Complete digestion of DNA with restriction enzyme that recognizes 4 bp site

Scenario 3
Partial digestion of DNA with restriction enzyme that recognizes 4 bp site

Source DNA (gDNA, cDNA, plasmid)

Restriction enzyme

Too small to clone in a bacteriophage vector and yield a recombinant DNA molecule of approximately 49Kb

Fragments too small to clone in a bacteriophage vector

Fragments of proper size to clone into vector and yield 49 Kb recombinant DNA

Fragments too large to clone into a vector and yield a 49 Kb recombinant DNA

figure 12.15 Comparison of complete and partial restriction digestion of genomic DNA. Complete digestion of genomic DNA produces a set of nonoverlapping fragments. Using a restriction enzyme that recognizes an 8 bp sequence (scenario 1), some of the generated DNA fragments will be either too large or too small to be cloned into a λ vector and produce a 49 Kb recombinant DNA. Those smaller and larger genomic fragments will not be present in the final library. The complete digestion with a restriction enzyme that cuts a 4 bp sequence (scenario 2) will produce nonoverlapping DNA fragments that will be too small for cloning into a λ vector. However, a partial digestion with a restriction enzyme that cuts a 4 bp sequence (scenario 3), will produce a series of overlapping genomic DNA fragments. While some DNA fragments will still be either too small or too large to clone into a λ vector, every genomic DNA sequence will be present in one or more clones that are the correct size to produce recombinant λ phage.

Maintaining the Library

The partially digested genomic DNA fragments are then cloned into a vector as described earlier. We now need a method to maintain the library so we can isolate specific recombinant clones. This is done in two different ways. First, for plasmid-based vectors (cosmids and BACs), we introduce the ligated (and packaged in the case of cosmids) DNA into *E. coli*. Each *E. coli* cell will accept a single recombinant DNA molecule (**fig. 12.16**). All of the cells are plated on medium containing an antibiotic. The antibiotic permits only the bacteria that have accepted a recombinant vector to grow. Each *E. coli* cell then begins to divide and produces a colony, and all the *E. coli* cells in the colony will have the exact same recombinant vector or genomic DNA insert. (We will discuss how a specific genomic DNA insert is identified later.)

For bacteriophage-based vectors, such as λ, each bacteriophage particle should contain a different recombinant genomic DNA insert (see fig. 12.16). Again, each *E. coli* cell is infected with only a single bacteriophage. The infected *E. coli* cells do not divide and form colonies, however. Rather, the recombinant bacteriophage genome replicates, makes more bacteriophage, and kills the *E. coli* cell by lysis, which releases the new phage. The infection–replication–lysis cycle produces a plaque (clearing) of lysed *E. coli* cells surrounded by growing *E. coli* cells. As the infection spreads, the size of the plaque also grows. However, all the bacteriophage within a single plaque contains the same recombinant DNA insert, and this phage can be isolated. Thus, instead of having live cells harboring the recombinant DNA clones, a solution of bacteriophage particles that are capable of infecting *E. coli* cells is maintained.

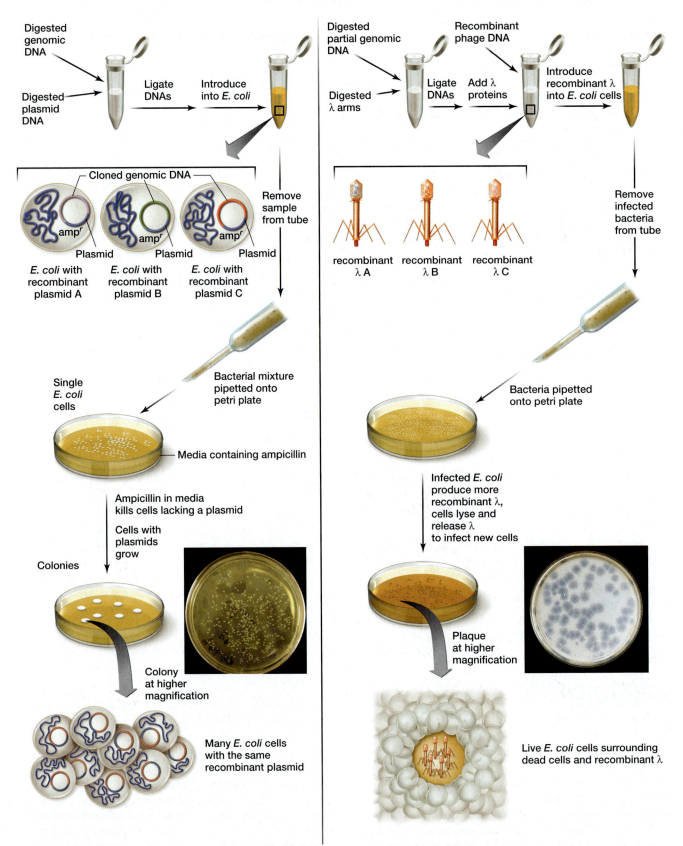

figure 12.16 Differences between colonies and plaques. Digested genomic DNA is ligated into a vector, such as a plasmid (left) or lambda (right). Plasmids can then be introduced (transformed) into *E. coli* cells. These cells can be pipetted onto solid media containing ampicillin. Colonies are then generated from the growth of single bacterial cells. If a cell contains a recombinant plasmid, then the cell will multiply into a colony that will contain many cells with the same recombinant plasmid. Recombinant λ DNA is mixed with λ coat proteins to produce recombinant λ phage. These phage are mixed with *E. coli* cells and then added to solid media. If a λ phage infects an *E. coli* cell, the λ will multiply, lyse the cell, and infect adjacent cells. This produces a plaque, which is a clearing of dead bacteria that are surrounded by living bacteria. All the λ within a single plaque will contain the same λ DNA.

Creating a cDNA Library

In higher eukaryotes, much of the genomic DNA represents sequences between genes, which may make trying to identify the transcribed genes more difficult. Genomic DNA also contains both exons and introns, and identification of the open reading frame (ORF) in genomic DNA may be problematic.

The process would be much easier if we could directly clone and sequence the mRNA. Although this is not possible due to the single-stranded nature of mRNA, we are able to isolate mRNA and convert it to double-stranded DNA. This **cDNA (complementary DNA)** can be useful in determining the ORF sequence (and deducing the amino acid sequence of the encoded protein), for expressing eukaryotic genes in prokaryotes, and in characterizing the expression of specific genes in a tissue. We will touch on some of these issues briefly and save the rest for chapter 13.

When isolating mRNA, it is important to use the proper tissue at the proper time, because not every gene is expressed at all times in all tissues. The extracted mRNA is purified based on the 3' poly(A) tail that this molecule possesses (see chapter 10). The RNA solution is passed over a resin that is covalently attached to oligo(dT), which is a polymer of 12 to 20 deoxythymidine residues that base-pairs with the 3' poly(A) tail of mRNA. The unbound RNAs (tRNAs, rRNAs, and others) are washed through the column, and the poly(A) mRNA is released from the oligo(dT) and collected.

In the first step of converting eukaryotic mRNA into cDNA, an oligo(dT) primer that is not covalently attached to a column is added to the mRNA. The 3' poly(A) tail of the mRNA base-pairs with the oligo(dT) to generate a short, double-stranded region that is now a primer for polymerase activity (**fig. 12.17**). Because DNA polymerase requires a DNA template for DNA synthesis, it is unable to use this mRNA template. Therefore, we use a polymerase that is encoded by an RNA tumor virus, **reverse transcriptase**, which will use an RNA template and will add deoxynucleotides to a primer in a 5' → 3' direction. The result is a DNA–RNA hybrid molecule (fig. 12.17).

This hybrid is now treated with the enzyme RNase H, which randomly nicks the RNA strand in an RNA–DNA hybrid. These nicks provide the primers for DNA polymerase I, which removes the RNA and replaces it with DNA; this process is similar to the removal of the RNA primers in Okazaki fragments during DNA replication (chapter 9). Finally, the short DNA segments of the second DNA strand are united with DNA ligase (fig. 12.17).

This blunt-ended, double-stranded cDNA can now be cloned to generate a cDNA library using the blunt-end linkers already described (fig. 12.6). The cDNA libraries are usually cloned into plasmids or λ vectors because of the size (0.5–10 kb) of the cDNA clones.

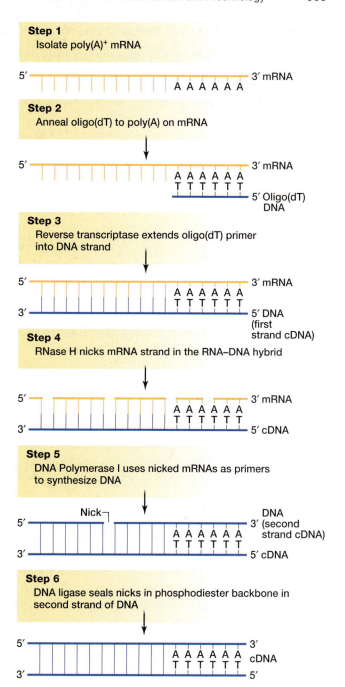

figure 12.17 Generating complementary DNA (cDNA) from mRNA. Single-stranded mRNA is isolated and mixed with oligo(dT) DNA, which acts as a primer by basepairing with the 3' poly(A) tail of the mRNA. Reverse transcriptase synthesizes DNA from the oligo(dT) primer in a 5' → 3' direction. RNase H nicks the RNA in the RNA–DNA hybrid. The RNA serves as primers for DNA polymerase I to synthesize the second strand of DNA. DNA ligase seals any phosphodiester nicks to generate the double-stranded cDNA.

table 12.2   Comparison Between Genomic and cDNA Libraries

Feature	Genomic DNA Library	cDNA Library
Average recombinant insert size	10–1000 kb	0.5–10 kb
Sequence representation	All genomic DNA sequences will be present.	Only the sequences corresponding to the mRNA exons will be present.
Sequence abundance	All genomic sequences will be present at nearly equal abundance.	The abundance of any specific cDNA will be proportional to the abundance of the mRNA.
Source of nucleic acid	Genomic DNA, will be the same from all tissues and developmental stages.	mRNA that is converted into double-stranded cDNA, will be different for each tissue and developmental stage.
Uses	Examine gene organization, promoter sequences, and intergenic regions.	Examine exons and open reading frames, easily deduce translated protein sequence, expression of eukaryotic genes in prokaryotes.

Comparing Genomic and cDNA Libraries

Genomic DNA libraries and cDNA libraries are created using similar approaches, but very significant differences are found between these two libraries (summarized in **table 12.2**).

Whereas genomic libraries contain essentially every DNA sequence that is present in a cell or tissue, the cDNA library contains only the sequences that correspond to the mRNAs in a specific cell or tissue. This means that promoter sequences, introns, and sequences between genes will be present in genomic libraries, but not cDNA libraries.

A genomic library from an organism should be essentially the same regardless of the tissue or developmental stage from which the genomic DNA was isolated. In contrast, a human liver cDNA library and a human brain cDNA library will have some clones in common, and some that are unique to the tissue and to the developmental stage at which the mRNA was isolated.

Furthermore, all the sequences in a genomic library will be present at nearly equal amounts, but the amount of each sequence in a cDNA library will be proportional to the amount of the corresponding mRNA in a cell. Some mRNAs are expressed at high levels (the corresponding cDNA will be plentiful), and other mRNAs will be expressed at low levels (the corresponding cDNA will be rare).

Both types of libraries can provide useful clones for further analysis. Genomic clones can reveal information on gene organization, promoter sequences, and regions between genes. In contrast, cDNA libraries and clones can provide information on the relative expression levels of specific mRNAs, clearly reveal the ORF of a gene, and allow the expression of eukaryotic proteins in prokaryotes (which cannot remove the introns from eukaryotic RNAs).

These differences become very important in setting up research projects, as discussed in later chapters.

Restating the Concepts

▶ Libraries contain clones that represent all the sequences from either the genomic DNA or the cDNA that is generated from a specific tissue.

▶ Genomic libraries are generated by partial digestion of genomic DNA and the cloning of large fragments.

▶ cDNA libraries are generated from mRNA using reverse transcriptase to make cDNA (complementary DNA), which is then cloned, usually without a restriction digest.

▶ The differences between genomic and cDNA libraries are primarily based on the differences between the genomic DNA, which includes all sequences, and the mRNA population, which contains only the transcribed sequences, in a given cell or tissue.

12.3 Identifying and Isolating Specific Sequences

After generating either a genomic or cDNA library, investigators are usually interested in identifying a specific recombinant clone within the library. You read earlier how bacteria with a recombinant vector can be distinguished from bacteria with a vector lacking an insert. But how do we know which of the hundreds of thousands of colonies or plaques are the ones that we desire?

To perform this task, researchers can utilize either *nucleic acid hybridization* or *polymerase chain reaction amplification*. The principles underlying these techniques can be applied to other nucleic acid populations and are not limited to only libraries. In addition, similar techniques have been invented to distinguish protein expression via labeled antibodies. This technique, *Western blotting*, or *immunoblotting*, is also discussed in this section.

These powerful techniques are widely used in the field of molecular biology.

Nucleic Acid Hybridization

Three main methods are discussed here by which nucleic acids may be identified: *Southern hybridization, colony* or *plaque hybridization,* and *Northern hybridization.*

Southern Hybridization

Southern hybridization, which was first developed by Edward M. Southern in the mid-1970s, is a procedure based on the ability of a nucleic acid to base-pair with its complementary sequence. It is also known as the *Southern blotting technique.*

Southern hybridization uses a nucleic acid with a known sequence, either a DNA or RNA, as a **probe.** The probe is labeled either radioactively so it can be detected later by autoradiography or with a small molecule (such as *biotin*) so that it can be detected with an antibody. In this latter approach, the antibody is usually covalently attached to an enzyme (usually alkaline phosphatase or horseradish peroxidase), which can cleave a compound to produce either a colored or a light-emitting product (*chemiluminescence*).

The **target** for the hybridization is a DNA sequence. As a specific example, if we wish to locate the rhodopsin gene in a library, we could use either radioactively labeled rhodopsin mRNA or a radioactively labeled rhodopsin cDNA clone as the probe. RNA–DNA or DNA–DNA hybrids would form between the radioactive probe and the specific gene sequence. Autoradiography would then locate the radioactive probe and in turn, the rhodopsin genomic clone.

The first step in a genomic Southern hybridization would be to digest human genomic DNA with a restriction enzyme (**fig. 12.18**). This digested DNA is then applied to an agarose gel, which is a gelatin-like substance. **Electrophoresis** is the process by which the DNA fragments are separated by applying an electrical current to the gel. Smaller DNA fragments move more quickly through the agarose gel than larger ones.

In a genomic DNA digest, there are usually so many fragments that the result is simply a smear of fragments, from very small to very large. After electrophoresis, we denature the double-stranded DNA by incubating the agarose gel in sodium hydroxide (NaOH). The resulting single-stranded DNA fragments must then be transferred from the gel to another medium for hybridization.

Nitrocellulose filters or nylon membranes are excellent for hybridization because the DNA fragments bind to these membranes and will not diffuse out. The transfer can be accomplished two different ways. In one approach, the DNA is electrophoresed

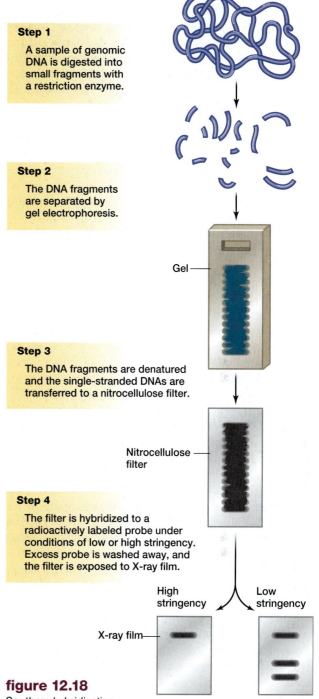

Step 1
A sample of genomic DNA is digested into small fragments with a restriction enzyme.

Step 2
The DNA fragments are separated by gel electrophoresis.

Gel

Step 3
The DNA fragments are denatured and the single-stranded DNAs are transferred to a nitrocellulose filter.

Nitrocellulose filter

Step 4
The filter is hybridized to a radioactively labeled probe under conditions of low or high stringency. Excess probe is washed away, and the filter is exposed to X-ray film.

High stringency   Low stringency

X-ray film

figure 12.18

Southern hybridization technique. Genomic DNA is digested with a restriction enzyme and the DNA fragments are separated by agarose gel electrophoresis. The DNA is denatured to single-stranded form and transferred to a nitrocellulose filter. The nitrocellulose is incubated with a radioactively labeled single-stranded probe that hybridizes with the single-stranded target DNA on the filter at either high stringency (high temperature or low ionic strength solution) or low stringency (low temperature or high ionic strength solution) conditions. The nitrocellulose is washed to remove unhybridized probe and exposed to X-ray film. At high stringency, only a single DNA fragment (corresponding to perfect base pairing between the target and probe) is detected, while additional DNA fragments are observed at low stringency.

out of the gel toward the nitrocellulose (see **fig. 12.19a**). The denatured DNA fragments will move toward the anode (positive electrode), and when it comes in contact with the nitrocellulose, it binds.

In the other approach, the NaOH-treated agarose gel is placed directly against a piece of nitrocellulose filter, and the resulting sandwich is placed agarose-side-down on a wet sponge. Dry filter paper placed against the nitrocellulose side wicks a salt solution from the sponge, through the gel, and past the nitrocellulose filter, carrying the DNA segments from the agarose to the nitrocellulose (fig. 12.19b). The nucleic acid hybridization can now take place on the filter.

The labeled cDNA or mRNA probe is denatured by heating and is added to the nitrocellulose in a salt solution (see fig. 12.18). We can control the hybridization conditions by regulating the **stringency** (specificity) of the base pairing between the probe and target. Increasing the hybridization temperature or decreasing the salt concentration results in a higher stringency, which requires more perfect base pairing between the probe and target. At high stringency, only a single DNA sequence hybridizes to the probe, whereas at lower stringency, multiple sequences may hybridize to the probe.

In the rhodopsin example, we can deduce that the single fragment at the higher stringency exhibits the greatest base pairing with the rhodopsin probe and therefore likely corresponds to the rhodopsin gene (see fig. 12.18). This fragment is also present at the lower stringency, as would be expected. However, the additional two fragments at the low stringency correspond to DNA fragments that possess less base pairing between the probe and target genomic DNA. These fragments could represent genes that are related, but not identical, to the rhodopsin gene. (These related fragments likely correspond to the cone opsin genes, which permit vision in bright light and of color.)

This example shows that Southern hybridization is a powerful way to identify a specific DNA fragment from a population of fragments. By adjusting the conditions of the hybridization, we can increase or decrease the specificity of the match between the probe and target, providing a method to identify related, but not identical, targets. We will use this idea in our next discussion.

Colony or Plaque Hybridization

Assume that we want to isolate a cone opsin cDNA. We do not have the cone opsin gene to use as a probe, but we do have the rhodopsin cDNA, which hybridizes under reduced stringency conditions. We also have the cDNA library cloned in a plasmid vector, so we would plate out the library colonies on a petri plate (**fig. 12.20**).

Because this plating separates the clones, we do not need to do a DNA digest or gel electrophoresis. We directly lay the nitrocellulose filter on the plate, which

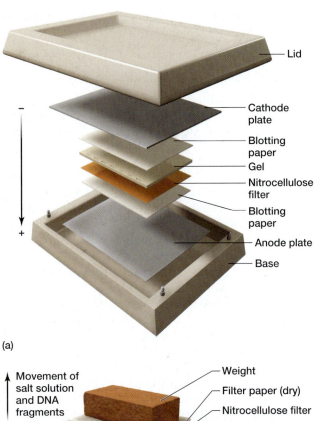

(a)

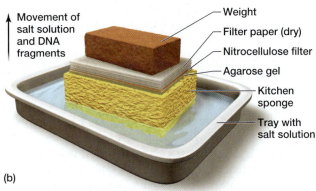

(b)

figure 12.19 Two methods to transfer DNA for a Southern blot. The electrotransfer method uses an electrical current to drive the DNA from the gel towards the anode (a). However, when the DNA reaches the nitrocellulose, it becomes bound. In the diffusion method, the salt solution is drawn upwards by the dry filter paper, transferring the DNA from the agarose gel to the nitrocellulose filter (b).

causes some of the cells from each colony to stick to the filter (see fig. 12.20). We can lift the filter, lyse the cells, and denature the DNA with NaOH. This action produces spots of single-stranded plasmid DNA on the filter that mimics the pattern of the colonies on the plate. We can then treat the nitrocellulose in the same manner as described before, by adding a radioactive probe and controlling the hybridization conditions.

If the hybridization is performed at high stringency, then the rhodopsin cDNA probe will base-pair with only rhodopsin cDNA clones. If the hybridization is performed at reduced stringency, then the probe will base-pair with both rhodopsin cDNA clones and

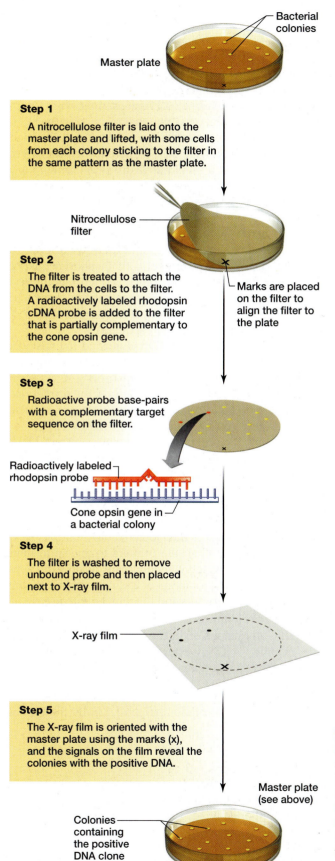

Bacterial colonies

Master plate

Step 1

A nitrocellulose filter is laid onto the master plate and lifted, with some cells from each colony sticking to the filter in the same pattern as the master plate.

Nitrocellulose filter

Step 2

The filter is treated to attach the DNA from the cells to the filter. A radioactively labeled rhodopsin cDNA probe is added to the filter that is partially complementary to the cone opsin gene.

Marks are placed on the filter to align the filter to the plate

Step 3

Radioactive probe base-pairs with a complementary target sequence on the filter.

Radioactively labeled rhodopsin probe

Cone opsin gene in a bacterial colony

Step 4

The filter is washed to remove unbound probe and then placed next to X-ray film.

X-ray film

Step 5

The X-ray film is oriented with the master plate using the marks (x), and the signals on the film reveal the colonies with the positive DNA.

Master plate (see above)

Colonies containing the positive DNA clone

clones containing a related sequence, such as the cone opsin cDNA.

The hybridized radioactive probe is detected on X-ray film that we can trace back to a specific colony (**fig. 12.21**). We can isolate a few cells that remain on the petri plate after the nitrocellulose lift, grow them in culture, and isolate the plasmid DNA. This yields a pure population of a single recombinant plasmid that contains the desired cDNA clone.

A similar approach, *plaque hybridization*, can be used with clones in a λ library. Because λ produces plaques instead of colonies, laying the nitrocellulose on the petri plate transfers phage particles that contain the λ clones. The remainder of the technique, denaturing the target DNA, hybridizing the probe to the target, and detecting the probe remains the same as for a colony hybridization. Once the probe is detected on the X-ray film, we can return to the corresponding plaque on the petri plate and isolate some of the recombinant phage.

As you can see, the ability to hybridize at reduced stringency is a powerful approach. We just described

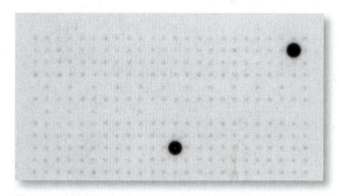

figure 12.21 Hybridization of a radioactively labeled probe with 288 bacterial colonies. The two dark spots indicate colonies containing the target sequence that is complementary to the probe. The weak background signals in most other spots is due to nonspecific hybridization of the probe which provides the spot pattern needed to orient the identity of the two positive clones.

figure 12.20 Screening a cDNA library by colony hybridization. The library is plated onto a petri plate and bacterial colonies with plasmid clones are grown. A nitrocellulose filter is laid on the plate and many of the bacterial cells from each colony are transferred to the nitrocellulose. The cells are lysed, the DNA is denatured, and the DNA is bound to the nitrocellulose. Radioactively labeled probe is incubated at the desired stringency. The nitrocellulose is washed and exposed to X-ray film. The film reveals spots where the probe hybridized to the cDNA clone. The X-ray film is aligned with the master plate. The colony that corresponds to the positive signal on the X-ray film contains the desired recombinant plasmid.

using one human cDNA to identify a related, but different, human cDNA, but a number of variations are possible. We can use a cDNA from one organism as a probe to hybridize to a related gene from a different organism. We can also reverse-translate the known amino acid sequence of a protein into oligonucleotide sequences (see box 11.1 in chapter 11). Because the probe only needs to be approximately 20 nucleotides long, the length of the nucleotide sequence deduced from a protein sequence is sufficient to be a probe. Thus, it is a common practice to compare the amino acid sequences of the same protein from several different organisms to identify the region that has the greatest similarity. Reverse translation of that similar sequence can often identify an oligonucleotide sequence that can be used as a probe for Southern hybridization or polymerase chain reaction (see later section).

Northern Hybridization

A third type of hybridization is **Northern hybridization,** the name being a takeoff on Southern hybridization. It differs from the Southern, colony, and plaque hybridizations in that RNA, not DNA, is the target.

In Northern hybridization, mRNA is selected on oligo(dT) purification, denatured to remove any secondary stem-loop structures, and then electrophoresed through an agarose gel. The mRNA is not digested with a restriction endonuclease, since these enzymes only function on double-stranded DNA. After separating the RNA molecules by molecular weight, they are transferred to nitrocellulose and incubated with a probe at the desired stringency.

The bands that appear on the X-ray film correspond to specific mRNAs. The size of the mRNA, in nucleotides, can be calculated by the distance that the mRNA migrated through the gel. By comparing mRNA isolated from different tissues, it is possible to determine when and where the mRNA is transcribed in the organism, the relative abundance or level of transcription of the mRNA, and alternative splice forms of the mRNA. This data cannot be extracted from a Southern hybridization, which reveals data only about DNA. Thus, Northern hybridization is a powerful complement to Southern hybridization (**table 12.3**).

Polymerase Chain Reaction

In a variety of instances, the isolation of nucleic acids (DNA or RNA) to perform the previously described procedures is extremely difficult. Often the DNA sample is available in very limited quantities (for example, crime scene evidence or very small tissue samples from biopsies) or in an ancient condition that resulted in the DNA being partially degraded (in museum specimens, dried specimens, and fossils). To circumvent the problem of having small amounts of tissue or nucleic acids, Kary Mullis, a biochemist working for the Cetus Corporation, devised the **polymerase chain reaction (PCR)** in 1983, for which he won the 1993 Nobel Prize in Chemistry.

PCR can be used to amplify whatever DNA is present in a sample, however small in quantity or poor in quality. The only requirement is that a short sequence of nucleotides on either side of the sequence of interest be known. These short sequences can be generated from known nucleotide sequences or from deduced, reverse-translated amino acid sequences.

This sequence information is needed to construct primers on either side of the sequence of interest. The primers are usually 15–25 nt in length. As with Southern and Northern hybridizations, the primers do not need to exactly base-pair with the target because the annealing conditions can be adjusted.

In the PCR technique, the template, primers, DNA polymerase, and dNTPs are combined. The following procedure then is followed:

1. The mixture is heated (for example, 92°C for 30 sec) to denature the DNA.

2. The temperature is then lowered (for example, 56°C for 30 sec) to allow the primers to anneal to their complementary target sequences. Adjusting this temperature allows for more- or less-perfect base pairing between the primers and the target sequences.

3. The temperature is then increased (for example, 72°C) for extension of the DNA strand. The length of time for DNA extension (replication) depends on the predicted length of the amplified product.

Then, a new cycle of denaturing, annealing, and DNA extension is initiated (**fig. 12.22**). By the end of

table 12.3   Comparison Between Various Hybridization Techniques

Feature	Southern Blot	Northern Blot	Western Blot
Target	DNA	RNA	Protein
Probe	DNA/cDNA/RNA	DNA/cDNA/RNA	Antibody
What Does It Reveal	Gene organization, number of gene copies	Size of mRNA, relative abundance of the mRNA, temporal and spatial pattern of the mRNA expression	Molecular weight of the protein, relative abundance of the protein, temporal and spatial pattern of protein expression

Cycle 3, we have generated the desired product that begins and ends with the oligonucleotide sequences on both strands. This DNA product is then amplified exponentially through subsequent cycles.

About 20 PCR cycles produce a million copies of DNA; 30 cycles make a billion copies. The various stages in the cycle can be controlled by temperature because denaturation, primer annealing, and DNA extension all require different temperatures.

In the original description of the technique, DNA polymerase had to be added just before each DNA extension step because the denaturing temperature also inactivated DNA polymerase. The technique became widely accepted when a DNA polymerase from a hot-springs bacterium, *Thermus aquaticus,* was identified that could withstand the denaturing temperatures. Using this polymerase, no additional enzyme had to be added to the reaction mixture after each cycle. Instead, the cycling could be continued without interruption in PCR machines, which could be programmed to automatically change temperatures after a specified length of time. Furthermore, the ability of these machines to handle hundreds of samples simultaneously has dramatically automated the isolation of DNA sequences without the need for making and screening genomic DNA and cDNA libraries.

Western Blotting, or Immunoblotting

The procedures just described can be used to identify and characterize a genomic DNA fragment, cDNA, or mRNA. A different approach needs to be employed to examine protein expression. Remember that the phenotype of an organism usually depends on the protein. Thus, a mutant phenotype, which is usually due to a change in the DNA, ultimately results from a change in the expression of a protein or an alteration in the protein's amino acid sequence. This suggests that nucleic acid hybridizations can be extremely informative for a variety of reasons, but analyzing protein expression can be equally important.

To examine the expression of a specific protein within a population of different proteins, we also need a probe. In this case, however, the probe is usually not a nucleic acid (unless the protein binds a specific DNA or RNA sequence, like a transcription factor binding a promoter). In the case of proteins, the probe is usually an antibody, a specialized protein that can recognize a precise, short amino-acid sequence in another protein. Antibodies can be very specific, in that each antibody may bind to only a single protein. This procedure is called either **Western blotting** (a takeoff on the Southern and Northern hybridizations) or **immunoblotting.**

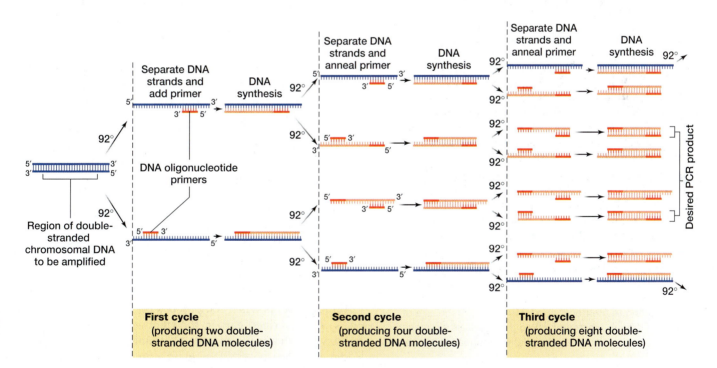

figure 12.22   Polymerase chain reaction. Double-stranded DNA is denatured by heating to 92°C. Primer oligonucleotides that are complementary to sequences flanking the desired product anneal to both single strands at 56°C. DNA polymerase synthesizes the DNA strands in a 5' → 3' direction at 72°C. This cycle of denaturation, annealing and polymerization is repeated 30 to 40 times by cycling the temperature. The original DNA template is shown as blue, the primers are red, and synthesized DNAs are pink. After three cycles, one double-stranded region of DNA becomes two desired double-stranded molecules (as shown). In subsequent rounds, the desired product increases exponentially, while the other products only increase linearly.

Denatured or intact proteins from a specific tissue or organism are electrophoresed through a polyacrylamide gel rather than an agarose gel (**fig. 12.23**). The proteins are separated by molecular weight as they move through the gel and are then electrotransferred to nitrocellulose. The nitrocellulose is then incubated with the desired antibody, which can bind to the specific protein on the filter. Sometimes the antibody is radioactively labeled and can then be detected on X-ray film. Alternatively, a second antibody is covalently attached to an enzyme (alkaline phosphatase or horseradish peroxidase), which is used in either a colorimetric reaction to cleave a colorless compound into a colored product or a chemiluminescent reaction. This secondary antibody will specifically bind the first (primary) antibody and the detected label would indicate the location of the desired protein. This can be used to determine the molecular weight of the desired protein, where it is expressed in the organism, or when it is expressed during development (see table 12.2). This immunoblot technique can also be performed on a cDNA library. In this case, the antibody identifies the colony or plaque that contains the cDNA that encodes the desired protein.

Restating the Concepts

▶ Southern hybridization uses a nucleic acid probe to base-pair with a target DNA within a population. The target DNA may be restriction digested and electrophoresed to separate the DNA fragments before adding the probe. Alternatively, the DNA may be represented as clones (as in a library). In this case, the distribution of colonies or plaques on a petri plate serves the purpose of separating the different clones before the probe hybridization.

▶ Northern hybridization uses a nucleic acid probe to anneal to a target mRNA within a population. The mRNA is denatured and then electrophoresed to separate the different mRNA molecules before hybridization with a probe.

▶ Southern hybridization can be used to determine features of the DNA, such as the size of the gene, intron–exon organization of the gene, and number of gene copies. Northern hybridization reveals features relating to the transcribed RNA, such as the size of the mRNA, when and where the mRNA is expressed in the organism, and the relative abundance of the mRNA.

▶ PCR utilizes a specialized DNA polymerase and different cycling temperatures to amplify a DNA sequence that is located between two known oligonucleotide primers.

▶ Western blotting utilizes an antibody to bind to a specific protein within a population. Proteins are

electrophoresed through a polyacrylamide gel and then transferred to a filter before incubating with an antibody that can bind a specific protein. Alternatively, the antibody can be incubated with an expression library to identify a cDNA clone that encodes the desired protein.

12.4 Restriction Mapping

The Southern hybridization technique identifies a particular DNA sequence in a population; however, the isolated genomic or cDNA clone must be further analyzed after it is isolated. We discuss two major methods for the characterization of a DNA clone. The first, **restriction mapping,** determines the location of restriction enzyme sites within the DNA. The second, **DNA sequencing,** is the most detailed type of analysis that we can perform on a DNA clone. In this section, we focus on restriction mapping; DNA sequencing is described in the following section.

The generation of a restriction map for either a genomic or cDNA clone can be extremely useful for several reasons. First, it can be used to determine the amount of identity or overlap between two different sequences. In this way we can compare a cDNA clone with a genomic clone and can localize the regions on the genomic clone that correspond to the gene exons.

Second, the restriction map provides useful landmarks that help gauge where in the genome a gene, mutation, or DNA sequence occurs. Imagine that you are driving west along Interstate-70 in Kansas, looking for the town of Hays. If you do not know where the Hays exit is, you might start feeling nervous after driving 3 or 4 h and think that you missed it. But if you know the Hays exit is between mile markers 159 and 158 going west, you will probably be fairly confident about finding the exit. A restriction map can provide a similar sort of confidence that you are in the proper region of the chromosome.

Third, a restriction map often allows researchers to correlate the genetic (recombination) map and the physical map of a chromosome. Certain physical changes in the DNA, such as deletions, insertions, or nucleotide changes at restriction sites, can be localized on the genetic map. These changes can be seen as changes in size or in the total absence of certain restriction fragments when compared with wild-type DNA.

As described earlier, the frequency of restriction enzyme sites in a segment of double-stranded DNA depends on the DNA sequence and the number of base pairs in the recognition sequence of the particular enzyme used. That is, a restriction enzyme with only 4 bp in its recognition sequence ($1/4^4$ or 1/256 bp) will cut more frequently than one with 6 bp in its sequence ($1/4^6$ or 1/4096 bp). These frequencies come into play when we construct a restriction map.

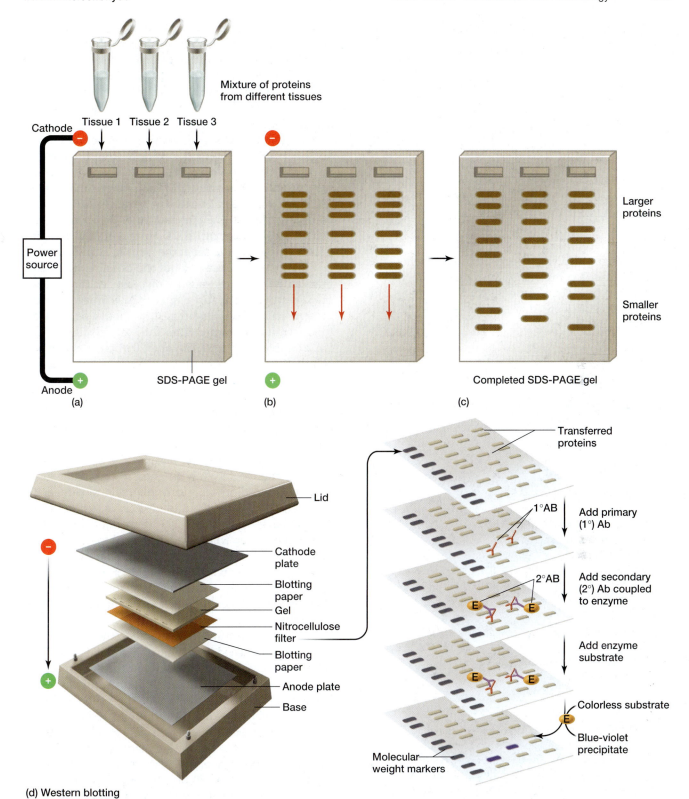

(d) Western blotting

figure 12.23   The Western blot technique is used to identify a protein. (a) Tissue or cells are lysed and the released proteins are separated by sodium dodecyl sulfate polyacrylamide gel electrophoresis (SDS-PAGE; b–d). All the proteins are then electrotransferred to a nitrocellulose filter. The filter is incubated with a primary antibody (1° Ab) that specifically binds the desired protein. A secondary antibody (2° Ab) that specifically binds the first antibody is added. The 2° Ab is coupled to an enzyme that converts a colorless substrate to a colored precipitate. The presence of the blue-violet precipitate on the membrane corresponds to the location of the desired protein. Western blotting can be used to determine what tissues express the protein and the molecular weight of the protein.

Constructing a Restriction Map

There are several different ways to generate a restriction map. The level of detail in the map depends on the type of restriction digests that are performed and how the DNA was treated prior to the digestion. We will go through a relatively simple example here, which demonstrates the types of data and how they are used in creating the map.

Figure 12.24a shows a piece of DNA with the location of the *Eco*RI restriction enzyme sites. Below this map are two restriction fragment profiles after gel electrophoresis. The restriction enzyme has three sites in the DNA, which generate four fragments that are 200, 50, 400, and 100 bp long.

The pattern of DNA fragments on the left lane of the gel in figure 12.24b is the result of electrophoresing the DNA after a complete *Eco*RI digestion. Each DNA fragment in the gel corresponds to the size of a single *Eco*RI fragment. Note that smaller fragments move faster than the larger ones. The sizes of the DNA frag-

ments are determined by comparison with standards of known size (which are not shown, although the computed size of each DNA fragment is shown). While the gel reveals the size of each restriction fragment, it does not reveal their relative order.

One method of determining the relative order is a *partial digest.* We already discussed partial digestion of genomic DNA in creating a genomic library. To partially digest the DNA, we can limit the amount of restriction enzyme, reduce the incubation temperature, or reduce the amount of time for the digestion to occur. This treatment results in some DNA molecules being digested at one site and other DNA molecules being digested at another site, producing the complex pattern shown in the gel at the right in figure 12.24b.

The pattern represents all the possible fragments that can be produced if zero, one, two, or three of the *Eco*RI restriction sites are cut. With only four fragments present in the complete digest, it should be possible to determine the order from this partial digest data. For example, only if the 50- and 200-bp fragments are adjacent to each other will you be able to generate the 250-bp partial digest fragment.

A method exists by which the partial-digest data can be simplified. This modification requires radioactively labeling the 5' ends of the DNA with ^{32}P using the enzyme polynucleotide kinase before either the complete or partial digest. Because the kinase acts on double-stranded DNA, both ends of the linear DNA become labeled. After electrophoresis of either the complete or partial DNA digestion in figure 12.24, the radioactively labeled 200-bp and 100-bp can be detected on X-ray film. The labeling of these two fragments in the complete digest reveals that they are the ends of the DNA, which means that there is an *Eco*RI site 100 bp from one end and 200 bp from the other end of the original DNA clone (it does not make any difference which fragment is at which end).

After determining that the 200- and 100-bp segments are on the ends, we know that the 50- and 400-bp segments must be on the inside. In the partial digest, a 250-bp labeled segment can be found, but not a 150-bp segment, which reveals that the 50-bp segment lies adjacent to the 200-bp terminus, not the 100-bp terminus (**fig. 12.25b**). There is also a 500-bp labeled segment, but not a 600-bp segment, which indicates that the 400-bp segment is adjacent to the 100-bp terminus (fig. 12.25c). A 450-bp segment confirms that the 400- and 50-bp segments are adjacent, and the absence of the ^{32}P label on this fragment confirms that it is an internal fragment.

In this way, the original DNA can be unequivocally reconstructed, creating a map of restriction enzyme sites that are separated by known lengths of DNA.

If a large number of restriction fragments occur after the complete digest, the partial-digest pattern

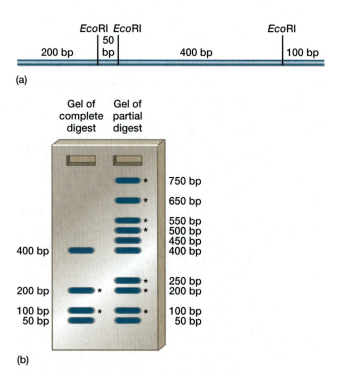

figure 12.24   Comparison of a restriction map with the gel electrophoresis pattern after a restriction endonuclease digest. (a) Shown is the *Eco*RI restriction map with the distances for each DNA fragment. (b) Schematic of an agarose gel showing the DNA fragments from both the complete and partial restriction digests (left and right, respectively). Asterisks mark radioactively labeled DNA fragments that were generated by end-labeling the DNA before the restriction digest. The complete digest produces four fragments with only the 200- and 100-bp fragments labeled. The partial digest yields five additional labeled fragments, which each must contain at least one of the two end fragments from the original DNA.

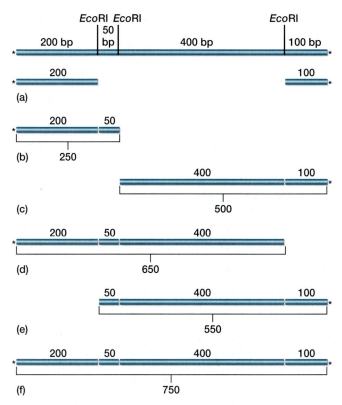

figure 12.25   Steps in the reconstruction of the DNA restriction map from figure 12.24. Asterisks show the location of the ^{32}P-labeled ends from the original DNA fragment. The complete digest establishes the 100- and 200-bp segments are the end segments because they are the only two fragments labeled (a). Because there are also 50- and 400-bp fragments within the DNA (revealed from the complete digest), only certain fragments are possible from the partial digest. The 50-bp fragment must be adjacent to the 200-bp fragment because of the presence of a labeled 250-bp fragment in the partial digest. Similarly, the 400-bp fragment must be adjacent to the 100-bp end segment (panels b and c). The occurrence of a labeled 650-bp and 550-bp labeled fragments in the partial digest (panels d and e) is consistent with the final map (e).

may get very complicated. To simplify the analysis, just one end of the DNA can be labeled with ^{32}P, or the DNA that is labeled at both ends can be cleaved with an enzyme that cuts once between the labeled ends. In this manner, the partial digest is built from a single end, and the size of the labeled fragments represents the distance each restriction site is located from that labeled end. (An example of this modification is presented in box 12.1.)

Double Digests

In the preceding example, a restriction map was generated using a single restriction enzyme. In practice, restriction mapping is usually done with several different restriction enzymes. **Figure 12.26** is a map of the

DNA in figure 12.24, with the sites for a second restriction enzyme, *Bam*HI, included. Using the end-labeling approach, we can determine that the 350-bp and 150-bp *Bam*HI fragments are on the outside, or the order of the *Bam*HI segments is 350 bp, 250 bp, and 150 bp. What we do not know is how to overlay the two maps. Do the *Bam*HI segments run left to right or right to left with respect to the *Eco*RI segments (fig. 12.26a and b)? We can determine the unequivocal order by digesting a sample of the original DNA with both enzymes simultaneously, thus producing a **double digest.**

The two orders shown in figure 12.26a and b are used to make different predictions about the double digest. From the first order (a), we predict that the 200-bp and 100-bp end segments would be radioactively labeled after the double digest. From the second order (b), we predict that the 150-bp and 100-bp fragments would be radioactively labeled after the double digest. The gel of the double digest shows 200-bp and 100-bp labeled fragments, indicating order (a) is correct. All other aspects of order (a) are consistent with the double digest.

Also notice that the double digest reveals the smallest end fragments in another way. The *Eco*RI digest shows the end fragments are 200 bp and 100 bp, whereas the *Bam*HI digest shows the 350-bp and 150-bp fragments. In the double digest, two of these four end fragments should be present, resulting from the first restriction site from each end. In the double digest data, the 200-bp and 100-bp fragments are present, and the 350-bp and 150-bp fragments are absent. We can deduce that the 200-bp and 100-bp *Eco*RI fragments are the closest to the ends.

Because the 150-bp *Bam*HI fragment is smaller than the 200-bp *Eco*RI fragment, we would see the smaller fragment (150 bp) if both of these fragments were on the same side. Because we failed to see the 150-bp fragment in the double digest, the 200-bp *Eco*RI fragment and 150-bp *Bam*HI fragment must be on opposite sides. (You will see this logic again in box 12.1.)

Restriction mapping thus provides a physical map of a piece of DNA, showing restriction endonuclease sites separated by known lengths of DNA. This technique gives us short DNA segments of known position that we can sequence, as well as a physical map of the DNA that can be compared with the genetic map and can be used to locate mutations and other particular markers.

Restriction Fragment Length Polymorphisms

A restriction map can be generated for any region of the genome in any organism, but differences will exist between individuals of a species. These differences sometimes create a new restriction enzyme site or delete one that was known to exist. Changes in restriction sites result in differences in the length of restriction fragments, a phenomenon that is termed **restriction fragment length polymorphisms (RFLPs).**

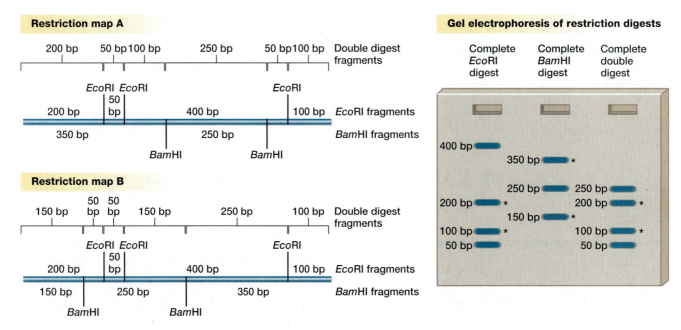

Restriction map A

Double digest fragments: 200 bp | 50 bp 100 bp | 250 bp | 50 bp 100 bp

EcoRI EcoRI   EcoRI
                50
EcoRI fragments: 200 bp | bp | 400 bp | 100 bp

BamHI fragments: 350 bp | 250 bp

BamHI   BamHI

Restriction map B

               50 50
Double digest fragments: 150 bp | bp bp | 150 bp | 250 bp | 100 bp

EcoRI EcoRI   EcoRI
                50
EcoRI fragments: 200 bp | bp | 400 bp | 100 bp

BamHI fragments: 150 bp | 250 bp | 350 bp

BamHI   BamHI

Gel electrophoresis of restriction digests

Complete EcoRI digest | Complete BamHI digest | Complete double digest

400 bp
              350 bp ∗
              250 bp   250 bp
200 bp ∗   200 bp ∗
              150 bp ∗
100 bp ∗   100 bp ∗
50 bp   50 bp

figure 12.26 Two possible restriction maps involving two different restriction endonucleases (*EcoRI* and *BamHI*). Two restriction maps are shown (A and B) with the location of the restriction sites. The DNA was radioactively end-labeled and then completely digested with either *EcoRI* alone, *BamHI* alone, or both enzymes together. The DNA fragments were separated by gel electrophoresis. Asterisks indicate the radioactive end-labeled DNA fragments. Restriction map A is consistent with all the fragments found in all the digests, whereas map B is not. For example, map B has an internal (unlabeled) 150-bp fragment that is not found in the double digest.

RFLPs can be valuable genetic markers in two areas of study: human gene mapping and forensics. Because a restriction digest of the whole human genome with a single restriction enzyme usually produces a smear, unique probes and Southern blotting are required to detect these RFLPs. Genetic variation lacking a restriction site, usually due to a mutation, will result in a larger DNA fragment (**fig. 12.27**).

Some probes have revealed **hypervariable loci** with many alleles. (Any one person has, of course, only two of the many possible alleles.) These hypervariable loci are the result of differences in the numbers of repeated DNA sequences that are called minisatellites and microsatellites. For example, a **microsatellite** is a short sequence (2–5 bp) that is repeated in a *tandem* array, that is, the repeated sequences are located adjacent to each other (**fig. 12.28**). The number of repeats that can exist in a given genomic location can be anywhere from one or two up to several hundred. As a result, hundreds of alleles may exist at a microsatellite location.

A **minisatellite** is also a tandem array of a repeat, but the unit sequence is 5-100 base pairs. Because of the variation in the number of repeated sequences at a given locus, minisatellites are also known as **variable number tandem repeats (VNTR).**

With such a large number of alleles that are phenotypically different, as shown by DNA sequencing or the size of the PCR amplification, a great likelihood exists that any given individual will be heterozygous

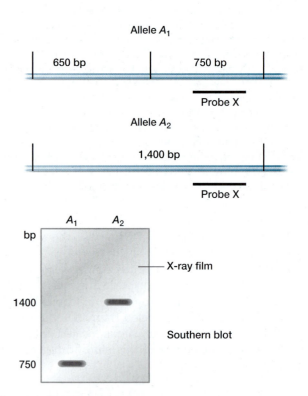

Allele A_1
650 bp | 750 bp
Probe X

Allele A_2
1,400 bp
Probe X

       A_1 A_2
bp

                        —— X-ray film

1400   ▬

                        Southern blot

750   ▬

figure 12.27 Restriction fragment length polymorphism (RFLP) analysis. Allele A_1 is a gene segment that is 750 bp long, which hybridizes to Probe X in a Southern blot. In allele A_2, the middle restriction site is lost. The same probe now recognizes a 1,400-bp fragment from allele A_2 instead of the 750-bp fragment on a Southern blot.

12.1 *applying genetic principles to problems*

Determining a Restriction Map

Several different approaches are possible when generating a restriction enzyme map, and there are also several different ways to examine the data. However, if the data are analyzed correctly, only a single correct restriction map will result. The problem and its solution, described later, shows a systematic method by which data can be examined and the correct solution realized. Not all the information described here will be in every mapping problem, and this approach may not always be the best way to work through the data, but you should see the basic logic that is applied.

The Data

Assume that you isolated a genomic DNA clone. You digested this DNA fragment with *Eco*RI and identified fragments of 9.5, 5.0, 4.5, and 3.0 kb. You also digested this genomic DNA fragment with *Bam*HI and identified fragments of 10.0, 6.0, 4.0, and 2.0 kb. You then performed a *Bam*HI/*Eco*RI double digest and identified DNA fragments of 8.0, 5.0, 4.0, 2.0, 1.5, 1.0, and 0.5 kb.

When you labeled the ends of the genomic DNA fragment and performed the *Bam*HI digest, you found that the radioactivity was in the 6.0- and 4.0-kb fragments. Double digestion of the labeled genomic DNA fragment with *Bam*HI/*Eco*RI revealed the 5.0- and 4.0-kb fragments were labeled. You then isolated the 10.0-kb *Bam*HI fragment from the earlier genomic DNA digest, end-labeled it, and digested it with *Xba*I. You identified an 8.0-kb fragment and a 1.0-kb fragment by ethidium bromide staining, and both fragments were labeled. When the 8.0-kb fragment was partially digested with *Hin*fI, 5.0-, 1.5-, 1.0-, and 0.5-kb labeled fragments were identified, whereas complete digestion of this 8.0-kb fragment with *Eco*RI yielded a 6.0-kb fragment and a 2.0-kb fragment.

To analyze this wealth of data, you utilize a six-step method:

1. Determine the total size of the DNA fragment with each restriction digest.
2. Determine the end fragments.
3. Attempt to place additional fragments.
4. Confirm that the map accounts for all the restriction fragments.
5. Include the partial-digest map information.
6. Confirm the final map with the data.

Each of these steps is worked through as follows.

The Analysis

STEP 1: Determine the size of the DNA fragment.

Determine the total size of the DNA fragment that is being studied. If you add the sizes of all the *Bam*HI fragments, you find that the undigested fragment is 22.0 kb. You also arrive at 22.0 kb if you add the sizes of all the *Eco*RI fragments or all the *Bam*HI/*Eco*RI double-digested fragments.

Sometimes, you may find that one of the digests produces a smaller total than the others. This difference may be due to either the error associated with calculating the size of DNA fragments from a gel (usually less than 5% of the total size) or to two or more different fragments migrating at the same size on the gel. In the latter case, you would need to add the missing fragments to the analysis.

STEP 2: Determine the end fragments.

The end-labeling experiment is often the best way to determine the end fragments. In this problem, the end-labeled *Bam*HI fragments are 4.0 kb and 6.0 kb, and the end-labeled *Bam*HI/*Eco*RI double-digest end fragments are 4.0 kb and 5.0 kb. Notice that the 6.0-kb *Bam*HI fragment appears to have decreased to 5.0 kb, which suggests that there is an *Eco*RI site 5.0 kb from that labeled end. Looking at the *Eco*RI digests, you find that there is a 5.0-kb fragment, which is consistent with this hypothesis.

Because the 4.0-kb *Bam*HI end-labeled fragment does not change in size with the double digest, there is either an *Eco*RI site and *Bam*HI site located both 4.0 kb from the same end, or no *Eco*RI site is located within the first 4.0 kb from that labeled end. Because there was not a 4.0-kb fragment in the complete *Eco*RI digestion, the latter hypothesis is correct. Furthermore, the only *Eco*RI fragments larger than 4.0 kb are the 4.5-kb and 9.5-kb fragments. (Remember that we already assigned the 5.0-kb *Eco*RI fragment to the other end). Thus, you have two possible maps at this time, shown in the following sketches. In this representation, E stands for *Eco*RI and B stands for *Bam*HI.

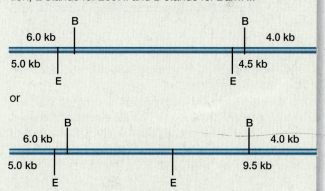

STEP 3: Attempt to place additional fragments.

It is always best to try to determine the simplest map first. If there were only three fragments for either single digest, you would have the map determined at this time. But in this problem, both single digests have four fragments. Therefore, it is best to try to place fragments into the map and compare the double-digest fragments that are produced with the double-digest information provided.

Continued

In the second map, a large area (5.5 kb) remains in the 9.5-kb *Eco*RI fragment that can accommodate the next *Bam*HI fragment; however, the two remaining *Bam*HI fragments are 2.0 kb and 10 kb. If you place the 10-kb *Bam*HI fragment next to the 4.0-kb *Bam*HI, that would produce a 5.5-kb *Bam*HI/*Eco*RI double-digest fragment. But because this size fragment was not observed in the double digest, the 10-kb *Bam*HI fragment cannot be adjacent to the 4.0-kb *Bam*HI fragment in the second map. Trying other arrangements reveals that there are no remaining possible arrangements for the 10-kb and 2.0-kb *Bam*HI fragments in the second map. (Can you see why?) Therefore, this map must be incorrect.

That means you only have one map to work on, which is the first map drawn earlier. Using the same logic, the largest region to work in is now the 1.0 kb remaining in the 6.0-kb *Bam*HI fragment. If you place the 2.0-kb *Bam*HI fragment adjacent to the 6.0-kb *Bam*HI fragment, that arrangement would produce two 1.0-kb *Bam*HI/*Eco*RI double-digest fragments. Because the data in the problem state that there is only one such fragment, and you do not have a missing 1.0-kb fragment (see step 1), then this tentative map would also be incorrect. This leaves only one possibility, which is:

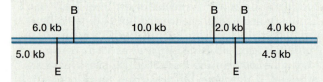

You can then use this map to determine the location of the final *Eco*RI site, which produces the following map:

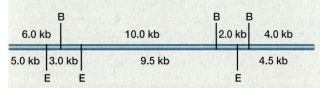

STEP 4: Confirm that the map properly accounts for all the restriction fragments.

As the map is drawn, you can easily see that all the fragments in the *Bam*HI and *Eco*RI single digests are present. You still need to confirm the *Bam*HI/*Eco*RI double-digest fragments. So, you can redraw the map to show each of the double-digest fragment sizes, which corresponds to:

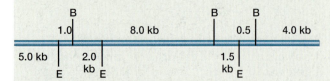

These fragments all correspond to the stated *Bam*HI/*Eco*RI double-digest fragments that you obtained experimentally. Thus, this restriction map is correct.

STEP 5: Include the partial-digest map information.

The problem also provides some data from a partial digest. The problem states that the 10.0-kb *Bam*HI fragment is end-labeled and digested with *Xba*I to produce two labeled fragments of 8.0 kb and 1.0 kb. However, these add up to only 9.0 kb, and the original fragment was 10.0 kb. Therefore, a second 1.0-kb fragment must exist. Because the 8.0-kb and 1.0-kb fragments are both end-labeled, they must correspond to the end fragments. This produces the following map of the 10-kb *Bam*HI fragment:

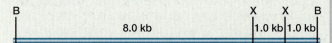

When the 8.0-kb fragment is isolated, it is labeled at only the *Bam*HI end. Thus, the *Hin*fI partial digest will show the location of each *Hin*fI site from the labeled *Bam*HI site. Note particularly that although the *Hin*fI partial digest fragments total 8.0 kb, they do not need to do so because the partial digest does *not* show the distance between adjacent *Hin*fI sites. It is a common mistake to think that it does.

Placing the *Hin*fI sites 0.5 kb, 1.0 kb, 1.5 kb, and 5.0 kb from the labeled end produces this composite map:

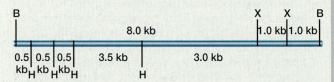

The only thing left to do is to orient this *Bam*HI fragment to the *Bam*HI and *Eco*RI map you just generated. This manipulation is possible because the problem states that digestion of the 8.0-kb *Bam*HI fragment with *Eco*RI produces a 6.0-kb and 2.0-kb fragment. This finding is also consistent with the *Bam*HI and *Eco*RI map if the *Eco*RI site is located 2.0 kb from the *Bam*HI site (rather than 2.0 kb from the *Xba*I site). Adding the data produces the following complete map:

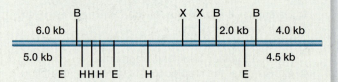

STEP 6: Confirm the final map with the data.

The most common mistake in map construction is failing to make sure that the final map reveals all the expected restriction fragments. As mentioned at the beginning of this problem, you may have to modify the order of these steps, based on the information you obtained. If you proceed in a logical fashion through the data and always look for the simplest piece of data to analyze first, you should be successful.

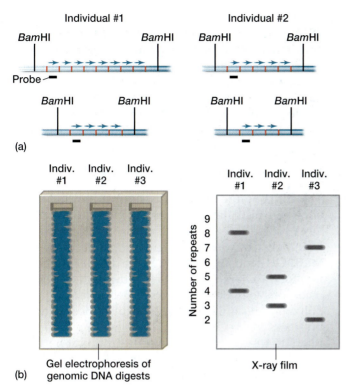

(a)

(b)

Gel electrophoresis of genomic DNA digests

X-ray film

figure 12.28 Variable number tandem repeat (VNTR) results from differences in the number of short repeats between restriction enzyme sites. (a) A typical VNTR locus is flanked by restriction enzyme sites *Bam*HI, which will digest the genomic DNA without disrupting the VNTR region. The repeat sequence will be used as a probe in the Southern blot. Different individuals will possess different numbers of repeats at the locus. Because each individual is diploid and the number of repeats can vary widely, it is likely that each individual will be heterozygous for the number of repeats. (b) A schematic of the gel containing digested genomic DNA (left) and the Southern blot X-ray film of the three heterozygous individuals in Panel a reveals the variability of the VNTR.

at each microsatellite or minisatellite locus. A population's genetic variation is presumably generated due to unequal crossing over at each locus (see chapter 7).

As a result, probing for one of these VNTR loci in a population reveals many alleles. Unlike the RFLP, where the variation is the presence or absence of a restriction site, the VNTR shows a change in the size of a restriction fragment length due to the number of repeated sequences between the restriction sites (fig. 12.28).

A Southern blot using a VNTR sequence as a probe creates a **DNA fingerprint,** which is the pattern of minisatellite alleles at different loci. Because of the hypervariable nature of the minisatellites, the DNA fingerprint may identify a single individual in a population or species, which makes it a powerful tool in forensics. DNA extracted from blood or semen samples left at a crime scene can be compared with DNA patterns of suspects (**fig. 12.29**). When a single probe recognizes a number of different loci, each individual will have many bands on a Southern blot, with most people producing unique patterns. In one system, developed by Alec Jeffreys, a single probe locates 50 or more variable bands per person. If the Jeffreys system probes are used, the likelihood that the patterns of two individuals would match randomly is infinitesimally small. This technique thus has greater power to identify individuals than using the prints from their fingertips.

figure 12.29 Forensic use of DNA fingerprinting. Southern blot of DNA from a defendant (D), evidence (jeans, shirt) and the victim (V) in a crime. Jeans and shirt refer to blood samples taken from the defendant's clothing. The DNA pattern from the blood on the shirt clearly matches the victim's blood, not the defendant's own blood. All of the other lanes of the blot contain controls and size standards. The probability that the blood stains were not from the victim was estimated at one in thirty-three billion, more than the number of people on earth. However, these probabilities are controversial, because they depend on statistical assumptions about variability within racial and ethnic subpopulations. The DNA in the blood sample from the genes was degraded, which prevented any PCR amplification from occurring.

It's Your Turn

While we discussed a logical method to determine a restriction map, it requires practice to become comfortable with the different analyses. You should now try problems 30, 32, and 34 to determine how well you understand restriction mapping and what aspects you still need to practice.

Restating the Concepts

▶ A restriction map is extremely valuable in characterizing a DNA molecule. The first step in generating the map is to identify the fragments at the end of the linear molecule. This is often done by end labeling and using X-ray film to detect the labeled end fragments.

▶ A partial digest is used to determine the order of the restriction enzyme sites from the labeled end. It is performed by limiting the amount of enzyme, reducing the length of time, or lowering the temperature of the restriction digest reaction.

▶ Most restriction maps involve more than a single restriction enzyme. By comparing the restriction digest patterns produced with a single enzyme with the pattern produced by two different enzymes simultaneously, it is possible to map the different restriction enzyme sites relative to each other.

▶ The restriction digest pattern at hypervariable regions, such as minisatellites and microsatellites, exhibits differences between individuals in a population due to the number of repeat sequences at each locus. These hypervariable loci can be used to generate a DNA fingerprint of an individual for forensic analysis.

12.5 DNA Sequencing

Although the restriction map provides much information about the physical nature of a DNA fragment, the nucleotide sequence yields the highest possible resolution of the DNA. The nucleotide sequence can be used to determine potential promoter sequences and the intron–exon relationships for a gene in a fragment of genomic DNA. It can also be used to determine the ORF of a cDNA strand and to deduce the amino acid sequence of the encoded protein.

Paul Berg, Walter Gilbert, and Frederick Sanger shared the 1980 Nobel Prize in Chemistry for their pioneering work in molecular biology. Berg won his prize for creating the first cloned DNA molecules when he spliced the SV40 viral genome into bacteriophage λ. Gilbert and Sanger were awarded their prizes

for independently developing methods of sequencing DNA. Gilbert, along with Allan Maxam, developed a chemical method of DNA sequencing. Sanger, who earlier won a Nobel Prize in Chemistry in 1958 for sequencing the insulin protein, developed an enzyme-mediated sequencing method with Alan Coulson. Sanger, Coulson, and Steve Nicklen later modified this method to use chain-terminating dideoxynucleotides.

The Dideoxynucleotide Method

The **dideoxynucleotide sequencing method** is an enzyme-based approach that is a modification of the DNA synthesis reaction. In chapter 9, you learned that DNA polymerase extends a growing DNA strand from the 3'–OH group (**fig. 12.30**). In the dideoxynucleotide method, a chain-terminating dideoxynucleotide is incorporated into the elongating DNA strand by DNA polymerase (fig. 12.30). Because this dideoxynucleotide lacks the 3'–OH group (the term *dideoxynucleotide* refers to the absence of the –OH groups at both the 2' and 3' carbons), it cannot be used for further DNA polymerization (see fig. 12.30).

Using each of the four chain-terminating dideoxynucleotides permits synthesis to be stopped at a known base. The approach is as follows:

1. The DNA to be sequenced is separated into four different reaction mixtures, each containing all four normal deoxynucleotides and one of the chain-terminating dideoxynucleotides. For example, if a reaction contains all four deoxynucleotide triphosphates (dATP, dCTP, dGTP, dTTP) and dideoxythymidine triphosphate (ddTTP), then one of two alternative events can occur on the growing strand when adenine appears on the template strand:
   - A deoxythimidine can be incorporated into the growing strand (see fig. 12.30), which permits synthesis to continue.
   - Alternatively, a dideoxythimidine can be incorporated into the growing strand (fig. 12.30), which prevents the continuation of DNA synthesis on that molecule.

2. Similar reactions are carried out in separate test tubes for each of the other three dideoxynucleotides, producing fragments in each tube that terminate at a specific dideoxynucleotide.

3. The resulting fragments from each reaction are electrophoresed, generating a pattern on the gel that reveals the sequence of the newly synthesized DNA.

Let's go through an example. In **figure 12.31**, we see a 9-bp DNA region to be sequenced. We take four aliquots (equal portions) of this DNA in solution and add all four deoxynucleotide triphosphates, an oligonucleotide primer, and DNA polymerase I to each reaction. Either the primer (fig. 12.31) or one of the deoxynucleotides triphosphates is radioactively labeled, usually with [32]P. This label allows us to identify newly synthesized DNA by autoradiography. To each of the four aliquots, we also add one of the dideoxynucleotides.

We see that the template has two adenines. Therefore, in the ddTTP reaction mixture, adenine's complement (thymine) is needed twice. There are two possible points for ddTTP to incorporate and terminate DNA synthesis, so two fragments could end in dide-

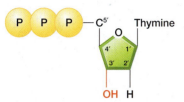

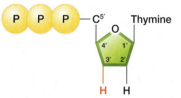

Deoxythymidine triphosphate (dTTP)

Dideoxythymidine triphosphate (ddTTP)

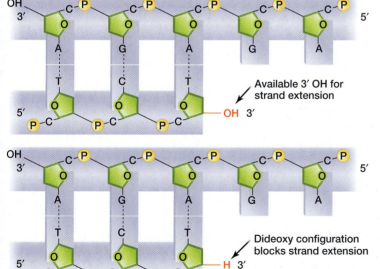

Available 3′ OH for strand extension

— OH 3′

Dideoxy configuration blocks strand extension

— H 3′

figure 12.30   Dideoxynucleotides cause chain termination during DNA replication. The deoxynucleotide contains a 3′-OH group on the deoxyribose sugar that is absent on the dideoxynucleotide. The 3′-OH is needed for DNA polymerase-mediated chain lengthening. Thus, incorporation of a dideoxynucleotide terminates the extension of the sequencing reaction due to the absence of a free 3′-OH to add another deoxynucleotide in a 5′ → 3′ direction.

oxythymidine, one with a length of two bases and one with seven bases, respectively. Similarly, three possible fragments could end in adenine, with lengths of one, three, and eight bases; three could end in cytosine, with lengths of four, five, and nine bases; and finally, one could end in guanine, of six bases (fig. 12.31).

After DNA synthesis is completed, the double-stranded DNAs in each reaction mixture are denatured, placed in separate slots of a polyacrylamide gel, and electrophoresed to separate the DNA strands by size. Because only newly synthesized DNA segments are radioactive, autoradiography reveals only the newly synthesized DNA.

The schematic of an autoradiograph in figure 12.31 shows that each aliquot produces fragments that end with a specific chain-terminating dideoxynucleotide. Furthermore, you can compare the four different aliquots and see the synthesized DNA strand increasing in length by a single nucleotide. By starting at the bottom of the gel and reading up, we can directly determine the exact sequence of the DNA segment. The smallest fragment is in the ddATP lane, which means that the synthesized strand had an A residue closest to the primer. The next larger fragment is in the ddTTP lane, which means that the next base in the synthesized strand is T.

We read the synthesized strand from the bottom to the top, corresponding to the 5′-to-3′ direction of the DNA strand. Because they have the appearance of stepladders in each lane (**fig. 12.32**), the gels are usually referred to as **sequencing ladder gels.**

Innovations to DNA Sequencing

Two significant modifications have simplified DNA sequencing. The first is the generation of plasmids that permit the rapid isolation of single-stranded DNA to serve as the sequencing reaction template. The second is the automation of reading the sequencing gels directly into a computer. These two innovations have led to the ability to produce massive amounts of sequencing data in a very short time.

Joachim Messing and his colleagues developed a very clever plasmid vector that could easily produce a single-stranded template. This plasmid is based on the circular bacteriophage M13, which replicates to produce single-stranded DNA that is packed into the infectious phage particles. Messing's plasmid contained both an *E. coli* plasmid origin of replication and the bacteriophage M13 origin of replication (**fig. 12.33**). Under normal conditions, this plasmid would use the plasmid origin of replication, which produces double-stranded DNA that can be cut with restriction enzymes. The M13 origin of replication (fl *ori*, fig. 12.7) can be induced, however, which produces

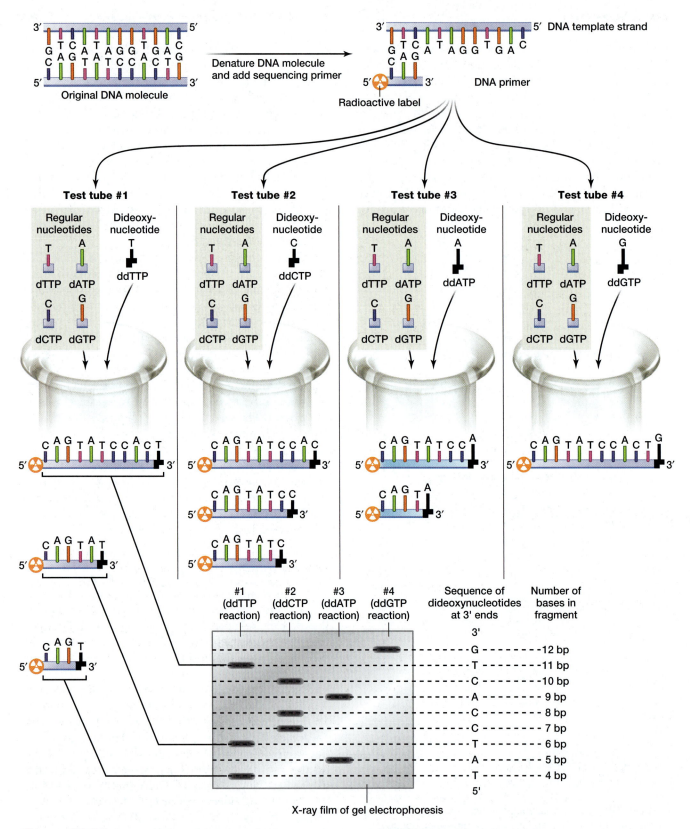

figure 12.31 The dideoxynucleotide DNA sequencing method. The DNA molecule to be sequenced is denatured and a radioactively labeled DNA primer is added. The template-primer pair is added to four tubes, along with all four deoxynucleotides and DNA polymerase I. Into each tube, a different dideoxynucleotide is added. Each reaction has the option of inserting either the deoxynucleotide or the dideoxynucleotide at a given location during DNA synthesis. When a dideoxynucleotide is added, synthesis of that DNA strand stops. Each reaction is loaded into a different lane on a gel and electrophoresed. The gel is then exposed to a piece of X-ray film to reveal the location of all the radioactively labeled DNA strands. Knowing what dideoxynucleotide was in each reaction reveals what dideoxynucleotide is at the 3' end of each DNA fragment on the X-ray film.

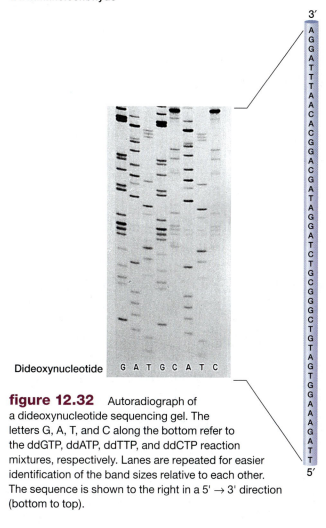

3′

AGGATTTAACACGGACGATAGGATCTGCGGGCTGTAGTGGAAAGATT

5′

Dideoxynucleotide G A T G C A T C

figure 12.32 Autoradiograph of a dideoxynucleotide sequencing gel. The letters G, A, T, and C along the bottom refer to the ddGTP, ddATP, ddTTP, and ddCTP reaction mixtures, respectively. Lanes are repeated for easier identification of the band sizes relative to each other. The sequence is shown to the right in a 5' → 3' direction (bottom to top).

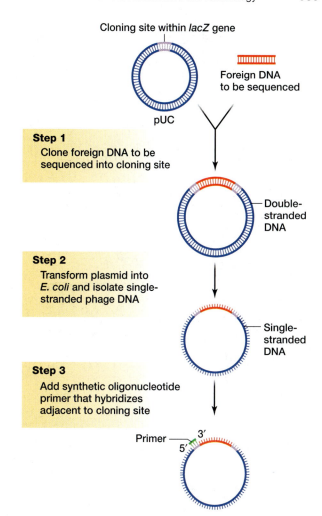

Cloning site within *lacZ* gene

pUC

Foreign DNA to be sequenced

Step 1
Clone foreign DNA to be sequenced into cloning site

Double-stranded DNA

Step 2
Transform plasmid into *E. coli* and isolate single-stranded phage DNA

Single-stranded DNA

Step 3
Add synthetic oligonucleotide primer that hybridizes adjacent to cloning site

Primer
3′
5′

figure 12.33 The pUC plasmid is a useful DNA sequencing vector. A DNA fragment is cloned into the double-stranded pUC plasmid (step 1) and then transformed into *E. coli*. Single-stranded DNA is produced from the *fl* replication origin (fig. 12.7). An oligonucleotide primer hybridizes adjacent to the cloning site to provide the primer for the DNA replication reaction in vitro.

single-stranded DNA that could easily serve as a DNA-sequencing template.

Messing also introduced the bacterial *lacZ* gene into this plasmid and placed a multiple cloning site in the *lacZ* gene (see fig. 12.33). The *lacZ* gene encodes the bacterial β-galactosidase enzyme, which acts to break lactose into its two components: glucose and galactose. β-Galactosidase can also cleave the artificial compound X-gal, which is colorless, into a molecule that forms a blue precipitate. Introducing a foreign DNA insert into the multiple cloning site inactivates this gene. Bacteria growing in the presence of X-gal will therefore be white if they contain a recombinant plasmid, which contains an insert in the multiple cloning site. The colonies will be white if the plasmid contains an insert. This straightforward method confirms that an insert is cloned in the multiple cloning site.

This insert could be easily sequenced by inducing the cell to produce single-stranded DNA and then using an oligonucleotide primer to the *lacZ* sequences that flank the multiple cloning site. This creates the template–primer configuration for dideoxynucleotide sequencing of the cloned DNA. Virtually any clonable

segment of DNA can be sequenced using this very general method. These plasmids, which are now named pUC, and their derivatives are still commonly used cloning vectors.

In recent years, PCR amplification has become the method of choice for DNA sequencing. Any single- or double-stranded DNA could be PCR-amplified for several cycles to generate the necessary amount of double-stranded DNA. This isolates a specific DNA fragment for DNA sequencing without cloning the fragment. The PCR-amplified DNA is purified and divided into four tubes, and a third primer, which is located between the two original PCR primers, is added along with a different dideoxynucleotide to each tube. This third primer would anneal to only one of the two DNA strands and serve as the sequencing primer. The PCR reaction can

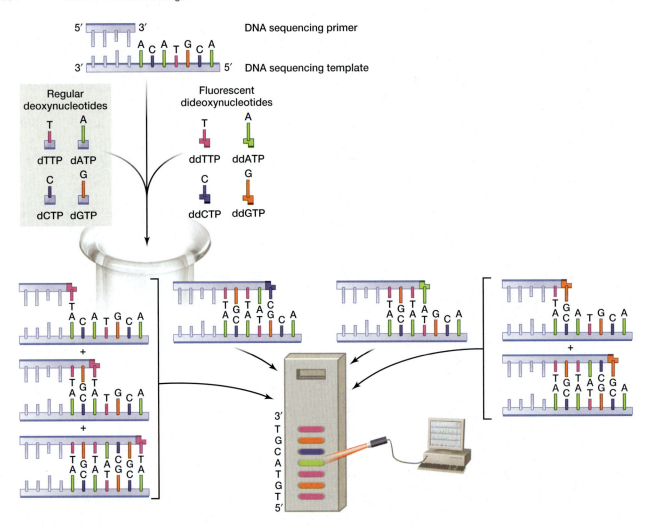

figure 12.34 Dideoxynucleotide DNA sequencing using fluorescent labels. The single-stranded DNA sequencing template is mixed with a DNA primer, DNA polymerase, and all four deoxynucleotides into a single reaction tube. The four dideoxynucleotides, each with a different fluorescent dye attached to it, are also added to the single reaction. Thus, at each sequencing position, either a deoxynucleotide or dideoxynucleotide can be incorporated. If a deoxynucleotide is inserted, the polymerase moves to the next position and the choice is made again. If a dideoxynucleotide is inserted, DNA synthesis of that strand terminates. The reaction is loaded into a single lane of a polyacrylamide gel. As the terminated DNA strands pass near the bottom of the gel, an argon laser excites the fluorescent dye attached to the dideoxynucleotide at the 3' end and it fluoresces at a specific wavelength. The wavelength of the fluorescence reveals the identity of the dideoxynucleotide at the 3' position of the terminated DNA strand.

then be continued for several additional cycles. If this third primer is labeled, then only the sequence produced from it will be revealed on the sequencing gel. This method also allows the DNA sequence to be generated from a very small amount of template DNA.

The most recent innovation in DNA sequencing involves attaching a different fluorescent dye to each of the four dideoxynucleotides (**fig. 12.34**). Each dye is designed to fluoresce at a different wavelength (505, 512, 519, and 526 nm). In this approach, all four sequencing reactions are performed simultaneously in a single tube and electrophoresed together in a polyacrylamide capillary gel (see fig. 12.34). As the gel runs, an argon laser at the bottom of the gel excites the dye molecules.

Because each dideoxynucleotide contains a dye that fluoresces at a different wavelength, the emitted fluorescence reveals the dideoxynucleotide at the 3' end of a specific fragment (**fig. 12.35**). This technique allows four times more sequencing reactions to be electrophoresed on a gel, because the four dideoxynucleotide reactions from a single template are in a single lane rather than four separate lanes (see fig. 12.32).

This automated method yields over 1000 bases of sequence per template, compared with the several hundred bases of sequence produced by ^{32}P-labeled sequencing. The combination of automation and increased read lengths has allowed the sequencing of genomes, including the human genome, to become a reality (chapter

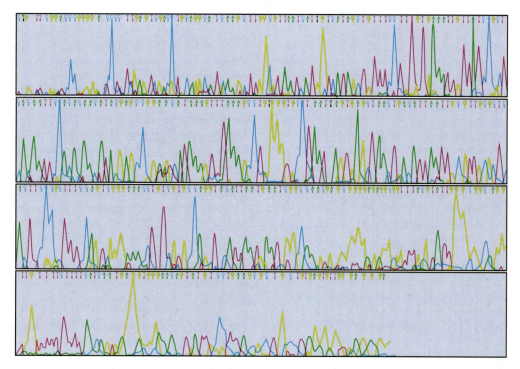

figure 12.35   Processed data from automated DNA sequencing analysis using fluorescent dyes. In this diagram, guanine is yellow, cytosine is blue, adenine is green, and thymine is red. The sequence is read left to right, top to bottom.

14). In fact, the ease of DNA sequencing has shifted the rate-limiting step from the actual sequence generation to the subsequent analysis of the sequence information, known as *bioinformatics* (chapter 14).

 Restating the Concepts

▶ DNA sequencing is primarily performed using DNA polymerase to incorporate dideoxynucleotides that would terminate the growing DNA strand at specific bases.

▶ By comparing the lengths of synthesized strands and knowing which of the four dideoxynucleotides is in each lane of the sequencing ladder, it is possible to read the sequence of the synthesized strand in the 5′ → 3′ direction from the bottom to top of the gel.

▶ DNA sequencing is simplified by using vectors that can produce the single-stranded template, such as pUC, or by using PCR amplification and a thermal cycler to amplify the template and produce multiple synthesized strands from the same template.

▶ The use of fluorescent tags on the dideoxynucleotides and an argon laser to excite the tags, which fluoresce at different wavelengths, allows DNA sequencing to become automated.

It's Your Turn

The ease of DNA sequencing has allowed the genome sequencing of numerous organisms to be done. Furthermore, it is more common to sequence a cDNA and deduce the encoded amino acid sequence than to determine the amino acid sequence directly from a protein. To test your ability to read a sequence ladder and deduce the sequence of the sequencing template strand, try problem 40. *Hint:* The sequence of the template strand is complementary and antiparallel to the sequence reactions.

genetic linkages

This chapter described many of the basic techniques that are part of recombinant DNA technology. Many of these techniques are built on standard biological processes, such as DNA replication, that we previously discussed in this book. Understanding that DNA synthesis proceeds in a 5' → 3' direction should help you remember that reading the sequencing ladder from the bottom of the gel to the top corresponds to the sequence on the synthesized strand in a 5' → 3' direction.

Restriction mapping is a powerful method of characterizing changes in the structure of chromosomes, such as deletions, insertions, inversions, and translocations (chapter 8). The ability to generate a restriction map of a normal chromosome region and compare it with a mutant using Southern hybridization can provide a fine level analysis of the chromosomal abnormality. In chapter 13, we discuss the process of chromosome walking and the creation of a contig (chapter 14). Both of these methods are significantly simplified by the ability to generate a restriction map. We also discuss the ability to generate transgenic animals and knock-out mice (gene replacement) in chapter 20. To identify these genetically altered individuals, it is essential to know the wild-type restriction map for the genomic region and to be able to compare the wild-type and mutant by Southern hybridization.

In chapter 13, we also discuss a variety of vectors that go beyond the scope of this chapter, including vectors that permit the expression of foreign genes in eukaryotes and vectors that are designed to assay promoter activity.

In this chapter, we discussed using Northern blots to examine details of gene expression. In chapter 14, we describe two additional methods for examining the broad global changes in gene expression between different tissues or different times in development. These techniques, termed gene microarray and SAGE are built on the nucleic acid hybridization and reverse transcriptase-mediated cDNA synthesis procedures that we discussed in this chapter.

In our discussion of restriction maps, we described restriction fragment length polymorphisms (RFLPs) and variable number tandem repeats (VNTRs) loci. These elements can also be used for recombination mapping. The ability to map human disease genes (chapter 13) based on the presence of VNTRs and microsatellites has significantly increased our ability to clone the genes that correspond to biological processes and inherited human diseases.

In chapter 14, we describe the relatively new field of bioinformatics. This research area was created when DNA sequencing reached a point that sequencing the entire genome of organisms was possible. Although much has been written about the Human Genome Project, genome projects have been initiated or completed in a large number of different bacteria, numerous viruses, the fruit fly, yeast, zebrafish, *Arabidopsis,* rice, corn, the mouse, and the rat. The analysis of these species' genomes may provide a deeper appreciation of biology and of the intricate processes of which living things are capable.

Chapter Summary Questions

1. For each term in column I, choose the best matching phrase in column II.

   Column I
   1. Northern hybridization
   2. Type I restriction enzyme
   3. DNA ligase
   4. BAC
   5. Western blotting
   6. λ
   7. DNA methylase
   8. Shuttle vector
   9. Type II restriction enzyme
   10. Southern blotting

   Column II
   A. Cuts DNA at sites other than the recognition site
   B. Nucleic acid hybridization technique used to analyze RNA
   C. Covalently modifies a single nucleotide in a restriction site
   D. Vector containing two origins of replication
   E. Technique used to identify specific proteins
   F. Cuts DNA at the recognition site
   G. Covalently links two nucleotides
   H. Vector that can accommodate inserts of 1000 kb
   I. Nucleic acid hybridization technique used to analyze DNA
   J. Bacteriophage used as vector

2. What purposes do restriction enzymes serve in bacteria that produce them? How do bacteria protect their own DNA from these enzymes?

3. What specific properties of type II endonucleases make them useful in gene cloning?

4. Under what circumstances is a restriction endonuclease unsuitable for cloning a piece of foreign DNA?

5. Compare and contrast plasmids and cosmids.

6. What methods exist to create sticky ends or create ends for joining two incompatible pieces of DNA? When is each method favored?

7. Identify three important features that cloning vectors possess.

8. What is the significance of the *lacZ* gene in a plasmid vector?

9. What are the steps by which mRNA can be converted into cDNA? How would we obtain radioactive cDNA? Radioactive mRNA?

10. Distinguish the differences between a genomic library and a cDNA library.

11. Define the following terms: **(a)** Electroporation; **(b)** Linkers; **(c)** Microsatellite.

12. Outline the three steps of PCR. Which step can be altered to change the stringency between the template and oligonucleotide primer?

13. What advantage does PCR have over gene cloning for amplification of DNA? Is there any disadvantage?

14. Arrange the following steps in the proper order in which they occur in Southern blotting:
    A. Transfer DNA from gel to nitrocellulose filter
    B. Cut DNA with restriction enzyme
    C. Expose filter to X-ray film
    D. Isolate DNA from organism
    E. Place filter in solution containing radiolabeled probe
    F. Separate DNA by gel electrophoresis

15. Explain the role of dideoxynucleotide triphosphates in the dideoxynucleotide method of DNA sequencing.

Solved Problems

PROBLEM 1: A piece of eukaryotic DNA is isolated using a restriction endonuclease that leaves blunt ends (*Hae*III). How could we get this piece of DNA into a *Bam*HI site in plasmid pcDNA2.1, and how would we know when the foreign DNA has been cloned?

Answer: Because the two pieces of DNA (the eukaryotic piece and the plasmid) have different ends, they must be made compatible before cloning. The simplest way to do this would be to attach blunt-ended linkers to the foreign DNA with phage T4 DNA ligase (see fig. 12.6). The linkers, of course, would have a *Bam*HI site within them. After the linkers are ligated to the foreign DNA, it would be digested with the *Bam*HI restriction enzyme, giving the foreign DNA *Bam*HI ends. The *Bam*HI-digested plasmid is then mixed with the *Bam*HI-digested eukaryotic DNA in the presence of *E. coli* DNA ligase, which seals up the plasmids, with or without cloned inserts (see fig. 12.8). Because both DNA pieces have compatible ends, some of the time, a piece of foreign DNA is inserted into a plasmid. The plasmids are then taken up by *E. coli* cells that are grown overnight in an incubator. The bacterial colonies are then replica-plated on media with X-gal and the antibiotic ampicillin. Colonies that are resistant to ampicillin, corresponding to *E. coli* that contain the plasmid, will grow on the medium. Of the colonies that grow, blue colonies will contain the pcDNA2.1 plasmid without a cloned insert; white colonies will contain the pcDNA2.1 plasmid with the cloned eukaryotic DNA fragment (see fig. 12.9).

PROBLEM 2: A piece of DNA has the sequence 3'-GGCGTATTC-5' immediately adjacent to a sequence that is complementary to a sequencing primer. This DNA is sequenced using the dideoxynucleotide method. How many bands are found on the ladder gel? How many bands and of what size are found for each reaction mixture?

Answer: Because the piece of DNA is nine bases long, the total number of bands in all four lanes of a sequencing gel ladder add up to nine (see fig. 12.31). Each lane contains a sequencing reaction that is composed of the template and primer DNAs, all four deoxyribonucleotides, DNA polymerase I, and one of the four dideoxynucleotides. In the reaction mixture with ddTTP, chain termination occurs at the adenine in the piece of DNA; that is, a DNA segment was synthesized that is six bases long (see the following figure). In the reaction mixture with ddATP, chain termination occurs opposite each of the thymines, producing DNA segments of five, seven, and eight nucleotides. In the reaction mixture

with ddGTP, chain termination occurs opposite the cytosines in positions three and nine. And, in the reaction mixture with ddCTP, chain termination occurs after synthesis of segments one, two, and four bases long. Note that the gel gives us the sequence of the complement strand of the original piece of single-stranded DNA.

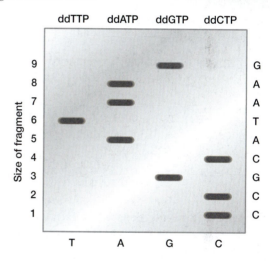

PROBLEM 3: A linear DNA molecule 1000 bp long is digested with the following restriction enzymes, producing the following results:

*Eco*RI	400 bp, 600 bp
*Bgl*II	250 bp, 750 bp
*Eco*RI + *Bgl*II	250 bp, 350 bp, 400 bp

Determine the restriction map.

Answer: Each enzyme alone produces two fragments, so the molecule has one site for each enzyme. Because we get different-sized fragments with each enzyme, the sites must be located asymmetrically along the DNA. Draw these sites:

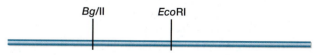

The *Eco*RI fragment that lacks a *Bgl*II site should appear in the double digest. If *Bgl*II cuts within the 400-bp *Eco*RI fragment, we would expect to see 150-, 250-, and 600-bp fragments. We don't see this, so the *Bgl*II site is not within the 400-bp *Eco*RI fragment. This means we must place the *Bgl*II site within the

600-bp *Eco*RI fragment. Similarly, the *Eco*RI site must be within the 750-bp *Bgl*II fragment. Thus, the map looks like this:

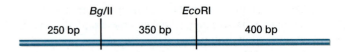

Exercises and Problems

16. Which of the following are palindromes?
    a. 1234567654321
    b. aibohphobia
    c. I'm a lasagna sang a salami
    d. 5'-TCGGCT-3'
       3'-AGCCGA-5'
    e. 5'-TCGCGA-3'
       3'-AGCGCT-5'

17. The following is a double helix of DNA. What, if any, are potential restriction enzyme recognition sequences?

    5'-TAGAATTCGACGGATCCGGGGCATGCAGATCA-3'

    3'-ATCTTAAGCTGCCTAGGCCCCGTACGTCTAGT-5'

18. The recognition sequences and cleavage sites of three restriction enzymes follow.

    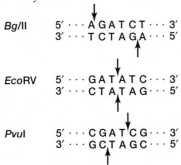

    State the type of end produced by each enzyme, and draw diagrams showing these cut ends.

19. Assuming a random arrangement and equal amounts of each nucleotide in DNA, what is the probability of finding the following restriction sites? What are the probabilities if the DNA is 65% GC?
    a. 5'-CGCG-3'
       3'-GCGC-5'
    b. 5'-AGNCT-3'
       3'-TCNGA-5'
    c. 5'-GTPyPuAC-3', where Py is a pyrimidine and Pu
       3'-CAPuPyTG-5'   is a purine
    d. 5'-GAATTC-3'
       3'-CTTAAG-5'
    e. 5'-GCGGCCGC-3'
       3'-CGCCGGCG-5'

20. Consider the hypothetical bacterium *Bacillus ignobleus* strain E1. If it is found to produce a single type II restriction endonuclease, what would you name this enzyme?

21. Diagram a possible heteroduplex between two phage λ vectors, one with and one without a cloned insert, created by DNA–DNA hybridization.

22. How would you isolate a human alanine tRNA gene for cloning? How would you locate a clone with a human alanine tRNA gene in a genomic library?

23. How would you develop a probe for a gene whose mRNA could not be isolated? How could an expression vector be used to isolate a cloned gene?

24. Exonuclease III is an enzyme that sequentially removes bases from the 3' end of double-stranded DNA. The following two molecules, each 100 bp long, are digested with exonuclease III. Molecule 1 is completely digested; molecule 2 is only partially digested. Explain these results.

    Molecule 1:   CGTTCAG...
                   GCAAGTC...

    Molecule 2:   AAAAAAAAAA...
                   TTTTTTTTTT...

25. The bacterium *Bdellovibrio bacteriovorus* has a genome that is 3.8 Mbp and 50% GC. Assume random distribution of nucleotides.
    a. How many *Not*I restriction sites (5' GC^GGCCGC 3') are likely to be found in this genome?
    b. What is the average spacing between these restriction sites?

26. A plasmid that contains an *Eco*RI site within a gene for ampicillin resistance is cut with *Eco*RI and then religated. This plasmid is used to transform *E. coli* cells, and the plasmid is reisolated from the ampicillin-resistant colonies. The reisolated plasmids from two different colonies are electrophoresed, and the results appear in the following figure.

    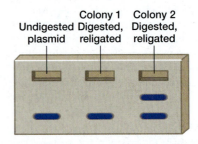

    How do you account for the two bands in colony 2?

27. Most human genes contain one or more introns. Because bacteria cannot excise introns from nuclear pre-mRNA (snRNPs are needed), how can bacteria be used to make large quantities of a protein encoded by one of these human genes?

28. The following segment of DNA is cut four times by the restriction endonuclease *Eco*RI at the places shown. Diagram the gel banding that would result from electrophoresis of the total and partial digests. Note the end-labeled segments (asterisks) and bands that contain two or more different DNA fragments.

    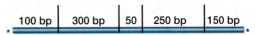

29. Three mutants of the DNA segment shown in problem 28 were isolated. They gave the following gel patterns when the total digests were electrophoresed. Asterisks denote the end-labeled segments. Can you determine the nature of each of the three mutations?

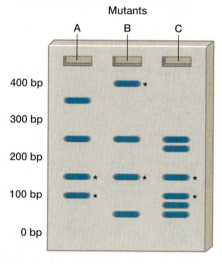

30. The following figure shows a gel and a total and partial digest of a DNA segment treated with *Hin*dII. End-labeled segments are noted by asterisks. Draw the restriction map of the original segment.

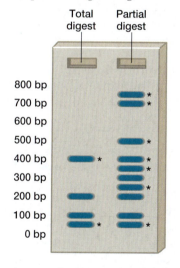

31. Restriction maps of a segment of DNA were worked out separately for *Bam*HI and *Taq*I. Two overlays of the maps are possible. The double-digest gel is shown in the following figure (asterisks denote end labels). Which overlay is correct?

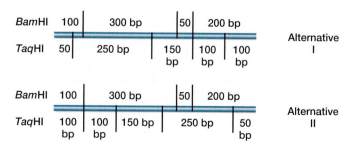

32. A linear DNA molecule, 1000 bp long, gives the following size fragments when treated with these restriction enzymes. Derive a restriction map.
*Eco*RI: 300 bp, 700 bp
*Bam*HI: 150 bp, 200 bp, 250 bp, 400 bp
*Eco*RI + *Bam*HI: 50 bp, 100 bp, 200 bp, 250 bp, 400 bp

33. A linear DNA molecule cut with *Eco*RI yields fragments of 3 kb, 4.2 kb, and 5 kb. What are the possible restriction maps?

34. You radioactively label the 5' ends of a double-stranded DNA molecule. Digestion of this molecule with either *Eco*RI or *Bam*HI yields the following fragments. The numbers are in kilobases (kb), and an asterisk indicates the fragments that are end-labeled.
*Eco*RI: 2.8, 4.6, 6.2*, 7.4, 8.0*
*Bam*HI: 6.0*, 10.0*, 13.0

If unlabeled DNA is digested with both enzymes simultaneously, the following fragments appear: 1.0, 2.0, 2.8, 3.6, 6.0, 6.2, 7.4. What is the restriction map for the two enzymes?

35. In a Northern blotting experiment, mRNA from human brain cells was run on a gel, and transferred onto a nitrocellulose filter. A cloned gene *z* from *Drosophila* was radiolabeled and hybridized to the filter under low- and high-stringency conditions. The X-ray films are shown below.

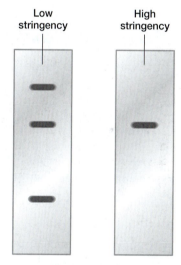

a. What does stringency refer to? What are the conditions that allow for low or high stringency?
b. Discuss the results of the X-rays.

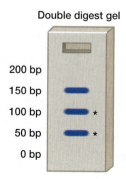

36. A 12-kb DNA molecule cut with *Eco*RI yields one 12-kb fragment. When the original molecule is cut with *Bam*HI, three fragments of 2 kb, 4.5 kb, and 5.5 kb are produced. When the fragment from *Eco*RI is treated with *Bam*HI, four fragments of 2 kb, 2.5 kb, 3.0 kb, and 4.5 kb are produced. Draw a restriction map.

37. A gene has the following *Eco*RI restriction map (in kilobases):

| 1.0 kb | 0.7 kb | 2.0 kb |

Draw the gel pattern expected from
a. a mutant that has lost the site between the 1.0- and 0.7-kb fragments.
b. a mutant that has a new site within the 2.0-kb fragment.

38. A DNA fragment 8 kb in size is labeled with ^{32}P at the 5' ends. It is then digested with *Eco*RI, *Bgl*II, or a mixture of both enzymes. The size of the fragments and the labeled fragments (*) appear as follows. Sizes are in kilobases.

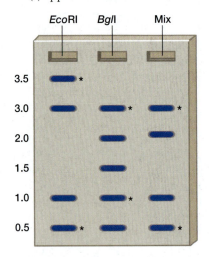

Which of the following two maps is consistent with the results?

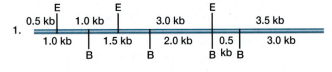

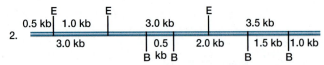

39. You now take an unlabeled molecule from problem 38, digest it with *Hin*dIII, and get two fragments, 5.5 and 2.5 kb in size. If *Hin*dIII does not cut within the 3.5-kb *Eco*RI

fragment, what size fragments do you expect in a double digest of *Hin*dIII and *Eco*RI?

40. The following diagram is of a dideoxynucleotide sequencing gel. What is the sequence of the DNA under study?

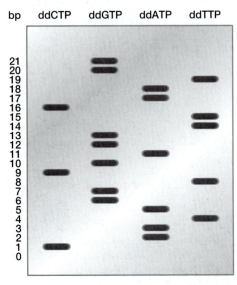

41. How can a particular piece of DNA be manipulated to be in the appropriate configuration for dideoxynucleotide sequencing?

42. Provide, if possible, DNA sequences that can mark the termination of one gene and the initiation of another, given that the genes overlap in one, two, three, four, five, six, or seven bases.

43. A Northern blot of human mRNA from four different cell types was probed with the *Drosophila* gene *z*. The X-ray film is shown here.

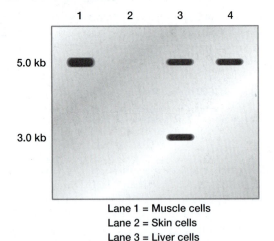

Lane 1 = Muscle cells
Lane 2 = Skin cells
Lane 3 = Liver cells
Lane 4 = Lung cells

Explain these results.

44. To study the expression of protein Z during the *Drosophila* life cycle, proteins are extracted from various samples, electrophoresed through a polyacrylamide gel, and transferred to a nitrocellulose filter. The filter was incu-

bated with a labeled anti-Z antibody. The results of the Western blot are shown here.

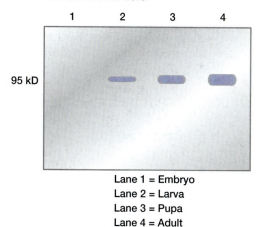

95 kD

1 2 3 4

Lane 1 = Embryo
Lane 2 = Larva
Lane 3 = Pupa
Lane 4 = Adult

Discuss these results.

45. Draw the expected gel pattern derived from the dideoxynucleotide sequencing method for a template strand with the following sequence:

5'-CAGCGAATGCGGAA-3'

46. A DNA strand with the sequence 3'-GACTATTCCGAAAC-5' is sequenced by the dideoxynucleotide method. If the reaction mixture contains all four radioactive deoxynucleotide triphosphates plus dideoxythymidine, what size labeled bands do you expect to see on the gel?

47. A plasmid 3 kb in length contains a gene for ampicillin resistance and a gene for tetracycline resistance. The plasmid has a single site for each of the following enzymes: *Eco*RI, *Bgl*II, *Hind*III, *Pst*I, and *Sal*I. If DNA is cloned into the *Eco*RI site, resistance to either antibiotic is not affected. DNA cloned into the *Bgl*II, *Hind*III, or *Sal*I sites abolishes tetracycline resistance, and DNA inserted into the *Pst*I site eliminates ampicillin resistance. If the plasmid is digested completely with enzyme mixes, the following fragments result:

Mixture	Fragment Size (kb)
*Eco*RI + *Pst*I	0.7, 2.3
*Eco*RI + *Bgl*II	0.3, 2.7
*Eco*RI + *Hind*III	0.08, 2.92
*Eco*RI + *Sal*I	0.85, 2.15
*Eco*RI + *Bgl*II + *Pst*I	0.3, 0.7, 2.00

Draw a restriction map of the plasmid and indicate the locations of the resistance genes and the sites of enzymatic cleavage.

48. In a certain animal, gene *R* encodes an enzyme that is involved in muscle development. Five mutants with muscle abnormalities have been discovered. You decide to investigate the expression of gene *R* in these mutants using Northern and Western blotting analysis. You isolate mRNA and protein from these mutants and you

probe them with labeled *R* DNA and anti-R antibody, respectively. The results follow:

Northern Blot

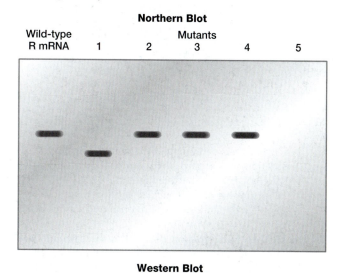

Wild-type
R mRNA   1   2   3   4   5
          Mutants

Western Blot

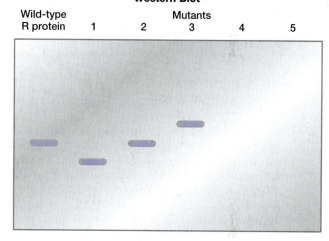

Wild-type
R protein   1   2   3   4   5
             Mutants

Describe as accurately as possible the molecular defect associated with each of these five mutations.

49. You isolated a genomic DNA clone. You digested this DNA fragment with *Eco*RI and identified fragments of 8.5, 7.0, 6.0, and 4.5 kb. You also digested this genomic DNA fragment with *Bam*HI and identified fragments of 8.0, 6.0, 4.5, 4.0, and 3.5 kb. You performed a *Bam*HI/*Eco*RI double digest and identified DNA fragments of 5.0, 4.5, 4.0, 3.5, 3.0, 2.5, 2.0 and 1.5 kb. You labeled the end of the genomic DNA fragment and performed the *Eco*RI digest and found that the radioactivity was in the 6.0- and 4.5-kb fragments. If the labeled genomic DNA fragment was digested with *Bam*HI, the 6.0- and 3.5-kb fragments were labeled.

a. Draw a restriction map for the genomic DNA fragment.

b. You isolated the 7.0-kb *Eco*RI fragment, end-labeled it, and digested it with *Xba*I. A 6.0- and 1.0-kb fragments were both radioactively labeled. When the 6.0-kb fragment was partially digested with *Hin*fI, 1.0-, 2.0-, 3.0- and 6.0-kb labeled fragments were identified. Digestion of this 6.0-kb *Eco*RI fragment with *Bam*HI yielded a 5.0 kb labeled fragment. Incorporate the *Xba*I and *Hin*fI sites into the restriction map that you generated in part (*a*).

Chapter Integration Problem

A hypothetical animal virus has a double-stranded DNA genome that is 50 kb in length and has a 70% GC content. Digestion of this molecule with *Bgl*II (which cuts at 5' A^GATCT 3'), *Eco*RI (5' G^AATTC 3') or both enzymes yields the following fragments (all in kilobasepairs):

*Bgl*II: 24, 22, and 4

*Eco*RI: 20, 18, and 12

*Bgl*II + *Eco*RI: 20, 17, 5, 4, 3, and 1

    a. Is this viral genome linear or circular? Justify your reasoning.

    b. Draw a restriction map of the genome. Label the restriction sites with B (*Bgl*II) and E (*Eco*RI), and indicate the distances between them.

    c. Is the number of *Bgl*II and *Eco*RI restriction sites compatible with random distribution of nucleotides in the DNA? Why or why not?

    d. How many restriction sites is this viral genome expected to have for each of the following enzymes? (N = any nucleotide; R = any purine; Y = any pyrimidine; K = G or T; and M = A or C)

*Mse*I	5' T^TAA 3'
*Hpa*II	5' C^CGG 3'
*Bsa*JI	5' C^CNNGG 3'
*Acc*I	5' GT^MKAC 3'
*Hae*II	5' RGCGC^Y 3'
*Pac*I	5' TTAAT^TAA 3'

    e. You decide to clone the viral DNA into the pEN plasmid, which has a single *Bgl*II (A^GATCT) restriction site and an overall GC content of 50%. You use the enzyme *Dpn*II (^GATC) to digest the viral DNA. You then ligate the fragments into the *Bgl*II-cut pEN plasmid. Could this cloned viral DNA be cut out of the plasmid with *Bgl*II? If not, why not? If so, what percentage of the time?

    f. The viral genome encodes 90 proteins with an average molecular weight of 25,000 daltons. Assume the average molecular weight of an amino acid is 125 daltons. What is the number of codons required to code for these 90 proteins? Explain how the viral DNA can contain this many codons.

Protein T, the smallest of these viral proteins, contains 45 amino acids and has the following primary sequence.

```
 1 Met Ala Thr Asp Gly Ser Trp Val Leu Cys Asn Leu Met Ile Tyr
16 Trp Ser His Leu Gly Glu Gln Trp Thr Ser Ile Leu Gly Trp Glu
31 Ile Met Arg Asp Pro Asn Leu Trp Trp Leu Gly Phe Phe Ile Ser
```

To find the location of the gene encoding protein T, you decide to perform a Southern blot analysis on the viral DNA that has been cut with *Bgl*II and *Eco*RI. The probes that you can use are 15 nucleotides long.

    g. What sequence of amino acids would provide the smallest number of degenerate probes? List the entire set of degenerate probes that would guarantee a perfect match to the gene in question.

    h. The Southern blot experiment using this mixture of degenerate probes produces the following results.

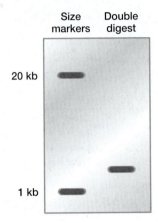

Determine as accurately as possible the location of gene *T*?

Two mutants of this virus are known. You isolate viral DNA for a Southern blot. The double restriction digest patterns for the two mutants are:

    Mutant 1:   *Bgl*II + *Eco*RI: 20, 17, 7, 5, and 1

    Mutant 2:   *Bgl*II + *Eco*RI: 20, 17, 5, 4, 2, and 1

The Southern blot of the wild-type virus and the two mutants is shown here.

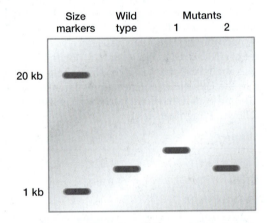

    i. Describe the molecular defect associated with each mutant, and localize gene *T* to one of the six restriction fragments of the wild-type viral DNA.

Do you need additional review? Visit **www.mhhe.com/hyde** for practice tests, answers to end-of-chapter questions and problems, interactive exercises, and animation tutorials all designed to enhance your understanding of key genetic concepts.

13

Application of Recombinant DNA Technology

Essential Goals

1. Understand how DNA markers are used in mapping human mutations.

2. Describe how mutant genes in humans are identified and cloned.

3. Discuss the different eukaryotic vectors and their unique features and applications.

4. Understand the various methods used to generate transgenic organisms.

5. Discuss the important considerations when developing gene therapy for a human disease.

Chapter Outline

Photo: A wild-type mouse (left) compared to a transgenic mouse (right) with a copy of the human growth hormone gene, which results in the increased size.

*I*n this chapter, we discuss the application of the recombinant DNA techniques introduced in chapter 12. In particular, we describe the methods used to clone wild-type and mutant genes and to introduce foreign genes into eukaryotic organisms.

Organisms that contain foreign genes are called **transgenic animals** or **transgenic plants.** These organisms are also known as **genetically modified organisms (GMOs).** The creation and uses of GMOs are currently topics of heated debate and controversy. Although a variety of concerns have been expressed about transgenic organisms, they have been shown to be useful for several human endeavors.

First, transgenic animals can provide information on basic cellular processes. In some organisms, it is possible to inactivate selected genes in the genome. This permits the analysis of loss-of-function mutations in a particular developmental, physiological, or biochemical pathway. Alternatively, a gene can be mutated in vitro such that the encoded protein is always in an active state. This mutant gene can then be introduced into an organism, and the dominant gain-of-function phenotype can be studied.

Second, transgenic animals can serve as models for a wide variety of human diseases. Either the mutant human gene is introduced into an animal (such as a mouse or fruit fly), or the animal version of the human gene is cloned, modified in vitro to contain the same mutation as the human gene, and then reintroduced back into the animal species. These transgenic animal models may reveal how underlying cellular processes are disrupted in diseases, providing clues to biochemical targets for therapy. The animal models can then serve to test pharmaceutical and genetic therapies. Using this approach, scientists are studying such wide-ranging human diseases as Alzheimer dementia, inherited blindness, obesity, mental retardation, and sickle cell anemia.

Third, transgenic insects are being explored as a way to eradicate particular insect-borne diseases, such as malaria and dengue fever. It may be possible to genetically modify the insects so that they are unable to sustain the disease-causing parasite. Releasing modified insects in selected regions to mate and spread the genetic modification throughout the population would reduce the pool of insects that could sustain the parasite, reducing the spread of the disease.

Transgenic plants are one of the more controversial subjects in modern genetics. Plants are being engineered for resistance to a variety of pests, agricultural chemicals (such as pesticides and herbicides), and environmental damage (such as frost). Some questions arise as to whether ingesting genetically modified plants is completely safe. Concern also exists that growing these plants in fields could lead to the uncontrolled spread of modified genes beyond the field site. To date, there is no data supporting these concerns.

13.1 Mapping Mutations in Eukaryotes

In chapter 12, you learned about the basic process of cloning genes by making genomic DNA and cDNA libraries. Libraries constructed from wild-type organisms contain the wild-type copies of all the genes. Similarly, libraries generated from mutants possess clones that correspond to particular mutations.

Trying to identify a specific gene within the library may not be straightforward, however. We discussed the method of hybridizing a probe at a reduced stringency to isolate a related sequence. But what if the related sequence is unknown? Alternatively, suppose we have an interesting mutant (or inherited human disease) and we want to isolate the corresponding mutation. In these cases, we need to localize the mutation to a chromosomal region. For this purpose, we can use DNA markers that can be mapped using the recombination techniques first discussed in chapter 6.

DNA Markers Used in Mapping

For most organisms, we do not have a sufficiently well-developed genetic map that contains a large number of evenly spaced mutations. DNA markers therefore provide the best alternative for recombination mapping.

Because of the large amount of variability in most DNA markers, the probability that an individual will be heterozygous at a specific DNA locus is relatively high. (Remember that recombination mapping depends on the individual being heterozygous.) This high probability of heterozygosity then increases the probability that any random cross can provide useful recombination data; designing specific genotypic crosses becomes unnecessary.

Four different types of DNA markers are described in this section: *RFLPs, VNTRs, microsatellites,* and *SNPs.*

RFLPs

A **restriction fragment length polymorphism (RFLP),** as you learned in the preceding chapter, is a nucleotide change that results in either the elimination or creation of a restriction enzyme site. An RFLP is characterized by either the division of one DNA restriction fragment into two smaller pieces, or the summation of two fragments into one.

An RFLP is detected by digestion of genomic DNA with a particular restriction enzyme, followed by gel electrophoresis and Southern hybridization (**fig. 13.1**).

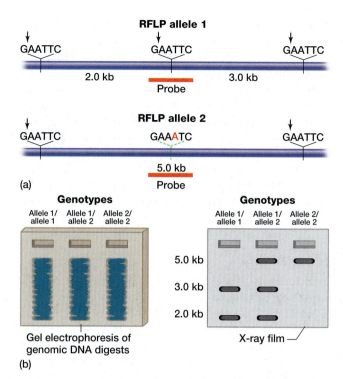

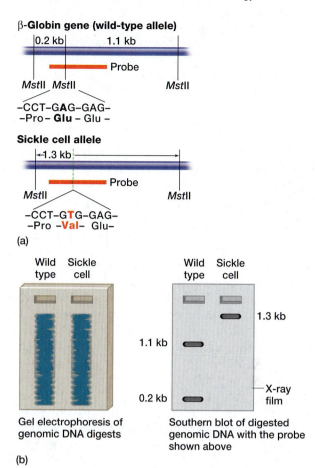

figure 13.1   Molecular characterization of an RFLP. (a) Two RFLP alleles are shown, with the top DNA molecule containing an internal *Eco*RI site (GAATTC) and the bottom molecule lacking the *Eco*RI site (GAAATC). The relative location of a probe used in a genomic Southern blot is shown. (b) A gel electrophoresis pattern of a genomic DNA digest from three genotypically different individuals (left) and the X-ray film after the genomic Southern blot (right). The fragment patterns for an individual who is homozygous for RFLP allele 1 (left lane), an individual who is heterozygous for the RFLP alleles (center lane), and an individual who is homozygous for RFLP allele 2 (right lane) are shown. Notice that all three genotypes can be distinguished from each other.

figure 13.2   Use of RFLP and Southern blot hybridization to distinguish between the wild-type and mutant β-globin allele, which causes sickle cell anemia. The location of the sickle cell mutation (A to T) and the resulting amino acid change in the β-globin protein (glutamic acid to valine) is shown in red. The genomic DNA digest patterns for a normal and sickle cell individual are shown (b, left). Southern blot is used to identify and distinguish the restriction digest fragments from the wild-type and sickle cell alleles (b, right).

A probe is used that will detect a single RFLP locus at a single site in the genome.

In some cases, the RFLP can be used to directly diagnose an inherited disease. For example, one allele of sickle cell anemia results from a DNA sequence change in the β-globin gene that eliminates an *Mst*II restriction enzyme site (**fig. 13.2**). This restriction enzyme recognizes the wild-type sequence 5'-CCTG**A**GG-3', which is absent in the sickle cell mutation due to the change of the A for a T (5'-CCTG**T**GG-3'). This single nucleotide difference results in a change of a glutamate in the wild-type β-globin amino acid sequence to a valine in the mutant protein.

Because the restriction site is eliminated, the *Mst*II-digested sickle cell genomic DNA produces one band during gel electrophoresis, whereas the wild-type DNA produces two (see fig. 13.2). Southern hybridization is required to identify either the wild-type or sickle cell β-globin gene fragments from the extremely large

number of DNA fragments that are produced from the genomic DNA digestion. The single band on the Southern blot is the molecular phenotype of the sickle cell DNA sequence detected by the probe, and the two bands on the blot represents the molecular phenotype of the wild-type DNA sequence. Thus, the molecular phenotype visible on a Southern blot can correspond to a specific clinical phenotype for an inherited disease.

Most RFLPs, however, are not directly associated with a visible phenotype. The RFLP can occur anywhere in the genome that a restriction enzyme site is lost or created, not only within the open reading frame (ORF) of a gene. We can use RFLPs, however, to reliably predict the probability of a particular disease allele through recombination mapping. One advantage of an RFLP for gene mapping is that RFLPs are scattered throughout the genome. The major disadvantage of using RFLPs is that only two alleles can be distinguished: one being the presence of the restriction site, and the other being the absence of the site (see fig. 13.1). The probability that any random individual is heterozygous for the RFLP is reduced with only two possible alleles.

VNTRs

A **variable number tandem repeat (VNTR),** or **minisatellite,** is a short sequence (10–100 bp) that is repeated in a tandem array (**fig. 13.3**). As we discussed in chapter 12, the number of repeats at a genome location can be anywhere from one or two up to several hundred. Thus, there can be hundreds of alleles at a VNTR locus. Minisatellites occur an average of once every 100 kb in the human genome.

With such a large number of phenotypically different alleles (as determined by Southern hybridization), any given individual is very likely to be heterozygous at each minisatellite locus. The high percentage of heterozygosity increases the likelihood that any mating will provide useful recombination mapping data.

A VNTR is usually detected by a restriction digest of genomic DNA. The restriction enzyme chosen will not cut within the repeated sequence; as a result, the size of the entire repeat and the flanking DNA can be detected by Southern hybridization. Because the probe detects the repeat, which is typically found at many different loci throughout the genome, the Southern hybridization pattern will be fairly complex (see fig. 13.3). It is still possible, however, to associate a particular VNTR band to the mutation of interest if they are linked and the pedigree is large.

Microsatellites

A **microsatellite** is also a short sequence that is repeated in a tandem array. However, unlike the VNTR, the microsatellite repeat is only 2–5 bp, and the total size of a microsatellite locus is usually 100–1,000 bp. The most common microsatellite repeat is $(CA)_n$ (**fig. 13.4**), which occurs an average of once every 30 kb in the human genome.

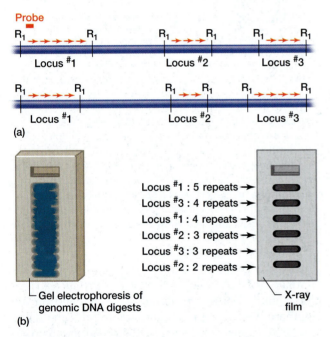

(a)

(b)

figure 13.3   Variable number tandem repeats and their analysis. (a) The number of VNTR repeats (represented as short arrows) at three different locations (Locus #1, Locus #2, and Locus #3) are shown for a diploid organism. (b) A VNTR analysis involves a restriction digest of genomic DNA with an enzyme that does not cut within the repeat locus (enzyme R_1). The digested DNA is gel electrophoresed (left) and analyzed by a Southern blot that uses the repeat sequence as the probe (right). The probe identifies restriction fragments that correspond to all the loci that contain the same repeat sequence. The size of each restriction fragment is roughly proportional to the number of tandem repeats and the distance from the repeat to the flanking restriction site.

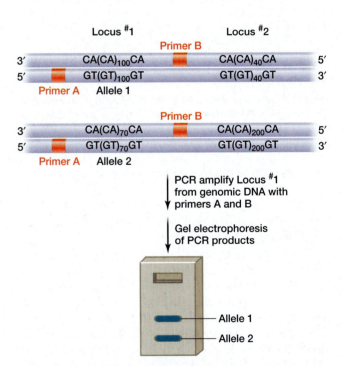

figure 13.4   PCR and gel electrophoretic analysis of microsatellites. The microsatellite repeat $(CA)_n$ in this example will be present at many locations in the genome, with two locations shown here (Locus #1 and Locus #2). The relative locations of two PCR primers (Primer A and Primer B) which are located outside the repeated sequence, are used to amplify only Locus #1. The PCR products are separated and sized by gel electrophoresis. Because the genome is diploid, two fragments should be generated by PCR amplification, one from each homologous chromosome.

Some microsatellites are directly associated with a class of inherited diseases called triplet repeat diseases. The **trinucleotide repeat diseases** result from an increased number of triplet nucleotides. If the repeat is within the ORF of a gene, it will cause a larger number of repeating amino acids in the translated protein. For example, Huntington disease, which is inherited as a dominant mutation, is characterized by involuntary movements and behavioral changes due to neurodegeneration. The symptoms first appear as early as 2 years of age and as late as 83. When the Huntington disease gene was cloned in 1993, it was found that the disease is due to an increase in the number of CAG repeats that encode the amino acid glutamine (**fig. 13.5**).

Unaffected individuals can have up to 34 of the CAG repeats in the gene, but individuals exhibiting the Huntington disease symptoms usually have more than 42 CAG repeats. More surprising, however, was that the age of onset corresponded with the number of repeats. Individuals who did not exhibit symptoms until later in life had between 50 and 75 repeats, but individuals who expressed the symptoms earlier had over 100 repeats. Thus, a Southern blot can be used to make clinical predictions, such as age of onset or severity associated with an inherited disease.

Microsatellites, like RFLPs, are only directly associated with a visible phenotype on a few occasions, such as Huntington disease and fragile-X syndrome. In the case of fragile-X, the microsatellite is located upstream of, rather than within, the ORF. The real power of microsatellites, however, lies in their use to map genes through recombination. Microsatellites provide the same advantages as VNTRs: they are scattered throughout the genome, a large number of alleles are present, and the probability that an individual is heterozygous is extremely high. Unlike VNTRs, microsatellites are usually examined by PCR amplification using primers that are complementary to sequences that flank the tandem repeats.

Although the repeat is located at a large number of different sites throughout the genome, the unique flanking primer sequences amplify the microsatellite at only a single site. Thus, PCR amplification usually produces a simple pattern that is straightforward to analyze (see fig. 13.4).

SNPs

One of the more powerful DNA sequences used in modern mapping is the **single-nucleotide polymorphism** (**SNP**, pronounced *snip*). A SNP is a single-nucleotide change in the DNA sequence, but it may not affect a restriction enzyme site (**fig. 13.6**). You can probably see that most RFLPs are SNPs, but most SNPs are not RFLPs. SNPs are more randomly and densely distributed throughout the genome. A SNP occurs approximately once every 1000 bp in the genome of most organisms. At the end of 2005, nearly 1.8 million SNPs were identified in the human genome. Although SNPs are very abundant, there are only four potential alleles at any given site. Thus, it is less likely that a random individual will be heterozygous for a specific SNP relative to the same individual being heterozygous at a specific VNTR or microsatellite locus.

SNPs can be identified by one of two methods. In the first method, the genomic DNA is digested with a

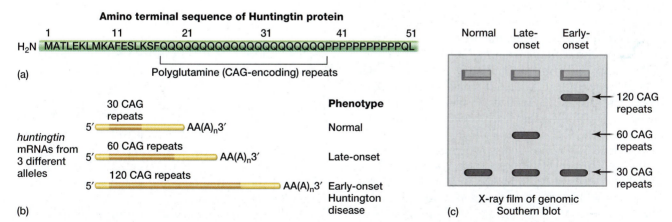

figure 13.5   Analysis of the triplet repeat associated with Huntington disease. (a) The first 51 amino acids of the Huntingtin protein encoded by the *huntingtin* gene. The polyglutamine region, which is encoded by the CAG repeats, is shown. (b) The number of CAG repeats correlates with the severity or onset of Huntington disease, with the greater number of repeats corresponding to an earlier onset of the symptoms. (c) Because Huntington disease is dominant, only one mutant allele is required to produce the clinical symptoms. A genomic Southern blot to determine the approximate number of CAG repeats shows individuals that express the Huntington symptoms are heterozygous, with one allele having 30 repeats and the second allele possessing 50 or more repeats.

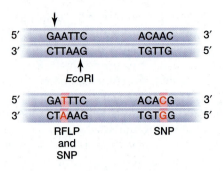

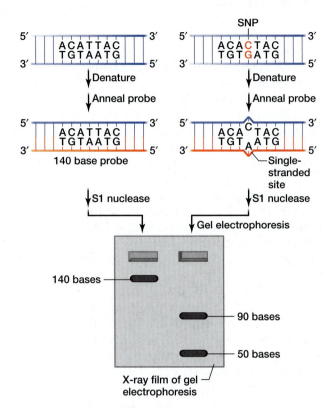

figure 13.6   The difference between an RFLP and a SNP. The RFLP (restriction fragment length polymorphism) is a mutation that creates or deletes a restriction enzyme site, while a SNP (single nucleotide polymorphism) is a single nucleotide change that may or may not affect a restriction enzyme site. The RFLP is a type of SNP (left sequence), while the rightmost SNP in this figure is not a RFLP because it fails to create or delete a restriction enzyme site. The vertical arrows show where the *Eco*RI restriction enzyme cuts.

figure 13.7   SNP characterization using S1 nuclease. Genomic DNA is denatured and annealed to a radioactively labeled probe, which is entirely complementary to only one of the SNP sequences. If the probe is completely complementary to the SNP (left), the probe lacks any single-stranded sequence and will not be cut by S1 nuclease, which leaves the probe (140 bases long) intact. If the probe is not completely complementary to the SNP, S1 nuclease cleaves the probe at the single-stranded site (A nucleotide). After gel electrophoresis, two fragments are observed (90 and 50 bases). The radioactive probe is detected by an X-ray film that is placed on the gel.

restriction enzyme or is randomly sheared to produce smaller DNA fragments (**fig. 13.7**). This DNA is denatured and annealed with a labeled single-stranded probe that corresponds to the desired SNP sequence. This mixture is treated with S1 nuclease, which cleaves single-stranded DNA. If the probe and the target are different at the SNP site, then both strands are cleaved at the site, and a shortened probe will be visible after electrophoresis. But if the probe and genomic DNA have the same sequence, the annealed double-stranded DNA is resistant to S1 nuclease, and the probe will remain unchanged in size.

The alternative method involves hybridization of the genomic DNA to oligonucleotides that correspond to the different SNP variant sequences. Genomic DNA is isolated from a small amount of tissue from an individual, and the region containing the SNP of interest is PCR-amplified (**fig. 13.8**). A third internal primer is used to synthesize a labeled single-stranded DNA probe. This probe is hybridized to four different short oligonucleotides (18–23 nt), each containing a different variation of the SNP sequence, which are attached to a solid support like nitrocellulose on a microscope slide.

By adjusting the hybridization temperature and salt conditions, the probe will hybridize only to the oligonucleotide with the perfect complementary sequence. The four oligonucleotides are arranged on the slide or membrane in a specific pattern (see fig. 13.8), and the pattern of hybridization will reveal the oligonucleotide sequence and, in turn, the probe sequence. Individuals who are homozygous for the SNP will exhibit hybridization to a single oligonucleotide, whereas heterozygous individuals will show hybridization to two different oligonucleotides. This technology can be miniaturized,

so that hundreds to thousands of SNPs can ultimately be examined on a single microscope slide!

Recombination Mapping of DNA Markers

We can now examine an example of mapping using microsatellites. We could use any of the previously mentioned DNA markers in this mapping discussion, as long as the individual being tested is heterozygous for both loci being mapped.

Let's assume that we are trying to map a rare dominant human disease, and we identified a family that exhibits the disease (**fig. 13.9**). We further assume that the disease is 100% penetrant, which suggests that individual I-1 is heterozygous (because she has unaffected children). Based on this assumption, we can assign a genotype to every individual in the pedigree (see fig. 13.9).

figure 13.8 SNP characterization using hybridization to oligos. The relative location of the PCR primers (green) used to amplify a specific SNP from genomic DNA are shown. Following amplification, a second primer (red) that is internal to the amplification primers, is used to synthesize and radioactively label only one DNA strand. This excess single-stranded DNA is hybridized to spots containing all four possible sequences for a specific SNP in a designated pattern. The hybridization pattern on the X-ray film reveals the complementary sequence to the probe at the SNP site. Using robotics and microarray technology (see chapter 14), thousands of SNPs can be placed on a single microscope slide, which allows the simultaneous analysis of thousands of SNPs throughout the genome.

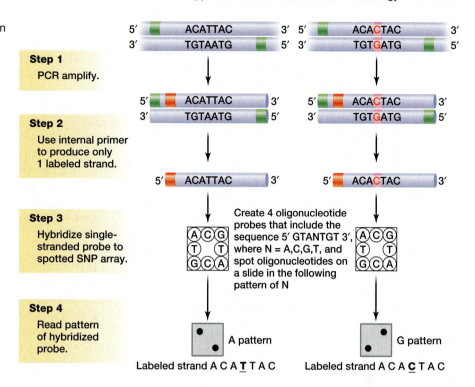

figure 13.9 Recombination mapping with a microsatellite. A pedigree with individuals exhibiting a dominant inherited trait is shown, with the analysis of two different microsatellites from genomic DNA shown below each individual. Microsatellite 1 fails to exhibit any linkage between the A or D fragments (both which are present in the affected mother) and the dominant trait in the offspring. Microsatellite 2 reveals that fragment A is observed in the affected mother and three of her four affected children. Further, all four of the unaffected children inherited fragment C, rather than fragment A, from their mother. The seventh child is the exception, where she has the dominant trait, but inherited fragment C (asterisk). This could be due to the recombination event that is shown at the bottom of the figure.

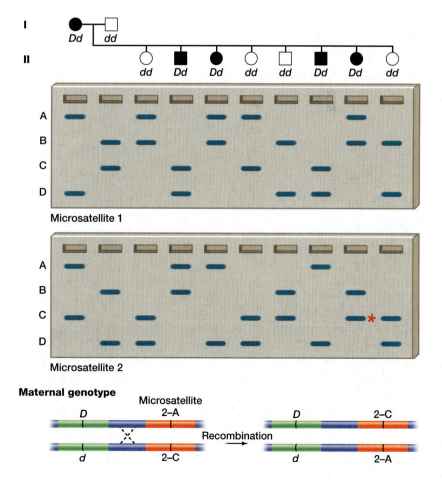

Analysis of PCR Data

Blood samples were taken from each individual in this pedigree, and two different microsatellites were analyzed by PCR amplification. A cursory examination of the PCR products reveals that for both loci, each child inherited one allele from each parent. We can look for a correspondence between the presence of the disease and a particular PCR amplification product.

Microsatellite #1 exhibits independent assortment, with half of the affected children receiving PCR product A from their mother, and the other half receiving product D (see fig. 13.9). An equal division of these two PCR products also occurred in the unaffected children. Therefore we cannot detect any linkage between a specific microsatellite allele inherited from the mother and the allele from the D/d gene.

A completely different situation is apparent with microsatellite #2. All of the unaffected children inherited PCR product C from their mother, and three of the four affected children inherited PCR product A from their mother (see fig. 13.9). The last affected child (II-7) however, possesses PCR product C. This result suggests that seven of the eight children inherited one of the two parental chromosomes from their mother, and child II-7 must have inherited a recombinant chromosome (see fig. 13.9).

We can use the production of recombinant offspring to determine the recombination distance between the two loci. As defined in chapter 6, the recombination frequency is the frequency of recombinant individuals among all the offspring, or $1/8 \times 100\% = 12.5\%$. This corresponds to 12.5 mu between microsatellite #2 and the dominant disease allele. But this family provides a small sample size, and the recombination can be greatly skewed by the presence or absence of a single extra recombinant child. Therefore, we need to employ a statistical analysis to this data. The chi square analysis described in chapter 2, however, requires a fairly large number of progeny.

Lod Score Analysis

In 1955, Newton E. Morton devised a different type of statistical analysis to deal with these smaller data sets in human pedigrees. The **logarithm of odds,** or **lod score** (pronounced as *load*) compares the probability that a particular set of births came from linked loci with a specific recombination frequency, with the probability that the same set of births came from unlinked loci. The resulting ratio is then expressed as a logarithm:

$$\text{lod score} = \log\left[\frac{\text{(probability of linkage at } x \text{ mu)}}{\text{(probability of no linkage)}}\right]$$

where x is the specified recombination distance in map units.

A positive lod score indicates some degree of linkage, and a negative lod score indicates independent assortment. By convention, a lod score of 3.0 or greater is considered to be very strong evidence for the possibility of linkage because the data are 1000 times more likely for a specific linkage, relative to no linkage (because the lod score is a logarithm).

In figure 13.9, if the D locus and microsatellite #2 are unlinked, then the genotype of each individual in the second generation occurs at a probability of 1/4 due to independent assortment (chapter 2). The probability of the specific outcomes on all eight individuals is therefore:

$$(1/4)^8 = 0.0000153$$

In contrast, if the two loci are linked, then individual II-7 represents a recombinant, and the other seven siblings represent nonrecombinants. For purposes of this calculation, assume that the two loci are 10 mu apart. (In actual practice, you will see that calculations are conducted for a number of different linkage values.) At 10 mu, recombinant gametes would occur at a frequency of 10%, and nonrecombinant gametes would occur at a frequency of 90%. Each of the two recombinant gametes would therefore occur at a frequency of 5%, and each of the two nonrecombinant gametes at 45%. The probability of seven nonrecombinant gametes and one recombinant gamete is:

$$(0.45)^7(0.05)^1 = 0.000187$$

The next step in calculating a lod score is to obtain the ratio of the probability of the 10 mu linkage to the probability of no linkage. This value is given by:

$$\frac{0.000187}{0.0000153} = 12.22$$

The logarithm of 12.22 is 1.087, which indicates some degree of linkage because the value is positive.

Table 13.1 shows the lod scores for a number of different linkage values for this pedigree. The highest lod score (1.09) is associated with a recombination frequency of either 10% or 15%. This result suggests that the best recombination frequency lies between 10% and 15%. Further calculations could be done to find which number between 10% and 15% produces the highest lod score.

table 13.1 Lod Score Calculations Using Different Recombination Frequencies

	Recombination Frequency			
	0.05	0.10	0.15	0.20
Odds (linked/unlinked)	8.94	12.24	12.31	10.7
Lod score	0.95	1.09	1.09	1.03

Linkage likely exists between the two loci, even though the lod scores do not meet the criterion of 3.0 for very strong linkage.

One of the powers of lod scores is that they are additive. A geneticist can identify and determine the lod scores of several different family pedigrees and then sum the individual lod scores to reach a composite lod score. If this composite lod score is 3.0 or more, the researcher can conclude that the two loci are linked.

Combining Lod Scores

Let's continue the example that we already started on the pedigree in figure 13.9. The calculated lod scores for this pedigree based on recombination frequencies from 5% to 45% are in **table 13.2**.

Assume that a second pedigree is identified that consists of six children: four who exhibit the nonrecombinant allele arrangements and two with the recombinant arrangement. The calculated lod scores for this pedigree are shown in table 13.2. The lod scores for a third pedigree consisting of four children, one of whom is a recombinant, are also included in table 13.2.

The sums of the lod scores from the three different pedigrees at each of the recombination frequencies are shown along the bottom in table 13.2. Based on the greatest lod score, we assume that the distance between the *D* locus and microsatellite #2 is between 20 and 25 mu.

As we identify additional pedigrees, we can continue to calculate the lod scores and add them to the current scores. When one of the lod score sums reaches or exceeds 3.0, the researcher can conclude the two loci are linked at the given recombination distance.

It's Your Turn

Mapping disease genes in pedigrees using DNA markers is a standard method before the gene can be cloned. Do not be nervous about using a molecular phenotype (such as a PCR band on a gel) compared to a morphological phenotype (such as hair color). To become comfortable with this type of analysis, you should try problems 25 and 32.

Restating The Concepts

▶ A variety of DNA markers exist, such as RFLPS, VNTRs, microsatellites, and SNPs. All of these markers can be used for recombination mapping. The latter three markers are especially useful because they are either scattered at a high density throughout the genome or exist as a large number of different alleles.

▶ DNA markers are analyzed by different methods that include restriction digestion, Southern hybridization, PCR amplification, S1 nuclease digestion, and nucleic acid hybridization. These techniques reveal a molecular phenotype that can be used during recombination mapping.

▶ Because of the small number of individuals in many human pedigrees, the lod score analysis was devised to statistically evaluate the potential linkage between two genes or DNA markers. One benefit of the lod score is that because it is a logarithm of the ratio between the probability of linkage and the probability of independent assortment, lod scores from different pedigrees can be added to produce a composite score.

▶ Linkage is thought to be demonstrated when the total lod score from one or more pedigrees reaches 3.0.

13.2 Cloning Eukaryotic Genes

Genes that correspond to a particular mutant phenotype can be localized to a chromosomal region by either recombination mapping or by characterizing chromosomal rearrangements (deletions, translocations, and insertions; see chapter 8) that are associated with the mutant phenotype. When the desired mutation is defined by a region of genomic DNA, we can proceed with identifying the mutation and corresponding gene. **Positional gene cloning** involves identifying a gene based on its location in the genome. However, several genes may occur in the identified location. Thus, additional approaches must be used to identify the correct gene.

table 13.2   Addition of Lod Using Multiple Pedigrees

	Recombination Frequency								
	0.05	**0.10**	**0.15**	**0.20**	**0.25**	**0.30**	**0.35**	**0.40**	**0.45**
Lod score for pedigree 1	0.95	1.09	1.09	1.03	0.93	0.80	0.64	0.46	0.24
Lod score for pedigree 2	−0.88	−0.37	−0.12	0.03	0.11	0.15	0.15	0.13	0.08
Lod score for pedigree 3	−0.16	0.07	0.17	0.21	0.23	0.22	0.19	0.14	0.08
Total lod score	−0.09	0.79	1.14	1.27	1.27	1.17	0.98	0.73	0.40

Chromosome Walking

Although recombination mapping can reveal that a mutation in a gene of interest is located near or between DNA markers, the gene may still be hundreds or thousands of kilobase pairs from the nearest marker. The intervening region could be so large that the DNA marker and the target gene may not reside on the same DNA fragment isolated from a genomic library (see chapter 12). In such a case, a series of overlapping genomic clones must be isolated that extends from the DNA marker to the target gene. **Chromosome walking** is the technique by which one genomic clone is used to isolate overlapping clones that extend from one or both ends of the original clone.

Assume that we are interested in cloning a gene that we already mapped to a location between two DNA markers. Because the VNTR and microsatellite repeats are located throughout the genome, these repeated sequences cannot be used as a probe. However, the unique sequences flanking the VNTR or microsatellite repeats, the RFLP probe, or the SNP probe can all be used to probe a genomic DNA library (**fig. 13.10**).

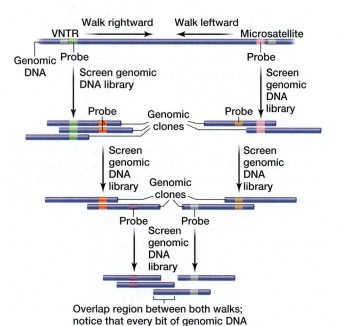

Overlap region between both walks; notice that every bit of genomic DNA between the two starting points is present on at least one of the genomic clones.

figure 13.10   Chromosome walking. Two DNA sequences are used as probes to independently screen a genomic DNA library. The isolated genomic DNA clones are compared by restriction fragment mapping (see chapter 12) or DNA sequencing. A region that extends further towards the other start site is used as a probe to screen the genomic DNA library again. This process is continued until the genomic clones that were isolated from both ends (walks) overlap, which signifies that the two walks have reached a point that covers all the genomic DNA between the original two start points.

For example, the unique DNA sequence flanking a VNTR can be used as a probe to screen a genomic DNA library. The positive clones from the screen can be isolated, and the genomic DNA insert can be isolated from each positive clone. A restriction map can be generated for each genomic DNA clone isolated as described in chapter 12. The restriction maps for the different clones should overlap because they all contain a sequence that is complementary to the probe and differ at their ends due to the partial digestion used to generate the library.

Comparing the different restriction maps will reveal the two clones that extend the furthest in both directions from the original probe. DNA fragments that correspond to these two ends are then used to screen the genomic DNA library again and isolate a new group of overlapping clones (see fig. 13.10). The restriction maps for the new clones can be generated, and a new probe can be identified for another round of library screening.

If the mutation maps to a region between two DNA markers, this genomic walking technique, which is similar to "walking" from an original point in the genome using DNA clones as "steps," can be performed from both DNA markers. This process of library screening, restriction mapping, and identification of the ends continues until the walks from both DNA markers overlap (see fig. 13.10). The end result is that the gene of interest that lies between the two DNA markers must be located within the genomic DNA clones that were isolated in the walk.

The recent genome projects in humans and a large number of different organisms have lessened the need to perform a genomic walk. As described further in chapter 14, these genome projects produce overlapping genomic DNA clones that serve as one source of the DNA that is sequenced. Additionally, the genomic sequence essentially produces the high-resolution detail necessary for identifying the locations of both the flanking DNA markers and the genes of interest. Genomic walking, however, remains a powerful method for organisms whose genomes have not been entirely sequenced.

It's Your Turn

The ability to organize clones with overlapping regions of identity is a fundamental tool that is used for a variety of different reasons in molecular genetics. You should be prepared to try this skill in problem 34.

Identifying Candidate Genes

The region defined by the molecular markers to contain the gene of interest may contain several genes. The smaller the genomic region that can be defined, the fewer

13.1 applying genetic principles to problems

Performing a Genomic Walk

We describe in this chapter the need to "walk" between two different DNA sequences to identify and close an intervening sequence containing a desired mutation. In chapter 14 you will learn how to generate a **contig,** which is a series of contiguous DNA clones or sequences. We may also have a genomic DNA clone and we may need to isolate only the adjacent genomic clone. In all of these cases, we start with a specific DNA sequence and we need to identify additional clones that extend in one or both directions.

Let's assume that you isolate a cDNA sequence and you wish to produce a genomic walk in both directions. Using this cDNA as a probe, you screen a genomic DNA library, and you isolate three different genomic clones. Here is how you proceed with a genomic walk:

1. Determine the restriction map for the genomic clones.
2. Extend the walk in both directions, using the two distal ends as probes.
3. Rescreen the library and extend the walk until the sequence of interest has been solved.

You can see how to work through each of these steps in the following discussion.

STEP 1: Determine the restriction map for the genomic clones.

The method you use to determine the restriction map for these three clones is slightly different from the method described in the application problem in chapter 12. In this example, you already know that all three genomic clones must contain overlapping sequences because they all hybridized to the original cDNA probe. Additionally, you are more interested in the ends of the genomic clones rather than the common regions between the clones because these are the starting points for extension.

Assume that the three clones contain the following *Bam*HI restriction fragments:

> Clone #1: 6.0 kb, 5.0 kb, 4.0 kb
> Clone #2: 6.0 kb, 4.0 kb, 3.0 kb
> Clone #3: 5.5 kb, 5.0 kb, 4.0 kb

Notice that all three genomic clones contain a common 4.0-kb *Bam*HI fragment. You can assume that this DNA fragment is the same in all three clones. Notice also that clone #1 and clone #2 share a common 6.0-kb *Bam*HI fragment and that clone #1 and clone #3 share a common 5.0-kb *Bam*HI fragment. You can then arrange

the three genomic clones in the following manner, based on the shared fragments:

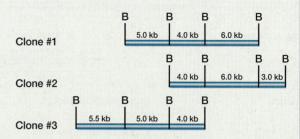

Southern blotting can confirm the identity of the shared fragments. You would digest the clones with *Bam*HI, electrophorese the digests, and probe with one of the three *Bam*HI fragments from clone #1. For example, if the 5.0-kb *Bam*HI fragment from clone #1 was used as a probe, it would detect the 5.0-kb fragment from both clone #1 and clone #3. This analysis should reveal that the 5.5-kb and 3.0-kb *Bam*HI fragments are the current ends of the walk because no other *Bam*HI fragment could be used as a probe to detect them.

STEP 2: Extend the walk in both directions.

In some cases we may know that we want to extend the walk in a particular direction, such as toward the telomere. In this case, however, you are currently unsure whether the 5.5-kb fragment or the 3.0-kb fragment is closer to the telomere. Therefore, you must take another step in the genomic walk in both directions. You would use the 5.5-kb fragment to screen the genomic DNA library again. When you do, let's say you isolate two clones that have the following *Bam*HI restriction enzyme fragments:

> Clone #3.1: 5.5 kb, 5.0 kb, 4.0 kb
> Clone #3.2: 6.0 kb, 5.5 kb, 4.0 kb

Notice that both these genomic clones contain a common 5.5-kb *Bam*HI fragment, which also corresponds to the size of your probe. This fragment is most likely the same in the original and in the two new clones.

Also notice that even though clone #3.2 contains a 6.0-kb fragment and 4.0-kb fragment that were in the original clones, it lacks the 5.0-kb fragment that lies between the 5.5-kb and 4.0-kb fragments. From this you can deduce that the 6.0-kb and 4.0-kb *Bam*HI fragments are new fragments that extend farther left.

continued

Finally, you can see that clone #3.1 contains a 5.0-kb fragment that is absent in clone #3.2, which means that it cannot be to the left. Rather, it must correspond to the 5.0-kb fragment that is to the right of the 5.5-kb fragment in the original clones. You now have a dilemma. This 4.0-kb fragment can correspond to the 4.0-kb fragment in the original clones, or it could be located immediately to the left of the 5.5-kb fragment. The three possibilities are shown here:

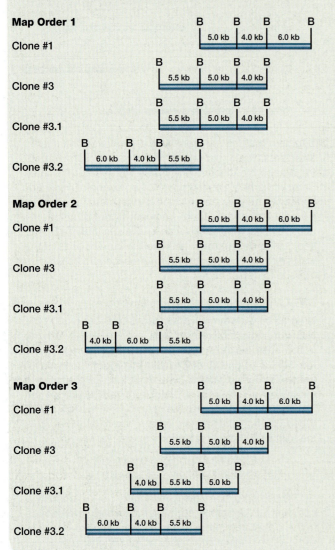

In any of these three cases, you can see that either the 6.0-kb fragment or the 4.0-kb fragment from clone #3.2 extends farthest to the left. You could screen a genomic library with both of these *Bam*HI fragments to take the next step in the walk. But, if you want to conclusively determine which fragment is farthest to the left, what would be the simplest approach?

One solution would be to perform a *partial digest* of clone #3.2 with *Bam*HI. If the 4.0-kb fragment is adjacent to the 5.5-kb fragment (map orders #1 and #3), you would see a 9.5-kb partial fragment. If the 6.0-kb fragment is adjacent to the 5.5-kb fragment (map order #2), you would see an 11.5-kb partial fragment. In the first case, the 11.5-kb partial *Bam*HI fragment would not be generated, and in the second case, the 9.5-kb partial *Bam*HI fragment would not be observed.

To distinguish between map orders #1 and #3, you would follow a similar procedure as detailed earlier. You would digest clones #3, #3.1, and #3.2 with *Bam*HI, electrophorese the fragments, and perform Southern blotting using the 4.0-kb *Bam*HI fragment from clone #3.1 as the probe. If map #1 is correct, then the probe will hybridize to the 4.0-kb *Bam*HI fragment in clone #3, but not clone #3.2. If map #3 is correct, then the probe will hybridize to the 4.0-kb *Bam*HI fragment from clone #3.2, but not clone #3.

You would also use clone #2 to extend the original walk to the right. The 3.0-kb *Bam*HI fragment would be used as the probe to screen the genomic library. The resulting clones would be mapped as described earlier. You are only interested in extending the walk with clone #2 to the right, because clone #3 is already extending the walk to the left.

STEP 3: *Continue extending the walk in both directions.*
You would then continue to extend the walk in both directions as described in step 2. You continue the procedure until you reach a point that signifies the completion of the walk. This point could be reaching another gene or DNA marker sequence, or a region that overlaps a walk coming from the other direction. Although this walk will contain the desired sequences, the identification of the gene or sequence within the walk sometimes requires more time and imagination.

genes will be present. However, the absence of informative DNA markers (due to the absence of heterozygosity) and limited numbers of pedigrees and recombinant individuals may result in the defined genomic region being relatively large and containing many genes.

The identification of candidate genes in a genomic region usually involves determining which genes exhibit an expression pattern that matches either the gene of interest or the disease symptoms. This can be performed using either Northern hybridization or a Western blot. If the disease affects a particular tissue, then the candidate gene's mRNA and protein should also be expressed in that tissue in normal individuals.

Expression Patterns to Identify Candidate Genes: Cystic Fibrosis

For example, cystic fibrosis is an autosomal recessive disease that causes the secretion of a viscous solution from the lungs, pancreas, and sweat glands. To clone the cystic fibrosis gene, researchers first mapped the gene between the *met* and *J3.11* markers on chromosome 7 and then to a small region between two DNA markers (*XV-2c* and *KM-19*) that were approximately 500 kb apart (**fig. 13.11**). Within this region were four genes, one of which was likely to be the cystic fibrosis gene.

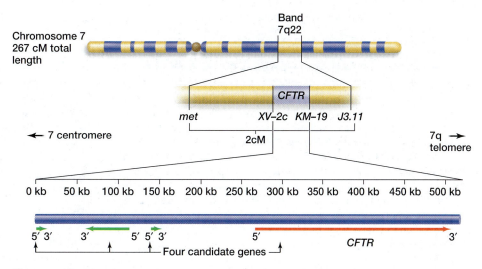

figure 13.11　The cystic fibrosis gene was first mapped between *met* and *J3.11* on chromosome 7, and later shown to be located between the DNA markers *XV-2c* and *KM-19* by recombination mapping. This region contains four candidate genes, which are marked by the three green arrows and the red arrow. The cystic fibrosis gene (*CFTR*) was identified because it was the only one of the four candidate genes that was expressed in the same tissues that exhibit the clinical symptoms: lungs, pancreas, and sweat glands.

To identify the cystic fibrosis candidate gene, the researchers determined which of the four genes were expressed in lungs, pancreas, and sweat glands. Using Northern blots that contained mRNA isolated from these and several other tissues (see chapter 12), the researchers applied probes that corresponded to each of the four different genes. Only one of the four genes was expressed in all three of the expected tissues. This became the major candidate for the cystic fibrosis gene.

This candidate gene was found to encode a chloride transporter that was named the cystic fibrosis transmembrane conductance regulator (CFTR). In normal cells, this CFTR channel is open to release chloride ions to the external environment, and water also leaves the cell by osmosis because of the ionic change. The result is moist, thin mucus that aids in protecting cells from pathogens. The nonfunctional version of this gene encodes a defective channel that yields a thick mucus that can clog the airways and the ducts in glands.

To confirm the identity of the cystic fibrosis gene, this candidate gene was cloned from both unaffected and affected individuals in several different pedigrees. In all cases that were tested, the affected individuals had mutations in both copies of the candidate gene. The unaffected individuals exhibited two patterns of sequences. Some possessed only the wild-type gene sequence. Others, because cystic fibrosis is a recessive disease, possessed both a wild-type and mutant copy of the candidate gene.

It is worth noting that the affected individuals could also possess the mutation in either the promoter, which would affect the transcription of the gene, or in an intron, which could affect the splicing of the final mRNA. In either case, however, a Northern blot would reveal an abnormality in the mRNA isolated from the affected individual. With mutations in the promoter, either greater or reduced levels of the mRNA would be produced relative to the wild-type, and mutations in the intron would produce mRNAs with an altered size relative to wild type.

Candidate genes can also be identified using the genomic DNA sequence information from several different organisms. As an example, assume that you are trying to identify the candidate gene for a human neurological disease. Recombination mapping has localized the disease gene to a small genomic region that consists of five genes. The Human Genome Project already has determined the wild-type sequence of all five genes. Of the five genes, one is related to a *Drosophila* gene that displays a neurological phenotype when mutated. This *Drosophila* mutant phenotype would strongly support the related human gene as the candidate for the neurological disease. Additional pieces of data, such as the DNA sequence of normal and affected individuals, would still be required to confirm the gene's identity.

Gene Cloning Using a Haplotype Map

The identification of nearly 1.8 million SNPs across the entire human genome provides a potentially powerful tool for recombination mapping. Recent analyses of these SNPs, however, revealed an unexpected level of organization. Clusters of SNPs, in regions up to 100- to 500-kb long, failed to exhibit recombination within a population (**fig. 13.12**). The absence of recombination within these regions suggests that identifying only a few SNPs in a given region will be sufficient to determine the remainder of SNPs in that region. These regions are flanked by very short regions (1–10 kb) that

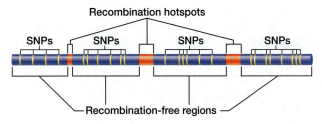

figure 13.12   Distribution of SNPs in the human HapMap. The human HapMap is built on SNPs (yellow lines) distributed approximately every 1000 base pairs throughout the genome. Analysis of the SNPs revealed regions that exhibit no recombination within one of the four test populations, flanked by short regions of high recombination frequency (red segments). This suggests that identifying only a few SNPs in each recombination free region will be sufficient to predict the remaining SNP alleles in the same region.

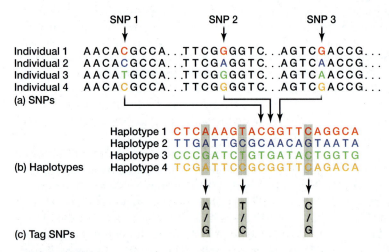

figure 13.13   SNP genotypes in the human HapMap. (a) Three different sequences from the same chromosomal region show the three SNPs in four different individuals. (b) The haplotype is the sum of all the SNPs for a given individual. (c) The presence of the recombination free regions within the genome permits the identification of the haplotype for a specific region after determining only a small number of SNPs in a region (tag SNPs, shown in shaded boxes). For example, an individual who has the tag SNPs "ACG" must possess haplotype 2.

Reprinted with permission from Macmillan Publishers Ltd: *Nature*, vol. 426, December 2003, copyright © 2003.

exhibit very high recombination frequencies. It is possible to determine the **haplotype** (*haplo*id geno*type*), the specific set of SNP alleles on a chromosome, for a given individual (**fig. 13.13**).

The human HapMap (haplotype map) project involves analyzing one SNP for every 5 kb across the human genome. Trios of individuals (father, mother, and child) were analyzed from four geographic populations: 30 trios of Yoruba people from Nigeria, 30 trios of Mormons in Utah, 15 trios of the Han population in Beijing, and 15 trios from Tokyo. In the near future, additional populations from Africa, Asia, Europe, and the Americas will be analyzed.

The haplotype of every individual in the initial phase of the human HapMap project has been determined. Comparing these haplotypes allowed researchers to identify *tag SNPs*, which are SNP alleles unique for a specific haplotype (see fig. 13.13). Determining the tag SNPs, called *genotyping*, for another individual reveals their haplotype. In figure 13.13, the tag SNPs "ATC" corresponds to haplotype 1, and "GTC" corresponds to haplotype 3. Thus, the haplotype can be determined by identifying only a subset of specific SNPs in an individual.

The development of the human HapMap will simplify the localization of human disease genes, particularly complex diseases. A complex disease, such as heart disease or manic depression, results from many genes, each contributing a part of the risk associated with displaying the disease. Researchers have already used the HapMap to identify new genes that contribute to heart disease, manic depression, type 2 diabetes, and prostate cancer.

Rather than looking for recombination between the disease genes and SNPs, the investigators look for associations when using the HapMap. Because of the regions that are largely devoid of recombination and the high density of SNPs throughout the human genome, it is highly likely that individuals who are related in a pedigree, or are culturally or ethnically related, have more similar haplotypes. Particular tag SNPs that are closely associated with a specific disease gene should be found in individuals that have the disease and are less likely to be found in individuals free of the disease. Tag SNPs that are not associated with the gene should show a random distribution between the affected and unaffected individuals.

An example of this association mapping is shown in **figure 13.14**. The tag SNPs for four individuals with heart disease and four individuals without heart disease are shown. If you examine each tag SNP, you will see that the two alleles are equally distributed between both groups. However, the tag SNP in red is not randomly distributed. All four individuals with heart disease have the C allele, and all four unaffected individuals have the T allele. This suggests that a gene that contributes to heart disease is associated with this tag SNP, and that the presence of the C allele suggests the presence of the disease allele.

Individuals with heart disease

Tag SNPs	
Individual 1	A C G A A G T C A A T G C G T C C T
Individual 2	T A T G C A G C C T G C C G C A G T
Individual 3	T C T A C A T C A T G G A T T C G A
Individual 4	A A G G A G G C C A T C A T C A C A

Individuals without heart disease

Tag SNPs	
Individual 5	T C G G A G T T A T G C A T C C C A
Individual 6	A A T A C G T T A A G G C G C A C T
Individual 7	T A G G C A G T C A T C C T T C G A
Individual 8	A C T A A A G T C T T G A G T A G T

figure 13.14   Use of SNPs in association mapping. The haplotypes of eight individuals, four with heart disease and four without, are shown. Of the four individuals with heart disease, only one SNP allele (C in red) is shared between them all. The four individuals without heart disease all show a different allele (T in red) for this SNP. This analysis shows that the C allele at this SNP is associated with a gene that contributes to heart disease.

Restating the Concepts

▶ Chromosome walking involves isolating a series of overlapping genomic clones between two different DNA markers. This process involves screening a genomic DNA library, producing a restriction map of the genomic DNA insert, isolating the unique end fragments, and beginning the library screen again.

▶ Candidate genes in a genomic region or walk are usually identified by a combination of expression patterns, using Northern and immunoblots and relatedness to genes previously identified and characterized in other organisms.

▶ The human HapMap examines the haplotypes in a population. This information revealed that recombination frequency is not constant across the genome and can be used to determine the location of complex disease genes by association with SNPs.

13.3 Eukaryotic Vectors

All of the vectors discussed in the preceding chapter were used to introduce foreign DNA into *E. coli*. For many reasons, geneticists would also want to introduce DNA into eukaryotic cells. First, *E. coli* is not capable of fully expressing some eukaryotic genes because it lacks the enzyme systems necessary for many posttranscriptional and posttranslational modifications, such as intron removal and glycosylation. Second, organization and expression of the eukaryotic genome may require study in vivo (in the living system), something that can only be accomplished

by working directly with eukaryotic cells. Finally, researchers may want to manipulate the genomes of eukaryotes for medical as well as economic reasons.

Several successful vectors have been developed for introducing genes into eukaryotes; we describe some of them in the following sections.

Yeast Vectors

Baker's, or brewer's, yeast, *Saccharomyces cerevisiae*, has a naturally occurring plasmid called the **2-micron (2-μm) plasmid,** named for its circumference. The yeast 2-μm plasmid contains an origin of replication that allows it to replicate autonomously in a yeast cell.

The **yeast episomal plasmid (YEp)** contains two origins of replication: an *E. coli* plasmid origin, and the yeast 2-μm origin (**fig. 13.15b**). This plasmid can therefore replicate in either yeast or *E. coli*. This is an example of a *shuttle vector,* which can replicate and be maintained in two different organisms. The advantage is that the plasmid and insert can be manipulated easily in *E. coli* and then transferred to yeast to examine the function or expression of the insert in a eukaryotic organism. The YEp also contains a selectable marker for *E. coli* (usually an antibiotic resistance gene) and a different selectable marker for yeast (usually a gene required for the biosynthesis of an amino acid or nucleotide base). These markers allow either *E. coli* or yeast containing the YEp to be easily selected.

The **yeast integrative plasmid (YIp)** is similar to a YEp, except that it lacks the yeast 2-μm origin (see fig. 13.15a). Thus, the YIp cannot autonomously replicate in yeast cells. The only way that the cloned insert can be stably expressed in yeast cells is if the plasmid or the insert recombines (integrates) into the host yeast genome.

The differences between YEp and YIp are shown in **figure 13.16**. *S. cerevisiae* is a haploid organism; introduction of a YEp into this yeast makes it diploid for only the genes that are cloned into the plasmid. This partial diploid (**merodiploid**) enables investigators to examine dominant–recessive relationships. Usually, the wild-type allele will be dominant to the mutant allele. The YEp can also be used to introduce a wild-type gene into a mutant yeast strain. If the resulting phenotype is wild-type, then the introduced wild-type gene must compensate for the mutant allele. This **rescue,** or restoration of the wild-type phenotype, is a common method to confirm the identity of the gene causing the mutant phenotype.

Because the YIp lacks a yeast origin of replication, it cannot be stably maintained as a plasmid in yeast. This requires that any gene cloned in the YIp plasmid must recombine with the genome to be stably expressed. This recombination can either result in the incorporation of the entire plasmid into the genome, or

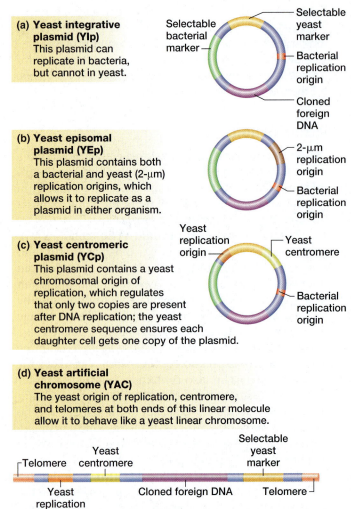

(a) Yeast integrative plasmid (YIp)
This plasmid can replicate in bacteria, but cannot in yeast.

Selectable bacterial marker — Selectable yeast marker — Bacterial replication origin — Cloned foreign DNA

(b) Yeast episomal plasmid (YEp)
This plasmid contains both a bacterial and yeast (2-μm) replication origins, which allows it to replicate as a plasmid in either organism.

2-μm replication origin — Bacterial replication origin

(c) Yeast centromeric plasmid (YCp)
This plasmid contains a yeast chromosomal origin of replication, which regulates that only two copies are present after DNA replication; the yeast centromere sequence ensures each daughter cell gets one copy of the plasmid.

Yeast replication origin — Yeast centromere — Bacterial replication origin

(d) Yeast artificial chromosome (YAC)
The yeast origin of replication, centromere, and telomeres at both ends of this linear molecule allow it to behave like a yeast linear chromosome.

Telomere — Yeast centromere — Selectable yeast marker
Yeast replication origin — Cloned foreign DNA — Telomere

figure 13.15 Different types of yeast vectors. The yeast integrative plasmid (a) lacks a yeast origin of replication, which prevents it from replicating in yeast. Thus, the yeast selectable marker is only stably expressed if the plasmid recombines into the yeast genome. The yeast episomal plasmid (b) contains the yeast 2 micron plasmid origin of replication, which allows it to replicate as a free plasmid vector in a yeast cell. The yeast centromere plasmid (c) contains a yeast chromosomal origin of replication and centromeric sequence. The origin of replication controls only two copies of the plasmid in a cell after DNA replication, while the centromere ensures that each daughter cell receives a copy of this plasmid. The linear yeast artificial chromosome (d) contains a yeast origin of replication, centromere, and telomeres at both ends so that it behaves like a yeast linear chromosome.

the replacement of the genomic copy of the gene with the cloned plasmid gene (see fig. 13.16b). Although this incorporation can also be used to confirm the identity of a mutation, another, more powerful use is possible.

We can disrupt a wild-type yeast gene that is cloned into a YIp by inserting a selectable gene inside it (**fig. 13.17**). This insertion is analogous to the cloning of genes into the *lacZ* gene present in many *E. coli* plas-

mids with the resulting inactivation of β-galactosidase activity (described in chapter 12). The resulting inactivated yeast gene then recombines from the YIp into the yeast genome, which replaces the wild-type genomic copy with the inactivated gene copy (see fig. 13.17). This approach allows us to rapidly produce an inactive allele of any yeast gene that we desire and determine the phenotype.

The **yeast centromeric plasmids (YCp)** possess a chromosomal origin of replication other than the 2-μm origin, plus a yeast centromere (see fig. 13.15c). These two features result in usually only one copy of the plasmid per cell. During genomic DNA replication, the plasmid also replicates, and the presence of a centromere ensures that each daughter cell gets only one copy of the plasmid.

Because the YEp can be lost during multiple rounds of cell division, the YCp is useful when you want to ensure that the copy number of the plasmid remains close to one in all the yeast cells. This is important when many copies of the plasmid may result in a higher level of expression of the cloned gene, which may prove lethal or deleterious to the yeast cell.

The **yeast artificial chromosome (YAC)** is a linear version of the YCp. In addition to the genomic origin of replication and centromere, it also contains telomeres (see fig. 13.15d). Because it is linear, larger inserts can be cloned. The ability to clone inserts of several thousand kilobase pairs into a YAC vector has made this vector a very useful tool in creating genomic DNA libraries from other species. Being able to isolate very large contiguous genomic fragments in YACs has been important in examining genomic DNA organization in the Human Genome Project and the genome projects of other organisms.

Plant Vectors

Recently, the production of genetically altered plants has received much public attention. The term genetically modified organisms (GMOs) has become a popular term to describe transgenic agricultural organisms in particular. Many individuals are concerned about the possibility of health risks associated with GMOs entering the human food supply. Some questions concerning this issue will be discussed later in this chapter.

The most studied system for introducing foreign genes into plants is the naturally occurring crown gall tumor system. The soil bacterium *Agrobacterium tumefaciens* can infect many dicotyledonous plants, causing tumors that are known as crown galls (**fig. 13.18**). In essence, the crown gall is made of plant cells that exhibit uncontrolled growth. The plant cells within the crown gall were transformed by a 200-kb bacterial plasmid called the **tumor-inducing (Ti) plasmid** (see fig. 13.18). Transformation occurs when a specific piece

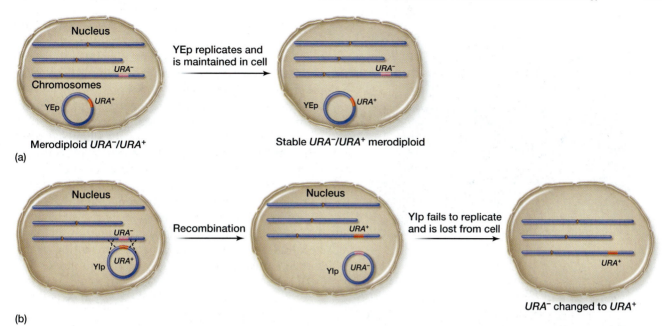

figure 13.16 Effects of YEp and YIp vectors in yeast. (a) The YEp is stably maintained in the yeast nucleus because it can replicate as a free episome due to the yeast 2-μm plasmid origin of replication. The stable maintenance of the YEp results in a partial diploid (merodiploid) for the genes that are cloned in the YEp. (b) The YIp lacks a yeast origin of replication, which requires the cloned genes to recombine with the genomic DNA to be stably expressed. This recombination exchanges the chromosomal allele (*URA*⁻) with the YIp allele (*URA*⁺). When the YIp is lost from the cell, the originally cloned allele (*URA*⁺) remains as the only copy of the gene in the chromosome.

of the Ti plasmid, called T-DNA (for *transferred* DNA), randomly integrates into the chromosome of the plant host. The T-DNA, which is approximately 15 kb in length, is flanked by a left border sequence and right border sequence that marks the region to move into the genome.

Because of the very large size of the Ti plasmid, it is very difficult to directly clone foreign DNA into it. To generate a transgenic Ti plasmid for plant infection, it is first necessary to clone the foreign DNA into a plasmid that contains a smaller region of the Ti plasmid. Introducing this smaller recombinant plasmid into *A. tumefaciens* cells that contain additional Ti plasmid sequences allows the insertion of the T-DNA into the plant genome through one of two major methods.

In the *integrating vector strategy*, a foreign gene, such as a gene conferring herbicide resistance, is cloned into

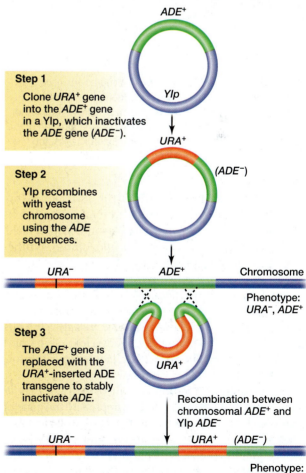

figure 13.17 YIp-mediated gene inactivation. Yeast genes can be inactivated by cloning a selectable marker (in this case *URA*⁺) inside the gene to be inactivated (*ADE*⁺) to generate a nonfunctional allele (*ADE*⁻). This recombinant YIp is transformed into an *ADE*⁺, *URA*⁻ yeast cell. Recombination between the *ADE* sequences and subsequent loss of the YIp will result in an *ADE*⁻ cell that was select by *URA*⁺. Notice that this also produces a stable *URA*⁻/*URA*⁺ merodiploid.

(a)

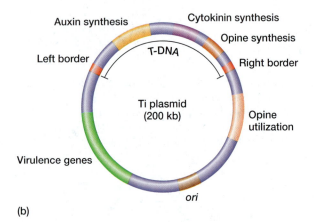

(b)

figure 13.18 (a) *Agrobacterium tumefaciens* cells that contain the Ti plasmid can infect tobacco plant (*Nicotiana tabacum*) cells and induce the formation of a crown gall tumor. (b) The organization of the Ti plasmid shows the location of the T-DNA that is transferred into the plant genomic DNA and the virulence genes that are required for this transfer.

a plasmid containing the T-DNA (**fig. 13.19**, left). This recombinant plasmid is introduced into *A. tumefaciens* cells that contain a defective Ti plasmid, which lacks the hormone (auxin, cytokinin, and opine) synthesis genes that are required for *A. tumefaciens* tumor growth in plants. The two plasmids recombine in the *A. tumefaciens* cell to generate a complete Ti plasmid that contains the foreign gene between the left and right borders of the T-DNA. These *A. tumefaciens* cells can infect damaged plant cells, such as leaf pieces, and transfer the recombinant T-DNA (containing the foreign gene) into the plant genome.

The alternative method is called the *binary vector method* (see fig. 13.19, right). In this approach, the gene of interest is cloned into a Ti plasmid that lacks the virulence genes, which are required for the insertion of the T-DNA into the plant genome. This plasmid is introduced into *A. tumefaciens* cells that contain a defective Ti plasmid that contains the virulence genes. Unlike the integrating

vector strategy, this second plasmid lacks the left border sequence. Thus, the two plasmids are unable to recombine and remain as two plasmids. When this *A. tumefaciens* cell infects a damaged plant cell, the virulence gene products on the second plasmid act on the right and left borders on the first plasmid to mediate the insertion of the recombinant T-DNA into the plant genome.

In both methods, infection of the damaged plant cells results in the growth of transformed leaf cells that form a body termed a *callus*. The callus can be induced with plant hormones to form roots and shoots and then placed in soil to generate a transgenic plant.

When this plant is crossed with others, the T-DNA (and herbicide resistance gene) will be inherited like any other plant gene. The resulting plants should exhibit some level of resistance to the desired herbicide.

Transposable Elements

Transposable elements, which are segments of DNA that are able to move from one DNA molecule or region to a new molecule or location, are widely distributed through all living organisms. They are actively used to create mutations and to move transgenes stably into organisms. (We will discuss the structures of transposable elements and their mechanisms of movement in chapter 18.) The basic structure of a transposable element is shown in **figure 13.20**.

The ends of this element are terminal inverted repeat sequences approximately 30 bp in length. **Transposase** is an enzyme encoded in the transposable element that acts on the terminal inverted repeats to catalyze the movement of the entire element. In the case of recombinant transposable elements that have a foreign gene cloned between the terminal inverted repeats (**figure 13.21**), the transposase can be transcribed from another copy of the transposable element in the cell.

One of the most common transposable elements used in generating transgenic organisms is the **P element,** which was identified in *Drosophila*. The general strategy for using the P element to generate a transgenic fly is shown in figure 13.21.

A recombinant P element containing the gene of interest is microinjected into fly embryos, along with a P element containing transposase but lacking the terminal repeats. Because these repeats are the targets for the transposase, this second P element is unable to transpose and is considered to be defective; it is also called a **helper element** because it supplies the transposase for the recombinant P element. The recombinant P element can then integrate into either a somatic or germ cell. Only an element that integrates into the genome of a germ cell will be passed on to subsequent generations.

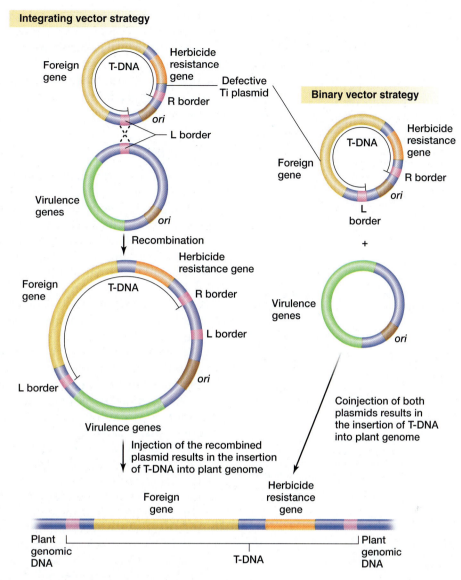

figure 13.19 Generation of a recombinant T-DNA insert in the plant genome. The integrating vector strategy involves cloning the foreign gene into a defective Ti plasmid that lacks the virulence and tumor forming genes. This plasmid is introduced into *Agrobacterium tumefaciens* cells that contain a defective Ti plasmid that lacks the R border, which prevents it from inserting into the plant genome. The two plasmids recombine through their common L border sequences to produce a recombinant Ti plasmid. In the binary vector strategy, the same defective Ti plasmid, containing the foreign gene, is introduced into *Agrobacterium tumefaciens* cells that contain a defective Ti virulence plasmid that lacks both the L and R border sequences. The virulence helper plasmid expresses the virulence genes that allows the recombinant T-DNA on the other plasmid to integrate into the plant genome.

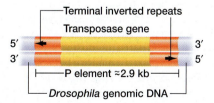

The injected embryos are then raised to adults. Often, the recombinant P element also contains a selectable marker, such as the wild-type *white*⁺ gene (see fig. 13.21). If the genotype of the injected embryo was *white*⁻, then the adults that possess an integrated recombinant P element will exhibit some pigmentation in their eyes (*white*⁺).

These adults, termed the G_0 *individuals* for *generation 0* could then be mated to *white*⁻ individuals, and the eye pigmentation of their progeny could be examined. Any *white*⁺ G_1 individuals would have inher-

figure 13.20 Structure of the P element. The *Drosophila* P element is approximately 2.9 kb in length and contains short inverted repeat sequences at its ends (arrows). Between the inverted repeats is a single gene that encodes the transposase enzyme (yellow) that is required for the movement of the P element.

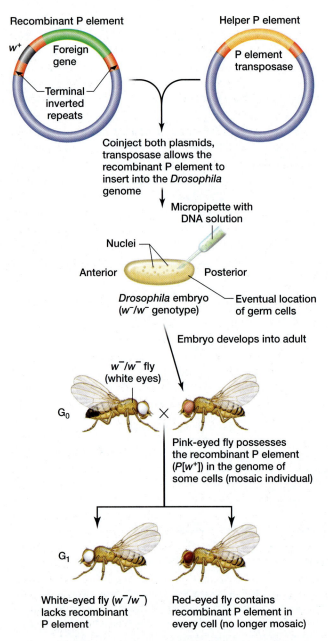

figure 13.21   General strategy of using a P element vector. A recombinant P element P[w^+], which contains the *white*$^+$ (w^+) gene and a foreign gene, are cloned between the terminal inverted repeats in place of the transposase gene. This DNA is coinjected into a *white*$^-$ embryo along with a helper P element that contains the transposase gene but lacks the terminal inverted repeats. The transposase enzyme allows the recombinant P element to insert into the genome, but the helper P element cannot transpose because it lacks the terminal inverted repeats. The resulting fly will have pink eyes if the P[w^+] element inserted into the genome (w^-/w^-). These flies are mosaic, because some cells will have the P[w^+] inserted in the genome and other cells will not. These pink-eyed flies can be mated to standard *white*$^-$ flies and the progeny will be either white-eyed (lacking the P [w^+] element) or red-eyed (containing the P [w^+] element in the genome).

ited the recombinant P element and would be able to stably pass it on to their progeny (G$_2$ individuals). Because the helper element could not transpose into the genome, the transposase source would be lost in the G$_0$ flies. Therefore, any recombinant P element insertions would be stably inherited and would not continue to transpose.

Viral Vectors

Viruses are one of the most commonly used vectors in higher organisms. Both DNA and RNA tumor viruses are often used, although other viruses can be utilized under certain circumstances. One of the advantages of using viruses is their ability to integrate their nucleic acid sequences into the host genome, which permits the foreign DNA to be stably inherited through cell divisions. (As with transposable elements, foreign DNA is not stably inherited through multiple generations of the host organism unless the integration event occurs in a germline cell.)

Because many viruses infect only specific cell types in a multicellular organism, the foreign DNA will only be integrated into the genome of certain cells.

Like λ (lambda) vectors, viral vectors permit the replacement of part of their DNA with foreign DNA. These viruses can then be used in recombinant DNA studies in one of two ways (**fig. 13.22**). They can replicate and complete their life cycle with the help of non-recombinant viruses; alternatively, they can replicate in the host without making active virus particles by existing as circular plasmids in the cytoplasm or by integrating into the host's chromosomes.

The SV40 virus, simian (monkey) virus 40, has become a valuable tool in mammalian genomic studies. For example, the rabbit β-globin gene was cloned in SV40, and enhancer sequences that control the rate of gene transcription were discovered in SV40. (Enhancers are described in chapter 17.) SV40 is a DNA tumor virus that infects and multiplies in a wide-range of cell lines, including humans. Thus, it is a common vector for introducing transgenes into mammalian cell lines and potentially for human gene therapy.

The use of many of these viral vectors carries with it two major concerns. First, the integration of the vector into the genome could result in the inactivation of a genomic gene, which may produce deleterious effects on the organism. The insertion will usually occur into only one copy of the gene in a diploid organism, which means that only the rare dominant mutations will affect the phenotype. However, a spontaneous mutation in the wild-type allele would then result in the expression of the recessive mutant phenotype. Second, because many of these viral vectors are derived from RNA and DNA tumor viruses, their use might result in the generation of a cancerous phenotype.

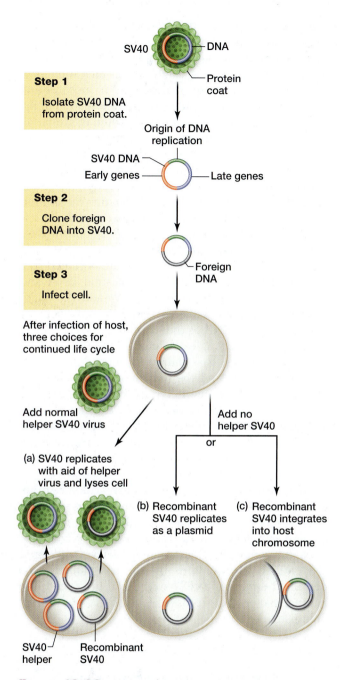

Step 1
Isolate SV40 DNA from protein coat.

Step 2
Clone foreign DNA into SV40.

Step 3
Infect cell.

After infection of host, three choices for continued life cycle

Add normal helper SV40 virus

Add no helper SV40

or

(a) SV40 replicates with aid of helper virus and lyses cell

(b) Recombinant SV40 replicates as a plasmid

(c) Recombinant SV40 integrates into host chromosome

SV40 helper   Recombinant SV40

figure 13.22 SV40 virus can be used as a gene cloning vector. Part of the virus is removed during the cloning of the foreign DNA, which inactivates SV40 normal function. The recombinant SV40 can replicate in the cell with the aid of normal helper viruses (lower left) and produce new viral particles. Without the aid of helper viruses, the recombinant SV40 can either replicate as a plasmid (lower center) or integrate into the host chromosome (lower right). However, neither of these two methods produces new SV40 particles, which prevents SV40 from infecting other cells.

With regard to this latter concern, a recombinant virus must first be grown in a cell culture before it can be introduced into an organism. But the recombinant virus is often unable to grow and replicate on its own. It must be grown in culture with another defective virus, much like the way two different types of defective P elements are used to achieve insertion into the genome DNA. Nevertheless, some risk may exist that the two different defective viruses can recombine to produce a normal DNA or RNA tumor virus. Great care must be exercised to ensure that wild-type DNA and RNA tumor viruses are not used to infect cells that are intended for later introduction into a human patient. We describe the seriousness of this complication later in this chapter during the discussion of severe combined immunodeficiency (SCID).

Use of Vectors to Express Foreign Genes

In many of the cases described earlier, a gene that has been isolated from one individual is being introduced into another of the same species. For example, we may use a P element to introduce a wild-type *Drosophila* gene into a *Drosophila* mutant. But it is sometimes informative to introduce a gene from one species into an individual from another species.

If a wild-type human gene is introduced into a *Drosophila* mutant, and the mutant phenotype is abolished, then the human gene must encode a protein that has a function either identical or very similar to the defective *Drosophila* protein. This type of analysis has proven very useful in determining the function of a wide range of proteins in different organisms.

Vectors must be capable of certain functions in order for the genes they carry to successfully be expressed. The first consideration is the promoter used to transcribe the foreign gene. Bacteria and eukaryotes utilize different promoters, so the expression of a eukaryotic gene in a bacterium usually fails unless a bacterial promoter is attached upstream of the gene. The situation can become more complicated when trying to express a foreign gene in a multicellular eukaryote. Quite often, a gene must be expressed only in a subset of cells or tissues and only at specific times in development. This requires the presence of numerous promoter elements that could be dispersed over several kilobases of genomic DNA (see chapter 17). Expression of the correct gene in the wrong tissues usually does not have the desired effect.

The next consideration is the posttranscriptional processing of the mRNA. For example, you may remember that a 5' cap, a 3' poly(A) tail, and splicing all must be present and be able to occur to produce the functional mRNA in eukaryotes. None of these functions occur in bacteria, so it is very rare that a eukaryotic gene is expressed in a bacterium. It is more common to express the cDNA that is produced from the eukaryotic mRNA, which has already undergone the splicing events.

For maximal protein expression, we also need to include the proper translation initiation signals. For a eukaryotic gene that is intended for expression in a bacterial cell, this may require introducing a Shine–Dalgarno sequence (chapter 11) immediately upstream of the translation AUG initiation codon.

Thus, the expression of a foreign gene requires significant forethought to: (1) find the proper vector to introduce the gene into the organism, (2) express the gene in the correct cells at the proper time, and (3) express the gene's product at the proper amounts within the cell.

Restating the Concepts

▶ Several different vectors are compatible with yeast. These vectors have varying uses, such as exchanging the plasmid copy of the gene with the genomic copy, creating partial diploids, and producing vectors that behave like chromosomes.

▶ Transgenic plants can be generated using the Ti plasmid that is present in *Agrobacterium tumefaciens*. The T-DNA (transferred DNA) region, along with any genes cloned within this region, is transferred from the plasmid to the host genome. They are then maintained and expressed like any other plant gene.

▶ Transposable elements are common vectors to introduce foreign genes into the genome. Their use requires the expression of the transposase enzyme and the presence of the foreign gene to be cloned between the terminal inverted repeats.

▶ Viruses are common vectors in higher organisms. They provide the advantage that they often infect only a subset of cells in an individual, which limits their integration and expression to only those specific cell types that are infected. Many of the viral vectors are DNA or RNA tumor viruses, and care must be taken to ensure that a wild-type virus is not used to infect a human cell that will later be introduced into a human patient.

▶ When expressing a foreign gene in an organism, the proper transcription (promoter) and translation signals must be present.

13.4 Site-Specific Mutagenesis

Although researchers may want to express the wild-type or normal copy of a transgene, in some cases a mutant version of the gene is more desirable. Two major types of mutations are usually generated. The first is an insertional mutation, in which a central portion of the wild-type gene is replaced with another selectable gene

or DNA sequence. We described this earlier with the yeast vectors, and we will revisit this type of mutation again when we discuss knockout mice.

The second type of mutation is alteration of a nucleotide sequence to change one or more amino acids in the encoded protein. This manipulation of the DNA sequence is called **site-specific mutagenesis.**

Site-specific mutagenesis is accomplished through a number of methods; here we describe the one most commonly used. An **oligonucleotide** of 17–25 bases is synthesized by a machine. This oligonucleotide sequence is designed such that it contains the desired mutation in the middle of the sequence (**fig. 13.23**). Notice that we are changing a single nucleotide to alter a single amino acid in the protein sequence.

The cDNA that we want to mutate is cloned into a plasmid vector. We purify the plasmid DNA and denature it by heating it to 92°C. We add the oligonucleotide, so that it will anneal to the complementary sequence in the cDNA as the solution cools (**fig. 13.24**), which generates a single-stranded DNA template with a complementary base-paired primer. Addition of DNA polymerase and the four deoxyribonucleotides then extends the sequence from the oligonucleotide primer in a 5'-to-3' direction until the circular plasmid is double-stranded. This DNA is then introduced into *E. coli* cells. As the bacteria grow, the plasmid replicates, such that half the plasmids contain the mutation and half possess the original wild-type sequence.

The bacteria are plated on agar, and colonies form. Plasmid DNA can be isolated from a number of different colonies and analyzed by DNA sequencing or by high stringency hybridization with the oligonucleotide that contains the mutation. In this way a plasmid is identified that contains the desired mutation. As described earlier, the cDNA can then be cloned into another vector and expressed in a variety of different organisms.

Oligonucleotide with wild-type sequence

5'CAC GGG GCA AAA ACC CCC CGA GAA 3'
His Gly Ala Lys Thr Pro Arg Glu

Oligonucleotide with site-specific mutation

5'CAC GGG GCA AAA GCC CCC CGA GAA 3'
His Gly Ala Lys Ala Pro Arg Glu

figure 13.23   Oligonucleotides are short DNA molecules. A wild-type oligonucleotide DNA sequence and the corresponding amino acid sequence that it encodes is shown. By changing a single nucleotide (in red), the oligonucleotide now encodes a peptide with a single amino acid change (threonine to alanine). The amino acid sequence is based on the standard genetic code.

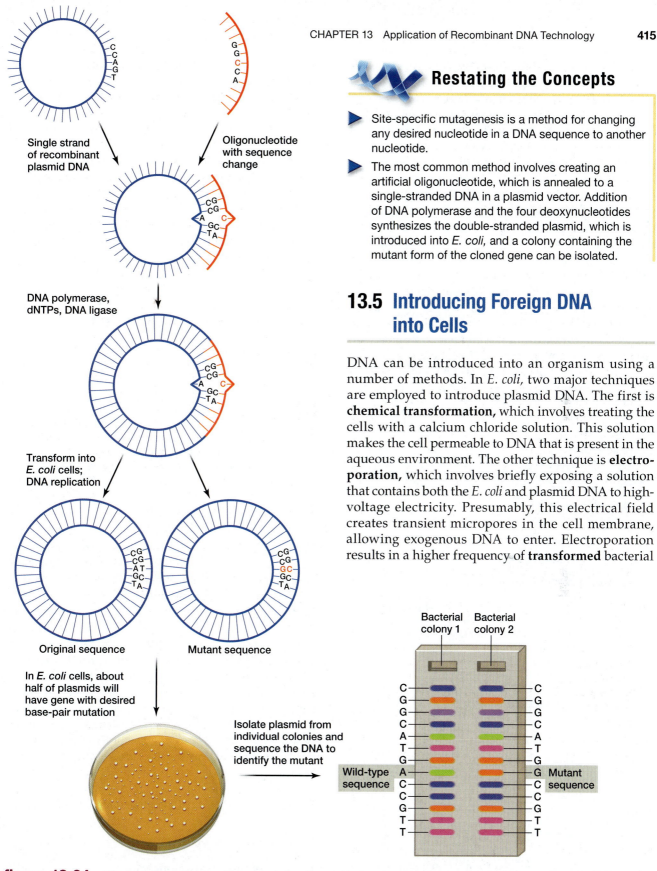

▶ Site-specific mutagenesis is a method for changing any desired nucleotide in a DNA sequence to another nucleotide.

▶ The most common method involves creating an artificial oligonucleotide, which is annealed to a single-stranded DNA in a plasmid vector. Addition of DNA polymerase and the four deoxynucleotides synthesizes the double-stranded plasmid, which is introduced into *E. coli,* and a colony containing the mutant form of the cloned gene can be isolated.

13.5 Introducing Foreign DNA into Cells

DNA can be introduced into an organism using a number of methods. In *E. coli,* two major techniques are employed to introduce plasmid DNA. The first is **chemical transformation,** which involves treating the cells with a calcium chloride solution. This solution makes the cell permeable to DNA that is present in the aqueous environment. The other technique is **electroporation,** which involves briefly exposing a solution that contains both the *E. coli* and plasmid DNA to high-voltage electricity. Presumably, this electrical field creates transient micropores in the cell membrane, allowing exogenous DNA to enter. Electroporation results in a higher frequency of **transformed** bacterial

figure 13.24   Site-specific mutagenesis protocol. An oligonucleotide that differs from the wild-type sequence by a single nucleotide (C rather than a T) is synthesized. Anneal the oligonucleotide to the single-stranded plasmid DNA. DNA polymerase and the four deoxyribonucleotides are added to extend the oligonucleotide and create a double-stranded DNA molecule, which contains a single nucleotide mismatch. This double-stranded DNA is transformed into *E. coli,* where the plasmid replicates to produce two different plasmids, one that has the site-specific mutation (CG pair) and one that contains the original wild-type sequence (TA pair). Plasmid DNA is isolated from individual bacterial colonies and sequenced. Approximately 50% of the plasmids will have the original wild-type sequence (Bacterial colony 1) and the rest will have the desired point mutation (Bacterial colony 2).

cells than the chemical method. Both of these methods contrast with the use of λ and other phage vectors that enter bacterial cells through infection.

Foreign DNA can also be introduced into eukaryotic cells using several methods. Animal cells, or plant cells with their cell walls removed (protoplasts), can be treated with calcium phosphate, allowing them to take up foreign DNA directly from the environment at a very low efficiency (fig. 13.25). This process is called **transfection,** because the term *transformation* in eukaryotes is used to mean cancerous growth. Eukaryotic organisms that take up foreign DNA are referred to as **transgenic.**

Electroporation can also be used to transfect cells at a higher frequency. Directly injecting the DNA into the cell using very small glass needles is another method to increase the frequency of transfection (see fig. 13.25).

Transfection can also be achieved by encapsulating the foreign DNA in artificial membrane-bounded vesicles called **liposomes.** By introducing different molecules to the liposome membrane, the liposomes will preferentially fuse with specific cell types. Fusion of the liposome and cell membrane results in the foreign DNA being introduced into the target cells. In one experiment, 50% of mice injected with DNA-containing liposomes were successfully transfected.

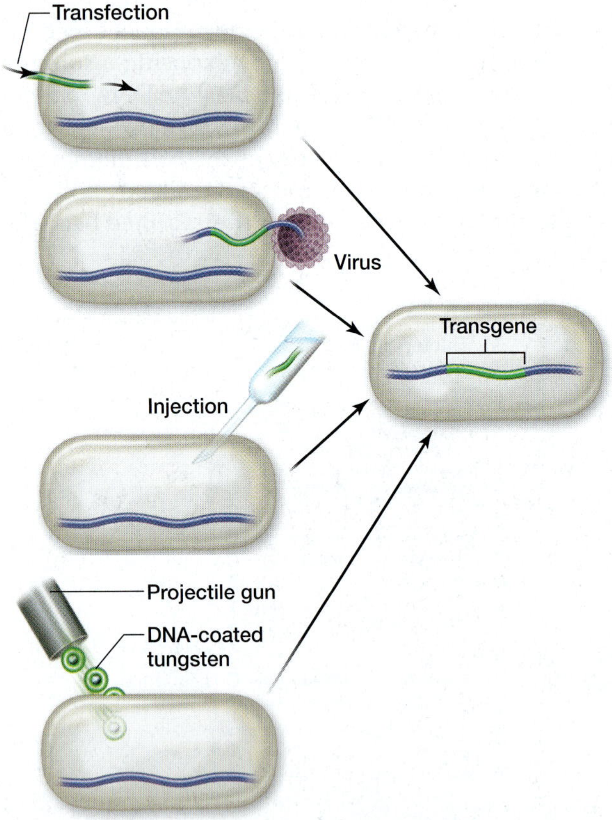

An alternative technique is required to deliver foreign DNA into mitochondria and chloroplasts. These organelles are difficult targets for genetic engineering because, among other reasons, they have double membranes that are not amenable to delivery of recombinant DNA. A successful approach to transfecting both mitochondria and chloroplasts involves a **biolistic** (biological ballistic) process, in which tungsten microprojectiles coated with the recombinant DNA are literally shot into these organelles (see fig. 13.25).

Restating the Concepts

▶ Introducing DNA into a bacterial cell is usually called transformation, whereas introducing DNA into an eukaryotic cell is called transfection.

▶ Other than via phage or viral infection, introduction of DNA can be accomplished with chemical treatment, electroporation, direct injection, artificially created liposomes, or biolistic processes by which DNA on tiny tungsten particles is shot into a cell.

13.6 Mouse Genetics

In recent years, the generation of genetically altered mice has revealed an enormous amount of information about development, physiology, and the underlying mechanism of many human diseases. Two different types of genetically altered mice have been created for use in these studies. The first, termed a **transgenic mouse,** results from the random insertion of DNA into the mouse genome. The second, called a **knockout mouse,** results from the physical exchange of the transgene with the endogenous genomic copy. The transgene is often a nonfunctional copy of the gene that replaces the endogenous wild-type gene; hence the term *knockout*. Crossing these mice can produce individuals homozygous for the defective transgene, which allows investigators to examine the recessive phenotype. Because both types of mice are considered extremely valuable in modern biology and genetics, we discuss them in some detail in the following sections.

Transgenic Mice

In generating a transgenic mouse, DNA is usually injected into the male pronucleus of newly fertilized eggs (fig. 13.26). After injection of about 2 picoliters (pL, 2×10^{-12} L) of DNA, the fertilized eggs are reimplanted into the uterus of a receptive female host. In about 15% of these injections, the foreign DNA (trans-

figure 13.25   Four different methods to introduce foreign DNA into an eukaryotic cell. Transfection involves a chemical modification of the cell to allow the passive diffusion of DNA into the cell. Virus is a biological carrier of DNA into the cell. Injection of DNA and DNA-coated projectiles are methods to manually introduce DNA into cells.

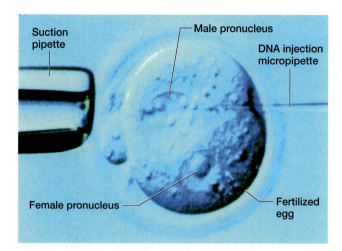

figure 13.26 Microinjection of DNA into the male pronucleus. A newly fertilized mouse egg is held by a suction pipette. DNA is injected through a micropipette into the larger male pronucleus. At this stage, the haploid male and female pronuclei are still separate. Shortly, they will fuse to generate the diploid nucleus.

gene) becomes incorporated into the genome of the embryo, usually as multiple copies (tandem repeats).

Identification of a transgenic mouse by its molecular phenotype usually involves either Southern blotting or PCR amplification. A small piece of the mouse's tail can be excised, and genomic DNA can be isolated from this tissue sample for use in these processes (**fig. 13.27**).

Gigantic Mice: An Example

As an example, let's look at the production of one of the first transgenic mouse lines. In this example, the rat growth hormone gene (*RGH*) was cloned so that it was under the transcriptional control of the mouse metallothionein promoter (**fig. 13.28**). The metallothionein promoter is inducible, which means that transcription from this promoter is controlled by environmental conditions. The metallothionein promoter is activated by adding heavy metals to the diet, such as the addition of zinc (in the form of zinc sulfate, $ZnSO_4$) to the water.

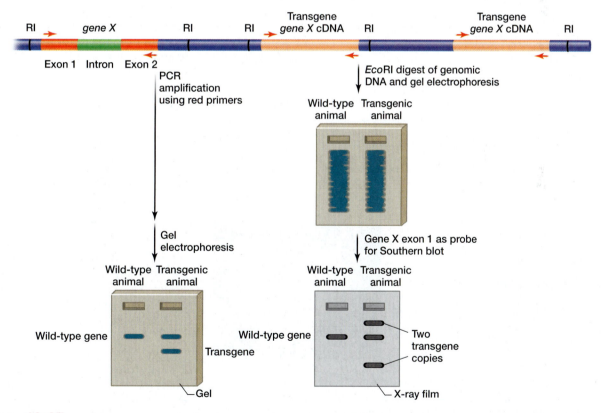

figure 13.27 PCR amplification and Southern hybridization of tail-clipped genomic DNA to identify transgenic insertions. A transgenic mouse contains both the endogenous wild-type *gene X* (containing an intron) and one or more copies of the *gene X* transgene (two in this example), which are derived from cDNAs and lack the intron. In the genomic Southern (right), DNA is isolated from a piece of the mouse's tail and digested with *Eco*RI. If exon 1 is used as the probe, the probe will hybridize to both the wild-type gene and the transgene. The wild-type gene will be present at the same location in both lanes. The transgene can be identified by the presence of additional DNA fragments, with each transgenic copy yielding a unique DNA fragment. In the PCR approach (left), primers located in exons 1 and 2 are used to PCR amplify the genomic DNA. The wild-type gene, with the intron, will produce a larger PCR product than the transgene. The transgenic mouse genomic DNA will also PCR amplify both the gene and transgene DNA fragments.

Richard Palmiter and his colleagues injected 170 eggs with this metallothionein:RGH transgene, which were then implanted into female mice. Of the 21 mice that were born, 7 showed the presence of the transgene on genomic Southern blots using a portion of the *RGH* gene as a probe.

After weaning, all 21 mice were given a solid-food diet along with water containing $ZnSO_4$. After approximately 6 weeks on the $ZnSO_4$ water, five of the seven RGH-positive mice were 32–87% larger than their siblings. In matings between a gigantic mouse and a wild-type mouse, approximately half the progeny would be gigantic and half normal sized (see fig. 13.28). The gigantic phenotype behaved as a dominant phenotype, and all the giant mice possessed the *RGH* gene.

This transgenic mouse line has many applications. First, these mice could serve as a model for the bio-logical consequences of gigantism, which is associated with pituitary and lung tumors. Second, expressing RGH in mice that contain a mutation in the mouse growth hormone gene (the *lit* mutant, which exhibits retarded growth) could serve as a model for gene therapy for human individuals who fail to express sufficient levels of growth hormone. Third, these transgenic mice could serve as a model for the production of growth hormone for other purposes, such as growth-hormone supplementation.

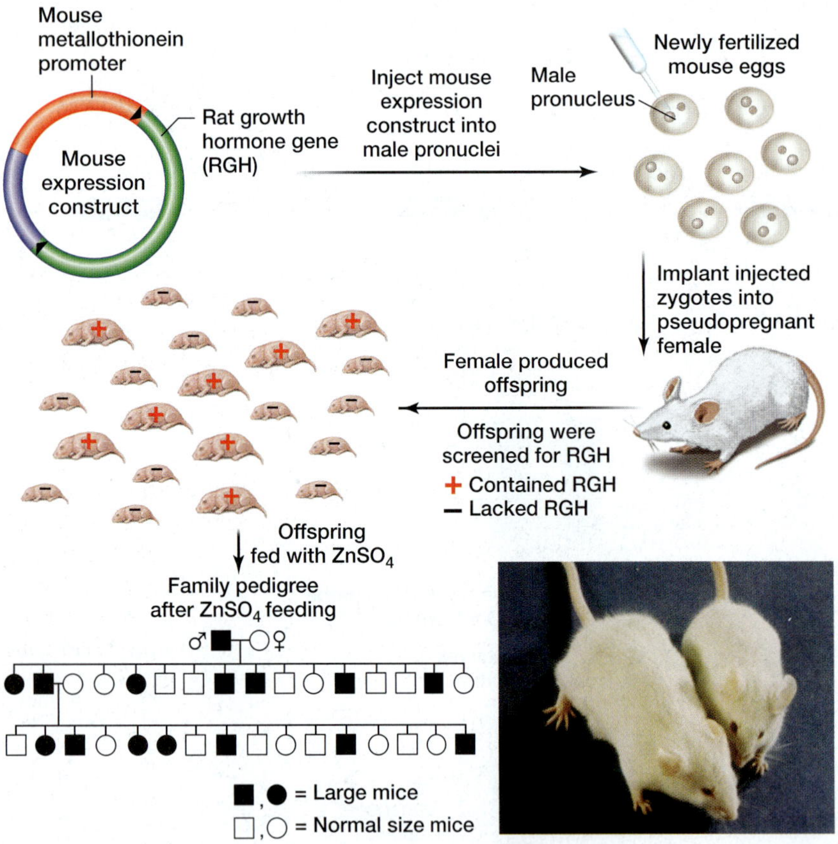

Knockout Mice

Normally, a gene used to transfect mice is incorporated randomly into the mouse genome. But in about one in one thousand insertions, the gene replaces the endogenous gene copy in the genome (that is, "knocks-out" the endogenous gene) by a process called **homologous recombination,** which requires the breakage and rejoining of related sequences in different DNA molecules. Because this event is rare relative to random insertion, geneticists had to devise a method to select for homologous recombination and the generation of *knockout mice.*

Creating a Vector

The first step in generating knockout mice is to create a vector (**knockout-targeting vector**) carrying the modified gene (**fig. 13.29**). First, the center of the gene to be modified is replaced with the gene conferring neomycin resistance (neo^r). The flanking gene regions are retained because they are required for homologous recombination to occur. The neo^r gene will serve as a selectable marker for insertion of the transgene into the genome.

Outside the flanking gene regions, the gene for thymidine kinase from the herpes simplex virus (tk^{HSV}) is cloned. This form of thymidine kinase phosphorylates the drug gancyclovir; the phosphorylated gancyclovir is a nucleotide analog that is incorporated during DNA synthesis, which kills the cell. Thus, the combination of the tk^{HSV} gene and gancyclovir is lethal; without the tk^{HSV} gene, gancyclovir is harmless.

figure 13.28   Generation of a transgenic mouse. The rat growth hormone (RGH) gene was cloned under the transcriptional control of the mouse metallothionein promoter. This transgene was microinjected into the male pronucleus of newly fertilized mouse eggs. The eggs were implanted into pseudopregnant females and the progeny were screened for the presence of the RGH transgene. A transgenic male produced approximately 50% giant sized progeny after being exposed to $ZnSO_4$ that induced expression of the RGH transgene. One of these large males also produced 50% larger progeny after mating with a female. This dominant pattern of inheritance is consistent with the large mice having only one copy of the RGH transgene. A giant sized transgenic mouse (left) and the smaller wild-type mouse (right) are shown.

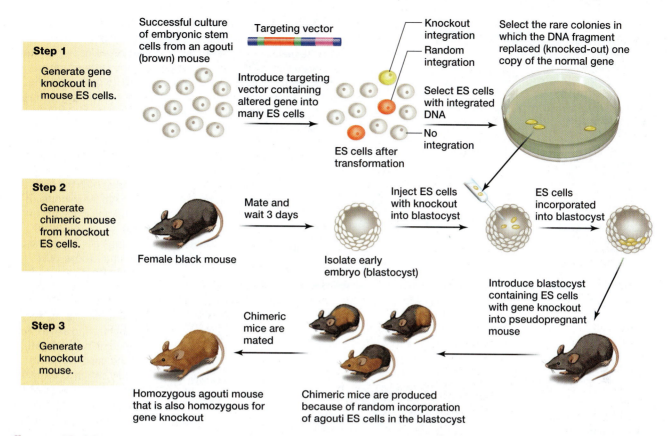

figure 13.29 Three major steps in making a knockout mouse. First, an in vitro altered version of the gene (targeting vector) is introduced into agouti ES cells. These cells are plated on special media to identify the colonies that contain the knockout copy of the gene. Next, these ES cells are introduced into an early embryo that should have black fur. If the ES cells are incorporated into the developing mouse, it will be chimeric and have both black and brown (agouti) fur. Finally, the chimeric mice are mated to produce agouti mice, which should be homozygous for the gene knockout.

Introducing the Vector into Embryonic Stem Cells

We next need to introduce the targeting vector into mouse **embryonic stem (ES) cells.** Stem cells are *totipotent,* which means they possess the ability to differentiate into any cell type in the organism. The ES cells will be derived from a mouse line with brown fur (agouti), which is phenotypically dominant to black fur color. The knockout targeting vector is introduced into the ES cells, usually by calcium phosphate precipitation.

The targeting vector can then undergo one of three events in the ES cell (**fig. 13.30**). First, the vector may not insert into the genomic DNA (fig. 13.30, right). In this case, the linear knockout-targeting vector will be degraded by nucleases in the cell. Second, the vector may insert randomly in the genome (fig. 13.30, center). Third, the vector may undergo homologous recombination and replace a genomic copy of the gene with the knockout copy of the gene (fig. 13.30, left).

The transfected ES cells are plated on medium that contains both neomycin and gancyclovir, which allows selection for the homologous recombination (see fig.

13.30). Cells that did not have the targeting vector inserted into the genome will die from the effects of neomycin. Cells that had the vector insert randomly in the genome will often contain both the *neo^r* and *tk^{HSV}* genes. These cells will also die due to the phosphorylation of gancyclovir and its incorporation into replicating DNA. Cells that underwent homologous recombination, however, will contain the *neo^r* gene but lack the *tk^{HSV}* gene. These cells will survive because they are resistant to neomycin and because the gancyclovir cannot be phosphorylated.

Analyzing Mice for the Presence of the Knockout Allele

The ES cells that underwent homologous recombination of the targeting vector are then injected into a blastocyst-stage mouse embryos, which are homozygous for the recessive allele that produces the black fur color phenotype (**fig. 13.31**). The ES cells will randomly become part of the developing mouse. The mice that develop will be **chimeric;** that is, they will be composed of two different genotypes. Cells derived from the injected ES

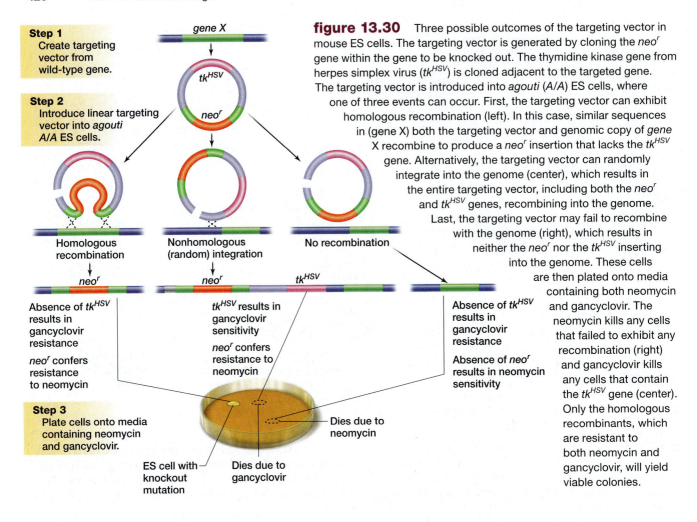

figure 13.30   Three possible outcomes of the targeting vector in mouse ES cells. The targeting vector is generated by cloning the *neo^r* gene within the gene to be knocked out. The thymidine kinase gene from herpes simplex virus (*tk^{HSV}*) is cloned adjacent to the targeted gene. The targeting vector is introduced into *agouti* (*A/A*) ES cells, where one of three events can occur. First, the targeting vector can exhibit homologous recombination (left). In this case, similar sequences in (gene X) both the targeting vector and genomic copy of *gene X* recombine to produce a *neo^r* insertion that lacks the *tk^{HSV}* gene. Alternatively, the targeting vector can randomly integrate into the genome (center), which results in the entire targeting vector, including both the *neo^r* and *tk^{HSV}* genes, recombining into the genome. Last, the targeting vector may fail to recombine with the genome (right), which results in neither the *neo^r* nor the *tk^{HSV}* inserting into the genome. These cells are then plated onto media containing both neomycin and gancyclovir. The neomycin kills any cells that failed to exhibit any recombination (right) and gancyclovir kills any cells that contain the *tk^{HSV}* gene (center). Only the homologous recombinants, which are resistant to both neomycin and gancyclovir, will yield viable colonies.

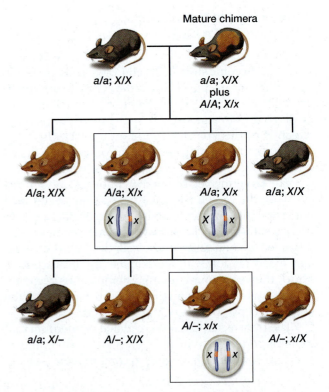

figure 13.31   Generating a knockout mouse. The *agouti* ES cells (*AA*) that contain the knockout mutation (*x*) are introduced into the blastocyst of a black mouse (*aa*) that lacks the knockout mutation (*XX*). The result is a chimeric mouse that has agouti (*AA; Xx*) and black fur (*aa; XX*). The chimeric mice are mated with black mice (*aa; XX*). The potential genotypes and phenotypes produced from the chimeric parent are shown. The agouti mice containing the knockout mutation (*Aa; Xx*) are identified from the agouti mice lacking the knockout mutation (*Aa; XX*) by either genomic Southern blot or PCR amplification of genomic DNA that is isolated from a clipped piece of the mouse tail. These *Aa; Xx* mice are mated together and 25% of the progeny will be homozygous for the gene knockout (*xx*), which can be determined by either morphological, behavioral, physiological, or molecular phenotypes (depending on the nature of the mutant allele).

cells will be homozygous for the dominant brown fur color allele and will also contain the knockout allele. The remainder of the mouse will be composed of cells that are homozygous for the recessive black fur color allele and lack the knockout allele. These chimeric mice will possess black fur with patches of brown.

The patches of brown fur confirm that the ES cells incorporated into the chimeric mouse. The critical question, however, is whether some of the knockout ES cells incorporated into the germ line of the chimeric mouse. Mating the chimeric mouse with a homozygous recessive black mouse would produce heterozygous brown mice, if any germ cells were derived from the knockout ES cells (see fig. 13.31). Some of these brown-furred mice would contain the knockout allele, and others would not. To identify the ones with the knockout allele, genomic DNA would be analyzed by either Southern blotting or PCR. After subsequent rounds of crossing and genetic analysis, mice that are homozygous for the knockout allele can be generated and carefully examined for a mutant phenotype.

Knockout mice are especially useful for studying a variety of processes, such as development, neuronal function, physiological processes, and immunology. Knockout mice have been generated to model human genetic diseases, ranging from cataracts and retinal degeneration to asthma and tumorigenesis.

Restating the Concepts

▶ Transgenic mice are produced by the random insertion of a transgene into the genome of an oocyte. The transgenic mouse contains the two endogenous copies of the gene and the transgene. Thus, the transgene must confer a dominant phenotype, which could be the wild-type allele in the homozygous mutant background or a dominant mutant allele.

▶ Knockout mice are produced by introducing a targeting vector into ES cells and selecting for homologous recombination to replace one copy of the endogenous gene. The ES cells are introduced into a blastocyst that develops into a chimeric mouse, which is often identified by different fur colors. Because the transgene replaces the endogenous gene in the knockout mouse, it is possible to generate a mouse that is homozygous for the transgene and observe the recessive phenotype.

13.7 Human Gene Therapy

Human gene therapy involves the expression of transgenes to correct an inherited disease. Unlike the transgenic and knockout mouse techniques described in the preceding sections, the transgenes are only introduced into human somatic cells. The children of these

individuals, therefore, do not inherit the transgene and retain a certain probability of exhibiting the disease.

Treatment of Severe Combined Immunodeficiency

The first human disease that was treated by gene therapy was **severe combined immunodeficiency (SCID),** which is sometimes known as *bubble-boy disease* because in a well-publicized case, an affected child was kept isolated in a plastic chamber (**fig. 13.32**). One autosomal recessive form of SCID results from an absence of a single enzyme, adenosine deaminase (ADA), which is required for purine metabolism. In the absence of adenosine deaminase, deoxyadenosine builds up in cells. This increased deoxyadenosine is especially toxic to B and T lymphocytes, essential components of the immune system. In affected individuals, the immune system is unable to fight infections, requiring them to live in a sterile environment, namely the enclosed chamber with filtered air. No drugs or other conventional therapies consistently work for these individuals.

Several features make this disease an attractive candidate for gene therapy. First, the disease is caused by a recessive mutation in a single gene. By introducing the wild-type *ADA* gene, it should behave as a dominant allele and mask the two recessive mutant alleles. Second, the defect is restricted to the lymphocytes. Thus, the wild-type *ADA* gene only needs to be introduced into the B and T cells. Because B and T lymphocytes are self-renewing cells, only a fraction of the lymphocytes need to be transfected with the wild-type gene. Finally, B and T lymphocytes are relatively straightforward to isolate from the individual and to reintroduce into the patient.

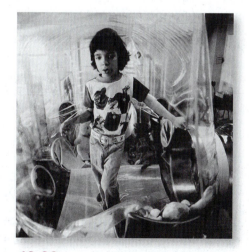

figure 13.32 A child with severe combined immunodeficiency (SCID) disease lives in a sterile environment to prevent a potentially life-threatening infection. Because SCID children often live in a sterile chamber, the disease is often called *bubble-boy disease.*

figure 13.33 The first person to receive gene therapy, Ashanti DeSilva (left), was treated for severe combined immunodeficiency (SCID) disease. R. Michael Blaese, M.D. (right) was the scientist at the National Institutes of Health that pioneered this gene therapy trial. A wild-type adenosine deaminase gene was introduced into Ashanti's T cells, which were then reinfused back into her circulatory system to partially restore her immune system.

On September 14, 1990, the first human gene therapy trial was approved for a SCID patient, Ashanti DeSilva. The procedure involved isolating her T cells and infecting them with a retrovirus that contained the wild-type human *ADA* gene. The T cells were grown in culture, and cells containing the transgene were selected and injected back into Ashanti. This procedure was performed by W. French Anderson, Michael Blaese, and their colleagues at the National Institutes of Health (NIH). Ashanti has maintained normal ADA enzyme levels in over 25% of her T cells, which has allowed her to lead a fairly normal life (**fig. 13.33**). This gene therapy treatment was less successful in a second patient, who is exhibiting normal ADA enzyme levels in a lower percentage of her T cells. However, the percentage of her T cells that are expressing normal ADA enzyme levels appears to be increasing.

Another form of SCID is a recessive X-linked mutation in the γ-subunit of the *interleukin-2 receptor* gene. Somatic gene therapy is performed by cloning the cDNA corresponding to this mRNA into a defective retroviral vector, which is derived from a RNA tumor virus. After the virus infects the patient's isolated bone marrow cells, the cells are reintroduced back into the individual. Using this approach on 10 candidates, 9 of the 10 patients exhibited significant long-term improvement in an otherwise lethal disease.

Unfortunately, two of the 10 patients exhibited T-cell leukemia due to the insertion of the virus into the genome near a protooncogene, *LMO2* (see chapter 22). Protooncogenes are required for normal cell division and growth, but lead to tumor development when they are misexpressed. It is believed that the insertion

of the retroviral vector in the genome of the bone marrow cells results in the abnormal expression of *LMO2*.

Because the insertion site of the retroviral vector cannot be controlled, further use of these vectors raises the potential concern of tumor development in other gene therapy patients. Current somatic gene therapy research involves trying to modify these retroviral vectors so that they either preferentially insert in "safe" genomic regions or are less likely to activate the transcription of genes adjacent to the insertion site. Improvements to these retroviral vectors is very important in the future development of somatic gene therapy.

Treatment of Cystic Fibrosis

The retroviral vectors have many potential problems, including the risk of causing an immune response or possibly inserting into the host genome and producing deleterious effects. Additionally, these retroviral vectors must be grown with a helper virus, which runs the risk of the two viruses recombining to produce a functional cancer-causing retrovirus. For these reasons, other types of viral vectors are being tested. One promising vector is adenovirus, which infects lung epithelial cells. On infection, the viral genome is maintained in the epithelial cells as extrachromosomal DNA, similar to a bacterial plasmid, rather than integrating into the host genome.

The ability of adenovirus to infect lung epithelial cells makes it an attractive vector for gene therapy of cystic fibrosis. As described earlier, cystic fibrosis is a recessive single-gene defect; it causes increased mucus buildup in the lungs, difficulty breathing, and lung infections due to defective chloride transport out of lung epithelium.

The normal cystic fibrosis gene has been cloned into adenoviral vectors, and a nasal spray is being tested to deliver the recombinant virus into the lungs. Because the adenovirus does not integrate into the epithelial genome, it will be necessary to continue adenoviral treatments for potentially the remainder of the individual's life. But this continual recombinant adenoviral reinfection is likely preferable to the physical disruption of mucus buildup and chronic lung infections that afflict cystic fibrosis patients.

Restating the Concepts

▶ Although gene therapy has been performed in humans, it has only been done in somatic cells and not germ-line cells. This therapy may reduce or eradicate the symptoms of a particular disease; however, the individual still possesses the disease allele(s) and has the same probability of passing them on to his or her offspring.

figure 13.35   Snuppy, the first cloned dog. (a) The Afghan hound that served as the somatic nucleus donor (left) and Snuppy (right). (b) Snuppy and his surrogate mother (right), a yellow Labrador retriever.

figure 13.34   Dolly is the first cloned mammal produced by the transfer of a nucleus from a somatic cell of an adult sheep.

13.8 Cloned Organisms

A cloned organism is genetically identical to another individual. This usually involves replacing the nucleus of an egg with the nucleus of a somatic cell, such as a skin cell. Fiction writers in the past have created stories in which scientists cloned one person to create numerous copies. The themes of these stories have varied from the cloning of Adolf Hitler to the cloning of a very busy man to help him fulfill his day-to-day obligations. In February of 1997, a group of scientists from the Roslin Institute and PPL Therapeutics, both in Edinburgh, Scotland, reported in *Nature* magazine that they had successfully cloned the first mammal from an adult somatic cell. The cloned lamb, which was derived from an udder cell taken from a 6-year-old ewe, was named Dolly (**fig. 13.34**).

The word *clone* is used here to mean the creation of a genetically identical individual. Dolly contained an exact copy of the 6-year-old ewe's nuclear genome. However, mitochondria also contain a genome that encodes several genes (see chapter 19). Because only the nucleus of the 6-year-old ewe's udder cell was transferred into the enucleated egg cell, Dolly's mitochondrial genome was derived from the egg. Thus, Dolly's total genotype was not completely identical to the 6-year-old ewe.

Dolly lived a celebrity's life, being featured on the news and numerous documentaries. She also gave birth to six lambs the natural way. On February 14, 2003, Dolly was put down by lethal injection. She had been suffering from lung cancer and crippling arthritis. At 6 years of age, her life span was significantly shorter than the 11

or 12 years of most Finn Dorset sheep. The postmortem examination failed to reveal any other abnormalities in Dolly besides the cancer and arthritis.

Since the birth of Dolly, cats, pigs, rabbits, cattle, mice, mules, a rhesus monkey, and several endangered species have been cloned. In 2005, the first report of a cloned dog was announced. Over 1000 nuclear transfers were required to produce Snuppy, a cloned Afghan hound (**fig. 13.35**). As expected, Snuppy possesses a nuclear genome that is identical to the Afghan hound donor, and a mitochondrial genome that is identical to the surrogate mother, a yellow Labrador retriever. Some consider cloning one mechanism for preserving endangered species, and cloning might also be a method to produce animal organs that can be more efficiently used for organ transplants in humans. The very large number of nuclear transfers required to produce a viable clone remains a significant hurdle before cloning mammals becomes acceptable. Although reports have surfaced about the cloning of humans, this has not been confirmed.

Some of the initial concerns about cloned animals have been their apparently short lifespan (or rapid aging), their propensity for disease, and the frequency with which they exhibit developmental and physiological abnormalities. In 2002, researchers from the Whitehead Institute for Biomedical Research reported that the genomes of cloned mice possessed defects, and that 4% of the genes were functioning abnormally. These problems were not caused by mutations, but rather by the disruption of the normal activation and repression of genes. These abnormalities could be a result of a differentiated cell being "reprogrammed" to produce the correct genomic expression pattern for different cell types in the cloned animal.

By contrast, groups at the University of Cincinnati and the University of Hawaii recently reported that the development, behavior, and physiology of cloned mice is indistinguishable from that of normal mice. Clearly, more work needs to be done to determine what problems and risks, if any, are associated with cloned animals.

Restating the Concepts

▶ Cloning animals involves replacing the haploid nucleus of an egg cell with the diploid nucleus of a somatic cell. This procedure has been done successfully in a wide range of animals.

▶ Cloned animals are not genetically identical to the donor animal because only the nuclear genome is transferred into the enucleated cell. The mitochondrial genome of the cloned animal is derived from the host cell and not the donor cell.

13.9 Practical Benefits from Gene Cloning

Throughout this chapter, we discussed a variety of applications of genetic engineering. Here we summarize some of the accomplishments and future directions in the medical, agricultural, and industrial arenas.

Medicine

In medicine, genetic engineering has had remarkable successes in several areas. First, basic knowledge about how genes work (and don't work) has advanced tremendously, and as a result we have a better appreciation of how an organism develops and functions. This in turn, increases our understanding of what causes developmental and physiological processes to be disrupted. This knowledge can be applied to a variety of diseases. The sequence of the human genome will further aid medicine by identifying genes that are associated with various diseases, a first step in discovering cures. So far, several genes of great importance have been located, cloned, and sequenced. Some human genetic diseases for which genes have been cloned include cystic fibrosis, Huntington disease, and some forms of retinitis pigmentosa (which result in a gradual loss of vision). A number of genes associated with certain cancers have also been cloned (*BRCA1*, *BRCA2*, *retinoblastoma*).

Recombinant DNA methodology has made available large quantities of biological substances previously in short supply. These include insulin for the treatment of diabetes, interferon (an antiviral agent), growth hormone, growth factors, blood-clotting factors, and vaccines for diseases such as hepatitis B, herpes, and rabies.

Recombinant DNA techniques are becoming a powerful method for expressing antigens (proteins) that can be used to immunize humans and animals. For example, the development of sera that can be used to combat or immunize against the flu virus, malaria, and HIV are just some of the more recent areas of investigation.

On another front, transgenic mice and cloned sheep have shown that genetic engineering can be applied to higher organisms. Transgenic animals can serve as models for a large number of inherited human diseases. Their usefulness stems in part from engineering the transgenic animal to contain the same defective gene that causes the human disease. These transgenic models can be used to develop and test pharmaceutical therapies for the disease or even test the potential of different gene therapy treatments.

The development of cloned animals also raises the possibility of producing organs that can be used in human transplants. With the severe shortage of liver, kidney, heart, and other organs that are compatible for transplantation, a serious effort is necessary to provide these organs to individuals in dire need of them.

Agriculture

Currently in the United States, approximately one quarter of farmland is planted with crops that are genetically modified. Most of the transgenic crops are resistant to certain insect pests because they contain genes from *Bacillus thuringiensis* (often referred to as Bt), which produce natural insecticidal proteins. For example, the proteins Cry1A and Cry1C from *B. thuringiensis* protect the plants against larval forms of lepidopterans such as the European corn borer. Cry3A protects against coleopterans such as the Colorado potato beetle.

In excess of 50 genetically altered crop plants have been approved for planting, including those protected against insect pests, frost, and premature ripening. Rice is being modified so that its vitamin A content is maintained even after the husks are removed, a procedure to increase storage life, since the husks become

figure 13.36 The genetically engineered Roundup Ready form of soybean contains a transgene for a glyphosate-resistant form of ESPS synthase. This field was sprayed with the herbicide glyphosate (Roundup) and the Roundup Ready soybean plants on the left continue to grow, while the nontransgenic plants on the right (arrows) are either dead or severely stunted in growth.

rancid. That change alone will improve the health of millions of people throughout the world.

Transgenic crops are also being created that are resistant to herbicides, such as glyphosate (commercially known as *Roundup*, manufactured by Monsanto Corporation). Glyphosate inhibits the enzyme ESPS synthase, which is important for amino acid biosynthesis in bacteria and plants. It is a broad-spectrum herbicide that efficiently kills a wide range of weeds and plants. A glyphosate-resistant form of ESPS synthase was isolated in *E. coli* and cloned into a variety of different plants using the *Agrobacterium* Ti plasmid. These transgenic plants are now resistant to the herbicide (**fig. 13.36**). This resistance allows farmers to spray their fields with glyphosate, without it killing the desired crops. The result is increased crop yields to help feed a growing population.

Industry

Industrial applications of biotechnology include engineering bacteria to break down toxic wastes, modifying yeast to use cellulose to produce glucose and alcohol for fuel, using algae in *mariculture* (the cultivation of marine organisms in their natural environments) to produce both food and other useful substances, and developing better food-processing methods and waste conversion.

As an example, the yeast *Saccharomyces cerevisiae* has been modified with a plasmid that contains two cellulase genes—an endoglucanase and an exoglucanase—that convert cellulose to glucose. The yeast are naturally capable of converting glucose to ethyl alcohol. These yeast strains can now digest wood (cellulose) and convert it directly to alcohol. The potential exists to harvest the alcohol the yeast produce as a fuel to replace fossil fuels that are in dwindling supply and increase pollution.

As you can see, biotechnology is advancing in many directions.

13.10 Ethical Considerations

Along with the many advantages of recombinant DNA technology comes significant responsibility. The scientific community is largely aware of the inherent dangers and the social, moral, and ethical ramifications of this work. This chapter would not be complete if we did not consider the potential effects of the application.

The Green Revolution and GMOs

The **Green Revolution** refers to altering agricultural practices to increase crop yields and creating new varieties of food plants to feed the growing human population of the world. The term genetically modified organisms (GMOs), which includes all genetically modified life forms, has come to be associated primarily with genetically modified plants. Although many

individuals argue that genetically modified plants pose little or no risk to animals and the environment, others demand much more caution. This debate is not simply about the safety of consuming GMOs, but also about the ethical and moral considerations in the generation and planting of GMOs.

Weed infestation destroys approximately 10% of all crops grown worldwide, and malnutrition among lower-income populations is prevalent throughout the world. As we discussed earlier, certain herbicides such as glyphosate can dramatically reduce the amount of weeds in fields, which could increase crop yield.

Several major questions may affect the development and use of GMOs. Are the genetically modified plants safe for human and animal consumption? Is it appropriate to market and plant GMOs that encourage the use of herbicides, such as glyphosate? Are we able to control the spread of the GMOs once they are planted in fields? We may not realize the potential dangers until later, and if GMOs have already spread and mixed with more conventional crops, we may be unable to remove their risk.

Cloning Organisms and Individuals

Probably many more ethical issues come immediately to mind when you think about the application of recombinant DNA technology and cloning to humans. As just one example, supposing that gene therapy becomes more practical in humans, can we introduce an inducible form of the growth hormone gene into embryos and create a generation of super athletes? Controversy already exists over whether use of injectible growth hormone should be banned in athletics. But if an athlete contains a growth hormone transgene, is the activation of this gene something that should also be banned—and how could it be tested and monitored?

Clearly, the ability to move genes at will brings up questions never before imagined.

Preimplantation genetic diagnosis (PGD) in humans is a newly devised medical technique primarily used by couples who have given birth to children suffering from a genetic disease, or by couples in which a high probability exists of having a child with a genetic disease. For the PGD procedure, eggs are removed from the woman's ovaries and sperm from the man to perform in vitro fertilization (**fig. 13.37**).

Once the embryo reaches the eight-cell stage, one cell is removed and grown in culture for genomic DNA isolation. (Removal of one cell at this stage apparently does not alter further embryonic development.) This cell's genomic DNA is analyzed for the presence of the mutant allele. If the allele is not present, then the embryo is transferred to the woman's uterus. This screening method is one way to reduce the likelihood that a child will be born with a particular genetic disease. It remains unclear, however, what the long-term effects, if any, are

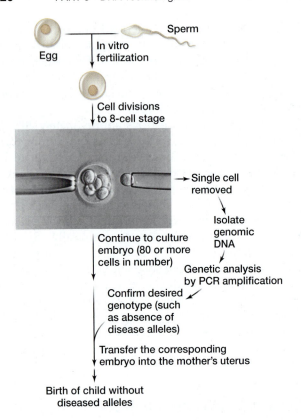

Egg — Sperm

In vitro fertilization

Cell divisions to 8-cell stage

→ Single cell removed

Isolate genomic DNA

Continue to culture embryo (80 or more cells in number)

Genetic analysis by PCR amplification

Confirm desired genotype (such as absence of disease alleles)

Transfer the corresponding embryo into the mother's uterus

Birth of child without diseased alleles

figure 13.37 Preimplantation genetic diagnosis (PGD). An egg is fertilized in vitro and grown in culture to the 8-cell stage. One cell is removed, while the remaining cells continue to grow in culture. Genomic DNA from the removed cell is PCR amplified and analyzed for the presence or absence of the genetic markers in question. If the desired genotype is identified, the corresponding embryo is implanted in the mother's uterus to continue development.

of an individual developing from a seven-cell embryo. It is possible that the consequences may not be apparent until individuals reach 40, 50, or even 70 years of age. At that point, large numbers of individuals may have been born through this testing method.

Ethical questions also arise when we consider what conditions are considered genetic diseases. Most people may agree that testing for untreatable diseases that result in early childhood death are a worthwhile use of this technology. But what about testing for a condition like Down syndrome? Although some individuals may feel that it would be very difficult to raise a child with Down syndrome, others would argue that their lives have been greatly enriched by these children.

A more difficult scenario is developing with PGD. In one reported case, a couple had a daughter who suffers from *Fanconi anemia*, a rare blood disorder that usually results in death by age 7. The only potential treatment is a blood transfusion of stem cells from an individual with a matching tissue type; however, this is not a cure for the disease. In 2000, the couple underwent the PGD procedure to produce a son with a matching tissue type. When he was born, stem cells isolated from his umbilical cord blood were used to treat their daughter.

This situation becomes more ethically complicated because the girl suffering from Fanconi anemia has a high risk of developing kidney problems, which might require a transplant. In this case, her younger brother is already known to be a good tissue match, and he could serve as a kidney donor. But who makes the decision about whether the brother, who could still be under the age of 12, should donate a kidney to save the life of his sister? What prevents parents from bearing a child with a desired tissue type to provide organs for transplant into another child?

Science is not the culprit in these questions. Science has simply provided the opportunity to choose between the different options, and all humans strive to possess options in their life. It is how society decides to use these options that determines our legacy.

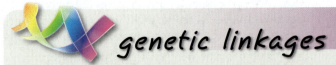

genetic linkages

We discussed the application of recombinant DNA techniques to several broad areas of modern biology, such as gene cloning, generation of transgenic organisms, altering gene expression, and medicine. Basic recombinant DNA techniques that were covered in chapter 12, such as restriction enzymes, PCR amplification, DNA cloning, and DNA sequencing, form the basis of these different applications.

In this chapter, we discussed the types of DNA sequences that are used for mapping and their molecular characterization. We also expanded our knowledge of statistical analysis to include the lod score analysis. Unlike chi square (see chapter 2), lod scores are better suited for the small numbers of offspring usually observed in human pedigrees.

We also described several different vectors used in eukaryotic organisms, including the transposable element, which can move between DNA locations or molecules. In chapter 18, we describe the different classes

(continued)

of transposable elements, the mechanisms they utilize to move to new DNA locations, and their potential function in both prokaryotic and eukaryotic organisms.

In chapter 10, we described the differences between the basic bacterial and eukaryotic promoters, which play an important role in expression of all genes, including transgenes. We will expand that discussion in chapter 17 to include a variety of eukaryotic promoter enhancer and control elements, the protein complexes that bind to these DNA sequences, and how they interact to properly regulate gene expression.

We concluded this chapter by raising some moral and ethical considerations in this research. With the creation of the Human Genome Project, the Ethical, Legal, and Social Implications (ELSI) Program was created to formally oversee these questions. We describe the ELSI Program in more detail in the following chapter.

Chapter Summary Questions

1. For each term in column I, choose the best matching phrase in column II.

   Column I
   1. S1 nuclease
   2. 2-μm plasmid
   3. GMO
   4. CFTR
   5. lod
   6. PGD
   7. Ti plasmid
   8. SNP
   9. SCID
   10. Transposase

   Column II
   A. Catalyzes movement of DNA elements
   B. Disorder of the immune system
   C. Changes in a single base in DNA
   D. Plasmid found in *Agrobacterium tumefaciens*
   E. Technique used in genetic studies of embryos
   F. Statistical tool used in genetic linkage studies
   G. Channel protein found in the cell membrane
   H. Transgenic animal or plant
   I. Plasmid found in *Saccharomyces cerevisiae*
   J. Catalyzes digestion of single-stranded DNA

2. Distinguish between a RFLP and an SNP.

3. What are minisatellites? What advantages do they provide over RFLPs for gene mapping studies?

4. Why is lod score analysis better suited than chi-square analysis for genetic linkage studies in human pedigrees?

5. Describe two ways in which plasmids are introduced into bacterial cells.

6. What methods are used to get foreign DNA into eukaryotic cells? What general term is used for these processes?

7. Compare and contrast a YCp and a YAC.

8. What is chromosome walking? When is it used?

9. What disease does the bacterium *Agrobacterium tumefaciens* cause in dicotyledonous plants? Describe how the disease is produced.

10. Outline the steps involved in site-specific mutagenesis.

11. Define the following terms: **(a)** haplotype; **(b)** YIp; **(c)** P element.

12. List the three ways by which an SV40 recombinant virus can continue its life cycle after infecting a mammalian cell.

13. What is a transgenic mouse? A knockout mouse? Why are they used in genetic studies?

14. What is gene therapy? How is the technique different from the ones used to create transgenic and knockout mice?

15. Was the lamb Dolly truly a genetic clone? Why or why not?

16. Describe some areas of practical benefit from genetic engineering.

17. Discuss the concerns that people may have with regard to the widespread use of genetically modified organisms.

Solved Problems

PROBLEM 1: Two normal individuals have a child with Down syndrome. RFLP analysis with a probe from chromosome 21 is performed on all three individuals, and the results of the gels appear as follows. Based on these results, what can you conclude about the origins of the number 21 chromosomes?

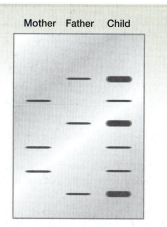

Mother   Father   Child

Answer: Two of the three number 21 chromosomes present in the child came from the father, not the mother. Because the probe produces different bands in the mother and the father, all of these bands must also be present in the child. The intensity of a band is proportional to the amount of DNA present. The bands that are of paternal origin are more intense than the maternal bands. This suggests that the father contributed more chromosomal 21 DNA than the mother, or that the father contributed two copies of chromosome 21 and the mother one copy.

PROBLEM 2: In humans, a VNTR locus contains the repeat $(TTCTATTGTA)_n$, where $n = 7-10$. How many possible alleles are there for this locus in the human population, and what are their lengths? How many possible genotypes exist?

Answer: The number or repeats (n) can be 7, 8, 9, or 10. Therefore, there are four alleles for this locus:

Allele 1 ($A1$) = $(TTCTATTGTA)_7$ and is 70 nucleotides long;

Allele 2 ($A2$) = $(TTCTATTGTA)_8$ and is 80 nucleotides long;

Allele 3 ($A3$) = $(TTCTATTGTA)_9$ and is 90 nucleotides long; and

Allele 4 ($A4$) = $(TTCTATTGTA)_{10}$ and is 100 nucleotides long;

With four alleles, 10 different genotypes (4 homozygous and 6 heterozygous) are possible: $A1A1$, $A2A2$, $A3A3$, $A4A4$, $A1A2$, $A1A3$, $A1A4$, $A2A3$, $A2A4$, and $A3A4$.

PROBLEM 3: The *Kpn*I restriction map of a 10-kb region of chromosomal DNA is shown here:

The letters indicate the sites where the restriction enzyme cuts. Shown are the distances (in kilobases) between these restriction sites.

Genomic DNA from four individuals is digested with *Kpn*I, and subjected to RFLP analysis using a labeled probe that is homologous to the region represented by the thick horizontal line. All four individuals are homozygous for this chromosome. They all possess restriction sites A, D, and E, and the presence of sites B and C are as follows:

Individual	Site B	Site C
I	Present	Present
II	Present	Absent
III	Absent	Present
IV	Absent	Absent

a. For each individual, draw the pattern of bands expected on the Southern blot.

b. Would these results change if site E is absent? Why or why not?

Answer: **a.** Individual I has all five restriction sites, and so *Kpn*I digestion will produce four DNA fragments: 2 kb, 1 kb, 3 kb, and 4 kb (from left to right). The genomic DNA region that is homologous to the probe overlaps the first three fragments. Therefore, the probe will hybridize to the 2-kb, 1-kb, and 3-kb fragments. Individual II lacks site C, and so *Kpn*I digestion will produce only three DNA fragments: 2 kb, 4 kb (1 + 3), and another 4 kb (from left to right). The probe will hybridize to the 2-kb fragment and one of the 4-kb fragments, although it will not be impossible to determine which of the 4-kb fragments the probe is hybridizing to on the gel. Individual III lacks site B, and so *Kpn*I digestion will produce only three DNA fragments: 3 kb (2 + 1), another 3 kb, and 4 kb (from left to right). Therefore, the probe will hybridize to the two 3-kb fragments, although it will not be possible to distinguish between these 3-kb fragments on the gel. Individual IV lacks sites B and C, and so *Kpn*I digestion will produce only two DNA fragments: 6 kb (2 + 1 + 3) and 4 kb (from left to right). Therefore, the probe will hybridize to the 6-kb fragment.

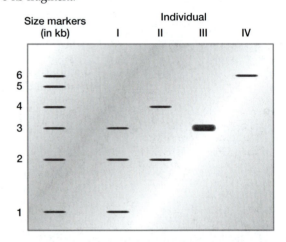

b. If site E is absent, digestion of the genomic DNA with *Kpn*I will yield a longer fragment to the right of site D. However, the probe will not detect this fragment (regardless of its size), because the probe is not homologous to that region of the DNA. Therefore, the results will not change.

Exercises And Problems

18. Describe several different ways in which RFLPs can be generated.

19. Shown here is a series of overlapping, contiguous regions of a chromosome generated by chromosome walking. Two cloned genes were found to hybridize to this region. Region D is the approximate location of the first gene; the second gene was mapped to region F. To which segment(s) (1–6) will each cloned gene hybridize?

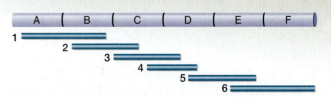

20. Jack and Jill have five children. They have conceived two together, each has one from a previous marriage, and they have adopted one child. VNTR analysis of a single locus is performed on all family members. From the results can you determine the identity of each child?

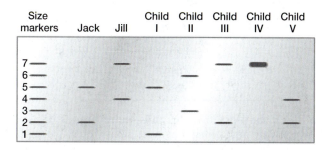

21. Are all RFLPs, SNPs, and VNTRs associated with a visible change in the cellular phenotype? Why or why not?

22. Restriction mapping of an 8-kb region of a chromosome reveals the four different patterns of *Hind*III sites shown:

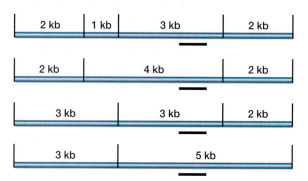

Genomic DNA is extracted from a random sampling of people and digested to completion with *Hind*III. The fragments are run on a gel, Southern blotted, and analyzed with a labeled probe that is homologous to the region represented by the thick horizontal line.
  a. How many different RFLP alleles are possible in this sample?
  b. How many genotypes are possible in this sample?

23. Most human genes contain one or more introns. Because bacteria cannot excise introns from nuclear messenger RNA (snRNPs are needed), how can bacteria be used to make large quantities of a human protein?

24. The restriction map of a particular region of a human chromosome is shown here.

| 3.3 kb | 1.1 kb | 2.6 kb |

Using a probe that is homologous to this region of the chromosome, RFLP analysis is performed on genomic

DNA obtained from a married couple. The results are shown in the following diagram:

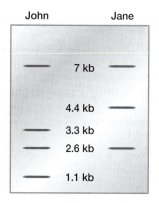

  a. Draw the chromosomal arrangement of the RFLPs in John and Jane.
  b. How many different RFLP patterns are possible in the DNA of this couple's children? List all of them.

25. The pedigree of a family affected with an autosomal dominant disorder is presented here, with the disease-locus RFLP pattern of each member of the family.

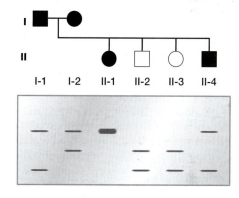

Which bands in I-1 and I-2 are associated with the disorder? For each individual in this pedigree, indicate their RFLP genotype as homozygous or heterozygous. Explain your reasoning.

26. Color blindness is an X-linked recessive disorder. Adonis and Beatrice, a phenotypically normal couple, have three children: Carla and Dora, two daughters, with normal eyesight, and Ethan, a color-blind son. Genomic DNA from all five is digested with a restriction enzyme, Southern-blotted, and probed with the cloned gene. The results are:

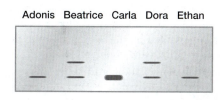

  a. Explain the presence of a single band in some individuals and a double band in others.
  b. What DNA fragment is associated with the recessive color blindness allele?
  c. Is Dora or Carla a carrier?

27. The normal DNA sequence of a 2-kb chromosomal locus has the following *Pvu*II restriction map. The sizes of the restriction fragments are given in base pairs.

The DNAs from three different individuals were isolated, digested with *Pvu*II, and Southern-blotted. The labeled probe used in the RFLP analysis is homologous to the entire locus. The banding patterns were as follows:

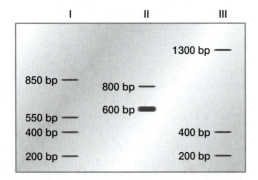

For each individual, indicate the location and nature of the change(s) leading to the different RFLP pattern?

28. You are a forensic science lab technician who is helping in a murder investigation. You perform a five-locus VNTR analysis of a blood sample found at the scene of the crime. The banding patterns you observe do not match those of the victim, but are an exact match to one of the suspects. The alleles are found in the general population in the following frequencies: allele A = 3%; allele B = 11%; allele C = 5%, allele D = 7% and allele E = 57%. What is the probability that these five alleles occur in an individual in the general population? Based on these five alleles, what are the odds that the suspect is innocent?

29. A restriction map of a 15-kb region of chromosomal DNA is shown here:

The letters E and N refer to the restriction sites of the enzymes *Eco*RI and *Not*I, respectively. The N sites are found in all individuals, but the E sites are polymorphic and may or may not be found in an individual.

Genomic DNA from hundreds of individuals is digested with *Eco*RI and *Not*I and subjected to RFLP analysis using a labeled probe that hybridizes to the region represented by the thick horizontal line. List all possible banding patterns expected.

30. A restriction map of another 15-kb region is shown here.

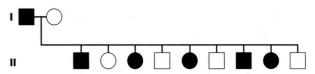

The letters A', A'', A''' and X', X'', X''' refer to restriction sites for the enzymes *Acc*I and *Xho*I, respectively.

Genomic DNA from three individuals is subjected to RFLP analysis using the probe indicated by the thick horizontal line. All three individuals are homozygous for this chromosome. They all possess restriction sites A', A'', X', X''', and the presence of sites A''' and X'' are as follows:

Individual	Site X''	Site A'''
I	Present	Present
II	Present	Absent
III	Absent	Present

a. If the DNA from the three individuals is digested with *Xho*I, what fragment sizes would the probe hybridize to?
b. If the DNA from the three individuals is double-digested with both *Xho*I and *Acc*I, what fragment sizes would the probe hybridize to?

31. In humans, how many different genotypes are possible for two SNPs that are located in the coding region of two genes?

32. The following is the pedigree of a family exhibiting a rare autosomal dominant illness that is 100% penetrant. Blood samples were taken from each individual and analyzed by PCR amplification with two different microsatellites. The results are given below each family member.

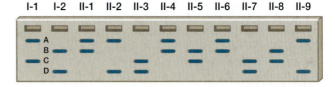

Microsatellite number 1

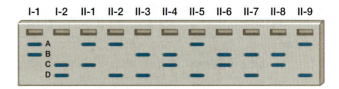

Microsatellite number 2

a. Using the symbols I/i, assign genotypes to each family member.
b. Which microsatellite appears to be linked to the illness locus? Provide a detailed analysis that accounts for any unusual genotypes.
c. Determine the lod score. For purposes of this calculation, assume that the two loci are 12 map units apart.

33. The motion picture *Jurassic Park* was based on the premise that DNA of dinosaurs could be extracted from the blood-meals of mosquitoes preserved in amber and inserted into the genome of a frog, which would then produce living dinosaurs. Is this premise reasonable? Please explain your logic.

34. You used a 1.0 kb cDNA to screen a genomic DNA library and identified four clones (A–D). You performed an *Eco*RI restriction digestion on these four clones and determined they contain the following *Eco*RI restriction fragments.

Clone A: 7.0 kb, 5.5 kb, 4.0 kb, 2.0 kb
Clone B: 8.0 kb, 5.5 kb, 3.0 kb, 2.0 kb
Clone C: 7.0 kb, 4.0 kb, 2.0 kb, 1.5 kb
Clone D: 9.0 kb, 7.0 kb, 4.0 kb, 2.0 kb, 1.5 kb

a. Arrange these four clones relative to each other based on the *Eco*RI restriction digest fragments.

b. What is the minimum *Eco*RI fragment(s) that is/are hybridizing to the 1.0 kb cDNA? Explain your reasoning.

c. What *Eco*RI fragments would you use as probes to extend your walk in both directions? Explain your reasoning.

d. What *Eco*RI fragment would you use as a probe to extend your walk specifically towards the centromere? Explain your reasoning.

Chapter Integration Problem

Yodeling, a hypothetical trait in humans, is under the control of a single gene that is completely penetrant. The *yodeling* gene is found on the p arm of a particular chromosome and is closely linked to two RFLP loci in this arrangement: *A – B – yodeling*. There are four RFLP alleles at locus *A* (*A1, A2, A3,* and *A4*) and four at locus *B* (*B1, B2, B3,* and *B4*).

The following is a pedigree of a family of yodelers. The RFLP genotype of a few individuals is also given. The slash "/" symbol indicates the chromosomal arrangement of the *A* and *B* loci. For example, the genotype of individual II-4 is *A2B3/A3B1*. This means that one chromosome carries the *A2B3* alleles and its homolog carries the *A3B1* alleles. Note that only one chromosomal arrangement is given for II-6.

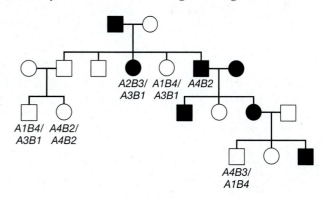

a. Which of the following modes of inheritance can be excluded: autosomal dominant, autosomal recessive, Y-linked, X-linked dominant, and X-linked recessive? Justify your answer.

b. What is the linkage arrangement at the *A, B,* and *yodeling* (*Y*) loci, in individuals I-1 and I-2? Explain your reasoning.

c. Determine to the extent that is possible, the linkage arrangement of the A B alleles for all members of this family. Assume that individuals II-6 and II-7 have identical genotypes and that IV-2 is homozygous.

One individual in the pedigree has an unusual genotype.

d. Who is it? What is unusual about their genotype?

e. What is the genetic mechanism underlying this unusual genotype? When does it occur? Provide a diagram of the chromosomes that will account for this individual's genotype. Be sure to indicate the *A, B,* and *Y* loci, as well as the position of the centromere.

A 50-bp segment of the *yodeling* gene is shown here. It includes the first 15 amino acids of the Yodeling protein.

ACACGGTATG GATCATACGG ATTGCTAATG ACTGTGATAG GCATGTACCA
TGTGCCATAC CTAGTATGCC TAACGATTAC TGACGGCGAT CGTACATGGT

f. Indicate the polarity of the DNA and the strand that serves as the template for transcription. Determine this 15-amino-acid sequence.

g. Suppose you want to explore how a particular change in the *yodeling* DNA sequence can affect the yodeling trait. You perform a site-specific mutagenesis to change the asparagine codon to a histidine codon. If the oligonucleotide you used is 21 bases long, determine the sequence of the oligonucleotide used in the mutagenesis, and indicate the 5' and 3' ends.

h. Suppose that you want to use site-specific mutagenesis to generate a nonsense mutation. Using an oligonucleotide that is 21 bases long, determine the sequence of the oligonucleotide, along with its 5' and 3' ends, used to change a single nucleotide in the *yodeling* gene.

 Do you need additional review? Visit **www.mhhe.com/hyde** for practice tests, answers to end-of-chapter questions and problems, interactive exercises, and animation tutorials all designed to enhance your understanding of key genetic concepts.

Genomics and Bioinformatics

Essential Goals

1. Describe the two different approaches used to sequence a genome.

2. Discuss how the different genetic maps (sequence, linkage, cytogenetic) correlate with one another and why this is important.

3. Describe the methods used to annotate the genomic DNA sequence.

4. Understand what a microarray is, how it is performed, and what information it provides.

5. Know what two-dimensional gel electrophoresis is, how it is performed, and what information it provides.

6. Outline a yeast two-hybrid experiment and the type of data it provides.

7. Compare and define genomics, transcriptomics, and proteomics.

Chapter Outline

Photo: Microarray chip that is used to examine the expression of thousands of genes simultaneously.

*I*n the spring of 2000, J. Craig Venter, CEO of Celera Genomics, and Francis Collins, director of the National Institutes of Health's Human Genome Research Institute, jointly announced that they and their colleagues had completed a draft sequence of the human genome. It was not until three years later, however, that the human genome sequence was considered "finished."

Although sequencing the human genome represented an enormous accomplishment, it is only a small part of the story. The first organism to have its genome sequenced was the bacterium *Haemophilus influenzae,* which was reported in 1995. Since that time, 679 different genomes have been completely sequenced in organisms that range from viruses and bacteria to plants to insects to mammals. As of November of 2007, the genomes of 2,315 different organisms (over half of which are bacteria) are currently being sequenced. During the intervening 12 years, the ability to sequence large and complex genomes has improved dramatically.

The field of **genomics,** the study of mapping and sequencing genomes, is now firmly established. To some, this field involves elucidating the very secret of life. To others, it opens the possibility of determining the genetic differences between various organisms.

Along with the power associated with genomics comes enormous responsibility. Significant moral and ethical considerations are associated with using this information to predict the susceptibility that an individual will develop certain genetic diseases. Society needs to discuss and consider how the government, employers, and insurance companies will utilize this information.

Genomics has allowed science to begin thinking about other broad-ranging questions as well. Because proteins are the molecules that usually confer the phenotypes to an organism, researchers have developed the field of proteomics. **Proteomics** is the study of all the proteins in an organism. This includes how the quantities of individual proteins change during a process (such as tumor progression), how proteins regulate gene expression, and how proteins interact to generate a biochemical or physiological process.

These analyses also produce large sets of data, which range from generating and analyzing genomic DNA sequences to all the potential protein interactions in an organism. To process these volumes of data, the field of **bioinformatics** has been created. In this chapter, we discuss some of these research areas and the types of analyses that are employed in their study.

14.1 Approaches to Sequencing a Genome

In chapter 12, we described the methodologies of DNA cloning and sequencing using dideoxynucleotides and DNA polymerase. These approaches were used until the late 1980s to clone and sequence individual genes. At the end of the 1980s, genome centers were beginning to sequence chromosomes or the entire genome of an organism. These centers used the same basic DNA sequencing technology employed to sequence individual genes. However, these centers possessed large numbers of DNA-sequencing machines to generate the raw sequence and computers to assemble the large amount of sequence data. In this chapter, we concentrate on the two major strategies used to generate and analyze the genomic DNA sequences: *ordered contig sequencing* and *shotgun sequencing.*

The Challenges of Sequencing

To illustrate the problem of sequencing the genome of an organism, let's consider the human genome. It is composed of approximately 3 billion base pairs (bp). If we could begin sequencing the DNA at the end of one chromosome, it would be a technical feat to sequence the entire chromosome.

The maximum sequence from an oligonucleotide primer would first need to be determined. This

sequence, which is called a **read,** corresponds to 600–1000 bp of unambiguous sequence. We could then generate a new oligonucleotide primer near the end of the read to initiate a new sequencing reaction and extend the sequence another 600–1000 bp. Thus, it would take an excruciating amount of time to systematically sequence along the chromosome.

The problem is somewhat aided by the presence of 22 autosomes and two sex chromosomes in humans. While we would need to identify the sequence immediately adjacent to the telomere on each chromosome, it would be possible to move along all the chromosomes at the same time. We could double the speed by sequencing from both ends of each chromosome simultaneously, but this would still be very slow.

To provide you with an idea of how long this approach would take, assume that an oligo primer can be synthesized in a few hours and the DNA-sequencing reactions can be performed and analyzed in several hours. This cycle—producing a primer, sequencing, and analyzing the sequence to identify the next primer—could be performed in 24 hours.

If we assume that the 24 human chromosomes (22 autosomes + X + Y chromosomes) were of equal length (which they are not), then the 3 billion genomic bases could be sequenced in:

3×10^9 bases/750 bases/read = 4×10^6 reads

4×10^6 reads/24 chromosomes read simultaneously = 16.67×10^4 reads/chromosome

16.67×10^4 reads/chromosome/2 ends per chromosome = 8.335×10^4 reads from each chromosome end

8.335×10^4 reads from each chromosome end/365 days/year = 228 years to sequence the human genome

Clearly, 228 years to sequence the human, or any other, genome is unacceptable for a variety of reasons, both scientific and economic.

To increase the rate of sequencing, the genome must be sequenced from multiple locations within each chromosome. Two different methods can generate these multiple sequencing start sites. In the **ordered contig sequencing** approach, genomic clones are identified and arranged in an overlapping series, similar to a genomic walk (see chapter 13). The minimal set of overlapping clones that encompasses a genomic region is called a **contig.** If we sequence a number of different contigs on each chromosome simultaneously, we would dramatically decrease the time needed to complete the sequencing of the entire genome.

The **shotgun sequencing** approach involves randomly sequencing genomic clones. The time saved by not generating the contig could be used to generate additional genomic DNA sequences. This approach requires more effort, however, to assemble the sequencing reads into a contiguous DNA sequence. This problem has been solved using computers and complex programs.

Ordered Contig Sequencing

The basic concept behind the ordered contig sequencing strategy is shown in **figure 14.1**. Briefly, a number of genomic clones are isolated and then arranged to generate a **physical map.** This physical map, which is a set of overlapping genomic DNA clones, represents all the genomic DNA sequences for a chromosome. The genomic clones are arranged relative to each other by comparing restriction maps and the presence of DNA markers, such as genes, minisatellites, and microsatellites.

From this physical map, a **minimal tiling path** is generated, which is a set of the fewest number of overlapping DNA clones that contains all of the genomic sequences. Each genomic clone within the minimal tiling path is then subcloned into smaller fragments and sequenced.

Because of the physical map, we know the location of each DNA sequence that is generated relative to the other genomic DNA clones. This simplifies the assembling of the large number of DNA sequences into a contiguous single sequence for each chromosome.

The vectors of choice for carrying very large genomic DNA inserts are either the bacterial artificial chromosome (BAC) or the yeast artificial chromosome (YAC). Both of these vectors can accommodate inserts

of several hundred to thousands of kilobase pairs (see chapter 13). A genomic DNA library is generated in the desired vector, and then each clone is individually isolated and arranged into a grid pattern. Usually, microtiter well plates that contain 96 or 384 wells in a grid pattern are used for saving each clone. The library can then be screened to identify a positive clone, and we can then return to a specific well that contains the desired clone in a bacterial or yeast cell.

Two general methods are used to produce the map of overlapping clones. First, we can take a specific DNA sequence, such as a microsatellite or previously cloned gene, and identify all of the genomic clones that contain this sequence. Alternatively, we can use restriction endonucleases to produce a pattern of DNA fragments unique for each clone.

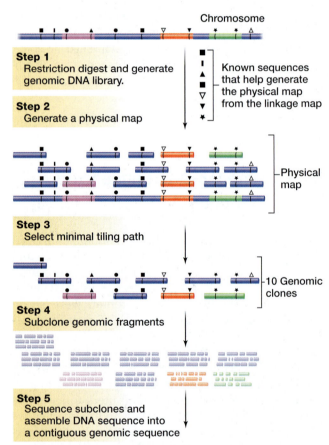

...ACAGGATACTAGCATACTAGGCATCGCCATACAGGATACTAGCATACTAGGCATCGCCATACAGGAT...

figure 14.1 General strategy of ordered contig sequencing. Clones from a genomic library are arranged using a variety of DNA sequences, such as ESTs, microsatellites, and other DNA markers (designated by circles, squares, triangles, and asterisks) to produce a physical map. The genomic clones are then organized into a minimal tiling path, which corresponds to all of the genomic sequences in the fewest number of genomic clones. These genomic clones are subcloned into smaller fragments that are sequenced. The contiguous DNA sequence is then assembled using the knowledge of the orientation and placement of each clone in the physical map.

Using a Known Sequence for Comparison

As an example of the first mapping method, we can arrange all of the clones in a grid pattern that mimics their arrangement in the microtiter well plates (**fig. 14.2**). Because the clones are separated during their spotting onto the membrane, this action is analogous to screening a library, as was discussed in chapter 12.

Screening the spotted array using gene *X* as the probe, reveals a number of positive clones, as shown in figure 14.2. Each of these clones must contain gene *X*. We can repeat this process and identify clones that contain gene *Y*, which we know is located on the same chromosome as gene *X*.

We could also take each individual genomic DNA clone and subject it to a PCR amplification reaction that uses oligo primers that will amplify a specific DNA marker, such as microsatellite #146 on chromosome 12 (see fig. 14.2). The clones that can amplify the PCR product will contain that specific microsatellite. We then compare the clones that were identified by the gene *X* probe, those that were revealed by the gene *Y* probe, and those that contain the microsatellite. In this example, we identified some clones that contain either one or two of the three DNA markers.

Of the clones that contain two markers, we found one clone that contains the microsatellite and gene *X* and another clone that contains the microsatellite and gene *Y*; however, we did not find a clone that contains both genes *X* and *Y*. The simplest explanation is that genes *X* and *Y* are far apart, and the microsatellite is located between them.

Random sequences can also be used in producing the physical mapping. **Sequence tagged sites (STSs)** are genomic DNA sequences that are generated from the ends of the genomic clones. Each read can be analyzed, and oligos can then be designed to PCR-amplify the read. We can then use these primers to try to PCR amplify the read from all of the genomic clones. Each genomic clone that produces an amplification product must contain the read sequence.

This information can be used to assemble the contig in the same manner as microsatellites. A major difference is that the STS is based solely on a random DNA sequence, and it may not be polymorphic like the microsatellite. The STS is also unlikely to be repeated at multiple locations in the genome as are the minisatellites.

Using Fragments Generated by Restriction Enzymes

An alternative method is to perform a restriction digestion using one or more restriction enzymes. Each restriction enzyme will produce a "fingerprint" pattern of DNA fragments that will be unique for each genomic clone (**fig. 14.3**). By comparing the fragment

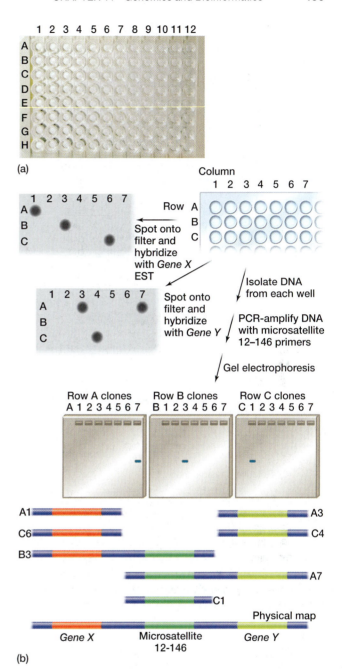

figure 14.2 Generating a physical map using either Southern hybridization or PCR amplification. a) A microtiter well plate with 96 wells is arranged in 8 rows (A–H) and 12 columns (1–12). b) Each well contains a different genomic DNA clone. DNA from each well can be spotted on a filter and hybridized with a probe (EST clone) to identify genomic clones that contain the same sequence. Alternatively, every clone can be subjected to PCR amplification with the same set of primers and only the clones containing the same sequence will exhibit a PCR product. Analyzing genomic clones that share common DNA markers allows the generation of the contiguous physical map shown at the bottom.

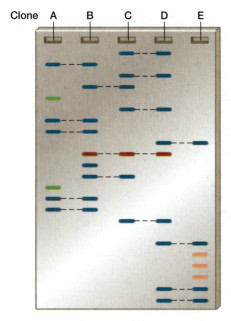

(a) DNA fingerprint

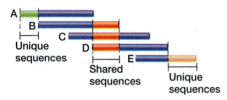

(b) Relative physical map

figure 14.3   DNA fingerprinting to generate a physical map. (a) Five different genomic clones are each digested with a restriction enzyme and then analyzed by gel electrophoresis. Common restriction fragments in different genomic clones are connected by dashed lines. (b) Comparing the common fragments between different genomic clones yields a relative arrangement of the genomic clones to each other and reveals the overlapping and unique regions.

patterns of several different genomic clones, it is possible to identify clones that contain overlapping regions and the relative amount of overlap between the clones.

In figure 14.3a, we find that clones A and B share common fragments, as do clones B, C, and D. However, clone A does not share any fragments with clones C, D, and E, which means that these clones do not overlap. Thus, we know that clone B must lie between clones A and C, D, and E. Continuing this type of comparison; we can generate a relative physical map of the clones (fig. 14.3b).

From the overlapping clones, we want to identify the minimal number of clones that contain all the genomic DNA sequences for a chromosome. This minimal tiling path reduces the amount of DNA sequence that must be generated to complete the entire genome (see fig. 14.1).

The genomic clones in the minimal tiling path are digested with restriction enzymes to produce smaller fragments. These fragments are then subcloned into a new vector. Because we want to confirm the sequence across the restriction enzyme sites in the genomic DNA, we independently digest the genomic DNA with at least two different restriction enzymes and sequence both sets of subclones (**fig. 14.4**). In this example, the enzymes *Eco*RI and *Bam*HI are used (chapter 12).

The fragments produced from the *Eco*RI restriction digest produces essentially all the genomic DNA information; however, we lack sufficient information to know how the different *Eco*RI fragment sequences join together. If we also sequence the *Bam*HI subclones, we can determine the sequences across the different *Eco*RI sites. This information allows us to determine how the different *Eco*RI fragment sequences join together.

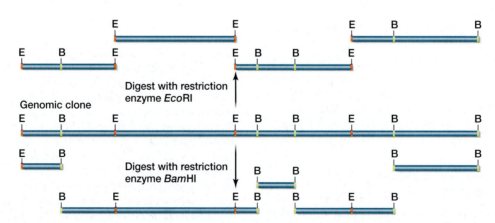

figure 14.4   Orienting DNA sequences across restriction enzyme sites. A genomic clone is digested individually with either *Eco*RI or *Bam*HI and then cloned for DNA sequencing. Sequencing the *Eco*RI clones reveals all of the sequence, but not how the different *Eco*RI fragments are joined together. Sequencing the *Bam*HI clones will produce the sequence across the *Eco*RI sites, which will reveal how the *Eco*RI fragments are joined together.

Because we know which subclone is derived from the larger genomic clones in the minimal tiling path, and we also know something about the overlap between different clones, we are able to fairly quickly assemble the DNA sequence that we generated.

Shotgun Sequencing

The map produced by ordered contig sequencing can be very useful in orienting the DNA sequence relative to a variety of DNA markers and genes, but it also requires a large amount of time and effort prior to starting the DNA sequencing. To circumvent this delay, the shotgun sequencing strategy was developed. Shotgun sequencing was first truly applied to a eukaryotic organism when Celera Genomics, a company headed by J. Craig Venter, completed the *Drosophila melanogaster* sequence.

Details of the Procedure in Drosophila

The shotgun sequencing approach employed by Celera Genomics is diagrammed in **figure 14.5**. The *Drosophila* genome was cloned into three libraries that each contained a different average insert size: 2 kb, 10 kb, and 150 kb.

Clones were randomly picked from all three libraries, and a single read (an average of 750 bp) was generated from both ends of each genomic DNA clone.

The sequence reads produced from both ends of a single genomic clone are known as **paired-end reads.** Generating a single read from both ends was simplified because the oligo primers used for the DNA sequencing reaction annealed to the vector sequence adjacent to the cloning sites. Thus, only two different oligos were required to generate the reads from both ends of all the clones in a single vector. The size of the genomic

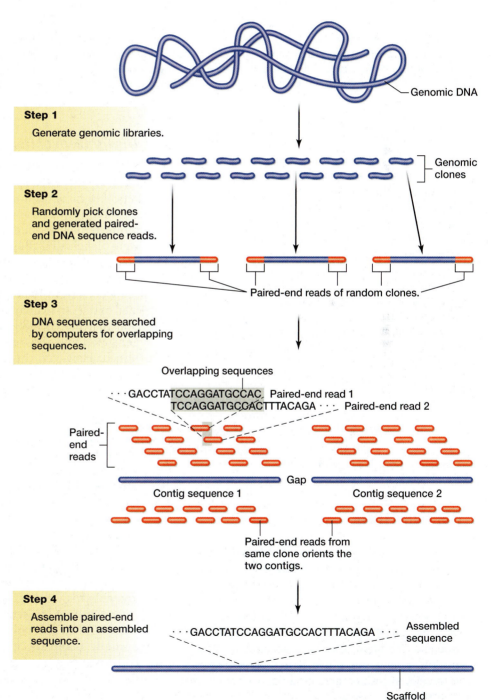

figure 14.5   Shotgun sequencing approach. DNA is sequenced from both ends of randomly isolated genomic clones to produce paired-end reads. Computer programs assemble the paired-end reads from thousands of genomic clones into contigs based on overlapping DNA sequences. Contigs that are separated by a gap can be joined to a common scaffold if a genomic clone has paired-end reads that align in both contigs. These paired-end reads also reveal the relative orientation of the two contigs.

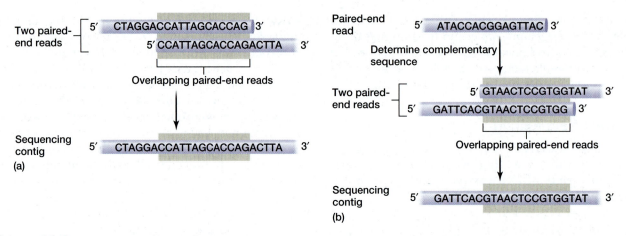

figure 14.6   Converting paired-end sequence reads into a sequencing contig. The two paired-end reads may exhibit the same overlapping sequence, which then requires only combining the two sequences (a). Alternatively, the complementary sequence of one paired-end read may match the paired-end read of another genomic clone (b). This requires taking the inverse complement of one of the sequences before combining them into the sequencing contig.

DNA clone also reveals approximately how far the two paired-end reads are separated from each other.

Celera Genomics automated this process, such that large numbers of machines could prepare the DNA for sequencing, run the sequencing reactions, and extract the sequence from the gel and store it in a computer file. This automation allowed Celera Genomics to generate the *Drosophila* sequence 24 h a day, 365 days a year. Automated sequencing machines can produce 96 reads in approximately 3 h, or nearly 600 kb of genomic DNA sequence per machine per day!

Sophisticated computer programs were developed to take the 750-bp average read and compare it with all the previously generated reads. When a region of significant nucleotide sequence identity or complementarity was found, the computer then stitched the two sequences together (**fig. 14.6**). In this way, the corresponding genomic clones became ordered relative to each other, while also producing a sequence contig. The **sequence contig** is the contiguous sequence through a region of a chromosome.

It's Your Turn

Computers are required to assemble the end reads from thousands of sequences in a genome project, however, you can try to manually align sequences from a limited number of reads in problems 14 and 15. Remember that either strand of the genomic DNA may serve as the template for the sequencing reaction. Therefore, you may need to convert the given sequence into the inverse (opposite strand polarity) complementary sequence to detect the overlapping sequence.

The Problem of Repeats

A problem was encountered that potentially could have doomed this approach, namely the presence of repeated nucleotide sequences. In some cases, these repeated sequences were tandemly arrayed (such as minisatellites and microsatellites), and in others they were dispersed throughout the genome (such as transposable elements, see chapter 18).

In the ordered contig approach, when repeated sequences were found, the region was already localized to the physical map. This simplified the confirmation that the correct flanking sequences were being generated and the total size of the repeat was determined. However, this information was not available initially during the shotgun sequencing.

These repeated sequences resulted in sequence contigs that were 150 kb on average, which were separated by gaps of unknown sequence and length. To cross these gaps, the researchers identified paired-end reads in which the two end reads resided in two different sequence contigs (**fig. 14.7**). This genomic clone then contained the sequence that was present in the gap; it joined two different sequence contigs and oriented them relative to each other. Because the size of the genomic clone was known, as was the location of the paired-end reads in the sequence contigs, it was possible to calculate the size of the gap between the sequence contigs.

The different sequence contigs that were joined by known gaps were called **scaffolds** (see fig. 14.5). The clones linking the sequence contigs could be further sequenced to the point of the repeated sequence, to partially fill in the gaps. The remaining gap would correspond only to the repeated sequence. Because the size of the gap and the sequence of the repeat were both known, it was straightforward to calculate the number of repeats that were present in the gap.

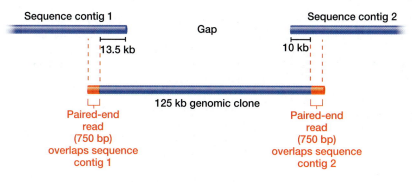

figure 14.7   Determining gap size between sequence contigs. A single genomic clone is identified that has its two paired-end reads residing in different contigs, which reveals the orientation of the contigs in a scaffold. Knowing the size of the genomic clone (125 kb) and how far each paired-end read is from the end of the contig, it is possible to determine the size of the gap between the contigs (100 kb).

One of the advantages of the shotgun sequencing approach was that the clones (and therefore the DNA sequences) were randomly selected and sequenced. This randomness led to the same sequence being determined multiple times from different clones, which produced a degree of confidence that the determined **consensus sequence** was correct. Do not confuse consensus sequence with wild-type sequence or the genomic sequence that all individuals must possess. The presence of repeats, single-nucleotide polymorphisms (SNPs, see chapter 13), and mutations throughout the genome will result in each individual possessing a different genomic DNA sequence.

Completing the Sequencing Project

In the absence of a complete minimal tiling path, the human genome was to be sequenced using the shotgun approach. The goal of the human genome **draft sequence,** completed in the year 2000, was to assemble sequence into scaffolds that covered at least 90% of the genome, with some ambiguities. The **finished sequence,** announced in April 2003, contained no gaps and was 99.99% accurate. Here we describe how the draft sequence was turned into the finished sequence.

The First Requirement: Increasing the Level of Accuracy

The first requirement was improving the level of accuracy in the finished versus the draft sequence. Each nucleotide position in the genome is sequenced from several different clones, and the position is sequenced on both strands. It is not uncommon for some ambiguity to exist at a particular position in a DNA sequencing read. The identity of the nucleotide, however, is often conclusively revealed on the subsequent reads.

Researchers feel that a draft sequence requires an average of 3 to 5 reads for each nucleotide position, whereas the finished sequence may require a total of 10 reads per nucleotide position. As the number of reads for each position increases, the accuracy of assigning a specific base to that position also increases. Therefore, the accuracy is largely a technical consideration, which requires additional time to ensure the correct sequence.

The Second Requirement: Eliminating the Gaps

The second requirement to move to a finished sequence was the removal of any gaps in the composite genomic sequence. This task can be more difficult to achieve than the accuracy because of the presence of repetitive elements throughout the genome.

Two major problems come up with repeats. First, when the genomic clone contains two or more repeats, they can recombine in either the *E. coli* or yeast cell that is the host for the genomic DNA library. This recombination can result in either the loss or the inversion of the sequences located between the repeats (**fig. 14.8a**).

Second, if the repeat is longer than the read, then the sequence produced from the read cannot be assigned to a unique location in the genome, or the sequences

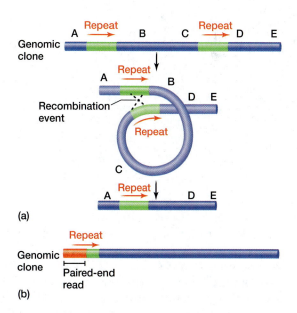

figure 14.8   Presence of repeats in a genomic clone can cause two problems. (a) The repeat sequences may recombine in the host cell. This will result in either a deletion of the sequences between direct repeats or an inversion of the sequence between the repeats (if the repeats are in an inverted orientation). (b) If the paired-end read contains a portion of the repeated sequence, then the paired-end read may not correspond to a unique location in the genomic sequence.

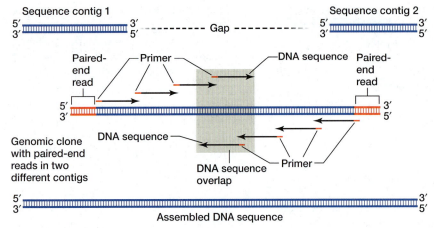

figure 14.9 Determining the gap sequence between two sequence contigs by primer walking. A single genomic clone, which contains two paired-end reads that join two sequence contigs, will serve as the DNA sequencing template. Oligonucleotide primers, generated from the two DNA sequence contigs, are used to produce sequence reads from the spanning clone. At the end of the sequence read, a new sequence is used to synthesize a new sequencing primer. New sequencing reads and primers continue to be generated from both ends of the spanning genomic clone until they overlap.

that flank the repeat cannot be determined (fig. 14.8b). Rather, the read sequence could be associated with any location that houses the repeat. This can result in the unexpected assignment of one DNA sequence to the wrong sequence contig or to the creation of an incorrect scaffold.

Another mechanism that generates gaps in the cloned DNA occurs because some sequences are not stable in *E. coli* (and sometimes yeast). These sequences may encode proteins that are lethal or harmful to the transgenic host cell or may not be properly replicated. The result is that clones containing these sequences are either absent or rare in the library, which results in a gap in the sequence contig.

To determine the sequence associated with gaps, two different approaches are often used. First, a genomic clone may be isolated that spans the gap region, even if it requires the clone to be several hundred kilobase pairs in length (see fig. 14.7). This large clone can be sequenced using a primer walking approach (**fig. 14.9**). DNA is sequenced from each end of the genomic clone. Near the end of the initial reads, a new primer sequence is selected, and oligo primers are synthesized in vitro (see fig. 14.9). These new primers are used to generate a new sequence read that extends further towards the middle of the genomic clone. The process is continued until the gap is entirely sequenced. This process can be slow if the gap is very large.

Alternatively, the genomic clone can be digested with a restriction enzyme, and the fragments subcloned and sequenced (see fig. 14.4). If the gap is small, PCR amplification can be used to isolate the genomic DNA fragment, which could be sequenced without cloning (fig. 14.10). This strategy would allow the DNA in the gap region, which is unstable in the host cell, to be sequenced without cloning.

Step 1

PCR amplify genomic DNA using primers to the A and D sequences.

Step 2

Use A and D primers to sequence PCR product.

Step 3

Generate new primers from the end of both sequence reads to sequence PCR product again.

Step 4

Repeat step 2 until sequence reads overlap.

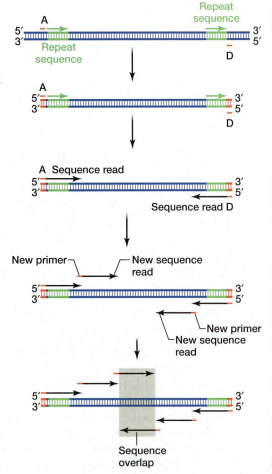

figure 14.10 Primer walking approach to sequence across small gaps in genomic DNA. Oligonucleotide primers (A and D) to regions outside the direct repeat sequences (green) are used to PCR amplify the genomic DNA region. Rather than clone this sequence and potential recombination-induced deletions, the primers are used to sequence the PCR product directly. At the end of the sequence read, new primers are designed to again sequence the same PCR product. This sequential generation of a primer for DNA sequencing from the previous sequence read is called primer walking. Primer walking continues until the sequence reads from both directions overlap, which yields the entire sequence between the repeats.

Another potential problem associated with repeats is the generation of sequence reads or scaffolds that join two noncontiguous regions through a common repeat that is dispersed throughout the genome. To reduce the possibility of these genome "jumping" errors, clones must be identified that contain the repeat sequence. If the paired-end reads from this clone flank the repeat, then it will join the correct sequence contigs. Because the repeats rarely extend more than several kilobase pairs, genomic clones that are 150 kb should easily span the repeat.

Time Considerations in Genome Sequencing

It should be obvious that the amount of time and effort increases as the sequence nears the "finished" status. Although initially any sequence information is useful in a sequencing project, only very certain sequences, and ultimately only specific bases, are required to move to the finished status.

Scientists are constantly reevaluating the standards for producing a "finished" project to keep the costs under control so that funds will be available to perform sequencing projects on other organisms. Recently, there seems to be a movement towards generating a draft sequence of a large number of organisms, rather than producing finished sequences for a smaller number of organisms. First, investigators feel that the sequences derived from an increased diversity of organisms will be useful for a variety of studies. Second, the draft sequence is often sufficient for researchers to identify genes and begin their analyses. Third, researchers will likely sequence particular areas of the genome that they are working on, which will ultimately increase the accuracy of the overall sequence.

Restating the Concepts

▶ Ordered contig sequencing involves producing a minimal tiling path of overlapping genomic clones that covers the entire genome. Each genomic clone is then subcloned into smaller fragments and then sequenced. The sequences are assembled into a complete genomic sequence based on the position of the genomic clones.

▶ Shotgun sequencing involves the sequencing of the ends of randomly selected genomic clones (sequence reads) and using massive computer power to assemble these reads into the genomic sequence.

▶ To move from a draft sequence to a finished sequence requires multiple sequencing of each base from many different clones to ensure the accuracy of the sequence and the completion of sequencing through gaps.

14.2 Correlating Various Maps

The DNA sequence alone is not of great use. We need landmarks within the sequence to provide an indication of relative location as we try to identify genes and mutations.

Imagine that this textbook lacked the chapter titles, chapter numbers, and any section headings. If all of the remaining text was then run together with no punctuation or spaces between words, it would be extremely difficult to read, let alone to understand. We need similar markers to correlate with the genomic DNA sequence to make the sequence useful. These markers can take a variety of forms, which are discussed in the following sections.

Linkage Maps

The most obvious useful markers are known mutations. You previously learned how to use recombination frequencies to generate a linkage map of either mutations or DNA markers (see chapters 6 and 13). When correlating linkage maps with DNA sequence, one significant point must be kept in mind: Recombination does not occur at the same rate throughout the genome.

Some genomic regions are "hot spots" for recombination, which means that recombination is more likely to occur in these regions relative to other areas. For example, 4.0 map units (mu) between two genes may correspond to a smaller physical distance than 4.0 mu between two other genes. Thus, the linkage map cannot simply be laid over the DNA sequence. The mutations that are placed on the linkage map, however, can serve as relative markers for comparison to the DNA sequence.

Linkage maps produced from DNA markers often exhibit a higher resolution than does a mutation map. We previously described these DNA markers, which include RFLPs, VNTRs, microsatellites, and SNPs. These markers usually occur at a higher density than the mapped mutations because of the greater number of DNA markers that are available to generate the map. Furthermore, the greater frequency of individuals who are heterozygous for these DNA markers relative to individuals heterozygous for mutations increases the number of meaningful matings and offspring, which tends to make the map more accurate (see chapter 6).

The locations of microsatellites on zebrafish chromosome 21 are shown in **figure 14.11**. Notice that the distribution of the microsatellites is not uniform throughout the chromosome, with some regions lacking any markers (between 73.2 and 110.0 cM), and other regions containing several markers (19 microsatellite markers at 3.7 cM). Regardless of the uneven distribution, the presence of a large number of micro-satellites along the length of this chromosome is much denser than the *seven* current mutations that have been mapped on this chromosome.

The high density of DNA markers is best exemplified in the SNPs, which are located approximately every 1000 bp throughout the genome of most organisms. This extremely large number of SNPs is the basis for the human HapMap (see chapter 13). It is anticipated that the human HapMap will permit the location of a large number of genes.

Unifying the Linkage and Cytogenetic Maps

To make the maps most useful, we would like to overlay the different maps. We can easily overlay parts of the linkage map (the DNA markers) onto the genomic sequence map by simply performing a computer search of the genomic sequence to identify the desired repeated sequences. This correlates the recombination map positions of the DNA markers to their physical location, which helps to approximate other recombination map data to the genomic DNA sequence.

We may also want to relate either the linkage map or the sequence map to the cytological map, or banding pattern, of the chromosome. This alignment requires the technique of **in situ hybridization.** Two basic types of in situ hybridization can be performed. The first involves annealing a nucleic acid probe to the chromosomes to identify the location of the complementary sequence within the chromosome. The second involves annealing a nucleic acid probe to the mRNA in a tissue to localize where the corresponding gene is expressed within an organism. We will discuss the first application here.

For chromosomal hybridization, the nucleic acid probe must be labeled so that its location can be easily detected. Currently, fluorescent molecules are attached to the probe, allowing the metaphase chromosomes (which are composed of two chromatids) and the label to be simultaneously observed under a microscope. This technique is called **fluorescent in situ hybridization,** or **FISH.** The technique is essentially like a Southern blot, but without the restriction digest and gel electrophoresis. An example of a fluorescent-labeled probe that is hybridized to human chromosome 11 is shown in **figure 14.12**.

The resolution of this map largely depends on the cytology of the chromosomes in an organism. For example, two probes that are 5 cM (or approximately 5000 kb) from each other in a human chromosome will likely appear as a single location in a FISH analysis. In contrast, a similar approach on the *Drosophila* polytene chromosomes can distinguish between two different probes that are less than 50 kb apart. **Figure 14.13** shows a non-fluorescent probe that is annealed to a *Drosophila* chromosome and the surrounding banding pattern of the polytene chromosome.

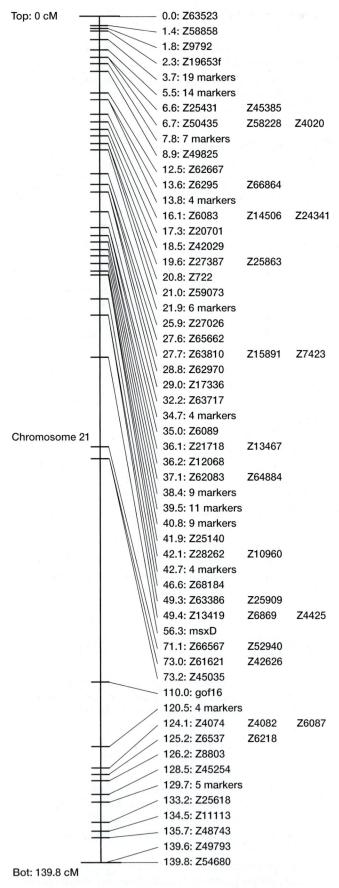

figure 14.11 Microsatellite map of zebrafish chromosome 21. The relative locations of the microsatellites on the chromosome are shown on the vertical line. To the right of the line is map position (in centiMorgans), and the name of the microsatellite present at that location.

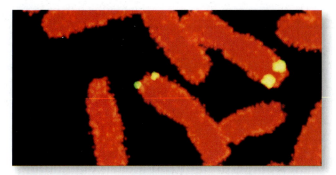

figure 14.12   Fluorescent in situ hybridization (FISH) involves hybridizing a probe to a specific DNA sequence on a metaphase chromosome. The location of the fluorescent-labeled probe is seen (bright yellow spots) using a laser scanning confocal microscope. The chromosomes are counterstained with propidium iodide, which makes them fluoresce red. In this case, the probe hybridized to a sequence on both human chromosome 11s (humans are diploid). The two yellow dots on each chromosome result from the presence of sister chromatids.

figure 14.14   Mapping inversion breakpoints in a polytene chromosome. An orcein-stain reveals the banding pattern of this *Drosophila* X chromosome containing an inversion. The inversion breakpoints, which are marked by the arrowheads, can be determined by comparison to the banding pattern of a wild-type *Drosophila* X chromosome.

figure 14.13   In situ hybridization of a cosmid clone to the *Drosophila* X chromosome. The arrowhead marks the location of the hybridization signal, which was pseudocolored red.

Chromosome Rearrangement Map

In chapter 8, we described a variety of chromosomal rearrangements, such as deletions, inversions, duplications, and translocations. A *cytogenetic map* is one that reveals the chromosomal banding pattern, which can be used to identify breakpoints and define the type of rearrangement.

One advantage of the cytogenetic map is that it is easily laid on the chromosome. The breakpoints are often defined initially by the change in the staining pattern of the chromosomal bands. In *Drosophila*, the cytogenetic map has a fairly high level of resolution because the polytene chromosomes contain hundreds of bands that have been well defined, and any disruption of that banding pattern can be specifically localized.

The distal half of a *Drosophila* X chromosome that contains an inversion (marked by arrowheads) shows the density of the bands (**fig. 14.14**). Comparing mutant chromosomes with the wild-type chromosome banding pattern, it is possible to define the breakpoints of a variety of different abnormalities.

It is possible to perform chromosomal in situ hybridization on either wild-type or rearranged chromosomes. For example, if a cDNA is located such that it spans an inversion breakpoint, we would expect that the cDNA probe would exhibit a single hybridization signal on a wild-type chromosome and two split hybridization signals on the inversion chromosome (**fig. 14.15**). When such a probe is hybridized to a *Drosophila* X chromosome that contains an inversion, investigators in fact do observe two hybridization signals. Because the probe sequence can be correlated to the genomic DNA sequence, the inversion breakpoint can be localized to anywhere from several hundred basepairs to an exact basepair.

Expressed-Sequence Tags

The linkage and cytogenetic maps can be very useful, but the genomic DNA sequence still lacks some features. Using our earlier text metaphor, we are still unsure when the "topics" change in the genome and how to read some of the words and sentences. We can get a picture of the gene locations using expressed sequence tags.

An **expressed-sequence tag (EST)** is a portion of a cDNA sequence, which corresponds to exons of transcribed genes. The EST sequence may be only 100–300 bp in length, and it may not correspond to the open reading frame, which prevents us from deducing the identity of the encoded protein. ESTs nevertheless serve as important landmarks that identify the location of a gene within the genomic DNA sequence.

We can generate ESTs by simply creating a cDNA library (see chapter 12), randomly picking clones, and

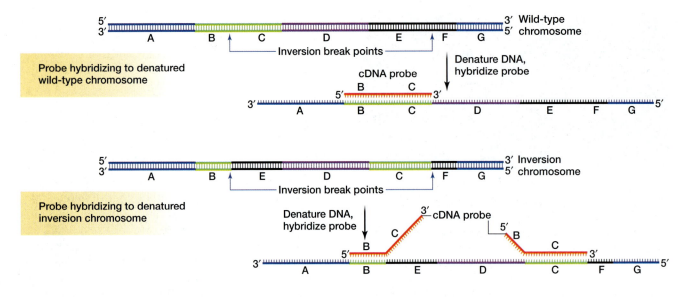

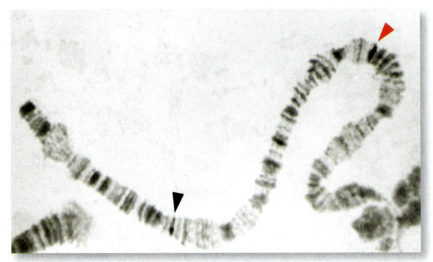

figure 14.15   In situ hybridization to an inversion chromosome. A cDNA (red) is complementary to a specific genomic DNA sequence (green) that is composed of two regions, B and C, in the wild-type *Drosophila* X chromosome. If an inversion breaks within this genomic sequence to separate B and C (middle panels), the same cDNA probe continues to hybridize to the green genomic DNA region that is now present as two sequences. The lower panel is an actual in situ hybridization of a cDNA probe to an inversion X chromosome in *Drosophila* (right). The red and black arrowheads mark the two in situ hybridization signals that correspond to the inversion breakpoints.

sequencing the ends of the individual cDNA clones. The sequences do not have to be as long as the genomic DNA sequence reads, because we are not trying to reassemble a contiguous sequence. Rather, we are only interested in generating sufficient sequence from the cDNA to produce a unique identifier.

Using computers to compare the EST sequences with the genomic DNA sequence, researchers can identify the genomic sequence that corresponds to the EST—thus revealing a gene location. Oligonucleotide primers that correspond to the ends of the EST sequence can also be used to PCR-amplify the EST from cDNA. This amplified EST can then be labeled and run through FISH to reveal the relative position of the gene within the chromosome.

The shotgun sequencing method makes it very difficult to correlate a specific genomic DNA sequence with a particular chromosome. Using the EST or another probe from the genomic DNA sequence for FISH provides an unequivocal demonstration of which chromosome contains that sequence.

It's Your Turn

Unifying different maps (sequence, physical, cytogenetic, and linkage) is not trivial because they all utilize different units and scales. However, estimations can be employed to generate these important unified maps. To examine what is involved when trying to generate a unified map, try Problem 42. You might also like to try Problem 19 to see how DNA markers can be used to assemble a contig from several genomic clones.

Unification of the Various Maps

The ultimate goal is to unify these various maps into a single map that allows us to move effortlessly between a linkage map location and the genomic DNA sequence and between chromosomal rearrangements and ESTs and potential genes.

Most of the genome-sequencing projects were originally performed on organisms that already had very strong linkage, cytological, or EST maps available. Many scientists felt that these other maps would be critical in assembling the sequence from a genome project into contiguous sequence per chromosome.

In contrast, J. Craig Venter and Celera Genomics believed that the shotgun sequencing approach could be completed without significant need for these other maps. It is now routine to initiate genome-sequencing projects without the other maps being available. Why have we spent so much time discussing these maps if they are not essential for the genome-sequencing projects?

The answer is that these other maps have proven extremely useful in the postgenomic sequencing era. As we quickly move to identifying and cloning disease genes, the ability to integrate information between the different maps becomes essential.

For example, if we are studying a gene associated with an inherited form of blindness, we can place it on the linkage map using lod score analysis (see chapter 13). This linkage map position can be compared with the genomic map to see whether the disease gene is located near a particular chromosomal rearrangement. The individuals used in determining the lod score can be checked to reveal if they also possess any chromosomal abnormalities in this region.

We can also use the genomic map to identify any DNA markers in the region in which the disease allele has been mapped, which may help to further define the location of the gene (**fig. 14.16**). These DNA markers can then be tested on the previously identified pedigrees to further map the location of the gene. Ultimately, we can use the genomic sequence to identify a candidate gene (see chapter 13) that may correspond to the mutated or disease allele.

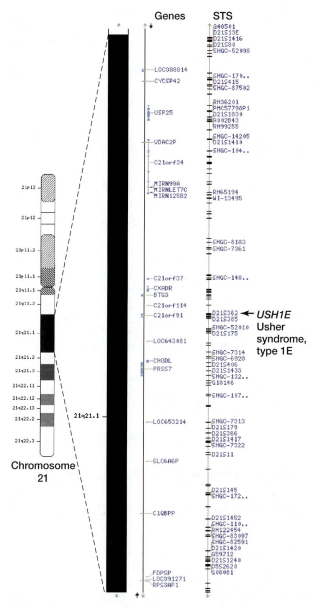

figure 14.16   Linkage map position of Usher syndrome, type 1E on human chromosome 21. The Usher syndrome gene is mapped relative to the genes that were identified in the genomic DNA sequence and SNPs.

Restating the Concepts

▶ To make the genomic sequence meaningful and useful, we must correlate landmarks with the genomic sequence.

▶ The linkage map can consist of DNA sequences or mutations (or both) that have been mapped through recombination. The linkage map is provided in a map unit or centiMorgan (cM) scale.

▶ The cytogenetic map is the banding pattern on a chromosome. A chromosomal rearrangement map is a variation of the cytogenetic map in which the banding pattern of rearranged chromosomes is examined.

▶ An EST map reveals the location of genes within either the cytogenetic map or the genomic DNA sequence.

14.3 Annotating the Genomic Sequence

We can think of the linkage, cytological, and EST maps as being analogous to the chapter headings of a textbook, the section headings within a chapter, and the paragraph breaks. All of this information would be helpful in starting to understand the meaning of the text. But we are still lacking the punctuation and spacing between words, namely where the genes are located and where the intron–exon boundaries occur in the genome.

Without this information, we may be able to identify the relative location of a gene, but we would have a hard time trying to piece together the open reading frame (ORF), especially if the gene is composed of several exons. The **annotation** of the genome involves trying to identify all of the genes and their corresponding ORFs for translation. (Remember that not every gene has an ORF; rRNAs, tRNAs, snRNAs, scRNAs, and miRNAs are not translated.)

Annotation requires a massive amount of computer analysis and the ability to learn the genomic nuances of the different organisms. We will focus on two major tasks in the annotation process, the identification of genes and the analysis of the genomic information.

Locating Genes Based on Sequence Data

Although EST mapping is a very powerful approach, it has several faults. First, some genes are expressed at only specific times in the life of the organism and in a particular tissue or subset of cells. This means that a complex organism, such as a human, would require a large number of different cDNA libraries that were generated from numerous tissues at a very large number of points in the lifetime of the organism to yield all the potential cDNAs.

Second, although cDNA libraries represent only the mRNA in the tissue, the frequency of each cDNA is relative to the abundance of the corresponding mRNA (some genes are transcribed at a high rate and others at a low rate). Thus, the cDNAs are present at levels that can differ by more than four orders of magnitude within a library. This implies that identifying every cDNA associated with an mRNA in a given tissue at a specific time would require sequencing an extremely large number of cDNAs to ensure that we identified all of the low-abundance cDNAs/mRNAs.

One final potential problem is that many genes exhibit alternative splicing, or the exons of a single gene are incorporated or removed during processing to produce different mRNAs. These alternative mRNAs may or may not encode different proteins. Each mRNA form for a given gene would correspond to a different cDNA, which may require the isolation of multiple cDNAs to account for all the exons in a single gene.

Computer Scan for ORFs

One approach to locate genes is to perform a computer search for ORFs. All the start and stop codons in all of the six potential reading frames can rapidly be identified (**fig. 14.17**). The problem becomes more difficult in eukaryotes, in which the exons may contain a portion of the ORF, and the flanking introns can be several tens of kilobase pairs in length. Because the exon may contain only a portion of an ORF, computer programs may often identify the region located

between two translation termination codons as an ORF. However, the true ORF in the mRNA begins with the translation initiation codon and ends with one of the three termination codons (see chapter 11).

Codon Bias

Another method to identify ORFs that likely correspond to a true exon is to look at **codon bias**, the preferential use of a particular codon for an amino acid. For example, *Drosophila* preferentially utilizes the GGC codon for glycine 43% of the time, whereas *E. coli* preferentially utilizes the GGU codon for this amino acid 59% of the time (**table 14.1**). In contrast, humans use the GGG codon for glycine much more frequently than either *Drosophila* or *E. coli*.

Codon biases also exist in humans. For alanine, which can be encoded by GCG, GCA, GCU, and GCC, humans utilize GCC approximately 41% of the time and GCG only 11% of the time.

These codon biases are likely due to differences in the relative abundances of the different tRNAs.

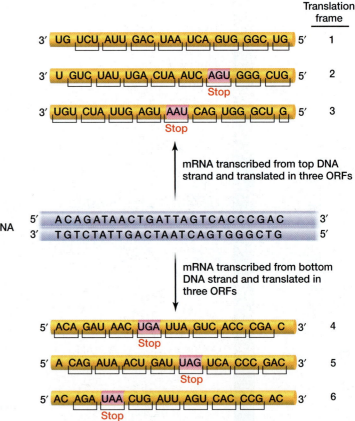

figure 14.17   Six potential reading frames are present in any segment of double-stranded DNA. Two different complementary mRNAs, labeled 5' to 3', can be transcribed from double-stranded DNA. The top mRNA is transcribed using the upper DNA strand as the template and the lower mRNA is transcribed using the lower DNA strand as the transcriptional template. Each mRNA has three different open reading frames, labeled 1, 2, and 3 for the top mRNA and 4, 5, and 6 for the lower mRNA. The translational stop codons in each reading frame are marked.

table 14.1 Codon Biases for Glycine in *Drosophila melanogaster, Escherichia coli,* and *Homo sapiens*

Codon	Percentage of Time Each Codon Is Used (%)		
	D. melanogaster	*E. coli*	*H. sapiens*
GGA	28	0	23
GGG	3	2	26
GGU	26	59	18
GGC	43	38	33

Knowing the codon biases for an organism helps to determine whether a putative ORF is more or less likely to encode a portion of a protein.

Gene Landmarks

Another method of gene identification is to use a computer to search for gene-related landmarks. A number of these landmarks were described in previous chapters.

For example, any gene that is transcribed must contain a promoter sequence that allows the RNA polymerase to properly initiate transcription. Although different promoters are recognized by the three eukaryotic RNA polymerases, common features exist for each promoter class. For example, many RNA polymerase II promoters recognize an AT-rich region at approximately 25 bp upstream of the transcription start site (**fig. 14.18**).

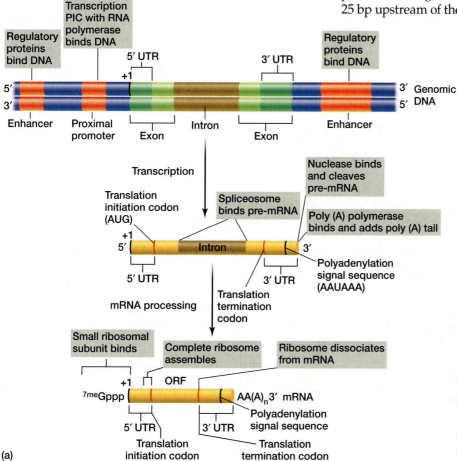

(a)

Gene structure	DNA example sequence	RNA example sequence
Enhancer element	variable	not present in RNA
Promoter TATA box	TATAAA	not present in RNA
Promoter CAT box	CAATT	not present in RNA
Translation start	ATG	AUG
Intron splice site	AG/GT··intron··AG/G	AG/GU··intron··AG/G
Translation stop	TAA, TAG, TGA	UAA, UAG, UGA
Polyadenylation site	AATAAA	AAUAAA

(b)

figure 14.18 Various sequences used to identify a gene. (a) Shown is a typical gene, the transcribed pre-mRNA and the processed mRNA. The DNA contains several functional sites, such as the promoter and enhancers. The pre-mRNA contains three functional sequences, the +1 nucleotide, the splice donor and acceptor sites (where the spliceosome binds), and the polyadenylation signal sequences. While these sequences are not functional in the DNA, their relative positions can be identified and are useful in defining a gene. For example, the 5' AAUAAA 3' polyadenylation signal sequence is not present in DNA, but the corresponding 5' AATAAA 3' is present and can be used to predict the potential 3' end of the gene. (b) A table with the consensus sequences for many of these elements is shown.

Additional landmark sequences that can be identified by computer programs include the various conserved promoter sequence elements that bind transcriptional activator and repressor proteins (see chapter 10), the conserved intron–exon boundary consensus sequences, and the conserved polyadenylation signal sequence. In bacteria, computer programs can be used to identify the –10 and –35 promoter elements and the Shine–Dalgarno sequence, which is bound by the small ribosomal subunit. The combination of multiple gene-related sequences associated with the same genomic DNA region increases the likelihood that a gene is truly present.

Computer programs (such as BLAST) can also use the previously identified genes in other organisms to locate related genes in a newly completed genomic DNA sequence. Two related proteins from different organisms should possess similar amino acid sequences. The computer can deduce the amino acid sequence of an ORF and compare this sequence with the deduced amino acid sequences from all the previously identified genes in large databases. Because of the redundancy in the genetic code, two related gene sequences are likely to possess less similarity than the two encoded amino acids (**fig. 14.19**).

EST Locations

An we already mentioned, the location of ESTs within the genomic DNA sequence is another powerful method for locating genes. Because ESTs are derived from sequencing cDNAs, they must correspond to the transcribed gene regions within the genomic DNA. It is relatively straightforward for a computer program to scan the genomic DNA sequence and identify all the matches

Comparison of crx nucleotide sequences

```
Human:    ATGATGGCGTATATGAACCCGGGGCCCCACTATTCTGTCAACGCCTTGGC
          ATGATG C TA AT AA C G    CCCCA TAT CTGT AACG  TT  C
Zebrafish:ATGATGTCCTACATAAAGCAG---CCCCATTATGCTGTGAACGGGTTAAC

Human:    CCTAAGGTGGCCCCAGTGTGGATCTGATGCACCAGGCTGTGCCTACCCAA
          CT    G C C  G  TGGA CTG T CAC   GC GT  CTACCCA
Zebrafish:ACTGTCCCGCCTCAGGAATGGACCTGCTCCACACCGCCGTCGCTACCCAG

Human:    GCGCCCCCAGGAAGCAGCGGCGGGAGCGCACCACCTTCACCCGGAGCCAA
          C C CC AGGAAGCAGCG CG GAGCGCACCACCTTCAC CG A CCA
Zebrafish:CCACTCCGAGGAAGCAGCGTCGAGAGCGCACCACCTTCACTCGCACCCAG

Human:    CTGGAGGAGCTGGAGGCACTGTTTGCCAAGACCCAGTACCCAGACGTCTA
          CTGGA    CTGGA GC TGTT  CCAA AC C  TA CCAGAC T T
Zebrafish:CTGGACATTCTGGAAGCTTTGTTCACCAAAACACGCTATCCAGACATATT

Human:    TGCCCGTGAGGAGGTGGCTCTGAAGATCAATCTGCCTGAGTCCAGGGTTC
          T   G GA GAGGT GCTCTGAA ATCAA CT CC GAGTCCAG GTTC
Zebrafish:TATGAGAGAAGAGGTAGCTCTGAAAATCAACCTTCCCGAGTCCAGAGTTC

Human:    AGGTTTGGTTCAAGAACCGGAGGGCTAAATGCAGGCAGCAGCGACAGCAG
          AGGT TGGTT AAGAACCG  G GCTAAATGC G CAGCAGC  CAGCAG
Zebrafish:AGGTGTGGTTTAAGAACCGCCGTGCTAAATGCCGCCAGCAGCAGCAGCAG
```

Comparison of CRX amino acid sequences

```
Human:    MMAYMNPGPHYSVNALALSGPSVDLMHQAVPYPSAPRKQRRERTTFTRSQLEELEALFAKTQ
          MM Y   PHY VN L LS    DL H AV YP  PRKQRRERTTFTR QL   LEALF KT
Zebrafish:MMSYIKQ-PHYAVNGLTLSASGMDLLHTAVGYPATPRKQRRERTTFTRTQLDILEALFTKTR

Human:    YPDVYAREEVALKINLPESRVQVWFKNRRAKCRQQRQQ
          YPD   REEVALKINLPESRVQVWFKNRRAKCRQQ QQ
Zebrafish:YPDIFMREEVALKINLPESRVQVWFKNRRAKCRQQQQQ
```

figure 14.19 Comparison of nucleotide and amino acid sequences. Approximately the first 300 nucleotides of the human and zebrafish (*Danio rerio*) *cone-rod homeobox* (*crx*) genes are shown (top). The sequence between the human and zebrafish sequence lines lists the location of identical nucleotides between the two sequences. Three dashes were placed in the zebrafish sequence to maximize the nucleotide match with the human sequence. The amino acid sequences of the CRX proteins are shown (bottom). The sequence between the human and zebrafish sequence lines shows the location of identical amino acids between the two sequences. One dash was placed in the *Danio* sequence to maximize the nucleotide match with the human sequence and corresponds to the three dashes in the nucleotide sequence above.

to EST sequences. These ESTs often identify the 5' or 3' untranslated sequence of the mRNA, because the EST sequence corresponds to one of the ends of the cDNA.

Although the 5' or 3' UTR (untranslated region, see chapter 10) may not reveal any significant information about the encoded protein, it does provide unambiguous proof that the genomic region contains a transcribed gene. The major difficulty with using ESTs to identify genes is that the correct cDNA libraries must be screened; otherwise, a gene expressed in a different tissue or another time in development will not be present. Therefore, the completeness of localizing genes based on the presence of ESTs depends on the number of cDNA libraries analyzed.

Because the EST usually corresponds to the 5' or 3' end of the mRNA, the location of exons that encode the internal mRNA sequences are usually not identified. Furthermore, alternative splicing of these central exons are not revealed. However, use of ESTs to confirm the location of a transcribed gene within genomic DNA sequence, coupled with the previously described methods, usually provides a good description of the gene organization.

Compiling Gene Identification in Genomic Sequences

Each of the methods just discussed usually locates only a subset of the genes in the genome. Looking at a given region of the genomic sequence, we see some genes are well defined by possessing a large number of different components, and other genes are defined by only one or two components. The final determination of the presence of a gene requires direct examination of the specific genomic DNA sequence for the presence of a gene.

One surprising result from the Human Genome Project was the small number of genes identified. Many investigators thought that the genome must contain a large number of genes to encode the complexity of the human organism. It was already known that *E. coli* contains approximately 4400 genes, *Drosophila* 16,000 genes and the nematode *C. elegans* approximately 20,000. Many scientists estimated that the human genome should contain between 100,000 and 200,000 genes. After sequencing, the human genome was found to contain closer to 30,000 genes—only 10,000 more than that of *C. elegans*.

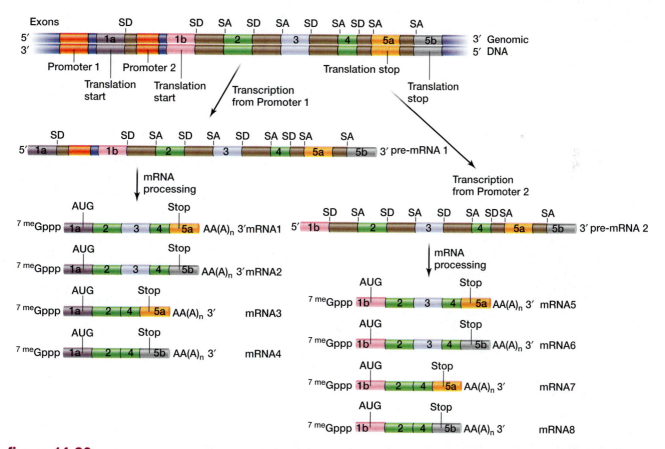

figure 14.20 Alternative splicing of a gene. In this example, a single gene has two different promoters (Promoter 1 and 2) that begin transcription at different locations (exon 1a and exon 1b, respectively) and produce two different pre-mRNAs. The pre-mRNAs are spliced by joining a splice donor site (SD) with a splice acceptor site (SA) and the removal of any sequences in between, including introns (brown). The SD at the 3' end of exon 4 can join with the SA at the 5' end of either exon 5a or 5b. Notice that exon 5a lacks a SD site, so it cannot be spliced to exon 5b. The result is eight alternative mRNAs that differ in either the 5' end, the 3' end, or within the middle of the open reading frame. Each of these eight different mRNAs contains a different open reading frame that will be translated into one of eight different proteins from a single gene.

Another surprise was the large amount of alternative splicing. The basic feature of alternative splicing is a gene that joins common exons with different exons to produce different mRNAs (**fig. 14.20**). In some cases, a single gene may have a number of promoters that control the transcription of a gene at either different times or different tissues during development. Each promoter may transcribe a different first exon that then splices into the common exons of the gene (see fig. 14.20, pre-mRNAs 1 and 2).

A single gene may also have different exons at the 3' end of the gene, which may have an effect on the stability of the mRNA (see fig. 14.20, mRNAs 1 and 2). Finally, the gene may use different exons within the ORF (see fig. 14.20, mRNAs 2 and 3). In this case, a single gene may encode different proteins that may have related, but different functions. This provides one mechanism for increasing the genetic complexity without increasing the number of genes.

In the case of humans, over half of the human genes are alternatively spliced, with an average of three splice variants per gene. We will explore one of the best examples of alternative splicing when we discuss *Drosophila* sex determination in chapter 21.

As the sequences of different genomes were completed, the impact of these data has reached beyond what was originally anticipated. By comparing the genes in the different genomes, the functions of the encoded proteins have been deduced for many of the genes (**fig. 14.21**). But for over 30% of the predicted proteins in the human genome, no function has yet been assigned.

In September of 2005, a consortium of researchers supported by the National Human Genome Research Institute of NIH published the genomic sequence of the chimpanzee, *Pan troglodytes*. This sequence permitted a direct comparison to the human genome sequence, which was found to be 96% identical. If DNA insertions and deletions are not considered, the two genomes share 99% identical sequences. However, this is still approximately 10-fold more differences than what is observed between two different humans.

If the putative protein sequences are compared, approximately 29% of the genes encode identical amino acid sequences in humans and chimps. On average, the typical human protein differs by only a single amino acid compared with the chimp protein.

Although the two genomes are very similar, it is the differences, which amounts to 35 million of the 3 billion nucleotides, that distinguish the two species. Analyzing these differences, researchers identified regions that very rapidly accumulated mutations in humans relative to the chimp genome. One of these regions corresponds to the human *FOXP2* gene, which is implicated in the ability to develop speech. Thus, studying the genomic differences between humans and chimps may reveal many of the genetic elements that differentiate us from our closest living relative.

Restating the Concepts

▶ Annotation of the genomic sequence involves trying to identify all the genes. This involves computer programs and the physical analysis of cDNAs.

▶ In addition to the number of genes, the genetic complexity can be increased through alternative splicing of a gene to produce multiple different mRNAs that can be translated into different proteins.

▶ Comparing the genomic sequences of different species will reveal related genes. Just as important, the differences between the genomic sequences, especially between closely related species, will reveal some of the genetic causes for the different traits and phenotypes between the species.

14.4 Transcriptomics

Genomics has become much more than simply determining the DNA sequence of an organism's genome. It has direct implications on the identification of genes, influences our diagnosis and treatment of genetic diseases in the future, and can deepen our understanding of evolution.

Furthermore, genomics has pushed science to think in much broader terms. Many scientists used to think about cloning single genes and studying the function of each gene's encoded protein. Now, many

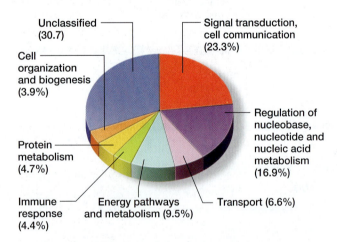

figure 14.21 The assigned molecular functions of the proteins encoded on the human X chromosome. These functions are based on either the known function of the human proteins or the similarity between the human protein's deduced amino acid sequence and proteins of known functions in other organisms. Notice that 30.7% of the proteins have not been assigned a function as of 2006.

Source: Reprinted by permission from Macmillian Publishers Ltd: *Nature Genetics*, vol. 37, no. 4, April 2005, copyright © 2005.

biologists are looking at how large groups of genes are regulated in response to a similar stimulus or disease, and how large numbers of proteins interact to form a signaling system or a structural component of a cell. Whereas genomics examines all the genomic sequence within an organism, **transcriptomics** is the study of the sequence and expression patterns of all the gene transcripts in an organism. Thus, transcriptomics is not concerned with the sequences between genes or in introns, which are both examined under genomics. In this section we explore this new area of biology.

SAGE Analysis

In chapter 12, we discussed Northern hybridization to reveal the expression pattern of a gene. As transcriptomics developed, researchers wanted to compare the expression patterns of thousands of genes between different tissues, different developmental stages, or between different conditions (normal versus diseased). Unfortunately, performing thousands of Northern blots would be slow.

Serial analysis of gene expression (SAGE) was an alternative that built on the high throughput DNA-sequencing efforts that were being used in the genome sequencing projects. The basic idea behind SAGE is that a short DNA sequence is sufficient to unambiguously identify a specific mRNA and gene. If a unique identification sequence could be generated from every mRNA in a tissue, the relative abundance of that sequence would be proportional to the relative abundance of the corresponding mRNA.

In SAGE, mRNA is isolated from the desired tissue and converted to cDNA using an oligo-dT primer that is attached to the small molecule biotin (see chapter 12). The cDNA is digested with a restriction enzyme (*Nla*III), which leaves a 3' nucleotide overhang of 5'-CATG-3'. This cDNA is mixed with beads that contain streptavidin, which binds the biotin (**fig. 14.22**). Because the oligo-dT is located in the region that corresponds to the 3' end of the mRNA, only the 3' end of the mRNA will be bound to the beads.

This bound mixture of *Nla*III-digested cDNA is divided in half. One of two different linkers is added to each half of the beads (see fig. 14.22). **Linkers** are short, double-stranded DNA molecules that are often synthesized with a specific sequence. Both linkers have a 5'-CATG-3' overhang, so they can base-pair with the *Nla*III-digested DNA. The remaining sequences in the two linkers are different for a later PCR amplification reaction.

The cDNA, still bound to the beads, is then digested with the *Bsm*FI restriction enzyme. *Bsm*FI cleaves 14 nucleotides after the 5'-GGGAC-3' recognition sequence. This sequence is generated when the linker ligates to the *Nla*II-digested cDNA. This digestion also releases a short DNA fragment from the bead. This 14 nucleotide

sequence that was derived from the cDNA represents the TAG.

The released DNA fragments are enzymatically treated to ensure that they have blunt ends. The two groups of DNA fragments, produced from the two linker reactions, are combined and ligated (see fig. 14.22). This ligation produces DITAGs that consist of two different TAG sequences. Oligonucleotide primers that are complementary to the linker sequences are added to the ligation reaction, and the ligated fragments are PCR-amplified. *Nla*III is then added, which again cleaves at the 5'-CATG-3' sequence and releases the linker sequences from the DITAG.

The DITAGs are ligated into a concatamer and cloned into a plasmid (see fig. 14.22). The DITAG concatamer is sequenced using primers that are located in the plasmid that flank the restriction enzyme site used to clone the DITAG concatamer. The sequence produces the 5'-CATG-3' sequence, followed by a 22-bp DITAG, another 5'-CATG-3' sequence, another 22-bp DITAG, and so on.

The DITAG sequences are then broken into the two TAG sequences and entered into the computer. A program can determine the number of times that a specific TAG sequence is present in a SAGE analysis (**table 14.2**). The frequency of each TAG sequence corresponds to the frequency of a specific mRNA in the original tissue. We can then compare the SAGE analysis between two different tissues or conditions and determine the change in gene expression.

One major advantage of SAGE is that all of the genes in the organism do not have to be identified prior to performing the analysis. SAGE will reveal a TAG sequence for a specific gene even if there is not a previously identified gene. We can subsequently identify the gene that corresponds to the TAG sequence. For example, table 14.2 contains two TAG sequences that are only identified as human ESTs rather than a gene encoding a known function.

Mass Processing: Microarrays

How can we examine all of the transcribed genes in a tissue simultaneously when there may be 20,000–30,000 genes? One approach would be to perform a Northern blot for every gene, in which a different cDNA would be used in each hybridization. This would be slow, and it would be difficult to maintain identical hybridization conditions for each of the 20,000–30,000 hybridizations. A more efficient approach is necessary.

The **microarray,** or **gene chip,** was developed to examine the expression of hundreds to tens of thousands of genes simultaneously. This method utilizes the standard nucleic acid hybridization, which has been slightly modified to work on a large number of samples at once. The basic idea is that a chip or microscope slide contains precisely arranged spots, each containing a single-stranded oligonucleotide or

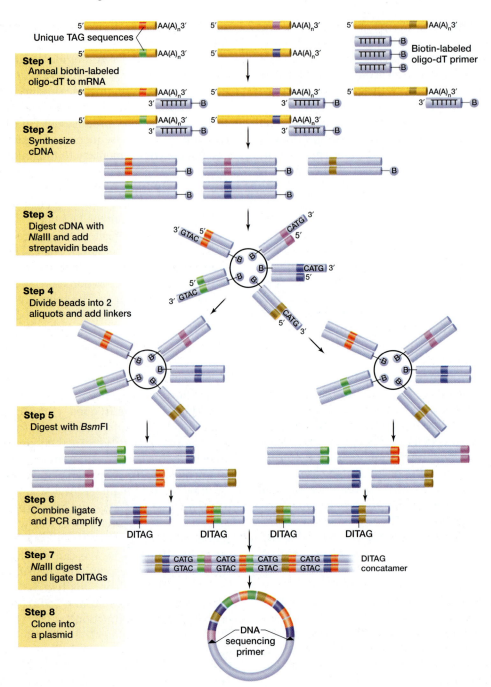

figure 14.22   General SAGE approach. Short sequences (TAG sequence, different colored segments) can serve as unique identifiers of each mRNA in an organism. The mRNA from a tissue or organism is converted to cDNA using oligo-dT primers that are linked to biotin (B). The cDNA is digested with the restriction enzyme *Nla*III, which removes the 5' ends of the cDNA, and the 3' end of the cDNA is bound to streptavidin beads through the biotin. Short DNA linkers are attached to the *Nla*III-digested cDNA ends to regenerate the *Nla*III restriction site and create a *Bsm*FI restriction enzyme site. Digestion of the DNA with *Bsm*FI cleaves the cDNA 14 or 15 bases downstream of the site, which releases a blunt-ended cDNA fragment from the streptavidin beads that contains the *Nla*III restriction site and the TAG sequence. Digestion of the cDNA fragment with *Nla*III creates a cDNA with a 3' overhang at one end and is blunt at the other end. These fragments are randomly ligated together to generate DITAG concatamers: two TAG seqences are ligated together through their common blunt ends and these DITAGs are joined to another DITAG through their common 3' overhang ends. This concatamer can be cloned into a plasmid and sequenced. Because the *Nla*III sequence is known, the DITAG sequences can be deduced and the frequency of any given TAG sequence in the concatamer population is directly proportional to the frequency of its mRNA in the tissue sample.

table 14.2   Twenty Transcripts with the Greatest Increase in Expression in Colon Cancer Identified Through SAGE Analysis

| TAG Sequence | Number of Occurences | | Corresponding Gene |
	Colon Tumor	Normal Colon	
CTTGGGTTTT	73	0	Insulin-like growth factor II splice form 1 (IGFII)
TACAAAATCG	42	0	Insulin-like growth factor II splice form 2 (IGFII)
GTGTGTTTGT	24	0	TGF-induced gene *Beta-igh3*
AAAAGAAACT	16	1	Human mRNA for poly(A)-binding protein
TGCTGCCTGT	15	1	*H. sapiens* HCG IV mRNA
CTGATGGCAG	14	0	EST 324128 3
GCCCAAGGAC	12	0	Human mRNA for actin-binding protein (filamin)
ACTCGCTCTG	12	0	EST 342926 3
ATCTTGTTAC	11	0	Human mRNA for fibronectin (FN precursor)
AAGCTGCTGG	10	0	Isoform 1 gene for L-type calcium channel, exons 41 and 41A
TGAAATAAAA	18	2	Human *hB23* gene for B23 nucleophosmin
TTATGGGATC	55	7	Human MHC protein homologous to chicken B complex
CAATAAATGT	60	9	Human mRNA for ribosomal protein L37
CTCCTCACCT	72	12	Human *Bak* mRNA
ACTGGGTCTA	29	5	*H. sapiens* RNA for *nm23-H2* gene
CTGTTGATTG	44	9	Human liver mRNA fragment DNA-binding protein UPI
TTCAATAAAA	56	12	Human acidic ribosomal phosphoprotein P1 mRNA
AAGAAGATAG	39	9	Human ribosomal protein L23a mRNA
CTGGGTTAAT	115	29	*H. sapiens* S19 ribosomal protein mRNA
CTGTTGGTGA	65	17	Human homolog of yeast ribosomal protein *S28*

The data represent the 20 human TAG sequences that exhibit the greatest increase in SAGE analysis comparing colon tumor tissue and normal colon tissue. HCG = human chorionic gonadotropin; MHC = major histocompatability complex.

Adapted with permission, from "Gene Expression Profiles in Normal and Cancer Cells," by Zhang et al., *Science*, 276: pp. 1268–1272. Copyright 1997 AAAS.

single-stranded cDNA that will be complementary to a different cDNA (**fig. 14.23**). Unlike SAGE, this microarray approach requires the identification of the genes prior to producing the oligonucleotides or cDNAs that are placed on the chip. Any gene that is not identified, will not have an oligonucleotide or cDNA synthesized and cannot be analyzed.

The probe for this hybridization will be synthesized from mRNA isolated from a particular tissue or developmental stage of the organism. The isolated mRNA is converted to single-stranded cDNA using reverse transcriptase and is then labeled with a fluorescent molecule (see fig. 14.23). When the labeled cDNA is applied to the chips, it anneals to the spot that contains the complementary sequence. This step is analogous to a Southern hybridization. Unhybridized probe can be washed away, and the chip or slide can be examined for the level of fluorescence in each spot. The brighter the fluorescent signal, the greater the amount of cDNA that is hybridized to the spot. This should correlate to a higher level of transcription of the gene in the tissue from which the mRNA was isolated.

Setting Up the Assays

Let's look at this methodology in detail, because it is becoming an increasingly more important tool in genetic analysis. The target sequences attached to the microarrays are either cDNAs or oligonucleotides. Many researchers avoid using cDNA as the target sequence for two reasons. First, the length of different cDNAs can vary, which results in each cDNA having a different optimal hybridization condition. Second, a cDNA for each potential gene in the genome must be isolated and confirmed to be correct. This could require a large amount of time due to some mRNAs being expressed at very low amounts and at very limited times or very limited tissues during development.

Because the oligonucleotides are short DNA sequences (typically 65 nucleotides long in microarray analysis), they can be synthesized using a machine if we have the annotated genomic sequence to predict every potential gene. We are also able to select oligonucleotide sequences that will all have an identical optimal hybridization condition. We can also select either a single oligonucleotide that will hybridize to all

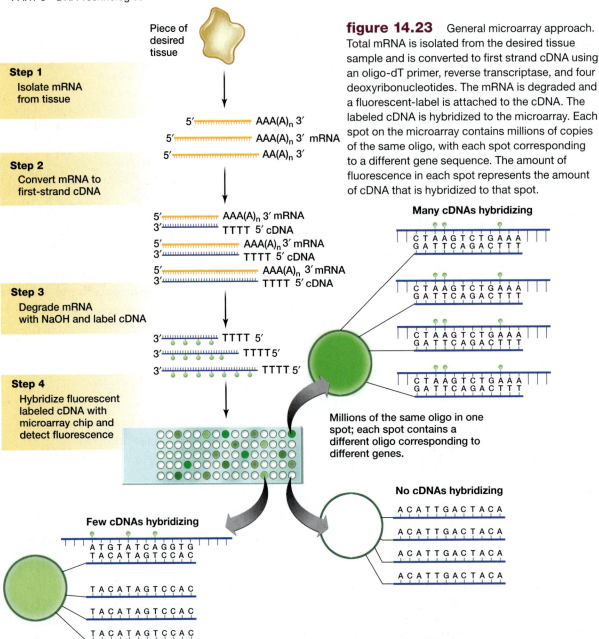

Piece of desired tissue

Step 1
Isolate mRNA from tissue

5'————— AAA(A)ₙ 3'
5'————— AAA(A)ₙ 3' mRNA
5'————— AA(A)ₙ 3'

Step 2
Convert mRNA to first-strand cDNA

5'————— AAA(A)ₙ 3' mRNA
3'————— TTTT 5' cDNA
5'————— AAA(A)ₙ 3' mRNA
3'————— TTTT 5' cDNA
5'————— AAA(A)ₙ 3' mRNA
3'————— TTTT 5' cDNA

Step 3
Degrade mRNA with NaOH and label cDNA

3'————— TTTT 5'
3'————— TTTT 5'
3'————— TTTT 5'

Step 4
Hybridize fluorescent labeled cDNA with microarray chip and detect fluorescence

figure 14.23 *General microarray approach. Total mRNA is isolated from the desired tissue sample and is converted to first strand cDNA using an oligo-dT primer, reverse transcriptase, and four deoxyribonucleotides. The mRNA is degraded and a fluorescent-label is attached to the cDNA. The labeled cDNA is hybridized to the microarray. Each spot on the microarray contains millions of copies of the same oligo, with each spot corresponding to a different gene sequence. The amount of fluorescence in each spot represents the amount of cDNA that is hybridized to that spot.*

Many cDNAs hybridizing

C T A A G T C T G A A A
G A T T C A G A C T T T

C T A A G T C T G A A A
G A T T C A G A C T T T

C T A A G T C T G A A A
G A T T C A G A C T T T

C T A A G T C T G A A A
G A T T C A G A C T T T

Millions of the same oligo in one spot; each spot contains a different oligo corresponding to different genes.

No cDNAs hybridizing

A C A T T G A C T A C A

A C A T T G A C T A C A

A C A T T G A C T A C A

A C A T T G A C T A C A

Few cDNAs hybridizing

A T G T A T C A G G T G
T A C A T A G T C C A C

T A C A T A G T C C A C

T A C A T A G T C C A C

T A C A T A G T C C A C

the alternatively spliced mRNAs from a single gene, or two different oligonucleotides that will each hybridize to a different alternatively spliced mRNA from the same gene (**fig. 14.24**).

Commercial companies have synthesized the set of oligonucleotides for different organisms, and they routinely add new oligonucleotides to their collections as new genes or alternatively spliced mRNAs are identified. These sets of tens of thousands of different oligonucleotides can be purchased and used to produce hundreds to thousands of microarrays. Microarray chips, which contain the oligonucleotides already attached to the glass and are ready for immediate hybridization, are available from several different companies. Affymetrix is one of the more well known companies that sell ready-made gene chips that cor-

respond to all the genes in an organism. Many of these companies produce chips for several different organisms, including humans.

If we decide to create our own microarray slides, rather than a commercially available slide, we must spot the thousands of oligonucleotides or cDNAs onto a microscope slide. Robots are available that accurately spot submicroliter amounts onto the slides (**fig. 14.25**). These slides can hold tens of thousands of different spots in an unambiguous pattern.

Selecting the Probes

Usually, two probes are selected that each originate as an mRNA population from two related tissues that differ in a specific manner. An example would be a piece of tissue from a normal prostate and a piece from a prostate tumor. The purpose is to identify the differences in gene expression between the two samples, as defined by the transcribed genes. By using tissues that differ in a specific way, such as the same tissue with and without the tumor, we will identify the changes in gene expression that either cause or result from the tumor.

Both mRNA populations are converted to single-stranded cDNA using reverse transcriptase (see chapter 12). A fluorescent molecule is then covalently attached to the cDNA, which will allow it to be easily detected (**fig. 14.26**). Alternatively, the cDNA can be reverse transcribed using three normal deoxyribo-

nucleotides and the fourth dNTP already covalently attached to the fluorescent molecule. Often one cDNA population is labeled to fluoresce green, and the other cDNA population to fluoresce red.

Equal amounts of each cDNA probe are mixed, however, individual cDNAs may be present in either greater or lesser numbers relative to the other cDNA probe. The mixed probe is then hybridized to the microarray (see fig. 14.26).

The excess probe is then removed by washing, and the microarray is excited by a laser. Every spot on the microarray that has probe bound will fluoresce. If an equal amount of cDNA from both probes is bound to the same spot, then the spot will fluoresce yellow (the color

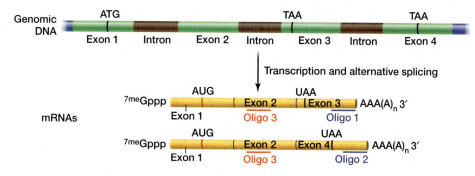

figure 14.24 Use of oligonucleotides to detect different alternatively spliced mRNAs. Two different oligonucleotides (oligo 1 and 2) each hybridize to different exons that will differentiate between the two mRNAs. Alternatively, one oligonucleotide (oligo 3) hybridizes to a common exon (exon 2), which will recognize both mRNAs.

(a)

(b)

figure 14.25 Equipment to generate a microarray. (a) A robot is used to spot the oligonucleotides on the microarray. (b) The microscope slides are arranged under a moving arm of the robot that contains pins that will place tens of thousands of individual spots (each containing a different oligonucleotide) on each microscope slide.

produced by equal amounts of red and green labels; see fig. 14.26). Note that this hybridization depends on a large excess of oligonucleotide target in each spot, relative to the cDNA in the probe population.

This microarray technology reveals whether the expression of every gene tested increases, decreases, or remains the same in the test condition (diseased tissue) relative to the control condition (normal tissue). The advantage of this approach is that every gene can be examined for a change in transcription in response to a specific and identical stimulus. The differences in the fluorescence in each spot correspond to differences in the level of gene expression between the two samples (**fig. 14.27**).

To illustrate an application of the technique, two individuals may have what appear to be similar tumors; but the cause or underlying molecular changes in the two tumors may be very different. Without completely understanding these molecular differences, doctors

may treat both tumors in an identical way. Yet the molecular differences may allow one tumor to respond to the treatment, while preventing the other tumor from responding. In the future, medicine may rely more on this molecular characterization to tailor specific treatments for individuals.

With the simultaneous analysis of tens of thousands of transcripts in a microarray experiment, enormous amounts of data are generated. The field of **bioinformatics** involves the use of computers to organize and analyze biological data.

Transcriptional Reporters

Geneticists may also be interested in studying the expression of a single gene in greater detail. Questions might include: What promoter sequences are essential for the expression of a gene? Which promoter elements activate transcription, suppress transcription,

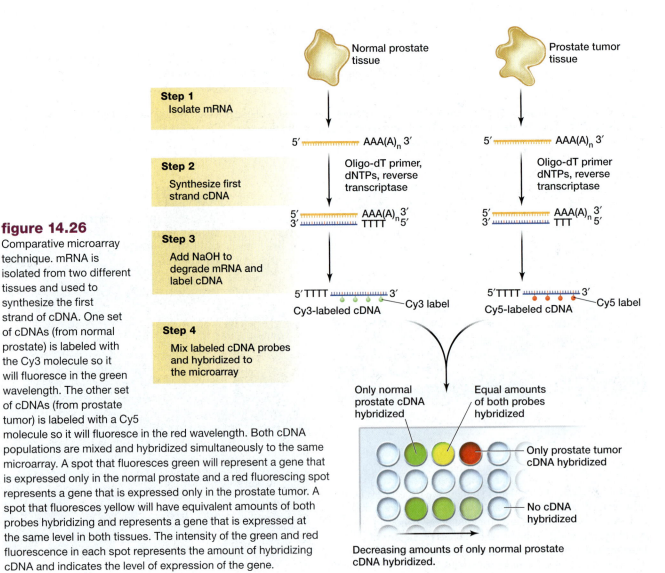

figure 14.26

Comparative microarray technique. mRNA is isolated from two different tissues and used to synthesize the first strand of cDNA. One set of cDNAs (from normal prostate) is labeled with the Cy3 molecule so it will fluoresce in the green wavelength. The other set of cDNAs (from prostate tumor) is labeled with a Cy5 molecule so it will fluoresce in the red wavelength. Both cDNA populations are mixed and hybridized simultaneously to the same microarray. A spot that fluoresces green will represent a gene that is expressed only in the normal prostate and a red fluorescing spot represents a gene that is expressed only in the prostate tumor. A spot that fluoresces yellow will have equivalent amounts of both probes hybridizing and represents a gene that is expressed at the same level in both tissues. The intensity of the green and red fluorescence in each spot represents the amount of hybridizing cDNA and indicates the level of expression of the gene.

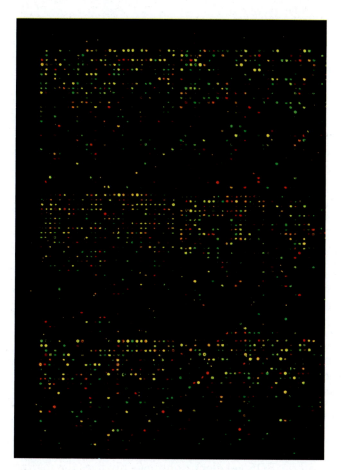

figure 14.27   A microarray result comparing normal and breast cancer tissue samples. The green spots represent genes that are expressed only in the normal breast tissue, while the red spots represent genes that are expressed only in the breast cancer tissue. The higher intensity of either red or green correlates with a higher level of expression. The yellow spots represent genes that are expressed at equivalent levels in both breast tissues.

and control the temporal and spatial expression of a gene? How does the transcriptional level of a gene change in response to different conditions? These questions are important in characterizing a promoter that may be used to control the expression of a transgene (chapter 13).

One method for examining the promoter of a gene is to generate a reporter. A **promoter reporter** is a transgene that is easily assayed to reveal the amount and location of its transcription from a foreign promoter. The most common reporters are β-galactosidase, green fluorescent protein (GFP), and luciferase.

A typical promoter analysis involves cloning several kilobase pairs of the promoter region from the gene of interest into a vector containing the reporter gene. Deletions are then systematically created from both ends of the promoter (**fig. 14.28a**). The wild-type promoter and each promoter deletion, all linked to

the reporter gene, are individually introduced into the organism from which the promoter was isolated. (Alternatively, the constructs can be transfected into a cell line derived from the organism.) The transgenic organisms are then tested for the level of reporter gene expression, as well as the temporal and expression pattern (fig. 14.28b).

The expression pattern of the transgene should match the expression of the endogenous wild-type gene. By comparing the different promoter deletions, promoter sequences can be identified that are essential for maximal level of expression and for the correct expression pattern (fig. 14.28a). This characterization is very useful to ensure that the essential promoter elements are present when trying to express other transgenes.

A transgenic animal can then be exposed to a variety of different conditions, such as a change in diet or a disease state. The expression of the reporter gene can then be compared with that of transgenic animals maintained under control conditions. This test will reveal how different conditions affect the transcription of a gene. One application may be testing various diets on the transgenic reporter organisms to determine if the diet alters the expression of specific genes that are involved in the progression of a disease. These dietary conditions can then be used to test potential drugs on the reporter animal to determine if they block the altered expression of the reporter gene.

Restating the Concepts

▶ Transcriptomics is the study of gene expression patterns in different tissues, developmental stages, or different conditions. SAGE analysis involves cDNA synthesis, followed by cloning and sequencing TAGs that are unique for every mRNA. The frequency of a TAG in the sequenced population corresponds to the relative abundance of the corresponding mRNA. SAGE requires no information about gene sequences in the organism prior to starting the analysis.

▶ Analysis of microarrays or gene chips is also used to simultaneously examine the expression of tens of thousands of genes. It involves the hybridization of one or two different fluorescent-labeled cDNA populations to gene-specific oligonucleotides or cDNAs that are spotted on a microscope slide or chip. Microarrays require sequence information for all the genes to synthesize the oligonucleotides prior to making the gene chip.

▶ Reporter genes allow the identification of the promoter elements that are essential for maximal expression and the correct temporal and spatial expression. The reporter transgene must be expressed in the species from which the promoter was derived or in a cell line generated from that species.

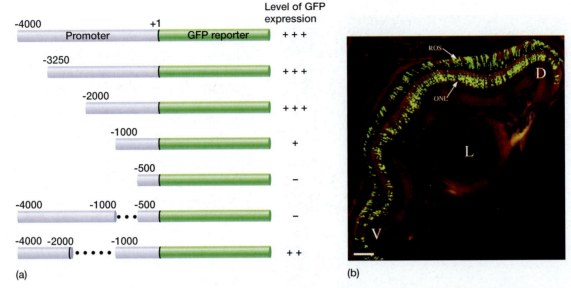

figure 14.28   Promoter analysis using a reporter gene. (a) A genomic DNA region (–4000 to +1) containing a putative promoter is cloned upstream of a reporter gene (GFP). A series of deletions from the promoter are made and each construct is introduced back into the organism to examine the specificity and level of expression. Three pluses (+++) represent normal expression, one plus (+) represents reduced expression, and a minus (–) represents no expression. Located between –1000 and –2000 is a sequence that is required for maximal expression and between –500 and –1000 is a sequence that is required for any expression. (b) A transgenic zebrafish expresses GFP from the rod opsin promoter. Sections from the adult transgenic eye reveals that the rod opsin promoter expressed GFP at a wild-type level in the rod photoreceptors in the adult retina.

> ▶ Animals containing a reporter transgene can be used to assay how different conditions, such as diet, affect the expression of a target gene through its cloned promoter. This could reveal how environmental conditions affect the progression of a disease.

14.5 Proteomics

Genomics provides us with information on the DNA sequence and gene organization within the genome of an organism, while transcriptomics can elucidate the changes in the levels and patterns of gene expression. Most phenotypes, however, result from the expression and function of proteins. Thus, the field of **proteomics,** the study of all the proteins in an organism, was created.

Within the proteome, researchers may be interested in studying the interactions between proteins (the *interactome*), or how proteins organize to produce signaling pathways in the cell (the *signalome*). We will discuss the study of interacting proteins here because it involves several interesting approaches, and because many different phenomena in a cell result directly or indirectly from the interaction of proteins.

Two-Dimensional Gel Electrophoresis

A variety of methods can be used to analyze proteins. Gel electrophoresis, described as a method to separate DNA restriction fragments in chapter 12, can also be used to separate proteins using different criteria. **SDS gel electrophoresis** involves incubating a protein mixture with sodium dodecyl sulfate (SDS) to equalize the charge on all the proteins. The proteins are then electrophoresed through a gel to separate them based entirely on their molecular weight. While this method is effective at separating proteins, many proteins will have similar molecular weights and cannot be distinguished from each other by this technique.

Another type of gel electrophoresis is isoelectric focusing. **Isoelectric focusing** involves separating proteins based on their charge. The isoelectric focusing gel uses an electrical current to drive proteins through a pH gradient until they reach their *isoelectric point* (pI), which is an indication of their total charge. Proteins of different molecular weights, but similar isoelectric points, cannot be distinguished from each other using this technique.

Both forms of gel electrophoresis can be combined in a technique called two-dimensional gel electrophoresis. **Two-dimensional gel electrophoresis** separates proteins based on both criteria, their isoelectric point and molecular weight. Two-dimensional gel electrophoresis involves several steps:

1. First, the protein mixture is electrophoresed in an isoelectric focusing gel (**fig. 14.29**).
2. The gel is then incubated in an SDS solution so the proteins bind the SDS and produce similar charges.

3. The SDS-soaked isoelectric focusing gel is laid on an SDS polyacrylamide gel and electrophoresed (see fig. 14.29).

4. The gel is then usually stained with a silver solution, which can detect as little as 0.5–1.0 ng of protein per spot.

The resulting proteins in the final gel are separated in one dimension based on their isoelectric point and in the second dimension by their molecular weight. This allows proteins of the same or very similar molecular weights to be clearly separated based on their charge.

This separates as many as 10,000 proteins from a tissue or organism into reproducibly distinct spots (**fig. 14.30**).

Besides clearly distinguishing a very large number of proteins, this technique allows the researcher to detect some proteins that are posttranslationally modified. For example, a protein that is phosphorylated, such as a histone protein (see chapter 7), will likely have a similar molecular weight, but a different pI and will shift positions along the IEF axis. Without two-dimensional gel electrophoresis, the two proteins would have appeared as a single molecular weight band on an SDS-polyacrylamide gel, or as two unrelated proteins in an isoelectric focusing gel.

Besides identifying proteins within a single sample, the proteins from two different samples can be compared to identify proteins that exhibit a change in expression level or charge. As with microarrays, the samples can be either normal tissue or a disease tissue, but here the focus is on proteins rather than transcripts.

In **figure 14.31**, the two samples are a *Drosophila* retina that was either maintained in the dark or exposed to light. In this case, three different proteins were identified that exhibit a change in their charge in response to light stimulation. This altered pI is due to phosphorylation of the proteins after light stimula-

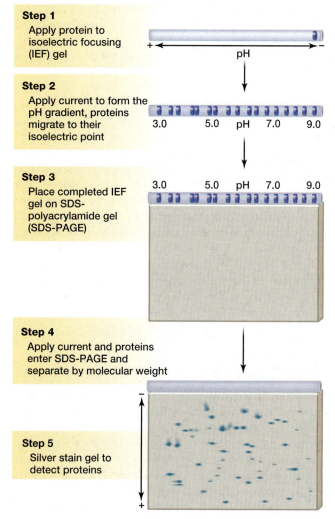

figure 14.29 Two-dimensional gel electrophoresis. Protein samples are first electrophoresed in an isoelectric focusing (IEF) gel, which separates proteins based on their isoelectric point (pI). After completion, the IEF gel is incubated in an SDS solution that allows all the proteins to bind SDS. The IEF gel is then placed on top of an SDS-polyacrylamide gel and electrophoresed to separate proteins based on their molecular weight. This results in each protein being separated by both its charge and molecular weight. The gel is incubated in a silver stain solution to detect as little as 0.5–1.0 ng of protein per spot.

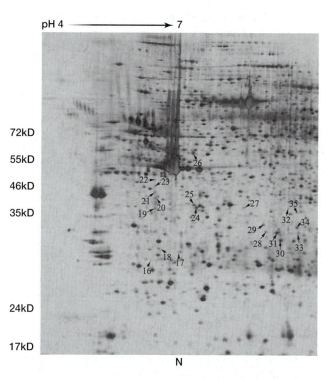

figure 14.30 Two-dimensional gel electrophoresis of proteins that were isolated from a colorectal carcinoma. The proteins were separated by charge (horizontal axis) and molecular weight (vertical axis). The relative locations of the molecular weight standards are shown to the left.

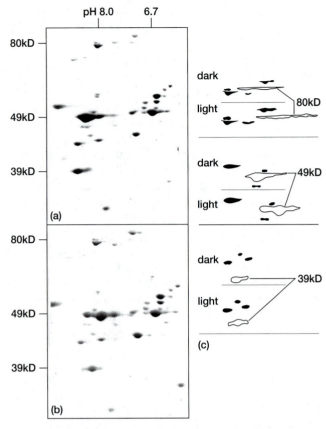

figure 14.31 Two-dimensional gel electrophoresis. *Drosophila* proteins were isolated from retinas that were either maintained in the dark (a) or immediately after turning on the light (b). The proteins were then separated by two-dimensional gel electrophoresis. Comparing the gel pattern of the light-stimulated proteins with the pattern of proteins isolated from a dark-treated retina, revealed three major protein changes upon light stimulation. A schematic of the three major proteins (80k, 49k, and 39k) that exhibit an altered pattern due to a change in their pI upon light stimulation is shown (c). These changes represent phosphorylation of these proteins.

tion. By comparing the two samples, we can determine that a greater percentage of these three proteins are phosphorylated in the light. The identification of these phosphorylation events revealed essential processes that regulate vision.

It's Your Turn

Two-dimensional gel electrophoresis is a powerful method to examine a variety of protein changes between two samples. Problems 31 and 32 will allow you to examine 2-D gels and observe how a variety of different posttranslational modifications can affect the protein in the gel.

Identifying Proteins by Mass Spectroscopy

Two-dimensional gel electrophoresis is a powerful method for separating thousands of different proteins and different forms of the same protein. However, the identities of the various spots, or proteins, in the gel are unknown. If we could determine the identities of the different proteins, we could use two-dimensional gel electrophoresis to determine the changes in the amount of protein expression and posttranslational modification between normal and diseased tissue.

The **mass spectrometer** is an instrument that measures the mass-to-charge ratio of proteins. The proteins to be analyzed are introduced into the spectrometer, where they are first ionized and then separated based on their mass-to-charge ratio through an electrical field (small molecules migrate faster than large ones). A detector measures the time between each protein. This time between proteins can be used to determine the mass of the protein.

A tandem mass spectrometer, or MS/MS, passes the proteins through a mass spectrometer to separate the proteins. The separated proteins are then fragmented by collisions with additional ions. The resulting smaller peptides are separated and their masses determined in the second mass spectrometer. This allows the amino acid sequence of the various peptides, and thus, the original protein to be determined.

The mass spectrometer can be used to analyze the different protein spots produced by two-dimensional gel electrophoresis. The individual proteins can be extracted from a single spot and introduced into the mass spectrometer to determine the amino acid sequence or posttranslational modification of a protein. Furthermore, the mass spectrometer can reveal the amino acid that possesses the posttranslational modification.

Mass Spectroscopy of Protein Complexes

Another important question that can be addressed using a mass spectrometer is the identity of the various molecules that interact to form a complex. This complex may be required for each of the individual proteins to be functional, or the complex may be required for proper regulation of the proteins. For example, one of the fastest biological processes known is the phototransduction cascade in *Drosophila*. The very rapid response after light activation results from the key proteins in this process being physically organized into a single protein complex.

How is the protein complex isolated before being analyzed by a mass spectrometer? Antibodies are proteins that recognize and bind a specific amino acid sequence or domain in a target protein. Because of this ability, they can be used to immunoprecipitate the protein (**fig. 14.32**). Under the proper conditions, the antibody will also *coimmunoprecipitate* any other proteins that interact strongly with the target protein, including a protein complex. We then need a way to identify the coprecipitated proteins. This immuno-precipitated complex of proteins, however, could be analyzed by the mass spectrometer to determine the sequence and posttranslational modifications of the different proteins.

Analyzing Interacting Proteins: The Yeast Two-Hybrid System

In some cases, an antibody against the desired protein may not exist. Thus, Stanley Fields at the University of Washington developed a genetic approach to complement the physical immunoprecipitation method to identify protein–protein interactions. The basic method described by Fields has been modified in recent years to study a variety of protein–protein and protein–nucleic acid interactions. We will describe the use of the yeast Gal4 protein (Gal4p), because it is commonly used today to identify protein–protein interactions.

The Gal4 protein is a transcriptional activator in yeast that binds a specific DNA sequence and assists in binding RNA polymerase II. The amino terminal region of Gal4p binds the 170-bp upstream activating sequence (UAS_G), and the carboxyl terminal region is required for activation of transcription (**fig. 14.33**).

These two domains are functionally distinct from each other. A mutation in the amino terminus of Gal4p prevents the protein from binding the UAS_G, while a mutation in the carboxyl terminus of Gal4p permits binding to the UAS_G but blocks transriptional activation (**fig. 14.34**). However, if the two functional domains are brought physically together, then the amino terminus of Gal4p binds the UAS_G and the carboxyl terminus of Gal4p activates transcription. This is the concept behind the yeast two-hybrid system.

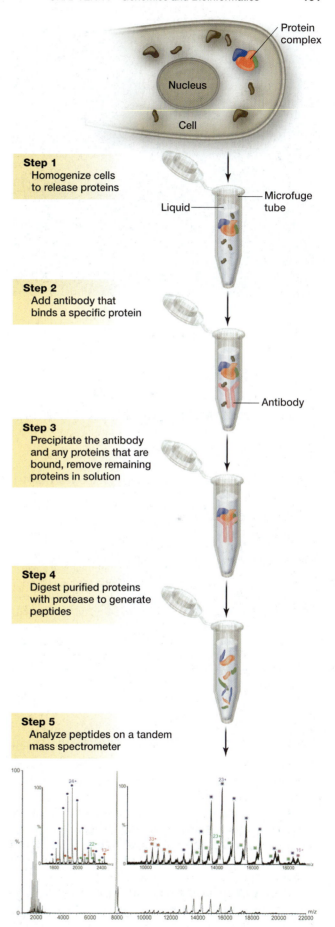

Step 1
Homogenize cells to release proteins

Step 2
Add antibody that binds a specific protein

Step 3
Precipitate the antibody and any proteins that are bound, remove remaining proteins in solution

Step 4
Digest purified proteins with protease to generate peptides

Step 5
Analyze peptides on a tandem mass spectrometer

figure 14.32   Identification of protein complexes by co-immunoprecipitation and tandem mass spectrometer. Cells or tissue containing the protein complex are homogenized to release the proteins. Proteins that are associated in a complex (red, green, orange, and blue proteins) remain bound together. An antibody to a desired (orange) protein is added. The antibody precipitates the protein that it binds, along with any proteins in the complex. The precipitated protein complex is purified, fragmented into peptides, and applied to a tandem mass spectrometer (MS/MS) to determine the mass of the various peptide fragments. If the amino acid sequence of all the proteins in the organism are known (usually through a genome sequencing project), the identity of the proteins corresponding to each peptide fragment can be determined.

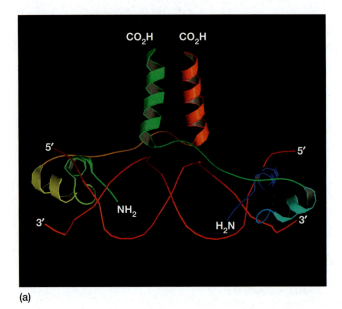

(a)

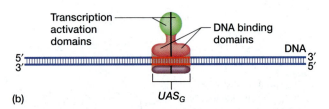

(b)

figure 14.33   Two representations of the yeast Gal4 protein binding DNA. (a) Ribbon structure of Gal4p based on the X-ray crystal structure of the protein. The alpha helices adjacent to the amino terminus of each Gal4p subunit binds in the major groove of the DNA (red is the sugar-phosphate backbone). Because Gal4p binds as a dimer, one amino terminal subunit binds the major groove on the right of the DNA and the other amino terminus binds the major groove behind the DNA to the left. Notice how the Gal4p transcription activation domain (carboxyl terminus) projects away from the DNA. (b) A schematic diagram of Gal4p is shown as two domains, the DNA binding domain (pink) and the transcriptional activation domain (green). Gal4p is shown binding to the UAS_G sequence (red) as a dimer.

Creating "Bait" and "Prey" Plasmids

The **yeast two-hybrid system** employs two different yeast episomal plasmids (YEps; see chapter 13) that each expresses a different fusion protein. A **fusion protein** is produced from two different ORFs that are ligated together to translate a single protein having an amino terminal portion identical to one protein and the carboxyl terminal portion identical to a different protein.

The first YEp is called the *bait* plasmid, as it will be used to "fish" for interacting proteins (**fig. 14.35**). The

bait YEp encodes a fusion protein that contains the amino terminus, or DNA-binding domain, of Gal4p fused to the protein of interest. The second YEp is often called the *prey* plasmid, since it contains the fusion protein target that the bait is trying to identify (see fig. 14.35). The prey fusion protein contains the Gal4p carboxyl terminus (transcriptional activation domain).

The prey (target) plasmid often contains a cDNA library from the organism of interest, because we may be unsure of the identity of the target—a library would contain all the potential targets (as well as every other cDNA from the source tissue). Each cDNA is cloned in the prey plasmid such that the fusion protein will contain the cDNA-encoded protein fused to the carboxyl or transcriptional activator domain of Gal4p. Only the cDNAs that are cloned in the proper orientation and with the correct alignment of the ORF between the cDNA and the Gal4p will produce the correct fusion protein.

The prey cDNA library and bait plasmid are transformed into the *leu2⁻ trp1⁻* yeast strain, which allows selection for transformants containing both the bait and prey plasmids (see fig. 14.35). This strain must also contain a reporter gene, such as β-galactosidase, that is under the transcriptional control of the UAS_G. Alternatively, the bait plasmid is tranformed into a *trp1⁻* strain, the cDNA is transformed into a *leu2⁻* strain, and the two strains are mated to bring the bait and prey plasmids together.

Testing for Interaction

The yeast cells are then plated on media lacking leucine and tryptophan. Only cells that contain both the bait and prey plasmids will leucine and tryptophan be synthesized.

The physical interaction between the two fusion proteins can be tested if the media contains X-gal. Only colonies expressing two interacting fusion proteins will transcribe the *lacZ* reporter gene and express β-galactosidase, which converts X-gal into a blue precipitate. Often, two or three additional reporters are in the yeast cell to ensure that the bait–prey interaction is real (see fig. 14.35).

The expression of the reporter gene reveals that these two different fusion proteins interact. The identity of the bait portion is already known. Sequencing the prey plasmid will reveal the identity of the prey portion. If the prey protein has a known function, the identified protein interaction may further reveal how a process works in the cell. Should the sequence of the prey protein not be sufficient to identify its function, its interaction with the bait protein suggests that it likely participates in a function related to the bait protein.

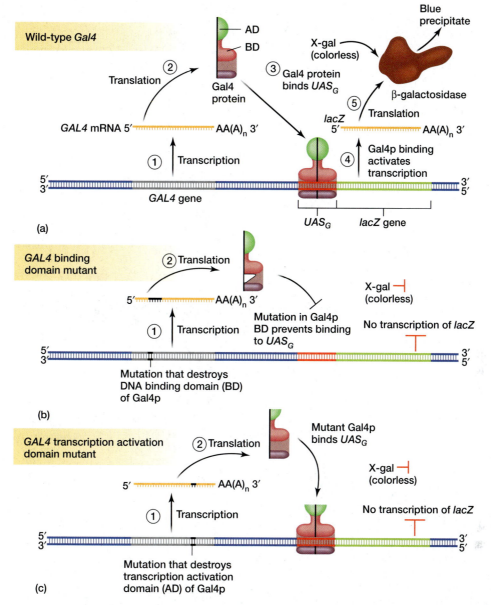

figure 14.34 Functional assay of two separate Gal4p domains. (a) Transcription of the wild-type *gal4* gene produces Gal4 protein that binds the UAS_G and activates transcription of the *lacZ* reporter gene. This leads to the expression of β-galactosidase, which hydrolyzes X-gal into a blue precipitate in the yeast cell. (b) A mutation near the beginning of the $gal4^-$ gene destroys the DNA binding domain (BD) without affecting the transcriptional activation domain. However, this mutant protein cannot bind the UAS_G, which prevents it from activating transcription of the *lacZ* gene. This results in the yeast failing to hydrolyze X-gal and remaining colorless. (c) A mutation near the end of the $gal4^-$ gene disrupts the transcription activation domain (AD), without affecting the DNA binding domain. The resulting protein specifically binds the UAS_G, but still cannot activate transcription of *lacZ*. Because X-gal is not hydrolyzed, the cells remain colorless.

Bioinformatics of Protein Relationships

The yeast two-hybrid approach was originally designed to identify if two specific proteins interacted. This approach was modified to screen libraries with a specific bait as described above, and more recently to provide a global picture of protein–protein interactions.

Using the sequence information from the genome projects, in particular the yeast and *Drosophila* genomes,

every potential ORF was cloned into both the bait and prey plasmids. Through transformation or cell matings, each bait plasmid is combined with each prey plasmid. All of the colonies that exhibit expression of the reporter gene are isolated, and the bait and prey cDNAs in each cell are sequenced. This approach should identify every protein–protein interaction and the identity of each protein.

This massive amount of information can be stored and sorted using computer programs. It is then possible

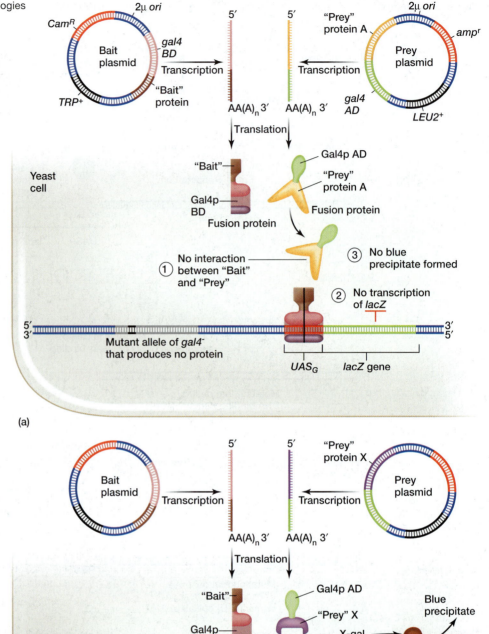

figure 14.35 The yeast two-hybrid system requires two YEp plasmids, the "Bait" plasmid and the "Prey" plasmid. Both plasmids contain a yeast 2μ origin of replication, an antibiotic resistance gene, and a yeast positive selection gene (*TRP⁺* and *LEU2⁺*). The Bait plasmid also contains a fusion gene that encodes the Gal4p DNA binding domain (BD) fused to a "Bait" open reading frame. The Prey plasmid contains a fusion gene that encodes a "Prey" open reading frame fused to the Gal4p transcription activation domain (AD). Both plasmids are introduced into a *gal4⁻* yeast mutant that does not encode a functional Gal4p and contains a reporter gene, the *Gal4-UAS* sequence upstream of the *lacZ* gene. If the Bait and Prey protein domains fail to produce a functional protein-protein interaction (a), the Gal4p AD will not be brought to the *UAS* sequence and the *lacZ* gene cannot be transcribed. The resulting yeast cell will be white when grown in the presence of X-gal. If the Bait and Prey protein domains interact with each other (b), the Gal4p AD is brought to the *UAS* element and activates transcription of the *lacZ* gene. The resulting β-galactosidase protein will convert the X-gal into a blue precipitate.

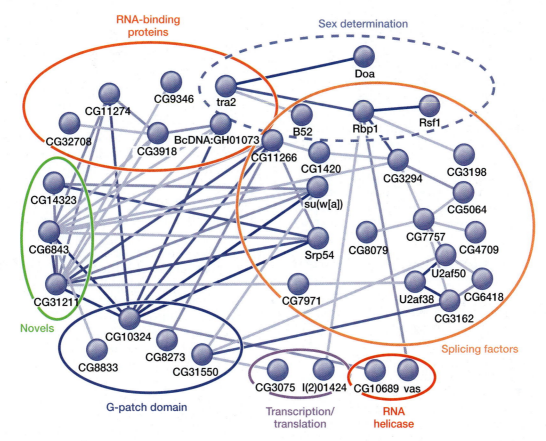

figure 14.36   An interactome map of the proteins involved in *Drosophila* sex determination. A yeast two-hybrid screen identified a large number of interacting *Drosophila* proteins that are associated with the splicing complex that is involved in *Drosophila* sex determination. Notice that the three novel proteins that are circled in green have no known functions. Because they physically interact with RNA-binding proteins and splicing factors, they are likely involved in the RNA splicing complex. (From Giot et al., Figure 5B, *Science* 302: 1727–1736, 2003. Reprinted with permission from the AAAS.)

to assemble a diagram that shows every protein–protein interaction in a particular organism. The proteins that were shown to interact in a splicing complex associated with *Drosophila* sex determination are illustrated in **figure 14.36**. Notice that three proteins with unknown functions (circled in green) are placed in this complex due to protein–protein interactions, which suggests that they must function in splicing. This *interactome map* has revealed not only that the network of protein interactions within a cell can vary in different tissues or different cells in an organism, but also that proteins do not function as independently from other proteins as previously expected. In some cases, these protein complexes can lead to faster processing of the substrate through a pathway or to more rapid regulation of the pathway.

Restating the Concepts

▶ Proteomics is the study of all the proteins in an organism's cells or tissues. We can concentrate on the proteins that interact with one another

(interactome) or all the proteins that are involved in signaling (signalome).

▶ Two-dimensional gel electrophoresis separates proteins based on their charge (first dimension) and on their molecular weight (second dimension). A two-dimensional gel can reveal 10,000 spots that each represent a different protein or a different form of the same protein, such as phosphorylated and dephosphorylated forms.

▶ The amino acid sequence of a protein can be rapidly determined using a mass spectrometer, which determines the mass-to-charge ratio of fragmented proteins. The mass spectrometer can be used to determine the sequence of proteins isolated after two-dimensional gel electrophoresis or coimmunoprecipitation.

▶ A yeast two-hybrid screen is used to identify proteins that interact with one another by regenerating the Gal4 protein, which then activates transcription of a reporter gene. Large-scale screens have recently been performed to examine all the potential protein–protein interactions in an organism and computer programs have been developed to diagram all these interactions.

14.6 Ethical Considerations

With the advances in molecular genetics described in the last three chapters, new and complex legal and ethical questions have arisen. To address these issues, the U.S. Department of Energy and the National Institutes of Health created the **ELSI Program** (Ethical, Legal, and Social Issues Program). This program seeks to clarify how science, society, corporations, and the law should deal with the information that is being produced from the Human Genome Project. A number of questions are being addressed under the ELSI program; here we provide a brief synopsis of some of the issues.

1. What is a fair use of the genetic information that is generated? In chapter 1, we considered some of the possible uses and abuses of this information. In the future, could insurance companies and employers demand that an individual submit to a battery of genetic tests, much like they already require drug testing? The major difference is that an individual who tests positive for a drug must purposely have taken the drug, but an individual who tests positive for a genetic variation was born with this condition and may not exhibit the clinical symptoms of a disease for decades, if at all. The argument becomes even more personal when we consider whether prospective parents have the "right" to learn the genetic background of a child they are considering for adoption. Would such a right lead to large numbers of children who are "unadoptable" because they are predisposed for perceived behavioral problems, such as drug addiction?

2. Is it right to develop tests to screen for inherited diseases if no reasonable treatment for the disease exists? Many of the genetic tests, in addition, are based on the detection of specific alleles that have already been identified. Therefore, an individual who tests negative is really only negative for the known alleles, and he or she could still have a disease allele and exhibit the clinical symptoms. If these tests are available, what are the societal, philosophical, and legal issues associated with parents having their children tested for adult-onset diseases? Should an individual have the freedom to choose whether they want knowledge of their own genetic makeup?

3. What are the potential reproductive issues associated with the new genetic technology? How will society and the scientific and legal communities deal with the ability to select the genotype or phenotype of children? And, what are the ethical and moral implications of creating children to provide matching tissue for their afflicted siblings? We described some these issues in chapter 13.

4. What portion of an individual's behavior is controlled by their genes versus their environment and their freedom of choice? What are the legal responsibilities of an individual who commits a crime and then argues that his actions were a result of a genetic predisposition to a behavioral condition, such as bipolar disease? What constitutes "normal" behavior?

5. We are increasingly able to create transgenic crops that are more nutritious or more resistant to environmental factors (such as drought, temperature extremes, insects), as discussed in chapter 13. At what level do we decide that GMOs are safe for human and animal consumption? Will these crops be affordable for less-developed countries to purchase and grow? These are the countries that potentially could reap the greatest benefits from these transgenic crops, but should their populations be exploited for profit? Does this situation also increase the dependency of the developing countries on the more technologically advanced nations?

These are only a small part of the issues that will have to be dealt with in the coming years. Genetics has advanced to a new level, where powerful techniques can be used to diagnose and treat a variety of human diseases. Genetics is also at the forefront of solving a number of complex questions in science that deal with development and neuroscience. The purpose of ELSI is to begin to discuss these issues so that opinions and thoughts are seriously considered before we reach the point of making these decisions.

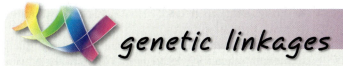

genetic linkages

In this chapter, we discussed some of the "omics" fields, such as genomics and proteomics. This work largely builds on the molecular genetic techniques that we previously described in chapters 12 and 13. These techniques are now applied on a very large scale and often automated to potentially allow the complete analysis of the DNA sequence of the entire genome or all the proteins in an organism.

Our discussion of genomics is based on the basic idea of DNA sequencing, which we first described in chapter 12, and the ability to generate recombination maps of mutations (chapter 6) and DNA markers (chapter 13). The major difference between the previous discussions and the genomics description in this chapter is the effort to sequence the entire genome of several organisms.

Ordered contig sequencing and shotgun sequencing differ in the approach of determining which clones to sequence and how the pieces of sequence are assembled. This sequencing of the genome requires the ability to amass and analyze a significantly large amount of data. Completing gaps in the genomic DNA sequence involves PCR amplification, which we originally described in chapter 12.

We described alternative splicing as a method for increasing the number of proteins that are encoded in a limited number of genes. Alternative splicing involves the choice between either two different splice donor sites and a common splice acceptor site, or a common splice donor site and two different splice acceptor sites (chapter 10). Alternative splicing can also help regulate the expression of a gene in different tissues. In chapter 21, we will describe the genetic basis of sex determination in *Drosophila*. This process involves alternative splicing of several genes in a sex-specific manner.

The identification of all of the genes in the genome allows for the analysis of the transcriptome, which is all the transcribed sequences in the organism. SAGE and microarray experiments are used to analyze the transcriptome in an organism. Oligonucleotides or cDNAs that correspond to each mRNA are spotted onto a microscope slide in an array. A fluorescent-labeled cDNA is hybridized to the microarray. Because the slide contains DNA targets, the hybridization reaction is essentially a Southern hybridization, which was discussed in chapter 12. This microarray can examine the global changes in gene expression, much like a Northern hybridization reveals the level of transcription of a single gene.

Proteomics is the study of all the proteins in an organism. We previously described an immunoblot as a way to examine the expression of a protein (chapter 12). In this chapter, we discussed two-dimensional gel electrophoresis for examining thousands of proteins simultaneously. Combined with a mass spectrometer, the amino acid sequence of each protein can also be determined.

One way to determine the potential function of a protein is to identify other proteins that physically interact with the protein of interest. We discussed the yeast two-hybrid method in this chapter for identifying interacting proteins. This approach involves transcriptional activation of one or more reporter genes. We previously discussed DNA sequences located in the eukaryotic promoter vicinity that are essential for transcriptional activation (chapter 10), and this topic is explored in more detail in chapter 17.

We completed this chapter with a brief introduction of the ELSI Program, which deals with the ethical, legal, and social issues that result from the fields of molecular genetics, genomics, and proteomics. These broad issues need to be considered and resolved by a community that includes specialists in fields far beyond the halls of genetics and industry.

Chapter Summary Questions

1. For each term in column I, choose the best matching phrase in column II.

Column I
1. Genomics
2. Expressed sequence tag
3. Annotation
4. Two-dimensional gel electrophoresis
5. Fusion protein
6. Bioinformatics
7. Shotgun sequencing
8. Contig
9. In situ hybridization
10. Proteomics

Column II
A. Product of two different open reading frames that are joined to encode a single protein
B. A method to separate proteins based on their isoelectric point and molecular weight
C. The minimal number of contiguous clones that cover a defined region
D. Use of computers to process and analyze large amounts of biological data, such as genomic sequence information
E. Random sequencing of genomic DNA clones
F. Annealing a nucleic acid probe to either tissue sections or chromosomes to determine where a gene is expressed or located
G. Portion of a cDNA sequence
H. Study of all the proteins in a cell or organism
I. Study of mapping and sequencing genomes
J. Identification of all the genes and open reading frames in a genome

2. What is genomics? What is proteomics? How do these two areas differ and why are they both important?

3. Why is there an interest in generating linkage maps instead of just determining the DNA sequence?

4. Why are repeat sequences a significant problem for shotgun sequencing? How is the problem overcome?

5. What is FISH? Why is it important for creating genomic maps?

6. Why are ESTs valuable in genomic mapping? Why aren't ESTs always useful in identifying ORFs in the genome?

7. What is the annotation of a genomic sequence? Why is this so important to understanding the sequence?

8. What is alternative splicing? Why does it complicate genomic analysis?

9. What is a sequence "read"? How is this different from a contig?

10. What are the differences between the ordered contig and shotgun sequencing techniques? What are the advantages and disadvantages of each?

11. What are the "tags" used when ordering contigs? What are the advantages of each? Why are several usually used together?

12. When is a genome project "complete"? Is it necessary to know every nucleotide of an organism to end a project? Why are scientists becoming more satisfied with "draft" genome sequencing projects?

Solved Problems

PROBLEM 1: In preparation for a sequencing project, a collection of YAC clones is assembled and arranged on grids. In an initial effort to order these clones, arrays are hybridized with four different short probes that correspond to different genes that were genetically mapped to the same chromosome. The following hybridization pattern is noted. You know that probe C is located closer to the centromere than the other three genes. Map the order of these genes on the chromosome and roughly identify the relative positions of the hybridizing YAC clones.

Answer: To begin this problem, we need to identify all the clones that contain overlapping regions. We can start by identifying any YAC clones that hybridize to probe A and any other probe. Only clone B7 has this type of hybridization pattern, hybridizing to both probes A and D. So we can put probes A and D next to each other on the map. We next examine if any YAC clones hybridize to probe D and another probe. We already determined that YAC B7 hybridizes to both probes A and D, so we will concentrate on YACs that hybridize to probe D and either B or C. We find that both YACs B1 and C5 hybridize with probes D and B. This places B next to D. Because none of the YACs hybridize to both B and A, we can assume that the order is: B–D–A. To determine if probe C is adjacent to A or B, we examine if any YAC hybridizes to both probes A and C or probes B and C. We can see that YAC A6 hybridizes to both probes B and C. So we can assume that the order is A–D–B–C. Because we know that the centromere is located closest to probe C, we can also add it to the map as such: A–D–B–C–centromere. We can also place all the YACs that hybridize to one or more of these four probes onto the map by specifying the location of the YAC DNA in relationship to the location of the four probes. (If the probe hybridizes to the YAC, then the YAC contains a sequence that is complementary to the probe. If the YAC did not hybridize to the probe, then they do not share a common sequence.)

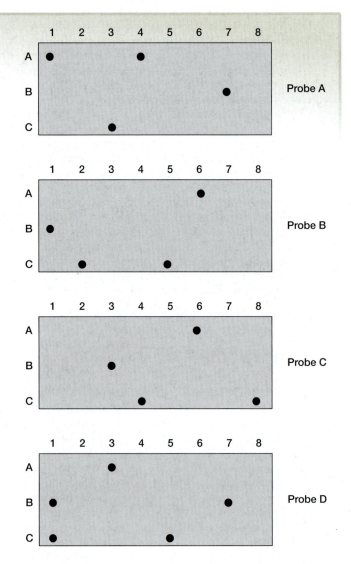

The final map, with relative positions of all the mapped YACs and the four probes is as follows:

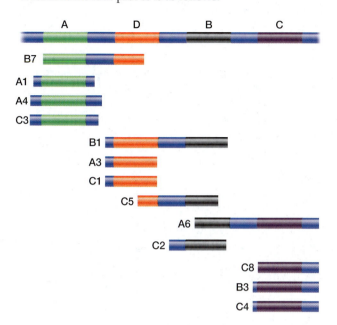

PROBLEM 2: DNA fingerprints were done on five different genomic clones to determine if they overlap. The following restriction fragment patterns were seen. Using these DNA fingerprints, map the relative positions of the five genomic clones.

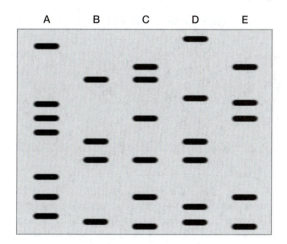

Answer: We first compare the restriction fragment patterns between the different genomic clones to identify what DNA fragments are shared between different genomic clones and what fragments are unique to only one genomic clone. The shared fragments correspond to overlapping regions between

genomic clones, but the unique fragments will correspond to the genomic clones that either contain regions located between the overlapping regions at both ends or the regions of clones that are located at the extreme ends of the contig. In lane A, we see that the second, third, and sixth bands are also present in lane E, with the remainder of the fragments in lane A not being present in any other clone. Because clone A contains restriction fragments that are not shared with any other clone, clone A must be located at one end of the contig, and the unique restriction fragments must correspond to the region of clone A that does not overlap clone E. Next, we can see what clone overlaps E. We see that the first, third, fourth, and fifth bands in lane E align with bands in lane C. We also see that the third and sixth bands in lane A align with both C and E. This represents a region in common with A, C, and E. Notice that all of the restriction fragments in clone E align with fragments in either clone A or clone C. This demonstrates that clone E overlaps both clone A and clone C, with the order being A–E–C. We can then determine which clone overlaps C. Two fragments in lane C, second and fourth bands, align with fragments in lane B, and only the fourth band in lane C aligns with D. Also three bands in B, second, third, and fourth bands, align with D. So the overall order is A–E–C–B–D as indicated below

PROBLEM 3: A series of short genomic fragments were generated from a single YAC clone. To place them in order, PCR is performed on each clone using four sets of two primers that will amplify sequences from sequence tagged site (STS) X, STS Y, STS Z and microsatellite sequence 21–11. The following pattern is seen.

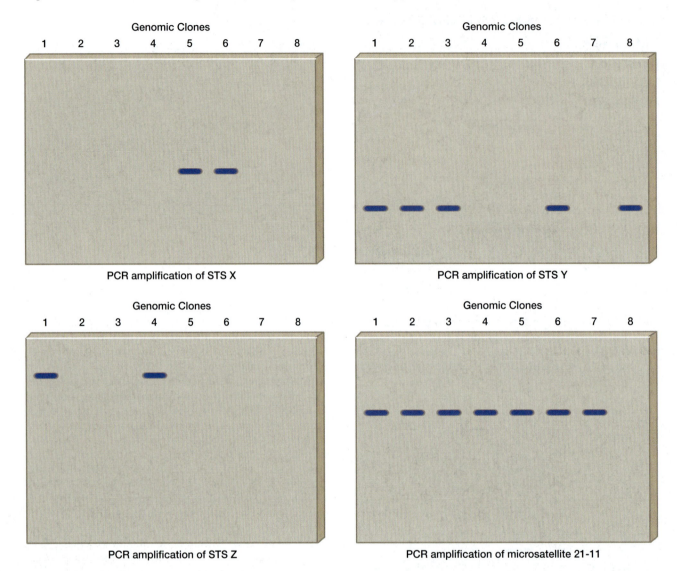

PCR amplification of STS X

PCR amplification of STS Y

PCR amplification of STS Z

PCR amplification of microsatellite 21-11

Order the three genes and the microsatellite sequence. What further experiments might be needed to resolve the final order?

Answer: Both the STS X and MS 21–11 primers amplify DNA sequences from clone 5 and clone 6, indicating that both of these clones must contain these two sequences (STS X and MS 21-11). Similarly, clones 2 and 3 both must share the DNA sequences that are amplified by the STS Y and MS 21–11 primers. Because clone 5 contains the MS 21–11 sequence and not STS Y, and clones 2 and 3 contain the MS 21-11 sequence and not STS X, the MS 21–11 sequence must be located between the STS X and STS Y sequences. This is verified by the absence of a clone that is amplified by both the STS X and STS Y PCR primers, and not the MS 21–11 primers. Clone 6, which contains sequences amplified by primers for STS X, STS Y, and MS 21–11, is also consistent with this order.

Clone 1 contains sequences that are amplified by the STS Y, STS Z, and MS 21-11 primers, which suggests the order is STS X, MS 21-11, STS Y, and STS Z. However, clone 4 contains sequences that are amplified by primers for STS Z and MS 21-11, but not STS Y. It is important to remember that microsatellite sequences may be repeated at different sites within the genome. If we ignore the MS 21–11 marker, the data are consistent with an order of STS Y, STS X, and STS Z, because none of the clones are amplified by the primers for both STS X and STS Z, but not STS Y. To account for the apparent descrepency of MS 21-11's location, we can assume that this repeat is located at two positions. The MS 21–11 sequences may be located on either side of STS Y, or between STS X and STS Y and a second MS 21–11 sequence on the end, next to STS Z.

To try to resolve the two different possible locations of MS 21–11 in the preceding maps, we can sequence the ends of all of the clones that have MS 21–11 sequences. We can then compare these sequences to see how many different MS 21-11 sites there are and use primers designed from these reads as STSs to map the different MS sites.

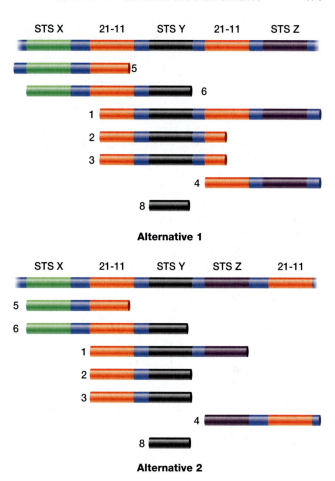

Exercises and Problems

13. A genomic DNA "fingerprint" is generated for several large, genomic fragments using a single restriction endonuclease. Using the gel pattern shown here, determine which of the following genomic DNA clones are likely to be related and why?

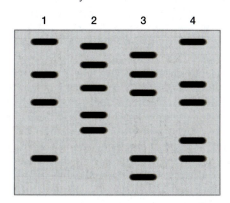

14. Four different runs are shown here from a sequencing project. Assemble them into a contig.

CTGGGTGGAAACCCATGTTGGGGAAGCACCCGGGGCCCTTTTTC

GGAAGCACCCGGGGCCCTTTTTCCTCTATCTCTATCGTGGACC

CCCTTTTTCCTCTATCTCTATCGTGGACCTGCCTTTTCCGAACCCC

CGGCAGGCAGTCTGGGTGGAAACCCATGTTGGGGAAGCACCC

15. Three different runs from a shotgun sequencing project follow. Assemble these three sequences into a contig. Is there more DNA sequence that needs to be generated for this contig? Based on this information, can you assign a polarity to the DNA sequence? Explain your answer.

GAGTGACGAGACCTATAGGATCCTGGATAGGCTGAACAAC

GATCCTGGATAGCTGAACAAGCAGAGTGCCGCTGATTGCCCTAC

TGATTGCCCTACATGGGCAAGGATAATGGCGGCAAGGGTGG

16. Genomics companies such as Celera may have 100 DNA sequencing machines working on a single genome project. Each machine can read 600,000 nucleotides of sequence a day, giving the entire company an output rate of 60,000,000 bp/day. At this rate, how long should it take to sequence the human genome? Why did it take much longer?

17. The animal *Fauna hypotheticus* (first encountered in Chapter 7's integration problem) has a nuclear genome of 2×10^9 bp. The normal karyotype is designated 28, XY. How long will it take to sequence this genome if one were to start at the ends of the chromosomes and move systematically along their length? Assume each read provides 700 nucleotides, and each cycle (producing a primer, sequencing, and analyzing the sequence to identify the next primer) is performed in 24 hours and that each chromosome is the same length.

18. A company called *Sequences R US* embarks on a shotgun sequencing project of the *Fauna* genome. First, scientists clone the genome into two libraries: one containing vectors with an average insert size of 2000 bp, and the other with vectors having an average insert size of 10,000 bp. To account for overlaps between the various inserts and to ensure a high accuracy rate, the scientists decide to sequence an amount of DNA equivalent to 50 times the *Fauna* genome. Suppose the company has 80 automated sequencing machines with each generating reads of 700 nucleotides every 2 minutes.
    a. How many different oligonucleotide primers are needed for this sequencing project?
    b. How long will this project take?

19. Six YAC clones of human DNA, designated A through F, were tested using primers for nine sequence-tagged sites (STS), numbered 1 through 9. The results are shown in the following table where the + symbol means that the YAC contains that STS.

					STS				
YAC	1	2	3	4	5	6	7	8	9
A	+	−	+	−	−	−	−	−	+
B	+	−	−	−	−	+	−	−	−
C	−	−	−	+	−	−	−	+	−
D	−	−	+	+	−	−	−	−	+
E	−	−	−	−	+	+	+	−	−
F	−	+	−	−	−	−	−	+	−

    a. Draw a physical map showing the order of the STS markers. Are there any markers whose order is ambiguous?
    b. Construct a contig map showing the alignment of the YAC clones.

20. The genome of *Mycobacterium tuberculosis* consists of a single circular chromosome that is 4.41 Mb in length and has a 66% GC content. You would like to construct a restriction map of this chromosome. You have the following five restriction enzymes at your disposal:

    *Asc*I which cuts at 5' GG^CGCGCC 3'

    *Mse*I which cuts at 5' T^TAA 3'

    *Nar*I which cuts at 5' GG^CGCC 3'

    *Psi*I which cuts at 5' TTA^TAA 3'

    *Sfi*I which cuts at 5' GGCCNNNN^NGGCC 3'

    a. How many different restriction sites are possible for *Sfi*I?
    b. Assuming a random distribution of nucleotides, which of the five enzymes would be the best choice for the restriction map? Justify your answer.
    c. If the *Mycobacterium* genome was first completely digested with *Asc*I and then exposed to *Mse*I, approximately how many pieces will each *Asc*I fragment produce?

21. Ideally, a genomic sequencing project will begin by cloning the genomic DNA into BAC or YAC vectors (or both), which can hold hundreds of thousands of bases. Yet the first step in shotgun sequencing is to subclone the BAC and YAC genomic fragments into lengths of 150 bp, 2 kb, and 10 kb fragments into sequencing vectors. If you ultimately subclone the large genomic fragments into smaller fragments, why not start by directly isolating the smaller genomic DNA fragments and skip the BAC and YAC cloning steps? Why do we use these different sizes of subclones instead of a single size?

22. Over half of the human genes are alternatively spliced, but far fewer *Drosophila* or *C. elegans* genes are. What does this imply about the evolution of complex organisms?

23. The human and chimpanzee genomes are 99% identical at the nucleotide sequence level. The typical human and chimp proteins differ by only a single amino acid. How can there be such a different outcome in form and intelligence?

24. The basic idea that makes SAGE analysis viable is that a very short sequence (20–22 bp) of a cDNA that might be thousands of nucleotides long is sufficient to definitively determine the identity of that transcript. Why is this a viable assumption?

25. Microarrays are usually created by synthesizing oligonucleotides typically containing around 65 nucleotides that are derived from genome projects. Why are these oligonucleotides preferable to using the full-length cDNAs that would more accurately represent the entire mRNA from the organism?

26. Microarrays are a very powerful technique because they allow the researcher to examine the complete gene expression pattern of two different states of a cell or organism. What are some comparisons that can be done and what can be learned from them?

27. A scientist is interested in determining the differences in gene expression between a normal and cancerous tissue. She performs a microarray analysis comparing healthy and cancerous tissues. Most of the genes on the chip are expressed at the same level in both tissues, but 75 genes show increased expression and 75 show a decrease. What should the scientist do next?

28. A scientific group is studying a gene they believe is involved in a critical kidney function in mice. They isolate 1000 bp of genomic DNA upstream from the transcriptional start site of the gene and clone it into a vector so it will control expression of a reporter gene. When this vector is used to generate a transgenic mouse, expression of the reporter is detected in the kidneys and several other tissues. Why would the scientists perform this experiment?

29. A scientific group is studying the mechanism by which a particular growth hormone alters cellular metabolism. They isolate a mutant cell line that no longer responds to this growth hormone. To try to identify the molecule that responds to the growth factor and is absent in the mutant, they compare two-dimensional gels of wild-type

and mutant cells. Propose a model for how each of the following changes could account for the desired protein.

a. A protein that has a lower molecular weight in the mutant relative to normal cell line

b. A protein with a higher molecular weight in the mutant relative to normal cell line

c. A protein that is the same molecular weight, but different isoelectric point, in the mutant relative to the normal cell line

d. A protein with the same molecular weight and isoelectric point in both mutant and wild-type, but is less abundant in the mutant cell line relative to the normal cell line

e. A protein with the same molecular weight and isoelectric point in both mutant and wild-type, but is more abundant in the mutant cell line relative to the normal cell line

30. Two-dimensional gels can detect small changes, such as phosphorylation or glycosylation, that are introduced into proteins after they are translated. Why is it very useful to have this ability when comparing proteomes?

31. A representation of a two-dimensional gel follows with three different sets of proteins circled. One of these sets of proteins shows multiple spots due to differences in phosphorylation, a second set is due to glycosylation, and the third set is simply unrelated proteins. Determine which set of proteins is associated with each posttrans-

lational modification and explain how you made this determination.

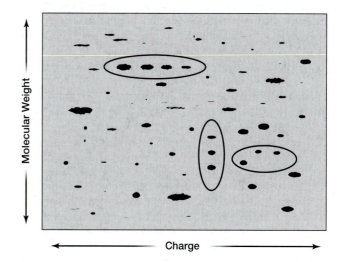

32. A scientist runs two two-dimensional gels comparing cells in a quiescent state versus cells treated with different compounds to induce rapid growth. If the first pattern is a part of the gel from the quiescent cells and the other three represent the corresponding region from gels of cells exposed to treatment A, treatment B, and treatment C, what are the most likely protein changes for each treatment?

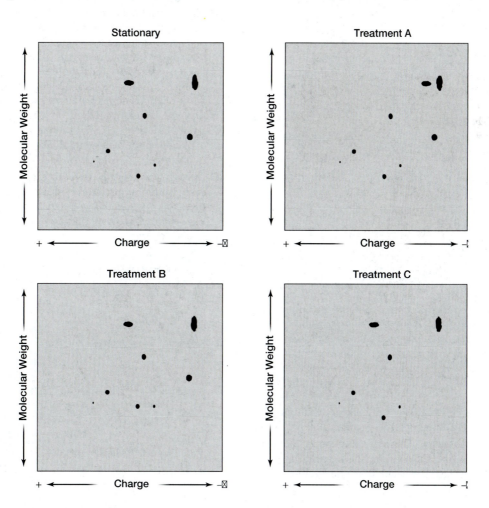

33. How does the yeast two-hybrid approach reveal potential protein–protein interactions in a cell? Will the yeast two-hybrid system always identify a positive interaction if two proteins are part of a multiprotein complex? Explain your reasoning.

34. There are a variety of different situations in which two proteins that are known to directly interact will not activate transcription of the reporter gene in the yeast two-hybrid system. Give two examples.

35. A scientist is interested in finding proteins that might interact with a new transcription factor she isolated. She uses a "bait" vector expressing her new transcription factor and the $TRP1^+$ selectable marker gene. A cDNA library is cloned in the "prey" vector that contains the $LEU2^+$ selectable marker gene. She cotransforms the bait vector and the cDNA library into a yeast strain that contains the $UAS–URA3^+$ reporter gene. She replica plates her colonies onto plates containing Trp, Leu and Uracil; Trp and Uracil; Leu and Uracil; Uracil alone or no additional nutrients.

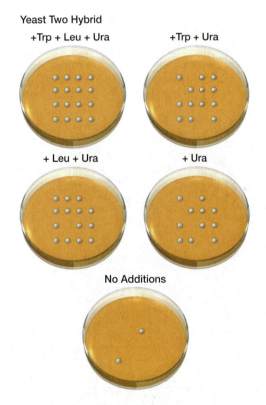

Yeast Two Hybrid

+Trp + Leu + Ura   +Trp + Ura

+ Leu + Ura   + Ura

No Additions

What are the properties of the colonies that grow on each plate? What plasmids, if any, are present in the different colonies? Does she have any candidate cDNAs that appear to interact with her transcription factor?

She decides to perform a positive control using TFIIA and TFIIB, two transcription factors that are known to physically interact. If she clones TFIIA into the bait vector and TFIIB into the prey vector, what pattern of colonies should she see on these same plates?

36. Two brothers who have familial colon cancer in their family have a DNA sequence analysis performed on themselves. The tests revealed that one brother pos-

sesses the allele that is associated with colon cancer in their family and the other brother lacks this allele. What conclusions can be drawn from this DNA test? What are the basic genetic principles that govern the expression of the disease-causing allele that was discovered?

37. A scientist is convinced that a serious disease is linked to a mutation in an unknown gene. What steps should be followed to try to identify the mutation?

38. While analyzing a collection of ESTs, you identify three ESTs that lie within 1500 bp of each other in the genomic sequence. Each EST was also shown to independently hybridize to an mRNA that is approximately 1000 nucleotides long. How is this possible and how would you test this hypothesis?

39. A rich philanthropist wishes to identify the gene that has caused a genetically linked kidney disease in his favorite nephew. He offers to pay a company to sequence the genomes of 10 individuals who have the disease and 10 individuals who are free of the disease. Why is this not likely to lead to identification of the mutation causing the disease? Devise an alternative sequencing approach that is more likely to identify the mutation causing the gene.

40. A group of scientists are trying to dissect a cell-signaling pathway. One of them develops an antibody that should specifically bind to the first protein in the pathway, a transmembrane receptor protein. However, when she immunoprecipitates the receptor, an additional protein is also precipitated. Another scientist develops an antibody that specifically detects the second protein in the cascade, an intracellular receptor-binding protein. When the second scientist immunoprecipitates the intracellular receptor-binding protein, he also precipitates the same additional protein as the first scientist. They both agree that they must have done something wrong and contaminated their antibody preparations. Is there another explanation for their results? How could they prove this hypothesis?

41. DNA restriction fragment fingerprints were done on seven different genomic clones to determine how they overlap. The following restriction fragment patterns were seen. Map the relative positions of the seven clones to one another.

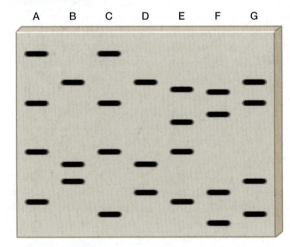

42. The following genetic map shows a region of the *Drosophila* X chromosome. Placed on this map is the location of two recessive loss-of-function mutations (*sme* and *nb*). The mutant *sme* has a smooth eye phenotype due to the fusion of the lenses from the individual ommatidia. The mutant *nb* has no bristles on the wing. Also shown on this map are the in situ hybridization locations for five different YAC genomic clones.

a. Place the five genomic clones in their proper relative positions based on the in situ hybridization data.

b. Predict the most likely positions of the *sme* and *nb* mutant genes on the genomic clones based on this information.

c. After you sequence the genomic clones, you identify two genes that are located near the deduced position of the mutant *sme* gene. Describe three different experiments you could perform that would determine which of these two genes actually corresponds to the mutant *sme* gene.

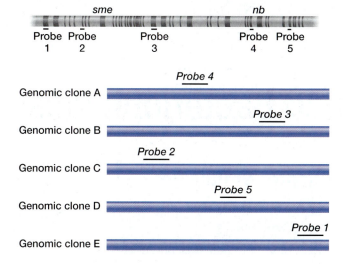

Chapter Integration Problem

A scientific group discovers a mutant *Drosophila* that has long bristles similar to those normally found growing out of the abdomen. When this fly is crossed to a wild-type fly, half of the offspring show this long-bristle phenotype and half are wild type. When two F_1 flies with the mutant long-bristle phenotype are crossed, one-quarter of the offspring are wild type and three-quarters have long bristles in the abdomen. One-third of long-bristle F_2 mutants have an additional phenotype that is the loss of bristles on the thorax.

a. Is each mutant phenotype dominant or recessive to wild type?

b. The scientists genetically map the long-bristle mutation to a position near the end of the third chromosome. They then use four different ESTs from this chromosomal region as probes in Northern blots of mRNA from homozygous mutants and wild-type insects. Which EST probe most likely corresponds to a mutation in this region?

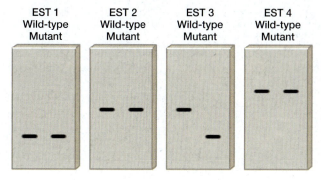

c. Using this EST for chromosomal in situ hybridization analysis of the mutant fly with both phenotypes showed the EST hybridized at a location near the centromere of the third chromosome. The EST sequence was only a few hundred base pairs long, and the short ORF it contained corresponded to the 3' end of the mRNA, which did not show significant homology with any known proteins. The scientists use the EST to screen a genomic library and isolate a large genomic fragment that also contains the 5' end of the ORF. When they use this genomic region in chromosomal in situ hybridization of the mutant, they find that this larger genomic clone hybridizes to both the location near the centromere and also near the end of chromosome 3. Interpret these results, including the location of the corresponding gene and its orientation of transcription along the chromsomsome.

d. When the scientists sequence the genomic fragment, they determine that the potential ORF encodes a protein that likely is a homeobox transcription factor. What is the most likely explanation for the cause of the mutant phenotypes? Be sure your model explains both mutant phenotypes.

e. An antibody is generated against the homeobox transcription factor and it detects a protein that has a slightly lower molecular weight in the mutant flies. How does this correlate with the slightly smaller mRNA on the Northern blot? What would the Western blot look like if the three lanes were (1) extract from wild-type flies; (2) extract from flies with long bristles on back and abdomen; and (3) extract from flies with long bristles on the abdomen, but none on the back?

Do you need additional review? Visit **www.mhhe.com/hyde** for practice tests, answers to end-of-chapter questions and problems, interactive exercises, and animation tutorials all designed to enhance your understanding of key genetic concepts.

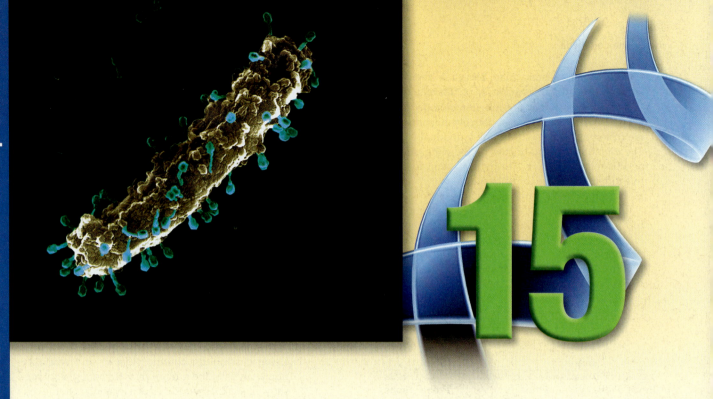

Photo: An electron micrograph of bacteriophage (colorized blue and green) infecting an *E. coli* cell.

Genetics of Bacteria and Bacteriophages

Essential Goals

1. Describe the different types of bacterial and phage mutants.

2. Identify the different forms of the F plasmid DNA, and explain how each form transfers bacterial genomic DNA and is used for gene mapping.

3. Understand how transformation is used for gene mapping.

4. Explain the two different types of transduction and how each can be used for gene mapping.

5. Describe how genes in the phage genome are mapped.

6. Summarize how Benzer assigned complementation groups in the *rII* gene and how he mapped mutations within an *rII* cistron.

Chapter Outline

*T*he sequence of the 4639 kilobase (kb) pairs in the *Escherichia coli* genome was published in the journal *Science* in 1997. It was found to contain 4289 open reading frames (ORFs). The bacteriophage lambda (λ), which consists of 48.5 kb, was sequenced in 1982 and found to contain 46 genes. **Bacteriophages,** also called *phages,* are viruses that infect and multiply within bacterial cells. Although the genomic sequence and gene annotation information are known for *E. coli* and bacteriophage λ, the roles of every protein in the life cycle of the respective organisms are not entirely understood.

This shows that the genome projects that we discussed in chapter 14 can provide the molecular details of the genetic makeup and identity of all the genes, but cannot provide all the information required to understand how an organism lives. The field of genetics is still critical for elucidating the function of a gene or its encoded protein. **Phenomics** is the term assigned to studying the phenotype produced by mutating every gene in the genome.

With the wealth of genomic information that is available in a variety of different eukaryotic organisms, and with the ability to perform genetic experiments on potentially every gene, why would we be interested in the genetics of bacteria and bacteriophages? There are three major answers to this question.

First, *E. coli* serves as one of the major host organisms for a variety of molecular genetic experiments. Having a bet-ter understanding of *E. coli* biology will allow us to further exploit it. For example, *E. coli* serves as an excellent host for the expression of large amounts of proteins from cloned foreign genes. These proteins cannot be posttranslationally modified; however, the recombinant proteins may still be useful as antigens for immunizing humans or livestock. The proteins can also be used for therapy in individuals who lack specific proteins or enzymes, such as human growth hormone.

Second, many of the initial findings in molecular genetics came from experiments in *E. coli* and their bacteriophages, including λ. It is likely that additional research in these "simple" organisms will continue to elucidate more complex processes in eukaryotes.

Third, many of these microorganisms are of medical relevance. *Bordetella pertussis, E. coli, Neisseria meningitidis, Salmonella, Shigella,* and *Staphylococcus* can cause a variety of diseases in humans. By continuing to study these microorganisms, we should further understand their biology within humans and develop new therapies to treat infections caused by them.

This chapter deals with the general methods used to genetically manipulate the bacterial genome. Using these approaches to generate linkage maps, the genetic map for *E. coli* was developed long before its genome was entirely sequenced.

15.1 Bacteria and Bacteriophages in Genetic Research

Several properties of bacteria and bacteriophages make them especially suitable for genetic research. First, bacteria and their phages generally have a short generation time. Some phages increase in number 300-fold in about a half hour; an *E. coli* cell divides every 20 minutes. (*E. coli,* the common intestinal bacterium, was discovered by Theodor Escherich in 1885.) In contrast, the generation time is 10 days in fruit flies, a year in corn, and 15 years or so in human beings. This rapid growth of bacteria and bacteriophages allows for the production of a large amount of DNA for analysis; furthermore, the large number of daughter cells allows the isolation of rare events, such as recombination between different allelic sites within a single gene.

Bacteria and phages are also well-suited for genetic research because they have much less genetic material than eukaryotes do, and the organization of this material is much simpler. The *E. coli* genome is 4639 kb containing 4289 ORFs. Compare these numbers to the *Saccharomyces cerevisiae* (yeast) genome of 12,069 kb and 6294 ORFs and the *Drosophila* genome of 180,000 kb and approximately 13,600 ORFs.

A third reason for the use of bacteria and phages in genetic study is their ease of handling. A researcher can handle millions of bacteria in a single culture with a minimal amount of work compared with the effort required to grow the same number of eukaryotic organisms such as fruit flies or corn. (Some eukaryotes, such as yeast or *Neurospora,* can also be handled using prokaryotic techniques.)

15.2 Morphological Characteristics

The term *prokaryote* arises from the lack of a true nucleus (*pro* means before, and *karyon* means kernel, or nucleus); they have no nuclear membranes (**fig. 15.1**) and usually have only a single, relatively "naked" circular chromosome, so they are haploid. As discussed in chapter 12, bacteria may, however, contain **plasmids** that are usually small, circular DNAs that replicate free of the genome.

Bacteriophages are even simpler. Although animal and plant viruses can be more complicated, the bacteriophages are exclusively genetic material surrounded by a protein coat. Bacteriophages are usually classified first by the type of genetic material (nucleic acid) they have (DNA or RNA, single- or double-stranded) and then by structural features of their protein surfaces (**capsids**). These features include type or symmetry and number of discrete protein subunits (**capsomeres**) in the capsid, as well as general size.

Most bacteriophages are complex, like T2 (**fig. 15.2**), or are made up of a head-like capsule like T2 without the tail appendages, or filaments. Most phages contain double-stranded DNA. Bacteriophages are obligate parasites; outside of a bacterial host, they are inert molecules. Once their genetic material penetrates a host cell, they can take over the metabolism of that cell and construct multiple copies of themselves.

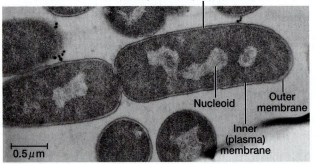

Periplasmic space and cell wall

Nucleoid

Outer membrane

Inner (plasma) membrane

0.5 μm

figure 15.1   Electron micrograph of an *E. coli* bacterium. The *E. coli* cell contains two membranes at the cell surface (inner and outer membranes) sandwiching a cell wall. The DNA is localized to the lightly stained, membrane-free nucleoid region.

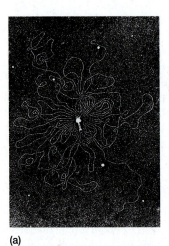

(a)

figure 15.2

Bacteriophage T2 and its chromosome. (a) An electron micrograph of the T2 phage chromosome, which is about 50 μm long, released from the bacteriophage protein head (visible in the center). (b) The intact phage consists of a protein exterior and DNA within the head. The phage attaches to a bacterium using its tail fibers and base plate and injects its genetic material into the host cell.

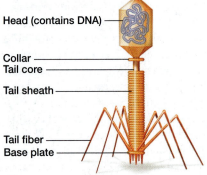

Head (contains DNA)

Collar
Tail core

Tail sheath

Tail fiber
Base plate

(b)

The smallest bacteriophages (for example, R17) have RNA as their genetic material and contain just three genes, one each for a coat protein, an attachment protein, and an enzyme to replicate their RNA. The larger bacteriophages (such as T2 and T4) have DNA as their genetic material and contain up to 130 genes.

15.3 Techniques of Cultivation

All organisms need a carbon source; an energy source; nitrogen, sulfur, phosphorus, several metallic ions; and water to grow. Those that require an organic form of carbon are termed **heterotrophs.** Those that can utilize carbon from carbon dioxide are termed **autotrophs.** All bacteria obtain their energy either by photosynthesis (autotrophs) or chemical oxidation (chemotrophs).

Bacterial Culture Media

Bacteria are usually grown in or on a chemically defined **synthetic medium,** either in liquid in flasks or test tubes, or on petri plates using an agar base to supply rigidity. When one cell is placed on the medium in the plate, it begins to divide, often as frequently as every 20 minutes. After incubation, often overnight, a colony, or clone, exists (**fig. 15.3**). This **colony** is composed of millions of cells that are genotypically

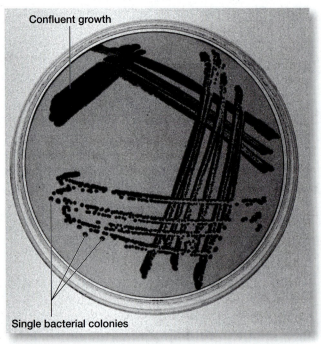

Confluent growth

Single bacterial colonies

figure 15.3   Growth of bacterial colonies on a petri plate. A bacterial culture was streaked on the petri plate, beginning at the upper left and continuing clockwise. With a heavy inoculation of bacteria, growth is confluent as the colonies cover the entire surface. At lower bacterial concentrations, single colonies are present (lower left).

table 15.1   Minimal Synthetic Medium for Growing *E. coli*, a Heterotroph

Component	Quantity
$NH_4H_2PO_4$	1 g
Glucose	5 g
NaCl	5 g
$MgSO_4 \cdot 7H_2O$	0.2 g
K_2HPO_4	1 g
H_2O	1000 ml

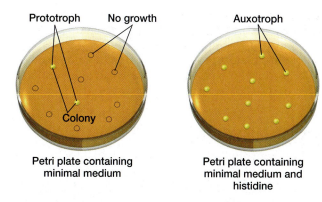

figure 15.4   Prototrophic and auxotrophic bacteria. Two petri plates containing either minimal medium (left) or minimal medium that is supplemented with histidine (right) are diagrammed. Colonies that grow on minimal medium are prototrophs, while colonies that require a supplement to grow on minimal medium are auxotrophs (mutants). The circles on the minimal medium plate represent auxotrophic colonies that did not grow.

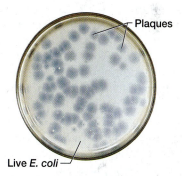

figure 15.5   Infection of a lawn of *E. coli* with bacteriophage T1 produces plaques. Each plaque corresponds to a region of lysed bacterial cells that contains phage.

identical to the original bacterial cell that was located at that position on the petri plate. Overlapping colonies form a confluent growth (see fig. 15.3).

A culture medium that has only the minimal components required by the wild-type bacterial species is referred to as *minimal medium* (**table 15.1**). This minimal medium requires a carbon source or sugar from which all organic compounds can be synthesized. Wild-type bacteria can utilize a variety of different sugars, although glucose is preferred. In the absence of glucose in the medium, wild-type bacteria will utilize other sugars, such as lactose or galactose, as their energy source.

Wild-type bacteria that can grow on minimal medium are referred to as **prototrophs** (**fig. 15.4**). By contrast, some mutant bacteria can only grow on minimal medium that is supplemented with more complex substances, including amino acids, vitamins, and other chemical components. These enriched media allow the growth of **auxotrophs,** mutant bacteria that have particular nutritional requirements. For example, an *E. coli* mutant strain that has an enzyme defect in the pathway that produces the amino acid histidine will not grow on minimal medium because it is unable to synthesize histidine; it requires histidine supplementation (see fig. 15.4). This type of mutant is called a **conditional-lethal mutant,** because these bacteria will be unable to survive under certain environmental conditions (absence of added histidine).

Mutants can also be defective in their ability to metabolize a specific sugar as their energy source. For example, some mutants cannot utilize lactose as an energy source (*lac⁻*). These *lac⁻* cells can grow on minimal medium because it contains glucose, which they can utilize as an energy source. However, if lactose replaces glucose in the medium, then the *lac⁻* bacteria are unable to grow. We will discuss the genetic control of these *lac⁻* conditional-lethal mutations in chapter 16.

Auxotrophic mutants can grow only on an **enriched** or **complete medium,** whereas prototrophs can grow on either minimal or complete medium. Media are often enriched by adding complex mixtures of organic substances such as blood, beef extract, yeast extract, or peptone, a product of protein digestion. Many media,

however, consist of a minimal medium plus only one other substance, such as a single amino acid or a vitamin. These are called **selective media;** we discuss their uses later in the chapter.

Bacteriophage Culture

The experimental cultivation of bacteriophages is somewhat different. Because phages can grow only in living cells, they are obligate parasites. Thus, to cultivate phages, petri plates of appropriate medium are inoculated with enough bacteria to form a confluent growth, or **bacterial lawn.** This bacterial culture serves as a medium for the growth of bacteriophages added to the plate. Because the phage attack usually results in rupture, or **lysis,** of the bacterial cell, addition of the phage usually produces clear spots, known as **plaques,** on the petri plates (**fig. 15.5**). Large quantities of bacteriophages can also be grown in flasks of bacterial suspensions.

15.4 Bacterial Phenotypes

Bacterial phenotypes fall into three general classes: colony morphology, nutritional requirements, and drug or infection resistance.

Colony Morphology

Colony morphology relates simply to the form, color, and size of the colony that grows from a single cell. Each wild-type species of bacteria will exhibit specific colony morphologies, which are usually under genetic control (**fig. 15.6**). Mutants that exhibit altered phenotypes can readily be identified.

Nutritional Requirements

Nutritional requirements are the compounds that must be added so that bacteria will grow on minimal media. Wild-type bacteria (prototrophs) have no nutritional requirements.

Nutritional requirements reflect mutations that disrupt one or more enzymes in the bacterium's biosynthetic pathways. For example, if an auxotroph has a requirement for the amino acid cysteine, then it most likely has a mutation that affects an enzyme in the cysteine biosynthesis pathway. **Figure 15.7** shows five steps in cysteine synthesis, with a different enzyme controlling each step. We discussed in chapter 11 how auxotrophs can be used to assemble the progression of a biochemical synthetic pathway.

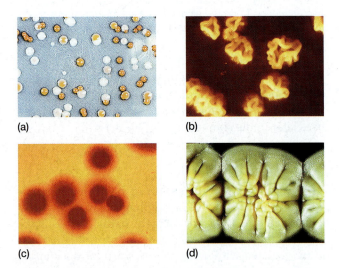

(a)   (b)

(c)   (d)

figure 15.6   Various bacterial colony phenotypes. Shown are the red and white colonies of *Serratia marcescens* (a), the irregular raised folds of *Streptomyces griseus* (b), the round colonies with concentrated centers and diffuse edges of *Mycoplasma* (c), and the irregularly folded raised colonies of *Streptomyces antibioticus* (d).

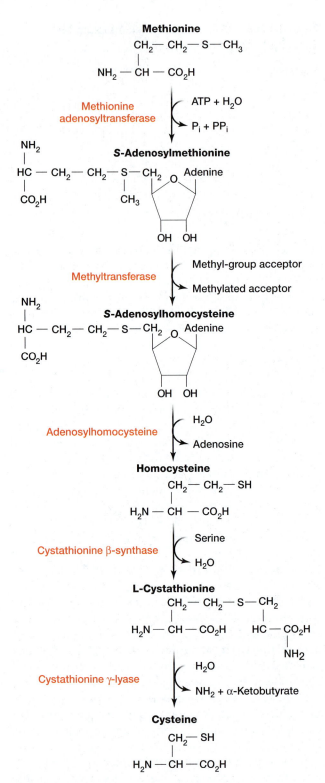

figure 15.7   Each of the five steps in the biosynthesis of cysteine from methionine is controlled by a different enzyme (red). The names of the different molecules are in boldface above the corresponding structure.

Replica Plating

To examine the specific nutrient needs of auxotrophic bacteria, geneticist Joshua Lederberg (**fig. 15.8**) developed a rapid screening technique known as **replica plating.** In this technique, a culture of bacterial cells are spread on a petri plate of complete medium. Both prototrophs and auxotrophs grow on the medium, and their colonies appear in a random arrangement.

figure 15.8   Joshua Lederberg (1925–) pioneered many fundamental techniques and principles in bacterial genetics, such as conjugation and transduction.

A piece of sterilized velvet is pressed on this plate of colonies, and the velvet picks up some cells from each colony. The velvet can then be pressed on a number of petri plates, which reproduces the same pattern of bacterial colonies. One of these replica plates contains minimal medium, and each of the remaining replica plates contain minimal medium supplemented with a different compound (**fig. 15.9**). A colony that grows on minimal medium represents a prototroph, whereas a colony that fails to grow on minimal medium represents an auxotroph. An auxotroph that grows on a supplemented minimal medium plate is unable to synthesize either the supplemented compound or a precursor to that compound (see chapter 11). The individual colonies of interest can then be isolated and grown out.

Designation of Mutants

The auxotroph in figure 15.9 that grows on medium containing methionine is designated genetically as *met⁻* (methionine minus or met minus). The *met⁻* designation refers to the fact that the auxotroph requires methionine to be added to the minimal medium for growth. It is assumed that the *met⁻* auxotroph possesses a mutation in a gene that encodes an enzyme that is required for one step in the methionine biosynthetic pathway.

In terms of energy sources, the plus or minus notation has a different meaning. For example, a strain of bacteria that can utilize the sugar galactose as an energy source is designated *gal⁺*. Bacteria that cannot utilize galactose would be called *gal⁻*. The latter strain will not grow on minimal media if galactose is the sole carbon source, but it will grow if a sugar other than galactose is present. Note that a *met⁻* strain needs methionine to grow, whereas a *gal⁻* strain needs a carbon source *other* than galactose for growth.

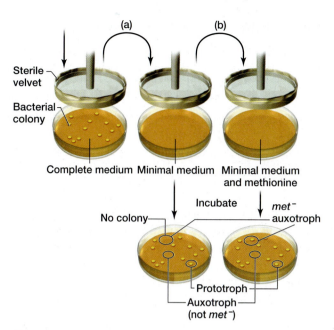

figure 15.9   Replica-plating technique. A pattern of colonies growing on a petri plate of complete medium is transferred (a) to a second plate of minimal medium, which lacks a variety of substances. The velvet is then touched (b) to a plate of minimal medium that is supplemented with methionine. The latter two plates are incubated to allow the bacteria to grow. Colonies that grow on the minimal medium plate are prototrophs (wild-type). Notice that the prototrophs grow on all the plates. Colonies that grow on the third plate, but not the second (minimal medium) plate are auxotrophs (*met⁻*). Colonies that grow on complete medium (first plate), but neither the second nor third, are also auxotrophs, but the nutrient defect is unknown.

Resistance and Sensitivity

The third common classification of bacterial phenotypes involves resistance and sensitivity to drugs, phages, and other environmental insults. We first described manipulation of these phenotypes in chapter 12 in the discussion of bacterial plasmids. For example, many plasmids contain a gene that allows a bacterial cell to be resistant to the antibiotic ampicillin (*ampʳ*). When ampicillin is present in the medium, most bacterial cells that lack the plasmid die; they are ampicillin-sensitive (*ampˢ*). The antibiotic resistance genes can be found on either plasmids or the bacterial genome. Numerous antibiotics are used in bacterial studies (**table 15.2**).

Because wild-type bacteria are typically sensitive to antibiotics, researchers usually screen for mutants that are resistant to antibiotics. This screening approach is relatively straightforward compared with the replica-plating screen for auxotrophs. Bacteria are grown in the presence of the desired antibiotic, which kills the wild-type bacteria. The only colonies that grow are the resistant mutants.

table 15.2   Some Antibiotics and Their Antibacterial Mechanisms

Antibiotic	Microbial Origin	Mode of Action
Penicillin G	*Penicillium chrysogenum*	Blocks cell wall synthesis
Tetracycline	*Streptomyces aureofaciens*	Blocks protein synthesis
Streptomycin	*Streptomyces griseus*	Blocks protein synthesis
Erythromycin	*Streptomyces erythraeus*	Blocks protein synthesis
Bacitracin	*Bacillus subtilis*	Blocks cell wall synthesis

Each wild-type bacterial species is also sensitive to specific types of bacteriophage. Infection of wild-type bacteria results in the lysis and death of the bacterial cell, which is visualized as a plaque on a petri plate. Screening for resistance to phages is similar to screening for drug resistance. When bacteria are placed in a medium containing phages, only those bacteria that are resistant to the phages grow and produce colonies.

Restating the Concepts

▶ The genetics of bacteria and bacteriophages can be studied because they both express distinct phenotypes.

▶ Bacterial phenotypes include colony morphology, nutrient requirements, and resistance to phage and antibiotics.

15.5 Bacteriophage Life Cycles and Phenotypes

Bacteriophages can only replicate within a bacterial cell, and their genetic material is dedicated to accomplishing the task of infection and replication. Phage phenotypes, therefore, affect their growth rate and plaque appearance on a bacterial lawn.

Phage Life Cycles

All phages infect bacterial cells by attaching to the cell surface and injecting their genetic material into the cell. Infection results in one of two events: immediate production of new phages and lysis of the cell, or maintenance of the phage genome in the bacterial cell for many generations. Phages are classified based on which course they take in the cell.

Virulent phages immediately begin producing new phages on entering a bacterial cell (**fig. 15.10**). The phage redirects the bacterial DNA replication, transcription, and translation to concentrate primarily on the phage genome and genes. Sometimes the bacterial genome

is degraded to accomplish this process. After hundreds of new phage particles have been generated, a phage-encoded lysozyme enzyme ruptures (*lyses*) the bacterial cell to release a **lysate** of phage particles to infect other cells.

Temperate phages have the ability to remain in the bacterial cell for some time before entering the lytic cycle (**fig. 15.11**). The phage DNA either replicates within the cell like a plasmid or integrates into the bacterial genome, in which case it is called a **prophage.** A bacterial cell that harbors a prophage is called a **lysogen.** Regardless of whether the quiescent phage resides as a prophage or plasmid, the majority of the phage genes are not transcribed until a

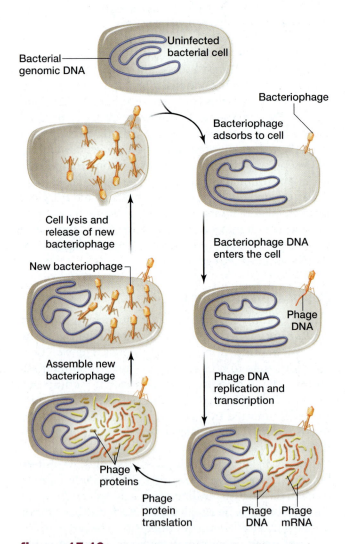

figure 15.10   The bacteriophage virulent life cycle. A wild-type phage infects a bacterial cell. The phage DNA is replicated (red), the phage genome is transcribed into mRNA (yellow), and phage proteins are translated. After the new phage are produced, the cell lyses to release the phage, which can infect new cells.

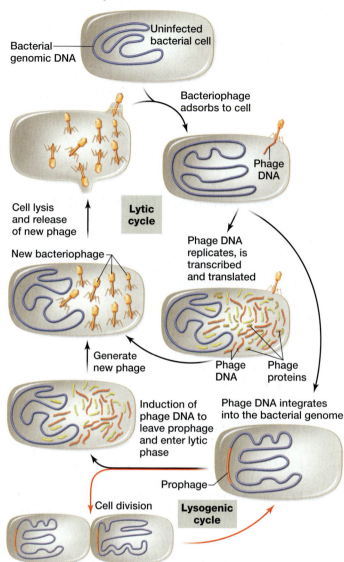

figure 15.11 Alternate life-cycles of a temperate phage. Upon infecting a bacterial cell, the phage may enter the lytic cycle, which results in the production of more phages and lysis of the cell. Alternatively, a temperate phage can enter a lysogenic cycle, where the phage genome integrates into the bacterial genome. The phage genome remains in the bacterial genome until conditions cause the phage to recombine out of the bacterial genome and enter the lytic cycle.

stimulus to the bacterial cell induces the phage to enter the lytic cycle (see fig. 15.11).

Lambda, which we discussed in chapter 12 as a cloning vector, is one of the best studied temperate phages. In chapter 16, we will discuss the gene regulation involved in λ's selection between becoming a prophage or entering the lytic cycle.

Phage Phenotypes

Phage phenotypes fall generally into two categories: plaque morphology and growth characteristics on different bacterial strains.

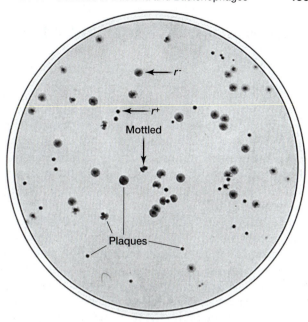

figure 15.12 Normal (r^+) and rapid-lysis (r^-) mutants of phage T2. The wild-type T2 phages produce small plaques, while the T2 r^- mutants produce large plaques. Mottled plaques occur when r^- and r^+ phages grow together.

The wild-type bacteriophage T2 infects *E. coli* to produce small plaques with fuzzy edges (T2 genotype r^+). This small plaque size results from a slow lysis of the bacterial cell. T2 mutants have been isolated that cause rapid lysis of infected *E. coli* cells (T2 genotype r^-) to produce large, smooth-edged plaques (**fig. 15.12**).

A related bacteriophage, T4, also has rapid-lysis mutants that produce large, smooth-edged plaques on *E. coli* strain B, but T4 will not infect and lyse a different *E. coli* strain, K12. If a bacterial lawn contains both B and K12 strains, then the T4 r^- mutant produces a turbid plaque due to the lysis of strain B but not strain K12. Here, the rapid-lysis mutant illustrates both the colony morphology phenotype and the growth restriction phenotype of phages. We return to these phenotypes later in this chapter when we describe recombination between and within phage genes.

Restating the Concepts

▶ A phage can have either a virulent life cycle, which causes lysis of the bacterial cell it infects, or temperate, which allows the phage to survive in a quiescent state in the bacterial cell (such as a prophage integrated in the genome) until it is induced to enter a virulent stage.

▶ Phage phenotypes that are encoded in the bacteriophage genome include plaque size, which is dictated by the rate of bacterial cell lysis, and the host range, which is the ability to grow on different bacterial strains.

15.6 Overview of Sexual Processes in Bacteria

Although bacteria and viruses are ideal subjects for biochemical analysis, they would not be useful for genetic study if they did not have sexual processes by which genotypes can be changed.

We previously defined a sexual process or sexual reproduction for a haploid organism as the fusion of two cells to form a diploid cell that undergoes meiosis (chapter 3). In this sense, haploid bacteria and phages fail to undergo sexual reproduction. If we define a sexual process more broadly, however, as the combining of genetic material from two individuals, then the life cycles of bacteria and bacteriophages include sexual processes. Bacteria have three different methods of gaining access to foreign genetic material: **transformation, conjugation,** and **transduction** (**fig. 15.13**).

Phages can also exchange genetic material when a bacterium is infected by more than one phage particle. During bacteriophage infection, the genetic material of different phages can exchange parts or recombine.

In the following sections, we first examine the exchange processes in bacteria and then in bacteriophages. We also discuss how to use these methods for mapping bacterial and viral chromosomes. Before we proceed, we need to refine the definition of chromosome: **Chromosome** refers to the structural entity in the cell or phage that contains the genetic material (DNA or RNA). In eukaryotes, a chromosome is double-stranded linear DNA complexed with proteins. In prokaryotes, the chromosome is usually a circular double-stranded DNA molecule. In bacteriophages, it is virtually any combination of linear or circular, single- or double-stranded RNA or DNA.

Restating the Concepts

▶ Bacteria and phages can undergo sexual reproduction without going through meiosis. In this case, sexual reproduction involves the recombination of genetic material between two different individuals or genotypes.

▶ Bacteria can undergo recombination with another cell through transformation, conjugation, or transduction.

▶ The term *chromosome* refers to the structural entity in a bacterial cell or a phage that contains the genetic material, either DNA or RNA.

15.7 Transformation

In **transformation,** a cell takes up extraneous DNA found in the environment and incorporates it into its genome through recombination. Transformation was first observed in 1928 by Frederick Griffith and was examined at the molecular level in 1944 by Oswald Avery and his colleagues, who used the process to demonstrate that DNA was the genetic material of bacteria. We discussed the details of these experiments in chapter 7.

Not all bacteria are competent to be transformed, and not all extracellular DNA is competent to cause transformation. To be competent to transform efficiently, the extracellular DNA must be double-stranded. To be competent to be transformed, some bacteria have a surface protein **competence factor,** which binds the extracellular DNA in an energy-requiring reaction. Other bacteria become competent by growing in an enriched medium. Bacteria that are not naturally competent, however, can be treated, usually with calcium chloride, which makes them more permeable or competent.

Mechanisms of Transformation

Transformation involves the binding of DNA by the cell and the digestion of one DNA strand by a membrane-

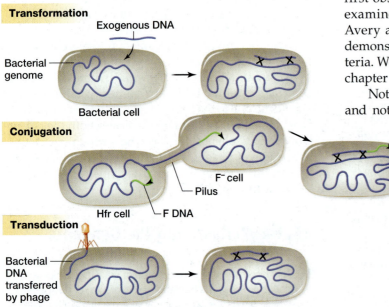

figure 15.13   Summary of three mechanisms to transfer foreign DNA into bacterial cells. Bacterial cells can take up free DNA from their environment (transformation), by mating (physical contact) with another bacterial cell (conjugation), or through infection by a bacteriophage (transduction). In transduction, the phage introduces bacterial DNA that was acquired from the previously infected cells. In all three cases, the foreign DNA must recombine with the recipient genome.

bound nuclease. Thus, only one strand of extracellular DNA enters the bacterial cell. The single-stranded linear donor DNA can recombine into the host genome by two crossover events (**fig. 15.14**). (We described the molecular mechanisms of crossing over in chapter 6.) These crossovers result in the exchange of one strand of genomic DNA with the single-stranded donor DNA in a region that is complementary with both molecules.

Unlike eukaryotic crossing over, bacterial crossing over is not a reciprocal process. The bacterial chromosome incorporates part of the foreign DNA to produce a **heteroduplex region,** which contains one host DNA strand and one donor DNA strand that may not be completely complementary. The resulting single-stranded DNA, originally part of the bacterial chromosome, is degraded by host enzymes called exonucleases; linear DNA is degraded rapidly in bacteria. When the bacterial chromosome is replicated, each strand of the chromosome serves as a template for a new strand. After cell division, one cell contains the original double-stranded chromosome, and the other cell contains a double-stranded DNA molecule with a double-stranded region of donor DNA.

A basic difference exists between recombination in bacteria and eukaryotes. Because eukaryotic chromosomes are linear, a single crossover event between the two linear molecules produces two recombinant linear molecules (**fig. 15.15**). In bacteria, the chromosome is circular and the foreign DNA is linear. A single crossover event between these two molecules

produces a linear molecule (see fig. 15.15); however, this linear molecule cannot replicate properly in a bacterial cell and the cell will die. Therefore, two crossover events are needed to regenerate a circular chromosome capable of proper replication. More generally, an even number of crossover events produces viable cells; an odd number does not.

For transformation to be detected, the donor DNA must contain a different allele from the host chromosome. In figure 15.14, the donor DNA contains the a^+ allele, and the host chromosome contains the a^- allele. Transformation and recombination produces a daughter cell that contains the a^+ allele. If we can detect a phenotypic difference between a^+ and a^-, such as a nutritional requirement, then we can detect the recombination event.

Cotransformation Mapping

The general idea of transformation mapping is to add DNA from a donor strain with a known genotype to a recipient strain, also with known genotype, but with different alleles at two or more loci. We then look for incorporation of the donor alleles into the recipient strain of bacteria.

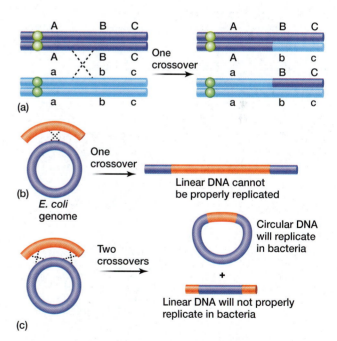

(a)
(b) *E. coli* genome
(c)

Linear DNA cannot be properly replicated

Circular DNA will replicate in bacteria

Linear DNA will not properly replicate in bacteria

figure 15.15   The consequences of different number of crossover events. (a) A single crossover between two linear nonsister chromatids produces two recombinant linear chromatids. (b) A single crossover between a linear DNA molecule and the circular *E. coli* chromosome produces a recombinant linear molecule that cannot properly replicate. (c) A double crossover between a linear DNA molecule and the circular *E. coli* chromosome produces a circular recombinant molecule and a linear recombinant molecule. In this case, the recombinant circular chromosome can properly replicate.

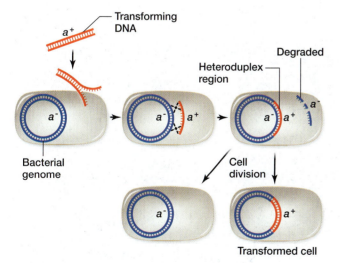

figure 15.14   Recombination of a transformed DNA fragment. A single strand of DNA (red with a^+ allele) enters a bacterial cell (blue chromosome with a^- allele). The single-stranded foreign DNA utilizes two crossover events to recombine into the bacterial genome. This recombination generates a heteroduplex region that consists of one host DNA strand and one transforming DNA strand. After DNA replication and cell division, one cell has the a^- allele and the other the transformed a^+ allele.

The more frequently alleles from two different loci are incorporated together into the recipient, the closer together these loci must be to each other. Thus, we can use an index of **cotransformation,** which is an *inverse* relationship to map distance. That is, the larger the cotransformation frequency of alleles at two different loci, the closer together the loci must be. This relationship results from the need for both alleles to enter the recipient cell on the same piece of DNA and then an even number of crossover events to recombine both alleles into the recipient genome. As discussed in chapter 6, loci that are close together have a low frequency of crossovers between them; conversely, there is an increased frequency of crossovers flanking them. By having the two crossover events flank the pair of loci, they are cotransformed into the recipient genome.

The next step is to select for recombinant cells. In fruit flies, every offspring of a mated pair represents a sampling of the meiotic tetrad, and thus a part of the total, whether or not recombination has taken place. In this case, however, many cells are present that do not take part in transformation. In a bacterial culture, for example, only one cell in a thousand might be transformed. Therefore, when working with bacteria, it is important to select for the transformed cells.

Example: Mapping in *Bacillus subtilis*

A recipient strain of *B. subtilis* is auxotrophic for the amino acids tyrosine ($tyrA^-$) and cysteine ($cysC^-$). We are interested in how close these loci are to each other on the bacterial chromosome. We isolate DNA from a prototrophic strain of bacteria ($tyrA^+\ cysC^+$), add this donor DNA to the auxotrophic strain, and allow time for transformation to take place (**fig. 15.16**).

If the experiment is successful, then some of the recipient bacteria will incorporate donor DNA that has either one of the donor alleles or, if the loci are close together, both donor alleles. The overwhelming majority of the cells, however, will retain the auxotrophic genotype, $tyrA^-\ cysC^-$. Either these cells failed to be transformed with the donor DNA (the most prevalent reason), or the donor DNA that entered the recipient cell either did not possess the $tyrA^+$ or $cysC^+$ alleles or did not recombine with the genome (a minor reason). To properly measure the cotransformation frequency, we need to count all of the transformed cells and not the nontransformed cells.

Distinguishing Transformants

After removing extraneous transforming DNA, samples of the transformation mixture are plated onto a

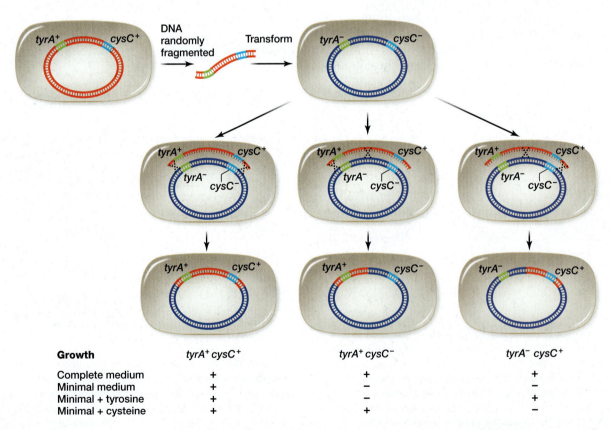

Growth	$tyrA^+\ cysC^+$	$tyrA^+\ cysC^-$	$tyrA^-\ cysC^+$
Complete medium	+	+	+
Minimal medium	+	−	−
Minimal + tyrosine	+	−	+
Minimal + cysteine	+	+	−

figure 15.16 Transformation experiment with *B. subtilis*. A $tyrA^-\ cysC^-$ strain is transformed with DNA from a $tyrA^+\ cysC^+$ strain. Three types of recombinant transformants, a $tyrA^+\ cysC^+$ prototroph and two different auxotrophs ($tyrA^+\ cysC^-$ and $tyrA^-\ cysC^+$) can be generated. Genotypes of the different transformants are determined by growth characteristics on four different types of petri plates (see fig. 15.17).

complete medium to determine the total number of cells, onto minimal medium to select for $tyrA^+$ $cysC^+$ transformants, minimal medium plus tyrosine to select for $cysC^+$ transformants, and minimal medium plus cysteine to select for $tyrA^+$ transformants. The cultures are allowed to grow overnight in an incubator at 37°C. **Figure 15.17** summarizes the plating.

The overwhelming majority of colonies grow on complete medium, but not on any of the three minimal medium plates. This majority is made up of the nontransformants ($tyrA^-$ $cysC^-$).

Checking for Reversion

Reversion is the spontaneous mutation of $tyrA^-$ to $tyrA^+$ or $cysC^-$ to $cysC^+$, which is unrelated to the generation of $tyrA^+$ or $cysC^+$ through transformation. Reversion is always a concern when we are detecting the generation of prototrophs from auxotrophs.

To test for reversion, several plates of auxotrophs that were not exposed to prototrophic DNA are grown on minimal medium and on minimal medium with tyrosine or cysteine added. The number of natural revertants are counted, and the experimental num-

bers can be corrected for the natural reversion rate. In this way we are sure that we are measuring the actual transformation rate, rather than just a mutation rate that we mistake for transformation. This control should *always* be carried out.

Data Analysis

From the experiment as shown in figures 15.16 and 15.17, we count 12 cotransformants ($tyrA^+$ $cysC^+$), 31 $tyrA^+$ $cysC^-$, and 27 $tyrA^-$ $cysC^+$. From these data, we calculate the **cotransformation frequency,** or **cotransfer index, (r)** as

$$r = \frac{12}{(12+31+27)} = 0.17$$

This is a relative number indicating the co-occurrence of the two loci, and thus their relative distance apart on the bacterial chromosome. Remember that this number increases for different pairs of loci as the loci are closer together, not farther apart.

By systematically examining many loci, we can establish their relative order. For example, if locus A is closely linked to locus B and B to C, we can establish the order A B C. It is not possible by this method to determine exact order for very closely linked genes. For this information we need to rely on transduction, which we will consider shortly. However, transformation has allowed us to determine that the map of *B. subtilis* is circular, a phenomenon found in most bacteria and phages. (The *E. coli* map is shown later.)

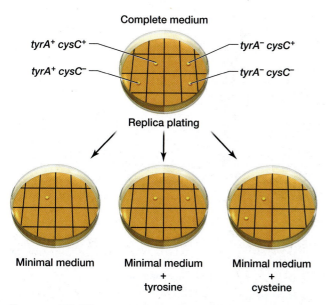

figure 15.17   Growth patterns on different media reveal bacterial genotypes. Only four colonies of different genotypes are shown, and a grid is added to simplify identification of the colonies on different plates. After transformation (see fig. 15.16), cells are plated on complete medium and then replica-plated onto minimal medium and minimal medium containing either tyrosine or cysteine. Only the prototroph ($tyrA^+$ $cysC^+$) grows on minimal medium. Only the colonies that are $cysC^+$ ($tyrA^+$ $cysC^+$ and $tyrA^-$ $cysC^+$) grow on the minimal medium plate with tyrosine. Similarly, only the $tyrA^+$ colonies grow on the minimal medium with cysteine plate. The double mutant ($tyrA^-$ $cysC^-$) will only grow on complete medium or minimal medium that also contains both tyrosine and cysteine.

In the figure: Complete medium; labels $tyrA^+$ $cysC^+$, $tyrA^-$ $cysC^+$, $tyrA^+$ $cysC^-$, $tyrA^-$ $cysC^-$; Replica plating; Minimal medium; Minimal medium + tyrosine; Minimal medium + cysteine.

Restating the Concepts

▶ Transformation occurs when a bacterial cell takes up foreign DNA molecules. Some bacteria possess a naturally high transformation rate; other bacteria (such as *E. coli*) must be chemically treated to increase their normally low transformation rate.

▶ Recombination of linear foreign DNA into a circular bacterial genome requires an even number of crossover events to regenerate the circular chromosome.

▶ Bacterial genes can be mapped based on their cotransformation frequency, which is inversely proportional to the distance between two genes. A high cotransformation frequency corresponds to a smaller distance between two genes.

15.8 Conjugation

In 1946, Joshua Lederberg and Edward L. Tatum, later to share the Nobel Prize in Physiology or Medicine (1958), discovered that *E. coli* cells can exchange genetic material through the process of conjugation. This finding was an important landmark in the history of genetics.

The Lederberg and Tatum Experiment

Lederberg and Tatum mixed two auxotrophic strains of *E. coli*. One strain required biotin and methionine (*bio⁻ met⁻*), and the other required threonine and proline (*thr⁻ pro⁻*). This cross is shown in **figure 15.18**.

Remember that a strain designated as *thr⁻ pro⁻* is wild-type for all other loci. Thus, the *thr⁻ pro⁻* auxotroph actually has the genotype of *thr⁻ pro⁻ bio⁺ met⁺*. Similarly, the *bio⁻ met⁻* strain is actually *thr⁺ pro⁺ bio⁻ met⁻*.

Lederberg and Tatum used auxotrophs with multiple mutations in order to rule out spontaneous reversion (mutation) in their experiment. About one in 10^7 *pro⁻* cells will spontaneously become a prototroph (*pro⁺*) every generation. With multiple mutations, however, the probability that several loci will simultaneously and spontaneously revert becomes vanishingly small. The product rule states that this probability is approximately $10^{-7} \times 10^{-7} = 10^{-14}$ (probability of *thr⁻ pro⁻* to revert to *thr⁺ pro⁺*). In reality, the control plates in the Lederberg and Tatum experiment showed no revertants for the parental double mutants. After mixing the two auxotrophic strains, however, Lederberg and Tatum found that about one cell in 10^7 was a prototroph (*thr⁺ pro⁺ bio⁺ met⁺*).

To determine if the prototrophs were produced by transformation, Bernard Davis repeated the experiment in 1950 with a slight modification. He placed one auxotrophic strain in complete medium in each arm of a U-tube with a sintered glass filter having known pore size at the bottom (**fig. 15.19**). The liquid and large molecules, including DNA, passed through the filter and were mixed by alternate application of pressure and suction to one arm of the tube; whole cells were too large to pass through the filter. The result was that the fluids surrounding the cells, as well as any large molecules, could be freely mixed while the cells were kept separate.

After cell growth stopped in the two arms, the contents were plated out on minimal medium. No prototrophs appeared in either arm. Therefore, cell-to-cell contact was required for the genetic material of the two cells to recombine; transformation of a cell by taking up free foreign DNA in the liquid culture was ruled out. This transfer of genetic material between two bacterial cells that must be in physical contact is called **conjugation.**

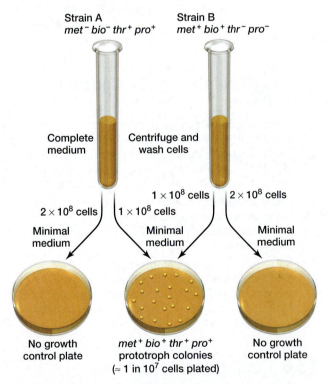

figure 15.18   Lederberg and Tatum's experiment demonstrating genetic recombination in *E. coli*. Two different auxotrophic strains are grown in liquid culture. Plating either bacterial strain independently on minimal media fails to produce any prototrophic revertants. Mixing the two strains before plating on minimal medium produces recombinant prototrophs at a frequency of approximately 1 in 10^7 cells.

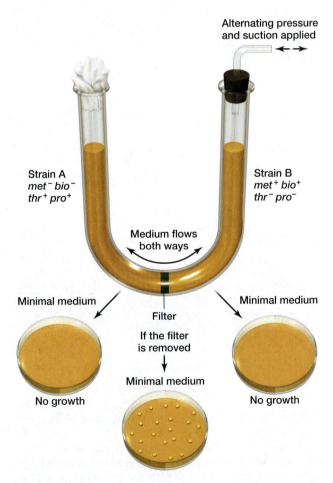

figure 15.19   The Davis U-tube experiment. Two different auxotrophic bacterial strains are placed on opposite sides of the U-tube and are separated by a filter. Alternating suction and pressure forces the liquid and macromolecules across the filter. Plating either bacterial strain independently on minimal medium fails to produce any prototrophs. When the filter, which prevents physical contact between the two bacterial strains, is removed, the mixed bacterial strains produces prototrophs. Thus, generating these prototrophs requires physical cell-cell contact.

F Factor

In bacteria, conjugation is a one-way transfer, with DNA being transferred from a donor strain to a recipient strain. In this one-way transfer, only the recipient cell can become the prototroph described earlier. This one-way transfer of genetic material is a key feature of conjugation.

F⁺ and F⁻ Bacteria

Sometimes the donor cells (F⁺), if stored for a long time, lose the ability to be donors, but they can regain the ability if they are mated with other donor strains. This discovery led to the hypothesis that a **fertility factor, F,** made any strain that carried it a male (donor) strain, termed **F⁺**. The strain that did not have the F factor, referred to as a female or **F⁻** strain, served as a recipient for genetic material during conjugation. The F factor turns out to be a *plasmid*, a term originally coined by Lederberg to refer to independent, self-replicating genetic particles. Plasmids are usually circles of double-stranded DNA (see chapter 12).

F⁺ × F⁻ Mating

Escherichia coli cells are normally coated with hairlike **pili.** F⁺ cells have one to three additional pili (singular: pilus) called **F-pili,** or sex pili, that the F⁻ cells lack. During conjugation, these sex pili form a connecting bridge between the F⁺ and F⁻ cells (**fig. 15.20**). Once a

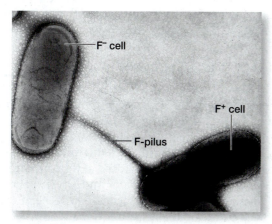

figure 15.20 F-pilus. Electron micrograph of conjugation between an F⁺ (upper right) and an F⁻ (lower left) cell with the F-pilus between them.

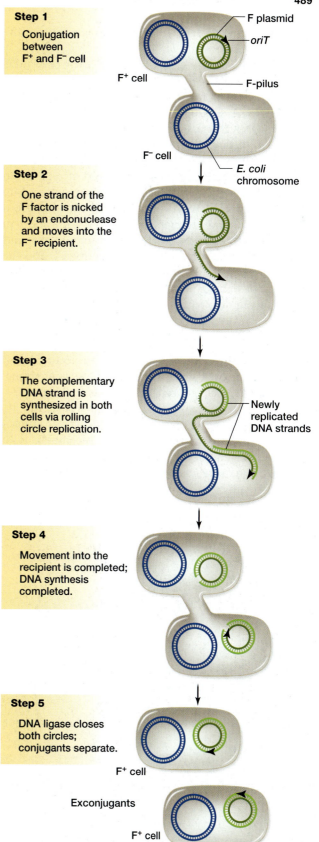

Step 1 Conjugation between F⁺ and F⁻ cell

F plasmid
oriT
F⁺ cell
F-pilus
F⁻ cell
E. coli chromosome

Step 2 One strand of the F factor is nicked by an endonuclease and moves into the F⁻ recipient.

Step 3 The complementary DNA strand is synthesized in both cells via rolling circle replication.

Newly replicated DNA strands

Step 4 Movement into the recipient is completed; DNA synthesis completed.

Step 5 DNA ligase closes both circles; conjugants separate.

F⁺ cell

Exconjugants

F⁺ cell

figure 15.21 Bacterial conjugation between a F⁺ and a F⁻ cell. The F⁺ cell contacts an F⁻ cell through the F-pilus and transfers one strand of the F plasmid into the recipient F⁻ cell. This single-strand transfer initiates at the origin of transfer (*oriT*) in the F DNA (arrowhead). Rolling circle DNA replication, which occurs as the DNA strand is transferred into the recipient F⁻ cell, begins at the origin of replication (*oriV*) that overlaps *oriT*. This conjugation results in the donor cell remaining F⁺ and the recipient becoming F⁺ after the transfer of the entire F plasmid.

connection is made, the sex pilus contracts to bring the two cells together.

After a F⁺ and F⁻ cell make contact, the F plasmid is nicked at the origin of transfer (*oriT*), which is the point that DNA transfer to the recipient and DNA replication both begin. A single strand of the double-stranded F plasmid DNA then passes from the F⁺ cell to the F⁻ cell across the cell membranes (**fig. 15.21**). This nick initiates the rolling circle mechanism of DNA replication (chapter 9), which begins at the origin of replication (*oriV*) that overlaps with *oriT*. This origin is therefore the first region of the F plasmid that transfers from the donor to the recipient cell. DNA replication in both cells reestablishes the double-stranded F plasmid DNA. In this manner, the F⁺ cell remains F⁺, and the F⁻ cell becomes an F⁺ cell (see fig. 15.21). The F factor itself carries the genes for sex pilus formation and DNA transfer to a conjugating F⁻ cell.

Transfer of Genomic DNA

This description tells how conjugation occurs, but it fails to explain how genes in the donor *genome* are transferred into the recipient cell. To understand this process, more information was needed about the F plasmid.

Although the F plasmid can exist as an autonomously replicating plasmid, it can recombine with the circular genome and integrate into it (**fig. 15.22**). This event requires a single crossover, because a single crossover between these two circular molecules

will result in a circular molecule. Double crossovers between the F plasmid and the genome can also produce a circular molecule, but a single crossover occurs more frequently than a double crossover.

This recombination event occurs between similar DNA sequences in both the F plasmid and bacterial genome that are called transposable elements or insertion sequences (see chapter 18). Because these transposable elements are scattered throughout the *E. coli* genome, the F plasmid may integrate at a variety of different sites. However, the number of these integration sites is limited, which reduces the probability of an entirely random integration pattern into the *E. coli* genome. Thus, an F⁺ population of cells contains a few cells that have the F plasmid integrated into their genome and the remainder of the cells contain the free F plasmid. When an integrated F plasmid starts to transfer from the donor to the recipient, it takes with it genomic DNA to which it is physically linked.

Hfr Conjugation

An *E. coli* strain was discovered that transferred genomic DNA at a rate about one thousand times higher than the normal F⁺ strain. This strain was called **Hfr,** for *high frequency of recombination.*

Several other phenomena occurred with this high rate of transfer. First, the ability to transfer the F factor itself dropped to almost zero in this strain. This results in the recipient (F⁻) cell remaining F⁻ after conjugation. Second, not all loci were transferred at the same rate. Some loci were transferred much more frequently than others. **Figure 15.23** shows a model that accounts for these data.

Details of Hfr × F⁻ Matings

An Hfr strain represents a population of cells in which all, or the vast majority, have the F plasmid integrated at the exact same location. (Different Hfr cultures can have different integration sites, but all the cells in a single Hfr culture share the same integration site.) The F plasmid sequence is nicked at the *oriT* site, from which transfer to the recipient and DNA replication both begin. Some of the F plasmid enters the recipient, followed by the donor's genomic DNA.

As the culture is conjugating, small disruptions (such as shaking) can occasionally break the F pilus and terminate the conjugation at a point when only some of the F plasmid has entered the recipient cell (see fig. 15.23). Because the DNA entering the recipient lacks all the essential F plasmid functions, which are the last sequences to transfer into the recipient, the recipient cell remains F⁻. (The donor remains Hfr.) Furthermore, the transferred piece of plasmid is unable to replicate autonomously in the recipient. The transferred DNA must recombine with the recipient genome to be stably

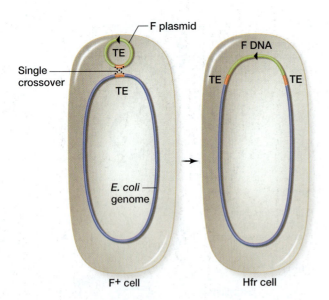

figure 15.22 Integration of the F plasmid by a single crossover. Recombination between the F plasmid and the *E. coli* genome can occur at specific sequences (transposable elements, TE) that are located at various sites in both DNA molecules. Recombination between these sequences in both circular molecules results in the F plasmid integrating into the *E. coli* genome to generate a circular Hfr molecule. The origin of transfer on the F plasmid is designated by the arrowhead.

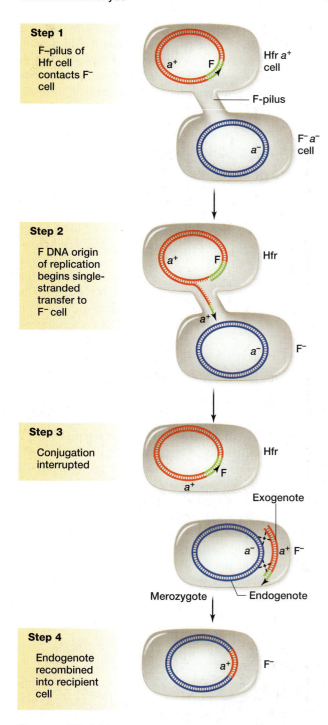

Step 1

F–pilus of Hfr cell contacts F⁻ cell

Step 2

F DNA origin of replication begins single-stranded transfer to F⁻ cell

Step 3

Conjugation interrupted

Step 4

Endogenote recombined into recipient cell

figure 15.23   Bacterial conjugation. The F-pilus draws an Hfr and an F⁻ cell close together. The *oriT* (arrowhead) of the F DNA begins the single-stranded DNA transfer of the donor genome into the F⁻ cell. The single strands in both cells are replicated through a rolling circle mechanism. Notice that only a portion of the F DNA enters the F⁻ cell at the start of conjugation, with the remainder of the F DNA transferring after the entire *E. coli* genome has entered the recipient cell. After conjugation is interrupted, two crossovers recombine the *a⁺* allele from the exogenote into the F⁻ *a⁻* endogenote chromosome. The donor cell remains Hfr because it retains an integrated copy of the F plasmid, while the recipient remains F⁻ because it still lacks a complete copy of the F DNA.

expressed. Because the foreign DNA is linear, an even number of crossover events are required for its insertion into the bacterial genome.

Note that for a short time, the conjugating F⁻ cell has two copies of whatever chromosomal loci were transferred from the Hfr cell: one copy of its own, and one transferred in (see fig. 15.23). With these two copies, the cell is a partial diploid, or a **merozygote.** The new foreign DNA is called the **exogenote,** and the host genome is called the **endogenote.** Whether recombination occurs or not, the piece of linear DNA is soon degraded by enzymes.

Interrupted Mating

To demonstrate that the transfer of genetic material from the donor to the recipient cell during conjugation is a linear event, François Jacob and Élie Wollman devised the technique of **interrupted mating.** In this technique, F⁻ and Hfr strains were mixed together in a food blender. After waiting a specific amount of time, Jacob and Wollman turned the blender on. The spinning motion separated conjugating cells and thereby interrupted their mating. Then the researchers tested the F⁻ cells for the transfer of various alleles from the Hfr cells.

To ensure that the alleles being tested were transferred to the F⁻ recipient cell, the F⁻ strain used is usually resistant to an antibiotic such as streptomycin, and the Hfr strain is sensitive to the antibiotic. After interrupting conjugation, the cells are plated onto a medium containing the antibiotic, which kills all the Hfr cells. Then the genotypes of the F⁻ cells can be determined by replica plating without fear of contamination by Hfr cells.

A mating was carried out between an Hfr strain and an F⁻ strain. The strains had the following characteristics:

Hfr Strain:	F⁻ Strain:
Streptomycin-sensitive (*str*ˢ)	Streptomycin-resistant (*str*ʳ)
Azide-resistant (*azi*ʳ)	Azide-sensitive (*azi*ˢ)
Phage T1-resistant (*ton*ʳ)	Phage T1-sensitive (*ton*ˢ)
Prototrophic for leucine, galactose, and lactose (*leu⁺*, *galB⁺*, and *lac⁺*)	Auxotrophic for leucine, galactose, and lactose (*leu⁻*, *galB⁻*, and *lac⁻*)

At various time intervals, ranging from 0 to 60 minutes, an aliquot of the culture was isolated and the conjugation was interrupted. The cells were plated on medium that contained streptomycin (to kill all the Hfr cells) and lacked leucine (to eliminate the F⁻ cells that did not conjugate). The only colonies that grew, therefore, were F⁻ conjugants that must have received the *leu⁺* allele from the Hfr. By replica plating onto specific medium, investigators were able to determine the percentage of *leu⁺* F⁻ cells that had also received the *azi*ʳ, *ton*ʳ, *lac⁺*, or *galB⁺* alleles from the Hfr strain.

Figure 15.24 shows that as time of mating increases, two things happen. First, each donor allele appears in the recipient cell at a specific time and order after the start of conjugation. The *azi*[r] allele first appears among the F⁻ conjugants after about 9 minutes of mating, whereas *tonA*[r] enters the F⁻ cells after about 11 minutes and *lac*⁺ and *galB*⁺ enter after 18 and 25 minutes, respectively. Repeating the experiment with the same Hfr and F⁻ strains produces the same time course of transfer. This observation suggests that a specific sequential transfer of loci from the Hfr into the F⁻ cells (**fig. 15.25**).

Second, after initial entry of each donor allele, increasing the conjugation time increases the percent-age of F⁻ cells that express the donor allele, up to a maximum. At 11 minutes, *ton*[r] is first found among F⁻ conjugants. After 15 minutes, about 35% of F⁻ conjugants have the *ton*[r] allele, and after 25 minutes, nearly 80% of the F⁻ conjugants have the *ton*[r] allele. This limiting percentage does not increase with additional time.

Mapping Genes in Conjugation by Time of Transfer

Jacob and Wollman conjugated several different Hfr strains of independent origin with an F⁻ strain. The results from these matings were quite striking at the time because they suggested that the bacterial chromosome was circular.

For each Hfr conjugation, the relative order of the transferred loci was always the same (**table 15.3**). The characteristics that differed with each independent Hfr were the point of origin (or the first bacterial gene transferred) and the direction of the transfer. This point is illustrated in **figure 15.26**, where the gene order for five of the independent Hfrs from table 15.3 have been laid out. The only difference between them is the location of the F plasmid origin and its orientation (arrowhead).

Jacob and Wollman proposed that normally the F factor is an independent circular DNA entity in the F⁺ cell, and that during conjugation the F factor is passed to the F⁻ cell. Because it is a small DNA plasmid, it can be passed entirely to the recipient cell in a high proportion of conjugations before the cells separate. This results in most of the recipient cells becoming F⁺.

Occasionally, however, the F factor randomly integrates into the bacterial host's genome, which then becomes an Hfr cell. Because the F plasmid directs the transfer of the bacterial genomic DNA, the integration

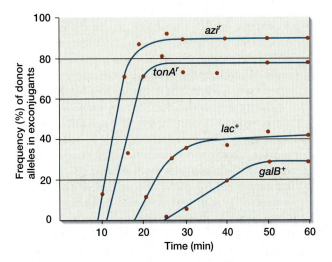

figure 15.24   Frequency of Hfr donor alleles among exonjugants after interrupted mating. As the length of conjugation time increases, new donor alleles appear in the recipient cell. This suggests that there is a specific temporal order of gene transfer from the Hfr to F⁻ cell. The frequency of any given allele reaches a limiting percentage in the exconjugants.

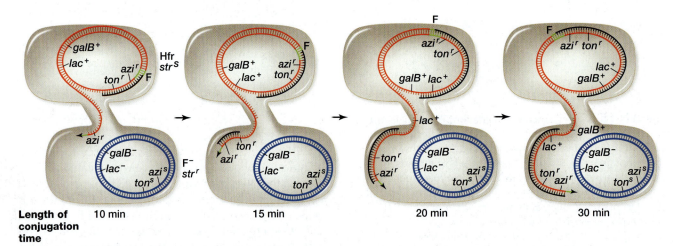

figure 15.25   Conjugation with an *E. coli* Hfr cell. The *E. coli* Hfr genome is red, F DNA is green, the F⁻ recipient genome is blue, and new DNA replication is black. As time proceeds, alleles from the Hfr enter the F⁻ cell in an orderly, sequential fashion. After conjugation is disrupted, two crossovers can bring the transferred Hfr alleles on the linear DNA into the F⁻ chromosome. The F DNA is the last part to transfer from the Hfr cell into the F⁻ cell.

table 15.3   Gene Order of Various Hfr Strains Determined by Means of Interrupted Mating

Types of Hfr	Order of Transfer of Genetic Characters*																	
HfrH	0	t	l	azi	t_1	pro	lac	ad	gal	try	h	s-g	sm	mal	xyl	mtl	isol	m	b_1
1	0	l	t	b_1	m	isol	mtl	xyl	mal	sm	s-g	h	try	gal	ad	lac	pro	t_1	azi
2	0	pro	t_1	azi	l	t	b_1	m	isol	mtl	xyl	mal	sm	s-g	h	try	gal	ad	lac
3	0	ad	lac	pro	t_1	azi	l	t	b_1	m	isol	mtl	xyl	mal	sm	s-g	h	try	gal
4	0	b_1	m	isol	mtl	xyl	mal	sm	s-g	h	try	gal	ad	lac	pro	t_1	azi	l	t
5	0	m	b_1	t	l	azi	t_1	pro	lac	ad	gal	try	h	s-g	sm	mal	xyl	mtl	isol
6	0	isol	m	b_1	t	l	azi	t_1	pro	lac	ad	gal	try	h	s-g	sm	mal	xyl	mtl
7	0	t_1	azi	l	t	b_1	m	isol	mtl	xyl	mal	sm	s-g	h	try	gal	ad	lac	pro
AB311	0	h	try	gal	ad	lac	pro	t_1	azi	l	t	b_1	m	isol	mtl	xyl	mal	sm	s-g
AB312	0	sm	mal	xyl	mtl	isol	m	b_1	t	l	azi	t_1	pro	lac	ad	gal	try	h	s-g
AB313	0	mtl	xyl	mal	sm	s-g	h	try	gal	ad	lac	pro	t_1	azi	l	t	b_1	m	isol

*The 0 refers to the origin of transfer.

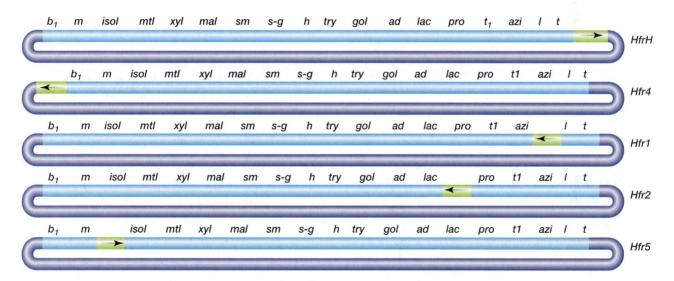

figure 15.26   Generating a circular genetic map in *E. coli*. Shown is the relative position and orientation of the integrated F DNA in five different Hfrs from table 15.3. Notice that the bacterial genes do not change order, only the position and orientation of the F DNA. By comparing the gene transfer order in the five Hfrs, it becomes apparent that the *E. coli* genome is circular.

site of the F plasmid into the genome determines the initiation point of transfer for the *E. coli* genome, as well as the direction of transfer.

As mentioned earlier, the origin of replication in the F plasmid is the point of transfer initiation and controls the direction of transfer into the recipient cell. The remainder of the F factor, however, is the last part of the *E. coli* chromosome to be passed from the Hfr cell. This fact explains why an Hfr, in contrast to an F⁺, rarely passes the entire F factor into the recipient. In the original work of Lederberg and Tatum, the one prototroph in 10^7 cells most likely came from a conjugation between an F⁻ cell and an F⁺ cell in which the F

plasmid had spontaneously integrated into the bacterial genome.

We can now diagram the *E. coli* chromosome and show the map location of several known loci (**fig. 15.27**). The map units would be in minutes, obtained by interrupted mating. Based on the time of transfer between genes, we know that the *E. coli* chromosome would require 100 minutes of conjugation to entirely transfer into the recipient. Interrupted mating is most accurate in giving the relative position of loci that are not very close to each other. But with this method alone, a great deal of ambiguity would arise as to the specific order of very close genes on the chromosome.

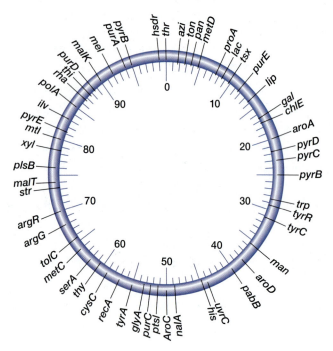

figure 15.27   Genetic map of the *E. coli* genome. Several different Hfrs were conjugated to F⁻ strains and the order and time of gene transferred were used to determine the "distance" between genes. The "distance" is measured as the time of gene transfer (inside scale) from the Hfr to F⁻ cells. Notice that the *E. coli* genome would take approximately 100 minutes to completely transfer by conjugation (in addition to approximately 20 minutes for the transfer of the F DNA).

Mapping Genes in Conjugation by Recombination

Stable expression of the donor alleles in the recipient cell after conjugation requires recombination of the alleles into the recipient genome. Time-of-transfer mapping, however, is not based on recombination between specific pairs of genes.

To map the recombination distance between pairs of genes, we must ensure that all the desired genes are transferred into the recipient cell. Therefore, for all the genes to be mapped, we select the last gene to enter the recipient. These recombinants can then be tested by replica plating for the presence of additional donor alleles that had entered the recipient cell earlier.

Determining Crossover Classes

Let's look at a diagram of this recombination mapping before we examine actual recombination data (**figure 15.28**). In this example, we conjugate an Hfr strain that is *strˢ, cysC⁺, tyrA⁻, glyA⁺* to a F⁻ strain that is *strʳ, cysC⁻, tyrA⁺, glyA⁻*. We know from time-of-transfer experiments that the *cysC⁺* gene is the last of the three genes to enter the recipient. After conjugation, we select

recombinants using minimal medium supplemented with streptomycin, tyrosine, and glycine. Streptomycin selects for the *strʳ* recipient F⁻ cells, and the absence of cysteine in the medium selects for recombination of the donor's *cysC⁺* allele into the recipient genome. This recombination event is shown as the leftmost crossover in all four recombination possibilities in figure 15.28.

Just as we discussed in chapter 6 on recombination, the rarest phenotypic class results from the greatest number of recombination events. Because we are recombining a linear DNA molecule with the circular *E. coli* genome, we must have an *even* number of crossovers to yield a circular molecule. In this case, the quadruple crossover will represent the rarest class. Identification of this class also reveals which of the three genes is in the middle (this represents the gene order). Although time of transfer alone may not be able to distinguish between two genes that are very close together, recombination mapping and the comparison of double and quadruple crossovers can reveal the correct gene order.

An Example of Data Analysis

Now consider some data generated from recombination mapping after conjugation. We will continue using the mating of the Hfr strain *strˢ, cysC⁺, tyrA⁻, glyA⁺* and the F⁻ strain *strʳ, cysC⁻, tyrA⁺, glyA⁻*. We already know that the *cysC⁺* allele was the last of the three genes to enter the recipient, so we selected for *strʳ cysC⁺* recombinants. We then replica plated the recombinants and found the following number of each genotype:

strʳ, cysC⁺, tyrA⁺, glyA⁻	30
strʳ, cysC⁺, tyrA⁻, glyA⁻	2
strʳ, cysC⁺, tyrA⁻, glyA⁺	78

Only three genotypes were identified and four were expected. The missing genotype (*strʳ, cysC⁺, tyrA⁺, glyA⁺*) must represent the rarest class, which would correspond to the quadruple crossover.

Comparing the quadruple crossover genotype with the Hfr genotype allows us to identify the middle gene of the three. The quadruple crossover will have the Hfr alleles for the flanking genes and the F⁻ allele for the middle gene. Comparing the Hfr *cysC⁺, tyrA⁻, glyA⁺* with the quadruple recombinant *cysC⁺, tyrA⁺, glyA⁺* reveals that the *tyrA* gene must be in the middle (see fig. 15.28).

We can now calculate the recombination distances. The distance from *cysC* to *tyrA* will be the percentage of all of the crossover events between these two genes (see fig. 15.28). This percentage corresponds to the 30 *strʳ, cysC⁺, tyrA⁺, glyA⁻* recombinants; the calculation reveals that the *cysC* to *tyrA* distance is 27.3 map units, mu ([100%] × 30/110). The distance from *tyrA* to *glyA* is 1.8 mu ([100%] × 2/110).

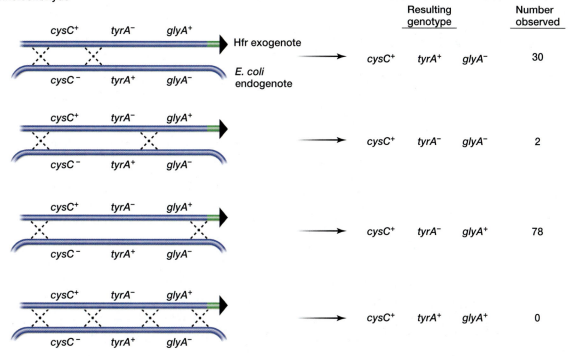

figure 15.28 Recombination mapping of linked genes after conjugation between a *cysC⁺ tyrA⁻ glyA⁺* Hfr cell and a *cysC⁻ tyrA⁺ glyA⁻* F⁻ cell. The F DNA's (green) origin of transfer (*oriT*) is shown as a black arrowhead in the exogenote. Because *cysC⁺* is the last gene to enter the F⁻ cell, we initially select for *cysC⁺* recombinants. The four different types of viable recombination events, their resulting genotypes in the F⁻ recipient cell, and the observed number of each phenotypic class are shown. Because the *cysC⁺ tyrA⁺ glyA⁺* phenotypic class was not observed, it must represent the quadruple crossover class.

Therefore, we can determine the distance between bacterial genes in a circular chromosome by interrupted conjugation mating, which produces a genetic map in either minutes or in map units.

F' Conjugation: F-duction or Sexduction

There is another form of donor cell besides the F⁺ and Hfr strains. As shown in **figure 15.29**, the F plasmid can spontaneously recombine between the transposable element sequences to integrate into the bacterial genome. This occurs in a small fraction of cells within an F⁺ culture. If a cell with an integration event is isolated and grown as a pure culture, these cells are designated as Hfr.

The F plasmid can also spontaneously excise itself from the bacterial chromosome. Usually, the excision process is a recombination event between the transposable element sequences (fig. 15.29, *left side*). This produces the intact F plasmid and an F⁺ cell. At a very low frequency, the recombination event is not between the flanking transposable element sequences, which results in an imprecise excision and the generation of an F plasmid that also contains some of the bacterial genomic sequences (fig. 15.29, *right side*). This does not produce an F⁺ cell, because the F plasmid contains some of the bacterial genomic DNA (the *lac⁺* gene in fig. 15.29). This new class of donor cell is called an F' (F prime).

The F' can carry up to 25% of the bacterial chromosome. The passage of this F' *lac⁺* factor to an F⁻ cell is called **F-duction**, or **sexduction**. (The *lac⁺* designation refers to the *E. coli* alleles that are carried on the F'.) Notice that after conjugation between an F' *lac⁺* and F⁻, the recipient cell usually becomes an F' cell that is also a merozygote (F' *lac⁺/lac⁻*) due to the two copies of the *lac* region (**fig. 15.30**). Unlike in the case of an Hfr, the *lac⁺* that was transferred to the recipient cell is stably maintained without recombination, because it is present in the autonomously replicating F plasmid. We will use the F' cell in examining the dominant/recessive nature of *E. coli lac* alleles in chapter 16.

The similarities and differences between the types of conjugating cells are summarized in **table 15.4**. The F⁺, Hfr, and F' cells all act as donors, and the F⁻ cell is the only recipient cell type.

It's Your Turn

We have now discussed two different ways to map bacterial genes using conjugation. Problem 25 will allow you to try mapping genes based on time of transfer. Make sure you designate the orientation of the integrated F plasmid correctly. Problem 31 will allow you to map genes by recombination after conjugation.

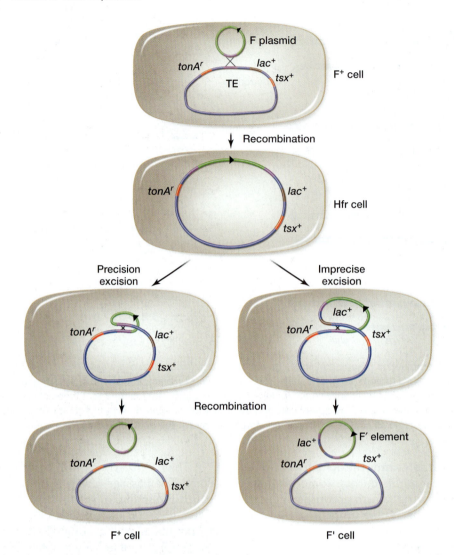

figure 15.29 Recombination events with F DNA. In an F$^+$ cell, a single crossover event between transposable elements (TE, purple) in the F plasmid (green) and the *E. coli* genome inserts the F DNA into the *E. coli* genome to form an Hfr cell. This usually occurs in less than 1% of the F$^+$ cells. The integrated F DNA can recombine out of the *E. coli* genome through two different mechanisms. First, precise excision involves recombination between the transposable elements, in an exact reversal of the reaction that produced the Hfr (left side). Second, imprecise excision happens at a low frequency when recombination occurs between unrelated sequences in the F DNA and bacterial genome. The excised F DNA removes part of the cell's genome to produce a free circular F' element (right side).

table 15.4   Similarities and Differences Between Different Types of F Strains

Phenotype	F$^-$ culture	F$^+$ culture	Hfr culture	F' culture
Donor/recipient	Recipient	Donor	Donor	Donor
F$^-$ cell after conjugation	–	F$^+$	F$^-$	F'
Transfer genomic DNA	–	No	Yes	Yes
Efficiency of genomic DNA transfer	–	Low	High	High
Recombination required to stably maintain transferred genomic genes?	–	Yes	Yes	No

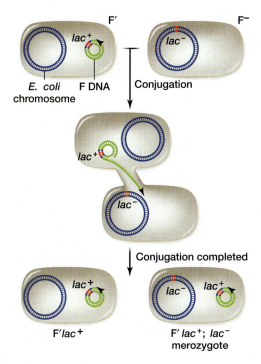

figure 15.30 Conjugation of an F' with an F⁻ cell. The F' element, which contains the *lac*⁺ allele (red DNA), transfers the genomic *lac*⁺ into the recipient cell. The donor cell remains F' and the recipient cell also becomes an F'. Notice that the donor cell has only one copy of the *lac* region, which is on the F'. In contrast, the recipient cell is a partial diploid (merozygote) for the *lac* region, with one copy in the genome and one in the F'.

Restating the Concepts

▶ Conjugation involves a cell containing F DNA (F⁺, Hfr, or F') transferring genetic information to an F⁻ strain. The F⁺, Hfr, and F' strains all contain the F DNA in different forms, transfer genomic DNA at different rates, affect the F⁻ strain differently, and have different requirements of recombination in the recipient cell to stably express the donor alleles.

▶ Conjugation allows genes to be mapped by time of transfer. This approach involves selecting the first allele that enters the recipient F⁻ cell from an Hfr cell and then determining the amount of time required for subsequent alleles to enter the F⁻ cell.

▶ Conjugation can also be used to map genes by recombination distance. This method involves selecting the last gene to enter the recipient, so that all the desired genes are present in the recipient cell. The recombination distance between pairs of genes can then be calculated based on the percent of expression of the donor alleles in the recipient.

15.9 Transduction

Before lysis, when phage DNA is being packaged into phage heads, an occasional error occurs that causes some bacterial genomic DNA to be incorporated into the phage head. This bacterial genomic DNA can be physically joined to either the phage DNA (*specialized transduction*) or a separate fragment of DNA (*generalized transduction*). When either of these two processes happens, bacterial genes are transferred via the phage coat to infect another bacterial cell, where the bacterial DNA can recombine into the recipient's genome. This phage-mediated transfer of bacterial genomic DNA, called **transduction,** has been of great use in mapping the bacterial chromosome.

Specialized Transduction

The process of **specialized,** or **restricted, transduction** was first discovered in phage λ by Lederberg and his students. Phage λ, unlike the F factor, recombines at a single site within the *E. coli* genome, termed *att*λ (**fig. 15.31**). This locus can be mapped on the *E. coli* chromosome; it lies between the galactose (*gal*) and biotin (*bio*) loci.

Phage Lambda

Normally, when the phage λ prophage is induced to enter the lytic cycle, it is recombined out of the bacterial genome in a precise manner to yield a complete λ genome and none of the flanking bacterial genomic DNA (fig. 15.31, left side). Occasionally, the λ is excised in a manner that takes some of the flanking bacterial genomic DNA in place of some of the lambda sequence (fig. 15.31, right side). Thus, specialized transduction is analogous to sexduction—it involves a mistake during a looping-out process. The resulting defective λ (λdgal) is not a fully functional λ—it lacks some λ sequences.

Method of Specialized Transduction

There are two methods by which the λdgal can transfer the *gal*⁺ gene into a new bacteria. First, it can coinfect a bacterial cell with a wild-type λ, which is called a λ *helper* because it possesses the necessary λ genes that encode the proteins that are required for both the λdgal and λ helper to integrate in the λ attachment site (see fig. 15.31). The result is a merozygote that is diploid for the *gal* locus. Unlike conjugation, however, both copies of the *gal* locus in this merozygote are present in the bacterial genome.

Second, only the λdgal may infect a bacterial cell. Although the λdgal cannot integrate into the bacterial genome at the λ attachment site, the *gal*⁺ gene itself can recombine with the endogenous *gal*⁻ gene. The result is the transfer of *gal⁺* from one cell into a new cell and the replacement of *gal⁻*.

Because λ only integrates between the *gal* and *bio* genes in the *E. coli* genome, and because only loci adjacent to the prophage can be transduced in this process, specialized transduction has not proven very useful for mapping the host genome.

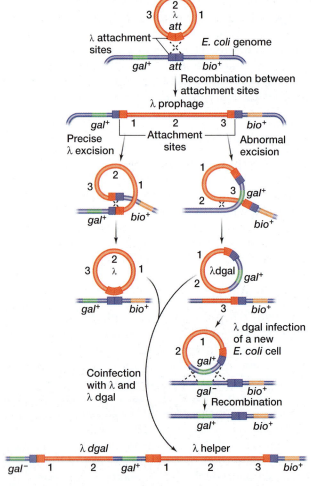

figure 15.31   Mechanism of specialized transduction. A single crossover between the lambda attachment sites (*att*) in both lambda and the *E. coli* genome inserts the lambda prophage between the *E. coli* *gal⁺* and *bio⁺* genes. Precise lambda excision results in the removal of lambda by recombination between the *att* sites. Alternatively, imprecise excision results in some of the flanking genomic DNA being removed, while some lambda DNA remains in the *E. coli* genome. This results in a defective lambda (λd) that is unable to generate a prophage in a newly infected cell by itself. The λdgal, however, can coinfect a bacterium with a wild-type lambda, which then provides the necessary functions to integrate λdgal into the *E. coli* genome at the *att* site (bottom). Alternatively, the λdgal can infect another *E. coli* cell and simply exchange the *gal* genes by recombination (right). The difference between these two λdgal events is the former produces a merozygote with two copies of *gal* and the latter yields only a single copy of *gal*.

Generalized Transduction

In 1952, Norton Zinder and Joshua Lederberg discovered **generalized transduction** while studying recombination in *Salmonella typhimurium*. Using the Davis U-tube apparatus, they made a surprising discovery: unlike Davis's experiment, which revealed that conjugation required physical contact, Zinder and Lederberg's experiment found that a *diffusible agent* could convert one of the auxotrophic strains to a prototroph (**fig. 15.32**).

The transfer was one-way, similar to conjugation. The *S. typhimurium* LA-22 auxotroph could become a prototroph, but the LA-2 auxotroph did not become a prototroph (see fig. 15.32). By changing the size of the filter pore, they demonstrated that the diffusible agent was in fact a bacteriophage.

Phage P22

The phage P22 resided as a prophage in the LA-22 strain. Occasionally, the prophage would be induced to enter the lytic cycle, releasing P22. These phages could not infect another LA-22 cell because the resident P22 prophage conferred immunity to subsequent infections. However, the P22 phage could diffuse across the filter and infect a LA-2 cell that lacked a P22 prophage.

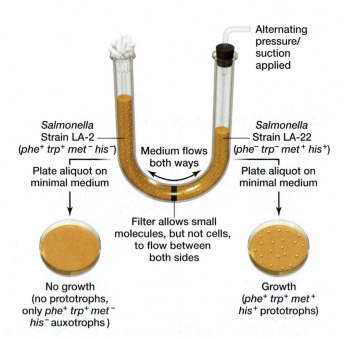

figure 15.32   The Lederberg-Zinder experiment of generalized transduction with *Salmonella*. Using the Davis U-tube apparatus, Lederberg and Zinder showed that strain LA-22, but not LA-2, could become a prototroph without physical contact between the strains. Furthermore, by adjusting the pore size of the filter, they demonstrated that the transfer required an agent that was the size of a bacteriophage.

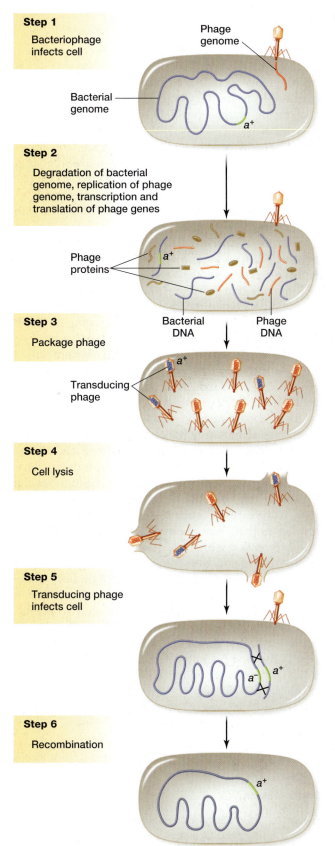

Step 1

Bacteriophage infects cell

Phage genome

Bacterial genome

a^+

Step 2

Degradation of bacterial genome, replication of phage genome, transcription and translation of phage genes

Phage proteins

a^+

Bacterial DNA

Phage DNA

Step 3

Package phage

a^+

Transducing phage

Step 4

Cell lysis

Step 5

Transducing phage infects cell

a^- a^+

Step 6

Recombination

a^+

During the generation of new P22 particles, the phage could package some of the bacterial genomic DNA instead of the phage DNA. If P22 packaged the phe^+ and trp^+ genes from LA-2 and then passed back through the filter and infected a LA-22 cell, the phe^+ and trp^+ alleles could recombine to convert the LA-22 auxotroph to a prototroph.

Mechanism of Generalized Transduction

The mechanism of generalized transduction does not depend on a faulty excision of a prophage, but rather on the random inclusion of a piece of the host genome inside the protein coat (**fig. 15.33**). Infection of some virulent phage, such as P22, results in the digestion of bacterial genomic DNA, which greatly reduces the replication of bacterial DNA, the transcription of bacterial mRNAs, and the translation of bacterial proteins. These processes can then focus on the replication, transcription, and translation of the phage.

When the new phage protein coats are synthesized, they are ready to package the phage DNA inside the head. However, this packaging does not depend on any specific phage DNA sequence, unlike the *cos* sequence in λ (see chapter 12). The only requirement is that the packaged DNA must be approximately the same size as the phage genome, which is 41.7 kb. This allows either the phage DNA or the fragmented bacterial genomic DNA to be incorporated into the phage head before lysis of the cell and release of all the packaged phage. Thus, virtually any bacterial locus can be transduced by generalized transduction.

A defective phage, one that carries bacterial DNA rather than phage DNA, is called a **transducing particle.** Transduction is complete when the genetic material from the transducing particle is injected into a new bacterial cell and recombines into the host's genome.

For P22, the rate of transduction is about once for every 10^5 infecting phages. Because a transducing particle can carry approximately only 2% of the host genome, or approximately 90 kb in the case of P22, only genes very close to each other can be transduced together (**cotransduction**). Cotransduction can thus help to fill in the details of gene order over short distances, after interrupted mating or recombination mapping is used to ascertain the general pattern. Transduction is similar to transformation in that cotransduction, like co-occurrence in transformation, is a relative indicator of map distance.

figure 15.33   The mechanism of generalized transduction. Phage infection induces digestion of the host genome and the production of new phage. During packaging of the new phage particles, either phage DNA or bacterial genomic DNA fragments may be incorporated into the phage particle. Phage carrying bacterial genomic DNA, transducing phage, can infect another bacterial cell and the donor genomic DNA (a^+ gene) can recombine with the recipient's genome (a^- genotype) and convert the cell's genotype from a^- to a^+. The frequency of these transducing phages is usually less than 0.01%.

Mapping with Cotransduction

Transduction can be used to establish both gene order and relative map distance. Methods include two-factor transduction and three-factor transduction.

Two-Factor Transduction

Gene order can be established by two-factor transduction. Remember that the greater the frequency that two alleles cotransduce, the closer they are relative to each other.

The data shown in **table 15.5** indicate that gene *a* is cotransduced with gene *b*, and *b* with gene *c*, but that *a* is never cotransduced with *c*. Genes *a* and *c* must be the farthest apart because they did not cotransduce. Therefore, the gene order must be *a b c*. Nothing is implied, however, about the relative distance between any two genes.

Three-Factor Transduction

Three-factor transduction allows us to simultaneously establish gene order and relative distance. Three-factor transduction is especially valuable when the three loci are so close that it is very difficult to make ordering decisions on the basis of two-factor transduction or interrupted mating.

For example, if genes *a, b,* and *c* are cotransduced, we can find the order and relative distances by taking advantage of the rarity of multiple crossovers. As an example, we will use a prototroph $a^+ b^+ c^+$ to make transducing phages that then infect the $a^- b^- c^-$ strain of bacteria.

We initially need to select cells in which some transduction has occurred. The simplest way is to select bacteria in which the wild-type allele has replaced at least one of the loci by recombination. For example, after transduction, the bacteria are grown on minimal medium with compounds B and C added; the result is that all the bacteria that are a^+ will grow. In this example, the *a* locus is the **selected locus.** Replica plating then allows us to determine the genotypes at the *b* and *c* loci for the a^+ bacteria.

Colonies that grow on minimal medium lacking compound A (compounds B and C are added) are replica-plated onto minimal medium with either compound B or C added. In this way, each transductant can be scored for the other two loci (**table 15.6**). The result is four classes of transductants in which the a^+ allele was selected: $a^+ b^+ c^+, a^+ b^+ c^-, a^+ b^- c^+,$ and $a^+ b^- c^-$. We can now compare the relative numbers of each of these four categories.

The rarest category is the quadruple crossover, which recombines only the outer two markers, but not the center one, into the recipient's genome (**fig. 15.34**). Therefore, by looking at the number of transductants in the various categories, we can determine that the gene order is *a b c* (**table 15.7**), because the $a^+ b^- c^+$ category is the rarest. **Figure 15.35** shows the recombination events that were required to generate each of the four different classes of transductants.

The cotransduction frequency is simply the percentage of total transductants that contain both of the stated alleles. If we want to determine the cotransduction frequency of a^+ and b^+, there are two phenotypic classes that show both a^+ and b^+ cotransduced: $a^+ b^+ c^+$ and $a^+ b^+ c^-$. Of the 426 a^+ transductants, 50 were of the first phenotype and 75 of the second phenotype. Thus the a^+ and b^+ cotransduction frequency is $(100\%)(50+75)/426 = 29.3\%$. The cotransduction frequency of a^+ and c^+ would be the frequency of transductants that are either $a^+ b^+ c^+$ (50) or $a^+ b^- c^+$

table 15.5   Gene Order Established by Two-Factor Cotransduction*

Transductants	Number
$a^+ b^+$	30
$a^+ c^+$	0
$b^+ c^+$	25
$a^+ b^+ c^+$	0

* An $a^+ b^+ c^+$ strain of bacteria is infected with phage. The lysate is used to infect an $a^- b^- c^-$ strain. The transductants are scored for the wild-type alleles they contain. These data include only those bacteria transduced for two or more of the loci. Since $a^+ b^+$ cotransductants and $b^+ c^+$ cotransductants occur, but no $a^+ c^+$ types, we can infer the *a b c* order.

table 15.6   Method of Scoring Three-Factor Transductants

| Colony Number | Minimal Medium | | | Genotype |
	With only Compounds B and C	With only Compound C	With only Compound B	
1	+	+	−	$a^+ b^+ c^-$
2	+	−	−	$a^+ b^- c^-$
3	+	+	+	$a^+ b^+ c^+$
4	+	−	+	$a^+ b^- c^+$

Note: The plus indicates growth, the minus lack of growth. An $a^- b^- c^-$ strain was transduced by phage from an $a^+ b^+ c^+$ strain.

(1). Thus, the a^+ and c^+ cotransduction frequency is $(100\%)(50+1)/426 = 12.0\%$. With the gene order $a\ b\ c$, it is logical that the distance between two adjacent genes (a and b) should be shorter (and therefore have a larger cotransduction frequency) than the distance between the outer two genes.

Comparison of Mapping Approaches

In conjugation, distances can be calculated based on either time of transfer or recombination distance. In the case of recombination distances, we select the last gene to enter from the donor and determine the recombination frequency between the adjacent pairs of genes.

In the case of either transformation or transduction, we are measuring the cotransformation or cotransduction frequencies. These represent the frequencies of two genes recombining into the recipient genome together. Thus, they represent the frequency of recombination events that flank the pair of genes, rather than the frequency of recombination events between the pair of

genes. Furthermore, a higher recombination distance or time of transfer represents genes that are farther apart, while a larger cotransformation or cotransduction frequency represents genes that are closer together.

The analysis in figure 15A shows details of how you can distinguish between mapping after conjugation and mapping in cotransduction.

table 15.7   Numbers of Transductants and Relative Cotransduction Frequencies

Class	Number
$a^+\ b^+\ c^+$	50
$a^+\ b^+\ c^-$	75
$a^+\ b^-\ c^+$	1
$a^+\ b^-\ c^-$	300
TOTAL	426

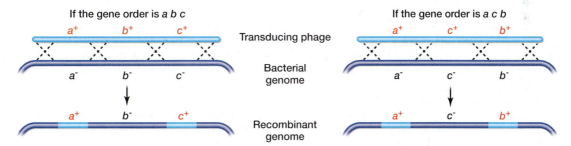

figure 15.34   Recombination mapping with transducing phage. When three genes are being mapped by cotransduction, the smallest recombinant class will correspond to the quadruple crossovers. Comparing the genotypes of the transducing phage with the rarest class will reveal the middle gene.

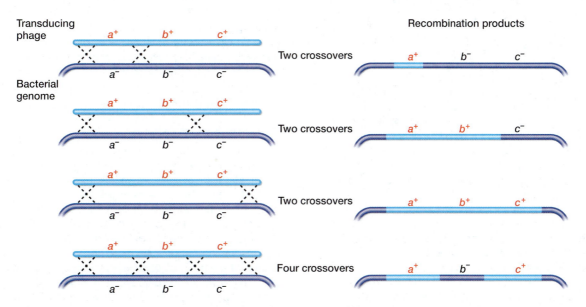

figure 15.35   The four different classes of recombinants produced from generalized transduction. One gene (a^+) is initially selected and cotransduction with each of the other two genes is tested. Of the four events shown, the second and third exhibit cotransduction of a^+ and b^+, while the third and fourth exhibit cotransduction of a^+ and c^+. Notice that the last event, which contains the bacterial genome's allele in the central position of the three genes, results from four recombination events.

15.1 applying genetic principles to problems

Distinguishing Between Mapping Through Conjugation and Transduction

The differences between recombination mapping after conjugation versus mapping in cotransduction are often difficult to discern and understand. In this section, we provide an opportunity to clarify the features and analysis of the two methods.

The mechanistic difference between recombination mapping after conjugation and cotransduction is the type of event that is measured by the crossovers. In both recombination mapping and cotransduction, the same crossover events are being analyzed. The difference lies in which events are meaningful and how they are interpreted.

Conjugation

In conjugation, DNA passes from one bacterium to another. The size of the exogenote that enters the recipient depends largely on the length of time that passes prior to interrupting the conjugation. Based on the time of transfer, you can be sure that all of the genes you want to map have had the potential to be transferred.

Let's say you are interested in mapping genes a, b, and c, and you know that gene a is the last of the three to enter the recipient cell using a specific Hfr prototroph. After mixing the cells and interrupting conjugation at appropriate times, you select for a^+ recombinants to be sure that the other two genes have also entered the recipient cell (fig. 15A).

Regardless of how far the three genes are from the F DNA insertion site, or how much distance lies between the genes, selecting recombinants for the last gene to enter the recipient guarantees that the other two genes also transferred into the recipient. Because all three genes are in the F⁻ cell, you can measure the recombination frequency between the genes.

Cotransduction

In contrast, cotransduction involves a phage infecting a bacterium, degrading the bacterial genome, generating new phage particles, and packaging fragments from the bacterial genome in a small percentage of the phage particles.

Let's say that you are interested again in genes a, b, and c. After infection of a prototrophic bacteria, you isolate the new phage particles. Using the phage lysate that contains transducing particles, you then infect the auxotrophic recipient cells and select for any one of the three wild-type genes, because order of entry into the recipient cell is not an issue—transduction is a random cleavage of the donor's genomic DNA to produce the transducing DNA fragments. The transducing phages on infection of a recipient cell, will introduce all of its DNA and will not be interrupted like occurs in conjugation.

In cotransduction, you calculate the frequency of two genes being recombined into the recipient's genome together. In figure 15A, you first selected that a^+ gene, which allows you to determine the cotransduction frequency of a^+ and b^+ and also a^+ and c^+.

Notice that in conjugation, you calculated distances between adjacent pairs of genes, but in cotransduction, you calculate the distance from the originally selected gene, which is randomly selected, to each of the other two genes. If you initially select the a^+ allele, the leftmost crossover event shown in figure 15A would be required.

Suppose you select the b^+ gene. Knowing that the phage requires approximately 90 kb of DNA to package, you know that there are a large number of potential genomic DNA fragments that are 90 kb in length, and they all contain the b^+ gene. But not all of these fragments contain the a^+ or c^+ genes.

Looking at figure 15B, you can see that the top line represents the sequence of genes in the bacterial genome. The fragments below are all 90 kb in length, but they have been broken at different places before being packaged into the phage particle. Gene d^+, as it turns out, is closest to gene b^+, but you may or may not know that. What you do know is that if you plate out transduced phages to select for gene b^+, and then select on different media for genes a^+, c^+, and d^+, you find that d^+ is present more often with b^+ than is either a^+ or c^+.

You would observe some transduced particles, revealed by plating, that contain a^+, d^+, and b^+, and some particles that contain d^+, b^+, and c^+. No particles, however, would appear to have genes a^+ and c^+, which indicates that these genes are farther than 90 kb apart.

The premise that the closer two genes are together, the more frequently they are located on the same 90-kb DNA fragment is fairly obvious. What is less obvious is that you are unable to examine the recombination distance between a pair of genes using cotransduction data. This is because all the fragments that contain the selected b^+ gene do not also have either the a^+ or c^+ genes. As you can see in figure 15B, some fragments that contain b^+ lack both a^+ and c^+. This result gives the appearance of recombination, which would erroneously increase the calculated recombination distance.

Differences in Analysis

The same crossover events are used in both recombination mapping in conjugation and cotransduction. But some of these events are meaningful in one process but not in the other. For example, in conjugation mapping, production of the a^+ b^- c^- genotype is essential in determining the a to b distance (fig. 15A), while the same genotype in cotransduction reveals nothing about the cotransduction frequencies (fig. 15A). Similarly, the generation of the a^+ b^+ c^- genotype provides information on the b to c distance in conjugation mapping (fig. 15A), but the same genotype instead provides information on the a to b distance in cotransduction (fig. 15A).

Thus, we are left with two different ways to map bacterial genes. In conjugation, we know the genes are

continued

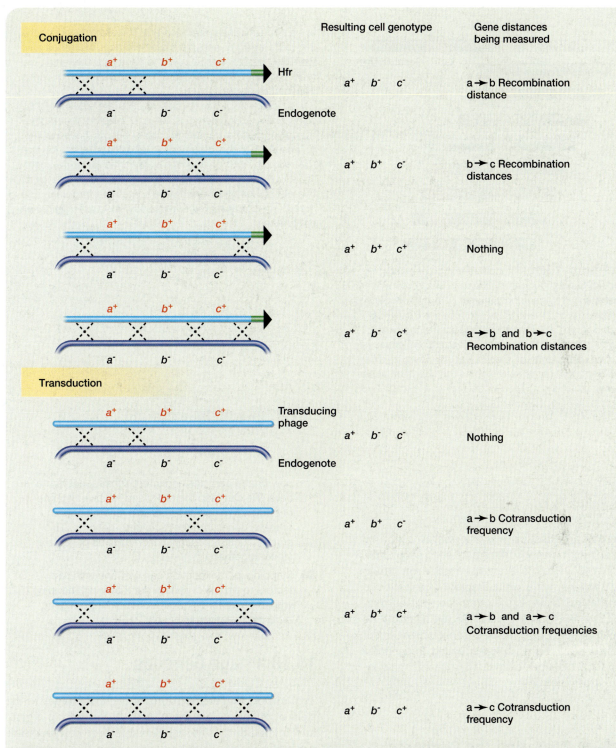

box figure 15A   Differences between conjugation and cotransduction mapping. Shown are the crossover events required to recombine alleles from either the Hfr exogenote (conjugation) or a transducing phage (transduction) into the recipient genome. The recombination events and the resulting genotypes are the same in both processes. However, different genotypes are used to measure either the recombination distances or cotransduction frequencies between the same pair of genes.

continued

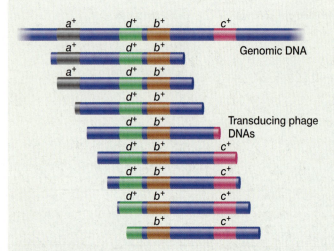

box figure 15B   Random digestion of a genomic region containing four genes. Infection of a bacteriophage can randomly fragment the genomic DNA that contains four genes (a^+ b^+ c^+, and d^+) to produce the 8 potential transducing phage DNAs shown. All the fragments are the same size and the presence of the different genes is shown on each fragment. Genes that are closer together (b and d) appear on more fragments together than genes further apart (d and c).

transferred in a linear arrangement. Therefore, selecting the last gene to enter ensures that the preceding genes also were transferred. Because all the genes are present in the recipient cell, we can measure the recombination distance between the genes.

In cotransduction, pieces of DNA are transferred by transducing phage particles. We can select for recombination of a specific transferred allele, but we have no way to ensure that the other genes of interest also got into the recipient cell on the same genomic DNA fragment. Thus, we use cotransduction to look at the frequency of two genes being present on the same DNA fragment. The greater the cotransduction frequency, the closer the genes are together.

It's Your Turn

Generalized transduction is a powerful approach for mapping mutations that are close to each other. Problem 43 utilizes generalized transduction to map three different genes, while Problem 44 asks you to map three mutations in the same gene (alleles) relative to two flanking genes. Both problems utilize the same concepts of generalized transduction: closer sites will have higher cotransduction frequencies and double crossovers are more frequent than quadruple crossovers.

Restating the Concepts

▶ Specialized transduction involves the production of defective λ genomes that possess flanking *E. coli* genomic DNA. Because λ integrates into a single location in the *E. coli* genome, the defective phage can carry either the *bio* or *gal* loci, which limits the usefulness of specialized transduction in mapping genes.

▶ Generalized transduction involves bacteriophage that infect a cell and cleave the bacterial genomic DNA. These genomic fragments can be packaged into

new phage particles and infect subsequent bacterial cells. We can map the order and relative distance of bacterial genes based on the recombination of these genes into the recipient's genome. This cotransduction frequency increases as genes are closer together.

▶ Mapping in transduction can be done with two factors to establish gene order; however, mapping with three factors allows deduction of both gene order and relative distances.

15.10 Phage Genetics

Two important contributions of bacteriophages to the field of genetics have been in the definition of a gene and in elucidating the relationship between recombination and genes. Their unique role in these contributions was based on the ability to produce hundreds of phage particles from a single bacterial cell infection, yielding tens of millions of phage particles in a single plaque. This huge number of phages permitted scientists to observe extremely rare events.

We first describe a basic method to map genes in the phage genome by recombination. We then explore Seymour Benzer's classic *rII* phage mapping experiments that helped to provide a definition for a gene and recombination.

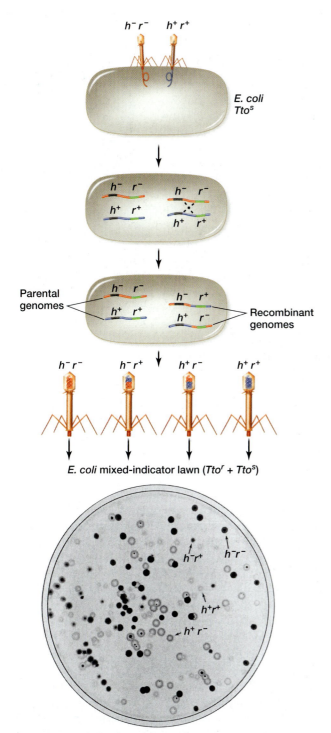

Recombination

Much of the genetic work on phages has been done with a group of seven *E. coli* phages called the T series (T-odd: T1, T3, T5, T7; T-even: T2, T4, and T6) and several others, including λ. To study recombination in a phage genome, a bacterial cell must be coinfected with two genetically distinct phage particles that possess different alleles.

Phage T2 Alleles

As an example of different phage alleles, consider the *rapid-lysis* locus. The *rapid-lysis* mutants (r^-) of the T-even phages produce large, sharp-edged plaques. The wild-type allele produces a smaller, more fuzzy-edged plaque (see fig. 15.12).

Another group of phage-encoded genes that exist as two distinct alleles are known as the *host-range* loci, which determine the bacterial strains the phage can infect. For example, the T2 phage can infect and lyse *E. coli* cells. These phages are designated as T2 h^+ for the normal host range. The *E. coli* cell is Tto^s, referring to their sensitivity to the T2 phage. Over time, a resistant *E. coli* mutant arose; this *E. coli* mutant strain is named Tto^r for T2 resistance. Subsequently, a T2 mutant was found that can infect and lyse the Tto^r strain of *E. coli*. These phage mutants are designated as T2 h^- for host-range mutant. Keep in mind that *host range* signifies a mutation in the phage genome, whereas *phage resistance* indicates a mutation in the bacterial genome.

Distinguishing Phenotypes

In 1945, Max Delbrück (one of the 1969 Nobel laureates in Physiology or Medicine) developed a method for distinguishing between the four phage phenotypes on the same petri plate (**fig. 15.36**, *bottom*). A lawn of mixed Tto^r and Tto^s bacterial strains is grown. On this lawn, the rapid-lysis phage mutants (r^-) produce large plaques, whereas the wild-type (r^+) produce smaller plaques.

Phages with the host-range mutation (h^-) lyse both Tto^r and Tto^s bacteria. They produce the plaques that are clear (but appear dark) in figure 15.36. Because phages with the wild-type host-range allele (h^+) can only infect the Tto^s bacteria, they produce turbid plaques, which appear light-colored in fig. 15.36. The turbidity results from the Tto^r bacteria that continue to grow in the plaque because they are resistant to the h^+ phage.

A Recombination Experiment

From the wild-type T2 phage stock, *host-range* mutants (h^-) can be isolated by looking for plaques on a Tto^r bacterial lawn. The *rapid-lysis* mutants (r^-) in these h^- mutant phages can be found by looking for large plaques. Once the double mutant strain ($h^- r^-$) is isolated, it can be mixed with the wild-type phage ($h^+ r^+$) and used to coinfect Tto^s sensitive bacteria (see fig. 15.36).

figure 15.36 Recombination mapping of phage genes. Two different phages, $h^- r^-$ (red) and $h^+ r^+$ (blue), are mixed with phage-sensitive bacterial cells (Tto^s) at a high moi (multiplicity of infection) to ensure coinfection of the bacterial cells. The h^+ phage will infect and lyse only Tto^s bacteria, while the h^- phage will infect and lyse both Tto^s and Tto^r bacteria. The r^+ allele causes slow bacterial lysis (producing small plaques) and r^- causes rapid lysis (large plaques). In the coinfected Tto^s cells, the different phage genomes may either remain intact to maintain the two parental genomes ($h^- r^-$, $h^+ r^+$) or may crossover to yield two recombinant genomes ($h^- r^+$, $h^+ r^-$). The phage that are released from the lysed cells are used to infect a lawn of mixed bacteria (Tto^s and Tto^r). Each of the four phage genotypes ($h^- r^-$, $h^+ r^+$, $h^- r^+$, and $h^+ r^-$) produces a phenotypically distinct plaque phenotype as shown on the plate, which allows both recombinant and parental phage genotypes to be distinguished.

The phages are added in a quantity that ensures a high **multiplicity of infection,** or **moi,** which corresponds to the ratio of phage per bacterium. The high moi ensures that each bacterium is infected by multiple phages, including at least one of each of the two phage types. Infecting the bacteria with two genotypically different phages creates the possibility of the phage genomes recombining within the bacterium.

After a round of phage multiplication, the phages are isolated and plated out on Delbrück's mixed-indicator stock (Tto^r and Tto^s cells) at a low moi, which ensures that a bacterium is infected with only a single phage. From this growth, the phenotype (and, hence, genotype) of each phage can be recorded. The percentage of recombinants can be read directly from the plate. For example, on a given petri plate (e.g., fig. 15.36, *bottom*) there might be

$$h^- r^- \ 46 \quad h^+ r^+ \ 52$$
$$h^+ r^- \ 34 \quad h^- r^+ \ 26$$

The $h^- r^-$ and $h^+ r^+$ represent the original, or parental, phage genotypes. The second two classes ($h^+ r^-$ and $h^- r^+$) result from recombination between the h and r loci on the parental phage genomes (see fig. 15.36).

A single crossover in this region produces the recombinants. The recombination frequency is the percentage of recombinant phage among the total number of phage produced:

$$\text{recombination frequency} = (34+26)/(46+52+34+26)$$
$$= 60/158$$
$$= 0.38, \text{ or } 38\%, \text{ or } 38 \text{ mu}$$

This percentage of recombinant phages is the map distance, which (as in eukaryotes) is a relative index of distance between loci: The greater the physical distance, the greater the amount of recombination, and thus the larger the map distance. One map unit (1 cM) is equal to 1% recombinant offspring.

Restating the Concepts

▶ Phage genetics can be performed on the rapid-lysis gene (*r*) and the host-range gene (*h*). It is possible to use a mixed culture of bacteria (Tto^s and Tto^r) to simultaneously determine the four different combinations of phage genotypes ($h^- r^-$, $h^+ r^+$, $h^+ r^-$, $h^- r^+$).

▶ Phages can infect at either a high moi, which increases the probability that a bacterium is infected with multiple phages, or at a low moi, which ensures that a single bacterium is infected with a single phage.

▶ Recombination can be detected between different phage genes based on plaque phenotype. The recombination distance or frequency between phage genes is the percentage of recombinant phage in the total number of phage produced.

Genetic Fine Structure

Phages exhibit recombination like the eukaryotes that we described in chapter 6; however, phages possess a limited number of genes that exhibit phenotypes that can be easily detected. Recombination distances for these genes have already been determined. Why would we then be interested in discussing phage genetics?

figure 15.37
Seymour Benzer (1921–), whose research with bacteriophage defined the cistron and the minimal recombination unit.

Phages played a very important role in answering some fundamental questions in genetics. Some of these questions include: How do we determine the relationship among several mutations that cause the same phenotypic change? What are the smallest units of DNA capable of mutation and recombination? The following discussion explores how Seymour Benzer (**fig. 15.37**) used phage genetics to answer these two questions.

Fine-Structure Mapping

In 1941 after Beadle and Tatum established that a gene controls the production of an enzyme that then controls a step in a biochemical pathway, Benzer used analytical techniques to dissect the fine structure of the gene. **Fine-structure mapping** involved examining the size and number of sites within a gene that are capable of mutation and recombination. In the late 1950s, when biochemical techniques were not yet available for DNA sequencing, Benzer used classical recombination and mutation techniques with bacteriophages to provide reasonable estimates on the details of fine structure and to give insight into the nature of the gene.

Before Benzer's work, genes were thought of as beads on a string. The analysis of mutational sites within a gene by means of recombination was hampered by the low rate of recombination and the limited number of progeny in most research organisms, such as fruit flies or corn. Although it certainly seemed desirable to map sites within a gene, the problem of finding an organism that would allow fine-structure analysis remained elusive until Benzer decided to use phage T4.

rII Screening Techniques

Benzer used T4 because of the growth potential of phages, in which a generation takes about an hour, and a single bacterial cell can yield hundreds of progeny phage on lysis. Benzer examined the rII^- mutants of T4, which produce large, smooth-edged plaques on *E. coli*, whereas wild-type (rII^+) phages produce small plaques.

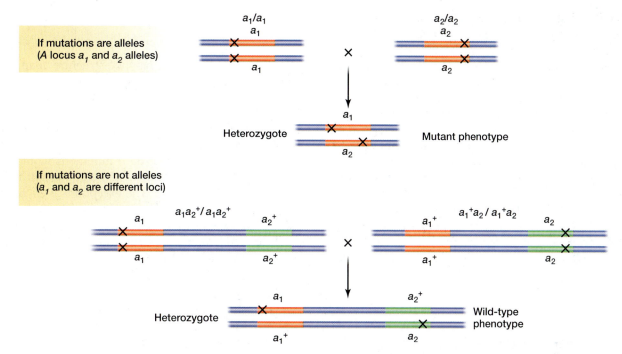

figure 15.38 The complementation test defines allelism. If the mutations are in the same gene (top), then there is no wild-type allele in the heterozygote, which results in the mutant phenotype. If the mutations are nonallelic (bottom), each DNA molecule contains a mutant allele in one gene and a wild-type allele in the second gene. In the heterozygote, there is a wild-type allele of each gene, which results in the wild-type phenotype.

Benzer's screening system used the fact that rII^- mutants did not lyse *E. coli* strain K12, whereas the wild-type rII^+ strain did. In contrast, both rII^- mutants and wild-type lysed the normal host strain, *E. coli* B. Thus, various rII^- mutants could be mixed to coinfect *E. coli* B cells at a high moi, and the progeny screened for wild-type recombinants by plating them at a low moi on *E. coli* K12, on which only wild-type recombinant phages produce plaques. Using this approach, it is possible to detect as few as one recombinant in over a billion phage particles in a single afternoon! In this way, recombination events occurring between mutations that are very close together, and thus happening at an extremely low frequency, could be detected.

Cis-Trans Complementation

If two recessive mutations arise independently and both have the same phenotype, how do we know whether they are both mutations of the same gene? That is, how do we know whether they are alleles? In chapter 5, we described the *cis–trans complementation test* in diploid organisms. Basically, if the a_1 and a_2 mutations are alleles, then the a_1/a_2 heterozygote will exhibit the mutant phenotype (**fig. 15.38**, top).

In contrast, if the mutations are not allelic, the gamete from the a_1 parent also contains an a_2^+ allele, and the gamete from the a_2 parent also contains the a_1^+ allele (fig. 15.38, bottom). Thus, the a_1/a_2 heterozygote

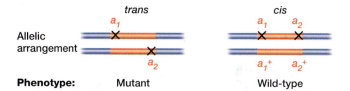

figure 15.39 The *cis* and *trans* arrangement. A heterozygote containing two recessive mutations can be in either the *trans* or *cis* arrangement. In the *trans* arrangement (left), the two mutations (black X's) are on different DNA molecules and produce a mutant phenotype. In the *cis* arrangement (right), both mutations are on the same DNA molecule, which always results in a wild-type phenotype because the second DNA molecule contains both wild-type sequences. In the *cis-trans* test, the two mutations must be in the *trans* configuration to determine complementation.

carries both the a_1^+ and a_2^+ wild-type alleles, which allows the two mutations to complement each other and produce the wild-type phenotype.

This test is only meaningful if the two mutant alleles are in the *trans* configuration (**fig. 15.39**). When the alleles are in the *cis* configuration, a wild-type phenotype will always be expected.

Identification of *rII⁻* Mutants. Benzer applied a similar test to examine his rII^- mutants, with a few modifications. Because there are multiple *r* mutant genes in the T4 phage, Benzer needed to identify only the *rII⁻* mutants, which was done by identifying *r* mutants that failed to complement. He correctly assumed that if two *r* mutant phages coinfected the same *E. coli* K12 cell, then a plaque would form if the *r* mutants were in different genes (**fig. 15.40**). Each phage genome would have one wild-type *r⁺* gene, which would encode a wild-type protein that could result in lysis.

To ensure that these plaques were not due to recombination, the lysate from the coinfected plaque could be used to infect *E. coli* B and K12 individually at a low moi. If the original plaque is due to recombination, then the subsequent infections would yield approximately equal number of plaques on both *E. coli* B and K12. But if the original plaque resulted from complementation, then the subsequent infections

would produce significantly more plaques on *E. coli* B than on K12.

The Cistron. To begin his fine-structure mapping, Benzer isolated a large number of independently derived *rII⁻* mutants and crossed them among themselves. The first thing he found was that the *rII* gene was composed of what appeared to be two genes, with almost all of the mutations belonging to one of two different complementation groups. The *A* mutations would not complement each other, but would complement the *B* mutations, and vice versa. The exceptions were mutations that seemed to belong to both groups, which were soon found to be deletions of both regions (**table 15.8**).

From this complementation data, Benzer coined the term **cistron** for the smallest functional genetic

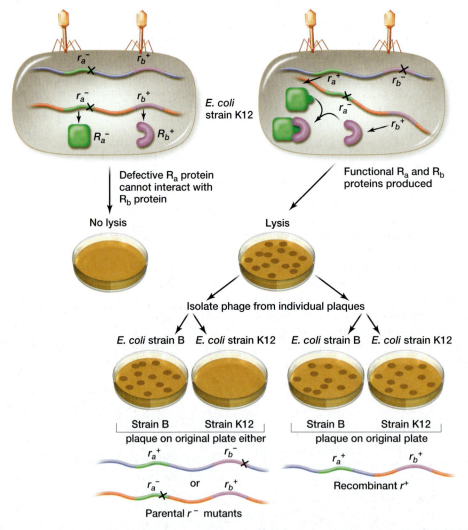

figure 15.40   Complementation mapping of *rII* mutations. Two different *rII* mutant phage are infected into *E. coli* strain K12 at a high moi (multiplicity of infection). If the mutations are in the same complementation group, no plaques will be produced (left). If the mutations are in different genes (right), many plaques will be produced (lysis). These plaques were produced by either complementation or recombination. To differentiate between these two possibilities, individual plaques were used to infect *E. coli* strains B and K12 individually. If plaques were only produced on strain B, then they contained one of the parental phage from complementation. If both strains B and K12 yield many plaques, then the original plaque contained a wild-type recombinant phage.

table 15.8 Complementation Matrix of Ten *rII⁻* Mutants

	1	2	3	4	5	6	7	8	9	10
1	–	–	+	–	–	–	+	–	–	–
2		–	+	–	–	–	+	–	–	–
3			–	–	+	+	–	+	–	+
4				–	–	–	–	–	–	–
5					–	–	+	–	–	–
6						–	+	–	–	–
7							–	+	–	+
8								–	–	–
9									–	–
10										–

Note: Plus sign (+) indicates complementation; minus sign (–) indicates no complementation. The two cistrons are arbitrarily designated *A* and *B*. Mutants 4 and 9 must be deletions that cover parts of both cistrons. Alleles: *A* cistron: 1, 2, 4, 5, 6, 8, 9, 10; *B* cistron: 3, 4, 7, 9.

unit (length of genetic material) that exhibits a *cis–trans* position effect. The cistron is usually equivalent to a gene. In essence, Benzer refined Beadle and Tatum's one-gene/one-enzyme hypothesis to a more accurate one-cistron/one-polypeptide concept.

Deletion Mapping

Benzer rapidly isolated large numbers of *rII⁺* mutants and collected *rII⁻* mutants from other labs. The complementation test placed about an equal number of mutations in each cistron. However, as the number of *rII⁻* mutations in a single cistron approached, and later surpassed, several thousand, it became obvious that making every possible pairwise cross for recombination mapping would entail millions of crosses. To overcome this problem, Benzer isolated mutants that had partial or complete deletions of each cistron.

Deletion mutations were easy to detect because they failed to complement mutations in cistron *A* and cistron *B* (see mutations 4 and 9 in table 15.8). Furthermore, deletion mutants failed to produce wild-type recombinant phage when *E. coli* was coinfected with phages having several independent mutations in one or both of the cistrons (**fig. 15.41**). Once a series of deletion mutations covering the *A* and *B* cistrons was isolated, a minimal number of crosses were required to localize a new mutation. A second series of even smaller deletions within each region was then isolated, further localizing each mutation (**fig. 15.42**).

Recombination Mapping

After each of the *rII⁻* mutations was localized into a subregion of one of the *rII* cistrons, Benzer crossed

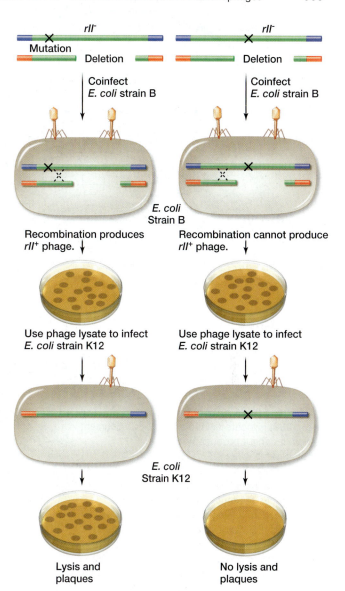

figure 15.41 Deletion mapping of *rII⁻* mutants. *E. coli* strain B cells are coinfected with an *rII⁻* point mutant and an *rII⁻* deletion mutant. If the point mutation is located outside the deleted region, then recombination can produce some wild-type phage (left). When this strain B lysate infects *E. coli* strain K12 at a low moi, only the wild-type recombinant phage will produce plaques. If the point mutation is located within the deleted region, then recombination cannot produce wild-type phage (right). These mutant phage will also lyse *E. coli* strain B and produce phage. However, this phage lysate cannot produce plaques on *E. coli* strain K12, because there are no wild-type recombinant phage.

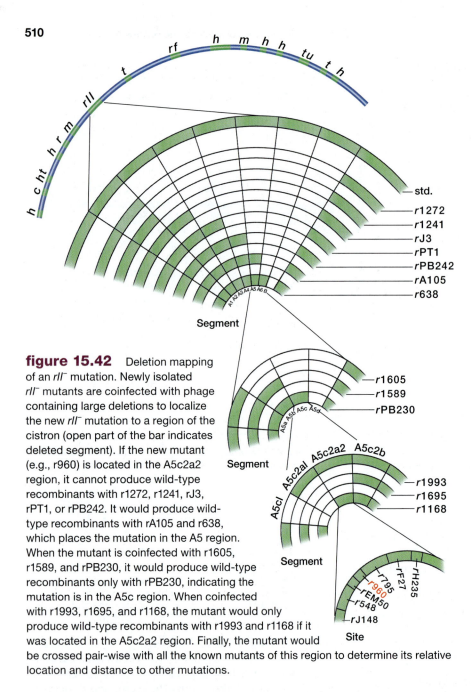

figure 15.42 Deletion mapping of an *rII⁻* mutation. Newly isolated *rII⁻* mutants are coinfected with phage containing large deletions to localize the new *rII⁻* mutation to a region of the cistron (open part of the bar indicates deleted segment). If the new mutant (e.g., r960) is located in the A5c2a2 region, it cannot produce wild-type recombinants with r1272, r1241, rJ3, rPT1, or rPB242. It would produce wild-type recombinants with rA105 and r638, which places the mutation in the A5 region. When the mutant is coinfected with r1605, r1589, and rPB230, it would produce wild-type recombinants only with rPB230, indicating the mutation is in the A5c region. When coinfected with r1993, r1695, and r1168, the mutant would only produce wild-type recombinants with r1993 and r1168 if it was located in the A5c2a2 region. Finally, the mutant would be crossed pair-wise with all the known mutants of this region to determine its relative location and distance to other mutations.

it with other mutations in the same region to test for recombination. When two *rII⁻* mutants were coinfected into *E. coli* strain B at a high moi, recombination could take place. If recombination occurred between the two mutant sites, then a phage with *rII⁺* would be produced along with a complementary phage having both mutations (**fig. 15.43**).

The lysate that was produced, which contained phage populations having the parental genomes and the two recombinant genomes, was then diluted and plated at a low moi on a single plate containing *E. coli* strain B, and on a different plate containing *E. coli* strain K12. The first plate revealed the **titer** of the phage, which is the concentration of all the phage particles per milliliter. The latter plate revealed the number of wild-type recom-

binant phage produced. The phage containing both mutations, similar to the two parental phage, would be unable to grow on *E. coli* strain K12.

Recombination Distance Calculations. To determine the recombination distance between the two mutant sites, we calculate the frequency of all the recombinants in the total population. Let's assume that the titer of phage determined on the *E. coli* strain B plate is 8×10^7 plaques (phage) per milliliter, and that we identified 14 plaques per milliliter on the *E. coli* strain K12 plate. The recombination frequency would then be:

$$(100\%) \times 2 \times (14/8 \times 10^7) = 3.5 \times 10^{-5}$$

We have to multiply by 2 in the preceding equation because there are two genotypic classes of recombinants. The first is the wild-type (*rII⁺*), which produces the plaques on *E. coli* strain K12. The second is composed of the phage containing both mutations (see fig. 15.43), which is phenotypically identical to the parental phage, unable to generate a plaque on *E. coli* strain K12. We assume that the number of wild-type phage particles equals the number of the double mutant recombinant class because they are the reciprocal products of the same crossover event. Multiplying the number of wild-type recombinants by two should then produce a good approximation of the total number of recombinant phages.

It's Your Turn

Problems 39 and 41 deal with recombination mapping between only two mutations at a time, much like the preceding discussion. However, it is also possible to do three-point crosses in phages. Problems 38 and 40 require you to extend the preceding discussion and what you learned in chapter 6 to perform three-point crosses in phage. Remember to account for all the recombinant phages in your calculations.

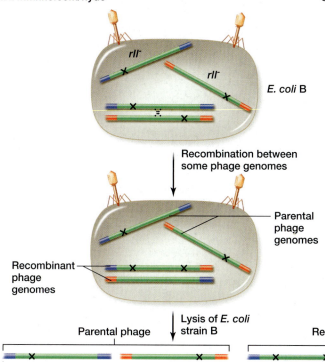

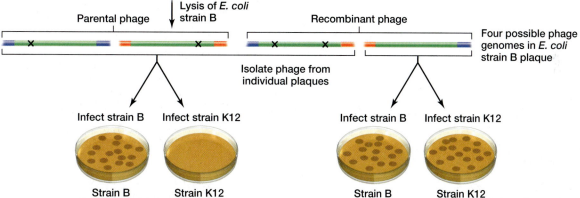

figure 15.43   Recombination between different *rII⁻* point mutations. *E. coli* strain B is coinfected with two different *rII⁻* mutant phage. Recombination between some of the phage genomes in the bacterial cell results in two different types of recombinant phage, one wild-type and one containing both mutations. The resulting phage lysate from the strain B coinfection, which contains both parental and both recombinant phage classes, is used to independently infect *E. coli* strains B and K12 at a low moi (multiplicity of infection). All the phage will produce plaques on *E. coli* strain B to reveal the total number of phage in the lysate. However, only wild-type phage will produce plaques on *E. coli* strain K12. Thus, the number of plaques on K12 reveals the number of wild-type recombinants, which corresponds to half of the total number of recombinants. The other half of the recombinants, the double mutants, are phenotypically identical to the original single mutant phage (they produce large plaques on strain B and no plaques on strain K12).

Localization of Mutation Position.   Next, all of the mutants in a subregion were crossed with one another to localize their relative positions. The exact position of each new mutation within the region was determined by the relative frequency of recombination between it and the known mutations of this region.

Benzer eventually analyzed about 350 mutations from 80 different subregions defined by deletion mutations. An abbreviated map is shown in **figure 15.44**. Because of the enormously large number of phages that could be examined in this process, extremely small recombination distances could be determined. Benzer found, however, that there was a minimum recombination frequency, which was several orders of magnitude larger than his minimum detection limit. This led him to postulate that a minimal recombination unit existed, which he named the **recon.** The recon, which we now know is a single base pair, was significantly smaller than what was previously thought to be the minimal recombination unit, namely the gene.

Mutation Uniformity.   It is obvious from figure 15.44 that the mutations were not evenly distributed through-

out the *rII* gene. Rather, some locations possessed a large number of independent mutations (note B4 and A6c in fig. 15.44). These regions were termed **hot spots.**

Presuming that all base pairs are either AT or GC, this lack of uniformity was unexpected. Benzer suggested that spontaneous mutation is not just a function of the base pair itself, but is affected by the surrounding bases as well. This concept still holds.

Conclusions from Benzer's Work

To recapitulate, Benzer's work supports the model of the gene as a linear arrangement of DNA whose nucleotides are the smallest units of mutation. The phosphodiester bond between any adjacent nucleotides can break in the recombinational process. The smallest functional unit, determined by a complementation test, is the cistron. Mutagenesis is not uniform throughout the cistron, but may depend on the particular arrangement of bases in a given region. These findings in the 1950s, which were accomplished without the aid of molecular genetic techniques and DNA sequencing, remain to be essentially correct to this day.

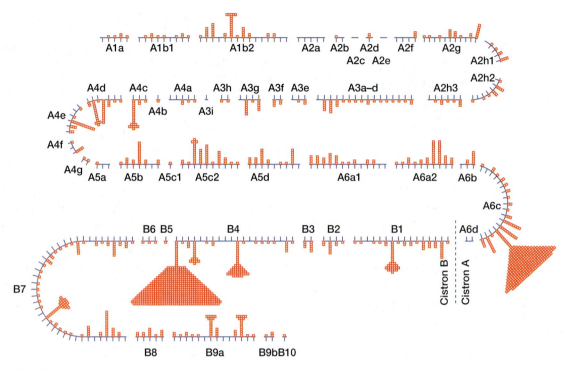

figure 15.44 *rII⁻* mutations. An abbreviated map of the spontaneous mutations that Seymour Benzer isolated and characterized in the A and B cistrons of the *rII* region of bacteriophage T4. Each red square represents one independently isolated mutation. Note the "hot spots" (large number of independent mutations) at A6c and B4.

 **Restating the Concepts**

▶ Phage genetics allows for the isolation of extremely rare events due to the large number of phages that can be easily examined in a short time. This permitted the identification of a single base pair as the minimal recombination unit.

▶ Recombination frequencies between mutations in a single gene can be determined if the number of progeny is very large.

▶ Regardless of the system being examined (prokaryotic or eukaryotic), the underlying principles of complementation and recombination between a deletion and a point mutation remain constant.

▶ Benzer provided working definitions for a cistron and a recon, and he found that the mutant sites within a gene were not evenly distributed.

 genetic linkages

In this chapter, we discussed the genetics of bacteria and bacteriophages. Because bacteria are haploid organisms, the dominant/recessive nature of an allele is not as straightforward to study as in diploids. However, we described the use of F' elements and conjugation to generate a partial diploid, a merozygote, to study this relationship. We visit this concept again in chapter 16 when we describe the *lac* operon and its various mutants.

We also discussed the life cycle of bacteriophages, their ability to transduce bacterial genes, and techniques used for mapping genes and mutations in the phage genome. We revisit λ in chapter 16, when we discuss how gene regulation controls its temperate life cycle.

The definitions of *cis* and *trans,* first introduced in chapter 5, were reintroduced in this chapter. The term *cis* refers to elements that are on the same DNA molecule, whereas *trans* elements are on different DNA molecules. The complementation test that Benzer employed with the *rII⁻* mutants of the phage T4 is essentially the same *cis–trans* complementation test that we described in chapter 5.

Recombination was first discussed in chapter 6, in the context of meiosis. You saw that single crossover events are more common than double crossovers, but that double crossovers can reveal the order of three genes (the three-point cross). In this chapter, we added a new wrinkle. We described how a linear molecule recombines

continued

with a circular genome. An odd number of crossovers produces a linear genome, which does not properly replicate in *E. coli*. Thus, an even number of crossovers is needed when recombining a linear Hfr exogenote or transducing phages into the *E. coli* genome.

Our previous discussions on recombination mapping dealt with measuring the frequency of crossovers between pairs of genes. We used the same definition when measuring map distances that result from recombination after conjugation. When mapping genes using either transformation or transduction, however, we examine the frequency of the crossovers that flank the pair of genes. This results in the cotransformation or cotransduction of genes.

We also looked at the recombination of the F plasmid and λ DNA into the *E. coli* genome. In both of these cases, we are recombining two circular molecules, which means that a single crossover can regenerate the circular molecule. Therefore, the minimal number of crossover events that can regenerate a properly replicating genome is the most common event in recombination.

Chapter Summary Questions

1. For each term in column I, choose the best matching phrase in column II.

   Column I
   1. Autotroph
   2. F$^+$ cell
   3. Virulent phage
   4. Hfr cell
   5. Prototroph
   6. Temperate phage
   7. F' cell
   8. Auxotroph
   9. F$^-$ cell
   10. Heterotroph

   Column II
   A. Cell with no nutritional requirement
   B. Cell with a fertility plasmid in its chromosome
   C. Virus that can only undergo a lytic cycle
   D. Cell with a fertility plasmid not in its chromosome
   E. Can utilize carbon from CO_2
   F. Requires an organic form of carbon
   G. Cell with no fertility plasmid
   H. Cell with a nutritional requirement
   I. Cell with chromosomal genes in its fertility plasmid
   J. Virus that can undergo a lytic or lysogenic cycle

2. What is the nature and substance of prokaryotic chromosomes and viral chromosomes?

3. What are the differences between a minimal and a complete medium? An enriched and a selective medium?

4. What properties of bacteria and phages make them very suitable for genetic research?

5. What are the three general classes of bacterial phenotypes and how do they differ?

6. What are the two general classes of phage phenotypes and how do they differ?

7. Explain the replica-plating screening technique.

8. What are the differences between a plaque and a colony?

9. What is a plasmid? How does a plasmid integrate into a host's chromosome? How does it leave?

10. Define the following terms: **(a)** prophage; **(b)** lysate; **(c)** lysogeny.

11. Distinguish between *specialized* and *generalized transduction*.

12. Indicate whether each of the following lettered items apply to: conjugation, transformation, transduction, a specific combination of these processes, or to none of them.
    a. moi
    b. Competence
    c. Increases genetic diversity
    d. Reciprocal DNA exchange between two bacteria
    e. Sensitive to extracellular DNases
    f. Can still occur in a U-tube apparatus
    g. Sexduction

13. Compare and contrast between the mapping approaches used in conjugation, transformation, and transduction.

14. Diagram the step-by-step events required to integrate foreign DNA into a bacterial chromosome in each of the three processes outlined in the chapter (transformation, conjugation, transduction).

Solved Problems

PROBLEM 1: A wild-type strain of *Baccilus subtilis* is transformed by DNA from a strain that cannot grow on galactose (*gal$^-$*) and also needs biotin for growth (*bio$^-$*). Transformants are isolated by plating the transformed cells on minimal medium with penicillin, killing the wild-type cells. Thirty of the penicillin-resistant transformants were replica-plated to establish their genotypes, which are shown here:

Class 1: *gal$^-$ bio$^-$*	17
Class 2: *gal$^-$ bio$^+$*	4
Class 3: *gal$^+$ bio$^-$*	9

What is the cotransformation frequency of these two loci?

Answer: The three classes of colonies represent the three possible transformant groups. Classes 2 and 3 are single transformants, and class 1 is the double transformant. We are interested in the cotransformation of the two loci. Therefore we divide the number of double transformants by the total: $r = 17/(17 + 4 + 9) = 0.57$. This is a relative value inverse to a map distance; the larger it is, the closer the loci are to each other.

PROBLEM 2: A $gal^- \ bio^- \ att\lambda^-$ strain of *E. coli* is transduced by P22 phages from a wild-type strain. Transductants are selected by growing the cells with galactose as the sole energy source. Replica-plating and testing for lysogenic ability gives the genotypes of 106 transformants:

Class 1: $gal^+ \ bio^- \ att\lambda^-$ 71
Class 2: $gal^+ \ bio^+ \ att\lambda^-$ 0
Class 3: $gal^+ \ bio^- \ att\lambda^+$ 9
Class 4: $gal^+ \ bio^+ \ att\lambda^+$ 26

What is the gene order, and what are the relative cotransduction frequencies?

Answer: We selected all transductants that are gal^+. Class 2 is in the lowest frequency (0 colonies) and therefore represents the quadruple crossover between the transducing DNA and the host chromosome. Comparing the quadruple crossover genotype with the transducing DNA genotype reveals that $att\lambda$ must be in the middle because this order produces the lowest frequency genotype by switching only the middle locus of the transducing DNA. In other words, the two end loci would recombine into the host genome, and the middle locus would have the host allele. We can calculate only two cotransduction frequencies because these frequencies are based on the gene that was initially selected (gal^+). Note that in class 1, there is no cotransduction between gal and either of the other two loci; class 2 would show the cotransduction of gal and bio; class 3 represents the cotransduction of gal and $att\lambda$; and class 4 represents the cotransduction of gal and both other loci. Therefore, cotransduction values are

$gal–att\lambda = (9 + 26)/106 = 35/106 = 0.33$
$gal–bio = (0 + 26)/106 = 26/106 = 0.25$.

PROBLEM 3: A series of overlapping deletions in phage T4 are isolated. All pairwise crosses are performed, and the progeny scored for wild-type recombinants. In the following table, + = wild-type progeny recovered; – = no wild-type progeny recovered.

	1	2	3	4	5
1	–	+	–	–	–
2		–	+	+	–
3			–	+	+
4				–	–
5					–

a. Draw the relative locations of each deletion.
b. A point mutation, 6, is isolated and crossed with all of the deletion strains. Wild-type recombinants are recovered only with strains 2 and 3. What is the location of the point mutation?

Answer: **a.** The main principle at work is that if two deletions overlap, then both DNA molecules are missing a common DNA region. Therefore, the deletions cannot recombine and generate a complete wild-type sequence. To solve this problem, we can draw overlapping gaps (no wild-type progeny produced) and nonoverlapping gaps (when the two mutants produce wild-type recombinants) to represent the deletions shown in the table.

Begin with deletions that yield mostly "–"s (failure to produce wild-type recombinants). These must be large deletions that cover most of the other deletions. Mutations 1 and 5 are such mutations. Because they give no wild-type, these deletions must overlap:

Now look at mutant 2. It gives wild-type recombinants with 1, but not 5. Therefore, the mutant 2 deletion must overlap the region deleted in 5, but not the deletion in mutant 1. The deletion in mutant 3, by similar logic, must overlap part of deletion 1, but not overlap the deletion in mutant 5. We can draw these results as follows:

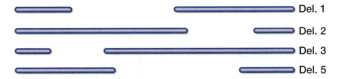

Now look at mutant 4. It does not produce wild-type recombinants with either mutant 1 or mutant 5. Therefore, the deletion in mutant 4 must be in the region that the deletions 1 and 5 overlap. If deletion 4 extended to and overlapped either deletion 2 or 3, we expect to see no wild-type recombinants produced between mutant 4 and either mutant 2 or 3. Because this prediction is not met (mutant 4 produces wild-type recombinants with both mutant 2 or 3), mutant 4 must be a small deletion spanning at least part of the overlap of 1 and 5, but not overlapping with both mutants 2 or 3. Therefore, the map of the five deletions is

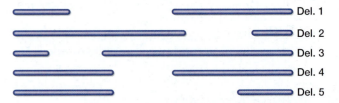

Note that the data do not provide the actual endpoints of the deletions, so other maps can be drawn as well. For example, deletion 3 could be a little shorter or longer (provided it does not overlap with either 4 or 5).

b. If wild-type recombinants are produced, the point mutation must have occurred in a region of the gene that is outside the deletion in question. Because mutant 6 gives no wild type with 1, 4, or 5, it must be within the common region deleted in all three strains.

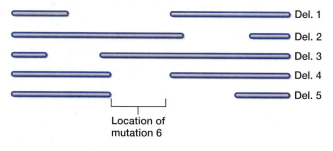

Location of mutation 6

PROBLEM 4: The following *E. coli* Hfr strains donate their markers to F⁻ cells in the order given:

Hfr Strain	Order of Gene Transfer					
	First					**Last**
1	*f*	*h*	*e*	*t*	*a*	*g*
2	*o*	*c*	*r*	*f*	*h*	*e*
3	*u*	*j*	*n*	*o*	*c*	*r*
4	*n*	*j*	*u*	*g*	*a*	*t*

Draw a map showing the relative order of these genes in the circular chromosome of the original F⁺ strain from which all four Hfr strains were derived. Be sure to indicate the sites of F plasmid integration and the orientations of the origins of transfer origin.

Answer: The best approach is to determine the overlapping genes in each sequence:

Hfr strain 1		*f h e t a g*
Hfr strain 2		*o c r f h e*
Hfr strain 3		*u j n o c r*
Hfr strain 4	*t a g u j*	

The gene sequence transferred by Hfr strain 4 overlaps the gene sequences of both strains 3 (genes *j* and *u*) and 1 (*g*, *a*, and *t*; shown in red). This should not be a surprise because the bacterial chromosome is circular. This simply indicates that the F plasmid in Hfr strain 4 is integrated in the opposite orientation to that of Hfr strains 1, 2, and 3. So the transfer of genes from Hfr strain 4 is in the opposite direction to that of the other three strains.

A consolidation of these gene sequences results in the following map, where the arrowheads represent the site of F plasmid integration and orientation of gene transfer:

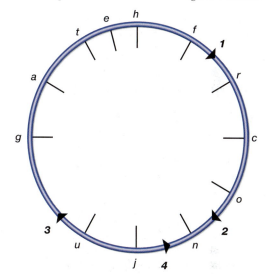

Exercises and Problems

15. What genotypic notation indicates alleles that make a bacterium
    a. resistant to penicillin?
    b. sensitive to azide?
    c. require histidine for growth?
    d. unable to grow on galactose?
    e. able to grow on glucose?
    f. susceptible to phage T1 infection?

16. An *E. coli* cell is placed on a petri plate containing λ phages. It produces a colony overnight. By what mechanisms might it have survived?

17. An *E. coli* lawn is formed on a petri plate containing complete medium. Replica-plating is used to transfer material to plates containing minimal medium and combinations of the amino acids arginine and histidine (see the figure). Give the genotype of the original strain as well as the genotypes of the odd colonies found growing on the plates.

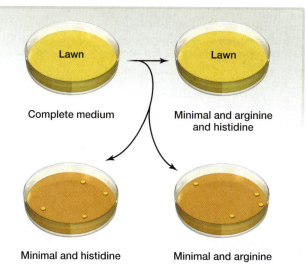

18. Give possible genotypes of an *E. coli*-phage T1 system in which the phage cannot grow on the bacterium. Give genotypes for a T1 phage that can grow on the bacterium.

19. Prototrophic Hfr *E. coli* strain G11, which is sensitive to streptomycin and can use maltose as an energy source (*malT*⁺), is used in a conjugation experiment. The *str*ˢ locus is one of the last genes to be transferred, whereas the *malT*⁺ locus is one of the first. This strain is mated to an F⁻ strain that is resistant to streptomycin, *malT*⁻ (cannot utilize maltose), and requres five amino acids (histidine, arginine, leucine, lysine, and methionine) to be added to minimal medium for growth. Recombinants are selected by plating on a medium containing streptomycin, with maltose as the sole carbon source, and all five amino acids present. Thus, all recombinant F⁻ cells will grow irrespective of their amino acid requirements. Five colonies are grown on the original plate with streptomycin, maltose, and all five amino acids in question (see the figure). These colonies are replica-plated onto minimal medium containing various amino acids. What are the genotypes of each of the five colonies?

Medium with amino acids, streptomycin, and maltose

Minimal medium +

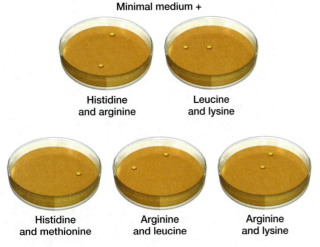

Histidine and arginine

Leucine and lysine

Histidine and methionine

Arginine and leucine

Arginine and lysine

20. A petri plate with complete medium has six colonies growing on it after a conjugation experiment. The colonies are numbered, and the plate is used as a master to replicate onto plates of glucose-containing selective (minimal) medium with various combinations of additives. From the following data, which show the presence (+) or absence (–) of growth, give your best assessment of the genotypes of the six colonies.

21. In conjugation experiments, the Hfr strain usually carries a gene for some sort of sensitivity (for example *azi*ˢ or *str*ˢ), whereas the F⁻ strain carries a resistance gene. Why is that so? Should this locus be near to or far from the origin of transfer point of the Hfr chromosome? What are the consequences of either alternative?

22. How does a geneticist doing interrupted mating experiments know that the locus for the drug-sensitivity allele has crossed from the Hfr into the F⁻ strain?

23. The DNA from a prototrophic strain of *E. coli* is isolated and used to transform an auxotrophic strain deficient in the synthesis of purines (*purB*⁻), pyrimidines (*pyrC*⁻), and the amino acid tryptophan (*trp*⁻). Tryptophan was used as the marker to determine whether transformation had occurred (the selected marker). What are the gene order and the relative cotransformation frequencies between loci, given these data:

trp⁺ *pyrC*⁺ *purB*⁺	86
trp⁺ *pyrC*⁺ *purB*⁻	4
trp⁺ *pyrC*⁻ *purB*+	67
trp⁺ *pyrC*⁻ *purB*⁻	14

24. Using the data in figure 15.24, draw a map of that region of the *E. coli* chromosome.

25. Three Hfr strains of *E. coli* (P4X, KL98, and Ra-2) are mated individually with an auxotrophic F⁻ strain using interrupted mating techniques. Using the following data, construct a map of the *E. coli* chromosome, including distances in minutes.

	Approximate Time of Entry		
Donor Loci	**Hfr P4X**	**Hfr KL98**	**Hfr Ra-2**
gal⁺	11	67	70
thr⁺	94	50	87
xyl⁺	73	29	8
lac⁺	2	58	79
his⁺	38	94	43
ilv⁺	77	33	4
argG⁺	62	18	19

How many different petri plates and selective media are needed?

26. Design an experiment using interrupted mating and create a possible data set that would correctly map five of the loci on the *E. coli* chromosome (see fig. 15.27).

	Colony					
On Minimal Medium Plus	**1**	**2**	**3**	**4**	**5**	**6**
Nothing	–	–	+	–	–	–
Xylose + arginine	+	–	+	+	–	–
Xylose + histidine	–	–	+	–	–	–
Arginine + histidine	–	+	+	+	–	–
Galactose + histidine	–	–	+	–	–	+
Threonine + isoleucine + valine	–	–	+	–	+	–
Threonine + valine + lactose	–	–	+	–	–	–

27. The following *E. coli* Hfr strains donate their markers to F⁻ cells in the order given:

Hfr Strain	Order of Gene Transfer						
1	z	i	a	d	e	l	r
2	m	f	b	p	t	o	n
3	y	j	k	r	l	e	
4	b	f	m	z	i		
5	t	o	n	y	j		

Draw a map showing the order of these genes in the circular chromosome of the original F⁺ strain from which all five Hfr strains were derived. Be sure to indicate the sites of F plasmid integration and the orientations of the origins of transfer within each of the integrated F plasmids.

28. Lederberg and his colleagues (Nester, Schafer, and Lederberg, 1963, *Genetics* 48:529) determined the gene order and relative distance between genes using three markers in the bacterium *Bacillus subtilis*. DNA from a prototrophic strain ($trp^+ his^+ tyr^+$) was used to transform the auxotroph. The seven classes of transformants, with their numbers, are tabulated as follows:

trp^+	trp^-	trp^-	trp^+	trp^+	trp^-	trp^+
his^-	his^+	his^-	his^+	his^-	his^+	his^+
tyr^-	tyr^-	tyr^+	tyr^-	tyr^+	tyr^+	tyr^+
2,600	418	685	1,180	107	3,660	11,940

Taking the loci in pairs, calculate cotransformation frequencies. Construct the most consistent linkage map of these loci.

29. In a transformation experiment, an $a^+ b^+ c^+$ strain is used as the donor and an $a^- b^- c^-$ strain as the recipient. One hundred a^+ transformants are selected and then replica-plated to determine whether b^+ and c^+ are present. What can you conclude about the relative positions of the genes, based on the following results?

$a^+ b^- c^-$	21
$a^+ b^- c^+$	69
$a^+ b^+ c^-$	3
$a^+ b^+ c^+$	7

30. In a transformation experiment, an $a^+ b^+ c^-$ strain is used as donor and an $a^- b^- c^+$ strain as recipient. If you select for a^+ transformants, the least frequent class is $a^+ b^+ c^+$. What is the order of the genes?

31. A mating between $his^+, leu^+, thr^+, pro^+, str^s$ cells (Hfr) and $his^-, leu^-, thr^-, pro^-, str^r$ cells (F⁻) is allowed to continue for 25 min. The mating is stopped, and the genotypes of the recombinants are determined. What is the first gene to enter, and what is the probable gene order, based on the following data?

Genotype	Number of colonies
his^+	0
leu^+	12
thr^+	27
pro^+	6

32. a. In a transformation experiment, the donor is $trp^+ leu^+ arg^+$, and the recipient is $trp^- leu^- arg^-$. You initially select for trp^+ transformants, which are then further tested. Forty percent are $trp^+ arg^+$; 5% are $trp^+ leu^+$. In what two possible orders could the genes be arranged?

b. You can do only one more transformation to determine gene order. You must use the same donor and recipient, but you can change the selection procedure for the initial transformants. What should you do, and what results should you expect for each order you proposed in part (a)?

33. DNA from a bacterial strain that is $a^+ b^+ c^+$ is used to transform a strain that is $a^- b^- c^-$. The numbers of each transformed genotype appear. What can we say about the relative position of the genes?

Genotype			Number
a^+	b^-	c^-	214
a^-	b^+	c^-	231
a^-	b^-	c^+	206
a^+	b^+	c^-	11
a^+	b^+	c^+	6
a^+	b^-	c^+	93
a^-	b^+	c^+	14

34. An Hfr strain that is $a^+ b^+ c^+ d^+ e^+$ is mated with an F⁻ strain that is $a^- b^- c^- d^- e^-$. The mating is interrupted every 5 min, and the genotypes of the F⁻ recombinants are determined. The results appear in the following table. (A *plus* indicates appearance; a *minus* the lack of the locus.) Draw a map of the chromosome and indicate the position of the F factor, the direction of transfer, and the minutes between genes.

Time	a	b	c	d	e
5	–	–	–	–	–
10	+	–	–	–	–
15	+	–	–	–	–
20	+	–	–	–	–
25	+	–	–	–	–
30	+	–	–	+	–
35	+	–	–	+	–
40	+	+	–	+	–
45	+	+	–	+	–
50	+	+	–	+	–
55	+	+	–	+	–
60	+	+	–	+	–
65	+	+	+	+	–
70	+	+	+	+	–
75	+	+	+	+	+

35. A bacterial strain that is $lys^+ his^+ val^+$ is used as a donor, and $lys^- his^- val^-$ as the recipient. Initial transformants are isolated on minimal medium + histidine + valine.

a. What gene is being selected and what genotypes will grow on this medium?

b. These colonies are replicated to minimal medium + histidine, and 75% of the original colonies grow. What genotypes will grow on this medium?

c. The original colonies are also replicated to minimal medium + valine, and 6% of the colonies grow. What genotypes will grow on this medium?

d. Finally, the original colonies are replicated to minimal medium. No colonies grow. From this information, what genotypes will grow on minimal medium + histidine and on minimal medium + valine?

e. Based on this information, which gene is closer to *lys*?

f. The original transformation is repeated, but the original plating is on minimal medium + lysine + histidine. Fifty colonies appear. These colonies are replicated to determine their genotypes, with these results:

val⁺ his⁺ lys⁺   0
val⁺ his⁻ lys⁺   37
val⁺ his⁺ lys⁻   3

Based on all the results, what is the most likely gene order?

36. Outline an experiment to demonstrate that two phages do not undergo recombination until a bacterium is infected simultaneously with both.

37. Consider the following portion of a phage genome:

r	2 mu	*t* 0.8 mu *z*

The cross $r^+ t^+ z^+ \times r\, t\, z$ is performed, and 50,000 progeny are scored.

a. How many $r^+ t\, z^+$ phages are expected?

b. If the progeny actually included 30 that are either $r^+ t\, z^+$ or $r\, t^+ z$, calculate the interference?

38. Doermann (1953, *Cold Spr. Harb. Symp. Quant. Biol.* 18:3) mapped three loci of phage T4: minute, rapid lysis, and turbid. He infected *E. coli* cells with both the triple mutant (*m r tu*) and the wild-type (*m⁺ r⁺ tu⁺*) and obtained the following data:

m	*m⁺*	*m*	*m*	*m⁺*	*m⁺*	*m*	*m⁺*
r	*r*	*r⁺*	*r*	*r⁺*	*r*	*r⁺*	*r⁺*
tu	*tu*	*tu*	*tu⁺*	*tu*	*tu⁺*	*tu⁺*	*tu⁺*
3,467	474	162	853	965	172	520	3,729

What is the linkage relationship among these loci? In your answer include gene order, relative distance, and coefficient of coincidence.

39. Wild-type phage T4 (r^+) produce small, turbid plaques, whereas *rII⁻* mutants produce large, clear plaques. Four *rII⁻* mutants (*a–d*) are crossed. (Assume, for the purposes of this problem, that *a–d* are four closely linked loci. Here, assume that $a \times b$ means $a^- b^+ c^+ d^+ \times a^+ b^- c^+ d^+$.) These percentages of wild-type plaques are obtained in crosses:

$a \times b$	0.3
$a \times c$	1.0
$a \times d$	0.4
$b \times c$	0.7
$b \times d$	0.1
$c \times d$	0.6

Deduce a genetic map of these four mutants.

40. A phage cross is performed between $a^+ b^+ c^+$ and *a b c* phage. Based on these results, derive a complete map:

$a^+ b^+ c^+$	1,801
$a^+ b^+ c$	954
$a^+ b\, c^+$	371
$a^+ b\, c$	160
$a\, b^+ c^+$	178
$a\, b^+ c$	309
$a\, b\, c^+$	879
$a\, b\, c$	1,850
	6,502

41. The *rII⁻* mutants of T4 phage will grow and produce large plaques on *E. coli* strain B, *but* will not grow on *E. coli* strain K12. Certain crosses are performed in strain B. (Assume that the three mutations affect three separate loci in the *rII* region.) By diluting and plating on strain B, it is determined that each experiment generates about 250×10^7 phages. By dilution, approximately 1/10,000 of the progeny are plated on K12 to generate these wild-type recombinants (plaques on K12):

1×2	50
1×3	25
2×3	75

Draw a map of these three mutants (1, 2, and 3) and indicate the distances between them.

42. Consider the following data from a generalized transducing phage.

Genes	Cotransduction
a – b	No
a – c	No
a – d	Yes
a – e	No
a – f	Yes
b – c	Yes
b – d	Yes
b – e	Yes
b – f	No
c – d	No
c – e	Yes
c – f	No
d – e	No
d – f	No
e – f	No

What is the order of these four genes on the phage chromosome?

43. In *E. coli*, the three loci *ara, leu,* and *ilvH* are within 0.5-min map distance apart. To determine the exact order and relative distance, the prototroph (*ara⁺ leu⁺ ilvH⁺*) was infected with transducing phage P1. The lysate was used to infect the auxotroph (*ara⁻ leu⁻ ilvH⁻*). The *ara⁺* classes of transductants were selected to produce the following data:

ara^+	ara^+	ara^+	ara^+
leu^-	leu^+	leu^-	leu^+
$ilvH^-$	$ilvH^-$	$ilvH^+$	$ilvH^+$
32	9	0	340

Outline the specific techniques used to isolate the various transduced classes. What is the gene order and what are the relative cotransduction frequencies between genes? Why do some classes occur so infrequently?

44. Consider this portion of an *E. coli* chromosome:

Three *ara* mutations, *ara-1⁻*, *ara-2⁻*, and *ara-3⁻*, are located in the *ara⁻* region. The order of the three *ara⁻* mutations with respect to *thr* and *leu* was analyzed by transduction. The donor was always *thr⁺ leu⁺*, and the recipient was always *thr⁻ leu⁻*. Each *ara⁻* mutant was used as a donor in one cross and as a recipient in another; *ara⁺* transductants were selected in each case. The *ara⁺* transductants were then scored for *leu⁺* and *thr⁺*. Based on the following results, determine the order of the *ara⁻* mutants with respect to *thr* and *leu*.

Cross	Recipient	Donor	Ratio: $\dfrac{thr^- \, ara^+ \, leu^+}{thr^+ \, ara^+ \, leu^-}$
1	$ara\text{-}1^-$	$ara\text{-}2^-$	48.5
2	$ara\text{-}2^-$	$ara\text{-}1^-$	2.4
3	$ara\text{-}1^-$	$ara\text{-}3^-$	4.0
4	$ara\text{-}3^-$	$ara\text{-}1^-$	19.1
5	$ara\text{-}2^-$	$ara\text{-}3^-$	1.5
6	$ara\text{-}3^-$	$ara\text{-}2^-$	25.5

45. An *E. coli* strain that is *leu⁺ thr⁺ azi^r* is used as a donor in a transduction with a strain that is *leu⁻ thr⁻ azi^s*. Either *leu⁺* or *thr⁺* transductants are selected and then scored for unselected markers. The following results are obtained:

Selected Marker	Unselected Markers
leu^+	48% azi^r
leu^+	2% thr^+
thr^+	3% leu^+
thr^+	0% azi^r

What is the order of the three loci?

46. Four different *rII⁻* deletion strains of phage T4 were crossed pairwise in *E. coli* strain B to test for their ability to produce wild-type recombinants. The results are shown in the following table. A + = recombinants produced; – = no recombinants produced.

	1	2	3	4
1	–	–	+	+
2		–	–	+
3			–	–
4				–

Draw a deletion map of these mutations.

47. Seven different *rII⁻* mutations (*m* to *s*) were tested in paired crosses with the four deletion mutants described in problem 46. The results are shown in the following table. A + = recombinants produced; – = no recombinants produced.

	Deletion			
Mutant	1	2	3	4
m	+	+	–	+
n	+	–	–	+
o	+	+	+	–
p	+	+	–	–
q	+	–	+	+
r	–	–	+	+
s	–	–	–	+

What is the map location of these mutants?

48. Why might transformation have evolved, given that the bacterium is importing DNA from a dead organism?

Chapter Integration Problem

The hypothetical bacterium *Dudeus maximus* has a genome of 2.5×10^6 bp of B-DNA. Interrupted mating experiments with various Hfr strains have revealed that the *D. maximus* chromosome is 70 min long. The F plasmid in this bacterium is a whopping 250 kbp.

Consider the following Hfr × F⁻ cross:

Hfr genotype: $a^+ \, b^+ \, c^+ \, d^+ \, e^+ \, str^s$ (streptomycin-sensitive)

F⁻ genotype: $a^- \, b^- \, c^- \, d^- \, e^- \, str^r$ (streptomycin-resistant)

This listing of genes does not imply their specific order, although the streptomycin-sensitivity gene is known to be transferred late in this cross. The a^+, b^+, c^+, d^+ and e^+ genes encode enzymes required for the biosynthesis of nutrients A, B, C, D and E, respectively. At various times after the mating is initiated samples were removed, and the matings were disrupted in a blender. The samples were then plated onto five types of agar plates. (A *plus* indicates presence; a *minus* the absence of the substance.)

		Nutrients				
Type	Streptomycin	A	B	C	D	E
1	+	–	+	+	+	+
2	+	+	–	+	+	+
3	+	+	+	–	+	+
4	+	+	+	+	–	+
5	+	+	+	+	+	–

The following table shows the number of colonies that formed on each type of agar at various sampling times. ($x =$ unknown number of minutes)

Number of Colonies on Agar Type

Time (min)	1	2	3	4	5
0	0	0	0	0	0
$x - 2.5$	0	0	0	0	0
x	0	0	10	0	0
$x + 2.5$	0	0	90	0	0
$x + 5$	10	0	170	0	0
$x + 7.5$	70	0	250	0	0
$x + 10$	130	0	330	0	10
$x + 12.5$	190	10	410	0	50
$x + 15$	250	30	410	0	90
$x + 17.5$	250	50	410	1	130
$x + 20$	250	70	410	2	130
$x + 22.5$	250	70	410	2	130
$x + 25$	250	70	410	1	130

a. Determine the order of genes *a, b, c, d,* and *e,* and indicate their approximate location relative to one another on the Hfr chromosome.

b. How many minutes is x? Assume the *oriT* site is in the middle of the F plasmid DNA in the Hfr chromosome. Also assume negligible lengths of time for DNA mobilization in the donor cell, and for DNA recombination in the recipient cell.

c. Approximately how many *micrometers* of chromosomal DNA will be transferred after 25 min of conjugation?

d. Very few colonies formed on agar type 4. Suggest two reasons for this.

e. Suppose *D. maximus* has the following reversion frequencies:

a^- to $a^+ = 1 \times 10^{-5}$
b^- to $b^+ = 2 \times 10^{-6}$
c^- to $c^+ = 3 \times 10^{-5}$
d^- to $d^+ = 1 \times 10^{-6}$
e^- to $e^+ = 1 \times 10^{-7}$

Indicate whether the following statements are true or false. Justify your answers.

 I. About 1 in 10 million e^- cells will spontaneously become prototrophic.

 II. $a^- b^-$ to wild-type revertants $> c^- d^-$ to wild-type revertants.

A 40-bp portion of the *oriT* sequence is shown here. The arrow indicates the position of the nick site.

5′ CAGGTTGGTG CTTCTCACCA CCAAAAGCAC CACACCGGTC 3′
3′ GTCCAACCAC GAAGAGTGGT GGTTTTCGTG GTGTGGCCAG 5′

f. This DNA segment has two interesting features. Can you spot them?

g. The transfer of the F plasmid sequence involves a rolling circle replication mechanism. What are the first 5 nucleotides that are transferred from the donor to the recipient cell during conjugation? What is your reasoning?

h. The Hfr cells were first grown in a medium containing ^{15}N until all cellular DNA was heavy. The Hfr × F⁻ cross was then carried out in a medium with only ^{14}N. Show this DNA region in the donor cell after DNA replication, indicating the heavy (H) and light (L) strands or sequences.

Do you need additional review? Visit **www.mhhe.com/hyde** for practice tests, answers to end-of-chapter questions and problems, interactive exercises, and animation tutorials all designed to enhance your understanding of key genetic concepts.

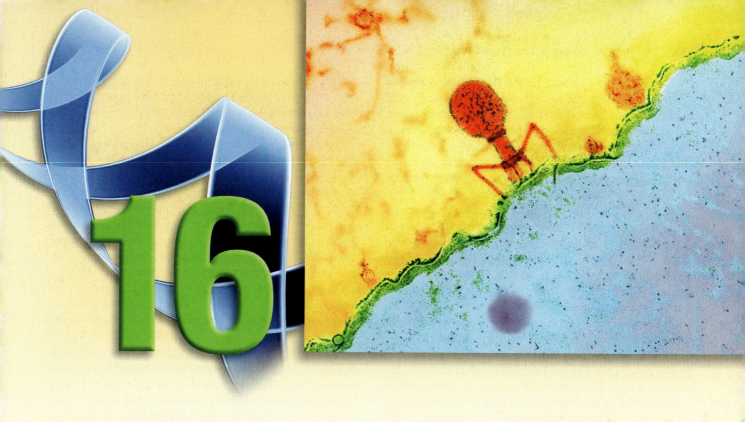

16

Gene Expression: Control in Bacteria and Phages

Essential Goals

1. Distinguish between inducible and repressible operons and describe the difference between positive and negative regulation.

2. Diagram how the *lac* operon is regulated by both *cis-* and *trans-*acting functions. Describe how expression of the *lac* operon will be affected by different genotypes in merozygotes.

3. Understand the similarities and differences in regulation of the *lac, ara,* and *trp* operons and how these differences relate to the biochemical pathway encoded by the structural genes.

4. Understand how phage λ regulates the choice between lysogeny and lysis.

5. Describe additional transcriptional, translational, and posttranslational regulatory mechanisms observed in bacteria and bacteriophages.

Photo: Pseudocolored electron micrograph of a bacteriophage attached to a bacterial cell wall.

Chapter Outline

521

*B*ecause of their simplicity, bacteria and bacteriophages are highly useful for understanding basic biological and genetic processes. For example, bacterial genes do not have to be regulated according to tissue or organ type, as we will see in higher eukaryotes. Nevertheless, not all bacterial genes in an organism are expressed at the same time and at the same level. In fact, regulation of gene expression for these simple organisms is very important.

Gene regulation plays an essential role in the life of any cell, including bacteria and bacteriophages. The bacterial cell needs to recognize whether the environment is conducive for large amounts of cell growth and then turn on specific genes and turn off other genes to stimulate rapid growth. The cell needs to recognize when nutrients are scarce and regulate the expression of the necessary genes to significantly slow growth.

As you will see in the discussion of bacteriophage λ, a temperate phage, the decision to become either a prophage (and integrate into the *E. coli* genome) or to enter the lytic cycle is the result of a simple competition between two λ proteins that bind to the same DNA sequences. If the first protein wins the DNA-binding competition, then λ becomes a prophage. If the second protein wins, then λ enters the lytic cycle. Thus, this binary decision is a result of gene regulation.

Transcriptional regulation involves altering the ability of RNA polymerase to bind the promoter of a gene and initiate transcription. Once transcription begins, the rate of RNA extension is relatively constant. There are three key components to transcriptional regulation. First is a regulatory protein that will either enhance or inhibit RNA polymerase binding to the promoter. Second is the specific DNA sequence that the regulatory protein will bind. In bacteria, this sequence is usually located either within or near the promoter. Third is a signal molecule that will be bound by the regulatory protein. This signal molecule will affect the ability of the regulatory protein to bind to the DNA sequence.

Regulation of expression does not stop once the gene is transcribed. Posttranscriptional levels of regulation also exist in bacteria, although they are more common in eukaryotes. One common point of eukaryotic posttranscriptional regulation is the transport of mRNA out of the nucleus to enable its translation. In bacteria, however, transcription and translation occur simultaneously, which precludes the need to transport the mRNA to the proper compartment for translation.

Posttranslational regulation can also be an important strategy; in general, posttranslational mechanisms result in the conversion of a protein to either an active or inactive form and the transport of the protein to the proper location within the cell.

Many biotechnology and pharmaceutical companies are interested in using bacteria to produce specific foreign proteins. These proteins could be used to immunize humans or animals against particular diseases, or they could be used as therapy, such as the expression of insulin in *E. coli* for human use in the treatment of diabetes. For these proteins to be properly expressed in bacterial cells and remain functional for later use, the expression of the protein must be properly regulated. Expressing too little protein can result in high production expense. Expressing too much protein can result in the bacterial cells either degrading the foreign protein or dying. Understanding the regulation of bacterial gene expression is therefore of immense importance to the health of eukaryotes, including humans.

16.1 The Operon Model

Gene expression involves a complicated interplay of signals and enzymes, geared to the needs of the bacterial cells and the environment in which they are living. Here we summarize two types of gene expression systems: namely inducible systems and repressible systems. We also briefly define the difference between positive and negative regulation. These topics are discussed in greater detail in the rest of this chapter.

Inducible Systems: Keyed to Substrates

Some genes are transcribed only under specific conditions, such as in the presence of a specific nutrient in the environment. These genes are known as **inducible genes** (fig. 16.1 left). Inducible genes often encode enzymes that catabolize (break down) a molecule, and that molecule also serves to induce or activate transcription of the inducible gene. The molecule that activates gene expression is called the **inducer.** The best-studied inducible system is the *lac* operon in *E. coli,* which encodes the enzymes that metabolize lactose.

Repressible Systems: Keyed to End Products

In contrast, enzymes in many synthetic pathways exist in low concentration or are completely absent when an adequate quantity of the end product of the pathway is already available to the cell. That is, if a bacterium encounters an abundance of the amino acid tryptophan in the environment, or if it is overproducing tryptophan, then the cell stops manufacturing additional tryptophan until a need arises again. A **repressible gene** is one whose transcription is stopped or slowed when its expression is no longer needed (fig. 16.1 right). Transcription of a repressible system is reduced or terminated by a molecule called the **repressor,** which is often the end product of the biosynthetic pathway that is controlled by the enzymes encoded by the regulated genes.

Positive and Negative Regulation

As you will see, genes can be under positive regulation, negative regulation, or both. **Positive regulation** occurs when a regulatory molecule activates tran-

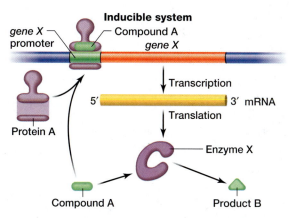

Inducible system

Compound A must be present and interact with protein A to bind the promoter and to transcribe *gene X* at the maximal rate.

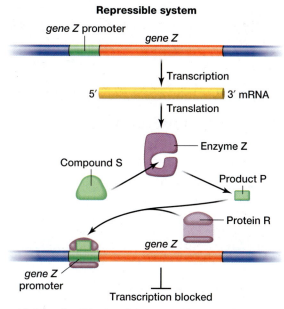

Repressible system

High levels of Product P interact with protein R to bind the promoter and block transcription of *gene Z*.

figure 16.1   General diagram of inducible and repressible systems. An inducible system (left) is maximally expressed when the substrate (compound A) is present. Protein A binds compound A to form a complex that binds the promoter to activate transcription of *gene X*. The enzyme encoded by *gene X* acts on compound A and converts it to product B. Thus, when compound A is present, transcription of *gene X* is induced, which produces more enzyme X to act on compound A. A repressible system (right) is one that is inactivated when increased amounts of product (product P) are present. Rather than continuing to make large amounts of product P, product P binds protein R that then binds the promoter to block transcription of *gene Z*, which reduces the amount of enzyme Z and reduces the geneation of product P.

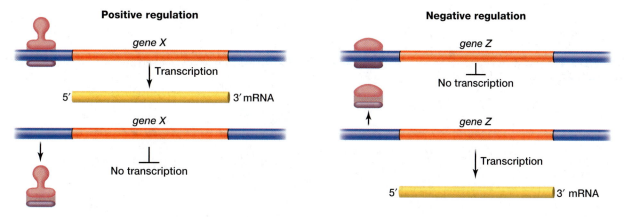

figure 16.2   Positive and negative regulation of transcription. Positive regulation (left) occurs when the binding of a protein (activator) to the DNA activates transcription of *gene X*. Negative regulation (right) occurs when the binding of a protein (repressor) to the DNA blocks transcription of *gene Z*. Releasing the repressor from the DNA leads to activation of *gene Z*.

scription (**fig. 16.2**, left). A **positive regulator** or activator is a protein that binds to DNA and, in turn, helps RNA polymerase bind to the promoter. The positive regulator may act by making the promoter sequence more accessible to RNA polymerase. Alternatively, the positive regulator may directly interact with the RNA polymerase to help it bind to the promoter.

Negative regulation occurs when the regulatory molecule blocks transcription (see fig. 16.2, right). A **negative regulator** or repressor is a protein that binds to DNA and blocks RNA polymerase binding to the promoter or reduces the ability of the RNA polymerase

to initiate transcription. The terms *positive* and *negative regulation* pertain to the effect that a regulatory protein has on transcription when it binds to DNA. As you will see with the *lac* operon, an inducible system can exhibit both positive and negative regulation.

The binding of either a positive or negative regulator to the DNA is controlled by a signaling molecule. The signaling molecule functions by binding to a specific region of the regulatory protein, which is called the **allosteric site.** Binding to the allosteric site causes the regulatory protein to change its shape or conformation (**fig. 16.3**), which affects the ability of the regulatory protein

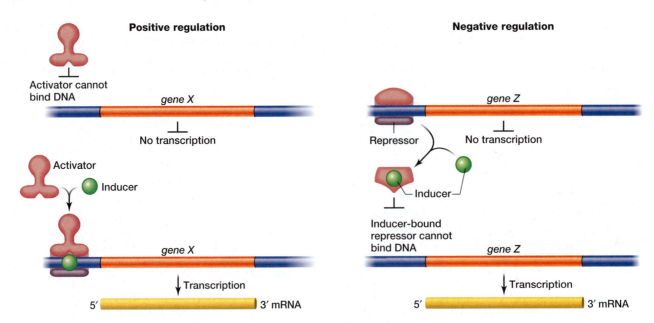

figure 16.3 General description of allostery for a DNA binding protein. The DNA binding protein has two sites, one binds the DNA and the second site binds a small molecule, the inducer. Binding the inducer causes a conformational or allosteric change in the DNA binding site. If the gene is under positive regulation (left), the DNA binding site only recognizes the DNA sequence when the inducer is bound. The activator-inducer complex binds the DNA and activates transcription. If the gene is under negative regulation (right), binding the inducer causes the DNA binding site to change conformation and the repressor-inducer complex releases the DNA and transcription occurs. Positive regulation still requires activator binding the DNA for transcription to occur, while negative regulation still requires repressor releasing the DNA for transcription to occur. In both models, however, the presence of the inducer leads to transcription.

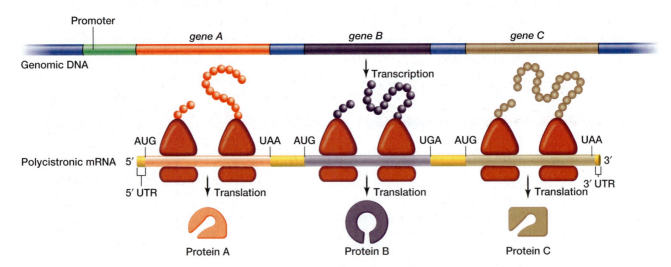

figure 16.4 General description of an operon. Multiple genes (A, B, and C) are expressed under a single regulatory mechanism (promoter) and transcribed on a single polycistronic mRNA. Each open reading frame is then translated into a different protein from the single mRNA.

to bind DNA and control transcription. Binding of some signaling molecules allow the regulatory protein to bind DNA, whereas other regulatory proteins bind the signaling molecule and are then unable to bind DNA.

It is important to remember that transcriptional regulation involves regulatory proteins binding to the DNA and affecting the binding of RNA polymerase to the promoter of a gene and initiating transcription. The effect of binding the signaling molecule determines if the regulatory protein will bind the DNA. Thus, the presence or absence of the signaling molecule will indirectly affect the transcription of the gene by affecting the binding of the regulatory protein to the DNA. The signaling molecule, however, does not change the inherent role of the regulatory protein; it cannot make a negative regulator a positive regulator or vice versa.

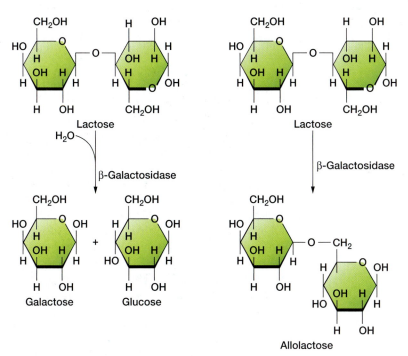

figure 16.5 The role of β-galactosidase. The enzyme β-galactosidase can either cleave lactose into glucose and galactose (left) or convert lactose to allolactose (right).

Definition of Operon

Because many enzymes are often required in a catabolic or biosynthetic pathway, and because a gene encodes only one polypeptide, regulation of these genes must occur together. For example, all the genes encoding enzymes that are involved in synthesizing tryptophan must be similarly regulated as the concentration of tryptophan changes. Bacteria often accomplish this regulatory coordination by grouping the genes together.

An **operon** is a group of genes that are coordinately regulated and are encoded on a single mRNA (**fig. 16.4**), which is called a *polycistronic mRNA* (see chapter 15 for a discussion of cistrons). An operon contains a promoter, DNA sequences that can bind regulatory proteins, and multiple genes that are transcribed onto a single mRNA molecule.

We begin by examining the *lac* operon as an example.

16.2 The Lactose Operon (Inducible System)

Lactose is a disaccharide that *E. coli* can use for energy and as a carbon source after it is broken down into glucose and galactose. The enzyme that cleaves lactose is β-**galactosidase** (**fig. 16.5**). (The enzyme can also convert lactose to allolactose, which, as we will see, is also

important.) A wild-type *E. coli* cell grown in the absence of lactose contains very few molecules of β-galactosidase. Within minutes after adding lactose to the medium, however, large amounts of this enzyme appear within the bacterial cell, which suggests that expression of this enzyme is induced.

The Structural Genes

When the synthesis of β-galactosidase, encoded by the *lacZ* gene, is induced, the production of two additional enzymes is also induced. The **lac permease** enzyme, encoded by the *lacY* gene, is involved in transporting lactose into the cell. The **transacetylase** enzyme, encoded by the *lacA* gene, is believed to protect the cell from the buildup of toxic products created by β-galactosidase acting on other galactosides. By acetylating galactosides other than lactose, the transacetylase prevents β-galactosidase from cleaving them.

Not only are these three *lac* genes (*lacZ, lacY, lacA*) induced together, but they are adjacent to one another in the *E. coli* genome; they are, in fact, transcribed on a single, polycistronic mRNA (**fig. 16.6**).

The Regulatory Gene

Inducing expression of these three genes involves the protein encoded by another gene, called the *lac* **repressor gene**, or *lacI* gene. The *lacI* gene encodes the repressor protein, which is a negative regulator of the transcription of the three structural *lac* genes. Although the *lacI* gene is located adjacent to the three structural *lac* genes (see fig. 16.6), it is a totally independent transcriptional entity. The *lacI* gene possesses its own promoter and is not transcribed in the same polycistronic mRNA as the three structural *lac* genes. Thus, the *lacI* gene is not a part of the *lac* operon, although it is essential for the proper regulation of the *lac* operon.

The Promoter and Operator

For the *lacI* repressor protein to exert its influence over transcription, a control element (DNA sequence) must be located near the beginning of the β-galactosidase (*lacZ*) gene to which the repressor can bind. This control element is referred to as the **operator,** or operator site (see fig. 16.6). When the repressor is bound to the operator, it either interferes with RNA polymerase binding or prevents the RNA polymerase from achieving the open promoter complex (see chapter 10). In either case, transcription of the operon is prevented

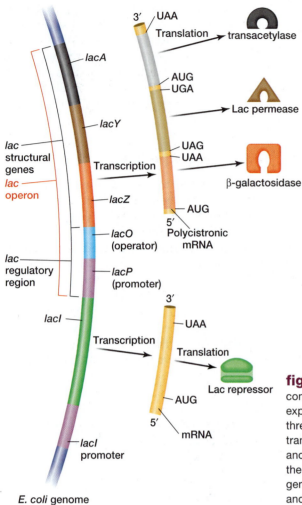

figure 16.6   Organization of the *lac* operon. The *lac* operon is composed of a promoter (*lacP*) and operator (*lacO*) that regulates the expression of the polycistronic *lac* mRNA that is transcribed through the three structural genes, *lacZ, lacY,* and *lacA*. This polycistronic mRNA is translated into three individual proteins, β-galactosidase, Lac permease, and transacetylase. Transcription of the *lac* operon is also regulated by the Lac repressor protein, which is encoded by the *lacI* gene. The *lacI* gene is not part of the operon because it contains a separate promoter and is not transcribed as part of the polycistronic mRNA.

or repressed (**fig. 16.7**) which represents a negative regulatory system.

The repressor is released from the operator when it combines with an *inducer,* which in this case is a derivative of lactose called allolactose (see fig. 16.5). In the laboratory, a synthetic analog of allolactose is used—isopropyl-β-D-thiogalactosidase (IPTG)—that also binds the *lac* repressor to induce expression of the *lac* operon.

The nucleotide sequence of the *lac* operator and promoter region is shown in **figure 16.8**. The operator in figure 16.7 is referred to as the primary operator, O_1, centered at +11 in figure 16.8. Two other operator sequences also exist. One, O_2, is centered at +412. The third, O_3, overlaps the C-terminal end of the *lacI* gene and is centered at −82. To achieve maximum repression of the *lac* operon transcription, all three operator sites must be bound by repressor. The *lac* promoter, which contains both −10 and −35 sequences (see chapter 10),

figure 16.7   General regulation of the *lac* operon. In the absence of lactose (left), the *lacI* gene is transcribed and the Lac repressor is translated. The Lac repressor binds the *lac* operator (*lacO*), which blocks transcription of the *lac* operon (negative regulation). In the presence of lactose (and the absence of glucose, right), the Lac permease permits lactose to enter the *E. coli* cell. Some of the lactose is converted to allolactose (inducer), which binds the Lac repressor and prevents it from binding *lacO*. This induces transcription of the *lac* operon as a polycistronic mRNA, which is then translated into the three proteins: β-galactosidase, Lac permease, and transacetylase.

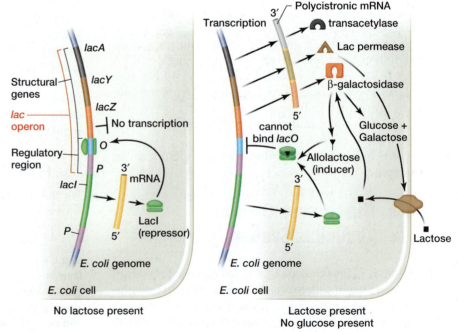

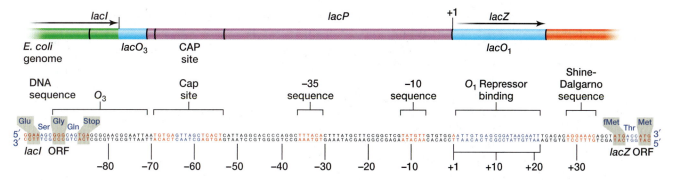

figure 16.8   The *lac* operon promoter and operator regions. A greater detailed view of the *lac* operon regulatory region is shown above the corresponding nucleotide sequence. The carboxyl terminal amino acids encoded in the *lacI* gene are shown on the left overlapping *lacO₃*, as well as the initial amino acids of the *lacZ* gene on the right. In addition, the Shine-Dalgarno sequence, the –10 and –35 sequences of the *lac* promoter (*lacP*), and two of the three Lac repressor binding sites (*lacO₁* and *lacO₃*) are shown. The CAP site is described later in this chapter.

and the catabolite activator protein (CAP) binding site are located between O_1 and O_3.

Repressor Binding

The structure of the repressor and its interaction with the operator sites was worked out in 1996 using X-ray crystallography. The functional repressor is a *homotetramer* of the LacI protein; that is, it is formed from four identical copies of the repressor protein. Because each operator site has two-fold symmetry, two repressor monomer proteins bind to each operator site. The monomer is shaped so that it fits into the major groove of the DNA to locate the exact base sequence of the operator; it then binds at that point through electrostatic forces.

A repressor tetramer can bind two operator sites at the same time, presumably O_1 and O_3 or O_1 and O_2. If O_1 and O_3 are bound by the repressor tetramer, the intervening DNA, which includes the *lac* promoter, is formed into a repressor loop (**fig. 16.9**). Positioning the promoter within the inside of this loop makes it inaccessible to RNA polymerase, which prevents transcription. Thus, transcription of the structural genes within the *lac* operon are under the control of a promoter and operator, which lie very close together in the regulatory region.

Induction of the *lac* Operon

Under conditions of repression, the repressor must be removed from the operator before the operon can be "turned on" to express the lactose-utilizing enzymes. The repressor is an **allosteric protein**: Binding one specific molecule changes the shape of the repressor protein, which changes its ability to interact with a second molecule. Here the first molecule is the inducer allolactose (or IPTG), and the second molecule is the operator DNA sequence. When allolactose is bound

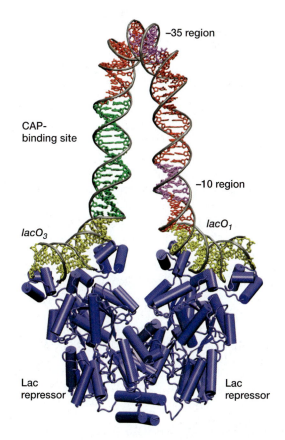

figure 16.9   Model of the functional Lac repressor binding to DNA. The Lac repressor forms a homotetramer with one dimer binding to *lacO₁*, and the second dimer binding *lacO₃*, which causes the DNA between the two operator sequences to form a loop. The N-terminus of each repressor subunit binds a portion of the operator sequence, while the C-terminal ends form tails that join the subunits together. Also indicated are the –10 and –35 *lacP* sequences and the CAP binding site that is discussed later in this chapter.

to the repressor, the repressor undergoes a change in shape that results in a decreased affinity for the operator sequences (see fig. 16.7).

With allolactose bound to the repressor, the **affinity,** or ability of the repressor to bind to the operator, is greatly reduced—by a factor of 10^3. Because the repressor is not bound to the operator by covalent bonds, the repressor-inducer complex simply dissociates from the operator. After the repressor leaves the operator, the repressor loop is lost (see fig. 16.9) and RNA polymerase is able to bind the promoter and begin transcription. The three *lac* structural genes are then transcribed and subsequently translated into their respective enzymes.

This system of control is very efficient. The presence of the lactose molecule permits transcription of the genes of the *lac* operon, which act to break down the lactose. After all the lactose is metabolized, the repressor dissociates from the allolactose, returning to its original shape and again binding the operator to block transcription. Using very elegant genetic analysis, the details of this system were worked out by François Jacob and Jacques Monod, who subsequently shared the 1965 Nobel Prize in Physiology or Medicine for their efforts.

Restating the Concepts

▶ An operon is a group of genes that are coordinately regulated by a single promoter and operator and are transcribed on a single polycistronic mRNA.

▶ The *lac* operon consists of a promoter, an operator, and three structural genes that encode the enzymes involved in lactose metabolism. A separate gene, *lacI*, encodes a repressor that binds the operator to negatively regulate transcription of the operon.

▶ An inducer, either allolactose (which is derived from lactose) or IPTG, can bind the Lac repressor protein, which causes a conformational change in the protein. This new shape can no longer bind the operator, so the repressor-inducer complex is released from the operator DNA sequence and transcription of the *lac* operon is initiated.

Lactose Operon Mutants

Discovery and verification of the *lac* operon system came about through the use of mutants and partial diploids of the *lac* operon, well before DNA sequencing techniques were developed. The structural genes of the *lac* operon all have known mutant forms in which the particular enzyme lacks its activity. These mutant forms are designated *lacZ⁻*, *lacY⁻*, and *lacA⁻*, and the wild-type alleles that encode the functional enzymes are designated *lacZ⁺*, *lacY⁺*, and *lacA⁺*.

Merozygote Formation

Partial diploids in *E. coli* can be created through sexduction using strains of *E. coli* that have the *lac* operon incorporated into an F' factor (see chapter 15). Because F' strains can pass the F' DNA into F⁻ strains, *lac* operon diploids (also called merozygotes, or partial diploids) are formed. By careful manipulation, various combinations of mutations can be examined in the diploid state.

Constitutive Mutants

Jacob and Monod identified a variety of mutants that revealed the regulatory mechanisms used by the *lac* operon. One class of mutants were constitutive mutants, which resulted in the three *lac* structural genes being transcribed at all times. In other words, they were transcribed even in the absence of lactose or an inducer molecule.

Inspection of figure 16.7 suggests two different mutations that would lead to constitutive production of the enzymes. First, a defective repressor would be unable to bind the operator and would fail to repress transcription. This inactive repressor mutation is designated *lacI⁻*. Second, a mutant operator sequence would not be recognized by the wild-type repressor, and, therefore, the repressor could not bind. This constitutive operator mutation is designated *lacOᶜ*. Both types of mutants can produce the same phenotype: constitutive expression of the three *lac* operon structural genes.

Distinguishing Between Constitutive Mutants

When a new constitutive mutant is isolated, researchers want to determine whether it is a *lac* repressor or *lac* operator mutation. The Jacob and Monod model predicts different modes of action for the two types of mutations. We can test those different modes by examining merozygotes in which the genomic copy of the operon contains wild-type alleles for one or two of the structural genes (e.g., *lacZ⁺ lacY⁻ lacA⁺*), and the F' copy contains the wild-type alleles for the remaining structural genes (e.g., *lacZ⁻ lacY⁺ lacA⁻*). The expression of a specific wild-type structural gene allows us to determine whether it is transcribed from the F' element or the genome.

In merozygotes, a constitutive operator mutation affects only the operon it is physically associated with. Thus, only the wild-type structural genes that are downstream of the *lacOᶜ* mutation are constitutively expressed. Operator mutations are therefore called **cis-dominant.** (To review, if two mutations are on the same piece of DNA, they are in the *cis* configuration. If they are on different pieces of DNA, they are in the *trans* configuration.)

In contrast, a constitutive *lacI⁻* mutation, which encodes an inactive repressor, is recessive to a wild-

figure 16.10 Effect of the *lacI⁻* mutation on *lac* operon expression. (a) The *lacI⁻* mutation encodes a defective repressor that cannot bind the *lac* operator (*lacO*), which results in the expression of the three *lac* structural genes in the absence or presence of the inducer. (b) In a merodiploid, the genomic copy of the *lacI⁻* gene encodes a defective Lac repressor, while the *lacI⁺* gene in the F' encodes a functional repressor. In the absence of an inducer (allolactose or IPTG), the functional repressor binds *lacO* in both the genome and F' and inhibits transcription of both copies of the operon. In the presence of an inducer, the functional repressor binds the inducer, which prevents it from binding *lacO* and results in transcription. The non-functional repressor from the genomic *lacI⁻* gene cannot bind the operator. A table showing the expression of the functional *lacZ⁺* (genomic copy) and the *lacY⁺* (F' copy) is shown.

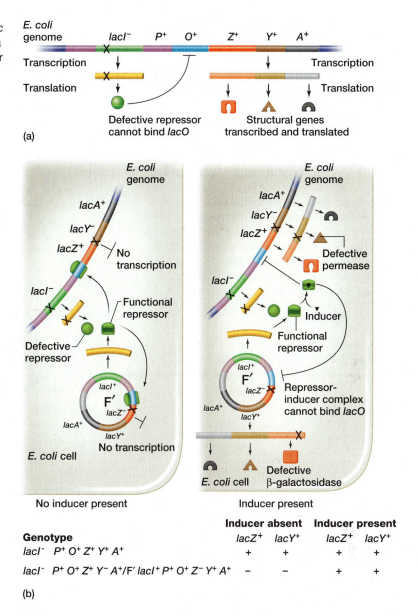

Genotype	Inducer absent		Inducer present	
	$lacZ^+$	$lacY^+$	$lacZ^+$	$lacY^+$
$lacI^-\ P^+\ O^+\ Z^+\ Y^+\ A^+$	+	+	+	+
$lacI^-\ P^+\ O^+\ Z^+\ Y^-\ A^+/F'\ lacI^+\ P^+\ O^+\ Z^-\ Y^+\ A^+$	–	–	+	+

(b)

type repressor gene in a merozygote, regardless of which operon (genomic or F' factor) the mutation is on. Thus, the wild-type LacI repressor is a ***trans*-acting** function. *Trans*-acting alleles usually work through a protein product that diffuses through the cytoplasm. ***Cis*-acting** mutants are changes in recognition sequences on the DNA.

Example of the Two Constitutive Mutations

In **figure 16.10a,** the bacterium has a constitutive repressor mutation (*lacI⁻*) that results in the cell constitutively transcribing the operon. If the wild-type *lacI⁺* allele is introduced on an F' plasmid, the normal (inducible) phenotype is restored (fig. 16.10b).

In this merozygote, the *lacI⁺* allele on the F' is dominant to the genomic *lacI⁻* mutation, and the wild-type repressor regulates both the genomic and F' oper-

ons. We know this because *lacZ⁺* is induced from the genomic copy of the operon, and *lacY⁺* is induced from the F'. Because both operons are inducible, the *lacI⁺* allele on the F' is *trans*-acting.

In **figure 16.11a**, the genomic copy of the operon carries the constitutive operator mutation (*lacOᶜ*) that results in the constitutive transcription of the operon. When a wild-type operator is introduced into the cell on an F' plasmid, the genomic operon remains constitutively expressed, but the operon on the F' remains inducible (fig. 16.11b). This difference in regulation is confirmed by the constitutive expression of the *lacZ⁺* allele from the genome and the inducible expression of *lacY⁺* from the F'. The *lacOᶜ* allele is *cis*-dominant because the constitutive phenotype is observed in a cell that also contains the wild-type *lacO⁺* allele, but only for the operon that is physically linked to the *lacOᶜ* mutant operator.

figure 16.11 Effect of the *lacO^c^* mutation on *lac* operon expression. (a) The *lacO^c^* mutation changes the *lac* repressor binding site, such that no form of the repressor can bind this mutant operator. This results in continual transcription of the three *lac* structural genes. (b) In a merodiploid, the genomic copy of *lacO^c^* encodes a mutant operator sequence, while the *lacO^+^* in the F' possesses a wild-type sequence that can be bound by the repressor. In the absence of an inducer (allolactose or IPTG), the functional repressor binds only the *lacO^+^* in the F' (*trans*-acting effect), which results in the transcription of only the genomic copy of the operon that is adjacent to the *lacO^c^* mutation (*cis*-acting effect). In the presence of an inducer, the functional repressor binds the inducer, which prevents it from binding *lacO^+^* in the F'. Thus, neither operator is bound and both operons are transcribed. A table showing the expression of the functional *lacZ^+^* (genomic copy) and the *lacY^+^* (F' copy) is shown.

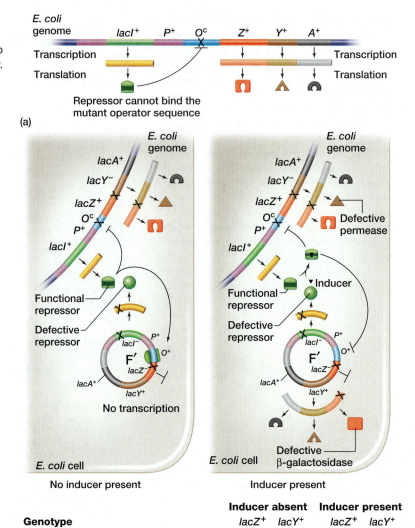

Genotype	Inducer absent		Inducer present	
	*lacZ^+^*	*lacY^+^*	*lacZ^+^*	*lacY^+^*
*lacI^+^ P^+^ O^c^ Z^+^ Y^+^ A^+^*	+	+	+	+
*lacI^+^ P^+^ O^c^ Z^+^ Y^− A^+^/F' lacI^− P^+^ O^+^ Z^− Y^+^ A^+^*	+	−	+	+

(b)

Other lac *Operon Control Mutations*

Other mutations have also been discovered that support the Jacob and Monod operon model. Two particularly interesting mutations are the *lacI^s^* superrepressor mutation and the *lacP^−^* promoter mutation. Both of these mutations are phenotypically similar in that they are unable to induce expression, but the mechanisms that they employ further demonstrate the concepts of *cis*- and *trans*-acting functions.

The *lacI^s^* **superrepressor mutation** represses the operon even in the presence of large quantities of the inducer (**fig. 16.12a**). Unlike the *lacI^−^* mutation that fails to bind the operator, the superrepressor has lost the ability to bind the inducer. Basically, the *lacI^s^* superrepressor acts as a constant repressor rather than as an allosteric protein.

The genetic interaction of the superrepressor with the wild-type repressor can be examined in a *lacI^s^/ lacI^+^* merozygote (fig. 16.12b). Although the wild-type repressor possesses the ability to bind the inducer and activate expression of the operon, the superre-

pressor will always be bound to the operator. Because the superrepressor is a *trans*-acting protein, it binds to the operator and constitutively represses transcription of both operons regardless of whether the inducer is present. Therefore, the *lacI^s^* mutation is dominant to the wild-type *lacI^+^* allele, and *lacI^+^* is dominant to the *lacI^−^* mutation. Notice that the *lacI^−^* and *lacI^s^* mutations are consistent with the LacI repressor protein possessing two binding sites; the inducer binding site and the DNA binding region.

Like the *lacI^s^* mutation, the *lacP^−^* **promoter mutation** blocks transcription of the *lac* operon either in the presence or absence of the inducer. Genetic crosses, however, revealed that the mechanism for this phenotype is quite different from that of the *lacI^s^* mutation. The *lacP^−^* mutation is due to changes in either the −10 or −35 region of the *lac* promoter, which results in RNA polymerase being unable to recognize and bind the mutant promoter sequence (**fig. 16.13a**).

figure 16.12 Effect of the *lacI*[s] mutation on *lac* operon expression. (a) The *lacI*[s] mutation (superrepressor) encodes a defective repressor that cannot bind the inducer, which results in repressor constantly being bound to *lacO*[+], regardless of the presence or absence of the inducer. Thus, the three *lac* structural genes are not transcribed. (b) In a merodiploid, the genomic copy of the *lacI*[s] gene encodes a superrepressor that cannot bind the inducer, while the *lacI*[+] gene in the F' encodes a functional repressor. In the absence of inducer, both the functional and superrepressors bind *lacO*[+] in both the genome and F' and inhibit transcription of both copies of the operon. In the presence of inducer, the functional repressor binds the inducer and releases from *lacO*[+]. However, the superrepressor, which cannot bind the inducer, binds both operators (*trans*-acting effect) and blocks transcription of both operons. A table showing the expression of the functional *lacZ*[+] (genomic copy) and the *lacY*[+] (F' copy) is shown.

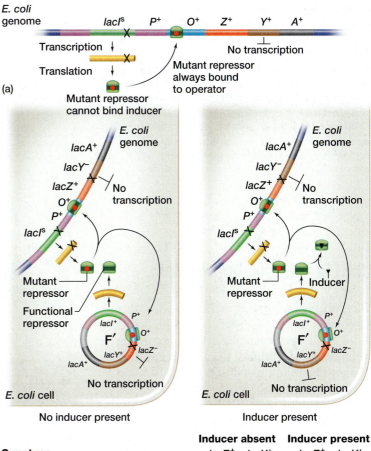

Genotype	Inducer absent		Inducer present	
	lacZ[+]	*lacY*[+]	*lacZ*[+]	*lacY*[+]
lacI[s] *P*[+] *O*[+] *Z*[+] *Y*[+] *A*[+]	–	–	–	–
lacI[s] *P*[+] *O*[+] *Z*[+] *Y*[−] *A*[+]/F' *lacI*[+] *P*[+] *O*[+] *Z*[−] *Y*[+] *A*[+]	–	–	–	–

(b)

We can again examine the genetic interaction of the *lacP*[−] promoter using the wild-type *lacP*[+] promoter in a *lacP*[−]/*lacP*[+] merozygote (fig. 16.13b). In this merozygote, the operon that contains the *lacP*[+] promoter can be induced to transcribe the operon, but the operon containing the *lacP*[−] promoter cannot. We see again, therefore, that a DNA sequence, the *lac* promoter, exhibits a *cis*-acting function on the expression of the operon in a merozygote.

More complex genotypes can be examined, and the expression of the *lac* structural genes can be predicted based on understanding the mutations described earlier. In **Box 16.1**, we take a more detailed look at how to examine *lac* merozygotes and deduce their expression.

It's Your Turn

We discussed various regulatory mutations in the *lac* operon and how they interact to affect transcription. You have also been presented with a methodical approach to determine how transcription is affected in a merozygote. You can now try to apply these tools in examining the merozygotes in Problems 16 and 17.

 Restating the Concepts

▶ Merozygotes provide a convenient way to examine the genetic properties of mutations in bacteria.

▶ *Cis*-acting functions are usually DNA sequences, such as the promoter and operator in the *lac* operon. *Trans*-acting functions are usually proteins that can diffuse throughout the cell, like the LacI repressor.

▶ The LacI[s] superrepressor, which lacks the inducer binding site, is dominant to the wild-type repressor. The wild-type repressor in turn is dominant to the LacI[−] repressor, which lacks the ability to bind the operator sequence.

▶ The *lacO*[c] mutation possesses an altered operator sequence that both the superrepressor and wild-type repressor are unable to bind. The *lacP*[−] mutation has an altered promoter sequence that RNA polymerase is unable to recognize and bind.

figure 16.13 Effect of the *lacP⁻* mutation on *lac* operon expression. (a) The *lacP⁻* mutation changes the promoter sequence so it is not recognized by RNA polymerase, which prevents transcription of the *lac* operon. (b) The genomic copy of the *lac* operon contains the *lacP⁻* mutation and the F' factor contains a wild-type *lacP⁺* sequence. The *lacP⁻* mutation in the bacterial genome results in no transcription of the genomic copy of the *lac* operon, while the *lac* structural genes in the F' element are expressed from the *lacP⁺* only under induced conditions (right panel).

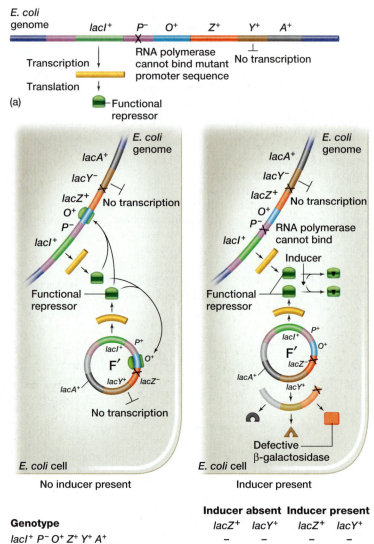

16.3 Catabolite Repression

You probably realized from the preceding discussion that the *lac* repressor is a negative regulator of *lac* operon transcription. According to the earlier definition, negative regulation occurs when binding of the regulatory protein to DNA blocks transcription. Another mechanism is involved in regulating the *lac* operon, one that is shared by other operons involved in catabolizing certain sugars (e.g., arabinose, galactose). This mechanism is called **catabolite repression,** and it involves **cyclic-AMP (cAMP)** (**fig. 16.14**).

Enzyme and Repressor Interactions

The rationale underlying this regulation is that catabolizing lactose produces glucose and galactose. The galactose must then be converted into glucose before it can be efficiently used by the cell. As a result, the cell must expend energy to convert lactose completely into glucose. Because glucose is the preferred energy source, the cell tries to repress transcription of certain operons that encode enzymes that catabolize other sugars if glucose is present—thus the term *catabolite repression.*

In eukaryotes, cAMP was known to be a *second messenger,* an intracellular signaling molecule that is regulated by certain extracellular hormones. Geneticists were surprised to discover cAMP also functions as a signaling molecule in bacteria.

When glucose is plentiful, the enzyme adenyl cyclase is inhibited in *E. coli* (see fig. 16.14). Because adenyl cyclase catalyzes the conversion of ATP to cAMP, the inhibition of this enzyme blocks the generation of additional cAMP and results in the eventual decline of cAMP levels. Thus, when glucose is abundant in the extracellular environment, the amount of intracellular cAMP is low; when

Genotype	Inducer absent		Inducer present	
	lacZ⁺	*lacY⁺*	*lacZ⁺*	*lacY⁺*
lacI⁺ P⁻ O⁺ Z⁺ Y⁺ A⁺	−	−	−	−
lacI⁺ P⁻ O⁺ Z⁺ Y⁻ A⁺/F' lacI⁺ P⁺ O⁺ Z⁻ Y⁺ A⁺	−	−	−	+

(b)

glucose is scarce in the environment, the intracellular cAMP concentration is high.

The Catabolite Repression Mechanism

As we already discussed, a signaling molecule will be bound by a regulatory protein to affect its binding to DNA. In catabolite repression, cAMP is the signaling molecule. The regulatory protein that binds cAMP is called the **catabolite activator protein (CAP).** Whereas the LacI repressor is a negative regulator for only the *lac* operon, the CAP protein is a positive regulator of several operons that encode enzymes that catabolize a variety of sugars, including lactose.

We can now examine how cAMP and CAP regulate expression of the *lac* operon. In the absence of glucose, the amount of cAMP is high, and cAMP combines with CAP. The CAP–cAMP complex binds to the distal part

16.1 applying genetic principles to problems

Deducing Expression of *lac* Structural Genes in Complex Merozygotes

A common genetics problem is to be given a genotype for an *E. coli* merozygote that is diploid for the *lac* operon and contains a variety of *lac⁻* gene mutations; your task is to determine under which conditions (induced or uninduced) each of the three structural genes will be expressed.

A complex genotype may mean complex reasoning, but the rationale for working through this type of problem is simple—especially if you understand the relationship between *cis*- and *trans*-acting mutations, and you know the basic function of the various mutations. Let's look at the following genotype:

lacI⁻ lacOᶜ lacZ⁺ lacY⁻ lacA⁺ / F' *lacIˢ lacP⁻ lacZ⁻ lacY⁺ lacA⁻*

You can move through this problem in a very orderly manner.

Step 1. Find the *cis*-Acting Mutations. First identify the *cis*-acting mutations, which are mutations in binding sites on the DNA. They affect only the operon to which they are physically linked. In this genotype, there are two *cis*-acting mutations, namely *lacP⁻* and *lacOᶜ*.

Remember that the *lacP⁻* mutation is an altered sequence in the promoter, so that it is not recognized by RNA polymerase, and the linked structural genes will never be expressed; the *lacOᶜ* mutation is an altered operator sequence that cannot be recognized by any form of the LacI repressor, and the linked structural genes are constitutively expressed.

Step 2. Find the *trans*-Acting Mutations. Second, identify the *trans*-acting mutations, knowing that these affect proteins, and they can diffuse throughout the cell and affect either of the operons. There are two *trans*-acting mutations in regulatory genes: *lacIˢ*, which encodes the superrepressor that lacks the inducer binding site and will always be bound to

the wild-type operator; and *lacI⁻*, which encodes an inactive repressor that cannot bind the operator. The *lacIˢ* mutation is dominant to *lacI⁻*, which means you can ignore the *lacI⁻* in the genotype.

Step 3. Associate the Wild-Type Alleles with Each DNA. Third, identify which wild-type *lac* structural gene is associated with each DNA molecule, and note which *cis*-acting sites are associated with them. The endogenote is *lacZ⁺* and *lacA⁺*, and it is associated with the *lacOᶜ cis*-acting mutation. (You decided in step 1 that the *lacOᶜ* mutation would constitutively express the linked structural genes). The exogenote has a *lacY⁺* allele, and it is linked to the *lacP⁻ cis*-acting mutation. (You decided in step 1 that the *lacP⁻* mutation would not express the linked structural genes regardless of the status of the operator).

Step 4. Create a Table. Make a table to clearly show the effects of gene expression for each structural gene under both induced and uninduced conditions (namely, the presence or absence of either allolactose or IPTG). The table headings should look like box table 16A.

Step 5. Fill in the Blanks. You can now methodically fill in the table.

a. You know that the only *lacZ⁺* and *lacA⁺* genes are in the endogenote, which is linked to a *lacOᶜ* mutation. Because these two genes are associated with a constitutive operator mutation that no form of the LacI repressor will bind, the linked genes will be constitutively expressed, regardless of the presence or absence of inducer. Therefore, your table should now look like the following (where + signifies expression and – signifies no expression) (see box table 16B).

box table 16A

Genotype	Without Inducer			With Inducer		
	lacZ⁺	*lacY⁺*	*lacA⁺*	*lacZ⁺*	*lacY⁺*	*lacA⁺*
lacI⁻ lacOᶜ lacZ⁺ lacY⁻ lacA⁺ / F' *lacIˢ lacP⁻ lacZ⁻ lacY⁺ lacA⁻*						

box table 16B

Genotype	Without Inducer			With Inducer		
	lacZ⁺	*lacY⁺*	*lacA⁺*	*lacZ⁺*	*lacY⁺*	*lacA⁺*
lacI⁻ lacOᶜ lacZ⁺ lacY⁻ lacA⁺ / F' *lacIˢ lacP⁻ lacZ⁻ lacY⁺ lacA⁻*	+		+	+		+

continued

b. Now you can examine the *lacY⁺* gene, which is only expressed from the exogenote. Because this gene is associated with a *lacP⁻* mutation, RNA polymerase cannot bind to this operon and the *lacY⁺* gene will not be expressed, regardless of the type of operator it is associated with and the forms of Lac repressor that are present in either DNA molecule. You can now complete box table 16C.

You can see that regardless of presence of the inducer (allolactose or IPTG), the expression of each structural gene does not change because each operon is under the control of a *cis*-acting mutation. In this example, the *lacIˢ* mutation has no effect on the expression of any of the three structural genes. This is because one operon has the *lacOᶜ* mutation, which is unable to be bound by any type of repressor (wild type or mutant). The other operon has the *lacP⁻* mutation, which cannot be transcribed because RNA polymerase cannot bind, regardless if repressor is bound or not to the operator sequence.

In working with genotypes, remember always to identify the *cis*-acting functions first because they exert their effects on the linked operon regardless of what *trans*-acting functions are present. Then, if an operon fails to have either of the *cis*-acting mutations, examine which *trans*-acting alleles are present in *both* genomes, because they are diffusible proteins that can interact with either genome.

box table 16C

Genotype	Without Inducer			With Inducer		
	lacZ⁺	*lacY⁺*	*lacA⁺*	*lacZ⁺*	*lacY⁺*	*lacA⁺*
lacI⁻ lacOᶜ lacZ⁺ lacY⁻ lacA⁺ / F' lacIˢ lacP⁻ lacZ⁻ lacY⁺ lacA⁻	+	−	+	+	−	+

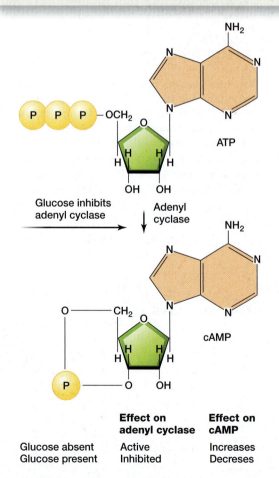

	Effect on adenyl cyclase	Effect on cAMP
Glucose absent	Active	Increases
Glucose present	Inhibited	Decreses

figure 16.14 Synthesis and structure of cyclic AMP (cAMP). The enzyme adenyl cyclase synthesizes cAMP from ATP. Adenyl cyclase is inhibited by glucose, so as glucose levels increase, cAMP levels decrease.

of the *lac* promoter (**figs.** 16.8 and **16.15**)—and also the promoters of other operons with CAP-binding sites.

This binding enhances the affinity of RNA polymerase for the promoter. Without the CAP–cAMP complex bound to the promoter, the transcription rate of the *lac* operon is very low. Because the presence of glucose reduces the amount of cAMP, which reduces the amount of the CAP-cAMP complex, glucose indirectly represses the transcription of operons with CAP sites (see fig. 16.15). The same reduction of transcription rates is noticed in mutant strains of *E. coli* when this part of the distal end of the promoter is deleted.

The binding of the CAP–cAMP complex to the CAP site causes the DNA to bend more than 90 degrees (**fig. 16.16**). This bending, by itself, may enhance transcription, making the DNA more accessible to RNA polymerase. In addition, at some point in the process of transcription initiation, the CAP protein is in direct contact with RNA polymerase. This has been shown by photo-crosslinking studies in which the CAP protein was treated with a cross-linking agent that covalently linked the CAP with the α-subunit of RNA polymerase. For the two proteins to cross-link, they must be in direct contact during the initiation of transcription.

Catabolite repression is an example of positive regulation: Binding of the CAP–cAMP complex at the CAP site enhances the transcription rate of the *lac* operon. Thus, the *lac* operon is both negatively and positively regulated; the LacI repressor exerts negative control, and the CAP–cAMP complex exerts positive control of transcription.

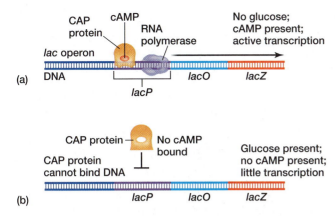

(a)

(b)

figure 16.15 Catabolite repression. (a) When glucose is absent in the environment, cAMP levels in the cell are high and it binds the CAP protein. The CAP-cAMP complex binds the CAP site in the *lac* operon and other sugar-metabolizing operons, which aids in RNA polymerase binding the promoter and inducing transcription. Thus, CAP functions as a positive regulator and cAMP is an inducer. (b) When glucose is present, cAMP levels are low. Thus no CAP-cAMP complex forms, and transcription of the operon is reduced.

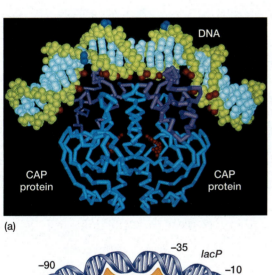

(a)

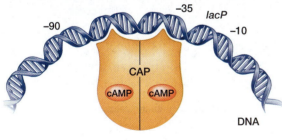

(b)

figure 16.16 Model of the CAP-DNA interaction. (a) The CAP binding site in the DNA has two-fold symmetry, like each operator site, which allows the CAP-cAMP complex to bind the DNA as a dimer. Binding of the CAP-cAMP complex causes a significant bend in the DNA molecule. The DNA-binding domain is purple and the cyclic AMP molecules within the protein are red. The DNA sugar-phosphate backbones are shown in yellow, the bases in light blue. Experiments demonstrated that the DNA phosphates in red (on the double helix) are physically close to the bound CAP protein. (Courtesy of Thomas A. Steitz.). (b) Binding of the CAP-cAMP complex causes a bending of the DNA in the region containing *lacP* (−10 and −35 sequences), which makes it more accessible for binding of RNA polymerase.

Restating the Concepts

▶ The *lac* operon is indirectly regulated by the presence of glucose, which inhibits the synthesis of cAMP. The cAMP forms a complex with CAP to bind to the distal promoter of the *lac* operon and activate transcription by binding RNA polymerase.

▶ The *lac* operon exhibits positive regulation through the CAP–cAMP complex and negative regulation with the *lacI*-encoded repressor.

16.4 The Arabinose Operon

The *lac* operon demonstrates the basic principles underlying the organization and regulation of bacterial operons. However, this basic regulatory machinery can exist in many variations. The **arabinose operon (*ara* operon)** is another *E. coli* operon that encodes enzymes that are involved in the catabolism of a different sugar, arabinose (**fig. 16.17**).

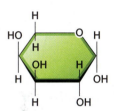

figure 16.17 Structure of the sugar arabinose.

Genes Involved in *ara* Operon Expression

The *ara* operon contains three structural genes: *araB*, *araA*, and *araD*. Because the cell would prefer to utilize glucose instead of arabinose as an energy source, the *ara* operon contains a CAP-binding site like the one discussed earlier. The operon also contains an inducer-binding site (*araI*) and an operator-binding site (*araO*). The organization of the *ara* operon is shown in **figure 16.18**. Transcription of the *ara* operon is controlled by a regulatory protein that is encoded by the *araC* gene, which is not part of the *ara* operon.

The Dual Action of the AraC Protein

The AraC protein can bind to the *araI* site and also *araO* under specific conditions. When arabinose is present in the medium, it is bound by the AraC protein. The AraC–arabinose complex binds to the *araI* site. When the CAP–cAMP complex is also bound to the promoter region, transcription of the *ara* operon is induced (**fig. 16.19** right). Because binding of the AraC protein to the inducer site activates transcription of the *ara* operon, the AraC protein functions as a positive regulator.

When arabinose is not present, the cell has no need to synthesize

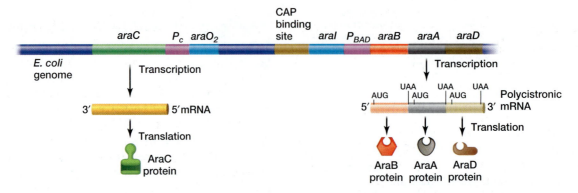

figure 16.18   Organization of the *ara* operon. The *ara* operon is composed of an inducer binding site (*araI*), an operator (*araO₂*), CAP binding site, promoter (P_{BAD}), and three structural genes (*araB, araA,* and *araD*). Transcription of the operon is regulated by the protein encoded by the *araC* gene, which is not part of the operon, and the CAP-cAMP complex. The *araC* gene is transcribed from its own promoter (P_C).

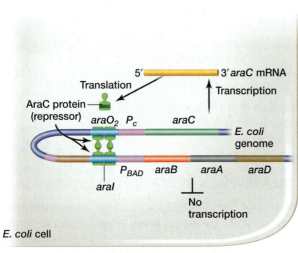

No arabinose (inducer) present

figure 16.19   Regulation of the *ara* operon. In the absence of arabinose (left), the AraC protein dimers bind to both *araI* and *araO₂* sequences and then interact to form a tetramer, which produces a tight DNA loop that represses transcription. In the presence of arabinose (right), the sugar binds the AraC protein, which still binds to *araI*, but not to *araO₂*. This prevents the AraC proteins from forming a tetramer and the tight DNA loop. The CAP-cAMP complex (formed under low glucose levels) can bind the CAP site and activate transcription of the *ara* operon from the promoter (P_{BAD}).

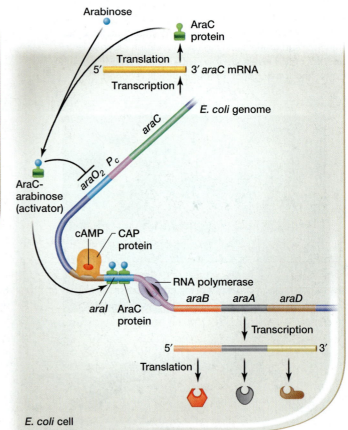

Arabinose (inducer) present

enzymes involved in catabolizing this sugar. The AraC protein, in the absence of arabinose, still binds to the *araI* sequences. Thus, binding of the inducer (arabinose) does not affect the ability of the AraC protein to bond *araI*. However, in the absence of the bound inducer, the AraC protein can also physically interact with other AraC proteins and assist in their binding to *araO*. Thus, in the absence of bound arabinose, the AraC proteins at *araI* and *araO* interact with each other to bend the DNA into a tight conformational loop,

which suppresses transcription of the *ara* operon (fig. 16.19 left).

Although the *lac* operon contains one protein that is a positive regulator and a second protein that is a negative regulator, the *ara* operon contains a single protein that exhibits both positive and negative control. In the presence of arabinose, the AraC protein binds the sugar and the AraC–arabinose complex binds only the inducer sequence to activate transcription. The AraC–arabinose complex cannot bind the operator sequence. In the absence of arabinose, the AraC protein binds to both the inducer and operator sequences, and the interaction between the proteins in these two regions forms a loop that negatively regulates *ara* operon transcription.

Restating the Concepts

▶ The *ara* operon is positively regulated by the CAP–cAMP complex that binds near the promoter.

▶ The *ara* operon is also regulated by the AraC protein, which exhibits positive regulation in the presence of arabinose and negative regulation in the absence of the sugar. In both cases, the AraC protein binds the inducer sequence. In the presence of arabinose, the AraC–arabinose complex cannot bind the operator, and transcription is induced. In the absence of arabinose, the AraC protein binds both the inducer and operator sequences, which form a repression loop that blocks transcription.

16.5 The Tryptophan Operon (Repressible System)

Catabolic operons are induced when the substrate to be broken down enters the cell. **Anabolic operons** function in the reverse manner: They are turned off (repressed) when the end product that their enzymes synthesize accumulates beyond the needs of the cell.

Two entirely different, although not mutually exclusive, mechanisms seem to control the transcription of repressible operons. The first mechanism follows the basic scheme of inducible operons, repressor proteins negatively regulating transcription, and involves the end product of the pathway; we describe this mechanism in this section. The second mechanism involves the secondary structure of the messenger RNA transcribed from an attenuator region of the operon, and it is discussed in the following section.

Tryptophan Synthesis

One of the best-studied repressible systems is the **tryptophan operon (*trp* operon)** in *E. coli*. The *trp* operon contains the five structural genes that encode the polypeptides that constitute the enzymes that synthesize tryptophan from chorismic acid (**fig. 16.20**). It has a promoter-operator sequence (*trpP* and *trpO*) as well as its own regulatory gene (*trpR*).

Operator Control

In this repressible system, the product of the *trpR* gene, the TrpR repressor, is inactive by itself; it does not bind the operator sequence of the *trp* operon. The repressor only becomes active when it combines with tryptophan, which is termed the **corepressor.** When the amount of tryptophan in the cell increases, enough becomes available to bind and activate the repressor. The corepressor-repressor complex then recognizes and binds the operator sequence, preventing transcription of the *trp* operon by RNA polymerase.

After the amount of tryptophan in the cell is reduced, the tryptophan that is bound to the repressor diffuses out of the complex. The repressor, now no longer bound with tryptophan, then detaches from the *trp* operator. The transcription of the *trp* operon is then possible and can proceed normally (the operon is now said to be **derepressed**).

Transcription continues until enough of the various enzymes have been synthesized to once more produce an excess of tryptophan. The operon is again shut off by the TrpR–tryptophan complex, and the process is repeated. This mechanism ensures that tryptophan is being synthesized as needed (**fig. 16.21**).

This type of regulation is modified, however, by the existence of the second mechanism for regulating repressible operons—attenuation.

16.6 The Tryptophan Operon (Attenuator-Controlled System)

Details of a second mechanism to repress the *E. coli* tryptophan operon were elucidated primarily by Charles Yanofsky and his colleagues. Controlling the expression of an operon by an **attenuator region** has been demonstrated for at least five other amino acid-synthesizing operons, including the leucine and histidine operons. This regulatory mechanism may be the same for most operons involved in the synthesis of an amino acid.

Leader Transcript

In the *trp* operon, an attenuator region lies between the operator and the first structural gene, *trpE* (**fig. 16.22**). The mRNA transcribed from the attenuator region, termed the **leader transcript,** has been sequenced, revealing two surprising and interesting facts. First, the mRNA has four subregions with base sequences that are complementary to one another, so that the mRNA can form three different stem-loop structures (**fig. 16.23**).

Depending on circumstances, regions 1–2 and 3–4 can form two stem-loop structures, or regions 2–3 can form a single stem-loop. If region 2 pairs with region 1, then it is unable to pair with region 3, and region 3 is able to basepair with region 4. Alternatively, if region 2 is prevented from pairing with region 1, then it will

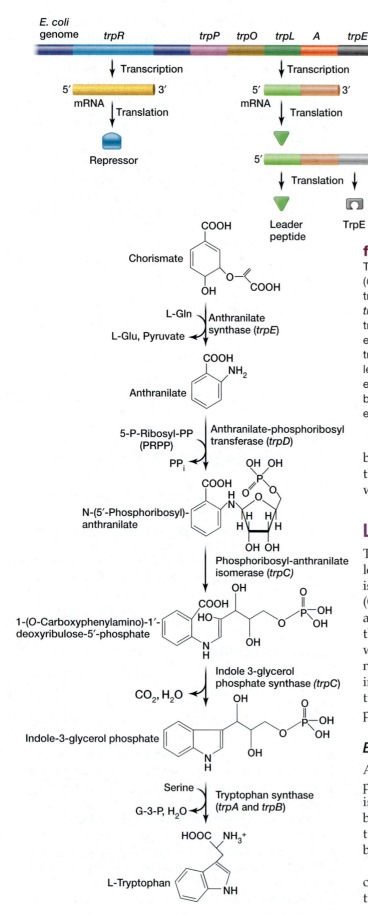

figure 16.20 Organization of the *trp* operon in *E. coli*. The *trp* operon is composed of a promoter (*P*), operator (*O*), leader sequence (*trpL*), attenuator (*A*) that regulates the transcription of the five structural genes (*trpE*, *trpD*, *trpC*, *trpB*, and *trpA*). The Trp repressor, which negatively regulates transcription of the operon by binding to the operator, is encoded in the *trpR* gene. Two different mRNAs can be transcribed from the *trp* promoter, one encodes only the leader and attenuator, while the other is polycistronic and encodes the five structural genes. The pathway for tryptophan biosynthesis, and the steps controlled by the five *trp*-encoded enzymes, is shown to the left.

base-pair with region 3. As you will see shortly, the particular combination of stem-loop structures determines whether transcription continues.

Leader Peptide Gene

The second surprising fact obtained by sequencing the leader transcript is that a small peptide-encoding gene is found from bases 27 to 68. This open reading frame (ORF) is referred to as the **leader peptide gene.** It encodes a peptide that consists of only 14 amino acids. Two of those amino acids, however, are adjacent tryptophans, which are highly unusual in a peptide (as tryptophan is not an abundant amino acid) and are critically important in attenuator regulation. Note that region 1 in the leader transcript is located within the leader peptide ORF. The proposed mechanism for this regulation follows.

Excess Tryptophan

Assuming that the operator site is available to RNA polymerase, (that is, the TrpR–tryptophan complex is not bound), transcription of the attenuator region begins. As soon as the 5' end of the mRNA has been transcribed, a ribosome attaches to this mRNA and begins translation of the leader peptide.

Depending on the levels of tryptophan in the cell, three outcomes can take place. If the concentration of tryptophan in the cell is abundant, then high

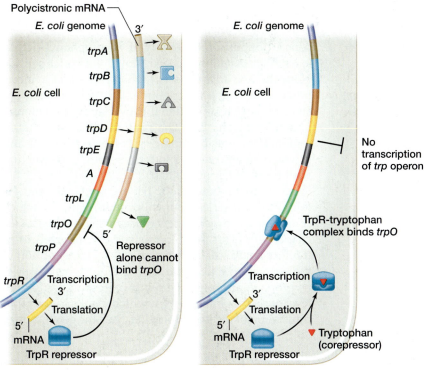

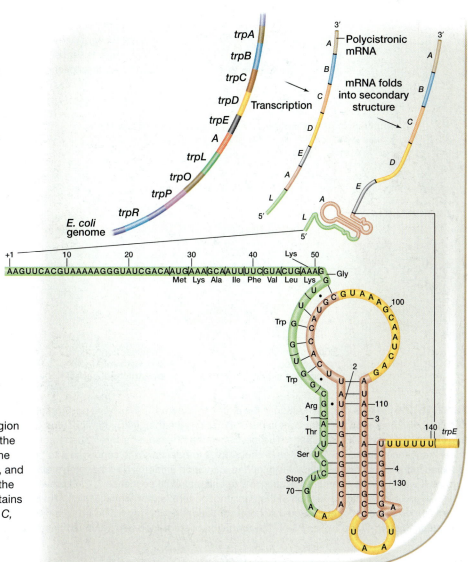

figure 16.21 Transcriptional regulation of the *trp* operon by the Trp repressor. Under low tryptophan levels (left), the TrpR repressor cannot bind a molecule of tryptophan (corepressor) and is unable to bind the *trp* operator. This permits transcription of the *trp* operon's polycistronic mRNA. Under high tryptophan levels (right), the Trp repressor binds a molecule of tryptophan and this repressor-corepressor complex binds the operator, which results in blocking transcription of the *trp* operon (negative regulation).

figure 16.22 Attenuator region of the *trp* operon, which contains the leader peptide gene (green) and the four attenuator sequences (1, 2, 3, and 4). This region is transcribed into the polycistronic mRNA that also contains the five *trp* structural genes (*E, D, C, B,* and *A*).

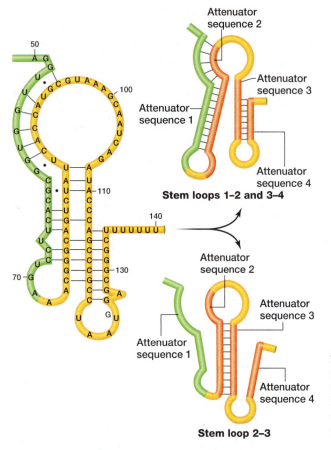

Stem loops 1–2 and 3–4

Attenuator sequence 2

Attenuator sequence 1

Attenuator sequence 3

Attenuator sequence 4

Attenuator sequence 2

Attenuator sequence 1

Attenuator sequence 3

Attenuator sequence 4

Stem loop 2–3

concentrations of the charged tryptophanyl-tRNAs exist. As translation proceeds down the leader peptide gene, the two adjacent tryptophans are quickly incorporated into the leader peptide. This permits the ribosome to move to a position that overlaps both regions 1 and 2 of the transcript. The location of the ribosome on region 2 prevents base pairing between regions 2 and 3, which allows region 3 to form a stem-loop structure with region 4 (fig. 16.24, left).

Note that the stem-loop between regions 3 and 4 is followed by a series of uracil-containing bases, which constitute a *rho-independent transcription terminator* (see chapter 10). This stem-loop structure, referred to as the **terminator,** or **attenuator,** causes transcription of the operon to be terminated. Therefore, when levels of tryptophan are high, there is a correspondingly high level of tryptophanyl-tRNA, which results in the rapid translation of the leader peptide gene, the generation of

figure 16.23   Nucleotide sequence and the potential base pairings in the *trp* attenuator region (bases 50 to 140) are shown (left). On the right (top), the simultaneous pairing of stem-loops 1–2 and 3–4 are shown. Alternatively, stem-loop 2–3 can form (right, bottom), which prevents sequence 4 from base pairing with sequence 3 and forming a transcriptional stop signal.

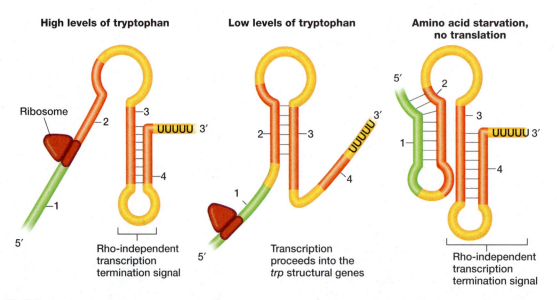

High levels of tryptophan

Low levels of tryptophan

Amino acid starvation, no translation

Ribosome

Rho-independent transcription termination signal

Transcription proceeds into the *trp* structural genes

Rho-independent transcription termination signal

figure 16.24   Model for attenuation in the *E. coli trp* operon. The relative position of the ribosome as it translates the leader peptide (fig. 16.22) is shown. Under conditions of excess tryptophan (left), the ribosome quickly translates through the two Trp codons in the leader peptide sequence. The location of the ribosome between attenuator sequences 1 and 2 blocks basepairing between sequences 2 and 3, which permits sequence 3 to basepair with sequence 4 to produce a rho-independent transcription termination signal. Under conditions of tryptophan starvation (middle), the ribosome stalls at the two adjacent Trp codons in sequence 1, which allows attenuator sequence 2 to basepair with sequence 3. This prevents sequence 3 from base pairing with sequence 4 and blocks the generation of the transcription termination signal, which allows transcription to continue into the five structural genes of the *trp* operon. Under general starvation (right), there is no translation of the leader peptide, which results in the formation of stem-loops 1–2 and 3–4. The formation of stem-loop 3–4 again terminates transcription.

the transcription terminator, and a failure to transcribe the five structural genes.

Tryptophan Starvation

If the quantity of tryptophan is reduced, there will be a correspondingly low level of the charged tryptophanyl-tRNA. The ribosome that is translating the leader peptide then must wait at the first tryptophan codon until it acquires a Trp-tRNA^{Trp}. The ribosome then pauses again at the second tryptophan codon. This situation is shown in the configuration in the middle of figure 16.24. The ribosome stalls on region 1, which prevents region 1 from base-pairing with region 2. As transcription continues, region 2 is then free to pair with region 3. Formation of the stem-loop between regions 2 and 3 precludes the formation of the terminator stem loop (between regions 3 and 4).

The stem-loop structure between regions 2 and 3 is referred to as the **antiterminator stem.** Note that the antiterminator stem is not followed by a string of uracils, which precludes it from being a rho-independent transcription terminator. In this configuration, transcription is not terminated, so that the entire operon is transcribed and translated, producing the enzymes that synthesize tryptophan to increase its level in the cell.

General Starvation

A final configuration is possible, as shown on the far right in figure 16.24. Here, no ribosome interferes with base pairing between regions 1 and 2, and, presumably, stem loops are formed between regions 1 and 2 and also between regions 3 and 4. This latter stem-loop configuration is a terminator that stops transcription of the *trp* operon.

This configuration is believed to occur if the ribosome is stalled at the beginning of the leader peptide, which is before the start of region 1 (see figure 16.22). This would occur when the cell is starved for other amino acids. Presumably, it makes no sense for the cell to manufacture tryptophan when other amino acids are also in short supply. In this way, *E. coli* can carefully raise the levels of the different amino acids in the most efficient manner.

TRAP Control

In the 1990s, Charles Yanofsky and his coworkers expanded their work on the regulation mechanisms of tryptophan operons by studying the *trp* operons in other organisms. The tryptophan operon in *Bacillus subtilis* is also controlled by attenuation; the secondary structure in the mRNA transcript is not induced by binding of the ribosome, however, but by a *trp* **RNA-binding attenuation protein (TRAP).**

This protein, which is composed of 11 identical subunits, attaches to the 5' leader sequence of the nascent mRNA. Before binding to the leader mRNA, each subunit must bind a tryptophan molecule, which then permits the protein to also bind to either a GAG or UAG triplet in the mRNA. Because there are 11 of these triplets in the leader mRNA, each separated by several bases, 11 protein subunits must each bind a tryptophan molecule before binding to the leader sequence. These 11 protein subunits bound to the mRNA leader sequence is called the TRAP structure (**fig. 16.25a**).

The result is when tryptophan levels are high, the TRAP structure forms and prevents regions A and B in the mRNA from base pairing. Because regions B and C overlap (fig. 16.25b), C is free to basepair with region D, creating a rho-independent transcription termination signal (notice the string of U's following region D, fig. 16.25b, right). In the absence of excess tryptophan, TRAP does not bind to the mRNA, and an antiterminator stem loop forms between regions A and B. The formation of the antiterminator prevents region C from base pairing with region D, allowing transcription to continue (fig. 16.25b, left).

Redundant Controls

Some amino acid operons are controlled only by attenuation, such as the histidine operon (*his* operon) in *E. coli*, in which the leader peptide gene contains seven histidine codons in a row, or the *trp* operon in *B. subtilis.* Redundant controls, which we saw with the *trp* operon in *E. coli*, allow the cell to test both the tryptophan levels (in the repression system, in which tryptophan is the corepressor) and the tryptophanyl-tRNA levels (in the attenuator control system). The attenuator system also allows the cell to regulate tryptophan synthesis on the basis of the shortage of other amino acids.

It's Your Turn

We finished our discussion on how the *trp* operon is regulated. Problem 25 presents you with three mutations that affect either the *trp* operon or its regulation. Based on several different genotypes you will need to determine which mutation affects *trpR*, *trpO*, and *trpE*.

Restating the Concepts

▶ In *E. coli*, the TrpR repressor protein must bind a molecule of tryptophan (corepressor) before the complex can bind to the operator and negatively regulate transcription of the *trp* operon.

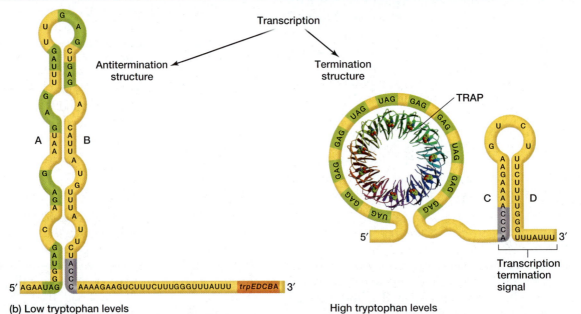

figure 16.25 Attenuation mechanism of the *trp* operon in *Bacillus subtilis*. (a) A ribbon diagram of the mRNA wrapped around the multisubunit Trp RNA-binding attenuation protein (TRAP). Each of the 11 TRAP subunits are shown in different colors binding tryptophan molecules (red spheres). (b) The triplets (GAG and TAG in DNA, or GAG and UAG in the mRNA) that the TRAP protein binds are highlighted in green. The label *trpEDCBA* refers to the structural genes of the *trp* operon. The sequences marked C and D can basepair and form the rho-independent transcription terminator stem, while the sequences marked A and B can basepair and form the anti-terminator stem. When tryptophan levels are low (left), the TRAP protein cannot bind Trp and, therefore, cannot bind the mRNA. This allows the anti-terminator stem to form, which includes part of the C sequence and prevents the C–D transcription termination signal from forming and transcription continues into the *trp* structural genes (red region). When Trp levels are high (right), the TRAP protein binds Trp, which then binds the mRNA. TRAP binding of the mRNA prevents sequences A and B from base pairing, which then allows sequences C and D to base pair and form the rho-independent transcription termination signal.

(Reprinted by permission from Macmillan Publishers Ltd. From Alfred A. Antson, et al., "Structure of the trp RNA-binding attenuation protein, TRAP, bound to RNA" *Nature* 40: 2347-237, 1999.)

▶ In *E. coli.* the *trp* operon is also under the control of attenuation. When tryptophan levels are high, translation of the *trp* leader sequence allows the mRNA to form a rho-independent transcription termination stem (between regions 3 and 4) to terminate transcription before the *trp* structural genes. When tryptophan levels are low, translation of the leader peptide stalls at two adjacent tryptophan codons, which allows the formation of the antiterminator stem (between regions 2 and 3) and permits transcription of the *trp* structural genes.

▶ *Bacillus subtilis* also utilizes attenuation to control transcription of the *trp* operon. However, it requires the 11 subunits of the TRAP protein to bind 11 molecules of tryptophan to form a pinwheel structure with the mRNA, which generates a rho-independent transcription termination signal (between regions C and D). When tryptophan levels are low, the TRAP structure does not form, and the antiterminator stem forms between regions A and B, which prevents region C from base-pairing with region D to generate the terminator signal, and transcription continues into the structural genes of the *trp* operon.

16.7 Lytic and Lysogenic Cycles in Phage λ

When a bacteriophage infects a cell, it must express its genes in an orderly fashion. Early genes usually control phage DNA replication; late genes usually express phage coat proteins and control the lysis of the bacterial cell. It could be disastrous to the phage if the late genes were transcribed first and the cell lyses before the phage DNA had completed replication and packaged into the phage particles. Also, temperate phages have the option of entering either a lysogenic or lytic cycle—here, too, control processes determine which path is taken. One generalization that holds true for most phages is that their genes are clustered into early and late operons, with separate transcriptional control mechanisms for each.

Phage λ is perhaps the best studied bacteriophage. It has a genome of about 48,500 bp. Because it is a temperate phage, it can exist either as a prophage, integrated into the host genome (lysogenic cycle), or as a plasmid-like entity, replicating in the bacterial cell prior to lysis (lytic cycle).

Briefly, the expression of one of the two life-cycle alternatives, lysogenic versus the lytic cycle, depends on which of two repressors, CI and Cro, gains access to the operator sites. The CI repressor acts to favor lysogeny and repress the lytic cycle. The Cro repressor favors the lytic cycle and represses lysogeny. Other control mechanisms are also involved in determining aspects of the λ life cycle, including antitermination and multiple promoters for the same genes. Entire books have been written on the life cycle and gene regulation mechanisms in λ; we only briefly discuss the most essential features here.

Phage λ Operons

The λ genes are grouped into four functional regions: left operon, right operon, late operon, and repressor region (fig. 16.26). The **left operon** contains the genes that are required for recombination and phage integration (lysogeny), and the **right operon** contains the genes that are required for DNA replication (lytic cycle). The **late operon** contains the genes that determine phage head and tail proteins and lysis of the host cell. The **repressor region** contains genes for repressor controls.

The Two Linear Forms of λ DNA

Shortly after infecting a bacterial cell, the λ genome exists as a circular molecule. However, the λ genome also has two linear stages in its life cycle. It is packed within the phage head in one linear form, and it integrates into the host chromosome to form a prophage in

another linear form. The two linear forms result from breaking the circular form of λ at two different locations so that they have different ends.

The mature DNA, which is packed within the phage heads before lysis of the cells, is flanked by *cos* sites (see chapter 12). These sites result from a break in the circular map between the *A* and *R* loci (see fig. 16.26). The prophage is integrated at the *att* site, and the circular map is thus broken there at integration into the bacterial genome. Both linear forms contain the same genes as the circular form; no DNA is lost in the conversion of circular to linear or linear to circular forms.

POP' and BOB'

The homologous integration sites (*att* sites) on both λ and the *E. coli* genome consist of a 15-bp core sequence (called "O" in both), flanked by different sequences on both sides in both the bacterium and the phage (**fig. 16.27**). In the phage, the region is referred to as POP', where P and P' (P for phage) are two different 3-bp sequences flanking the O core on the phage DNA. In the bacterium, the region is called BOB'; B and B' (B for bacterium) are the 3-bp sequences flanking the O core in the *E. coli* genome. The O core sequence is identical in both the λ and bacterial sites, but the flanking bacterial and λ sequences are different.

Integration and Excision

Integration, which is a part of the lysogenic life cycle, requires the product of the λ *int* gene, a protein known as *integrase*, in a process that is referred to as **site-specific recombination** because the recombination event involves two specific sequences (POP' and BOB'). Later excision of the prophage, when the phage leaves the host chromosome to enter the lytic cycle during induction, requires both the integrase protein and the protein product of the neighboring *xis* gene, *excisionase*.

After infection of the *E. coli* cell by λ, the phage DNA circularizes, using the complementary sequences of the single-stranded *cos* sites. Transcription begins, and within a very short time the phage is guided toward either entering the lytic cycle and producing virus progeny or entering the lysogenic cycle and integrating into the host chromosome. The choice of entering the lysogenic phase requires the expression of the *int* gene. What events, however, lead up to this "decision" on which path to take?

Early and Late Transcription

When the phage first infects an *E. coli* cell, transcription of the left and right operons begins at the left (P_L)

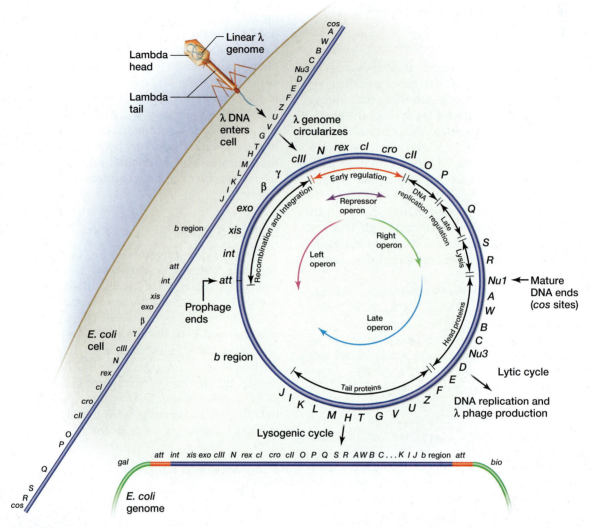

figure 16.26 Different forms of lambda DNA. The bacteriophage lambda (λ) infects an *E. coli* cell and injects its linear genome into the cell. The ends of the linear DNA, *cos* sites, are single-stranded and complementary, which allows the λ DNA to circularize after entering the cell. The λ genome is composed of four operons, the right, left, repressor and late operons (marked in the circular genome). Because λ is a temperate phage, it can enter either a lytic or lysogenic cycle. The early regulation region (red in the circular genome) determines which life cycle λ will enter. The lytic cycle requires replication of the circular λ genome and the generation of more λ bacteriophage particles. Entering the lysogenic cycle requires the circular λ genome to recombine with the *E. coli* genome. This recombination involves a specific sequence in the λ genome (*att*) and a similar sequence in the *E. coli* genome that is located between *gal* and *bio*. Thus, the lysogenic decision causes an insertion of the λ genome (prophage) at a specific location in the *E. coli* genome.

and right (P_R) promoters. The left promoter transcribes the *N* gene, and the right promoter transcribes the *cro* gene (**fig. 16.28**). Transcription from both promoters stops at the rho-dependent terminators t_{R1} and t_{L1}. Transcription cannot continue until the protein product of the *N* gene is translated. This protein is called an **antiterminator protein.**

The antiterminator protein binds at sites upstream from the t_{R1} and t_{L1} terminators, called *nutL* and *nutR* (*nut* stands for *N* utilization; *L* and *R* stand for left and right). When the N protein is bound to the transcript, RNA polymerase reads through the terminators and

continues on to transcribe the left and right operons. Although it is not completely clear why antitermination has evolved here, it seems to give the phage better control over the timing of events.

Transcription then continues along the left and right operons through the *cIII* and *cII* genes, respectively (see fig. 16.28). Later, if the lytic pathway is followed, the *Q* gene (see fig. 16.26), which codes for a second antiterminator protein in the right operon, is transcribed. Without the *Q* gene product, transcription of the late operon proceeds about 200 nucleotides and then terminates. With the *Q* gene product, the late operon is completely transcribed.

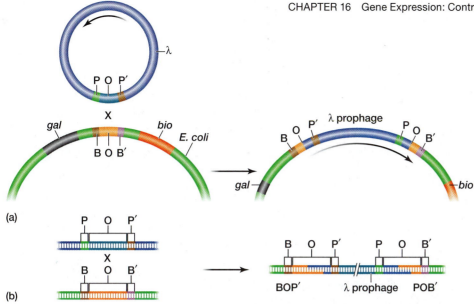

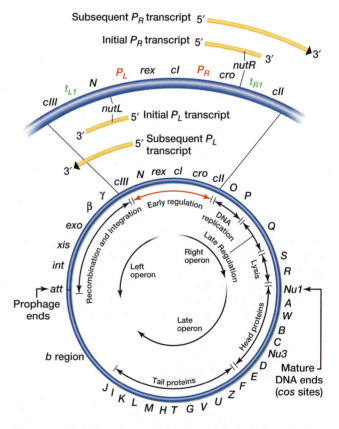

figure 16.27   The lambda attachment sites. (a) Integration of the λ phage DNA into the *E. coli* genome requires recombination between the two attachment (*att*) sites, called POP' (in λ) and BOB' (in *E. coli*). (b) The recombination sites contain identical 15 nucleotides sequences (O) that are flanked by 3 nucleotide phage (P and P') or bacterial-specific (B and B') sequences.

In λ, then, proteins that allow RNA polymerase to proceed past terminators mediate general control of transcription. The pathway just described always leads to the lytic pathway; however, a complex series of events can also take place in the repressor region that may lead to a "decision" to follow the lysogenic cycle instead.

Repressor Transcription

The CIII protein, which is encoded in the P_L mRNA, inhibits an *E. coli* protease called FtsH that degrades the CII protein (**fig. 16.29**). Thus, expression of the CIII protein allows the stable expression of the CII protein, which is encoded in the P_R mRNA, that binds at two promoters: P_{III} and P_{RE}. The CII protein is a positive regulator of both of these promoters, which enhances their availability to RNA polymerase—just as the CAP–cAMP product enhances transcription of the *lac* operon.

The P_{III} promoter transcribes the *int* gene, which is required for λ integration into the genome, and the P_{RE} promoter (*RE* stands for *repression establishment*) transcribes the *cI* repressor gene. Note that although the mRNA transcribed from P_{RE} moves through the *cro* gene, it fails to encode the Cro protein because the wrong strand of the gene is transcribed from P_{RE} (see chapter 10).

At this point, the phage can still "choose" between either the lytic or the lysogenic cycles. The integrase and CI repressor proteins favor lysogeny, and the Cro protein, also called the *antirepressor*, is transcribed from the P_R promoter and favors the lytic pathway. We can now focus further on the repressor region with its operators and promoters.

figure 16.28   Transcription of the lambda early regulation region. Transcription begins at the left and right promoters (P_L, P_R) and proceeds in opposite directions to the left and right terminators (t_{L1}, t_{R1}). The *N* gene, encoded in the initial P_L transcript, encodes a transcription anti-terminator. Binding of the N protein to the *nutL* sequence on the P_L RNA permits transcription to continue through t_{L1} and express the longer RNA that transcribes through the *cIII* gene. Similarly, the N protein binds the *nutR* sequence on the P_R transcript to permit transcription through t_{R1} to the *cII* gene. The expression of the *cII* and *cIII* genes are essential for the lysogenic, but not the lytic, cycle in lambda.

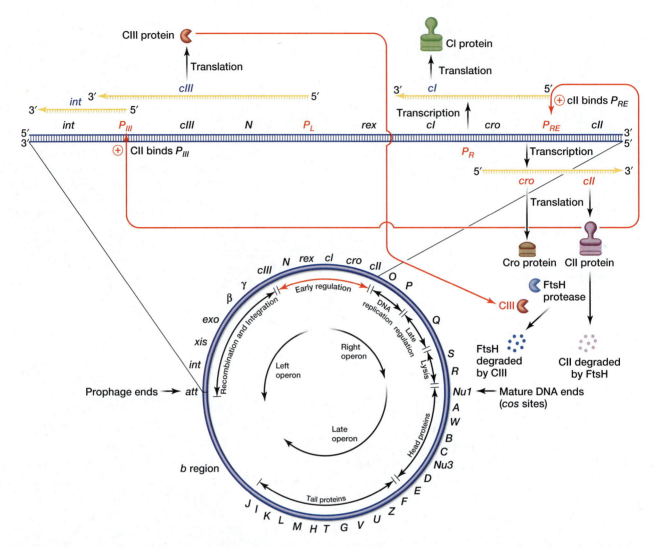

figure 16.29 Initial transcriptional control of the lambda early regulation region. Lambda transcription originates from P_L to express the N protein and P_R to express the Cro protein. After the N protein suppresses transcription termination from P_L and P_R, the *cIII* and *cII* genes are transcribed. The CII protein, expressed with Cro from the P_R promoter, binds and activates transcription from the *cI* repressor establishment promoter (P_{RE}) and the *int* promoter (P_{III}). The CIII protein, in contrast, acts to degrade the *E. coli*-encoded FtsH protease, which normally degrades the CII protein. Thus, stable expression of FtsH suppresses transcription from P_{RE} and P_{III}. Stable expression of the CII protein increases expression of the *cI* and *int* genes, which is required for λ recombination into the *E. coli* genome.

Maintenance of Repression

The CI repressor binds the left and right operators, O_L and O_R, to terminate transcription of the left and right operons. There are several ramifications of this repression. First, lysogeny can be initiated because the *int* gene has been transcribed at the early stage of infection from the P_{III} promoter. Second, the late operon is not transcribed, which prevents the expression of the late lytic genes. Third, transcription of the *cII* and *cIII* genes are terminated, which ends CII-dependent transcription of the *cI* gene from the P_{RE} promoter. The *cI* gene can still be transcribed, however, because a second promoter, P_{RM} (*RM* stands for *repression maintenance*), allows low levels of *cI* transcription.

Transcription from P_{RM}, however, requires the CI protein to bind to an operator to activate transcription.

Details of Operator Structure

When the right and left operators were sequenced, each was discovered to have three DNA-binding sites for the CI repressor; for the right operator, these sites were named O_{R1}, O_{R2}, and O_{R3} (**fig. 16.30**). As described earlier for the *lac* operator, O_{R1}, O_{R2}, and O_{R3} each have two-fold symmetry. The λ repressor is a dimer of two identical subunits (**fig. 16.31**). Each subunit is composed of two domains, or "functional regions." The carboxyl and amino terminals are separated by a relatively open region, susceptible to protease attack. The α-helical

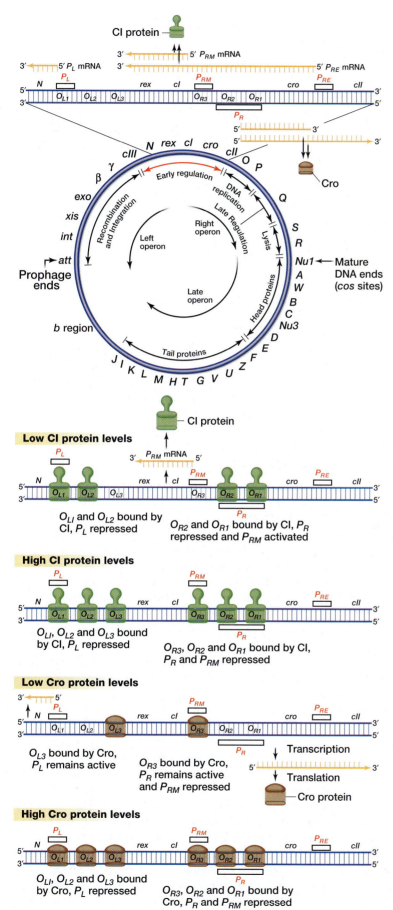

regions of the amino terminals interdigitate into the major groove of the DNA to locate the specific sequences making up the left and right operator sequences.

On the right operator, for example, the right-most site (O_{R1}) was found to be the most efficient at binding repressor. When the repressor was bound to O_{R1}, it aided the binding of another dimer of CI to the O_{R2} site. This **cooperativity,** in which one protein helps a second one to bind, results from the second domain in the CI protein. This protein–protein interaction domain is located in the carboxyl terminus of CI protein and allows one CI dimer that is bound to O_{R1} to also interact with a second dimer of CI and improve its ability to bind O_{R2}.

When O_{R1} and O_{R2} are bound by CI, the right operon was found to be repressed (i.e., P_R was negatively regulated), and transcription of cI was enhanced (P_{RM} was positively regulated). But excess CI repressor, when present, also bound the third site within O_R (O_{R3}), which repressed transcription of the cI gene itself. This result indicated that maintenance levels of CI can be kept within very narrow limits.

Promoter Differences

The promoters for repressor establishment and maintenance differ markedly in their control of repressor gene expression. When P_{RE} is active, a very high level of CI repressor is produced, whereas P_{RM} expresses only a low level of repressor. The level of repressor translation is due to the length of the leader RNA transcribed on the 5′ side of the cI gene. The P_{RE} promoter transcribes a very long leader RNA and is very efficient at translation of the cI ORF. In contrast, the P_{RM} promoter begins transcription at the translation initiation codon of the cI gene. This leaderless mRNA is translated into CI very inefficiently.

figure 16.30 The effect of CI and Cro proteins on transcription of the early regulatory RNAs. The five mRNAs, four promoters (P_L, P_R, P_{RM}, and P_{RE}) and protein binding sites (operators) are shown. The right operator (O_R) overlaps the P_{RM} and P_R promoters, while the left operator (O_L) overlaps P_L. There are three repressor recognition sites within each operator with CI and Cro having different affinities for each sequence. Preferential binding by the Cro repressor to O_{R3} and the CI repressor to O_{R1} and O_{R2} determines whether transcription occurs to the right (from the P_R promoter) or the left (from the P_{RM} promoter). At high protein levels both CI and Cro block transcription from all the promoters.

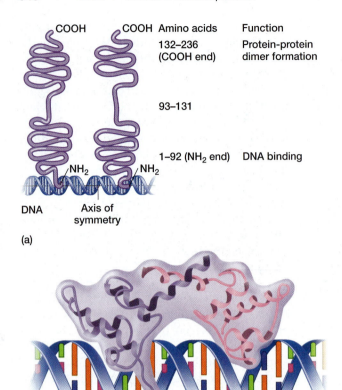

figure 16.31 The λ CI repressor. (a) The λ CI repressor is a dimer with each subunit having helical amino- and carboxyl-terminal ends. One alpha helix of each amino-terminal end binds in the major groove of DNA. (b) Diagram of the interaction of amino acid residues 1–69 (purple) with O_{R1} (blue) in phage 434, which is very similar to lambda. The two repressor subunits are shown in purple and magenta. The amino terminal tails of the repressor subunits are shown fitting in the major grooves of the DNA.

Lysogenic Versus Lytic Response

We just described the mechanism by which λ establishes lysogeny, the binding of CI to O_{R1} and O_{R2} to terminate transcription of P_R and activate transcription of P_{RM}. How then does λ turn toward the lytic cycle? In this case, control is exerted by the Cro protein, which is another repressor molecule that also works at the left and right operators but in a manner antagonistic to the way the CI repressor works. Put another way, using the right operator as an example, the Cro protein preferentially binds O_{R3}, which represses expression of the *cI* gene and enhances the transcription of *cro* from P_R (see fig. 16.30). The Cro protein can direct the cell toward a lytic response if it occupies the O_R and O_L sites before the CI repressor, or if the CI repressor is removed.

Effect of Bacterial Growth on Lysis Versus Lysogeny

Initially, however, when the phage first infects an *E. coli* cell, the "decision" for lytic versus lysogenic growth is probably determined by the CII protein. This protein, as mentioned, is susceptible to a bacterial protease, which, in turn, is an indicator of cell growth. When *E. coli* growth is limited, its proteases tend to be limited, a circumstance that would favor lysogeny for the phage. If bacterial growth is limited and the lytic cycle were preferred, then the phage would rapidly infect and lyse all the bacterial cells, which would also lead to the loss of the phage. Thus, lysogeny is preferred when bacterial cell growth is limited and the lytic cycle is preferred when the bacteria are actively growing.

Under active bacterial growth, the CII protein is more readily destroyed by the bacterial FtsH protease, and therefore it fails to enhance *cI* transcription; lysis follows. When bacteria are not growing actively, the CII protein is not readily destroyed. Therefore, it enhances *cI* transcription, and lysogeny results. During initial infection, the choice between lysogeny or the lytic cycle depends primarily on the CII protein, which gauges the health and activity of the bacterial host. After lysogeny is established, it can be reversed by processes that inactivate the CI protein.

Evolution of the Timing of Lysis

From the point of view of phage λ, when would be a good time for the CI repressor to be removed? In evolutionary terms, a prophage might be at an advantage if it excised from a host's genome and began the lytic cycle when it "sensed" damage to the host. In fact, one of the best ways to induce a prophage to enter the lytic cycle is to direct UV light at the host bacterium.

UV radiation damages the DNA and induces several repair mechanisms (see chapter 18). One, called SOS repair, makes use of the protein encoded in the *recA* gene. The RecA protein is normally involved in DNA recombination. When RecA encounters damaged DNA, however, it is converted into a protease, called RecA* (RecA-star). RecA* cleaves specific repressor proteins that allows transcription of genes that encode enzymes that repair damaged DNA. The RecA* activity also cleaves the λ CI repressor in the susceptible region between the DNA-binding and protein–protein interaction domains. The cleaved repressor loses the cooperative binding of CI to the O_{R2} site and the efficient dimer binding of CI to the O_{R1} site. This loss of the CI repressor makes the operator sites available for the Cro protein. The lytic cycle then follows.

Not all the details regarding the CI–Cro competition are known, but an understanding of the relationship of lytic and lysogenic life cycles and the nature of DNA–protein recognition has emerged (**fig. 16.32** and **table 16.1**).

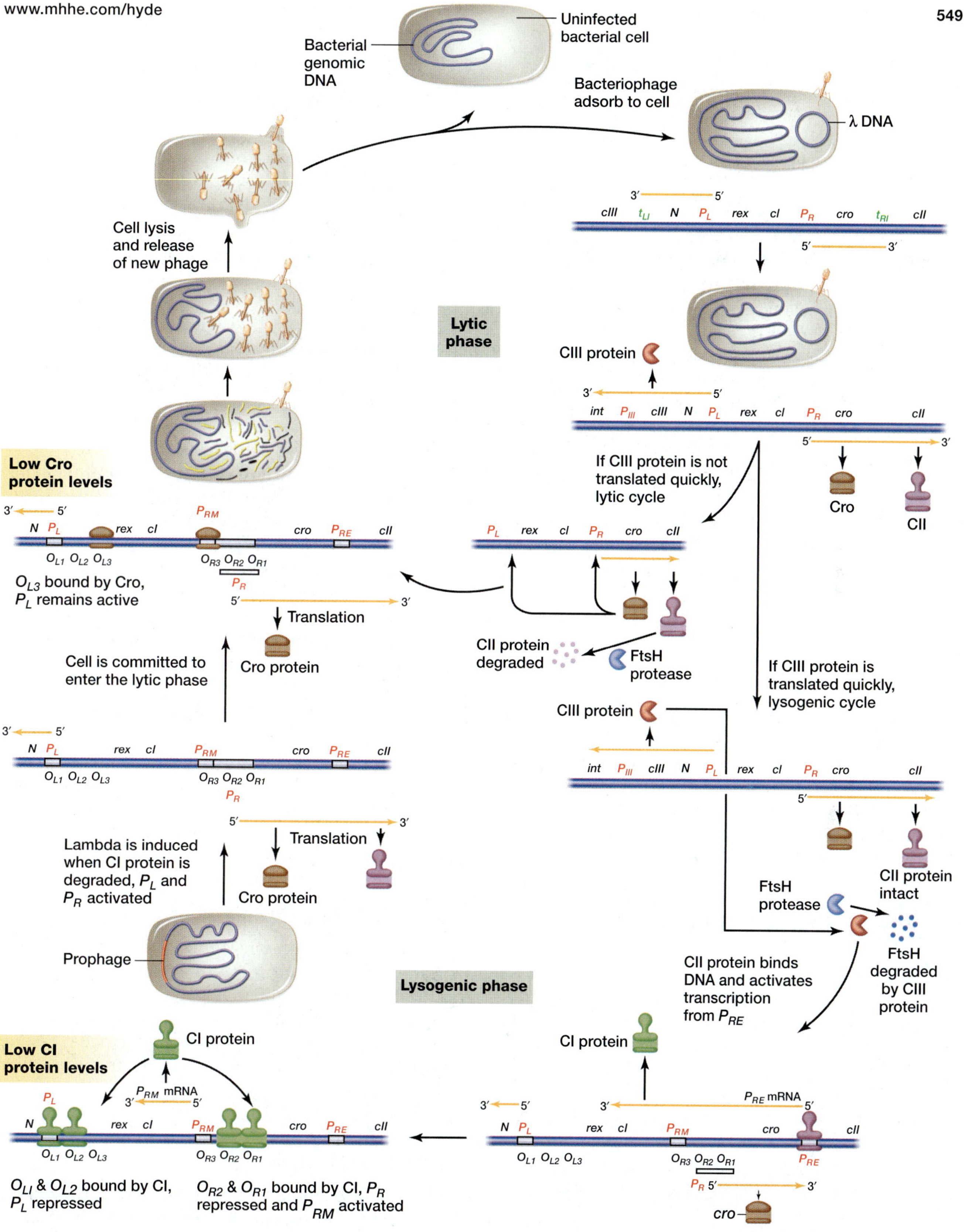

figure 16.32 Summary of regulation of phage lambda life cycles. Upon infecting a cell, transcription begins from P_R to express the *cro* gene and from P_L to express the *N* gene, but both RNAs terminate at left and right terminators (t_{L1} and t_{R1}). The N protein allows transcription to proceed through the initial terminators and permits transcription of additional genes. The lytic cycle will prevail if the Cro protein gains access to the right and left operators (O_L and O_R), which leads to continued transcription from P_L and P_R and no transcription from P_{RM}. Lysogeny will occur if the CII protein is not degraded by FtsH, which leads to CI expression. CI protein can gain access to the right and left operators and block transcription from P_R and activate transcription from P_{RM}.

table 16.1 Elements in Phage λ Infection

Gene Products	
cI	Repressor protein whose function favors lysogeny
cII	Enhances transcription at the P_I and P_{RE} promoters
cIII	Inhibits the FtsH protease
cro	Antirepressor protein that favors lytic cycle
N	Antiterminator acting at *nutR* and *nutL*
rex	Protects bacterium from infection by T4 *rII* mutants
int	Integrase for prophage integration
Q	Antiterminator of late operon
FtsH	Bacterial protease that degrades CII protein
Promoters of	
P_R	Right operon
P_L	Left operon
P_{RE}	Establishment of repression at repressor region
P_{RM}	Maintenance of repression at repressor region
P_R	Late operon
P_I	*int* gene
Terminators	
t_{R1}	Terminates after *cro* gene
t_{L1}	Terminates after *N* gene
Antiterminators	
nutR	In *cro* gene
nutL	In *N* gene

Restating the Concepts

▶ Proteins that suppress transcription termination (such as the N and Q proteins in λ) play a general role in regulating gene expression. Without these antiterminator proteins, the mRNA transcript never reaches downstream genes.

▶ The CI repressor is required for "choosing" the lysogenic state, whereas Cro is required for the lytic state.

▶ The CI repressor binds to three different sites in O_R. Binding to O_{R1} cooperatively binds a second CI dimer to O_{R2}, which represses transcription of P_R and activates transcription of P_{RM}. Higher levels of CI will bind to O_{R3}, which then blocks further transcription from P_{RM}.

▶ The Cro protein acts antagonistically relative to the CI repressor to induce the lytic cycle. Cro preferentially binds O_{R3} to repress transcription of P_{RM}. Unlike CI, Cro does not bind cooperatively to O_R and O_L, such that increased amounts of Cro are required to bind to O_{R2} and O_{R1}, which then represses P_R.

16.8 Other Transcriptional Control Systems

Transcription of bacterial and phage genomes may also be controlled by alteration of transcription factors under certain conditons and by changes in promoter efficiency.

Transcription Factors

Some phages, notably phage T4, have highly specific transcription factors that ensure proper timing of transcription. In both bacteria and eukaryotes, stresses such as heat causes production of proteins that help protect the cell.

Phage T4

Phage T4, a relatively large phage with approximately 300 genes, has transcription controlled by different RNA polymerase subunit proteins. Like phage λ, phage T4 has early, middle, and late genes, which need to be expressed in a particular order.

Early T4 gene promoters are recognized by the bacterial sigma factor (σ^{70} of *E. coli*). Early transcription of the T4 phage expresses the AsiA and MotA proteins. The AsiA protein binds the same –35 promoter sequence as σ^{70} but does not bind the RNA polymerase. Thus, binding of the AsiA protein to the –35 promoter sequence blocks the binding of the σ^{70} factor and prevents the transcription of the bacterial and early phage genes. AsiA is thus called an **anti-sigma factor,** because it interferes with σ^{70} binding to the –35 promoter sequence.

Unlike the early phage gene promoters, the middle phage gene promoters lack the –35 recognition region and are not affected by AsiA. However, the middle phage gene promoters contain a binding site for the MotA protein. Bacterial RNA polymerase bound with the σ^{70}–AsiA complex recognizes these MotA-bound promoters. As a result, MotA functions as a positive regulator for the transcription of the phage middle genes.

Finally, late phage genes have promoters that depend on the phage-encoded sigma factor σ^{gp55}. Some proteins are needed both early and late in the infection process; they are specified by genes that have promoters recognized by different specificity factors.

Heat Shock Proteins

Heat shock proteins are produced by both bacteria and eukaryotes in response to elevated temperature (see chapter 10). In *E. coli*, elevated temperatures cause the general shutdown of protein synthesis at the same time as at least 17 heat shock proteins are expressed. These proteins help protect the cell against the consequences of elevated temperature; some are molecular chaperones that stabilize folding of other proteins (see chapter 11).

The production of these *E. coli* proteins is the direct result of the gene product of the *htpR* gene, which encodes an alternative sigma factor (σ^{32}) rather than the normal sigma factor, σ^{70} (encoded by the *rpoD* gene). The heat shock genes have promoters that are recognized by σ^{32} rather than σ^{70}. Heat activates transcription of the *htpR* gene, as well as stabilizing the σ^{32} protein.

From DNA sequence data, the difference in promoters between normal genes and heat shock genes seems to lie in the –10 consensus sequence (Pribnow box). In normal genes, it is TATAAT; in heat shock protein genes, it is CCCCATXT, in which *X* is any base. Thus, the σ^{70} protein recognizes the normal A–T-rich –10 promoter sequence, whereas the σ^{32} protein recognizes the alternative –10 promoter sequence.

Promoter Efficiency

Other mechanisms exist to regulate the transcription of messenger RNA. One is to control the efficiency of various processes. For example, different genes in *E. coli* have different promoter sequences. Because the affinity for RNA polymerase is different for these sequences, the rate of initiation of transcription also varies. The more efficient promoters are transcribed at a greater rate than are the less efficient promoters.

An example of different promoter efficiencies is the *lacI* promoter. The wild-type *lacI* promoter is constitutively expressed at a low level, usually producing only about one mRNA per cell division. Mutations in the *lacI* promoter sequence, which increases the efficiency of RNA polymerase binding to the promoter, are known that produce up to 50 mRNAs per cell division. This increased level of *lacI* expression results in significantly more LacI repressor protein being present in the cell, which requires significantly higher concentrations of the inducer (allolactose or IPTG) to activate transcription of the *lac* operon.

Efficiency can be controlled by the direct sequence of nucleotides (i.e., differences from the consensus sequence) or by the distance between consensus regions. For example, promoters vary in the number of bases between the –35 and –10 sequences. Seventeen seems to be the optimal number of bases separating the two. Presumably, more or fewer than 17 reduces the efficiency of transcription.

Restating the Concepts

▶ Expression of genes may be regulated by either affecting the activity of the sigma factor or replacing the sigma factor. This results in an inability to recognize some promoters, while also recognizing some new promoter sequences.

▶ The efficiency of initiating transcription is another way to regulate gene expression. Increasing the initiation frequency results in more mRNA molecules present in the cell.

16.9 Translational Control

When considering control of gene expression, it is important to remember that all control mechanisms are aimed at exerting an influence on either the amount or the activity of the gene product. In the previous examples, we discussed ways to regulate transcription of a gene. As the amount of mRNA is increased, more protein can be translated (and conversely, less mRNA results in less protein). In addition to these transcriptional controls, translational controls exist that could affect how efficiently the mRNA is translated.

In bacteria, translational control is of less importance than transcriptional control for two reasons. First, mRNAs are extremely unstable; with a lifetime of only about 2 minutes, there is little room for controlling the rates of translation of existing mRNAs—they simply do not last very long. Second, although some indications of translational control have been found in bacteria, such control is inefficient—energy is wasted synthesizing mRNAs that may never be used.

Translational Polarity

Translational control can be exerted on a gene if the gene occurs further from the promoter in a polycistronic operon. In an operon, the gene transcribed last is usually translated at a lower rate than the gene transcribed first. This **translational polarity** is seen in the three *lac* operon structural genes, which are translated roughly in a ratio of 10:5:2 (with *lacZ* translated at the highest rate and *lacA* at the lowest).

Polarity is evident in bacteria, where transcription and translation are coupled—an mRNA can have ribosomes attached to it well before transcription ends. Genes at the beginning of the operon are therefore available for translation before genes at the end, which means the first gene usually has more ribosomes translating it during the lifetime of the mRNA.

Additionally, exonucleases seem to degrade mRNA more efficiently from the 3' end. Thus, the last gene in the polycistronic mRNA may be degraded while the first gene is still being translated. Presumably, natural selection has ordered the genes within operons so that those producing enzymes needed in greater quantities are found at the beginning of an operon.

RNA–RNA Hybridization

Translation can also be regulated by RNA–RNA hybridization. RNA complementary to the 5' end of an mRNA can prevent the translation of that mRNA. The regulating RNA is called an **antisense RNA.**

In **figure 16.33**, the mRNA from the *ompF* gene in *E. coli* is prevented from being translated by complementary base pairing with an antisense *micF* RNA (*mic* stands for *mRNA-interfering complementary RNA*). This complementary base pairing covers the Shine–Dalgarno sequence and the AUG translation initiation codon, which blocks translation initiation of the *ompF* open reading frame. The *ompF* gene codes for a channel protein called *porin,* which provides pores in the outer membrane for the transport of materials. Surprisingly, a second porin gene, *ompC,* seems to be the source of the *micF* RNA. Transcription of one DNA strand produces the *ompC* mRNA, while transcription of the other strand in the same region yields the antisense RNA. One porin gene (*ompC*) therefore seems to regulate the expression of another porin gene (*ompF*), for reasons that are not completely understood.

Antisense RNA has also been implicated in such phenomena as the control of plasmid number and the control of transposon transposition (see chapter 18). Control by antisense RNA is a fertile field for gene therapy because antisense RNA can be artificially synthesized and then introduced into eukaryotic cells.

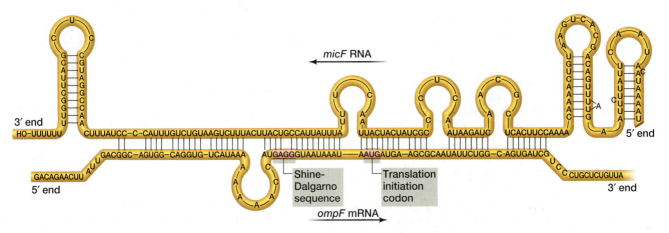

figure 16.33   Model of antisense RNA regulation of translation. Complementarity between the *ompF* mRNA and the antisense *micF* RNA allows them to base pair as diagrammed. The *ompF* Shine-Dalgarno sequence and translation initiation codon, lie within this double-stranded region, which effectively prevents ribosome binding and translation of the *ompF* RNA. Thus, expression of the *micF* mRNA represses translation of the *ompF* mRNA.

Reprinted, with permission, from the *Annual Review of Biochemistry,* Volume 55 © 1986 by Annual Reviews. www.annualreviews.org.

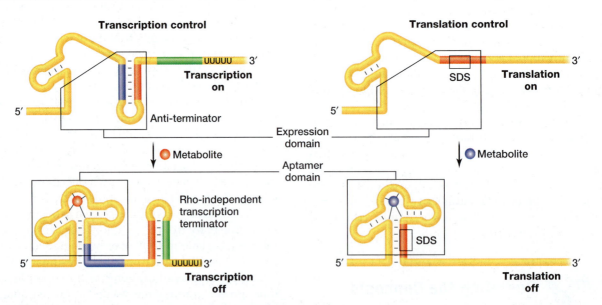

figure 16.34   Riboswitches can exert either transcriptional or translational control on an mRNA. In transcriptional control (left), binding of the metabolite causes the mRNA to alter its base pairing from an anti-termination structure (top) to a rho-independent transcription terminator (bottom). In translational control (right), the Shine-Dalgarno sequence (SDS, ribosome binding site) is accessible to the ribosome (top) for translation to occur. Binding of the metabolite alters the base pairing in the mRNA so the Shine–Dalgarno sequence is masked within a double-stranded RNA region (bottom). In both processes, the switch between the two RNA conformations is controlled by the binding or absence of a small molecule in the aptamer domain of the RNA.

Riboswitches

We already discussed situations in which the secondary structure of the RNA directly affects the expression of a gene. In attenuation, the translation of a leader peptide or the binding of the TRAP molecule dictates if a rho-independent transcription termination stem-loop signal is produced in the mRNA. In both of these cases, the RNA conformation is an indirect measurement of tryptophan levels, through the presence of either the Trp-tRNA^{Trp} or the tryptophan-bound TRAP. Some RNA molecules, however, can directly bind the small molecule that must be measured.

Riboswitches are RNA sequences that regulate gene expression by directly binding small molecules to control their secondary structure (**fig. 16.34**). Riboswitches possess two different domains. An **aptamer domain** is a folded region of the RNA that can bind specific small molecules, such as amino acids or purines. The **expression domain** is a RNA sequence that exists in two different conformations that controls gene expression.

The riboswitch can regulate either transcription or translation of a gene or operon. To control transcription, one conformation of the expression domain yields a rho-independent transcription termination signal, and the second conformation encodes an antiterminator (see fig. 16.34, left side). To control translation, one conformation of the expression domain has a freely accessible Shine-Dalgarno sequence for ribosome binding, and the second conformation masks the Shine–Dalgarno sequence within a base-paired RNA stem, which cannot be bound by the small ribosomal subunit (see fig. 16.34, right side).

Riboswitches can be used to either inactivate or activate gene expression. In the two examples in figure 16.34, the binding of the small molecule either terminates transcription or blocks translation initiation. Both of these represent riboswitches that inactivate gene expression. However, the adenine-sensing riboswitch is an example of a riboswitch that activates gene expression. When the aptamer binds adenine (**fig. 16.35**, top), the expression domain assumes a conformation that lacks a transcription termination stem signal. In the absence of adenine, a rho-independent transcription termination stem signal is formed, and transcription of the following genes is stopped.

As you might predict based on your knowledge of attenuation and the ability of some riboswitches to regulate translation initiation, the riboswitches are often found in the 5' untranslated region of the RNA. However, riboswitches differ from our previous gene control mechanisms because the RNA molecule directly binds the small regulatory molecule.

Ribosome-Binding Efficiency

A fourth translational control mechanism consists of the efficiency with which the ribosome binds to the mRNA. This binding is related to some extent to the sequence of nucleotides at the 5' end of the mRNA that is complementary to the 3' end of the 16S ribosomal RNA segment in the ribosome (the Shine–Dalgarno sequence, chapter 11). The Shine–Dalgarno sequence is usually located just upstream of the translation initiation codon. Variations in this consensus sequence demonstrate different efficiencies of binding, and, therefore, the initiation of translation occurs at different rates.

Redundancies and Codon Preference

The redundancy in the genetic code can also play a part in translational control of some proteins because different tRNAs occur in the cell in different quantities. Genes with abundant protein products may have codons that specify the more common tRNAs, a concept called **codon bias.**

Each species has certain codons that are preferred; they specify tRNAs that are abundant. Genes that code for proteins not needed in abundance could have sev-

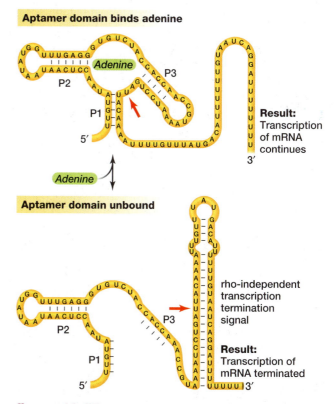

figure 16.35   The adenine-sensing riboswitch. When the aptamer domain binds a molecule of adenine (top), the RNA lacks a rho-independent transcription termination signal and transcription proceeds. When adenine is not bound by the aptamer domain, the RNA assumes a different conformation that produces the rho-independent transcription termination signal. The uracil that is critical for binding adenine in the aptamer domain is designated with a red arrow.

eral codons specifying the rarer tRNAs, which would slow down the rate of translation for these genes. As an example, although the phage MS2 uses every codon except the UGA stop codon (**table 16.2**), the distribution of codons is not random. (The numbers in the table refer to the incidence of a particular codon in the phage genome.) For example, the amino acid glycine has two common codons and two rarer codons. The same holds for arginine but not, for example, valine.

Amino Acid Levels

Finally, translational control can be exerted at the ribosome in a mechanism called the **stringent response.** When a bacterial cell has very low levels of several amino acids (termed *amino acid starvation*), an uncharged tRNA can enter the A site of the ribosome. This uncharged tRNA stimulates an **idling reaction.** The **stringent factor protein,** which is associated with the ribosome, then converts the nucleotide GDP to guanosine tetraphosphate (5'-ppGpp-3'; **fig. 16.36**).

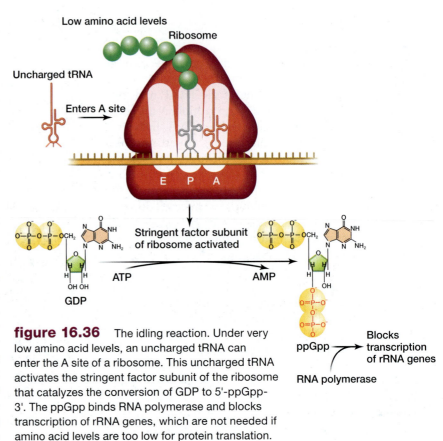

figure 16.36 The idling reaction. Under very low amino acid levels, an uncharged tRNA can enter the A site of a ribosome. This uncharged tRNA activates the stringent factor subunit of the ribosome that catalyzes the conversion of GDP to 5'-ppGpp-3'. The ppGpp binds RNA polymerase and blocks transcription of rRNA genes, which are not needed if amino acid levels are too low for protein translation.

table 16.2   Codon Distribution in MS2, an RNA Bacteriophage

First Position	Second Position								Third Position
	U		C		A		G		
U	Phe	10	Ser	13	Tyr	8	Cys	7	U
	Phe	13	Ser	10	Tyr	13	Cys	4	C
	Leu	11	Ser	10	stop	1	stop	0	A
	Leu	4	Ser	13	stop	1	Trp	14	G
C	Leu	10	Pro	7	His	4	Arg	13	U
	Leu	14	Pro	3	His	4	Arg	11	C
	Leu	13	Pro	6	Gln	10	Arg	6	A
	Leu	6	Pro	5	Gln	16	Arg	4	G
A	Ile	8	Thr	14	Asn	11	Ser	4	U
	Ile	16	Thr	10	Asn	23	Ser	8	C
	Ile	7	Thr	8	Lys	12	Arg	8	A
	Met	15	Thr	5	Lys	17	Arg	6	G
G	Val	13	Ala	19	Asp	18	Gly	17	U
	Val	12	Ala	12	Asp	11	Gly	11	C
	Val	11	Ala	14	Glu	9	Gly	4	A
	Val	10	Ala	8	Glu	14	Gly	4	G

This guanosine tetraphosphate serves as a signaling molecule and interacts with RNA polymerase to represses transcription of rRNA, so no energy is wasted synthesizing ribosomes when translation is not possible. However, many amino acid-synthesizing operons require ppGpp for transcription. Thus, ppGpp inhibits rRNA synthesis and enhances the transcription of genes that encode enzymes that synthesize amino acids, when the cell is starved for amino acids.

Restating the Concepts

▶ Expression of proteins can be regulated at the translational level. In bacteria, the major translational control mechanisms include polarity of the operon, efficiency of ribosome binding of mRNA, and codon preference in the ORF. These mechanisms are all fixed and depend on the mRNA sequence.

▶ Riboswitches control gene expression through RNA sequences (aptamers) that can directly bind small molecules and invoke a conformational change in an adjacent RNA structure that can affect either transcription or translation of the RNA.

▶ Binding of an antisense RNA to an mRNA can regulate translation of the mRNA. This is a control mechanism that can be modulated based on the expression of the antisense RNA (or the amount of mRNA relative to the antisense RNA).

▶ The stringent response is a translational control mechanism that responds to the level of available amino acids. When the amount of amino acids becomes very low, the stringent factor produces 5'-ppGpp-3' in the ribosome. This molecule interacts with RNA polymerase to suppress transcription of rRNA genes and activates expression of operons that encode enzymes involved in biosynthesis of some amino acids.

16.10 Posttranslational Control

Even after a gene has been transcribed and the mRNA translated, a bacterial cell can still exert some control over the functioning of the enzymes produced if the enzymes are allosteric proteins. We already discussed the activation and deactivation of operon repressors (e.g., Lac, Trp) owing to their allosteric properties. Similar effects occur with other proteins.

The need for posttranslational control is apparent because of the relative longevity of proteins relative to mRNA. When an operon is repressed, it no longer transcribes mRNA; however, the protein that was previously translated from the mRNA remains and is still functional. Thus, during operon repression, it would also be efficient for the cell to control the activity of existing proteins.

Feedback Inhibition

An example of posttranslational control occurs with the enzyme aspartate transcarbamylase, which catalyzes the first step in the pathway of pyrimidine biosynthesis in *E. coli* (**fig. 16.37**). An excess of one of the end products of the pathway, cytidine triphosphate (CTP), inhibits the activity of aspartate transcarbamylase. This method of control is called **feedback inhibition** because an end product of the pathway is the molecule that inhibits one of the first enzymes in the pathway.

Aspartate transcarbamylase is an allosteric enzyme. Its active site is responsible for the condensation of carbamyl phosphate and L-aspartate (**fig. 16.38**). However, it also has regulatory sites that bind CTP. When CTP is bound in a regulatory site, the conformation of the enzyme changes (see fig. 16.38), and the enzyme has a lowered affinity for its normal substrates. In this way, abundant CTP inhibits the condensation reaction the enzyme normally carries out. A number of allosteric enzymes (usually the first enzyme in a pathway) bind the final product in a biochemical pathway to regulate enzyme activity after the protein has been synthesized.

Notice that in feedback inhibition, an end product acts to regulate the activity of an enzyme and not on the transcription of the gene that encodes the enzyme. It is therefore different from the repressible systems described earlier, which slow or halt gene transcription.

Protein Degradation

A final control point at which the amount of gene product in a cell can be affected is the rate at which proteins degrade. The normal life spans of proteins vary greatly. For example, some proteins last longer than a cell cycle, whereas others may be broken down in minutes.

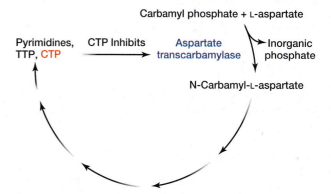

figure 16.37 Mechanism of feedback inhibition. Aspartate transcarbamylase catalyzes the first step in pyrimidine biosynthesis. One of the end products, cytidine triphosphate (CTP), inhibits the aspartate transcarbamylase enzyme. Thus, when CTP levels get high, the CTP represses the synthesis of more CTP.

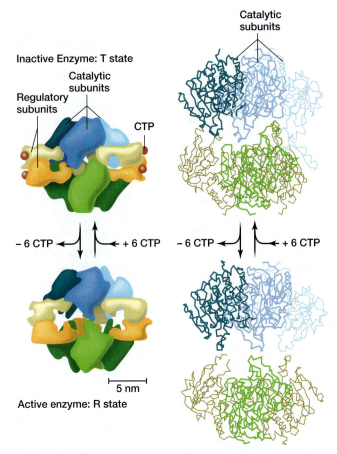

Inactive Enzyme: T state

Catalytic subunits

Regulatory subunits

CTP

Catalytic subunits

− 6 CTP ⇌ + 6 CTP   − 6 CTP ⇌ + 6 CTP

5 nm

Active enzyme: R state

figure 16.38 Two views of the allosteric enzyme aspartate transcarbamylase. Left: A space filling cartoon shows the enzyme is made up of six identical subunits, each containing a regulatory domain (CTP-binding) and catalytic domain. The top panel shows the tightly closed (inactive) protein when each subunit has a cytidine triphosphate (CTP) molecule bound, while the lower panel shows the open (active) protein when CTP is not bound. Right: A line diagram of only the catalytic domains of aspartate transcarbamylase in the inactive (top) and active (bottom) conformations. The regulatory subunits are not shown to simplify the diagram and better reveal the difference between the catalytic subunits. Notice the difference in the accessibility of the region between the catalytic subunits when CTP is bound and unbound.

Several models have been suggested for control of protein degradation, including the **N-end rule** and the **PEST hypothesis.**

The N-End Rule

According to the N-end rule, the amino acid at the amino-, or N-terminal, end of a protein is a signal to proteases that control the average lifetime of a protein. Although nearly every protein initially has a methionine at the amino terminus because of the requirement

table 16.3 Relationship Between Modified N-Terminal Amino Acid and Half-Life of *E. coli* β-Galactosidase

N-Terminal Amino Acid	Half-Life
Met, Ser, Ala, Thr, Val, Gly	> 20 h
Ile, Glu	30 min
Tyr, Gln	10 min
Pro	7 min
Phe, Leu, Asp, Lys	3 min
Arg	2 min

of the translation initiation codon, this amino terminus can be cleaved to reveal a new N-terminal amino acid in the functional protein (see chapter 11).

In recent experiments, the life span of the β-galactosidase protein was determined with almost complete predictability based on its modified N-terminal amino acid. Protein life spans range from 2 minutes for those with an N-terminal arginine to greater than 20 hours for those with N-terminal methionine or one of five other amino acids (**table 16.3**). Therefore, a delay in removal of the N-terminal methionine, or other N-terminal amino acids, is one way to confer a longer life span.

The PEST Hypothesis

According to the PEST hypothesis, protein degradation is determined by regions rich in one of four amino acids: proline, glutamic acid, serine, and threonine. (The single-letter abbreviations of these four amino acids are P, E, S, and T, respectively.) Proteins that have these regions tend to degrade in less than 2 hours. In one study of 35 proteins with half-lives of between 20 and 220 hours, only three contained a PEST region.

Restating the Concepts

▶ Posttranslational control of enzyme activity is a final mechanism to regulate biochemical pathways in a cell.

▶ Allosteric enzymes often exhibit feedback inhibition, in which the first enzyme in a biosynthetic pathway can usually bind the final product to inhibit the activity of the enzyme. This limits the generation of more product when sufficient amounts are already available.

▶ Protein degradation is another way to limit the amount of protein in a cell. This degradation cannot be easily regulated or changed by the cell, however; it depends upon the amino acid sequence of the functional protein.

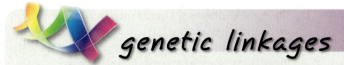
genetic linkages

In this chapter, we discussed several different ways in which bacterial and bacteriophage genes and their products are regulated. In chapter 17, we begin to examine mechanisms that eukaryotes employ to regulate gene expression. Many of the concepts that we discussed in this chapter, such as DNA-binding proteins, positive regulation, and negative regulation, still apply. Furthermore, many of these gene regulation concepts are relevant when we discuss eukaryotic development and cancer (see chapters 21 and 22).

The bacterial and bacteriophage −10 and −35 promoter regions were previously described in chapter 10. In this chapter, we discussed how proteins that bind in this region can affect the binding of RNA polymerase. The importance of the Shine–Dalgarno sequence for the binding of the 3' end of the 16S ribosomal RNA to initiate translation was described in chapter 11; you now know that alterations of this sequence can serve as a regulatory mechanism.

Some of the topics described in this chapter, such as polycistronic mRNAs and operons, are rarely seen in eukaryotes. These characteristics result from the coupling of transcription and translation in bacteria and their phages. Because transcription occurs in the eukaryotic nucleus and the mRNA is then transported to the cytosol for translation, it is usually not possible to simultaneously transcribe a eukaryotic mRNA and translate it.

In chapter 12, we considered the use of phage λ as a cloning vector. In this chapter, we discussed how λ chooses between the prophage and lytic forms. The choice of the lytic cycle is largely based on the ability of the bacteria to multiply and grow (which favors the lytic cycle). In chapter 18, we further describe the mechanism by which UV light damages DNA and how this induces the SOS repair system and the excision of the λ prophage.

Short complementary RNAs and their ability to either degrade or block the translation of the complementary mRNA were mentioned in chapter 10. In this chapter, you were reintroduced to this concept and its use in bacterial cells. The following chapter describes how they operate as a mechanism to regulate eukaryotic gene expression.

Finally, in chapter 17 we describe many of the common protein motifs (structures) that are found in eukaryotic DNA-binding proteins. Some of these motifs, in particular the helix-turn-helix motif, are common to both bacteria and eukaryotes.

Chapter Summary Questions

1. For each term in column I, choose the best matching phrase in column II.

   Column I
   1. Inducible
   2. Leader sequence
   3. Allostery
   4. Antisense RNA
   5. Aptamer domain
   6. Repressor
   7. Operator
   8. *Trans*-acting factors
   9. Lysogen
   10. Operon

   Column II
   A. Molecule that binds the DNA to stop transcription
   B. Phage integrated into the host genome
   C. A group of genes encoded on a single mRNA, resulting in coordinate regulation
   D. Transcription controlled in response to the presence of a compound in the environment
   E. Complementary RNA molecule that regulates translation
   F. Proteins that can diffuse within a cell and bind to a DNA sequence
   G. Change in shape of a molecule due to binding another molecule
   H. RNA region that can bind a specific molecule
   I. Binding site of repressor molecule
   J. Short coding region regulating expression of the rest of the mRNA

2. Describe the role of cyclic-AMP in transcriptional control in *E. coli*.

3. The *lac* operon is negatively controlled by the repressor binding to the operator site. The CAP system exerts positive control because it acts through enhancement of transcription. Describe how an inducible operon could work if it was dependent only on positive control.

4. Describe the interaction of the attenuator and the operator control mechanisms in the *trp* operon of *E. coli* under varying concentrations of tryptophan in the cell. How does the attenuator control react to shortages of other amino acids?

5. Assume we can construct a merozygote of the *trp* operon in *E. coli* with two forms of the first two structural genes (*trpE⁺* and *trpE⁻*, *trpD⁺* and *trpD⁻*) in the operon so we can differentiate expression from each copy of the operon. Describe the types of *cis* and *trans* effects that are possible, given the existence of mutants of any component of the operon (*trpR*, *trpP*, *trpO*, and *trpL*). Can this repressible system work for any type of operon other than those that control amino acid synthesis?

6. What is antisense RNA? How does it work? What is the obvious source of this regulatory RNA? How could this RNA be used to treat a disease clinically?

7. What is a riboswitch and why is it such an effective method for controlling gene expression?

8. What is the fate of a λ phage entering an *E. coli* cell that contains quantities of λ repressor? What is the fate of the same phage entering an *E. coli* cell that contains quantities of the *cro*-gene product?

9. Describe the fate of phages during the infection process with mutations in the following genes: *cI, cII, cIII, N, cro, att, Q.*

10. What is the fate of λ phages during the infection process with mutations in the following sequences: O_{R1}, O_{R3}, P_L, P_{RE}, P_{RM}, P_R, t_{L1}, t_{R1}, *nutL, nutR?*

11. What three different physical forms can the phage λ chromosome take?

12. How does ultraviolet light (UV) damage induce the lytic life cycle in phage λ?

13. What is feedback inhibition? What other roles do allosteric proteins play in regulating gene expression?

14. What is one mechanism controlling the rate of protein degradation?

Solved Problems

PROBLEM 1: How could you determine whether the genes for the breakdown of the sugar arabinose are under inducible control in *E. coli*?

Answer: Inducible means that the genes are not transcribed in the absence of the inducer molecule. In this example, the genes that encode the enzymes that break down arabinose, a five-carbon sugar, would not be transcribed in the absence of the inducer (again, arabinose). Therefore, when arabinose is absent in the cells' environment, the arabinose utilization enzymes should not be present within the bacterial cells. When arabinose is added to the medium, the enzymes should be present. We thus need to assay the contents of the cells before and after arabinose is added to the medium, performing the assay after the cells are broken open and the DNA destroyed so as not to confound the experiment. Using a standard biochemical analysis for arabinose, the bacterial cell should be incapable of metabolizing arabinose before induction but capable of metabolizing it afterward. If the cells metabolized arabinose in both cases, the expression of arabinose would be constitutive. If the cells were incapable of utilizing arabinose in both cases, we would conclude that the bacterium is incapable of using the sugar arabinose as an energy source. If the cells only metabolized arabinose after it was added to the medium, then we could conclude that it was inducible. An alternative to the standard biochemical analysis, we could directly examine the transcription of the genes by either Northern blot hybridization or PCR amplification of mRNA. In the inducible situation, the mRNA should be present when arabinose is available in the environment and absent when arabinose is missing.

PROBLEM 2: Why would the RecA protein of *E. coli* cleave the λ repressor?

Answer: Since the cleaving of the λ repressor is a signal to begin the lytic phase of the life cycle of the phage, it seems odd that the lysogenized bacterial cell would be an accomplice to its own destruction. However, the phenomenon makes much more sense if we realize that the RecA protein has several functions critically important to the bacterial cell. One of those functions is to recognize when the bacterial genomic DNA is damaged. This DNA damage activates the RecA* activity, which is a protease that cleaves specific transcriptional repressors. Loss of these repressors activates the

transcription of several genes that encode enzymes that act to repair damaged DNA. However, this repair is not always effective, which still results in the cell dying. If this cell is a lysogen, the integrated λ phage would also be lost. Thus, the λ repressor has evolved the ability to be sensitive to the RecA* activity. Thus, DNA damage results in the generation of RecA* activity, the cleavage of λ repressor, the induction of the λ lytic cycle and the release of λ particles to infect other cells. Evolutionary biologists view this as "coevolution"— two interacting organisms evolving to take advantage of a common property, UV-damaged DNA. This is one plausible explanation as to why RecA liberates phage λ.

PROBLEM 3: What are the differences in action of the λ promoters P_{RE} and P_{RM}?

Answer: P_{RE} and P_{RM} are both promoters of the repressor operon of phage λ. Transcription from both of these promoters allows production of the CI repressor protein, the repressor that favors lysogeny. Initially, transcription from the P_{RE} promoter is activated by the CII protein, which is transcribed from the P_R operon. This promoter, P_{RE}, produces an mRNA with a long leader that is translated very efficiently. Once the repressor binds at the operators of the left and right operons, the *cII* gene is no longer transcribed, and therefore P_{RE} is no longer a site for transcription. However, the repressor gene can still be transcribed from the P_{RM} promoter, which does not need the product of the *cII* gene. This promoter produces a transcript with no leader and thus is translated very inefficiently. At that point, however, only a very small quantity of repressor is needed to maintain lysogeny. Thus, the two promoters are the sites for the initiation of the repressor operon under different circumstances: one early in the infection stage and one after lysogeny is under way.

PROBLEM 4: What are the phenotypes of the following partial diploids for the *lac* operon in *E. coli* in the presence and absence of lactose?

    a. *lacI⁻ lacO⁺ lacP⁺ lacZ⁺ /F′ lacI⁺ lacO⁺ lacP⁺ lacZ⁻*
    b. *lacI⁺ lacO⁺ lacP⁻ lacZ⁺ / F′ lacI⁺ lacOᶜ lacP⁺ lacZ⁺*

Answer: Consider one DNA molecule at a time. If one DNA molecule can never make the enzyme, it can be ignored. In (a), the plasmid DNA (*F′*) will never make enzyme (it is *lacZ⁻*), so we are only concerned with the *lacZ* gene that is present in the genome. The genomic DNA will never make repressor (it is *lacI⁻*). However, the functional repressor expressed from the *F′* plasmid (*lacI⁺*) will bind to the operator in both DNAs, and hence the genomic operon will be induced, it will not be transcribed in the absence of lactose and will transcribe the *lacZ⁺* in the presence of lactose. In (b), the *F′* element will always transcribe the *lacZ* operon on the *F′* because the repres-

sor can never bind the *lacOᶜ* operator mutation (constitutive operator); hence, the operon and *lacZ⁺* will be transcribed all the time. There is no reason to examine the other operon if one is constitutively transcribed. However, the genomic operon can never be transcribed because it is associated with a *lac* promoter mutation (*lacP⁻*).

Exercises and Problems

15. Are the following *E. coli* cells constitutive or inducible for the *lacZ* gene?
    a. *lacI⁺ lacO⁺ lacZ⁺*
    b. *lacI⁻ lacO⁺ lacZ⁺*
    c. *lacI⁻ lacOᶜ lacZ⁺*
    d. *lacI⁺ lacOᶜ lacZ⁺*
    e. *lacIˢ lacO⁺ lacZ⁺*
    f. *lacI⁻ lacO⁺ lacZ⁺*

16. Determine whether the following *lac* operon merozygotes are inducible or constitutive for the *lacZ* gene.
    a. *lacI⁺ lacO⁺ lacZ⁺ / F′ lacI⁺ lacO⁺ lacZ⁺*
    b. *lacI⁻ lacO⁺ lacZ⁺ / F′ lacI⁺ lacO⁺ lacZ⁻*
    c. *lacI⁺ lacO⁺ lacZ⁺ / F′ lacI⁻ lacO⁺ lacZ⁻*
    d. *lacI⁻ lacOᶜ lacZ⁻ / F′ lacI⁺ lacO⁺ lacZ⁺*
    e. *lacI⁻ lacO⁺ lacZ⁻ / F′ lacI⁺ lacOᶜ lacZ⁺*
    f. *lacIˢ lacO⁺ lacZ⁻ / F′ lacI⁺ lacOᶜ lacZ⁺*

17. For each of the following genotypes, determine if β-galacotsidase and permease activities are present.

	no IPTG present		IPTG present	
	β-Galactosidase	Permease	β-Galactosidase	Permease
lacZ⁺ Y⁻ / F′ lacZ⁻ Y⁺				
lacIˢ Z⁺ Y⁻ / F′ lacI⁻ Z⁻ Y⁺				
lacIˢ Oᶜ Z⁺ Y⁻ / F′ lacP⁻ Z⁻ Y⁺				
lacZ⁺ Y⁻ / F′ lacP⁻ Z⁻ Y⁺				
lacIˢ Z⁺ Y⁻ / F′ lacOᶜ I⁻ Z⁻ Y⁺				
lacI⁻ Oᶜ Z⁺ Y⁻ / F′ lacZ⁻ Y⁺				
lacP⁻ Oᶜ Z⁺ Y⁻ / F′ lacI⁻ Z⁻ Y⁺				

18. What would be the response of *E. coli* to the addition of lactose if it contained a *lac* operon with a mutant, completely nonfunctioning *lacY* (permease) gene?

19. A mutant *E. coli* is isolated that does not produce *lacZ*. A merodiploid is formed by introducing an F′ with a wild-type *lac* operon, except that the *lacZ* gene is nonfunctional. The merodiploid also does not produce *lacZ*. What are the possible causes of this mutation?

20. You isolated two *E. coli* mutants that synthesize β-galactosidase constitutively.
    a. If these mutants affect different functions, in what two functions could they be defective?
    b. You can make a partial diploid of the mutants with the wild type. What result do you expect for each mutant?

21. A hypothetical operon has a sequence of sites, *Q R S T U*, in the promoter region, but the exact location of the operator and promoter consensus sequences have not been identified. Various deletions of this operator region

are isolated and mapped. Their locations appear as follows, with a gap representing a deleted region.

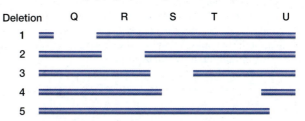

Deletions 3 and 4 are found to produce constitutive levels of RNA from the operon, and deletion 1 is found to never make RNA. Where are the operator and promoter consensus sequences probably located?

22. Jon Beckwith isolated point mutations that were simultaneously uninducible for the *lac, ara, mal,* and *gal* operons, even in the absence of glucose. Provide two different functions that could be missing in these mutants.

23. Two mutations in the arabinose operon are recovered that do not affect the induced expression in the absence of glucose and presence of arabinose, but which do not show full repression in the absence of arabinose. Merodiploid studies show that one mutation is *cis*-acting and the other is *trans*-acting.
    a. What are the two mutations?
    b. If a merodiploid is generated using these two mutant operons, describe the expression in the presence of arabinose (and absence of glucose) and the absence of glucose.

24. A double mutant is generated in the *lac* operon of *E. coli* that is both operator constitutive and produces a super-repressor. What will be the phenotype of this mutant for *lacZ* inducibility?

25. The tryptophan operon is under negative control; it is on (transcribed) in the presence of low levels of tryptophan and off in the presence of excess tryptophan. The symbols *a*, *b*, and *c* represent the gene for tryptophan synthetase, the operator region, and the repressor—but not necessarily in that order. From the following data, in which superscripts denote wild type or mutant, determine which letter is the gene, the repressor, and the operator (+ is tryptophan synthetase activity; – is no activity).

Strain	Genotype	Tryptophan Absent	Tryptophan Present
1	$a^- b^+ c^+$	+	+
2	$a^+ b^+ c^-$	+	+
3	$a^+ b^- c^+$	–	–
4	$a^+ b^- c^+/a^- b^+ c^-$	+	+
5	$a^+ b^+ c^+/a^- b^- c^-$	+	–
6	$a^+ b^+ c^-/a^- b^- c^+$	+	–
7	$a^- b^+ c^+/a^+ b^- c^-$	+	+

26. What advantage, if any, does the attenuator system have over the TRAP system for regulating amino acid synthesis?

27. The histidine operon is a repressible operon. The corepressor is charged His-tRNA^{His}, and the repressor gene is not part of the operon. For the following mutants, determine whether the genes of the operon will be transcribed; then determine whether each mutant would be *cis*-dominant in a partial diploid.
    a. RNA polymerase cannot bind the promoter.
    b. The repressor-corepressor complex cannot bind the operator DNA (the operator has the normal sequence).
    c. The repressor cannot bind charged His-tRNA^{His}.

28. What might be the advantage to the bacteria in having two systems—repression and attenuation—simultaneously acting to control operons that encode enzymes for amino acid biosynthesis?

29. What would happen to a λ phage that contained a Cro protein with a high affinity for O_{R1}?

30. Plaques produced by wild-type λ phage tend to have a turbid appearance due to the presence of a low level

of "survivor" bacteria growing. There are mutants of λ phage, however, that only produce clear plaques. What is the most likely site for these mutations in the λ phage genome and why?

31. What mutations would cause a shift so that λ phage followed the lytic cycle more frequently? What mutations would make the phage always go into the lytic pathway?

32. A temperature-sensitive mutant in the λ *cI* gene was isolated. At 30°C the CI repressor binds λ DNA, but it cannot bind DNA at 42°C (the CI protein denatures). What is the consequence of incubating *E. coli* that are lysogenic for this λ mutant at 42°C?

33. The mutant in problem 32 is heated to 42°C for 5 min, cooled to 30°C, and grown for 1 h so that the cells divide several times. The temperature is then raised to 42°C, and you wait for lysis. Many of the cells are not lysed and are in fact able to form colonies. Explain these results.

34. What are the advantages of transcriptional control over translational control? What are the advantages of translational control over transcriptional control?

35. How is the expression of a heat shock protein induced? Why would a GC-rich promoter sequence be used for heat shock genes instead of an AT-rich promoter?

36. In phage T4, the genes *rIIA* and *rIIB* lie adjacent to each other on the T4 chromosome. During the early phase of infection, the rIIA and rIIB proteins are present in equimolar amounts. In the late phase of infection, the amount of rIIB protein is 10–15 times greater than the rIIA protein. Nonsense mutations (mutations to a stop codon) in the *rIIA* gene eliminate early, but not late, *rIIB* transcription. In mutants that contain small deletions near the end of the *rIIA* gene, the amount of rIIA proteins is always equal to the amount of rIIB protein, regardless of the time of infection. Based on this information, devise a map of the *rII* region, showing the location of the *rIIA* and *rIIB* genes, the location(s) of the promoter(s), and the direction of transcription.

37. A mutant bacteria is found that produces excess levels of thymidine and cytosine. The RNA and protein levels of the genes involved in thymidine and cytosine synthetic pathways appear normal. What is the most likely mutation that would cause this phenotype?

38. From an evolutionary perspective, why do you think *E. coli* evolved a CAP system of positive control of gene expression? Why not just metabolize any and all sugars in the environment as they appear?

39. Why are most mRNAs produced in *E. coli* degraded in a matter of minutes instead of being used over and over for several hours? Doesn't this seem like a wasteful system?

40. A mutant bacteria is found that synthesizes reduced amounts of the enzymes of the *trp* biosynthetic pathway when the bacteria is starved for tryptophan. This mutant produces full amounts of the enzymes if the bacteria is starved for both tryptophan and cysteine. If the mutation was mapped genetically, what region of the *trp* operon would it be found in? What is the most likely nature of this mutation?

41. A new strain of bacteria is found that uses a leader control system similar to the *trp* operon of *E. coli* to regulate transcription of the glycine pathway. Six identical codons for glycine are found in the leader region reading frame. Mutants are found that still produce significant levels of the full-length RNA of the glycine operon when glycine is present. The mutation is mapped to the leader region, but when the leader peptide is analyzed, it still contains the wild-type number of glycines? What might be the explanation for this mutation?

42. Are inducible systems always positively regulated? Are repressible systems always negatively regulated?

43. Why might allolactose, which is produced by β-galactosidase about 1% of the time when cleaving lactose, used as the inducer in the *lac* system instead of lactose?

44. The *lac* operon has three operator sites, all of which must be bound to produce maximum repression. How do these three sites work together to reduce transcription? What are the advantages to using three different sites that are widely separated from one another?

45. The hypothetical *luv* operon in *E. coli* encodes two enzymes designated 1 and 2. Symbols *v, w, x, y,* and *z* represent the two structural genes, the repressor gene, the operator region, and the promoter sequence—although not necessarily in this order. Eleven strains are presented in the table at the bottom of the page. Symbols: + = wild type, – = mutant, A = active enzyme, I = inactive enzyme, N = no enzyme.
  a. Is the *luv* operon inducible or repressible? Why?
  b. Deduce the identities of *v, w, x, y,* and *z*. Explain your reasoning.
  c. Complete the table by using the symbols A, I, and N.

46. The biosynthetic phenylalanine (*phe*) operon in *E. coli* is regulated by a leader–attenuator mechanism similar to that of the *trp* operon. The attenuator sequence contains four subregions with base sequences that are complementary to one another, so that the transcribed mRNA can form three different stem-loop structures, similar to that shown in figure 16.24. The DNA sequence of the leader peptide is shown below.

5' ATG AAA CAC ATA CCG TTT TTC TTC GCA TTC TTT TTT ACC TTC CCC TGA 3'
   Met Lys His Ile Pro Phe Phe Phe Ala Phe Phe Phe Thr Phe Pro STOP

  a. Is this the template or nontemplate strand of DNA?
  b. What is the minimum expected length of a random polypeptide chain that will contain the tripeptide Phe-Phe-Phe? Assume equal proportions and random distribution of amino acids.
  c. The leader peptide of the *trp* operon is 14 amino acids long and contains two codons for tryptophan. The leader peptide of the *phe* operon is about the same length (15 amino acids), yet it contains 7 codons for phenylalanine. Suggest a reason for this discrepancy.
  d. Will attenuation occur under the following conditions? If yes why, and if not why not? Assume the given mutations disrupt the stem-loop-forming ability of the region in question.
    i. Wild-type leader sequence; charged Phe-tRNAs are scarce; abundance of other amino acids.
    ii. Wild-type leader sequence; general amino acid starvation.
    iii. Mutant in region 1; tryptophan levels are high; starvation for other amino acids.
    iv. Mutant in region 2; tryptophan levels are low; abundance of other amino acids.
    v. Mutant in region 4; general amino acid starvation.

Strain	Genotype	Luv not found Enzyme 1	Luv not found Enzyme 2	Luv all around Enzyme 1	Luv all around Enzyme 2
1	$v^+ w^+ x^+ y^+ z^+$	N	N	A	A
2	$v^- w^+ x^+ y^+ z^+$	N	N	A	I
3	$v^+ w^- x^+ y^+ z^+$	A	A	A	A
4	$v^+ w^+ x^- y^+ z^+$	N	N	N	N
5	$v^+ w^+ x^+ y^- z^+$				
6	$v^+ w^+ x^+ y^+ z^-$	A	A	A	A
7	$v^- w^+ x^+ y^+ z^+/$ F' $v^+ w^+ x^+ y^+ z^+$				
8	$v^+ w^- x^+ y^+ z^+/$ F' $v^+ w^+ x^+ y^+ z^+$				
9	$v^+ w^+ x^- y^+ z^+/$ F' $v^+ w^+ x^+ y^+ z^+$				
10	$v^+ w^+ x^+ y^- z^+/$ F' $v^+ w^+ x^+ y^+ z^+$				
11	$v^+ w^+ x^+ y^+ z^-/$ F' $v^+ w^+ x^+ y^+ z^+$	A	A	A	A

The table has a two-level header: "Luv not found" spanning Enzyme 1 and Enzyme 2, and "Luv all around" spanning Enzyme 1 and Enzyme 2.

e. A new strain of *E. coli* is isolated and found to have two point mutations in the leader DNA sequence: a deletion of the middle T of the first Phe codon, and an insertion of an A immediately in front of the TGA stop codon. What is the amino acid sequence of this mutant's leader peptide? What are the conditions that would prevent attenuation in this mutant? Justify your answer.

Chapter Integration Problem

Scientists decide to map a bacterial chromosome using conjugation. They use a tetracycline resistant (*tet^r*), *leu^−*, *trp^−*, *met^−*, *bio^−* F^− recipient. The *tet^r* is used to select against the donor cells. After 3 min, *leu^+* colonies are recovered. After 5 min, *leu^+ met^+* colonies are recovered. They do not recover any *trp^+* or *bio^+* colonies and note that after 60 minutes of conjugation, the number of *leu^+* colonies they recover drops dramatically.

a. What could be causing the blockage of mapping the two markers?

b. The scientists then attempt the conjugation at the elevated temperature of 42°C. All of the markers are transferred in the order *leu^+*, *met^+*, *trp^+* then *bio^+*. Why would the experiment have worked at the higher temperature?

c. They next switch to a different recipient strain that is kanamycin-resistant. Again, the transfer works with the order *leu^+*, *met^+*, *trp^+* then *bio^+*. What besides the kanamycin resistance might be responsible for the experiment working?

d. They next pick individual donor bacterial colonies and test for their ability to transfer genes. Most of them have the same inability to transfer either *trp^+* or *bio^+*. One of the individually picked strains (strain F) does transfer genes, however, the gene order is now *leu^+*, *met^+*, then *bio^+*, with no colonies that are prototro-

phic for *trp^+* being found. If the scientists plated out this strain on plates that contain all 20 amino acids but lack leucine, methionine, or tryptophan, what pattern of growth would be expected?

e. What is the most likely event that resulted in the characteristics of this strain?

f. Another of the individually picked donor strains (strain M) is found to transfer *leu^+*, then *trp^+* and no other markers. What is the most likely cause of this variant?

g. To try to understand the basis of these changes, DNA is prepared from the original donor strain and variant strains F and M. The DNA is digested with *Eco*RI, electrophoresed through 1% agarose, denatured, and transferred to nitrocellulose (Southern blot). The nitrocellulose is then probed with a radioactively labeled 10 kb genomic fragment that contains the *trp* region. The following pattern is observed.

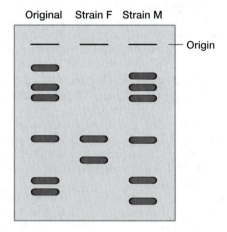

What does this tell us about the molecular events that generated strains F and M?

Do you need additional review? Visit **www.mhhe.com/hyde** for practice tests, answers to end-of-chapter questions and problems, interactive exercises, and animation tutorials all designed to enhance your understanding of key genetic concepts.

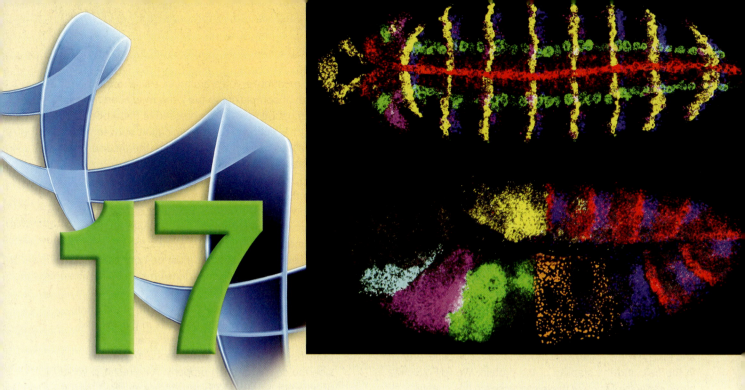

Gene Expression: Control in Eukaryotes

Essential Goals

1. Understand the organization of the *cis*-acting control elements and their various roles in regulating transcription.

2. Understand the different types of *trans*-acting elements, the common motifs that they share, and how they function to regulate gene expression.

3. Describe a complex regulatory network and identify the steps that could be involved in regulating gene expression.

4. Enumerate the different methods used to modify chromatin that result in changes in gene expression.

5. Describe the posttranscriptional and translational regulatory mechanisms that affect protein expression.

6. Diagram the common themes in regulating gene expression in bacteria and eukaryotes.

Photo: Top panel: *Drosophila* embryo stained for the expression of the *sog* (red), *ind* (green), *msh* (magenta), *wg* (yellow), and *en* (blue) mRNAs.

Lower panel: *Drosophila* embryo stained for the expression of seven homeobox genes: *lab* (light blue), *Dfd* (magenta), *Scr* (green), *Antp* (orange), *Ubx* (dark blue), *abd-A* (red), *Abd-B* (yellow).

Chapter Outline

The images of the *Drosophila* embryos at the beginning of this chapter reveal the expression patterns of several different genes in the same embryo. These precise expression patterns highlight one of the important concepts of this chapter. Genes and their encoded proteins must be properly expressed in specific cells and tissues at the correct time for multicellular eukaryotic organisms to develop and thrive.

Gene regulation is often thought of as the group of processes that control the amount of mRNA that is produced from a gene. In the previous chapter, you saw that many of the regulatory events occurred at the promoter, which involved positive or negative regulation of transcription. In eukaryotic cells, however, a wider variety of other processes can be involved. These include the rate of RNA transcription, the splicing of the primary RNA transcript into mRNA, the stability of the mRNA, the rate of protein translation, the posttranslational processing and folding of the protein, the proper localization of the protein in the cell, and the stability of the protein. We will describe some aspects of each of these processes in this chapter.

Five essential differences must be accounted for when we discuss regulating gene expression in eukaryotes versus bacteria:

- First, eukaryotes have three different RNA polymerases that each recognize different promoters and transcribe different classes of genes (see chapter 10). As a result, the proteins that regulate transcription are likely to be more diverse in nature and able to accommodate a complex regulatory network.
- Second, eukaryotic mRNA is spliced and transported from the nucleus prior to translation. Transcription and translation are therefore not linked, which is an important aspect of transcribing a polycistronic mRNA from bacterial operons. The scarcity of eukaryotic operons requires more extensive transcriptional control mechanisms to ensure appropriate levels of enzyme production for a single biochemical pathway.

- Third, most eukaryotes are multicellular, adding another level of complexity to the gene regulation model. A gene must be correctly expressed in the proper cells, and at the right time. Communication between cells and tissues is required to generate the precise gene expression patterns observed.
- Fourth, most eukaryotes have multiple chromosomes, unlike the single chromosomes that we discussed in *E. coli* and phage λ. The expression of genes requires the alteration of the chromatin arrangement on these chromosomes. Furthermore, the expression of some genes on different chromosomes must be coordinated, but the expression of adjacent genes may require different expression patterns.
- Fifth, eukaryotic cells contain a variety of different organelles and functional domains in the cell and membrane that are absent in bacteria. Eukaryotic proteins must find their correct location within or outside the cell to function properly. As a result, a variety of different protein modifications are required for the expression of functional eukaryotic proteins.

In this chapter, we describe each of these five differences and their effect on eukaryotic regulatory mechanisms, beginning with the *cis-* and *trans-*acting functions that regulate transcription. We show an example of a complex regulatory network in *Drosophila,* namely the action of the Dorsal protein. The roles of chromatin and histone proteins, not present in bacteria, and of methylation of DNA are described; both these control mechanisms use the addition of chemical groups to regulate trascription. Finally, posttranscriptional, translational, and posttranslational controls are reviewed. The chapter concludes with an example of how these different regulatory mechanisms interact to yield the correct gene expression pattern and how disruption of these mechanisms can lead to inherited human diseases.

17.1 The *cis* Regulatory Elements: Control of RNA Polymerase II Action

You learned in chapter 16 that the promoter and operator sequence in bacteria is only a few hundred nucleotides in length at most. This length of DNA is sufficient for the DNA-binding proteins and RNA polymerase to recognize and bind. In the multicellular eukaryotic organism, however, the complexity of controlling gene expression requires the *cis-*acting regulatory sequence to be significantly larger. In the following discussion, we concentrate on the RNA polymerase II *cis-*acting elements, which we divide into the promoter, proximal promoter, enhancers, and silencers (**fig. 17.1**).

The RNA Polymerase II Promoter

As described in chapter 10, the RNA polymerase II promoter is composed of two elements. An initiator region (*InR*) is located in the region that flanks the transcriptional start site (+1). The second element is the TATA, or Hogness, box, which is similar in sequence to the bacterial −10 region and is located at approximately −25. (A second class of RNA polymerase II promoters contains a downstream promoter element instead of the TATA box.) Although both the Hogness box and *InR* sequences are sufficient for assembly of the preinitiation complex and initiating RNA polymerase II transcription at the correct site (+1) in vitro, they cannot provide basal levels of transcription in vivo.

RNA polymerase II is unable to efficiently identify and bind the promoter. Binding of RNA polymerase

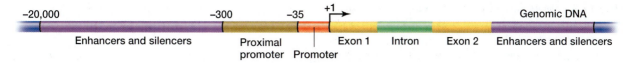

figure 17.1   The transcription of a gene by RNA polymerase is controlled by three different genomic DNA domains, the promoter, the proximal-promoter element, and enhancers and silencers. While the locations of the promoter and proximal promoter regions are fixed, enhancers and/or silencers may be located many kilobases either upstream or downstream of a gene, and even within the intron of a gene.

II to the promoter region requires the formation of a **preinitiation complex (PIC).** The PIC is a group of six different general transcription factors (see table 10.4) that bind to the promoter region and assist in RNA polymerase II binding before transcription begins. One of those transcription factors is **transcription factor IID (TFIID),** which is composed of over 10 different proteins, including the **TATA-binding protein (TBP).** The TBP is the protein that binds the TATA (Hogness) box. Because TBP binds the minor groove of DNA, it causes significant bending of the DNA and permits the binding of other proteins, called TBP-associated factors (TAFs).

The $TAF_{II}150$ and $TAF_{II}250$ proteins of the TFIID complex bind the initiator region of the promoter. As we discussed in chapter 10, the $TAF_{II}250$ protein is a histone acetyltransferase. This enzyme binds DNA that is associated with acetylated histones and then adds additional acetyl groups on the amino terminal lysine residues that are found on histones H3 and H4. It is possible that TFIID recognizes the Hogness box and initiator region that are associated with acetylated histones. Binding of TFIID results in dramatic bending of the DNA and further acetylation of histones H3 and H4, which may enhance the binding of additional transcription factors.

The additional transcription factors bind the RNA polymerase II promoter in a specific sequence (see fig. 10.24). The association of RNA polymerase II and these general transcription factors results in the formation of the preinitiation complex that permits basal level transcription to occur. For the maximal transcription level of any particular RNA polymerase II gene, however, the proximal-promoter elements and enhancers must bind their required proteins.

Proximal-Promoter Elements

The **proximal-promoter elements** are required for directing the proper level of transcription for a gene. They are usually found in the first 200 nucleotides upstream of the promoter (−35 to −200), although they occasionally extend further upstream. These DNA sequences bind specific proteins that either activate or repress transcription. A single gene may contain over 10 different DNA-binding sites in this region. The combination of activators and repressors that bind to this region determines whether the gene is transcribed and

at what level. Two general classes of promoter-proximal elements have been identified: generic, and cell- or tissue-specific.

Generic Proximal-Promoter Elements

Two generic proximal-promoter elements are the CCAAT box and the GC box. The relative positions of these two sequences upstream of the β-globin gene are shown in **figure 17.2**. Mutations can be introduced in these sequences at desired nucleotide positions using site-directed DNA mutagenesis (see chapter 13). The location of mutations can be plotted against the relative rate of transcription for that mutant sequence (see fig. 17.2). Mutations in three different regions significantly decreased the level of β-globin gene transcription (GC box, CCAAT box, and Hogness box), whereas mutations outside of these conserved sequences did not have much of an effect. Thus, in this 300-nucleotide upstream region of the β-globin gene, these three sequences represent the only critical *cis*-acting elements that are required for transcription.

Furthermore, if either the CCAAT or GC boxes are moved different distances from the TATA box, they lose their ability to maximally activate transcription. These generic elements are therefore required for regulating the amount of transcription, rather than controlling where and when the gene is expressed.

The proteins that bind to the GC and CCAAT boxes are expressed in all cells, which limits the ability of these sequences to direct cell- or tissue-specific transcription. The SP1 protein binds to the GC box, and the CCAAT-binding transcription factor (CTF) binds to the CCAAT box.

How could these proteins help to influence the level of transcription? Let's use the SP1 protein and GC box as an example. As mentioned earlier, TFIID is composed of several different proteins (**fig. 17.3a**), including the TBP, $TAF_{II}150$, and $TAF_{II}250$. An additional component of TFIID is the $TAF_{II}110$ protein, which interacts with the SP1 protein.

We can imagine a model by which TFIID binds a RNA polymerase II promoter through the TBP subunit's binding of the Hogness box and the $TAF_{II}150$ and $TAF_{II}250$ subunits' binding of the initiator region (fig. 17.3b). When a GC box is present and is at the proper distance from the TATA box, the SP1 protein will bind

figure 17.2 Mutations in the β-globin gene promoter and proximal-promoter region affect the level of transcription. Single nucleotide mutations were introduced throughout the upstream region of the β-globin gene and the level of transcription for each mutation was determined. The relative location of each mutation in the genomic DNA is represented as a bar in the graph. The height of each bar represents the relative level of transcription, with the wild-type promoter being 1.0. Mutations in three different regions (Hogness, GC and CCAAT boxes) exhibited significantly reduced levels of transcription. Nearly all the mutations outside of these sequences had only minimal effects on transcription. The only two exceptions were two mutations adjacent to the CCAAT box that increased transcription.

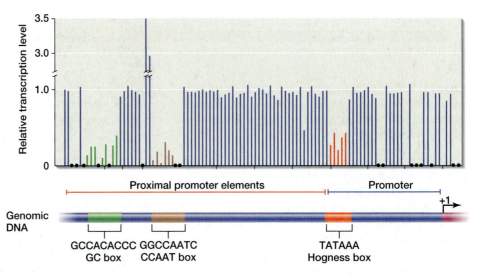

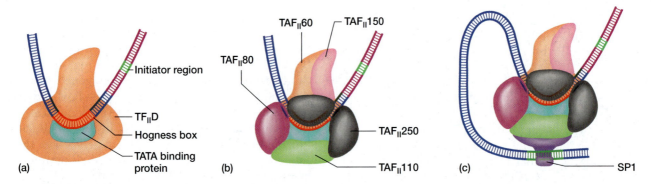

figure 17.3 A model for the role of the generic proximal-promoter element. (a) A model of the organization of the TATA binding protein (TBP) and the TF$_{II}$D protein complex bound to the promoter region. Binding of TBP to the Hogness box causes a significant bend in the DNA. (b) The TAF$_{II}$150 and TAFI$_{II}$250 subunits interact with the initiator region. Additional subunits in the TF$_{II}$D complex are shown. (c) A GC box, which is located upstream of the Hogness box, binds the SP1 protein. The SP1 protein can also interact with the TAF$_{II}$110 protein in TF$_{II}$D to either help bind TF$_{II}$D or stabilize it at the promoter before transcription begins.

the GC box and assist in the binding or stabilizing of the TFIID complex through an interaction with the TAF$_{II}$110 subunit (fig. 17.3c). This action increases the efficiency of transcription initiation, resulting in a higher level of mRNAs produced from the gene.

Cell- or Tissue-Specific Proximal-Promoter Elements

The cell- or tissue-specific proximal-promoter elements function in a similar manner as the generic elements. Two major differences exist, however, between these elements and the generic elements. First, the number of these cell-specific elements can be significantly greater in the upstream region of a gene than the number of generic elements. For example, the zebrafish rhodopsin gene has seven different proximal-promoter elements that are required to properly

transcribe the gene in the rod photoreceptors of the retina (fig. 17.4). Furthermore, the rhodopsin genes in humans, mice, chickens, and zebrafish all have a similar organization of these protein-binding sites. The conservation of these sequences in the upstream region in a variety of different organisms suggests that they play a very important role in properly controlling the transcription of the rhodopsin gene.

Second, the proteins that bind these tissue-specific elements are expressed in a subset of cells in the organism and only at specific times during development. The pattern of expression of the DNA-binding proteins therefore dictates the expression of the rhodopsin gene. This allows for a wonderful mechanism to coordinate gene expression: For example, one proximal-promoter element may bind a transcription factor that is expressed in all neurons; a second proximal-promoter element may bind a transcription factor that is

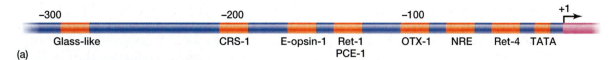

(a)

(b)

figure 17.4   Diversity of proximal-promoter elements. (a) Several putative protein binding sites are located in the proximal-promoter region (−50 to −300) of the zebrafish rhodopsin gene. (b) When this proximal-promoter region and TATA sequence were placed upstream of a GFP reporter gene and reintroduced into zebrafish, GFP was expressed in the retinal rod photoreceptors (arrows). This expression pattern is the same as the endogenous rhodopsin. The dorsal (D) and ventral (V) regions of the head are labeled, as is the lens (L) of one of the eyes.

present in all neurons that are located in the retina; and finally, a third element may bind a transcription factor that is present in only retinal neurons that are photoreceptors. If transcription of the rhodopsin gene requires all three of these different transcription factors to bind their proximal-promoter element, then the rhodopsin gene will be transcribed only in photoreceptors.

How Cell- or Tissue-Specific Protein Activation Produces Complex Regulation

The *combinatorial model of gene expression* suggests that the transcription of a gene depends on which transcription factors are present and absent in a particular cell. This ability to distinguish a particular cell or tissue based on the expression of DNA-binding proteins or transcription factors is diagrammed in **figure 17.5**. In this example, we have three different transcription activators (protein G, protein R, and protein P). By expressing these three proteins in different patterns, we can generate seven unique classes that each express at least one of these proteins. Thus, these three proteins would permit the production of seven different transcription expression patterns, which could correspond to patterns of different cells or tissues. This combined generation reduces the need for having a different transcription activator for each cell or tissue in the organism. (We come back to this idea when we discuss the generation of a pattern during development in chapter 21.)

It is important to note that a transcription factor may appear in one tissue during development, then vanish, appear in a different tissue at a later point in development, and then appear in a third tissue at a time that overlaps with its appearance in the second tissue. Expressing a single transcription factor at different times in development allows the protein to interact with other

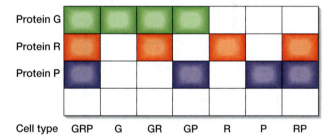

figure 17.5   Combinatorial model of regulating gene expression. Shown is a model of the hypothetical expression of three different DNA binding proteins (proteins G, R, and P). A filled square represents the expression of the corresponding protein in a specific cell. Based on the different combinations of these three proteins, seven different cell types can be determined. In this model, each cell type must express at least one of the three DNA binding proteins.

transcriptional activators in diverse cell types or tissues. This complexity of expression is one mechanism for generating the elaborate pattern of transcriptional regulation required in a multicellular organism.

Enhancers

Enhancers are DNA sequences that bind proteins to control the transcription of a gene in a particular cell or tissue. Although most proteins that bind enhancers are called **activators** because they increase transcription, **repressors,** which suppress transcription, can also bind some enhancers (see the section Interactions Within and Between Enhancers). Two features distinguish an enhancer from the previous class of *cis*-acting elements. First, the distance of an enhancer from the gene does not affect its ability to regulate transcription.

For example, an enhancer can be moved up to tens of kilobases upstream or downstream of the gene, and it would continue to exert its effect on transcription (**fig. 17.6**). In contrast, moving a GC box 50 nucleotides farther upstream of its normal location could prevent it from activating transcription of the gene. Second, the orientation of the enhancer is not critical; reversing the orientation of the enhancer sequence does not affect its ability to activate transcription.

Enhancers differ from the promoter and proximal-promoter elements in one other significant way. The

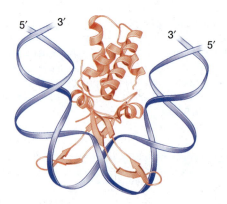

figure 17.6   Effect on DNA by binding an architectural protein. The binding of an architectural protein (red) into the minor groove of DNA creates a significant bend in the DNA molecule (blue) that it binds.

promoter is required for the assembly of the preinitiation complex and determining the transcription start site, and loss of these DNA sequences usually results in virtually no transcription occurring. The proximal-promoter elements control the level of transcription of a gene, and loss of these DNA sequences usually results in a significant decrease in the level of transcription. In contrast, enhancers affect the maximal level of transcription and control the temporal (time) and spatial (tissue) expression pattern. Thus, loss of an enhancer means transcription of the associated gene will be reduced and the gene may be expressed in the wrong tissue or at the wrong time.

If an enhancer is moved near a different gene, that gene will be transcribed in a pattern that is directed by the enhancer. For example, a number of lymphomas are due to chromosomal translocations involving chromosome 8. All of these translocations place the *c-myc* protooncogene (see chapter 22) near the enhancer of an immunoglobulin gene. In this location, the *c-myc* gene is inappropriately expressed in lymphoid cells, which results in the lymphoma phenotype.

Enhancers contain protein-binding sites that are recognized by transcriptional activators and repressors, similar to the proximal-promoter elements. But enhancers must function in a manner different from simply binding a protein that directly aids or blocks the binding of additional proteins in the transcriptional complex because its location can be dramatically

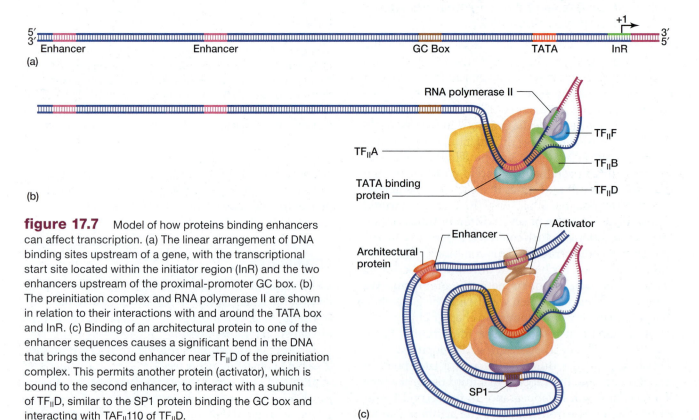

figure 17.7   Model of how proteins binding enhancers can affect transcription. (a) The linear arrangement of DNA binding sites upstream of a gene, with the transcriptional start site located within the initiator region (InR) and the two enhancers upstream of the proximal-promoter GC box. (b) The preinitiation complex and RNA polymerase II are shown in relation to their interactions with and around the TATA box and InR. (c) Binding of an architectural protein to one of the enhancer sequences causes a significant bend in the DNA that brings the second enhancer near $TF_{II}D$ of the preinitiation complex. This permits another protein (activator), which is bound to the second enhancer, to interact with a subunit of $TF_{II}D$, similar to the SP1 protein binding the GC box and interacting with $TAF_{II}110$ of $TF_{II}D$.

altered without altering its effect on the expression of a gene. Two major models suggest how proteins that bind to enhancers can exert their effect over tens of kilobases either upstream or downstream of a gene.

The first model is that the enhancer binds proteins that are capable of bending the DNA, that is, they are **architectural proteins** (fig. 17.6). Bending the DNA can potentially bring additional enhancer sequences near the promoter, which allows other proteins bound to the enhancers to physically interact with the RNA polymerase II or transcription factors in this region (**fig. 17.7**). This model suggests that as the enhancer region is moved different distances from the promoter, the size of the DNA loop varies. Because this loop does not interact with any of the proteins located in the promoter region, the size of the loop is irrelevant.

The second model suggests that the enhancer changes the conformation of the chromatin in the region of the gene, which makes it more accessible for the binding of transcription factors and RNA polymerase II. The mechanism by which this conformational change occurs is still unclear.

Silencers

Enhancers can have a profound effect on where or when the gene will be transcribed. As you saw with the *lac* operon, there are times when utilizing both positive and negative control can provide a tighter

regulation of transcription. In eukaryotes, **silencers** provide a negative form of regulation (**fig. 17.8**); they can either exert a gene-specific or a global transcription repression.

The gene-specific silencers bind proteins, which change the conformation of the DNA. For example, the lymphoid-specific zinc finger protein subfamily 1A,1 (ZFPN1A,1 or IKAROS) binds specific silencers and alters the chromatin for a given gene from euchromatin to heterochromatin.

Global silencing involves a sequence that is found in many promoters that are bound by the same protein. For example, the neural restrictive silencer element (NRSE) is found in the promoter of many genes that are only expressed in neurons. The RE1-silencing transcription factor (REST), which is expressed in all cells other than neurons, binds the NRSE. The REST factor is recognized by additional proteins (histone deacetylases or histone methyltransferases), which modify the histone proteins. These modified histones alter the chromatin conformation and make the DNA transcriptionally inactive (see the section The Role of Chromatin in Gene Regulation). The result of the NRSE being bound by REST, which is not expressed in neurons, makes the chromatin transcriptionally inactive in all nonneuronal cells. Another global silencer protein is called Polycomb, which represses the expression of embryonic genes during later times in development.

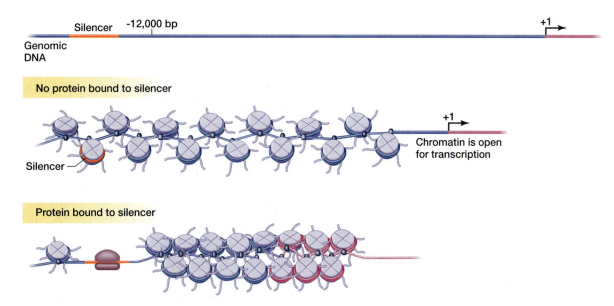

figure 17.8   Effect of a silencer on transcription. A silencer can be located up to tens of kilobases upstream or downstream from the gene it controls. For a gene that is being transcribed, the silencer is not bound by a protein because it is located within a nucleosome. However, when a protein binds the silencer, it could suppress transcription by increasing the compactness of the chromatin. A protein bound to a silencer may also destabilize the preinitiation complex or block the binding of an activator. Other mechanisms of chromatin silencing are discussed later in this chapter.

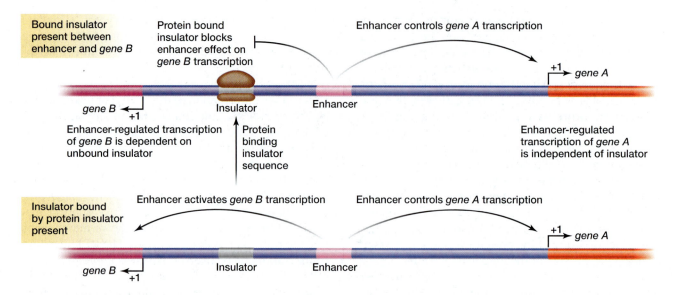

figure 17.9   A model for insulator action. An insulator located between *gene B* and an enhancer will prevent the enhancer from activating transcription of *gene B* when the insulator is bound by a protein. Loss of protein binding to the insulator will allow the enhancer to control transcription of *gene B*. In either case, *gene A* transcription is regulated by the enhancer.

Insulators

One of the interesting features of enhancers and silencers is their ability to function either upstream or downstream of a gene and over long distances. You may wonder why an enhancer that is located upstream of one gene does not influence the transcription of a gene on the other side (**fig. 17.9**). **Insulators** are DNA elements that can shield a gene from the effects of a neighboring enhancer or silencer.

We know that insulators are DNA sequences that bind proteins that recognize these specific sequences. For example, *Drosophila* insulators have a conserved GAGA sequence that is bound by the Trl protein. If either the GAGA sequence or the Trl protein is mutated, then the insulator does not function. The mechanism by which insulators block the effects of enhancers and silencers remains unclear, however.

Interactions Within and Between Enhancers

Note that many genes contain multiple enhancers, silencers, or both, that in combination produce the final level of transcription. The *Drosophila even-skipped* (*eve*) gene is an example. The *eve* gene is required for the proper development and identity of the segments in the adult fly. It is expressed as a pattern of seven stripes in the early *Drosophila* embryo.

The eve Gene and Its Enhancers

The *eve* gene is composed of a transcribed region and five different enhancers, with each enhancer being required for the transcription of one or two specific stripes (**fig. 17.10a**). Notice that two of the enhancers are upstream of the transcribed region, and the other three enhancers are downstream.

We can determine the relative role of each enhancer by cloning the enhancer and a weak promoter upstream of a reporter gene (for example, *lacZ*) and then introducing this transgene into flies using the *P* element vector (see chapter 13). We can then compare the expression of the reporter gene simultaneously with the expression of the even-skipped protein using an antibody. In this case, we find that a specific enhancer directs β-galactosidase expression in a pattern that precisely overlaps one or two of the Eve protein stripes (fig. 17.10b–d).

We could expand this analysis and look at a single enhancer, such as the stripe 2 enhancer. Analysis has revealed that four different proteins bind to this enhancer region to regulate *eve* gene expression. By mutating the predicted binding sites in the transgene, we can examine what happens to the *lacZ* transgene expression pattern.

For the stripe 2 enhancer, the Bicoid and Hunchback proteins bind as activators (**fig. 17.11**). Mutation of the five Bicoid-binding sites and the three Hunchback-binding sites results in the absence of expression in stripe 2. In contrast, the Krüppel and Giant proteins are repressors of stripe 2 expression. Loss of the seven Krüppel- and four Giant-binding sites results in a stripe 2 that is wider than normal.

This loss of repression can also be observed when the reporter transgene is expressed in a *giant* mutant. In this example, the *eve* stripe 5 is also repressed by the Giant protein. The embryonic expression of a *lacZ* transgene from the *eve* stripe 5 enhancer is shown in figure 17.10d. If this transgene is introduced into a *giant* mutant embryo, which fails to express the Giant protein, a wider β-galactosidase-expressing stripe 5 is present in a normal Eve protein stripe pattern (fig. 17.10e). Thus, the interaction between different enhancers, and the

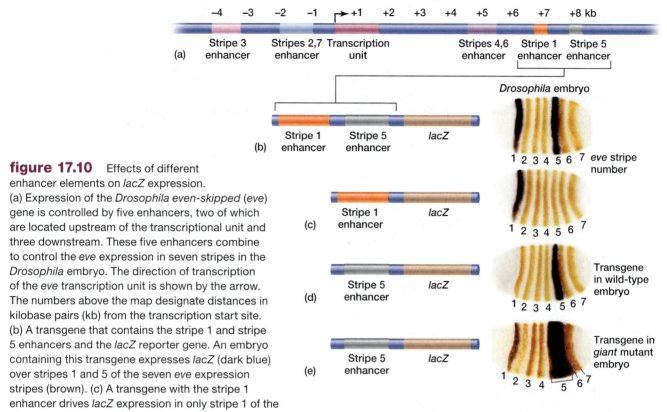

figure 17.10 Effects of different enhancer elements on *lacZ* expression.
(a) Expression of the *Drosophila even-skipped* (*eve*) gene is controlled by five enhancers, two of which are located upstream of the transcriptional unit and three downstream. These five enhancers combine to control the *eve* expression in seven stripes in the *Drosophila* embryo. The direction of transcription of the *eve* transcription unit is shown by the arrow. The numbers above the map designate distances in kilobase pairs (kb) from the transcription start site. (b) A transgene that contains the stripe 1 and stripe 5 enhancers and the *lacZ* reporter gene. An embryo containing this transgene expresses *lacZ* (dark blue) over stripes 1 and 5 of the seven *eve* expression stripes (brown). (c) A transgene with the stripe 1 enhancer drives *lacZ* expression in only stripe 1 of the embryo. (d) A transgene with the stripe 5 enhancer expresses *lacZ* in only stripe 5 of the embryo. (e) Putting the same transgene in a *giant* mutant embryo results in stripe 5 getting wider. This reveals that the wild-type Giant protein normally restricts the *eve* stripe 5 expression to a narrow stripe.

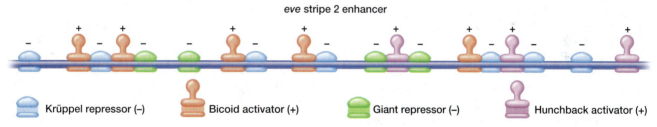

figure 17.11 Multiple protein binding sites in an enhancer. The *Drosophila even-skipped* stripe 2 enhancer is composed of multiple binding sites for four different DNA-binding proteins, Bicoid, Giant, Hunchback and Krüppel. Giant and Krüppel are negative regulators (−), while Bicoid and Hunchback are activators (+) of *eve* transcription in stripe 2. The arrangement of an activator (+) and repressor (−) binding site adjacent to each other forces the proteins to compete for DNA binding and to either activate or repress *eve* transcription.

activators and repressors within each single enhancer, all combine to yield the final expression pattern of the gene.

The Thyroid Hormone Response Element (THRE)

Another example of interaction within an enhancer is a DNA-binding site that can either activate or repress transcription, depending on what form of the protein is bound. For example, the thyroid hormone response element (THRE) is found upstream of many genes and serves as the binding site for the thyroid hormone receptor (**fig. 17.12**).

When the thyroid hormone receptor binds a molecule of the thyroid hormone and then binds to the THRE, it activates transcription. When the receptor binds to the THRE without being bound to the thyroid hormone, however, transcription is repressed.

In this manner, a gene transcribed at a low basal level can exhibit increased transcription in the presence of the hormone, or it can undergo further reduction in expression in the absence of the hormone. Thus, enhancers can affect the temporal and spatial expression of a gene (*eve* enhancers) or the level of transcription (THRE).

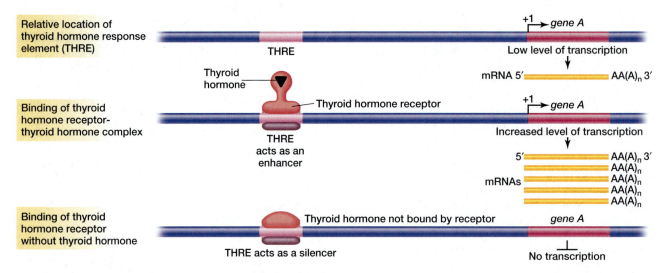

figure 17.12 A DNA binding site can function as either an enhancer or silencer. In the absence of thyroid hormone receptor binding the thryroid hormone response element (THRE), there is a low level of gene transcription. When the thyroid hormone receptor binds thyroid hormone, this complex can bind the THRE and activate transcription of the gene. Binding of only the thyroid hormone receptor to the THRE represses transcription of *gene A*. Thus, the THRE can function as either an enhancer or silencer depending on what form of the thyroid hormone receptor binds.

 Restating the Concepts

▶ The important *cis*-acting sequences for regulating transcription of a gene are the promoter, proximal-promoter, enhancer, silencer, and insulators.

▶ The promoter and proximal-promoter elements control the basal level of transcription of a gene. The location of these elements cannot be changed significantly relative to the transcription start site without severely affecting transcription.

▶ The enhancer and silencer elements regulate the maximal level of transcription of a gene temporally and spatially. They are position-independent, in that they can be located upstream or downstream of a gene at distances of up to tens of kilobases from the transcription start site. The interaction of different proteins binding within an enhancer and between different enhancers and silencers determines the final expression pattern of the gene and the level of expression.

▶ Insulators act to ensure that the enhancers and silencers affect only the proper gene. Because enhancers and silencers can function over large distances, they could affect the expression of unintended genes. Insulators function, often by binding a protein, to direct the effects of enhancers and silencers in only one direction.

17.2 The *trans* Regulatory Elements: Proteins

You have seen how a variety of DNA sequences can activate or repress expression of a gene. In all of these cases, proteins bind to these sequences to carry out the desired action on transcription. What common features among these proteins can help us understand their action?

Each of these proteins possesses a DNA-binding domain. Although each protein recognizes very different sequences, some common **motifs,** or tertiary structures, can be found in the DNA-binding regions.

Most of these proteins also bind to DNA as either dimers or tetramers, which means that they must contain a protein–protein interaction domain to recognize their binding partner. Protein–protein interaction domains must also be present if the activator binds or helps to stabilize the binding of TFIID or RNA polymerase II to the promoter.

Finally, some of these proteins bind a small molecule, for example, the thyroid hormone receptor binds thyroid hormone. In these cases, the protein must have a ligand-binding domain. A *ligand* is a small molecule that is bound and either activates or inactivates the molecule to which it is bound, usually called a receptor.

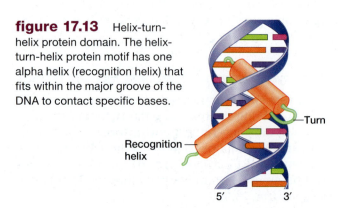

figure 17.13 Helix-turn-helix protein domain. The helix-turn-helix protein motif has one alpha helix (recognition helix) that fits within the major groove of the DNA to contact specific bases.

DNA-Binding Motifs

The DNA-binding motifs described here are found in several proteins. The ability of the protein to bind to a specific sequence lies within the amino acid sequence *in* the motif, rather than the motif itself. The general motif within a protein confers the ability to bind DNA, however, the specific amino acids within the motif dictate what DNA sequences will be bound.

Helix-Turn-Helix

The **helix-turn-helix** motif was the first DNA-binding motif to be identified. It is found in a variety of bacterial DNA-binding proteins (Lac repressor, Cro repressor, Trp repressor) and a wide array of eukaryotic transcriptional regulators. It is characterized by two α-helices that are connected by a "turn" that consists of several amino acids (**fig. 17.13**). The arrangement of the two α-helices allows one helix to bind in the major groove of DNA and the other helix to interact with the sugar–phosphate backbone.

One class of eukaryotic proteins that contains the helix-turn-helix domain is the **homeobox** protein. As discussed in chapter 21, the homeobox proteins were first identified in *Drosophila* by Walter Gehring and his colleagues. They were studying a class of genes identified in a series of mutants that had one or more of their body segments transformed into a different fly segment. Examples of *homeotic mutations* include the transformation of a haltere wing into a full wing in the *Bithorax* mutant or the transformation of the antennae into legs in the *Antennapedia* mutant (**fig. 17.14**).

All of these homeotic genes contain a conserved 180-bp region in the open reading frame (ORF) that encodes a homologous 60-amino-acid sequence. Not only was this sequence similar in all the homeotic genes, it also encoded the helix-turn-helix motif.

As with the other helix-turn-helix domains, an α-helix of the homeobox domain makes contact with the bases in the major groove of DNA. Examination of other eukaryotes found homeobox-encoding genes in a wide variety of organisms, including humans. In mammals, the homeobox proteins are required for the development of the brain.

Zinc Fingers

A second DNA-binding motif is the **zinc finger,** which was first identified in the *Xenopus* TFIIIA protein. There are several different variations of the zinc finger; one of the most common consists of two cysteine and two histidine residues that covalently bind an atom of zinc, which folds the amino acids into a loop (**fig. 17.15a**). The covalent bond with zinc forms the "finger" from two antiparallel β-sheet strands that contain the two cysteine residues and an α-helix that contains both histidine residues.

As with the helix-turn-helix proteins, it is the α-helix of the zinc finger that makes contact with the bases in the major groove of the DNA. A zinc finger protein often has multiple fingers—from 2 to 13—with the α-helix of each finger making contact in the major groove (fig. 17.15b).

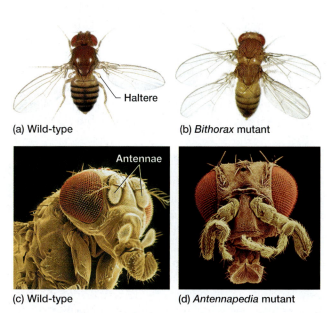

(a) Wild-type   (b) *Bithorax* mutant

(c) Wild-type   (d) *Antennapedia* mutant

figure 17.14 Homeotic transformations in *Drosophila*. (a) The wild-type *Drosophila* has one large pair of wings and one small pair of haltere wings. (b) The *Bithorax* mutation transforms the small haltere wings into a large pair of wings. (c) The wild-type *Drosophila* head has a pair of antennae. (d) The *Antennapedia* mutation transforms the antennae into a pair of legs.

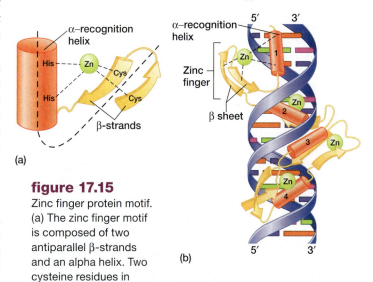

figure 17.15

Zinc finger protein motif. (a) The zinc finger motif is composed of two antiparallel β-strands and an alpha helix. Two cysteine residues in the β-strands and two histidine or cysteine residues in the alpha helix covalently bind a zinc atom to fold the motif into a finger structure (dashed line). (b) The zinc finger's alpha helix contacts specific bases in the major groove of DNA. Shown are four zinc fingers (alpha helices numbered) bound in the major groove of DNA.

Basic Helix-Loop-Helix and Basic Leucine Zipper

Another type of DNA-binding domain is characterized by a region of basic amino acids at the end of an α-helix. This basic region is a subdomain of two larger motifs, the **basic helix-loop-helix (bHLH)** and **basic leucine zipper (bZIP).** The basic region of the bHLH protein is an extension of an α-helix, which contacts the bases in the major groove of the DNA (**fig. 17.16**). Although the basic region in the bZIP protein also is an α- helix that contacts the bases in the major groove, it is not an extension of either the leucine zipper or a helical domain (**fig. 17.17**). As you can see, the α-helix provides a common mechanism among all of these motifs for binding DNA.

Protein–Protein Interaction Domains

The leucine zipper portion of a bZIP protein is composed of four or more leucine residues that are seven amino acids apart (see fig. 17.17). Because these leucines are located within a helix, they are found along one side and at every turn of the helix. This spacing provides a convenient mechanism to bring together two protein monomers. Each leucine on one helix fits between adjacent leucines on the other helix. This fit brings the two helices together, which then coil around each other to form a **coiled coil** structure (see fig. 17.17). The bHLH proteins also form a dimer, which is mediated through an interaction in the α-helices (see fig. 17.16).

bZIP Dimerization

The bZIP proteins will often form either homodimers (protein complex composed of two identical protein subunits) or heterodimers (protein complexes composed of two different subunits). The basic amino acid domain of each bZIP subunit recognizes and binds a specific DNA sequence. A homodimer binds one DNA sequence composed of two copies of the same subsequence, whereas a heterodimer binds a DNA sequence that is composed of two different subsequences. This mechanism increases the diversity of DNA-binding proteins without increasing the number of subunits or genes.

Two of the better characterized bZIP proteins are called Jun and Fos. Jun can form either a homodimer or a heterodimer with Fos, but Fos can form only the heterodimer. Because neither monomer alone can bind DNA to regulate transcription, only two potential DNA-binding proteins exist: Jun–Jun and Jun–Fos.

The Jun–Jun homodimer and Jun–Fos heterodimer bind similar DNA sequences, but with different affinities. As a result, the presence of the Fos protein affects the level of the Jun–Jun homodimer and the amount of dimer that is available to bind to the DNA site.

The Basic Helix-Loop-Helix/Leucine Zipper

A variation of the bZIP is the **basic helix-loop-helix/leucine zipper,** which contains a DNA bind-

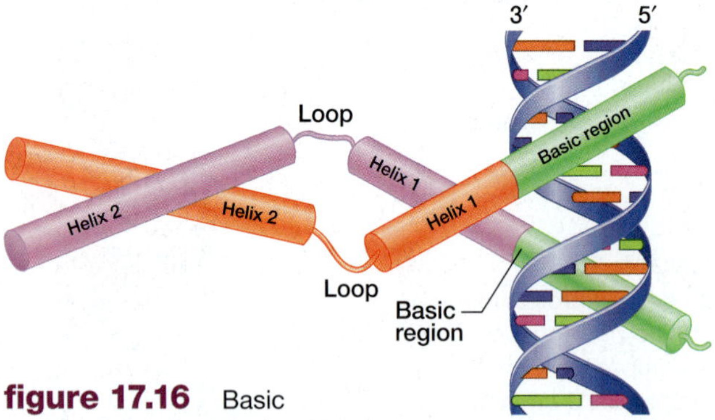

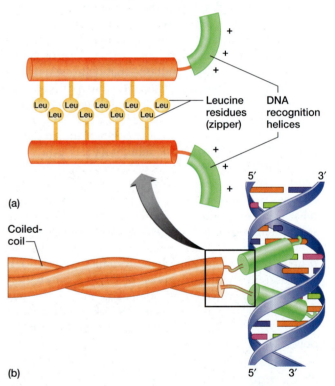

(a)

(b)

figure 17.17   Models of the structure of the leucine zipper. (a) The leucine zipper is an alpha helix with leucine residues every seventh amino acid, which places all the leucines along one side of the helix (red cylinder). The leucines from two different coils can interdigitate to form a protein dimer. The leucine zipper helix is followed by an alpha helix containing basic amino acids (green cylinder). (b) The interdigitating of the leucines from two alpha helices (on different proteins) causes the two helices to coil (called a coiled-coil). The charged alpha helices lie in the major groove of the DNA and contact specific bases.

figure 17.16   Basic helix-loop-helix protein motif. The basic helix-loop-helix motif is composed of two amphipathic alpha helices with all the charged amino acids on one side of the helix. The protein forms a dimer through protein-protein interactions between the charged amino acids in the alpha helices. One helix has an extended region that contains basic amino acids that contact specific bases in the major groove of DNA.

ing domain rich in basic amino acids along with the protein-protein dimerization domain in the leucine zipper and in the adjoining helix-loop-helix domain (**fig. 17.18**). It is unclear why these proteins utilize two different protein–protein interaction motifs.

Transcription Activation Versus Repression

The enhancer sequences previously discussed possess DNA sequences that bind proteins, which serve either to activate or to repress transcription of a gene. This choice depends on the protein that binds to the enhancer and the interaction with other proteins. It is possible for a protein that binds to a specific enhancer to serve as an activator in one cell and a repressor in another cell. We now examine the Myc–Max transcription factor system as an example of this complexity.

The Myc Protein

Myc is a DNA-binding protein that activates transcription of genes encoding proteins required for cell division. The Myc protein is therefore often expressed in cells that are actively dividing and not in cells that are terminally differentiated and have stopped dividing.

The *myc* gene is also a protooncogene; misexpression of Myc causes a cell to begin to proliferate (actively divide, as in a tumor). Earlier we mentioned the lymphomas produced when *c-myc* is involved in a translocation

near the enhancer of an immunoglobulin gene. In this location, the *c-myc* gene is inappropriately expressed in lymphoid cells, which results in the persistent cell division of the lymphoid cells.

When the *myc* gene was isolated, it was determined that the Myc protein is a member of the basic helix-loop-helix/leucine zipper protein family and also possesses a transcription activation domain at the amino terminus (**fig. 17.19**). But work with purified Myc protein failed to identify either a DNA-binding or a transcriptional activation activity. All data, therefore, except for the physical data, suggested that Myc was a transcriptional activator. This discrepancy in these data was resolved when the Max protein was identified.

The Max Protein

Max is also a basic helix-loop-helix/leucine zipper protein, but it lacks a transcription activation domain (see fig. 17.19). Like Jun and Fos, the Max protein can form a homodimer, or a heterodimer with Myc, but Myc is unable to form a homodimer. Max also has a higher affinity to interact with Myc than with itself.

Because Max is expressed in all cells at all times, it is usually found as a homodimer that can bind to DNA. Because it lacks the transcription activation domain, binding of the Max homodimer represses gene transcription (**fig. 17.20**).

The Myc–Max System: Competition in Dimer Formation

When Myc is expressed, the Max homodimer falls apart and preferentially forms the heterodimer, which then binds to the same DNA sequences as the Max homodimer. However, the heterodimer has a transcription activation domain that is supplied by the Myc subunit, permitting the Myc–Max heterodimer to activate transcription (**fig. 17.21**). Thus, the Max protein is

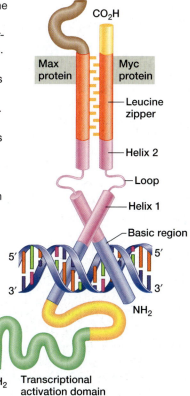

figure 17.18   A model for the dimerization of the Myc and Max proteins, which are leucine zipper-helix-loop-helix proteins. The leucine zipper and helix-loop-helix domains allow the Myc and Max proteins to form a dimer. The basic region (blue) on both molecules binds the major groove of DNA. However, only the Myc protein contains a transcriptional activation domain (green).

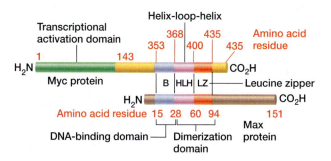

figure 17.19   A linear representation of the Myc and Max DNA-binding proteins. The location of the leucine zipper (LZ, red), helix-loop-helix domain (HLH, pink), and basic DNA-binding domain (B, blue) in both the Myc and Max proteins are shown. Notice that the Myc protein (top), but not Max (bottom), has a transcriptional activation domain at the amino terminus (green). The numbers represent the location of particular amino acid residues in both proteins.

a subunit of both a repressor and activator that binds to the same enhancer sequences. The difference lies in whether Max is a homodimer or heterodimer with Myc.

In the case of the Myc–Max system, the repressor (the Max homodimer) functions by binding to the same enhancer as the activator (the Myc–Max heterodimer), but it lacks the ability to activate transcription (**fig. 17.22a**). Because Max preferentially forms the heterodimer with Myc, the repression by the Max homodimer decreases when the Myc protein is expressed.

Repressors often function in two additional general mechanisms. These differ according to which domain of the activator is bound and masked by the repressor. In the first mechanism, the repressor can bind the activator protein, which results in masking of the DNA-binding site on the activator (fig. 17.22b). As a result, the activator is unable to bind the enhancer sequence as a heterodimer. In the second mechanism, binding of repressor to the activator subunit blocks the transcription activation domain of the activator. As a result, the

heterodimer can bind the enhancer sequence, but is unable to activate transcription (fig. 17.22c).

The ability of DNA-binding proteins to form dimers allows another level at which gene expression can be regulated, namely competition between forming a homodimer or a heterodimer.

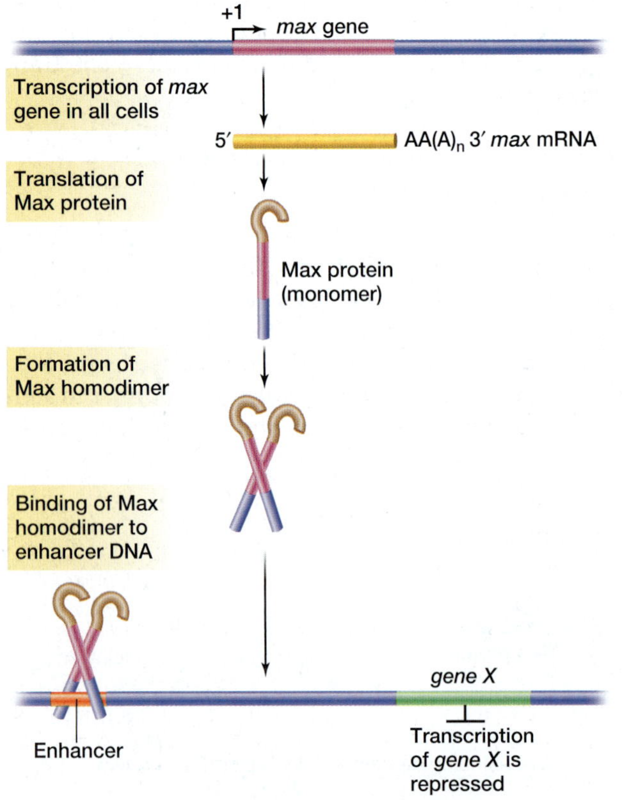

Restating the Concepts

▶ DNA-binding proteins have common motifs that mediate the DNA-binding activity. These motifs include helix-turn-helix, zinc finger, and helices that contain basic amino acids. In all three cases, an α-helix fits in the major groove of DNA to interact with the bases of the DNA.

▶ Most DNA-binding proteins function as dimers. The leucine zipper and helix-loop-helix are two motifs that allow proteins to interact and form a dimer. The formation of a dimer permits the generation of a larger number of DNA-binding molecules and allows another level of regulating gene expression through the formation of either a heterodimer or homodimer.

▶ A regulatory protein may function as either an activator or repressor, depending on the sequence to which it binds and the other proteins that are present. Repressors can bind to a site and prevent the formation of the transcriptional complex or can bind to an activator and prevent it from either binding the DNA or activating transcription.

17.3 The *trans* Regulatory Elements: RNAs

The *trans*-acting factors described to this point are proteins that have the ability to diffuse through the cell to reach their targets—either DNA sequences or other proteins. But another class of *trans*-acting factors exists, and these can either diffuse through the cytosol and recognize and bind specific complementary RNA sequences or enter the nucleus and recognize specific complementary DNA sequences. These factors are RNAs that fall into two different classes: miRNAs and antisense RNAs.

miRNAs

We first discussed microRNAs (miRNAs) in chapter 10. These miRNAs regulate the expression of a gene by disrupting translation through either the cleavage of the corresponding mRNA or by base-pairing with the mRNA to block translation.

The miRNA, which is transcribed as a pri-miRNA (primary miRNA) by RNA polymerase II, folds into a

figure 17.20 A model for repressing gene transcription by the Max homodimer. The *max* gene is transcribed in all cells and the Max protein is translated. The Max homodimer is formed through the leucine zippers and helix-loop-helix domains. The Max homodimer binds specific enhancer DNA sequences and blocks transcription of several genes because the Max protein lacks a transcriptional activation domain (see fig. 17.19).

stem-loop structure (**fig. 17.23**). The Drosha protein binds the pri-miRNA and cleaves it into an approximately 70-nucleotide pre-miRNA, which retains the stem-loop structure. The pre-miRNA is exported from the nucleus.

The Dicer protein binds the stem region adjacent to the loop of the pre-miRNA and cleaves the double-stranded stem from the loop (see fig. 17.23). After the loop region is removed, one 22-nucleotide RNA strand is preferentially transferred to the Argonaute protein to form a microribonucleoprotein (miRNP). Argonaute is one protein component of the RNA-induced silencing complex (RISC).

The miRNA of the RISC has the potential to base-pair with a specific mRNA. If the miRNA completely base-pairs with the mRNA, the mRNA is cleaved at a specific site (see fig. 17.23). This cleavage prevents the translation of the mRNA. In contrast, if there are limited number of mismatches between the miRNA and mRNA, the double-stranded region blocks translation without cleaving the mRNA. Either of these events results in a decreased amount of protein that is translated from the mRNA.

Antisense RNAs

The second class of regulatory RNAs, called antisense RNAs, also controls the translation of a mRNA, but are functionally different from the miRNAs. First, this class of RNAs does not form a stem-loop structure that is processed into a short RNA. Second, these RNAs do not form a ribonucleoprotein.

These antisense RNAs are complementary to a sequence within a target RNA. The base pairing of the

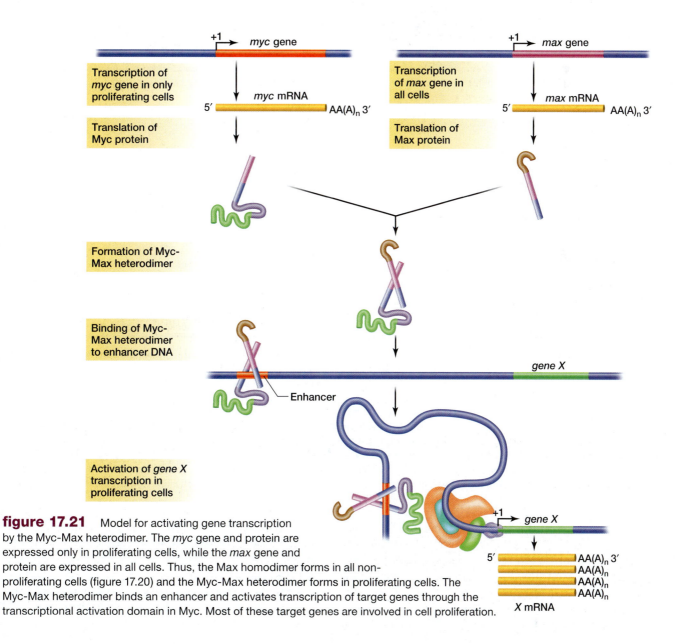

figure 17.21   Model for activating gene transcription by the Myc-Max heterodimer. The *myc* gene and protein are expressed only in proliferating cells, while the *max* gene and protein are expressed in all cells. Thus, the Max homodimer forms in all non-proliferating cells (figure 17.20) and the Myc-Max heterodimer forms in proliferating cells. The Myc-Max heterodimer binds an enhancer and activates transcription of target genes through the transcriptional activation domain in Myc. Most of these target genes are involved in cell proliferation.

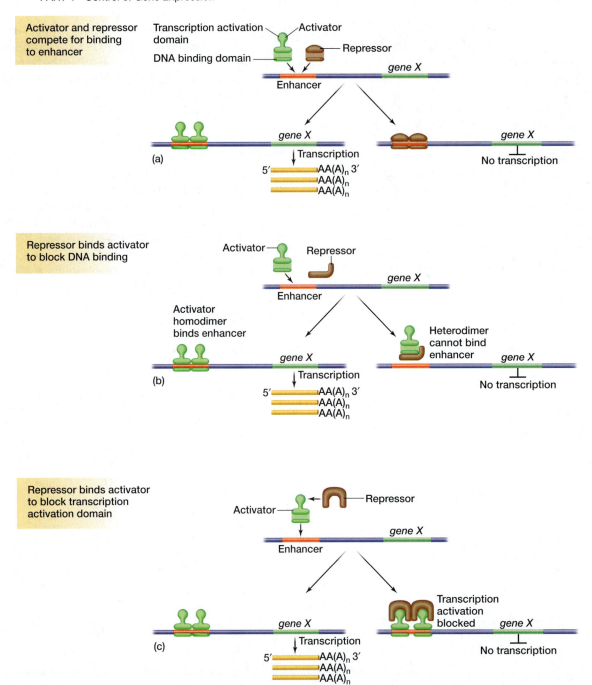

figure 17.22 Mechanisms of repressor action on transcription. (a) The repressor and activator proteins may compete for the same or overlapping enhancer binding site. Because the repressor lacks a transcription activation domain, repressor binding suppresses transcription. This model is similar to the Max homodimer repressor competing with the Myc-Max heterodimer for binding to the enhancer. Alternatively, the repressor can bind the activator to prevent transcription by two different mechanisms. (b) The repressor-activator complex is unable to bind the enhancer site because the repressor blocks the activator's DNA binding domain. (c) The repressor-activator complex binds the enhancer, but the repressor masks the transcription activation domain to block transcription.

regulatory RNA to the target RNA has one of three consequences (**fig. 17.24**). First, this base pairing can block translation of the mRNA. Second, base pairing at an intron–exon junction can block the splicing of the pre-mRNA. In our discussion of Prader–Willi and Angelman syndromes later in this chapter, we will see one example of a small antisense RNA affecting the transcription or translation of one gene and an example of a different antisense RNA affecting the splicing of a pre-mRNA. Third, the base pairing can affect the stability of the mRNA. In chapter 21, we will discuss how the complementary base pairing between the *XIST* and *TSIX* RNAs leads to the decreased stability of the *XIST* RNA during X chromosome inactivation.

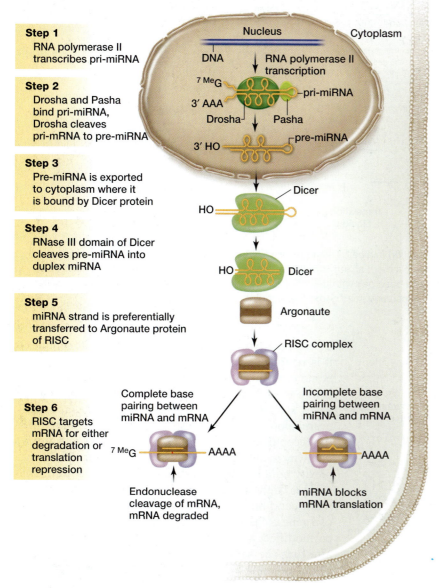

Step 1
RNA polymerase II transcribes pri-miRNA

Step 2
Drosha and Pasha bind pri-miRNA, Drosha cleaves pri-mRNA to pre-miRNA

Step 3
Pre-miRNA is exported to cytoplasm where it is bound by Dicer protein

Step 4
RNase III domain of Dicer cleaves pre-miRNA into duplex miRNA

Step 5
miRNA strand is preferentially transferred to Argonaute protein of RISC

Step 6
RISC targets mRNA for either degradation or translation repression

figure 17.23   Mechanisms of miRNA action in a cell. RNA polymerase II transcribes the pri-miRNA, which is then cleaved by the Drosha protein to produce the pre-miRNA. After being exported from the nucleus, the pre-miRNA is bound by the Dicer protein, which cleaves the pre-miRNA into a double-stranded RNA. One of the two strands is preferentially transferred to the Argonaute protein in the RISC complex, which can also bind a mRNA. Perfect basepairing between the miRNA and the mRNA results in the endonuclease cleavage of the mRNA followed by mRNA degradation (left). Limited basepair mismatches between the miRNA and the 3' UTR of the mRNA results in a stable miRNA-mRNA duplex, which suppresses mRNA translation.

17.4 Example of a Complex Regulatory Network: The Dorsal Protein

The preceding discussion on the role of *cis-* and *trans-*acting factors suggests that the mechanism of regulating transcription lies only in whether the protein (or miRNA) binds to the target sequence. This is a gross oversimplification of the regulatory network in most cases. In actuality, the *trans-*acting factor itself is often regulated between an active and inactive form—often through a **signal transduction pathway,** in which signals pass from the external environment through the cytoplasm, into the nucleus.

As an example, let's consider the Dorsal protein in *Drosophila.* Dorsal plays a critical role in the development of the dorsal–ventral axis in the *Drosophila* embryo. Dorsal can be either an activator or repressor of gene expression, depending on the gene it is regulating.

Spätzle Is Cleaved and Binds to Toll to Activate Pelle

To activate Dorsal so that it can regulate gene expression, an extracellular protein, Spätzle, first must be cleaved in a specific manner by a protease (**fig. 17.25**). The cleaved Spätzle protein can now bind to a receptor, Toll, which is located on the membrane of the *Drosophila* embryo. Spätzle binding activates the kinase activity of the Toll protein, which then phosphorylates another kinase called Pelle. The Tube protein, which is associated with Pelle, is also involved in the activation of Pelle in an unknown manner.

Pelle Phosphorylates Cactus to Release Dorsal

The activated Pelle, which is a protein kinase, can phosphorylate the Cactus subunit of the Dorsal-Cactus heterodimer. Phosphorylation of Cactus disrupts the dimer, which releases the Dorsal protein (see fig. 17.25). The phosphorylated Cactus protein is also degraded to prevent its reassociation with Dorsal. As a component of the Dorsal–Cactus dimer, Dorsal is inactive and present in the cytosol of the cell. However, the free Dorsal subunit is now able to enter the nucleus and regulate the expression of genes. Thus, Dorsal exists in both an inactive (heterodimer) and active states.

In reality, at least three other proteins must be cleaved outside of the cell prior to the cleavage of Spätzle, which is required for the activation of Toll and Pelle.

580

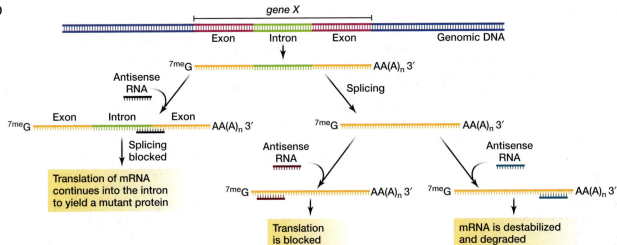

figure 17.24 Mechanisms of antisense RNA action in a cell. The antisense RNA can basepair with the intron-exon junction of the pre-mRNA (left) and block splicing. Failing to splice out an intron within an open reading frame results in the translation of a mutant protein. Alternatively, the antisense RNA can base pair with the mRNA and either block translation of the open reading frame (*center*) or decrease the stability of the mRNA resulting in its degradation (right).

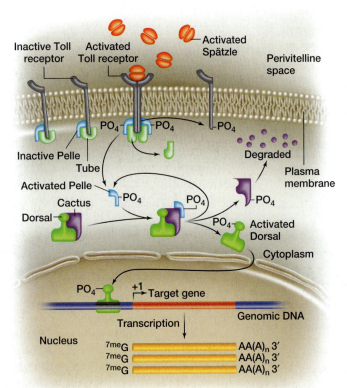

figure 17.25 The Dorsal signal transduction pathway in *Drosophila*. The cleavage of the Spätzle protein in the perivitelline space of the embryo produces an active Spätzle ligand that binds the Toll transmembrane receptor protein. Spätzle binding causes the Toll receptor to form a dimer, which activates the Pelle-Tube heterodimer. The activated Pelle, which is a kinase, phosphorylates itself, Tube, and Toll. The activated Pelle is released and then phosphorylates the Cactus and Dorsal proteins. This phosphorylation releases Cactus, which is then degraded. The phosphorylated Dorsal then enters the nucleus and binds specific enhancers to activate transcription of specific target genes.

Although this pathway appears to be quite complex, it is relatively simple compared with other pathways. Some pathways involve many more intracellular steps before the DNA-binding protein is activated or inactivated, as required. Some pathways require several steps even before a cell can release a signal to be bound by the receptor on a different cell, which then activates a pathway to modify the DNA-binding protein. The complexity built into these pathways allows many different steps to be the subject of regulation, or put another way, many conditions that must be met before the regulation of a gene is changed. Alternatively, these pathways may be modified, which could result in a different signal affecting the same DNA-binding protein, or a single signal affecting different DNA-binding proteins.

It's Your Turn

You previously learned that DNA binding proteins can function as activators and repressors of transcription. In this chapter you also learned that the activity of these DNA binding proteins can be regulated through a variety of different mechanisms. To test your understanding of how DNA binding proteins can regulate transcription, try problems 28, 30, and 34.

Restating the Concepts

▶ MicroRNAs (miRNAs) can regulate expression at the translational level. The mechanisms of regulation are based on the 22-nucleotide miRNA of the RISC that base-pairs with the target mRNA. The miRNA may base-pair with the mRNA and block translation of a mRNA or may target the cleavage of the mRNA.

▶ Antisense RNAs are usually longer RNAs that are complementary to a target RNA sequence. Base pairing of the antisense RNA and the target RNA can block the splicing of the target RNA, repress the translation of the target mRNA, or affect the stability of the target mRNA.

▶ Regulation of expression is often more complex than a single protein binding to genomic DNA. Often, the regulation is in response to a stimulus, either environmental or from another cell, which requires a complex signaling pathway to activate or inactivate the DNA-binding protein.

17.5 The Role of Chromatin in Gene Regulation

We have been discussing the binding of proteins to different types of DNA sequences to regulate transcription. For the proteins to bind, however, the genomic DNA must be accessible to these activators and repressors. Similarly, RNA polymerase II and its promoter-recognizing TFIID need access to the promoter sequence in the genomic DNA.

We know that the eukaryotic genomic DNA is present in chromatin, a complex of DNA and a variety of proteins (see chapter 7). The tight association of genomic DNA with these proteins, especially the histone proteins in the nucleosome, is a major factor in repressing gene transcription. Transcription initiation therefore requires modifying the chromatin state so that the genomic DNA is accessible to the transcription machinery.

Several different mechanisms can modify the nucleosomes, which results in remodeling the chromatin. We mentioned in chapter 7 that chromatin remodeling can be achieved by acetylation of histones, by the incorporation of variant histones into the nucleosome, and by the enzyme-mediated movement of nucleosomes. Before we discuss these three mechanisms in more detail, we take a quick look at how chromatin remodeling near a gene is detected.

Detection of Nucleosome Position

To examine the status of the chromatin, we use the enzyme DNase I, which cuts any DNA sequence that is accessible to the enzyme. At low DNase I concentrations, genomic DNA that is free of nucleosomes is digested, and genomic DNA associated with nucleosomes is protected from digestion. **DNase I hypersensitive sites** represent DNA that is cut at low DNase I concentrations.

To examine DNase I hypersensitivity, nuclei are isolated from the tissue being tested and from a control tissue. The nuclei are treated with DNase I to digest the accessible genomic DNA (**fig. 17.26**). The DNA is then treated to remove the DNase I and the histone proteins.

The protein-free genomic DNA is then digested with a restriction enzyme, electrophoresed in an agarose gel, and transferred to a membrane. A Southern blot is then performed using probes for the gene of interest. DNase I hypersensitive sites are revealed as fragments that decrease in size in response to DNase I treatment in transcriptionally active tissues relative to the control tissues that do not express the gene.

From this type of analysis, we learn that DNase I hypersensitive sites are often found upstream of genes that are transcriptionally active. These sites correspond to the absence of nucleosomes in tissues where the gene is transcribed. In tissues in which the gene is not transcribed, nucleosomes are present at these sites. The location of nucleosomes along the length of the genomic DNA must change to permit the binding of transcription factors and the RNA polymerase II transcription complex. Furthermore, these nucleosomes when present often overlap DNA sequences that are also bound by activator proteins. Thus, the interaction between the nucleosome and the activator protein is one level of regulating transcription.

Histone Acetylation and Deacetylation

In chapter 7, we described the analysis of nucleosome structure. In eukaryotes, the nucleosome consists of two copies of each of four histone proteins (H2A, H2B, H3, and H4). A fifth histone, H1, binds to the linker DNA that is located between nucleosomes. Because these histones are functionally the same in all eukaryotic organisms, their role in regulating gene expression must also be conserved in all eukaryotes.

Histones can exist in either an acetylated or deacetylated form. The acetyl groups are added to the amino-terminal lysines of histones H3 and H4, which decreases the interaction between the histone core and the negatively charged sugar–phosphate backbone of the genomic DNA. The acetylated histones, therefore, are associated with transcribed genes, and deacetylated histones are found near the genes that are transcriptionally repressed.

Acetylation of histones is performed by a group of enzymes called histone acetyltransferases (HATs). Type B HATs are cytosolic enzymes that acetylate the newly synthesized H3 and H4 histones. These histones are then transported into the nucleus and incorporated into nucleosomes (**fig. 17.27**).

Type A HATs are nuclear enzymes that bind acetylated lysines and add additional acetyl groups to the amino-terminal lysines on H3 and H4. It is interesting to note that the $TAF_{II}250$ subunit of TFIID, which binds the initiator region of the promoter during the formation of the transcription preinitiation complex, also has the capacity to bind acetylated lysines. It is possible that $TAF_{II}250$ recognizes those initiator regions that are associated with acetylated histones. The bound $TAF_{II}250$

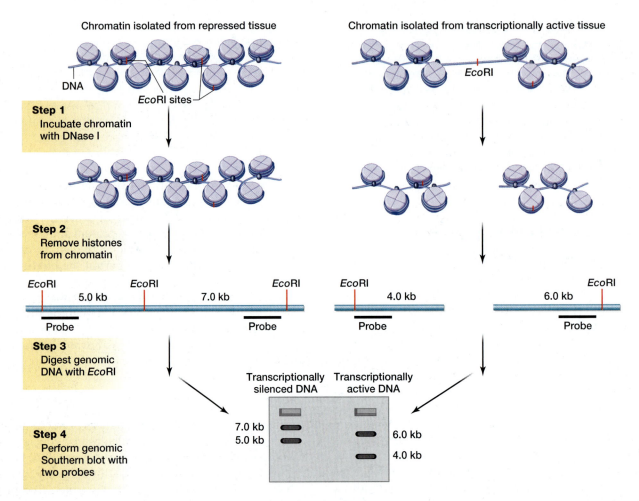

Chromatin isolated from repressed tissue

Chromatin isolated from transcriptionally active tissue

DNA

*Eco*RI sites

*Eco*RI

Step 1
Incubate chromatin with DNase I

Step 2
Remove histones from chromatin

*Eco*RI   5.0 kb   *Eco*RI   7.0 kb   *Eco*RI

Probe   Probe

*Eco*RI   4.0 kb   *Eco*RI   6.0 kb   *Eco*RI

Probe   Probe

Step 3
Digest genomic DNA with *Eco*RI

Transcriptionally silenced DNA   Transcriptionally active DNA

Step 4
Perform genomic Southern blot with two probes

7.0 kb
5.0 kb

6.0 kb
4.0 kb

figure 17.26   DNase I hypersensitivity assay. Nuclei are isolated from a tissue that does not express the gene being tested (left) and a tissue that expresses the gene (right). The nuclei are treated with a small amount of DNase I, which degrades unprotected DNA (right) but not compact chromatin (left). The histone proteins are removed from the chromatin and the DNA is digested with a restriction enzyme (*Eco*RI). A genomic DNA Southern blot reveals that the repressed genomic DNA has the expected *Eco*RI genomic DNA fragments near the gene of interest (left) because the histones protected the DNA from DNase I degradation. The actively transcribed gene will have smaller DNA fragments due to the DNase I degradation (right). The regions that are accessible to DNase I upstream of transcribed genes are called DNase hypersensitive sites.

figure 17.27   Acetylation and deacetylation of histones. The newly translated H3 and H4 histone proteins may be acetylated by the histone acetyltransferase-B (HAT-B) protein in the cytosol, which then move into the nucleus and incorporate into nucleosomes. The H3 and H4 histones can be deacetylated by the histone deacetylase (HDAC). The nuclear HAT-A enzyme can acetylate H3 and H4 histones that are already incorporated in a nucleosome.

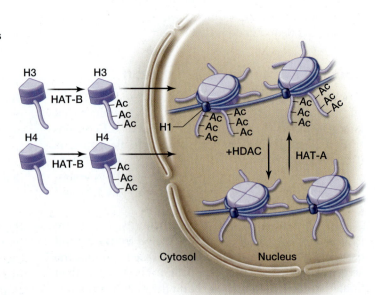

H3

H3

HAT-B

- Ac
- Ac
- Ac

H1

- Ac   - Ac
- Ac   - Ac
- Ac

- Ac
- Ac   - Ac
- Ac   - Ac
- Ac

H4

H4

HAT-B

- Ac
- Ac
- Ac

+HDAC   HAT-A

Cytosol   Nucleus

may also acetylate the nearby histones to make the DNA further accessible for transcription initiation.

Cells also express histone deacetylases (HDACs), which reduce the number of acetyl groups present on the H3 and H4 histones (see fig. 17.27). HDACs are necessary to reduce the levels of histone acetylation, which are required to repress transcription.

Recent evidence suggests that DNA-binding proteins may recognize not only a specific DNA sequence, but also the acetylation state of the histones. Binding of some activators to a DNA sequence may recruit HATs that would increase the acetylation of the histones, and thereby increase the accessibility of the DNA to other activators and to the RNA polymerase complex (**fig. 17.28**). Similarly, binding of a repressor to DNA may recruit HDACs, which reduce the histone acetylation and further repress transcription by making the DNA less accessible.

Histone Methylation

Acetylation is not the only modification that histone proteins undergo. The lysines and arginines on all four nucleosome histone proteins can be methylated by histone methyltransferase (HMTase) and HMTase-associated protein (HP1). Whereas acetylation only activates transcription, methylation of histones can either activate or repress transcription.

Effects of Histone Methylation

Methylation of lysine-4 on histone H3 enhances transcription through two different mechanisms (**fig. 17.29a**). First, this methylated lysine inhibits a HDAC from binding and reducing acetylation of the H3 tail. Repressing deacetylation maintains the enhanced state of transcription. Second, the methylation of lysine-4 inhibits the methylation of lysine-9 on histone H3. Because methylation of lysine-9 reduces transcription, methylation of lysine-4 inhibits the repressive effect of lysine-9, again enhancing transcription.

In fact, lysine-9 reveals that an intimate relationship exists between acetylation and methylation of histone H3 that results in a regulatory switch. Methylation of lysine-9 blocks the phosphorylation of the adjacent serine-10 (fig. 17.29b). Phosphorylation of serine-10 enhances the acetylation of lysine-14, which is required for the transcription of some genes. Thus, methylation of lysine-9 represses transcription by blocking acetylation of lysine-14.

In contrast, acetylation of lysine-14 and phosphorylation of serine-10 represses the methylation of lysine-9 (fig. 17.29c). Similarly, acetylation of lysine-9 prevents methylation of the same residue (fig. 17.29d) to activate transcription.

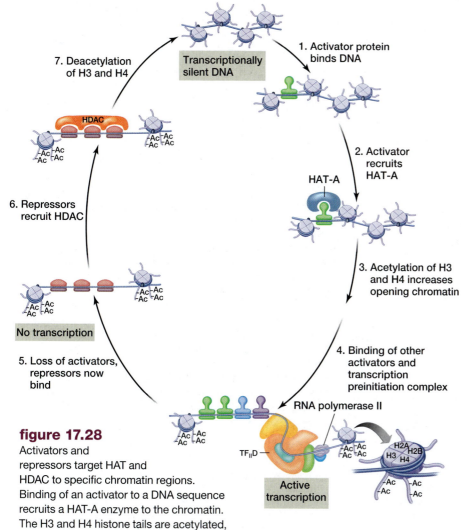

figure 17.28

Activators and repressors target HAT and HDAC to specific chromatin regions. Binding of an activator to a DNA sequence recruits a HAT-A enzyme to the chromatin. The H3 and H4 histone tails are acetylated, which opens the chromatin conformation to permit the binding of additional activators and formation of the transcription preinitiation complex (TFIID and RNA polymerase II). Over time, the activators may be lost and replaced with repressors, which terminate transcription of the target gene. The repressors may recruit a HDAC, which deacetylates the H3 and H4 histones. This causes the chromatin to become more compact and further suppresses transcription.

Spreading of Methylation Effects

The effects of methylation can also spread from nucleosome to adjacent nucleosome. Methylation of lysine-9 aids the binding of HP1 to the nucleosome, which then recruits a HMTase that methylates lysine-9 on the adjacent nucleosome (**fig. 17.30**). The result is the spreading of the repressive chromatin state.

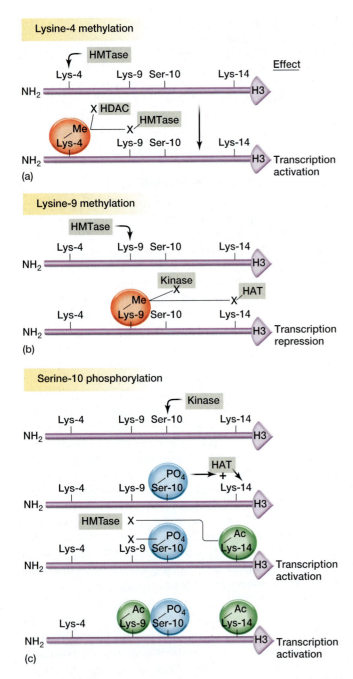

figure 17.29   Variations of the histone H3 modification code. (a) Methylation of Lysine-4 by the histone methyltransferase (HMTase) blocks the methylation of Lysine-9 and any deacetylation, which combine to activate transcription. (b) Methylation of Lysine-9 blocks methylation of Serine-10 and acetylation of Lysine-14, which represses transcription. (c) Phosphorylation of Serine-10 activates acetylation of Lysine-14. Both of these modifications block methylation of Lysine-9. In the absence of Lysine-9 methylation, Lysine-9 may be acetylated, both leading to transcription activation.

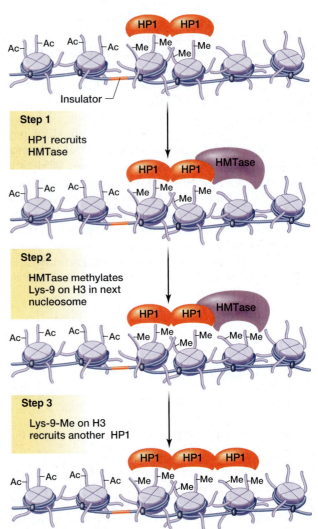

figure 17.30   Expansion of a transcriptionally silent region. The insulator region (red) prevents the rightward expansion of the transcriptionally active region (acetylated nucleosomes on the left) and the leftward expansion of the transcriptionally silent region (methylated nucleosomes on the right). However, the transcriptionally silent region can expand to the right. The HP1 protein binds the methylated Lysine-9 on histone H3 and recruits a histone methyltransferase (HMTase), which then methylates the histone H3 tail on the next nucleosome. This methylated nucleosome can then bind a newly recruited HP1 and the transcriptionally repressed region expands one nucleosome further to the right.

Notice that the repressive methylation cannot proceed across an insulator region (see fig. 17.30). This region protects the acetylated state of the nucleosome to the left that functions to activate transcription.

Enzyme-Mediated Movement of Nucleosomes

As you have seen, increasing the acetylated state of the histones reduces the association between the nucleosome and the genomic DNA, which permits increased transcription of nearby genes. An alternative approach to increase transcription is to change the location of the nucleosomes so that the promoter and enhancer sequences are available for protein binding. Either the **ISWI** or the **SWI–SNF protein complexes** can mediate the change in the nucleosome location. (The SWI–SNF complex was originally discovered in yeast, where it was found to facilitate *swi*tching of mating types. ISWI stands for imitation SWI. Homologs are found in mammals.) How do these protein complexes function to make the genomic DNA more accessible for transcription?

Very little is actually known about the mechanisms that underlie the nucleosome movement. It is known that the ISWI protein complex hydrolyzes ATP to produce energy to slide the DNA across the surface of the histone core (**fig. 17.31a**) This action repositions the nucleosome and can make a promoter region accessible for transcription initiation. In contrast, the SWI–SNF protein complex appears to utilize ATP to dissociate the DNA from the histone core, followed by reassociation of the histone core to a new location on the DNA and the rewrapping of the DNA to regenerate the nucleosome (fig. 17.31b).

Certain genes have been identified that are activated by the SWI–SNF protein complex, followed by HAT-mediated acetylation of histones (fig. 17.31c). These

two chromatin-remodeling mechanisms therefore are not mutually exclusive, but may be required in a specific sequence to properly initiate transcription. This is not much different from the requirement of a specific sequence of transcription factors binding to the DNA to produce the preinitiation transcription complex.

Incorporation of Variant Histone Proteins into Nucleosomes

As described in chapter 7, the nucleosome is composed of two subunits of H2A, H2B, H3, and H4 histones, and the H1 histone is associated with the linker DNA between the nucleosomes. Several variant histone proteins also exist in addition to these five standard histone proteins. These variant histones share 40–80% amino acid identity with their respective common histone.

The function of these variant histones differs significantly from their common histone and from each other. Some H1 histone variants are likely involved in repressing transcription, and others are likely involved in activating transcription.

There are several H3 histone variants, with the functions of H3.3 and CENPA (Centromere Protein A) being the best defined. The H3.3 histone is associated with genes that are actively transcribed. In contrast, CENPA is essential for the generation of the chromatin in the centromeric region of the chromosome.

Different H2A variants also appear to be associated with specific functions. One particular histone H2A variant (mH2A) appears to replace the standard H2A histone during X chromosome inactivation in female mammals (see chapter 4). This mH2A variant may be involved in the generation or maintenance of the Barr body. In contrast, the H2AZ variant is associated with genes that are transcriptionally active.

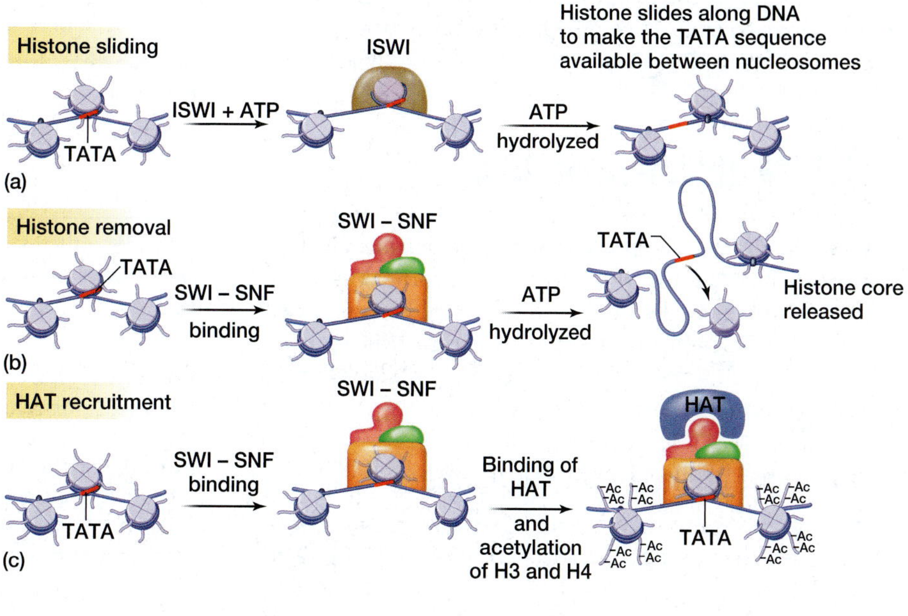

figure 17.31 Potential mechanisms of protein-mediated movement of nucleosomes. (a) The ISWI protein complex binds a nucleosome and hydrolyzes ATP to slide the histones along the genomic DNA. This shifts a TATA sequence (red) from being present on a nucleosome to being located between nucleosomes and available for transcription initiation. (b) The SWI-SNF protein complex binds a nucleosome region and hydrolyzes ATP to remove the histone proteins. Again, a promoter region that is initially located on a nucleosome (red) is subsequently located between nucleosomes. (c) The binding of the SWI-SNF complex can recruit a HAT, which acetylates the adjacent nucleosomes to enhance transcription initiation.

The mechanism by which the histone variants are deposited into the chromatin remains somewhat unclear. In the case of H2AZ, a complex that includes a H2AZ and H2B histone dimer and the Swr1 and Bdf1 proteins is required. The Swr1 protein is a member of the SWI–SNF family of chromatin-remodeling proteins, whereas Bdf1 binds acetylated H4 histones. It is believed that the Bdf1 protein targets the complex to transcriptionally active chromatin and the Swr1 protein exchanges the H2AZ–H2B dimer for the H2A–H2B dimer in the nucleosome, which further enhances the transcriptionally active state of the chromatin.

Restating the Concepts

▶ Chromatin remodeling is the change in nucleosome location or in the interaction between the nucleosome core and genomic DNA. Chromatin remodeling, which is required for repression and activation of gene expression, can be detected by DNase I hypersensitive sites.

▶ The level of acetylation of histones is controlled by the HAT and HDAC complexes. In general, increased acetylation is associated with transcription activation.

▶ Histone methylation can either activate or repress transcription. Complex regulatory networks exist in which the acetylation pattern affects the methylation of the histone and the methylation status affects the acetylation. This complex situation permits the dynamic regulation of transcription of some genes.

▶ The ISWI and the SWI–SNF protein complexes have the ability either to slide nucleosomes to a new location relative to a promoter or to disassemble and reassemble nucleosomes. The modification of histones and the movement of nucleosomes often work in a sequence to regulate transcription.

▶ Variant forms of histones exist that can be incorporated into nucleosomes to regulate the chromatin state. In some cases, the chromatin-remodeling proteins are required to deposit the histone variant into the nucleosome.

17.6 Methylation of DNA

Methylation of the DNA molecule can also play an important role in DNA–protein interactions. We already discussed that methylation of histones can repress transcription of some genes. In this section, we will see that methylation of DNA can also repress transcription, resulting in gene silencing.

CpG Islands and Gene Silencing

In eukaryotic organisms, a small percentage of cytosine residues are methylated by the enzyme DNA methyl-ase (**fig. 17.32a**). Most of the methylated cytosines are present in CpG sequences, with up to 80% of the cytosines in CpG sequences methylated in human DNA. (Often, when we refer to a sequence of two bases on the same strand of DNA, we put a "p" between them—CpG—to indicate that they are on the same strand connected by a phosphodiester bond, and not on two different strands as a hydrogen-bonded base pair.)

CpG islands are regions that are highly enriched in the CpG sequence; they are 1–2 kb in length and usually are located from several hundred base pairs upstream of the transcription start site to several hundred base pairs downstream. Thus, many CpG islands are located near the promoters of eukaryotic genes, which places them in an ideal location to regulate transcription.

The degree of DNA methylation is related to the silencing of a gene. In housekeeping genes, which encode proteins that are expressed at all times in all cells of a multicellular organism, the CpG islands are methylated at a very low level (fig. 17.32b). In contrast, genes that are not expressed in a particular cell or tissue possess a high degree of methylation in the CpG islands. Thus, methylation is involved in silencing transcription.

An extreme example of silencing transcription is the inactivation of one of the two X chromosomes in every somatic cell of a female mammal (see chapter 4). Analysis of the inactive X chromosome reveals a higher level of methylation than the active X chromosome.

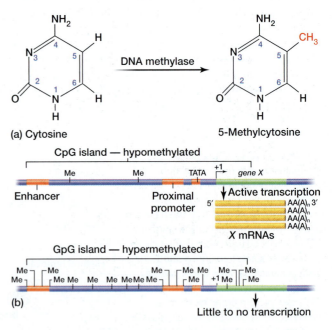

figure 17.32   DNA methylation. (a) The C-5 position of cytosine is methylated by a DNA methylase. This methylation does not affect base pairing with guanine. (b) The CpG island is located upstream of a gene and includes the promoter and enhancer. Hypomethylation of the CpG sequences in this region permits active transcription, while hypermethylation suppresses transcription.

Transcription Repression

How can methylation of the CpG sequence repress transcription? The most likely mechanism involves extension of the methyl group, which is located on C-5 of cytosine, into the major groove of the DNA, where most activators bind (**fig. 17.33**). The methylation of the CpG sequences may therefore actually block the regulatory proteins from binding their target sequence.

Another mechanism would involve the identification of proteins that bind methylated DNA. These **methyl-CpG-binding proteins** recognize the 5-methyl-cytosine, regardless of the flanking nucleotide sequences (see fig. 17.33). It is possible that these binding proteins may recruit the binding of HDACs, which would lead to the further repression of transcription by remodeling the chromatin.

The discovery of Z-DNA has generated further interest in the role of methylation. As mentioned in chapter 7, Z-DNA can be stabilized by methylation. This observation has led to a model of transcriptional regulation based on alternative DNA structures. Sequences (such as CpG repeats) could exist as Z-DNA when repressed and as B-DNA when being transcribed. If the gene is to be silenced, the CpG sequences are converted to stable Z-DNA by cytosine methylation, which then represses transcription. This possibility has gained some interest because of the recent discovery of an enzyme, double-stranded RNA adenosine deaminase (ADAR1), which binds to Z-DNA sequences.

It's Your Turn

Transient changes to the DNA, such as the chromatin organization or DNA methylation, can have dramatic effects on gene expression. We discussed different ways that these changes can be examined. In questions 10, 12, and 14, and problem 22, you have the opportunity to utilize your knowledge to examine how these DNA modifications affect transcription.

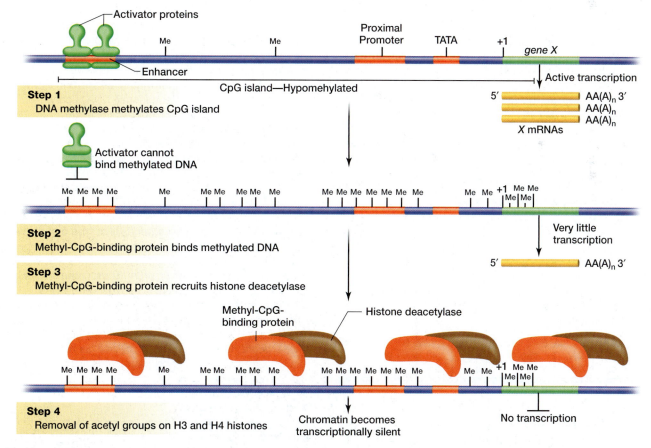

figure 17.33   Models of transcription repression by DNA methylation. The promoter region and CpG island are shown for *gene X*. An activator protein is bound to an enhancer, which activates transcription of *gene X*. Methylation of various sites in the CpG island leads to two effects that each repress transcription. First, methylation results in methyl groups protruding into the major groove of DNA, which prevents activator binding. Second, methylation of additional cytosines recruits a methyl-CpG-binding protein to the DNA. The methyl-CpG-binding protein recruits a histone deacetylase (HDAC) that deacetylates neighboring H3 and H4 histones. This decreased H3 and H4 acetylation compacts the chromatin to a transcriptionally silent state.

17.1 *applying genetic principles to problems*

Determining the Pattern of DNA Methylation

The methylation of CpG islands in eukaryotic DNA is an important mechanism in repressing or silencing gene expression. The ability to identify the location of the methylated sequences and determine if they are methylated under different conditions is an important analysis in studying gene expression.

Principles of the Technique

The analysis of DNA methylation patterns is based on the specificity of DNA restriction endonucleases. As you remember from chapter 12, restriction endonucleases cut DNA at specific sequences. Thus, it is possible to use the restriction enzyme *Hpa*II or *Msp*I to cut the DNA at the 5'-CCGG-3' sequence. However, these two enzymes possess a slight difference in their specificity for the DNA sequence that is cleaved. Although *Hpa*II only cleaves the 5'-CCGG-3' sequence if both cytosines are not methylated, *Msp*I can cleave the 5'-CCGG-3' sequence whether or not the second cytosine is methylated. The second cytosine is part of a CpG sequence.

After the genomic DNA is digested with either the *Hpa*II or *Msp*I enzyme, it is analyzed by a Southern blot to determine the sizes of the DNA fragments cut by the enzymes. If the Southern blot reveals that the two enzymes produce the same size fragments, then the 5'-CCGG-3' sites are not methylated. However, if the Southern blot reveals that the *Hpa*II fragment is larger than the *Msp*I fragment, then the second cytosine in the 5'-CCGG-3' sequence is methylated.

Analysis of Methylation Pattern from Genomic Southern Blot

Let's apply the basic information that we just discussed to determine the methylation pattern in the promoter region of a gene. The locations of the 5'-CCGG-3' sequences in the promoter region are shown in the following diagram. The vertical lines represent the location of the 5'-CCGG-3' sequences. The distances between restriction enzyme sites are shown as are the probes used for the genomic Southern blots.

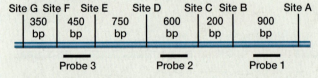

Genomic DNA is isolated from both the liver and kidney, knowing that the gene is transcribed in the liver and not in the kidney. The genomic DNA from each tissue is divided in two, with half being digested with *Hpa*II and half with *Msp*I. The digested DNA is analyzed by Southern blots with each of the three probes.

The genomic Southern blots with the three probes are illustrated as follows.

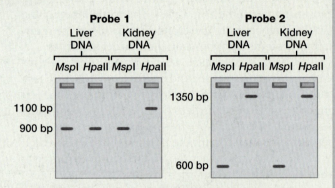

When analyzing these genomic Southern blots, remember that the restriction endonuclease *Hpa*II only cleaves the 5'-CCGG-3' sequence if it is not methylated, whereas *Msp*I cleaves the 5'-CCGG-3' sequence whether the second cytosine is methylated or not.

Probe 1 detects a 900-bp *Msp*I fragment as would be expected from the restriction map. The *Hpa*II digest reveals a 900-bp fragment in liver genomic DNA and a 1100-bp fragment in the kidney genomic DNA. This difference between the 1100-bp *Hpa*II fragment and the 900-bp *Msp*I fragment in kidney genomic DNA reveals that site B is methylated. The methylation of site B in kidney genomic DNA and not in liver genomic DNA is consistent with the gene not being transcribed in the kidney.

Probe 2 detects a 600-bp *Msp*I fragment and a 1350-bp *Hpa*II fragment in both liver and kidney genomic DNA. This reveals that the 5'-CCGG-3' sequence in site D is methylated in both genomic DNAs. Because it is methylated in both genomic DNAs, it is not likely critical in regulating the tissue-specific expression of the gene.

If site F, but not site E, is methylated in silencing the transcription of the associated gene, what type of pattern would you expect on the genomic Southern blot using probe 3? As discussed, *Msp*I cleaves the 5'-CCGG-3' sequence regardless of the methylation pattern. Thus, probe 3 would detect a 450-bp *Msp*I fragment in both genomic DNAs.

Because site E is not methylated in either kidney or liver genomic DNA, then *Hpa*II would cleave it in both DNAs. If site F is methylated in the silenced kidney gene, then it would not be methylated in liver genomic DNA. This would produce a 450-bp fragment in liver genomic DNA. The kidney DNA would have site E, and site D (see preceding paragraph) would be methylated. Thus, probe 3 would detect a *Hpa*II fragment that is cleaved at sites E and G, which would be 450 + 350 = 800 bp in length. The Southern blot with probe 3 would then appear as:

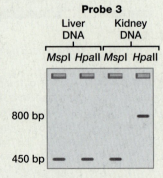

17.7 Imprinting

The regulatory mechanisms that we already discussed are transient; they will change in different tissues and at different times in development. This is critical because some genes must be expressed in a specific tissue at a particular time, then become transcriptionally repressed before they are transcribed again in a different tissue at a later time. Furthermore, these regulatory mechanisms affect both copies of the gene in a diploid organism. Both of these features are critical for a cell or tissue to respond to different cellular and environmental signals.

In contrast, the expression of some genes exhibits a very different pattern. In this class of genes, only one of the two copies of the gene is transcribed. The copy of the gene that is transcribed depends on whether it is inherited from the male or female parent. Thus, for some genes, the phenotype is determined by the allele that is received from a particular parent. DNA methylation is the mechanism used to inactivate the gene inherited from a particular parent. This phenomenon is called **imprinting** (or *molecular imprinting* or *parental imprinting*)—the differential expression of the alleles at a locus depending on which parent contributed the gene.

Imprinting falls under the general classification of **epigenetics,** which is an effect resulting from an environmentally induced change in the genetic material that does not cause a change in the DNA sequence. Because the DNA sequence is unaffected, it is not a permanent change. Although an epigenetic effect can be fixed through the lifetime of an individual, it can change in the next generation. Here we look at an example of imprinting so you can see the role of methylation and the pattern of allele expression from only one parent.

The *Igf-2* Gene in Mice

The mouse *Igf-2* gene encodes the insulin-like growth factor 2. The wild-type allele (*Igf-2*) produces a normal-sized mouse, and the mutant allele (*Igf-2m*) yields a small mouse (**fig. 17.34**). However, two normal-sized heterozygous mice can produce heterozygous offspring who are phenotypically different (**fig. 17.35**). Because of imprinting, the paternal allele is expressed in an individual, and the maternal allele is silenced through DNA methylation. A heterozygous individual who received the *Igf-2* allele from its father is normal size, but a heterozygous mouse who received the *Igf-2m* (mutant) allele from its father will be small—as though it were homozygous for *Igf-2m* (see fig. 17.35). In other words, the maternally inherited allele is silenced in the progeny and the paternally inherited allele is not, which makes the paternally inherited allele responsible for the offspring's phenotype.

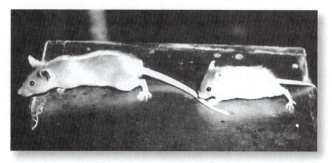

figure 17.34 The phenotypes associated with two *Igf-2* alleles in mice. The normal-sized mouse on the left expresses the wild-type *Igf-2* allele. The smaller mouse on the right expresses the *Igf-2m* mutant allele. However, both mice have the same genotype, *Igf-2/Igf-2m*. The phenotypic difference is due to the mouse on the left receiving the *Igf-2* allele from its father and the mouse on the right receiving the *Igf-2* allele from its mother. The maternal allele is silenced in both cases.

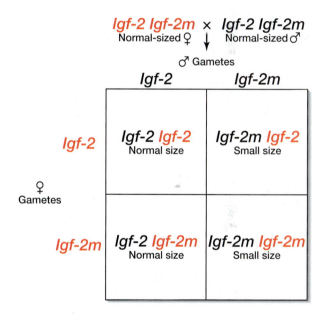

Genotypic ratio:

1	:	2		:	1
Igf-2 Igf-2 Normal size		*Igf-2 Igf-2m* Normal size	*Igf-2m Igf-2* Small size		*Igf-2m Igf-2m* Small size

Phenotypic ratio:

1 normal size : 1 small size

figure 17.35 Imprinting the *Igf-2* gene in mice. In the cross, both heterozygous parents are normal size. The ova contain imprinted *Igf-2* alleles (red), while the sperm contain *Igf-2* alleles (black) that will be expressed in the progeny. Fertilization results in the expected 1:2:1 genotypic ratio, but not the expected 3:1 phenotypic ratio. This unexpected 1:1 phenotypic ratio is due to imprinting, where the maternal allele (red) is silenced in all the progeny. Notice that the heterozygous mice produce two different phenotypes that depends on what allele is inherited from the male parent.

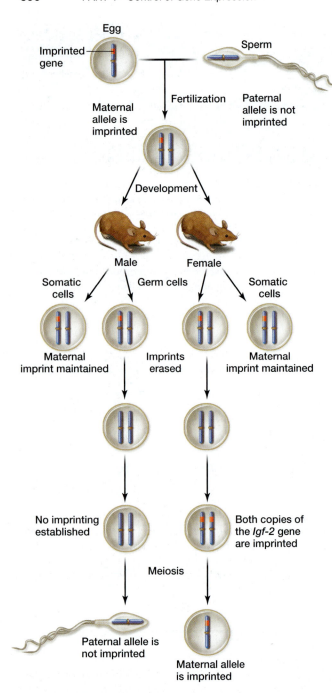

figure 17.36   Diagram of *Igf-2* gene imprinting. The *Igf-2* gene is imprinted in the female gametes, but not in the male. As the mouse, either male or female, develops, the maternally inherited *Igf-2* allele remains imprinted in all somatic cells. However, the imprint is erased in both male and female germ cells. In the male (left), all the sperm that are produced lack the imprinted *Igf-2* gene. In the female germ cells (right), the imprint is reset on the *Igf-2* gene on both homologous chromosomes, which results in all the eggs containing an imprinted *Igf-2* gene.

The imprint (DNA methylation) that is associated with the maternal *Igf-2* gene is inherited from the egg to the zygote (**fig. 17.36**), where it is passed on through multiple rounds of cell division (mitosis). In the germ cells, however, the imprint is removed briefly, such that neither copy of the *Igf-2* gene is methylated. At this point, both copies of *Igf-2* then become methylated in a female, but neither copy is methylated in a male. Thus, all the eggs will again contain an imprinted (methylated) *Igf-2* gene and all the sperm will contain an unmethylated *Igf-2* gene.

The *H19* Gene in Mice

The mouse *H19* gene is located approximately 70 kb downstream from the *Igf-2* gene in mice. The *H19* gene, which is found as the common alleles *H19* and *H19hc*, confers a histocompatibility phenotype on the mice.

The *H19* gene is also imprinted in mice. In contrast to the *Igf-2* imprinting, the *H19* gene is imprinted in the male rather than the female. This results in the progeny expressing the *H19* phenotype that corresponds to the maternally inherited allele.

Mechanism of Imprinting the *Igf-2* and *H19* Genes

The relative locations of the *Igf-2* and *H19* genes are shown in **figure 17.37**. Also located in this region is an insulator element between the two genes and an enhancer element that is located downstream of the *H19* gene.

The CCCTC-binding factor CTCF, which contains 11 zinc fingers, binds the DNA sequence CCCTC that is located in the insulator region between the *Igf-2* and *H19* genes. On the maternally inherited chromosome, the insulator is not methylated, which allows the CTCF protein to bind (see fig. 17.37). Binding of the CTCF protein to the insulator prevents the enhancers downstream of the *H19* gene to activate transcription of the *Igf-2* gene. The enhancers, however, do activate the transcription of the *H19* gene in females. Thus, the absence of DNA methylation in this region on the maternally inherited chromosome results in the repression of transcription of the *Igf-2* gene and activation of the *H19* gene expression.

On the paternally inherited chromosome, the insulator and the *H19* promoter region are methylated (see fig. 17.37). The methylation of the insulator prevents the binding of the CTCF protein. This removes the insulator effect and allows the enhancers to activate transcription of the *Igf-2* gene. Methylation of the *H19* promoter also silences the expression of the *H19* gene. Therefore, methylation results in the prevention of CTCF protein binding, activation of the *Igf-2* gene expression, and silencing of the *H19* gene.

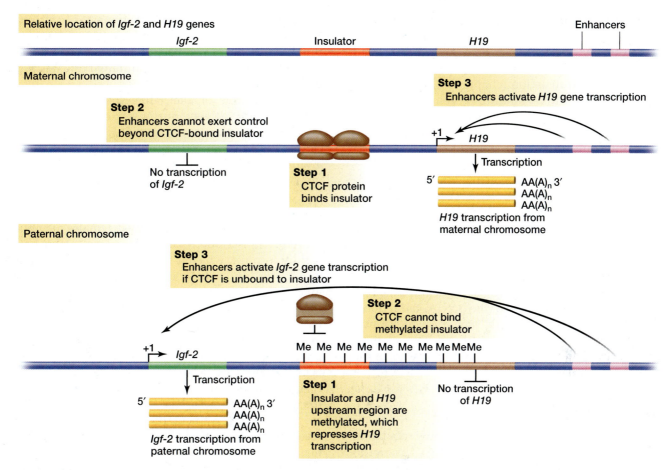

figure 17.37   *Mechanism of imprinting the Igf-2 and H19 genes in mice. The location of the Igf-2 and H19 genes are shown relative to an insulator and downstream enhancers. The maternal chromosome lacks methylation in the insulator region. The CTCF protein binds the unmethylated insulator, which prevents the enhancers from activating transcription of the Igf-2 gene. Thus, the enhancers only activate transcription of the H19 gene and the Igf-2 gene is transcriptionally inactive on the maternal chromosome. Thus, maternal imprinting of Igf-2 is due to the absence of methylation. The insulator and the H19 promoter are both methylated on the paternal chromosome, which silences expression of the H19 gene and prevents the binding of CTCF. The absence of the CTCF protein bound to the insulator allows the enhancers to activate transcription of the Igf-2 gene, but not the H19 gene, on the paternal chromosome. Thus, paternal imprinting of the H19 gene is due to DNA methylation.*

Restating the Concepts

▶ DNA is methylated on the cytosine of CpG sites in many organisms. Methylation of DNA can repress transcription by preventing activators from binding to particular DNA sequences and by binding methyl-CpG-binding proteins that can recruit repressors or HDACs to deacetylate histones. Methylation can prevent the binding of other proteins, such as the CTCF protein, which affects the expression of genes.

▶ Imprinting occurs when a gene is reproducibly silenced in either the maternal or paternal gamete and is not expressed in the somatic cells of the progeny. Imprinting is erased in the germ cells of the progeny and reestablished in a gender-specific manner. Imprinted genes fail to exhibit the expected Mendelian phenotypic ratios due to the silencing of the allele in one of the parents.

17.8 Posttranscriptional Regulation

We already encountered one mechanism of posttranscriptional regulation. The miRNAs, which are a component of the miRNP, can basepair with a specific mRNA and either repress translation or induce the cleavage of the mRNA. This type of mechanism depends on the nucleotide sequences of both the miRNA and the mRNA and on the level of complementarity between the two RNAs. Several additional posttranscriptional regulatory mechanisms exist; here we consider three of them.

Proteins Binding to 3' Untranslated Regions

One posttranscriptional regulatory mechanism involves the binding of a protein to the mRNA to exert an effect. Although we discussed many DNA-binding proteins in this chapter, a class of proteins also exists that specifically recognizes and binds mRNA molecules. Often these molecules bind the 3' untranslated region (3' UTR) of the mRNA and modify the mRNA or repress its translation (chapter 11). Let's look at one way that an RNA-binding protein can regulate expression by modifying the mRNA (we will look at a second way that involves a RNA-binding protein affecting translation in the section on translational control).

The Nanos protein is known to inhibit the translation of a *Drosophila* mRNA called *hunchback.* In the 3' UTR of the *hunchback* mRNA is a sequence that binds the Pumilio protein (**fig. 17.38**). Binding of Pumilio does not affect the translation of the *hunchback* mRNA; however, the Pumilio that is bound to the 3' UTR can simultaneously bind the Nanos protein. This Pumilio–Nanos complex then deadenylates the 3' poly(A) tail of the *hunchback* mRNA. Loss of the poly(A) tail reduces the efficiency of translating the *hunchback* mRNA and greatly decreases the stability of the mRNA. Both of these mechanisms result in reduced expression of the Hunchback protein.

Alternative Splicing

Another mechanism of posttranscriptional regulation is the alternative splicing of the primary transcript to produce multiple mRNAs. Alternative splicing uses various exons to produce different mRNAs from the same primary RNA transcript. One result of alternative splicing is the ability to encode different pro-

teins from the same gene. As we discussed in chapter 14, alternative splicing is fairly common in humans, with over half of the human genes being alternatively spliced and an average of three different mRNAs produced from each gene.

One example of alternative splicing is seen in the *Sex lethal* (*Sxl*) gene during sex determination in *Drosophila.* There are two classes of *Sxl* mutations. The first is homozygous lethal to females, but the males are unaffected. The second is dominant and deleterious to males, without affecting females. The *Sxl* gene is required for all aspects of sexual dimorphism in *Drosophila,* including dosage compensation (see chapter 4). Complete loss of *Sxl* activity in females results in the hypertranscription of the X-linked genes from both X chromosomes, which is too great and the *Sxl* mutant female dies. In contrast, expression of *Sxl* activity in males is lethal, because it prevents the hypertranscription of the X-linked genes, resulting in insufficient transcription of X-linked genes.

Early in embryonic development, a female fly transcribes the *Sxl* gene from the P_e promoter; but a male is unable to transcribe the *Sxl* gene (**fig. 17.39**). Female flies, therefore, translate the Sxl protein, but males lack the protein.

Later in development, the *Sxl* gene is transcribed from the P_L promoter in both males and females. In males, the primary *Sxl* transcript is spliced to include exon 3, which contains a translational stop codon. This splicing leads to a truncated and nonfunctional Sxl protein. In females, the Sxl protein that was made earlier in development binds to the splice acceptor site

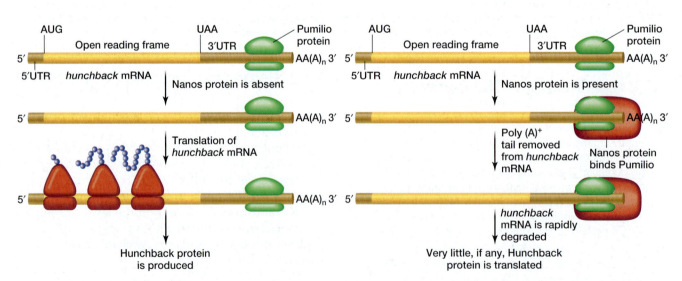

figure 17.38 Posttranscriptional regulation of transcription. The Pumilio protein binds the 3' untranslated region (3' UTR) of the *hunchback* mRNA. In the absence of the Nanos protein, the 3' poly(A) tail of the *hunchback* mRNA is retained and the mRNA is translated (left). In the presence of the Nanos protein (right), Nanos binds the Pumilio protein, which induces the cleavage of the 3' poly(A) tail. Loss of the poly(A) tail, decreases the stability of the *hunchback* mRNA and reduces the efficiency of translation. Both of these effects result in a failure to translate Hunchback protein.

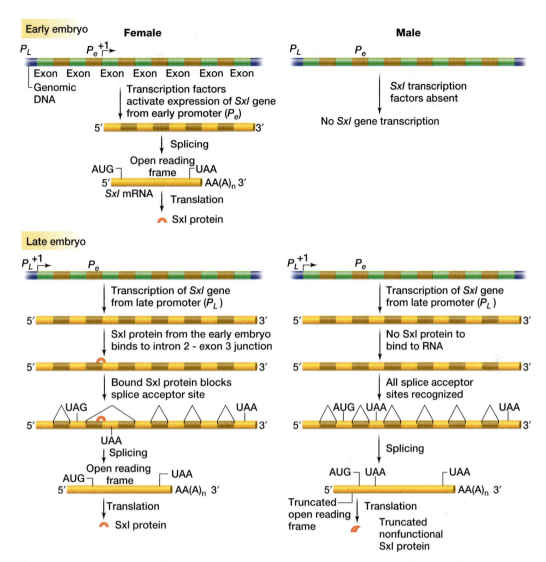

figure 17.39 Alternative splicing regulates gene expression. In early embryonic development, the *Sex lethal* (*Sxl*) gene is transcribed from the P_e promoter in the female (left), but not in the male (right). This results in the Sxl protein being present in the female and not in the male. Later in development, the *Sxl* gene is transcribed from the P_L promoter in both males and females. The Sxl protein, which is already present in the female, binds to the exon 3 splice acceptor site. Masking this site prevents exon 3 from being spliced into the final mRNA. This mRNA translates the wild-type Sxl protein. In the male, the absence of the Sxl protein leads to exon 3 being incorporated into the mature mRNA. Because exon 3 contains a premature translational stop codon, males express a truncated and nonfunctional Sxl protein.

on exon 3. This Sxl binding keeps exon 3 from being recognized during splicing. Thus, the mature mRNA in females lacks exon 3 and its premature translational stop codon. The *Sxl* mRNA in females is then translated to produce a full-length and functional Sxl protein.

Alternative splicing provides a mechanism to produce two different forms of the Sxl protein in the different genders, one a functional Sxl protein in females and the other a nonfunctional Sxl protein in males. Similar scenarios are present for a large number of genes. In some cases, the two protein forms (one functional and the other nonfunctional) are expressed in different cells or tissues. In other cases, the different

mRNAs produced through alternative splicing each encode a functional protein, although they may have slightly different properties. Thus, alternative splicing can regulate the expression of different forms of the protein encoded in the same gene.

Nonsense-Mediated mRNA Decay

In recent years, a new way of regulating the stability of the mRNA transcript has been identified. **Nonsense-mediated mRNA decay (NMD)** is a process that eliminates mRNAs that possess either a premature translational termination codon or a translation termination codon that is too far from the poly(A) tail of the

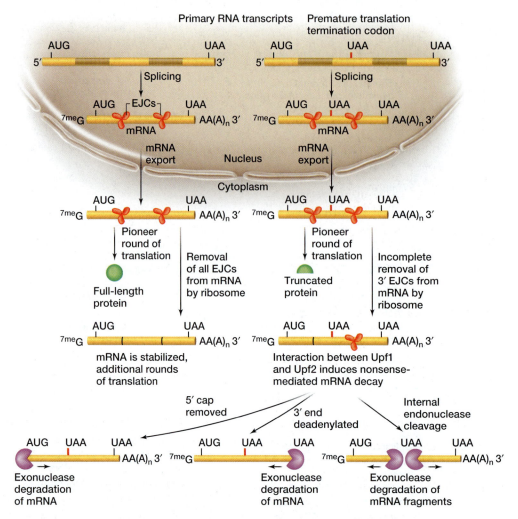

figure 17.40 Mechanism of nonsense-mediated mRNA decay. Transcription of a wild-type and mutant gene produces primary RNAs that are either wild-type in sequence (left) or containing a nonsense mutation (in red, right). The primary RNAs are spliced in the nucleus and exon junction protein complexes (EJCs) are deposited at the exon junctions on the mRNAs. Upon reaching the cytosol, the pioneer (first) round of translation occurs, where the ribosomes encounter and displace EJCs from the mRNA. In the wild-type mRNA, all the EJCs are removed and the mRNA is stabilized for additional rounds of translation. In the mutant mRNA, if the nonsense mutation occurs greater than 50 nucleotides upstream of an EJC, the ribosome cannot remove the EJCs that are 3' of the nonsense mutation. Upf proteins (Upf1 in the translation termination complex and Upf2 in the remaining EJC) interact on the mRNA and induce nonsense-mediated mRNA decay through one of three different mechanisms: decapping and 5' → 3' exonuclease, 3' deadenylation and 3' → 5' exonuclease, or internal endonuclease cleavage and exonuclease. The end result is mRNA degradation.

mRNA. NMD acts on a variety of mRNAs, including those that are transcribed from a mutant gene, those that incorporate a mutation during transcription, and incorrectly spliced mRNAs. Besides serving as a mechanism for monitoring mutations, NMD also acts on some wild-type mRNAs to regulate their abundance.

The basic premise of NMD is that after a primary RNA transcript is spliced, a complex of proteins called the *exon-junction complex* (*EJC*) becomes associated with the mRNA at approximately 20 nucleotides upstream of the exon–exon junctions (**fig. 17.40**). One model suggests that the initial mRNA translation, called the *pioneering translation event,* occurs during or immediately after the mRNA is transported across the nuclear pore. The translation machinery removes the EJCs along the

ORF. If any EJCs remain associated with the mRNA after the first translation event, then NMD is activated. In many cases, a premature termination codon (PTC) that is greater than 50–55 nucleotides upstream of an EJC will not remove the downstream EJCs. It is believed that a component of the translation termination complex (Upf1) may interact with a component of the EJC (Upf2) to initiate NMD.

The mechanism of the mRNA decay occurs through three different pathways. First, the 5' cap of the mRNA is removed and the mRNA is degraded in a 5' → 3' direction by an exonuclease. Second, the 3' end is deadenylated and the mRNA is then degraded in a 3' → 5' direction. *Drosophila* utilizes the third mechanism, which involves an endonucleolytic cleavage of the mRNA near the site of the PTC. The two fragments are then degraded in a 3' → 5' direction toward the 5' cap and 5' → 3' toward the poly(A) tail.

Nonsense-mediated mRNA decay plays an important role in the degradation of mRNAs that contain mutations, which could encode proteins that exert a dominant phenotype. It also plays a critical role in the regulation of wild-type mRNAs, however. Based on sequence and microarray analysis, approximately one-third of the naturally occurring, alternatively spliced mRNAs would be targeted for degradation. This suggests that NMD is used to regulate the correct levels of normal gene expression. An indication that NMD is an essential process for normal gene expression in mammals is that loss of the Upf1 protein in mice results in an embryonic lethal phenotype.

 Restating the Concepts

- Several different mechanisms can exert posttranscriptional regulation of expression, including miRNA suppressing translation.

- Proteins that bind to the mRNA can regulate expression by removing the poly(A) tail, which leads to reduced translation efficiency and stability of the mRNA.

- Alternative splicing of the primary transcript can yield different mRNAs that can encode different protein products. Therefore, regulating the production of the different mRNAs produced from the same gene controls the expression of different forms of the protein.

- Nonsense-mediated mRNA decay is a process that regulates gene expression of both alternatively spliced mRNAs and mutant mRNAs with premature termination codons. NMD occurs when a translation termination codon is greater than 50–55 nucleotides upstream of an exon–exon junction in the mRNA. NMD results in degradation of the mRNA from the 5' end, the 3' end, or from a cleavage within the mRNA.

17.9 Translational Control

When considering control of gene expression, it is important to remember that all control mechanisms are aimed at exerting an influence on either the amount or the activity of the gene product. Up to this point, we primarily concentrated on mechanisms that control the amount of mRNA. This is the most likely step in regulating expression because it is the first step. If this step is properly controlled, the cell does not waste energy in transcribing mRNA that is not translated or converted into a functional protein. We now turn to examples of mechanisms in which the control is exerted at the translational level.

General Translation Regulation: Phosphorylation

Translation can be regulated in a general manner by modifying the translation initiation factors. You have seen how modifications to the histone proteins and DNA methylation can affect transcription. In a similar way, phosphorylation of eukaryotic initiation factor 2 (eIF2; see chapter 11) can affect the general translation machinery.

Phosphorylation of eIF2 prevents the initiation tRNA^{Met} from binding the 40S ribosomal subunit. Because this modification is directed at all eIF2 molecules, it has a general effect on translation, blocking the translation of all mRNAs. Therefore, it must be utilized under conditions when all translation must be blocked, such as in nutrient deprivation when the cell needs to conserve its available energy. This serves as a mechanism of global regulation.

An Example of Specific Translation Regulation: The Nanos Protein

Translation of specific mRNAs can also be subject to regulation. Previously in this chapter, you saw how the Nanos protein can cleave the poly(A) tail of the *hunchback* mRNA, a posttranscriptional event. In contrast, regulating the expression of the Nanos protein occurs at the translational level.

The *nanos* mRNA in *Drosophila* is transcribed in the female and deposited in the egg before fertilization. The mRNA accumulates in the posterior portion of the egg, where it will be translated early in embryonic development. This posterior localization of the *nanos* mRNA allows it to interact with two proteins that regulate *nanos* translation. A protein called Smaug specifically binds to the 3' UTR of the *nanos* mRNA (**fig. 17.41**). The Cup protein then binds Smaug.

The bound Cup protein can also bind the eukaryotic translation initiation factor eIF4E. For translation to initiate in eukaryotes, eIF4E must interact with eIF4G. With eIF4E bound to the 3' UTR of *nanos* mRNA by the Cup protein, it is unable to interact with eIF4G. Translation of the *nanos* mRNA cannot begin. Repression of translation initiation by binding the eIF4E protein and preventing its interaction with eIF4G is seen in other species as well.

Another Example of Specific Translation Regulation: Transferrin and Ferritin

Another example of translational control of specific mRNAs is observed in regulating the levels of iron.

How Mammals Deal with Iron

In mammals, iron is acquired through the diet, from which it is absorbed in the bloodstream and bound by the protein transferrin. The transferrin–iron complex

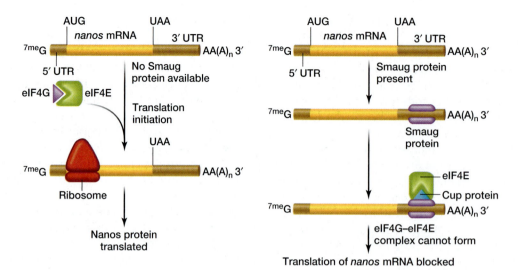

figure 17.41   Translational regulation of the *nanos* mRNA. In the absence of the Smaug protein, eIF4G and eIF4E interact and participate in translation initiation of the *nanos* mRNA (left). In the presence of the Smaug protein, Smaug binds the 3' untranslated region (3' UTR) of the *nanos* mRNA (right). Smaug also binds the Cup protein (blue), which then binds the eIF4E protein. The absence of free eIF4E to interact with eIF4G prevents translation initiation of the *nanos* mRNA.

is bound by the transferrin receptor on the surface of cells (**fig. 17.42**). The transferrin–iron complex is then moved into cells by endocytosis, where it can interact with a variety of proteins that require iron for activity.

Iron levels that are too high can be toxic to the cell or organism. To prevent this situation, iron is stored in ferritin, a protein within cells that sequesters the iron.

To summarize, transferrin and the transferrin receptor are essential for getting the excess iron into the cell, where it can complex with the ferritin. The cell needs to regulate the expression of transferrin, the transferrin receptor, and ferritin in response to levels of iron.

Action of the Iron Regulatory Protein (IRP)

Regulation of the transferrin receptor and ferritin occurs through the iron regulatory protein (IRP), which binds iron as a cofactor (see fig. 17.42). The IRP is also an RNA-binding protein that recognizes the iron response element (IRE), which is a sequence present on the ferritin and transferrin receptor mRNAs. But the IRP exerts different effects on these two mRNAs.

The ferritin mRNA has its IRE in the 5' UTR. Normally, the IRP binds to the ferritin IRE to prevent translation of the mRNA. But when iron levels are high, the iron binds the IRP, which is now unable to bind the IRE; now the ferritin mRNA is translated.

In contrast, the IRE on the transferrin receptor mRNA is located in the 3' UTR. Binding of the IRP to the transferrin receptor mRNA increases the stability of the mRNA by suppressing the activity of mRNA degradation enzymes. The result is increased amounts of translation. When iron levels are high, the IRP–iron complex again forms; the complex cannot bind the IRE,

and the transferrin receptor mRNA is more rapidly degraded, which yields less protein.

In this instance, one RNA-binding protein has different effects on the translation of two different mRNAs to regulate a physiological process in a coordinated fashion. At low iron concentrations, the translation of transferrin receptor is increased and ferritin is decreased. The result is the increased amount of free iron within the cell. When iron levels are high, the translation of transferrin receptor is decreased and ferritin is increased. This reduces the amount of iron that is endocytosed into the cell and the sequestering of the large amount of free iron, by ferritin, that is already in the cell.

17.10 Posttranslational Control

Posttranslational control of expression deals with the modification of a protein. These modifications may affect the activity, localization, and the stability of the protein.

Protein Modifications

By this point you understand that proteins can be modified in a variety of ways. In this chapter, we discussed the methylation, phosphorylation, and acetylation of histones and the phosphorylation of the Pelle and Cactus proteins. These **posttranslational modifications** are covalent changes that occur to proteins that modify their activity within the cell.

In the case of the histone proteins, the posttranslational modifications affect the activation or repres-

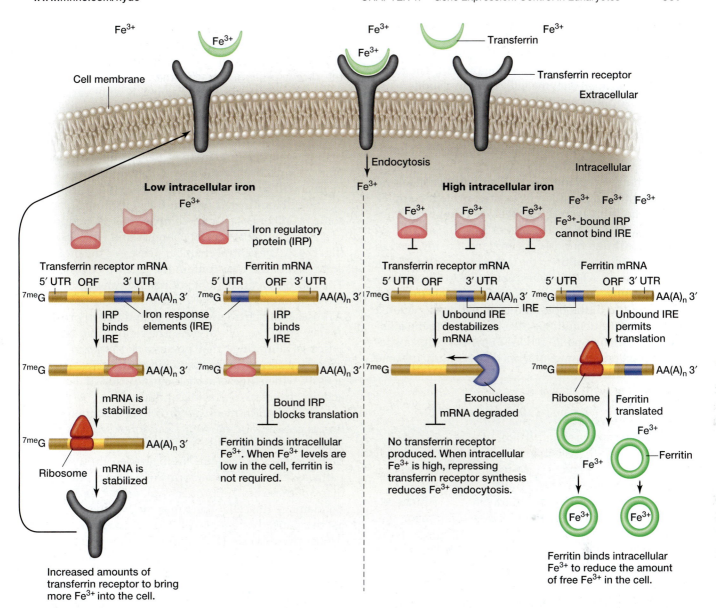

figure 17.42   Role of iron in translation regulation. Extracellular iron (Fe^{3+}) is bound by transferrin and this complex is bound by the transferrin receptor and endocytosed into the cell. When intracellular Fe^{3+} concentration is low (left), there are large amounts of iron regulatory protein (IRP) not bound with Fe^{3+}. The free IRP binds the iron response element (IRE) in the 3' untranslated region (3' UTR) of the transferrin receptor mRNA. The bound IRP prevents the mRNA from degradation, which permits high levels of transferrin receptor to be translated. This leads to increased endocytosis of Fe^{3+} into the cell. The IRP also binds the IRE located in the 5' untranslated region (5' UTR) of the ferritin mRNA, which blocks translation of the mRNA. Ferritin, which binds intracellular Fe^{3+}, is not required when intracellular Fe^{3+} levels are low. When intracellular iron levels are high (right), the Fe^{3+} binds the available IRP, which cannot bind the IRE. The unbound transferrin receptor mRNA is degraded and the resulting decrease in transferrin receptor reduces the endocytosis of more Fe^{3+}. The unbound ferritin mRNA is translated into high levels of ferritin protein. This ferritin binds much of the excess free intracellular Fe^{3+}.

sion of gene transcription. Phosphorylation of eIF2, in contrast, blocks the initiation of translation. The *Drosophila* Dorsal protein is bound by an inhibitory protein, Cactus; phosphorylation of Cactus releases Dorsal, which is now able to enter the nucleus and regulate gene transcription. In this case, the posttranslational modification of one protein frees another protein that is then able to control gene expression.

In chapter 11, we described two additional forms of posttranslational modification. First is the cleavage of the leader sequence, which is composed of the first 15–25 hydrophobic amino acids at the amino terminus. This sequence must be cleaved if the protein is be secreted from the cell or inserted into the membrane. A second class of posttranslational modification is the addition of sugars and lipids to proteins.

Protein Degradation: Ubiquitination

Another form of posttranslational regulation is manifested through the stability of the protein. The rapid degradation of an enzyme will result in a sudden loss of the associated activity. One form of protein degradation involves a group of enzymes called *proteases.* Another form of regulated protein degradation involves *ubiquitination,* which is the process of attaching ubiquitin, a peptide of 76 amino acids, to a protein as a signal for degradation. Ubiquitinated proteins are targeted for degradation by a large protein complex called the **proteasome.** The average human cell contains 20,000–30,000 proteasomes.

Three different classes of enzymes are involved in attaching the ubiquitin to the target protein (**fig. 17.43**). Hydrolysis of ATP attaches ubiquitin to the ubiquitin-activating enzyme E1 (UAE-E1), which transfers the ubiquitin to the ubiquitin-conjugating enzyme E2 (UCE-E2). The ubiquitin is then transfered to the ubiquitin ligase enzyme E3 (ULE-E3), which binds a specific target protein and attaches the ubiquitin. In some cases, the ubiquitin ligase enzyme E3 binds the target protein and delivers it to the ubiquitin-conjugating enzyme E2, which directly transfers the ubiquitin to the target protein. Because there could be possibly hundreds of different E3 proteins, a wide variety of target proteins could be recognized. The ubiquitinated target protein remains bound to the ubiquitin ligase enzyme E3, which adds additional ubiquitin peptides to the target protein. Generally, the greater the number of ubiquitins that are attached to the target protein, the faster it will be degraded.

The ubiquitinated protein then enters the proteasome, a 2400-kDa protein complex that is composed of a 20S cylindrical central chamber and two 19S caps (fig. 17.43). The central chamber possesses the protease activity, and the caps contain the regulatory function. The cap only permits ubiquitinated proteins to enter the central chamber. The proteasome also possesses ATPases that help to unfold the ubiquitinated proteins as they enter the central core. The central core is composed of four rings that possess *chymotrypsin* activity, which cleaves at hydrophobic amino acids, *trypsin* activity, which digests at basic amino acids, and *caspase* activity that cleaves at acidic amino acids.

The ubiquitin-mediated protein degradation pathway is critical for removing a variety of proteins from within the cell. In the context of genetics, two particular classes of proteins are degraded through this pathway. First are mutant forms of proteins. Although many of these mutant proteins are inactive and would not damage the cell, others may be detrimental, such as the causative agents of Huntington disease or Alzheimer disease. Degradation of these potentially dominant-acting mutant proteins allows the cell to continue functioning in the presence of the reduced amount of wild-type protein. The second class of proteins that must be regulated through the ubiquitin-mediated degradation pathway are proteins that are needed for a short time, such as transcription factors. If transcription factors stably persisted in a cell, it would be difficult to alter levels of gene transcription.

Many of the cell cycle proteins that we discussed in chapter 3 are another group of proteins that must be regulated by the ubiquitin-mediated degradation pathway. The anaphase-promoting complex/cyclosome (APC/C) functions as the ubiquitin ligase enzyme E3 that binds many of these cell cycle proteins that regulate the G_1/S and M/G_1 transitions.

Recently, it was found that the APC/C also plays a very important role in neurons, which have left the cell cycle. The APC/C is present in the synapse, where it regulates the amount of different proteins. By targeting specific synaptic proteins for degradation, the APC/C plays an important role in the formation, activity, and plasticity of the synapse.

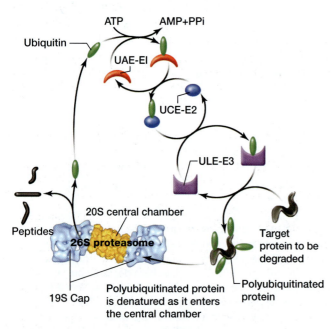

figure 17.43   Ubiquitin-mediated protein degradation. Hydrolysis of ATP provides the energy to attach ubiquitin to the ubiquitin-activating enzyme E1 (UAE-E1). The ubiquitin is then transferred to the ubiquitin conjugating enzyme E2 (UCE-E2). The ubiquitin is transferred to the ubiquitin ligase enzyme E3 (ULE-E3), which attaches multiple ubiquitin molecules to the target protein. The polyubiquitinated protein is targeted to the proteasome, where it is denatured by the 19S Cap protein complex. The polyubiquitinated protein enters the 20S central chamber of the proteasome, where it is degraded into peptides by three different protease activities.

Restating the Concepts

▶ A variety of mechanisms exist that affect the translation of the mRNA, which results in a change in the amount of protein that is produced in a cell.

▶ Proteins can be covalently modified after translation, which may affect their activity. In some cases, these changes affect the ability of the protein to regulate another step in the gene regulatory pathway.

▶ An important aspect of posttranslational regulation is the degradation of proteins. One common mechanism is the ubiquitin-mediated pathway, which adds the 76 amino acid ubiquitin peptide to proteins, which targets the protein to the proteasome where the protein is cleaved. Misfolded and mutant proteins are degraded through this pathway. Proteins that are expressed for a relatively short time, such as transcription factors or cell cycle regulatory proteins, are also degraded by the ubiquitin-mediated pathway.

17.11 Interaction of Multiple Regulatory Mechanisms

We have discussed a number of different mechanisms that lead to the regulated amount of protein in an eukaryotic cell, from transcriptional regulation to translational control to posttranslational modifications of the protein. It is important to realize that these different mechanisms do not act alone. For a given protein, the amount present in a cell at a given moment is a consequence of the amount of its mRNA that is being transcribed, processed, and translated into protein. To appreciate the complexity of these different regulatory mechanisms acting on two genes, we describe the genes that are associated with two human genetic diseases: Prader–Willi and Angelman syndromes.

The Genetics of Prader–Willi and Angelman Syndromes in Humans

A striking example of genetic imprinting in human beings involves two medical syndromes. In Prader–Willi syndrome (PWS), infants feed poorly. This improves until about 18 months when the individual begins to exhibit an insatiable appetite. Additional features of PWS includes reduced sensitivity to pain, skin picking, poor gross motor skills, and mild mental impairment. Individuals suffering from Angelman syndrome (AS) exhibit loss of muscle coordination; high frequency of seizures; severe mental deficiency and usually a lack of speech, and hyperactivity. Both syndromes are found in approximately 1 in 15,000 live births.

Examining the pedigrees of families that have either PWS or AS individuals, geneticists often find that two unaffected individuals can have an affected child (**fig. 17.44**). This suggests that either these syndromes result from a dominant mutation that has reduced expressivity or from a recessive mutation. Furthermore, males and females are equally affected (except for one form of AS), which suggests that both syndromes are autosomal.

figure 17.44   Pedigree of an Angelman syndrome family. Below the individuals in the pedigree are their haplotypes. The boxed haplotype is associated with the Angelman syndrome mutation in the imprinting center.
From Espiritu et al., "Molecular Mechanism of Angelman Syndrome in Two Large Families Involves an Imprinting Mutation," *American Journal of Human Genetics*, February 1999. Copyright © 1999 The University of Chicago Press. Reprinted with permission.

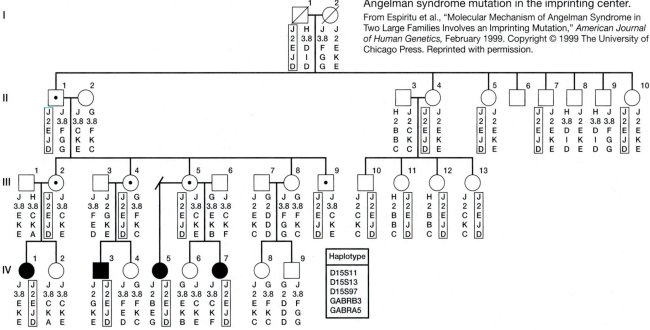

A surprising result was that approximately 70% of all individuals that suffer from either syndrome possess a deletion of the 15q11-q13 region that spontaneously arose in the germline of one of the parents. Thus, the genes associated with these two syndromes, with strikingly different developmental and neurobehavioral symptoms, must be located near each other. A greater surprise occurred when the inheritance of the deletion chromosome was characterized in the parents and their affected children. Using haplotype analysis (see chapter 14), researchers found that individuals with PWS inherited their deletion chromosome from their father, whereas AS individuals inherited their deletion chromosome from their mother (fig. 17.45). This strict division of the two syndromes based on which parent contributed the deletion chromosome is a clear indication of imprinting.

The Molecular Nature of Imprinting the 15q11-13 Region

Current evidence suggests that AS results from lack of expression of the *UBE3A* gene, which encodes the ubiquitin protein ligase 3E. The *UBE3A* gene is imprinted in the male, so that an individual only expresses the maternal allele of this gene. Thus, if a female passes the 15q11-q13 deletion on to her children (fig. 17.45b), they exhibit AS because the only copy of the *UBE3A* gene in the children is the paternal copy that is silenced.

Approximately 10–15% of the AS individuals have a mutation in the *UBE3A* gene, rather than the 15q11-q13 deletion. As shown in figure 17.45c, this *UBE3A* mutation can be maintained in a pedigree without exhibiting the AS symptoms as long as it is only passed through the paternal lineage (left side). When this mutation enters a daughter, she has a 50% chance of passing the mutation on to her children, who would then exhibit the AS symptoms (fig. 17.45c, right side).

PWS most likely is not the result of a single imprinted gene, but several that are located in the same region. The gene that appears to have the greatest contribution to the PWS phenotype is the *SNRPN* gene, which encodes a small ribonucleoprotein polypeptide N that controls mRNA splicing and is required for normal brain development and function. Unlike the *UBE3A* gene, the *SNRPN* gene is maternally imprinted (see fig. 17.46a). A second component of PWS corresponds to the first three exons of the *SNRPN* gene and is called *SNURF* (<u>SN</u>RPN <u>u</u>pstream <u>r</u>eading <u>f</u>rame). It is possible that *SNURF* corresponds to the imprinting center (IC). The region that corresponds to the *UBE3A*, *SNRPN*, and IC is shown in figure 17.46a.

The imprinting of this genomic region appears to be under the control of the IC. The IC on the paternal chromosome is not methylated and neither are several CpG islands. The absence of DNA methylation allows several genes to be transcribed (fig. 17.46b). Additionally, an antisense RNA transcript is transcribed from the

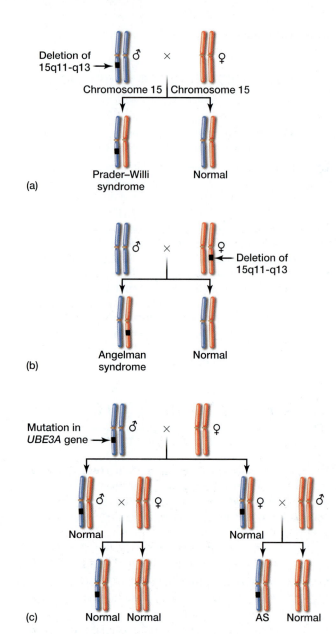

figure 17.45   Inheritance patterns of Prader-Willi and Angelman syndromes. (a) The inheritance pattern of Prader-Willi syndrome (PWS) is shown. The 15q11-q13 deletion is marked in one of the paternal chromosomes as a black bar. Offspring who inherit the deletion chromosome from their father will exhibit PWS. (b) The 15q11-q13 deletion is shown on one of the maternal chromosomes as a black bar. Offspring who inherit the deletion chromosome from their mother will exhibit Angelman Syndrome (AS). (c) A mutation in the *UBE3A* gene is seen in one of the paternal chromosomes. When the *UBE3A* mutation is inherited from the father, all the offspring will be normal (left). When the *UBE3A* mutation is inherited from the mother (right), the offspring will exhibit AS.

strand opposite to that used to transcribe the *UBE3A* gene (*UBE3A-AS*). It is possible that transcription of the *UBE3A-AS* RNA interferes and blocks either the transcription or translation of the *UBE3A* mRNA, which prevents expression of the ubiquitin protein ligase 3E from the paternally-inherited chromosome. Thus, when the maternally-inherited chromosome contains either

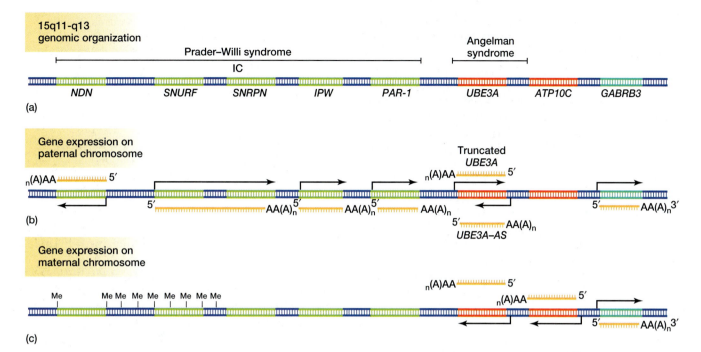

figure 17.46 Gene expression in the 15q11-q13 imprinting region. (a) The genetic organization of the 15q11-q13 region shows some of the genes and the imprinting center (IC). Some genes are expressed from only the paternal chromosome (green), other genes are only expressed from the maternal chromosome (red), and the *GABRB3* gene is expressed from both chromosomes (light blue). The regions important in Prader-Willi syndrome and Angelman syndrome are marked. (b) The absence of significant methylation on the paternal chromosome permits the transcription of the genes in the orientations shown by arrows. Transcription of the *UBE3A* antisense RNA (*UBE3A-AS*) prevents complete transcription of the *UBE3A* gene, which results in silencing of the *UBE3A* gene. (c) On the maternal chromosome, the imprinting center and several other sequences are methylated (Me) This silences the expression of the *NDN, SNURF, SNRPN, IPW, PAR-1,* and *UBE3A-AS* transcripts. The loss of *UBE3A-AS* transcription permits the transcription and expression of the *UBE3A* and *ATP10C* genes. The designated genes are *NDN,* Necdin; *SNURF,* SNRPN upstream reading frame; *SNRPN,* small nuclear ribonucleoprotein polypeptide N; *IPW,* unknown; *PAR-1,* unknown; *UBE3A,* ubiquitin ligase enzyme 3A; *ATP10C,* probable calcium-type P-type ATPase; *GABRB3,* GABA neurotransmitter receptor subunit.

the 15q11-q13 deletion or *UBE3A* mutation, the UBE3A protein is not present and AS occurs.

On the maternal chromosome, the CpG islands in the IC are methylated, which leads to the silencing of the genes (fig. 17.46c). In the absence of expressing the *UBE3A-AS* RNA, the *UBE3A* mRNA can now be translated. However, the *SNRPN* and several nearby genes are not expressed. If the paternally-inherited chromosome contains the 15q11-q13 deletion, then the resulting individuals will not express these genes and PWS results.

Consequences of the 15q11-q13 Imprinted Region

In addition to the *UBE3A* gene on chromosome 15, the *methyl-CpG-binding protein-2 (MECP2)* gene on the X chromosome can also cause Angelman syndrome when it is mutated. The *MECP2* gene is not imprinted, however. It is likely that the MECP2 protein binds the methylated IC to either expand the methylation pattern on the CpG islands or recruit HMTases to methylate the histones and further silence the paternally imprinted genes. This is consistent with the finding that the maternally inherited chromosome possesses

methylated histones that are associated with the *SNRPN* promoter and IC.

One additional feature in the 15q11-q13 region is the presence of several **small nucleolar RNAs (snoRNAs)** that are transcribed downstream of *SNRPN* (between *SNRPN* and *UBE3A*). One of these snoRNAs, *snoRNA HBII-52*, is complementary to the *serotonin receptor 5-HT$_{2C}$R* RNA. Recent studies show that the *snoRNA HBII-52* RNA can basepair with the *serotonin receptor 5-HT$_{2C}$R* primary RNA, which directs the alternative splicing of the RNA. These alternatively spliced mRNAs encode different forms of the serotonin receptor that are expressed in Prader–Willi individuals. This may explain why these individuals respond to certain serotonin reuptake inhibitors.

You should appreciate that the phenomenon of imprinting utilizes the same basic gene regulatory mechanisms that we already described. Silencing of the maternal chromosome requires DNA and histone methylation. It also requires the MECP2 protein to recognize and bind the methylated DNA.

Angelman syndrome results from the loss of one aspect of the ubiquitin-mediated protein degradation, specifically the loss of ubiquitin protein ligase

3E. Thus, the inability to properly degrade the specific protein targets that are recognized by this UBE3A protein results in AS.

Silencing of the paternal chromosome simply requires transcription of the *UBE3A-AS* RNA. This antisense RNA likely interferes with the transcription of the *UBE3A* RNA or base pairs with the *UBE3A* mRNA to disrupt its translation. We also described how the *SNRPN* mRNA encodes a ribonucleoprotein that regulates mRNA splicing that is required for brain development and function. In contrast, it is the *snoRNA HBII-52* RNA that is required for the alternative splicing of the serotonin receptor RNA. Without *snoRNA HBII-52*, one specific form of the serotonin receptor mRNA and protein are not produced, which affects brain function in PWS individuals. Thus, a variety of different regulatory systems are required to correctly control gene expression and produce a normal phenotype.

Restating the Concepts

▶ Prader–Willi and Angelman syndromes result from a defect in imprinting of the same genomic region. PWS is due to methylation of the paternal chromosome and the subsequent failure to express several transcripts, including *SNURF/SNRPN*. AS results from the failure to methylate the DNA in the imprinting center, which allows expression of the *UBE3A-AS* transcript that prevents expression of the *UBE3A*-encoded ubiquitin protein ligase 3E.

▶ Production of a normal phenotype requires multiple levels of gene and protein expression. In the case of the 15q11-q13 region, we discussed DNA and histone methylation, proteins that bind methylated DNA, antisense RNA, alternative splicing, and ubiquitin-mediated protein degradation.

17.12 Eukaryotic Versus Bacterial Gene Regulation

Some fundamental similarities and differences exist between bacterial and eukaryotic regulation of gene expression. Some of these are summarized in **Table 17.1**.

As for similarities, both bacteria and eukaryotes have activator and repressor proteins that bind DNA sites to regulate transcription. Some of the basic DNA-binding motifs, such as helix-turn-helix, are common in all organisms. The activity of a DNA-binding protein can be modified by binding coregulatory molecules.

The differences, however, easily outnumber the similarities:

• The bacterial DNA-binding sites are clustered near the promoter, whereas the eukaryotic enhancers and silencers can be tens of kilobases upstream or downstream of the transcription unit.

• A larger variety of *cis*-acting functions, such as promoters, enhancers, silencers, and insulators, are present in eukaryotes.

table 17.1   Comparison of Eukaryotic and Bacterial Gene Regulation

Regulatory Method	Eukaryotes	Bacteria
Gene organization	Single gene transcriptional units, complex promoters with enhancers several kilobase pairs from transcription start site	Operons with polycistronic mRNAs, small promoters with nearby operator sequences
DNA-binding proteins	Both activators and repressors, can exert their control over large distances	Both activators and repressors, only exert their control over short distances
RNAs	mRNAs encode proteins, miRNAs can control cleavage of mRNA, translation and transcription, and antisense RNA can control gene transcription, RNA splicing, and protein translation	mRNAs encode proteins, antisense RNAs control transcription and translation of target genes, miRNAs have yet to be identified
Chromatin organization	Chromatin organization essential for transcription, chromatin organization controlled by covalent modifications to histone proteins, location of nucleosomes, and DNA methylation	Absence of chromatin in bacteria, supercoiling of circular DNA can affect the level of gene expression, DNA is methylated but not to regulate transcription
Posttranscriptional Regulation	mRNA stability, alternative splicing, and RNA editing significantly affect the expression of a gene	Absence of mRNA splicing and RNA editing, mRNA stability has only a minimal effect because transcription and translation are coupled
Translation control	Proteins are covalently modified to affect their activity in the cell, these modifications may activate or repress transcription	Proteins are not covalently modified after translation

- In eukaryotes, but not bacteria, covalent modifications of the DNA-binding proteins (phosphorylation, methylation, acetylation) affect the activity of these proteins to bind DNA and regulate transcription.
- In eukaryotes, but not bacteria, methylation of the DNA has a profound effect on repressing transcription.
- Chromatin organization plays a critical role in eukaryotic gene regulation. The absence of chromatin eliminates this possibility as a regulatory mechanism in bacteria.
- Although DNA-binding proteins and antisense RNAs serve as *trans*-acting factors in both bacteria and eukaryotes, the latter also employ miRNAs to regulate gene expression through several different mechanisms.

Some of the differences result from fundamental biological differences between bacteria and eukaryotes. One example is the role of chromatin, which is only present in eukaryotes. Another example is the coupling of transcription and translation in bacteria, which is not possible with the eukaryotic nuclear membrane separating transcription and translation. The requirement that the small ribosomal subunit attaches to the 5' cap of the eukaryotic mRNA to begin translation, makes polycistronic mRNAs an inefficient regulatory mechanism in eukaryotes. Another difference can be found in the processing of the eukaryotic mRNA, which allows alternative splicing and maintenance of the poly(A) tail as steps that can regulate eukaryotic gene expression. The absence of mRNA processing in bacteria eliminates these events as possible regulatory steps.

In summary, bacteria and eukaryotes have developed similar, yet different, mechanisms to regulate gene expression. One feature remains in common among all organisms, however, namely that expression of genes and the corresponding proteins must be regulated in response to the environment and the needs of the cell and organism.

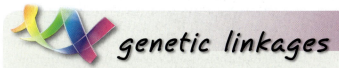

genetic linkages

In this chapter, we explored the variety of methods that can be used to regulate gene expression in eukaryotic organisms. Differences between bacteria and eukaryotes result from the fundamental processes of genomic DNA organization and chromatin conformation (chapter 7), transcription (chapter 10), and translation (chapter 11).

Information on the covalent modification to the histone proteins and the ability to slide and remove histones makes it clear that chromatin is significantly more dynamic than the figures depicted in chapter 7 might suggest.

In chapter 7 we also examined the three major forms of DNA (A-DNA, B-DNA, and Z-DNA). You learned that Z-DNA is primarily found in alternating CpG sequences and could be stabilized by methylating the DNA. In this chapter, we specifically discussed methylation of CpG sequences within the CpG islands and how this represses transcription. The methylation of CpG sequences provides a mechanism by which Z-DNA can regulate gene expression.

Chromosomal rearrangements were examined in the context of genetic diseases. First, you saw how translocations could bring an enhancer for one gene near a protooncogene to induce a lymphoma (discussed again in chapter 22). You also saw how a deletion could not only reveal the pattern of genetic imprinting, but also affect the expression of two different syndromes.

Continuing with imprinting, you saw how phenotypic ratios deviate from standard Mendelian frequencies (chapter 2) due to the fact that one parental chromosome is silenced. Because of imprinting, the standard pedigrees for autosomal and X-linked dominant and recessive traits (chapter 4) are not apparent. Thus, imprinting makes the study of the underlying genetic phenomenon more difficult.

We previously presented information on RNAs, their transcription, and their processing in eukaryotes (chapter 10). We expanded on that material in this chapter, spending more time examining the proteins, including TFIID and other basal transcription factors that regulate the ability of RNA polymerase II to recognize the promoter. We also discussed one mechanism by which alternative splicing can result in the regulation of the final gene product. The earlier views on microRNAs (chapter 10) and their increasingly more important role in regulating gene expression were restated and emphasized.

We also saw how some of the components and events required in translation (chapter 11) can serve as control points in regulating the expression from a gene. You should now appreciate that translation is not a process that occurs immediately upon the generation of a mRNA. Translation, like transcription, is regulated to respond to a variety of intracellular and extracellular signals to express the correct proteins at the correct times. Furthermore, posttranslational control mechanisms are critical for the proper expression of the active form of proteins and their corresponding phenotypes.

Chapter Summary Questions

1. For each term in column I, choose the best matching phrase in column II.

   Column I
   1. H1
   2. H19
   3. Ligand
   4. Motif
   5. GC box
   6. CpG island
   7. Antisense RNA
   8. Architectural protein
   9. Silencer
   10. Insulator

   Column II
   A. A tertiary structure of a protein
   B. A DNA element that can shield a gene from the effects of *cis*-elements
   C. Binds to DNA and bends it.
   D. Substrate for DNA methylation
   E. Imprinted in mice
   F. A *cis*-regulatory element that suppresses transcription
   G. Small molecule that activates or inactivates proteins it binds to
   H. Associated with the linker DNA between nucleosomes
   I. A *trans*-regulatory element that suppresses translation
   J. A genetic proximal-promoter element

2. What alterations occur in the DNA surrounding the TATA region as the transcription preinitiation complex (PIC) forms? How do these changes affect the rate of transcription?

3. What are the differences between a general transcription factor and a cell-specific transcription factor?

4. List the characteristics of enhancers.

5. Name and describe two DNA-binding motifs and two motifs that are used for protein dimerization.

6. How can transcriptional repressors inhibit the function of transcriptional activators?

7. Many mechanisms are commonly used to regulate the activity of transcription factors. Name three.

8. What are the functional differences between miRNA and antisense RNA?

9. Discuss the relationship between the Nanos and Hunchback proteins of *Drosophila*.

10. Describe the role of each of the following enzymes and its effect on gene expression: **a.** HAT-A; **b.** HAT-B; **c.** HDAC; **d.** HMTase

11. Discuss the role of alternative splicing during *Drosophila* sex determination.

12. Compare and contrast the ISWI and SWI-SNF protein complexes.

13. What is nonsense-mediated mRNA decay? How is the mRNA decay actually accomplished?

14. What are the three mechanisms currently believed to explain how methylation of cytosine blocks transcription?

15. What are the main differences in the control of gene expression between bacteria and eukaryotes?

16. What is eIF2? Describe the role it plays in the control of gene regulation.

17. The protein Dorsal plays a very important role in the development of the dorsal–ventral axis in the *Drosophila* embryo. Arrange the following events in the proper order in which they occur in the Dorsal signal transduction pathway:
    1. Phosphorylation of Pelle
    2. Binding of Spätzle to the Toll receptor
    3. Phosphorylation of Cactus
    4. Entry of Dorsal into the nucleus
    5. Cleavage of Spätzle
    6. Release of Dorsal from its heterodimer partner

18. Consider the following statements about the Myc-Max system of gene expression. Indicate which of the following statements is true. If the statement is false, change a few words to make it true.
    1. The *myc* gene is an oncogene that may cause cancer when inappropriately expressed.
    2. The *myc* gene is expressed in cells that are terminally differentiated.
    3. The *max* gene is continuously expressed in all cells.
    4. The Myc homodimer represses gene transcription.
    5. The Myc-Max heterodimer is preferentially formed over the Max homodimer.
    6. The Max protein contains a transcription activation domain.
    7. The Myc-Max heterodimer activates gene transcription.
    8. Both the activator and repressor bind to the same enhancer sequence.

19. Consider the following statements about the proteasome. Indicate which of the following statements is true. If the statement is false, change a few words to make it true.
    1. The addition of the sugar ubiquitin to proteins targets them for destruction by the proteasome.
    2. Attachment of ubiquitin to target proteins is mediated by three different classes of enzymes.
    3. $ATP + ubiquitin + UAE\text{-}E1 \rightarrow UAE\text{-}E1\text{-}ubiquitin + ADP + P_i$
    4. $UAE\text{-}E1\text{-}ubiquitin + UAE\text{-}E2 \rightarrow UAE\text{-}E2\text{-}ubiquitin + UAE\text{-}E1$
    5. Enzyme 3 is referred to as the ubiquitin-ligase enzyme.
    6. The proteasome complex is 26S and consists of a 20S cylindrical chamber and a 6S cap.
    7. The proteins are degraded in the central core by three protease activities: chymotrypsin, trypsin, and ATPase.
    8. The proteasome is involved in the degradation of mutant proteins, as well as proteins that are needed for long periods of time.

Solved Problems

PROBLEM 1: One way to identify enhancers and silencers associated with a gene is to clone a large region upstream of that gene in front of a reporter gene. Large deletions can then be generated in the promoter region; the resulting transgene can then be introduced into the organism, and the expression of the reporter relative to the endogenous gene can be analyzed. Deletion of an enhancer will result in either decreased expression or improper spatial expression of the transgene. Additionally, silencers can also be identified because deletion of a silencer will cause either increased or broader spatial expression of the transgene.

You cloned a 15-kb region upstream of a *Drosophila* gene. The following figure shows several deletion constructs that were all fused to the β-galactosidase reporter. You then introduce each construct individually into *Drosophila*. The 15-kb region exactly reproduces the expression pattern of this gene in the embryo, which corresponds to three stripes. For each construct, the embryonic expression pattern is diagrammed to the right.

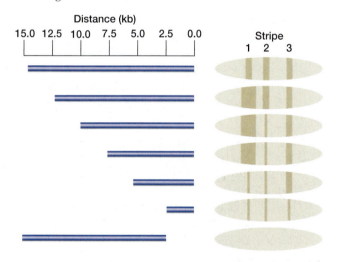

State where each element is located (between what kilobase distances), which stripe it affects, and if it is involved in activating or repressing the wild-type stripe expression pattern.

Answer:

1. The region from 12.5 to 15.0 kb contains a silencer element that is required to restrict stripe 1 to the proper width. In the absence of this silencer, stripe 1 is wider than it should be.
2. The region from 10.0 to 12.5 kb contains an enhancer element that is required for maximal expression of stripe 2.
3. The region from 7.5 to 10.0 does not contain an element that controls the expression of this gene based on the reporter gene expression.
4. The region from 5.0 to 7.5 kb contains an enhancer element that is required for maximal expression of stripe 1.
5. The region from 2.5 to 5.0 kb contains an enhancer element that is required for maximal expression of stripe 3.
6. The region from 0 to 2.5 kb contains the basal promoter that is required for any expression in the three stripes.

PROBLEM 2: Fill in the following two tables dealing with the transferring-ferritin translational regulation of gene expression in mammals.

Name of effector protein	
Name of DNA sequence	
Sequence location in ferritin mRNA	
Sequence location in transferrin receptor mRNA	

For the following table, use the terms: *increases, decreases,* and *no change* to accurately describe what happens under low iron levels (middle column) and high iron levels (right column).

	Low Levels of Iron	High Levels of Iron
Uptake of iron		
Storage of iron		
Concentration of transferrin receptor mRNA		
Synthesis of transferrin receptor		
Concentration of ferritin mRNA		
Synthesis of ferritin		

Answer:

Name of effector protein	Iron-regulatory protein (IRP)
Name of DNA sequence	Iron response element (IRE)
Sequence location in ferritin mRNA	5' untranslated region (5' UTR)
Sequence location in transferrin receptor mRNA	3' untranslated region (3' UTR)

Note that the regulation of gene expression in response to changes in iron levels occurs at the level of translation of mRNA rather than transcription of DNA. For example, when levels of iron are low, IRP binds to IREs, thereby (1) blocking the translation of ferritin mRNA and (2) suppressing the degradation of transferrin receptor mRNA. With this in mind, let us complete the table.

	Low Levels of Iron	High Levels of Iron
Uptake of iron	*Increases*	*Decreases*
Storage of iron	*Decreases*	*Increases*
Concentration of ferritin mRNA	*No change*	*No change*
Synthesis of ferritin	*Decreases*	*Increases*
Concentration of transferrin receptor mRNA	*Increases*	*Decreases*
Synthesis of transferrin receptor	*Increases*	*Decreases*

PROBLEM 3: Outline the genetic defect in Prader–Willi syndrome (PWS) and Angelman syndrome (AS)? Assuming that all individuals with PWS and AS are fertile, predict the phenotypic ratios in the offspring of the following crosses.
1. Female with AS × Normal male
2. Male with AS × Normal female
3. Female with PWS × Normal male
4. Male with PWS × Normal female

Answer: Prader–Willi and Angelman syndromes are a result of genomic imprinting. The vast majority of cases in either syndrome are due to a deletion of the 15q11-q13 region. Individuals with PWS inherited the deletion chromosome from their father, whereas AS individuals inherited the deletion chromosome from their mother. The AS gene is imprinted in the male, and so an individual only expresses the maternal allele of this gene. On the other hand, the PWS genes are imprinted in the female, and so an individual only expresses the paternal alleles of these genes.

1. Female with AS × Normal male
$$\downarrow$$
$$\frac{1}{2} \text{ AS} : \frac{1}{2} \text{ Normal}$$

The affected female would be heterozygous for the 15q11-q13 deletion. Therefore, her offspring have a 50% chance of inheriting the deletion chromosome. Children with the deletion chromosome would exhibit AS because the deletion was obtained from their mother.

2. Male with AS × Normal female
$$\downarrow$$
$$\frac{1}{2} \text{ PWS} : \frac{1}{2} \text{ Normal}$$

The affected male would be heterozygous for the 15q11-q13 deletion. Therefore, his offspring have a 50% chance of inheriting the deletion chromosome. Children with the deletion chromosome would exhibit PWS because the deletion was obtained from their father.

3. Female with PWS × Normal male
$$\downarrow$$
$$\frac{1}{2} \text{ AS} : \frac{1}{2} \text{ Normal}$$

The affected female would be heterozygous for the 15q11-q13 deletion. Therefore, her offspring have a 50% chance of inheriting the deletion chromosome. Children with the deletion chromosome would exhibit AS because the deletion was obtained from their mother.

4. Male with PWS × Normal female
$$\downarrow$$
$$\frac{1}{2} \text{ PWS} : \frac{1}{2} \text{ Normal}$$

The affected male would be heterozygous for the 15q11-q13 deletion. Therefore, his offspring have a 50% chance of inheriting the deletion chromosome. Children with the deletion chromosome would exhibit PWS because the deletion was obtained from their father.

Exercises and Problems

20. What are the similarities and differences between eukaryotic basal transcription factors and bacterial sigma factors?

21. What would be the effect on transcription of genes normally regulated by the thyroid hormone response element if the thyroid hormone receptor were mutated so it could not bind DNA?

22. A scientist is studying the expression of gene *T* in three different tissues. He isolates DNA from the tissues and treats it with a low concentration of DNase I. He then removes the histone proteins from DNA and subjects it to restriction endonuclease digestion, followed by electrophoresis and Southern blot hybridization using a genomic fragment containing gene *T* as a probe. The results are shown in the following figure.

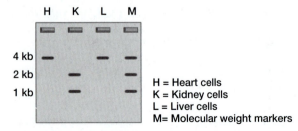

Southern Blot

H = Heart cells
K = Kidney cells
L = Liver cells
M= Molecular weight markers

In which tissue is gene *T* actively expressed? Provide a detailed explanation.

23. Predict the genotypic and phenotypic ratios in the offspring of the following crosses in mice:
    Cross A: *Igf-2 Igf-2m* female × *Igf-2 Igf-2* male
    Cross B: *Igf-2 Igf-2m* male × *Igf-2 Igf-2* female
    Cross C: *Igf-2m Igf-2m* female × *Igf-2 Igf-2* male
    Cross D: *Igf-2m Igf-2m* male × *Igf-2 Igf-2* female

24. The 3' untranslated region of the transferrin receptor mRNA contains five iron response elements (IREs) termed TfR-IRE-a, b, c, d, and e. These five sequences are highly conserved. The 29 nucleotide sequence of TfR-IRE-b is

    5' AAUUAUCGGAAGCAGUGCCUUCCAUAAUU 3'.

    Can this sequence assume any secondary structure? If so, draw it out. How could you test whether the secondary structure or the nucleotide sequence is important for protein binding?

25. Why is it likely that nucleosome regulation of gene expression occurs in all eukaryotes?

26. Jack is heterozygous for a mutation in the *UBE3A* gene. His wife Jill is homozygous normal. Their daughter Jasmine is married to Jethro who is also homozygous normal. Jasmine and Jethro are expecting their first child. What is the probability that their child will have Angelman syndrome if
    a. it is a boy?
    b. it is a girl?

27. Predict the phenotype of a newly fertilized *Drosophila* XY embryo in which the *Sxl* gene is deleted?

28. The fur of a certain animal could be white or black. The expression of the genes for black pigmentation is under the control of two transcriptional activators whose loci are on two different chromosomes. A cross of two double heterozygotes is carried out. Predict the genotypic and phenotypic ratios in the offspring in each of the following cases:
    a. Both transcriptional activators are required for black fur.
    b. Either transcriptional activator is enough for black fur.

29. You are examining the expression of *gene Z* in three tissues using Northern and Western blotting analysis. You isolate mRNA and protein from an animal, and you probe them with labeled Z DNA and protein Z antibody, respectively. The following results occur.

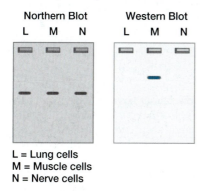

L = Lung cells
M = Muscle cells
N = Nerve cells

Is *gene Z* regulated at the transcriptional or translational level? Justify your answer.

30. A scientist creates a cell line that no longer produces the endogenous Myc or Max proteins. She then introduces into this cell line a construct that produces a Max protein containing a transcription activation domain and a Myc protein that lacks its activation domain. What would happen to the expression of genes in this cell line that are normally regulated by Myc-Max?

31. Why are operons common in prokaryotes but not in eukaryotes?

32. Consider the following primary mRNA transcript, where the yellow regions represent exons (E) and the brown regions are introns (I):

If a functional protein requires the presence of at least three different exons, how many proteins are possible from the alternative splicing of this molecule?

33. One way to identify enhancers and silencers associated with a gene is to clone a large region upstream of that gene before a reporter gene. Large deletions can then be generated in the promoter region, the resulting transgene can be introduced into the organism, and the expression of the reporter relative to the endogenous gene can be analyzed. Deletion of an enhancer will result in either decreased expression or improper spatial expression of the transgene. Additionally, silencers

can also be identified because deletion of a silencer will cause either increased or broader spatial expression of the transgene.

You cloned a 20 kb region upstream of the *Drosophila* gene Z to characterize the gene control elements. The following figure shows several deletion constructs that were all fused to the β-galactosidase reporter. You then introduce each construct individually into *Drosophila*. The 20 kb region exactly reproduces the expression pattern of this gene in the embryo, which corresponds to four stripes. For each construct, the embryonic expression pattern is diagrammed to the right.

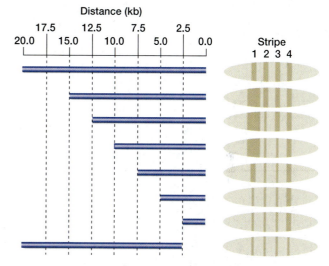

How many different gene expression control elements were identified by this analysis? State where each element is located (between what kilobase distances), which stripe it affects, and if it is involved in activating or repressing the wild-type stripe expression pattern.

34. The Myc-Max system regulates the proliferation of eukaryotic cells. However, a variety of different mutations in either the *myc* or *max* gene can disrupt this regulation. For each of the following mutations, determine its effect on (i) nonproliferating cells and (ii) proliferating cells.
    a. A *max* mutation that prevents Max from binding DNA
    b. A *myc* mutation that prevents Myc from binding DNA
    c. A *max* mutation that prevents Max from binding Myc
    d. A *myc* mutation that prevents Myc from binding Max
    e. A *myc* mutation that abolishes the transcriptional activation domain of Myc

35. The *myc* gene is a proto-oncogene (see chapter 22), whose expression is misregulated in a variety of different cancers. Provide a model for how *myc* mutations could be related to cancer development or progression (or both).

36. Below is a restriction map for a genomic DNA region that contains a gene that is expressed in the liver, but not in the kidneys.

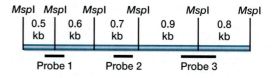

To identify sites of DNA methylation, you isolated genomic DNA from both liver and kidney, digested both DNA samples with *Hpa*II and *Msp*I individually, and then perform a genomic Southern blot. The results of the blot are shown below.

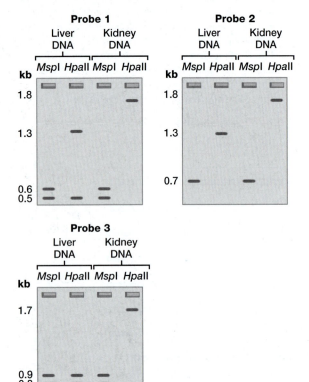

a. Determine what sites in the map are methylated.
b. Are these the only methlyated CpG sequences in this genomic region?
c. What methylation sites are most likely associated with repressing expression of the gene?
d. If the DNA methylase that acts in this region was nonfunctional, what would the genomic Southern blot look like and what would you predict would happen to the expression of this gene in the liver and kidney?

Chapter Integration Problem

You are studying the methylation pattern in a 4 kb region of a mouse (*Mus musculus*) gene with the following *Msp*I partial restriction map (distances in kilobase pairs):

Previous studies revealed that the restriction site marked with an asterisk is methylated in all tissues.

You isolated genomic DNA from kidney and liver tissue. You divided each sample into two halves, which you subjected to restriction endonuclease digestion with *Msp*I and *Hpa*II, respectively. The digested DNA was run on an agarose gel and analyzed by Southern blot hybridization using a probe that spans the entire 4 kb region. The X-ray film is shown here:

a. A few bands have been left out of the gel. Which ones? Justify your reasoning.
b. A single restriction site for *Hpa*II and/or *Msp*I (or both) has been left out of the restriction map. Where is it? Explain.
c. Indicate which restriction sites are likely to be methylated. Justify your answer.
d. From the results of this experiment, can you determine in which tissue the gene is transcriptionally active. Why or why not?
e. How would you determine the relative location of the introns and exons for this gene?
f. How would you determine the direction of transcription of this gene or the relative position of the promoter?
g. The restriction site for the enzyme *Msp*I is 5' C^CGG 3'. What type of ends are generated by *Msp*I? Draw them.

The hypothetical restriction enzyme *Ren*I has the restriction site 5' AYC^CGGRT 3' (where R = any purine and Y = any pyrimidine). *Ren*I cuts at the position indicated whether the second cytosine is methylated or not. The mouse genome has an overall GC content of 42%. Assume the nucleotide sequence is random.

h. Is every *Ren*I site in the mouse genome a *Msp*I site? If not, then what fraction is?
i. Is every *Msp*I site in the mouse genome a *Ren*I site? If not, then what fraction is?

Do you need additional review? Visit **www.mhhe.com/hyde** for practice tests, answers to end-of-chapter questions and problems, interactive exercises, and animation tutorials all designed to enhance your understanding of key genetic concepts.

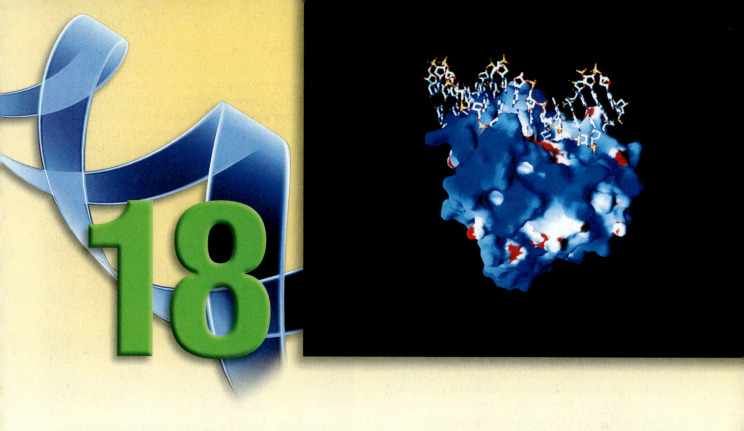

18

DNA Mutation, Repair, and Transposition

Essential Goals

1. Learn the major mechanisms by which spontaneous mutations are induced.

2. Describe the various types of mutagens and how they produce mutations in DNA.

3. Understand the various mechanisms by which a cell detects mutations and repairs them.

4. Describe the various classes of transposons and their mechanisms of transposition.

5. Diagram how transposable elements can produce mutations.

Chapter Outline

Photo: Computer-generated space-filling model of an enzyme repairing damaged DNA.

When we discussed DNA replication (chapter 9), we talked about the accuracy of the DNA polymerase and its 3' → 5' exonuclease proofreading activity to minimize the potential incorporation of an incorrect base. A variety of mechanisms, however, are capable of spontaneously introducing a mutation into the DNA sequence. Let us consider for a moment an example of how somatic mutations can lead to a human disease.

Retinoblastoma is an inherited disease that results in tumors of the retina (**fig. 18.1**). It is the most common eye cancer in children, with nearly 350 new cases occurring in the United States and 5000 cases worldwide each year. In approximately 90% of the cases, there is no history of retinoblastoma in the family. In the 10% of the cases where there is a family history, the pedigree suggests that retinoblastoma is inherited as an autosomal dominant disease. Medical investigators know, however, that retinoblastoma results from either the loss or mutation of a specific gene (*Rb*) on chromosome 13 and that it is a recessive phenotype. Why, then, does it appear like a dominant disease in pedigrees? As you will see, the answer lies in the rate of spontaneous mutation at this locus and the high rate of cell division in the fetal retina.

We will return to the topic of retinoblastoma when we discuss spontaneous mutations, and we will examine the details of how somatic mutations contribute to this disease. In addition, this chapter presents the different categories of mutations and the mechanisms by which they are produced. We also take a closer look at DNA repair mechanisms—enzymes that have evolved to detect and correct problems in DNA strands. The chapter concludes with information about transposable genetic elements and their ability to generate mutations.

figure 18.1   A child with retinoblastoma exhibits an abnormal white pupil reflex in his left eye. Instead of a camera flash producing a normal black or red pupil, an abnormal white color is observed. This white pupil indicates a variety of retinal problems, one of which is retinoblastoma.

18.1   Mutation

The concept of *mutation*, a term coined by Hugo de Vries, who was a rediscoverer of Mendel's work, is pervasive in genetics. **Mutation** is both the process by which a gene (or chromosome) changes structurally and the end result of that process. Without mutations producing alternative forms of genes (*alleles*), the biological diversity that exists today could not have evolved. Without different alleles, it would also have been virtually impossible for geneticists to determine which of an organism's characteristics are genetically controlled.

Studies of mutation provided the background for our current knowledge in genetics. In some cases, the mutations are induced to generate the desired alleles. In other cases, the mutations occur spontaneously. A specific allele will exhibit the same phenotype, regardless of whether the mutation was induced or spontaneous.

Phenotypic Classes

Mutations result in a change in the DNA sequence, which may produce a different phenotype in the organism. We can classify the different phenotypes that can be affected by mutations, including *morphological, nutritional* or *biochemical,* and *behavioral.* Many morphological and behavioral phenotypes, however, result from defects in biochemical pathways.

Morphological phenotypes are the observable traits of the organism, such as eye color and wing shape in *Drosophila.* In humans, hair color and height represent morphological phenotypes.

Nutritional or **biochemical phenotypes** are the description of substances that are required to support growth of the organism. Most often, we think of bacteria and fungi as having nutritional phenotypes. A *his⁻* auxotroph, for example, only grows on minimal medium if histidine is added; the *his⁻* mutation prevents the cell from synthesizing histidine. However, nutritional phenotypes also occur in humans. For example, phenylketonuria is a genetic disorder that results in the inability to break down phenylalanine. The nutritional phenotype is elevated phenylalanine levels in the blood, which requires the reduction of phenylalanine in the diet.

Behavioral phenotypes affect the patterns of behavior for an organism. Several different behavioral mutants have been identified in *Drosophila* that affect courtship and mating behavior, including mutations that result in a male fly preferring to court other males rather than females. Wild-type flies prefer to climb up the sides of a bottle (geotaxis), whereas behavioral mutants would either remain sedentary or prefer to climb down.

Behavioral phenotypes in humans can be more complex. However, several behavioral diseases have been shown to have a genetic component, such as manic depression.

A **lethal phenotype** results in the death of an organism, whereas a **deleterious phenotype** results in reduced viability of the organism. A mutation that produces a lethal phenotype is often called a lethal mutation, and the same is true for deleterious mutations. We described a variety of lethal mutations in previous chapters, including the *Yellow* allele (A^Y) in mice and Huntington disease in humans.

Although investigators could observe phenotypic mutations, the details of what a mutation represents genetically had to wait until the middle of the twentieth century to be revealed. One large question had to do with the nature of phenotypic differences in bacteria.

The Fluctuation Test: A Turning Point for Bacterial Genetics

Prior to the 1940s, very little work had been done in bacterial genetics because of the feeling that bacteria did not have "normal" genetic systems like the systems of fruit flies and corn. Rather, bacteria were believed to respond to environmental change by physiological adaptation, a non-Darwinian view. In 1943, Salvador Luria and Max Delbrück published a paper entitled "Mutations of Bacteria from Virus Sensitivity to Virus Resistance." This paper ushered in the era of bacterial genetics by demonstrating that the phenotypic variants found in bacteria are actually the result of mutations rather than induced physiological changes.

The Luria and Delbrück Experiment: Induction Versus Mutation

Luria and Delbrück studied the generation of ton^r (phage T1-resistant) mutants from a wild-type ton^s (phage T1-sensitive) *Escherichia coli* strain. When wild-type *E. coli* were infected with the T1 phage, the bacterial cells were lysed as the T1 phage replicated and were released. Thus, a petri plate that was spread with wild-type *E. coli* bacteria and T1 phages would exhibit no bacterial colonies because all the bacteria would be lysed. But if one of the bacterial cells was resistant to T1 phages, it would produce a bacterial colony with all the cells in this colony being T1-resistant.

Two explanations were possible for the appearance of T1-resistant colonies:

1. Any *E. coli* cell may be induced to be resistant to phage T1, but only a very small number actually are. That is, all cells are genetically identical, each with a very low probability of exhibiting resistance when exposed to the T1 phages. When resistance is induced, the cell and its progeny remain resistant.

2. In the culture, a small number of *E. coli* cells exist that are already resistant to phage T1; in the presence of phage T1, only these cells survive.

If the presumed rates of physiological induction and mutation are the same, determining which of the two mechanisms is operating is difficult. Luria and Delbrück, however, developed a means of distinguishing between these mechanisms. They assumed that if T1 resistance was physiologically induced, the relative frequency of resistant *E. coli* cells in a wild-type culture (ton^s) should be constant, and it should be independent of the number of cells in the culture or the length of time that the culture has been growing.

In contrast, if resistance was due to random mutation, the frequency of producing mutant (ton^r) cells would depend on when the mutations occurred (**fig. 18.2**). A mutation occurring early in the growth of the culture would produce many cells descended from the mutant cell, and therefore many resistant colonies would develop. But a mutation occurring late in the growth of the culture would produce a low number of resistant cells.

If the mutation hypothesis was correct, therefore, considerable fluctuation from culture to culture should be observed in the number of resistant cells present (fig. 18.2).

Results of the Fluctuation Test

To distinguish between these hypotheses, Luria and Delbrück developed what is known as the **fluctuation test.** They counted the number of resistant cells both in small ("individual") cultures and in subsamples from a single large ("bulk") culture.

All subsamples from a bulk culture should have the same number of resistant cells, differing only because of random sampling error. But if a mutation occurs, the number of resistant cells in the individual cultures should vary considerably from culture to culture; the number would be related to the time at which the mutation occurred during each culture's growth (**fig. 18.3**). Early mutations would produce many resistant cells. Late mutations would result in relatively few resistant cells.

Under physiological induction, the distribution of resistant colonies should not differ between the individual and bulk cultures. All cells would be identical, and the same number of cells in each individual culture should undergo induction, subject only to random error.

Luria and Delbrück inoculated 20 individual cultures and one bulk culture with *E. coli* cells and incubated them in the absence of phage T1. Each individual culture was then spread out on a petri plate containing a very high concentration of T1 phages; 10 subsamples from the bulk culture were plated in the same way. We can see from the results (**table 18.1**) that there was

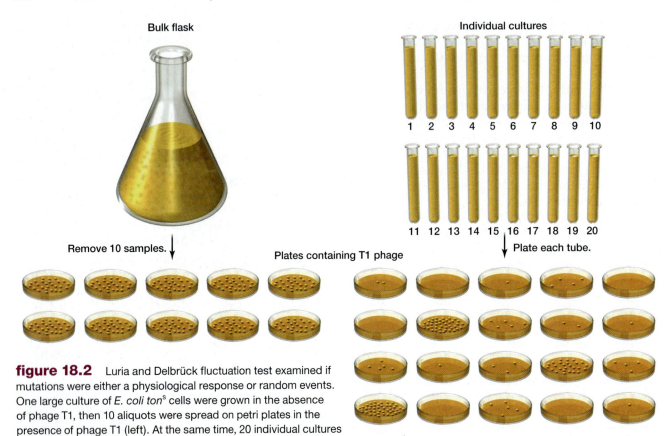

figure 18.2   Luria and Delbrück fluctuation test examined if mutations were either a physiological response or random events. One large culture of *E. coli ton*ˢ cells were grown in the absence of phage T1, then 10 aliquots were spread on petri plates in the presence of phage T1 (left). At the same time, 20 individual cultures of *E. coli ton*ˢ cells were grown in separate test tubes in the absence of phage T1, then spread on petri plates in the presence of phage T1 (right). The *ton*ʳ resistant cells grow into colonies on the plates. A uniform distribution of resistant cells is expected on all the plates if the physiological induction hypothesis is correct. If mutation is a random event, there should be a great fluctuation in the number of resistant colonies on the plates from individual tubes because each tube may or may not have experienced the mutation. These data demonstrated that mutation is a random event.

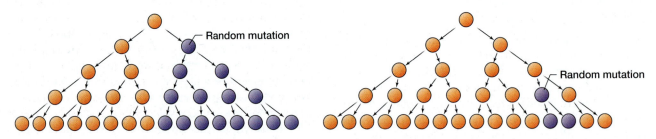

figure 18.3   The number of mutant cells is a consequence of when the mutation is produced. If the mutation occurs earlier in the cell division (left), then more mutant cells (purple) will be derived from the original mutant. If the mutation occurs later (right), then fewer mutant cells will be present.

minimal variation in the number of resistant cells among the bulk culture subsamples. However, a very large amount of variation, as predicted for random mutation, was observed among the individual cultures.

This finding demonstrated that bacteria have "normal" genetic systems that undergo mutation. As a result, bacteria could then be used, along with higher

organisms, to answer genetic questions. As we have seen in past chapters, the modern era of molecular genetics began with the use of bacteria and phages in genetic research. In the next section, we turn our attention to several basic questions about the gene, questions whose answers were found in several instances only because bacterial systems were available.

table 18.1   Results from the Luria and Delbrück Fluctuation Test

Individual Cultures*		Samples from Bulk Culture*	
Culture Number	*Ton*^{r} Colonies Found	Sample Number	*Ton*^{r} Colonies Found
1	1	1	14
2	0	2	15
3	3	3	13
4	0	4	21
5	0	5	15
6	5	6	14
7	0	7	26
8	5	8	16
9	0	9	20
10	6	10	13
11	107		
12	0		
13	0		
14	0		
15	1		
16	0		
17	0		
18	64		
19	0		
20	35		
Mean (*n*)	11.4		16.7
Standard deviation	27.4		4.3

Source: From E. Luria and M. Delbrück, *Genetics*, 28: 491. Copyright © 1943 Genetics Society of America.

* Each culture and sample was 0.2 mL and contained about 2×10^7 *E. coli* cells.

Restating the Concepts

▶ Some of the different phenotypes that can be affected by mutations are classified as morphological, nutritional or biochemical, and behavioral.

▶ Luria and Delbrück developed the fluctuation test and demonstrated that bacteria underwent random spontaneous mutagenesis rather than physiological induction.

▶ Bacteria, like eukaryotes, have phenotypes that result from the presence of different alleles, which are attributable to mutations.

18.2 Types of Mutations and Their Effects

Luria and Delbrück's work on the generation of *E. coli* mutants resistant to the T1 phage was based on the generation of **spontaneous mutations,** which is the natural production of mutations in the normal life cycle of the organism. Earlier, in 1927, Hermann J. Muller demonstrated that exposing fruit flies to varying doses of X-rays caused mutations in their offspring. **Induced mutations** are produced by treating the organism with either high-energy radiation or a chemical to generate mutations at a significantly higher frequency than spontaneous mutations. At about the same time, Lewis J. Stadler showed that he could also use X-rays to induce mutations in barley.

The basic impetus for Muller and Stadler's work was the fact that spontaneous mutations occur infrequently, and genetic research was significantly slowed by the inability to obtain mutants. Their studies led to two major conclusions. First, X-rays greatly increased the occurrence of mutations. Second, the inheritance patterns of X-ray-induced mutations and the resulting phenotypes were similar to those that resulted from natural, or spontaneous, mutations.

We return to the topic of mutation rate a little later on; first, we review some of the categories of mutations. Definition of these categories required the understanding of DNA as being the genetic material in which mutation takes place.

Point Mutations

The mutations of primary concern in this chapter are **point mutations,** which consist of single-nucleotide changes in the DNA sequence. (We already discussed *chromosomal mutations,* changes in the number and visible structure of chromosomes, in chapter 8.) If the mutation replaces one nucleotide with another within an open reading frame (ORF), then a new codon is created. In many cases, this new codon will specify a different amino acid, which results in a change in the translated protein sequence. These new proteins can alter the morphology or physiology of the organism and result in mutant phenotypes or lethality.

Some point mutations may not produce a detectable phenotypic change, in which case they are called **silent mutations.** Silent mutations can be found in the intergenic regions, in introns, in the 5' and 3' untranslated regions (UTR) of a gene, and in the wobble position of a codon.

Missense mutations are nucleotide changes in the ORF that result in one amino acid becoming a different amino acid (see chapter 11). For example, a conservative mutation may change the codon for glutamate to a codon for aspartate. **Nonsense mutations** occur when the single-nucleotide change alters a codon encoding an amino acid into a translation termination codon. The effect of a nonsense mutation is a prematurely terminated protein.

Frameshift Mutation

A point mutation may consist of replacement, addition, or deletion of a nucleotide in a DNA sequence (**fig. 18.4**). A **frameshift mutation** is one that adds or sub-

tracts a base. Because a frameshift mutation within a gene changes the reading frame from the site of mutation onward (**fig. 18.5**), it is potentially more devastating to the cell than a nucleotide change.

A frameshift mutation causes two problems. First, all the codons from the frameshift on will be different and thus will yield (most probably) a useless protein. Second, translation termination signals will be misread. One of the new codons may be a nonsense codon, which causes translation to stop prematurely. Or, if the translation apparatus reaches the original stop codon, it will no longer be recognized as such because it is in a different reading frame; therefore, translation will continue beyond the end of the gene.

Back Mutation and Suppression

It is possible that a second mutation can occur within a mutant gene. In this case, there are three possible outcomes (see fig. 18.4). First, the second mutation can occur at a different nucleotide position relative to the first mutation, which results in the same or worse mutant phenotype. Alternatively, the second mutation may change the mutant nucleotide back to the original wild-type nucleotide. This second mutation is known as a **back,** or **reverse, mutation**: the mutant sequence is restored to the original sequence and function. Third, the second mutation may occur at a different site than the first and still restore the wild-type function without restoring the original sequence. This type of mutation, an **intragenic suppressor,** occurs when the second mutation masks the effect of the origi-

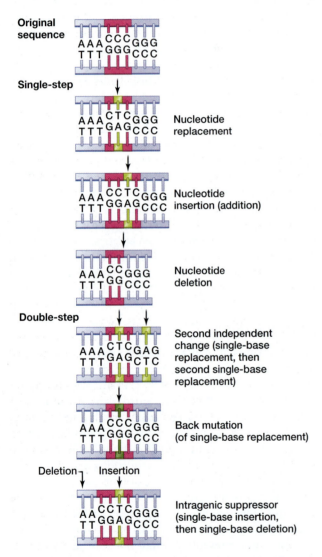

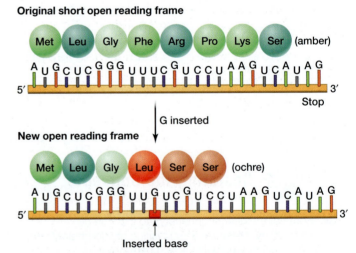

figure 18.4   Types of DNA point mutations. The original DNA sequence (top) can be mutated in either a single-step or double-step process. Single-step changes are nucleotide replacements, insertions (additions), or deletions. A double-step process involves two independent mutations in the same gene, which can result either in a double mutation, reversion to the original sequence, or intragenic suppression. In the last case, intragenic suppression is illustrated by the addition of one base followed by the nearby deletion of a different base.

figure 18.5   Possible effect of a frameshift mutation. The wild-type mRNA and short peptide sequence are shown at the top. The translation termination codon (UAG) is labeled as (amber). The insertion of a single base (G) shifts the reading frame, which produces a new amino acid sequence in the translated protein (red circles) after the inserted base and a new premature translation stop sequence (UAA, ochre).

nal mutation in the same gene. The new sequence is a double mutation that appears to have the wild-type phenotype.

In figure 18.4, an insertion of a T results in a frameshift mutation, which alters the entire reading frame following this mutation. A second mutation, deletion of an A, occurs in the same gene and also alters the reading frame. However, the combination of a loss of a nucleotide and an insertion results in an altered reading frame between the two mutations, followed by the wild-type reading frame after the downstream mutation. If the region between the two mutations is sufficiently small, then the encoded protein will retain sufficient activity to produce the wild-type phenotype.

Intragenic suppression can also occur if the first mutation was a nucleotide change that results in the change of a single amino acid in the encoded protein. The second mutation, or reversion mutation, is a change in a different codon that results in a second amino acid change in the protein. Suppression occurs when the second amino acid change counteracts the first to restore the phenotype of the organism to approximately wild type. Suppressed mutations can be distinguished from true back mutations either by subtle differences in phenotype, by genetic crosses, by changes in the amino acid sequence of a protein, or by DNA sequencing.

Conditional Mutations

A class of mutants that has been very useful to geneticists is the **conditional mutant,** an organism that is mutant under one set of circumstances but normal under another. All conditional mutants have a **permissive condition,** in which the wild-type phenotype is observed, and a **restrictive condition,** in which the mutant phenotype is observed. Conditional mutants have been very informative in examining time-dependent processes, such as when a protein must function in development to produce the desired phenotype, and in studying lethal mutations.

Nutrient-requiring mutants are good examples of conditional mutants; they survive (wild-type phenotype) when specific nutrients are added to the minimal medium and fail to survive (mutant phenotype) when grown on only minimal medium.

Another important class of conditional mutants is **temperature-sensitive mutants.** This class of mutant exhibits the wild-type phenotype at one temperature and the mutant phenotype at a different temperature. Temperature-sensitive lethal mutants have made it possible for geneticists to work with genes that control vital functions of the cell, such as DNA replication. Many temperature-sensitive lethal mutants are completely normal at 20–25°C (permissive temperature), but are not viable, due to the mutant protein, at 37–42°C (restrictive temperature).

Presumably, temperature-sensitive mutations encode enzymes with amino acid substitutions that result in the mutant protein forming an abnormal structure at the restrictive temperature. Thus, the organism can be grown at the permissive temperature and then heated to the restrictive temperature to examine the mutant (lethal) phenotype in more detail.

Let's look at one example of how studying conditional-lethal mutants revealed important information about a genetic process. Certain *E. coli* mutants cannot synthesize DNA at the restrictive temperature; however, they have a completely normal DNA polymerase I. These mutants revealed that DNA polymerase I is not the only enzyme *E. coli* normally uses for DNA replication. This led to the identification of the function of DNA polymerase III (chapter 9).

When an organism with a conditional mutation in the DNA polymerase II gene was isolated, it was found to be able to replicate its DNA normally, but could not repair damaged DNA. This observation led to the conclusion that DNA polymerase II is primarily involved in DNA repair rather than replication. Conditional-lethal mutants, therefore, permit the genetic analysis of mutant genes that would otherwise be impossible to study because of their lethal phenotype.

Mutation Rates

The **mutation rate** is the number of mutations that arise per gene for a specific length of time in a population. For bacteria and single-cell organisms, the mutation rate is usually measured as the number of new mutations in a gene per cell division. In eukaryotic organisms, the mutation rate is usually measured as the number of mutations that arise in a gene per gamete.

Mutation rates vary tremendously, depending on the length of genetic material, the kind of mutation, and other factors. Luria and Delbrück found that in *E. coli* the mutation rate per cell division of *ton^s* to *ton^r* was 3×10^{-8}, whereas the mutation rate of the wild-type prototroph to the histidine-requiring auxotroph (*his$^+$* to *his$^-$*) was 2×10^{-6}.

The **reversion rate,** which corresponds to the change from the *his$^-$* mutant phenotype to the *his$^+$* wild-type phenotype, was found to be 7.5×10^{-9}. The mutation and reversion rates differ because many different mutations in the *his$^+$* gene can cause the *his$^-$* phenotype, whereas reversion requires specific, and hence less probable, mutations to change the *his$^-$* allele back to the wild-type *his$^+$* phenotype.

The lethal mutation rate in *Drosophila* is about 1×10^{-2} per gamete for the total genome. This number is relatively large because many different mutations in many different genes can produce the same phenotype (lethality, in this case).

table 18.2   Spontaneous Mutation Rates Differ Between Genes and Organisms

Organism	Trait	Gene	Rate	Units
Bacteriophage T2	Plaque size	r^+	1×10^{-8}	per gene replication
	Host range	h^+	3×10^{-9}	
E. coli	Streptomycin sensitivity	str^r	1×10^{-8}	per cell division
	Leucine prototroph	leu^-	7×10^{-10}	
D. melanogaster	White eye	w^+	4×10^{-5}	per gamete per generation
	Ebony body	e^+	2×10^{-5}	
	Yellow body	y^+	1.2×10^{-6}	
M. musculus	Piebald coat	s^+	3×10^{-5}	per gamete per generation
	Brown coat	b^+	8.5×10^{-4}	
H. sapiens	Retinoblastoma	Rb^+	2×10^{-5}	per gamete per generation
	Huntington disease	Hu^+	5×10^{-6}	

table 18.3   Induced Mutation Rates Are Higher than Spontaneous Mutation Rates

Mutagen	Exposure Time (min)	Percent Survival	Number of *ad-3* mutants per 10^6 survivors
None (spontaneous)	—	100	0.4
Amino purine (1-5 mg/mL)	During growth	100	3
Ethyl methanesulfonate (1%)	90	56	25
Nitrous acid (50 mM)	160	23	128
X-rays (2000 r/min)	18	16	259
UV rays (600 erg/mm²/min)	6	18	375
Nitrosoguanidine (25 mM)	240	65	1500

The spontaneous mutation rate can differ dramatically from gene to gene within an organism and from organism to organism (**table 18.2**). It is interesting to note that bacterial genes have a mutation rate of approximately 1×10^{-9} to 1×10^{-8}. In contrast, eukaryotes typically have spontaneous mutation rates of 1×10^{-6} to 1×10^{-5}. Some factors that affect the spontaneous mutation rate include the size of the gene, the rate that mismatched bases are incorporated into the DNA, and the efficiency of repairing damaged DNA in the organism.

The induced mutation rate can be several orders of magnitude larger than the spontaneous mutation rate (**table 18.3**). For this comparison to be meaningful, we must look at a single gene in a specific organism (*ad-3* and *Neurospora*, respectively, in table 18.3). The factors that can affect the induced mutation rate are the size of the gene, the base composition of the gene, the efficiency with which the mutagen can modify the base or DNA in the organism, and the efficiency with which the organism repairs the specific type of damage.

Restating the Concepts

▶ Mutations can occur naturally or spontaneously. The spontaneous mutation rate is low, which led to the identification of mutagens that induce a much higher mutation rate.

▶ Point mutations can be a base change, base insertion, or base deletion. An insertion or deletion produces a frameshift mutation that changes the reading frame.

▶ A back, or reverse, mutation changes the original mutation back to the original sequence. An intragenic suppressor is a second mutation that fails to restore the original sequence, but does restore the original function back to the encoded protein.

▶ Conditional mutations exhibit the wild-type phenotype under permissive conditions and the mutant phenotype under restrictive conditions. Conditional mutants permit the analysis of mutant genes that otherwise would be lethal.

18.1 *applying genetic principles to problems*

Calculating Mutation Rate and Mutation Frequency

Mutation rate represents the number of mutations that occurs in a specific gene per unit of time, such as generation, cell division, or round of DNA replication. This represents the likelihood that a new mutation will be generated in a specific gene. Spontaneous mutation rates vary from 1×10^{-5} to 10^{-9} per cell division. Mutagens are used to increase the mutation rate so that there is a greater chance of isolating the desired mutation.

You must be careful not to confuse mutation rate with mutation frequency. **Mutation frequency** for a given gene is the number of mutant alleles per the total number of copies of the gene within a population. Thus, mutation frequency represents the prevalence of a specific mutation, both old and newly generated, in the population. Different factors affect the mutation rate and mutation frequency in a population.

Mutation rate is affected by factors that will influence the generation of new alleles in a specific gene. First, different organisms will exhibit different spontaneous mutation rates, different repair rates, and also different susceptibilities to mutagens. Second, mutagens, which produce higher mutation rates than spontaneous mutations, will yield different mutation rates for a given gene based on the effectiveness of the mutagen at modifying the DNA, how effectively the DNA modification is repaired, and how the mutagen is metabolized by the organism. Third, the size of the gene will affect the mutation rate. Larger genes have more bases that can be mutated than small genes. Thus, a variety of factors affect the mutation rate and result in a large variation for any given gene.

Mutation frequency depends more on the effect the mutation has on the organism. Mutations that are not deleterious or lethal are likely to be present in a larger percentage of the population, which relates to a higher mutation frequency.

To get a different perspective of mutation rate and mutation frequency, let's look at an example to see how they are calculated.

Neurofibromatosis 1 (NF1) is an autosomal dominant human disease that is completely penetrant, but exhib-

its variable expressivity. Individuals with *NF1* mutations exhibit discrete tumors along the peripheral and central nervous system, freckling on areas of skin not exposed to sunlight (called café-au-lait spots), learning disabilities, and increased risk for additional tumors.

In 1989, a study was published on the presence of NF1 among the 668,100 residents of southeast Wales. The researchers found 135 cases of NF1 in this population. Because 41 of these individuals did not have ancestors who exhibited the NF1 symptoms, they were determined to be new mutations. (It is possible that a new mutation is present in one or more of the remaining 94 individuals. However, we cannot detect any new alleles in the background of an existing dominant mutation.)

The mutation rate of the *NF1* gene in this population is the number of new mutations per gamete (two gametes are required to produce an individual), which corresponds to:

$$\frac{41}{668,100 \times 2} = 3.1 \times 10^{-5}$$

This is one of the highest spontaneous mutation rates that exist in humans. This is consistent with the *NF1* gene corresponding to approximately 350 kb of genomic DNA! Even the 11 to 13 kb mRNA is relatively large for human genes (which average approximately 2 kb).

The mutation frequency corresponds to the number of mutant alleles divided by the total number of gene copies in the population. To calculate the mutation frequency, we do not need to differentiate between new and old mutations. Because each individual has two copies of the gene, there are a total of $2 \times 668,100 = 1,336,200$ gene copies. The mutation frequency is then:

$$\frac{135}{1,336,200} = 1.0 \times 10^{-4}$$

Thus, mutation rate is the frequency of generating new mutations and should be less than the mutation frequency in the population, unless the mutation is very deleterious or causes death at a relatively young age.

It's Your Turn

To test your understanding of the difference between mutation rate and frequency, try problem 40.

18.3 Mutagenesis: Origin of Mutation

As mentioned earlier, mutations may arise spontaneously in an organism's DNA. Mutations may also be induced by a variety of environmental factors, such as exposure to X-rays and other radiation or to certain chemicals. Rates of mutation are site- and organism-dependent.

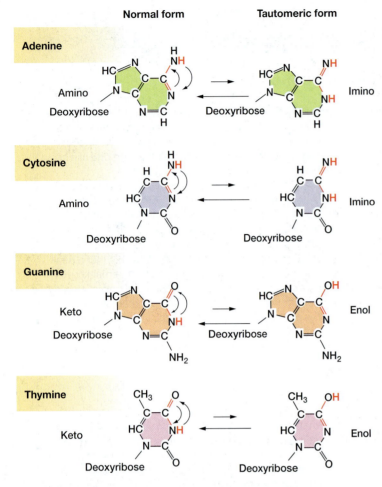

figure 18.6 Normal and tautomeric forms of DNA bases. Adenine and cytosine can exist in either the normal amino form (left) or the rare imino form (right), while guanine and thymine can exist in either the normal keto form (left) or the rare enol form (right).

Spontaneous Mutagenesis

Watson and Crick originally suggested that mutation could occur spontaneously during DNA replication if base-pairing errors occurred. Although DNA polymerase possesses a 3' → 5' exonuclease (proofreading) activity to reduce the possibility of incorporating the wrong base during replication, some mechanisms can yield a mismatched base pair.

Transition Mutations

The four DNA bases can exist in two different conformations: adenine and cytosine can exist in either the normal amino form or the rare imino form, and guanine and thymine can exist in either the normal keto form or the rare enol form (**fig. 18.6**). The normal form of each base can undergo a proton shift to produce its rare tautomeric form (tautomeric shift).

When any of the bases are in their rare tautomeric form, an inappropriate base pairing occurs. For example, the rare imino forms of adenine and cytosine can base-pair with the normal forms of cytosine and adenine, respectively (**fig. 18.7**). Similarly, the rare keto forms of guanine and thymine will base-pair with the normal forms of thymine and guanine, respectively (see fig. 18.7).

During DNA replication, a tautomeric shift in either the incoming base (*substrate transition*) or the base in the template strand (*template transition*) results in mispairing. The mispairing will produce a new base pair after an additional round of DNA replication, with the original strand remaining unchanged (**fig. 18.8**).

The replacement of one base pair with a different base pair through a tautomeric shift maintains the same purine–pyrimidine relationship: an A–T pair is replaced by G–C, and G–C by A–T. In both examples, a purine replaces another purine and a pyrimidine replaces a pyrimidine. This type of

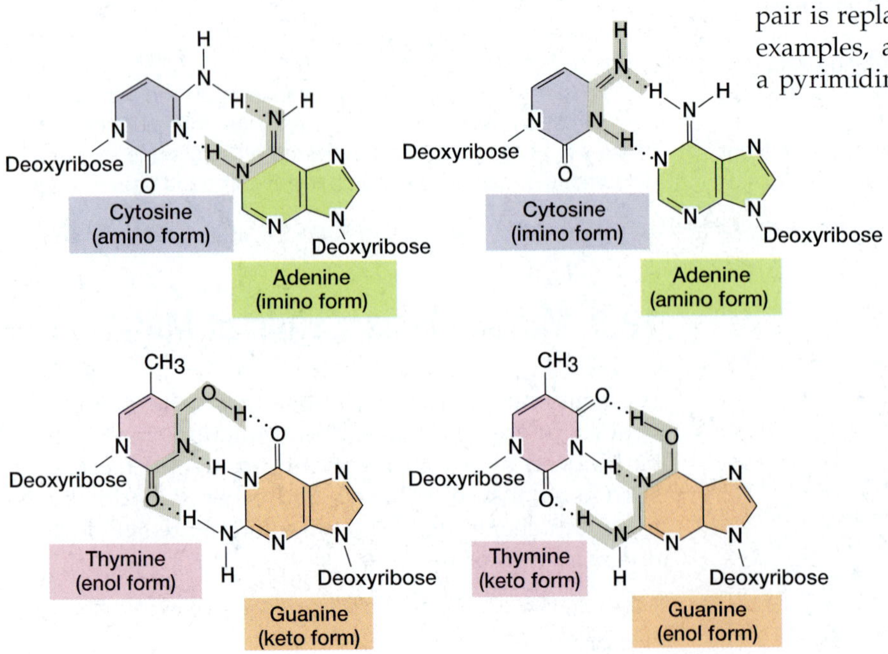

figure 18.7 Rare tautomeric forms of the bases lead to unexpected basepairs. In the rare imino form, adenine can base pair with the normal amino form of cytosine (upper left). Similarly, the imino form of cytosine can base pair with the normal amino form of adenine (upper right). The rare enol form of thymine can base pair with the normal keto form of guanine (lower left) and the rare enol form of guanine can base pair with the normal keto form of thymine (lower right).

mutation in which one base is replaced by a similar type of is referred to as a **transition mutation.** In contrast, when a purine replaces a pyrimidine, or vice versa, the mutation is termed a **transversion mutation.**

Transversion Mutations

Transversion mutations may arise by two different mechanisms. First, a tautomeric shift and a base rotation must both occur. (We discussed base rotations in the formation of Z-DNA in chapter 7.) The normal configuration of the glycosidic (base–sugar) bond is referred to as the *anti* configuration; the rotated form is the *syn* configuration. An example of this type of transversion occurs when one adenine undergoes a tautomeric shift and the other rotates to the *syn* configuration (**fig. 18.9a**). In this case, an A–T base pair is converted to a T–A base pair through an A–A pairing intermediate (fig. 18.9b).

The second mechanism of transversion is through base modification. For example, if guanine is oxidized to 8-oxoguanine (**fig. 18.10a**), it can base-pair with adenine. This converts a G–C base pair to a T–A base pair through an 8-oxoguanine–adenine intermediate

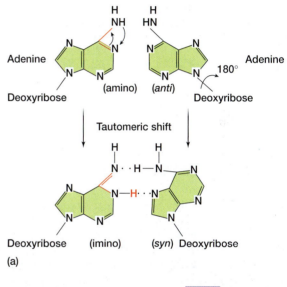

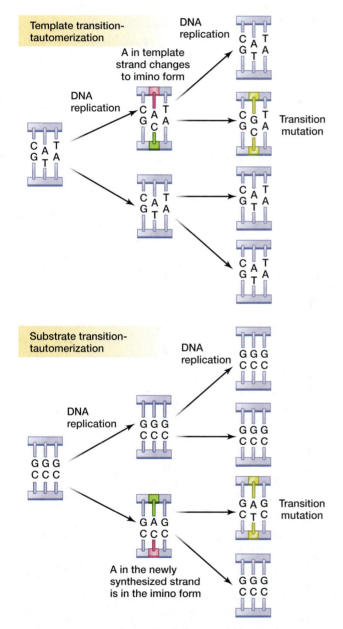

figure 18.8 Tautomeric shifts result in transition mutations. The tautomerization of bases can occur in either the template strand (top) or in the newly synthesized strand (bottom) during DNA replication. The rare tautomeric forms are shown in red; the normal base pairing with the rare tautomer is in green, and the resulting transition is in yellow. The transition is completed after a second round of DNA replication.

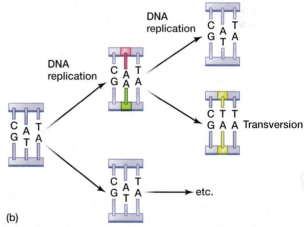

figure 18.9 A model for transversion mutagenesis. (a) An AA base pair can form if one base undergoes a tautomeric shift and the other rotates about its glycosidic (sugar) bond (from the normal *anti* configuration to *syn*). In this situation, two hydrogen bonds are formed in the AA base pair. (b) A schematic of how an AT base pair can be converted to a TA base pair (transversion) by way of an intermediate AA base pair. The red base is in the rare tautomeric form, while the green base is in the *syn* configuration. After a second round of DNA replication, one DNA duplex will have a transversion (yellow) at that point.

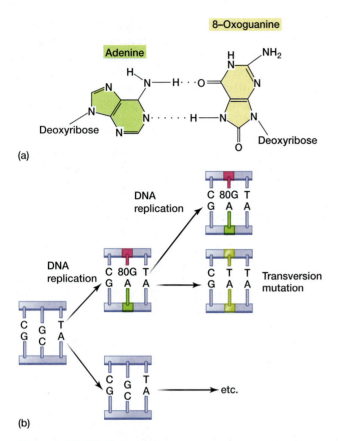

(a)

(b)

figure 18.10 Transversion through a base modification. A GC base pair can undergo a transversion to a TA base pair by modification of the guanine to 8-oxoguanine. (a) The base pairing between adenine and 8-oxoguanine is shown with two hydrogen bonds. (b) A schematic of how a GC base pair can be converted to a TA base pair (transversion) by way of a modified 8-oxoguanine (8OG) base pairing with adenine. Notice that the 8-oxoguanine-adenine base pair persists after DNA replication (top line) due to the covalent modification of the guanine.

(fig. 18.10b). This transversion mechanism has been found to be common in cancers.

Note that transversions are usually more serious mutations than transitions. The reason is apparent when we look at the genetic code (see table 11.4). For most codons, mutating a purine to purine (or pyrimidine to pyrimidine) in the wobble position fails to change the encoded amino acid. In contrast, the purine-to-pyrimidine change and pyrimidine-to-purine change at the wobble position is more likely to change the encoded amino acid. As a specific example, the CAU transition to CAC still encodes a histidine, but the CAU to CAA transversion changes the histidine codon to a glutamine codon.

Other Spontaneous Mutations

Three other types of spontaneous mutations occur to DNA. During the normal cell cycle, purines are randomly lost from the DNA. This **depurination** is fairly common, with approximately 10,000 purines lost

from the mammalian genome per cell in a 20h cell cycle. Depurination produces a point in the double-stranded DNA called an *apurinic (AP) site* where the purine base is missing, but the deoxyribose to which it was attached remains. Although the general information remains in the DNA because of the presence of the complementary pyrimidine, DNA replication and RNA transcription cannot proceed across the AP site without knowing what purine should be present.

Deamination occurs when the amine group is lost from either cytosine (to produce uracil) or 5-methylcytosine (to produce thymine). Both of these deaminations result in transition mutations.

The final class involves the interaction of active oxygen species, such as superoxide radicals (O_2^-), hydrogen peroxide (H_2O_2), and hydroxyl radicals (OH^-) acting on primarily thymine and guanine to produce a variety of mutations, including thymine glycol and 8-oxo-7-hydroguanine.

Retinoblastoma: An Example of Spontaneous Mutagenesis

To demonstrate the potential effect of spontaneous mutations in generating an abnormal phenotype, we will examine the human retinoblastoma gene (*Rb*). A mutation in the *Rb* gene can result in the development of an eye tumor (also called retinoblastoma) in children, usually before 18 months of age. The retinoblastoma phenotype was documented in children more than 2000 years ago. If not properly treated, the tumor can potentially move up the optic nerve and enter the brain.

The majority of the time (90%), there is no previous history of retinoblastoma in the family before a child is diagnosed. When retinoblastoma is present in a family pedigree, however, it appears to be inherited as an autosomal dominant allele—and yet it is known to result from the loss of function of the *Rb* gene on chromosome 13.

One important piece of information that helped to solve this puzzle is that the tumors may be either bilateral (affecting both eyes) or unilateral. Most cases of retinoblastoma (approximately 75%) are unilateral. This unilateral phenotype and the absence of retinoblastoma in the family history are both consistent with the disease resulting from a spontaneous mutation that produces the tumor.

The *Rb* gene encodes a protein that is required to control cell division in the developing retina. The spontaneous mutation rate of the *Rb* gene is known; if a single mutation occurs, there is no significant effect on the eye. But if the *Rb* gene on the homologous chromosome also experiences a mutation in an eye cell, then the cell containing two *Rb* mutations may begin uncontrolled cell division and produce a tumor.

This double mutation accounts for the 90% of the individuals who do not have a history of retinoblastoma in the family, and it also accounts for the majority of the unilateral tumors. Most individuals will have two wild-type *Rb* alleles. This requires the generation of a mutation in both *Rb* alleles in a single cell before that cell can develop into a tumor. The probability that this double mutation will occur in a cell in both retinas is very low, which accounts for the lower percentage of bilateral tumors in an individual whose family lacks a previous history of retinoblastoma.

When an individual's family has a history of retinoblastoma, there is a 50% chance that a mutant *Rb* allele will be passed on to their children. In these cases, only a single mutation (in the wild-type *Rb* allele) is required to produce the tumor. Because of the rapid cell division that occurs in the fetal eye, many opportunities exist for individuals who inherited a defective copy of the *Rb* gene to experience the second mutation. In this case, there is a greater chance that a single spontaneous mutation can occur in a cell in both retinas, compared with the two spontaneous mutations in the same cell in both retinas, to produce the bilateral tumors.

Retinoblastoma can potentially be lethal if the tumor progresses up the optic nerve and invades the brain. Because of the potential severity of this cancer, the original treatment for retinoblastoma was simply removal of the entire eye. Now the treatment varies from cryotherapy (applying a very cold probe to the eye to kill the tumor), to application of radiation to the surface of the tumor, to chemotherapy and external beam radiation, or removal of the eye in only the most extreme cases. Over 95% of the retinoblastoma patients in the United States are cured, with over 90% of the individuals retaining at least one eye and more than 80% maintaining 20/20 vision.

Chemical Mutagenesis

Certain chemicals can cause mutations at rates that are much higher than spontaneous mutations (see table 18.3). Determining the mode of action of various chemical mutagens has provided insight into the mutational process. Because many forms of cancer are chemically induced, such as lung cancer resulting from compounds in cigarette smoke, studying chemical mutational mechanisms may increase our understanding of the process of carcinogenesis. In addition, knowing how chemical mutagens act has allowed geneticists to produce large numbers of specific mutations at will.

The Ames Test

In 1975, molecular biologist Bruce Ames developed a routine screening test for identifying mutagens. Ames used several different auxotrophic strains of *Salmonella typhimurium* that require histidine to grow. Applying a mutagen to this strain caused a reversion mutation,

and the resulting prototroph would grow on minimal medium. Different *his⁻* auxotrophic strains are used to detect reversion of different classes of mutations, such as transition or transversion point mutations and frameshift mutations. Under normal circumstances, there is a spontaneous reversion rate for each of the auxotrophs; however, treatment with a mutagen increases the frequency of revertants.

Ames reasoned that the human body metabolized compounds, which could convert an inert compound into a mutagen. To improve this test's ability to detect mutagens, Ames added a supplement of rat liver extract to the medium containing the *Salmonella* and the mutagen (**fig. 18.11**). This liver supplement could break down noncarcinogenic compounds into potentially carcinogenic substances. Because the rat liver enzymes act on a biochemical substance the same way the human liver does, the conversion of a noncarcinogenic substance into a carcinogen by the rat liver extract would be a good approximation of what would happen to the noncarcinogenic compound in the human body. You should also realize that the liver enzymes could also inactivate a carcinogen.

The Ames test therefore has the capacity to measure the mutagenicity of a compound and how the liver affects the mutagenicity of the compound. Because many compounds that cause cancer (carcinogens) are mutagens, the Ames test can be a rapid, inexpensive, and easy initial test for mutagens and potential carcinogens.

Chemicals That Produce Transitions

As discussed earlier, spontaneous mutations can produce transition mutations. Transitions can also be routinely produced by **base analogs,** which are chemicals that mimic the structure of a base and can be incorporated into DNA.

Two of the most widely used base analogs are the pyrimidine analog 5-bromouracil (5-BU) and the purine analog 2-aminopurine (2-AP). The mutagenic mechanisms of the two are similar. The 5-bromouracil is incorporated into DNA in place of thymine and basepairs with adenine. But the bromine atom causes 5-bromouracil to tautomerize more readily than thymine does, to shift from the keto form to the enol form (**fig. 18.12**). Transitions frequently result when the enol form of 5-bromouracil pairs with guanine.

The 2-aminopurine is mutagenic because like adenine, it can exist in two forms. The major form of 2-aminopurine basepairs with thymine, and the rare tautomer basepairs with cytosine (**fig. 18.13**). Thus, 2AP can generate a transition mutation by replacing an A–T base pair with a G–C.

Nitrous acid (HNO_2) also readily produces transitions by a different mechanism. Rather than acting as a base analog, nitrous acid acts by modifying bases.

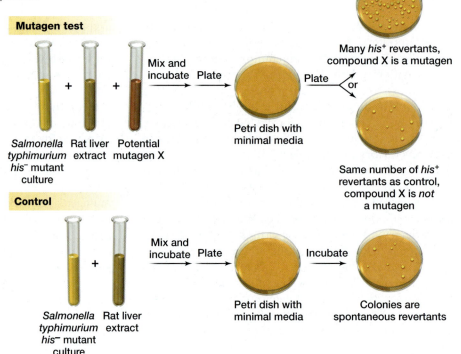

figure 18.11   The Ames test to identify mutagens. Different *his⁻* strains of *Salmonella typhimurium* (either missense or frameshift mutants) are mixed with the potential mutagen and a rat liver extract. The mixture is plated on minimal medium to select for *his⁺* revertants. The control reaction (bottom) contains the same *his⁻* strains and rat liver extract, but lacks the potential mutagen. When these control cells are plated on minimal medium, spontaneous *his⁺* revertants are selected. If the mutagen plate (top) has more revertants than the control plate (bottom), then the compound tested is a mutagen. If the test plate (center) has the same number of revertants as the control plate, then the tested compound is not a mutagen because it failed to generate more revertants than spontaneous mutations.

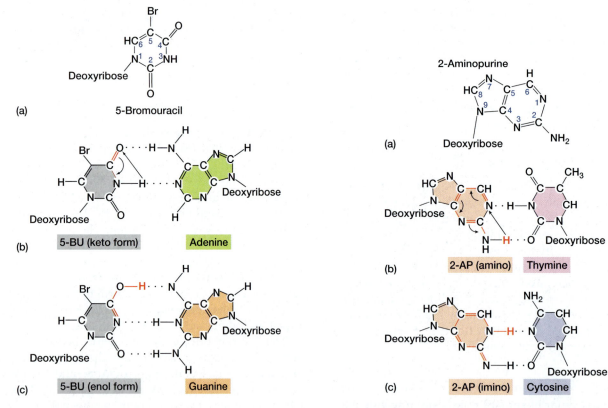

figure 18.12   Structure and base pairing with 5-bromouracil. (a) Structure of 5-bromouracil (5-BU). (b) In the preferred keto form, 5-bromouracil acts like thymine and pairs with adenine through two hydrogen bonds. (c) In the rare enol form, 5-bromouracil acts like cytosine and basepairs with guanine through three hydrogen bonds.

figure 18.13   Structure and base pairing with 2-aminopurine. (a) Structure of 2-aminopurine (2-AP). (b) In the preferred amino form, 2-aminopurine acts like adenine and basepairs with thymine through two hydrogen bonds. (c) In the rare imino form, 2-aminopurine acts like guanine and basepairs with cytosine through two hydrogen bonds.

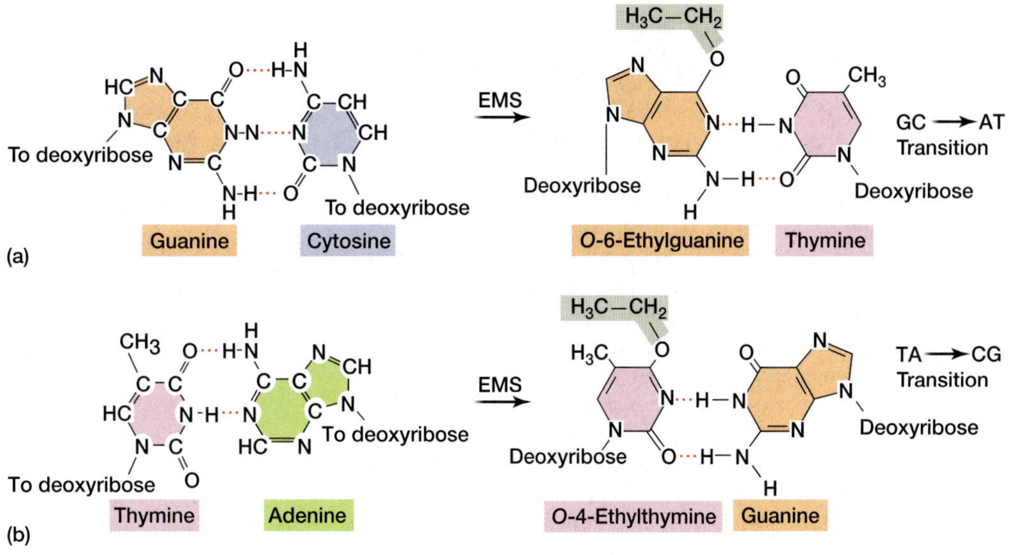

figure 18.14 Alkylation-induced mutations. (a) Ethyl methanesulfonate (EMS) can add an ethyl group to the oxygen at position 6 of guanine, which produces *O*-6-ethylguanine. The *O*-6-ethylguanine acts like adenine and basepairs with thymine to produce a GC to AT transition mutation. (b) EMS can also add an ethyl group to the oxygen at position 4 of thymine, which produces *O*-4-ethylthymine. The *O*-4-ethylthymine basepairs with guanine to produce a TA to CG transition.

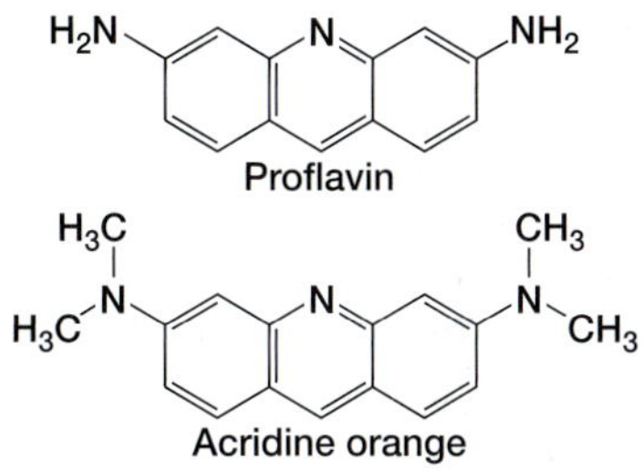

figure 18.15 Structure of two acridine dyes: proflavin and acridine orange.

Specifically, nitrous acid replaces amino groups on nucleotides with keto groups ($-NH_2$ to $=O$), which converts cytosine to uracil, adenine to hypoxanthine, and guanine to xanthine.

Like nitrous acid, heat can also deaminate cytosine to form uracil and thus bring about transitions (C–G to T–A). Apparently, heat can also bring about transversions by an unknown mechanism.

Two chemicals that can generate transversion mutations are ethyl methanesulfonate (EMS, $CH_3SO_3CH_2CH_3$) and ethyl ethanesulfonate (EES, $CH_3CH_2SO_3CH_2CH_3$). Both of these chemicals add an ethyl group to any of the four bases. However, mutations most often occur when the ethyl group is added to the oxygen at position 6 of guanine, which produces the *O*-6-ethylguanine, or to the oxygen at position 4 of thymine, which produces *O*-4-ethylthymine (fig. 18.14). The *O*-6-ethylguanine can base-pair with thymine to produce a G–C to A–T transition mutation, and the *O*-4-ethylthimine can base-pair with guanine to produce a T–A to C–G transition mutation. EMS is one of the most common chemicals used to induce mutations in *Drosophila* in research laboratories.

Insertions and Deletions

Insertions and deletions can be very strong mutations because the frameshift alters the ORF and changes multiple codons. Acridine dyes, such as proflavin and acridine orange (fig. 18.15), are flat molecules that often cause insertion and deletion mutations in the DNA. These dyes *intercalate*, or insert, themselves between the stacked bases of a DNA molecule. This intercalation causes the helix to buckle in the region of insertion, leading to the addition or loss of bases during DNA replication. Crick and Brenner used acridine-induced mutations to demonstrate both that the genetic code was read from a fixed point and that it was triplet (chapter 11).

It's Your Turn

As we discussed, there are a variety of different methods and chemicals to induce mutations. Each treatment, however, primarily generates a specific type of mutation through a particular mechanism. Problems 29, 37, and 38 explore the various choices of mutagen to induce specific types of mutations.

Misalignment Mutagenesis

Additions and deletions in DNA can also be produced by misalignment of a template strand and the newly synthesized (nontemplate) strand in a region containing a repeated sequence. For example, in **figure 18.16** we expect the nontemplate strand to contain six adjacent adenines because the template strand contains six adjacent thymines. Misalignment of the nontemplate strand results in seven consecutive adenines: six thymines replicated, plus one already replicated but misaligned. Misalignment of the template strand results in five consecutive adenines because one thymine is not available in the template. Regions with long runs of a particular base may be very prone to such mutation.

18.2 applying genetic principles to problems

Applying Mutagenic Mechanisms to Generate Nonsense Mutations

A geneticist who is trying to generate mutations in an organism, such as *Drosophila,* will often ask two initial questions. First, what is the most efficient mutagen that can be used in this organism? Second, what types of mutations am I interested in generating? Here we will examine the second question in more detail.

As we already discussed, a variety of different mutagens can be used to induce various types of mutations in the DNA. Let's look at a specific region of the ORF of a gene. The following sequence corresponds to the nontemplate strand of the DNA (see chapter 10). If the ORF corresponds to the first three nucleotides in the nucleotide sequence, which is 5'-GUU-3' the resulting amino acid sequence is located below the nucleotide sequence:

5′-GTTCCTTGCTGGAATAGGACAATGAGTTAGCCC-3′
  Val Pro Cys Trp Asn Arg Thr Met Ser stop

Based on this sequence, how could a transition, transversion, and frameshift mutation each produce a premature translational termination codon in the preceding sequence? Additionally, how would each of these mutations affect the encoded amino acid sequence?

To generate a simple nonsense mutation, we need to change a single codon into a translational termination codon. The possible codons in the mRNA (read 5'-to-3') correspond to: GUU, CCU, UGC, UGG, AAU, AGG, ACA, AUG, and AGU. The next codon on the mRNA, 5'-UAG-3', is the natural translational termination codon. How can we change one of these nine amino acid-encoding codons into a translational termination codon (UAA, UGA, or UAG)?

We could change the UGC codon to UGA by replacing the cytosine to adenine, which corresponds to a transversion. This can occur spontaneously by oxidizing the DNA. If the guanine on the template strand (5' GCA 3') is oxidized to 8-oxoguanine, it could basepair with adenine

to convert a G–C base pair to a T–A base pair after DNA replication (see fig. 18.10b).

We could also change the UGG codon to UGA by replacing the second guanine with adenine, a transition mutation. This could be done by treating the DNA with ethyl methanesulfonate, which alters guanine to *O*-6-ethylguanine (or thymine to *O*-4-ethylthymine). This causes a transition of a G–C base pair to A–T, with the adenine replacing the guanine (*O*-6-ethylguanine) as diagrammed in figure 18.14.

Unfortunately, with this type of chemically-induced mutagenesis, as with spontaneous mutations, we are unable to control which specific nucleotide is altered. However, the induced mutations will occur at a greater frequency than the spontaneous mutations. As we also see in this example, the EMS treatment will preferentially produce transition mutations, so the production of a transversion will be quite low.

Both of these mutations correspond to point mutations, in which a single nucleotide is altered. The resulting mutant gene will encode a protein that is the wild-type amino acid sequence up to the premature translational termination codon.

A final type of mutation that would produce the nonsense mutation is a frameshift mutation. We could produce this mutation by treating the organism (and its DNA) with either proflavin or acridine orange. An insertion of a single base in the third position of the previous sequence would produce the following nucleotide and amino acid sequence:

5′-GT**N**TCCTTGCTGGAATAGGTCAATGAGTTAGCCC-3′
  Val Ser Leu Leu Glu stop

This exercise should provide an example of some of the rationale used when trying to generate a mutation. The type of mutagen used will influence the types of mutations that will preferentially be generated.

Triplet Diseases

Misalignment mutagenesis appears to be a significant source of generating an abnormal phenotype for a particular class of diseases in humans. The **triplet diseases** occur when certain genes that contain a 3-nucleotide repeat exhibit an increase in the number of repeats. These repeats can be present anywhere within the gene.

For each gene, there is a range in the number of repeats that a normal individual possesses. In subsequent generations, the number of repeats can increase—and above a certain number (which varies for each gene/disease) the disease symptoms are exhibited. For example, Huntington disease is associated with a CAG repeat in the ORF. Normal individuals usually

have from 6 to 37 copies of the CAG sequence (and a corresponding number of adjacent glutamines in the protein). When the number of CAG triplets increases above 40, individuals begin to exhibit the Huntington disease neurodegeneration phenotype.

Features of Triplet Diseases

Most of the triplet diseases have two interesting features in common. First, they usually affect neurons, producing symptoms such as mental retardation or neurodegeneration. It is unclear whether neurons are more susceptible to proteins and RNAs that possess these repeats, or the genes containing these repeats are primarily expressed in neurons. Second, these dis-

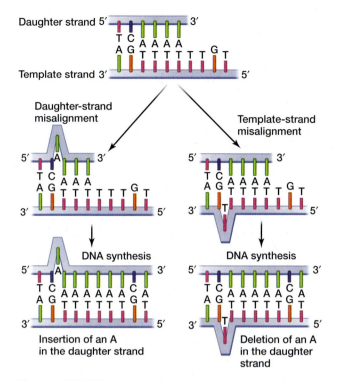

Daughter strand 5′   3′

Template strand 3′   5′

Daughter-strand misalignment

Template-strand misalignment

DNA synthesis

DNA synthesis

Insertion of an A in the daughter strand

Deletion of an A in the daughter strand

figure 18.16 Misalignment of a template or daughter strand during DNA replication. If the daughter strand becomes misaligned during DNA replication, the resulting daughter strand will have an additional base (left). If the template strand is misaligned during DNA replication, the resulting daughter strand will have a deleted base (right). These changes will show up after another round of DNA replication.

eases exhibit **anticipation,** which means the severity of the disease increases in subsequent generations due to the increasing number of triplet repeats. This characteristic is in contrast to many diseases that either exhibit approximately the same severity of symptoms in every generation, or the severity of symptoms fluctuates from one generation to the next.

Effects of Triplet Expansion

The triplet nucleotides associated with these diseases exert their effect through three different mechanisms, depending on the location of the triplet repeat.

Fragile-X syndrome is associated with a CCG repeat that is located upstream of the fragile-X mental retardation 1 gene (*FMR1*). Therefore, the CCG repeat is not present in the mRNA. Rather, the fragile-X syndrome results from the increased methylation of the second cytosine, which also corresponds to a CpG sequence. As the number of CCG repeats increase, the amount of CpG methylation increases. This results in silencing the expression of the *FMR1* gene, which encodes a possible component of the RNA-induced silencing complex (RISC) that is involved in miRNA-dependent suppression of gene expression (see chapters 10 and 17).

Huntington disease is associated with the expansion of a CAG repeat within the ORF. **Polyglutamine diseases** refer to the collection of diseases that result from an increase of either CAG or CAA repeats in the ORF, which produces an increasing polyglutamine stretch within the encoded proteins. It appears that the increased length of the polyglutamine region causes misfolding of the protein and its abnormal aggregation in the cell.

Myotonic dystrophy is associated with an expansion of a CUG repeat in the 3′ UTR of an mRNA encoding a protein kinase. A variety of experiments suggest that this trinucleotide expansion affects a variety of processes in the cell. First, CTG expansion is associated with hypermethylation of the adjacent genomic DNA, which suppresses the transcription of at least one gene in the area. Second, the CUG expansion in the mRNA appears to bind various transcription factors and RNA splicing factors, which represses the expression of several different mRNAs and prevents the generation of specific alternatively spliced mRNA isoforms.

Intergenic Suppression

When a mutation results in a mutant protein and phenotype, several routes can still produce a wild-type phenotype. We already discussed simple reversion (back mutation) and intragenic suppression in this chapter. A third route is through **intergenic suppression,** which is the restoration of the normal function of a mutated gene by changes in a different gene, called a **suppressor gene.** Suppressor genes usually encode either a tRNA or interacting protein. When mutated, intergenic suppressors change the way in which a codon is read.

Types of Mutations Suppressed

Suppressor genes can restore proper reading to missense, nonsense, and frameshift mutations. As we discussed earlier in this chapter, missense mutations change a codon from one amino acid to a different amino acid, while frameshift mutations alter the reading frame of codons by inserting or deleting bases.

Some missense mutations encode a mutant protein that is unable to interact with another protein in a multiprotein complex. The inability to form this complex results in the mutant phenotype. A suppressor may be a missense mutation in a second gene that encodes a different protein in the same multiprotein complex. In this scenario, the suppressor protein can interact with the first mutant protein to generate a functional multiprotein complex and restore the wild-type phenotype.

A frameshift mutation can be suppressed by a mutation in a tRNA that has four, rather than three, bases in the anticodon (**fig. 18.17a**). Binding of the suppressor tRNA to the mutant mRNA reads four bases as a codon and thus restores the original reading frame of the mRNA.

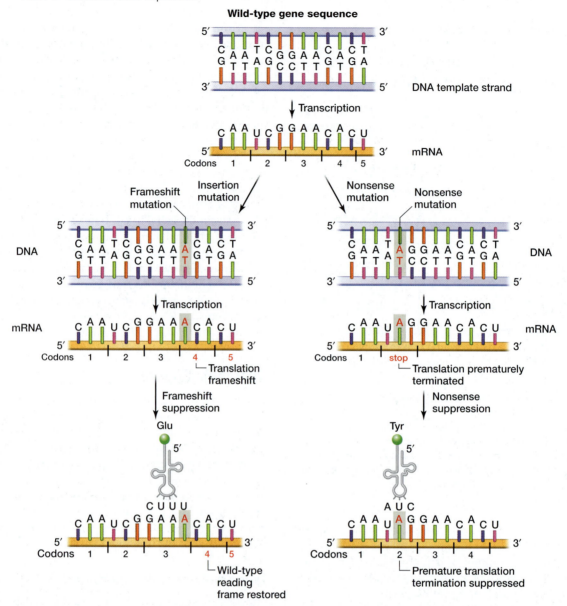

figure 18.17   Frameshift and nonsense suppression by mutant tRNAs. A portion of a wild-type gene's open reading frame and its mRNA are shown (top). An AT base pair (left) was inserted into the DNA (blue) to produce a frameshift mutation that affects the open reading frame beginning at codon 4. However, a mutant tRNA with four bases in the anticodon reads the inserted base as part of the third codon in the mRNA. The translation frameshift occurs and all the subsequent codons are translated in the wild-type reading frame. A nonsense mutation (right) results in premature translation termination. This UAG nonsense codon can be read by a mutant tyrosine tRNA, which is mutated in the anticodon sequence to be complementary to the UAG codon. This prevents premature translation termination and the subsequent codons are unaffected. The inserted nucleotide in the frameshift mutation and the base change in the nonsense mutation are shown in red.

A nonsense suppressor gene is a tRNA gene that has a mutation in the anticodon that is complementary to the nonsense codon. The suppressor tRNA therefore inserts an amino acid into the protein at the point that corresponds to the nonsense mutation in the mRNA.

Suppression of Amber Codon Mutants: An Example

E. coli contains at least three suppressors of the mutant *amber* stop codon (UAG). These suppressors insert tyrosine, glutamine, or serine into the protein chain at the point of an *amber* codon. Normally, the tyrosine tRNA has the anticodon 3'-AUG-5'. The tyrosine-charged suppressor tRNA that reads the *amber* termination codon (5'-UAG-3') as a tyrosine codon has the complementary anticodon 3'-AUC-5' (fig. 18.17).

If the *amber* nonsense codon is no longer read as a stop signal, then won't all the genes that normally terminate in an *amber* codon fail to properly terminate translation and produce a large number of different mutant proteins? In the tyrosine case, two genes for tyrosine tRNA exist; one contributes the major fraction of the tRNAs, and the other, the minor fraction. The minor-fraction tyrosine tRNA gene is the one that mutates into the suppressor. As a result, the *amber* codon is suppressed only a small percentage of the time. Wild-type mRNAs that naturally contain an *amber* termination codon are usually terminated properly.

Although the *amber* mutations also usually terminate prematurely at the mutant codon, a small percentage are suppressed and express the wild-type-length protein. This wild-type sized protein is often present in sufficient amounts to restore the wild-type phenotype to the cell.

In general, intergenic suppressor mutants would be eliminated quickly in nature because they are inefficient—the cells are not healthy. In the laboratory, geneticists can provide artificial conditions that allow them to be grown and studied.

Mutator and Antimutator Mutations

Whereas intergenic suppressors represent mutations that "restore" the normal phenotype, **mutator** and **antimutator mutations** cause an increase or decrease in the overall mutation rate of the cell. They are frequently mutations of DNA polymerase, which, as you remember, not only polymerizes DNA nucleotides in a $5' \rightarrow 3'$ direction, but also checks to be sure that the correct base was incorporated using a $3' \rightarrow 5'$ exonuclease (proofreading) activity. Mutator and antimutator mutations sometimes involve changes in this exonuclease activity.

Phage T4 has its own DNA polymerase with known mutator and antimutator mutations. Mutator mutants are very poor proofreaders, and thus they permit mutations to remain in the replicated DNA throughout the phage genome. Antimutator mutants, however, have exceptionally efficient proofreading ability, and therefore a very low mutation rate is observed for the whole genome.

Restating the Concepts

▶ Transitions are mutations of a purine to the other purine or a pyrimidine to another pyrimidine. Transversions are mutations of purine to pyrimidine or vice versa. Frameshift mutations result from either insertion or deletion of one or more bases.

▶ A spontaneous transition is produced by a tautomeric shift of a base. A spontaneous transversion mutation is generated either by a tautomeric shift of one base and a rotation of the other base to the *syn* configuration, or by the oxidation of guanine to 8-oxoguanine.

▶ Transitions can be induced by base analogs, by compounds that deaminate bases, or by compounds that alkylate bases.

▶ Insertions and deletions can be induced by intercalating agents. They also occur spontaneously,

especially in regions with a long run of a repeating sequence. An increase in the number of trinucleotide repeats is a cause of several human diseases.

▶ Suppressor tRNAs are mutations in tRNA genes that can correct for missense mutations, nonsense mutations, and frameshift mutations.

▶ Mutator and antimutator mutations modify the $3' \rightarrow 5'$ exonuclease (proofreading) function of a DNA polymerase.

18.4 DNA Repair

Radiation, chemical mutagens, heat, enzymatic errors, and spontaneous decay constantly damage DNA. For example, it is estimated that several thousand DNA bases are mutated each day in every mammalian cell due to spontaneous decay.

In the long evolutionary challenge to minimize mutation, cells have evolved numerous mechanisms to repair damaged or incorrectly replicated DNA. Many enzymes, acting alone or in concert with other enzymes, repair DNA. Repair systems are generally placed in four broad categories: damage reversal, excision repair, double-strand break repair, and postreplicative repair.

Enzymes that process repair steps have been conserved during evolution. That is, enzymes found in *E. coli* have homologs in yeast, fruit flies, and humans. However, eukaryotic systems are almost always more complex.

Damage Reversal

One mechanism for repairing mutated DNA is through **damage reversal,** which attempts to directly restore the original DNA sequence. Cells can perform damage reversal on only specific types of mutations, such as pyrimidine dimers and alkylated bases.

Repair of Dimerization

Ultraviolet (UV) light causes cross-linking, or **dimerization,** between adjacent pyrimidines in DNA. For example, two adjacent thymines can form a *cyclobutane ring* that incorporates two carbons from each thymine (**fig. 18.18a**). This cyclobutane ring distorts the DNA, resulting in a bulge in the phosphodiester backbone that is linked to the dimer and a failure of the thymines to basepair with their complementary adenines.

Although thymine–thymine dimers are the principal products of UV irradiation, cytosine–cytosine and cytosine–thymine dimers are occasionally produced. When cytosine–cytosine or cytosine–thymine dimers form, they generate a 6–4 product between the carbon-6 of one pyrimidine and the carbon-4 of the other pyrimidine (fig. 18.18b). This 6-4 product also

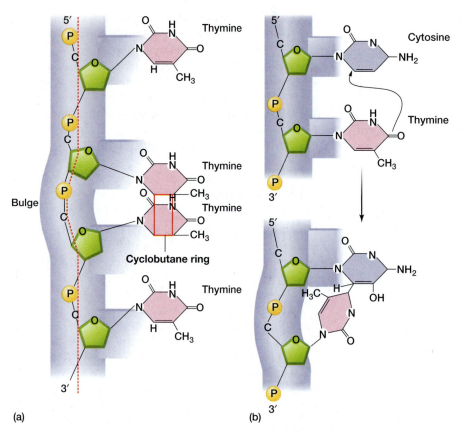

figure 18.18   UV-induced pyrimidine dimers in DNA. (a) UV light induces dimerization of adjacent thymines, which is represented by the formation of a cyclobutane ring (solid red lines). The red dashed line represents the bulge in the phosphodiester backbone that results from the formation of the cyclobutane ring. (b) UV light can also generate a cytosine-thymine 6-4 product between carbon-6 of cytosine and carbon-4 of thymine.

causes a disruption in the phosphodiester backbone and the inability of the pyrimidines to base-pair with the complementary purines on the other strand. These pyrimidine dimers can be repaired in several different ways. The simplest is to reverse the dimerization process and restore the original unlinked thymines.

In *E. coli*, an enzyme called DNA photolyase, the product of the *phr* gene (for **photoreactivation**), binds to dimerized thymines (**fig. 18.19**). When light shines on the cell, the enzyme is activated by absorbing light in the UV-to-blue range. The activated photolyase breaks the cyclobutane bonds between the thymines. The enzyme then falls free of the DNA. Photolyase exists in a wide range of bacterial and eukaryotic organisms, but it has not been identified in mammals.

Repair of Alkylation

Another example of an enzyme that directly repairs DNA damage is O^6-methylguanine DNA methyltransferase. This enzyme repairs the major damage caused by DNA-alkylating agents. For example, it can accept

either the methyl group from O^6-6-methylguanine (**fig. 18.20a**) or the ethyl group that is added to O^6-6-ethylguanine by EMS (see earlier discussion). This enzyme catalyzes the transfer of the methyl or ethyl group from the base to the sulfur atom on a cysteine in the enzyme (fig. 18.20b). This results in the covalent modification and inactivation of the methyltransferase enzyme. Thus, this enzyme is classified as a *suicide enzyme* because once it completes a transfer it is inactivated and unable to perform a subsequent transfer.

Excision Repair

Excision repair is the general mechanism that *removes* the damaged portion of a DNA molecule, rather than repairing the damage. Various enzymes sense the damage or distortion in the DNA double helix and then remove the bases and nucleotides from the damaged strand. The gap is then repaired using complementarity with the remaining strand.

We can broadly categorize these systems as base excision repair and nucleotide excision repair, which includes mismatch repair. Although many repair pathways exist, we will discuss only the major repair mechanisms here. Presumably, selection for redundancy in repair has occurred because of the critical need to keep DNA intact and relatively mutation-free.

Base Excision Repair

Base excision repair involves the removal of the damaged base, followed by nicking the phosphodiester backbone, which then allows DNA polymerase to add nucleotides in a 5' → 3' direction. The base is removed from a nucleotide within DNA by a **DNA glycosylase,** which is an enzyme that senses damaged bases and removes them. Currently, at least five DNA glycosylases are known. For example, uracil-DNA glycosylase, the product of the *ung* gene in *E. coli*, recognizes uracil within DNA (where it does not belong) and cleaves it out at the glycosidic bond.

The resulting site is called an AP (apurinic or apyrimidinic) site, because it lacks either a purine or

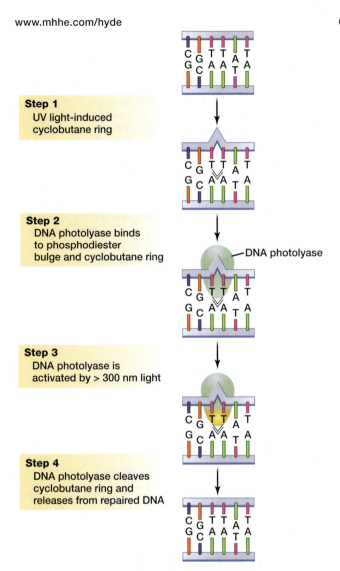

Step 1
UV light-induced
cyclobutane ring

Step 2
DNA photolyase binds
to phosphodiester
bulge and cyclobutane ring

— DNA photolyase

Step 3
DNA photolyase is
activated by > 300 nm light

Step 4
DNA photolyase cleaves
cyclobutane ring and
releases from repaired DNA

figure 18.19   DNA photolyase repairs thymine dimers. UV light induces a cyclobutane (TT) dimer between adjacent thymines. DNA photolyase recognizes the bulge in the phosphodiester backbone and binds to the cyclobutane dimer. Absorption of light in the UV to blue range activates the DNA photolyase, which cleaves the cyclobutane bonds to restore the two thymines and then releases the DNA.

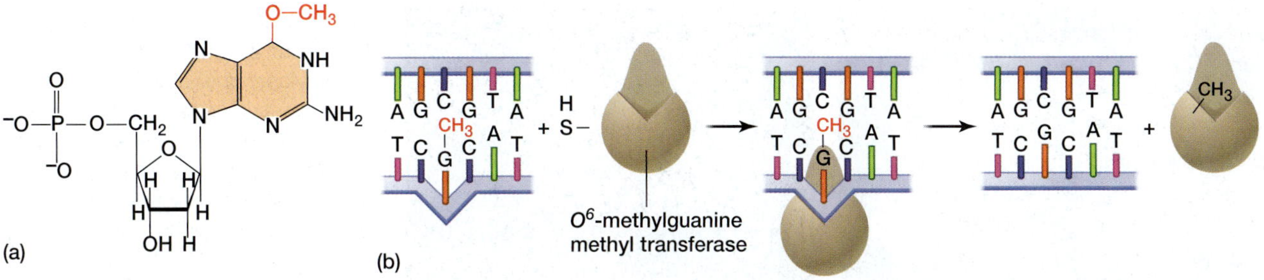

pyrimidine at the site (**fig. 18.21**). An **AP endonuclease** then senses the minor distortion of the DNA double helix and nicks the DNA at the 5' side of the base-free AP site. A DNA polymerase then inserts a nucleotide at the 3' end of the nicked strand (which corresponds to the AP site). The replacement of just one base occurs 80–90% of the time, with several bases being replaced in 10–20% of cases (see fig. 18.21). The number of replaced nucleotides probably depends on which DNA polymerase (I or III) first repairs the site.

An exonuclease, lyase, or phosphodiesterase enzyme then removes either the base-free nucleotide or the single-stranded DNA. (Lyases are enzymes that can break C–C, C–O, and C–N bonds.) DNA ligase then catalyzes the formation of a phosphodiester bond to seal the nick (see fig. 18.21). In mammals, DNA polymerase β performs two roles in base excision repair: It both inserts a new base at the AP site and eliminates the AP nucleotide residue by exonuclease activity.

One question that concerned scientists was how the glycosylases gain access to the inappropriate or damaged bases within the double helix. Recently it was shown that these enzymes remove the bases by first flipping them out of the interior of the double helix in a process called **base flipping.** For example, the enzyme in humans that recognizes 8-oxoguanine in DNA (see fig. 18.10), 8-oxoguanine DNA glycosylase, flips the base out to excise it (**fig. 18.22**). Base flipping seems to be a common mechanism in repair enzymes that need access to bases within the double helix.

Nucleotide Excision Repair

Whereas base excision repair usually involves the replacement of only one nucleotide residue, **nucleotide excision repair** excises and replaces a short stretch of nucleotides.

As an example, six enzymes in *E. coli* excise a short stretch of DNA containing thymine dimers if the dimerization is not reversed by photoreactivation. Two

(a)   (b) O^6-methylguanine
                                                                                                        methyl transferase

figure 18.20   Methylation of guanine and its repair. (a) The structure of O^6-methylguanine. The methylation of guanine occurs on the oxygen (red), which is normally a double-bonded oxygen (keto form, fig. 18.6). (b) Repair of O^6-methylguanine involves the binding of O^6-methylguanine methyl transferase to an O^6-methylguanine. The methyl group on the guanine is transferred to the sulfur atom in a cysteine in the protein. While the DNA sequence is restored, the enzyme is covalently modified and unable to perform any subsequent methyl transfers.

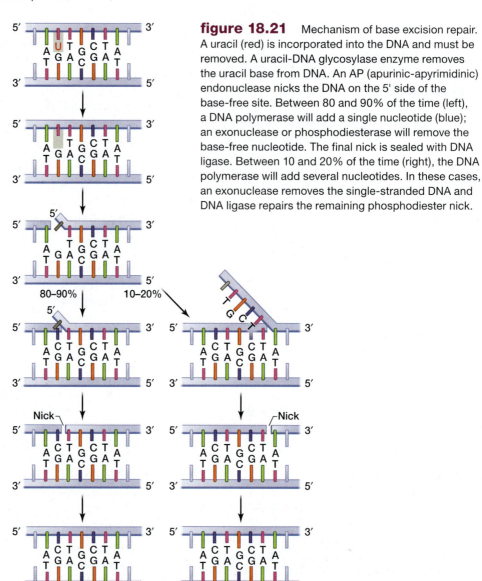

Step 1

Uracil-DNA glycosylase removes uracil base

Step 2

AP endonuclease cleaves phosphodiester backbone on the 5′ side of AP site

Step 3

DNA polymerase adds one (left) or more bases (right)

Step 4

Single-stranded DNA is cleaved off

Step 5

DNA ligase seals phosphodiester bond (nick)

figure 18.21 Mechanism of base excision repair. A uracil (red) is incorporated into the DNA and must be removed. A uracil-DNA glycosylase enzyme removes the uracil base from DNA. An AP (apurinic-apyrimidinic) endonuclease nicks the DNA on the 5′ side of the base-free site. Between 80 and 90% of the time (left), a DNA polymerase will add a single nucleotide (blue); an exonuclease or phosphodiesterase will remove the base-free nucleotide. The final nick is sealed with DNA ligase. Between 10 and 20% of the time (right), the DNA polymerase will add several nucleotides. In these cases, an exonuclease removes the single-stranded DNA and DNA ligase repairs the remaining phosphodiester nick.

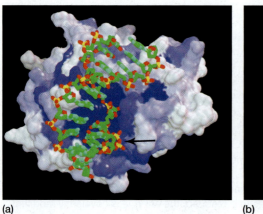
(a)

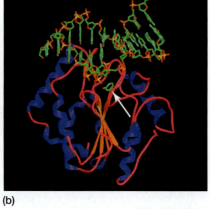
(b)

figure 18.22 Two views of the human enzyme, uracil-DNA glycosylase, bound to DNA and the uracil-containing residue flipped out of the DNA before cleavage. (a) The enzyme is shown as a molecular surface, with the flipped uracil-containing nucleotide marked with the arrow. (b) The uracil-containing nucleotide residue in the DNA has been flipped out and the uracil cleaved. Note the cleaved uracil is bound to the protein (arrow). In both panels, DNA is a green stick figure with oxygen being red, phosphorus being yellow, and nitrogen atoms being blue.

molecules of the UvrA protein (UV-repair-A) combine with one molecule of the UvrB protein to form a UvrA$_2$–UvrB complex that moves along the DNA, looking for damage (**fig. 18.23**). (The complex has 5'-to-3' helicase activity.) When the complex finds damage such as a thymine dimer, the UvrA$_2$ dimer dissociates, and the UvrB subunit remains bound to the region near the lesion. This causes the DNA to bend and attracts the UvrC protein.

The UvrB subunit first nicks the phosphodiester backbone 4–5 nucleotides on the 3' side of the lesion. The UvrC subunit then nicks the DNA 8 nucleotides on the 5' side of the lesion. (The three components, UvrA, UvrB, and UvrC, are together called the *ABC excinuclease,* for excision endonuclease.) The UvrD helicase removes the 14- to 15-base oligonucleotide and UvrC. DNA polymerase I fills in the gap and removes UvrB. DNA ligase finally seals the remaining nick (see fig. 18.23).

Like base excision repair, nucleotide excision repair is present in all organisms. In yeast, approximately 12 proteins are involved, many in what is called the RAD3 gene group. In humans, 25 proteins are involved. In contrast to the 14–15 nucleotides removed in *E. coli,* the human nucleotide excision repair system removes 27–29 nucleotides.

Transcription and nucleotide excision repair are linked in eukaryotes. For example, transcription factor TFIIH, which possesses a helicase activity like UvrD mentioned earlier, is involved in both transcription and repair of UV damage (see chapter 10). We know that actively transcribed genes are preferentially repaired. It is likely that when transcription is blocked by a DNA lesion like a thymine dimer, the formation of a repair complex is induced. This eukaryotic *repairosome* is a protein complex that contains several proteins that are involved in transcription (like TFIIH) and the

figure 18.23 Nucleotide excision repair. UV light induces a cyclobutane thymine dimer (red) in the DNA. The thymine dimer is recognized by a protein complex made of two copies of UvrA (purple) and one of UvrB (green). After binding the DNA, the UvrA subunits detach, and UvrC (orange) attaches on the 5' side of the lesion. UvrB nicks the DNA on the 3' side and UvrC on the 5' side of the lesion; UvrD helicase (blue) unwinds the DNA and releases the oligonucleotide containing the lesion and UvrC. DNA polymerase I (brown) fills in the single-stranded gap and releases UvrB. DNA ligase (red) binds and seals the phosphodiester nick.

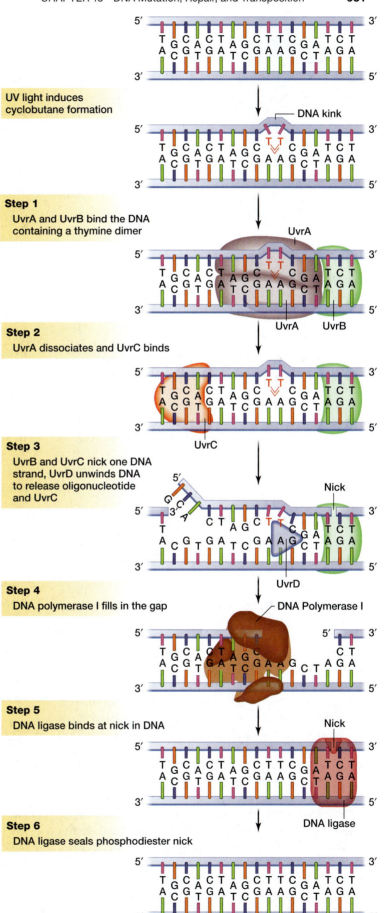

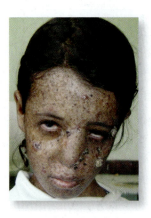

figure 18.24 A person with *xeroderma pigmentosum* exhibits an unusually high number of freckles when exposed to sunlight. The freckles develop into severe skin lesions as the disease progresses and the symptoms worsen.

DNA repair enzymes. Researchers believe that when eukaryotic RNA polymerase II encounters a DNA lesion, it backs up until the lesion is repaired, without losing the RNA transcript.

In humans, the autosomal recessive disease *xeroderma pigmentosum* is caused by an inability to repair thymine dimers induced by UV light. Persons with this disease freckle heavily when exposed to the UV rays of the sun (**fig. 18.24**), and they have a high incidence of skin cancer. Seven different complementation groups (loci *XPA–XPG*) each cause the xeroderma pigmentosum phenotype. These seven complementation groups encode different proteins that are involved in the first steps of nucleotide excision repair. Consistent with the previous discussion, the *XPD* complementation group encodes a component of TFIIH.

Mismatch Repair

In base excision repair, the damaged base is removed by glycosylases, and usually only one nucleotide residue is replaced. In some circumstances, however, the mutation is not a damaged base, but rather simply a mismatched base pair. **Mismatch repair** encompasses about 99% of all DNA repairs.

As DNA polymerase replicates DNA, some errors are made that the 3' → 5' exonuclease proofreading activity does not correct. It is thought that *E. coli* DNA polymerase III incorporates the wrong base approximately once every 100,000 bases. The 3' → 5' proofreading exonuclease activity corrects approximately 99% of those errors. Thus, DNA polymerase III has a mutation rate of 1 in every 10^7 bases replicated. The mismatch repair system, which follows behind the replication fork, recognizes these mismatched errors.

The *E. coli* mismatch repair system consists of proteins encoded by the *mutH, mutL, mutS,* and *mutU* genes, which are responsible for the removal of the incorrect base by an excision repair process. (The genes are called *mut* for mutator because mutations of these genes cause high levels of spontaneous mutation. The *mutU* gene is also known as *uvrD,* which we described under nucleotide excision repair.)

You might wonder how the mismatch repair system recognizes which base is the wrong one. After all, in a mismatch, there are no defective bases, so either partner in the complementary base pair could be the "wrong" base. The cell machinery assumes that the parental DNA strand is correct, and that the error occurred during the last round of DNA replication in the daughter strand.

But how does the cell recognize the daughter strand? In *E. coli,* the answer lies in the methylation state of the DNA. DNA methylase, the product of the *dam* locus, methylates the adenine in 5'-GATC-3' sequences, which are relatively common sequences in the *E. coli* DNA. Because the mismatch repair enzymes follow the DNA replication fork, they usually reach the mismatch site before the methylase does. Template strands were already methylated in the previous round of replication, whereas newly synthesized daughter strands are not (**fig. 18.25**). Put simply, the methylation state of the DNA cues the mismatch repair enzymes to eliminate the daughter-strand base for repair. After the methylase passes by, both strands of the DNA are methylated, and the methylation cue is gone.

A Model of Mismatch Repair in E. coli

In **figure 18.26**, we present one model of mismatch repair. The MutS homodimer detects and binds the mismatched base pair. MutL, also in the form of a homodimer, binds at or near the methylation signal. MutL also activates the endonuclease MutH, which then nicks the *unmethylated* strand at the 3'-CTAG-5' recognition site, which can be one thousand to two thousand bases away from the mismatch.

At the recognition site, the MutS–MutL tetramer loads the MutU helicase (UvrD), which unwinds the nicked strand. Any one of at least four different exonucleases degrades the unwound oligonucleotide. DNA polymerase III then repairs the gap, and DNA ligase seals the final phosphodiester nick.

This sequence of events highlights a common theme in DNA repair: Once a lesion is found, the damaged DNA has a protein bound to it until the repair is finished.

Double-Strand Break Repair

Some damage to DNA, such as that caused by ionizing radiation, is capable of breaking both strands of the double helix. When that happens, the cell uses one of two mechanisms to repair the broken ends: It can simply bring the ends back together (a process called *non-homologous end-joining*), or it can use a mechanism that relies on the nucleotide sequences of a homologous piece of DNA, such as a sister chromatid or a homologous chromosome. That method is called *homology-directed recombination.*

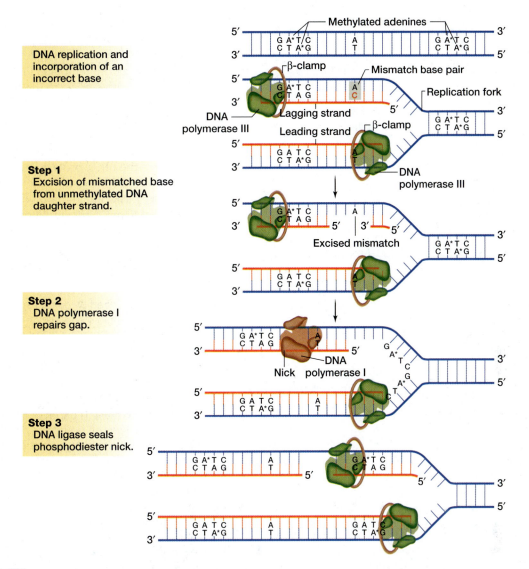

figure 18.25 Nucleotide excision repair of mismatched base pairs. The mismatch repair system follows the replisome at the DNA replication fork. When a mismatched base is encountered, the repair system determines the DNA template strand based on the methylated adenine (A*) in the 5′ GATC 3′ sequence. The base on the unmethylated daughter strand is then removed. This process preserves the base in the parental DNA strand in mismatch repair. DNA polymerase I replaces the removed bases and DNA ligase seals the final phosphodiester nick.

Nonhomologous End-Joining

In nonhomologous end-joining, a protein called Ku, a heterodimer of Ku70 and Ku80, binds the broken chromosomal ends and recruits a protein kinase (**fig. 18.27**). The complex trims the single-stranded DNA ends to produce blunt ends that can be properly ligated by DNA ligase IV. No particular sequence information is used in the trimming or ligation reactions, and if more than two broken ends are present, incorrect attachments can take place (such as translocations).

Although the trimming can result in the loss of DNA sequences, the cell has a better chance of surviving if some DNA is lost rather than an entire portion of the chromosome because of the double-strand break.

Homology-Directed Recombination

In contrast to nonhomologous end-joining, the homology-directed recombination mechanism usually restores the entire DNA sequence by using the sister chromatid sequence. The double-stranded break is recognized by several enzymes, including an exonuclease that degrades the DNA in a 5′ → 3′ direction (**fig. 18.28**). This produces a single-stranded region at the 3′ end that is bound by several proteins, including RAD51. RAD51 is related to the *E. coli* RecA protein, which is also involved in DNA repair. The fundamental importance of RAD51 is seen by its presence in all eukaryotes.

RAD51 protein coats the single-stranded DNA to form a helical filament, which then invades the

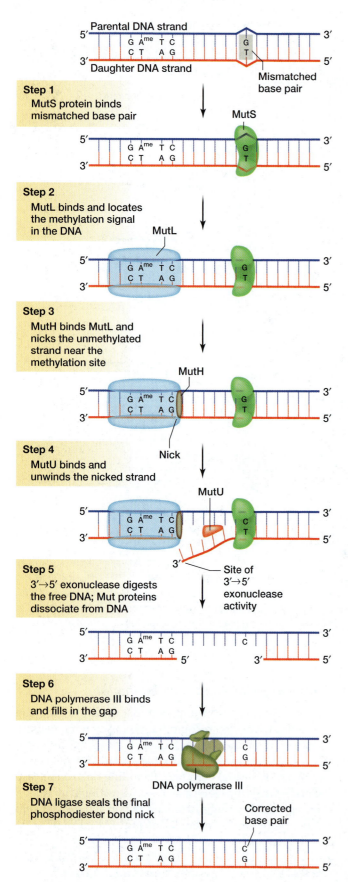

Parental DNA strand
5′ G A^{me} T C ... G ... 3′
   C T   A G ... T ...
3′ **Daughter DNA strand** 5′

Mismatched base pair

Step 1
MutS protein binds mismatched base pair

MutS

Step 2
MutL binds and locates the methylation signal in the DNA

MutL

Step 3
MutH binds MutL and nicks the unmethylated strand near the methylation site

MutH

Nick

Step 4
MutU binds and unwinds the nicked strand

MutU

3′
Site of 3′→5′ exonuclease activity

Step 5
3′→5′ exonuclease digests the free DNA; Mut proteins dissociate from DNA

Step 6
DNA polymerase III binds and fills in the gap

DNA polymerase III

Step 7
DNA ligase seals the final phosphodiester bond nick

Corrected base pair

complementary sequence in the sister chromatid. A heteroduplex region is formed, and DNA polymerase synthesizes new DNA from the broken 3′ end using the sister chromatid as the template. (You may notice that this structure is very similar to the recombination structure used to generate the Holliday structure described in chapter 6.)

DNA polymerase can then utilize either this new strand or the other strand of the sister chromatid as a template to restore the other broken end. In this manner, the sister chromatid provides the missing information to restore the double-stranded broken DNA without any loss of genetic information.

Restating the Concepts

▶ Mutations can be repaired by damage reversal, which involves an enzyme-mediated removal of the damage from the base without removing the base.

▶ Base excision repair involves the removal of the damaged base to generate an AP site; the AP site is then nicked, and the correct nucleotide is incorporated by DNA polymerase.

▶ Nucleotide excision repair is performed on short damaged regions, such as pyrimidine dimers. The damaged strand is nicked and a short sequence of nucleotides is removed. The remaining strand serves as a template for DNA polymerase to replace the missing nucleotides.

▶ Mismatch repair removes a mismatched nucleotide from the newly synthesized daughter strand. The template (parental) strand is recognized in *E. coli* by the methylation of the adenine in 5′-GATC-3′.

▶ Double-stranded DNA breaks can be repaired by a nonhomologous end-joining mechanism, which results in the loss of some sequence information, or by a homology-directed recombination process, which uses the RAD51 protein and the complementary sequence of the sister chromatid as a template to restore DNA sequence.

figure 18.26   Mismatch repair of recently replicated DNA. The MutS protein dimer binds mismatched base pairs; the MutL protein dimer binds the 5′-GA^{me}TC-3′ sequence. The MutH endonuclease binds MutL and nicks the daughter strand at the 3′-CTAG-5′ sequence. MutU helicase unwinds the nicked DNA strand containing the mismatched base. Exonuclease digestion cleaves the single-stranded DNA. DNA polymerase III fills in the gap and DNA ligase seals the final phosphodiester bond nick.

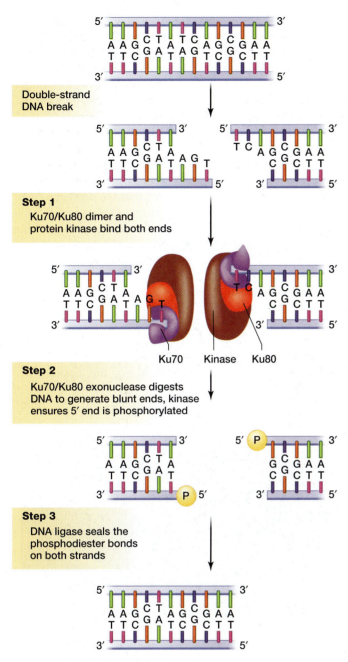

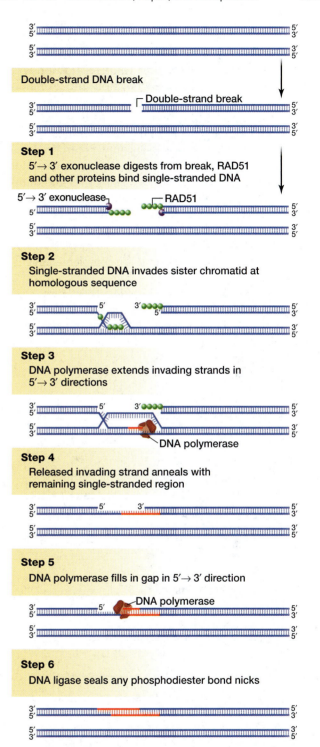

figure 18.27   Non-homologous end-joining of a double-stranded break. A double-strand break occurs in the DNA. The Ku70/Ku80 dimer and a protein kinase bind the ends of the DNA. The Ku70/Ku80 nuclease digests the single-stranded ends to generate blunt ends, while the kinase ensures the 5' end contains a phosphate. DNA ligase binds and seals the phosphodiester bonds on both DNA strands. Notice that this repair mechanism results in the loss of some DNA sequence by the exonuclease activity (3 base pairs in this example relative to the original sequence).

figure 18.28   Homology-directed recombination repair of a double-strand break. The double-strand break is digested by an exonuclease that removes bases in a 5' → 3' direction, which leaves single-stranded DNA at the 3' end. Several proteins, including RAD51, bind the single-stranded DNA, which then invades the complementary sequence in the sister chromatid. The sister chromatid sequence is used by DNA polymerase as a template to synthesize a DNA strand that traverses the break. This newly synthesized strand can then be used as a template by DNA polymerase to complete the double-stranded DNA synthesis across the break. DNA ligase is required to seal the final phosphodiester nick.

Recombination Repair

Recombination repair is a critical process that occurs if damaged DNA persists after DNA replication. We already discussed photoreactivation and excision repair as two postreplicative repair mechanisms. A third class of postreplication repair utilizes recombination to repair gaps in one of the DNA strands.

When DNA polymerase III encounters certain types of damage in *E. coli,* such as thymine dimers, it cannot proceed. Instead, the polymerase skips over the damage, leaving a gap, and resumes replication as far as 800 or more bases away (**fig. 18.29**). This gap cannot simply be filled in by DNA polymerase because the thymine dimer remains and DNA polymerase is unable to properly incorporate nucleotides that are complementary to this dimer.

While one daughter DNA molecule contains this gap opposite the thymine dimer, the second daughter DNA molecule is completely double stranded. Thus, the second DNA molecule contains an undamaged copy of the gap region that exists on the first daughter DNA molecule (fig. 18.29). A group of enzymes, one of which is specified by the *recA* locus and has central importance, repairs the gap. Because the repair takes place at a gap created by the failure of DNA replication, the process is called **postreplicative gap repair.**

The *recA* locus was originally discovered due to its role in the recombination process. Thus, RecA-mediated repair is sometimes called recombinational repair because it shares many enzymes and mechanisms with recombination. The eukaryotic RAD51 enzyme that was required for homology-directed recombination repair of double-stranded DNA breaks is also related to RecA, which is consistent with the repair mechanism being similar to the recombination models.

The RecA protein coats single-stranded DNA (**fig. 18.30a**) and assists the single-stranded DNA to invade double-stranded DNA (fig. 18.30b). By invasion, we mean that the single-stranded DNA displaces one strand of double-stranded DNA to base-pair with the complementary strand. A mechanism for this activity, assuming two sites on the RecA enzyme, appears in **figure 18.31**.

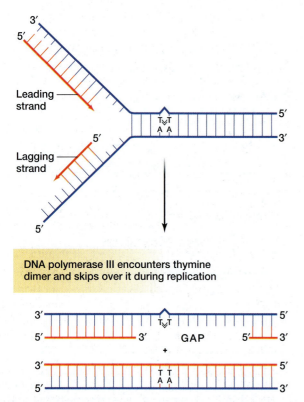

figure 18.29 Effect of a thymine dimer on DNA polymerase III. As DNA polymerase III is synthesizing the leading strand, it encounters a thymine dimer on the DNA template strand (blue). Because DNA polymerase III cannot incorporate nucleotides in the leading strand that will basepair with the thymine dimer, it skips a region and begins the leading strand again. This results in a gap, that consists of a single-stranded region containing the thymine dimer, in the final replicated DNA. The lagging strand, on the other hand, can replicate across the corresponding two adenines.

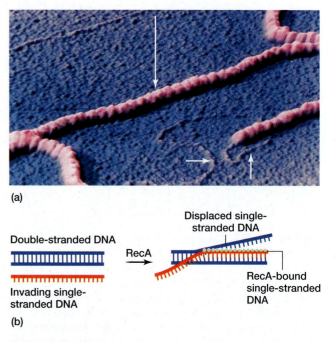

figure 18.30 Role of RecA protein in *E. coli* postreplicative DNA repair. (a) Scanning tunneling microscope image of single-stranded DNA coated with RecA protein (large arrow). The small arrows indicate double-stranded DNA that is not coated with RecA protein. (b) Diagrammatic representation of the invasion of the RecA-coated single-stranded DNA. The single-stranded DNA is red, the double-stranded and displaced single-stranded DNAs are blue, and the RecA protein is depicted as the green balls.

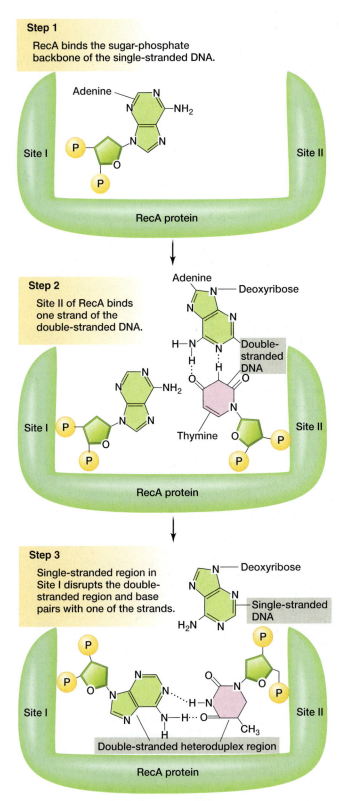

Step 1

RecA binds the sugar-phosphate backbone of the single-stranded DNA.

Adenine

Site I   RecA protein   Site II

Step 2

Site II of RecA binds one strand of the double-stranded DNA.

Adenine — Deoxyribose

Double-stranded DNA

Thymine

Site I   RecA protein   Site II

Step 3

Single-stranded region in Site I disrupts the double-stranded region and base pairs with one of the strands.

Deoxyribose

Single-stranded DNA

Double-stranded heteroduplex region

Site I   RecA protein   Site II

figure 18.31   Model of RecA-mediated single-stranded invasion. The RecA protein (green) binds the sugar-phosphate backbone of single-stranded DNA. One strand of the double-stranded DNA is bound to site II of RecA. When both sites are filled, RecA rotates the bases so that the single-stranded DNA (adenine) forms a complementary base pair with the thymine in the double-stranded DNA. This rotation and new base pairing displaces the other strand of the original duplex and results in the single strand invasion of the double-stranded DNA (see fig. 18.32).

RecA continues to move the single-stranded DNA along the double-stranded DNA until a region of complementarity is found.

The RecA protein is responsible for filling a postreplicative gap in newly replicated DNA with a strand from the undamaged sister duplex. **Figure 18.32** shows a replication fork with a gap in the progeny strand in the region of a thymine dimer. The RecA protein binds to the damaged single strand to form a filament that invades the sister duplex (fig. 18.32). Endonuclease activity then frees the double helix containing the thymine dimer (fig. 18.32). DNA polymerase I and DNA ligase return both daughter helices to the intact state (fig. 18.32). The thymine dimer still exists, but now its duplex is intact, and another cell cycle is available for photoreactivation or excision repair to remove the dimer.

The SOS Response

In *E. coli*, the final postreplicative repair mechanism is part of a reaction called the **SOS response.** When an *E. coli* cell is exposed to excessive quantities of UV light, other mutagens, or agents that damage DNA, or when DNA replication is inhibited, gaps are created in the DNA. When RecA interacts with single-stranded DNA, a protease activity is stimulated that normally is silent in the RecA protein. This RecA* protease activity has the ability to cleave specific proteins, one of which is the LexA protein.

The LexA protein normally represses about 18 genes, including itself. The other genes include *recA*, *uvrA*, *uvrB*, and *uvrD*; two genes that inhibit cell division, *sulA* and *sulB*; and several others. Each of these genes has a consensus sequence in its promoter called the **SOS box:** 5'-CTGX$_{10}$CAG-3' (where X$_{10}$ refers to any 10 bases), to which the LexA protein normally binds to repress transcription of these genes.

When the RecA* activity cleaves the LexA protein, the LexA protein fragments are released from the SOS box and all of the normally repressed genes are transcribed (**fig. 18.33**). The two inhibitors of cell division, the SulA and SulB proteins, presumably increase the amount of time the cell has to repair the damage before the next round of DNA replication.

Several of these other derepressed genes were known to encode proteins that facilitated the replication of DNA with lesions. These proteins were thought to interact with DNA polymerase to allow the damaged DNA to be used as a template. We now know that these proteins are, in fact, polymerases that can replicate damaged DNA. In *E. coli*, polymerase V can copy damaged DNA. In yeast, polymerases η and ζ, also called REV3/7 and RAD30 polymerases, respectively, can also copy damaged DNA.

These SOS repair polymerases have a higher error rate during DNA replication relative to the standard

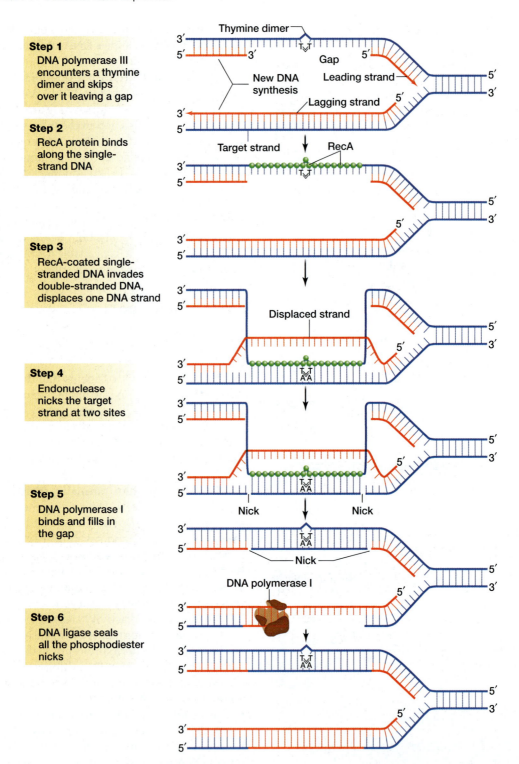

Step 1
DNA polymerase III encounters a thymine dimer and skips over it leaving a gap

Step 2
RecA protein binds along the single-strand DNA

Step 3
RecA-coated single-stranded DNA invades double-stranded DNA, displaces one DNA strand

Step 4
Endonuclease nicks the target strand at two sites

Step 5
DNA polymerase I binds and fills in the gap

Step 6
DNA ligase seals all the phosphodiester nicks

figure 18.32   RecA-dependent postreplicative DNA repair. DNA polymerase III skips past a thymine dimer on the leading strand during DNA replication. The RecA protein (green) coats the single-strand DNA containing the thymine dimer, which then invades the sister duplex. One DNA strand is displaced so the invading strand with the thymine dimer can base pair with its complementary target strand. An endonuclease nicks the target strand at either side of the thymine dimer site. These nicks free the new duplex with the thymine dimer and leaves a single-stranded gap in the sister duplex. DNA polymerase I (brown) fills any gaps in both molecules and DNA ligase seals any nicks in the phosphodiester backbone.

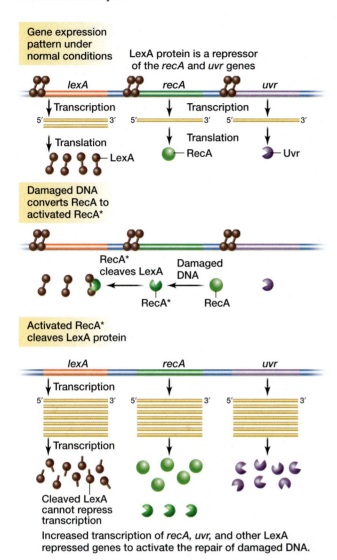

Gene expression pattern under normal conditions

LexA protein is a repressor of the *recA* and *uvr* genes

Damaged DNA converts RecA to activated RecA*

RecA* cleaves LexA

Damaged DNA

RecA*   RecA

Activated RecA* cleaves LexA protein

Cleaved LexA cannot repress transcription

Increased transcription of *recA*, *uvr*, and other LexA repressed genes to activate the repair of damaged DNA.

figure 18.33 LexA-mediated gene regulation. The *lexA* gene encodes a repressor that suppresses transcription of itself, *recA*, and several other loci (*uvrA*, *uvrB*, *uvrD*, *sulA*, and *sulB*) by binding the SOS box DNA sequence at each gene. LexA repressor binding allows low level of transcription from each of these genes. When DNA damage occurs, RecA becomes activated RecA*, which is a protease that cleaves the LexA protein. The cleaved LexA cannot function as a repressor. This results in increased transcription of the *recA* and *uvr* genes, which encode proteins that function to repair damaged DNA.

polymerases. This is due to their ability to replicate damaged DNA without requiring completely correct base pairing. This increased error rate is preferred to failing to complete DNA replication.

When the DNA damage is repaired, no single-stranded DNA is left to activate RecA, and, therefore, LexA is no longer destroyed. Newly translated LexA again represses the transcription of the suite of genes involved in the SOS response, and the SOS response is

over. **Table 18.4** summarizes some of the enzymes and proteins involved in DNA repair.

RecA* and the Lambda Repressor

Another protein that is cleaved by RecA* is the λ CI repressor (chapter 16). As we described, λ prophage can be induced to enter the lytic cycle by UV light. This UV light damage of DNA induces the activation of the RecA* protease, which cleaves λ repressor. Loss of λ CI repressor permits transcription of the λ *cro* gene in the prophage. Expression of the Cro protein induces the λ prophage to enter the lytic cycle.

From an evolutionary point of view, it makes sense for phage λ to have evolved a repressor protein that the RecA* protein inactivates. As a prophage, λ depends on the survival of the host cell. When that survival might be in jeopardy, such as through UV damage of the genomic DNA, the prophage would be at an advantage if it could sense the danger and make copies of itself that could leave the host cell. The SOS response is a signal to a prophage that the cell has received that damage. From an evolutionary perspective, the *E. coli* cell has not created an enzyme (RecA) that seeks out the λ repressor for the benefit of λ. Rather, the λ repressor has evolved for its own advantage to be sensitive to RecA*.

Restating the Concepts

▶ The RecA protein can mediate postreplicative repair of DNA using a recombination mechanism where the replicating strand invades the complementary sequence on the sister DNA molecule.

▶ The SOS response is induced by DNA damage that stimulates a protease activity in the RecA protein (RecA*), which cleaves the LexA transcriptional repressor. The inactivation of LexA allows transcription of up to 18 different genes in *E. coli* that are involved in DNA repair.

▶ Postreplicative repair can be achieved by a specific group of DNA polymerases that exhibit a high error rate while randomly inserting nucleotides opposite of a damaged template DNA strand.

18.5 Transposable Genetic Elements

Up to this point, we discussed mechanisms by which radiation, chemicals, and base analogs can generate spontaneous and induced mutations. Another mutagen exists within the genome itself. **Transposable elements,** or **mobile genetic elements,** are segments of DNA that can move (transpose) from one place to another within a genome. Because they can insert into a gene when

table 18.4 Some of the Enzymes and Proteins Involved in DNA Repair in *E. coli*, Not Including DNA Polymerase I and III, DNA Ligase, and Single-Stranded Binding Proteins

Enzyme	Gene	Action
Damage Reversal		
DNA photolyase	*phr*	Undimerizes thymine dimers
DNA methyltransferase	*ada*	Demethylates guanines in DNA
Base Excision Repair		
Uracil–DNA glycosylase	*ung*	Removes uracils from DNA
Endonuclease IV	*nfo*	Nicks AP sites on the 5' side
Exonuclease, lyase, or phosphodiesterase	*several*	Removes base-free nucleotide
Nucleotide Excision Repair		
UvrA	*uvrA*	With UvrB, locates thymine dimers and other distortions
UvrB	*uvrB*	Nicks DNA on the 3' side of the lesion
UvrC	*uvrC*	Nicks DNA on the 5' side of the lesion
UvrD (helicase II)	*uvrD*	Unwinds oligonucleotide
Mismatch Repair		
MutH	*mutH*	Nicks DNA at recognition sequence
MutL	*mutL*	Recognizes mismatch
MutS	*mutS*	Binds at mismatch
MutU (UvrD)	*mutU*	Unwinds oligonucleotide
Exonucleases	*recJ, xseA, sbcB*	Degrades unwound oligonucleotide
DNA methylase	*dam*	Methylates 5'-GATC-3' DNA sequences
Double-Stranded Break Repair		
Ku	*Ku70, Ku80*	Binds to broken chromosomal ends
PK$_{CS}$	*PK$_{CS}$*	Protein kinase
DNA ligase IV	*LIG4*	Ligates broken ends of DNA
XRCC4	*XRCC4*	Stabilization protein
Postreplicative Repair		
Polymerase IV	*DinB*	DNA polymerase
Polymerase V	*UmuC, UmuD*	DNA polymerase
Polymerase +RAD30+DNA polymerase		
Polymerase +	*REV3, REV7*	DNA polymerase
RecA	*recA*	Single-stranded DNA invades double-stranded DNA; causes LexA to autocatalyze; protease
LexA	*lexA*	Represses SOS proteins
SulA, SulB	*sulA, sulB*	Inhibit cell division

they transpose, they possess the ability to disrupt the expression of a gene and produce a mutant phenotype.

Transposable elements were discovered by Barbara McClintock in the 1940s. She proposed the existence of mobile DNA elements to account for the genetic phenomena that she observed in maize. Into the 1960s, most geneticists insisted that the genome was static, and the idea that segments of DNA could move seemed unreasonable. It was not until the 1970s, when transposable elements were discovered in bacteria, that

McClintock's work became recognized as being applicable to a variety of organisms. McClintock's work on mobile genetic elements in maize earned her the Nobel Prize in Physiology or Medicine in 1983.

We now know that transposable elements exist in all organisms that have been studied. They serve important roles in the transmission of antibiotic resistance in bacteria, regulation of gene expression in eukaryotes, and as a DNA vector by which scientists can efficiently introduce foreign DNA into an organism.

Bacterial Transposable Elements

The study of transposable elements increased significantly when they were identified in bacteria in the 1970s. Researchers identified three different classes of bacterial elements: insertion sequences, composite transposons, and complex transposons. Analyzing these classes revealed three different mechanisms by which they move or transpose to new sites. We briefly describe each of the three classes of bacterial elements and the major transposition mechanisms before we move on to discussing the eukaryotic elements.

IS Elements

The first transposable elements discovered in bacteria were called **insertion sequences,** or **IS elements.** IS elements are the simplest transposons, with a total length of about 700 to 2000 bp. Like all transposable elements, IS elements contain terminal inverted repeat sequences of about 15 to 40 bp, the length depending on the specific IS element (**fig. 18.34**).

Between the terminal inverted repeats is a gene that encodes **transposase,** the enzyme that catalyzes the movement of the element. The transposase recognizes the terminal inverted repeats to define the sequence that will be transposed. Unlike other transposons, the relatively small IS elements carry no bacterial genes between the terminal inverted repeats.

The target site to which the transposable element moves is not a specific sequence, as with the *att* site of phage λ (see chapter 16). When a transposable element moves to a new location, we find that a short sequence of 3 to 7 bases has been duplicated in a directly repeated orientation on both sides of the transposable element (see fig. 18.34). The length of this repeat is specific for each IS element.

The creation of this direct repeat of the target sequence suggests a model of insertion (**fig. 18.35**). The target site is cut in a staggered fashion, leaving single-stranded ends. The IS element is then inserted between the single-stranded ends. DNA polymerase converts the two single-stranded regions to double-stranded segments and, hence, to direct flanking repeats.

Like all transposable elements, IS elements exist in multiple copies in the genome. For example, IS1 is present in five to eight copies in the *E. coli* genome, but IS4 is found in only one or two copies. The IS elements also exist in multiple copies in the F plasmid. It is likely that the insertion of the F plasmid into the *E. coli* genome is due to recombination between IS elements that are present in both DNA molecules. Furthermore, an IS element on the F plasmid can be transferred into an F⁻ cell through conjugation (see chapter 15).

Once the IS element is in the recipient cell, it can move to a new location either in the recipient's genome or to another plasmid in the recipient. This provides a mechanism by which the transposable elements can spread throughout a bacterial population.

Composite Transposons

After the discovery of IS elements, a more complex type of transposable element, a **composite transposon (Tn),** was discovered. A composite transposon consists

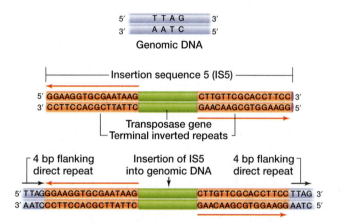

figure 18.34 General organization of a bacterial insertion sequence (IS5). The IS5 element contains a 16 bp terminal sequence (red) that is in an inverted orientation. Between the terminal inverted repeats is the transposase gene (green), which encodes an enzyme that mediates the insertion of IS5 into a new DNA location. Insertion of IS5 into the chromosome results in a duplication of the sequence flanking the IS5 element (5' TTAG 3').

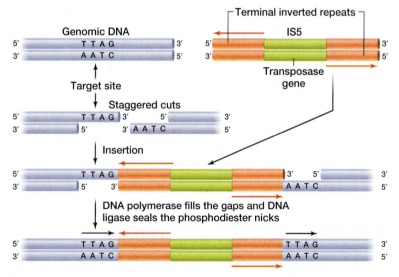

figure 18.35 Mechanism that generates direct repeat sequences flanking an inserted transposable element. Transposition of IS5 involves generating a staggered cut in the genomic DNA at the insertion site. The IS element can then be ligated to the free ends at the target site. The single-stranded regions are repaired by DNA polymerase to produce two directly repeated sequences flanking the inserted IS element.

of a central region flanked by two similar IS elements. The central region usually contains bacterial genes, frequently antibiotic resistance loci. For example, the composite transposon Tn10 contains the gene for transposase in one of the two IS10 elements (IS10R) and the gene for bacterial resistance to tetracycline (tet^r) in the region between the two IS10 elements (**fig. 18.36**). The IS10R is a fully functional IS element, but the IS10L is a nonfunctional element because it does not express transposase.

If the movement of the transposon is not properly regulated, it can potentially insert into essential bacterial genes and kill the cell. This result is not to the transposon's advantage, since the death of the cell would also mean that the inserted element would also be lost. Thus, transposons move to new locations at only a modest frequency.

Transposons have developed complex schemes to regulate their movement (transposition). The Tn10 element encodes functional transposase in only the IS10R sequence. There are two promoters in IS10R; P_{in} is used to transcribe the transposase gene and P_{out} transcribes an RNA in the opposite direction (fig. 18.36). The two RNAs overlap by 36 bp and are therefore complementary in this region.

The purpose of the P_{out} RNA is to hybridize to the P_{in} transposase mRNA. This double-stranded region at the 5' end of the P_{in} mRNA blocks translation initiation of transposase, ensuring that the amount of transposase protein does not become too great.

The second level of regulation involves transposase binding to the terminal inverted repeat DNA sequences. Located within the terminal inverted repeats is an adenine residue that can be methylated. Because transposition of the Tn10 requires DNA replication (to be explained shortly), the transposase preferentially binds hemimethylated terminal inverted repeats, in which only one strand is methylated, relative to the fully methylated repeats. This is similar to the MutS and MutL proteins binding the hemimethylated sequence during mismatch repair. This mechanism ensures that the element has just been replicated, and the replication machinery is nearby and available to assist in the transposition of the Tn10 element.

Two IS elements can transpose virtually any region between them. In fact, composite transposons most likely originated when two IS elements became located near each other. We can see this very clearly in a simple experiment. **Figure 18.37** illustrates a small plasmid constructed with transposon Tn10 in it. The "reverse" transposon, consisting of the two IS elements and the plasmid genes, or the normal transposon, could each transpose. The ability of transposons to move any sequence that is located between the terminal inverted repeats has made them a useful system for introducing foreign DNA into an organism's genome (see P elements).

Complex Transposons

The final class of transposable elements in bacteria are called **complex transposons**. One member of this class is Tn3, which is diagrammed in **figure 18.38**.

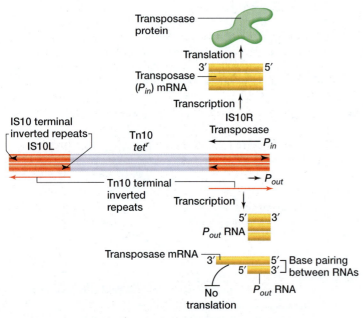

figure 18.36   Organization of the composite Tn10 transposon. The composite Tn10 transposon contains two terminally inverted IS10 elements flanking the tetracycline resistance gene (tet^r). Only IS10R produces functional transposase from the P_{in} promoter. The P_{out} promoter makes a RNA that has 36 base pairs that are complementary to the P_{in} mRNA. Base pairing between these two mRNAs blocks translation of the transposase enzyme. The small black triangles represent the terminal inverted repeat sequences in the IS10 elements.

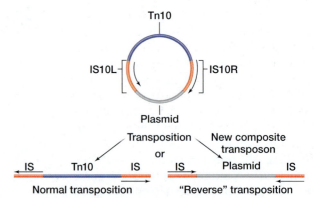

figure 18.37   Two IS elements in a plasmid can transpose either of the two regions between them. In the example shown, either the Tn10 transposon or the "reverse" transposon (Plasmid) is transposed. Both of these transposition products have terminal inverted IS10 elements, although they are in different orientations in the two transposed elements (see orientation of black arrows). Because each IS10 has short terminal inverted repeats, both transposed elements have the same short terminal inverted repeats.

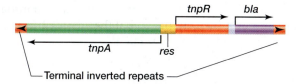

figure 18.38 Organization of the complex Tn3 transposon. The Tn3 transposon encodes the transposase (*tnpA*) and resolvase (*tnpR*) genes that are required for complete transposition. The *res* site is required for recombination of the cointegrate transposition intermediate. The *bla* gene encodes resistance to β-lactam antibiotics, such as ampicillin. The horizontal arrows represent the direction of transcription for the three genes. The small black triangles represent the short terminal inverted repeat sequences.

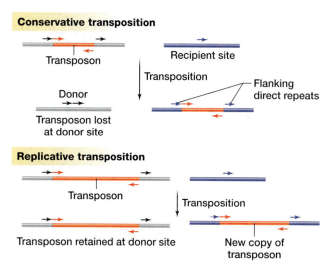

figure 18.39 Two models of transposition. In conservative transposition, double-stranded breaks flanking the transposon at the donor site release the transposon. A staggered double-stranded break at the recipient site (figure 18.35) creates the flanking duplication of the target sequence. Notice that the donor site retains the duplicated direct repeats that previously flanked the transposon, but not the transposon. In replicative transposition, the transposon is replicated so a copy is retained at the donor site and a copy is inserted in the recipient site. While examination of the donor site can reveal if transposition was either conservative or replicative, both models produce the same transposition product at the recipient site.

Like all transposable elements, Tn3 contains terminal inverted repeat sequences and a transposase gene (*tnpA*). However, complete transposition of Tn3 also requires the resolvase gene (*tnpR*) and a specific internal sequence called the internal resolution site (*res*). The Tn3 transposon also contains a gene that confers resistance to the β-lactam antibiotics (such as ampicillin). Thus, the complex transposons contain genes other than transposase, unlike the IS elements, and they do not contain flanking IS elements, as the composite transposons do.

Mechanisms of Transposition

Transposition can come about by two major mechanisms: conservative and replicative transposition. Both mechanisms generate the flanking direct repeats of the target sequence at the insertion site. Each type of transposon utilizes a specific mechanism.

Conservative Replication

In **conservative replication,** the transposon moves to a new location without copying itself. The transposon is liberated from the donor site by double-stranded breaks in the DNA that flank the element at the donor site (**fig. 18.39**). Staggered double-stranded breaks at the recipient site allow the excised transposon to ligate and generate the flanking direct repeats.

Notice that even though the transposon has left the donor site, it has left behind a sign of the former presence of the transposon, the direct repeat sequence (fig. 18.39). Comparing this donor sequence with a wild-type sequence, prior to the presence of the transposon, will reveal the short direct repeat sequence. This is often used to reveal the former presence of a transposon. Furthermore, if the repeat sequence is not a multiple of 3 bp, then the loss of the transposon will create a frameshift mutation if it is located within an ORF.

Replicative Transposition

Replicative transposition involves the duplication of the transposon; the copy is inserted in the recipient site, while the original transposon remains at the donor site (see fig. 18.39). Unlike the conservative mechanism, which required double-stranded breaks, the replicative mechanism involves only single-stranded breaks of the DNA at the donor site.

One model of replicative transposition, described by James Shapiro, suggests that during transposition from one plasmid to another, a **cointegrate** intermediate is formed, made up of both plasmids and two copies of the transposon (**fig. 18.40**). Then, through a process called *resolution*, the cointegrate is reduced back to the two original plasmids, each now containing a copy of the transposon. In this model, the first step is to generate staggered cuts in the donor and recipient DNA molecules (fig. 18.40). Then nonhomologous ends are joined so that only one strand of the transposon connects them (fig. 18.40). DNA replication now takes place to fill in the single-stranded segments. The result is a cointegrate of the two plasmids joined by two copies of the transposon.

The last step is a recombination event at a homologous site within the two transposons (*res* site in Tn3). This is catalyzed by a *resolvase* enzyme (encoded by the

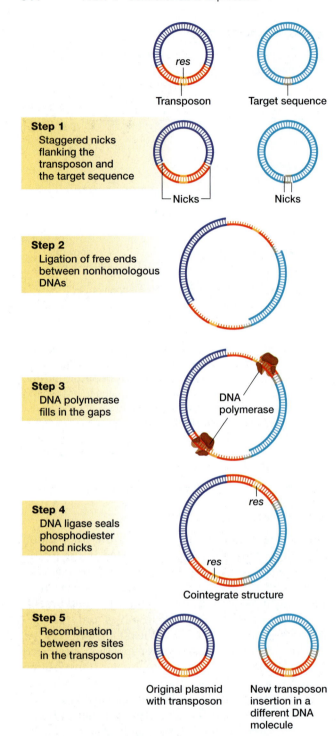

Step 1
Staggered nicks flanking the transposon and the target sequence

— Nicks —   Nicks

Step 2
Ligation of free ends between nonhomologous DNAs

Step 3
DNA polymerase fills in the gaps

DNA polymerase

Step 4
DNA ligase seals phosphodiester bond nicks

res

res

Cointegrate structure

Step 5
Recombination between *res* sites in the transposon

Original plasmid with transposon

New transposon insertion in a different DNA molecule

figure 18.40   The Shapiro model of replicative transposition. Staggered double-strand cuts are made at the target site and on either side of the transposon at the donor site. Nonhomologous single strands are ligated together, resulting in two single-stranded copies of the transposon joining the plasmid and donor DNA. DNA polymerase then fills in the gaps of the single-stranded regions to produce the cointegrate structure that contains two copies of the transposon. A crossover between the *res* site in both transposons resolves the cointegrate into two plasmids, each with a copy of the transposon.

tnpR gene in Tn3), which recombines the cointegrate molecule into the original two plasmids, each with a copy of the transposon (fig. 18.40).

Phenotypic and Genotypic Effects of Transposition

Transposition can have several effects on the genotype and corresponding phenotype of an organism. If the transposable element inserts into a bacterial gene it can produce an aberrant mRNA with an altered ORF. If the insertion is within the promoter of a gene, it could affect the level of expression of the bacterial gene. We discussed in previous chapters how either increased or decreased expression of a gene could result in a mutant phenotype.

A transposon can also produce genomic rearrangements, such as deletions, inversions, and translocations (**fig. 18.41**). Through a replicative mechanism, two copies of a transposable element can be present in a directly repeated orientation on a chromosome. If recombination then occurs between these two transposons, then one transposon and the region between the transposons is deleted. If the transposons are present in an inverted orientation, pairing followed by recombination between the elements can result in an inversion of the region between the repeats. Finally, if the transposons are present on separate eukaryotic chromosomes, recombination between the elements results in a reciprocal translocation.

Transposable elements can therefore have a significant effect on the phenotype by their actions in transposition and by the fact that they may carry genes valuable to the cell. But they can also exist without any noticeable consequences, especially if they are located between genes and are not actively transposing. This fact has led some evolutionary geneticists to suggest that transposons are an evolutionary accident that, once created, are self-maintaining. Because they may exist without a noticeable benefit to the host cell, transposons have been referred to as **selfish DNA.** In recent theoretical and experimental studies, however, some scientists have suggested that transposons improve the evolutionary fitness of the bacteria that have them (see chapter 25).

Restating the Concepts

▶ Bacteria contain three types of transposable elements. IS elements are the simplest and are composed of terminal inverted repeats and encode a transposase enzyme. Composite transposons contain IS elements at their ends and usually encode bacterial genes, such as antibiotic resistance, between the IS elements. Complex transposons are similar to the composite transposons, except they lack the IS elements at the end.

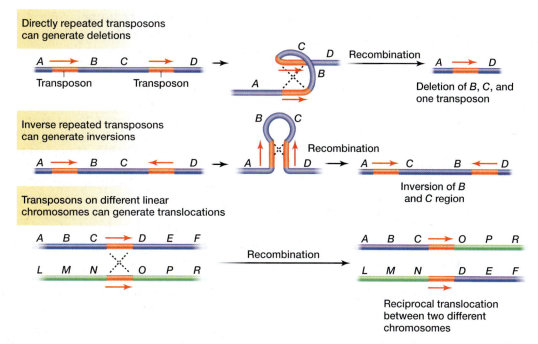

figure 18.41   Recombination between transposable elements produces chromosomal rearrangements. Transposable elements that are in a directly repeated orientation (top) can recombine, which results in a deletion of the region between the elements and the loss of one transposable element. Transposable elements in an inverted repeat orientation (middle) can recombine to yield an inversion of the region between the elements. Two transposable elements on different chromosomes can recombine (bottom), which results in a reciprocal translocation due to a single crossover between the two elements.

▶ Transposons contain terminal inverted repeat sequences and duplicate the insertion site sequence that flanks the element in a directly repeated orientation.

▶ Transposons can replicate through either a conservative mechanism or a replicative mechanism. In conservative transposition, the donor site loses the transposon, but retains the duplicated flanking sequence. In replicative transposition, the donor site retains the transposon after transposition.

▶ Transposons can generate mutations by their insertion, their excision during transposition, and their recombination between different elements.

Eukaryotic Transposable Elements: Corn *Ac–Ds* Elements

The *Ac–Ds* system in corn was first described by Barbara McClintock to explain the phenotypes that resulted from the interaction between the *Dissociation* (*Ds*) and *Activator* (*Ac*) mutations (**fig. 18.42**). McClintock carefully examined the phenotypes produced from different crosses and correlated them with the cytology of the chromosomes. She deduced that the *Ds* gene was located on chromosome 9, and the presence of the *Ac* allele caused a chromosome break adjacent to *Ds*.

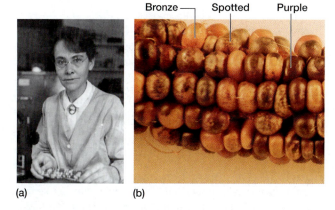

figure 18.42   The *Ac-Ds* transposon system in corn. (a) A photograph of Barbara McClintock (1902–1992) taken in 1947, at about the time that she began her work on the *Ac-Ds* transposon system in corn, which earned her the 1983 Nobel Prize in Physiology or Medicine. (b) Shown is a portion of an ear of corn with purple and bronze kernels. The purple kernels have no transposons. The bronze kernels (light-colored) lack the purple pigment because they have a *Ds* element in both copies of the *Bz2* locus, which disrupts pigment production. In the presence of the *Ac* element, the *Ds* element can leave its position, restoring the wild-type allele and producing a purple spot in a bronze kernel. Spots differ in size based on when the *Ds* element was excised during the development of the kernel: early excision yields large bronze spots; late yields small bronze spots.

Phenotypes Produced by the Ac–Ds Elements

In figure 18.42, you see three kinds of corn kernels: purple, bronze (light-colored), and bronze with purple spots. The purple kernels result from dominant functioning alleles that provide enzymes in the pathway for purple pigment. In these kernels, the *Ds* element is not located within the *Bz2* locus (**fig. 18.43a**).

In the kernels that are bronze without spots, the *Ac* element has produced transposase that allows *Ds* elements to transpose into both copies of the *Bz2* locus, disrupting the pigment pathway. Without the *Ac* element, the *Ds* elements (that lack a transposase gene) remain in place, and the kernels are a uniform bronze color.

In the bronze kernels with purple spots, the *Ac* element produces transposase that allows the *Ds* elements to leave the *Bz2* locus in some of the cells. This restored activity creates purple spots in those cells and their progeny with the functioning *Bz2* allele. The cells that continue to have the *Ds* element in the *Bz2* locus remain bronze. This demonstrates the mutagenic potential of transposable elements.

Nature and Relationship of the Ac and Ds Elements

Ds and *Ac* elements have been cloned and sequenced. They are typical transposons that are very similar to each other (fig. 18.43b). The *Ac* element is 4563 bp long and contains 11 bp terminal inverted repeats. It also contains two ORFs, one of which encodes a transposase. The *Ac* element is considered an autonomous element, as it contains all the essential functions (transposase and terminal inverted repeats) that are required for transposition.

Several different *Ds* elements appear to have arisen from the *Ac* element by internal deletions that remove at least a portion of the transposase gene. Each of these *Ds* elements must have both terminal inverted repeats, however, in order for them to transpose. Therefore, *Ac* must provide the transposase for *Ds* to transpose.

Although *Ds* is unable to transpose without *Ac* (*Ds* is a nonautonomous element), *Ds* can directly produce a phenotype by blocking expression of the genes it transposes into, as well as by causing the loss of alleles in acentric chromosomal fragments lost when *Ds* breaks its chromosome. The ability of a wild-type transposon to allow transposase-defective elements to move is a common theme in transposition.

Restating the Concepts

▶ The *Ac–Ds* system in corn is composed of two related classes of transposons. The *Ac* elements are complete transposons and encode transposase, whereas the *Ds* elements contain internal deletions of at least some of the transposase gene. The *Ds* elements cannot move to a new location unless the transposase is provided by an *Ac* element.

▶ Transposons can generate mutant phenotypes by inserting in genes.

Drosophila Transposable Elements

Drosophila contains several different classes of transposable elements. For our discussion, we will concentrate on two different classes. The **P element** is of interest because of some unusual biology and its common use in generating transgenic flies. The *P* element is also similar to the corn *Ac–Ds* system in many ways.

The second class is called *copia,* which is of interest due to its transposition mechanism through a RNA intermediate. *Copia* also serves as a model for several classes of transposable elements that are present in humans.

P Elements

The *Drosophila P* element was first identified through a phenomenon called **hybrid dysgenesis,** which occurs

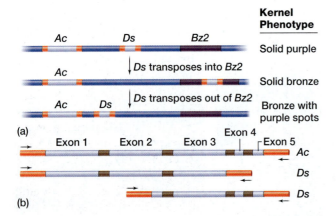

figure 18.43 The *Ac-Ds* system in maize. (a) Shown are the molecular details associated with the three different kernel colors in maize. The purple kernel color is due to an intact *Bz2* gene. If the *Ac* element provides transposase to insert the *Ds* element into both copies of the *Bz2* gene, the *Bz2* gene is inactivated and the kernel has a bronze color. If *Ac* again provides transposase, the *Ds* element can move out of the *Bz2* gene and restore the purple pigmentation. If *Ds* moves out early in kernel development, large purple spots appear, while a late excision of *Ds* produces small purple spots. (b) The *Ac* element contains two terminal inverted repeats (black arrows) flanking the transposase gene (5 exons and 4 introns) that transcribes a 3.5 kb in mRNA. Deletion of any part of the transposase gene without affecting the terminal inverted repeats yields the defective *Ds* element. Shown are two internal deletions that produce different forms of *Ds*. However, all *Ds* elements can transpose if an *Ac* element produces transposase in the same cell.

when a male of a P stock is crossed to a female of an M stock. The resulting hybrid F_1 progeny are sterile. However, if a male of an M stock is crossed to a female of a P stock, the F_1 progeny are normal.

Analysis of the stocks revealed the P stock contains one or more copies of a transposable element, the *P* element, but the M stock either completely lacked the *P* element or contained shorter versions of the *P* element (much like the *Ds* elements in corn). The sterility in the F_1 hybrid resulted from transposition of the *P* element within the genome of the germ cells, causing chromosomal rearrangements and insertions into essential genes. The two major questions from the hybrid dysgenesis observation were: Why are the germ cells preferentially affected? And why did the *P* element have to be introduced through the paternal lineage?

Before we can address those questions, we must examine the organization of the *P* element (**fig. 18.44**). Like all transposable elements, the *P* element contains terminal inverted repeats (31 bp long) and generates an 8-bp flanking direct repeat of the insertion sequence. The *P* element contains four exons that are called ORF0, ORF1, ORF2, and ORF3.

The primary transcript from the *P* element is not spliced the same way in all tissues. In somatic tissues, only the first two introns are spliced out of the primary transcript. In the somatic tissues, translation of the mRNA proceeds into the third intron, which contains a premature translation termination codon to yield a protein of 66 kDa (fig. 18.44, left). In germ cells, all three introns are spliced out of the primary transcript to produce a mRNA that is translated into the 87 kDa transposase protein (fig. 18.44, right).

This difference in splicing of the third intron is due to the presence of a protein in somatic tissues

that recognizes and binds a sequence in the third intron. Binding of this protein to the primary transcript blocks the removal of the third intron during splicing. Because this factor is not found in germ cells, the third intron is spliced out to generate the fully processed mRNA.

We now know that the 66 kDa protein is a repressor of transposase. This repressor possibly binds to the same DNA sequences in the *P* element as transposase, without catalyzing the transposition of the element. Because the *Drosophila* eggs are loaded with proteins, they contain the *P* element repressor (**fig. 18.45**). A zygote produced from an M male and a P female therefore has the repressor protein to prevent *P* element transposition. A zygote produced from a P male and M female (who lacks a full-length *P* element) would not have the repressor. The *P* elements passed on through the sperm would transpose in the germ cells of this hybrid F_1 individual. This accounts for the relationship of the P and M stocks.

Scientists use *P* elements routinely to generate transgenic flies using two plasmids (**fig. 18.46**). The first plasmid contains the terminal inverted repeats of a *P* element, without the transposase gene. A foreign gene that is to be introduced into the fly genome is cloned between the repeats in this plasmid. The second plasmid (*helper plasmid*) contains a transposase gene that lacks one or both of the terminal inverted repeats. Both plasmids are injected into *Drosophila* embryos near the region where the germ cells are located. The helper plasmid produces transposase, but cannot transpose because it lacks one or both of the terminal inverted repeats. The transposase, however, can act on the two terminal inverted repeats on the other plasmid and transpose the transgene from the other plasmid into the genome. The use of transposons has proven to be an efficient method of introducing foreign DNA into the genome of a wide range of organisms.

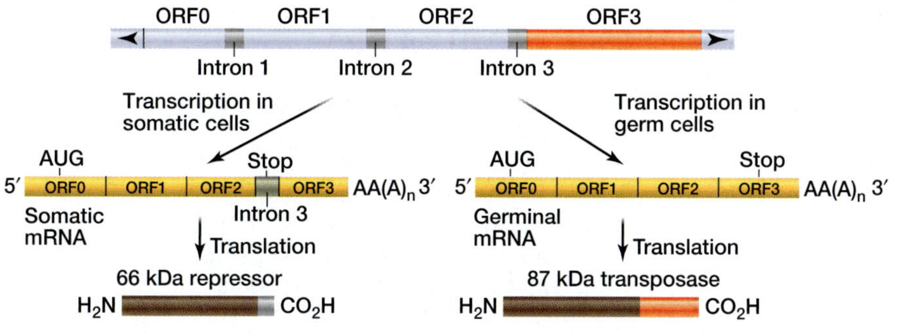

figure 18.44 Organization of the *Drosophila* P element. The 2.9 kb full-length *P* element contains terminal inverted repeats (black arrowheads) that flank four exons (ORF0, ORF1, ORF2, and ORF3). When the *P* element RNA is transcribed in somatic cells (left), the third intron is not spliced out. Because intron 3 contains a translation termination codon, this somatic mRNA encodes a 66 kilodalton (66 kDa) repressor protein. When the *P* element is transcribed in germ cells (right), all three introns are removed to produce an mRNA that encodes the 87 kDa transposase protein. The 66 kDa and 87 kDa proteins differ in only their carboxyl termini, with the 66 kDa protein containing sequences corresponding to intron three (gray) and the 87 kDa protein possessing sequences corresponding to ORF3 (red).

Copia *Elements*

The *Drosophila copia* element serves as the model for several related classes of transposable elements, which are collectively called *copialike elements*. The *copia* element is approximately 5 kb long. The ends of *copia* are defined by long terminal repeats (LTRs) that are nearly 300 bp in length (**fig. 18.47**). Each LTR contains terminal inverted repeats, similar to the IS elements in bacterial composite transposons.

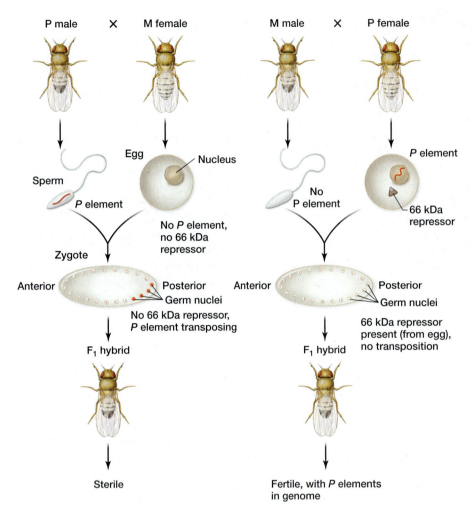

figure 18.45 Mechanism of hybrid dysgenesis. Flies can either have *P* elements in their genome (P type) or lack *P* elements (M type). When a P male and M female mate (left), the sperm provides *P* elements to the zygote, while the egg lacks both *P* elements and *P* element-encoded repressor protein. The *P* element's transposase gene is transcribed in the germ cells, transposase is translated, and the *P* element transposes in the germ cells. The resulting F$_1$ hybrid fly is sterile due to the transposition of the *P* element into essential genes in the germ cells. When an M male and P female mate (right), the egg contains the *P* element in its genome and the repressor (66 kDa) protein, while the sperm provides none of these. The presence of the maternally-supplied repressor protein prevents transposition of the *P* element in the germ cells of the zygote. The resulting F$_1$ hybrid fly is fertile due to the absence of additional transposition, although the *P* elements remain in both its somatic and germ cells.

The insertion site of *copia* creates a 5 bp flanking direct repeat of the target insertion sequence. But unlike the other transposable elements that we have discussed, *copia* does not encode a conventional transposase gene. Instead, *copia* contains one ORF that is homologous to the *gag, int, env,* and *pol* genes in RNA retroviruses. The *pol* gene encodes a reverse transcriptase that converts mRNA into double-stranded DNA.

The mechanism of *copia* transposition is outlined in figure 18.47. The *copia* element is transcribed as a single mRNA, which is converted into double-stranded DNA by the *pol*-encoded reverse transcriptase. The associated integrase activity creates a staggered cut at the target site that permits insertion of the *copia* DNA. Repair of the single strands by DNA polymerase generates the flanking direct repeats.

As a further similarity between *copia* and the RNA retroviruses, virus-like particles have been identified in *Drosophila* cells that contain *copia* RNA and reverse transcriptase. The class of transposons that move through a RNA intermediate are termed **retrotransposons** due to their similarity with RNA retroviruses.

Retrotransposons have been identified in yeast (Ty element), humans, and a variety of other organisms.

Restating the Concepts

▶ The regulation of *P* element transposition involves alternative splicing of the transposase mRNA so that transposase is encoded in the germline mRNA and a repressor is encoded in the somatic tissue mRNA. This results in the *P* element only transposing in germ cells under normal circumstances.

▶ Use of the *P* element and other transposons has proven to be an effective method to introduce foreign genes into the genome of a wide assortment of organisms.

▶ The *Drosophila copia* element represents a retrotransposon. These elements encode a reverse transcriptase enzyme to convert the mRNA into double-stranded DNA. The *copia* RNA is often found in *Drosophila* cells in a virus-like particle with reverse transcriptase.

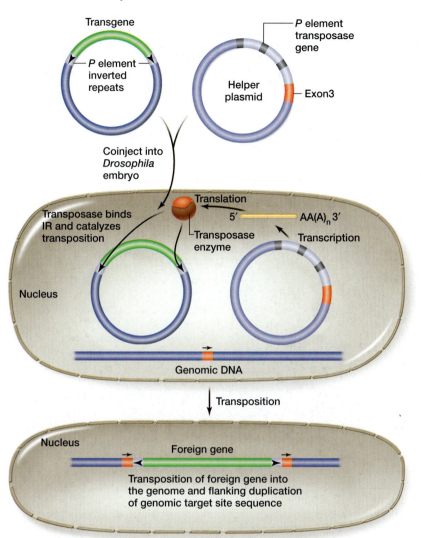

figure 18.46   Generation of flies with transgenic *P* elements. A foreign gene (transgene) is cloned between the *P* element terminal inverted repeats (black arrowheads). The helper plasmid contains the *P* element transposase gene, but lacks both terminal inverted repeats, which prevent it from transposing. Both plasmids are microinjected into *Drosophila* embryos. The helper plasmid expresses transposase protein in the germ cells, which acts on the first plasmid to transpose the transgene, flanked by the inverted repeats into the genome. The resulting fly contains the transgene in the genome of the germ cells. Mating this fly produces progeny that contain the transgene in both the somatic and germ tissues.

Transposable Elements in Humans

Transposable elements are also found in the human genome. Two of the more common classes of human transposons are called **long interspersed elements (LINEs)** and **short interspersed elements (SINEs).** Both of these elements are believed to transpose through RNA intermediates, but there are clear differences between them.

LINEs

There are at least two different classes of human LINEs, named L1 and L2. LINEs are 1 to 6.5 kb in length and found in 100,000 to 500,000 copies in the human genome. Together, these two elements constitute over 10% of the human genome! The 6.5 kb sequence is thought to correspond to a full-length element, and the shorter ones are deleted, nonautonomous versions. The full-length LINE contains an ORF that appears to encode a reverse transcriptase.

Unlike retrotransposons, the LINEs lack the LTRs. Whereas the LTR contains the retrotransposon promoter to transcribe the mRNA, the LINEs must contain an internal promoter to synthesize the mRNA.

SINEs

The SINEs in the human genome are primarily composed of the *Alu* sequence, which is 150–500 bp in length. The *Alu* sequence is present in close to 1,000,000 copies in the human genome, which represents approximately 10% of the genomic sequence!

Like the LINEs, the SINEs lack the LTR sequences, but none of the SINEs contain an ORF that could encode a reverse transcriptase. How do researchers know that the SINEs represent a transposable element? The data are based on the presence of the 7 to 21 bp direct repeats of the insertion sequence that flanks each SINE.

The *Alu* sequence is also very similar to a small cytoplasmic RNA called 7SL. The *Alu* sequences are thought to be derived from 7SL RNAs that were converted to DNA by a reverse transcriptase enzyme in the cell (possibly from a LINE) and then inserted into the genomic DNA. Other SINEs are related to small nuclear RNAs and tRNAs. In each case, the SINE sequence in the genome is flanked by direct repeats.

Restating the Concepts

▶ The human LINEs are considered retrotransposons due to the presence of a reverse transcriptase gene, even though they lack the LTR sequence. Like other transposable elements, one class of LINE is considered to be full length and another class has internal deletions.

▶ The SINEs appear to be short RNAs that were converted to DNA through a reverse transcriptase reaction and inserted into the genome. Unlike LINEs, SINEs do not encode the reverse transcriptase and depend completely on the reverse transcriptase and integrase being provided by another element.

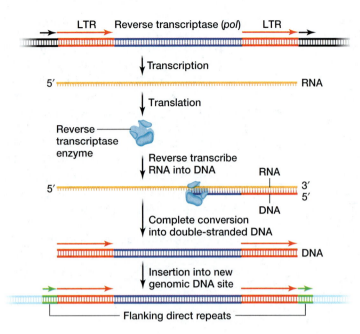

figure 18.47   Organization and transposition of a retrotransposon. The retrotransposons contain long terminal repeats (LTRs) that flank a reverse transcriptase gene. The LTR (red) contains a promoter that transcribes a mRNA that contains the LTRs and reverse transcriptase. Translation of the mRNA yields the reverse transcriptase enzyme, which converts the mRNA into double-stranded DNA. Insertion of this DNA between staggered double-stranded breaks at the target DNA site (green) produces the inserted retrotransposon, flanked by duplicated direct repeats of the target sequence.

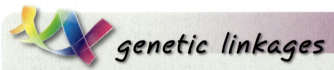
genetic linkages

In this chapter, we discussed the variety of mechanisms, both spontaneous and inducible, that can be used to generate mutations. Without mutations, the diversity of alleles that allow us to examine the function of a gene would not be possible. The dominant and recessive alleles that we discussed in chapters 2, 4, and 5 were the result of these mutations.

A basic knowledge of the structure of both the DNA bases and the DNA molecule allows for a better understanding of how spontaneous and induced mutations are produced. For example, the rotation of bases that is important in generating transversions was first discussed in the formation of Z-DNA (chapter 7). Knowing the functional groups in the four bases that are involved in hydrogen bonding of the base pairs allows the prediction of how particular base modifications affect base pairing. In a related issue, understanding the relationship between the different codons and the underlying genetic redundancy (chapter 11) provides the basis for realizing that transition mutations are usually less severe than transversion mutations.

In this chapter, you saw how DNA repair is linked to a variety of molecular processes that were already described. For example, some of the repair mechanisms, such as mismatch base repair, depend on DNA replication (chapter 9). The cell utilizes DNA methylation as a mechanism for knowing which strand has been newly synthesized, and therefore, which base needs to be repaired. We previously described the role of DNA methylation in the context of the bacterial modification/restriction system (chapter 12), genetic imprinting (chapter 17), and in the regulation eukaryotic gene expression (chapter 17).

You learned the similarities between the recombination mechanism and formation of the Holliday structure (chapter 6) and the mechanism of recombination repair of double-stranded breaks. These similarities indicate that the cell utilizes some basic functions and mechanisms to perform different processes in the cell. Not only does this provide a method to conserve cellular functions, but it also ties together functions that might appear to be unrelated.

We described how chromosomal alterations (chapter 8) can be generated by transposable elements. We discussed the role of overlapping RNAs in IS10R in blocking translation of another RNA, which is not very different from the general idea behind miRNA and antisense RNA (chapters 10 and 17).

We also discussed how alternative splicing of the *P* element RNA can produce either the transposase or repressor protein. This is another demonstration of how a single gene can encode multiple proteins through alternative splicing (chapters 10 and 17). We also saw how retrotransposons utilized a RNA intermediate and reverse transcriptase to move. Reverse transcriptase was previously described in chapter 10. You should now have a basic understanding of how transposable elements serve as mutagens and how this has been exploited to introduce genes into organisms.

Chapter Summary Questions

1. For each term in column I, choose the best matching phrase in column II.

   Column I
   1. Biochemical mutant
   2. *Ac* element
   3. Xeroderma pigmentosum
   4. DNA glycosylase
   5. *Copia* element
   6. Morphological mutant
   7. Retinoblastoma
   8. DNA photolyase
   9. IS element
   10. Huntington disease

   Column II
   A. Removes damaged bases
   B. Disease caused by an inability to repair thymine dimers
   C. Corn transposon
   D. An *E. coli* cell that is *his⁻*
   E. *Drosophila* with an abnormal wing shape
   F. *Drosophila* transposon
   G. Disease caused by a triplet expansion
   H. *E. coli* transposon
   I. Breaks bond between two thymines
   J. Disease caused by a mutation in a gene controlling cell division

2. What are silent mutations? Where do they occur in the genome?

3. What test was developed by Salvador Luria and Max Delbrück in the 1940s? What did it accomplish?

4. Distinguish between intragenic suppression and intergenic suppression.

5. What is a tautomeric shift? How does it cause mutations?

6. How is the mutation rate different from the mutation frequency?

7. Define triplet diseases and name their two main features.

8. Ethyl ethanesulfonate and nitrous acid are chemical mutagens. What does each do?

9. What is the difference between a substrate and a template transition mutation?

10. What types of mutations does UV light induce? List three mechanisms that can be used to repair the damage.

11. Discuss the mechanism of mismatch repair in *E. coli*.

12. Describe the role of the RecA protein in the SOS response.

13. Can double-stranded DNA breaks be repaired? If so how?

14. Why are IS elements sometimes referred to as "selfish DNA"?

15. Why are transposons flanked by direct repeats?

16. How do transposons induce deletions? inversions?

17. Name the three different classes of transposable elements in bacteria? How do they differ in structure and mode of transposition?

18. What is hybrid dysgenesis? Describe the role of *P* elements in it.

19. Compare and contrast LINEs and SINEs.

20. What properties are shared by bacterial and eukaryotic transposable elements?

Solved Problems

PROBLEM 1: A wild-type polypeptide consists of 35 amino acids. Many single-base substitution mutants have been discovered that produce polypeptides that are longer than normal.

  a. Do you expect the first 35 amino acids in the mutant polypeptides to have the same sequence as that in the wild type?

  b. What amino acids could occupy the 36th position in the mutant polypeptides? Consult the genetic code (table 11.5)

Answer:

  a. The mutants have all undergone base substitutions that produced longer polypeptides. This indicates that the mutations converted the stop codon in each mutant into an amino acid-specifying (sense) codon. Therefore, the first 35 amino acids in each mutant polypeptide should have the same sequence as the wild-type polypeptide.

  b. There are three stop codons: UAA, UAG, and UGA. The possible single-base mutations are as follows:

UAA can be converted to

AAA *Lys*	UCA *Ser*	UAC *Tyr*
CAA *Gln*	UGA *stop*	UAG *stop*
GAA *Glu*	UUA *Leu*	UAU *Tyr*

UAG can be converted to

AAG *Lys*	UCG *Ser*	UAA *stop*
CAG *Gln*	UGG *Trp*	UAC *Tyr*
GAG *Glu*	UUG *Leu*	UAU *Tyr*

UGA can be converted to

AGA *Arg*	UAA *stop*	UGC *Cys*
CGA *Arg*	UCA *Ser*	UGG *Trp*
GGA *Gly*	UUA *Leu*	UGU *Cys*

Therefore, 10 possible amino acids can occupy the 36th position of the mutant polypeptides: *Lys, Gln, Glu, Ser, Leu, Tyr, Trp, Arg, Gly,* and *Cys.*

PROBLEM 2: What are the similarities and differences between mismatch repair and AP repair?

Answer: Both processes are similar in that they entail removal of an incorrect base in a DNA double helix by an excision process followed by a repair process. The processes differ in the event that triggers them. Mismatch repair is triggered by a base pair that does not occupy the correct space in the double helix—that is, by a non-Watson and Crick pairing (neither AT nor GC). AP repair is triggered by enzymes that recognize a missing base.

PROBLEM 3: Four chemically induced mutants, *w, x, y,* and *z,* are treated with the following mutagens to determine if revertants can be produced: 2-aminopurine (2-AP), 5-bromouracil (5-BU), acridine orange (AO), and hydroxylamine (HA). *Note:* HA is a base-modifying agent that adds a hydroxyl group to cytosine allowing it to pair with adenine and thereby producing a *one-way* transition mutation from a GC basepair to AT. In the following table, + = revertants and − = no revertants. For each mutation, determine the probable base change that occurred to convert the wild type to the mutant.

	Chemical			
Mutant	**2-AP**	**5-BU**	**AO**	**HA**
w	−	−	+	−
x	+	+	−	+
y	+	+	−	−
z	−	−	−	−

Answer: Let us first review the effects of these four mutagens. 2-Aminopurine is a base analog that induces mainly AT to GC transition mutations, but could also cause GC to AT transitions. 5-Bromouracil is a base analog that induces mainly GC to AT transition mutations, but could also cause AT to GC transitions. Acridine orange is an intercalating agent that causes insertions or deletions of bases and therefore induces frameshift mutations in the DNA. HA is a base-modifying agent that causes GC to AT transitions only.

Mutant *w*: Only acridine orange was capable of producing revertants. Therefore, the original mutation that converted *w⁺* to *w* was a frameshift mutation caused by the addition or deletion of a base.

Mutant *x*: Revertants were produced by 2-AP, 5-BU, and HA. The critical point here is the response to HA, which only causes GC to AT transitions. This mutation is reverted by HA, therefore *x* must be a GC transition mutation, and the wild-type *x⁺* was AT.

Mutant *y*: This mutation is reverted by 2-AP and 5-BU, but not by HA. Therefore, *y* must be an AT transition mutation, and the wild-type *y⁺* was GC.

Mutant *z*: This mutation could not be reverted by any of the four chemical mutagens. Because 2-AP, 5-BU, and HA cause transition mutations and AO causes frameshift mutations, the mutation must be of another type. Therefore, a transversion is the most likely cause of the *z* mutation.

Exercises and Problems

21. Suppose a single-base substitution occurs in a codon that specifies arginine. List all amino acids that would replace arginine if the mutation was a
    a. transition.   b. transversion.

22. Which of the following double-stranded DNA molecules is expected to be the most sensitive to UV light damage? The least sensitive? Explain.

Molecule	% GC content
I	50
II	40
III	30

23. The first 21 bases of the coding region of a gene are shown here along with the specified amino acids. For each of the following mutants, indicate whether the mutation is a transition, transversion, insertion, or deletion, and whether it is a silent, missense, nonsense, or frameshift mutation.

    5' ATG CCG ACT AAC TAC AGG AAA . . . 3'
       Met Pro Thr Asn Tyr Arg Lys

    a. 5' ATG CCG ACT AAC TAT AGG AAA . . . 3'
    b. 5' ATG CCG ACT AAC TAA AGG AAA . . . 3'
    c. 5' ATG CCG ACT AAC TAC AAG AAA . . . 3'
    d. 5' ATG CCG ACT CAA CTA CAG GAA . . . 3'
    e. 5' ATG CCG ACT AAC TAA GGA AAC . . . 3'

24. In general, which point mutations are expected to have a greater deleterious effect on an organism: frameshift or base substitutions? Why?

25. Assuming base-substitution mutations occur at the same frequency, what is the expected ratio of transitions to transversions?

26. Would a genetic system that correctly repaired every damaged base pair be an advantage to an organism?

27. A point mutation occurs in a particular gene. Describe the types of mutational events that can restore a functional protein, including intergenic events. Consider missense, nonsense, and frameshift mutations.

28. A new IS element was discovered in bacteria. Which of the following pairs of DNA sequences could be found at its two ends? Why?
    a. 5' GAGACTCTAC 3' and 5' GAGACTCTAC 3'
    b. 5' GAGACTCTAC 3' and 5' CATCTCAGAG 3'
    c. 5' GAGACTCTAC 3' and 5' CTCTGAGATG 3'
    d. 5' GAGACTCTAC 3' and 5' GTAGAGTCTC 3'

29. Which of the following amino acid replacements could result from single-base substitution mutations induced by 2-aminopurine?
    a. Pro → Gln,   b. Ala → Thr,   c. Lys → Arg,
    d. Ser → Tyr,   e. Gly → Glu

30. You isolated a new histidine auxotroph, and, despite all efforts, you cannot produce any revertants. What probably happened to produce the original mutant?

31. A nonsense suppressor is isolated and shown to involve a tyrosine tRNA. When this mutant tRNA is sequenced, the anticodon turns out to be normal, but a mutation is found in the dihydrouridine loop. What does this finding suggest about how a tRNA interacts with the mRNA?

32. Which of the following amino acid changes could result from a single transversion event?
    a. Glu → Asp,   b. Ile → Val,   c. Leu → Pro,
    d. Ala → Gly,   e. Met → Thr

33. Can hydroxylamine induce nonsense mutations in wild-type genes? Can it reverse nonsense mutations? Explain.

34. During transcription of a gene by RNA polymerase, a ribonucleotide is deleted from the mRNA molecule. Would this event be considered a mutation? Why or why not?

35. Is the term *"jumping" gene* an accurate description of all transposable elements?

36. You perform an Ames test on three different chemicals and obtain the following results: (numbers represent revertant colonies)

	Tube		
	I	**II**	**III**
Compound	*his⁻* Culture Without Chemical	*his⁻* Culture plus Chemical	*his⁻* Culture plus Chemical plus Rat Liver Extract
A	7	6	8
B	10	12	107
C	15	213	17

Indicate which of the following statements about this experiment are true. If the statement is false, change a few words to make it true.

1. The *his⁻* culture used in the Ames test is that of the bacterium *Staphylococcus aureus.*
2. The revertant colonies were formed on agar containing histidine.
3. Tube I serves as control.
4. Tube II approximates what would happen in the human body.
5. Compound A is not mutagenic, but its metabolites are.
6. Compound B is mutagenic, and so are its metabolites.
7. Compound C is mutagenic but not its metabolites.
8. Compound C is carcinogenic.

37. Consider the following DNA sequence:

5' ATCG 3'

3' TAGC 5'

a. Draw the sequence as it would appear immediately after treatment with nitrous acid. Use H for hypoxanthine and U for uracil. Assume guanine bases are unmodified.

b. If the nitrous acid is removed and the DNA is replicated for two generations, what final sequences of DNA will be produced?

38. The original DNA sequence of problem 37 is incubated with 2-aminopurine and 5-bromouracil for one round of replication. The base analogs are then removed and the DNA is replicated for two more rounds. What final DNA sequences will result? Use the symbols P for 2-aminopurine and B for 5-bromouracil.

39. Construct a data set that Luria and Delbrück might have obtained that would prove the mutation theory wrong.

40. Achondroplasia is a form of dwarfism that is inherited as an autosomal dominant trait. A detailed survey of all hospital records in the hypothetical country of Mendelania revealed that among 420,316 live births in 2001-2003, there were 21 infants with achondroplasia. In nine cases, the affected children had one parent with the disorder.

a. Calculate the mutation rate and mutation frequency for achondroplasia among these newborns?

b. Propose two reasons why your estimate may not be accurate.

41. One of these achondroplastic children was born to Klaus, a man who had been working in Mendelania's nuclear power plant for 10 years. Neither Klaus nor his wife are achondroplastic. There is also no history of achondroplasia in either family. Franz, Klaus' co-worker for 7 years, becomes the father of a boy with Duchenne muscular dystrophy (an X-linked recessive disorder). There is no evidence of this disorder in him, his wife, or anyone in their extended families. Klaus and Franz sue their employer for damages, claiming that their working environment was responsible for the diseases that their sons were born with. A genetics expert, Karl-Heinz, was asked to testify in a pretrial hearing. What do you think was his take on the validity of the two lawsuits?

42. A wild-type *Drosophila* male is subjected to a heavy dose of X-rays, and then crossed to a wild-type female with no mutant genes. Assume the radiation exposure induced an X-linked recessive lethal mutation. What is the expected ratio of male to female in the

a. F_1?

b. F_2?

43. The DNA polymerase III holoenzyme of *Escherichia coli* is the major chromosome replication enzyme. DNA pol III is composed of 10 different subunits. It contains two cores, each consisting of an α, ε, and θ subunit that are arranged in a linear fashion: α–ε–θ (ε binds to both α and θ, which are not bound to each other). The *dnaE, dnaQ,* and *holE* genes encode the α, ε, and θ subunits, respectively. Genetic analysis of the DNA pol III core has yielded insight into the roles of these three subunits. For example, a number of *dnaE* temperature-sensitive mutants have been discovered. These mutants encode subunits that are functional at the lower permissive temperature and defective at the higher temperature, which results in conditional lethality. Other *dnaE* mutants exhibit mutator or antimutator effects. On the other hand, many *dnaQ* mutants display strong mutator phenotypes (more than 10^4 times higher than the spontaneous rate), including transitions, transversions, and even frame-shift mutations. Lastly, deletion mutants of *holE* are fully viable, showing no difference in cell health, shape, and mutation rates when compared with wild-type cells.

Based on this information, indicate the most likely functions of the α, ε, and θ subunits.

44. The Tn10 transposase preferentially binds hemimethylated DNA. What advantage does this have for the transposon?

45. Spontaneous depurination is fairly common in the diploid human genome (6×10^9 bp). It is estimated that a diploid cell loses about 19,500 purines every 18 h. In light of this information, calculate the rate of spontaneous depurination per purine per minute.

46. The wild-type mRNA and specified polypeptide sequences of a gene are:

5' CCGAC <u>AUG UGG ACA AGU GAA CCG UCA GCA UAA</u> GCACG 3'
          Met Trp Thr Ser Glu Pro Ser   Ala Stop

A single point mutation was induced within this gene by proflavin. Analysis of the wild-type and mutant DNA revealed a single-base difference in the region that encodes the underlined sequence in the preceding diagram. However, this point mutation was silent, producing a polypeptide with an identical amino acid sequence as that of the wild type. Determine, to the extent that is possible, the identity and location of this point mutation.

47. An IS element is found to contain a large deletion in its transposase gene. Will this element be still capable of transposition? If not, why, and if so, how?

48. In a particular organism, the mutation frequencies of genes T and Z are 7×10^{-5} and 3×10^{-6} per generation, respectively. What is the likelihood that
   a. *either* gene will undergo a mutation in a single generation?
   b. *neither* gene will undergo a mutation in a single generation?

49. Why is postreplicative repair a necessity for multicellular organisms?

50. Suppose two consecutive spontaneous point mutations occurred in a codon specifying arginine. These mutations are:

          Transition   Transition
Arginine ——▶ Amino acid ——▶ Stop codon

Determine, to the extent that is possible, the codons and missing amino acid(s) in this sequence.

51. A scientist decides to create a mosaic patch by injecting DNA encoding a gene into embryonic *Drosophila* cells. He borrows a construct containing the gene between the repeats of a *P* element and another construct containing the *P* element transposase gene. He injects individual cells of over 1000 embryos, but does not see any expression of the added gene in even a single adult. What was wrong with the experimental design?

52. The amino acid sequence of a particular polypeptide in a wild-type bacterium and three single point mutant strains are shown in the following table:

	Codon									
	1	2	3	4	5	6	7	8	9	10
Wild type =	Met	Tyr	Ile	Thr	Trp	Asp	Glu	Pro	Val	Lys
Mutant 1 =	Met	Tyr	Ile	Thr	Trp	Asp	Glu	Pro	Val	Ile
Mutant 2 =	Met	Tyr	Ile	His	Val	Gly				
Mutant 3 =	Met	Tyr	Ile	Thr	Trp	Met	Asn	Leu		

   a. For each mutant indicate the type of point mutation (transition, transversion, addition, or deletion) and the codon in which it occurred.
   b. Determine, to the extent that is possible, the ribonucleotide sequence of the wild-type mRNA molecule?

Chapter Integration Problem

Inhabitants of the planet Humouranus are quite funny! Sense of humor is under the control of an autosomal gene called *fb* (for "funny bone"). The gene locus has three alleles: *fb*⁺, which is wild type and dominant, and *fb*^Ok and *fb*⁻, which are both rare and recessive.

The allele *fb*⁺ gives individuals the ability to tell funny jokes and to quickly "get" the jokes of others. Individuals that are homozygous for *fb*^Ok have a so-so sense of humor: they tell jokes that are lame or have no punch lines, and they do get jokes, but only after considerable time. Finally, *fb*⁻*fb*⁻ individuals have no sense of humor whatsoever; they cannot tell jokes and simply don't get the jokes of others.

Assume that the molecular biology and gene expression of Humouranusians is the same as that of humans.

The *fb*⁺ gene is 3751 bp long and is organized as follows:

| A | B | C | D | E | F | G |

where A is the 5' UTR = 153 bp
       B is exon 1 = 375 bp (including the start codon)
       C is intron 1 = 111 bp (does not interrupt any codon)
       D is exon 2 = 1971 bp
       E is intron 2 = 290 bp (does not interrupt any codon)
       F is exon 3 = 672 bp (including the stop codon)
       G is the 3' UTR = 179 bp

The following double-stranded sequence of DNA is obtained from codons 136–145 of the *fb*⁺ gene:

```
GAC CTT CGA GTT TGA AAA AAA AAT CTA ATG
CTG GAA GCT CAA ACT TTT TTT TTA GAT TAC
```

   a. Label the 5' and 3' ends of the DNA, as well as the template and coding strands. Justify your answers.
   b. Write the mRNA and polypeptide sequences corresponding to this gene fragment.
   c. How many different single-base substitution mutations are possible in the above DNA sequence?
   d. What type of mutations are most likely to occur in the previous DNA sequence and why?
   e. Consider the following frameshift mutations:
      i. Deletion of the base pair at the 3' end of codon 145;
      ii. Insertion of a G–C base pair immediately following codon 145.
   What is the probability that mutation **(i)** will generate a translation termination triplet in codon 145? What is the probability that mutation **(ii)** will generate a translation termination triplet in codon 146? In codon 147? Assume the genome of Humouranusians has a 50% GC content and random sequence.

f. Predict the changes that the following mutations will induce on the fb^+ gene at the levels of DNA, RNA, and encoded polypeptide. All mutations occur in the fb^+ gene region. Assume that the DNA damage is not repaired and that it will eventually affect both DNA strands.

	Mutation	Location
i.	Tautomerization to the rare imino form	First A in the template strand of codons 136–145
ii.	Tautomerization to the rare enol form	Second T in the nontemplate strand of codons 136–145
iii.	Deamination	Third C in the coding strand of codons 136–145
iv.	A → G	Nucleotide number –173 (template strand)
v.	Insertion of TGA	After nucleotide number + 208 (template strand)
vi.	Deletion	Nucleotides + 529 to + 538 (template strand)
vii.	TTG → CTG	Nucleotides + 1840 to + 1842 (template strand)
viii.	T → C	Nucleotide + 3570 (template strand)

g. The following table displays data from five different homozygous mutants:

	Mutation	Nucleotides (Template Strand)	Sense of Humor
Mutant 1	Deletion	– 35 to – 17	Absent
Mutant 2	Deletion	+ 760 to + 768	Okay
Mutant 3	CTC → CTA	+ 1005 to + 1007	Absent
Mutant 4	AAG → GAG	+ 2440 to + 2442	Absent
Mutant 5	CGA → TGA	+ 3558 to + 3560	Okay

For each mutant provide a (reasonable) explanation for why the particular change caused a mutation and why it specifically produced no sense of humor instead of an OK sense of humor or vice versa.

h. Zorkan Tarazan is an individual who is homozygous for a transversion mutation at nucleotide + 692. What is his genotype for the sequence in question? Is Zorkan's sense of humor very good, okay, or absent? Justify your answers.

i. Zorkan's wife, Zelda, is homozygous for the wild-type fb^+ gene. They have a son named Zultan. What is the probability that Zultan's son will inherit the transversion mutant allele from Zorkan? If the transversion mutant allele is present in 1 in 12,500 in the Humouranusian population, what is the probability that Zultan's son will be homozygous for this allele?

Do you need additional review? Visit **www.mhhe.com/hyde** for practice tests, answers to end-of-chapter questions and problems, interactive exercises, and animation tutorials all designed to enhance your understanding of key genetic concepts.

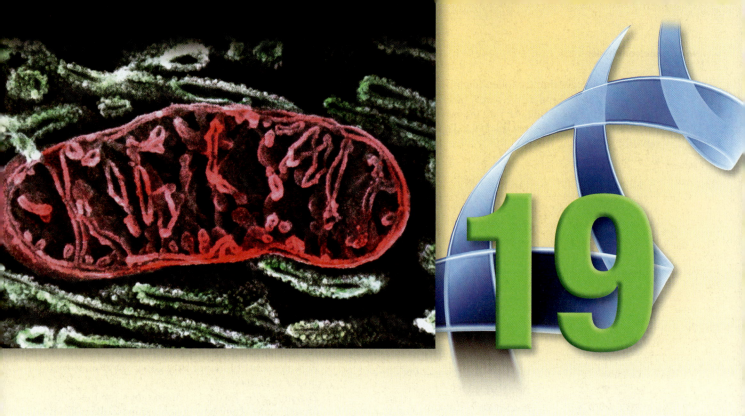

Extranuclear Inheritance

Essential Goals

1. Describe the various functions of chloroplasts and mitochondria in the eukaryotic cell.

2. What is the endosymbiotic theory? Compare the similarities and differences in the origin of chloroplasts and mitochondria according to this theory.

3. Compare and contrast the genomes associated with chloroplasts and mitochondria. How do these organellar genomes compare with the nuclear genomes in the same cell? Why do these organelles require their own genome?

4. What are the different patterns of organellar inheritance? How can you distinguish between these different patterns of inheritance?

5. What are the general features of inherited human mitochondrial diseases?

Chapter Outline

Photo: Pseudocolored scanning electron micrograph of a mitochondrion (red) in the cytoplasm of an intestinal epithelial cell.

Y ou may remember from chapter 4 that Y-linked traits are passed from father to son and do not appear in females. X-linked traits are passed from a mother potentially to all of her children, whereas males pass X-linked traits only to their daughters. A group of inherited human diseases exists, however, that fails to follow any of these pedigree patterns. This new pattern exhibits strict **maternal inheritance,** which means that the mother passes it on to all her children, and the father never passes it on to his children. As you will see in this chapter, one major source of maternal inheritance is through the genomes that are found in organelles, such as the mitochondria and the chloroplasts.

Studying the human mitochondrial genome is interesting in two ways. The first is the study of genetic diseases that result from mutations in the mitochondrial genome. We discuss some of these diseases and their pattern of inheritance at the end of this chapter. The second interesting analysis pertains to the identification of the origins of the human race. Mutations in the mitochondrial DNA of animals are repaired at a low rate relative to the nuclear DNA. This results in a relatively high mutation rate in the mitochondrial genome that has allowed scientists to draw conclusions about human origins.

The **mitochondrial Eve hypothesis** suggests that all human mitochondrial genomes evolved from a "single" original genome approximately 200,000 years ago. Because the mitochondrial genome is maternally inherited, the original genome must have been present in the first *Homo sapiens* female (which accounts for the name "Eve").

This dating is in remarkably close agreement with a similar analysis performed on the human Y chromosome. Molecular analyses suggests that the "original" human Y chromosome, present in the first *H. sapiens* male (which correspondingly would be named "Adam"), appeared approximately 180,000 years ago. Both estimates also correspond very well with the isolation of human remains in Ethiopia that were determined to be nearly 160,000 years old.

Of course, these findings and the names do not suggest that only two humans existed at that time. Instead, these discoveries indicate that the DNA humans have today has come down from a small group of ancestral humans. Over time, events that drastically reduced populations may have also limited the variability that was originally present in female mitochondria and the male Y chromosome—a type of bottleneck effect.

The presence of traits that can be transmitted outside of nuclear DNA is yet another example of how inheritance can be much more complex than the simple dominant and recessive model envisioned by Mendel.

19.1 Physiological Roles of Mitochondria and Chloroplasts

Mitochondria and chloroplasts are **organelles,** membrane-bounded compartments within eukaryotic cells. All eukaryotes, excluding some protists, possess mitochondria, which serve as the site of high levels of ATP production in the cell. Chloroplasts, in which photosynthesis occurs, are found in plants and algae.

Unlike many eukaryotic organelles, mitochondria and chloroplasts contain small, and usually circular, genomes. These organellar genomes exhibit **extranuclear inheritance** because they are passed on to progeny in a manner that is separate from the nuclear genome. We discuss the inheritance of these organelles later in this chapter. First, however, we discuss the structure and function of these organelles, followed by their genome organization.

Structure and Function of Mitochondria

A mitochondrion (singular) is usually viewed as a bean-shaped organelle composed of an outer and inner membrane (**fig. 19.1**). The inner membrane is highly folded into structures termed cristae (**fig. 19.2**). The inner membrane also surrounds the central matrix, which is the location of the mitochondrial genomic DNA, RNA, and ribosomes for protein translation.

The mitochondrion is the site where *electron transport* takes place, which is the process of producing high levels of ATP via oxidative phosphorylation. The actual number of mitochondria per cell depends on the energy requirements of the cell, with cells that utilize high levels of ATP possessing the most mitochondria. Estimates range between 10 and 10 thousand mitochondria per cell, depending on the organism and cell type.

Mitochondria contain several features that make them ideal for ATP synthesis. First, ATP production occurs on membranes, and the presence of the membrane-enriched cristae makes an ideal environment for large amounts of ATP production. Second, the inner membrane is impermeable to ions and small molecules. ATP is produced through the metabolism of sugars to pyruvate, which enters the mitochondria and is degraded to CO_2 in the Krebs cycle. The Krebs cycle also produces reduced NADH and $FADH_2$ molecules that are used to generate the electrons that pass through the respiratory electron-transport chain (**fig. 19.3**), with oxygen as the final electron acceptor. This electron transfer creates a proton gradient across the mitochondrial inner membrane that drives the production of ATP.

In addition to ATP production, portions of a number of biosynthetic pathways occur within mitochondria, including lipid metabolism, nucleotide metabolism, amino acid metabolism, heme synthesis, and ubiquinone synthesis.

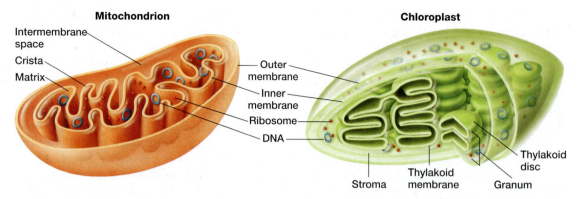

figure 19.1 Comparison of mitochondrial and chloroplast structures. Shown are the two double-membrane bounded DNA-containing organelles of the eukaryotic cell: the mitochondrion and the chloroplast. The mitochondrial inner membrane is continuous, forming the inner folds or cristae. The matrix is the soluble portion of the organelle that is within the inner mitochondrial membrane. In contrast to the two mitochondrial membranes, the chloroplast contains a third membrane that forms the thylakoid disks. Both organelles contain DNA molecules and ribosomes, which allows transcription and translation to be carried out within the organelle.

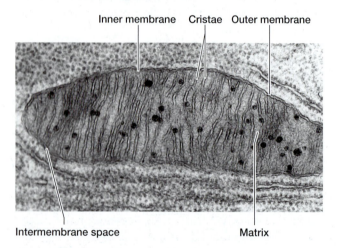

figure 19.2 Electron micrograph of a mitochondrion from bat pancreas. The outer membrane and inner membrane, which contains many folds that are called cristae, are shown.

Structure and Function of Chloroplasts

Chloroplasts are one of several different types of **plastids,** organelles found in all plants and some protists. Chloroplasts are found in the green plant tissues, such as the leaves and stems, where photosynthesis occurs. In nonphotosynthetic plant tissues, such as flower petals and roots, plastids are also present, but they have not differentiated into chloroplasts that carry out photosynthesis. In these tissues, the plastid is called either a chromoplast (colored plastid), elaioplast (oil-containing plastid), or leucoplast (starch-containing plastid). These other plastids are active in certain essential plant functions, such as amino acid synthesis, fatty acid synthesis, and the production of pigments and growth regulators. These functions are probably the reason that even root and floral cells of plants are not viable unless they contain plastids.

The plastids develop from **proplastids,** which are small immature organelles found in the plant meristems. A proplastid is about the size and shape of a mitochondrion. When grown in the dark, however (and under some other circumstances), plastids do not develop into chloroplasts, but remain reduced in size and complexity.

Like mitochondria, chloroplasts also contain two peripheral membranes, but both the inner and outer membranes are smooth, unlike the highly folded inner membrane in mitochondria (see fig. 19.1). The chloroplasts also contain a third membrane, which is arranged in flattened disks called thylakoids. The thylakoids are arranged in stacks called grana. The inner fluid environment surrounding the thylakoids is called the stroma.

The energy from sunlight drives electron transfer, similar to what goes on in mitochondria to produce ATP. In chloroplasts, this process transports ions from the stroma into the thylakoid space within the thylakoid disks (**fig. 19.4**). The proton gradient drives the production of ATP, and both the ATP and the NADPH are used for carbon fixation.

Photosynthesis occurs in green plant tissues, such as the leaves and stems, and the cells of those tissues contain many chloroplasts.

Restating the Concepts

▶ The mitochondrion is a double-membraned organelle found in most eukaryotes. It is the site where oxidative phosphorylation occurs to generate large amounts of ATP for the cell.

▶ Chloroplasts, found in plants and algae, are composed of three membranes: an inner and outer membrane like mitochondria, and one additional membrane that surrounds the internal thylakoid disks. Chloroplasts are the site of photosynthesis in the cell. Plastids are the precursor organelle of chloroplasts.

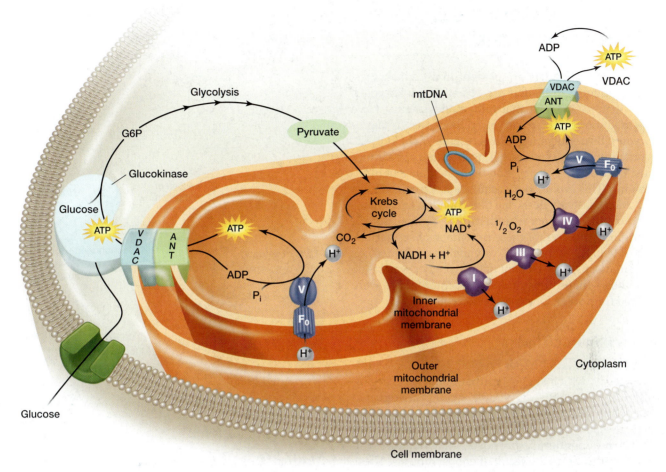

figure 19.3   Overview of mitochondrial metabolism. The breakdown of sugars, such as glucose into pyruvate through glycolysis, occurs in the cytoplasm. Pyruvate is then transported into the mitochondrial matrix, where it enters the Krebs cycle and is metabolized into CO_2. This metabolism also produces the cofactors NADH and $FADH_2$ that are then oxidized to NAD^+ and FAD, which produce electrons that enter the electron transport chain (purple structures I, III, and IV) to reduce O_2 to H_2O. This ion gradient across the inner mitochondrial membrane then allows for the aerobic production of ATP through Complex V (ATP synthase) in the mitochondrion.

figure 19.4   Overview of chloroplast metabolism. Light activates Photosystem I and Photosystem II to generate NADPH and a proton gradient by hydrolyzing H_2O to $2H^+$ and $\frac{1}{2}O_2$, respectively. The ATP synthase uses the proton gradient to generate ATP. The NADPH, ATP, and CO_2 enter the Calvin cycle to generate sugars. The sugars are then transported out of the chloroplast and proceed through glycolysis to produce pyruvate in the cytosol. The pyruvate can enter the mitochondria to generate ATP.

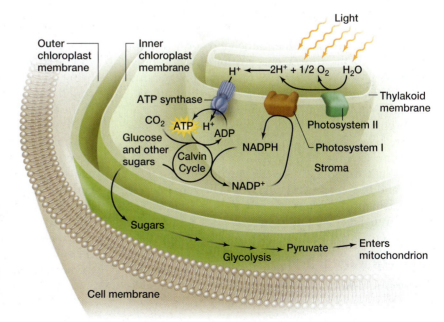

19.2 Origin of Mitochondria and Chloroplasts

Because mitochondria are present in all eukaryotes, except some protists, and are absent in prokaryotes, investigators deduced that mitochondria must have appeared shortly after eukaryotic cells evolved from prokaryotes. By a similar logic, the presence of chloroplasts in only some eukaryotes suggests that their appearance must have occurred later than the appearance of mitochondria. This evolutionary acquisition of mitochondria and chloroplasts suggests that mitochondria and chloroplasts should share some similarities with their source of origin. Searching for these fundamental similarities should help reveal how these organelles originated. As we will discuss, the current data strongly support the hypothesis that mitochondria and chloroplasts were derived from an endocytosed prokaryotic cell.

Endosymbiotic Theory

In 1883, Andreas Schimper proposed the **endosymbiotic theory,** which suggested that eukaryotic organelles were originally derived by a cell that engulfed (phagocytosed) either DNA or a prokaryotic cell. In the early 1900s, Konstantin Mereschkowsky and Ivan Wallin independently proposed that plastids and mitochondria were the result of endosymbiosis. These theories were quickly discounted because these organelles were thought to lack DNA. However, the discovery of DNA in mitochondria and chloroplasts in the 1960s led to the reconsideration of this theory.

In the 1980s, Lynn Margulis and her collaborators collected a variety of data that supported the endosymbiotic theory for the origin of mitochondria and chloroplasts. This theory proposes that an archaean phagocytosed an aerobic α-proteobacterium, which evolved into a mitochondrion (**fig. 19.5**). In this model, the inner membrane of the mitochondrion would be derived from the α-proteobacterium, and the outer membrane would be derived from the phagocytosing archaean. The resulting cell could then proceed through evolution into an animal cell.

A chloroplast could be generated if this cell (or its descendents) subsequently phagocytosed a cyanobacterium (see fig. 19.5). The inner membrane of the chloroplast would be derived from the cyanobacterium, and the outer membrane would be derived from the endocytosing eukaryotic cell. This cell could then evolve into either a plant or an alga.

Origin of Mitochondria

The endosymbiotic event that produced mitochondria would have occurred over 2 billion years ago. However,

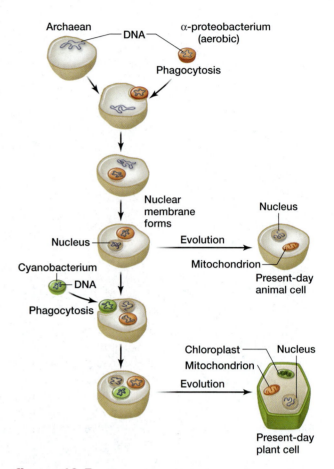

figure 19.5   The endosymbiont theory of generating plant and animal cell organelles. An archaeal cell engulfed an aerobic α-proteobacterium that evolved into a mitochondrion. This cell later phagocytosed a cyanobacterium, which evolved into a chloroplast. The model depicted here shows the nucleus evolving before the mitochondrion became established.

the mitochondrion should still retain features of its bacterial ancestry, and evidence of this ancestry has been found. For example, the great majority of mitochondrial genes, with the yeast mitochondrial genome being the exception, lack introns. This makes the mitochondrial genes more similar to bacterial genes than to the eukaryotic nuclear genes.

With progress in genome-sequencing projects, it is possible to perform a variety of DNA sequence comparisons. For example, the mitochondrial gene sequences can be compared with a variety of different bacterial genomic sequences. This DNA sequence analysis confirms the similarity of mitochondrial gene sequences with related genes from α-proteobacteria, a group that includes *Rickettsia prowazekii* and *Rhodospirillum rubrum*. Many of the α-proteobacteria, furthermore, form symbiotic relationships with eukaryotes, such as *Agrobacterium tumefaciens, Bradyrhizobium japonicum,* and *Sinorhizobium meliloti.*

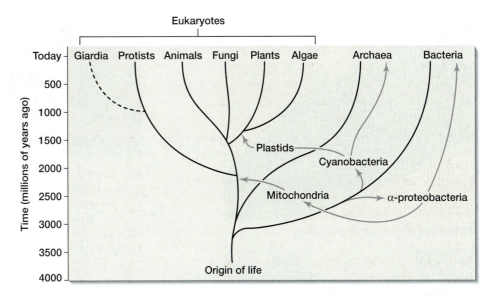

figure 19.6 Phylogenetic tree showing the acquisition of mitochondria and chloroplasts in relation to when the major life forms were thought to have originated. Placed within this tree is the estimated time and location when the aerobic α-proteobacterium and cyanobacterium were phagocytosed to yield the mitochondrial and chloroplast organelles, respectively.

Another important DNA sequence comparison can be made between the mitochondrial genome sequences from different organisms. If the endosymbiotic event that created mitochondria occurred before the acquisition of chloroplasts (**fig. 19.6**), then the mitochondrial genomes would have had more time than the chloroplast genomes for their sequences to diverge from the original bacterial ancestor. As discussed later in this chapter, mitochondrial genomes do exhibit much more variability than the chloroplast genomes.

Further support that mitochondria arose from a bacterial cell stems from the basic genetic functions of this organelle. For example, translation initiation of mitochondrial mRNAs employs tRNA$_f$^{Met} (tRNA charged with *N*-formyl methionine, see chapter 11), as does bacterial translation initiation. Remember that eukaryotes also utilize a special tRNA for translation initiation, tRNA$_i$^{Met}, although the methionine lacks the *N*-formyl group.

Additionally, the mitochondrial ribosomes contain some features that are more similar to bacterial ribosomes than they are to eukaryotic ribosomes. First, the mitochondrial rRNAs are more similar to bacterial rRNAs than to eukaryotic rRNAs. Second, even though the mitochondrial ribosome is constructed of imported cellular proteins, it is sensitive to particular antibiotics, such as streptomycin and chloramphenicol, which can inhibit translation with bacterial ribosomes but not with eukaryotic ribosomes. These and other similarites between mitochondria and bacteria strongly support the endosymbiotic origin of mitochondria.

Origin of Chloroplasts

Chloroplasts appear to be related to cyanobacteria at several different levels. For example, chloroplasts and cyanobacteria possess many similar structural features,

such as internal membranes (**fig. 19.7**). Additionally, the cyanobacteria, formerly called blue-green algae, carry out photosynthesis as do chloroplasts. Finally, the ribosomal RNA of chloroplasts share significant DNA sequence similarities with the cyanobacterial DNA. Taken together, it appears that chloroplasts and cyanobacteria share a common ancestor.

Another similarity that exists between plastids (and mitochondria) and bacteria occurs during cell division. When a eukaryotic cell divides, the plastids are not generated de novo; they arise from pre-existing plastids. Both plastids and mitochondria divide by fission (**fig. 19.8**) in a process that is very similar to bacterial fission. In fact, organellar division relies on many proteins that can be tracked back to a bacterial origin. As you will also see shortly, further similarities can be found between the chloroplast and the cyanobacterial genomes.

Restating the Concepts

▶ The endosymbiotic theory proposes that eukaryotic organelles were generated by an archaean phagocytosing a bacterial cell. Mitochondria are thought to be evolved from an engulfed aerobic α-proteobacterium, and chloroplasts from a phagocytosed cyanobacterium.

▶ Support that mitochondria are related to aerobic α-proteobacteria comes from DNA sequence comparison, the similar use of *N*-formyl methioine and tRNA$_f$^{Met} for translation initiation, and the ribosome structure and function.

▶ Support that chloroplasts were evolved from cyanobacteria includes the structure of the organelle, the ability to perform photosynthesis, and similarities in rRNA sequence.

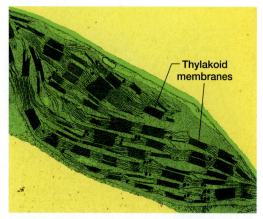

(a) Electron micrograph of chloroplast

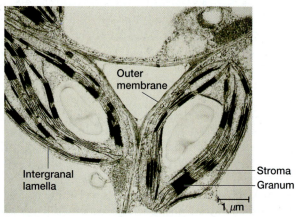

(b) Electron micrograph of cyanobacterium

figure 19.7   Structural similarities between a chloroplast and cyanobacterium. An electron micrograph reveals that the stacked grana or thylakoids in the lettuce chloroplast (a) are similar to the organization of grana and stroma in the cyanobacterium *Microcoleus rushforthii* (b).

19.3 Structure and Organization of the Organellar Genomes

If mitochondria and chloroplasts were derived from endosymbiotic bacteria, then their genomes should also possess some similarities with the bacterial genomes. The genomic DNAs in most mitochondria and chloroplasts are circular and essentially free of chromosomal proteins, like bacterial genomes. In some protozoa, the mitochondrial DNA is a double-stranded linear molecule. Furthermore, both organellar genomes replicate using bacterial-like mechanisms, such as a D-loop mechanism (**fig. 19.9**; see chapter 9).

Mitochondrial Genome

In animal and plant cells, the **mitochondrial DNA (mtDNA)** is located within the nucleoid region of the

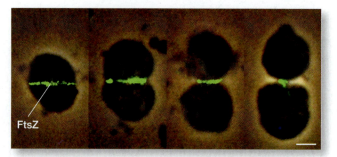

figure 19.8   Immunolocalization of the FtsZ protein during chloroplast division. Shown are four different stages of cell division. The FtsZ protein, which is detected by an antibody that fluoresces green, forms a ring at the site of division of the chloroplast. This process is similar to the fission of bacterial cells.

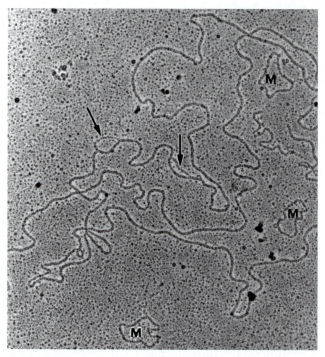

5 kb

figure 19.9   Chloroplast DNA replicates through a D-loop structure. The circular chloroplast DNA from *Oenothera* is shown in this electron micrograph. Two D-loops are marked with arrows. The small circular bacteriophage genomes (M) were included for size purposes.

mitochondria (**fig. 19.10**). Usually one to three nucleoids occur per mitochondrion, with several copies of the circular genome in each nucleoid. Because every cell contains many mitochondria (**fig. 19.11**), there are numerous copies of the mitochondrial genome in every cell.

Among the mitochondrial DNAs that have been sequenced from over 1,280 different organisms (as of November, 2007), we see great variation in content and

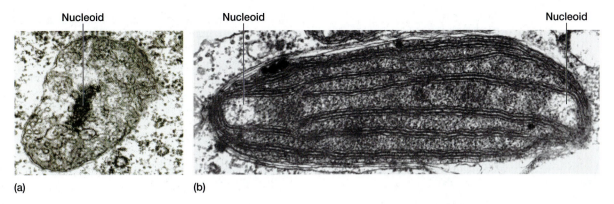

figure 19.10   Electron micrograph of the nucleoids in a mitochondrion and chloroplast. Within the mitochondrion (a) is the electron dense nucleoid, which is the location of the mitochondrial DNA. A chloroplast (b) shows two nucleoids, which are where the chloroplast DNAs are located.

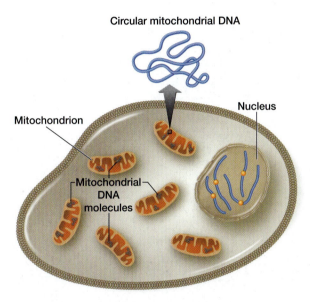

figure 19.11   Relationship between mitochondria and mtDNA in a cell. A eukaryotic cell may contain multiple mitochondria, which each contain one or more identical circular mitochondrial DNA molecules. The nuclear genome consists of one or more linear DNA molecules, which may all be different.

table 19.1   Comparison of Mitochondrial DNA Genome Sizes

Organism	Size of Mitochondrial Genome (in kb)
Mouse (*Mus musculus*)	16.2
Human (*Homo sapiens*)	16.6
Fruit fly (*Drosophila melanogaster*)	18.4
Yeast (*Saccharomyces cerevisiae*)	75.0
Pea (*Pisum sativum*)	110.0
Liverwort (*Marchantia polymorpha*)	186.0
Mustard plant (*Arabidopsis thaliana*)	367.0

gene organization (**fig. 19.12**). The sizes of the mitochondrial genomes range from less than 4 kb to more than 360 kb and encode from 3 to 97 genes (**table 19.1**). With this wide range of sizes and number of genes, one of the few generalities we can make about mtDNA is that the large and small mitochondrial ribosomal RNAs, as well as most of the mitochondrial transfer RNAs, are usually encoded in the mitochondrial genome. Several proteins in respiratory complexes III and IV (cytochrome *c* oxidase and cytochrome *c* oxidoreductase), which are required for ATP synthesis in the mitochondrial oxidative phosphorylation pathway, are also encoded in the mitochondrial genome.

The human mitochondrial genome is present in 2 to 10 copies per organelle. The mtDNA is 16,569 bp long and represents a model of economy, with very few noncoding regions and no introns (see fig. 19.12). Each strand of the double-stranded DNA is transcribed into a single RNA product that is then cut into smaller pieces, primarily by freeing the 22 tRNAs interspersed throughout the genome. A 16S rRNA and a 12S rRNA are also formed, along with 13 mRNAs that encode polypeptides.

Although proteins and small molecules such as ATP and tRNAs can move in and out of the mitochondrion, large RNAs cannot. Thus, the mitochondrion must be relatively self-sufficient in terms of the RNAs needed for translation of the 13 mitochondrial-encoded polypeptides.

The mitochondrial oxidative phosphorylation process requires at least 69 polypeptides. The human mitochondrion has the genes for 13 of these (see fig. 19.12): cytochrome *b*, two subunits of ATP synthase, three subunits of cytochrome *c* oxidase, and seven

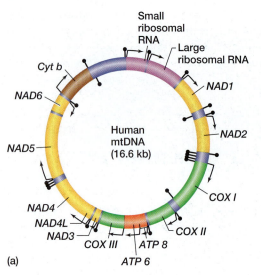

(a)

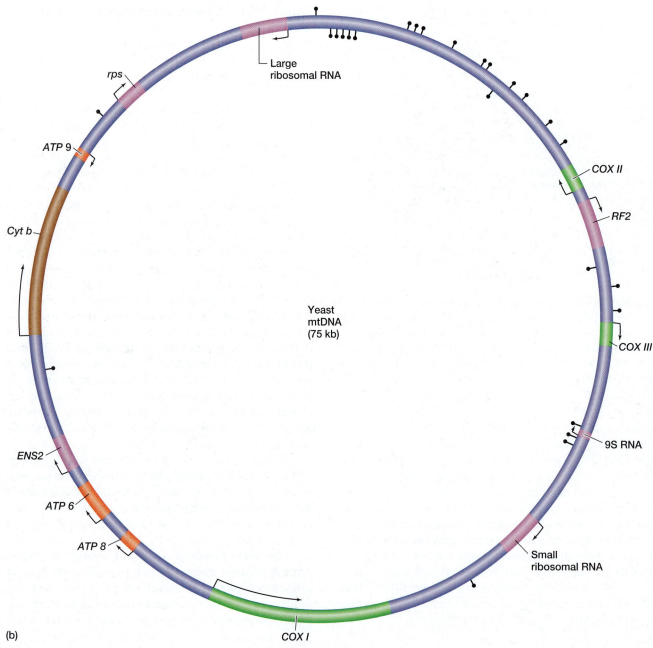

(b)

subunits of NADH dehydrogenase. The remaining polypeptides needed for oxidative phosphorylation are transported into the mitochondrion; they are transcribed from nuclear genes and are then translated in the cytoplasm.

Mitochondrial Gene Transfer to the Nucleus

The nuclear genes that encode proteins required by the mitochondria represent an interesting aspect of mitochondrial evolution. Both mitochondria and plastid genomes encode only a limited number of the proteins that they require to perform their physiological functions. Many of the genes of the original symbiotic bacteria were lost, as they acquired metabolic intermediates from the host, and a number of genes were transferred into the host genome over a long evolutionary period.

Mitochondria show great variability in the genes that have been transferred to the nuclear genome. Animal mitochondira are among those that have retained the fewest genes (**fig. 19.13**). Fungal mitochondria have retained a few more genes, but the plant mitochondria have the largest genomes, having kept far more functions encoded within their own DNAs. When researchers analyzed the relative sizes of the mitochondrial genomes relative to the number of genes they retained, they found that animals have some of the smallest known mitochondrial genomes (**fig. 19.14**). A few protists have similarly small genomes, whereas other protists, such as *Reclinomonas*, have larger genomes due to their retention of many genes.

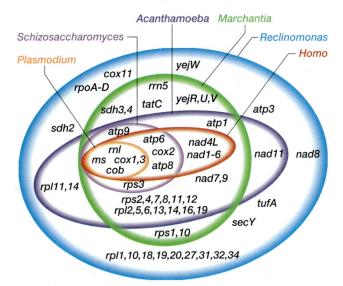

figure 19.13   Relationship of the genes found in the mtDNA in a variety of organisms. All the genes found in the *Reclinomonas* mtDNA are shown. Within each of the colored circles are the genes found in the mtDNA of the different organisms. *Plasmodium* mtDNA contains only five genes (orange circle), which are found in all the other mtDNAs. The human (*Homo*) mtDNA genes are within the red circle.
(From M. W. Gray et al., *Science*, 1999, 283: pp. 1476–1481, figure 1. Reprinted with permission from AAAS.)

Although a large number of genes appear to have been transferred from the mitochondrial genome to the nuclear genome, this type of transfer is only likely to continue happening in plants. The reason is that fungi and animals use a slightly different genetic code for mitochondrial genes relative to the nuclear-encoded genes (**table 19.2**). These five codons do not even encode the same amino acid in the various mitochondrial genomes.

For example, the mitochondrial codon 5'-AGA-3' is a translational termination codon in vertebrates, but it encodes serine in invertebrates, glycine in ascidians, and arginine in yeast. Thus, if a mitochondrial gene from any of these organisms were transferred to the nucleus, the 5'-AGA-3' codon would then encode an arginine. In vertebrates, this would result in the failure to correctly terminate translation at the 5'-AGA-3' and the production of an altered (longer) protein from the nuclear gene.

Gene transfer between a mitochondrion and the nucleus is still possible in plants because the genetic code used by both the nucleus and mitochondria in plants is universal.

How does this gene transfer from the mitochondria to the nucleus occur? Jeffrey Palmer and his colleagues studied the *cox2* gene in plants, which encodes the subunit II of the cytochrome *c* oxidase complex that is required for the last step of oxidative electron transfer. They found that the *cox2* gene is encoded in the mtDNA of most flowering plants, with a few notable exceptions, such as the bean lineage of the legumes. In pea, soybean, and the common bean, a copy of the *cox2* gene can be found in both the nuclear and the mitochondrial genomes. It was known that the mitochondrial *cox2* mRNA is posttranscriptionally edited. Comparison of the mitochondrial and nuclear *cox2* genes revealed that the nuclear gene possesses the same sequence as the edited mitochondrial *cox2* mRNA. Palmer suggested that the *cox2* gene was transcribed in the mitochondria and some of the edited mRNA was reverse-transcribed into cDNA and integrated into the nuclear genome.

In some legumes, such as the pea, only the mitochondrial copy of the *cox2* gene is transcriptionally active. In other legumes, such as soybean and the common bean, only the nuclear copy appears to be transcribed. Thus, either the nuclear copy or the mitochondrial copy of the *cox2* gene could be mutated or inactivated in the different legume lines. Based on this data, Palmer concluded that if transcriptionally active copies of the gene were present in both the nucleus and the mitochondria, either copy could attain dominance and the other copy could be lost due to mutation. Although either copy of the gene could remain transcriptionally active, the transfer of the gene is probably unidirectional, from the mitochondria to the nucleus.

In addition to transferring genes from the mitochondrial genome to the nuclear genome, examples

α-proteobacterial genome

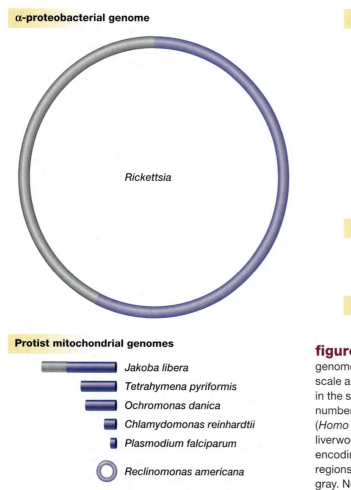

Rickettsia

Protist mitochondrial genomes

Jakoba libera

Tetrahymena pyriformis

Ochromonas danica

Chlamydomonas reinhardtii

Plasmodium falciparum

Reclinomonas americana

Chlamydomonas eugametos

Plant mitochondrial genomes

Arabidopsis thaliana

Marchantia polymorpha

Fungus mitochondrial genomes

Spizellomyces punctatus

Saccharomyces pombe

Animal mitochondrial genome

Homo sapiens

figure 19.14   Relative sizes of the *Rickettsia* α-proteobacterial genome and 12 mitochondrial genomes. The genomes are drawn to scale and show if the genome is linear or circular. The wide variation in the size of the mitochondrial DNAs does not correspond to the number of genes in the molecule. For example, human mtDNA (*Homo sapiens*) contains only a fraction of the genes found in liverwort (*Marchantia polymorpha*) mtDNA. The amount of DNA encoding genes with a known function is in blue, while intergenic regions and open reading frames with no known function are in gray. Notice that the *Spizellomyces punctatus* mtDNA is actually composed of three different circular DNA molecules.

table 19.2   Codons That Encode Different Amino Acids in Mitochondrial Genomes

Codon	Universal Code	Amino Acid Specified by Mitochondria of:				
		Vertebrate	Invertebrate	Ascidian	Yeast	Plants
UGA	stop	Trp	Trp	Trp	Trp	stop
AUA	Ile	Met	Met	Met	Met	Ile
AGA	Arg	stop	Ser	Gly	Arg	Arg
AGG	Arg	stop	Ser	Gly	Arg	Arg
CUN	Leu	Leu	Leu	Leu	Thr	Leu

were also found in which mitochondrial genes were simply lost when they were no longer required. The gut parasite *Giardia intestinalis* lacks mitochondria and was thought to be derived from the eukaryotic ancestor prior to the mitochondrial endosymbiosis. However, this protist contains organelles called mitosomes. The mitosome contains a genome that is similar to mtDNA, except for the absence of several genes that encode enzymes that are required for oxidative phosphoryla-

tion. It is now apparent that *Giardia intestinalis* underwent degenerative evolution; it lost these genes because it lived primarily in an anaerobic environment where it was unable to utilize oxidative phosphorylation.

Many transcripts in both mitochondria and plasmids are modified (**fig. 19.15**). Some RNAs, as you have learned, are spliced to remove introns. Whereas organellar genes do not tend to have as many introns as do nuclear genes, the presence of introns is a trait

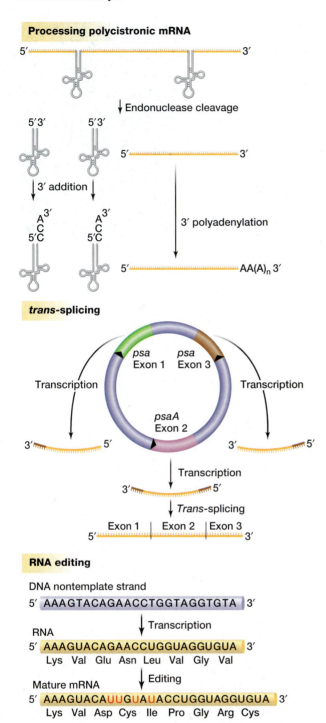

Processing polycistronic mRNA

↓ Endonuclease cleavage

↓ 3′ addition

3′ polyadenylation

trans-splicing

psa Exon 1   *psa* Exon 3

Transcription   Transcription

psaA Exon 2

Transcription

Trans-splicing

Exon 1 | Exon 2 | Exon 3

RNA editing

DNA nontemplate strand
5′ AAAGTACAGAACCTGGTAGGTGTA 3′

↓ Transcription

RNA
5′ AAAGUACAGAACCUGGUAGGUGUA 3′
   Lys Val Glu Asn Leu Val Gly Val

↓ Editing

Mature mRNA
5′ AAAGUACAUUGUAUACCUGGUAGGUGUA 3′
   Lys Val Asp Cys Ile Pro Gly Arg Cys

figure 19.15   Modifications to organellar RNAs. Mitochondrial RNAs undergo splicing, polyadenylation, *trans*-splicing, and RNA editing. In *trans*-splicing, the three *psa* RNAs that will be spliced together have an exon (yellow) and splice donor and/or acceptor sites (brown). The mechanism of *trans*-splicing is described in chapter 10 (see figure 10.44). In RNA editing, specific nucleotides in the RNA, uracils (red U) are inserted at specific locations in the open reading frame of the mRNA to alter the encoded amino acids.

that is shared between the organellar genomes and their closest ancestors: the aerobic α-proteobacteria and cyanobacteria. In some cases, those introns are found in exactly the same position as in the organelle's genome, providing a strong argument for their existence in the mutual ancestor. Some organellar RNAs also undergo RNA editing, where bases are replaced, modified, removed, or inserted (see chapter 10). And in some instances, RNA exons transcribed from opposite DNA strands are spliced together, through a process called *trans*-splicing (see chapter 10).

Chloroplast Genome

The chloroplast genome is generally similar between different algae and plants. Although some rearrangements of portions of the chloroplast genome have been detected among algal lines, most are very similar in gene content and size (**fig. 19.16**). In the case of some parasitic plants, such as the beechdrop (*Epifagus virginiana*), the need to perform photosynthesis has been lost, and their plastids lack many of the genes that encode proteins required for photosynthesis (**fig.19.17**).

As further support that chloroplasts were derived from endocytosed cyanobacteria, Jeffrey Palmer and his colleagues studied the leucine tRNA gene that possesses the 5′-UAA-3′ anticodon. In the chloroplast genome, this gene contains a single intron located within the UAA anticodon sequence. Palmer found that the related gene from five diverse cyanobacteria also contained an intron in the same location. Furthermore, there was significant sequence similarity between the chloroplast intron and the cyanobacterial intron. This suggests that they shared a common origin over one billion years ago!

It was a great surprise when the malarial parasite *Plasmodium falciparum* was found to contain an organelle that has a degenerate plastid genome (**fig. 19.18**). If plastids represent an organelle that participates in photosynthesis, then why would a parasite that exists in mosquitoes and mammals need a genome that encodes these components of photosynthesis? As it turns out, *Plasmodium* and other opportunistic pathogenic protists, such as *Toxoplasma gondii* and *Cryptosporidium parvum*, contain very small, degenerate plastid *and* mitochondrial genomes. This group of protists is thought to be related to dinoflagellates (a large class of unicellular algae). Evolutionary biologists have hypothesized that these protists were originally pathogens on marine worms, before vertebrates ever existed. Their pathogenic lifestyle evolved as eukaryotes evolved. As they developed the ability to move between mosquitoes and mammals, they lost the need to participate in photosynthesis. Thus, their plastid genomes underwent degenerative evolution with the loss of the genes required for photosynthesis, while retaining other essential organelle functions.

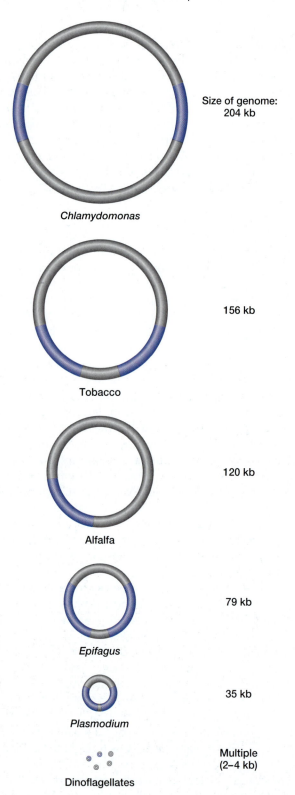

figure 19.16 Relative sizes of the chloroplast genomes from several different plants and algae. The blue represents a large inverted repeat that contains similar genes, while *Plasmodium* and Dinoflagellates possess smaller versions of the repeat.

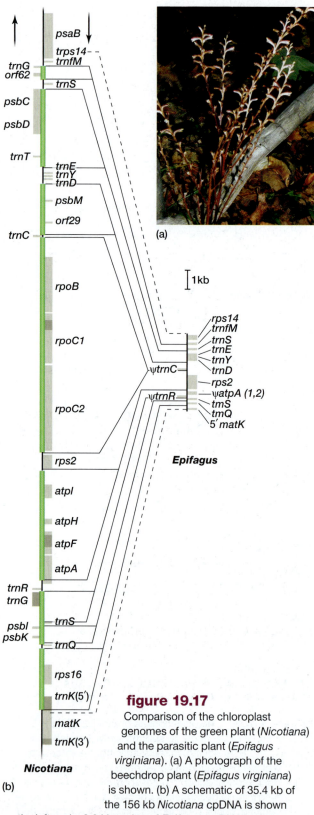

figure 19.17
Comparison of the chloroplast genomes of the green plant (*Nicotiana*) and the parasitic plant (*Epifagus virginiana*). (a) A photograph of the beechdrop plant (*Epifagus virginiana*) is shown. (b) A schematic of 35.4 kb of the 156 kb *Nicotiana* cpDNA is shown on the left and a 3.9 kb region of *Epifagus* cpDNA is shown on the right. The green portions of the *Nicotiana* genome represent genes that were lost in the degenerative evolution of the *Epifagus* genome. The main losses represent genes that encode proteins required for photosynthesis, which are not required in the parasitic *Epifagus*.

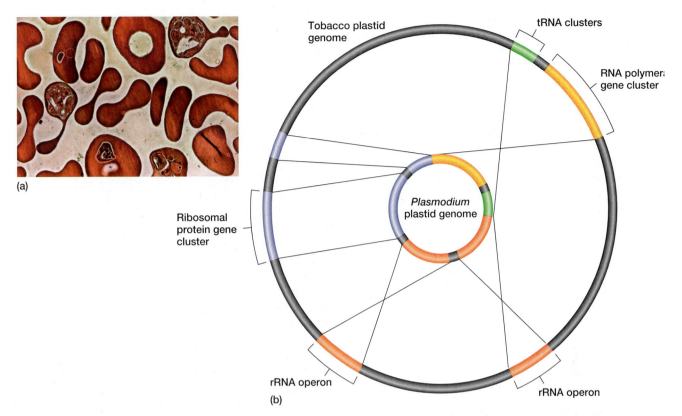

figure 19.18 The animal malarial parasite has a degenerate chloroplast. (a) A colored transmission electron micrograph of the malarial parasite *Plasmodium* in its trophozoite stage, when it inhabits the red blood cell. (b) Comparison of the tobacco (outer circle) and *Plasmodium falciparum* (inner circle) plastid genomes. The color-coded segments correspond to conserved regions between the two genomes. Notice that the *Plasmodium* plastid lacks a large amount of genetic information that is present in the tobacco plastid genome. Furthermore, both genomes have lost all the genes that are relevant for photosynthesis, which are found in their chloroplast genomes.

 ## Restating the Concepts

▶ The circular mitochondrial genomes are located in the nucleoid, with several copies per organelle. Mitochondrial genomes range from 4 kb to 367 kb and possess from 3 to 97 genes that encode proteins that are required for oxidative phosphorylation.

▶ Many genes have been transferred from the mitochondrial genome to the nuclear genome over time. Most of these genes were likely mitochondrial RNAs that were reverse-transcribed into DNA that were then inserted into the nuclear genome.

▶ Mitochondrial genes use a slightly different genetic code from that of nuclear genes. This difference in the genetic codes reduces the ability to continue transferring genes from the mitochondria to the nucleus.

▶ Some animal parasites, such as *Plasmodium falciparum*, contain mitochondria and degenerate plastids. This suggests that they evolved from parasitic unicellular algae that may have required the photosynthetic function long ago.

19.4 Mechanisms of Organellar Inheritance

The organelles are governed by rules of heredity that differ so greatly from those that we previously described for nuclear genes that they can be considered principles of *non-Mendelian* genetics. Unlike the nuclear genome, the organellar genome is not subdivided with the precision by which chromosomes are segregated in mitosis and meiosis. Rather, the organelles may be randomly partitioned into daughter cells at cytokinesis. The two main principles that govern organellar inheritance are (1) that they are preferentially inherited from only one of the two parents, and (2) that the organelles exhibit somatic segregation.

Inheritance of Organelles from a Single Parent

In sexual reproduction, parents usually contribute an equal number of nuclear chromosomes. In contrast, the parents usually contribute unequal volumes of cytoplasm and cytoplasmic components to their off-spring, including the organelles. In many instances, sexual dimorphism has so minimized the size of the sperm that very little male cytoplasm is contributed to the fertilized egg (**fig. 19.19**).

For example, human sperm is primarily composed of a haploid number of nuclear chromosomes and very little cytoplasm, whereas the egg contains a haploid number of nuclear chromosomes and a large amount of cytoplasm and cytoplasmic components, including organelles. In some ferns, as the motile sperm makes its way to the egg, it discards a cytoplasmic bundle with many organelles, in order to become more streamlined (fig. 19.19b).

These unequal ratios of cytoplasm between the sperm and egg are reflected in the organelle genotypes of the progeny. Hence, the organelles in animals and many plants are inherited almost exclusively from the female parent.

Geneticists have developed two new terms to deal with the presence of multiple mitochondria in a single cell and more than one genome per organelle. **Homoplasmy** is the presence of a single common genotype in the organellar genome (**fig. 19.20a**). **Heteroplasmy** refers to the presence of a mixture of organellar genomes in a cell (fig. 19.20b).

Heteroplasmy can occur when an egg contains an organelle type, such as mitochondria, that is composed of two different genotypes. This condition can occur two different ways. First, a mutation may occur within the organellar DNA. If this mutant DNA and the organelle persist in the cell, the cell is heteroplasmic for both the wild-type and mutant genomes. In some cases, such as when the mutation is a deletion, the mutant genome may replicate faster than the wild-type genome and increase the likelihood that it will persist in the daughter cells.

Second, heteroplasmy can occur if both the male and female parents contribute cytoplasmic organelles to the fertilized egg. Even when the female contributes most of the cytoplasm to the fertilized egg, heteroplasmy may occur if even a few cytoplasmic organelles are contributed by the male. For example, heteroplasmy occurs in approximately 30% of the angiosperm species because of plastids being inherited from both parents. This biparental inheritance of plastids is exemplified by the geranium (**fig. 19.21**). Even in organisms such as mice, humans, and tobacco plants, which are considered

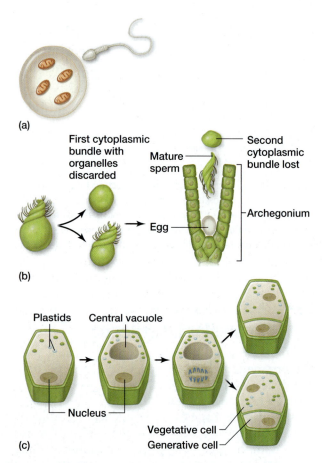

figure 19.19 Sexual dimorphic mechanisms of producing predominantly maternal organelle inheritance. (a) In most animals, including humans, the large egg supplies most of the cytoplasm and organelles to the zygote, while the highly motile sperm contains primarily a haploid genome. (b) In some organisms, such as ferns, the sperm matures as it moves toward the egg in the archegonium. The sperm loses two cytoplasmic bundles containing most of the organelles. The mature sperm, containing the nuclear contents and mitochondria in a flagellated coil, move towards the egg for fertilization. (c) In many higher plants, unequal cell division of the microspore produces a vegetative cell, from which the pollen tube will grow, and a generative cell that lacks organelles and acts as the male gamete. The unequal cell division results in all the organelles localizing to the vegetative cell.

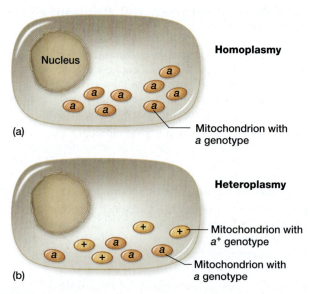

figure 19.20 Cellular definition of homoplasmy and heteroplasmy. (a) Homoplasmy refers to a cell that contains mitochondria (or plastids) that all have the same genotype. (b) Heteroplasmy is a cell that contains mitochondria (or plastids) with two or more different genotypes.

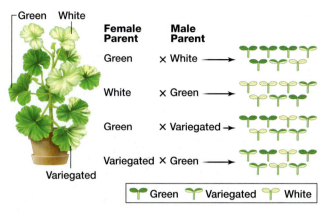

figure 19.21 Biparental inheritance in the geranium. This geranium plant could produce a flower with a homoplasmic germline of either green or white stems, or a heteroplasmic germline if the flowers arose on the variegated branch. Because the pollen also contributes plastids to the progeny, the seedling phenotypes would reflect the relative contributions from both gametes.

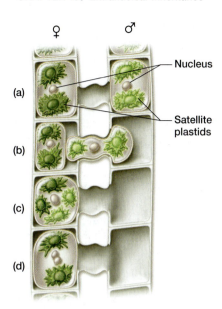

figure 19.22 Uniparental inheritance of female chloroplasts in *Zygnema*. A conjugation tube forms between cells of different mating types (a) which leads to transfer of the male protoplast into the female cell (b). Fusion of the protoplasts (c) is followed by nuclear fusion. This zygote clearly contains two satellite plastids from each parent. The two paternally contributed plastids then degenerate (d), as the nucleus undergoes meiotic division.

to have solely maternal inheritance of organelles, some paternal organelles have been found to be transmitted to the offspring when progeny are carefully examined.

In some cases, however, the sperm may possess a relatively large amount of cytoplasm and would then contribute many organelles to the progeny. This situation occurs in many gymnosperm plants (such as pine and fir trees) and in the occasional angiosperm, such as alfalfa. In these plants, many more plastids are inherited from the male parent than from the female parent.

Both the male and female parent produce gametes that possess cytoplasmic organelles, but the organelles from one of the two parents are preferentially destroyed after fusion of the gametes. For example, several related filamentous green algae, including *Zygnema* and *Spirogyra,* have chloroplasts in both of the gametes. Shortly after the gametes fuse, the chloroplasts donated by the "male" parent disintegrate (**fig. 19.22**).

In contrast to the previous examples, some gymnosperm plants, such as coastal redwoods, exhibit paternal inheritance of their mitochondria. Biparental and paternal inheritance patterns, however, are the exceptions to the general rule of maternal inheritance of mitochondria.

Complex Patterns of Inheritance

Chlamydomonas reinhardtii has two different mating types that are referred to as "mating-type plus" (mt^+) and "mating-type minus" (mt^-). Only *Chlamydomonas* cells of the opposite mating types can fuse. The resulting diploid zygote can then undergo meiosis and produce four haploid cells, two mt^+ and two mt^-. This pattern of inheritance is consistent with the mt gene being located in the nuclear genome.

Streptomycin resistance can be selected in *Chlamydomonas* by several ways. Normal cells, which are

sensitive to the antibiotic, are killed in its presence. If cells are plated on medium containing high levels of streptomycin (500–1600 g/mL), some resistant colonies grow. When these resistant cells are crossed with the wild type, however, a 1:1 ratio in offspring does not ensue. This higher level of resistance, therefore, is unlikely due to be controlled by a nuclear locus. Streptomycin's target is the chloroplast's small ribosomal subunit. Thus, it is likely that the parental origin of the chloroplast determines the sensitivity or resistance to streptomycin.

If the mt^+ parent is resistant to streptomycin, then both the mt^+ and mt^- haploid progeny will be resistant (**fig. 19.23**). It is as though the mt^+ parent is contributing the cytoplasm to the diploid zygote in a manner similar to maternal plastid inheritance in plants. The mt^- parent acts like a pollen parent by making a chromosomal contribution, but not a cytoplasmic one.

In actuality, both the mt^+ and mt^- cells contribute a chloroplast during mating. After cell fusion, the two chloroplasts fuse. But the *Chlamydomonas* cell preferentially digests the chloroplast DNA from the mt^- parent, such that 0.4–11.0% of the offspring in crosses of the type shown in figure 19.23 have the streptomycin phenotype of the mt^- parent. This case is therefore an example of biparental contribution of chloroplasts to the diploid zygote, and preferential uniparental inheritance of the organellar genome in the haploid meiotic products.

Interestingly, *Chlamydomonas* also exhibits uniparental inheritance of mitochondrial phenotypes. In contrast to the chloroplast, the mitochondria appear

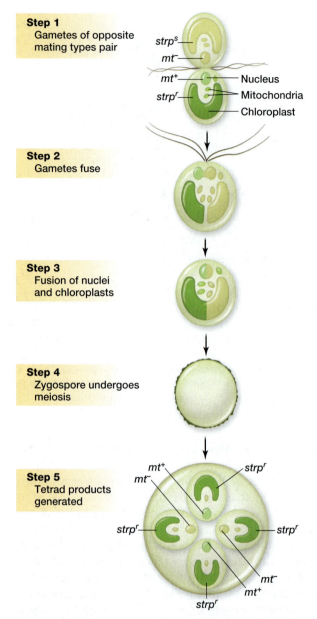

Step 1
Gametes of opposite mating types pair

strp^s
mt$^-$
mt$^+$ —————— Nucleus
strp^r —————— Mitochondria
—————— Chloroplast

Step 2
Gametes fuse

Step 3
Fusion of nuclei and chloroplasts

Step 4
Zygospore undergoes meiosis

Step 5
Tetrad products generated

mt$^+$ strp^r
mt$^-$
strp^r —————— strp^r
mt$^-$
mt$^+$
strp^r

figure 19.23   Uniparental inheritance of organelle genes in *Chlamydomonas reinhardtii*. Gametes of opposite mating types pair and fuse, followed by fusion of the haploid nuclei and chloroplasts (mitochondria do not fuse). The zygospore enters meiosis, yielding four products that show a 1:1 segregation pattern of the nuclear markers. However, uniparental organellar inheritance results in chloroplasts being inherited predominantly from the *mt$^+$* parent, while the mitochondria primarily come from the *mt$^-$* parent.

to be inherited from the *mt$^-$* parent. Because the cells of the two mating types fuse, both cells contribute both mitochondria and a chloroplast (each cell has only one chloroplast). It remains unclear how the cell recognizes the chloroplast from the *mt$^+$* cell and the mitochondria from the *mt$^-$* cell (see fig. 19.23). It is also not completely clear why *Chlamydomonas* inherits one organelle type from each parent.

Yeast Petite Mutants

Under anaerobic conditions, yeast colonies are small, and the structures of the mitochondria are reduced. Anaerobic conditions prevent oxidative phosphorylation from occurring in the mitochondria, and without this mechanism to generate ATP, the yeast cells grow more slowly. Under aerobic conditions, however, the mitochondria can produce more energy, which allows the colonies to grow larger. This permits the yeast colonies to exhibit distinctive morphologies (**fig. 19.24a**).

Boris Ephrussi and his colleagues studied *Saccharomyces cerevisiae*, and in the late 1940s identified mutants that appeared as small, anaerobic-like colonies when growing aerobically. These colonies are caused by **petite mutations.**

Categories of Petite Mutations

When petite mutants are crossed with wild type, three modes of inheritance emerge (fig. 19.24b). The **segregational petite** is caused by a mutation of a nuclear gene and exhibits a Mendelian pattern of inheritance in the meiotic progeny (2 petite:2 wild type). We know that the nuclear mutation is in a gene that encodes a protein required for mitochondrial function.

In contrast, when the **neutral petite** (*rho^0*) is crossed to a wild type, the petite phenotype is lost among all the progeny (0 petite:all wild type). Crossing a **suppressive petite** (*rho$^-$*) with a wild type shows some variability in expression. Highly suppressive petites produce almost all petite progeny when crossed with wild type, whereas less suppressive petites produce a smaller fraction of progeny with the petite phenotype. Both the neutral and suppressive petites are due to mutations in the mitochondrial genome.

Analysis of the mitochondrial genomes in the neutral and suppressive petites have revealed the mechanisms that produce these petite phenotypes. To fully understand these inheritance patterns, it is important to remember that *Saccharomyces cerevisiae* exhibits a biparental form of mitochondrial inheritance.

Neutral Petites Versus Suppressive Petites

Neutral petites seem to have mitochondria that entirely lack DNA. When neutral petites are crossed with the wild type to form diploid cells, mitochondria from both parents are inherited in the progeny. The mitochondria can fuse, and in some cases, the mitochondria lacking mtDNA (petite mitochondria) fuse with the wild-type mitochondria. When these fused mitochondria divide, they all will contain at least one copy of the mtDNA. Thus, only wild-type mitochondria will be present in these cells. The resulting meiosis produces large numbers of haploid meiotic products (spores) that all contain normal mitochondria. Because wild-type mitochondria are inherited, they can produce the wild-type

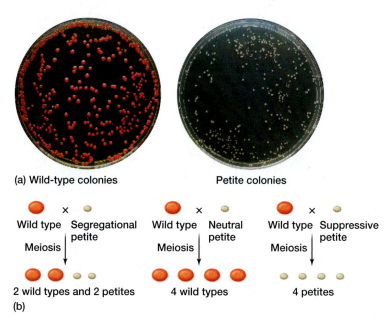

(a) Wild-type colonies   Petite colonies

(b)

Wild type × Segregational petite — Meiosis → 2 wild types and 2 petites

Wild type × Neutral petite — Meiosis → 4 wild types

Wild type × Suppressive petite — Meiosis → 4 petites

figure 19.24   Yeast petites. (a) The large wild-type and small petite colony morphologies are shown (left and right, respectively). (b) Crosses between a wild type and one of three different petites (segregational, neutral, and suppressive) yields three different phenotypic ratios in the meiotic progeny. The segregational petite crossed to wild type produces a Mendelian ratio of 2 wild type:2 petites (left). The neutral petite crossed to wild type produces 4 wild-type progeny (center), while the suppressive petite crossed to wild type produces 4 petites (right). Both of the latter are non-Mendelian ratios, as only one phenotypic class of progeny is produced.

aerobic colony morphology, even when derived from a cross with neutral petites.

Unlike the neutral petites, the suppressive petites contain small deletions in the mitochondrial genome. The presence of both wild-type and mutant mitochondrial genomes makes defining the mechanism of the suppressive petites more difficult to understand.

Suppressive petites could exert their influence over normal mitochondria in one of three ways. First, the suppressive mitochondria might replicate more rapidly than the normal mitochondria. In this way, the relative number of suppressive petite mitochondria would increase within the cells.

Second, fusion of the suppressive petite mitochondria with the wild-type mitochondria would produce mitochondria with both types of genomes. If the petite mtDNA replicates faster than the wild-type mtDNA, then a greater percentage of mtDNA in the fused mitochondria will be of the petite nature. When a fused mitochondrion divides, it will produce organelles that contain both the petite and wild-type mtDNA. The relative abundance of the wild-type mtDNA in each mitochondrion will help to determine if the cell will have a petite or wild-type phenotype.

Third, some data suggest that recombination occurs between the wild-type mtDNA and the suppressive petite mtDNA when the mitochondria fuse.

Presumably, recombination in mtDNAs occurs when two or more mitochondria fuse, bringing the two different sets of mtDNA in contact within a single organelle. Recombination between the different mitochondrial genomes could presumably alter the relative abundance of the mutant petite mtDNA. When this mitochondrion divides, the frequency of petite versus wild-type mtDNA will affect the phenotype of the cell.

Somatic Segregation of Organelles

When somatic cells divide in a eukaryotic organism, mitosis ensures that each daughter cell receives equal and exact copies of all the nuclear chromosomes. The organelles, by contrast, do not necessarily undergo DNA replication during S phase and then equally divide their duplicated genomes during cell division. Rather, organelles tend to increase in number as the cell volume increases. When the cell divides, the organelles and their genomes are not separated by a spindle apparatus. On the contrary, they are divided randomly into the daughter cells, carried along in the part of the cytoplasm in which they happen to be located.

When a heteroplasmic cell divides, the daughter cells may be either heteroplasmic or homoplasmic, depending on the random segregation of the organelles. This random segregation will more likely yield a heteroplasmic cell, however, than a homoplasmic cell (**fig. 19.25**).

These different outcomes in random somatic cell segregation can be observed in an individual. If a heteroplasmic cell has organelles with different genotypes, the resulting daughter cells may exhibit a variety of phenotypes. First, if the heteroplasmic cell generates a homoplasmic cell during cell division, then all the subsequent cells will also be homoplasmic, resulting in the generation of a phenotypic clone within the individual (**fig. 19.26**). Alternatively, the number of wild-type genomes relative to mutant genomes in a heteroplasmic offspring cell may affect the phenotype that is observed.

It's Your Turn

As you saw from the discussion in this section, the pattern of organellar inheritance can range from straightforward (strict maternal inheritance) to more complex (one organelle inherited maternally and a different organelle inherited paternally). To practice determining the underlying modes of organelle inheritance based on phenotypes, try problems 30 and 31.

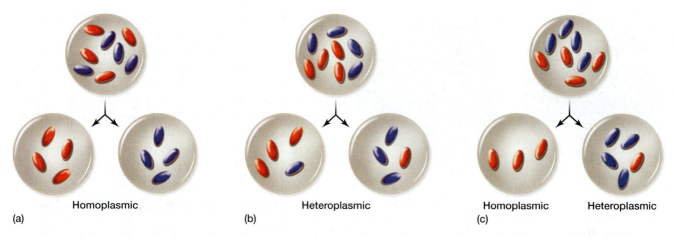

(a)   Homoplasmic
(b)   Heteroplasmic
(c)   Homoplasmic   Heteroplasmic

figure 19.25   Vegetative segregation in heteroplasmic cells. (a) Vegetative segregation can produce homoplasmic cells from a heteroplasmic cell if the two organellar genotypes segregate from each other when the cell divides. (b) Alternatively, random segregation of the organelles will likely continue to produce heteroplasmic cells. (c) Finally, random segregation of the organelles can yield both a homoplasmic and a heteroplasmic daughter cell.

 Restating the Concepts

▶ Mitochondria can be inherited in a maternal, paternal, or biparental manner depending on the organism. Maternal inheritance is the most common and is the method observed in animals.

▶ A cell can have a single mitochondrial genotype (homoplasmy) or multiple genotypes (heteroplasmy). Because mitochondria randomly segregate during cell division, a heteroplasmic cell can produce either a homoplasmic daughter cell or a daughter cell that contains a different percentage of the two mitochondrial genomes relative to the parental cell. The generation of homoplasmy from a heteroplasmic cell can produce phenotypic clones in an individual.

▶ In *Chlamydomonas,* we observe biparental contribution of organellar genes to the diploid zygote, but preferential uniparental inheritance of the genes in the haploid meiotic progeny. The chloroplast genes are preferentially inherited from the mt^+ parent, but the mitochondrial genes are preferentially inherited from the mt^- parent.

▶ The yeast petite mutants are due to the inability to properly use oxidative phosphorylation to produce increased amounts of ATP. This results from mutations in both the nuclear (segregational petite) and mitochondrial (neutral and suppressive petites) genomes.

▶ The neutral petites lack the entire mitochondrial genome, and these organelles are lost in heteroplasmy to wild-type mitochondria. The suppressive petites lack regions of the mitochondrial genome, which permits a shorter length of time to replicate, giving them an advantage in heteroplasmy with wild-type mitochondria.

Mosaic leaf

figure 19.26   Plastid variegation in the evening primrose. White sectors of this evening primrose plant represent the somatic segregation of a mutant plastid in the leaves. The mosaic leaf still has many heteroplasmic cells (green colored).

19.5 Inherited Human Mitochondrial Diseases

In human beings, certain diseases trace their dysfunction to mitochondrial pathologies. The first human disease that was reported to be maternally inherited was Luft disease in 1962, which is characterized by excessive sweating and general weakness. Since 1962, over a hundred diseases, including some of the general symptoms of aging and cancer, have been attributed to mitochondrial pathology.

Leber Hereditary Optic Neuropathy

In 1988, Douglas Wallace and his colleagues showed that **Leber hereditary optic neuropathy (LHON)** is a cytoplasmically inherited disease. This disease causes blindness, with a median age of onset of approximately 24 years. The age of onset and phenotype vary, depending on the degree of heteroplasmy in the individual.

Because the optic nerve requires a great amount of energy, defects in mitochondrial oxidative phosphorylation are not tolerable. Some damage to the heart is also part of the disease. Pedigrees have shown that LHON is transmitted only maternally (**fig. 19.27**). Mutations have been identified in four different genes that lead to LHON, with the majority of the cases resulting from a single specific mutation, a change in guanine to adenine within codon 340 of the NADH dehydrogenase subunit 4 gene. This mutation results in an arginine to histidine change in the protein product. This was the first human disease traced to a specific mitochondrial DNA mutation.

Interestingly, many cases of LHON appear spontaneously, with no previous history of the disease in the family (fig. 19.27). This may be due to the loss of normal mtDNAs in heteroplasmic individuals through several generations, until an individual falls below a specific threshold of mitochondrial function and exhibits the symptoms. Alternatively, the identification of four different mitochondrial genes being associated with LHON suggests that the disease may simply result from a general loss of mitochondrial function, rather than the loss of a specific protein function. If this is the case, then spontaneous mutations could be the cause of the spontaneous LHON appearance.

Myoclonic Epilepsy and Ragged Red Fiber Disease (MERRF)

Another human mitochondrial disease is **myoclonic epilepsy and ragged red fiber disease (MERRF),** which primarily affects the central nervous system and leads to the loss of skeletal and cardiac muscle function (**fig. 19.28**). MERRF is a progressive disease that usually starts with vision problems and myopathy, which progresses through loss of hearing, uncontrolled jerky movements (myoclonic epilepsy), dementia, cardiac failure, renal failure, and, ultimately, death.

MERRF is maternally inherited and results from a mutation in the mitochondrial tRNA^{Lys} gene. Two things can be seen in a MERRF pedigree (fig. 19.28). First, it is inherited in a strictly maternal manner. An affected male (II-6) does not pass the disease on to his children. Second, the percentage of mutant mitochondrial DNAs in the individual correlates with the severity of the clinical symptoms.

Because most individuals are heteroplasmic, it is interesting to examine what percentage of mutant mitochondrial genomes are required to produce the MERRF

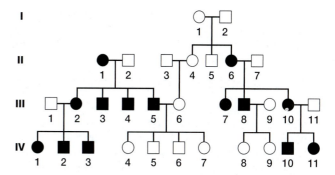

figure 19.27 Pedigree of LHON exhibits strict maternal inheritance. Affected females (II-1, II-6, III-2, and III-10) pass the disease on to all their children, while affected males (III-5 and III-8) do not pass the disease on to any of their children. The disease spontaneously appeared in individual II-6, whose mother was unaffected.

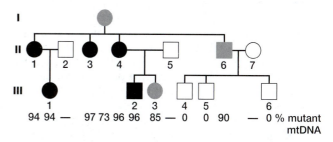

figure 19.28 The MERRF pedigree exhibits strict maternal inheritance. Affected females (I-1, II-1, and II-4) pass the disease on to all their children, while an affected male (II-6) does not pass the disease on to his children. The black and gray symbols represent individuals with severe and mild MERRF symptoms, respectively. The percentage of mutant mitochondrial genomes was determined for each person in the pedigree and is shown at the bottom. Greater than 90% of the mitochondria must be mutant to develop the severe MERRF symptoms.

symptoms. In one large MERRF family, an individual with only 15% wild-type mitochondrial genomes showed no obvious symptoms (except for some muscle abnormalities that were detected only by using a microscope). In the same family, two individuals having only 2% wild-type mitochondrial genomes showed severe MERRF symptoms. Thus, a small percentage (between 2% and 15%) of wild-type mitochondrial genomes is sufficient to yield a clinically normal individual.

General Features of Human Mitochondrial Diseases

Four general features are observed in mitochondrial diseases in humans.

- They are inherited in a maternal manner. An affected woman passes the disease on to all of her children, but an affected man does not pass the disease on to any of his children.

19.1 applying genetic principles to problems

Inheritance of Mitochondrial-Associated Metabolic Disease in Humans

We previously discussed both X-linked and Y-linked patterns of inheritance (see chapter 4). In nearly all cases, human mitochondria are inherited through the maternal lineage, except in rare instances that one or two mitochondria are inherited from the male through the sperm. It is important to have a good understanding of these different mechanisms of inheritance and how they would appear in a pedigree. Look at the three different pedigrees shown here and see whether you can determine the mode of inheritance.

In figure 19A (below), you have a pedigree for an inherited human disease. You can determine the mode of inheritance by studying the pedigree carefully. Notice that the disease does not skip any generations, and that affected individuals have an affected parent. This suggests that it is a dominant disease. Second, affected women can pass the disease on to either their daughters or sons. Affected males, however, only pass the disease on to their daughters. As you know, males have only a single X chromosome and always pass that X chromosome on to their daughters and never to their sons. Therefore, you can conclude that the following pedigree appears to show a dominant X-linked disease.

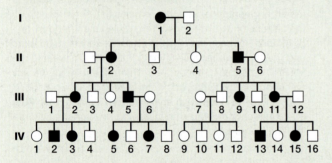

In figure 19B (top of next column), you see an entirely different-looking pedigree, even though the female in the first generation is affected as in the previous pedigree. The first difference that you should notice is that many fewer individuals are affected in the second pedigree compared with the first. Looking at the pedigree more carefully, you can notice that only males are affected in generations II through IV. You can also see that every affected male is the son of a female who is directly related to the original affected female. This suggests that these females are carriers of an X-linked recessive disease. The males are primarily affected because they are hemizygous for X-linked genes, and they express the recessive phenotype if they inherit the X chromosome with the recessive disease allele.

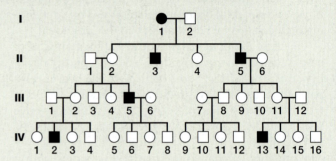

In figure 19C (below), you see still another different-looking pattern of inheritance, even though it shows the same arrangement of individuals within the pedigree. Notice that an affected female produces only affected children, but an affected male (II-5) never produces affected children. Even more striking is that all of the descendents of an affected male do not exhibit the disease symptoms. Individual II-5 is affected, but none of his 12 descendants in the next two generations have the disease. This is very different from what you saw for the inheritance of the X-linked recessive disease in figure 19B (above), in which the disease skipped generations.

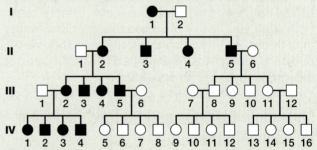

You can conclude that the pedigree in figure 19C is an example of maternal inheritance. A female that has a mutation in her mitochondrial DNA passes those mitochondria on to all of her children, thereby causing them to exhibit the recessive mitochondrial disease. If the child is a female, as is individual II-2, then all of her children also inherit their mitochondria from their mother and have the disease. If the affected child is a male, as is individual II-5, then all of his children inherit his wife's mitochondria and do not exhibit the disease.

Occasionally, a mitochondrial disease appears in a female whose mother does not exhibit the disease. This outcome would result from the generation of a spontaneous mutation in the mitochondria of the mother's gametes. Because the daughter has inherited mitochondria with the mutation, she and all of her children will exhibit the clinical symptoms of the disease.

If the mitochondrial disease appears in a male, the mutation again was likely generated in his mother's gametes. However, his descendants will not exhibit the disease.

- Heteroplasmy makes it very likely that an individual contains some percentage of wild-type mitochondrial genomes, which will affect the severity of the disease.

- Random segregation of mitochondria during cell division suggests that some tissues will contain a lower percentage of wild-type mitochondrial genomes than other tissues. Because of this random segregation, some individuals may exhibit more severe symptoms in one tissue, and siblings may show more severe symptoms in a different tissue. This effect has even been observed in identical twins, where the random segregation of mitochondria prior to the separation of the fertilized egg into two zygotes could result in one twin possessing more mutant mitochondrial genomes than the other twin, and therefore exhibiting more severe symptoms.

- Disease-causing mutations in the mitochondrial genome basically affect the energy-producing oxidative phosphorylation capacity of the mitochondria. Some tissues, such as neurons and muscles, have very high energy requirements and are likely to be affected to a greater extent by any decrease in the ATP production capacity. Consistent with this thought, muscles and neurons are more sensitive to mitochondrial diseases than other tissues.

It's Your Turn

The strict maternal inheritance of mitochondria helps to make the identification of human mitochondrial diseases obvious. However, spontaneous mutations and random segregation in heteroplasmic individuals creates a significantly more complex situation. Try problems 19 and 23 to check your reasoning skills.

Restating the Concepts

▶ Mutations in the mitochondrial genome affect the expression of proteins required for oxidative phosphorylation or their translation. Thus, mitochondrial diseases primarily affect the energy production of a cell. More severe phenotypic effects occur in tissues that have a higher demand for energy, such as neurons and muscles.

▶ Human diseases due to mitochondrial mutation exhibit strict maternal inheritance, can exhibit some differences in expression due to the random segregation of mitochondria during cell division, and primarily affect the nervous system and muscles.

genetic linkages

In this chapter we discussed ways in which some traits are inherited through the genomes found in organelles other than the nucleus. Mitochondria are found in almost all eukaryotes, and plastids are present in all plants and algae. The genomes and genetics associated with both of these organelles share some common features, however. First, they are inherited from generation to generation. Second, they fail to exhibit a standard Mendelian pattern of inheritance. Third, they are randomly distributed to the daughter cells during cell division.

We previously described other situations in which we failed to observe the standard Mendelian inheritance patterns. For example, in genomic imprinting, the alleles of a particular gene are silenced in the gametes from a particular parent and remain silenced in the progeny (chapter 17). Thus, we observe either a maternal or paternal pattern of inheritance. However, the imprinted genes are nuclear compared with the organellar genes that we discussed in this chapter. And, as discussed in chapter 18, the *Drosophila P* element transposes in the germ cells that result from a cross between an M cytotype egg and a P cytotype sperm. The phenotype is not observed, however, until the subsequent generation. Although this effect also exhibits a non-Mendelian inheritance pattern, it nevertheless represents nuclear-encoded genes.

In chapter 12, we discussed bacterial plasmids that confer antibiotic resistance to the cell. Because these circular plasmids usually do not integrate into the bacterial genome, they are inherited as extrachromosomal elements. They are usually present in several copies in the cell, and their random segregation usually still results in each daughter cell receiving at least one plasmid. Similarly, the organellar genomes are circular and present in many copies in each cell (and within each organelle). They are also randomly distributed into the daughter cells. The major difference between bacterial plasmids and the organellar genomes is that the plasmid is freely distributed within the cytoplasm, but the organellar genome is within the membrane-bounded environment of the mitochondria or plastid.

We will discuss maternal effect mRNAs in chapter 21. For example, the *Drosophila hunchback* and *nanos* mRNAs are transcribed in the female and deposited in the cytoplasm of the unfertilized egg. The Pumilio protein is translated in the female and also deposited in the unfertilized egg. In these cases, the *hunchback, nanos,* and *pumilio* are nuclear-encoded genes. Because they are transcribed in the female and deposited in the egg, the maternal genotype determines the phenotype of the offspring during early embryonic development. This is in contrast to maternal inheritance of organelles, where the organellar genes are transcribed and expressed in the individual rather than the mother.

These non-Mendelian inheritance mechanisms clearly play important roles in the development and energy pro-duction of multicellular diploid organisms. The expression of genes in an individual's nuclear genome alone is clearly not sufficient for an organism's normal development and function.

Chapter Summary Questions

1. For each term in column I, choose the best matching phrase in column II.

Column I	Column II
1. Mitochondrion	A. Stacks of thylakoid membranes in chloroplasts
2. Chloroplast	B. Organelle where the Krebs cycle occurs
3. *Giardia intestinalis*	C. Possesses mt^+ and mt^- mating types
4. *Plasmodium*	D. Organelle where carbon fixation occurs
5. Cristae	E. Possesses petite mutants
6. Grana	F. Contains mitosomes
7. *Chlamydomonas reinhardtii*	G. Starch-containing plastid
8. *Saccharomyces cerevisiae*	H. Contains a degenerate plastid genome
9. Leucoplast	I. Immature plastid
10. Proplastid	J. Infoldings of the inner mitochondrial membrane

2. What is the "mitochondrial Eve" hypothesis?
3. Describe the structure and organization of human mito-chondrial DNA.
4. Discuss the role of mitochondria in energy production.
5. What basic functions occur within chloroplasts?
6. What similarities do mitochondria and chloroplasts share?
7. How does gene expression in animal mitochondria dif-fer from that of bacteria and eukaryotic nuclei?
8. What is the endosymbiotic theory? Describe some of the data that support it.
9. Compare and contrast the segregational, neutral, and suppressive petite mutants of *Saccharomyces cerevisiae*.
10. Distinguish between homoplasmy and heteroplasmy.
11. A petite yeast strain is crossed with a wild-type strain. What phenotypic ratio do you expect after meiosis if the petite is
    a. segregational?
    b. highly suppressive?
    c. neutral?
12. The mitochondrial genomes of many organisms have gotten smaller since their bacterial origin. How did this shrinking occur?
13. List the four general features of human mitochondrial diseases.

Solved Problems

PROBLEM 1: When chloroplast DNA from *Chlamydomonas* is digested with a particular restriction enzyme and then hy-bridized with a specific probe, two bands are detected on the Southern blot. Some *Chlamydomonas* strains (type 1) yield bands of 1.5 and 3.7 kb; other strains (type 2) yield bands of 2.5 and 6.0 kb. For the following crosses, predict the progeny:
   a. mt^+, strain 1 × mt^-, strain 2
   b. mt^+, strain 2 × mt^-, strain 1

Answer: In *Chlamydomonas*, both parents contribute chlo-roplasts to the diploid zygote. However, there is preferen-tial uniparental inheritance of the chloroplast genome in the haploid meiotic products. Only the mt^+ parent contributes cpDNA to the progeny because the cpDNA from mt^- cells is digested and lost.

Therefore, the progeny of cross **(a)** will all be of type 1 and show bands of 1.5 and 3.7 kb, and all those of cross **(b)** will be type 2 and produce bands of 2.5 and 6.0 kb.

PROBLEM 2: What results would you obtain by making all possible pairwise crosses of the three types of yeast petites?

Answer: The rule of thumb is that suppressive petite mito-chondria will dominate a cell, whereas neutral petite mito-chondria will be lost in a competitive situation. Therefore,

   segregational petite × segregational petite →
   segregational petites

   segregational petite × neutral petite → segregational petites

   segregational petite × suppressive petite → suppressive petites

   neutral petite × neutral petite → neutral petites

   neutral petite × suppressive petite → suppressive petites

   suppressive petite × suppressive petite → suppressive petites

PROBLEM 3: Suppose you are a botanist who discovered mutant plants that are chlorophyll-deficient and have yellow leaves. Explain how you would determine if this trait exhibits autosomal nuclear inheritance or non-Mendelian cytoplasmic inheritance.

Answer: Traits that exhibit cytoplasmic inheritance will show differences in reciprocal crosses, but autosomal traits do not. Therefore, you would first make reciprocal crosses between strains with normal (green) leaves and strains with yellow leaves. Let G = green, g = yellow, and assume that G is dominant to g.

If the yellow trait is controlled by nuclear genes, the reciprocal crosses

female GG × male gg

and female gg × male GG

will produce the same results: offspring with genotype Gg (green).

If on the other hand the yellow trait is controlled by a cytoplasmic (organellar) gene, the reciprocal crosses will produce different results.

Cross	Progeny	
	If Maternally Inherited	If Paternally Inherited
Female G × male g	All green (G)	All yellow (g)
Female g × male G	All yellow (g)	All green (G)

Exercises and Problems

14. What data indicate that it is not absolutely essential, in an evolutionary sense, for mitochondria to have genes for specific components of oxidative phosphorylation?

15. Suppose you identified a person who has introns in his or her mitochondrial DNA. What would you deduce about the origin of this DNA?

16. A mutation identified in the human mitochondrial genome causes an inherited form of blindness. If reciprocal matings between affected and normal individuals occur in a family pedigree, what types of children would you expect from each cross?

17. You just noticed a petite yeast colony growing on a petri plate under aerobic conditions. What type of petite is it?

18. In *Drosophila*, sensitivity or resistance to carbon dioxide displays maternal inheritance. Predict the outcomes of reciprocal crosses involving sensitive and resistant flies.

19. A 40-year-old man has been diagnosed with myoclonic epilepsy and ragged red fiber disease (MERRF). His wife is phenotypically normal, with no history of MERRF in her extended family. What is the probability that one of the couple's four children will develop the disease? Why?

20. A male from a true-breeding strain of nervous zeebogs is crossed to a calm female, also from a true-breeding strain. The F_1 and F_2 offspring are all calm. Propose an explanation for these findings. If the F_2 generation information were unavailable, what conclusions would you have drawn?

21. Jesse James, the notorious bank robber, was allegedly killed in 1882. Since that time, rumors that the dead man was not Jesse James circulated widely and were supported by several men who claimed to be the dead legendary outlaw. In 1995, a group of scientists exhumed the remains of the person believed to be Jesse James in an effort to confirm his identity. The research team obtained DNA from two teeth and two hairs. They also obtained DNA samples from the descendants of James' sister Susan. Why were these samples chosen to confirm the identity of Jesse James? Would a similar analysis have been useful using the descendants of James' brother? Why?

22. In humans, both X-linked inheritance and cytoplasmic inheritance produce different results from reciprocal crosses. So how can these two modes of inheritance be distinguished?

23. Complete the following pedigree to suggest maternal inheritance.

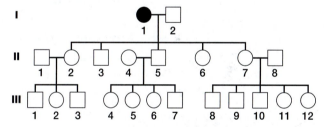

24. In plants, twisted leaves, a hypothetical mutant phenotype, may be caused by either a chloroplast gene (C^T = normal; C^t = twisted) or a recessive nuclear gene (N^T = normal; N^t = twisted). Assume chloroplast DNA is maternally inherited. Predict the genotypic and phenotypic ratios in the offspring of the following crosses.
   a. $C^T N^t N^t$ female × $C^t N^T N^T$ male
   b. $C^t N^t N^t$ female × $C^T N^T N^T$ male
   c. F_1 female from cross (a) × F_1 male from cross (b)
   d. F_1 female from cross (b) × F_1 male from cross (a)

25. Restriction mapping of an 11-kilobasepair region of an animal's mitochondrial DNA reveals the two different patterns of *Pst*I sites shown here:

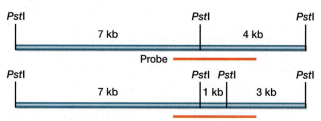

Mitochondrial DNA is extracted from a large number of individuals and digested to completion with *Pst*I. The fragments are run on a gel, Southern blotted, and analyzed with a labeled probe that is homologous to the region represented by the thick horizontal line. How many different patterns of hybridization are possible in this sample? Explain how each pattern is obtained.

26. In the white campion plant, *Silene latifolia*, sex is determined by the XX-XY chromosomal mechanism. Individuals with an intact Y chromosome develop male reproductive organs, and those without the Y develop into females. How can Y-linked inheritance and paternal inheritance be distinguished in this species?

27. The mitochondrial DNA from a particular organism is isolated and digested with the restriction enzymes *Pst*I, *Eco*RI, and *Bgl*II individually and all three enzymes simultaneously. The DNA fragments are then electrophoresed, Southern blotted and hybridized with a radioactive cDNA probe derived from a ribosomal RNA gene that was isolated from a related species. The results are as follows, with the Southern blot bands indicated by an asterisk (*):

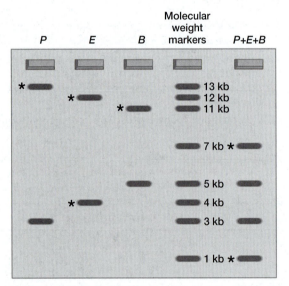

a. Is this mitochondrial DNA circular or linear? Explain.
b. Draw a restriction map of this mitochondrial genome. Indicate the approximate location of the rRNA gene.

28. Consider an animal cell that has only two mitochondria: one with DNA encoding the wild-type trait *A*, and the other containing a mutation encoding trait *a*. Assume that each mitochondrion replicates once during a cell division, and that mitochondrial segregation is random so that each daughter cell receives exactly two mitochondria. Predict the genotypes and the probabilities of the two cells that would arise after one cell division.

29. Consider the following cross in *Chlamydomonas reinhardtii*:

$mt^+, a^+ b^+ c^+ d^- \times mt^-, a^- b^- c^- d^+$

where *a* and *b* are nuclear genes, and *c* and *d* are chloroplast genes. What is the genotype of the
a. zygote?
b. meiotic products?

30. Prescottine (*pr*), eduardomycin (*ed*), and brownicillin (*br*) are three hypothetical antibiotics. You isolated three strains of *Chlamydomonas reinhardtii* that each display resistance to one of these antibiotics. From the following reciprocal crosses, deduce the nature of antibiotic resistance in each strain?

Strain 1: $mt^+, pr^r \times mt^-, pr^s \rightarrow$ all offspring are pr^r
$mt^+, pr^s \times mt^-, pr^r \rightarrow$ all offspring are pr^s
Strain 2: $mt^+, ed^r \times mt^-, ed^s \rightarrow$ offspring are $1/2\ ed^r : 1/2\ ed^s$
$mt^+, ed^s \times mt^-, ed^r \rightarrow$ offspring are $1/2\ ed^r : 1/2\ ed^s$
Strain 3: $mt^+, br^r \times mt^-, br^s \rightarrow$ all offspring are br^s
$mt^+, br^s \times mt^-, br^r \rightarrow$ all offspring are br^r

31. In a particular plant, leaves could be green, white, or variegated (white and green). Flowers can develop on different parts of the plant, and the flower color corresponds to the leaf color. Flowers from branches of each type were used to perform all possible crosses (outlined in the following table). Predict the progeny phenotypes expected if inheritance is **(a)** biparental (with codominance); or **(b)** maternal.

	Parents		Inheritance	
	Female	Male	(a) Biparental	(b) Maternal
1.	green	× green		
2.	green	× white		
3.	green	× variegated		
4.	white	× green		
5.	white	× white		
6.	white	× variegated		
7.	variegated	× green		
8.	variegated	× white		
9.	variegated	× variegated		

32. Ribulose-1,5-bisphosphate carboxylase/oxygenase (commonly referred to as RuBisCO) is one of the most abundant proteins on earth. This enzyme catalyzes CO_2 fixation in photosynthesis. In most photosynthetic organisms, the RuBisCO holoenzyme consists of 16 subunits, 8 identical large and 8 identical small subunits, which are encoded by the genes *rbcL* and *rbcS*, respectively. The DNA from the nucleus and chloroplast of a plant are isolated, digested to completion by restriction enzymes, electrophoresed, Southern blotted, and hybridized to labeled *rbcL* and *rbcS* probes. The results are shown in the following diagrams.

a. What conclusions can be drawn about the nature and location of the *rbcL* and *rbcS* genes?

b. If *E. coli* cells are transformed with plant *rbcL* and *rbcS* genes, the large- and small-subunit polypeptides are synthesized, however the RuBisCO holoenzyme fails to form. Propose an explanation.

Chapter Integration Problem

The following diagram is the pedigree of a rare neurodegenerative disease in humans.

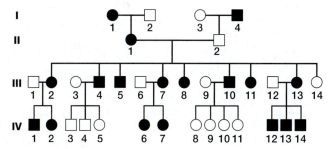

a. Based on the pedigree, characterize each of the following modes of inheritance as likely, unlikely, or impossible: autosomal dominant, autosomal recessive, Y-linked, X-linked dominant, X-linked recessive, sex-limited, sex-influenced dominant in males, and sex-influenced dominant in females. Justify your reasoning.

Further scrutiny of members of this family reveals the following: (1) The disease's degree of expression varies among individuals: some are mildly affected, others moderately affected, and still others severely affected; and (2) Individual III-14 was previously married. Her ex-husband is phenotypically normal, with no history of the disease in his family going back five generations. He has custody of the couple's two children, a girl and a boy, both of whom are affected with the disease. The partial pedigree follows (the "slash" represents separation or divorce)

b. Do these findings eliminate any of the remaining "likely or unlikely" modes of inheritance listed in part **(a)**? Are any of these modes of inheritance still possible? If so, which one(s) and why?

c. Is mitochondrial inheritance a likely, unlikely, or an impossible mode of inheritance for this disease? Does

it provide an explanation for the disease's variable expressivity?

d. How can mitochondrial inheritance explain individual III-14?

The mitochondrial DNA of several family members was analyzed by digestion with a type II restriction endonuclease that recognizes a 6 bp sequence. Agarose gel electrophoresis of the restriction digests produced the following results.

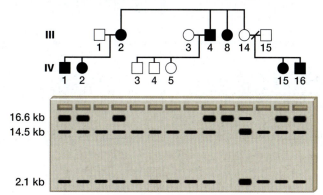

e. The wild-type human mitochondrial DNA sequence of bases 6901–6960 is shown here in the 5' to 3' direction:

```
6901            6911            6921
 |               |               |
AATGATCTGC  TGCAGTGCTC  TGAGCCCTAG

6931            6941            6951
 |               |               |
GATTCATCTT  TCTTTTCACC  GTAGGTGGCC
```

This sequence includes one of the restriction sites for the 6 bp cutter, as well as three different restriction sites for three 4 bp cutters. Locate these four restriction sites. Justify your choices.

f. Do the results of the restriction digest validate your answers to parts **(c)** and **(d)**? Explain in detail.

g. What type of mutation may have caused this disease? Be as specific as possible.

h. Assume individual III-8 marries a normal male. What is the probability that their first child will be affected with the disease?

Do you need additional review? Visit **www.mhhe.com/hyde** for practice tests, answers to end-of-chapter questions and problems, interactive exercises, and animation tutorials all designed to enhance your understanding of key genetic concepts.

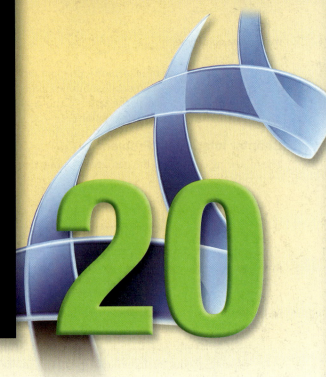

Mutational Analysis

Essential Goals

1. Describe the various classes of mutations and how they differ.

2. Understand the genetic crosses used to generate different types of mutants.

3. Describe the differences between forward and reverse genetics and the advantages of both approaches.

4. Discuss the various approaches used to isolate mutants in forward and reverse genetic approaches.

5. Understand the different ways that enhancers and suppressors are isolated and how they may operate.

6. Discuss the advantages of isolating recessive mutations in a mosaic screen and how it is performed.

7. Describe two different ways to phenocopy a gene and the advantages of each.

Chapter Outline

Photo: A wild-type zebrafish (*Danio rerio*) and five different fin mutants isolated in a large-scale forward genetic screen.

*I*n this chapter, we examine in greater depth the ways in which mutant organisms are generated, isolated, and characterized. The isolation of specific mutants can serve several important uses. First, they can be used to study general cellular processes, such as DNA replication. Second, we can generate or identify mutations in laboratory organisms (mice, fruit flies) that correspond to mutations in inherited human diseases. These laboratory mutants can serve as models for either studying the progression of the disease or developing potential therapies.

As an example, one dog breed, the Briard, has a recessive form of blindness. The blindness is due to a defective gene named *RPE65*. Mutations in the human *RPE65* gene also cause a genetic disease, Leber's congenital amaurosis (LCA), which results in early childhood blindness like that seen in the Briards. In 2001, Dr. Jean Bennett and coworkers cloned the wild-type canine *RPE65* gene into a virus and introduced the recombinant virus into the eye of a 3-month-old Briard, named Lancelot, who had been blind since birth. Within 3 months, Lancelot's vision had been restored. In the subsequent years, Lancelot has appeared several times at U.S. Congressional hearings to demonstrate his continued

ability to see and show the promise of gene therapy. Currently, clinical trials using viral vectors to introduce the wild-type human *RPE65* gene into the eyes of LCA patients are moving forward. Thus, the *RPE65* mutation in the Briard has served as a model for identifying a potential cure for this human disease.

The third reason to isolate mutants is to identify genes and their corresponding proteins that interact with a previously identified and characterized mutation. These interactions may be physical, as in two different proteins forming a heterodimer, or they may be in different points in a biochemical pathway. This type of study may help to further reveal the function of the gene or the biological process in which the encoded protein is involved.

The ability to generate desired mutations and characterize the mutant phenotypes is a critical tool for a geneticist. Isolating mutants requires knowledge of the organism and the available techniques, a keen insight into the problem being studied, and an ability to accurately predict potential phenotypes. The search for specific genetic mutants often allows the geneticist to fully express his scientific creativity.

20.1 Types of Mutations

Genetics researchers often begin by thinking about a type of mutant that they would like to isolate and study, and then consider the DNA alteration that would produce that mutant.

As an example, suppose you are interested in a mutation that produces no functional protein. You could use X-rays to generate a deletion or a chromosomal break within the gene. However, if you wanted a mutant that expressed some functional protein, but at less than normal levels, you would likely use a chemical mutagen rather than X-rays. For this reason, the type of mutagen used becomes important (chapter 18).

When a researcher isolates a mutant, a series of genetics tests can be performed to define the level of protein activity present in the mutant relative to wild type. For example, is the mutant phenotype resulting from a loss of protein activity, increased protein activity, or the expression of the protein in the wrong location? Therefore, the ability to identify the correct mutant requires first using the proper mutagen, and then analyzing the mutants for the one with the desired expression.

Genetic Description of Mutational Classes

Several different descriptions are used to categorize mutations. These descriptions are based on the following characteristics of mutations:

- The effect on the encoded protein; that is, what sort of change occurs in the amino acid sequence.
- The phenotype produced under different environmental or nutritional conditions.
- How the mutant behaves in certain genetic backgrounds.

We next consider these descriptions in detail.

Mutational Classes Based on Changes in Protein Sequence

In chapter 18, we described a **missense mutation** as causing a change from one amino acid in the encoded protein sequence to a different amino acid. If the mutation introduces a premature translational termination codon in the open reading frame (ORF) and produces a truncated protein, it is called a **nonsense mutation.** A **frameshift mutation,** due to the insertion or deletion of one or more nucleotides (but not a multiple of three), will affect all the codons following the mutation. The mutant protein will possess many amino acid differences relative to the wild-type protein and often results in a protein that is altered in length relative to the wild-type protein.

Because missense mutations produce a full-length protein with only one altered amino acid, they usually retain some of the protein's original activity. In contrast, the closer a nonsense mutation is located to the 5' end of the ORF, the smaller the encoded protein will be and the greater chance it will lack any of the protein's

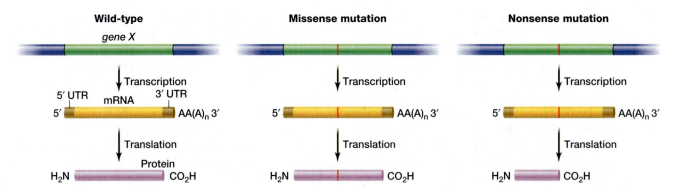

figure 20.1   Comparison of missense and nonsense alleles. A wild-type gene (left) transcribes an mRNA (yellow) that is translated into a functional protein (purple). The 5′ UTR and 3′ UTR are in brown. A missense mutation (center, red line) produces a wild-type sized protein that contains a single amino acid difference (red line) relative to the wild-type protein. A nonsense mutation (right, red line) introduces a premature translation termination codon in the mRNA that is translated into a protein that is shorter than the wild-type protein.

original activity. Thus, nonsense mutations are usually more severe than missense mutations. **Figure 20.1** illustrates the difference between missense and nonsense mutations. Similarly, the closer a frameshift mutation is to the 5′ end of the ORF, the less wild-type amino acid sequence will be present in the encoded protein and potentially the more severe the mutant phenotype.

Mutational Classes Affected by Environmental Conditions

Conditional mutants express the wild-type phenotype under certain environmental conditions and the mutant phenotype under different environmental conditions. For a conditional mutant, the **permissive condition** allows the wild-type phenotype to be expressed, and the **restrictive condition** yields the mutant phenotype.

In **nutritional mutants** (sometimes called nutrient-requiring mutants), the rich medium is the permissive condition that allows cell growth. The restrictive condition is the minimal medium, which lacks numerous compounds that are present in the rich medium. One of the compounds missing in minimal medium is required for the growth of the mutant. The mutants (auxotrophs) exhibit the wild-type phenotype, growth, on the rich medium and the mutant phenotype, inability to grow, on minimal medium (restrictive condition).

Temperature-sensitive mutants are conditional mutants that express the wild-type phenotype at some temperatures and the mutant phenotype at others. Often, temperature-sensitive mutants exhibit the wild-type phenotype at low temperatures (*permissive temperature*) and exhibit the mutant phenotype at higher temperatures (*restrictive temperature*). This occurs because the conformation of the mutant protein is stabilized at lower temperatures and is functional (**fig. 20.2a**). The mutant amino acid is unable to maintain the structure of the protein as the temperature

increases, which results in the protein losing some conformation and activity.

An example of a temperature-sensitive mutation is the Himalayan rabbit (fig. 20.2b). These rabbits contain a specific *albino* allele that causes pink eyes and white fur over most of the body. Unlike a true *albino* mutant, however, the Himalayan rabbit has dark fur at its extremities, ears, snout, feet, and tail. The specific *albino* allele in Himalayan rabbits is a temperature-sensitive mutation. At lower temperatures, which are found at the extremities of the body, the mutant protein is able to produce melanin and the dark pigment. At higher temperatures, such as in the body core, the mutant protein cannot fold properly and is inactive, which prevents melanin production. Similar *albino* mutants with pigmented extremities have also been identified in the Siamese cat and mouse.

Defining Mutational Classes Based on Genetic Interactions

We can also classify mutations based on how they interact with other alleles. In the case of a recessive mutation, for example, we could compare the homozygous mutant phenotype with the phenotype of a heterozygote that contains one copy of the mutant allele and a deletion that removes the second copy (**fig. 20.3**). Because the chromosome containing the deletion does not encode any of the desired protein, this heterozygote contains half of the corresponding protein activity relative to the homozygous mutant.

If the phenotype of the homozygous recessive mutant is identical to the deletion heterozygote (see fig. 20.3), then the mutation is called a null mutation. A **null allele** (sometimes called an **amorph**) is one that produces the same phenotype as a deletion of the gene. When the deletion heterozygote and mutation homozygote are phenotypically identical, the

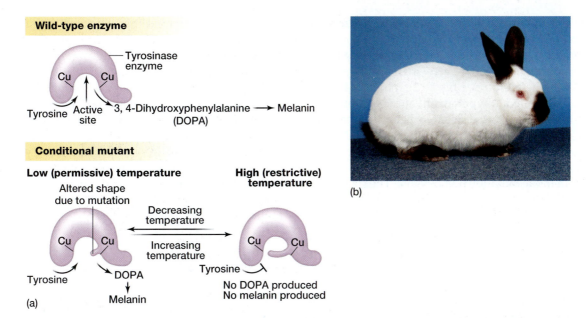

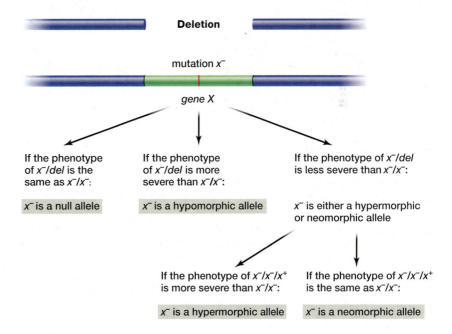

figure 20.2   A model for the mechanism of a conditional mutation. (a) The wild-type tyrosinase enzyme has an active site that catalyzes the production of 3,4-dihydroxyphenylalanine (DOPA) from tyrosine (top). The DOPA is subsequently converted to the pigment melanin. A temperature-sensitive conditional mutation encodes an altered form of the tyrosinase enzyme (bottom). At a low (permissive) temperature the mutant tyrosinase maintains a relatively normal active site that continues to convert tyrosine to DOPA. At the elevated (restrictive) temperature, the mutant tyrosinase changes shape and cannot produce DOPA, which prevents pigment production. (b) A Himalayan rabbit has a temperature-sensitive mutation in the *albino* gene, which encodes the tyrosinase enzyme. At the cooler temperatures in the body extremities, the mutant protein is functional and the dark pigment is produced. In the warmer body core, the mutant tyrosinase enzyme is inactive, melanin is not produced, and the *albino* phenotype is observed.

figure 20.3   Use of a deletion to define the allele. The phenotype of an individual with a mutation over a deletion is compared to the homozygous mutant phenotype. If the x^-/del phenotype is the same as the homozygous mutant (x^-/x^-), then the mutation is a null allele. If x^-/del is more severe than the homozygote, the mutation is a hypomorph. If the phenotype is less severe, the mutation is either a hypermorph or neomorph. To distinguish between these two possibilities, an $x^-/x^-/x^+$ individual is generated. If the phenotype of this individual is more severe than x^-/x^-, the mutation is a hypermorph. If the phenotype is the same as x^-/x^-, the mutation is a neomorph.

mutation in the homozygote has the same effect as the deletion in the heterozygote and produces an equivalent phenotype.

This definition of a null allele is based on the resulting phenotype. Any mutation that produces the same phenotype in both a homozygous state or heterozygous with a deletion is genetically defined as a null allele.

Clearly, a mutation that does not encode the protein will be a null allele, which is also called a *molecular null*. A mutation encoding a protein that possesses a small amount of activity, however, may be a genetic null if the homozygote yields the same phenotype as the heterozygote with a deletion. In this situation, a certain threshold of protein activity must be reached to produce a phenotype that is not as severe as the homozygous null mutant.

Alternatively, the deletion heterozygote may have a phenotype that is more severe than the homozygous mutant phenotype (see fig. 20.3). In this case, the mutant allele, called a **hypomorph,** encodes a protein with some activity. In the deletion heterozygote, the amount of expressed protein activity is half that of the homozygous mutant. This may produce a mutant phenotype in the deletion heterozygote that is more severe than in the mutation homozygote. Thus, a hypomorphic allele encodes more protein activity than does a null allele, but less than a wild-type allele.

To help understand this relationship, assume that each copy of a wild-type allele encodes 50% of the protein activity in a diploid cell (**fig. 20.4**). Assume that a mutant allele encodes a protein with 10% of the total wild-type activity. If the mutant homozygote (total 20% activity) has the same phenotype as the deletion heterozygote (total 10% activity), then the mutation is classified as a null allele. In this case, a threshold of greater than 20% wild-type protein activity must be achieved before a less severe phenotype is produced. However, if the phenotype of the mutant homozygote is less severe than the deletion heterozygote, then the mutation is a hypomorphic allele. Notice that the phe-

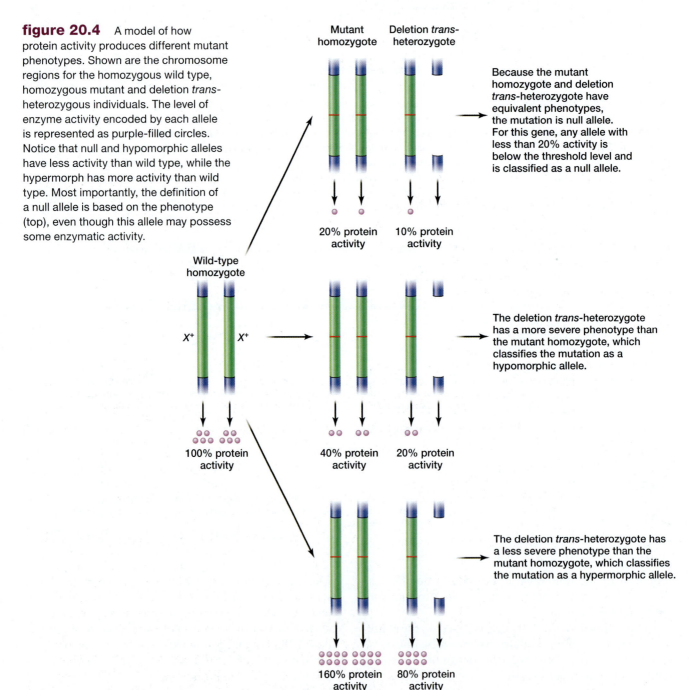

figure 20.4 A model of how protein activity produces different mutant phenotypes. Shown are the chromosome regions for the homozygous wild type, homozygous mutant and deletion *trans*-heterozygous individuals. The level of enzyme activity encoded by each allele is represented as purple-filled circles. Notice that null and hypomorphic alleles have less activity than wild type, while the hypermorph has more activity than wild type. Most importantly, the definition of a null allele is based on the phenotype (top), even though this allele may possess some enzymatic activity.

Mutant homozygote   Deletion *trans*-heterozygote

Because the mutant homozygote and deletion *trans*-heterozygote have equivalent phenotypes, the mutation is null allele. For this gene, any allele with less than 20% activity is below the threshold level and is classified as a null allele.

20% protein activity   10% protein activity

Wild-type homozygote

X^+ X^+

100% protein activity

The deletion *trans*-heterozygote has a more severe phenotype than the mutant homozygote, which classifies the mutation as a hypomorphic allele.

40% protein activity   20% protein activity

The deletion *trans*-heterozygote has a less severe phenotype than the mutant homozygote, which classifies the mutation as a hypermorphic allele.

160% protein activity   80% protein activity

notype of the mutant homozygote relative to the deletion heterozygote reveals whether the mutation is a null or hypomorphic allele, and not simply the amount of encoded protein activity.

Both null and hypomorphic alleles are examples of loss-of-function mutations. **Loss-of-function** mutations encode less protein activity than a wild-type allele. This may result from either reduced expression (transcription or translation) of the encoded protein or the encoded protein possessing less activity than the wild-type protein.

A third genetic class of mutation is one that encodes increased levels of protein activity relative to wild-type, which is termed a **hypermorphic allele.** If a single wild-type allele encodes 50% of the protein activity in a diploid cell, then a hypermorphic allele would encode greater than 50% of the activity. Hypermorphic mutations are also called **gain-of-function alleles,** because they possess increased amounts of protein activity. It is important to realize that increased wild-type activity can, but does not always, produce a mutant phenotype.

Assume that a recessive hypermorphic allele encodes 80% of the wild-type protein activity (see fig. 20.4). The hypermorphic homozygote would possess 160% of the wild-type activity, which results in the recessive mutant phenotype. The deletion heterozygote would express 80% of the wild-type protein activity, because it contains only one hypermorphic allele. This deletion heterozygote would exhibit a less severe phenotype, even possibly a wild-type phenotype, relative to the mutant homozygote (see fig. 20.3).

Neomorphic Alleles

Another type of mutation is a **neomorphic allele,** which produces a phenotype that results from the expression of a "new" protein activity. Neomorphic alleles are also classified as gain-of-function mutations because they represent the addition of a different protein activity. A neomorphic mutation is usually produced through one of three different mechanisms. First, the mutation may cause the misexpression of a gene in the wrong tissue or at the wrong time to produce a "new" activity in those cells. This can occur through the joining of a promoter for one gene with the ORF of a second gene. Second, the mutation may join or fuse the ORFs of two different genes, producing a fusion protein. In this situation, an authentic new activity, which functions as a hybrid between the two original proteins, is produced in the organism. Third, a missense mutation may encode a mutant form of the protein that possesses a new function or activity, rather than a reduced activity (hypomorph) or no activity (amorph).

For a recessive neomorphic mutation, the mutant phenotype requires two copies of the mutant alleles, and the presence or absence of the original wild-type

alleles does not affect the phenotype. Thus, the deletion heterozygote, which has only a single copy of the recessive neomorphic allele, would express the wild-type phenotype (see fig. 20.3). The recessive hypermorphic allele and the recessive neomorphic allele produce the same result in this test.

How can we distinguish between a hypermorphic and neomorphic allele? Remember that a hypermorphic allele has excess wild-type activity, and that the neomorphic allele is unaffected by the presence of wild-type activity. Thus, the simplest way to differentiate between these two classes of alleles is to introduce wild-type alleles into the mutant background (see fig. 20.3).

Let's look at an example of this type of analysis. Assume that we are studying the pigmentation of the *Drosophila* compound eye, which is a deep red in wild-type flies. We identified a recessive mutation (*p*) that exhibits a pink eye color (**fig. 20.5**). We cross the homozygous recessive mutant with a fly that contains a deletion (*del*) for the chromosomal region that corresponds to the mutation and observe the eye pigmentation of the heterozygous progeny (*p/del*).

If the deletion heterozygotes have pink eyes that are phenotypically equivalent to the homozygous mutant, then the mutation is classified as a null allele. However, if the deletion heterozygotes have light pink eyes, then the mutation is classified as a hypomorphic

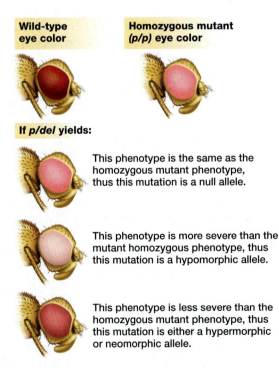

figure 20.5 Use of a deletion to define the type of mutant allele. The eye color phenotypes of wild type (upper left), homozygous mutant (upper right), and three different deletion *trans*-heterozygotes (*p/del*). The level of pigmentation (phenotype) in the *p/del* fly, relative to the homozygous mutant (*p/p*), is used to define the mutant allele.

allele. If the deletion heterozygotes have either wild-type or dark pink eyes, then the mutation is either a hypermorphic or neomorphic allele.

We can distinguish between these latter two possibilities by examining the $p/p/+$ flies. If these flies have lighter pink eyes than the p/p flies, the mutation is a hypermorphic allele because adding the wild-type allele (+) makes the mutant phenotype more severe. Alternatively, if the $p/p/+$ flies have the same pink eyes as the p/p flies, the mutation is a neomorphic allele because the extra wild-type allele did not affect the phenotype.

It is important to remember that in this type of genetic analysis, we always compare the homozygous mutant phenotype to either the deletion heterozygote's phenotype or the phenotype of the homozygous mutant carrying an extra wild-type allele. We want to test what happens to the phenotype when a copy of the mutant allele is removed (deletion), which is essentially decreasing the encoded protein activity of the homozygous mutant by half.

Chemical-Induced Mutations

Chemicals are a commonly used mutagen because they can generate a diversity of loss-of-function and gain-of-function mutations, as well as conditional mutations. Chemicals may be easy to introduce into an organism; they can often be simply added into food or water. Some chemicals produce very high mutation rates, which increases the likelihood of creating a mutation in the desired gene.

Chemicals can mutate DNA in specific ways. The chemical mutagens ethyl methanesulfonate (EMS, discussed in chapter 18) and *N*-ethyl-*N*-nitrosourea (ENU) are both alkylating agents. EMS adds an ethyl group to primarily guanine or thymine (**fig. 20.6**), which produces transitions from either a G–C base pair to an A–T base pair or a T–A base pair to a C–G base pair. ENU primarily produces A–T to T–A transversions and A–T to G–C transitions. As discussed in chapter 18, transversions are usually more severe mutations than transitions.

Base analogs, such as 5-bromouracil and 2-aminopurine, are also common mutagens in some organisms. Both primarily produce A–T to G–C transitions.

All four of these chemical mutagens primarily produce single-base changes in the DNA, generating both missense and nonsense mutations. The missense mutations can include conditional mutations, null alleles, hypomorphs, hypermorphs, and neomorphs. In contrast, nonsense mutations usually correspond to the loss-of-function alleles, hypomorphs and nulls. Thus, when a full range of mutations is desired, chemical mutagens are preferred.

Energy-Induced Mutations

A second type of mutagen is high levels of energy, which is most often in the form of X-rays (see chapter 18). X-ray-mediated mutagenesis has been well studied for nearly a century. Hermann Muller began studies on the mutagenic effect of X-rays on *Drosophila* in 1918.

The effects of X-rays on DNA are due to their ability to transfer high levels of energy into the DNA molecule, which breaks the phosphodiester bond. As discussed in chapter 8, single and double breaks in the DNA molecule can produce inversions, translocations, and deletions. These chromosomal rearrangements usually result in loss-of-function alleles, either because of the absence of protein expression (deletion) or because of the expression of a truncated protein (**fig. 20.7**). At a lower frequency, gain-of-function alleles can also be generated by chromosomal rearrangements that join either the ORFs of two different genes to create a fusion protein or the promoter of one gene with the ORF of a second gene (see fig. 20.7).

One major advantage of chromosomal rearrangements is that they enable the rapid identification of the location of the DNA breakpoints. If a chromosomal rearrangement affects a particular gene and produces a phenotype, then the localization of the breakpoints by genomic Southern hybridization can simplify the isolation of the mutant allele (see chapter 12).

DNA Insertional Mutations

A third way to mutate DNA is by inserting a piece of foreign DNA randomly into the genome. These insertions are usually the result of a transposable element moving to a new location in the genomic DNA (chapter 18).

Insertions can have profound effects on the expression of a gene. They often result in the introduction of a premature translation termination codon into the ORF (nonsense mutation), or the prevention of transcription altogether, by insertion into the promoter region.

The insertion of DNA has provided a powerful method for identifying genes and their patterns of transcription, as will be described shortly. Furthermore, the location of an insertion can be easily identified by genomic Southern hybridization, which greatly simplifies the cloning of a mutant allele.

It's Your Turn

We have discussed a variety of different ways that mutations can be classified, from the molecular defect in the DNA to the phenotype. Problems 18, 24, and 35 examine these differences and provide you with an opportunity to reinforce these concepts.

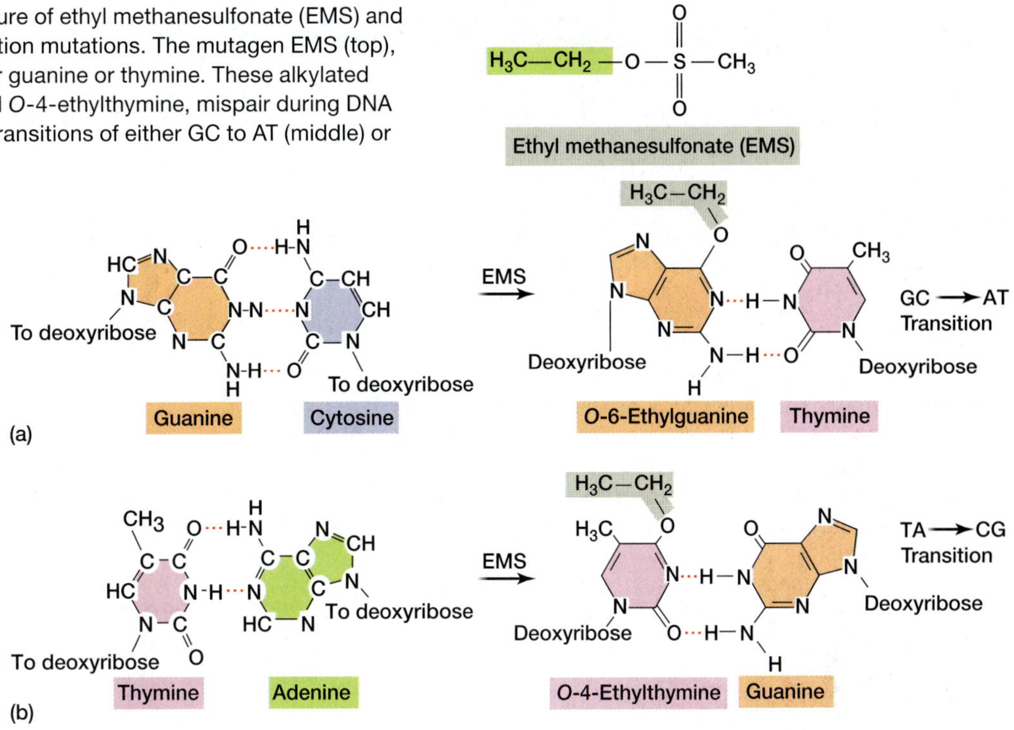

figure 20.6 The structure of ethyl methanesulfonate (EMS) and its mode of generating transition mutations. The mutagen EMS (top), alkylates the oxygen of either guanine or thymine. These alkylated bases, O-6-ethylguanine and O-4-ethylthymine, mispair during DNA replication, which results in transitions of either GC to AT (middle) or TA to CG (bottom).

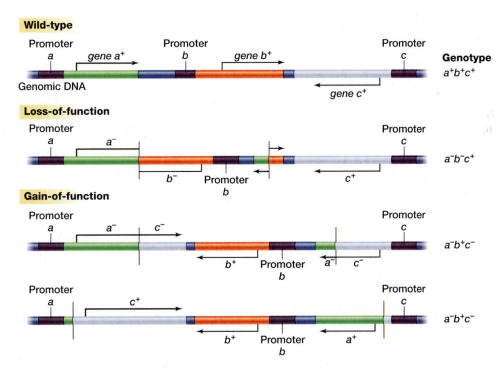

figure 20.7 The generation of loss-of-function and gain-of-function mutations by inversions. Three genes (a, b, and c), their corresponding promoters (purple), and their direction of transcription (arrows) are shown. Loss-of-function mutations result in proteins with either reduced or no activity. The a⁻ and b⁻ mutant alleles are loss-of-function because they both express only part of the open reading frame and will produce truncated proteins. Gain-of-function mutations are generated by either the expression of a fused gene (genes a⁻ and c⁻, top) and the corresponding fusion protein with a new activity or the expression of a gene from a new promoter (c⁺ mRNA is expressed from the a promoter and the a⁺ mRNA is expressed from the c promoter). These fusions result in the misexpression of the wild-type gene, which results in the mutant phenotype.

20.1 applying genetic principles to problems

Mapping Chromosomal Breakpoints Associated with a Mutant Phenotype

Mutations generated by X-rays are the result of breaks in the DNA, and one advantage of X-rays as a mutagen is that the mutation and the associated gene can be quickly identified. This analysis requires analysis of chromosome-banding patterns and techniques such as chromosome in situ hybridization and genomic Southern hybridization, which were discussed in earlier chapters.

One approach is in situ chromosome hybridization, which is the hybridization of a specific DNA sequence (probe) to the metaphase chromosomes. Fluorescent in situ hybridization (FISH) utilizes a fluorescent molecule to label the probe. The location at which the probe hybridizes reveals the chromosomal location of the complementary sequence.

Figure 20.A shows a diagram of two different chromosomes from an organism: a metacentric chromosome (blue) and a submetacentric chromosome (green). The red mark on the metacentric chromosomes represents the FISH pattern using a specific probe that hybridizes to the metacentric chromosome.

The purpose of the FISH analysis is to determine what chromosome has been affected by the X-rays. It can also localize the breakpoints to a general region of the chromosome and provide information about the type of chromosomal rearrangement that occurred. Genomic Southern hybridization is the preferred method to define the location of the breakpoints to a narrower region.

Here you will apply some of these fundamental concepts of chromosome rearrangements to interpreting a genomic Southern hybridization. First, the restriction map for the region is determined (see below).

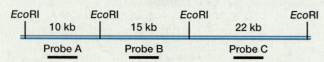

You would then perform Southern hybridizations using genomic DNA that was isolated from wild-type and X-ray-induced mutants. The *Eco*RI-digested genomic DNA is hybridized with each of the three probes individually, and the results of the genomic Southern blots are shown as follows.

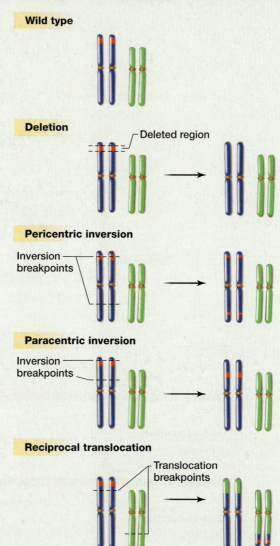

box figure 20.A   The effect of chromosomal rearrangements on the FISH signal. The locations of the probe hybridizing to a wild-type chromosome and four different chromosomal rearrangements are shown in red. The breakpoints associated with each rearrangement are marked with horizontal lines.

Mutants

Wild-type #1   #2

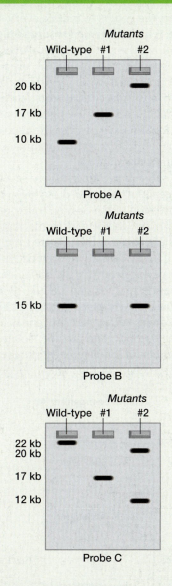

20 kb

17 kb

10 kb

Probe A

Mutants

Wild-type #1   #2

15 kb

Probe B

Mutants

Wild-type #1   #2

22 kb
20 kb

17 kb

12 kb

Probe C

To determine the chromosomal aberration associated with each of the mutants, you need to compare the restriction fragment that each probe detects in both wild-type and the individual mutant genomic DNAs.

Mutant #1 shows a different restriction fragment than wild-type for all three probes. First, probe B fails to detect a complementary sequence in mutant #1. This suggests that a deletion removes the complementary sequence from mutant #1. And yet, probes A and C also detect bands of different sizes than the wild type. Additionally, both probes are detecting the same 17-kb band in mutant #1. The sequences complementary to probes A and C must be present in this mutant, because these probes are detecting bands. Probes A and C are detecting the same fragment because the second and third EcoRI sites are missing.

The location of the beginning and end of the deletion are a little unclear. The deletion must begin before the

second EcoRI site, but also it must leave at least some of the sequence that is complementary to probe A. The deletion must also end after the third EcoRI site, but leave at least some of the sequence that is complementary to probe C. You can therefore draw this deletion as follows, with the gray line representing the region deleted. The only other condition is that the distance remaining in the deletion between the two EcoRI sites must be 17 kb.

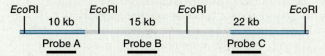

EcoRI   EcoRI   EcoRI   EcoRI
      10 kb   15 kb   22 kb
   Probe A   Probe B   Probe C

Mutant #2 is different from mutant #1 in three respects. First, all three probes detect DNA fragments, so mutant #2 is not a deletion. Second, probe C detects two DNA fragments (20 and 12 kb in length) instead of one fragment. Third, the flanking probes, A and C, detect different size DNA fragments in mutant #2 compared with wild type, whereas the central probe (B) detects the same size fragment in both mutant #2 and wild type. These observations suggest that the DNA rearrangement affects the flanking probes, but not the central region.

Remember, an inversion breaks the DNA at two sites and inverts the central region (chapter 8). This could account for the differences observed with the flanking probes. The central region, furthermore, would have the same sequence and restriction enzyme sites, but just in the opposite orientation. If mutant #2 is an inversion, how could probe C detect two DNA fragments?

If an inversion breaks within the sequence that is complementary to the probe, the probe now has the ability to hybridize to two different regions. If restriction sites are located between these two regions, then the probe will hybridize to two different restriction fragments.

By the same logic, mutant #2 must also be an inversion, and the right breakpoint must be within the sequence that is complementary to probe C. Because probe A does not detect two restriction fragments, the sequence complementary to probe A is not broken by the inversion breakpoint. We are unable to determine, however, if the sequence that is complementary to probe A is outside the inversion or within the inversion. Thus, two maps are possible for mutant #2.

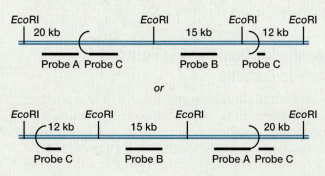

EcoRI   EcoRI   EcoRI   EcoRI
   20 kb   15 kb   12 kb
Probe A Probe C   Probe B   Probe C

or

EcoRI   EcoRI   EcoRI   EcoRI
   12 kb   15 kb   20 kb
Probe C   Probe B   Probe A Probe C

Restating the Concepts

▶ We can define mutations based on their effect on the encoded protein. A missense mutation results in the change of an amino acid. A nonsense mutation results in the introduction of a premature translation termination codon.

▶ Conditional mutants exhibit the wild-type phenotype at the permissive condition and a mutant phenotype at the restrictive condition.

▶ Mutations can be classified based on their behavior in different genetic backgrounds. Null (amorphic) alleles genetically behave as a deletion, with no apparent encoded protein activity. Hypomorphs and hypermorphs produce mutant phenotypes as a result of reduced and increased wild-type levels of protein activity, respectively. Neomorphs genetically behave as a new protein activity, which may represent a common activity that is simply expressed at either a new location in the organism or at a new time in its life cycle.

▶ Common types of mutagens include chemicals, X-rays, and DNA insertions. The type of mutagen used depends on the type of mutation sought.

20.2 Forward Genetics

In the last 50 years, genetics has been widely used to isolate mutants in a variety of model organisms, ranging from bacteria to mice. The power of genetic analysis is that a mutation in a gene should help to reveal the role of the gene and its encoded product (usually a protein) through the mutant phenotype.

There are two common ways to isolate mutants. **Forward genetics** involves randomly mutagenizing the genome to generate a wide array of mutants (**fig. 20.8**). The mutants are screened to reveal the ones that exhibit the desired phenotype, and the mutated genes are then isolated to determine the identity of the encoded gene product. **Reverse genetics,** described in the following section, involves the targeted mutagenesis of a specific gene. With the advancement of molecular biology and the completion of genome sequencing projects, the reverse genetics approach has become more popular and powerful.

Genetic Crosses Used in Forward Genetic Screens

Once the desired mutagen is selected, it is applied to the organism. To be efficient, the mutagen must produce a large number of mutations in each individual. This mutagenized individual (F_0) is then mated to produce progeny. The individual (F_1) progeny will possess a subset of the mutations that were generated in the mutagenized individual. Thus, generation of progeny serves as a way to reduce the total number of mutations, and phenotypes, in a mutagenized individual.

Quite often, the goal is to mutate every single gene in the genome, which is termed **saturation.** Saturation theoretically allows the identification of every mutation that affects a specific process in the organism. This is a powerful approach when trying to understand a process with multiple components, such as how a neuron transmits an excitatory signal to a muscle. In the following sections, we discuss the types of crosses that are performed in eukaryotic organisms to isolate a variety of different mutants.

Reaching saturation depends on several factors. First, the mutagen must be applied at a high enough dose to ensure the creation of the desired mutations, without generating too many lethal and sterile mutations that would make the recovery of the desired mutation difficult. Second, all of the desired genes must be mutable to roughly the same efficiency by the mutagen. For example, if the mutation rate for one gene is 10-fold greater than another gene (see chapter 18), roughly 10 mutations will need to be generated in the first gene before a single mutation in the second gene may be isolated. Third, a very large number of mutagenized progeny must be generated and screened for the mutant phenotype.

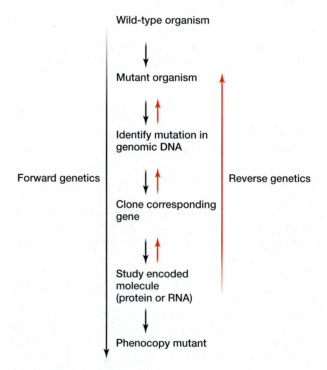

figure 20.8 Forward versus reverse genetics. In forward genetics, a mutant is isolated based on a phenotype, followed by the cloning of the gene and the characterization of the encoded protein's function. In reverse genetics, a protein or gene is first identified, a mutation is subsequently generated in the gene and the resulting mutant organism is characterized.

Isolation of Dominant Mutations

If we are interested in isolating mutations that were generated in the germ cells that can be stably inherited, we can mate the mutagenized individual with a wild-type individual to produce F_1 progeny (**fig. 20.9**). The F_1 progeny can then be examined for dominant mutant phenotypes in an **F_1 screen.** Only a small number of the F_1 individuals will exhibit the desired mutant phenotype.

Often, an F_1 individual may exhibit a more complex phenotype than might be expected. In isolating mutant flies with curly wings, for example, you might identify some F_1 individuals that contain both curly wings and abnormal bristles on the body. It is possible that a single dominant mutation can cause both phenotypes (pleiotropy, chapter 5) or the F_1 individual may possess two dominant mutations: one affecting the wing and the second mutation affecting bristles. How do you determine if there is one or two mutations?

The F_1 individual is again mated with a wild-type individual. If a single dominant mutation causes both phenotypes, then half of the F_2 progeny will exhibit both mutant phenotypes and half will be wild-type. None of the F_2 progeny will exhibit only curly wings or only abnormal bristles (see fig. 20.9). If each dominant mutant phenotype is caused by a different mutation, then some of the F_2 progeny will be wild-type, some will have curly wings, some abnormal bristles, and some both curly wings and abnormal bristles. Some of these F_2 progeny, therefore, reveal the phenotypes associated with each single mutation.

Isolation of Recessive Viable Mutations

Producing and screening F_1 progeny for dominant mutants is fairly straightforward. But many more mutations exhibit recessive phenotypes than dominant ones. The loss-of-function mutations are also critical for deducing the function of the gene's encoded product. Therefore, it is desirable to screen for recessive mutations (**fig. 20.10**). Rather than screening the F_1 progeny as we discussed for dominant mutations, each F_1 individual is mated with a wild-type individual to produce separate families of F_2 progeny. Approximately, 50% of the F_2 progeny from an F_1 individual containing induced mutations will also be heterozygous for any given recessive mutation. Thus, several siblings within an F_2 family will possess the same allele that was inherited from their common F_1 parent. If these F_2 siblings mate with each other, approximately 25% of the F_3 progeny will exhibit the recessive mutant phenotype. In this **F_3 screen,** mating separate F_1 individuals with wild-type individuals produced several heterozygous F_2 siblings that all possess the same mutation, and who are then mated with their siblings to generate homozygotes that exhibit the recessive mutant phenotype.

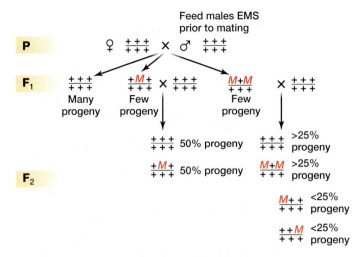

figure 20.9   Genetic scheme to isolate a dominant mutation. Males are mutagenized with EMS and then mated to wild-type females. The resulting F_1 progeny are analyzed to identify the few individuals with a dominant mutation (red M). To distinguish if the mutant possesses one or more dominant mutations, the F_1 mutant is crossed to a wild-type individual and the frequency of mutants in the F_2 generation is determined. If half the progeny have the same dominant phenotype as the parent, then it is a single mutation. If some of the progeny have the same mutant phenotype as the parent and others have only a portion of the dominant phenotype, then the parent had multiple mutations. However, some of the F_2 progeny will now have only a single mutation.

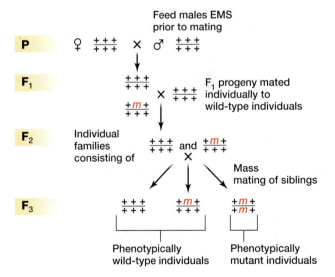

figure 20.10   A typical F_3 screen to identify recessive mutants. Males are mutagenized with EMS and mated to wild-type females. The resulting F_1 progeny are individually mated to wild-type individuals to generate independent lines. The F_2 progeny are randomly mated with their siblings. Some random matings will occur between heterozygous mutant siblings, which will produce the homozygous mutant F_3 progeny that express the recessive mutant phenotype. Notice that the recessive phenotype is not observed until the F_3 generation, while dominant mutations will produce a mutant phenotype in the F_1 generation (Figure 20.9).

In figure 20.10, it is possible that the F_1 individual who is mated may lack the mutagenized chromosome. If this occurs, none of the F_2 progeny will have the desired mutation. Therefore, it is common to mutagenize an individual in the P generation that contains a recessive mutation to mark the mutagenized chromosome. This allows the researcher to identify and mate individuals that contain the mutagenized chromosome, even though it is heterozygous, which increases the chances of producing progeny that are homozygous for the induced mutation.

When mutagenizing a marked chromosome, it is possible that the induced mutation may recombine off the marked chromosome. To decrease the likelihood that the marked chromosome will lose the induced mutation, balancer chromosomes were developed in *Drosophila*. **Balancer chromosomes** contain multiple overlapping inversions that cover most of a single chromosome and suppress recombination (see chapter 8). Often, the balancer chromosome contains a dominant morphological mutation to easily identify progeny

that contain the balancer chromosome. The dominant mutation also often has a recessive lethal phenotype to prevent the generation of an individual that is homozygous for the balancer chromosome. These features make it possible to select heterozygotes expressing the dominant balancer mutant phenotype and know the origin of the homologous chromosome. Balancer chromosomes, which exist and are extensively used for the *Drosophila* X chromosome and the second and third autosomes, are also being developed and used in mouse genetics.

A common balancer in *Drosophila* is *SM1*, which stands for **s**econd chromosome **m**ultiple inversions **1**. This balancer, which was generated through a series of overlapping inversions, contains the dominant *Curly wing* mutation (*Cy*). The *Cy* mutation also has a recessive lethal phenotype. A second mutation on *SM1* is *cinnabar* (*cn*), which causes the eye to be bright scarlet red relative to the deep red wild-type eye color. The *SM1* balancer would be used in a mutagenesis as diagrammed in **figure 20.11**.

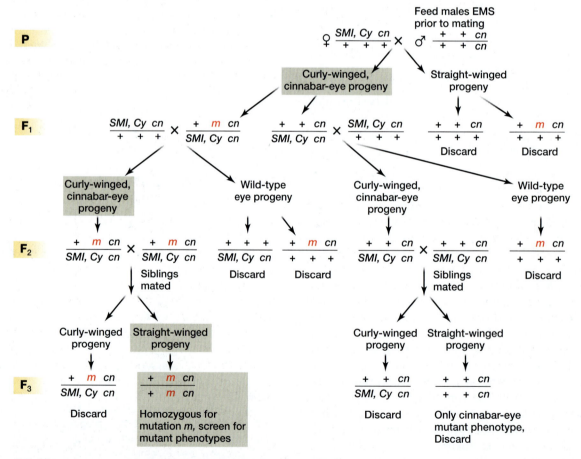

figure 20.11 The use of a balancer chromosome in genetic screens. The balancer chromosome, *SM1*, contains both a dominant (*Cy*, curly wings) and a recessive visible marker (*cn*, cinnabar-eye color). A *cn* mutant male is mutagenized to generate recessive mutations on the *cn*-marked chromosome. The inversions on the balancer chromosome, which is marked by the dominant mutation (*Cy*), prevent the newly generated recessive mutation (*m*) from recombining onto the homologous chromosome (*SM1*). Isolating progeny in the various generations based on the expression of the curly wing and cinnabar eye phenotypes ensures the presence of specific unrecombined chromosomes in the individual.

In this scheme, *Drosophila cinnabar* (*cn*) males are mutagenized. The *cn* allele serves as a marker for the mutagenized second chromosome. Because the *SM1* balancer prevents recombination on the second chromosome, the *cn* mutation and whatever mutation is induced on the second chromosome will remain linked throughout the crosses.

The curly-winged, cinnabar-eyed F_1 progeny are selected, which should be of two genotypes: one possessing induced mutations, and the other lacking mutations. These F_1 progeny are individually backcrossed to the *SM1, Cy cn* flies. The curly-winged, cinnabar-eyed F_2 progeny are mated to their siblings. The cinnabar phenotype ensures that a chromosome from the mutagenized male (either + *m cn* or + + *cn*) is present in the F_2 individuals. Maintaining the mutagenized chromosome with the *SM1* balancer chromosome ensures that the induced mutation remains on the same chromosome as the *cn* marker. The straight-winged, cinnabar-eyed F_3 progeny are then examined for a recessive mutant phenotype.

Isolation of Recessive Lethal Mutations

Occasionally, geneticists are interested in isolating recessive lethal mutations. This task is a little more difficult than isolating recessive viable mutants for two reasons. First, if the lethality occurs early in development, such as during embryonic development, the dead individuals may not be apparent. Second, if a dead individual is identified, it is no longer able to mate and propagate the chromosome that contains the recessive lethal mutation. Therefore a variation of the F_3 recessive viable mutation cross just discussed must be employed.

In this variation, cinnabar-eyed males are mutagenized and mated as diagrammed (see fig. 20.11). In the F_3 generation, the phenotype of interest is lethality, which may not always be obvious if the individuals die early in development. If a lethal mutation exists, however, then the F_2 mating will fail to produce straight-winged F_3 progeny. The production of only curly-winged F_3 progeny reveals that a recessive lethal mutation exists on the second chromosome.

How do we recover a recessive lethal mutation if all the straight-winged F_3 progeny are dead? If you look at the cross, you can see that the curly-winged F_3 progeny are heterozygous for the recessive lethal mutation (labeled as *m*) with the *SM1* balancer chromosome. The curly-wing phenotype of the dominant *Cy* mutation on the balancer chromosome, which is also homozygous lethal, is used to identify heterozygous individuals containing the chromosome with the induced recessive lethal mutation. This technique allows lethal and deleterious mutations to be isolated, maintained as heterozygotes, and studied.

Eric Wieschaus and Christiane Nüsslein-Volhard used a similar F_3 screen to identify recessive lethal mutations on the *Drosophila* X chromosome and the second and third autosomes that disrupted embryonic development. This work, which earned them the Nobel Prize in Physiology or Medicine in 1995, identified a large number of genes that affect the development of either the embryonic axes (dorsal–ventral and anterior–posterior) or the identity of regions and cells within the embryo. The genes that correspond to these mutations were later shown to be present in a wide variety of organisms, including humans. These *Drosophila* mutations also revealed that similar proteins and mechanisms are utilized in the proper development of the fly embryo and humans.

In 1996, Christiane Nüsslein-Volhard and Wolfgang Driever independently announced their initial results from F_3 screens for mutants that affect embryonic development in zebrafish. Over 500 different genes were mutated in these two large-scale screens for recessive lethal mutations. Mutants were identified that affect the earliest stages of embryonic development, within 15 h after fertilization (**fig. 20.12**). Additional mutants were identified that affected the embryonic development of the fins (see opening figure in this chapter), eye, brain, and heart. Many of these genes have already been shown to be related to the genes required for *Drosophila* embryonic development. It is hoped that these zebrafish mutants will expand our knowledge about the mechanisms that are involved in embryogenesis of a vertebrate.

Isolation of Mutants Resulting from Insertion of Transposable Elements

The previous examples work for chemically induced and radiation-induced mutations. If we are interested in isolating insertion mutations due to transposable elements, we must make a small change in the procedure.

The *Drosophila* P element (see chapter 18) is used as the transposon that randomly inserts within and inactivates genes throughout the *Drosophila* genome. The recombinant P element used in these experiments (**fig. 20.13**) cannot transpose on its own because the transposase gene, which encodes the enzyme that mediates P element movement, was replaced with a version of the wild-type *white* gene to generate $P[w^+]$. If every individual in the cross is homozygous w^-, then the red eye phenotype confirms the presence of the $P[w^+]$ element in the genome. In the P generation, males containing the modified $P[w^+]$ elements are mated to females that provide the transposase gene in the form of a Δ2–3 transgene. Because the Δ2–3 transgene lacks the intron between exons 2 and 3, it does not encode the 66 kDa repressor and the transposase is expressed throughout the embryo (chapter 18). The transposase

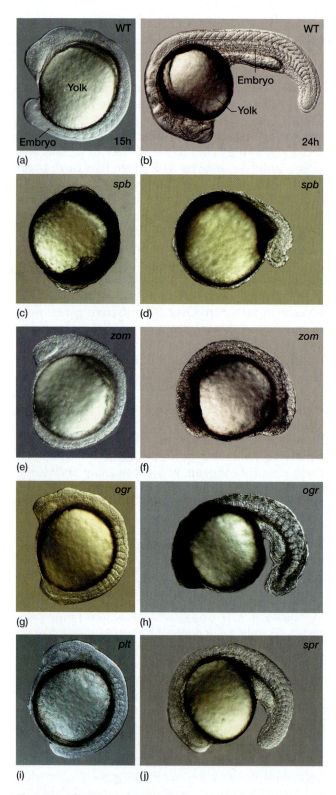

figure 20.12   Early embryonic recessive lethal zebrafish mutants identified in an F$_3$ screen. Embryos were examined at 15 and 24 hours after fertilization (left and right columns, respectively). Photos a, b, wild-type embryos; c, d, *speed bump* (*spb*) mutant; e, f, *zombie* (*zom*) mutant; g, h, *ogre* (*ogr*) mutant; i, *poltergeist* (*plt*) mutant; j, *specter* (*spr*) mutant.

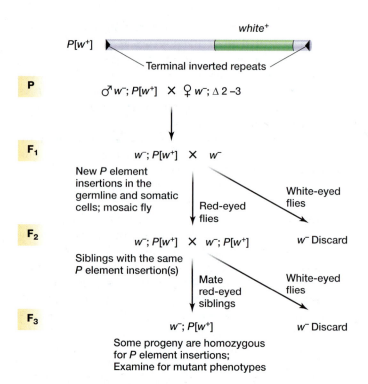

figure 20.13   *Drosophila P* element mutagenesis. Shown are the sample crosses to generate mutants by transposing a recombinant *P* element. The *P[w$^+$]* element contains the terminal inverted repeats of the *P* element flanking the *white$^+$* gene, which confers the red-eyed phenotype. Continually mating the *P[w$^+$]* flies to *w$^-$* flies and selecting the red-eyed (*w$^+$*) phenotype in the progeny identifies the flies with the mutagenized chromosome. The Δ2–3 element in the P generation provides the transposase that mediates the movement of the *P[w$^+$]* element.

then drives the movement of the *P* element to new locations in the genome.

The F$_1$ progeny contain germ cells in which the *P* element has transposed to new locations (see fig. 20.13). These F$_1$ progeny are mated to homozygous *w$^-$* individuals and the red-eyed F$_2$ progeny are isolated. Each red-eyed F$_2$ individual represents one or more different *P[w$^+$]* insertions that may correspond to new mutations.

The red-eyed F$_2$ progeny are again individually mated with *w$^-$* flies (see fig, 20.13). The resulting red-eyed F$_3$ siblings, which are heterozygous for the same insertional mutations, are mated. The resulting red-eyed F$_4$ progeny are then examined for either dominant or recessive mutant phenotypes that resulted from the *P[w$^+$]* insertion. Because insertions are more likely to generate loss-of-function alleles, most of these insertions will be recessive mutations.

The genes that are disrupted by the insertion of the *P* element can be easily cloned through two different methods. First, we can generate a genomic library from the progeny of these mutant flies (see chapter 12). The library is screened using the *P* element as the probe, and a series of genomic clones containing the recombinant *P* element is identified. A restriction map of these genomic clones is created and compared with a restriction map of the recombinant *P* element to identify the genomic DNA flanking the *P* element insertion. These flanking sequences, which correspond to the disrupted gene, are cloned and used as a probe of a wild-type genomic DNA library to clone the wild-type gene. Alternatively, this flanking genomic DNA can be sequenced and compared with the genomic DNA sequence database to identify the gene.

A second approach to identify the disrupted gene is based on a method called **inverse circular PCR.** The genomic DNA, which contains the recombinant *P* element insert, is digested with a restriction enzyme (**fig. 20.14**). The genomic DNA fragments are diluted and ligated under conditions that preferentially cause each fragment to circularize. Two different oligonucleotide primers, which are complementary to the known recombinant *P* element sequence, are added to the circular molecules. They will only base-pair with the *P* element DNA and can be used to PCR-amplify the circular molecule. This results in the production of a linear DNA product that corresponds to the genomic DNA flanking the recombinant *P* element insertion. This PCR product can be sequenced and compared with the genomic DNA sequence database to identify the gene that is disrupted by the *P* element insertion.

It's Your Turn

Any successful forward genetic screen requires designing the proper genetic crosses. The crosses will vary depending on the type of mutation. To get a feel for the critical issues in designing crosses, try problems 27 and 28.

Approaches Used to Screen for Mutants

A forward genetic screen may in some cases require mutagenizing hundreds to thousands of individuals to generate hundreds of thousands to hundreds of millions of progeny. Many of these progeny will contain mutations. The number that contain a desired mutation, however, depends on the mutagen, the size of the target gene, and the number of target genes. In some cases, the frequency of the desired mutant in the population of progeny may be as small as one in a million. How do we expect to identify these rare individuals?

The key to any successful forward genetics project is the **genetic screen,** the method used to identify the

rare individuals with the desired phenotype in the large population. The design of the genetic screen is an art, and it can reveal the creativity and insight of the geneticist. Using the right screen easily identifies all the desired mutants; an incorrect screen discards desired mutants, isolates undesirable mutants, and takes a great deal of time.

Genetic screens have been performed in a variety of different organisms, from bacteria and bacteriophages, to molds and yeast, to flies and worms, to zebrafish and mice. These genetic screens identified genes that are required for basic cellular functions, developmental processes, viability, physiological processes, behaviors, and longevity, to name a few. Genetic screens have also been performed to identify mutants that mimic forms of inherited abnormalities in humans. For example, mutations that cause degenerative blindness in flies, zebrafish, and mice have been isolated that mimic forms of human blindness such as retinitis pigmentosa or night blindness, macular degeneration, and cataracts. In some cases, the mutant genes helped to identify the corresponding gene in humans. In other cases, the mutants helped to reveal the underlying cause of the human abnormality.

Mutant Selection Versus Detection

In any mutagenesis, one important consideration is how the desired mutants will be identified in the overall population. Two general approaches are used. **Selection** involves segregating the desired mutants from the remainder of the population. For example, to identify wild-type bacteria (prototrophs) from a population of mutagenized auxotrophs, we can use minimal medium to select the prototrophs. Auxotrophs will only grow on minimal medium that has specific compounds added, such as *ala*⁻ auxotrophs requiring alanine to be added to minimal medium for growth. If the mutagenized auxotrophs are plated on minimal medium, only the prototrophs will grow. Thus, they are selected from the auxotrophs.

Detection improves the identification of the desired mutants from the remainder of the population rather than separating the desired mutants from all the other individuals. For example, to isolate *E. coli* mutants that cannot utilize lactose as an energy source (*lac*⁻ mutants), we could mutagenize wild-type bacteria and plate them on medium that contains IPTG and X-gal. The IPTG will induce expression of the *lac* operon, and X-gal is a colorless compound that will turn blue if it is cleaved by the β-galactosidase enzyme (see chapter 16). This detection method allows us to identify the desired *lac*⁻ mutants, which are present as white colonies because they are unable to cleave the X-gal, within a population of wild-type bacteria (blue colonies).

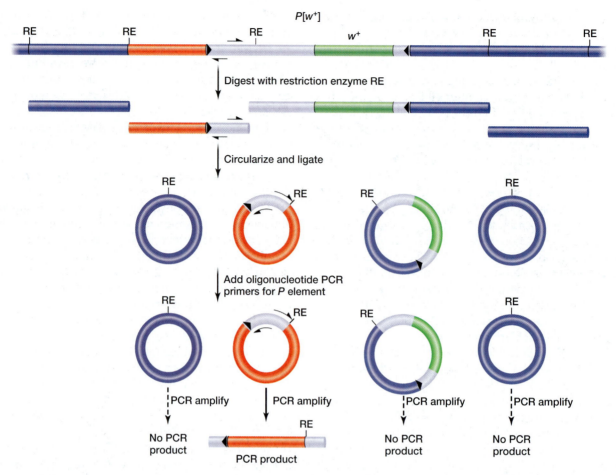

figure 20.14 Inverse circular PCR amplification to identify the insertion site location of a *P* element. Shown is a *P[w⁺]* element inserted within the genomic DNA and several restriction enzyme sites (RE). The genomic DNA is digested with the restriction enzyme and the genomic DNA fragments are ligated to circularize them. Two oligonucleotide primers within the *P* element (arrows) anneal to the specific *P* element fragment. PCR amplification with these primers produces a fragment from only the circular DNA molecule that contains the correct portion of the *P* element. The PCR product has some *P* element sequences flanking the genomic DNA that is adjacent to the inserted *P* element. Sequencing the PCR product will reveal the DNA sequence that is adjacent to the inserted *P[w⁺]* element.

Visible Mutant Phenotypes

Mutants exhibiting a visible phenotype are the most obvious type to select. A visible phenotype may be obvious to the naked eye, such as the coat color in mice; it may require a simple observation under a microscope, such as wing shape in flies; or it may require a more prolonged examination, such as an altered behavior.

For the more obvious phenotypes, examining the individuals one at a time is usually the simplest and most effective way of identifying the desired mutants. An example of this is the detection of zebrafish fin mutants. Using a standard F₃ screen, geneticists identified several mutants that possessed abnormal fins. Some of the mutants had unusually long fins, others had very short fins, and others even lacked some fins (see the opening figure in this chapter). As with these fin mutants, the detection of mutants with unusual size or pigmentation is fairly easy.

The detection of mutants with an abnormal behavior can be more difficult. In many cases, not every individual in the population will exhibit exactly the same phenotype all the time. An example of a behavioral phenotype is the escape response in zebrafish. Wild-type zebrafish normally swim in circles in a circular tank. If a dark stripe is rotated around the outside of the tank, however, the fish perceive it as a threatening object and quickly change their swimming direction to avoid the stripe (**fig. 20.15**). If wild-type fish are given 10 encounters with the stripe, they will exhibit the escape response 8–10 times. Mutants can then be detected as exhibiting the escape response 7 or fewer times out of 10 encounters.

The complexity underlying the behavior can be seen in the variety of mutants that exhibit an abnormal escape response. First, eye mutants that cannot see the stripe will not exhibit an escape response. Second, a

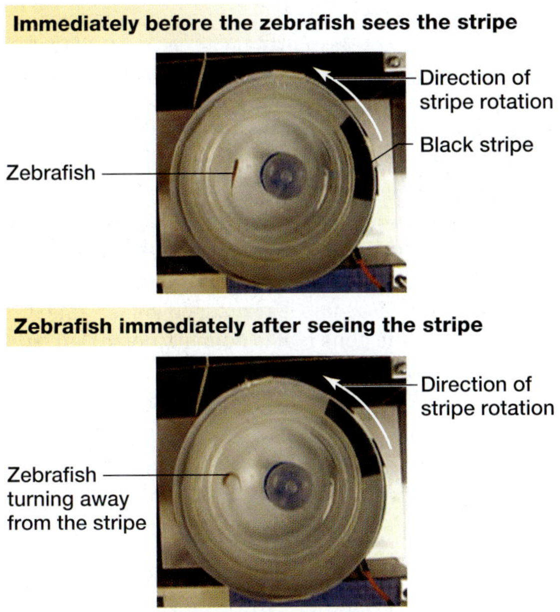

Immediately before the zebrafish sees the stripe

- Direction of stripe rotation
- Black stripe
- Zebrafish

Zebrafish immediately after seeing the stripe

- Direction of stripe rotation
- Zebrafish turning away from the stripe

figure 20.15   The escape response measures a zebrafish visual behavior. The fish swims around the perimeter of a circular tank. If the fish approaches a black stripe that rotates around the outside of the tank, it is perceived as a predator and the fish switches its swimming direction to "escape" the stripe. By reducing the intensity of the ambient light, it is possible to measure the relative ability of the fish to see in both dim and bright light.

variety of brain mutants, which are unable to process the visual information as a threatening object, will not attempt to escape. Third, some mutants that have muscle defects may be unable to rapidly change their direction when they encounter the stripe. Thus, defects that affect a variety of different tissues can lead to the same mutant behavior. This suggests that the geneticist must either use a very well characterized behavior or apply another level of analysis to detect the desired mutants from a behavioral screen.

Enhancer Trap Expression Screens

The visible phenotype screens require the researcher to make a prediction about the potential phenotypes that might be observed. In the previous example, we would predict that mutations affecting vision would prevent the mutant zebrafish from seeing the stripe, and they would therefore fail to show the escape response. But the phenotype of the desired mutant may not be so obvious.

For example, you may be interested in identifying genes that are expressed in the brain. You could imagine that mutations in these genes would affect learning, memory, or perception of different sensory signals. In some cases, a single mutation may affect zero, one, two,

or all of these traits. If we screened for only learning mutants, we could miss important mutants that affect memory and not learning.

For this reason, expression screens were developed. *Expression screens* identify genes based on when and where the genes are expressed in the organism. When a gene with a desirable expression pattern is identified, a mutant can be generated through a reverse genetics approach or by phenocopy (described later in this chapter).

A clever way to identify expression patterns and simultaneously generate mutations in those genes was pioneered in bacteria and then modified for eukaryotic organisms. Many years before the bacterial genomes were completely sequenced, researchers sought to identify genes through genetic methods. Transposable elements were constructed that contained a selectable marker that lacked a promoter. Only when this transposable element inserted downstream of a promoter, which would be within a gene, would the selectable marker be expressed. Those colonies expressing the selectable marker could be isolated and analyzed to reveal the location of the transposable element and the gene that it had inserted within.

In eukaryotes, the problem was simplified a little. We mentioned in chapters 10 and 17 that eukaryotic genes often contain enhancer elements that can control when and where a gene is expressed. More important, these enhancers function regardless of their orientation relative to the gene and regardless of whether they are located upstream or downstream of a gene. Because of these factors, the identification of enhancers should provide an approximate location of a gene.

These **enhancer trap screens** have been widely used in *Drosophila*. They employ the transposable *P* element and a transposase source to move the *P* element into new locations throughout the genome.

The *P* element is constructed to contain a reporter gene, such as the bacterial *lacZ* gene, that is under the transcriptional control of a weak promoter from the *P* element (**fig. 20.16**). If this *P* element fails to insert near an enhancer, only very low levels of *lacZ* expression are observed. When this *P* element inserts near an enhancer, the expression of *lacZ* increases significantly. When and where the β-galactosidase is expressed reveals the expression pattern that is normally controlled by the enhancer.

These enhancer screens in *Drosophila* have revealed a wide array of different expression patterns (**fig. 20.17**). Many of these insertions have led to the identification of genes. However, flies that contain an interesting *P* element insertion often fail to have a mutant phenotype. This is because the enhancer has the ability to control the expression of both its natural gene and the *lacZ* reporter in the *P* element, which may be separated by tens of kilobases.

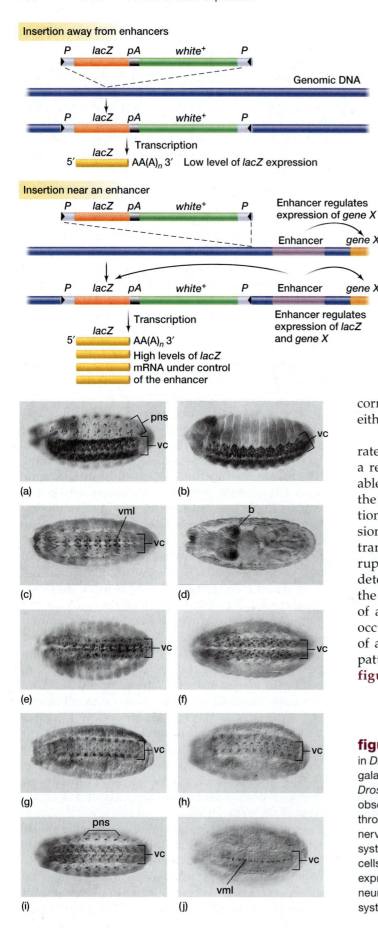

Insertion away from enhancers

P   lacZ   pA   white⁺   P

Genomic DNA

P   lacZ   pA   white⁺   P

↓ Transcription

5′ lacZ AA(A)ₙ 3′   Low level of *lacZ* expression

Insertion near an enhancer

P   lacZ   pA   white⁺   P   Enhancer regulates expression of *gene X*

Enhancer   *gene X*

P   lacZ   pA   white⁺   P   Enhancer   *gene X*

↓ Transcription   Enhancer regulates expression of *lacZ* and *gene X*

5′ lacZ AA(A)ₙ 3′

High levels of *lacZ* mRNA under control of the enhancer

(a)   pns / vc
(b)   vc
(c)   vml / vc
(d)   b
(e)   vc
(f)   vc
(g)   vc
(h)   vc
(i)   pns / vc
(j)   vc / vml

figure 20.16   Enhancer trap mutagenesis. The enhancer trap mutagenesis uses a modified *P* element that contains the *Drosophila white⁺* gene, the bacterial *lacZ* gene under the transcriptional control of the weak *P* element promoter and a polyadenylation signal (*pA*). If the *P* element inserts in a genomic region that lacks enhancer elements (top), only very low levels of β-galactosidase are produced from the *lacZ* gene. If the *P* element inserts near an enhancer (bottom), transcription of *lacZ* increases and is expressed in the same pattern as the gene that is normally controlled by the enhancer.

To generate a mutation in a specific enhancer trap line, small deletions are generated from one end of the *P* element into the flanking genomic DNA. If these deletions extend far enough, they may delete either the enhancer or even the corresponding gene. Either of these two possibilities may yield the desired phenotype in the corresponding gene. These deletions usually result in either null or hypomorphic alleles.

Two modifications to this scheme were incorporated into an approach to isolate genes in mice. First, a recombinant virus is used to integrate the selectable reporter gene into the mouse genome. Second, the viral reporter was altered so that only viral insertions within a gene were able to activate the expression of the reporter. These viral insertions within the transcribed region of the gene would also likely disrupt the mouse gene. Thus, the system allows for the detection of reporter expression patterns, much like the *P* element, but also the simultaneous generation of an insertional mutation, which does not always occur with the enhancer trap approach. An example of a mouse that exhibits an eye-specific expression pattern from an inserted reporter virus is shown in **figure 20.18**.

figure 20.17   Embryonic expression of β-galactosidase in *Drosophila* enhancer trap lines. Shown are the β-galactosidase expression patterns (black) in 10 different *Drosophila* enhancer trap lines (see fig. 20.16). The different observed β-galactosidase patterns include: expression throughout the central nervous system and peripheral nervous system (a), expression only in the central nervous system (b), expression only in the non-neuronal glial support cells (c), expression only in specific brain regions (d), and expression only in a subset of the central nervous system neurons (e–j). b, posterior brain; pns, peripheral nervous system; vml, ventral midline; vc, ventral chord.

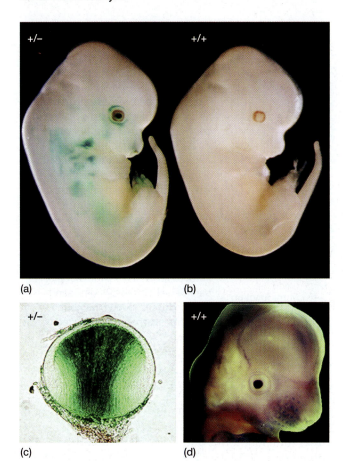

(a)   (b)

(c)   (d)

figure 20.18   Mouse gene trap mutagenesis. β-galactosidase expression of a *lacZ* gene trap that inserted in the mouse *aquaporin* gene. The β-galactosidase expression (blue) is seen in the lens, around the eye, snout, and rib regions of 12–13 day-old embryos that are heterozygous for the insertion (a), but not in similarly aged wild-type embryos (b). β-galactosidase expression is detected in the lens of the heterozygote (c). An in situ hybridization of a wild-type embryo with the aquaporin cDNA shows the gene is expressed in the lens but not around the eye like the *lacZ* gene trap (d).

Restating the Concepts

▶ Forward genetics is an approach that involves random mutagenesis of an organism, followed by screening the individuals to identify those that exhibit a particular phenotype.

▶ Screens can be designed to identify either dominant or recessive mutations. Balancer chromosomes, which contain multiple inversions, can be used to suppress recombination and mark the homologous chromosome that was not mutated.

▶ Mutants can be isolated based on a visible phenotype, a behavior, or a reporter gene expression pattern.

20.3 Reverse Genetics

In recent years, reverse genetic approaches have become more popular for studying the role of genes in a variety of processes. This popularity results from the abundance of genes that have been identified in genome-sequencing projects and is based on homologies to genes in other organisms. As genes are identified, researchers are interested in determining the function of the encoded proteins in the organism. Thus, genetic approaches must allow this "reverse genetic" analysis of first identifying a gene, then generating mutations within the gene, and then determining the mutant phenotype.

Screening-Targeted Mutations

One of the most direct methods in reverse genetics is to generate specific mutations in a desired gene. This is most efficiently done by creating mutations in a cloned gene in vitro and reintroducing the mutant form of the gene into the organism. We previously discussed this approach in the generation of knockout and transgenic mice (chapter 13). We will briefly summarize those approaches here.

To generate a knockout mouse (see figure 13.29), a selectable marker gene, usually encoding neomycin resistance (neo^r), is cloned into a genomic copy of the gene. This recombinant target gene, now in a targeting vector, is introduced into mouse embryonic stem (ES) cells where it recombines to replace the wild-type gene in the genome. The neomycin resistant ES cells, which contain the transgene in their genome, are introduced into a blastocyst that develops into a chimeric mouse, which contains cells from both the blastocyst genotype and the ES cell genotype. Through crosses, we can generate knockout mice that are derived entirely from the ES cell genotype. Because the knockout mouse contains a disrupted version of the gene, it usually expresses a loss-of-function phenotype.

In generating a transgenic mouse (see figure 13.28), specific mutations are introduced into the target gene in vitro. The altered target gene is then introduced into the male pronucleus of a recently fertilized egg. The altered target gene randomly inserts into the genome, without necessarily affecting the endogenous wild-type copies of the target gene. Because the wild-type alleles remain, the altered target gene must produce a dominant (gain-of-function) mutant phenotype for it to be observed. The production of transgenic animals can occur in a wide range of organisms, including bacteria, yeast, flies, zebrafish, mice, and even sheep, to name a few.

It is also possible to generate a transgenic animal that expresses a dominant loss-of-function phenotype. For example, extracellular ligands must be bound

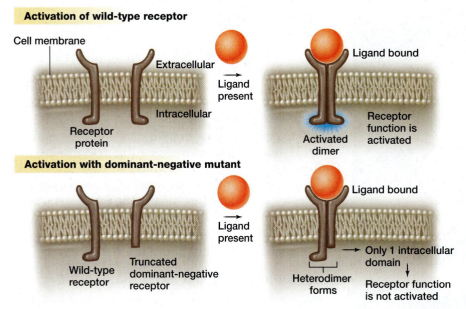

Activation of wild-type receptor

Cell membrane

Receptor protein

Extracellular

Intracellular

Ligand present

Ligand bound

Activated dimer

Receptor function is activated

Activation with dominant-negative mutant

Wild-type receptor

Truncated dominant-negative receptor

Ligand present

Ligand bound

Heterodimer forms

Only 1 intracellular domain

Receptor function is not activated

figure 20.19   Rationale behind a dominant-negative receptor mutant. The wild-type transmembrane receptor exists as a monomer. The receptor forms a dimer when it binds the ligand, which activates the intracellular function of the receptor, such as a protein kinase or protein binding. The dominant-negative receptor lacks the intracellular domain. Binding of the ligand by either the wild-type or dominant-negative receptor induces the formation of the dimer. If there is an excess of the dominant-negative receptor, the heterodimer will primarily be produced. Because this heterodimer lacks one of the intracellular domains, the receptor cannot be activated. Thus, ligand binding fails to induce the normal intracellular response produced by the receptor.

before transmembrane receptors can form the homodimer (**fig. 20.19a**). The formation of this homodimer activates the intracellular domain of the receptor. If a truncated receptor that lacks the intracellular domain binds the ligand, it will form either a homodimer or a heterodimer with the wild-type receptor (fig. 20.19b). This heterodimer would lack one of the two intracellular domains and would be unable to carry out its intracellular function, thus behaving as a loss-of-function mutation. These **dominant-negative mutations** exhibit a loss-of-function (negative) phenotype in the presence of the wild-type allele (dominant effect).

Screening Mutations Uncovered by Deletions

Although the knockout and transgenic animal approaches target the mutation to a specific gene, it is also possible to randomly mutagenize the genome in reverse genetics. However, we must screen for specific mutations, such as recessive mutations, in only the desired gene. The use of deletions and pseudodominance is consistent with this idea (chapter 8). Diploid individuals possessing a single recessive mutant allele will express the mutant phenotype if the homologous chromosome contains a deletion of the wild-type allele. We already discussed how pseudodominance is used to map the relative location of recessive alleles.

In the reverse genetics approach, the gene of interest is localized to a specific region in the chromosome. One or more strains that contain overlapping deletions of this region are identified. Wild-type males are randomly mutagenized and then crossed to females that possess this deletion chromosome (**fig. 20.20**). The F$_1$

progeny can then be examined for the recessive mutant phenotype.

Because the mutations are randomly generated, the sperm from the mutagenized male contains many different mutations. If the sperm that fertilizes the egg contains mutations that are outside of the deleted region, these mutations will not contribute to the recessive mutant phenotype (see fig. 20.20). But if the sperm contains a mutation within the deleted region, a recessive phenotype could be observed.

It is important to have an understanding of the potential mutant phenotypes associated with the gene of interest. If the mutagen creates both recessive and dominant mutations, then the desired recessive mutants will need to be selected from the dominant mutants. Knowing when and where the target gene is expressed may help determine whether the observed mutant phenotype corresponds to a mutation in the gene of interest. For example, if we know that the target gene is expressed only in the adult eye, we would screen for mutants that exhibit either an abnormal eye structure or function. We could distinguish the recessive mutants from the dominant ones by crossing them with a wild-type individual. Because the wild-type individual lacks the deletion, pseudodominance will not occur. Thus, if half the progeny from this cross are mutant, the corresponding mutation is dominant. However, if all the progeny are wild type, then the mutation must be recessive and located within the deletion encompassing the target gene.

This approach is limited by several factors. First, a deletion must exist for the region that contains the target gene. Second, the deletion should be relatively small. If the deletion is large, it will exhibit pseudodominance

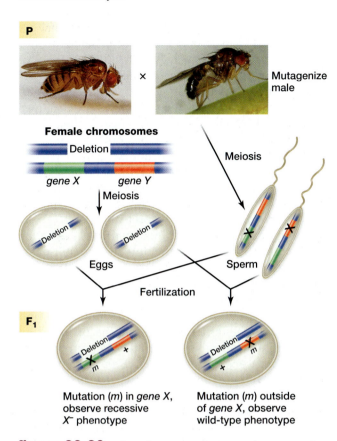

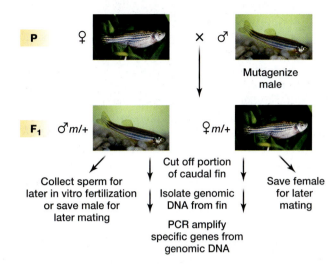

figure 20.21 Mutagenesis for TILLING. Males are mutagenized and mated to wild-type females. The F_1 progeny, which are heterozygous for random mutations, are produced and a tissue sample (such as a zebrafish fin) is collected for genomic DNA isolation. The desired gene is PCR amplified from the genomic DNA and analyzed for a mutation. The F_1 progeny with the desired mutation are either mated or their frozen sperm are used for in vitro fertilization.

figure 20.20 Genetic crosses to screen for a recessive mutation over a deletion. A *Drosophila* male is mutagenized and mated to a female that has a deletion for *gene X*, but not *gene Y*, on one of the chromosomes. The male produces sperm that may contain a mutation in either gene, while the female produces eggs with either the wild-type or deletion chromosome (only the eggs with the deletion chromosome are shown for simplicity). The resulting F_1 genotypes are shown. If the male contributed a sperm with a mutation in *gene X*, the F_1 individual will express the mutant phenotype due to pseudodominance. If the male contributed a sperm with a mutation in *gene Y*, the F_1 individual will express the wild-type phenotype.

for the target gene and several nearby genes. This could complicate identifying the mutant phenotype for only the target gene. Third, the mutagenized male must be mated to a female with the correct deletion before we can even determine if a mutation was generated. If the mutation rate for the target gene is low, this will require a large number of crosses to ensure that we generate a mutation in the target gene. Fourth, it might be difficult to predict the recessive mutant phenotype, which would result in potential mutants being discarded because they did not exhibit the predicted phenotype. For example, we might screen for a mutation in a gene that is expressed in the brain that would cause short-term memory defects. However, if the mutant phenotype actually caused a defect in long-term memory or only visual-based memories, we would not isolate the

mutants based on a short-term memory screen. Because of these, and other concerns, a different approach was devised to screen the randomly mutagenized population for mutations in the target gene.

TILLING

To get around these limitations, a new screening method was developed. **TILLING,** which stands for **t**argeting-**i**nduced **l**ocal **l**esions **in** **g**enomes, relies on the ability to detect F_1 individuals that are heterozygous for the mutation of interest and mate them to generate homozygous progeny that can be analyzed for the mutant phenotype. This approach has been warmly accepted by large groups of researchers working in both plants (*Arabidopsis thaliana*) and animals (*Caenorhabditis elegans* and *Danio rerio*).

Initially, an individual is mutagenized and mated with a wild-type individual to produce heterozygous F_1 progeny (**fig. 20.21**). Genomic DNA is isolated from each F_1 individual. This can be easily performed in organisms that either have dispensable somatic tissue (such as plants) or the ability to regenerate lost tissues (such as the caudal fin in zebrafish or the tail in the mouse). These F_1 individuals are then either maintained for later mating, or in the case of males, sacrificed after collecting their sperm. The isolated genomic DNA from each F_1 individual is analyzed for the desired mutation, and a mechanism to later recover the mutant organism is preserved.

How do we analyze the genomic DNA for the desired mutations? The easiest way is to use PCR amplification of the exons and analyze the PCR products. To minimize the number of PCR reactions and concentrate efforts on the potentially more relevant parts of the ORF, a computer program named CODDLE was developed (codons optimized to discover deleterious lesions). This program analyzes the genomic DNA sequence of the target gene and identifies the exons in which the mutagen will most likely produce nonsense and missense mutations (fig. 20.22). PCR primers are

then designed to amplify the selected exons that are then analyzed by automated DNA sequencing.

One significant advantage of TILLING is the generation of a series of missense and nonsense alleles without any preconceived idea about the corresponding phenotypes. This has the potential of producing an allelic series, which represents a range of alleles with different phenotypes. Null, hypomorphic, hypermorphic, and neomorphic alleles may all be generated using TILLING. The allelic series is important because sometimes null and hypomorphic phenotypes need to be compared with gain-of-function alleles to fully appreciate the function of the encoded protein.

Because TILLING does not require a prediction of the mutant phenotype, mutants that exhibit unexpected phenotypes can be identified. The mutants are identified by their DNA sequence and can then be analyzed by a battery of tests to determine the mutant phenotype. The testing of a known mutant to find a phenotype is similar to the analysis of knockout and transgenic mice; however, TILLING can produce a wide array of different alleles. Even though these mutant phenotypes may not be what was expected, they can often be informative about the function of the encoded protein, and they may not have been identified in normal screening approaches. Thus, TILLING serves several useful advantages as a method for identifying mutations in a specific gene.

Conditional Gene Mutations

Another way to generate mutations in a gene of interest is to create specific dominant mutations in the gene in vitro and then introduce the transgene into the organism. However, these dominant mutant transgenes must be expressed at specific times to examine the desired phenotype. Expression of a dominant mutation during embryonic development may result in death, which would prevent us from examining its mutant phenotype in the adult organism. Thus, controlling the expression of transgenes is an important consideration.

Tet-Off Regulated Gene Expression

In chapter 13, we described the ability to generate transgenic mice that contain an extra copy of a gene. This may not seem very useful, because a cell usually contains two endogenous copies of the gene to begin with. For the transgene to produce a phenotype, it must be a dominant allele to the two endogenous copies.

We already discussed the potential for generating dominant-negative transgenes to reveal the potential loss-of-function phenotype in an organism. Many dominant-negative mutations encode proteins that form heterodimers with the wild-type protein and produce an inactive complex. For example, most transcription factors must form a dimer to bind to the DNA

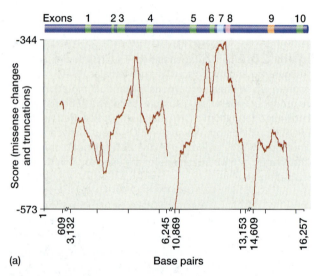

(a)

MQNSHSGVNQLGGVFVNGRPLPDSTRQKIVELAHSGARPCDI
SRILQTHADAKVQVLDNENVSNGCVSKILGRYYETGSIRPRAIG
GSKPRVATPEVVSKIAQYKRECPSIFAWEIRDRLLSEGVCTNDN
IPSVSSINRVLRNLASEKQQMGADGMYDKLRMLNGQTGSWG
TRPGWYPGTSVPGQPTQDGCQQQEGGGENTNSISSNGEDSD
EAQMRLQLKRKLQRNRTSFTQEQIEALEKEFERTHYPDVFARE
RLAAKIDLPEARIQVWFSNRRAKWRREEKLRNQRRQASNTPS
HIPISSSFSTSVYQPIPQPTTPVSSFTSGSMLGRTDTALTNTYSA
LPPMPSFTMANNLPMQPPVPSQTSSYSCMLPTSPSVNGRSY
DTYTPPHMQTHMNSQPMGTSGTTSTGLISPGVSVPVQVPGSE
PDMSQYWPRLQ

(b)

figure 20.22   The CODDLE (codons optimized to discover deleterious lesions) software analyzes the DNA sequence of a gene and determines the nucleotides and exons that will most likely yield missense or nonsense mutations. This analysis is based on the DNA sequence, the mutagen being used and the codon preference for the organism being mutagenized. (a) The top of the figure shows the exons (green) and introns (blue) in the zebrafish *Pax6* gene. Below is the relative score of generating a deleterious mutation in this gene using *N*-ethyl-*N*-nitrosourea (ENU) as a mutagen. The less negative the score, the greater the likelihood of producing a deleterious mutation. (b) The amino acid sequence of the Pax6 protein is shown, with the region corresponding to exon 7 in blue, exon 8 in magenta, and exon 9 in orange. The 12 potential amino acids in this region that could be mutated by ENU to a nonsense codon are in red and underlined.

Reprinted with permission from MacMillan Publishers Ltd: *Nature Review Genetics*, Stemple, vol. 5, figure 3, page 4, copyright 2004.

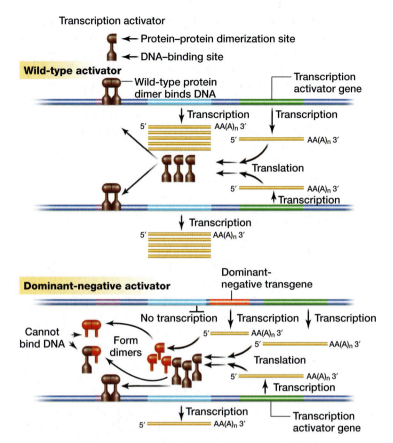

Transcription activator
← Protein–protein dimerization site
← DNA–binding site

Wild-type activator

Wild-type protein dimer binds DNA

Transcription activator gene

↓ Transcription   ↓ Transcription

5′ ——— AA(A)n 3′
         5′ ——— AA(A)n 3′

← Translation

         5′ ——— AA(A)n 3′
              ↑ Transcription

↓ Transcription

5′ ——— AA(A)n 3′

Dominant-negative activator

Dominant-negative transgene

Cannot bind DNA

Form dimers

No transcription   ↓ Transcription   ↓ Transcription

5′ ——— AA(A)n 3′
         5′ ——— AA(A)n 3′

← Translation

         5′ ——— AA(A)n 3′
              ↑ Transcription

↓ Transcription

5′ ——— AA(A)n 3′

Transcription activator gene

figure 20.23   Model of action for a dominant-negative transcription activator. The wild-type transcription activator contains a protein-protein dimerization domain and a DNA binding domain and must form a dimer before it binds DNA and activates transcription of a target gene (top). A dominant-negative transgene, which encodes a truncated protein that lacks the DNA binding site, is introduced into a cell (bottom). This truncated protein can form a dimer with either another truncated protein or the wild-type protein. In both cases, the dimer lacks two DNA binding sites and cannot bind DNA and activate transcription. The small amount of wild-type protein dimer that is produced will yield reduced transcription of the target gene and a loss-of-function phenotype.

(**fig. 20.23**). If you express a transgene that has the dimer-binding site, but not the DNA-binding site, this truncated protein would bind the wild-type protein and prevent it from binding to the DNA. In this way, expression of the transgene acts in a dominant manner to effectively yield a loss-of-function phenotype.

Suppose that you are interested in studying the function of a particular transcription factor in the development of the mouse eye. Unfortunately, this transcription factor is also required very early in embryonic development, and the null and hypomorphic alleles cause lethality much earlier in development than the eye development you want to study. These null and hypomorphic mutants therefore cannot be easily used for this purpose. You could create a transgene that expresses a truncated form of this transcription factor, which would act as a dominant-

negative mutation; however, its expression during early embryonic development would lead to the same embryonic lethality that the hypomorphic mutants yield. You would need to develop a method to express this transgene just before eye development, so that it does not affect early development.

The tetracycline-regulated systems, Tet-Off and Tet-On, provide a method to control the expression of a transgene through environmental conditions. This system is based on the bacterial tetracycline repressor binding to the tetracycline response element (TRE) sequence. By adding tetracycline or its derivative, doxycycline, to the drinking water of mice, it is possible to either repress transcription of a transgene that is linked to a TRE sequence with the Tet-Off system or activate transcription with Tet-On.

The **Tet-Off system** represses transcription of specific transgenes in the presence of tetracycline. The bacterial tetracycline repressor is a negative regulator of transcription. In the Tet-Off system, the tetracycline repressor gene is fused with a portion of the VP16 transcriptional activator from herpes simplex virus to create the *tTA* transgene (**fig. 20.24**, top). The encoded tTA protein binds to the TRE and activates transcription of a downstream gene. Addition of either tetracycline or doxycycline binds the tTA protein and changes its conformation so it cannot bind the TRE and transcription of the gene downstream of the TRE is terminated.

In the previous example, you would place the dominant-negative transcription factor transgene downstream of the TRE and introduce this into the mouse genome using standard transgenic techniques (chapter 13). Crossing this transgenic mouse with one that contains the *tTA* transgene allows expression of the dominant-negative transgene in the progeny mice whenever tetracycline is not present in the drinking water. Because the dominant-negative transgene would be lethal during early development, we would need to ensure that tetracycline is constantly available to the developing fetus. Only when the mouse has reached the desired age would we remove the tetracycline-containing water to examine the expression of the transgene.

This is a rather cumbersome approach for repressing the expression of the dominant-negative transgene until a desired time. The Tet-Off system is primarily used when we want to terminate the expression of a specific transgene for a specified period.

For example, assume we had a null mutation in the transcription factor gene described before. We cannot

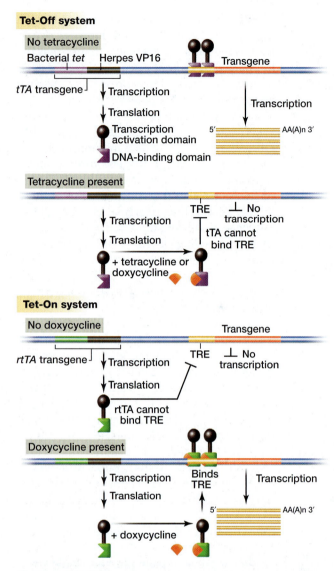

Tet-Off system

No tetracycline
Bacterial *tet*   Herpes VP16
tTA transgene

Transgene

↓ Transcription
↓ Translation
Transcription activation domain
DNA-binding domain

↓ Transcription
5′ ‖‖‖‖‖ AA(A)n 3′

Tetracycline present

TRE   ⊥ No transcription
↓ Transcription   tTA cannot bind TRE
↓ Translation
+ tetracycline or doxycycline

Tet-On system

No doxycycline
rtTA transgene

Transgene

TRE   ⊥ No transcription
↓ Transcription
↓ Translation
rtTA cannot bind TRE

Doxycycline present

↓ Transcription   Binds TRE   Transcription
↓ Translation
5′ ‖‖‖‖‖ AA(A)n 3′
+ doxycycline

figure 20.24   Tet-Off and Tet-On systems. The Tet-Off system (top) encodes a hybrid tTA protein, with the bacterial Tet repressor fused to the Herpes Simplex Viral VP16 transcriptional activator domain. This fusion protein binds the TRE sequence and activates transcription of the downstream transgene. When tetracycline or doxycycline is added, it binds the tTA fusion protein and prevents it from binding the TRE, which prevents transcription. The Tet-On system (bottom) encodes a hybrid rtTA, which is a mutated tTA that must bind doxycycline before it can bind the TRE. Addition of doxycycline permits the binding of the rtTA to the TRE, which activates transcription of the transgene.

study the function of this allele in eye development because the homozygous recessive embryos die early in development. However, we could clone the wild-type transcription factor gene downstream of the TRE and introduce this into the Tet-Off-expressing mouse that is homozygous for the null mutation. In the absence of the tetracycline-containing drinking water, transcrip-

tion of the wild-type transgene will occur. After we reach the desired point in development, we could add tetracycline to the drinking water, and the tTA would not bind the TRE, and transcription of the transgene would be terminated. Because the mouse is homozygous for the null mutation, it would now exhibit the loss-of-function phenotype in eye development. Thus, the Tet-Off system is primarily used to induce termination of transgene transcription.

Tet-On-Regulated Gene Expression

The **Tet-On system** was developed as a method for inducing transcription by the addition of tetracycline. In the Tet-On system, the *tTA* gene was mutated to encode a protein (rtTA) with four altered amino acids that now must bind doxycycline before the rtTA can bind the TRE (fig. 20.24, bottom).

In our dominant-negative transcription factor example, you would clone the dominant-negative transgene downstream of the TRE and introduce this into the mouse genome using standard transgenic techniques (chapter 13). This transgenic mouse would be crossed with one that contains the *rtTA* transgene. The progeny would develop normally in the absence of doxycycline. At the desired time, doxycycline would be added to the drinking water. The doxycycline would bind the rtTA, which would then bind the TRE and activate transcription of the dominant-negative transgene. Later moving the mice to doxycycline-free water would once again prevent the rtTA from binding the TRE and repress transcription of the dominant-negative transgene.

Thus, the tetracycline system allows for the conditional expression of a transgene. This is a powerful reverse genetic approach for controlling the expression of transgenes at specific times or in specific tissues of an organism. This permits the dissection of complex expression patterns of a single gene and simplifies the study of its function.

Restating the Concepts

▶ Reverse genetics is an approach used to generate a mutation in a target gene.

▶ Random mutations generated by any mutagen can be screened by crossing the mutated individual with an individual that is heterozygous for a deletion of the target gene. The progeny that contain the deletion will express the mutant phenotype if a new allele was generated and present in the F_1 progeny.

▶ TILLING is another approach for screening randomly generated mutations in a target gene. A computer program (CODDLE) is used to identify the exons of the target gene that are most likely to yield missense and nonsense mutations.

▶ Conditional expression of transgenes using the tetracycline-inducible systems, allows the analysis of the individual temporal and spatial requirements within a gene's complex expression pattern. The tetracycline-inducible system employs a transcriptional activator protein, a form of the tetracycline repressor fused to the VP16 transcriptional activator domain, and a specific DNA-binding sequence, TRE.

▶ In the Tet-On system, transgene expression is activated in the presence of doxycycline. In the Tet-Off system, transgene expression is repressed in the presence of either tetracycline or doxycycline.

20.4 Advanced Genetic Analysis

Mutations are often generated in a gene to reveal the function of the encoded protein or RNA in the organism. Sometimes, the mutant phenotype associated with the allele may not be completely informative. The mutant phenotype may be nearly normal, or the allele may have a strong general effect, such as lethality. For these reasons, it is often desirable to perform additional genetic analyses.

Isolating Suppressors and Enhancers

We have discussed several different ways to generate and identify mutants. A number of different experiments can be performed after a mutant is isolated, such as a detailed characterization of the phenotype, cloning the gene that is mutated, and studying how the mutation affects the normal cellular processes to produce the mutant phenotype. A different approach is to isolate mutations in other genes that modify the phenotype of the first mutant. As discussed in chapter 5, a **suppressor** is a mutation in one gene that makes the phenotype of another gene "less mutant." An **enhancer** is a mutation in one gene that makes the phenotype of another gene "less wild-type" or "more mutant."

Suppressors and enhancers may function through a direct protein–protein interaction with the mutant protein of the gene of interest. In this model, the mutation is usually a hypomorphic mutation, such that its mutant phenotype is not as severe as a null mutation. For example, two proteins may form a dimer to generate a functional protein complex in a cell (**fig. 20.25**). A mutation in the first gene reduces the interaction between the two proteins, but the complex still retains some activity. If a mutation is generated in the gene that encodes the second protein, the resulting altered protein may either make the phenotype more like wild-type (suppressor) or further worsen the mutant phenotype toward the null phenotype (enhancer).

Suppressors and enhancers may also identify genes that encode proteins that function in a related

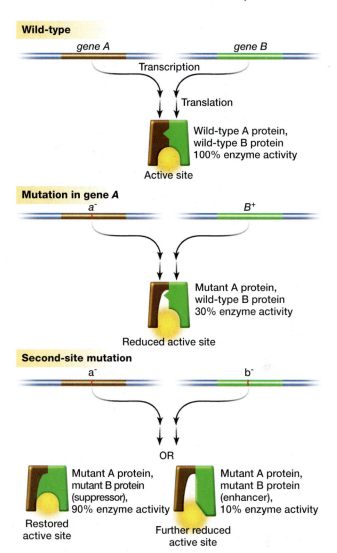

figure 20.25 Suppressors and enhancers can act through direct protein-protein interactions. Genes *A* and *B* encode proteins that interact in a specific manner to produce a functional enzyme (top). A mutation in *gene A* produces a protein that interacts less favorably with Protein B to generate an enzyme with reduced activity (30% of wild-type activity). A mutation in *gene B* that restores the favorable interaction with the mutant protein A (lower left) and increases enzyme activity (90% relative to 30%) is considered a suppressor (suppresses the mutant phenotype). A mutation in *gene B* that further reduces the interaction with mutant protein A or further disrupts the active site (lower right) will further reduce the enzyme activity (10% relative to 30%) and is considered an enhancer (enhances the mutant phenotype).

biochemical pathway. For example, a loss-of-function mutation in a gene may produce a mutant phenotype because of reduced activity through a biochemical pathway, with less of the final product being produced (**fig. 20.26**). A suppressor mutation in a second gene could be a gain-of-function mutation that increases the production of one of the intermediates in the pathway.

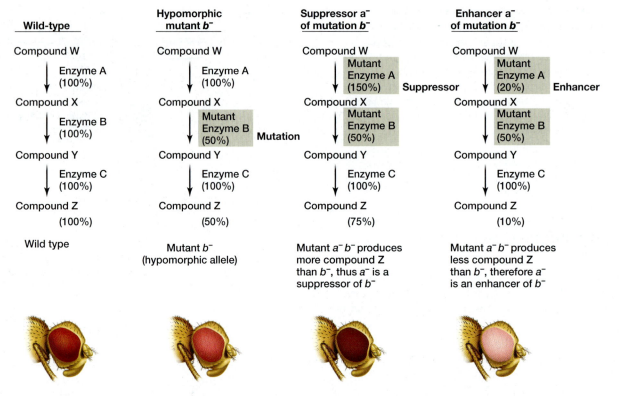

figure 20.26   Suppressors and enhancers function through a biochemical pathway. Genes *A*, *B*, and *C* encode enzymes that function sequentially in a biochemical pathway that converts substrate W to product Z, which confers the eye color phenotype. A hypomorphic mutation in Gene *B* (b^-) encodes an enzyme with reduced activity (50%) relative to wild type, which results in a mutant eye color phenotype due to the reduced amount (50%) of product Z relative to wild type. A different mutation in Gene *A* (a^-) results in either increased (suppressor) or decreased (enhancer) activity of enzyme A, which alters the amount of Compound X. This altered amount of Compound X changes the amount of Compound Z produced and alters the mutant *B* phenotype, with the suppressor producing a phenotype closer to wild-type and the enhancer resulting in a more severe mutant phenotype.

This increase could lead to the production of more final product, which would reduce the effect of the mutant phenotype.

In contrast, if the second mutation is also a loss-of-function allele, it could result in a further reduction in the amount of final product. In this case, the second mutation would be an enhancer.

To isolate suppressors and enhancers, males that are homozygous for the first mutation are mutagenized and then mated to homozygous mutant females (**fig. 20.27**). The F_1 progeny, which are homozygous for the first mutation and potentially heterozygous for the second mutation, are individually mated with homozygous mutant individuals. The F_2 siblings are mated, and the F_3 progeny are examined. All the F_3 progeny would be homozygous for the first mutation, so screening is done to identify individuals that exhibit either a more severe or less severe mutant phenotype.

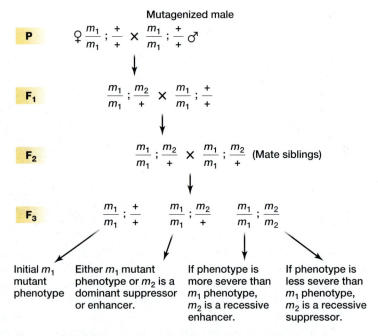

figure 20.27   Genetic crosses used to isolate either a recessive suppressor or enhancer. Homozygous recessive mutant males, m_1, are mutagenized and mated with homozygous mutant females. In the F_1 (and subsequent generations) dominant enhancers and suppressors in the m_2 gene can be identified. In the F_3 generation, recessive m_2 suppressors or enhancers can be isolated.

One example of how enhancers and suppressors were particularly useful is in the identification of the signaling pathway during the development of the *Drosophila* compound eye. The compound eye is composed of approximately 750 ommatidia (seen as bumps on the surface of the eye), each being a simple eye that contains eight photoreceptor cells that detect light (**fig. 20.28a**). One of these photoreceptor cells, called R7, is located at the same position in every ommatidia, and it projects its light-detecting organelle (called a rhabdomere) into the center of the ommatidium (fig. 20.28b).

A *Drosophila* mutant was isolated and was called *sevenless* (*sev*), because every ommatidium lacked the R7 cell (fig. 20.27c). Cloning the *sevenless* gene revealed that the encoded Sev protein is a transmembrane receptor that contains a tyrosine kinase domain on the intracellular side of the receptor. However, many questions remained about the Sev protein and the signaling pathway required to generate the R7 cell. What ligand did the Sev protein bind? What proteins directly interacted with the Sev protein? What proteins were in the intracellular signaling pathway that the Sev protein activated? To answer these questions, screens for enhancers and suppressors were initiated.

These screens used two tricks to simplify the identification of enhancers and suppressors. First, the researchers used known mutations in other tyrosine kinase proteins to generate mutations in vitro that would produce a temperature-sensitive form of Sev. This in vitro-synthesized sev^{B4} mutation was introduced as a transgene into a *sevenless* null mutant. The resulting flies and their progeny expressed only the temperature-sensitive form of Sev^{B4}. The researchers carefully studied the temperature sensitivity of the transgenic sev^{B4} mutant flies. They determined that at 22.7°C (permissive temperature) the transgenic flies had a wild-type phenotype, and at 24.3°C (restrictive temperature) the transgenic flies exhibited the *sevenless* mutant phenotype.

Second, the researchers developed a sensitized screen based on the temperature-sensitive sev^{B4} allele and the temperature that the flies were raised. A **sensitized screen** involves adjusting the conditions of the mutant so that it is very close to balancing between the mutant and wild-type phenotypes. Small changes in any protein in the same pathway as the mutant protein would be sufficient to switch the mutant phenotype to wild-type or the wild-type phenotype to mutant.

Thus, the researchers mutagenized the sev^{B4} flies and raised the progeny at 22.7°C. Normally, this temperature would produce a wild-type phenotype. However, a mutation that further reduces the signaling through the Sev pathway (enhancer) would be sufficient to cause

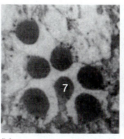
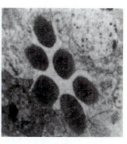

(a)   (b)

figure 20.28 The *Drosophila* compound eye (a) is composed of approximately 750 precisely arranged ommatidia, each corresponding to a smooth bump on the surface of the eye. (b) Each wild-type ommatidium contains eight photoreceptor cells, although only seven are seen in any section. Each photoreceptor cell contains a rhabdomere (dark circle) that is the membrane-rich organelle that detects light. In the wild-type ommatidium (left), the R7 photoreceptor cell's rhabdomere is located in the center of the other six rhabdomeres. In the *sevenless* mutant (right), the central R7 rhabdomere is missing.

a mutant phenotype. Alternatively, mutagenized sev^{B4} flies that are raised at 24.3°C would normally exhibit a mutant phenotype (**fig. 20.29a**), but if a mutation increases the signaling through the pathway (suppressor), then a wild-type phenotype would be produced (fig. 20.29b). Using this approach, a wide variety of genes that encode components of the Sev-signaling pathway were identified. Many of these genes encode molecules that are also used in intracellular signaling pathways in other organisms, including humans!

One advantage of this sensitized screen is that the enhancers and suppressors only need to alter the signaling pathway slightly. Thus, they could be isolated as dominant, rather than recessive, mutations. Some of the suppressors and enhancers identified in the sensitized sev^{B4} screen are lethal in a homozygous state. This suggests that the identification of these genes through normal genetic screens would not have identified them as being involved in the Sev-signaling pathway due to their recessive lethal phenotype.

Mosaic Expression Analysis

As we discussed in several of the examples earlier in this chapter, many of the genes that exhibit a recessive lethal phenotype may also be required later in development. However, the early lethal phenotype makes the analysis of the mutation later in development difficult. We already described how to use a conditional gene expression system (Tet-On or Tet-Off) to express a wild-type copy of the gene as a regulated transgene until the developmental stage in which we are interested.

Another approach to examining the phenotype associated with a recessive lethal mutation later in development is through mosaic individuals. A **mosaic**

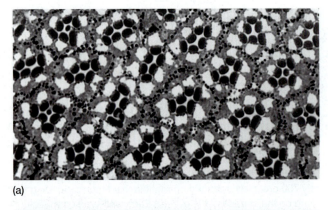

(a)

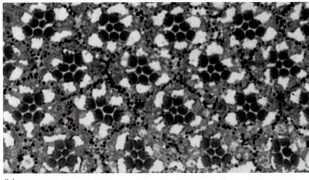

(b)

figure 20.29   Isolation of suppressors in a sensitized genetic background. (a) A section of a *Drosophila sentless* (*sev^{B4}*) mutant retina that was raised at the restrictive temperature (24.4°C), which shows that most ommatidia exhibit the *seventless* mutant phenotype of only six rhabdomeres (dark circles) in an ommatidium. (b) A suppressor of *sev^{B4}* at the restrictive temperature restores the wild-type phenotype of seven rhabdomeres to most ommatidia.

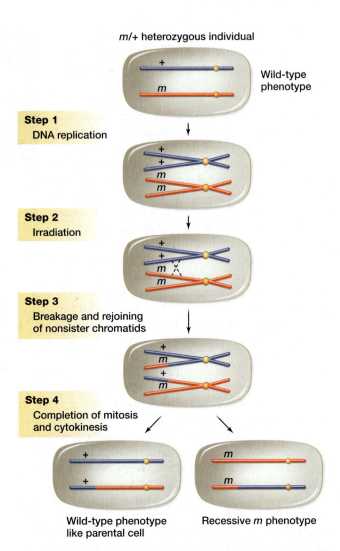

figure 20.30   Somatic recombination can be used to generate homozygous recessive mutant cells in a background of heterozygous cells. A heterozygous cell (*m/+*) undergoes DNA replication and is then irradiated, which induces breakage and reunion between nonsister chromatids. When this cell completes mitosis and cytokinesis, one daughter cell is homozygous for the recessive *m* mutation (lower right) and the other cell is homozygous for the dominant wild-type allele (lower left).

individual is composed of more than one genotype and may express different phenotypes in different tissues. In this type of analysis, the two genotypes are the heterozygous and homozygous recessive. If the homozygous recessive phenotype is death, the individual is generated as a heterozygote and then at some point in development, the homozygous recessive tissues or cells are generated. There are two general methods for producing this homozygous recessive genotype. The first is irradiation-induced mitotic recombination. The second is site-specific recombination.

Irradiation-Induced Mitotic Recombination

We first described X-ray-induced mitotic recombination in chapter 6. Mitotic recombination apparently occurs when two homologous chromosomes come to lie next to each other (after DNA replication and before entering mitosis) and irradiation induces a breakage and reunion event between nonsister chromatids. When mitotic recombination occurs in a cell that is heterozygous for a given gene, then one daughter cell becomes homozygous for the dominant allele and the other cell becomes homozygous for the recessive allele (**fig. 20.30**).

Because mitotic recombination is a rare event, irradiation of the organism will only induce mitotic recombination in a limited number of cells, unless mitotic recombination occurs early in development. In most cases, however, the majority of the cells in the organism will remain heterozygous. Only half of the small number of cells that are induced to recombine will become homozygous recessive. Thus, most of the cells in the mosaic individual are heterozygous, and half of the remaining cells are homozygous recessive and the remainder are homozygous dominant. The tissue of interest can then be examined for a mutant phenotype. The major disadvantage of the irradiation-induced mitotic recombination, however, is that the cells involved in the recombination are largely random.

For example, if we isolated a suppressor of the sev^{B4} mutant phenotype in a sensitized screen (see previous section), we may wish to examine the recessive phenotype of this mutation in the adult eye. Because the recessive phenotype is early embryonic lethality, we cannot raise homozygous recessive adults to examine their eyes. We can, however, irradiate an individual who is heterozygous for this mutation. In many of the resulting individuals, the induced homozygous recessive tissue may not include the eyes. Thus, we may have to generate a large number of irradiation-induced mitotic recombinants before we create a homozygous recessive region, known as a *patch*, that includes the eye.

Site-Specific Recombination

One method of efficiently inducing the recombination event in the proper tissue is through site-specific recombination. **Site-specific recombination** involves enzyme-mediated recombination between two specific DNA sequences. One common site-specific recombination system involves the yeast Flippase enzyme (FLP) and the DNA sequence recognized by the FLP recombinase, called FRT sequences.

The FLP–FRT system has been extensively used to create mosaics in the *Drosophila* eye. To begin, *Drosophila* stocks were generated that contain an FRT sequence (within a *P* element), either near the centromere on the X chromosome or one of the arms of the second and third chromosomes. On the same chromosome as the FRT is a second *P* element that contains the wild-type *white* (w^+) transgene.

The lethal mutation (*l*) is introduced onto the FRT-containing chromosome arm using standard meiotic recombination (**fig. 20.31**). It is essential, however, that the lethal mutation be located between the FRT and the w^+ transgene. This individual is mated with a fly that contains the *ey-FLP* transgene on the X chromosome and the parental FRT chromosome that also contains the $P[w^+]$ transgene. The *ey-FLP* transgene expresses flippase in the developing eye, which is where we want the recombinant clone to be generated.

Flippase catalyzes the recombination between FRT sites on nonsister chromatids, which results in the generation of cells that are homozygous for the lethal mutation (see fig. 20.31). The clone that is homozygous for the lethal mutation is identified by the white (w^-) patch (**fig. 20.32**). Because the rest of the fly is heterozygous for the recessive lethal mutation, the fly survives. If the recessive lethal mutation affects a general cellular function that is required for the survival of all

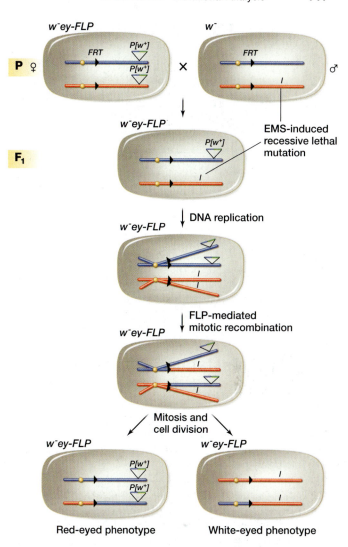

figure 20.31 Genetic crosses to screen for recessive lethal mutations in a mosaic background. A *Drosophila* male, which is homozygous for a chromosome with an FRT sequence (black arrowhead), is mutagenized to generate a recessive lethal mutation (*l*). This male is then mated to a female who is homozygous for the FRT chromosome that also contains a $P[w^+]$ transgene (inverted triangle). The F_1 progeny containing the recessive lethal mutation will undergo flippase (*FLP*)-mediated recombination between the FRT sequences in the eye tissues (where the *eyeless* promoter transcribes the *flippase* transgene, *ey-FLP*). This recombination occurs between nonsister chromatids after DNA replication. Mitosis and cell division yield two genotypically different cells, one of which is homozygous for the l^- mutation. If this cell continues to multiply, it will produce a white-eyed patch in the eye, because it lacks the w^+ transgene on the *P* element.

cells, then there will not be a w^- clone in the eye. In contrast, if the recessive lethal mutation affects a process that permits the retinal cells to survive, the phenotype associated with the recessive lethal mutation can now be examined in this region of the eye.

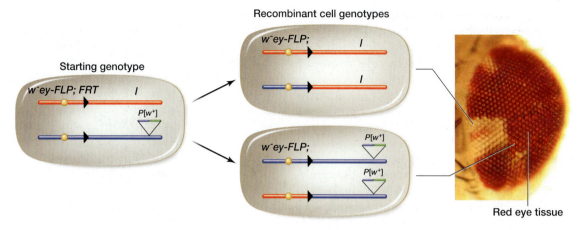

figure 20.32   Flippase-mediated recombination in the *Drosophila* eye. The crosses in figure 20.31 produce a *Drosophila* eye that contains a white patch. The cells within this homozygous *w⁻* patch are also homozygous for any mutations (*l*) that were also on the FRT chromosomal arm. The red-eyed tissue contains cells that either failed to undergo the flippase-mediated recombination, which corresponds to the original genotype (heterozygous for the recessive *l* mutation), or underwent the recombination event and lack the recessive mutation (*l*).

Restating the Concepts

▶ Suppressors and enhancers modify the phenotype of another mutation. They usually correspond to genes that encode proteins that either directly interact with the original mutant protein or act in the same biological pathway.

▶ A mosaic analysis involves generating an individual with three different genotypes, usually heterozygous, homozygous for a recessive mutation, and homozygous for the dominant mutation. The heterozygous genotype is often required to allow the organism to survive.

▶ If the mosaic individual was induced by irradiation, then individuals that contain the homozygous recessive patch in the desired tissue must be identified. If the FLP–FRT system was used to generate the mosaic individual, the expression pattern associated with the promoter that controls the FLP expression determines when and where the homozygous recessive patch will be located.

20.5 Phenocopying

So far we have dealt with methods for generating and identifying mutations in genes. A genetic analysis can be extremely powerful in determining the role of the encoded molecule in the organism, but this can be a time-consuming endeavor. Additionally, some genetic techniques, such as generating knockout individuals, cannot be efficiently produced in many species. To get around these two problems in studying the loss-of-function phenotype in a specific gene, phenocopy methods were developed. **Phenocopy** is the produc-

tion of a mutant phenotype in a genetically wild-type individual by introducing molecules or altering the environment. We describe here two different ways to phenocopy mutant phenotypes.

Phenocopying via RNAi

In the early 1990s, a phenomenon was discovered in plants, fungi, and the nematode *Caenorhabditis elegans.* This phenomenon, which was named **RNAi (RNA interference)** in *C. elegans,* occurred when the worms were injected or fed a double-stranded RNA (dsRNA). If this dsRNA contained a sequence that was complementary to a specific mRNA, translation of that mRNA was blocked, and transcription of the corresponding gene was suppressed in both the injected worm and its F_1 progeny. The reappearance of the gene product in subsequent generations, however, suggested that this phenomenon was not due to a mutation that permanently modified the genome.

RNAi employs the same Dicer enzyme and RISC machinery as the miRNAs that were described in chapters 10 and 17. Whereas the miRNAs are endogenous transcripts that are involved in gene regulation, RNAi involves double-stranded foreign RNAs that are introduced into the cell. The dsRNA is processed into a 22-nucleotide, single-stranded, **short interfering RNA (siRNA)** by the Dicer protein. The siRNA is then transferred to the Argonaute protein, and the RISC can either block the translation of an endogenous mRNA or cleave a specific mRNA to prevent its translation (**fig. 20.33**).

The siRNA can also drive the RISC to perform one additional function that has not been clearly demonstrated by miRNAs. The siRNA can direct the RISC to the nuclear genome where transcription of the cor-

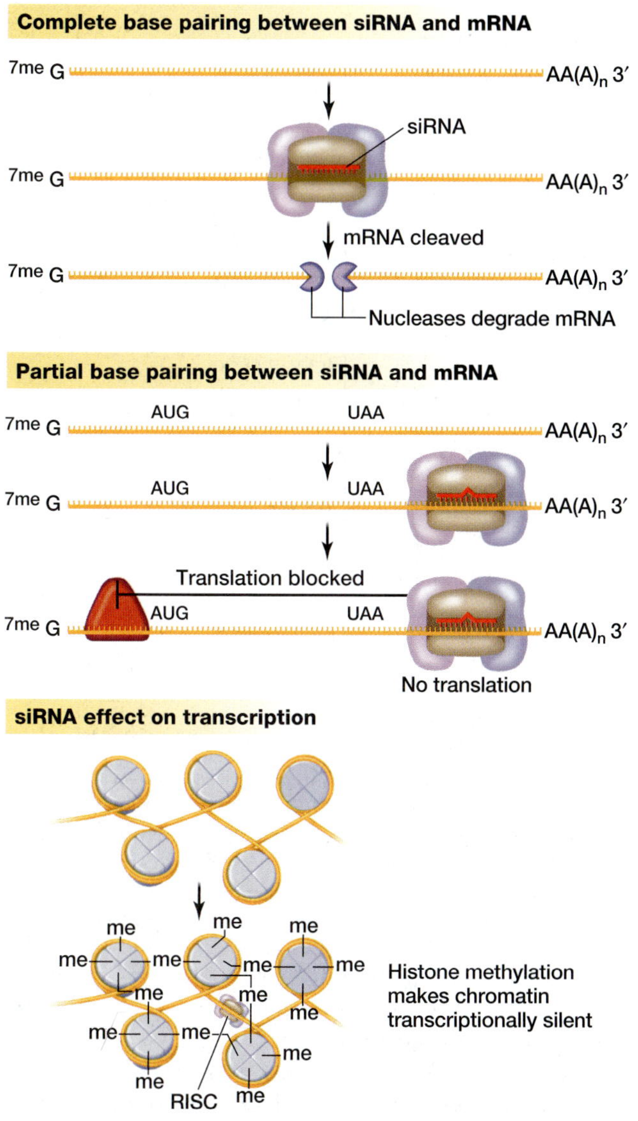

Complete base pairing between siRNA and mRNA

Partial base pairing between siRNA and mRNA

siRNA effect on transcription

figure 20.33   Small interfering RNAs (siRNAs) can affect protein expression three ways. The siRNA in the RISC complex can completely basepair with a complementary mRNA and induce cleavage of the mRNA, which is then degraded by nucleases (top). If the siRNA is not completely complementary to the target mRNA (middle), the siRNA-RISC complex will basepair with the mRNA and block translation. The siRNA-RISC complex can also reenter the nucleus and bind specific genomic DNA sequences and alter the chromatin conformation by histone methylation to block transcription of the gene (bottom).

responding gene can be silenced by targeted histone methylation and deacetylation (fig. 20.33c).

It is possible to harness this cellular process to temporarily block the expression of any desired gene in a wide variety of organisms. This is accomplished in three general ways. First, two short complementary siRNAs are synthesized in vitro. They are annealed to produce a double-stranded siRNA that is injected into a cell line

(**fig. 20.34a**). In many cases, the phenotypes are studied in these cell lines. Companies are currently marketing these complementary siRNAs for most of the genes that have been identified in yeast, flies, and humans.

Alternatively, a region of the gene of interest is cloned between two bacterial or phage promoters (fig. 20.34b). This region of the gene can be transcribed from both promoters in vitro, which results in two complementary siRNAs. The siRNAs are combined and coinjected into the cell or organism.

The final method involves cloning a DNA segment, which corresponds to either the 5' or 3' UTR of an mRNA, in an inverted-repeat orientation into a vector along with a promoter (fig. 20.34c). The promoter can either direct expression in the same tissues or cells as the targeted gene, or it could be an inducible system (such as Tet-On). This vector can be introduced into a cell, and the transgene will express a long stem-loop pri-siRNA that is processed by Drosha into a pre-siRNA. Dicer then further processes the transcript into an siRNA that is transferred to Argonaute. This approach allows for the stable expression of the RNAi from the transgene in an organism and a persistent phenotype rather than the temporary phenotype that results from the injection of double-stranded RNA into cells.

Recently, the development of morpholinos allows for translation suppression in a variety of organisms. **Morpholinos** are modified deoxyribonucleotides (**fig. 20.35a**) that are 20–25 nucleotides in length. Unlike standard oligonucleotides, morpholinos are more stable in the cell and less toxic. The morpholinos do not require the Argonaute protein or RISC complex. The morpholinos are designed to be complementary to the 5' UTR of a specific mRNA. The morpholino will then basepair with the mRNA and prevent translation of the mRNA (see fig. 20.35).

The siRNAs and morpholinos have allowed researchers to reduce the expression of any desired protein and examine the resulting loss-of-function phenotype. While these two approaches often fail to produce the equivalent of a null phenotype, they have permitted genetic analysis of specific genes without the need for generating mutants. Furthermore, siRNAs and morpholinos have allowed the identification of loss-of-function phenotypes in any targeted gene, even in organisms that cannot be manipulated by any of the other genetic approaches described earlier in this chapter.

Phenocopying via Chemical Genetics

With the RNAi technique, a phenotype is produced by reducing the amount of protein through the silencing of transcription, specific degradation of a mRNA, or the blocking of translation. In all three cases, a phenotype is produced that should mimic the loss-of-function phenotype. But the full spectrum of potential

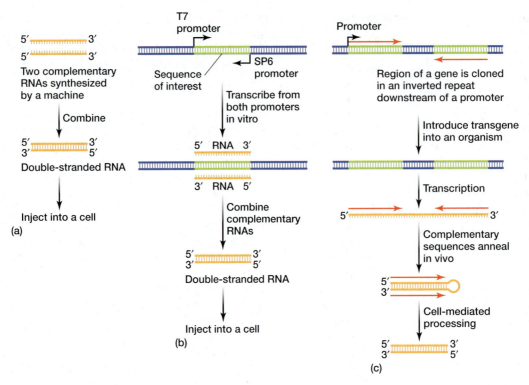

figure 20.34 Three major methods used to generate siRNAs. (a) Two complementary RNA sequences are synthesized in vitro using a machine. The two RNAs, which spontaneously basepair, are coinjected into the cell. (b) A region of the gene is cloned between two prokaryotic promoters (T7 and SP6) and both RNAs are transcribed in vitro. Because the RNAs are complementary, they spontaneously basepair and are coinjected into the cell. (c) A region of the desired gene is cloned in an inverted repeat orientation (red arrows) downstream of a cell or tissue-specific promoter. This transgenic construct is introduced into the organism, where the promoter transcribes an RNA that forms a stem-loop structure in vivo. This stem-loop RNA is processed by the endogenous cellular enzymes (such as Dicer) to yield the double-stranded siRNA.

phenotypes that can be generated through an allelic series is not produced.

Furthermore, it is difficult to mimic the conditional phenotypes that can be isolated by mutations. Although it is possible, using an inducible promoter, to activate transcription of the dsRNA that can be processed by the RNAi machinery, it is very difficult to induce the reexpression of the targeted protein.

Both of these problems can be circumvented by **chemical genetics,** which is the identification and use of small molecules that can interact with specific proteins so as to generate an altered phenotype.

Chemical genetics is based on the availability of large collections of small molecules, either small organic molecules or **peptide aptamers,** which are short amino acid oligomers. The small organic molecules can readily cross the cell membrane; they potentially have a high level of complexity, with different stereocenters, that could provide additional interaction specificity.

Large pharmaceutical companies have libraries of millions of different small molecules, which they have collected for over a hundred years as the numerous

variants of each drug candidate were being synthesized. Additionally, tens of thousands of natural products have also been isolated from marine life, fungi, bacteria, and plants.

These libraries have become more useful in recent years with the ability to create assays that are rapid, require small volumes, and allow the analysis of large numbers of small molecules in a short time (termed high throughput). Two basic types of screens are used: a phenotype-based screen, and a target-based screen.

Phenotype-Based Screen

A **phenotype-based screen** is similar to a forward genetics approach, with the small molecules being added to the cells or organism to generate a specific phenotype (fig. 20.36, left). These screens are usually performed on single-cell organisms or cell cultures that are derived from multicellular organisms because of the ease in introducing the small molecules into cells. The phenotype-based screen has successfully identified small molecules that affect mitosis, cell cycle

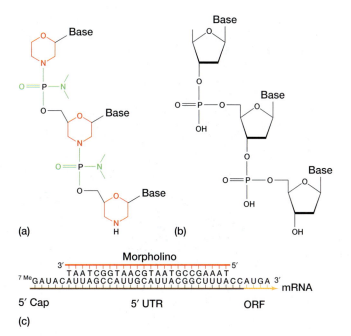

(a)   (b)

Morpholino

3′   TAATCGGTAACGTAATGCCGAAAT   5′
7 Me GAUACAUUAGCCAUUGCAUUACGGCUUUACCAUGA 3′
                                                → mRNA

5′ Cap   5′ UTR   ORF

(c)

figure 20.35   Morpholino. Comparison of the structure of a single-stranded morpholino (a) and a single-stranded deoxyribonucleotide oligo (b). The morpholino contains a morpholine ring (red) in place of deoxyribose. Furthermore, the morpholine rings are connected by non-ionic phosphorodiamidate linkages (green) that cannot be cleaved by endogenous cellular enzymes. (c) The location of a morpholino (red) annealed to the 5' untranslated region (UTR, brown) of an mRNA. The relative locations of the 5' CAP, 5' UTR and the start of the open reading frame (ORF, yellow) are shown below the mRNA. Base pairing of the morpholino to the 5' UTR of an mRNA blocks movement of the small ribosome subunit to the translation initiation codon, which results in the inhibition of translation.

progression, and programmed cell death (apoptosis). Once the desired phenotype is observed, the target protein must be identified.

Target-Based Screen

The **target-based screen** is similar to a reverse-genetics approach, where the library of small molecules are tested for the ability to specifically bind to a single desired target protein (see fig. 20.36, right). The target protein is purified and tested against all of the small molecules in the library. Alternatively, the target protein is expressed in a foreign cell, such as a human protein in bacterial cells, and the mixture of cellular proteins, cell lysate, is tested against all the molecules in the library. Once a specific interaction is observed between the target protein and a small molecule, potential phenotypes are studied in either cell culture or in a multicellular organism.

The potential specificity of these small molecules gives them another advantage over conventional

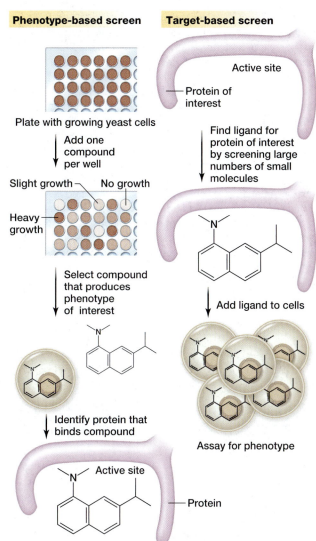

figure 20.36   General description of chemical genetics. In the phenotype-based screen (left), cells are plated into microtiter wells and an aliquot of different small molecules are individually added to separate wells of the plate. Each well is examined for a potential cell phenotype, such as the extent of cell growth, which suggests the small molecule perturbs a specific cellular function. The identity of the specific protein affected by the small molecule (depicted as the molecule binding in the protein's active site) remains to be identified. In the target-based screen (right), a specific protein is purified and aliquotted into individual wells of a microtiter plate. Different small molecules are added to each well and tested for their ability to bind the protein. The cellular phenotype that is generated by this protein-small molecule interaction is determined by adding the small molecule to cells and carefully testing a wide variety of phenotypes.

mutant alleles—different molecules may interact with different portions of a single protein. This characteristic has two possible advantages. First, some proteins may have multiple functions in the cell or in different tissues. For example, a kinase may be activated by two

figure 20.37 Comparison of a mutational and a chemical-genetic analysis. A hypothetical protein kinase (PKX) is activated by two different receptors (top), which leads to two different phenotypes (change in cell shape and cell adhesion). In the classical genetics approach (lower left), a mutation in the *PKX* gene (*PKX⁻*) will produce a loss-of-function allele that exhibits both mutant phenotypes. This mutant does not reveal if the kinase has a single function in the cell that produces both phenotypes or if the kinase has two different roles in the cell. In the chemical-genetics approach (lower right three columns), different small molecules can be added to the cells and the resulting phenotype can be determined. Small molecule 1, which binds and inactivates the PKX protein, phenocopies both mutant phenotypes because it acts at a common point in both pathways. In contrast, small molecules 2 and 3 each phenocopy a different phenotype. The results from small molecules 2 and 3 suggest that either PKX has two different roles in the cell (one affecting cell shape and one affecting cell adhesion) or it has a single role that can be separated into two different domains of the enzyme. If we learn that small molecules 2 and 3 bind different receptors, then PKX likely has two different functions in the cell, each regulated by a different transmembrane receptor.

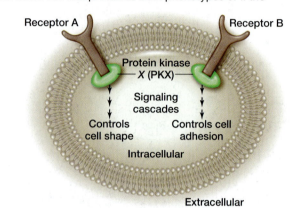

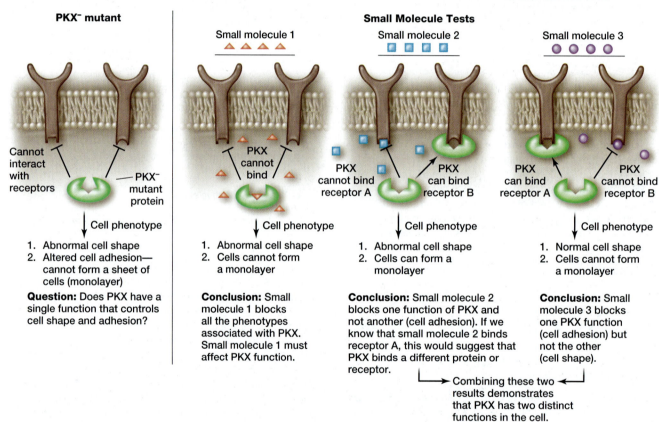

or three different receptor proteins (**fig. 20.37**). One small molecule may prevent the interaction of receptor A with the kinase, and a different small molecule may block the interaction between receptor B and the kinase. Thus, the different small molecules could be useful in examining the different functions of a single protein. In contrast, a genetic mutation in the kinase may block the response of both receptors equally and simultaneously, which may not reveal all the functions of the kinase.

The second advantage of small molecule genetics is their ability to function as either suppressors or

activators of the target protein's activity. As an example, the Hedgehog pathway is essential for invertebrate and vertebrate development. Binding of the Hedgehog ligand to the Patched receptor activates an intracellular signaling pathway and leads to increased transcription of the *patched* gene. In a phenotype-based small molecule screen, the Hh-Ag1.1 small molecule was shown to be an activator of the Hedgehog pathway. When Hh-Ag1.1 was delivered orally into pregnant mice, which contain the *patched* promoter (*ptc1*) upstream of the *lacZ* reporter gene, the embryos showed an increased level of β-galactosidase expres-

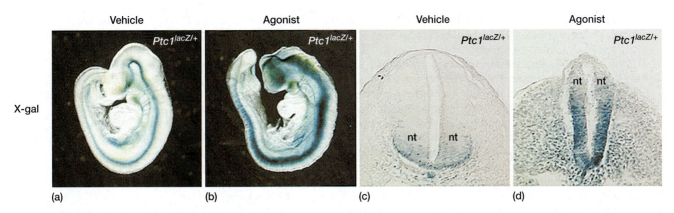

figure 20.38   The effect of a small molecule on the organism. Transgenic female mice containing the *patched* gene promoter fused to the bacterial *lacZ* gene were fed either the Hedgehog agonist Hh-Ag1.1, which activates transcription from the *patched* promoter, or a control solution (vehicle). An embryo receiving the vehicle treatment showed β-galactosidase expression (blue staining) in the neural tube that mimicked the expression pattern of the *patched* gene (a). An embryo receiving the Hh-Ag1.1 molecule exhibited an increased amount of β-galactosidase expression (b). A transverse section of the embryos fed Hh-Ag1.1 showed the increased β-galactosidase expression was throughout the ventral neural tube (nt) and the adjacent mesoderm (d) relative to the control (c).

sion throughout the ventral neural tube and surrounding mesoderm (**fig. 20.38**).

Hh-Ag1.1 was shown to bind the Smoothened protein, which functions after the Patched receptor in the signaling pathway. Previously, the small molecule Cur61414 was shown to be a repressor of this pathway by binding the Smoothened protein. Thus, a single protein target (Smoothened) can be activated by one small molecule (Hh-Ag1.1) and inactivated by another small molecule (Cur61414).

Restating the Concepts

▶ RNAi and morpholinos can be used to reduce the levels of gene expression in an organism without deleting or mutating the gene of interest.

▶ Chemical genetics employs small molecules to bind and either activate or inactivate a specific target protein without altering the underlying gene. Libraries containing millions of natural and synthetic compounds, most of which are organic, can be tested to determine if any molecule can bind and modify the activity of a target protein.

▶ Both RNAi and chemical genetics can be used to phenocopy a mutant phenotype in organisms that are not readily accessible to other genetic approaches described in this chapter.

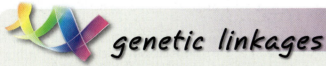

genetic linkages

This chapter began with an overview of different methods that are used to generate mutations in DNA, such as chemicals, energy, and transposable elements (chapter 18). We next discussed the various ways that these mutations are classified.

The forward genetics approach is based on the identification of abnormal phenotypes, and it represents historically the most common strategy for identifying mutants. The different crosses used to isolate dominant and recessive mutations require a solid understanding of general transmission genetics (chapters 2, 4, and 6), chromosome segregation during meiosis (chapter 3), recessive lethal mutations (chapter 5) and the role of inversions in balancer chromosomes (chapter 8). As you have seen throughout this chapter, the expression of transgenes (chapter 13) has become a powerful way to examine changes in gene expression.

The reverse genetics approach involves generating mutations in a specific target gene. It has been sparked by advances in molecular biology, such as gene cloning (chapter 12) and the explosion of genomic DNA-sequencing information from a variety of different organisms (chapter 14). The principle of pseudodominance with deletions (chapter 8) and TILLING employ a random mutagenesis and a selection of mutations within a targeted gene. TILLING has become a powerful way to perform a random mutagenesis and allow large numbers of individuals to selectively identify mutations in a gene of interest.

We discussed two different genetic approaches to further analyze a mutation. The isolation of suppressors, which were first described as mutations that modify the phenotypic ratio in a dihybrid cross (chapter 5), and enhancers are important in studying how the encoded protein interacts with other molecules, either directly or indirectly in a pathway, to generate a phenotype. The second method, mosaic analysis, was first discussed as a result of mitotic recombination (chapter 6), which occurs at a very low frequency.

The last section dealt with phenocopying, which we first described in chapter 5. In this chapter, we presented two different strategies for generating a phenocopy, the first involved nucleotide sequences, RNAi and morpholinos, and the second used chemical genetics. Understanding that RNAi depends on the processing of double-stranded

RNAs (chapter 10) should help demonstrate the important features that are necessary for this process.

The field of chemical genetics is relatively new. We discussed two different strategies, which are analogous to a forward and reverse genetics approach, for identifying small molecules that specifically interact with proteins. We also demonstrated that these small molecules may function to mimic either gain-of-function (activators) or loss-of-function (repressors) phenotypes.

The classical genetic approaches, the recent reverse genetic methodologies, and the phenocopy techniques have made genetics a critical component for nearly all areas of biological sciences. In the following chapters, you will see how these genetic approaches are applied to important areas of study, such as the various mechanisms involved in the development of a multicellular organism (chapter 21) and the development of a disease, namely cancer (chapter 22).

Chapter Summary Questions

1. For each term in column I, choose the best matching phrase in column II.

Column I	Column II
1. Mutagen	A. Mutation in a gene that encodes a protein with a function not present in the wild-type protein
2. Hypomorph	B. A second mutation that further increases the mutant phenotype of the first mutation
3. Hypermorph	
4. Neomorph	
5. Phenocopy	C. A second mutation that decreases the mutant phenotype caused by the first mutation
6. Mutant selection	
7. Saturation mutagenesis	D. A mutation in a gene that encodes a protein possessing a reduced level of wild-type activity
8. Suppressor	
9. Enhancer	E. A mutation in a gene that encodes a protein with increased wild-type activity
10. Genetic mosaic	
	F. An individual who possesses two different genotypes
	G. A substance that causes changes in the genetic code
	H. Environmental factors that cause a similar phenotype as a genetic mutation
	I. The process of screening enough mutants to generate mutant alleles in every gene
	J. Identification of mutant alleles

2. Why might scientists want to generate or isolate specific mutants?

3. Differentiate between a molecular null allele and a genetic null allele.

4. The terms *gene first* and *mutant first* could be used to describe forward and reverse genetics. Which phrase applies to which approach and why?

5. What characteristics would make an organism amenable to saturation mutagenesis?

6. What are the essential characteristics of a balancer chromosome?

7. Would the FM7 balancer be useful in transposon-mediated mutagenesis using P[w^+]? Why or why not?

8. For a large-scale mutagenesis project, would you prefer to use a selection screen or a detection screen? Explain.

9. For what reasons might a researcher prefer a *P* element-mediated mutagenesis screen rather than a chemical mutagenesis?

10. What component of the *Drosophila P* element used in enhancer trap screens allows a single enhancer to direct expression from both the endogenous and the reporter gene?

11. In the Tet-On conditional expression system, expression of the dominant-negative transgene is controlled only by the tetracycline response element (TRE) rather than tissue-specific enhancers. How might this be a problem?

12. Compare a dominant-negative mutation to a neomorphic mutation.

13. How does mutant screening using the TILLING approach differ from forward genetics?

14. Describe an individual that is a genetic mosaic. How might such an individual arise?

15. Differentiate between siRNA and antisense RNA techniques.

Solved Problems

PROBLEM 1: In studying flower color of a strain of petunias, you generated a range of mutants in the *PIP* (*Pigment In Petals*) gene, which produces the pigment that determines the color of each petunia flower. You designate your alleles *PIP¹* through *PIP⁴*. The table below shows how the different *PIP* alleles interact with one another, wild-type *PIP⁺* allele, and a deletion of the *PIP* gene (*PIP^del*). Fill in the remainder of the table for *PIP¹* through *PIP⁴*. (Assume that a 25% change in activity level produces a visible change in phenotype.)

Allele of Interest	Second Allele	Flower Color	Estimated % of wild-type activity level	Interpretation
PIP^del	*PIP⁺*	blue	50	
	PIP^del	white	0	
	PIP⁺/PIP⁺	blue	100	*PIP^del* is a complete null allele.
	PIP⁺/PIP⁺/PIP⁺	dark blue	150	
PIP¹	*PIP⁺*	blue		
	PIP^del	white		
	PIP¹	white		
PIP²	*PIP⁺*	blue		
	PIP^del	light blue		
	PIP²	blue		
PIP³	*PIP⁺*	dark blue		
	PIP^del	blue		
	PIP³	dark blue		
	PIP²	blue		
PIP⁴	*PIP⁺*	purple		
	PIP^del	red		
	PIP⁴	red		
	PIP³	purple		

Answer:

Allele of Interest	Second Allele	Flower Color	Estimated % of wild-type activity level	Interpretation

We can see that the *PIP¹* homozygote produces the same phenotype as the *PIP¹/PIP^del* heterozygote, thus the *PIP¹* allele is equivalent to *PIP^del*, leading us to the interpretation that *PIP¹* is a complete null allele.

PIP¹	*PIP⁺*	blue	50%	
	PIP^del	white	0%	*PIP¹* is a complete null allele.
	PIP¹	white	0%	

PIP² must still retain some activity, since, in combination with the deletion, a small amount of pigment is still produced. If we estimate this amount as 25%, two *PIP²* alleles will together generate 50%, and *PIP²* along with *PIP⁺* will generate 75%. Because *PIP²* possesses less activity than wild-type, but more than a null allele, *PIP²* must be a hypomorphic allele.

PIP²	*PIP⁺*	blue	75%	
	PIP^del	light blue	25% **(Start here!)**	*PIP²* is a hypomorphic allele.
	PIP²	blue	50%	

Two copies of *PIP³* generate the same phenotype observed with three copies of the wild-type allele. If we estimate this to be about 150%, each *PIP³* allele must contribute about 75%. Because the *PIP³* allele possesses greater activity than the wild-type allele (50% activity), *PIP³* must be a hypermorphic allele.

PIP³	*PIP⁺*	dark blue	125%	
	PIP^del	blue	75%	*PIP³* is a hypermorphic allele.
	PIP³	dark blue	150% **(Start here!)**	
	PIP²	blue	100%	

PIP⁴ shows a new phenotype not observed in previous alleles. In the two cases where no wild-type *PIP* function is present, the phenotype is red, so the *PIP⁴* mutant must produce red pigment. Because *PIP⁴* generates a phenotype that is not observed by either increasing or decreasing the wild-type *PIP⁺* activity, *PIP⁴* must be a neomorphic allele.

PIP⁴	*PIP⁺*	purple	50% + new function	
	PIP^del	red	0% + new function	*PIP⁴* is a neomorphic allele.
	PIP⁴	red	0% + new function	
	PIP³	purple	75% + new function	

PROBLEM 2: If the normal function of PIP^+ is as shown in the following figure, what types of mutations might be likely to produce PIP^4?

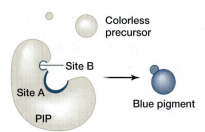

Answer: The PIP^4 neomorph still produces a pigment molecule, but with a different color. Specific mutations in the active site of the protein could either convert the substrate into a different pigment or alter the substrate that is bound, which would produce a different pigment.

PROBLEM 3: FM7 is an X-chromosome balancer marked with recessive mutations *y* (*yellow*; produces yellow body color) and *w* (*white*, causes white eyes). The dominant mutation *B* (*Bar*, bar eyes) is also present. Heterozygous *B*/+ flies have kidney-shaped eyes with about half the normal number of facets; *B* flies without a wild-type *Bar* allele show a "bar"-shaped eye with about 1/8 the normal number of facets.

  a. What would be a logical phenotype/genotype of male flies that were to be mutagenized to isolate mutations on the X chromosome?

  b. Diagram a cross using the FM7 balancer to recover recessive mutants on the *Drosophila* X chromosome.

Answer:

  a. The FM7 balancer includes recessive *y* and *w* alleles, so we will want to begin by mutagenizing males with one of those mutations on their X chromosome. The example shown here uses *y*.

  b.

(Figure: Drosophila crosses using the FM7 balancer)

P ♀ FM7 — Bar and white eyes, yellow females (*B y w* / *B y w*) × ♂ Feed EMS mutagen — Yellow males (*y*)

F₁ ♀ FM7 Mutagenized — Kidney eyes, yellow females (*B y w* / *m y*) × ♂ Bar, yellow, white males (*B y w*)

Offspring from single F₁ females

F₂
- Bar, yellow, white (*B y w*)
- (*B y w* / *B y w*) Kidney eye, yellow females
- (*B y w* / *m y*) Kidney eye, yellow females
- (*m y*) Wild-type eyes, yellow males

Random mass mating of F₁ progeny

- (*m y* / *m y*) Wild-type eye, yellow females and males. These flies express the recessive mutant phenotype. If the mutation is a recessive lethal, then there will be no non-Bar eyed F₂ progeny.
- (*m y*)
- (*m y* / *B y w*) Kidney eye, yellow females. These females can maintain the recessive lethal mutation as a heterozygote.
- (*B y w*) Bar, white eye, yellow males

Exercises and Problems

16. Imagine you were assigned the task of selecting a single mutagen to generate both null and hypomorphic alleles in a strain of *Drosophila*. You have EMS, 5-bromouracil, and an X-ray source. State if each is potentially useful and why or why not.

17. While screening the F₃ progeny of flies that were mutagenized by EMS, you discover an interesting mutant that you name *captain hook* (*chk*), because the ends of the legs are malformed. You realize that the *chk* phenotype is very similar to a mutant that you have been studying for several years called *bent legs* (*bl*). You determine that *chk* fails to complement *bl*. You do a Western blot using an antibody to the Bl protein (shown here). You also PCR-amplify the *bl* gene from the *chk* mutant and determine the sequence, part of which is shown here. From these data, determine the cause of the *chk* mutation.

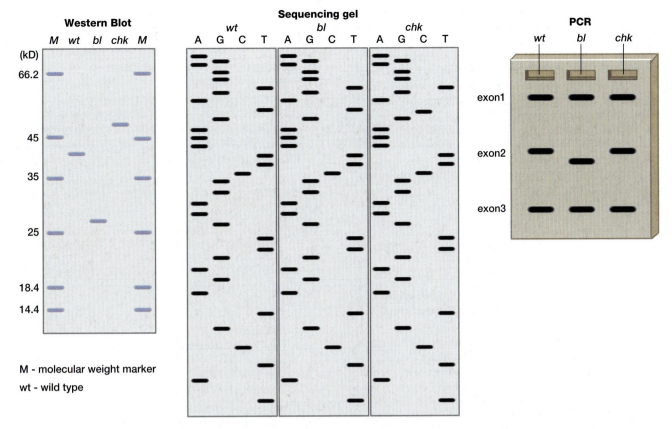

M - molecular weight marker

wt - wild type

18. You are analyzing a group of alleles (alleles 1–5) that were created by saturation mutagenesis. These alleles fail to complement a known genetic null mutation (*del*). You analyze five of the mutants using a genomic Southern blot, Northern blot, Western blot, and DNA sequencing. The data follow. Determine as accurately as possible the nature of each of the five mutations.

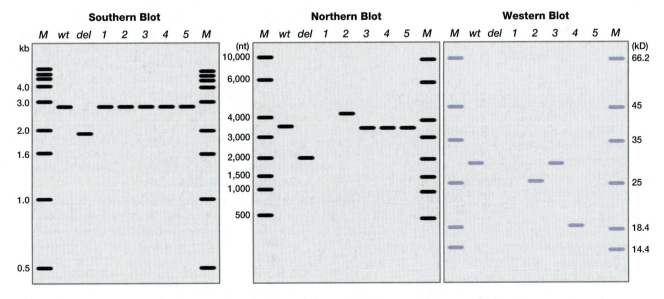

Promoter/mRNA start

wt	ataaattttagtatattttgtgggcaattcccagaaattaatggccttct
allele 1	ataaattttagtgtattttgtgggcaattcccagaaattaatggccttct
allele 2	ataaattttagtatattttgtgggcaattcccagaaattaatggccttct
allele 3	ataaattttagtatattttgtgggcaattcccagaaattaatggccttct
allele 4	ataaattttagtatattttgtgggcaattcccagaaattaatggccttct
allele 5	ataaattttagtatattttgtgggcaattcccagaaattaaaggccttct

Exon 2

wt	ttttggtatagtggtatagtcgaaaagccagatcccccaaagctcttggag
allele 1	ttttggtatagtggtatagtcgaaaagccagatcccccaaagctcttggag
allele 2	ttttggtatagtggtatagtcgaaaagccagatcccccaaagctcttggag
allele 3	ttttcgtatagtggtatagtcgaaaagccagatcccccaaagctcttggag
allele 4	ttttggtatagtggtatagtcgaaaagccagatcccccaaagctcttggag
allele 5	ttttggtatagtggtatagtctaaaagccagatcccccaaagctcttggag

19. After EMS mutagenesis of male *Drosophila,* single mutagenized males are mated to wild-type females. An F_1 offspring from male #75 displays a complex phenotype that includes reduced bristle number and uneven distribution of ommatidia in the eye. When this F_1 individual is mated to a wild-type female, 517 wild-type offspring and 498 offspring with reduced bristle number and eye abnormalities are produced.

   a. Are bristle number and ommatidia distribution affected by the same mutation or different mutations? What allowed you to make this inference?

   b. Are the mutation(s) described here dominant or recessive? How do you know?

20. List all the possible genotypes and corresponding phenotypes for the offspring from the following cross.

$$\frac{m \quad\quad cn}{\text{SM1} \quad Cy \quad cn} \times \frac{+ \quad\quad +}{\text{SM1} \quad Cy \quad cn}$$

21. What is the minimal number of inversion events that could have generated the SM1 balancer from the original chromosome?

22. Scientists believe that Himalayan coloration in rabbits is due to a temperature-sensitive *albino* allele. To study the production of pigment in your pet Himalayan rabbit, you shave patches of fur from its back. You put a cold pack on one of the patches while the fur is growing back, but otherwise treat the rabbit normally. What results do you expect?

23. Give the phenotype you expect for each of the following coat color genotypes at the *albino* locus in rabbits.

   A = wild type a = albino a^h = Himalayan, temperature-sensitive allele

Genotype	Expected Phenotype	Genotype	Expected Phenotype
AA		*Aa^h*	
Aa		*aa^h*	
aa		*a^h a^h*	

24. When analyzing mutant alleles, the phenotypes of homozygous mutants are often compared with the phenotype of a "deletion heterozygote."

   a. What is a deletion heterozygote, and why is it useful?

   b. You identify four new mutants that all exhibit a recessive curly wing phenotype. You cross each of the new mutants to a recessive *curled* mutant and to a fly that possesses a deletion of the *curled* locus. You generate progeny for each cross that exhibit the following phenotypes:

   mutant 1/mutant 1—curly wing phenotype

   mutant 1/curled—curly wing phenotype

   mutant 1/deletion—slightly curly wing phenotype

   mutant 2/mutant 2—curly wing phenotype

   mutant 2/curled—curly wing phenotype

   mutant 2/deletion—curly wing phenotype

   mutant 3/mutant 3—curly wing phenotype

   mutant 3/curled—straight wing phenotype

   mutant 3/deletion—straight wing phenotype

   mutant 4/mutant 4—curly wing phenotype

   mutant 4/curled—curly wing phenotype

   mutant 4/deletion—very curly wing phenotype

Describe as accurately as possible the nature of each of the four new mutations.

25. In *Drosophila,* the protein D-Raf is normally activated only when brought to the cell membrane on activation of the Ras/Raf signaling cascade. Activation of D-Raf results from phosphorylation at the site indicated, causing a change in protein shape.

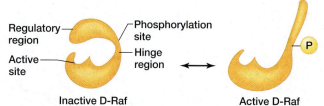

What result would you expect from each of the following mutants?

   a. Conservative missense mutation of the phosphorylation site

   b. Nonconservative missense mutation in the hinge region

   c. Missense mutation in the active site

   d. Deletion of the regulatory region

26. Ras mutations are unable to activate the Ras/Raf pathway described in the previous question. Suppose you have a *ras* hypomorphic mutation that results in weak Ras/Raf signaling. Suggest what type of mutations might be suppressors or enhancers of this *ras* hypomorphic mutation.

27. When screening for new mutants following an EMS mutagenesis, individual F_1 flies are mated to wild-type flies.

   a. Under what circumstances can new mutants be directly identified in the resulting F_2 flies, and when would they be intercrossed to generate F_3 flies?

   b. Under what circumstances can recessive mutants be identified in fewer generations?

28. The *white* gene (*w*), which is required to produce the dark red pigment in the compound eyes of *Drosophila,* is located on the X chromosome. Explain why *P* element mutagenesis using $P[w^+]$ can be used to identify mutants on any chromosome, rather than only the X chromosome containing the w^- allele.

29. Imagine you have a chimeric mouse containing cells in which your favorite gene (*YFG*) has been knocked out. The blastocyst came from an *albino* strain, and the ES cells were derived from a strain with wild-type (agouti) coloration.

   a. How do you know your mouse is chimeric?

   b. Diagram the necessary crosses that will be necessary to create mice that are homozygous for the *YFG* knockout allele.

   c. The following is a restriction map for the wild-type *YFG* gene and the targeting vector. Also shown is a genomic Southern blot of several individual mice using a probe in the *YFG* gene. The genomic DNA was digested with *Eco*RI before gel electrophoresis. Based on the restriction maps and Southern blot, determine the genotype of each mouse.

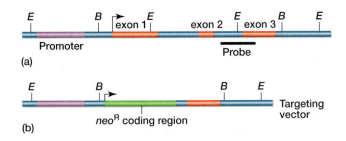

(a)

(b) neo^R coding region — Targeting vector

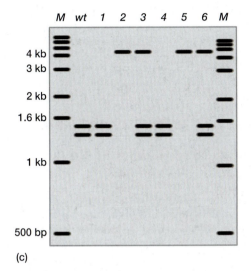

(c)

the following diagram. The CRE recombinase enzyme catalyzes recombination between two *loxP* sites to excise the sequences between the sites, which results in a deletion of the gene. These "floxed" alleles will function normally, and a homozygous line can be easily maintained. A second line of mice expresses CRE recombinase in a specific temporal and spatial pattern. Mating the two lines will generate an animal in which the floxed gene is deleted in any cells where CRE is expressed. As long as CRE is not expressed in the germline (and as long as the conditional knockout does not preclude reproduction), backcrosses can generate a mouse in which both copies of the floxed gene are knocked out in the cells expressing CRE. Compare this with the Tet-On and Tet-Off systems for generation of conditional knockouts.

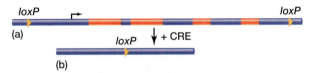

(a)

(b)

30. Although most mutagenesis/screening methods allow isolation of mutants in many different genes, TILLING identifies many mutations in only a single gene of interest. Give a situation where each might be preferable.

31. You identified a gene that is essential for correct embryonic mouse development. However, you are interested in studying the function of this gene in the adult. Expression of the gene is limited to a few key tissues. How could you use antisense technology to study the adult function of this gene?

32. The *CRE-lox* system is used to conditionally knockout a gene of interest that is flanked by *loxP* sites as shown in

33. During formation of the *Drosophila* eye, the ligand Boss (Bride of sevenless) binds the Sevenless receptor tyrosine kinase, which activates the Ras signaling pathway. Mutations that interfere with either the activation or correct signaling of this pathway result in a failure to specify the R7 photoreceptor cells. Design a mutagenesis/detection screen to produce mutants in the process of R7 specification. (Do not use the sensitized screen described in the text.)

34. You identified four new mutants that interfere with R7 specification. Describe your plan to determine if these mutations are located in the *sevenless* gene, or if they represent other genes involved in the process.

35. For each of the following examples, state whether you think the encoded mutant protein would represent a hypomorph, hypermorph, neormorph, or null, and explain your reasoning. The translation initiation codon corresponds to the first ATG codon. Substitutions are shown by giving the substituted base; deleted bases are shown with a .; inserted bases indicated with ^.

```
          ATGATCATGAGCTCGTATTTGATGGACTCTAACTACATCGATCCGAAATTTCCTCCATGCGAA
mutation 1 G
mutation 2                                                          ...
          GAATATTCGCAAAATAGCTACATCCCTGAACACAGTCCGGAATATTACGGCCGGACCAGGGAA
          TCGGGATTCCAGCATCACCACCAGGAGCTGTACCCACCACCGCCTCCGCGCCCTAGCTACCCT
mutation 3 .................................................................
          CGTCAGTATAGCTGCACCAGTCTCCAGGGGCCTGGCAATTCGCGAGCCCACGGGCCGGCCGCG
          CCTCTCTCAGGCACCTCTGCCTCCCCGTCCCCAGCCCCGCCAGCCTGCAGCCAGCCAGCCCCT
          GACCATCCCTCCAGCGCCGCCAGCAAGCAACCCATAGTCTACCCTTGGATGAAAAAAATTCAC
mutation 4                                T
          AGCACGGTGAACCCCAATTATAACGGAGGGGAGCCCAAGCGCTCGAGGACAGCCTACACCCGG
          GCTGCGCCCAGCACCCTCTCGGCAGCCACCCCCGGCACTTCTGAAGACCACTCCCAGAGCGCC
          ACGCCGCCGGAGCAGCAACGGGCAGAGGACATCACCAGGTTATAA
mutation 5                                                    ^   (insert A)
```

36. You are analyzing *fuzzy* (*fz*), a recessive mutation affecting bristle formation in *Drosophila*. You recovered this allele from a saturation mutagenesis screen using a chromosome 2 deletion that removed the *fz*$^+$ gene. What genotype(s) of fly will you want to generate in order to characterize the nature of this mutation? What molecular technique(s) will you use to further describe the mutation?

37. You discover that the *fz* mutation is homozygous embryonic lethal, although the *fz*/*del* genotype is viable. You hope to study the functions of the Fz protein at later stages in development, so you decide to generate mosaic individuals using X-rays. Describe how you will do this, and diagram the relevant crosses.

38. After limited success studying *fz* in X-ray-induced recombinants, you decide to use the FLP-FRT system instead. You generate female flies with the following genotype for chromosome 2:

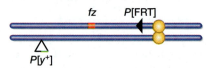

a. You will need to cross these flies to homozygous males to generate mosaics. Diagram the chromosome 2 genotype of these males.
b. Describe any other specific genotypes that must be present in your males, and explain why.

39. You are studying the Tailless protein, a transmembrane receptor that specifies the ends of the *Drosophila* embryo. You know that binding of the ligand causes the Tailless receptor to form homodimers and autophosphorylate both intracellular tails of the receptor. Each receptor protein of the homodimer must have a functional intracellular kinase domain in order to carry out the autophosphorylation. Instead of screening for new mutations, you decide to generate transgenic animals that carry various mutant versions of this receptor. Describe the phenotype you would expect for each of the following transgenic proteins in a wild-type background.
a. Truncated receptor with no intracellular domain
b. Truncated receptor with no extracellular domain
c. Insertion of a domain that allows dimerization in the absence of ligand binding
d. Missense mutation changing the autophosphorylated tyrosine to cysteine

40. To further characterize the function of Tailless signal, you use the mutants from the previous question in a rescue experiment. You generate embryos that are a complete null for *tailless* (no Tailless protein expressed at all) and embryos that produce receptor protein with a missense mutation in the ligand binding domain. Which of the mutants do you expect to demonstrate rescue, and why?

Chapter Integration Problem

"Sprite" is a hypothetical signal transduction pathway in cats. Activation of this pathway (by as yet unidentified factors known as 'cat sprites') leads to the hyperactive play behavior known as the cat crazies. Mellow cats rarely or never exhibit cat crazies, but spritely cats exhibit cat crazies during all waking periods. Assuming you have access to all the necessary resources, design experiments to accomplish each of the following.
a. Identify the pattern of inheritance of the sprite trait in cats.
b. Identify all the genes involved in this pathway.
c. Most signal transduction pathways have a transmembrane receptor protein that must bind an extracellular ligand and activate intracellular components. The original *sprite* mutation is a loss-of-function mutation in this receptor protein. To map the location of this mutation, you cross the homozygous *sprite* mutant with a cat that is heterozygous for the dominant *Wiskerless* and *Tailless* mutations. You know that the *Wiskerless* mutation maps to the third chromosome and *Tailless* maps to the fifth chromosome. You cross the *Wiskerless*, *Tailless* female progeny with homozygous *sprite* males. The resulting number of progeny in each phenotypic class are:

wild type	4
sprite, Wiskerless, Tailless	5
Wiskerless	6
sprite, Tailless	3
Tailless	19
sprite, Wiskerless	20
sprite	22
Wiskerless, Tailless	21

Based on this information, map the *sprite* mutation as accurately as possible.

d. Your efforts paid off in the following sequences that you believe represent the 5' end of the *sprite* gene. The first sequence is from a genomic library, and the second is from a cDNA library. Identify the transcriptional start site, the translational start codon, and the location and length of the first intron.

genomic sequence

```
CACCCATTTGGTATATTAAAGCTCTTCTGGTCCCCACAGACTCAGAGAGAACCCACC
ATGGTGCTGTCTCCTGCCGACAAGACCAACGTCAAGGCCGCCTGGGGCAAGGTTGGC
GCGCACGCTGGCGAGTATGGTGCGGAGGCCCTGGAGAGGATGTTCCTGTCCTTCCCC
ACCACCAAGACCTACTTCCCGCACTTCGACCTGAGCCACGGAGACAGAGAAGACTCT
TGGGTTTCTGATAGGCACTGACTCTCTCTGCCTATTGGTCTATTTTCCCACCCTTAG
GGCTCTGCCCAGGTTAAGGGCCACGGCAAGAAGGTGGCCGACGCGCTGACCAACGCC
GTGGCGCACGTGGACGACATGCCCAACGCGCTGTCCGCCCTGAGCGACCTGCACGCG
CACAAGCTTCGGGTGGACCCGGTCAACTTCAAGCTCCTAAGCCACTGCCTGCTGGTG
ACCCTGGCCGCCCACCTCCCCGCCGAGTTCACCCCTGCGGTGCACGCCTCCCTGGAC
AAGTTCCTGGCTTCTGTGAGCACCGTGCTGACCTCCAAA
```

cDNA sequence

```
CTCTTCTGGTCCCCACAGACTCAGAGAGAACCCACCATGGTGCTGTCTCCTGCCGAC
AAGACCAACGTCAAGGCCGCCTGGGGCAAGGTTGGCGCGCACGCTGGCGAGTATGGT
GCGGAGGCCCTGGAGAGGATGTTCCTGTCCTTCCCCACCACCAAGACCTACTTCCCG
CACTTCGACCTGAGCCACGGCTCTGCCCAGGTTAAGGGCCACGGCAAGAAGGTGGCC
GACGCGCTGACCAACGCCGTGGCGCACGTGGACGACATGCCCAACGCGCTGTCCGCC
CTGAGCGACCTGCACGCGCACAAGCTTCGGGTGGACCCGGTCAACTTCAAGCTCCTA
AGCCACTGCCTGCTGGTGACCCTGGCCGCCCACCTCCCCGCCGAGTTCACCCCTGCG
GTGCACGCCTCCCTGGACAAGTTCCTGGCTTCTGTGAGCACCGTGCTGACCTCCAAA
```

e. DNA sequence analysis of this gene from the *sprite* mutant reveals that it contains a missense mutation. How would you definitively demonstrate that the missense mutation in this gene results in the mutant phenotype?

f. How would you use TILLING to analyze the function of the sprite receptor? Describe the variety of mutations that you could isolate from TILLING and the expected effect on the receptor protein.

g. You discover a colony of cats that don't seem to fit your expectations regarding *sprite* behavior. These cats may be mellow, spritely, or an intermediate phenotype wherein they exhibit cat crazies only within an hour of eating. How would you determine if this new *sort-of-spritely* (*sos*) phenotype acts in the *sprite* pathway or not? If not, what other explanations are possible?

 Do you need additional review? Visit **www.mhhe.com/hyde** for practice tests, answers to end-of-chapter questions and problems, interactive exercises, and animation tutorials all designed to enhance your understanding of key genetic concepts.

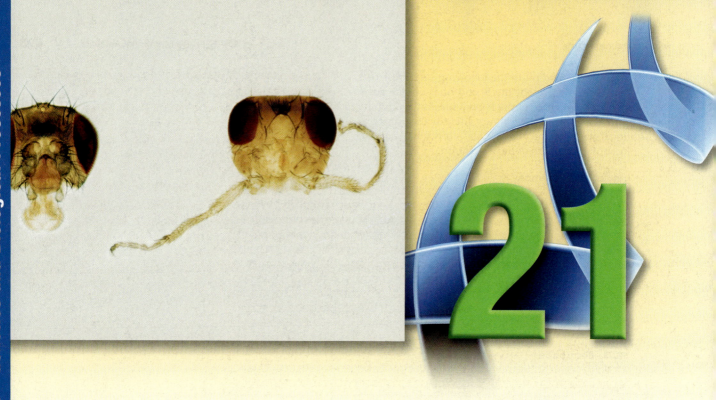

Developmental Genetics

Essential Goals

1. Distinguish between determination and differentiation.

2. Understand maternal-effect genes, and describe the roles of cytoplasmic determining factors in pattern formation.

3. Explain the levels at which zygotic gene activity can be regulated.

4. Gain a basic knowledge of the regulatory molecules and their roles in cell–cell signaling during development.

5. Discuss the role of homeotic genes in development and their conservation between invertebrates and vertebrates.

6. Explain the differences between master regulator genes and the genes required for development of the R7 photoreceptor.

7. Describe the differences between necrosis and apoptosis and the roles of apoptosis in development.

Photo: A wild-type *Drosophila* head (left) and an *Antennapedia* mutant (right) with a pair of legs growing in place of the antennae.

Chapter Outline

istorically, the study of animal development began with the limited observation of developmental processes, a field called embryology. As the understanding of sterile techniques improved and better microscopes became available, some manipulations of developing embryos were attempted. Results from these early studies provided an essential basis for our understanding of the basic processes of development.

As the field of genetics gained popularity, biologists began to study how mutations and the expression of genes affected the development of an organism. This type of analysis created the field of **developmental genetics,** which lies at the interface between embryology, genetics, and molecular biology.

Genes are not the sole source of an organism's phenotype. You learned in chapter 5 that the environment can also influence the characteristics of an organism, which can result in incomplete penetrance or reduced expressivity. In this chapter, we concentrate on the general developmental processes that are controlled by genetic factors.

The central question of developmental biology begins with a single cell: the fertilized egg, or zygote. By what mechanisms does that single cell give rise to all the variety of cell types and functions in a complex, multicellular adult? We do not yet have complete answers to this question, but we do know that the key processes begin even before fertilization, and that they continue throughout the organism's life span. Furthermore, because many of these processes are conserved in all multicellular organisms, we will primarily focus our discussion on model organisms that can be easily manipulated genetically.

21.1 Basic Concepts in Animal Development

To fully appreciate this discussion on developmental genetics, it is important that some new terminology be introduced. We will not concern ourselves in this chapter with the specifics of development or the developmental differences across a wide range of organisms, but rather we will look at a few examples to demonstrate some of the general concepts.

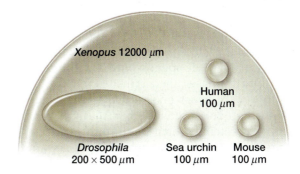

figure 21.1 The relative sizes of several animal oocytes are shown.

Fusion of Gametes at Fertilization

For diploid organisms, sexual reproduction relies on the process of meiosis to produce haploid cells called gametes (chapter 3). A male-produced gamete, sperm, fuses with a female-produced gamete, the ovum, or egg, to produce a zygote. Correct structure and function of gametes are critical to the success of the developmental program.

An animal sperm cell must carry its haploid complement of chromosomes to the oocyte. As a result, all animal sperm share some critical characteristics: they are motile, they possess limited cellular materials, and they have the ability to penetrate the protective layers and cell membrane surrounding the ovum.

In contrast, the ovum must contribute both a haploid chromosome complement and the cytoplasm and organelles that are initially required for the zygote. As a result, ova are among the largest cells in the animal kingdom (**fig. 21.1**). For example, a chicken egg is a single ovum, complete with large extracellular supporting structures.

To generate these very large cells, the process of meiosis in female animals usually includes asymmetrical cell divisions. Thus, a precursor cell entering meiosis produces three polar bodies, which contain only a haploid complement of chromosomes, and a single functional ovum, which contains a haploid complement of the chromosomes and the majority of the cytoplasmic material. In many cases, meiosis of oocytes is not completed until after fertilization has occurred (**fig. 21.2**; see also chapter 3).

Gene Expression in the Developing Embryo

The ova in many species are precisely organized; RNA and proteins produced by the female parent are distributed asymmetrically in the cytoplasm. The genes that express these RNAs and proteins are called **maternal-effect genes.** Cell divisions then compartmentalize those gene products into specific cells, leading to differing cell identities. These maternal-effect RNAs and proteins must carry the zygote through many rounds of cell division at the beginning of development.

Cell Division in the Zygote

The initial rounds of cell division are extremely rapid. The cells move between S phase and M phase with very little time spent in either the G_1 or G_2 phase. The rate

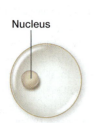

Young primary oocyte	**Fully grown primary oocyte**	**Metaphase I**	**Metaphase II**	**Female pronucleus**
The annulate worms *Dinophilus* and *Saccocirrus* The flatworm *Otomesostoma*	The round worm *Ascaris* The mesozoan *Dicyema* The sponge *Grantia* The clam *Spisula*	The mollusk *Dentalium* Many insects Starfish and ascidians	The lancelet *Branchiostoma* Amphibians Most mammals	Cnidarians (e.g., sea anemones) Sea urchins

figure 21.2   Animal oocytes are fertilized at specific meiotic stages in different species. Most female animals pause meiosis (oogenesis) at a particular stage until the female is ready to produce oocytes for fertilization. Additionally, the sperm often cannot fertilize the oocyte until it has reached a particular developmental stage.

of cleavage divisions varies from as little as 8 min per cycle in *Drosophila* to 12–24 h per division in mammals.

Because the cell spends only a short period in either G_1 or G_2 phases, there is insufficient time for the transcription of genes and subsequent translation of mRNAs to produce proteins. Thus, the maternal-effect RNAs and proteins, which are the only gene products that are present in the early embryo, must carry out the initial steps in development.

Zygotic Induction

At some point in development, the rate of cell division slows. This point varies from one organism to another. At this time, transcription of the zygotic genome begins. This **zygotic induction** marks the first point at which the paternally contributed genome has an opportunity to affect the development and phenotype of the embryo.

Because early development is controlled by the maternal-effect genes, the phenotype during this period depends solely on the maternal genotype. Only after zygotic induction does the zygote's genotype control the phenotype. Therefore, if the female parent is homozygous recessive for a specific gene and the male parent is homozygous dominant, the heterozygous embryo exhibits either the recessive phenotype if it is controlled by a maternal-effect gene or the dominant phenotype if the gene is not expressed until after zygotic induction.

Differential Gene Activity

Cell division by itself produces only a multicellular mass, not a complex multicellular adult organism. Additional mechanisms controlled by key developmental genes establish the initial body plan and, along with growth, continue to act to produce a fully formed adult.

Many embryonic cells are **undifferentiated**—they do not yet display the characteristics of any specific adult cell type. Some cells, called **embryonic stem cells,** retain this undifferentiated nature throughout their life. A cell that possesses the capacity to become any cell in the organism is called **totipotent.** As the embryo develops through cell divisions, however, most cells undergo a progressive restriction of possible fates until they are committed to become a particular adult type. This process is called **determination.**

Using *C. elegans* embryonic development as an example, the fertilized egg is totipotent (**fig. 21.3**). But after the first cell division, the AB cell loses the ability to become gut, muscle, and germ cells. The P_1 cell has lost the ability to become neurons. Thus, neither cell is totipotent, but they are **pluripotent,** meaning that they possess the ability to become several different cell types. Through several additional cell divisions, the P_4 cell is produced (see fig. 21.3). All the cells that are produced from P_4 will become germ cells. Thus, the P_4 cell is determined, or committed, to the germ cell fate.

Although a determined cell can be recognized by a specific pattern of gene activity, the genes in question have not yet led to a physical change in that cell to distinguish it from another cell. Each determined cell must ultimately take on the characteristics of its adult fate in a process called **differentiation.** A fully differentiated cell, such as a neuron, has undergone dramatic changes in its physical structure as well as function. These changes are mediated by differential gene expression.

Genetic control of determination probably results from many interacting genetic factors. Expression of cell surface molecules that mediate cell–cell communication is one likely example. Cells can respond only to signals for which they have specific receptor proteins. The control of the expression of genes encoding such receptors, therefore, results in differential cellular

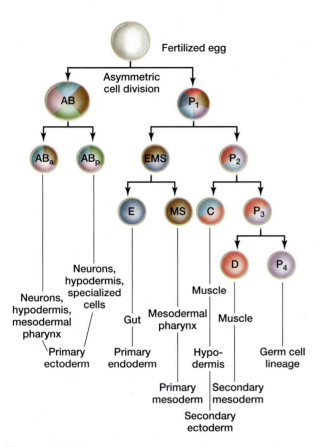

figure 21.3 The gradual determination of different cell fates in *Caenorhabditis elegans.* The first several cleavages of the *C. elegans* fertilized egg produces a fixed pattern of development. The first cell division is an unequal cleavage producing large (AB) and small daughter cells, with the smaller P₁ cell already committed to produce the germ cells. However, other cell types, such as the gut, muscle, and hypodermis, are also produced from the P_1, P_2, and P_3 cells. It is not until the P_4 cell is generated that it and all if its descendants are determined to become germline cells. The cells are color-coded to help demonstrate their commitment through the first several cell divisions. Hypodermis, light blue; germline, purple; gut, navy blue; mesodermal pharynx, brown; muscle, red; neurons, pink; and specialized cells, green.

responses to signaling factors and causes a totipotent cell to become determined. Other classes of gene products that control determination include transcription factors and cell cycle control proteins.

Internal and External Determination Cues

Determination comes about via two primary methods: **intrinsic** methods, in which factors present within each cell control the process, and **extrinsic** methods, in which external signals from neighboring cells provide cues. Although most organisms rely more heavily on one or the other, both methods must work in concert for proper development.

Intrinsic Cues

We already mentioned that maternal-effect gene products are distributed asymmetrically in the oocyte. During the early rounds of cell division, these asymmetrically arranged factors become unevenly distributed into the resulting daughter cells (**fig. 21.4**). The presence of some factors and the absence of others determine the future identity of the cell.

Because many of the maternal-effect gene products are either transcription factors or regulators of transcription factors, they control zygotic gene activity and determine which genes are transcribed—and thus they determine the identity of the cell.

Extrinsic Cues

Some cells contain identical mRNA and protein content, and yet they become very different types of cells. In this case, extrinsic cues such as diffusible proteins allow a cell to communicate with neighboring cells and identify its relative location within the embryo and to adjust its fate accordingly.

Comparison of Intrinsic and Extrinsic Mechanisms

The difference between these two mechanisms of development can be easily observed. If the developmental process is entirely intrinsic, then the death of

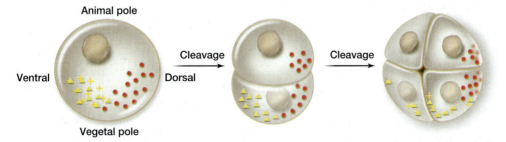

figure 21.4 Asymmetric distribution of cytoplasmic factors. Red circles and yellow triangles represent asymmetrically distributed cytoplasmic determining factors. After two rounds of cell division, each cell may contain neither, only one, or both determining factors. These different combinations of determining factors may result in unique cell fates.

Experiment demonstrating intrinsic determination cues

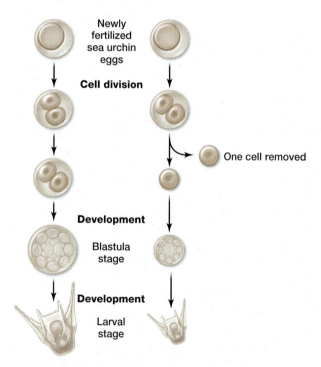

Experiment demonstrating extrinsic determination cues

figure 21.5 Difference between intrinsic and extrinsic determination. An example of intrinsic signals is the early development of *C. elegans* (top). If the P₁ cell is destroyed after the first cell division (left), the remaining AB cell can continue to divide to produce about 100 cells. None of these cells, however, will differentiate into muscle or gut cells. If the AB cell is killed (right), the P₁ cell can continue to divide and yield gut and muscle cells. An example of extrinsic cues is seen with the early development of the sea urchin embryo (bottom). Removing one of the first two cells results in the development of a complete sea urchin larva that is only reduced in size. Because a complete larva develops from half the number of cells, there were no essential cues in the cell that was removed and all the required information comes from the remaining cells.

a cell will result in the loss of its intrinsic information. The cells that were to be determined from the dead cell would then never appear. This loss could have profound effects on the development or function of the adult. These effects are observed when one of the cells of a *C. elegans* zygote is killed at the two-cell stage (see fig. 21.3). Neither surviving cell can develop into a worm, but either cell can continue to divide and produce approximately 100 cells. If the smaller P₁ cell survives, it can produce muscle and gut cells (**fig. 21.5a**). If the AB cell survives, the resulting cells cannot differentiate into muscle or gut cells.

When the cues are external, no information is lost when a cell dies. For example, removing one of the cells of a sea urchin embryo at the two-cell stage produces a larval sea urchin that contains all the cell types, but that is smaller in size than an undisturbed embryo (fig. 21.5b). The reduced size results from development proceeding with half the number of cells (one instead of two) after the first cell division of the fertilized egg.

Restating the Concepts

▶ The egg contains maternal-effect products—RNAs and proteins—which are produced by the female and deposited in the egg. The maternal-effect products must control the development of the embryo for many of the initial rounds of cell division until transcription of the zygotic genome begins.

▶ Cells that can differentiate into any cell type are called embryonic stem cells, or totipotent cells. Determined cells have become limited in their range of potential fates, and differentiated cells have made the physical changes necessary to adopt a specific fate. Both of these processes are mediated by differential gene expression.

▶ Determination can be controlled by either intrinsic cues or extrinsic (external) cues.

21.2 Pattern Formation During *Drosophila* Embryonic Development

We now apply the basic concepts and terminology used in development to specific examples. To begin, we explore the role of both maternal-effect and zygotic factors in generating a pattern of determined cells during *Drosophila* embryonic development.

An adult fruit fly, like other insects, has a body organized into three major regions: head, thorax, and abdomen (**fig. 21.6a**). Each of these regions can be subdivided into segments. For example, the abdomen is

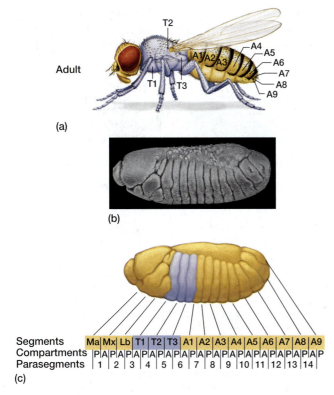

Adult

(a)

(b)

Segments	Ma	Mx	Lb	T1	T2	T3	A1	A2	A3	A4	A5	A6	A7	A8	A9
Compartments	P A	P A	P A	P A	P A	P A	P A	P A	P A	P A	P A	P A	P A	P A	P
Parasegments	1	2	3	4	5	6	7	8	9	10	11	12	13	14	

(c)

figure 21.6 *Drosophila* segmentation pattern. (a) The adult *Drosophila* is composed of a segmented body plan with three head segments, three thoracic segments (T1–T3), and nine abdominal segments (A1–A9). (b) The segments are first observed in the scanning electron micrograph of an embryo. (c) The identities of the *Drosophila* segments are shown. The two maxillary and one labial segments (Ma, Mx, and Lb) will produce the adult head. The three thoracic segments in both the adult and the embryo are colored purple.

composed of nine segments that are termed A1 through A9. Each of these adult segments is produced from segments initially generated in the embryo (fig. 21.6b and c). Significant research has been devoted to learning how this segmented embryo is generated from the *Drosophila* egg.

Maternal-Effect Factors and Generation of the Initial Axis Polarities

The first pattern element that must be generated in the *Drosophila* embryo are the polarity axes: dorsal–ventral and anterior–posterior. These axes are essential for the cells in the embryo to recognize their relative position so that the maxillary segments are located in the most anterior part, the abdominal segments are located posterior to the thoracic segments, and the adult legs are located on the dorsal side of the fly instead of the ventral, to name just a few examples. To generate these axes within the developing zygote, gradients of maternal-effect factors are created.

Generation of the Anterior–Posterior Axis

Four genes are essential in producing the anterior–posterior axis: *bicoid, hunchback, nanos,* and *caudal.* Because all four genes are maternal-effect genes, their expression in the female is essential for the development of the embryo; the translation products of these genes, in contrast, must function within the embryo to generate the anterior–posterior axis.

The *bicoid* mRNA is deposited in the anterior end of the egg (**fig. 21.7a**). After the egg is fertilized, the *bicoid* mRNA is translated. The Bicoid protein then diffuses from the site of translation. Because the Bicoid protein has a relatively short half-life (approximately 30 min), it produces a steep gradient, with the highest Bicoid concentration at the anterior end of the fertilized egg and the concentration decreasing toward the posterior pole (fig. 21.7b). Thus, the Bicoid protein gradient is generated from the initial localization of the *bicoid* mRNA at only the anterior end of the embryo and the degradation of the protein.

How do we know that the developing embryo requires the Bicoid protein? The answer can be seen in embryos that are produced from *bicoid* mutant females

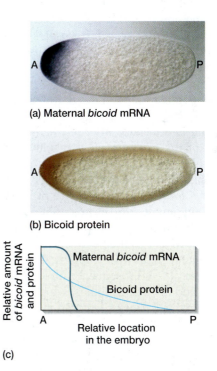

(a) Maternal *bicoid* mRNA

(b) Bicoid protein

(c)

figure 21.7 Generating the Bicoid protein gradient in the embryo. (a) The maternal *bicoid* mRNA is deposited in the anterior pole of the egg (blue). (b) After fertilization, the *bicoid* mRNA is translated into Bicoid protein (brown), which diffuses from the anterior pole to produce a concentration gradient. (c) The relative amounts of the *bicoid* mRNA and protein along the anterior–posterior axis of the embryo is shown. Notice that the Bicoid protein diffusion produces a broader gradient than the anteriorly packed mRNA.

Wild-type larva

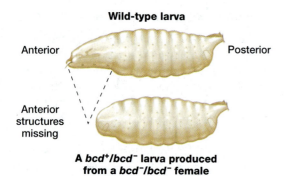

figure 21.8   Larva produced from a *bicoid* mutant female lacks anterior structures. A wild-type larva is drawn with the anterior and posterior regions labeled (top). The lower larva is heterozygous for the recessive *bcd⁻* allele (*bcd⁺/bcd⁻*). This larva lacks anterior structures, however, because the maternal parent was a *bicoid* mutant (*bcd⁻/bcd⁻*). The early developmental defect shown here was due to the absence of *bicoid* mRNA being deposited in the egg.

(*bcd⁻/bcd⁻*). The *bcd⁻/bcd⁻* females are normal because their female parents are heterozygotes (*bcd⁻/bcd⁺*). But the *bcd⁻/bcd⁻* females are unable to deposit *bicoid* mRNA in their ova. When these ova are fertilized, they become embryos that fail to develop into larvae because they lack the anterior structures—a lethal condition (**fig. 21.8**). The *bicoid* mRNA must therefore be required for the proper development of the anterior region of the embryo.

An analogous situation occurs for the *nanos* gene. The *nanos* mRNA is deposited in the posterior pole of the egg and is translated after fertilization (**fig. 21.9**). Diffusion of the Nanos protein results in a gradient with the highest concentration at the posterior pole and decreasing toward the anterior pole. Analogous to the *bicoid* mutant, larvae that are produced from a *nanos* mutant female lack the posterior region.

The 3' untranslated regions (UTRs) of the *bicoid* and *nanos* mRNAs contain sequences that bind proteins associated with microtubules. The protein that binds the *bicoid* 3' UTR sequence also associates with the minus ends of microtubules, which are located at the anterior pole of the egg. In contrast, the protein that binds the *nanos* mRNA associates with the plus ends of microtubules in the posterior pole.

How do we know that these 3' UTR sequences control the localization of the mRNAs in the egg? We can generate a transgene that contains the *nanos* mRNA and the *bicoid* 3' UTR. In this event, this mRNA is localized to the anterior pole like the *bicoid* mRNA, rather than the posterior pole as with wild-type *nanos* mRNA. In addition, deletion of the 3' UTR sequence from either the *bicoid* or *nanos* mRNA results in their uniform distribution throughout the unfertilized egg. Finally,

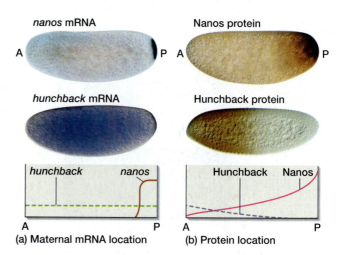

figure 21.9   Generation of the Nanos and Hunchback protein concentration gradients. (a) The maternally-deposited *nanos* mRNA is localized to the posterior pole (P) of the egg (blue), while the *hunchback* mRNA is evenly distributed throughout the egg's cytoplasm. The relative amounts of the mRNAs are plotted against the relative distance along the egg (bottom). (b) After fertilization, the Nanos protein (brown) is translated at the posterior pole and diffuses to generate a concentration gradient that is high at the posterior pole. Because the Nanos protein suppresses translation of the *hunchback* mRNA, the translated Hunchback protein concentration gradient is high at the anterior (A) end of the embryo. The relative amounts of the two proteins are plotted against the relative distance along the embryo (bottom).

adding either the *bicoid* or *nanos* 3' UTR to the *hunchback* mRNA results in the localization of this hybrid mRNA to either the anterior or posterior pole, respectively.

In contrast to the *bicoid* and *nanos* mRNA localization, the *hunchback* mRNA is uniformly distributed throughout the unfertilized egg (see fig. 21.9). If the *hunchback* mRNA is uniformly distributed throughout the unfertilized egg, how does it participate in generating the anterior–posterior axis? The answer is that the Nanos protein acts to suppress the translation of the *hunchback* mRNA, which produces a Hunchback protein gradient that is high at the anterior pole and low at the posterior pole (see fig. 21.9).

The fourth maternal-effect factor, the *caudal* mRNA, is also uniformly expressed throughout the unfertilized egg. The Bicoid protein acts to suppress translation of the *caudal* mRNA. This results in a Caudal protein gradient that is high at the posterior pole and low at the anterior pole.

By sequestering two of the maternal effect mRNAs (*bicoid* and *nanos*) to opposite poles, and through the ability of their proteins to diffuse and block the translation of two evenly distributed mRNAs (*hunchback* and *caudal*), four protein concentration gradients are generated (**fig. 21.10**).

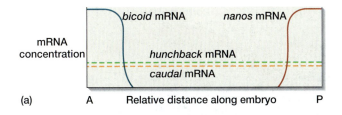

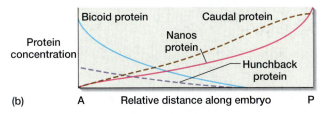

figure 21.10 Expression concentration of the four maternal-effect mRNAs and proteins. (a) The relative concentrations of the *bicoid*, *hunchback*, *caudal*, and *nanos* mRNAs are shown relative to the anterior (A) and posterior (P) ends of the unfertilized egg. (b) After fertilization, the maternal mRNAs are translated. Due to the location of the translation, protein diffusion, and the suppressed translation of *caudal* and *hunchback* mRNAs by the Bicoid and Nanos proteins (respectively), four protein concentration gradients are generated along the anterior–posterior axis.

Generation of the Dorsal–Ventral Axis

The generation of the dorsal–ventral axis also requires maternal factors to be deposited in the unfertilized egg; however, in this case the deposition of a protein, rather than mRNAs, initially determines the axis polarity.

Drosophila embryonic development involves one rather unusual feature. After the egg is fertilized by a sperm to create the zygotic diploid genome, a nuclear division occurs without a cell division. This produces a one-cell embryo with two diploid nuclei. The nuclei continue to proceed through rounds of mitosis, without a cell division. Ultimately, the embryo consists of several thousand nuclei within a single cell membrane.

The nuclei then migrate to the periphery of the plasma membrane to generate the *syncitial blastoderm*. The membrane then invaginates between each nucleus and generates a cell membrane around each nucleus to generate the *cellular blastoderm*. The continuous cytoplasm that is shared by all the nuclei until formation of the cellular blastoderm simplifies the generation of the later axis polarity gradients.

With this in mind, we can describe the generation of the dorsal–ventral axis. The protein Spätzle is deposited in the unfertilized egg. It is attached to and evenly distributed throughout the vitelline membrane, which lies just external to the egg's plasma membrane. Additional proteins are also deposited in the unfertilized egg that will be able to cleave the Spätzle protein

and release the activated Spätzle ligand. In contrast to Spätzle, these proteases are concentrated along the ventral midline of the egg.

After fertilization, the Spätzle protein is activated by protease cleavage to release it from the vitelline membrane along the ventral midline. This freed Spätzle ligand forms a gradient within the perivitelline space, with the highest concentration at the ventral midline and decreasing toward the dorsal midline (**fig. 21.11**).

The Spätzle ligand can bind the Toll receptor protein, which crosses the plasma membrane in the syncitial blastoderm (see fig. 21.11). When Toll binds Spätzle, the Pelle protein is activated (**fig. 21.12**). Pelle, which is a kinase located throughout the cytoplasm of the syncitial blastoderm, phosphorylates the Cactus and Dorsal proteins. Prior to phosphorylation, the Cactus-bound Dorsal is maintained in an inactive state in the cytoplasm. After phosphorylation, Dorsal is released from Cactus, Cactus is degraded, and the phosphorylated Dorsal enters the nucleus and activates transcription of zygotic genes.

The activated Spätzle is in a gradient with its highest concentration along the ventral midline, but the Toll and Dorsal proteins are evenly distributed in and adjacent to the plasma membrane (see fig. 21.11). All of the Toll receptors at the ventral midline bind the activated Spätzle, which in turn activates the maximal amount of Dorsal. Dorsal then enters the nuclei at the ventral midline. Moving away from the ventral midline, in a dorsal direction, fewer Toll receptors bind the activated Spätzle, because the concentration of Spätzle is reduced. The result is that less Dorsal protein is phosphorylated and less activated Dorsal enters the adjacent nuclei.

A gradient of Dorsal protein now exists in the nuclei that mimics the activated Spätzle gradient that had been present in the perivitelline space. The amount of phosphorylated Dorsal protein in the nuclei determines which zygotic genes are transcribed.

Restating the Concepts

▶ The axes in the *Drosophila* embryo are initially generated by maternal-effect gene products, either mRNAs or proteins. Females that are mutant for a maternal-effect gene produce larvae that lack the regions corresponding to where the maternal-effect protein normally would be localized in the egg.

▶ Gradients within the unfertilized egg can be generated with either maternal mRNAs that will be translated or with maternal-effect proteins.

▶ In generating the gradients, some proteins act as transcription factors (Bicoid), repressors of translation (Nanos), or activators of receptor proteins (activated Spätzle).

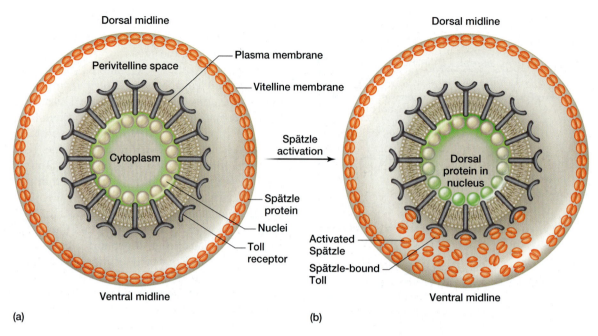

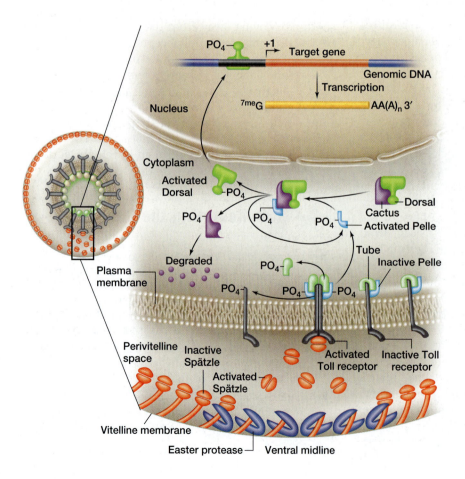

figure 21.11 Generation of activated Spätzle and Dorsal gradients. (a) The Spätzle protein is evenly distributed on the vitelline membrane before fertilization. The Toll receptor is also evenly distributed on the plasma membrane, and the Dorsal protein (green shading) is evenly distributed in the cytoplasm under the plasma membrane before fertilization. b) After fertilization, maternal proteins located at the ventral midline cleave Spätzle from the vitelline membrane to produce a soluble activated Spätzle gradient in the perivitelline space. When Toll binds activated Spätzle, Toll activates Dorsal, which then allows Dorsal to enter the nucleus. Thus, the activated Spätzle gradient is translated into an activated Dorsal gradient that can be seen by the localization of Dorsal in the ventral nuclei.

figure 21.12 Summary of the Spätzle-Dorsal signaling cascade. At the ventral midline, the Easter protease cleaves Spätzle from the vitelline membrane. This active Spätzle binds the Toll transmembrane receptor and induces Toll to form a homodimer. The activated Toll receptor is bound by Pelle, a kinase protein, which phosphorylates itself, Tube, and the Toll receptor. These phosphorylation events allow the three proteins to separate from each other. The activated Pelle phosphorylates Cactus and Dorsal, which frees Dorsal and targets Cactus for degradation. The freed and phosphorylated Dorsal can now enter the nucleus and activate transcription of target genes that are required for the development of the ventral side of the *Drosophila* embryo.

Generation of the Segmented Pattern in the *Drosophila* Embryo

We have already discussed how maternal-effect gene products are involved in generating the initial polarity axes in the *Drosophila* syncitial blastoderm. This information must be transmitted on to the individual cells in the cellular blastoderm and to the daughter cells that are produced after cell division begins. Because the zygotic genes are first expressed during the syncitial blastoderm stage, the positional information generated by the maternal gene products must be translated into the expression of specific zygotic genes. Control of zygotic gene activation and regulating the function of the gene products is thus critical for correct development.

Direct Control of Gene Expression: Activation of Gap Genes

Anterior–posterior pattern formation in *Drosophila* is governed by four main classes of genes that are transcribed in the embryo: gap, pair-rule, segment polarity, and homeotic selector genes. The first three groups function in a linear cascade that establishes the termini and a series of unspecified segments. The specific identities of each segment are determined by the homeotic genes.

We already discussed the mechanism by which the Bicoid protein forms an anterior–posterior gradient in the embryo. Proteins that function to specify two or more embryonic cell fates, such as Bicoid, are called **morphogens.** The Bicoid protein is a transcription factor that controls expression of several **gap genes,** which are required for defining regions within the embryo along the anterior–posterior axis. The gap genes are transcribed during the syncitial blastoderm stage.

Two such gap genes are *hunchback* and *Krüppel,* each activated by a different level of Bicoid protein. We previously mentioned *hunchback* as a maternal-effect gene; it is also one of the first zygotic genes to be expressed in *Drosophila.* Activation of zygotic transcription of *hunchback* occurs at moderate concentrations of Bicoid, and embryonic Hunchback protein soon replaces the initial maternally produced gradient. Thus, Hunchback is expressed in two regions in the developing embryo: a large anterior region, and a smaller posterior area (**fig. 21.13**). One of the functions of Hunchback is to repress the expression of *Krüppel,* which is activated at relatively low concentrations of Bicoid. The interaction of Bicoid and Hunchback activity thus limits Krüppel function to a stripe near the middle of the embryo (see fig. 21.13).

You can see in figure 21.13 that the Hunchback and Krüppel proteins define three regions in the anterior half of the embryo. The anteriormost region contains only the Hunchback protein (stained green). This is followed by a small region that contains both proteins

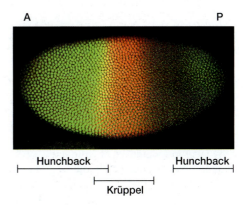

figure 21.13 The Hunchback (green) and Krüppel (red) proteins were localized in the *Drosophila* embryo with antibodies. The Hunchback protein is found in both the anterior (A) and posterior (P) ends, while Krüppel is detected in one central region. The yellow stripe corresponds to a region where both the Hunchback and Krüppel proteins are expressed.

(red + green = yellow stripe). Posterior to this is a region that contains only the Krüppel protein (stained red). To date, four gap genes (*hunchback, Krüppel, giant,* and *knirps*) and one terminal gene (*tailless*) have been well characterized in *Drosophila* and are known to be expressed in unique and overlapping patterns along the anterior–posterior axis (**fig. 21.14**).

The function of the gap genes is apparently to define regions that will ultimately consist of several segments. The expression of each gap gene is controlled by the relative expression of the different maternal-effect protein gradients and by interactions between the different gap genes. After these gap regions are defined, the location of the actual segments must be determined, which is the function of the pair-rule and segment polarity genes.

Organization of the Drosophila Segments

As was diagrammed in figure 21.6, the *Drosophila* embryo consists of 15 segments: three that go on to produce the adult head (Ma, Mx, and Lb), three that generate the thorax (T1, T2, and T3), and nine that yield the abdomen (A1–A9). When we discuss embryonic gene expression, however, we must consider parasegments. A **parasegment** is composed of the posterior half of one segment and the anterior half of the following segment (fig. 21.6c). A parasegment is defined by the expression pattern of genes, whereas the segments are defined by the physical or morphological structures in the organism.

The Pair-Rule Genes

Each **pair-rule gene** is expressed in seven stripes that correspond to every second parasegment. This can be easily observed by looking at the expression of two

21.1 applying genetic principles to problems

Identification of Genes Involved in a Signaling Pathway

You have seen that several genes are involved in determining the dorsal–ventral axis in *Drosophila*. To clarify this pathway and how it may interact with other signaling pathways, you would want to identify all the genes that encode components in this pathway. For identified mutants that affect dorsal specification, you would need to determine whether they are involved in this particular pathway, and if so, where.

The first step is to identify other mutants that disrupt dorsal-ventral patterning. A quick search of the appropriate databases would reveal that *snake, gurken, pelle,* and other mutations all cause a dorsalized phenotype (where the ventral surface has acquired a dorsal phenotype) similar to *dorsal, spätzle,* and *Toll.* This common mutant phenotype suggests that they are also involved in the same processes and perhaps even the same pathway. To confirm their action, you can look for interactions between these new mutations and the known molecules in the cascade.

For example, an early step in determining the ventral phenotype is the binding of Spätzle to the Toll receptor. If activated (soluble) Spätzle is injected in the ventral surface of a mutant embryo, Spätzle would "rescue" ventral development—that is, restore the wild-type phenotype—if a mutation affects a step *prior* to the production of active Spätzle. If a mutation affects a step *after* the normal activation of Spätzle, then the injection of soluble Spätzle would not bypass the defective step.

A later step in the pathway is the movement of Dorsal from the cytoplasm into the nucleus. If large amounts of Dorsal protein are injected into the ventral side of the embryo, it overwhelms the normal mechanisms that maintain it in the cytoplasm prior to phosphorylation (activation). Thus, injecting additional Dorsal protein rescues ventral development if the mutation disrupts a step before the production of active Dorsal.

To examine whether *snake, gurken,* and *pelle* are members of the same pathway, you can inject either active Spätzle or Dorsal into the ventral side of these mutant embryos and then determine whether the mutant phenotype is rescued. The results from an experiment using both of these treatments on all three mutants are shown in **table 21.A**.

Based on these results, the Snake protein acts in this pathway at an earlier point than Spätzle activation, because the mutant phenotype is rescued by the injection of active (soluble) Spätzle. Snake may function in the generation of the maternal inactive Spätzle protein, in its proper localization in the egg, or in the production of the active Spätzle from the inactive maternal protein. Although these two experiments do not differentiate between these possibilities, separate experiments have shown that Snake is a protease that cleaves the inactive Spätzle to produce the active form. The identification of this function required cloning and sequencing the *snake* gene as well as specific biochemical experiments.

The *gurken* mutant phenotype is not rescued by either of the two treatments. This finding suggests that the Gurken protein functions in a step after the movement of Dorsal from the cytoplasm to the nucleus. Alternatively, Gurken may function in a different pathway. Other experiments have demonstrated that Gurken binds to a different transmembrane receptor, Torpedo, to specify the cells that have the ability to adopt the ventral fate.

The Pelle protein is also a member of this pathway, because the mutant phenotype is rescued by the addition of Dorsal protein. Unlike Snake, Pelle must function at a point between Spätzle activation and movement of Dorsal into the nucleus. As would be expected, injection of active Spätzle is unable to block the mutant defect that occurs later in the pathway. The injection of additional Dorsal, however, functions later than the mutant defect, and therefore it is able to rescue the phenotype. We know that Pelle is the kinase that phosphorylates Cactus and Dorsal, which leads to Dorsal activation.

table 21.A Determining the Potential Role of Three Mutations in Dorsal–Ventral Patterning

Treatment	Mutant		
	snake	*gurken*	*pelle*
Active Spätzle	Rescue	No rescue	No rescue
Excess Dorsal	Rescue	No rescue	Rescue

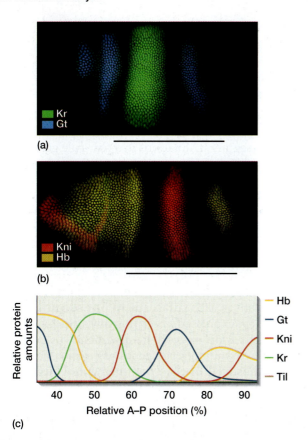

(a)

(b)

(c)

figure 21.14 Expression patterns of gap and terminal genes. (a) The location of the Krüppel (green) and Giant (blue) proteins in the *Drosophila* embryo were determined using antibodies specific for the two proteins. (b) An embryo is similarly stained for the location of the Knirps (red) and Hunchback (yellow) proteins in the embryo. The region that expresses both Knirps and Hunchback is stained orange. Anterior is to the left in both panels. (c) The relative amount of expression for each of the four gap proteins and the terminal Tailless protein between 35 and 90% along the length of the embryo (region corresponding to the lower bar in panels a and b).

different pair-rule genes: *even-skipped* and *fushi tarazu* (**fig. 21.15**). Each gene is expressed in seven stripes in the embryo, and there is no overlap between the two expression patterns. Thus, a total of 14 stripes, or parasegments, are defined by this expression pattern.

The expression of the pair-rule genes, which are first transcribed during the syncytial blastoderm stage, is controlled by the relative expression of the gap gene proteins and the maternal-effect proteins. Each pair-rule gene contains multiple transcriptional control regions that regulate the expression of the gene in a different stripe. In chapter 17, we discussed how these enhancers were experimentally defined by expressing a reporter transgene (see fig. 17.10). We then discussed the *even-skipped* stripe 2 (parasegment 3) enhancer and the relative locations where the gap gene proteins bind the enhancer element (see fig. 17.11). We saw that

Anterior   Posterior

figure 21.15 The location of two pair-rule proteins, Even-skipped (blue) and Fushi tarazu (brown), were determined in a developing *Drosophila* embryo using antibodies. The two proteins are expressed in alternating parasegments, with Even-skipped expressed in the odd-numbered parasegments. There is no overlap in the expression of the two proteins.

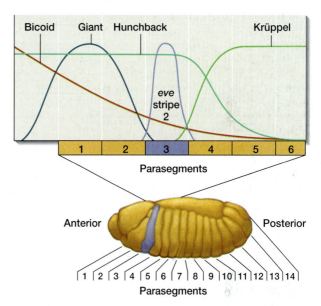

figure 21.16 Relative expression patterns of four different gap proteins define the location of the second stripe. The relative expression levels of the Bicoid (red), Giant (blue), Hunchback (turquoise) and Krüppel (green) proteins in parasegments 1 through 6 are shown on the graph. Notice that each parasegment along the embryo can be identified by the unique combined expression patterns of these four proteins. Bicoid and Hunchback are activators of *even-skipped* expression in stripe 2, while Giant and Krüppel are suppressors (Figure 17.11). The relative positions of parasegments 1–6 and the location of the *even-skipped* stripe 2 in parasegment 3 are shown.

Bicoid and Hunchback bind to this enhancer to activate *even-skipped* expression, whereas the Krüppel and Giant proteins bind to repress transcription.

The actual decision on whether this enhancer is utilized to drive *even-skipped* transcription depends on the concentration of the four gap gene proteins. Examining the relative concentrations of the four proteins in the embryo reveals that they have a collective expression pattern at parasegment 3 that is unique (**fig. 21.16**;

high Hunchback, low Bicoid, and little to no Giant and Krüppel). From this we can imagine that each parasegment has a unique expression profile of the four gap gene proteins (at least for the six parasegments shown in fig. 21.16). If every pair-rule gene has an enhancer for each stripe it produces in the embryo, then the activation of each enhancer should be determined by the relative gap gene proteins that bind to it.

Most of the pair-rule genes are also transcription factors and contain a helix-turn-helix region (described in chapter 17). These transcription factors act on three different classes of genes. First, they further regulate the expression of other pair-rule genes. This helps to produce the very finely organized expression pattern that is only one or two cells wide in each stripe (see fig. 21.15). Second, they control the expression of the segment polarity genes. Third, in conjunction with the gap gene proteins, they control the expression of the homeotic genes.

Defining the Segment Boundaries: Segment Polarity Genes

The pair-rule genes define the number of parasegments and the location of the anterior boundary of each parasegment. Mutations in a pair-rule gene result in the loss of alternating segments in the embryo (**fig. 21.17a**). The **segment polarity genes,** in contrast, determine the identity and polarity of each segment. Mutations in a segment polarity gene produce an embryo having either mirror images or tandem duplications of either the anterior or posterior portion of each parasegment (fig. 21.17b).

To define the polarity across the entire parasegment, the segment polarity genes must be expressed in specific cells in each parasegment. The segment polarity genes are expressed in very narrow stripes, only one or two cells wide, in every parasegment. For example, the *engrailed* and *hedgehog* genes are expressed in the anteriormost cell of each parasegment, which corresponds to the posterior margin of every segment (**fig. 21.18**). The *wingless* gene is expressed in the posterior margin of every parasegment. Finally, the *patched* gene is expressed in every cell that does not express either *engrailed* or *hedgehog*. This permits *patched* and *wingless* to define the posterior margin of every parasegment.

A tight regulatory mechanism must be used to prevent mixing cells from different parasegments and to ensure the correct size of a parasegment. How is the expression of a segment polarity gene regulated in

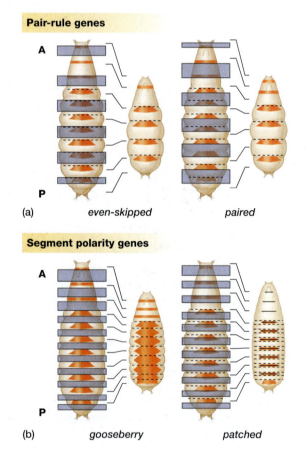

(a)   *even-skipped*   *paired*

(b)   *gooseberry*   *patched*

figure 21.17   The phenotypes of two pair-rule mutants (*even-skipped* and *paired, panel a*) and two segment polarity mutants (*gooseberry* and *patched, panel b*) at the first instar larval stage. The orange trapezoids represent the bristle pattern on the ventral surface that is unique for each segment. The blue rectangles represent the regions that are deleted in the pair-rule mutants and the regions that are mirror imaged in the segment polarity mutants. The boundary of each segment is marked by the dashed line. In each case, the wild-type pattern is to the left, the mutant is to the right, and anterior (A) is to the top.

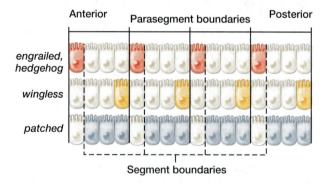

figure 21.18   Relative expression of four segment polarity genes in the parasegment and segment. The top of the figure shows the boundaries of the four-cell parasegment, while the bottom of the figure shows the position of the four-cell segment boundaries (anterior is to the left). The expression of a segment polarity gene in a cell is marked by coloring the cell. The top row reveals the expression pattern of the *engrailed* and *hedgehog* genes, the middle row represents *wingless* expression, and the bottom row shows *patched* gene expression. The four cells in the segment (or parasegment) are defined by three expression patterns, with two adjacent cells having the same expression pattern.

such a very tight pattern, only one or two cells wide? The segment polarity genes encode proteins that function to regulate other segment polarity genes. The four genes just described function in a regulatory loop.

An example of the regulatory mechanism can be seen at the parasegment boundary. The *engrailed* gene

encodes a transcription factor that activates the transcription of the *hedgehog* gene. The Hedgehog protein is secreted from this anteriormost cell, and it is bound by a transmembrane receptor, the Patched protein (**fig. 21.19**). The *patched* gene is transcribed in a single row of cells that lie anterior to the *engrailed*-expressing cells.

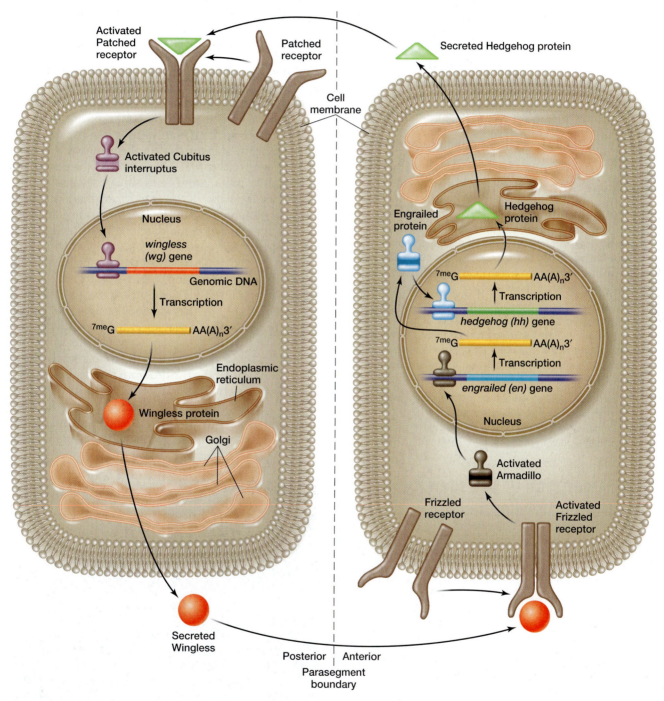

figure 21.19   The signaling pathways involved in defining the posterior/anterior cells at the parasegment boundary. The Hedgehog protein is secreted from the anterior-most cell in each parasegment and binds the Patched receptor in the posterior-most cell. The Patched receptor activates the Cubitus interruptus protein, which enters the nucleus and activates transcription of the *wingless* gene. The Wingless protein is translated, secreted, and binds the Frizzled receptor in the anterior-most cell. The Frizzled receptor activates the Armadillo protein, which activates transcription of the *engrailed* gene. The Engrailed protein activates transcription of the *hedgehog* gene, which produces the secreted Hedgehog protein. Thus, gene expression in one cell activates transcription of genes in adjacent cells.

Binding of Hedgehog activates the Patched receptor, which stabilizes the Cubitus interruptus transcription factor that activates transcription of the *wingless* gene. The Wingless protein is secreted from the cell and it binds and activates the Frizzled receptor on the adjacent cell, which stabilizes the Armadillo transcription factor. Armadillo activates transcription of the *engrailed* gene to produce additional Engrailed protein in the same cell in which we started this discussion.

As you can see, many proteins are involved in what is basically a reinforcing loop that stabilizes the parasegment boundary. Hedgehog and Wingless are secreted ligands that act as signals between the two rows of cells. Patched and Frizzled are receptors. Cubitus interruptus and Armadillo are transcription factors that ultimately act in the cell nucleus.

Restating the Concepts

▶ The gap genes are the first transcribed zygotic genes and are required for generating the segmentation pattern. Expression of the gap genes is controlled by the relative expression of the maternal-effect genes at specific points in the embryo.

▶ The pair-rule genes are expressed in seven stripes in the embryo that correspond to every other parasegment. Their expression is controlled by the relative amounts of the different gap gene proteins at any particular location in the embryo. The pair-rule genes are transcription factors.

▶ The segment polarity genes are required for individual cells to recognize their relative position within a parasegment. The segment polarity genes encode receptors, signaling proteins, and transcription factors.

21.3 Generation of Cell Identity During Development

The maternal and zygotic genes combine to generate the 14 parasegments, but they are unable to assign unique identities to cells in each parasegment. The **homeotic genes** are essential for assigning the identity to a particular cell. Mutations in the homeotic genes cause one body part to develop as a different part. A classic example of this type of mutation is *Antennapedia* (**fig. 21.20**), where the antennae are replaced with legs. In this section, we look more closely at the actions of homeotic genes in *Drosophila* development.

Homeotic Genes and Segment Identity

The *Drosophila* homeotic genes are located in two major clusters: the *Antennapedia* and the *Bithorax* gene clusters. The *Antennapedia* complex, which is composed

of the *labial, proboscipedia, Deformed, Sex combs reduced,* and *Antennapedia* genes, is required for the correct development of the adult head and the first two thoracic segments. The *Bithorax* complex, composed of the *Ultrabithorax, abdominal A,* and *Abdominal B* genes, is required for the proper development of the posterior region of the second thoracic segment, the third thoracic segment, and all the abdominal segments in the adult.

Evidence from Mutant Studies

We know that these homeotic genes determine the identity of segments and, ultimately, the adult structures from two types of mutants. First, gain-of-function mutations cause the gene to be expressed in a different segment than normal. For example, the *Antennapedia* gene is normally expressed in the segment that produces the second thoracic segment. In the dominant *Antennapedia* mutant, the gene is expressed both in its normal location and in the anterior segment that produces the antennae. Thus, the antennae are converted into the appendage normally found in the second thoracic segment, the legs (see fig. 21.20).

The second type of informative mutation is the loss-of-function mutation, where the gene is not expressed at all. When this occurs, the segment that is normally defined by the homeotic gene is now under the control of the gene that specifies the *adjacent* anterior segment.

For example, the wing and haltere, a small wing-like appendage on the third thoracic segment, consists of both an anterior and posterior compartment (**fig. 21.21a**). The original *bithorax* mutation causes the anterior compartment of the haltere to become an anterior wing compartment, or the anterior compartment of the

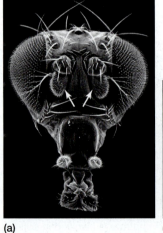
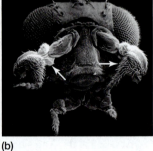

(a)   (b)

figure 21.20   The *Antennapedia* phenotype. A wild-type fly head possesses a pair of antenna (a, arrows). In the *Antennapedia* mutant, a pair of thoracic legs replaces the antenna (b, arrows).

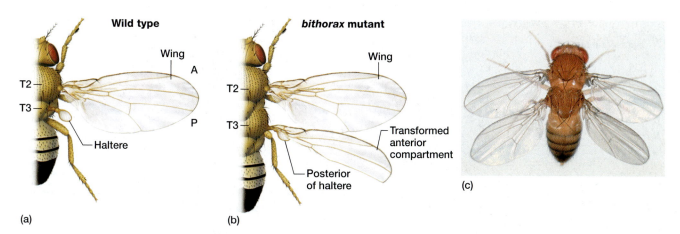

figure 21.21   Homeotic transformation of the haltere into a *Drosophila* wing. (a) A drawing of the right side of an adult *Drosophila* showing the wing (located on the second thoracic segment, T2) and haltere (located on the third thoracic segment, T3). (b) The original *bithorax* mutation transforms the anterior compartment of the haltere into the anterior compartment of a wing. However, the posterior compartment remains a haltere. (c) Introduction of a second mutation into the *bithorax* mutant causes the transformation of the haltere into an entire wing.

third thoracic segment to become a duplication of the anterior compartment of the second thoracic segment (fig. 21.21b). Introducing a second mutation to the *bithorax* mutant causes the halteres to be entirely converted into the second thoracic wings (fig. 21. 21c).

Homeotic mutations, both gain-of-function and loss-of-function, result in the adult structures assuming a new identity without changing the number of segments.

Regulation of Homeotic Genes

The gap and pair-rule genes regulate the establishment of the expression of the homeotic genes. In addition, two classes of genes are important in maintaining homeotic gene expression. The *Polycomb* genes repress the expression of the homeotic genes in the segments where they are not to be expressed. In contrast, the *trithorax* genes are required to maintain the expression of the homeotic genes in the correct segments.

Both of these gene classes regulate homeotic gene expression by controlling chromatin conformation. The *Polycomb* genes encode proteins that remodel the chromatin in a closed conformation to prevent transcription of the desired homeotic genes; *trithorax* proteins remodel the chromatin to an open conformation to allow transcription of the desired genes.

This regulation of transcription by chromatin remodeling is simplified by the organization of the genes in the *Antennapedia* and *Bithorax* complexes. The genes are linearly arranged within both complexes in the same order in which they are expressed along the anterior-posterior axis in the embryo (**fig. 21.22**). Within each complex, the genes located at the 3' end of the complex are expressed more anteriorly than the

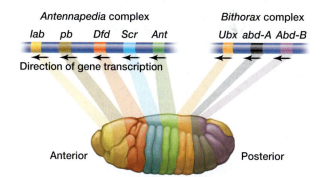

figure 21.22   The organization of the genes in the *Antennapedia* and *Bithorax* complexes correlates with their expression pattern in the embryo. Shown at the top is the arrangement of the genes in the *Antennapedia* and *Bithorax* complexes along with their orientation of transcription. The relative location of each embryonic segment that is determined by a homeotic gene is color-coded to match its corresponding gene.

genes located more toward the 5' end in the complex. Thus, the *Polycomb* and *trithorax* genes can regulate expression within each complex by simply remodeling the chromatin along the chromosome.

For the *Bithorax* complex, the situation is further simplified because the three genes are not uniquely expressed in different segments. Rather, they are expressed in overlapping groups of segments (**fig. 21.23**). The *Ultrabithorax* gene is expressed in parasegments that produce the third thoracic segment through the eighth abdominal segment; *abdominal A* is expressed in parasegments that generate the second through eighth abdominal segments; and *Abdominal B* is expressed in parasegments that generate the fifth through eighth abdominal segments.

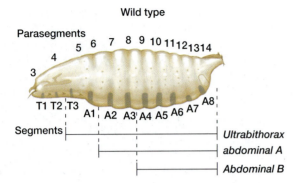

Wild type

figure 21.23   Expression patterns of the three *Bithorax* complex genes. Shown is a diagram of the segments and parasegments in a wild-type *Drosophila* embryo. The segments of the embryo that express each of the three *Bithorax* complex genes (*Ultrabithorax*, *abdominal A*, and *Abdominal B*) are marked under the embryo. Notice that the three genes share a common expression region at the posterior end of the embryo.

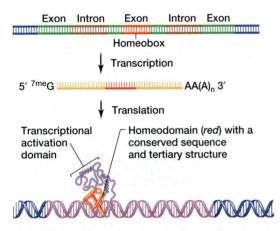

figure 21.24   Relationship between the homeobox and homeodomain. The homeobox is a 180 basepair sequence that is within the open reading frame of several genes that encode the homeodomain, which is a 60-amino-acid sequence that forms a helix-turn-helix motif in a DNA-binding protein. The homeodomain binds a conserved DNA sequence to regulate gene expression in organisms ranging from the fruit fly to humans.

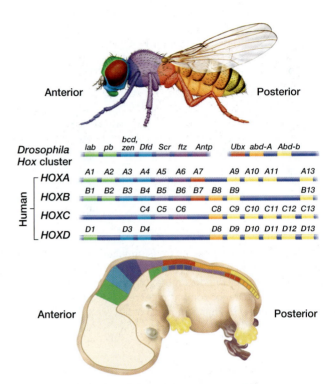

figure 21.25   Homeotic genes in fruit flies and humans are organized into complexes that display colinearity of gene location with expression pattern. The relative gene positions within the *Drosophila Antennapedia* and *Bithorax* gene clusters are shown. The genes are color coded to correspond with the regions of the adult fly that they determine. The gene order along the chromosome corresponds to the region along the anterior-posterior axis of the fly that they specify. The genes in the four human *HOX* gene clusters are aligned with their related *Drosophila* genes. Shown at the bottom is the inferred relative expression patterns of the *HOX* genes in the stage 19 human embryo based on the corresponding *HOX* gene expression patterns in the mouse embryo. Notice that the expression pattern along the anterior-posterior axis again matches the gene order along the chromosome. Thus, the homeotic genes show a conserved DNA sequence, gene order, and expression pattern from fruit flies to humans.
(Reprinted with permission from MacMillan Publishers Ltd: *Molecular Genetics Metabolism*, Veraksa, Del Campso, and McGinnis, vol. 69, figure 1, pp. 85–100, copyright 2000.)

Comparing the amino acid sequences of the proteins encoded by the genes in the *Antennapedia* and *Bithorax* complexes has revealed a surprising similarity. Within each gene is a 180-bp gene sequence that is called the **homeobox** (**fig. 21.24**). The homeobox is in the ORF of each gene, and it encodes a 60-amino-acid sequence termed the **homeodomain,** which is a helix-loop-helix DNA-binding motif. Although it is not surprising that the homeotic genes encode proteins with DNA-binding domains, it is surprising that the homeodomain sequences are so similar in all of these proteins.

Homeobox Genes and Vertebrate Development

When the homeobox sequence was identified in the *Drosophila* homeotic genes, it led to the identification of related genes in organisms from worms to mice to humans. Mammalian homeobox genes (called **Hox genes**) share more than just sequence similarity with the *Drosophila* genes. First, *Hox* genes exist in carefully regulated complexes, and they display *colinearity*, in which expression of the gene in time and space correlates with its position in the complex (**fig. 21.25**). Second, *Hox* genes control specification of body segments or regions along primarily the anterior–posterior axis, but also the proximal–distal axis in vertebrate limb development.

Seeing how the homeotic genes control segment identity in *Drosophila* is a straightforward matter, but vertebrates are not similarly segmented. There are times during vertebrate development, however, when discrete compartments are present within a larger structure. For example, development of the hindbrain involves the temporary formation of rhombomeres, which are a series of restricted compartments, each of

which receives sensory neural connections and extends motor neural connections in a specific pattern.

The restricted expression of the *Hox* genes is particularly obvious in the rhombomeres (**fig. 21.26**). Each rhombomere has a specific pattern of *Hox* gene expression. The tight spatial and temporal regulation of *Hox* gene expression generates the **Hox code,** by which segments of the animal are specified based on their unique pattern of *Hox* gene expression.

Based on observations in *Drosophila,* we would expect changes in expression of a single *Hox* gene will lead to the cells within the rhombomere assuming a new identity. Using zebrafish, we can easily manipulate the level of expression of a specific gene (see chapter 20). For example, we can increase the amount of Hoxb1b protein by injecting *hoxb1b* mRNA into the newly fertilized egg. Alternatively, we can inject a morpholino that is complementary to the *hoxb1b* mRNA into the newly fertilized egg and reduce the expression of the Hoxb1b protein.

In **figure 21.27**, rhombomeres r3 and r5 are visualized by the expression of the *krox20* mRNA. The location

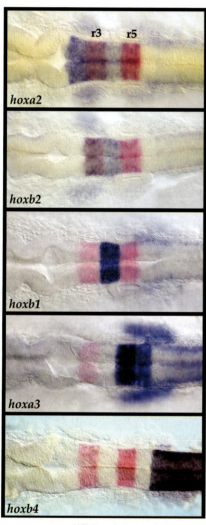

figure 21.26 The "Hox code" is demonstrated in the zebrafish hindbrain. During development of the zebrafish hindbrain, temporary structures known as rhombomeres form in a segmental pattern. Each rhombomere displays a unique *Hox* gene expression pattern. Pink staining in each picture indicates the positions of a marker (*krox20*) for rhombomeres 3 and 5, while blue staining indicates expression of the specific *Hox* gene indicated in each panel. The *hoxa2* is expressed in rhombomeres 2, 3 and 4, *hoxb2* is present in rhombomeres 3 and 4, *hoxb1* is expressed in rhombomere 4, *hoxa3* is found in rhombomeres 5 and 6, and *hoxb4* is expressed in rhombomere 7 and further posteriorly.

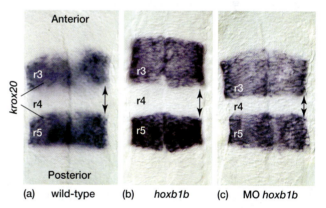

figure 21.27 Changes in the zebrafish Hoxb1b expression affects rhombomere size and position. Based on the *Drosophila* homeotic mutants, changes in Hox protein expression should alter the identity of the developing rhombomeres. The expression of *krox20* gene expression is used to mark the location of rhombomeres r3 and r5. (a) A wild-type zebrafish embryo shows the normal size and position of rhombomeres r3, r4, and r5. The double-headed arrow represents the width of rhombomere r4. (b) Injection of wild-type *hoxb1b* mRNA into the newly fertilized egg produces a gain-of-function mutant phenotype. The increased amount of *hoxb1b* causes rhombomere r3 to become wider and extend further anteriorly (towards r2). There is no change in the width or position of either r4 or r5. (c) Injection of a *hoxb1b* morpholino (MO *hoxb1b*) into the newly fertilized egg reduces Hoxb1b protein and generates a loss-of-function phenotype. Loss of Hoxb1b causes rhombomere r3 to become wider than wild type and extends further posteriorly toward r4, which decreases the width of rhombomere r4. This suggests that extra Hoxb1b causes the posterior r2 cells to adopt an r3 fate, while loss of Hoxb1b causes the anterior cells of r4 to adopt a r3 cell fate.

and size of rhombomeres r3, r4, and r5 are shown in the wild-type zebrafish hindbrain at 20 h after egg fertilization (fig. 21.27a). Overexpression of Hoxb1b, which corresponds to a gain-of-function phenotype, results in the expansion of r3 in the anterior direction (fig. 21.27b). There is no change in the size or location of either r4 or r5, which means that cells that should be in r2 are adopting an r3 identity. The morpholino-induced repression of Hoxb1b expression results in the reduction of the size of r4 and the expansion of r3 in the posterior direction (fig. 21.27c). Thus, cells that should be in r4 are adopting an r3 identity. Thus, the vertebrate *Hox* genes follow the same general rules as we discussed for the *Drosophila* homeotic genes.

It's Your Turn

We have described several different ways that a pattern is generated in the newly fertilized egg and then leads to unique cell identities in the developing embryo. Problems 21, 22, and 23 explore these fundamental processes in development.

Restating the Concepts

▶ The homeotic genes define the identity of individual segments during *Drosophila* embryonic development. Each homeotic gene contains a 180-bp homeobox, which encodes a homeodomain, a 60-amino-acid sequence corresponding to a helix-loop-helix DNA-binding motif.

▶ *Drosophila* contains two homeotic gene clusters: *Antennapedia* and *Bithorax.* The genes are arranged in each cluster such that the genes are expressed in an anterior to posterior direction based on their 3' to 5' position within the cluster.

▶ Vertebrate *Hox* gene clusters also control aspects of development. Similar to *Drosophila*, the relative position of the *Hox* gene within the cluster corresponds to the relative position where the gene is expressed. The vertebrate *Hox* genes are also required for determining the identity of specific compartments, such as the rhombomeres.

▶ The Hox code states that the identity of a segment or compartment of a developing animal can be determined based on the combination of *Hox* genes that are expressed.

21.4 Sex Determination

Up to this point, we discussed how segments and cells adopt a specific identity within an organism. A second

decision that must be made for each cell is related to gender determination. The general mechanisms of sex determination, such as the X/A ratio in *Drosophila* and the presence or absence of the Y chromosome in mammals was covered in chapter 4; we now discuss the molecular details behind these decisions. Related to this decision is the maintenance of gene balance in the two sexes. Differences in the number of sex chromosomes in the two genders require a mechanism to equalize the level of expression of the sex-linked genes. Failure by a cell to correctly determine the number of sex chromosomes it possesses will lead to gene imbalance and cell death.

Sex Determination in *Drosophila*

The "decision" to be either a *Drosophila* male or female depends on the ratio of X chromosomes to the number of sets of autosomes. An X/A ratio of 0.5 produces a male, and an X/A ratio of 1.0 yields a female. After this ratio is determined early in development, the *Sex lethal* gene is transcribed. Alternative splicing of several genes, beginning with the *Sex lethal* primary transcript, leads to the establishment of a specific sex.

Expression of the Sex lethal Gene

Expression of the *Sex lethal* gene involves transcription from two different promoters (**fig. 21.28**). Only females transcribe the *Sex lethal* gene from the establishment promoter (*PE*), which generates a transcript that is spliced and translated to generate functional protein. After establishing the initial expression, *Sex lethal* is transcribed from the late promoter (*PL*) in both males and females.

Transcription of the *Sex lethal* gene from *PE* occurs only during the syncitial and cellular blastoderm stages. Transcription from *PE* depends on the same proteins that determine the X/A ratio. In female embryos, the amount of these proteins is sufficiently high that transcription from *PE* occurs. The *PE*-derived *Sex lethal* transcript is spliced to produce functional Sex lethal protein (see fig. 21.28). In male embryos, the amount of X/A proteins is low, which represses transcription from *PE*. Later in development, and throughout the remaining life of the fly, the cells transcribe the *Sex lethal* gene from *PL* in both males and females (see fig. 21.28).

The initial Sex lethal protein that was generated from the *PE* transcript in females functions to prevent exon L3 from splicing to L2 in the *PL* transcript. Instead, L2 splices to exon 4, which can be translated to generate additional Sex lethal protein. The absence of Sex lethal protein in males results in exon L2 splicing to L3 on the *PL* transcript. This male transcript introduces a translation termination codon into the ORF, which again prevents the production of functional Sex lethal protein. In this manner, Sex lethal protein is initially produced only in females, and it continues to be expressed only in females. This system is an example of a positive feedback system in a biological setting—presence of Sex lethal protein in females permits continual production of Sex lethal protein!

The productive splicing of the *PL Sex lethal* transcript depends on the prior presence of Sex lethal protein, which only occurs in females. This is typical of later steps of the sex-determination pathway: The default pathway in *Drosophila* leads to male development, whereas regulated activity leads to female-specific gene activation.

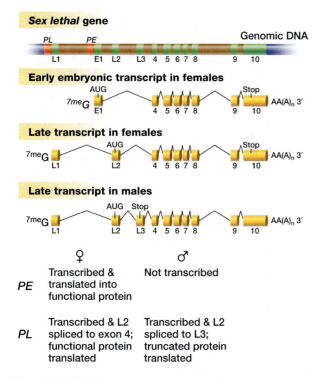

figure 21.28 Transcription and alternative splicing of the *Drosophila Sex lethal (Sxl)* gene. Shown is a schematic of the *Drosophila Sxl* gene where the exons are green and the introns are brown. The gene contains two promoters (red), *PE* that drives transcription early in female development and *PL*, which is transcribed in both genders later in development. The location of the translation initiation codons (AUG) are shown in exons E1 and L2, while the translation termination codons (Stop) are in exons L3 and 10. The black lines designate the various splices that are used. In the early embryo, a single mRNA is transcribed from the *PE* promoter and is only found in females. This mRNA encodes a functional Sxl protein. Later in development, the *Sxl* gene is transcribed from the *PL* promoter. In females, the pre-mRNA is spliced from exon L2 to 4 to generate the complete open reading frame. In males, the pre-mRNA is spliced from exon L2 to L3, which produces a truncated open reading frame that encodes a non-functional protein. The lower table summarizes the transcript that is produced from both promoters in both genders. Only in the female is functional Sex lethal protein ever produced.

The Action of the transformer Gene

The second key gene in the sex-specification cascade is *transformer* (*tra*). Like *Sex lethal*, the *transformer* primary transcript is alternatively spliced (**fig. 21.29**). The *transformer* primary transcript is spliced to one mRNA form in the presence of the Sex lethal protein in females. This female-specific *tra* mRNA can be translated to yield a functional Tra protein. In the absence of Sex lethal protein (in males), the *transformer* primary transcript is spliced into an alternative form. Translation of this male *tra* mRNA is terminated early due to the presence of a translational termination codon in an exon that is absent in the female-specific *tra* mRNA. This yields a truncated and nonfunctional Tra protein in males.

In females, the Tra protein interacts with the Transformer-2 protein, which regulates the splicing of the *doublesex* (*dsx*) and *fruitless* (*fru*) primary transcripts (see fig. 21.29). The female-specific *dsx* and *fru* mRNAs are translated to generate female-specific Doublesex and Fruitless proteins. In males, the Transformer-2 protein, with the truncated Tra protein, generates different spliced male-specific *dsx* and *fru* mRNAs that are translated to produce male-specific and functional Doublesex and Fruitless proteins.

The Msl-2 Protein

One final important control in sex determination is the production of the Male-specific lethal-2 (Msl-2) protein (fig. 21.29). In this case, the absence of the Sex lethal protein splices the *msl-2* primary transcript into an mRNA that is translated into a functional Msl-2 protein. In females, the Sex lethal protein alters the splicing of the *msl-2* primary transcript to produce a female *msl-2* mRNA that is translated into a truncated and nonfunctional Msl-2 protein.

The Msl-2 protein is essential for properly balancing expression of X-linked genes in the two genders. Because females have two X chromosomes relative to one in males, they have the possibility of expressing twice as much mRNA from each X-linked gene compared to males. As we discussed in chapter 8, these differences in chromosome number can be lethal or deleterious. This is resolved in *Drosophila* males by the Msl-2 protein functioning to double the transcription of the X-linked genes. This increased transcription results in both genders exhibiting similar levels of transcription of X-linked genes. In the following section, you will see how mammals solve this problem.

Results of the Splicing Cascade

The end result of this cascade of alternative splicing is sex-specific control of gene expression. Whereas the *Sex lethal* gene is the overall control gene, the Doublesex,

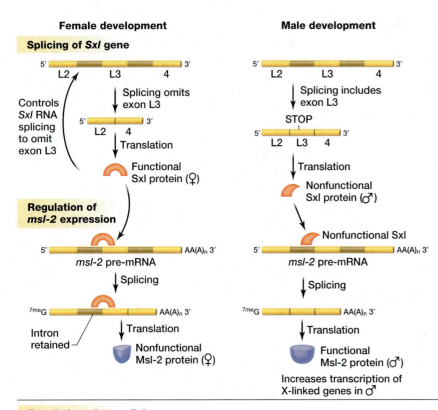

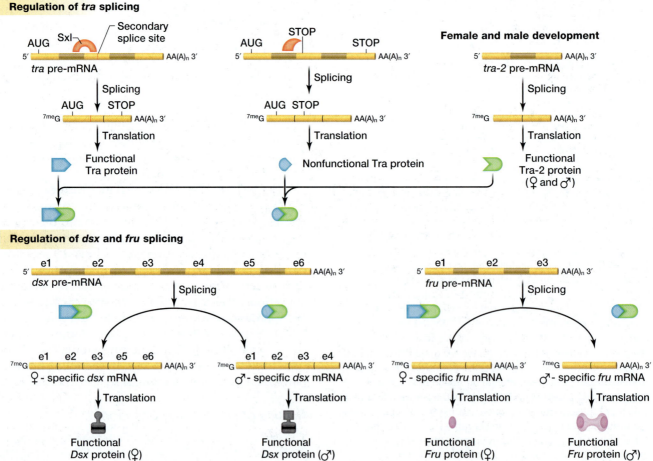

figure 21.29 Role of alternative splicing in *Drosophila* sex determination. Alternative splicing of the *Sex lethal* (*Sxl*), *male-specific lethal-2* (*msl-2*), *transformer* (*tra*), *doublesex* (*dsx*), and *fruitless* (*fru*) RNAs are required for correct sex determination in *Drosophila*. Only three of the *Sxl* exons are shown, which correspond to the late *Sxl* transcript in both males and females. The L3 exon is alternatively spliced to produce functional Sxl protein in females and truncated, nonfunctional Sxl protein in males. In females, functional Sxl protein affects splicing of both *msl-2* (yields a nonfunctional Msl-2 protein) and *tra* (to yield a functional Tra protein). In males, the truncated, nonfunctional Sxl protein yields an *msl-2* mRNA that encodes a functional Msl-2 protein and a nonfunctional Tra protein. In females, the Tra/Tra-2 complex affects *dsx* splicing to produce a female-specific functional Dsx protein and affects *fru* splicing to produce a functional Fru protein. In contrast, Tra-2 and the nonfunctional Tra in males result in the expression of a male-specific functional Dsx protein and a functional Fru protein.

table 21.1 Role of Different Proteins in *Drosophila* Sex Determination

	Female	Male
Sex lethal	Functional protein translated from *PE* transcript; correct splicing of the *PL* pre-mRNA results from earlier Sex lethal protein expression to produce a *PL* mRNA that can also be translated to yield functional Sex lethal protein. Required for the alternative splicing of the *tra* and *msl-2* pre-mRNAs.	*PE* is not active; *PL* pre-mRNA is incorrectly spliced in the absence of Sex lethal protein to yield truncated and nonfunctional Sex lethal protein.
Transformer	Sex lethal protein regulates splicing of *tra* pre-mRNA to produce *tra* mRNA that can be translated to yield functional Transformer protein. Required to interact with the Tra-2 protein for the alternative splicing of the *dsx* and *fru* pre-mRNAs.	Absence of Sex lethal protein results in alternative splicing of *tra* pre-mRNA that produces an mRNA that is translated into a truncated and nonfunctional Transformer protein. Required to interact with the Tra-2 protein for the alternative splicing of the *dsx* and *fru* pre-mRNAs.
Transformer-2	Functional protein is expressed. Required in either the presence or absence of the Tra protein to splice the *dsx* and *fru* pre-mRNAs.	Functional protein is expressed.
Doublesex	Transformer and Transformer-2 proteins work together to regulate *dsx* pre-mRNA splicing to produce an mRNA that is translated into a functional female-specific Dsx^F protein. Required for the expression of genes involved in female-specific differentiation.	Transformer-2, in the presence of truncated Transformer, regulates *dsx* pre-mRNA splicing to produce an mRNA that is translated into a functional male-specific Dsx^M protein. Required for the expression of genes involved in male-specific differentiation.
Fruitless	Transformer and Transformer-2 proteins work together to regulate *fru* pre-mRNA splicing to produce an mRNA that is translated into a functional female-specific Fru^F protein. Required for the expression of genes involved in female-specific behavior	Transformer-2, in the absence of Transformer, regulates *fru* pre-mRNA splicing to produce an mRNA that is translated into a functional male-specific Fru^M protein. Required for the expression of genes involved in male-specific behavior
Male-specific lethal-2	Sex lethal protein regulates splicing of *msl-2* pre-mRNA to produce *msl-2* mRNA that is translated into a truncated and nonfunctional Msl-2 protein.	Absence of Sex lethal protein results in alternative splicing of *msl-2* pre-mRNA to produce an mRNA that is translated into a functional Msl-2 protein. Required for hyperactivation of X-linked gene transcription.

Fruitless, and Male-specific lethal-2 proteins regulate specific functions associated with the two sexes. As summarized in **table 21.1**, the Doublesex protein regulates the expression of genes that control the development of the somatic and germ-line tissues in both sexes. The Fruitless protein regulates the expression of genes that allow differences in the central nervous system between the two sexes, which results in sex-specific behaviors. This difference is very important in the male- and female-specific mating behaviors. The Msl-2 protein is critical for the hyperactivation of X-linked gene expression in males.

Note that the *doublesex* and *fruitless* genes encode functional proteins in both sexes. Aberrant splicing of either primary transcript may therefore result in the gender of the individual switching. In contrast, the *msl-2* gene encodes a functional protein only in males. If improper splicing of the *msl-2* primary transcript occurs

in males, hyperactivation of the X-linked genes does not occur, and the fly dies due to genetic imbalance. In contrast, if incorrect splicing of the gene occurs in females to produce functional Msl-2 protein, the fly also dies due to hyperactivation of transcription of both X chromosomes and genetic imbalance again. Mutants can be isolated in which the sex of the fly has been switched relative to its X/A ratio, but mutations that affect the level of X chromosome transcription are lethal.

Sex Determination in Mammals

As we saw with *Drosophila*, two issues must be resolved in sex determination. First, differences must occur during the development of the two genders, including both the somatic and germ tissues. Second, differences in the number of sex chromosomes between the two genders must be compensated for by altering the level of transcription of the sex-linked genes in one of the genders. In *Drosophila,* this is achieved by increasing the level of transcription of the X-linked genes in males (one X chromosome) relative to females (two X chromosomes). These two issues must also be resolved in mammalian sex determination.

Mechanism of Mammalian Sex Determination

As described in chapter 4, mammalian sex is determined by the presence or absence of the Y chromosome. The initial gene in the sex determination cascade is the *SRY* gene, which is located on the Y chromosome. Evidence that the *SRY* gene is sufficient to determine the gender of mammals was observed in mice and humans who lack this gene on the Y chromosome. These XY(*SRY* deleted) individuals develop as females. Similarly, when the *SRY* gene is translocated from the Y chromosome to an autosome, XX individuals develop as males if they possess the *SRY* gene translocation.

The *SRY* gene encodes a DNA-binding protein with an HMG domain. The HMG domain is a 79-amino-acid DNA binding motif that is composed of three α-helices (**fig. 21.30**). The HMG domain binds in the minor groove of DNA, which causes an approximately 80-degree bend in the DNA. The result is that the DNA becomes significantly unwound and even nucleosomes that are associated with some specific genes are displaced.

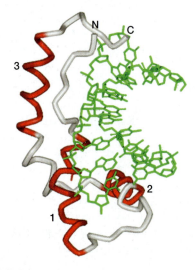

figure 21.30   Structure of the HMG box of the SRY transcription factor that is required for male differentiation in mammals. The HMG box is a 79-amino-acid domain in the SRY protein that consists of three alpha helices. The first alpha helix binds the minor groove of the DNA, which results in the bending and unwinding of the DNA. Shown is the structure of the HMG box domain, with the three alpha helices in red and numbered. The amino (N) and carboxyl (C) ends of the domain are labeled. The bound DNA (green) is bent nearly 80° to the right due to the HMG box binding the minor groove.

The end result of HMG binding is usually transcriptional activation of specific genes. Many of the missense mutations in the *SRY* gene that cause human sex reversal are located in the HMG domain, which suggests that this region of the protein is essential to its function.

Sex determination in mammals begins with the initial formation of an indeterminate gonad within the urogenital system. If the *SRY* gene is present within these cells, then it is expressed, and the resulting SRY protein performs two critical functions (**fig. 21.31**). First, it activates expression of the *SOX9* gene, which also encodes a HMG-containing protein. Unlike the Y-linked *SRY* gene, the *SOX9* gene is an autosomal gene. Transgenic mice that express *SOX9* develop as males, even if they lack the *SRY* gene. Thus, either *SRY* or *SOX9* is sufficient to determine maleness in mice. The SOX9 protein is involved in the activation of genes that encode testis-determining factors and AMH (anti-Müllerian hormone). Expression of AMH causes the Müllerian ducts to regress, which in combination with the expression of the testis-determining factors, commits the indeterminate gonad to accept the male fate.

The second significant function of the SRY protein is to block the expression of the *WNT4* and *DAX1* genes (see fig. 21.31). The WNT4 protein activates the expression of genes that encode the ovary-determining factors; the DAX1 protein blocks the expression of proteins that activate expression of the testis-determining factors. In this manner WNT4 and DAX1 combine to induce female gonad development, while also blocking male gonad development. When SRY represses *WNT4* and *DAX1* expression, it is preventing female gonad development in concert with the SOX9-induced expression of AMH.

Mechanism of Dosage Compensation in Mammals

Mammalian dosage compensation, as you learned in chapter 4, is resolved by inactivation of one of the X chromosomes in females. In most mammals, the decision of which X chromosome is inactivated is fairly random, which can lead to the generation of mosaicism in females.

The process of X-chromosome inactivation can be divided into three major steps. First, the number of X chromosomes in a cell must be counted. If there is only one X chromosome, then inactivation must not occur. If more than one X chromosome is present, then all but one must be inactivated. Second, a choice must be made as to which X chromosome(s) will be inactivated. Third, the selected X chromosome(s) must actually be inactivated.

The mechanism of X inactivation begins with a locus near the center of the X chromosome (Xq13) that is called *XIC* (**X**-**I**nactivation **C**enter). The major gene in this locus is called ***XIST*** (**X**-**I**nactivation **S**pecific **T**ranscript). The *XIST* RNA is approximately 15 kb in length, is

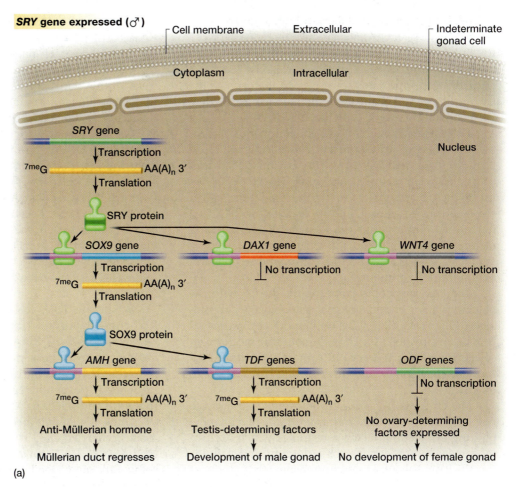

(a)

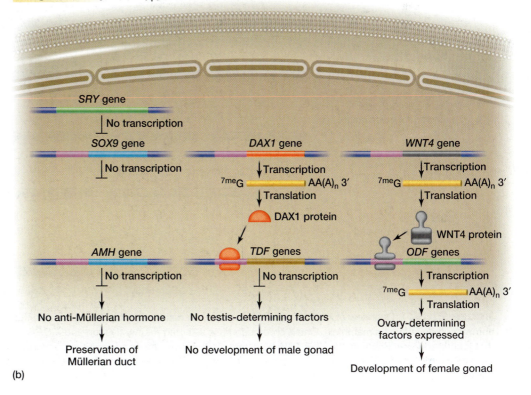

(b)

figure 21.31

Sex determination in mammals begins with either the presence (a) or absence (b) of *SRY* expression in the indeterminate gonad. Expression of *SRY* results in the subsequent expression of the *SOX9* gene and the repression of the *DAX1* and *WNT4* genes. The SOX9 protein induces expression of the anti-Müllerian hormone, which results in the regression of the Müllerian duct, and the testis-determining factors, which are required to form the male gonads. In the absence of *SRY* expression, *DAX1* and *WNT4* are expressed. DAX1 represses expression of the testes-determining factors and WNT4 induces expression of the ovary-determining factors. In the absence of SOX9 protein expression, the anti-Müllerian hormone is also not expressed. The sum effect is the development of the female gonad.

polyadenylated, very unstable, and does not encode a protein. During X chromosome inactivation, the *XIST* RNAs bind to the chromosome from which they were transcribed (**fig. 21.32**), initially to regions that flank the *XIST* gene and later spreading to the ends of the chromosome. The X chromosome is inactivated when it is completely covered with the *XIST* RNAs.

The current model for how the number of X chromosomes in a cell are counted is shown in **figure 21.33**. A molecule called *blocking factor* is transcribed and translated from an autosomal gene. The blocking factor binds to a region adjacent to the *XIST* gene, called the counting element (CE). The location of the CE is fairly well defined, but the identity of the blocking factor remains unknown. Sufficient blocking factor is synthesized in the cell so that it can bind to one, and only one, CE in each nucleus.

The X chromosome that contains the CE with the blocking factor bound becomes the active X chromosome (X_a) in the cell. All the unbound X chromosomes become inactivated (X_i). Thus, the counting mechanism selects the sole X_a in every cell, rather than counting all the X chromosomes and determining the number to inactivate.

The choice of which X chromosome to inactivate partially involves the counting mechanism, since the binding of blocking factor determines which X chromosome is X_a. A second factor in determining choice is the *XIST* RNA itself. The X chromosome that produces the greater amount of *XIST* RNA is usually selected

as X_i. Investigators think that the higher level of *XIST* RNA corresponds to a greater amount of *XIST* RNA bound to the region flanking the *XIST* gene. Because the CE is adjacent to the *XIST* gene, binding of *XIST* RNA in this flanking region can reduce the ability of the blocking factor to bind to the CE, which results in that chromosome becoming X_i.

One additional factor involved in choice is the **TSIX** RNA. The *TSIX* (reverse of *XIST*) RNA is transcribed from the DNA strand that did not serve as the transcription template for *XIST* RNA (**fig. 21.34**). It also lacks an ORF, which makes the RNA the functional unit analogous to *XIST*. Because transcription always goes in a 5'-to-3' direction, these two RNAs are transcribed toward each other. In fact, their transcription overlaps, which results in the two RNAs having complementary sequences for the entire *XIST* locus, allowing them to base-pair.

When the *TSIX* RNA base-pairs with the *XIST* RNA, the stability of the *XIST* RNA decreases and results in less *XIST* RNA associated with the X chromosome. The decreased amount of *XIST* RNA then frees the CE to bind the blocking agent. Thus, the *TSIX* RNA acts to positively regulate blocking factor binding to the CE and selecting X_a.

Once the X_i has been selected, the inactivation must be carried out. The first step is the spreading of the *XIST* RNA toward the ends of the chromosome. Binding of the blocking factor to CE appears to prevent that spreading, which further reinforces the

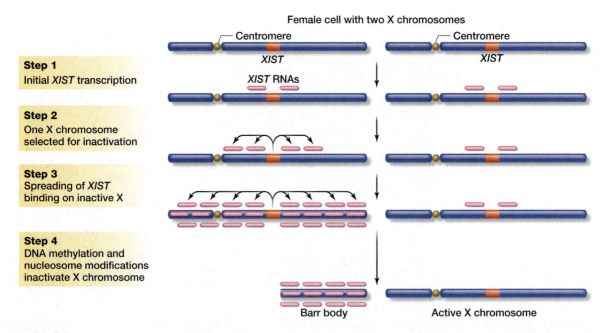

Female cell with two X chromosomes

Step 1
Initial *XIST* transcription

Step 2
One X chromosome selected for inactivation

Step 3
Spreading of *XIST* binding on inactive X

Step 4
DNA methylation and nucleosome modifications inactivate X chromosome

Centromere

XIST

XIST RNAs

Barr body   Active X chromosome

figure 21.32   Overview of X chromosome inactivation in mammals. The *XIST* gene is initially transcribed from both X chromosomes in female mammals. One X chromosome is selected for inactivation and *XIST* RNA continues to be expressed from that X chromosome. Because the *XIST* RNA binds only to the X chromosome that it is transcribed from, the *XIST* gene on the X chromosome selected for inactivation continues to be transcribed. X chromosome inactivation is subsequently stabilized by chromatin modifications to produce the Barr body.

selection of X_a to remain active. Because the spreading requires larger amounts of *XIST* RNA, transcription of the *TSIX* gene is suppressed on X_i. This suppression also increases the stability of the *XIST* RNA. As *XIST* RNA spreads along X_i, the *XIST* gene promoter on X_a

is methylated to terminate *XIST* transcription on the active X chromosome.

To maintain the inactive nature of X_i, several events occur. The X_i histones, particularly H3, undergo increased methylation and decreased acetylation.

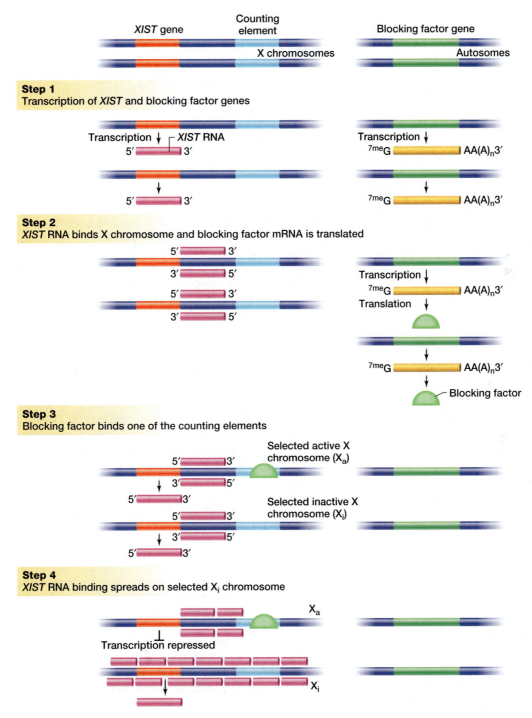

figure 21.33   X chromosome inactivation begins with transcription of *XIST* RNA from the X chromosome and binding of the *XIST* RNA to the X chromosome from where it was transcribed. The autosomal blocking factor gene is transcribed and the translated blocking factor protein binds the counting element on one of the two X chromosomes. Binding of the blocking factor selects that X chromosome as the active X chromosome (X_a). *XIST* RNA continues binding to regions flanking the *XIST* gene, but cannot spread past the bound blocking factor. Spreading of bound *XIST* RNA on the inactive X chromosome (X_i) is the commitment for inactivation and repression of gene expression on X_i.

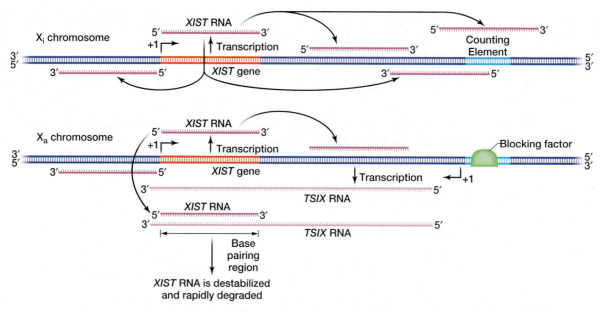

figure 21.34 Relative location of the overlapping and complementary *XIST* and *TSIX* transcription units involved in X chromosome inactivation. The directions of transcription of both RNAs are shown as arrows above (*XIST*) and below (*TSIX*) the double-stranded DNA. The top inactive X_i chromosome lacks the bound blocking factor and transcribes little *TSIX* RNA, which allows spreading of the bound *XIST* RNA along the chromosome. The lower active X_a chromosome contains the bound blocking factor, which prevents the spreading of the bound *XIST* RNA to the X chromosome and activates transcription of the *TSIX* RNA. Base pairing in the complementary region between *XIST* and *TSIX* destabilizes the *XIST* RNA and leads to its rapid degradation. The 5' and 3' ends of the *XIST* and *TSIX* RNAs are labeled, as well as the putative location of the Counting Element.

Recent data suggest that a *Polycomb*-related protein, which we described as being important in repressing expression of the homeotic genes in *Drosophila,* may be involved in the methylation of the X_i chromosome.

Transcription from X_i

Originally, mammalian X-chromosome inactivation was thought to result in the silencing of all the genes on the X_i chromosome, except for the 18 genes in the pseudoautosomal region. Because these 18 genes are present on both the X and Y chromosome, males possess two copies of each of these 18 genes. This would imply that females express both copies also, to maintain genetic balance.

Recent technical advances have permitted the analysis of 471 of the 1098 X-linked genes in human female (XX) fibroblast cells. Two very surprising results were observed. First, 15% of the X-linked genes (excluding the pseudoautosomal genes) were not completely silenced on the inactive X_i chromosome. This finding suggests that a significant difference exists in the level of expression for many X-linked genes between the two genders. Second, an additional 10% of the genes were expressed to varying extents from different X_i chromosomes. This suggests that the level of mosaic expression in females may be even greater than imagined. The expression from the X_i chromosome may increase the overall expression of this group of X-linked genes from 0% to 100% from one cell to another throughout the female.

It's Your Turn

Gender determination and balancing the expression from sex chromosomes are fundamental processes that must be accomplished by all animals. As we have seen, there are different ways in which these can be achieved. In problems 29, 30, and 31, you have the opportunity to compare these differences.

Restating the Concepts

▶ Sex determination in *Drosophila* involves the alternative splicing of the *Sex lethal* gene, which sets in motion a cascade of alternatively spliced mRNAs that ultimately produce male- and female-specific proteins encoded by the same gene.

▶ Sex determination in mammals involves the Y-linked *SRY* gene. The presence or absence of the SRY DNA-binding protein controls a cascade of proteins that regulate the expression of testis- or ovary-specific genes, respectively.

▶ X-chromosome inactivation in mammals balances X-linked gene expression between males and females. It involves the *XIST* RNA, which binds along the entire length of the X chromosome that will be inactivated, followed by several modifications to the chromatin of X_i.

▶ Recent data reveal that inactivation of genes along X_i is not complete, with up to 25% of the genes showing some variability in inactivation.

21.5 *Drosophila* Eye Development

Previously, we have discussed how large-scale changes are produced during early development, such as the generation of polarity in the embryo, the determination of gender for the entire organism, and the inactivation of X chromosomes throughout the mammalian female. Another important aspect of development is the generation of specific organs or structures. The development of the heart, limbs, gut, and nervous system are all under genetic control. We now discuss two aspects of *Drosophila* eye development as an example of the genetic control of organ development.

Master Regulator Genes

During development, cells have the ability to adopt a variety of different cell fates until they become committed to either a specific tissue/organ or cell type. A **master regulator gene** is a gene that commits a cell to a particular cell fate, and its action usually involves the initiation of a cascade of gene activations. A mutation in a master regulator gene should prevent the formation of the corresponding organ or tissue. Additionally, a master regulator gene that is expressed in an incorrect location in the organism should create aberrant cells or tissue in this new location.

In *Drosophila*, mutations in five different genes produce flies that lack compound eyes. The complete absence of eyes in these mutants suggest that one or more of these genes are master regulator genes that are absolutely required to generate an eye. These genes are called *eyeless, twin of eyeless, sine oculis, eyes absent,* and *dachshund.* Four of these genes encode transcription factors, with the *eyes absent* gene encoding a protein phosphatase that regulates transcription. The *eyeless* phenotype that is associated with the *eyeless* mutant is shown in **figure 21.35b.**

To prove that the *eyeless* and *twin of eyeless* genes are master regulators, Walter Gehring expressed the wild-type genes in the wrong parts of the developing fly. The resulting mutant flies had small compound eyes in the antenna, wing, and leg (fig. 21.35c and d). The ability of either the *eyeless* or *twin of eyeless* gene to induce eye formation in the wrong parts of the fly confirmed that both genes are master regulators. By comparing which transcription factor binds each promoter, a regulatory network has been described for these genes (**fig. 21.36**).

The potential importance of these *Drosophila* genes in eye development is further suggested by the identification of related genes in both humans and mice (**table 21.2**). *Drosophila* contains two homeobox-containing genes, *eyeless* and *twin of eyeless;* in contrast, mammals have only a single gene, called *Pax6.* Mutations have been identified in most of these genes in either humans or mice.

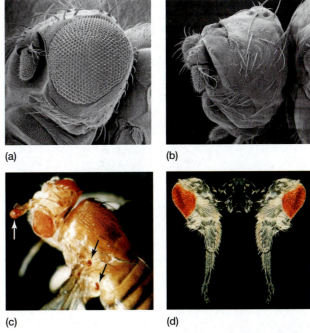

(a)   (b)

(c)   (d)

figure 21.35

Expression of master eye regulator genes is required and sufficient to form compound eyes. A scanning electron micrograph of the wild-type compound eye (a). The *eyeless* mutant lacks the compound eye (b). Misexpression of the wild-type *eyeless* gene induces small compound eyes in the antenna and wing margins (c, arrows). Similarly, misexpression of the wild-type *twin of eyeless* gene in the legs produces compound eyes on the legs (d). Expression of the wild-type mouse *Pax6* gene in the antennae of wild-type flies produces compound eyes on the antenna (e, arrow).

(e)

As would be expected if these genes are critical for eye development, most of the mammalian mutants have either no eyes, very small eyes, or no lenses. Walter Gehring further demonstrated the universal importance of these homeobox genes in eye development by expressing the wild-type mouse *Pax6* gene in wild-type *Drosophila.* These flies possessed extra compound eye tissue on the antenna (fig. 21.35e). This experimental finding demonstrates that many of the genetic pathways and encoded proteins have similar functions in very different organisms.

Generation of the R7 Photoreceptor Cell

The *Drosophila* compound eye is composed of approximately 750 facets or ommatidia. Each ommatidium, which is represented as a "bump" on the surface of

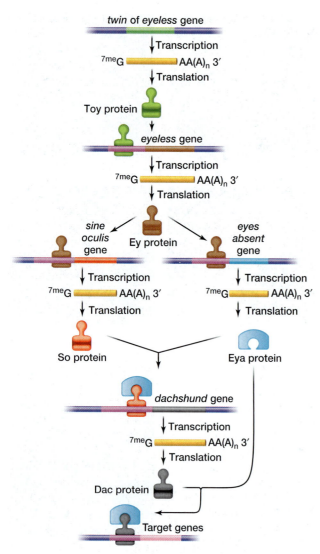

figure 21.36 The regulatory network of the five master regulator genes involved in *Drosophila* eye development. The *twin of eyeless* (*toy*), *eyeless* (*ey*), *sine oculis* (*so*), and *dachshund* (*dac*) genes all encode transcription factors, while *eyes absent* (*eya*) encodes a transcription cofactor with protein phosphatase activity. The Toy protein activates *ey* gene transcription and the Ey protein activates transcription of both the *so* and *eya* genes. The So and Eya proteins interact, with So providing the DNA binding activity and Eya contributing the transcription activation domain, to express the *dac* gene. Similarly, the Dac and Eya proteins function together to activate the transcription of target genes that are required for eye development.

the eye, is a simple eye, with the exterior lens overlying eight photoreceptors (**fig. 21.37**, R1–R8). Each photoreceptor contains a tightly packed microvillar region, the rhabdomere, which contains the proteins that are involved in light reception. Each photoreceptor projects its rhabdomere into the center of the ommatidium, with the R1–R6 rhabdomeres forming a trapezoid, and the R7 and R8 rhabdomeres located in the center (see fig. 21.37). However, at any specific depth in the ommatidium, only the R7 or R8 rhabdomere is present since the R7 rhabdomere lies directly over the R8 rhabdomere.

The sevenless *Mutant*

The *sevenless* mutant was isolated as a fly that was unable to respond to ultraviolet light. Examining the mutant eye revealed that the only defect was the absence of the R7 photoreceptor in every ommatidium (**fig. 21.38b**). In the *sevenless* mutant, the cell that should become the R7 cell becomes a cone cell, which is an example of a homeotic cell transformation.

The *sevenless* gene encodes a receptor tyrosine kinase (RTK) molecule. A RTK is a transmembrane protein with an extracellular domain that binds a ligand. On ligand binding, the intracellular domain of the RTK phosphorylates itself or another RTK molecule (or both). The phosphorylated protein can interact with other proteins to initiate an intracellular signaling cascade. Quite often the result of such a cascade is activation of a DNA-binding protein that enters the nucleus to regulate transcription of other genes.

This general scheme suggested to investigators that other molecules must be required for the determination of the R7 photoreceptor. At a minimum, a ligand protein must exist that binds the Sevenless protein. Also, intracellular signaling molecules and at least one DNA-binding protein must be required. How could these other molecules in the Sevenless signaling cascade be identified?

Determining the Components of the Signaling Cascade

The first approach involved identifying additional mutants with the *sevenless* phenotype. These mutants would be unable to respond to ultraviolet light and should lack the R7 photoreceptor. One mutant that was

table 21.2 Relationship of *Drosophila* and Mammalian Genes Involved in Eye Development

Drosophila gene	Mammalian homolog	Mutant Mammalian Phenotype
eyeless, twin of eyeless	*Pax6*	Aniridia (human), small eye (mouse)
eyes absent	*Eya1, Eya2, Eya3*	Cataracts and anterior eye defects (human), no eye phenotype (mouse)
sine oculis	*Six3*	Very small eyes (human)
dachshund	*Dach1*	Not determined

identified in this manner was *bride of sevenless* (*boss,* fig. 21.38c). The *boss* gene also encodes an integral membrane protein. Unlike Sevenless, the Boss protein is expressed on the membrane of the R8 photoreceptor cell. The Boss protein serves as a ligand that binds the Sevenless receptor and signals the Sevenless-containing cell to become an R7 photoreceptor.

The second approach involved the sensitized screen that we discussed in chapter 20. A temperature-sensitive mutation was generated in the *sevenless* gene that corresponded to the tyrosine kinase domain of the Sevenless protein. At the permissive temperature, sufficient Sevenless protein activity is available to allow the R7 cell to be generated. At the higher, restrictive temperature, the Sevenless protein is inactive, and the R7 cell is absent. Researchers raised the flies at an intermediate temperature that was just sufficient to permit R7 determination. If a different mutation affects another component in the sensitized Sevenless signaling pathway, then it should further reduce the activity of the pathway and prevent the R7 cell from developing. Because only small changes in these other genes are required to reduce the overall signaling activity below the threshold and prevent R7 development, these additional mutations could be identified as heterozygotes.

This sensitized screen identified a *ras* gene, which encodes a small protein that is active when it binds GTP and inactive when GTP is hydrolyzed to GDP. A second mutation, called *Son of sevenless* (*Sos*) affects a guanine exchange factor. The Sos protein removes the GDP from the inactive Ras and replaces it with a GTP to reactivate the Ras protein. A third mutation affects a protein that is an adapter protein (GRB2) that binds to the phosphorylated Sevenless protein and also binds the Sos protein.

These secondary mutations were identified in heterozygotes, but the homozygous recessive phenotype of these mutations is lethal. This is because the genes identified in the sensitized screen are essential in many other cells during *Drosophila* development. The proteins encoded by these genes are activated by different receptors and stimulate different signaling components in the various tissues. Therefore, loss of these genes and their proteins will block the development of many critical organs and cause death. Using the sensitized screen revealed these genes and their proteins, which would not have been identified as being involved in R7 cell development using a standard forward genetic screen.

The initial steps of the Sevenless signaling cascade can be assembled from these proteins (**fig. 21.39**). The Boss protein, expressed on the surface of the R8 cell, binds to the Sevenless receptor on the surface of the

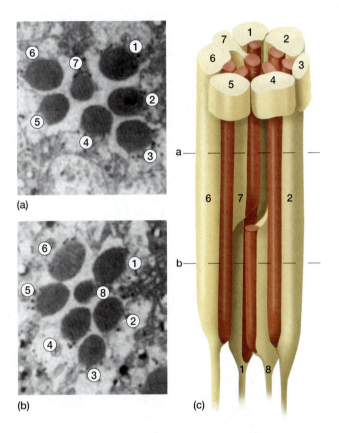

(a)

(b)   (c)

figure 21.37   Organization of the *Drosophila* ommatidium. A transmission electron micrograph through a distal (a) and proximal (b) section of a *Drosophila* ommatidium. At both locations, six peripheral rhabdomeres (r1–6), the dark staining circular structures toward the center of the ommatidium, form a trapezoid around the smaller central rhabdomere. The central rhabdomere is R7 in the distal retina and R8 in the proximal retina. The relative locations of the two sections are shown in the longitudinal view of the ommatidium (c).

figure 21.38   The *sevenless* mutant phenotype. (a) A transmission electron micrograph of a wild-type ommatidium shows six large rhabdomeres arranged in a trapezoid around the smaller central R7 rhabdomere. The *sevenless* (b) and *bride of sevenless* (c) mutants exhibit only the six large rhabdomeres. The missing central R7 rhabdomere causes the outer six rhabdomeres to collapse on the center and lose the trapezoidal arrangement.

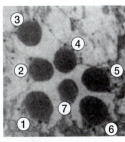

(a) Wild type

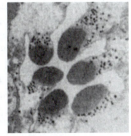

(b) *sevenless*

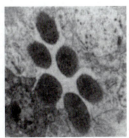

(c) *bride of sevenless*

1 µm

cell that is to become the R7 cell. This binding results in the activation of the Sevenless protein, which phosphorylates itself on the carboxyl terminus located on the cytoplasmic side of the membrane. The phosphorylated Sevenless can be bound by a specific region of the GRB2 protein (SH2 domain). A second region of the GRB2 protein (SH3 domain) binds the Sos protein.

Thus, Sevenless binding the Boss protein results in the Sos protein moving to the intracellular surface of the membrane, where the Ras protein is located. Sos then removes the GDP from the inactive Ras protein (Ras$_i$) and exchanges it for a GTP. This results in the Ras protein becoming active (Ras$_a$).

To confirm that the initial steps to generate an R7 cell require the activation of Ras, the *ras* gene was in vitro mutagenized (see chapter 13) to produce a dominant form of Ras that was always active. This mutant *ras* (*ras^D*) was cloned into a *P* element and transposed into the genome of a *sevenless* mutant (*sev$^-$*) fly. These transgenic flies (*sev$^-$*; *P*[*ras^D*]) contained the R7 photoreceptor cell. This experiment demonstrated that the purpose of the Sevenless protein is to generate an active form of Ras to produce the R7 cell. However, the steps that are required after Ras activation were still unclear.

Clarification of the Steps after Ras Activation

Additional genetic screens were performed to isolate further components in the Sevenless signaling cascade. Again, sensitized screens proved to be effective. One sensitized screen used the *P*[*ras^D*] transgenic flies. In these flies, the *ras^D* transgene is expressed in several different cells in the retina, which results in extra R7 cells being produced in each ommatidium. The extra R7 cells produce a rough eye phenotype (**fig. 21.40b**). If this transgenic fly is mutated so that the extra R7 cells are not present, then a smooth eye phenotype is produced (fig. 21.40c). This **suppressor mutation** makes the mutant transgenic phenotype more similar to the wild-type phenotype.

If the transgenic fly is mutated so that even more R7 cells are present in each ommatidium, then the

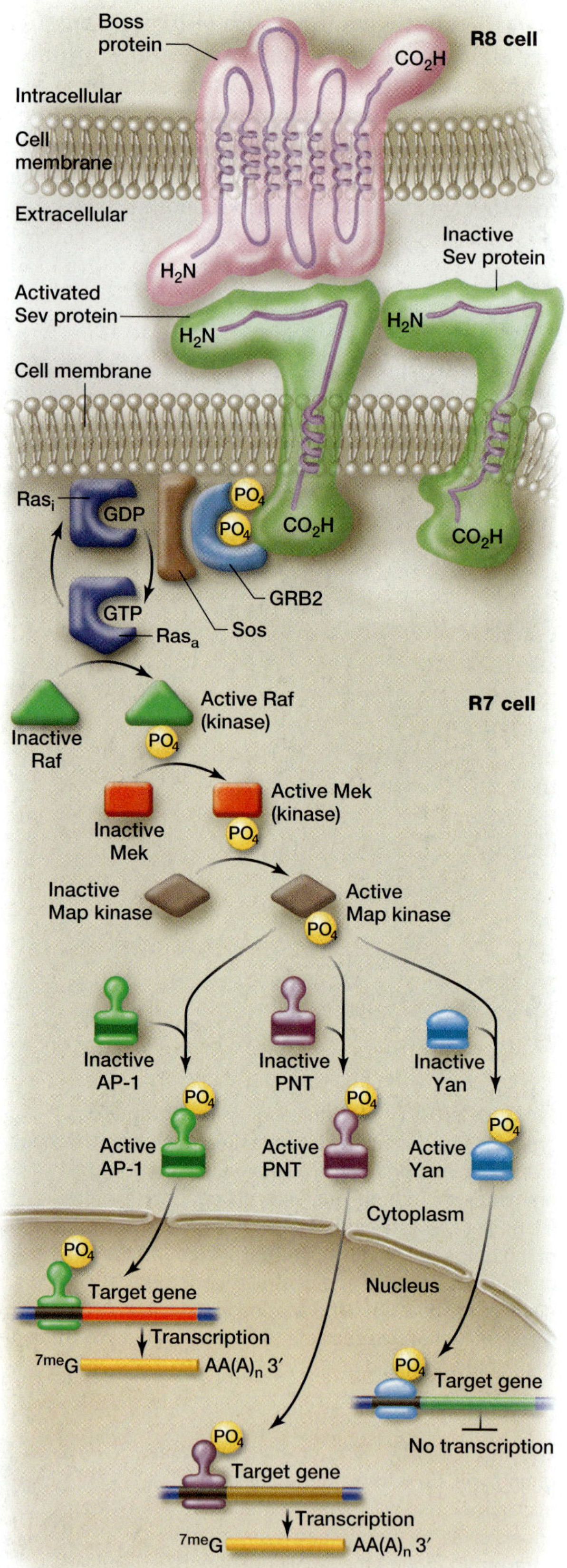

figure 21.39 *Sevenless* signaling cascade. The *Sevenless* signaling cascade begins with the Boss protein, which is expressed on the surface of the R8 cell, binding the Sevenless receptor on the surface of the future R7 cell. This binding stimulates the phosphorylation of the Sevenless protein (PO$_4$ near the carboxyl terminus). The GRB2 protein binds the phosphorylated Sevenless, which is then bound by the Sos protein. Sos catalyzes the exchange of GDP for GTP in the Ras protein. The active GTP-bound Ras (Ras$_a$) initiates a cascade of phosphorylating kinases (Raf, Mek, and Map kinase). Map kinase phosphorylates the transcription factors AP-1, PNT, and Yan. These proteins enter the R7 nucleus, with AP-1 and PNT activating transcription of target genes required for R7 determination and differentiation and Yan repressing a set of target genes.

eye appears rougher (fig. 21.40d). This **enhancer mutation** makes the transgenic mutant phenotype even less like the wild-type phenotype. These suppressor and enhancer mutations should be located in genes that are required for the activation and repression of the R7 signaling cascade.

By screening for suppressor and enhancer mutations in a sensitized background, additional components in the Sevenless signaling cascade were identified (see fig. 21.39). The Ras protein activates a series of kinases (Raf, Mek, and Map kinase). Map kinase activates two transcriptional activators (AP-1 and PNT) and a transcriptional repressor (Yan) all through phosphorylation events.

Therefore, activation of the Sevenless receptor initiates the activation of a series of kinases that in turn affect several DNA-binding proteins. The proteins then activate or repress the transcription of a specific set of genes. Expression of the correct set of genes results in the generation of a R7 photoreceptor cell. In the absence of expressing this set of genes, a homeotic transformation occurs and the cell becomes a cone cell rather than a photoreceptor cell.

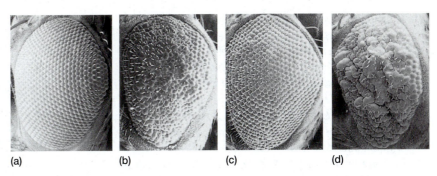

(a)   (b)   (c)   (d)

figure 21.40   Sensitized transgenic background to isolate suppressors and enhancers in the *sevenless* signaling cascade. (a) A transmission electron micrograph of a wild-type compound eye reveals the orderly and smooth packing of the ommatidia. (b) A *P[ras^D]* transgenic fly has a compound eye with a rough surface due to the extra R7 cells in each ommatidia that disrupt the orderly packing. A suppressor of the *P[ras^D]* fly (c) has a smooth looking appearance similar to wild-type, while an enhancer of the *P[ras^D]* fly (d) has an even more extreme rough appearance.

Restating the Concepts

▶ *Drosophila* eye development involves first the determination of the eye tissue and then the determination and differentiation of the different cell types within each ommatidium.

▶ Master regulator genes control the ability of cells to become a specific tissue. Expression of master regulator genes, such as *eyeless*, in the wrong cells can change the future identity of those cells.

▶ Many master regulator genes are expressed in a wide variety of organisms and perform related functions. For example, the *PAX6* gene is required for eye development in humans and can induce eye development in flies.

▶ Determination of specific cell types can involve communication between neighboring cells and a series of intracellular proteins that regulate gene expression. Sensitized genetic screens for enhancers and suppressors can identify genes required for either activation or repression of the signaling cascade.

▶ The *Drosophila* Sevenless signaling cascade is required for the determination of the R7 photoreceptor cell identity. Many of the molecules in this cascade are utilized in other receptor tyrosine kinase cascades.

21.6 Programmed Cell Death in Development

Most cells are determined to become a specific cell type, and in the absence of specific cues, they adopt a different cell type. These homeotic transformations—one cell type into another—are under genetic control. A different path that cells can undertake leads to cell death. **Apoptosis** is the form of cell death that is under genetic control.

Discovery of Apoptosis

The first genes involved in apoptosis were identified in *Caenorhabditis elegans*. Studies of *C. elegans* development revealed that 131 of 1090 cells die in the generation of the 959 somatic cells in the adult hermaphrodite, and 147 of 1178 cells die during the generation of the 1031 cells in the male.

Apoptosis is characterized by condensation of the cytoplasm and nucleus, blebbing of the plasma membrane (a bubbled or blistered appearance), fragmentation of the nuclear DNA, and, ultimately, engulfment of the cell by macrophages (**fig. 21.41**). This is in contrast to necrosis, which exhibits cell swelling, fragmentation of organelles, and rupturing of the plasma membrane (see fig. 21.41). Whereas apoptosis is the result of a specific cascade of proteins that results in the engulfment of a dying cell, necrosis results in cell injury and then death, without a specific series of proteins mediating the process.

The Process of Apoptosis

The "decision" to enter apoptosis usually involves two groups of genes. The first is *bcl-2* and its related genes. These genes encode proteins that associate with the mitochondrial membrane, which is disrupted during apoptosis. Some of the Bcl-2 related proteins (Bcl-2 and Bcl-x$_i$) suppress apoptosis, and other members (Bax, Bad, and Bak) promote apoptosis. The second group of genes involves *p53* and its related genes. The p53 protein is a transcription factor that controls cell proliferation and DNA repair. The role of these proteins in apoptosis is described in greater detail in chapter 22 (see fig. 22.14).

Once the cell is committed to enter apoptosis, a series of cysteine proteases, called *caspases,* are activated. Some caspases act to cleave cytoskeletal proteins, which allow the cell to condense. Other caspases cleave cell adhesion proteins, which allow the macrophage to move around and engulf the dying cell. Still another caspase cleaves an inhibitor of a specific DNase, which then allows the DNase to digest the genomic DNA in the cell. The outcome of apoptosis is to safely remove the cell without significant damage to the surrounding cells.

Many examples exist of the role of apoptosis during development. We now look at two situations: neuronal development, and the generation of digits.

Patterning of Neural Connections

During formation of the vertebrate nervous system, many more neurons are produced than will ultimately be needed. When neurons synapse to the proper target, the remaining neurons are not required and become expendable.

The mechanisms involved in identifying a properly connected neuron versus a neuron that has not generated a correct synapse are unclear; however, when motor neurons make the correct connection to muscles, the muscle produces neurotrophic factors that are required for the survival of the connected neuron. Neurons that do not connect to targets or that are selected against then undergo apoptosis.

Formation of Digits in the Tetrapod Limb

During embryogenesis, tetrapod vertebrate limbs begin by forming a limb bud that develops into a paddle-like extension. Most tetrapod adults have individual digits, however, rather than paddle-shaped limbs.

Two different mechanisms are likely to contribute to the formation of the adult *autopod* (foot, hand, wing, and so on). First, regions that will give rise to individual digits probably have a higher rate of cell proliferation than the interdigital zones, leading to selective outgrowth of the future digits. Second, cells located in the interdigital zones are selectively removed by apoptosis.

Regulation of apoptotic cell death in specific regions during vertebrate limb development leads to formation of different types of adult autopod structures. For instance, in the chick hindlimb, bone morphogenetic proteins (BMPs) act as ligands to stimulate a signaling cascade that initiates apoptosis in the interdigital zone

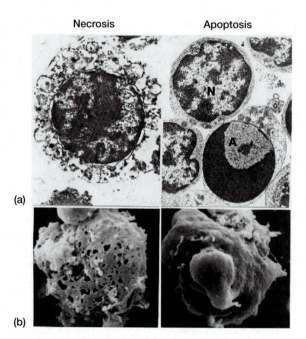

Necrosis   Apoptosis

(a)

(b)

figure 21.41 Necrosis and apoptosis are distinctly different types of cell death. (a) Shown are transmission electron micrographs of cells undergoing necrosis (left) and apoptosis (right). Necrotic cells display swelling, breakdown of organelle function, and fragmentation of the plasma membrane. In contrast, apoptotic cells (labelled A) maintain their plasma membrane integrity and organelle function while rearranging chromatin in a clearly distinguishable pattern. (b) In these scanning electron micrographs, breakdown of the plasma membrane is obvious in the necrotic cell (left), while characteristic 'blebbing' of the membrane, in preparation of producing vesicles, is present in the apoptotic cell (right).

(**fig. 21.42**). In the duck hindlimb, an inhibitor of the BMP signaling cascade leads to lack of apoptosis and formation of interdigital webbing in the adult. If the BMP signaling pathway is blocked in chick development, then interdigital webbing occurs.

Restating the Concepts

▶ Apoptosis is a mechanism of cell death that is controlled by the action of a specific group of genes. Apoptosis involves the condensation of the cytoplasm, fragmentation of the nuclear genome, and engulfment of the cell by macrophages.

▶ Necrosis is a form of cell death that results from injury to the cell. Necrosis does not require the action of a series of proteins.

▶ Apoptosis is a normal process in the development of an organism. It is involved in a number of different developmental events, such as the patterning of connections in the vertebrate neural system and establishing independent digits during autopod development.

figure 21.42 Control of interdigital webbing on duck and chicken hindlimbs. Expression of BMP proteins is similar during embryonic development of both chickens and ducks (a). Gremlin expression apparently suppresses BMP signaling specifically in the duck (b). BMP signaling leads to increased rates of apoptosis in the interdigital regions (c), producing separate digits in the chicken (d). Absence of BMP signaling in the duck prevents apoptosis, leading to interdigital webbing in the adult.

Chicken hindlimb

Duck hindlimb

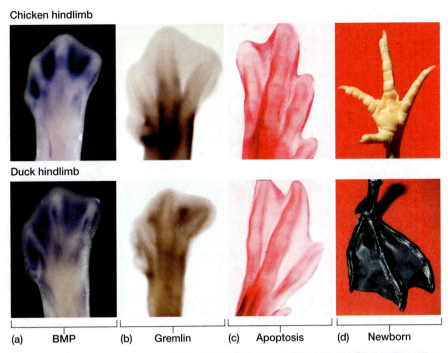

(a)   BMP   (b)   Gremlin   (c)   Apoptosis   (d)   Newborn

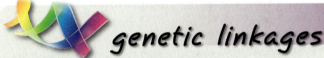

genetic linkages

In this chapter, we discussed a variety of mechanisms that are involved in determining the identity of a cell and ultimately the pattern of cells to generate a tissue and organism. These mechanisms often depend on regulating the expression of a specific set of genes in the proper pattern.

In embryonic development, a set of genes are expressed in the female and deposited in the egg in a highly organized fashion. These maternal mRNAs and proteins must carry the development of the embryo until zygotic gene transcription begins. These maternal-effect mRNAs and proteins generate the anterior–posterior axis and dorsal–ventral axis in the *Drosophila* embryo.

The generation of the *Drosophila* segments and their identity are also controlled by a cascade of DNA binding proteins. We discussed how the expression of a protein, Hedgehog, in one cell acts as a ligand to signal a neighboring cell to initiate a protein expression cascade. We also described the role of homeobox proteins, which contain a helix-loop-helix DNA-binding domain (chapter 17). The universality of these signaling cascades in both invertebrates and vertebrates suggests a fundamental conservation of these processes in development and throughout evolution (chapter 25).

The molecular mechanism for sex determination was described in both flies and mammals. We previously noted that the gender of a fly was a consequence of the X/A chromosome ratio (chapter 4). In this chapter, we discussed the events subsequent to the determination of the X/A ratio, including alternative splicing of genes to produce male- and female-specific mRNAs. The mechanism of splicing was described in chapter 10. These gender-specific mRNAs are then translated into gender-specific proteins that can affect the splicing of subsequent genes.

In contrast, gender determination in mammals initially depends on the presence of the Y chromosome, and more specifically, the *SRY* gene on the Y chromosome.

A third gene regulatory mechanism is observed in dosage compensation in mammals, which requires X-chromosome inactivation in females (chapter 4). This process initially involves the expression of the *XIST* RNA, whose stability is partially regulated by the binding of a complementary *TSIX* RNA. The stable expression of *XIST* RNA results in its binding to the X chromosome from which it was transcribed. This initially inactivates the X chromosome. To stably maintain this inactivation, chromatin modifications must occur to silence gene expression (chapter 17).

We also discussed mechanisms involved in the development of the *Drosophila* compound eye. Eye development begins with the expression of master regulator genes that encode transcription factors. After the eye tissue is determined, then genetic pathways are activated to determine the identity of the various cell types. In the case of the R7 photoreceptor cell, determination requires a signal from the R8 cell to the putative R7 cell and the initiation of a signaling cascade. Using sensitized screens (chapter 20), enhancers and suppressors (chapter 5) can be isolated that have identified different components in the Sevenless signaling cascade.

Development also involves programmed cell death. Apoptosis is a form of cell death that requires the expression of a series of genes. It is a critical component of development because it removes unnecessary cells, such as neurons and the interdigital cells, which have no role in the fully differentiated organism. We will revisit aspects of apoptosis in a discussion of cancer in chapter 22.

Chapter Summary Questions

1. For each term in column I, choose the best matching phrase in column II.

   Column I
   1. Zygote
   2. Maternal-effect genes
   3. Zygotic induction
   4. Zygotic genome
   5. Differentiation
   6. Determination
   7. Intrinsic molecule
   8. Extrinsic molecule
   9. Morphogen
   10. Master regulator gene

   Column II
   A. Factor from outside a cell that determines the cell's fate
   B. Factor within a cell that determines the cell's fate
   C. The process of establishing a cell's fate
   D. Specifies two or more identities along an axis
   E. Controls a program of development
   F. Activation of zygotic gene expression
   G. Genes that contribute to oocyte production
   H. Combined genetic material from both parents
   I. Single-cell embryo
   J. Gene expression influences a cell's phenotype

2. Describe the chromosomal complement of a sperm cell.

3. What is the function of a maternal effect gene?

4. Describe the polarity axes.

5. Describe the differences and similarities between a syncitial and a cellular blastoderm.

6. How does the syncitial blastoderm stage facilitate the use of gradients in axis determination in *Drosophila*?

7. What is a morphogen? How does the Bicoid protein of *Drosophila* function as a morphogen?

8. In *Drosophila*, what four classes of zygotic genes function in anterior–posterior pattern formation? Give the basic role of each.

9. What gene has roles both as a maternal effect gene and as a gap gene in *Drosophila* development?

10. What four genes play a role in controlling *even-skipped* expression in stripe 2 and what are their roles?

11. Would you expect a loss-of-function mutation in the *Krüppel* gene to affect *even-skipped* expression in stripe 2? Why or why not?

12. Describe the role(s) of the Sex lethal protein in *Drosophila* sex determination.

13. Why is hyperactivation of X-linked gene transcription an important part of *msl*-2 function?

14. How does *TSIX* RNA affect the choice of which X chromosome will be inactivated?

15. What are the characteristics of a master regulator gene?

16. In order to respond to a signal, a cell must be competent, or able to recognize the signal and respond appropriately. What cell or cells are competent to receive the Boss signal, and what is the appropriate response?

17. Define apoptosis.

Solved Problems

PROBLEM 1: Relate the homeobox, homeodomain, and homeotic mutation concepts.

Answer: A homeotic mutation is one that causes a cell or compartment of cells to adopt a different fate or identity. The corresponding gene contains a homeobox, which is a 180-basepair sequence that is conserved from flies to humans. The homeobox (DNA motif) encodes the 60-amino acid homeodomain (protein motif), which is a DNA-binding element.

PROBLEM 2: If drugs that inhibit transcription are injected into fertilized eggs, early cleavage divisions and protein synthesis will still occur. Explain.

Answer: Maternal-effect genes are transcribed in the mother, and either the mRNA or protein is deposited in the oocyte to be used during the first several cell divisions after fertilization. Maternal-effect gene products are necessary to allow these early cell divisions to be very rapid, and also allow early cleavage divisions to occur without additional transcription. Likewise, because maternal mRNAs along with cytoplasmic organelles such as ribosomes are present in the oocyte, translation of those mRNAs can occur in the absence of any additional transcription.

Exercises and Problems

18. For a hypothetical maternal-effect gene located on the *Drosophila* X chromosome, embryos affected by the recessive phenotype are unable to hatch and die after 24 h of development. Describe the possible offspring that can be produced by a heterozygous female. For each type of offspring, predict the results of a testcross.

19. To examine intrinsic versus extrinsic signals, one cell of a *Xenopus* zygote is killed at the two-cell stage. After several rounds of cell division, only one side of the embryo develops. Additional experiments show that separation of the two cells allows each cell to produce a small but otherwise normal embryo. Explain this apparent contradiction.

20. You decide to perform a further experiment with dividing early frog embryos.

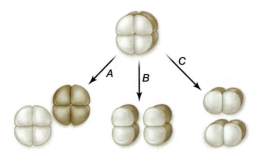

Division of the eight-cell embryo in the plane indicated in *A* results in two small, but otherwise normal embryos. Division in either the plane indicated by *B* or *C* results in two disorganized lumps of tissue.

Discuss what this tells you about extrinsic and intrinsic cues in frog development.

21. Exuperantia (Exu) is a protein required for transportation of *Bicoid* mRNA to the anterior pole of the *Drosophila* oocyte. Under normal conditions, the Exu protein binds both the *Bicoid* mRNA and microtubules in nurse cells, and is subsequently transported to the anterior pole of the oocyte cytoplasm. The Exu activity requires that it be phosphorylated by Par-1, a serine/threonine kinase. Describe the phenotype you would expect for the following *Exuperantia* mutations.

H_2N ▬▬▬▬▬▬▬▬▬▬▬▬▬▬▬▬▬ CO_2H
　　　　A　　*B*　　　*C*　　　*D*　　*E*

Protein domain	Functional significance
A	microtubule binding domain
B	spacer region
C	BLEI: *Bicoid* mRNA binding region
D	Par-1 phosphorylation site
E	binding sites for mRNAs unrelated to *Bicoid*

   a. *Exu* promoter mutation, which prevents the female sex-specific splicing of the *Exuperantia* transcript.
   b. Conservative missense mutation in region *A*.
   c. Nonconservative missense mutation in region *D*.
   d. Nonsense mutation in region *E*.

22. What effect would you expect from a loss-of-function mutation in *cactus*? Would the *cactus* mutant phenotype be rescued by constitutively active Toll? Why or why not?

23. What effect would you expect a *patched* null mutation would have on establishment and maintenance of parasegment boundaries? What effect would a hypermorphic *armadillo* allele have?

24. You are studying *engrailed* gene regulation in *Drosophila* and discover a recessive mutation in a second gene, which results in *engrailed* being expressed in seven stripes rather than the usual 14.
   a. What regulatory gene(s) might control gene expression in seven stripes? How could a mutation in that gene(s) cause this pattern of expression of *engrailed*?

   b. What is the probable mechanism of action of the gene affected in your mutant?
   c. What next step(s) would you take to get more information about the gene in which your mutation is located?

25. A centipede is similar to an insect that has one kind of segment, which includes legs. Based on your understanding of the roles of the homeotic genes in *Drosophila* pattern formation, what pattern(s) of expression might you expect to see if you look at *bithorax* complex gene expression in centipedes?

26. Dragonflies and other four-winged insects are considered to be an entirely separate family of insects from two-winged insects. What specific pattern of homeotic gene mutations could change a dragonfly into a two-winged insect?

27. Your investigations into pattern formation in *Drosophila* lead you to a strain of flies that show apparent homeotic transformations in several segments at once, rather than only one. List at least two possible explanations, and explain one experiment that would allow you to distinguish between them.

28. Mutations in the *Drosophila* homeotic genes cause dramatic structural changes known as homeotic transformations. What factors would affect the likelihood of similar homeotic transformations, which result from deletions of individual homeotic genes, from occurring in mammals?

29. Compare the mechanisms of balancing transcription of X-linked genes in males and females between *Drosophila* and mammals.

30. List three different genes essential for correct X-chromosome inactivation in mammals. What effect would you expect a null mutation in each of these genes to produce? A hypermorph?

31. How do you expect the genes and gene products for blocking factor, CE, and *XIST* will behave in male mammals?

32. You want to study X-chromosome inactivation in mice in more detail. Describe a plan to use knockout technology to disrupt CE function. Describe the cell level phenotype you expect if you are successful.

33. Do you expect to be able to generate a line of mice carrying the mutation mentioned in problem 32? Why or why not?

34. You identify a gene in zebrafish that may be a master regulator of fin development. What experiment(s) would allow you to confirm (or exclude) the possibility that the gene is a master regulator?

35. How could you demonstrate whether *eyeless* and *twin of eyeless* function in the same regulatory pathway or if they operate in different pathways? State expected results for both possibilities.

36. Walter Gehring demonstrated that the mammalian Pax6 protein can function in *Drosophila*. Do you think *eyeless* would be able to rescue *Pax6* mutants in mammals? Why or why not?

37. During the specification of photoreceptor cells in *Drosophila* ommatidial development, the precursors for five cells, R1, R3, R4, R6, and R7, are competent to respond to the Boss ligand. What other factors might determine the fates of these cells?

38. In addition to essential functions during development, apoptosis is thought to play a major role in the pathogenesis of brain injury and neurodegenerative diseases. Identify potential genes or gene products that could be used to combat undesired apoptosis. Comment on the potential limitations for this type of therapy.

Chapter Integration Problem

You want to study the development and genetics of an animal species that you recently discovered. This is a germline cell from this species.

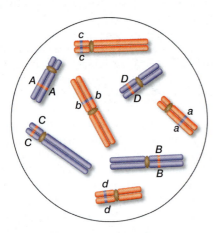

a. Your first hypothesis is that gamete formation follows the same rules as in other animal species with which you are familiar. Diagram the arrangement of chromosomes you expect to see from this cell at metaphase I of meiosis.

b. How many chromosomal combinations are possible in the resulting gametes?

c. Diagram an arrangement of chromosomes in metaphase I of meiosis that would result in a different chromosomal combination in gametes than you showed in part a. Continue your diagram through meiosis II and show the gametes produced. Include any potential crossing over that could occur.

d. Assume these gametes will be generated in a male. Make a similar diagram for female gamete formation. Select one male and one female gamete. Color coding your new chromosomal arrangements appropriately, diagram the chromosomal complement in the resulting zygote after fertilization.

e. Gene *A* represents a maternal effect gene, with the dominant *A* allele possessing a recessive embryonic lethal phenotype. Will your zygote be able to develop into an adult? Explain.

f. Assuming your zygote is female, will she be able to produce viable offspring? Explain.

g. You want to understand the structure and function of gene *A*. Assume this species is similar to fruit flies as far as housing/maintenance goes. Describe a mutagenesis screen(s) that you could use to collect mutants in gene *A*. Explain why you chose the technique(s) that you did.

Do you need additional review? Visit **www.mhhe.com/hyde** for practice tests, answers to end-of-chapter questions and problems, interactive exercises, and animation tutorials all designed to enhance your understanding of key genetic concepts.

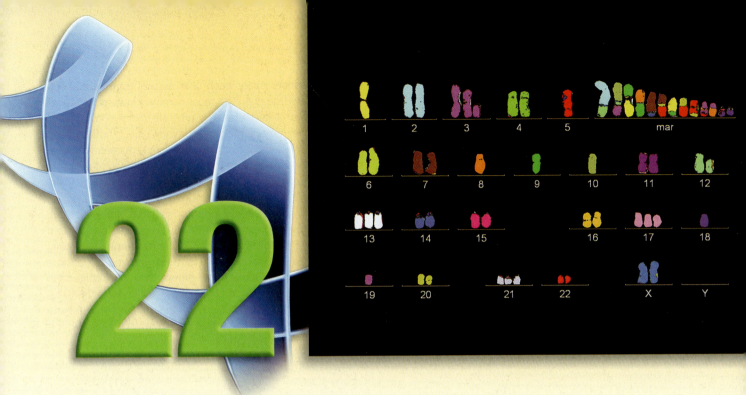

Cancer Genetics

Essential Goals

1. Understand the various stages of cancer progression and the cellular basis of each stage.

2. Describe different types of carcinogens and how they function to modify DNA.

3. Explain how the cell cycle is regulated at checkpoints and understand how various types of mutations disrupt the regulation of the cell cycle.

4. Diagram the various pathways involved in cell death (apoptosis) and describe how apoptosis can be altered during cancer progression.

5. Describe the similarities and differences between oncogenes and tumor suppressor genes, the types of mutations associated with each class, and their roles in cancer development.

6. Compare and contrast the development and types of mutations associated with sporadic cancers and familial cancers.

5. Explain the cellular events involved in tumor progression, the reason behind the two-hit hypothesis, and how this relates to colorectal cancer progression.

Chapter Outline

Photo: Chromosome painting of a karyotype of a cancerous biopsy. The karyotype shows either the loss or addition of several chromosomes, along with many complex chromosome translocations. Chromosome painting results in each chromosome having a unique color.

*I*n the preceding chapters, we discussed a variety of genetic diseases, each of which results from a mutation in a single, specific gene. This simplifies the genetic analysis, because a Mendelian pattern of inheritance (or modified Mendelian ratio) is observed.

Many traits, however, such as height in humans, are the result of many genes acting together. **Quantitative genetics** is the study of how several different genes each contribute to a final, single phenotype. We discuss how quantitative genetic traits are analyzed in chapter 24.

Many genetic diseases in humans are also the result of a combination of mutations in many different genes. Examples of such diseases include heart disease, manic depression, and cancer. To illustrate the genetic complexity of this type of disease, we focus on cancer in this chapter.

Cancer remains one of the most feared diseases in the world. In the United States alone, more than a million new cases of cancer occur each year as well as over 500,000 cancer-related deaths. The most common form of cancer in males in the United States is prostate cancer; in females, breast cancer remains the most common type. In both of these cancers, early detection can permit a variety of treatments that are effective at eliminating the cancer. In contrast, other forms of cancer, such as pancreatic or lung cancer, have limited treatment options and most frequently lead to a very rapid death.

Cancer is not usually inherited within a family. Instead, a **predisposition,** or likelihood, of developing cancer can be inherited for some cancers. Cancer development usually requires the accumulation of several somatic mutations. Some are spontaneous, and others may be induced by environmental factors, such as X-rays or ultraviolet light. As you study this chapter, notice how the genetic mechanisms and analysis of cancer differs from that of a single-gene disease, such as hemophilia (chapter 4).

22.1 What Is Cancer?

Cancer is a group of diseases characterized by uncontrolled cell growth and the spread of abnormal cells. The size of adult tissues is determined by the balance between the number of newly generated cells (requiring *mitosis* and *cytokinesis*) and the number of dying cells (*apoptosis*). The normal balance of cell division and cell death, a type of homeostasis, is tightly regulated by genes that control cell division or cell death.

These genes can be subdivided into two groups: **protooncogenes** encode proteins that stimulate cell division, and **tumor suppressor genes** encode proteins that either slow the cell cycle or induce apoptosis. Mutations that constitutively activate protooncogenes or inactivate tumor suppressor genes act to increase cell numbers and disrupt homeostasis. This increase in cell number, termed **hyperplasia,** is the first step in tumor development and leads to the formation of a primary tumor (**fig. 22.1**).

These initial mutations result in tumor growth, but the tumors are usually benign and normally do not kill the organism. The largest tumor ever removed from a human was a benign ovarian tumor that weighed 303 pounds. The patient, who weighed 180 pounds after surgery, made a full recovery.

These primary tumors often acquire additional mutations, however, either from inappropriate chromosomal segregation or from environmental carcinogens. **Carcinogens** are environmental agents that cause DNA mutations—either in specific genes or changes in chromosomal number or structure, which further affect cell division and lead to cancer progression. This stage of tumor progression varies from cell to cell within the tumor, and whereas most cells die as a consequence of these mutations, a few cells acquire the ability to escape from the primary tumor and colonize distant sites. This process of cell migration is known as **metastasis.** If these cancerous cells adopt the ability to invade healthy tissues, the cancer is defined as **malignant,** and the disease can become life-threatening (see fig. 22.1).

Restating the Concepts

▶ Carcinogens are agents that can create a mutation in DNA that results in cancer development.

▶ Cancer progresses through several different stages. The first mutation results in cancer initiation. Subsequent mutations permit a cell to migrate or metastasize from its original location to a different site. Additional mutations allow the migrating cells to adopt the ability to invade healthy tissue, which is termed malignancy.

22.2 Carcinogens

Less than 10% of all human tumors are hereditary or familial (**fig. 22.2**). Although individuals with a family history of cancer have a *predisposition* or higher risk of developing cancer than the general population, not all members of the family will develop tumors.

In contrast, more than 90% of all human tumors arise spontaneously (see fig. 22.2). Most of these cancers are induced by human-created and natural carcinogens found in air pollution, tobacco smoke, and our food, often as by-products of cooking. A small percentage of these spontaneous tumors are also the result of mutations that arise through the normal DNA replica-

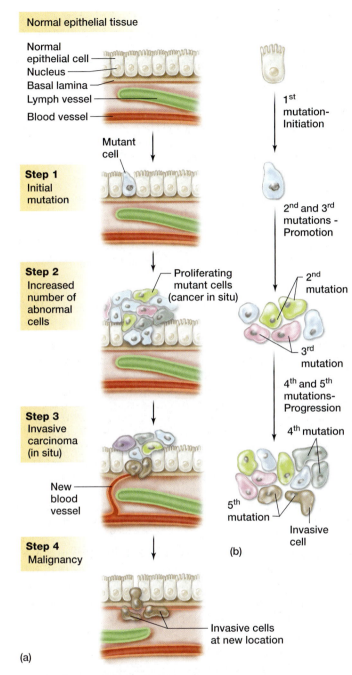

Normal epithelial tissue

Normal epithelial cell
Nucleus
Basal lamina
Lymph vessel
Blood vessel

Mutant cell

Step 1
Initial mutation

Step 2
Increased number of abnormal cells

Proliferating mutant cells (cancer in situ)

Step 3
Invasive carcinoma (in situ)

New blood vessel

Step 4
Malignancy

Invasive cells at new location

(a)

1st mutation-Initiation

2nd and 3rd mutations - Promotion

2nd mutation

3rd mutation

4th and 5th mutations-Progression

4th mutation

5th mutation

Invasive cell

(b)

figure 22.1   Model for cancer progression. (a) A mutation occurs in a cell to initiate cancer progression. The mutant cell possesses an advantage to actively divide (proliferate). As the cells proliferate new blood vessels must grow into the cancer to feed the cells. The proliferating cells may ultimately break through the basal lamina, which removes their physical restriction. These invasive cells may migrate through the lymph system. If they ultimately invade and proliferate at a new site, malignancy is achieved. (b) Each step in the progression of cancer requires the introduction of new mutations. The accumulation of mutations in the cell increases the aggressiveness of the cancer cells to invade and become malignant.

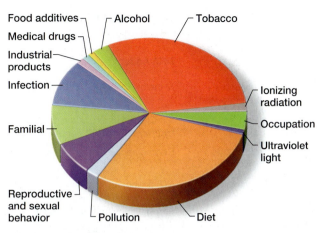

Food additives
Medical drugs
Industrial products
Infection
Familial
Reproductive and sexual behavior
Pollution
Diet
Ultraviolet light
Occupation
Ionizing radiation
Tobacco
Alcohol

figure 22.2   Causes of cancer. A pie chart shows the percentage of cancers that are due to different causes. The major carcinogens are tobacco and the diet.

tion and repair processes, radiation damage, or viral infection. It has been estimated that the human body undergoes approximately 10,000 DNA lesions per day, which are usually repaired. When these mutations are not repaired, the risk of developing cancer increases.

Chemical Carcinogens

Chemical carcinogens fall into three general classes: alkylating agents, aralkylating agents, and arylhydroxylamines (**table 22.1**). These compounds are either intrinsically reactive with DNA or are metabolized in the body to form electrophiles that react with the electron-sharing atoms of the nucleotides in DNA—particularly with the ring nitrogens and oxygen atoms—to form stable altered nucleotides, or **DNA adducts.**

Action of Enzymes on Chemical Carcinogens

Most of these carcinogens are absorbed through the lungs and intestines and travel through the circulatory and lymphatic systems to the liver. In the liver they are metabolized by a group of enzymes, the P450 mixed-function oxidases. These enzymes are members of the cytochrome P450 (CYP) family (**fig. 22.3**).

Action of the P450 oxidases alters the chemical structure of the carcinogens to generate the *proximate carcinogens* that are more water-soluble and, therefore, more readily excreted by the kidneys. Unfortunately, hepatic metabolism also makes the proximate carcinogens more likely to enter target cells, where they can be further metabolized by tissue-specific CYPs to form very active *ultimate carcinogens*. These modified ultimate carcinogens readily form covalent bonds with nucleotides in DNA, creating DNA adducts.

The action of these enzymes to produce ultimate carcinogens explains why, in addition to causing lung

table 22.1 Mutations Caused by DNA-Damaging Agents

Modifying Agent	DNA Adduct	Old Base Pairing	New Base Pairing	Consequence
Alkylating agents (N-methyl-N-nitrosourea)	N-methyl-7-methyl-deoxyguanosine (small adduct)	(Me-)G/C	A–T	Transition
Aflatoxin B1	8,9-Dihydro-8-(N7-deoxyguanyl)-9-hydroxyaflatoxin B1 (large adduct)	(Afl)-G/C	T–A	Transversion
Aralkylating agents (Benzo[a]pyrene)	Benzapyrene-N2-deoxyguanosine (large adduct)	(BP)-G/C	T–A	Transversion
Arylhydroxylamines (2-naphthylamine)	N(2) aminonaphthalene-deoxyguanosine (large adduct) and/or 8-oxo-7,8-dihydroxy-deoxyguanosine (small adduct)	(NA)-G/C (8-OH)-G/C	T–A T–A	Transversion Transversion
Base oxidation	8-hydroxydeoxyguanosine (small adduct)	(8-OH)-G/C	T–A	Transversion
UV light	Thymine dimers (TT)	—	—	Single-base deletion (frame shift)
Ionizing radiation	Strand breaks	—	—	Deletion

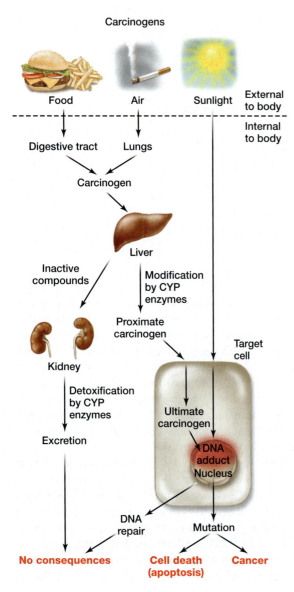

Carcinogens

Food   Air   Sunlight   External to body

Internal to body

Digestive tract   Lungs

Carcinogen

Liver

Inactive compounds

Modification by CYP enzymes

Kidney

Proximate carcinogen

Target cell

Detoxification by CYP enzymes

Excretion

Ultimate carcinogen

DNA adduct
Nucleus

DNA repair   Mutation

No consequences   Cell death (apoptosis)   Cancer

figure 22.3 Metabolic generation of a carcinogen. Environmental carcinogens (food, air, sunlight) can enter the body through a variety of mechanisms, including the lungs and digestive tract. Many carcinogens are detoxified by enzymes, including cytochrome P450 and related enzymes. If properly detoxified, the substance is excreted without consequence to the organism. In some cases, the modified carcinogen (proximate carcinogen) is taken up by cells, where it is further modified into the ultimate carcinogen that generates DNA adducts. The DNA adducts create mutations in the DNA that are either repaired or create a stably inherited mutation. A mutation can lead to either cancer or apoptosis (cell death). While cell death does not appear to be a desirable consequence, the loss of a small number of cells is often preferable compared to those cells proliferating and developing a tumor.

cancer, smoking increases the risk of other cancers, such as breast and prostate cancers. The proximate carcinogens from smoking are transported in the circulation from the liver to target tissues, where they are converted to ultimate carcinogens.

Consequences of DNA Adduct Formation

Formation of DNA adducts significantly alters the secondary structure of the DNA and can lead to abnormal base pairing (chapter 18). The nature of the adducts varies depending on the carcinogen involved. The adduct may be very large (as in the case of the carcinogen benzo[a]pyrene, described in the following section) or simply the addition of a single methyl group (as in the case of N-methyl-N-nitrosourea).

These adducts are usually recognized by DNA repair enzymes and excised before DNA synthesis occurs; however, if the adduct is not repaired, DNA polymerase misreads the modified base and inserts an incorrect nucleotide into the newly synthesized DNA strand. In most cases this produces a point mutation that is inherited by the daughter cell (see table 22.1).

The consequence of the mutation depends on where it occurs within the genome. The vast majority of point mutations do not affect the phenotype of the cell because they may occur in intergenic regions, within the introns of a gene, within the third nucleotide of a codon (wobble position), or within a codon resulting in a conservative amino acid substitution. Mutations that occur in exons and that alter the amino acid sequence of the protein frequently alter the activity of the protein. If the protein is one that regulates either mitosis or apoptosis, an increase in cell number and the hyperplasia phenotype may result.

Environmental Carcinogens

Many, but not all, carcinogens are human-created. Exposure to environmental carcinogens increases the incidence of many different kinds of cancer. Benzo[a]pyrene, for example, was first identified as the carcinogen that induced scrotal cancer in chimney sweeps (**fig. 22.4**). This chemical was also found, along with dimethylbenzanthracene and other polyaromatic hydrocarbons (PAH), in cigarette smoke and charcoal-grilled meats. In addition to the increased risk of lung cancer, exposure to benzo[a]pyrene and dimethylbenzanthracene increases the risk of breast and prostate cancer. Tobacco smoke, which accounts for approximately 30% of all human cancer deaths, is the most important carcinogen in our environment.

Arylhydroxylamines, such as 2-naphthylamine (see fig. 22.4), also found in tobacco smoke, are known to produce bladder cancer predominantly. In this case, the liver conjugates the 2-naphthylamine with glucoronate, forming a soluble proximate carcinogen compound that is excreted into the urine. In the bladder, however, the glucoronide conjugate enters the epithelial cells that line the bladder and forms both large and small DNA adducts that both result in G–C to T–A transversions.

Some naturally occurring compounds also form DNA adducts. Aflatoxin B1 (see fig. 22.4) is a natural product found in the common mold, *Aspergillus flavus*, which contaminates grains, nuts, and vegetables in some areas of the world, particularly southeast Asia and central Africa. Aflatoxin B1 is metabolized by at least three CYP enzymes. It forms DNA adducts by reacting with the nitrogen at position 7 in guanine, forming 8,9 dihydro-8-(N7-deoxyguanyl)-9-hydroxyaflatoxin B1, which leads to hepatocellular carcinoma (liver cancer).

Consumption of red meat and animal fat have also been implicated in specific forms of cancer, such as colon, breast, and prostate cancers. It is unclear how these foods facilitate cancer progression.

Not all compounds that produce tumors form DNA adducts. Asbestos, for example, induces tumors primarily after being taken up by the lung epithelial cells after inhalation. Asbestos then physically disrupts normal chromosome segregation, producing chromosomal aberrations that lead to mesothelioma, a particularly aggressive form of lung cancer.

Radiation

Radiation is also known to induce tumor formation. Ultraviolet radiation, particularly UV-B (280–320 nm), induces the formation of pyrimidine cyclobutane dimers (predominantly thymine dimers) between adjacent pyrimidines in DNA (**fig. 22.5**). These dimers substantially distort the DNA double helix, and if not repaired by excision repair (see chapter 18), induce either G–C to A–T transitions or frameshift mutations.

Because UV radiation does not penetrate beyond the first few layers of the skin, tumor formation induced by ultraviolet light is limited to skin cancers. UV radiation has been implicated in both squamous cell carcinoma and basal cell carcinoma, both of which are relatively benign tumors and readily treatable.

A third form of skin cancer, malignant melanoma, is also induced by UV radiation and is particularly aggressive and difficult to treat. The development of the malignant melanoma also requires other factors, including the intensity of light exposure and degree of sunburn, probably due to the release of cytokines from immune cells in the skin that are stimulated by severe sunburn.

Ionizing radiation, including X-rays and γ-radiation (as one would see in a nuclear detonation) produce oxygen-based free radicals in cells and oxidized nucleotides, such as 8-hydroxyguanine. If not repaired, 8-hydroxyguanine is mistaken for thymine by DNA polymerase, producing a transversion (see chapter 18).

Ionizing radiation also produces single-strand breaks (SSBs) and double-strand breaks (DSBs) in DNA.

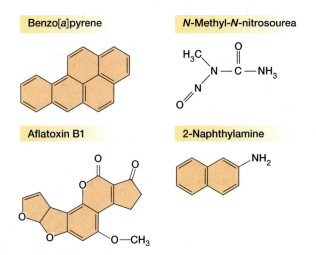

figure 22.4 Structure of four carcinogens, benzo[a]pyrene, aflatoxin B1, N-methyl-N-nitrosourea, and 2-naphthylamine.

Although SSBs and DSBs can be repaired within the cell, the accuracy of these repair systems depends on a number of factors, including the number of DSBs induced at one time and the stage of the cell cycle. DSBs also often result in chromosomal rearrangements, such as deletions, during cell division. Most cells with severely damaged genomes undergo apoptosis, with only a small percentage surviving with compromised genomes. Cells with large chromosomal aberrations are often tumorigenic (tumor causing) because the regulation of the cell cycle is significantly disrupted.

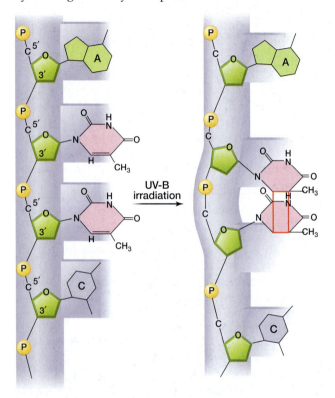

figure 22.5 Exposure of cells to ultraviolet light results in the formation of cyclobutane dimers between adjacent thymines. Notice how the cyclobutane dimer (red structure on right) results in a structural change (bulge) in the phosphodiester backbone. The repair of cyclobutane dimers is described in chapter 18.

Viral Carcinogens

Viral infections probably account for 5–10% of all cancers worldwide. In many cases, after the viral genome integrates into the host genome, viral proteins are produced that interfere with cell cycle control in the infected cell (**table 22.2**). The infected cell then has a growth advantage over the neighboring cells, inducing hyperplasia and enhancing the propagation of the infected cell. Some viruses, such as Epstein–Barr virus, are associated with genomic DNA translocations that result in the overproduction of proteins that stimulate the cell cycle. Other viruses, such as the hepatitis B virus, induce chronic inflammation that changes the growth factor environment of the cell, which inappropriately stimulates the cells to divide.

Human Papilloma Virus (HPV)

There are more than 100 strains of human papillomavirus (HPV), most of which produce nothing more offensive than warts. Two strains of HPV—HPV 16 and HPV 18—have been implicated in cervical cancer, however. These two strains are sexually transmitted, and they infect the cells of the cervix. Their transmission method explains why one of the risk factors for cervical cancer is the number of sexual partners.

The mechanism by which HPV 16 and HPV 18 induce cervical cancer involves two viral proteins and two cellular proteins. The HPV genome encodes the E6 and E7 proteins (**fig. 22.6**), whereas the retinoblastoma protein (Rb1) and p53 are encoded in the nuclear genome. (The Rb1 and p53 proteins are critically important for the regulation of the cell cycle and are described in detail later on. You may recall that the disease retinoblastoma was discussed in chapter 18.) The E6 protein binds and inactivates the p53 protein, and E7 binds and inactivates the Rb1 protein. Loss of p53 and Rb1 activity in the HPV 16- or HPV 18-infected cell initially results in hyperplasia and subsequently in dysplasia and cancer, because the viral proteins interfere with cell cycle regulation in the host cell.

table 22.2   Interaction Between Viral Gene Products and Host Proteins That Alter Cell Cycle Regulation

Virus	Viral Gene Product	Binds Host Protein	Tumor	Class of Virus
HPV 16	E6 and E7	p53 and Rb1, respectively	Cervical cancer	Papillomavirus
Adenovirus	E1A and E1B	Rb1 and p53, respectively	Various rodent tumors, but no human tumors	Adenovirus
Simian virus 40 (SV40)	Large T antigen	p53 and Rb1	Mesothelioma (associated with asbestos?)	Polyomavirus
Epstein–Barr virus (EBV)	EBNA-2	Transcription factors that induce c-*myc*	B cell leukemias, Burkitt lymphoma	Herpesvirus
Hepatitis B virus (HBV)	?	?	Hepatocellular carcinoma	Hepadnavirus

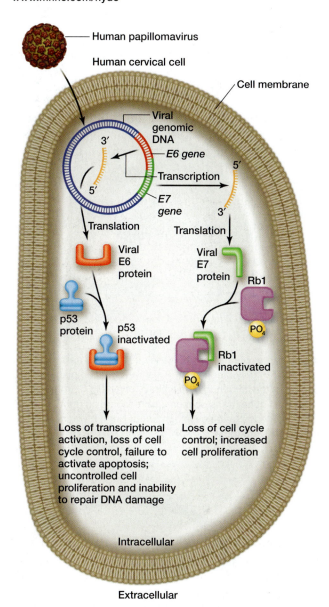

figure 22.6 Human papillomaviruses cause cancer by disrupting the function of human-encoded proteins. Some forms of human papillomaviruses (HPV 16 and HPV 18) encode the E6 and E7 proteins. The E6 protein binds the p53 tumor suppressor protein, while E7 binds the phosphorylated Rb1 (Retinoblastoma) tumor suppressor protein. The p53 protein is a transcription activator, regulator of the cell cycle, and activator of apoptosis. Thus, E6 binding of p53 blocks cell death and increases cell proliferation. Because the Rb1 protein also regulates the cell cycle, E7 binding of Rb1 leads to increased cell proliferation.

Adenovirus and Polyomavirus

Several other oncogenic DNA viruses, including adenoviruses and polyomaviruses, such as simian virus 40 (SV40), also encode proteins that disrupt cell-cycle regulation and initiate tumor formation. The adenoviruses encode two proteins—E1A and E1B—which bind and interfere with Rb1 and p53 activity, respectively.

In SV40, a single protein, large T antigen, binds to both p53 and Rb1 and blocks their regulation of the cell cycle. Like HPV 16 and 18, this initially induces hyperplasia, but it also leaves the cells more susceptible to further genetic damage because inactivation of p53 also affects the DNA repair processes (discussed later in this chapter).

Epstein–Barr Virus

Epstein–Barr virus (EBV) predominantly infects primary B cells and induces hyperplasia. Infection of epithelial cells with EBV increases the expression of BCRF-1, a growth factor for B cells that also inhibits the activation of cytotoxic T cells. This growth factor stimulates the cell division of B cells and protects them from immune surveillance by the cytotoxic T cells.

EBV also encodes two proteins—EBNA-1 and EBNA-2—which are required for infection and survival of the B cells. In particular, EBNA-2 interacts with several transcription factors to increase the expression of cellular Myc (c-Myc), a protein that controls the expression of many of the regulatory proteins in the cell cycle (discussed later in this chapter).

In areas of the world where malaria is endemic, B-cell proliferation is chronically stimulated in individuals infected with the malarial parasite. Occasionally, these infected individuals will also undergo a reciprocal translocation between the *MYC* locus on chromosome 8 and the immunoglobulin gene locus on chromosome 14 in the B cells that are already being stimulated to divide. This translocation places the *MYC* gene under the transcriptional control of the powerful immunoglobulin heavy-chain promoter (**fig. 22.7a**). The result is the production of very high levels of c-Myc protein throughout the cell cycle. Because c-Myc regulates the transcription of the cyclin genes that control progression through the cell cycle, the cells divide very rapidly and produce a large, benign tumor called Burkitt lymphoma (fig. 22.7b), named after the man who first described the disease in East Africa.

Hepatitis B and C Viruses

Another cancer associated with viral infection is liver cancer. A very strong correlation has been observed between hepatitis B and hepatitis C infection and hepatocellular carcinoma. This disease is usually preceded by *cirrhosis,* or scarring of the liver, which can be caused by viral infection, alcohol abuse, or diseases that result in chronic inflammation of the liver.

When the liver epithelial cells, hepatocytes, are exposed to cytokines (growth factors secreted by inflammatory cells) over a long period, this excess mitogenic signaling drives the hepatocytes through the cell cycle and overrides the normal cell-cycle control mechanisms. Data have also indicated that p53

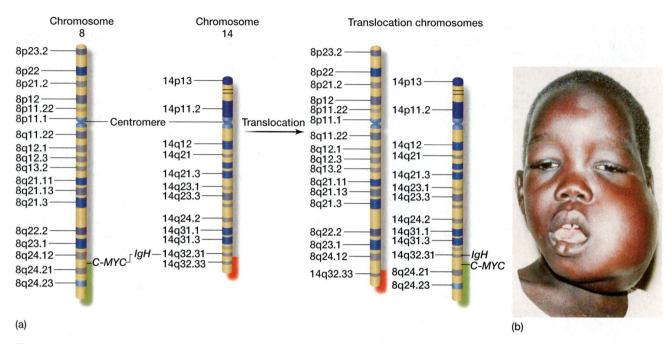

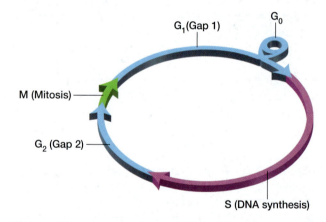

(a)

(b)

figure 22.7   Chromosomal translocation that leads to Burkitt lymphoma. (a) Burkitt lymphoma is occasionally associated with a reciprocal translocation between 8q24 (highlighted in green) and 14q32 (highlighted in red). This translocation places the human *MYC* gene from chromosome 8 under the transcriptional control of the immunoglobulin heavy chain (*IgH*) promoter region on chromosome 14, which results in increased transcription of the *MYC* gene (translocation chromosome on right). (b) The increased expression of the Myc protein increases the rate of cell division and leads to the development of Burkitt lymphoma. Shown is a boy with a large, benign facial tumor that is associated with Burkitt lymphoma.

is inactivated at a later stage in many hepatocellular carcinomas, suggesting that there are several stages of tumor progression in the liver and that the cells accumulate genetic damage (mutations) over time if the DNA repair processes are compromised.

 Restating the Concepts

▶ Chemicals, both natural and human-made, can induce tumor formation. These chemicals are metabolized in the liver into proximate carcinogens and then in target cells into ultimate carcinogens. DNA adducts are formed by these carcinogens that result in the incorporation of a mutation.

▶ Examples of environmental carcinogens are chemicals in tobacco smoke and in charred meat, and ultraviolet light (radiation).

▶ Viral infection can also lead to tumor formation. Often, virus-encoded proteins that inactivate the cellular Rb1 and p53 proteins are involved. Rb1 and p53 regulate the cell cycle, and loss of their activity results in increased cell proliferation.

22.3 The Cell Cycle

Cell proliferation is normally a tightly regulated process that involves an orderly progression of steps during

figure 22.8   The stages of the cell cycle. The cell cycle is composed of a DNA replication phase (S), two gap phases (G_1 and G_2), and mitosis (M). Some cells exit the cell cycle into a phase termed G_0.

which the cell integrates internal and external signals while monitoring cell size and DNA integrity. The cellular processes and biochemical events that constitute the cell cycle are highly conserved in eukaryotic cells.

Stages of the Cell Cycle

M phase is the portion of the cell cycle in which cell division actually takes place (**fig. 22.8**). Mitosis and cytokinesis are responsible for the appropriate segregation of chromosomes, cytoplasm, and other organelles into the daughter cells. This is the phase of the cell cycle during which changes in cellular structure are visible.

The remaining phases of the cell cycle are often referred to collectively as **interphase** (see fig. 22.8). During G_1 **phase** (first gap phase), the cell synthesizes the components required for cell division.

In many tissues, terminally differentiated cells exit the cell cycle and do not reinitiate DNA synthesis or cell division. This state is often designated as the G_0 **phase** of the cycle (see fig. 22.8). Most somatic cells in the body are in either G_0 or G_1 phase and are not actively involved in cell division. It is during these two phases of the cell cycle that cells perform specialized, tissue-specific functions.

DNA is replicated during **S phase** (synthesis phase). During G_2 **phase** (second gap phase), the cell completes the synthesis of nucleosomes and the packaging of the newly synthesized DNA into chromatin, prior to the initiation of mitosis.

DNA replication and chromosomal segregation are confined to discrete parts of the cell cycle (the S and M phases, respectively). The transitions from the gap phases to either S or M phase are tightly regulated to ensure that the processes are not initiated unless they can be completed without complication. For example, cells typically respond to proliferative and antiproliferative signals (growth factors and cytokines) during the G_1 and G_2 phases. These signals tell a cell whether to stop within the G_1 or G_2 phase without interrupting the critical events of DNA replication or chromosome segregation.

The orderly execution of cell-cycle events results from a series of interdependent relationships that require the completion of one event is required for the beginning of the next.

Regulation of the Cell Cycle

The timing of a cell's division depends on a complex array of proteins in the nucleus and cytoplasm, which are modified as a result of intracellular signaling. Two major mechanisms are responsible for regulating the cell cycle. The first involves the formation of heterodimers between cyclins and the cyclin-dependent kinases. The second responds to extracellular growth factors that activate the Ras protein and an intracellular signaling cascade.

Cyclin/Cyclin-Dependent Kinase Heterodimers

The *cyclins* are a class of proteins that increase and decrease their level of expression during specific stages of the cell cycle (**fig. 22.9**). There are four classes of cyclins: A, B, D, and E. Each functions in a specific stage of the cell cycle (**fig. 22.10**). Each cyclin activates a specific **cyclin-dependent kinase** (**CDK, table 22.3**), which is expressed at nearly constant levels throughout the cell cycle.

Monomeric CDK subunits are devoid of enzymatic activity and require forming a heterodimer with a spe-

cific cyclin protein to become activated. Binding of the regulatory cyclin subunit also determines the specificity of the substrate that is phosphorylated. Degradation of the cyclin protein leaves the CDK partner as a nonfunctional monomer until the expression of the cyclin increases at the same point in the next cell cycle.

Transit through interphase is regulated at several checkpoints:

- A restriction point in G_1, where the cell decides to enter the quiescent G_0 phase or continue through the cell cycle,

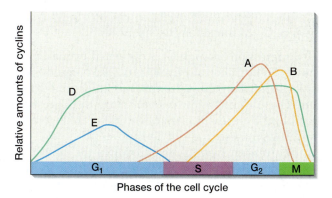

figure 22.9 Progression through the cell cycle is regulated by the transient expression of the cyclin proteins. Each of the cyclin proteins (cyclins A, B, D, and E) accumulates at specific stages of the cell cycle and then is degraded. The four phases of the cell cycle are shown along the bottom.

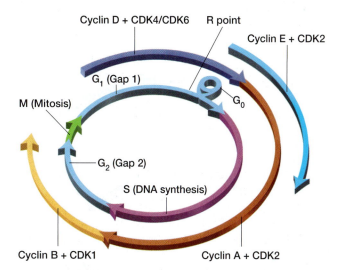

figure 22.10 Cyclins and cyclin-dependent kinases regulate the progression through the cell cycle. Each of the transiently expressed cyclin proteins (A, B, D, and E) form heterodimers with specific cyclin-dependent kinases at different points in the cell cycle. The formation of each specific heterodimer is required to regulate the progression through the G1/S transition (Cyclin E and CDK2), through S phase (Cyclin A and CDK2), the G2/M transition (Cyclin B and CDK1), and through the G1 restriction (R) point (Cyclin D and CDK4/CDK6) of the cell cycle.

table 22.3 Cyclin and Cyclin-Dependent Kinase Heterodimeric Partners

Cell Cycle Checkpoints	Cyclin	CDK Partner
G_1 restriction	Cyclin D	CDK4 and/or CDK6
G_1/S transition	Cyclin E	CDK2
S phase	Cyclin A	CDK2
G_2/M transition	Cyclin B	CDK1

- The G_1/S transition, and
- The G_2/M transition (see fig. 22.10).

Progression through G_1 is regulated primarily by the Cyclin D/CDK4 and Cyclin D/CDK6 heterodimers that respond to signaling from extracellular growth factors. These cyclins allow cells to transit the G_1 restriction point if the extracellular conditions are favorable for DNA synthesis and cell division.

Progression from G_1 to S requires the participation of the active Cyclin E/CDK2 heterodimer, which commits the cell to DNA synthesis in cooperation with Cyclin A/CDK2.

The final checkpoint, in metaphase, is regulated by Cyclin B/CDK1, which ensures that chromosome segregation is not initiated until all of the daughter chromosomes are aligned appropriately at the metaphase plate and spindle formation is complete. If these criteria are met, chromosomes then segregate to the opposite poles to form the daughter cells.

The kinase activity of each cyclin/CDK complex is also modulated by cyclin kinase inhibitory (CKI) proteins, particularly p21 and p27, which are induced in response to DNA damage. The CKI proteins bind reversibly to the activated cyclin/CDK complexes to arrest the cell cycle before the start of S phase. The expression of the CKIs is regulated by p53, a transcription factor that is elevated in response to cellular stress and DNA damage. Once the damage has been repaired, p53 is degraded and the transcription of CKIs ceases, allowing the cell cycle to resume.

Elevated p53 expression for prolonged periods (in the event of substantial DNA damage) leads to the induction of apoptosis, and the death of the damaged cell. Because of its fundamental importance in ensuring the integrity of the DNA template prior to the initiation of DNA synthesis, p53 is often referred to as the guardian of the genome.

Regulation of the transition through G_1 is also accomplished by action of the Rb1 (retinoblastoma) protein (**fig. 22.11**). The progression through the G_1 phase and through the G_1/S transition is regulated by the sequential phosphorylation of Rb1 by the active Cyclin D/CDK4 and Cyclin E/CDK2 heterodimers. This releases the E2F protein from Rb1. E2F is a transcription factor that is required for the transcription of 20–30 genes (including *cyclin A*) that are necessary for progression through the G_1/S transition. Mutations in Rb1 that cannot bind E2F

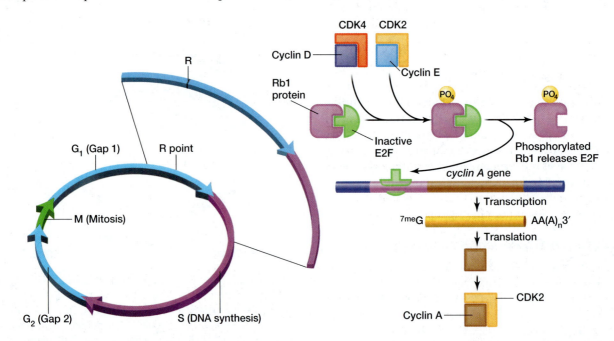

figure 22.11 Role of Rb1 protein in cell cycle progression. In the G_1 phase of the cell cycle, the retinoblastoma protein (Rb1) binds and inactivates E2F. At this same time, the Cyclin D/CDK4 and Cyclin E/CDK2 heterodimers are present. Both of these heterodimers can phosphorylate the Rb1 protein, which releases the E2F transcription factor. The freed E2F activates transcription of a set of genes (including *cyclin A*) that are required for progression through the cell cycle. The E2F-activated expression of the Cyclin A protein leads to the binding of CDK2, which drives the cell through S phase.

lead to the continuous transcription of these genes, and rapid progression through the cell cycle.

Ras-Mediated Intracellular Signaling Cascade

In most tissues, cells respond to growth factors such as IGF (insulin-like growth factor), EGF (epidermal growth factor), FGF (fibroblast growth factor) and PDGF (platelet-derived growth factor), which may exist in their immediate extracellular environment. These growth factors are synthesized by neighboring cells in the tissue and provide a means of communication between different cell types. This mechanism is referred to as **paracrine signaling.**

Epithelial cells may also respond to cytokines, extracellular signaling molecules released by cells of the immune system during inflammation. The tissue-specific response to growth factors and cytokines is determined by which receptors are present on the surface of the cells. Each growth factor and cytokine is bound by a different **transmembrane receptor;** on binding with the factor on the outside of the cell, the receptor transmits a signal into the interior of the cell. Many of these growth factor receptors are tyrosine kinases that trigger intracellular kinase cascades.

One of the most prominent intracellular signaling pathways is the Ras-mediated signal transduction cascade (**fig. 22.12**). There are three related Ras proteins: H-Ras, K-Ras and N-Ras. These Ras proteins are **GTPases,** low-molecular-weight proteins that bind and then hydrolyze guanosine triphosphate (GTP). The GTP-bound form of the GTPase is active and then becomes inactive after the GTP is hydrolyzed to GDP.

Through the steps of the cascade, Ras becomes bound to GTP and activates a protein called Raf kinase. The Raf kinase phosphorylates a variety of downstream signaling molecules, including a series of **mitogen-activated protein kinases (MAPKs).** The phosphorylated MAPKs enter the nucleus and activate a variety of transcription factors that are required for the transcription of genes that regulate cell-cycle progression, including Cyclin D (which subsequently phosphorylates Rb1) and Myc.

Ras activation also stimulates a second pathway, the Akt/PKB (Protein kinase B) pathway, which is essential for cell survival. We describe this pathway later in this chapter.

The *MYC* gene encodes a member of the *basic helix-loop-helix leucine zipper* (bHLH-Zip) family of transcription factors that regulate the transcription of several genes that are essential for cell survival and cell-cycle progression. We previously discussed the action of the *c-myc* gene in chapter 17. The Myc protein forms heterodimers with several other proteins, including Mad and Max (see chapter 17). Depending on the partner that Myc forms a dimer with, transcription is either activated

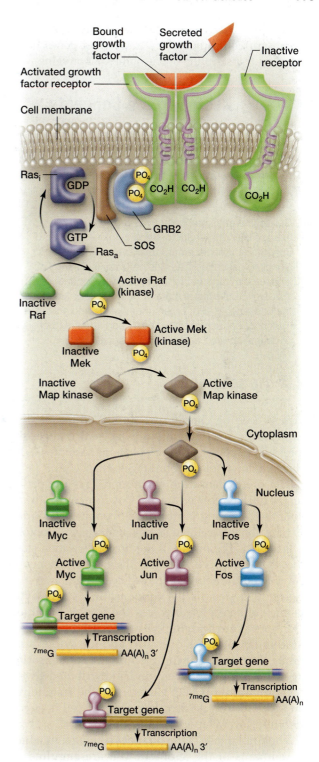

figure 22.12 The Ras intracellular signaling cascade. Binding of growth factors to specific transmembrane receptors initiates a series of protein interactions that cause Ras to bind to GTP. The activated GTP-bound Ras initiates the sequential phosphorylation and activation of three kinases, Raf, Mek, and Map kinase. The final tyrosine kinase, Map kinase, then enters the nucleus and phosphorylates several transcription factors, including Myc, Jun, and Fos, which are required for the expression of cell cycle genes.

or repressed. In the context of the cell cycle, Myc induces the transcription of the *cyclin D* and *cyclin E* genes, while repressing transcription of one of the CKI proteins.

Restating the Concepts

▶ The cell cycle is regulated through a series of checkpoints to ensure that the cell is ready to proceed to the next stage. These checkpoints are controlled by a group of proteins, cyclins, which interact with cyclin-dependent kinases to activate the kinase activity.

▶ Cell proliferation can be activated by a variety of extracellular growth factors and cytokines that bind transmembrane receptors to activate intracellular signaling cascades. The Ras pathway is a series of kinases that respond to the active (GTP-bound) and inactive (GDP-bound) state of Ras. The final kinase in the cascade, MAPK, enters the nucleus and activates transcription factors that in turn bring about expression of genes required for cell cycle progression.

22.4 Apoptosis

In the preceding chapter, we described the process of **apoptosis,** a cell death pathway that requires the expression of specific cellular proteins. Apoptosis can be broken down into several stages (**fig. 22.13**).

The **precondensation stage** represents the period after the cell has received a signal to induce cell death and before there are any visible signs of apoptosis. During this time, the intracellular signals that trigger apoptosis are activated. The length of the precondensation stage varies from cell to cell, reflecting the growth factor environment and the nature of the cell death signal.

The next stage is the **condensation stage.** Cytoplasmic condensation involves the loss of the interactions between the dying cell and its neighbors as the extracellular matrix is degraded and the cytoplasmic volume decreases. During the **nuclear condensation stage,** the DNA is cleaved and redistributed to the nuclear margins. This is followed by the **fragmentation stage,** during which the apoptotic cell is subdivided into several apoptotic bodies.

The final phase is the **phagocytic stage,** in which the remnants of the apoptotic cell are engulfed by neighboring cells. This results in the death of a cell without producing cellular debris that can negatively affect the remaining cells.

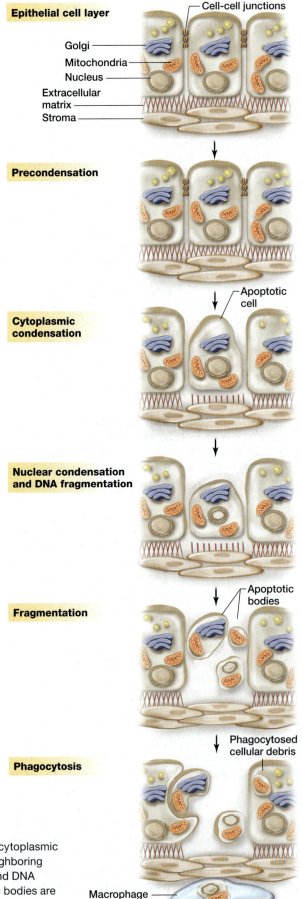

figure 22.13   Stages of apoptosis. First, the cell undergoes cytoplasmic condensation, which causes it to lose contact with most of the neighboring cells. The apoptotic cell then experiences nuclear condensation and DNA fragmentation. After DNA fragmentation, the fragmented apoptotic bodies are phagocytosed by either macrophages or neighboring cells.

Induction of Apoptosis

Cells initiate apoptosis because either paracrine survival signals (from growth factor signaling) are lost or extrinsic or intrinsic cell death signals appear. Extrinsic factors are those that come from outside the cell, whereas intrinsic factors occur internally (chapter 21). The increase in cell number seen in many slow-growing cancers, such as prostate cancer, is caused by defects in the activation of the intrinsic cell death pathway that is regulated primarily by the mitochondria.

The mitochondrion serves as a very effective sensor of the general health of the cell. The intrinsic pathway can be triggered by the loss of extracellular growth factor signaling or an increase in intracellular stress or DNA damage (fig. 22.14).

First, electrical potential is lost across the inner mitochondrial membrane, which is referred to as the **mitochondrial membrane permeability transition.** This is triggered by a Caspase-8-mediated cleavage of the monomeric Bax protein to t-Bax, or truncated Bax (see fig. 22.14). The t-Bax protein inserts into the outer mitochondrial membrane, forming pores that release cytochrome c from the mitochondria.

In the cytoplasm of the cell, cytochrome c binds Procaspase-9 and the Apaf-1 protein to form the *apoptosome*, leading to the proteolytic activation of Caspase-9. The activated Caspase-9 in the apoptosome cleaves Procaspase-3 to produce active Caspase-3, which activates DNA fragmentation.

Regulation of Apoptosis

Two biochemical mechanisms regulate programmed cell death. One involves the cleavage of the Bax protein. The other mechanism utilizes growth factor signaling through receptor tyrosine kinases.

Formation of t-Bax

In normal cells, the cleavage of monomeric Bax to t-Bax and its subsequent insertion into the mitochondrial membrane is blocked to ensure cell survival. This blockage is achieved by the anti-apoptotic proteins, Bcl-2 and Bcl-X_1, which form heterodimers with Bax to block the formation of t-Bax (see fig. 22.14). Bcl-2 also preferentially heterodimerizes with another protein, Bad. When Bad levels increase in the cytoplasm, Bcl-2 preferentially heterodimerizes with Bad and releases Bax. Bax is then exposed to proteolytic cleavage by Caspase-8 to generate t-Bax.

Toggling Bcl-2 between heterodimer formation with Bax and Bad essentially determines whether the cell will live or die. The level of Bcl-2 in the cytoplasm determines the sensitivity or resistance of cells to apoptosis. One of the unfortunate side effects of many chemotherapies is that they induce cellular stress and

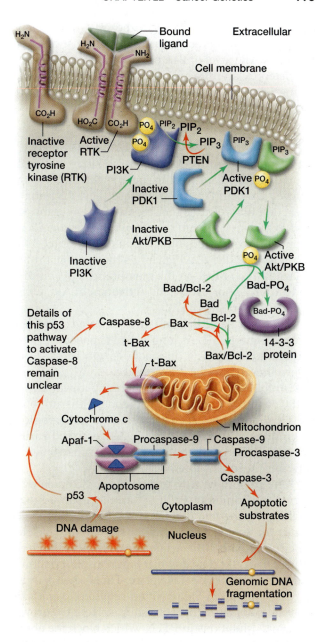

figure 22.14   Intracellular signaling pathways that regulate cell death and cell survival. Apoptotic cell death occurs when the Bax protein is cleaved into t-Bax by Caspase-8. The t-Bax protein then inserts into the outer mitochondrial membrane, which releases cytochrome c into the cytoplasm. Cytochrome c is required for the production of Caspase-3, which activates the DNA fragmentation pathway. Production of t-Bax can be blocked by the Bcl-2 or Bcl-X_1 proteins binding to the Bax monomer. Because Bcl-2 can also dimerize with Bad, the amount of Bad protein determines how much Bcl-2 protein is available to bind Bax and repress apoptosis. The amount of Bad is regulated by the ligand binding to specific receptor kinases, through the action of the Akt and 14-3-3 proteins. The receptor kinase can activate Akt, which then phosphorylates Bad. Phosphorylated Bad is bound by the 14-3-3 protein and unable to bind Bcl-2, which is then available to bind Bax and allow cell survival.

increase the levels of Bcl-2 in the cells that survive apoptosis, making them more resistant to subsequent therapies. The level of Bad protein in the cytoplasm is regulated by an intracellular signaling pathway emanating from the growth factor receptors, particularly the receptor tyrosine kinases.

Receptor Tyrosine Kinase Action

Signaling through receptor tyrosine kinases also can regulate apoptosis. Binding specific receptor kinases induces activation of phospholipase C, which initiates a reaction that releases the second-messenger inositol trisphosphate (IP_3) from the membrane (see fig. 22.14). Then, IP_3 is converted into phosphatidylinositol-3,4,5-trisphosphate (PIP_3), which binds the enzyme Protein kinase B (Akt/PKB) on the membrane. This binding leads to phosphorylation and activation of Akt, which then phosphorylates Bad.

Phosphorylated Bad is then sequestered by binding to the protein 14-3-3. As long as Bad is phosphorylated and bound to 14-3-3, it does not disrupt the Bcl-2/Bax heterodimers, and the cell survives.

Inhibiting this survival signaling pathway results in the dephosphorylation of Bad and the disruption of the equilibrium between the pro- and anti-apoptotic heterodimeric partners of Bcl-2. As described earlier, monomeric Bax is then exposed to Caspase-8, which activates the intrinsic cell death pathway.

One important way this pathway is regulated is through the inhibition of Phosphatidylinositol 3-kinase (PDK1), the enzyme that activates Akt. PTEN, which stands for "phosphatase and tensin homolog deleted on chromosome 10," metabolizes PIP_3, blocking the activation of Akt. The downstream phosphorylation of Bad therefore does not take place. Thus, when cells are triggered to die, PTEN activity is required to block the survival pathway signals transmitted from the growth receptors. Mutations in PTEN, which block the ability of the enzyme to metabolize PIP_3 and block survival pathway signals are common in slow-growing cancers, such as prostate cancer.

Restating the Concepts

▶ Apoptosis is a cell death pathway that requires the activation of an intracellular signaling pathway. Apoptosis involves four stages: precondensation, condensation, fragmentation of the genomic DNA, and engulfment of debris by neighboring cells.

▶ Apoptosis can be induced by either extrinsic or intrinsic signals. Intrinsic signals induced by cell stress or DNA damage involve the Caspase-8-mediated cleavage of Bax to t-Bax, which inserts into the mitochondrial membrane and releases cytochrome c.

▶ Anti-apoptotic proteins, such as Bcl-2, form heterodimers with Bax to prevent the formation of t-Bax. Increasing the amount of Bad causes the formation of a Bcl-2/Bad heterodimer, which frees Bax so that it can be cleaved to t-Bax.

▶ Lipid second messengers, such as PIP_3, help to regulate apoptosis in a cell by activating Akt, which then phosphorylates Bad. The PTEN protein negatively regulates this pathway by metabolizing PIP_3.

22.5 Spontaneous Cancers

Tumor formation can be initiated by a wide variety of mutations. As mentioned earlier, 90% of tumors occur as a result of spontaneous random mutations. What these various mutations have in common is that they alter the homeostatic regulation of either the cell cycle or cell death, which leads to hyperplasia.

Because somatic cells contain two copies of each gene, the result of spontaneous mutations depends on whether the mutations generate a gain-of-function or a loss-of-function allele. A gain-of-function mutation needs to occur in only one of the two copies of the gene because it acts in a dominant manner. In contrast, a loss-of-function mutation is usually recessive and requires two copies of the mutant allele to produce an altered phenotype.

Two classes of proteins can be mutated to induce tumor formation. **Protooncogenes** are genes that encode proteins that stimulate cell-cycle progression. When these genes are mutated, they may result in the creation of an **oncogene**. An oncogene produces a mutant form of the protein that induces tumor formation. Most mutations that create an oncogene are gain-of-function mutations. These may result from either increased expression of the wild-type gene or increased or constitutive activity of the mutant protein.

In contrast, **tumor suppressor genes** usually encode proteins that regulate cell-cycle checkpoints or stimulate cell death. Tumor suppressor mutations in cancer are primarily loss-of-function mutations that result in suppression of apoptosis or unregulated progression through the cell cycle. Because the progression through the cell cycle is not regulated in the tumor suppressor mutants, mutations necessary for cancer progression accumulate in the DNA (see fig. 22.1).

Activation of Oncogenes

With more than 100 oncogenes currently identified, we have many examples of gain-of-function mutations to choose from. **Table 22.4** shows a few examples of known oncogenes, demonstrating that mutations affecting a variety of proteins in the intracellular signaling pathways can lead to tumor growth.

Most tissue and cell types have unique regulatory pathways that can be disrupted, leading to tumor

table 22.4   Selective List of Activated Oncogenes

Gene	Location	Tissue	Function
ABL	9q34	Chronic myeloid leukemia	Nonreceptor tyrosine kinase
BCL-2	18q21	Leukemia, follicular lymphoma	Cytoplasmic/mitochondrial antiapoptotic protein
ERB-B	17q21	Breast, lung, colon, ovarian glioblastoma, medulloblastoma	Receptor tyrosine kinase (EGF receptor)
ETS	11q23	Lymphoblastic leukemia, lymphoma, breast	Transcription factor
FES	15q26	Promyelocytic leukemia	Nonreceptor tyrosine kinase
FOS	14q24	Osteosarcoma, skin	Transcription factor (heterodimerizes with Jun)
INT-2	11q13	Breast	Fibroblast growth factor-3 (FGF-3)
MET	7q31	Kidney	Receptor tyrosine kinase binds hepatocyte growth factor (scatter factor)
MYC	8q24	Leukemia, breast, colon, stomach, lung, neuroblastoma, glioblastoma	Transcription factor (heterodimerizes with Mad or Max)
RAF	3p25	Liver, lung	Cytoplasmic serine/threonine kinase
H-RAS	11p15	Bladder, breast and skin breast, lung, head and neck, ovarian, pancreatic	GTP-binding protein
K-RAS	12p12	Pancreas	GTP-binding protein
N-RAS	1p13	Multiple myeloma	GTP-binding protein
REL-B		Multiple	Nuclear protein (NF6B subunit)
SIS	22q12	T-cell leukemias and lymphomas	Growth factor receptor (PDGF, B chain)
SRC	20q12	Multiple tumor types	Nonreceptor tyrosine kinase

progression in rare tumors. In contrast, mutations in the Ras signaling pathway (see fig. 22.12) are common to most cells and play a critical role in regulating cell-cycle progression and tumor progression in many different cancers. Some of the known tumor progression mutations in the Ras pathway include the constitutive expression of growth factors, irreversible activation of the growth factor receptors, continuous production of individual second messengers, and persistent activation of transcription factors.

Cyclins as Oncogenes

As described previously, the expression of the cyclin proteins increases and decreases at specific points in the cell cycle, whereas the cyclin-dependent kinases are expressed at roughly the same amount throughout the cell cycle. The cyclin genes act as oncogenes when they are overexpressed. For example, overexpression of the cyclin D gene is associated with lung, breast, and bladder cancer.

The overexpression of the cyclin genes is often achieved through DNA amplification. If a cell has made hundreds of copies of the wild-type cyclin D gene, for example, very high levels of the Cyclin D protein are expressed. The Cyclin D protein is normally degraded during the cell cycle, but a very high level of protein allows Cyclin D to remain in the cell for prolonged periods. This in turn leads to an unregulated progression from G_1 to S phase.

In some cancers, the cyclin D gene is not amplified, but rather undergoes a translocation that places it under the transcriptional control of a very strong promoter. The result is still abnormally high levels of cyclin D mRNA and, in turn, Cyclin D protein. This situation is similar to the MYC gene translocation in Burkitt lymphoma that we discussed earlier in this chapter.

The cyclin E gene can also function as an oncogene when it is overexpressed. Overexpression of Cyclin E is observed in breast cancer, colon cancer, and some leukemias.

Growth Factor Receptor Genes as Oncogenes

In most tissues, growth factors are expressed by one cell type (often stromal cells), and the receptors are present on a second cell type (such as epithelial cells). This facilitates cell–cell (paracrine) interaction between different cells in the same tissue. Constitutive overexpression of the growth factor by stromal cells, or inappropriate expression of the growth factors by the epithelial cells can both lead to continuous stimulation of cell division.

An example of the latter is the expression of the int-2 oncogene, which encodes FGF-3 (fibroblast growth factor-3) and is increased in nearly 25% of all breast cancers. In this case, amplification of the wild-type gene in the epithelial cells establishes an autocrine loop. The FGF-3 protein is expressed in the epithelial cells and, when secreted, binds to the FGF receptors on the epithelial cell membrane, leading to persistent stimulation and continuous proliferation.

RAS as an Oncogene

Generation of *RAS* mutations by *N*-methyl-*N*-nitrosourea (MNU) and 7,12-dimethylbenz[*a*]anthracene (DMBA) are associated with a variety of tumors. MNU induces a G-to-C transversion in codon 12 of *RAS*, causing a glycine-to-alanine substitution in the Ras protein. In contrast, DMBA produces an A-to-G transition in codon 61 of *RAS*, which results in a glutamine-to-arginine substitution. These two events take place frequently in epithelial cells of the lungs, breast, and prostate because the CYP enzymes needed to generate the ultimate carcinogen are expressed at reasonably high levels in these cells, but at lower levels in most other tissues.

As you learned earlier, Ras proteins are GTPases. The mutation in codon 12 results in a conformational change in the GTP-binding pocket of Ras that prevents the release of the phosphate that is hydrolyzed from GTP. This keeps the mutant Ras, called K-Ras, in a constitutively active form. In contrast, the mutation at codon 61 alters the active site to block the hydrolysis of GTP. This keeps the mutant Ras, called N-Ras, also in a constantly active state.

These mutations have two important consequences. First, constitutively activated (oncogenic) Ras continually stimulates the Akt pathway, which maintains Bad in a phosphorylated state and bound to protein 14-3-3.

This blocks the induction of cell death through the intrinsic apoptotic pathway, even if p53 levels increase in response to DNA damage.

Second, oncogenic Ras increases the level of cyclin D, even when there is no paracrine signaling from neighboring cells. As a result, the cells are continually stimulated to divide and progress rapidly through G_1 and enter S phase.

Stimulation of this pathway by other oncogene products, including Raf and Jun, can result in the loss of cell-cycle control in a number of different tumor types, including liver and lung tumors, osteosarcomas (bone cancers), and skin cancer.

Myc as an Oncogene

The overexpression of Myc, either through amplification of the wild-type gene or translocation of the gene downstream of a new promoter, as occurs in Burkitt lymphoma, also leads to rapid cell-cycle progression. The tumor progression is largely due to the persistent elevated transcription of the Myc-responsive genes required for cell division.

The Myc protein has also been shown to act cooperatively with mutated (constitutively activated) Ras to transform normal cells by up-regulating cyclin expression and down-regulating CKI expression. It is important to note that although Ras is mutated in this context, the *c-myc* gene is not mutated. Rather, the effect is due to the overexpression of wild-type Myc protein (see fig. 22.7).

22.1 applying genetic principles to problems

Detection of *K-RAS* Mutations in Pancreatic Cancer.

Pancreatic cancers are very aggressive, with only 5% of the affected individuals surviving five years after the first diagnosis. In many cases, the individual has only several months to live after being diagnosed. Because of the severity and speed at which these tumors develop and spread, early identification of pancreatic tumors is critical for the potential success of any treatment.

The gene most commonly mutated in pancreatic cancers is the *RAS* gene. As we discussed, *RAS* is an oncogene that encodes a GTP-hydrolyzing enzyme. The most common *RAS* mutations in pancreatic cancer are the *K-RAS* mutations, which prevent the release of the hydrolyzed phosphate from K-Ras. The three most common types of *K-RAS* mutations change codon 12 (5'-GGT-3'), which encodes glycine (Gly), into 5'-GAT-3' (Asp), 5'-GTT-3' (Val) or 5'-AGT-3' (Ser). The loss of glycine at amino acid 12 in all these mutants results in the constitutive activation of the K-Ras enzyme and disruption of cell-cycle regulation, which leads to tumor progression.

Because the goal is early detection of the *K-RAS* mutation, a sensitive method must be used to identify the muta-

tion from a small amount of biopsy material. The polymerase chain reaction (PCR) is used to amplify a region of the *RAS* gene and then determine if one of the *K-RAS* mutations is present (fig. 22.A). Because codon 12 is located within a *Mva*I restriction enzyme site (5'- GGTACC-3'), it is possible to digest the PCR product with *Mva*I and detect the presence or absence of this restriction enzyme site, which will produce a restriction fragment length polymorphism (RFLP).

The standard procedure is to isolate genomic DNA from a portion of the biopsied pancreas. A 157 bp DNA fragment is PCR-amplified that contains the codon 12 sequence. The PCR product is then digested with the *Mva*I restriction endonuclease and separated on an agarose gel. The bands can then be detected by ethidium bromide staining of the DNA in the gel. A *Mva*I site is located near the 3' end of the PCR product that produces 143 bp and 14 bp fragments after digestion (see fig. 22.A).

The presence of the *Mva*I site in the wild-type codon 12 will result in the 143 bp fragment being digested into 114 bp and 29 bp fragments (see fig. 22.A). Thus, the wild-type *RAS* allele will produce 114 bp, 29 bp, and 14 bp fragments

RAS gene organization

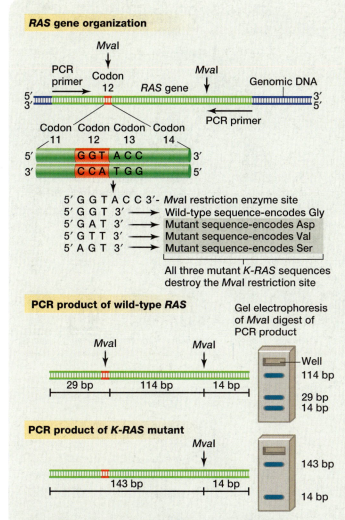

5′ G G T A C C 3′- *Mva*I restriction enzyme site
5′ G G T 3′ ⟶ Wild-type sequence-encodes Gly
5′ G A T 3′ ⟶ Mutant sequence-encodes Asp
5′ G T T 3′ ⟶ Mutant sequence-encodes Val
5′ A G T 3′ ⟶ Mutant sequence-encodes Ser

All three mutant *K-RAS* sequences
destroy the *Mva*I restriction site

box figure 22.A   PCR analysis of wild-type and mutant *RAS* alleles. The wild-type *RAS* gene contains two *Mva*I sites, with one located in codon 12, which encodes a glycine. Mutation of codon 12 to GAT, GTT, or AGT destroys the *Mva*I site and results in a *K-RAS* mutation. PCR amplification of the wild-type *RAS* gene, followed by *Mva*I restriction digestion, produces 114, 29, and 14 bp fragments. PCR amplification of the *K-RAS* allele (which lacks the *Mva*I site at codon 12), followed by *Mva*I restriction digestion, produces a 143 and 14 bp fragments. Thus, the wild-type and *K-RAS* alleles can be distinguished by the presence of either a 114 or 143 bp *Mva*I fragment.

after *Mva*I digestion of the PCR product, whereas the *K-RAS* mutations will yield only 143 bp and 14 bp fragments under the same conditions. In practice, identification of only the 114 bp and 143 bp fragments is sufficient to determine the presence or absence of the *K-RAS* mutations.

Let us interpret some actual data. Biopsies were collected from two different individuals with pancreatitis, two individuals with pancreatic neuroendocrine tumors, two individuals with pancreatic carcinomas, and a normal pancreatic epithelium. Genomic DNA was collected from all seven biopsy samples, PCR-amplified, and digested

with *Mva*I. The restriction digests were electrophoresed through an agarose gel, stained with ethidium bromide, and photographed (fig. 22.B).

Notice that all the samples contain the 114 bp fragment, which represents the wild-type *RAS* allele. Samples 1 and 2 are the only ones that contain the 143 bp fragment, which represents the *K-RAS* allele. If the *K-RAS* allele is an indicator of pancreatic carcinoma, then samples 1 and 2 must be biopsies from pancreatic carcinomas.

The absence of the 143 bp fragment in the remaining five samples is consistent with this mutation being specific for pancreatic carcinomas. The absence of the 143 bp fragment in the two pancreatic neuroendocrine tumor samples is a good indication that not every cancer of the pancreas possesses the same mutation and supports the specificity of the *K-RAS* mutations for pancreatic carcinomas. In fact, we have no way of determining that the two pancreatic neuroendocrine tumor biopsies correspond to lanes 3 and 4 or that the pancreatitis biopsies were analyzed in lanes 5 and 6.

Why do samples 1 and 2 contain both the 143 bp and 114 bp fragments? There are two possibilities. First, as we discussed earlier, the *K-RAS* mutation is a dominant oncogenic allele. Samples 1 and 2, therefore, may represent carcinomas that contain cells that are heterozygous for the *RAS* alleles. Alternatively, every biopsy will contain a mixture of carcinoma and normal human cells (either neighboring epithelial or stromal cells) that are probably required for tumor cell survival or progression. These normal cells will be homozygous for the wild-type *RAS* allele. In all likelihood, the presence of both bands is due to both the presence of carcinoma cells that are heterozygous for the dominant gain-of-function *K-RAS* allele and the presence of some homozygous wild-type *RAS* cells. This does not, however, affect our identification of the pancreatic carcinoma biopsy tissues. We are able to correctly identify a pancreatic carcinoma biopsy potentially early in its development. However, our understanding of the RFLP pattern allows us a more complete understanding of the composition of the pancreatic carcinoma biopsy.

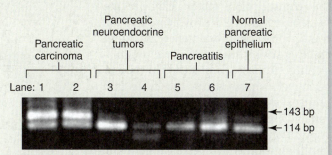

box figure 22.B   An ethidium bromide stained gel of the *Mva*I RFLP analysis of PCR-amplified *RAS* gene from genomic DNA. Each lane represents the analysis of a pancreatic biopsy from a different individual with either pancreatic carcinomas (lanes 1 and 2), pancreatic neuroendocrine tumors (lanes 3 and 4), pancreatitis (lanes 5 and 6), or a normal pancreatic epithelium (lane 7). The relative location of the 143 bp and 114 bp restriction fragments are shown.

Mutations in Tumor Suppressor Genes

Mutations in tumor suppressor genes also lead to an increase in cell number. In this case, both alleles of the tumor suppressor gene must be mutated before there is an effect on cell number.

Tumor suppressor genes fall into two general categories: those that regulate cell-cycle progression, and those that regulate cell survival and cell death.

Mutation of *RB1*

The human *retinoblastoma 1 (RB1)* gene was originally identified during the analysis of retinoblastoma, which, as you learned in chapter 18, results in the development of tumors in the eyes of children. Individuals who inherit the mutant $RB1^-$ allele have greater than an 80% chance of developing the eye tumor and an increased chance of developing other tumors later in life.

In heterozygous individuals who do develop the tumors, a second mutation was found to have developed in the wild-type copy of the *RB1* gene (**fig. 22.15** top). Thus, tumor formation depends on having two mutant $RB1^-$ alleles. Individuals who inherit two wild-type $RB1^+$ alleles rarely develop the retinoblastoma, because the spontaneous generation of two mutant $RB1^-$ alleles is very rare (**fig. 22.15** bottom).

As we described earlier, cyclin/CDK complexes phosphorylate RB1 during the G_1 phase of the cell cycle, which releases the transcription factor E2F. In the nucleus, released E2F protein induces the transcription of several genes, including *cyclin A*, which are required for progression into S phase. Loss of the Rb1 protein effectively shortens the cell cycle and reduces or eliminates the opportunity to repair DNA damage. The accumulation of additional mutations with each round of cell division is referred to as **genomic instability** and leads to more and more aggressive tumor behavior. $RB1^-$ mutations are associated with a variety of cancers, including breast, lung, and bladder cancers.

The sporadic form of retinoblastoma occurs when an individual who has

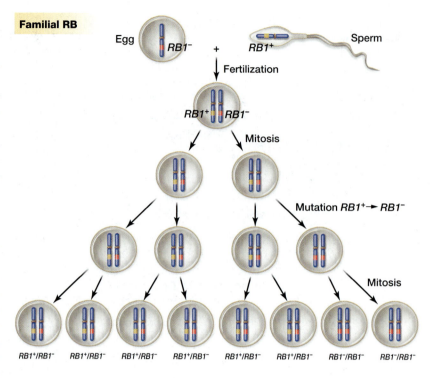

Familial RB

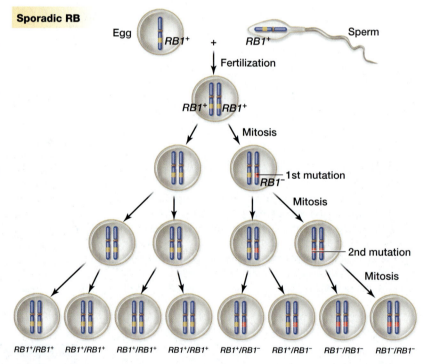

Sporadic RB

figure 22.15 Mechanisms of familial and sporadic retinoblastoma. Cell divisions that are associated with the progression of familial (top) and sporadic (bottom) forms of retinoblastoma are shown. Because the familial form already possesses an inactive $RB1^-$ allele from the egg, only a single mutation is required to inactivate the remaining wild-type $RB1^+$ allele. In contrast, the sporadic form involves the inheritance of two wild-type $RB1^+$ alleles, and needs two independent mutations to yield two $RB1^-$ alleles.

inherited two wild-type $RB1^+$ alleles undergoes two mutations (fig. 22.15 bottom). Two events are required to inactivate both wild-type alleles. The likelihood of the second mutation occurring in the same cell that already possesses the first mutation is very rare, explaining why the affected individuals are not prevalent in a family pedigree. Furthermore, because both of these mutations occur in the somatic tissue, none of the offspring of the affected individual will have an increased risk of retinoblastoma. In the inherited form, however, the first $RB1^-$ allele is present in the germ-line cells, and offspring will be predisposed to developing retinoblastoma.

Mutation of p53

The *p53* gene is mutated in over 50% of all cancers. The p53 protein is a transcription factor that controls the expression of over 50 different genes. The level of p53 in the cell is low during the cell cycle unless there is cellular stress or DNA damage. Single- or double-strand DNA breaks activate the ATR and ATM proteins that phosphorylate p53. This posttranslational modification stabilizes p53 and increases its abundance in the cell, which leads to the activation of the CKIs and cell-cycle arrest prior to entry into S phase. ATM also phosphorylates the BRCA-1 protein (for the DSB breast cancer susceptibility gene 1). This protein along with BRCA-2 and several other proteins initiate repair by homologous recombination (chapter 18).

The p53 protein also regulates apoptosis in the cell by activating transcription of the *BAX* gene and repressing transcription of the *BCL-2* gene (see fig. 22.14). Because Bax forms a heterodimer with Bcl-2, increasing p53 expression results in an increase in Bax and thus potentially a greater amount of t-Bax. The result, as described earlier, is apoptosis.

The accumulation of *p53* mutations in tumors of the colon, prostate, and breast is thought to occur at later stages of disease progression and coincides with the substantial increase in genomic instability that is the hallmark of untreatable cancer.

Mutation of BRCA-1

BRCA-1 mutations are responsible for about 50–70% of the cases of familial breast cancer. These mutations and mutation of *BRCA-2*, which cooperates with the BRCA-1 and Rad51 proteins in the identification and repair of double-strand DNA breaks, result in the loss of DNA repair and the accumulation of DNA double-strand breaks in the surviving daughter cells. This produces genomic instability and tumor progression. Since the failure to repair strand breaks rapidly leads to genomic instability, women with *BRCA-1* and *BRCA-2* mutations usually develop breast cancer in

their mid-thirties, considerably sooner than women who acquire sporadic forms of breast cancer that are most often diagnosed after 50 years of age.

Mutation of PTEN

Cancer, or uncontrolled cell growth, can also occur if cells incurring significant amounts of DNA damage fail to die. Spontaneous mutations in the *PTEN* gene that prevent the enzyme from metabolizing PIP_3 result in the constitutive phosphorylation of Bad and its continual binding to the 14-3-3 protein (see fig. 22.14). As described earlier, this prevents activation of the intrinsic apoptotic pathway and ensures cell survival, even when the intrinsic pathway is stimulated. Cells then accumulate DNA damage and proliferate. Because this *PTEN* mutation represents a loss-of-function mutation, both copies of the *PTEN* tumor suppressor gene must be inactivated. *PTEN* mutations are very common in prostate cancer. However, since only a small proportion of prostate cancer cells are in the cell cycle, the increase in cell number due to failure of the cells to die occurs very slowly. These cells continue to accumulate mutations over time, some of which, including mutations in *RAS* and *MYC*, increase the rate of cell division. These tumors start to grow more rapidly and may become clinically relevant 20–30 years after their initiation.

Restating the Concepts

▶ Protooncogenes encode proteins that stimulate cell-cycle progression. Protooncogenes that undergo gain-of-function mutations to become oncogenes lead to cancer. Mutations in growth factor receptors, *RAS* and *MYC*, are examples of oncogenes. Overexpressed cyclins can also function as oncogenes.

▶ Tumor suppressor genes encode proteins that regulate cell-cycle checkpoints or activate apoptosis pathways. Loss-of-function mutations in tumor suppressor genes can also lead to cancer. The *RB1*, *p53*, and *PTEN* genes are examples of tumor suppressor genes.

22.6 Cancer Predisposition and Familial Cancers

Like any other gene in the genome, the genes that regulate the cell cycle and cell death can be mutated or inherited in the germ line. Because embryonic cell division must be very tightly regulated, most gain-of-function oncogenic mutations that directly affect the regulation of the cell cycle are lethal during embryonic development. Those that are not lethal, such as the

RET oncogene that underlies multiple endocrine neoplasia type 2 (MEN-2), do not have any serious consequence in embryonic development, largely because the affected gene is not expressed until after birth. Furthermore, such genes may be expressed in only a very small number of cells within a tissue.

In contrast, embryonic development in individuals who carry one nonfunctional copy of a tumor suppressor gene is usually unaffected by this loss-of-function mutation; the wild-type allele compensates for the inactive allele. Furthermore, not all individuals that are heterozygous for a tumor suppressor gene develop cancer—because for cancer to occur, the wild-type allele must be inactivated through a spontaneous second mutation, which is a relatively rare occurrence.

Nevertheless, individuals who inherit a mutant tumor suppressor gene have a genetic predisposition to cancer, and they have a higher risk of developing tumors, relative to individuals who inherit two wild-type alleles. A number of well-characterized instances of cancer predisposition are known that involve tumor suppressor genes (**table 22.5**).

Retinoblastoma

Retinoblastoma is one of the most thoroughly studied familial cancers, even though it is relatively rare, affecting 1 in 16,000 to 24,000 live births. To review what we described in chapter 18, pedigrees of individuals that exhibit tumors in both eyes suggest a dominant autosomal pattern of inheritance (*familial retinoblastoma*), whereas pedigrees of individuals with the tumor restricted to a single eye appear to have an autosomal recessive pattern of inheritance (*sporadic retinoblastoma*, **fig. 22.16**).

During embryonic development of heterozygous individuals, the wild-type allele functions to regulate cell-cycle progression, ensuring normal development. In those who manifest the retinal tumors (retinoblasto-

mas), the wild-type allele likely mutated or was deleted late in the developmental process, and therefore it did not disrupt embryonic development. In many of these individuals, both eyes are affected (*bilateral disease*). Because inheritance of an *RB1* mutant allele is not an absolute determinant that cancer will develop, this mutant allele simply increases the likelihood, or predisposition, of developing a cancer.

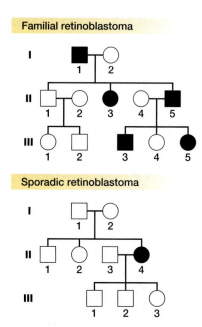

figure 22.16   Retinoblastoma pedigrees. Pedigrees showing the inheritance of either the familial form of retinoblastoma (top) or sporadic retinoblastoma (bottom). The familial form of retinoblastoma appears to be inherited as a dominant mutation, while the sporadic retinoblastoma appears to be inherited as a recessive mutation. However, both forms require mutations in both copies of the *RB1* gene. The apparent dominant inheritance in the familial retinoblastoma is due to the inheritance of one recessive mutant allele and the need to generate only one mutation, rather than two as in the sporadic retinoblastoma.

table 22.5   Tumor Suppressor Genes That Are Associated with Cancer Predisposition.

Syndrome	Tumor Suppressor Gene	Chromosomal Location	Tumor Types
Li–Fraumeni	*p53*	17p13.1	Brain tumors, leukemia, breast cancer
Familial retinoblastoma	*RB1*	13q14.1	Retinoblastoma, osteogenic, sarcoma
Wilms tumor	*WT1*	11p13	Pediatric kidney cancer
Familial adenomatous polyposis	*APC*	5q21-q22	Colon cancer
Familial breast cancer	*BRCA1*	17q21	Breast cancer, ovarian cancer
Familial breast cancer	*BRCA2*	13q12.3	Breast cancer, male breast cancer, ovarian cancer
Multiple endocrine neoplasia type 1	*MEN1*	11q13	Parathyroid adenomas, pituitary adenomas
Deleted in pancreatic carcinoma 4	*DPC4*	18q21.1	Pancreatic carcinoma, colon cancer
Deleted in colorectal carcinoma	*DCC*	8q21.3	Colorectal cancer

The loss of the remaining functional allele is referred to as **loss of heterozygosity (LOH),** because the wild-type allele is lost and the cell is now either homozygous for *RB1* mutant alleles—or hemizygous for the mutant allele if the wild-type allele was deleted. Loss of the wild-type allele initially results in hyperplasia.

The sporadic retinoblastoma phenotype appears in pedigrees as an autosomal recessive pattern of inheritance, and the tumor is usually restricted to a single eye (*unilateral disease*). Individuals who exhibit the sporadic form usually inherited the two wild-type *RB1* alleles, one from each parent. In this case, both wild-type alleles must be mutated to generate the retinoblastoma phenotype. This "double-hit" is much less frequent than only the single mutation that is required for the familial retinoblastoma. This difference explains why tumors skip one or more generations or only occur in one eye.

Li-Fraumeni Syndrome and *p53* Mutations

Like retinoblastoma, homozygous mutations that inactivate both alleles of *p53* in an individual are usually lethal because of the profound effect the loss of cell-cycle regulation has on embryonic development. Cancer predisposition due to mutations in *p53* is therefore usually transmitted through heterozygotes, and it is called Li-Fraumeni syndrome. There is one well-documented report, however, of a patient who inherited mutant maternal and paternal *p53* alleles and nevertheless had normal embryonic development, which raises some interesting issues about the role of *p53*.

The majority of *p53* mutations are missense mutations, mainly confined to the DNA-binding domain of the p53 protein, which effectively eliminates the protein's ability to function as a transcription factor. Such

defects prevent p53 from inducing transcription of the *p21* and *p27* genes in response to DNA damage. The loss of p53 activity also blocks the expression of Bax and the activation of cell death. As a result, cells rapidly proceed through the cell cycle, even though DNA damage has not been repaired. This leads to chromosome deletions, missense and nonsense mutations, and other aberrations that should normally trigger the apoptotic response, but now do not. The number of cells increases, and tumor progression occurs because the cells are unable to repair the DNA damage or to induce cell death.

In a heterozygous individual, the inactivation of the functioning wild-type *p53* allele, resulting in the loss of heterozygosity, occurs as a result of random mutagenesis, and it may occur essentially in any tissue. This explains why patients with Li-Fraumeni syndrome are predisposed to many kinds of tumors, including sarcoma; breast cancer; leukemia; osteosarcoma; melanoma; prostate cancer; and cancers of the colon, pancreas, adrenal cortex, and brain. In many cases these individuals develop tumors at more than one site (**fig. 22.17**).

PTEN Mutations

Mutations that affect the cell death pathways can also lead to uncontrolled cell growth. Individuals who are heterozygous for a mutation in the *PTEN* gene and who subsequently lose the wild-type allele also develop tumors. This is not because the cells are dividing more rapidly; rather, it is because the cells are failing to die. In this case, the loss of the functioning *PTEN* allele means that the dephosphorylation of Bad is not blocked (see fig. 22.14). The apoptotic removal of cells with substantial DNA damage or excessive cellular stress does not take place, and their proliferation leads to tumor progression.

In this regard, the regulation of the intrinsic cell death pathway by PTEN actually mirrors the oncogenic activation of the Ras pathway. Overexpression of the wild-type Bcl-2 protein, the principal anti-apoptotic protein in the cytoplasm, also blocks the intrinsic pathway by preventing Bax cleavage, which prevents the insertion of t-Bax into the mitochondrial membrane.

Follicular lymphoma, one of the most frequent hematological malignancies, is associated with a reciprocal translocation that places the *BCL-2* coding sequence under the control of the heavy-chain immunoglobulin promoter (**fig. 22.18**). The result is overexpression of Bcl-2, which blocks normal apoptosis of B lymphocytes.

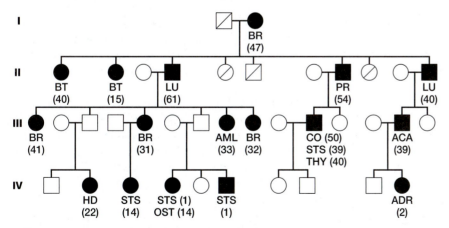

figure 22.17 Pedigree of four generations of a family that exhibits Li-Fraumeni syndrome. The dominant nature of this inherited cancer is seen by individuals in every generation exhibiting one or more different forms of cancer. Notice that affected individuals in the same family can exhibit multiple forms of cancer, including acute myeloid leukemia (AML), astrocytoma (ACA), adrenal (ADR), breast (BR), brain (BT), colon (CO), Hodgkin's disease (HD), lung (LU), osteosarcoma (OST), prostate (PR), soft tissue sarcoma (STS), and thyroid (THY). The age that each individual was diagnosed with each form of cancer is shown in parentheses after the cancer designation.

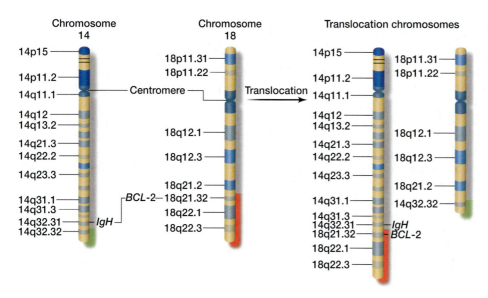

figure 22.18 One common form of follicular lymphoma is due to a reciprocal translocation between chromosomes 14 and 18. The *BCL-2* gene located on chromosome 18, is translocated to chromosome 14, which places the expression of the *BCL-2* gene under the transcriptional control of the immunoglobulin heavy chain (*IgH*) gene. This results in very high levels of *BCL-2* expression and leads to follicular lymphoma.

In many ways, this action is analogous to the activation of Myc in Burkitt lymphoma, except that in this case the translocation produces a protein that blocks cell death.

Restating the Concepts

▶ Familial (heritable) forms of cancer are not usually inherited as dominant mutations because these would be lethal during embryonic development. Rather, individuals inherit mutant alleles of genes that provide a predisposition to causing cancer; the accumulation of additional somatic mutations dictate whether cancer develops.

▶ Loss of heterozygosity occurs when a mutation occurs in the single wild-type tumor suppressor allele—a change of one or more nucleotides, deletion of the wild-type allele, or loss of the chromosome containing the wild-type allele. Loss of heterozygosity then leads to cancer because of the unregulated cell cycle or the suppression of apoptosis. Loss of heterozygosity often makes a loss-of-function phenotype appear as a dominant phenotype in a pedigree.

22.7 Tumor Progression

Most of the protooncogenes and tumor suppressor genes identified to date function in signal transduction pathways that stimulate cell proliferation. The initial mutations result in hyperplasia, or uncontrolled cell growth, rather than metastatic disease. Although there is still considerable debate as to the minimum number of mutations required for the initiation of tumor progression, statistical analysis of sporadic and hereditary cancers suggests that two genetic "hits" or mutations—the activation of one oncogene, and inac-

tivation of one tumor suppressor gene—is the minimum requirement. This idea is often referred to as Knudson's **multiple-hit hypothesis.** After the initial mutations initiate the formation of a benign hyperplasia, many additional mutations are required to generate the malignant metastatic disease.

Virtually all of the activated oncogenes that are initially mutated accelerate the cell cycle by decreasing or eliminating the G_1/S checkpoint, which reduces the opportunity for DNA repair. Additional genetic abnormalities, such as point mutations and chromosomal alterations, then increase. Cells therefore continue to accumulate mutations that affect regulation of mitosis, DNA repair, and cell death. Loss of both alleles of a tumor suppressor gene has essentially the same effect.

Analysis of many cases of advanced cancers suggests that although tumors may acquire a similar repertoire of mutations, the order in which they acquire the mutations is not fixed. The exception to this general observation may be colorectal cancer, in which the multistep progression appears to follow a fairly predictable sequence (**fig. 22.19**).

The development of colorectal cancer involves an initial mutation in the *adenomatous polyposis coli* (*APC*) gene, which regulates the cytoplasmic levels of β-catenin. In the absence of functional APC, β-catenin is translocated into the nucleus where it forms a heterodimer with LEF1 and induces transcription of several genes, including *MYC*.

The next step in colorectal cancer development involves mutation of the *RAS* gene, which overstimulates downstream signaling pathways from growth factor receptors, causing the cells to divide continuously while also stimulating the Akt-mediated survival pathways. These two events correlate with the formation of early and intermediate adenomas, which are relatively benign early stages of cancer.

The next mutation is in the *SMAD4* gene (also known as *DCC* for "deleted in colon cancer"). This gene encodes an intracellular signaling molecule that is required for a specific growth factor-mediated (TGF-4) cell-cycle arrest in the colon epithelium. This mutation removes an additional brake on the cell cycle and further reduces the time available at the G₁/S transition for DNA repair.

The transition to carcinoma is marked by the loss of p53 activity, further disrupting cell-cycle regulation and, more importantly, blocking repair of DNA adducts and double-stranded breaks. The loss of double-stranded break repair leads to multiple chromosomal aberrations and deletions, which results in marked aneuploidy (through loss of heterozygosity) in cells of late-stage cancers.

Restating the Concepts

▶ The initial stage of cancer development, hyperplasia, requires two mutations according to Knudson's multiple-hit hypothesis. One involves a gain-of-function mutation in a protooncogene. The second involves two loss-of-function mutations in a tumor suppressor gene, or loss of heterozygosity if a loss-of-function allele has been inherited.

▶ The progression of most tumors involves additional mutations that affect the G₁/S checkpoint, which results in an accumulation of even more mutations, along with a loss of regulation of the cell death pathway.

▶ The progression of colorectal cancer appears to be different from other cancers, in that it involves a series of mutations that occur in a specific sequence. However, the mutations still occur in protooncogenes and tumor suppressor genes.

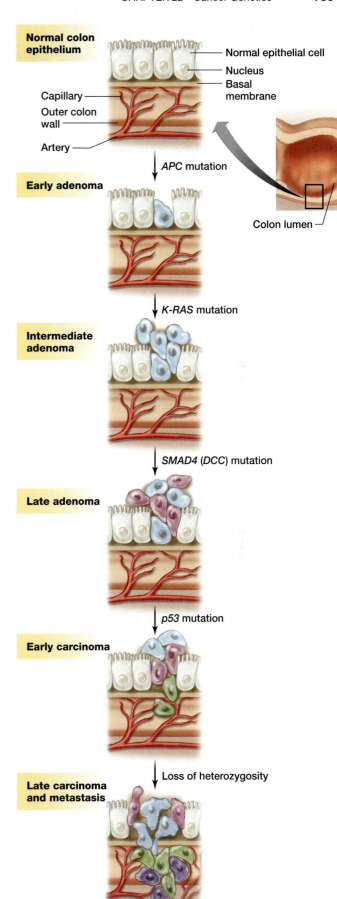

figure 22.19   The molecular progression of colorectal cancer. Nearly all cancers exhibit a similar cellular progression (fig. 22.1). Unlike most cancers, however, development of colorectal cancer exhibits a specific sequence of mutations that correspond to the cellular changes. The molecular mutations associated with colorectal tumor progression are shown, along with the designated stages of tumor progression and a schematic of tumor development.

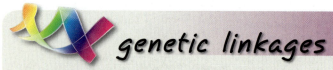

genetic linkages

In this chapter, we utilized many of the concepts that were addressed in earlier chapters. We previously discussed various aspects of loss of chromosomes and changes in the structure of chromosomes (chapter 8). You have now seen how these issues relate to loss of heterozygosity, translocations, producing gain-of-function mutations in oncogenes, and just a general accumulation of mutations as the cell progresses through the phases of tumor progression.

In chapter 18, we discussed mechanisms employed by mutagens to modify the genomic DNA sequence. In this chapter you learned how many of these mutagens may also act as carcinogens. Many of the DNA repair systems described in chapter 18 are not able to function properly during cancer progression because they act at the G_1/S transition, which is dramatically shortened or lost in cancerous cells.

The different phases of the cell cycle and how they are regulated were first described in chapter 3. Loss of cell-cycle regulation through mutations that encode the proteins acting at the cell-cycle checkpoints, are critical for cancer development. Loss of these regulatory mechanisms results in a shorter cell cycle and the inability to properly repair DNA damage before entering S phase. This results in an accumulation of mutations and an increased number of DNA strand breaks. Growth factors and the Ras signaling cascade play roles in the regulation of the cell cycle and in tumor progression.

Chapter 21 introduced the genetic complexity required for normal development of an organism and the way in which mutations have been used to determine how the normal process occurs. In this chapter, we revisited this complexity and saw how mutations affected normal processes to produce the abnormal phenotype of cancer. Through the study of these processes, we hope to learn how to prevent cancer's development.

Chapter Summary Questions

1. For each term in column I, choose the best matching phrase in column II.

Column I
1. DNA adduct
2. Cyclin
3. CDK
4. Checkpoint
5. Apoptosis
6. Paracrine signaling
7. Genomic instability
8. Predisposition
9. Hyperplasia
10. Carcinogen

Column II
A. Signals sent from nearby cells in a tissue
B. Protein that mediates the effects of cyclins
C. Excess growth of tissue that is dividing rapidly
D. Programmed cell death
E. Stable altered nucleotides within a strand of DNA
F. An increased risk for a particular event
G. A substance that can cause changes in the genetic material
H. Protein that controls movement of a cell through the cell cycle
I. Accumulation of additional mutations with each round of cell division
J. Specific points in the cell cycle that control rate of cell division

2. Define *cancer*.
3. Differentiate between protooncogenes and tumor suppressor genes.
4. Describe metastasis and the steps required to reach it.
5. Describe the basic mechanism by which most chemical carcinogens cause mutation.
6. Give two reasons mutations might not affect a cell's phenotype.
7. List at least four examples of environmental carcinogens.
8. Why does viral inactivation of Rb1 or p53 result in hyperplasia?
9. What do Epstein–Barr virus and Burkitt lymphoma have in common?
10. How does expression of the cyclins differ from that of CDK's?
11. Describe the three checkpoints that regulate a cell's ability to move through interphase.
12. List and describe the events that occur in the different stages of apoptosis.
13. How does genomic instability relate to cancer progression?
14. What is the typical sequence of mutations in colorectal cancer? Why is this different from other types of cancer?
15. How is a loss of heterozygosity related to tumor progression?

Solved Problems

PROBLEM 1: For each class of cyclin, what type of mutation would be likely to lead to cancer and why?

Answer: Loss-of-function mutations in *cyclin D, E,* or *A* would prevent any future cell divisions and would not be expected to lead to cancer. However, gain-of-function mutations in any of these three cyclins would likely increase the rate of cell division and could lead to cancer. In contrast, loss-of-function mutations in *cyclin B* may allow cells to move too quickly through the anaphase spindle checkpoint and increase the G_2 to M phase transition, leading to chromosomal errors and potential cancer formation.

PROBLEM 2: Assume a person is exposed to an environmental carcinogen that causes new mutations at a rate of 250 mutations per cell per day, and the repair rate is 95%. You expect about 0.2% of these mutations to be oncogenic. In a healthy liver, about 0.15% of cells die and are replaced by cell division daily; in this patient a subpopulation of cells is dividing at a rate of 5% per day.

   a. Calculate the risk that one of this person's healthy liver cells will become oncogenic in the next day.

   b. Calculate the risk that one of this person's rapidly dividing liver cells will accumulate another oncogenic mutation in the next day.

Answer:

   a. Of the 250 new mutations per cell per day, the 5% that are unrepaired correspond to about 12. If 0.2% of these 12 mutations per cell is oncogenic, then there would be 0.24 oncogenic mutations per cell per day. However, only 0.15% of cells in the liver are actually dividing, so relative to the total liver, the risk will be $0.24 \times 0.0015 = 0.00036\%$ per cell per day.

   b. For the more rapidly dividing cells, the risk will be $0.24 \times 0.05 = 0.012\%$ per cell per day.

PROBLEM 3: From the name of the disorder, you know that Burkitt lymphoma primarily affects tissues of the lymphatic system. Explain how this disorder is connected to the lymphatic system.

Answer: In Burkitt lymphoma, a translocation places the *MYC* gene under the transcriptional control of the immunoglobulin heavy-chain (*IgH*) promoter. This will misexpress the Myc protein, which will affect cells such as lymphocytes where the *IgH* promoter is normally active. Lymphocytes normally circulate through the body and spend a lot of time in the lymph nodes and other structures of the lymphatic system, so it is not surprising that cancer resulting from this specific mutation would primarily affect tissues of the lymphatic system.

Exercises and Problems

16. For tumor suppressor genes and protooncogenes, could each of the following types of mutations lead to growth of cancerous cells? If so, explain how.
   a. hypomorphic or null mutation
   b. hypermorphic mutation

17. Assume that a particular oncogene produces a growth factor.
   a. How could a retrovirus affect the oncogene so that the cell becomes cancerous?
   b. How could you test your hypothesis?

18. You isolated a cDNA that corresponds to a protooncogene. Genomic DNA is isolated from normal cells and a clone of cells infected with a retrovirus that lacks the oncogene (clone 1). Both genomic DNA samples are digested with a particular restriction enzyme and analyzed by a genomic Southern blot using the cDNA as a probe. The results appear in the following figure.

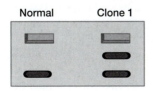

Interpret these results by describing the location of the retroviral insertion.

19. The *E1B* gene of adenovirus produces a protein that binds p53, which inactivates p53 and allows the virus to multiply in the cell. Given that more than 50% of cancer cells lack p53

activity, how might you engineer the adenovirus to attack only cells without p53 activity? That is, can you engineer adenovirus to attack a large proportion of cancerous cells?

20. Certain oncogenic viruses produce proteins that interfere with the Rb1 protein in a cell. How is this similar to and different from a loss-of-function mutation in the *RB1* gene?

21. Individuals that inherit a mutation in *RB1* have a greater than 80% risk of developing retinal tumors as children (see chapter 18), and their risk of developing other types of tumors later in life is also increased. What factors might influence this difference in risk for different tissues?

22. Most cells in the body display two key characteristics: anchorage dependence and contact inhibition. Anchorage dependence means the cell cannot survive unless connected to a substrate, such as the basal lamina (see figures 22.1 and 22.13). Most cells cling to a basement membrane or to other cells, and will commit apoptosis if this is not possible. Contact inhibition means cells stop dividing as they become crowded, and this is thought to be mediated by signals from adjacent cells. What mechanisms described in this chapter allow cells to overcome these two limitations and become cancerous?

23. Many viruses cause infected cells to continue proliferating. From an evolutionary standpoint, how might this be an adaptive trait for a virus?

24. What factors will affect the cell cycle of a tumor cell relative to a nearby cell that does not display cancerous changes? Give an example of a situation that would lead to a slow-growing tumor, and an example that would lead to a rapidly growing tumor.

25. Would you expect loss-of-function mutations in cyclins to be *more* or *less* common in tumor cells than cyclin gain-of-function mutations? Can you make the same statement for all cyclins? Explain.

26. Notice in figure 22.9 that Cyclin D levels are very low at the beginning of G_1. Cyclin D levels increase in response to extracellular signals. Briefly describe the mechanism that mediates the transcriptional activation of Cyclin D genes in response to these extracellular signals.

27. What effect on protein function and what cellular phenotype would you expect from the following *RB1* mutations? Exon 5 encodes the serine and threonine residues that are phosphorylated by CDKs.

```
Exon 1     ACCGCCGCGG AAAGGCGTCA TGCCGCCCAAA ACCCCC CGAA AAACGGCCGC

Allele 1   ACCGCCGCGG AAAGGCGTCA TGCCGCCCAAA ACCCCCCCGAA AAACGGCCGC

Exon 1     CACCGCCGCC GCTGCCGCCG CGGAACCCCCG GCACCGCCGC CGCCGCCCCC

Exon 1     TCCTGAGGAG GACCCAGAGC AGGACAGCGGC CCGGAGGACC TGCCTCTCGT CAG

Allele 2   TCCTGAGGAG GACCCAGAGC AGGACAGCGGC CCGGAGGACT TGCCTCTCGT CAG

Allele 3   TCCTGAGGAG GACCCAGAGC AGGACAGCGGC CCGGAGGACC TGCCTCTCGT CAC

Exon 2     GCTTGAGTTT GAAGAAACAG AAGAACCTGA TTTTACTGCA TTATGTCAGA

Exon 2     AATTAAAGAT ACCAGATCAT GTCAGAGAGA GAGCTTGGTT AACTTGGGAG

Exon 2     AAAGTTTCAT CTGTGGATGG AGTATTG

Exon 3     GGAGGTTATA TTCAAAAGAA AAAGGAACTG TGGGGAATCT GTATCTTTAT

Exon 3     TGCAGCAGTT GACCTAGATG AGATGTCGTT CACTTTTACT GAGCTACAGA


Exon 4     AAAACATAGA AATCAGTGTC CATAAATTCT TTAACTTACT AAAAGAAATT

Exon 4     GATACCAGTA CCAAAGTTGA TAATGCTATG TCAAGACTGT TGAAGAAGTA

Exon 4     TGATGTATTG TTTGCACTCT TCAGCAAATT GGAAAG

Exon 5     GACATGTGAA CTTATATATT TGACACAACC CAGCAGTTCGT

Allele 4   GACATGTGAA CTTATATATT TGACACAACC CAGCAGTTGGT

Allele 5   GACATGTGAA CTTATATATT TGACCCAACC CAGCAGTTCGT
```

28. Overexpression of the *MYC* gene is seen in many lymphoid malignancies and other cancers. The oncogenic potential of Myc is thought to relate to the ability of Myc to regulate expression of up to 15% of the genes in the human genome. You are studying a recessive mutant form of *MYC* in mice. This mutant, when homozygous, is embryonic lethal due to reduced levels of cell division. You generate mice heterozygous for a high-expressing variant of this mutant allele and find that even though no normal Myc is present, these mice display the wild-type phenotype. Using this as the background for a sensitized mutagenesis screen, you identify recessive mutants in a second gene that increase levels of cell division when homozygous. Animals homozygous recessive for both genes display a wild-type phenotype. Despite this apparent interaction, your studies are unable to identify any physical interaction between the two proteins.
    a. Describe the nature of the two mutations.
    b. Describe your next step in studying the function of Myc and this unknown gene in regulation of cell division.

29. What is cytochrome c, and why would its removal from mitochondrial membranes be an early step in the apoptotic path?

30. Growth factor binding to cell surface receptors simulates a survival signal that is mediated through Akt phosphorylation of Bad (figure 22.14).
    a. How would you expect increased expression of Bcl-2 to interfere with Bad/Bax equilibrium? Would this be an oncogenic mutation?
    b. Would a promoter mutation that decreases expression of Bax protein by 30% be likely to have an oncogenic effect? Explain.

31. One source of cellular stress is the presence of reactive oxygen species (ROS, or free radicals). It is estimated that ROS cause 20,000 lesions per cell per day in humans! Since most cells are in G_1/G_0, DNA repair must also happen in G_1/G_0. How does activation of p53 by cellular stress facilitate this repair?

32. Assume your cells are 99% efficient at repairing ROS damage in the G_1 phase of the cell cycle. How many unrepaired lesions will be present if a cell is in G_1 for 3 days prior to entry into S phase?

33. We described two different translocations that both contain a breakpoint near the heavy-chain immunoglobulin promoter (figures 22.7 and 22.18). Do you expect to find cancers due to misexpression of genes under this promoter in all tissues? Why or why not?

34. Describe the factors affecting the tissues that are more likely to be affected by mutations in specific tumor suppressor genes (table 22.5) or oncogenes (table 22.4).

35. A common urban legend claims that breast cancer is caused by use of deodorant. One line of "evidence" is the fact that a high proportion of breast cancers are found in the quadrant near the armpit. This quadrant is also a location of a large concentration of lymph nodes.

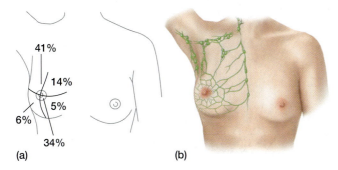

(a)   (b)

Based on your understanding of carcinogenesis, what is a potential *biological* explanation for the increased percentage of breast cancers originating in the upper, outer quadrant of the breast?

Chapter Integration Problem

Imagine you are a cytotechnician. You received a tissue sample from a biopsy of a suspicious epidermal growth and were asked to analyze it.

a. Your first step is to prepare a section of the tissue on a microscope slide to examine the tissue structure. Describe the signs you will look for to determine if this tissue represents a benign growth or if it is or is likely to become malignant. For simplicity, use the following simplified sketch of a normal epidermis as a reference.

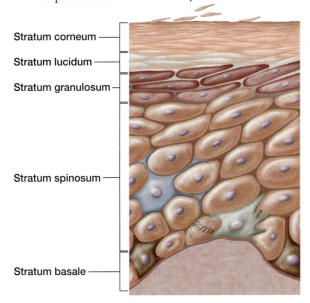

Stratum corneum

Stratum lucidum

Stratum granulosum

Stratum spinosum

Stratum basale

b. As part of your analysis, you need to know if there have been any changes to the amount and organization of genetic material in these cells. Explain the procedure(s) you will use.

c. The expression pattern of a family of genes known as *SUNBURN* can help you identify the stage of this tumor, as some genes in this family are activated early in tumor development and others only later. What technique(s) will you use to study gene expression in this biopsy specimen? Explain why you chose as you did.

d. The following chart shows the typical patterns of expression of *SUNBURN* genes in epithelial tissue under several different conditions. Which genes are likely to be involved in the growth factor response? Which are likely to be tumor suppressors? Explain your answers.

SUNBURN gene family member	Healthy Skin	Acute UV Exposure	Early Tumor (benign)	Late Tumor (malignant)	Metastatic Growth
1				Active	Highly active
2		Active	Active		
3	Active	Active	Active	Active	
4					Active
5		Highly active			
6	Active				
7	Active	Active			
8			Active	Active	Highly active

e. You are working with a researcher, trying to prevent damage caused by excessive UV exposure. Your colleague wants to identify compounds that prevent activation of *SUNBURN 2* and *5* after acute UV exposure. What effects do you expect treatment with such compounds will have on cancer risk following UV exposure?

f. Do you think inhibition of *SUNBURN* gene expression is a viable mechanism for treatment or prevention of UV-induced skin cancers? Explain.

g. The patient wants to know if her children have a higher risk of getting cancer because she has this growth. What other information would you need to have in order to answer this question? What factors probably influenced development of this growth, and do you think those factors are heritable? Does the stage of development of type of tumor affect your reasoning?

h. The patient's mother was recently diagnosed with breast cancer at age 62. Does this increase the likelihood that this family has an inherited predisposition to cancer? Explain.

Do you need additional review? Visit **www.mhhe.com/hyde** for practice tests, answers to end-of-chapter questions and problems, interactive exercises, and animation tutorials all designed to enhance your understanding of key genetic concepts.

Population Genetics

Essential Goals

1. Learn the forces that determine the amount and kind of genetic variation found in a population.

2. Describe the way the five major evolutionary forces change the amount and kind of genetic variation found in a population.

3. Understand the principle behind the Hardy–Weinberg equilibrium and apply it in a problem.

4. Use the different types of departures from Hardy–Weinberg equilibrium to predict the causes of disequilibrium.

5. Discuss the differences between the effects of random genetic drift and inbreeding.

Chapter Outline

Photo: A colony of ants represents a population that is under several different genetic principles.

*I*n previous chapters, we discussed genetics in relatively small groups, such as the F_1 and F_2 offspring from a pair mating. This type of analysis is useful for deducing potential genetic interactions, recombination mapping, or creating a specific genotype, but it fails to mimic what occurs in the real world. Outside of the laboratory, mating between individuals varies from completely random to biased for or against certain traits, and the expression of some alleles may affect the viability of the individual.

The study of the patterns of genetic variation in and among populations of individuals and the forces that shape those patterns is called **population genetics.** Population genetics provides the theoretical and conceptual underpinning necessary to understand quantitative genetics (chapter 24) and evolutionary genetics (chapter 25). These interrelated subjects are of great importance to real-world problems, such as the mapping of genes associated with complex diseases in humans and their domesticated plants and animals, the conservation of the world's precious biodiversity, the evolution of antibiotic resistance in bacteria and other pathogens, and the rapidly expanding field of biotechnology. Modern population geneticists require skills ranging from the ability to understand mathematical models to familiarity with the latest molecular genetics techniques.

The theory of population genetics was established by Ronald Fisher, J. B. S. Haldane, and Sewall Wright in the 1920s and 1930s. With the explosion of molecular techniques starting in the late 1960s, the types of questions being asked changed, and the data needed to answer long-standing disputes in population genetics became available. The result has rejuvenated population genetic studies, with a more applied focus. At its core, population genetics is the study of how differences in genetic constitution among individuals in a population are reflected in differences in behavior, morphology, and physiology among those individuals and their populations.

23.1 What Is Population Genetics?

Population genetics is the study of genetic variation at the population level that seeks to explain changes in allelic, genotypic, and phenotypic frequencies in a population through time and the analysis of divergence among populations. In particular, it explores why and how genetic variation and allelic frequencies are affected in a population in response to four evolutionary forces: mutation, selection, random genetic drift, and migration.

Definition of Population

Although defining a population is actually very difficult, as a general definition we can say that a **population** is an interbreeding group of individuals that exist together in time and space. Within a population, the spatial position of individuals does not affect their mate choice. From a genetic viewpoint, all the alligators in the Florida Everglades during a given year might constitute a population.

The important thing in defining a population is gene flow. All the alligators in the Everglades have the potential of interbreeding, and this is an important part of their being considered a population. Individual alligators may occasionally leave the Everglades, however, and successfully breed in other wetlands in Florida. The genetic constitution, or sum of all the genetic information, of a population is referred to as its **gene pool.**

Factors Influencing Genetic Variation

Five major forces affect the amount and distribution of genetic variation within and among populations:

Mutation.   **Mutation** is the source of all novel genetic variation. The vast majority of mutations either decreases fitness or has no effect on fitness at all. **Fitness** is the relative ability of individuals to pass on their alleles to future generations. Types of mutation and their analysis were covered in chapters 18 and 20; in this chapter, you'll see how alleles, which ultimately result from mutation, can be evaluated in populations.

Selection.   In combination with mutation, **selection** is the most important force in evolution. Natural selection has been responsible for the amazing array of biodiversity on earth and the fit of organisms to their environment. Selection occurs when some phenotypes (and underlying genotypes) are favored in an environment.

Random Genetic Drift.   Random sampling of gametes, in finite populations, results in a change in allelic frequencies from one generation to the next. This change is termed **genetic drift.** The smaller the population, the stronger the effects of genetic drift. Random genetic drift will lead to genetic divergence between populations.

Migration.   Movement of individuals in and out of populations, or **migration,** causes populations that would otherwise become genetically differentiated to become more genetically homogeneous.

Nonrandom Mating.   **Nonrandom mating** rearranges the structure of genetic variation without causing genetic diversity to be lost or gained. For example, inbreeding, per se, causes individuals to become more homozygous, while the population-level genetic diversity remains unchanged.

table 23.1   Factors Influencing Genetic Variation Within a Single Population

Factor	Influence on Genetic Diversity	Summary
Mutation	Increases genetic diversity	Provides the building blocks for evolution
Selection	*Directional* selection decreases genetic diversity, *stabilizing* selection decreases the number of homozygotes without changing the allelic frequency, *disruptive* selection decreases the number of heterozygotes	Primary force driving changes in the form and function of organisms
Random genetic drift	Decreases genetic diversity	Change in allelic frequencies due to chance
		Can decrease fitness and cause nonadaptive evolution
		Effect of drift increases with either a decreasing number of breeders or population size
Migration	Increases genetic similarity of populations linked by gene flow	Can introduce novel genetic variation to a population
		Under the proper conditions, migration can change allelic frequencies much more rapidly than mutation
Nonrandom mating	Restructures patterns of existing genetic diversity at the individual level	Changes phenotypic frequencies, but not allelic frequencies

We will explore these influences on the genetics of populations in the rest of this chapter. They are summarized in **table 23.1**. Next, we look at the mathematical treatment of genotypes in populations, which allows us to find the frequency with which an allele is present in a population.

Restating the Concepts

▶ A population is an interbreeding group of individuals that exist together in time and space. The gene pool is the collective sum of genetic information in a population.

▶ The five major forces that affect the genetic variation in a population are mutation, selection, random genetic drift, migration, and nonrandom mating.

23.2 Frequencies of Genotypes and Alleles

To describe the genetic constitution of a population, we begin by specifying the different genotypes in the population and determine the frequency of each genotype. Let's start with a simple case, an autosomal locus (A), with two different alleles A_1 and A_2.

There are three possible diploid genotypes: the A_1A_1 homozygote, the A_1A_2 heterozygote, and the A_2A_2 homozygote. The genetic constitution of the group could be fully described by the proportion of individuals that belong to each genotype. Naturally, the frequencies of all the genotypes must add up to 1.0.

Calculating Genotypic Frequencies

The proportion of individuals in a population that have a particular genotype can be calculated directly, as long as a method exists for identifying each genotype. Genotype can refer to one locus, two loci, or the whole genome, depending on the context.

Assume that we identify the following number of individuals with each corresponding genotype:

$$55 \text{ individuals are } A_1A_1$$
$$85 \text{ individuals are } A_1A_2$$
$$\underline{35 \text{ individuals are } A_2A_2}$$
$$= 175 \text{ Total individuals}$$

The frequency of each genotype is simply the proportion of individuals with each genotype. Thus, we can calculate the genotypic frequencies as:

$$p_{A_1A_1} = 55/175 = 0.31$$
$$p_{A_1A_2} = 85/175 = 0.49$$
$$p_{A_2A_2} = 35/175 = 0.20$$

Notice that the sum of the three genotypic frequencies equals 1.0.

Calculating Allelic Frequencies

In a diploid organism, each individual has two alleles that are either the same (homozygous) or different (heterozygous). To determine the total number of A_1 alleles in the previous example, we look at the number of A_1 homozygous individuals and the number of heterozygous individuals. Because the homozygotes have two A_1 alleles, this corresponds to 110 A_1 alleles (55 individuals × 2). Because the heterozygotes have one A_1 allele, this represents another 85 alleles (85 individuals × 1). The total number of all alleles is twice the total number of individuals of all genotypes, or 350 (175 × 2).

We can then calculate the frequency of the A_1 allele in this population as:

$$p_{A_1} = \frac{(110 + 85)}{350} = 0.557$$

The frequency of the A_2 allele, similarly, is the probability that any allele (gamete) drawn from that population at random is an A_2 allele. Stated another way, it is simply the proportion of all the gametes in that population that carry the A_2 allele. We know that 35 individuals are homozygous for A_2, so

$$p_{A_2} = \frac{(70 + 85)}{350} = 0.443$$

A summary of the calculations for determining the genotypic and allelic frequencies in this problem is shown in **table 23.2.**

Because the sum of the frequencies must total 1.0, we can also calculate p_{A_2} as:

$$p_{A_2} = 1.0 - p_{A_1} = 1.0 - 0.557 = 0.443$$

Restating the Concepts

▶ The genotypic frequency is the proportion of individuals in a population that possess the same genotype. To determine the genotypic frequency in a diploid organism, the heterozygous individuals must be distinguishable from each of the homozygous classes.

▶ The allelic frequency is the proportion of each allele in the genetic pool. This is analogous to the frequency of gametes that have a specific allele.

23.3 Hardy–Weinberg Equilibrium

Imagine a very large population in which the individuals mate randomly and all individuals and genotypes survive to the same extent. In this ideal situation, the population exhibits a simple relationship between the genotypic and allelic frequencies: If the frequencies of two alleles in the parental groups are p and q, then the genotypic frequencies of the two homozygous genotypes are p^2 and q^2 and the heterozygote is $2pq$. These genotypic frequencies correspond to individuals homozygous for the p allele (denoted P), heterozygous individuals (denoted H), and individuals homozygous for the q allele (denoted Q), respectively. The genotypic frequency equation

table 23.2   Calculating Genotypic and Allelic Frequencies.

	Genotype			
	A_1A_1	A_1A_2	A_2A_2	Total
Number of individuals	55	85	35	175
Genotypic frequency	55/175 = 0.31	85/175 = 0.49	35/175 = 0.20	
Number of A_1 alleles	110	85	0	195
Number of A_2 alleles	0	85	70	155
Total number of alleles				350
A_1 allelic frequency	(110 + 85)/350 = 0.557			
A_2 allelic frequency	(70 + 85)/350 = 0.443			

$$p^2 + 2pq + q^2 = 1.0$$

is referred to as the **Hardy–Weinberg equilibrium.** A defining feature of the Hardy-Weinberg equilibrium is that there are no changes in the allelic frequencies within or between generations.

Necessary Conditions for Hardy–Weinberg Equilibrium

The Hardy–Weinberg equilibrium is based on several assumptions about the organisms making up a population:

1. The organism is diploid and engages in sexual reproduction.
2. Generations are nonoverlapping.
3. Mating is random.
4. The size of the population is infinite (that is, there is no random genetic drift).
5. No mutation, migration, or selection occurs.

These assumptions are almost never true for a real-world population. For example, no population lacks mutation; selection is probably acting on at least some loci in a genome; and no population is infinite in size. Nevertheless, the Hardy–Weinberg equilibrium remains a very useful "null model"—that is, a baseline condition in which the allelic and genotypic frequencies are not changing in a population.

The way in which genotypic frequencies depart from Hardy–Weinberg expectations can tell us a lot about what has caused those departures:

Mutation.   Introduces new alleles (changes the allelic frequencies) into a population. Effects can usually only be measured over very long periods of time or with very large numbers of individuals.

Selection.   *Directional* selection increases the frequency of the favored allele, thus changing allelic and genotypic frequencies. Selection at a single locus is usually weak, but can potentially cause rapid departures from Hardy–Weinberg equilibrium. Other forms of selection have variable effects on genetic variation.

Random genetic drift.   May be confused with selection, as one allele increases in frequency. However, the allele that increases in one replicate population (or at another locus) will not be the same that increases in a different population. The changes in allelic frequencies are not consistent over time. Thus, drift increases the differences (variance) among populations.

Migration.   Introduces new alleles (changes allelic frequencies) in the population. Effects can usually only be measured over very long periods of time or with very large numbers of individuals.

Nonrandom mating.   Depends on the type of mating. Inbreeding leads to a number of homozygotes in excess of that expected under Hardy–Weinberg equilibrium.

Although few of the assumptions hold exactly, many hold approximately (especially over short periods of time)—so that many populations are approximately in Hardy–Weinberg equilibrium, for at least some selected loci. For example, mutation can be safely ignored over the span of a few generations; in addition, large populations undergo very little drift. We discuss these deviations in more detail later on.

Application of the Hardy–Weinberg Equilibrium

In the previous section, we counted the number of individuals with each genotype and then calculated the genotypic and allelic frequencies (see table 23.2). In contrast, the Hardy–Weinberg equilibrium is based on allelic frequencies (p and q) that are then used to predict the genotypic frequencies.

Genotypic Frequencies

Assuming that all of the conditions of Hardy–Weinberg equilibrium are met, the expected genotypic frequencies can be calculated by the simple rules of probability governing the random sampling of gametes from the population. Assume that the frequencies of the two alleles are as follows:

Frequency of allele $A_1 = p$

Frequency of allele $A_2 = q$

Then the frequencies of the genotypes would be:

Frequency of genotype $A_1A_1 = P$

Frequency of genotype $A_1A_2 = H$

Frequency of genotype $A_2A_2 = Q$

We know from the rules of probability that the sum of the allelic frequencies is equal to 1 ($p + q = 1$), and also that the sum of the genotypic frequencies is equal to 1 ($P + H + Q = 1$). Because each individual, being diploid, contains two alleles, the frequency of p can be shown as:

$$p = P + (0.5)H$$

Keep in mind that p includes all A_1A_1 homozygous individuals (p^2) and half of the A_1A_2 heterozygotes ($2pq$).

Following the same logic, the frequency of q can be shown as:

$$q = Q + (0.5)H$$

The frequency of the second allele, q, includes all A_2A_2 homozygotes plus half the heterozygotes ($2pq$).

We can also express $P + H + Q = 1$ in terms of the allelic frequencies. Imagine a gene with two alleles (A and a), with the frequency of $A = 0.5$. If we cross two heterozygous individuals (Aa), we can diagram the following Punnett square:

		Paternal Gametes	
		A ($p = 0.5$)	a ($q = 0.5$)
Maternal Gametes	A ($p = 0.5$)	AA ($p^2 = 0.25$)	Aa ($pq = 0.25$)
	a ($q = 0.5$)	Aa ($pq = 0.25$)	aa ($q^2 = 0.25$)

This is nothing new—you first learned in chapter 2 that there are four potential genotypic contributions, and that the probability of each is based on the probability of the parent producing a gamete with the designated allele. The total probability of producing one of these four genotypes is $p^2 + 2pq + q^2 = 1.0$. This is the basis for the 1:2:1 genotypic frequency produced in a monohybrid cross.

Phenotypic Frequencies

In population genetics, we are not setting up a single cross, but instead examining random matings in the population. The probability of producing the two different gametes is not based on the genotype of each parent, but rather on the frequency of each allele in the population. Knowing the allelic frequency allows a researcher to determine the frequency of the AA homozygote (p^2), the Aa heterozygote ($2pq$), and the aa homozygote (q^2). We can also apply these allelic frequencies to calculate the phenotypic frequencies in the population.

Let's look at an example of how allelic frequencies can be used to calculate the phenotypic frequencies. Assume that the frequency of alleles determining the M-N blood group in a certain population is as follows:

$$M = 0.6$$

$$N = 0.4$$

If the population is under Hardy–Weinberg equilibrium, we can calculate the phenotypic frequencies as follows:

$$MM = P = p^2 = (0.6)^2 = 0.36$$

$$MN = H = 2pq = 2 \times 0.6 \times 0.4 = 0.48$$

$$NN = Q = q^2 = (0.4)^2 = 0.16$$

Alternatively, if we only know the phenotypic frequencies, we can use our knowledge of the inheritance of the phenotype to determine allelic frequencies, provided the population is in Hardy–Weinberg equilibrium.

In humans, the presence or absence of dimples is governed by one gene with two alleles. The allele for dimples is completely dominant to the allele for no dimples. The frequency of each phenotype is:

Dimples present: 36%

Dimples absent: 64%

To solve for the allelic frequencies we note that individuals lacking dimples (recessive phenotype) must be homozygous for the recessive allele. Because 64% of the population possesses the recessive phenotype ($q^2 = 0.64$), the frequency of the q allele must equal 0.8 (the square root of 0.64) under Hardy–Weinberg expectations. The frequency of the p allele would then be equal to $1 - q$, or 0.2.

Knowing the allelic frequencies ($p = 0.2$ and $q = 0.8$) allows us to calculate the genotypic frequencies:

Homozygous (dimples) = $p^2 = (0.2)^2 = 0.04$

Heterozygous (dimples) = $2pq = 2 (0.2) (0.8) = 0.32$

Homozygous (no dimples) = $q^2 = (0.8)^2 = 0.64$

Total = $0.04 + 0.32 + 0.64 = 1.00$

We conclude that approximately 89% of the people that have dimples are expected to be heterozygotes (0.32 / 0.36 = 0.889).

You can see from this example that if a population is in Hardy–Weinberg equilibrium, the genotypic and phenotypic frequencies can be determined from the frequency of one of the two alleles in the population. This is possible because $p + q = 1.0$ and $p^2 + 2pq + q^2 = 1.0$; therefore, the two expressions are equivalent.

We can diagram this relationship between the allelic frequency and the genotypic frequencies as

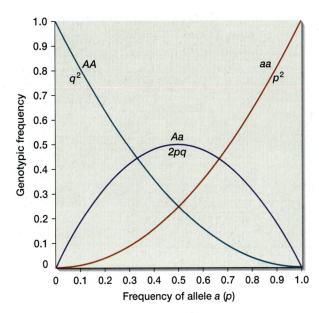

figure 23.1 The Hardy–Weinberg equilibrium predicts the different genotypic frequencies in a population based on the frequency of one allele. When only two alleles (*A* and *a*) are present at a given locus that is under Hardy–Weinberg equilibrium, the frequency of one allele (*p*) is used to calculate the genotypic frequencies in the population based on the equation $p^2 + 2pq + q^2 = 1$. In this graph, knowing the frequency of the *a* allele, you can determine the frequencies of the three different genotypes (*AA*, *Aa*, and *aa*) by reading up the vertical axis.

shown in **figure 23.1.** If we know the frequency of p, we can read straight up on the graph and determine the frequency (read from the vertical axis) of both homozygotes (p^2 and q^2) and the heterozygote ($2pq$). For example, if $p = 0.55$, then $q^2 = 0.203$, $p^2 = 0.303$, and $2pq = 0.494$.

Testing Whether the Genotypic Frequencies in a Population Are in Hardy–Weinberg Equilibrium

As mentioned earlier, the conditions for Hardy–Weinberg equilibrium are usually never entirely satisfied in a population; however, they can be nearly satisfied over a short period of time. In this latter case, how can we determine whether the genotypic frequencies in a population are in Hardy–Weinberg equilibrium?

Let's examine a case study using the wolf spider, *Rabidosa rabida* (**fig. 23.2**). Assume that we determined the genotypes of 139 individual spiders at two different loci (*S* and *W*). First, let's look to see if the *S* locus is under Hardy–Weinberg equilibrium. The *S* locus contains two alleles, S_1 and S_2, and the genotypes of the 139 individual spiders are shown in **table 23.3.**

figure 23.2   A wolf spider (*Rabidosa rabida*).

The frequency of the S_1 allele is 91/278, or 0.327, and that of the S_2 allele is 187/278, or 0.673. Based on these allelic frequencies, we can calculate the expected genotypic frequencies:

$$(S_1S_1) = p^2\,(N) = (0.327)^2 \times 139 = 14.9$$

$$(S_1S_2) = 2pqN = 2\,(0.327)\,(0.673)\,(139) = 61.2$$

$$(S_2S_2) = q^2\,(N) = (0.327)^2 \times 139 = 62.9$$

where N is the number of individuals sampled.

Although these expected genotypic frequencies differ from the observed values, we do not know whether they are statistically different or not. To perform this analysis, we employ the chi-square test (see chapter 2). The null hypothesis is that the number of individuals is equal to those expected under Hardy–Weinberg equilibrium, given our observed allelic frequencies. To conduct the statistical analysis we can arrange the data as in **table 23.4.**

Recall that a chi-square value is calculated by using the formula:

$$\chi^2 = \sum \frac{(O-E)^2}{E}$$

where O is the observed value and E is the expected value.

Because we are examining three different genotypes and we have to use the observed data to calculate one of the parameters (p), there is only one degree of freedom in the chi-square analysis. The calculated chi-square value (0.654) is far less than the chi-square critical value ($p < 0.05$) for one degree of freedom, which is 3.84. Thus, we fail to reject the null hypothesis. This population appears to be in Hardy–Weinberg equilibrium at this locus.

We can perform a similar analysis for the W locus, which also consists of two alleles (W_1 and W_2). **Table**

table 23.3   Genotypes of 139 Wolf Spiders at the *S* Locus

	Genotype			
	S_1S_1	S_1S_2	S_2S_2	Total
Number of individuals	17	57	65	139
Number of S_1 alleles	34	57	0	91
Number of S_2 alleles	0	57	130	187
Total number of alleles				278

table 23.4   Chi-Square Analysis of the Numbers of Individuals with Different Genotypes at the *S* Locus

Genotype	Expected	Observed	χ^2
S_1S_1	14.9	17.0	0.296
S_1S_2	61.2	57.0	0.288
S_2S_2	62.9	65.0	0.070
Totals	**139**	**139**	**0.654**

table 23.5   Genotypes of 139 Wolf Spiders at the *W* Locus

	Genotype			
	W_1W_1	W_1W_2	W_2W_2	Total
Number of individuals	45	86	8	139
Number of W_1 alleles	90	86	0	176
Number of W_2 alleles	0	86	16	102
Total number of alleles				278

23.5 shows the genotypes for the same 139 wolf spiders. The observed frequency of the W_1 allele is 176/278, or 0.633, and of the W_2 allele, 102/278, or 0.367. The expected genotypic frequencies are therefore:

$$(W_1W_1) = p^2\,(N) = (0.633)^2 \times 139 = 56$$

$$(W_1W_2) = 2pqN = 2\,(0.367)(0.633)(139) = 64$$

$$(W_2W_2) = q^2\,(N) = (0.367)^2 \times 139 = 19$$

Again, the null hypothesis states that the expected genotypic frequencies under Hardy–Weinberg equilibrium are the same as those observed. The chi-square analysis of this data is shown in **table 23.6.**

For the W locus, the calculated chi-square value (16.092) far exceeds the chi-square critical value ($p < 0.05$) for one degree of freedom (3.84). This result is highly unlikely to occur if the null hypothesis is true.

table 23.6 Chi-Square Analysis of the Numbers of Individuals with Different Genotypes at the *W* Locus

Genotype	Observed	Expected	χ^2
W_1W_1	45	56	2.161
W_1W_2	86	64	7.563
W_2W_2	8	19	6.368
Totals	139	139	**16.092**

Therefore we reject the null hypothesis and conclude that this population is not in Hardy–Weinberg equilibrium at the *W* locus.

In fact, this population consists of a notable excess of W_1W_2 heterozygotes and a deficit of both homozygous genotypes. One possible explanation for this observation is better survival of heterozygous individuals.

This example shows that it is possible for a population to be in Hardy–Weinberg equilibrium at one locus but not at another locus.

Hardy–Weinberg Equilibrium with More Than Two Alleles

In the discussion with two alleles, the genotypic frequency of $p^2 + 2pq + q^2 = 1.0$ is derived from the binomial expansion, $(p + q)^2 = 1.0$. For many genes, more than two alleles need to be evaluated under Hardy–Weinberg expectations. In these cases, the same probability rules apply.

Let's look at a three-allele system, with alleles A_1, a_2, and a_3. In this example, A_1 is completely dominant to both a_2 and a_3. Assume we identify a population in which 1 in 2500 individuals have the recessive phenotype. Furthermore, we know that the a_2 allele represents 30% of all the recessive alleles. What are the frequencies of all three alleles?

Assuming the frequencies of the A_1 allele is p, of the a_2 allele is q, and of the a_3 allele is r, then the allelic frequencies must total:

$$p + q + r = 1.0.$$

Based on these allelic frequencies, the genotypic frequencies must be:

$$(p + q + r)^2 = 1.0.$$

The equation resolves into:

$$p^2 + q^2 + r^2 + 2pq + 2pr + 2qr = 1.0.$$

Individuals with the recessive phenotype must contain two recessive alleles (a_2a_2, a_3a_3, and a_2a_3), which correspond to the genotypic frequencies of q^2, r^2, and

$2qr$. Because the frequency of the recessive phenotype is 1 in 2500, we can state that:

$$q^2 + r^2 + 2qr = \frac{1}{2500}$$

If the a_2 allele represents 30% of all the recessive alleles (q), then the a_3 allele frequency (r) must be 70%, which is 2.33 times greater than q. Another way of saying this is $r = 2.33q$. We can then replace r in the preceding equation with $2.33q$ and solve for the frequency of the q allele:

$$q^2 + r^2 + 2qr = 1/2500$$
$$q^2 + (2.33q)^2 + 2(q)(2.33q) = 0.0004$$
$$q^2 + 5.4\,q^2 + 4.66q^2 = 0.0004$$
$$11.09q^2 = 0.0004$$
$$q^2 = 0.0004/11.09$$
$$q = 0.006$$

Because $r = 2.33q$, $r = 0.014$. Returning to the original allelic frequency equation, $p + q + r = 1.0$, we can solve for p, obtaining a value of 0.980. Thus, the A_1 allelic frequency (p) is 0.98, the a_2 allelic frequency (q) is 0.006, and the a_3 allelic frequency (r) is 0.014.

Using these allelic frequencies, you can calculate any genotypic frequency using the Hardy–Weinberg equilibrium equation. For example, the frequency of the homozygous dominant individuals (A_1A_1) can be calculated as p^2, or $0.98 \times 0.98 = 0.96$. The frequency of the heterozygous individuals with the dominant phenotype (A_1a_2 and A_1a_3) would be $2pq + 2pr = 0.012 + 0.027 = 0.039$.

The genotypic frequency of a homozygote is the square of the allelic frequency, and the genotypic frequency of a heterozygote is double the product of the two allelic frequencies.

Restating the Concepts

▶ Hardy–Weinberg equilibrium defines a condition in which the allelic and genotypic frequencies in a population are not changing. This requires no random genetic drift, no gene flow from other populations, no mutation, random mating, and equal fitness of genotypes.

▶ For a gene with two alleles, the sum of the allelic frequencies ($p + q$) must equal 1.0 and the sum of the genotypic frequencies ($p^2 + 2pq + q^2$) must also equal 1.0.

▶ Hardy–Weinberg equilibrium can occur at a subset of loci within a population. It can also occur at loci that possess more than two alleles.

23.1 applying genetic principles to problems

Application of the Hardy–Weinberg Equilibrium to a Human Population

Let's look at an example of three alleles, the ABO blood type, with alleles I^A, I^B, and i. The blood type of 500 individuals from Massachusetts will serve as our population (table 23.A). Clearly this is not an infinite population, but a relatively large number of randomly selected individuals may serve to represent the larger population. We can see whether this population is under Hardy–Weinberg equilibrium.

Assume that p is the frequency of I^A (f_{I_A}), q is the frequency of I^B (f_{I_B}), and r is the frequency of i (f_i), such that $p + q + r = 1.0$. What are the allelic frequencies?

Under Hardy–Weinberg equilibrium, if you can determine the genotypic frequency of a homozygote, then the square root of that value gives you the allelic frequency. The only blood type phenotype that is composed of only homozygotes is blood type O, which can only be produced by the ii genotype. Using the observed numbers from table 23.A, the frequency of this blood type is 231/500, or 0.462. Under Hardy–Weinberg equilibrium, this homozygous genotypic frequency would therefore correspond to r^2. If r^2 is equal to 0.462, then $r = 0.680$.

Now you must use the frequency of the i allele to calculate one of the remaining two alleles. In table 23.A, you can see that blood types A and O include only genotypes $I^A I^A$, $I^A i$, and ii, which is similar to the two-allele problems discussed in this chapter.

Because these two blood types do not account for all 500 individuals, we must first determine the proportion of the population that contains these two blood types. Under Hardy–Weinberg equilibrium:

$$(p + r)^2 = \frac{(199 + 231)}{500}, \quad \text{or } 0.860$$

If you then take the square root of both sides, $(p + r)$ is equal to 0.927. Using the value of r that you calculated earlier, 0.680, p is then equal to $0.927 - 0.680$, or 0.247.

You can calculate the frequency of the third allele based on the equation $(p + q + r) = 1.0$. Knowing that $r = 0.680$ and $p = 0.247$, you can easily deduce that q must be 0.073. Thus, the frequency of the I^A allele (p) is 0.247, that of the I^B allele (q) is 0.073, and finally that of the i allele (r) is 0.680.

To determine whether this group of 500 individuals is in Hardy–Weinberg equilibrium, you need to calculate

table 23.A ABO Blood Type Distribution in 500 Individuals from Massachusetts

Blood Type	Genotype	Observed Number
A	$I^A I^A$ or $I^A i$	199
B	$I^B I^B$ or $I^B i$	40
AB	$I^A I^B$	30
O	ii	231
Total		500

the expected number of individuals with each blood type based on the allelic frequencies. Blood group O, produced only by the ii genotype, will occur at a frequency of $r^2 \times 500$ individuals, or $(0.680)^2 \times 500 = 231$ (table 23.B).

Blood group A can be produced by either $I^A I^A$ or $I^A i$. The frequency of $I^A I^A$ is $(0.247)^2 \times 500$, or 31 individuals. The frequency of the $I^A i$ heterozygote is $2pr$, or $2 \times 0.247 \times 0.680 = 0.336$, which corresponds to 168 out of 500 individuals. Thus, a total of 199 individuals (31 + 168) should exhibit blood group A.

Similarly, blood group B will be produced by either $I^B I^B$ or $I^B i$. The frequency of $I^B I^B$ is $(0.073)^2$, or 3 individuals. The frequency of $I^B i$ is $2 \times 0.073 \times 0.680$, or 50 individuals, for a total of 53 individuals exhibiting blood group B.

Finally, the $I^A I^B$ frequency is $2 \times 0.247 \times 0.073$, or 18 individuals having AB blood type.

You can now apply a chi-square analysis to this data, with your null hypothesis being that the expected genotypic frequencies are under Hardy–Weinberg equilibrium given our observed allelic frequencies are the same as those observed genotypic frequencies. In this case, we have four genotypes and we used the observed data to calculate two of the phenotypes, which yields one degree of freedom. You can see that the calculated chi-square value, 11.19, far exceeds the chi-square critical value ($p < 0.05$) for one degree of freedom, which is 3.84. This result is highly unlikely to occur if your null hypothesis is true. Therefore, you would reject the null hypothesis and state that this population is not in Hardy–Weinberg equilibrium at the *ABO* blood type locus.

table 23.B ABO Blood Type Distribution in 500 Individuals from Massachusetts

Blood Type	Genotype	Observed Number	Expected Number	$(O - E)^2/E$
A	$I^A I^A$ or $I^A i$	199	199	0
B	$I^B I^B$ or $I^B i$	40	53	3.19
AB	$I^A I^B$	30	18	8
O	ii	231	231	0
Total		500	500	$\chi^2 = 11.19$

23.4 Selection: Shifting Allelic Frequencies

Natural selection works because of reproductive excess, which means that more offspring are produced than actually survive to sexual maturity. In the wolf spider, an average of only 1% of the offspring survive to sexual maturity. Some females will have many offspring reaching sexual maturity, but most females may have none. These offspring differ in any number of phenotypic traits, and some of these differences will contribute to an individual's ability to pass their genes on to the next generation—that is, the offspring differ in fitness.

Fitness results from the presence of certain alleles in an individual's genotype. Individuals with greater fitness leave more offspring, and the alleles these individuals carry therefore increase in frequency. This is the essence of selection, whether it is natural selection or selection imposed in the farm or laboratory. Because some alleles change frequency during selection, the Hardy–Weinberg equilibrium cannot be achieved.

The Concept of Relative Fitness

Absolute fitness refers to the average reproductive rate of individuals with the same genotype. At an absolute fitness of 1, the population size is relatively constant. **Relative fitness** refers to a genotype's ability to survive and reproduce relative to other genotypes in the population. The relative fitness of different individuals is reflected in their ability to pass on their alleles to future generations. It is easier to calculate the effects of selection if we use relative, rather than absolute, fitness of genotypes.

Let's look at a hypothetical genotype of wolf spider (**table 23.7**). *Fecundity* equals the number of offspring born to an average female with a specific genotype, and *survival* equals the proportion of those offspring that, on average, survive to sexual maturity. Fitness is expressed as the product of fecundity times survival.

Fitness for a particular gene is standardized relative to the genotype with the highest fitness (A_1A_2 in

this example). The A_1A_1 homozygote has 95% of the fitness of the heterozygote ($2.66 / 2.80 = 0.95$) and the A_2A_2 homozygote has 43% of the fitness of the heterozygote ($1.20 / 2.80 = 0.43$).

This example helps to demonstrate the importance of both fecundity and survival in determining fitness. The A_1A_1 genotype has the greatest fecundity, but not the highest fitness because a lower proportion of the offspring reach sexual maturity. In contrast, the heterozygote has the greatest fitness even though it has a lower fecundity, because a greater proportion of the offspring reach sexual maturity.

Simple Models of Selection

Selection is the process by which one allele or allele combination is favored over another and results in the allelic frequencies changing away from Hardy–Weinberg equilibrium. The selection may result from one allele producing a phenotype that provides more protection from a predator than another phenotype, such as the coloring on butterflies. The selection may also pertain to one allele producing a phenotype that yields increased viability in a specific environment relative to a second allele.

An example of environmental selection is the presence of organophosphate pesticide resistance in the mosquitoes *Anopheles gambiae* and *Culex pipiens,* which transmit malaria and West Nile virus, respectively. Organophosphate resistance can be generated by a single amino acid substitution in the mosquito's acetylcholinesterase gene (*ace-1*). Researchers isolated resistant mosquitoes across Africa, America, and Europe and found them all to contain the same nucleotide difference in the *ace-1* gene relative to sensitive mosquitoes. This difference in a single nucleotide resulted in the acetylcholinesterase being either resistant or sensitive to the organophosphate pesticide. The presence of this same *ace-1* mutation in resistant mosquitoes on three different continents demonstrates the power of selecting an allele that confers an advantage to an organism.

Selection can act in three main ways (**fig. 23.3**). **Directional selection** acts to continuously remove individuals from one end of the phenotypic distribution. This results in a shift in the phenotypic mean and a decrease in the frequency of one of the alleles in a two-allele system. The *ace-1* resistant allele is an example of directional selection, where the *ace-1*-sensitive allele decreased in frequency in areas that are exposed to the organophosphate pesticides.

Stabilizing selection acts to remove individuals from both ends of the phenotypic distribution. This selection does not change the phenotypic mean or the allelic frequency, but it does decrease the phenotypic frequency of both homozygous groups.

table 23.7   Fitness of a Genotype, Standardized to a Reference Genotype

Genotype	Fecundity	Survival	Fitness	Relative Fitness
A_1A_1	266	0.0100	2.66	0.95
A_1A_2	250	0.0112	2.80	1.00
A_2A_2	150	0.0080	1.20	0.43

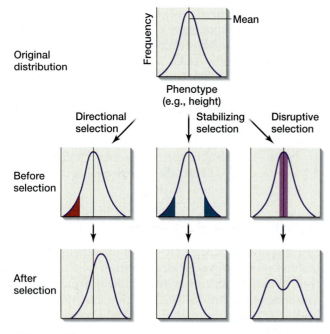

figure 23.3 The effects of directional, stabilizing, and disruptive selection. The top graph shows a phenotypic distribution and the mean phenotype. In all the subsequent graphs, the vertical line represents the original mean phenotype. Colored areas in the middle row show the phenotypic groups being selected against in each selection mechanism. The bottom row shows the final phenotypic distributions after selection. In directional selection (left column), one homozygous trait is at a disadvantage, which shifts the phenotypic distribution towards the other homozygous trait. In stabilizing selection (center column), both homozygous traits are selected against, which increases the frequency of the mean trait. In disruptive selection (right column), the intermediate trait is selected against, which increases the frequency of both homozygous traits without changing the mean.

Disruptive selection works by favoring the individuals at the phenotypic extremes over the individuals in the middle of the phenotypic distribution. This usually results in the loss of heterozygous individuals and an increase in the number of both homozygous genotypes.

Let's start with the simplest case of selection in a diploid: directional selection operating at one locus, consisting of two alleles with one allele completely dominant to the other. In this case, we make several assumptions:

1. Fitness differences are due only to differences in viability.
2. Fitness differences are due to genotypic differences at a single locus.
3. Random mating occurs.
4. Selection is identical in both sexes.
5. The **selection coefficient,** which is a measure of fitness disadvantage, is constant through time.

6. Population size is infinite.
7. No mutation occurs at the locus.

In this example, the two alleles being examined are A_1 and A_2. The relative fitness of the A_1A_1 homozygote and the A_1A_2 heterozygote are equivalent and equal to 1 and the A_2A_2 homozygote has reduced relative fitness. The genotypic frequencies correspond to the allele frequencies (p and q).

Genotype	A_1A_1	A_1A_2	A_2A_2
Frequency (0)	p^2	$2pq$	q^2
Relative Fitness	1	1	$1 - s$

The reduced relative fitness of the A_2A_2 homozygote can be defined as $1 - s$, where s is the *selection coefficient* and corresponds to a number between zero and one. The selection coefficient represents the difference in the relative fitness between two genotypes or a measure of fitness disadvantage of one genotype relative to another genotype. In this example, if s is 0.33 then individuals with the A_2A_2 genotype produce only 67% as many viable offspring as the other two genotypes, or put differently, their relative fitness is 0.67. In the absence of mutation and genetic drift, the A_1 allele should increase in frequency, in a deterministic manner but not at a constant rate, and it will eventually become fixed (that is, reach a frequency of 1.0). The A_2 allele will be lost.

How rapidly will the population change through time? If the allele frequencies are $A_1 = 0.7$ and $A_2 = 0.3$, we can determine the genotypic frequencies (**table 23.8**). Multiplying the genotypic frequencies by the relative fitness yields the contribution of each genotype in the next generation. However, the total fitness of the population, *mean fitness,* does not add up to 1.0 because of the reduced fitness of the A_2A_2 genotype.

If we divide each of the genotypic contributions by the mean population fitness (0.97), we have the new genotypic frequencies (table 23.8). In this example, nearly one-third of the A_2A_2 genotype was lost in a single generation (from 9% of the population to 6.2%).

The change in allelic frequencies from one generation to the next can now be calculated. Once the proportion of each genotype is known, it is a simple matter to get the new frequency of the A_1 allele: $p' = 0.505 + 0.5 (0.433) = 0.7215$. Thus, the frequency of the A_1 allele increased from 0.7000 to 0.7215, or a change, Δp, of +0.0215. Because $p + q = 1.0$, the A_2 allele decreased (Δq) by −0.0215 to a frequency of 0.2785.

Additive Effects

In the case just described, the A_1 allele was totally dominant to the A_2 allele in its effects on fitness. In this case, the heterozygote's fitness was equal to that of the A_1 homozygotes. Now we'll consider what happens with directional selection on the same allele, under the same circumstances, except that the effects of the alleles are additive. **Additive effects** occur when the phenotypic value for the heterozygote is the mean of the phenotypic value for each of the homozygotes.

In this case, the relative fitness of A_1A_1 remains 1.00 and of A_2A_2 remains 0.67. However, the relative fitness of A_1A_2 becomes the *mean* relative fitness, or 0.835 (**table 23.9**). Again, we scale the relative contribution of each genotype by the mean fitness of the population (0.90) to get the new genotypic frequencies. The new frequency of the A_1 allele, $p' = 0.544 + 0.5 (0.389) = 0.7385$. Thus, the frequency of the A_1 allele has increased from 0.7000 to 0.7385, or $\Delta p = +0.0385$.

Notice that the change in the allelic frequency is larger for the additive case (+0.0385) relative to the case with dominance (+0.0215). That is because in the additive case, selection acts against the deleterious allele in both homozygotes and heterozygotes. In contrast, dominance shields the recessive allele against selection in the heterozygotes and weakens selection's ability to weed out deleterious mutations. This phenomenon explains the observation that in natural populations, deleterious alleles are almost always recessive. Deleterious mutations that are dominant are generally eliminated quickly, and researchers seldom ever sample them.

An example of an additive effect is seen with the *actin-2* (*act2*) gene in the weed *Arabidopsis thaliana*. Researchers examined two alleles, the wild-type A_1 allele and a mutant allele, A_2. Starting with a single heterozygous plant (A_1A_2), the researchers selfed the plant and the progeny for five generations. As expected, approximately 50% of the F_1 plants were heterozygous, 25% were A_1A_1 homozygotes and 25% were A_2A_2 homozygotes (**fig. 23.4,** *solid line and filled symbols*). By the F_2 generation, it is apparent that the A_1A_1 homozygote is much more frequent than the A_2A_2 homozygote. By the F_5 generation, the A_1A_1 homozygote is approximately four-fold more frequent than the A_2A_2 homozygote. The researchers determined that this change in allelic frequency can be predicted (fig. 23.4, *dashed line and unfilled symbols*) if the fitness of the A_1A_1 homozygote is 1.0, the A_1A_2 heterozygote is 0.85, and the A_2A_2 homozygote is 0.73.

Response to Selection

The speed with which allelic frequencies change depends on the strength of selection, the level of

table 23.8   Computing Changes in Genotypic Frequencies, over One Generation, with Directional Selection

Genotype	Genotypic f	Relative Fitness	Contribution	New Genotypic f
A_1A_1	$(0.7)^2 = 0.49$	1.00	$(0.49)(1.00) = 0.49$	$0.49/0.97 = 0.505$
A_1A_2	$2(0.7)(0.3) = 0.42$	1.00	$(0.42)(1.00) = 0.42$	$0.42/0.97 = 0.433$
A_2A_2	$(0.3)^2 = 0.09$	0.67	$(0.09)(0.67) = 0.06$	$0.06/0.97 = 0.062$
Total			Mean fitness = 0.97	1.000

table 23.9   Computing Changes in Genotypic Frequencies, over One Generation, with Directional Selection on Alleles with No Dominance (Additive Case)

Genotype	Genotypic f	Relative Fitness	Contribution	New Genotypic f
A_1A_1	0.49	1.00	$(0.49)(1.000) = 0.49$	$0.49/0.90 = 0.544$
A_1A_2	0.42	0.835	$(0.42)(0.835) = 0.35$	$0.35/0.90 = 0.389$
A_2A_2	0.09	0.67	$(0.09)(0.670) = 0.06$	$0.06/0.90 = 0.067$
Total			Mean fitness = 0.90	1.000

dominance, and the allelic frequencies themselves. In general, selection at any single locus tends to be very weak, and selection coefficients (s) are generally much less than 0.01. For these reasons, changes in allelic frequencies from selection alone are, usually, very slow. However, many excellent examples are known of very rapid changes in allelic frequencies in cases where strong selection is applied.

One example of rapid change is seen with the coloration of the peppered moth (*Biston betularia*) in Europe and North America. The peppered moth has two major color phenotypes: a white-speckled coloration and a dark pigmented (melanic) phenotype (**fig. 23.5**). It is known that the melanic phenotype is controlled by a single locus, with the dark allele being dominant to the white-speckled.

Before the Industrial Revolution in England, the white-speckled peppered moth was camouflaged from its major predator, birds, as it rested on the trunks of trees. Industrialization led to severe pollution and an accumulation of soot on trees, which led to the death of lichen and the darkening of the previously light-colored trees. At approximately the same time, the melanic peppered moth increased in frequency. By the 1950s, a sharp change in the frequency of the melanic phenotype existed across England, with 90% or more of the peppered moths in the northeast having the melanic phenotype.

The Clean Air Act of 1956 in England resulted in significantly less air pollution, and the trees returned to a light color. This led to a significant decrease in the proportion of the melanic peppered moth. By the

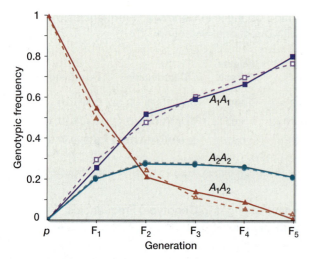

figure 23.4 The change in genotypic frequencies under additive selection. A single plant (*Arabidopsis thaliana*), which was heterozygous for the *actin-2* gene (A_1A_2), was selfed. The progeny were also selfed for four additional generations. The observed genotypic frequencies (filled symbols and solid line) are plotted against the generation. The open symbols and dashed lines represent the predicted genotypic frequencies using a fitness of 1.0 for the A_1A_1 genotype, 0.85 for the A_1A_2 genotype, and 0.73 for the A_2A_2 genotype. The good correspondence between the observed and expected curves suggests that the A_1 allele confers a higher fitness than the A_2 allele.

(From Lui, et al., *Genetical Research*, pp. 41–53, figure 6, © 2003. Reprinted with the permission of Cambridge University Press.)

(a)   (b)

figure 23.5 Two coloration phenotypes of the peppered moth (*Biston betularia*). (a) The white-speckled (top) and melanic (bottom) peppered moths on a light-colored tree. An arrow shows the location of the white-speckled peppered moth. (b) The white-speckled (top) and melanic (bottom) peppered moths on a dark colored tree. An arrow shows the location of the melanic moth.

1990s, the maximum frequency of the melanic peppered moth had dropped to less than 50% and in most places it has decreased to less than 10%. Because the two phenotypes are controlled by two alleles at a single locus, the changes in phenotype are attributed

to changes in the allelic frequencies. Similar rapid changes in the frequency of peppered moth coloration have been described in Europe and North America.

An even more rapid shift between phenotypes has been observed in the guppies (*Poecilia reticulata*) in the northern mountains of Trinidad. These guppies were relatively small in size and produced large numbers of small progeny. The dominant guppy predator on the southern slopes is a pike (*Crenicichla alta*) that eats guppies of any size. The large numbers of progeny were required to ensure that the guppy population was maintained in the presence of the pike. Upstream of the waterfalls, the killifish (*Rivulus hartii*) is a general predator that feeds on small guppies. Guppies were transplanted from a downstream location that contained a high concentration of predators to an upstream location that had a low concentration of killifish predators. Within four years in this new environment, guppies were selected that grew larger in size and produced fewer and larger progeny. These changes were estimated to be 10,000–10,000,000 times faster than the average rates of change as determined from the fossil record.

The strength of selection against a particular allele obviously has a huge effect on how quickly the frequency of that allele will change. Let's look at two scenarios to see the difference (**table 23.10**). In the first, selection is very strong ($s = 0.99$). In the second case, selection is much weaker ($s = 0.01$). If the genotypic frequencies are equivalent at the start, the change in allelic frequencies (Δp) is ten-fold greater when selection against the allele is stronger.

Selection also changes allelic frequencies faster when allelic frequencies themselves are intermediate. This is especially true when one allele is dominant to the other. If selection occurs only against individuals homozygous for the recessive allele, and that allele is rare, then homozygous individuals are going to be very rare (assuming random mating). Selection cannot "see" the allele to act against it—because it is so rarely expressed in the phenotype. You can test these hypotheses for yourself by plugging different allelic frequencies (or dominance levels) into the preceding example tables.

Maintaining Genetic Variability Through Selection

Selection is usually thought of as a "purifying process." It is the force that drives the favorable allele towards fixation, thus reducing genetic variation. But there are a number of scenarios in which selection does not fix one single allele, usually because there is no single best allele.

Heterozygote Advantage (Stabilizing Selection)

Heterozygote advantage, or stabilizing selection, occurs when the heterozygote is the fittest genotype (see fig. 23.3). Although it appears that factors other than heterozygote advantage are responsible for most of the observed genetic diversity, many convincing cases of heterozygote advantage are known. Most cases are specific to a particular environment.

The classic example of heterozygote advantage is β-hemoglobin in humans. There are many alleles at the β-globin locus, but we will define them in two broad categories: $Hb^A(A)$ and $Hb^S(S)$. The Hb^A allele is the wild-type alleles; Hb^S is the sickle cell anemia allele.

The ability of the malarial parasite to infect and proliferate in humans is affected by the individual's genotype. *AA* individuals are sensitive to malarial infection, whereas *SS* individuals are resistant to malarial infection but exhibit sickle cell anemia. The *AS* heterozygotes are resistant to malaria but exhibit only a slight sickling of their red blood cells. The relative fitness of the three genotypes, as measured in the West African malarial area, are 0.89 for *AA*, 1.00 for *AS*, and 0.2 for *SS*.

Let's consider in a general way how selection can maintain a polymorphism when the heterozygote is fitter than either homozygote. Below is the familiar table showing genotypes, their frequencies, and the relative fitness of each genotype.

Genotype	AA	AS	SS
Frequency (0)	p^2	$2pq$	q^2
Relative Fitness	$1-s$	1	$1-t$

In this case, *s* is the selection coefficient for the *AA* genotype and *t* is the selection coefficient for the *SS* genotype.

Because the heterozygote has the highest fitness, the proportion of heterozygous individuals should increase in the population. However, the mating of two heterozygotes produces homozygotes, which prevents either allele from being fixed through selection alone. Instead, the two alleles reach an equilibrium allelic frequency when this condition is met:

$$(p-1) \times [(s+t)\,p - t] = 0$$

Three cases give this result: $p = 0$, $p = 1$, or $p = t/(s + t)$. The first two cases are trivial solutions; however, the third case provides a solution for a single locus with two alleles.

Using the relative fitness data provided earlier (0.89 for *AA*, 1.00 for *AS*, and 0.2 for *SS*), we can solve for the equilibrium frequency of the two alleles. In this case the selection coefficients against the *A* and *S* homozygotes, respectively, are $s = 0.11$ and $t = 0.80$. The equilibrium frequency for the *A* allele is $[0.80 / (0.11+0.80)]$

table 23.10   Computing Changes in Allelic Frequencies, over One Generation, with Directional Selection

Genotype	Genotypic f	Relative Fitness (s = 0.99)	Relative Fitness (s = 0.01)
A_1A_1	0.49	1.00	1.00
A_1A_2	0.42	1.00	1.00
A_2A_2	0.09	0.01	0.99
Δp		+0.0685	+0.0063

= 0.879, which means the equilibrium frequency for the *S* allele is 0.121. These allele frequencies maximize population fitness. (Try plugging in other values and see what happens to mean population fitness.)

Under Hardy–Weinberg expectations, approximately 1.5% of the population ($[0.121]^2$) would have sickle cell anemia, and 77.3% of the population ($[0.879]^2$) would be susceptible to malarial infection. Another way of thinking about this is that the approximately 1.5% of the population would exhibit sickle cell anemia to allow 21.2% of the population to be immune to malaria.

Other forms of selection that can maintain genetic variation are frequency-dependent selection, density-dependent selection, and genotype–environment interactions.

Frequency-Dependent Selection

Frequency-dependent selection occurs when the fitness of an allele (or genotype) depends on the frequency of the other alleles (or genotypes) in the population. The most common examples of frequency-dependent selection have often been observed in sexual selection and in viability, such as predators primarily killing specific genotypes.

One example of frequency-dependent selection is the scale-eating fish, *Perissodus microlepis* (**fig. 23.6**). These fish feed on the scales of other fish. They

figure 23.6   Photograph of the scale-eating fish, *Perissodus microlepis.*

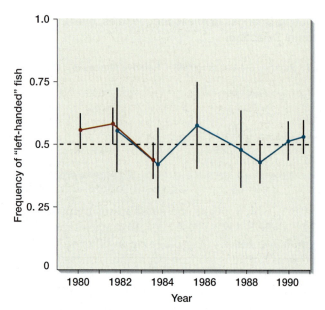

figure 23.7   The frequency of "left-handed" scale-eating fish (*Perissodus microlepis*) over an 11-year period at two different sites. Notice that the percentage of "left-handed" fish in the population fluctuates above and below the 50% frequency (dashed horizontal line). As the prey fish learn from which side most of the *Perissodus* approach, they become adept at avoiding the predator. This confers an advantage for *Perissodus* with the alternative phenotype, which drives the fluctuation seen on the graph. The 95% confidence intervals for each point are represented as the vertical lines.

(Hedrick, *Genetics of Population*, 2005: Jones and Bartlett Publishers, Sudbury, MA. www.jbpub.com. Reprinted with permission.)

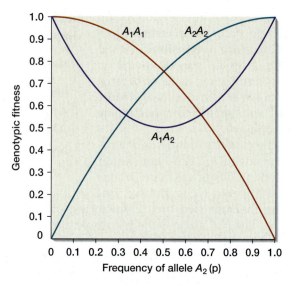

figure 23.8   A plot showing the relative fitness for a frequency-dependent hypothetical model. The fitness of all three genotypes (A_1A_1, A_1A_2, and A_2A_2) are plotted as a function of the frequency of the A_2 allele (p). The population reaches an equilibrium when $p = 0.5$, where the fitness of the two homozygous genotypes are 0.75 and the heterozygous genotype is 0.5.

(Fig. 12.8a, p. 502 from INTRODUCTION TO POPULATION GENETICS by Richard Halliburton. Copyright © 2004 by Pearson Education. Reprinted by permission.)

approach their prey from behind and then remove some scales on one side or the other. To perform this maneuver, *Perissodus* opens its mouth to either the left or the right. An asymmetrical joint in its jaw confers either a "left-handed" or "right-handed" phenotype on the fish that allows it to open its mouth to only one side. There is a complete correspondence of "right-handed" fish attacking the left side of prey and "left-handed" fish attacking the right side. This phenotype appears to be due to a single locus, with the "right-handed" allele being dominant over the "left-handed" allele.

The prey becomes alert to this behavior and swims to defend against the attack. Thus, if either "handed" allele became fixed, the prey would be able to avoid the majority of the attacks. The *Perissodus* exhibit frequency-dependent selection at this locus, so that neither allele becomes fixed. As the proportion of individuals increase with the "right-handed" allele, selection for the "left-handed" allele occurs (**fig. 23.7**). When the "left-handed" allele becomes more common, selection for the "right-handed" allele develops. In this way, the frequency of one allele dictates the selection that is exerted on the *Perissodus*. This ensures that enough *Perissodus* exist of both phenotypes, so that the species can continue to exist even when prey defend themselves.

We can use this frequency-dependent selection to plot the fitness of all three genotypes against the frequency of one of the alleles (**fig. 23.8**). As the frequency of the A_1 allele decreases (p), its fitness increases. This inverse relationship between allele frequency and fitness appears to be an important mechanism to maintain genetic variation in natural populations. In this case, selection pushes the population to reach an equilibrium balance of $p = 0.5$, when the fitness of the two homozygotes (A_1A_1 and A_2A_2) is 0.75 and the A_1A_2 heterozygote is 0.5.

Density-Dependent Selection

Another form of selection is **density-dependent selection,** in which the relative fitness of a genotype depends on the population density. This selection has been witnessed less often in nature than frequency-dependent selection.

Genetic variation is maintained by stochastic, or randomly occurring, changes in the environment. These changes make the environment more or less suitable to the organism, causing the population density of the organism in the habitat to change through time. If one genotype does better under crowded conditions and another under low-density conditions, and if the environment changes frequently enough that neither genotype becomes fixed, a polymorphism can be preserved.

An example of density-dependent selection is the foraging locus in *Drosophila melanogaster*. Larvae with

the *rover* allele (*for^R*) exhibit significantly longer foraging paths on yeast plates than larvae that are homozygous for the *sitter* allele (*for^s*). A population must
maintain both alleles, but in different ratios depending
on the density of the population and the amount of
food. In high-density populations, the rover phenotype is selected possibly because the larvae must travel
farther to find food when there are a greater number
of larvae feeding. In low-density populations, the sitter
phenotype is selected.

Spatial and Temporal Variation in the Environment

Spatial variation, temporal variation, or both in an
environment are also thought to be major factors in
maintaining genetic variation. But it is difficult to demonstrate this effect in natural populations. The effect
is similar to that of frequency- or density-dependent
selection; one genotype does not have the highest fitness across all environments, temporal and spatial.

In this case, as the environment changes through
time or is heterogeneous in space (or both), genetic
variation can be maintained because some genotypes
are favored in one time period (or habitat type), and
another genotype in another. This is especially true,
theoretically, if genotype-specific habitat selection
occurs—that is, individuals preferentially settle in
habitats where they have higher fitness.

This type of *genotype–environment interaction* for
fitness has been heavily explored theoretically and in
the laboratory. There is little doubt that it does contribute to the large amount of genetic variation witnessed
in most populations, however, how large a contribution it makes is still controversial.

Mutation–Selection Balance

As discussed earlier, mutation is more likely to generate a deleterious allele, whereas selection usually
decreases the presence of deleterious alleles (**fig. 23.9**).
Because mutation and selection are both crucial for a
population to adapt to new environments and evolve,
a careful balance must be maintained between these
two processes.

We can examine the equilibrium between muta-
tion and the selection for a mutant allele. If the deleterious allele is completely recessive, then the equilibrium
frequency, q_e, depends on two factors, namely the forward mutation rate (μ) and the selection coefficient (s).
The relationship can be expressed in this way:

$$q_e = \left(\frac{\mu}{s}\right)^{0.5}$$

As the mutation rate increases, so does the equilibrium
frequency; as the selection coefficient increases, the
equilibrium frequency decreases. This equation works

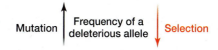

figure 23.9   The relationship between mutation and
selection on the generation/preservation of a deleterious
allele. While mutation increases the frequency of deleterious
alleles in a population, selection decreases the frequency of
deleterious alleles in the population.

because very deleterious alleles or recessive lethal
alleles exhibit primarily only the recessive phenotype.
In a heterozygote, the recessive allele has only a very
small deleterious effect on the phenotype or expression of the dominant allele.

But if an allele exhibits an additive effect, then the
equilibrium allelic frequency can be estimated as:

$$q_e = \frac{\mu}{hs}$$

where the factor h represents the dominance level. The
dominance level refers to the amount by which the
recessive allele alters the phenotype or fitness of the heterozygote, relative to the homozygous dominant genotype. Usually when the homozygous recessive genotype
does not have a very strong phenotype, the recessive
allele exhibits a stronger effect on the heterozygous
genotype.

As an example, consider that the relative fitnesses
of three genotypes, A_1A_1, A_1A_2, and A_2A_2, are 1.00, 0.80,
and 0.20, respectively.

Genotype	A_1A_1	A_1A_2	A_2A_2
Frequency	p^2	$2pq$	q^2
Relative Fitness	1	$1 - hs = 0.8$	$1 - s = 0.2$

We know from the A_2A_2 genotype that $s = 0.8$.
Inserting 0.8 as the selection coefficient for the A_1A_2
heterozygote produces a value of $h = 0.25$. In other
words, the effect of the A_2 allele in the heterozygous
state is 25% of that in the homozygous state.

We can now apply this to a problem. Assume that
the mutation rate for a particular disease allele is $1 \times
10^{-5}$ and that the selection coefficient $s = 0.80$. If the
A_2 allele is completely recessive, the equilibrium frequency is $q_e = \mu/s^{0.5} = 0.0035$.

If the A_2 allele is additive, however, then we
must calculate the dominance level as shown earlier.
Assuming that $\mu = 1 \times 10^{-5}$, $s = 0.80$, and $h = 0.25$, then
the equilibrium frequency becomes approximately
0.00005. This frequency is 70 times lower than in the
completely recessive case, although only 25% of the
selection coefficient occurs in the heterozygotes. This
result agrees with the earlier discussion regarding
directional selection. It is harder to eliminate a strong
deleterious allele that is not additive, because it has a
higher equilibrium frequency.

Restating the Concepts

▶ Natural selection is generally believed to be the dominant force shaping the course of evolution.

▶ Many types of selection actually work to maintain genetic variation. Directional selection removes individuals from either end of the phenotypic distribution or one of the homozygous genotypes. Directional selection works fastest when selection coefficients are larger, allele frequencies are intermediate, and when alleles have additive effects.

▶ Stabilizing selection refers to a case in which the heterozygous genotype has greater fitness than either of the homozygous genotypes. This drives the genetic variation primarily into the heterozygous genotype. In contrast, disruptive selection favors the individuals at the phenotypic extremes over the heterozygotes.

▶ A population is exposed to two opposing forces that are crucial for adapting to new environments and evolving. Mutation is likely to generate a deleterious allele, whereas selection acts to reduce the presence of a deleterious allele in a population. Thus, a balance is achieved in a population between mutation and selection.

23.5 Random Genetic Drift

Changes in allele frequency are termed evolution, whether they have positive, adaptive value or not. As mentioned earlier in this chapter, **random genetic drift** refers to changes in allelic frequencies that occur by chance alone; as with any chance event, drift can actually have effects that are negative, as well as positive or neutral. Furthermore, random genetic drift does not exhibit a steady, directional change in allelic frequency. In this section we look at the possible effects of genetic drift and how it interacts with selection and mutation.

Effect of Random Genetic Drift

In finite populations, sampling variance during fertilization can alter allelic frequencies. Imagine a hypothetical diploid population that consists of a single male and single female. We can assign each individual two unique alleles, with the female being A_1A_2 and the male being A_3A_4. We are starting with maximum heterozygosity and allele diversity, therefore, for a population of this size.

To simplify this example, each mating pair is allowed two offspring per generation, and these offspring happen to be a male and a female. In this example, there is no mutation, none of the alleles have a selective advantage over the other, and there is no migration.

What happens to these alleles over time? We can mimic the random assortment of chromosomes during meiosis by flipping a coin. Denoting each allele in the parents for each generation as being a head or a tail, we use the results of coin tosses to determine which allele segregates into the haploid gamete that produces each diploid zygote.

In the worked-out example shown in **table 23.11**, the A_4 allele has already been randomly lost in the first generation, because the male parent by chance gave both offspring a copy of his A_3 allele. Thus, allele diversity is reduced after only one generation. Because both offspring are heterozygotes (A_2A_3 and A_1A_3), heterozygosity remains 100%.

If drift continues long enough, it eventually fixes one allele at each locus, as the table shows. One allele is fixed, reaching a frequency of 1.0 in the population, and the other alleles are lost (reach a frequency of 0.0). Even without selection, heterozygosity has declined to zero, when no mutation or migration takes place to introduce new alleles.

Drift over Time

Random genetic drift can continue until one allele is fixed or lost. Without selection, mutation, or migration, eventually every allele will be either fixed or lost. The probability of an allele being fixed only by random genetic drift is equal to p (its beginning allelic frequency).

It is impossible to determine precisely how much an allelic frequency in a population will change due to genetic drift. We can, however, calculate the probability of any given change, based on the allelic frequencies and the size of the population. The mean change in allelic frequency (Δq) is

$$\Delta q = \frac{pq}{2N}$$

In this equation, N is the size of the population. Therefore, the amount of drift is inversely proportional to population size. Larger populations would be expected to exhibit less drift.

Differences Between Genetic Drift and Selection

Random genetic drift and selection are different in a number of crucial ways. The allele that is fixed with genetic drift is random. If we repeated the same experiment again with new coin tosses, any of the four alleles would be fixed. In fact, the probability for this same pattern to occur again is very tiny. If we performed this exercise 100 times, we would expect each allele would be fixed in about 25 of those populations. The allele is fixed or lost in a stochastic way, not because it is advantageous. In contrast, selection provides an advantage for a specific allele or genotype and is not random.

The effect of drift increases as the number of effective breeders decreases. This can occur with a decrease in the population size. As shown in the example in table 23.11, drift is very strong in a tiny population, and it affects all loci in the genome. Although there is no way to predict the effect of genetic drift for each allele, individual populations will eventually fix a particular allele at every locus.

The concern over random genetic drift stems from the fact that it can sometimes overwhelm selection and fix a specific allele, despite the allele being deleterious. Since the beginning of population genetics, controversy has existed concerning the importance of random changes in allelic frequencies due to finite population size. Part of the controversy has resulted from the large numbers observed in many natural populations, numbers large enough to make chance effects small in comparison with the effects of other factors, such as selection. As populations of some species have become smaller, however, the importance of random genetic drift has become almost irrefutable.

Random genetic drift is therefore an important consideration in the conservation of endangered species of plants and animals. One well-known example is the Florida panther (**fig. 23.10**), *Puma concolor coryi,* which reached a population of 50 to 70 wild animals in southwestern Florida. Unlike other mountain lions, the Florida panther is characterized by a flatter skull, a 90-degree kink near the end of its tail, and lowered fitness due to a high frequency of sperm abnormalities, cryptorchidism (undescended testicles) and congenital abnormalities. In the Florida panther population, 88% of the animals had a kinked tail and 68% had cryptorchidism. The frequency of the alleles producing these deleterious traits was increasing and could ultimately become fixed.

To reduce the frequency of the deleterious alleles in the Florida panther population and increase the population size, eight female Texas panthers (cougars) were captured and relocated to Florida in 1995. Within three generations, only 7% of the panthers that contained some Texas ancestry had kinked tails and none of the males with Texas ancestry had cryptorchordism.

As populations become endangered, random genetic drift increases the probability that deleterious alleles will be fixed in the population. The introduction of additional breeders into the population provides the

table 23.11   Decline in Genetic Variation (Number of Different Alleles) with Random Genetic Drift

Generation	Female Genotype	Male Genotype	H	Number of Alleles
0	A_1A_2	A_3A_4	100%	4 (A_1, A_2, A_3, A_4)
1	A_2A_3	A_1A_3	100%	3 (A_1, A_2, A_3)
2	A_2A_3	A_1A_3	100%	3 (A_1, A_2, A_3)
3	A_2A_3	A_2A_3	100%	2 (A_2, A_3)
4	A_2A_2	A_2A_3	50%	2 (A_2, A_3)
5	A_2A_3	A_2A_3	100%	2 (A_2, A_3)
6	A_2A_3	A_2A_2	50%	2 (A_2, A_3)
7	A_2A_2	A_2A_2	0%	1 (A_2)

figure 23.10   The Florida panther, *Puma concolor coryi.*

opportunity to decrease the frequency of the deleterious alleles and reduce the effect of random genetic drift.

Interactions Between Selection and Random Genetic Drift

When no differential selection occurs at a locus, the fate of an allele is largely the result of random genetic drift. The probability of fixation of a favorable allele, by contrast, is a function of the initial frequency of the allele, the strength of selection on the allele, dominance relationships among alleles at the locus, and the population size. Alleles are classed as **effectively neutral** when random genetic drift overrides selection, and the fate of an allele becomes primarily stochastic rather than deterministic.

The interaction between selection and random genetic drift can be observed in different-sized populations. Four populations of the red flour beetle, *Tribolium castaneum,* were studied in the laboratory (**fig. 23.11**). Twelve replicates were studied for each population size. Each population started with an initial frequency of the b^+ allele of 0.50 (the *b* allele confers a black body color).

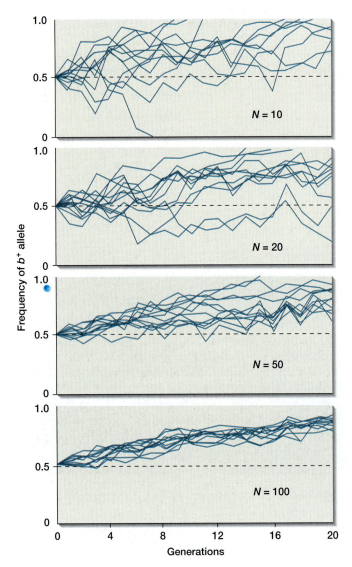

figure 23.11 Effect of population size on natural selection and random genetic drift. Four different-sized populations (N = 10, 20, 50, and 100) of *Tribolium castaneum* were maintained in the laboratory. Each population started with a b^+ allelic frequency of 50%. Random mating occurred in the population and the b^+ allelic frequency was determined and plotted against the generation number. Each line represents an independent population of the same size as the other populations on the same graph. As the population size increases, random genetic drift decreases and selection predominates.

In the top panel of figure 23.11 the population size is 10 individuals, with each line representing a different population. Although the top panel may appear to exhibit random genetic drift, that is not entirely correct. Under random genetic drift, the variation in allele frequencies between populations becomes larger—and, half the populations will fix one allele and half will fix the other allele (if both alleles started at a frequency of 50%).

The populations in the top panel do diverge from each other in allelic frequencies, but 11 of the 12 populations are increasing to fix the b^+ allele. The exception is a single population that rapidly fixed the deleterious allele (*b*). This panel demonstrates what happens when an allele is very close to the limit for effective neutrality: The allele behaves in a semideterministic manner, reflecting the almost equal strength of selection and drift at this population size.

The fixation of a deleterious allele in one population may seem trivial. But keep in mind that the same thing is happening across the genome of each of these populations. We might estimate, therefore, that 1 out of every 12 deleterious mutations across the genome is being fixed in a population. This might cause a substantial reduction in fitness in all the populations.

As we move through the other panels, the population size increases to 20, 50, and 100 individuals. As you can see, selection begins to exert a more powerful effect than random genetic drift as the population size increases. In the population of 20 individuals, 9 of the 12 populations have either fixed the same allele or are moving toward fixing that allele. The remaining three populations have nearly the same allelic frequency as when the experiment started. In contrast, a population size of 100 is sufficiently large to further reduce random genetic drift, and all 12 populations are primarily exhibiting selection of the same allele at a uniform rate.

Mutation–Drift–Selection Balance

None of the evolutionary factors we have examined acts in isolation. The amount and type of genetic variation maintained in a population, for a particular locus, depends on the interactions among mutation rates, the amount of random genetic drift (population size), and the strength and type of selection acting on alleles at that locus. In fact, these three forces acting together can explain much of the genetic variation that exists in natural populations.

 Restating the Concepts

▶ In the absence of other evolutionary forces (such as selection), the allelic frequency in a population will not be exactly reproduced in the next generation because of sampling error.

▶ Random genetic drift is the stochastic force that results in the random fixation of an allele at a given locus. Because this is a random process, both advantageous and deleterious alleles can be fixed. However, if the alleles are either advantageous or deleterious, then the system will be under selection, which is the deterministic loss of deleterious alleles.

▶ Two different populations that are isolated from each other may both undergo random genetic drift to fix different alleles. Thus, the heterozygosity within a population will approach zero, but the genetic variation between populations will increase.

▶ Random genetic drift is stronger in smaller populations.

23.6 Migration and Population Structure

Gene flow occurs when individuals migrate from one population into another, interbreed, and reproduce. The result is the movement of alleles from one population into the second. The major effect of gene flow between populations is to make those populations more genetically similar.

Isolated populations become genetically different from each other by a variety of methods.

1. **Mutation:** Mutations are rare events and not likely to be duplicated in isolated populations. Thus, the accumulation of mutations in separate populations after genetic isolation increases the genetic differences between the populations.
2. **Random genetic drift:** Drift is a stochastic process and may fix different alleles in different populations, which will also increase the genetic variation between populations.
3. **Differential selection:** Isolated populations experience at least slightly different environments. This leads to different alleles being favored, perhaps fixing different alleles.

Gene flow among populations slows or halts the process of genetic differentiation and makes populations more genetically homogeneous.

The Continent-Island Model

In many ways, migration resembles mutation. Migration introduces new alleles at some (assumed) constant rate just as mutation does; however, migration rates can be much larger than mutation rates, and selection for or against migrant alleles might differ substantially from that for a typical new mutation.

A large number of "standard" models for migration have been devised, and a great variety of nonstandard models can be imagined. We examine here only the continent-island model that assumes *one-way* gene flow from a large population ("continent") to a small population ("island") (**fig. 23.12**). We assume in this discussion that gene flow occurs in the absence of mutation, random genetic drift, or selection.

Let the proportion of migrants moving to the small population each generation be m, and the proportion

of natives in that population be $1 - m$. The frequency of a given allele among migrants is p_m, and the allelic frequency among natives is p_0. After one generation of migration, the frequency of this allele is expressed as:

$$P_1 = (1-m)\,p_0 + mp_m$$

To see how migration affects allelic frequency, assume that the frequency of the A_1 allele among all 1000 native people of a certain Polynesian island is 0.76. Among continental Americans, the frequency of the A_1 allele is 0.13. If 20 Americans are shipwrecked on the island, marry natives, and produce children; what is the new frequency of the A_1 allele in the next generation?

$$P_1 = (1-m)\,p_0 + mp_m$$
$$P_1 = (0.98)\,0.76 + (0.02)\,0.13 = 0.7474$$
$$\Delta A_1 = 0.7474 - 0.7600 = -0.0126.$$

The rate of change of allelic frequency in a small population subject to immigration depends on the rate of individuals moving into the small population and the difference in allelic frequencies, ΔA_1, between the natives and the immigrants. The greater the immigration rate, the faster allelic frequencies change. To see for yourself, assume 20% migration instead of 2% in the previous example. If this is the case, $P_1 = 0.634$ and the change in A_1 frequency is -0.126.

A little less obvious is the fact that the more the two populations initially differ from each other, the faster the allelic frequencies will change. In the original

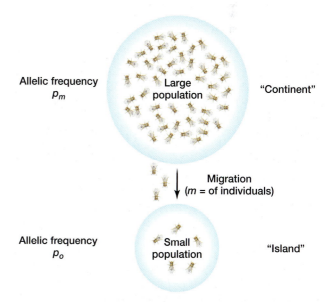

figure 23.12 The Continent-Island Model of migration. This model is based on a particular number of individuals migrating one-way from a large population (continent) to a small population (island). A particular allele that is being studied is at one frequency (p_m) in the large population and a different frequency (p_o) in the small population.

example, if we increase the A_1 allelic frequency in the continental Americans to 48% (which reduces the difference between the initial frequencies), then $P_1 = 0.7544$, and the change in the A_1 frequency is only -0.0056.

Migration, Selection, and Drift

Migration was the only force acting in the previous example, but in a natural population, other forces (drift, mutation, and selection) are operating as well. Because mutation and drift are probably small relative to gene flow, they can often be ignored. Selection on alleles introduced into the small population, however, may be very strong. For example, alleles introduced into a small, previously isolated population might be under strong positive selection, because these immigrants might ameliorate inbreeding depression (described in the next section). The immigrant alleles may spread more rapidly than otherwise expected for this reason.

Immigrant alleles may also face strong negative selection. Environments can be heterogeneous at many spatial scales, and each patch of habitat is different from any other patch. A population may exhibit a high frequency of a certain allele because that allele is locally adaptive.

Evidence for local adaptation includes **clines**, which are directional changes in allelic frequencies across a geographic region, an altitude, or a number of subpopulations. An example of a cline is the frequency of the sickle cell allele (Hb^S) across a linear array of tribes in Liberia (**fig. 23.13**). In this example, the allelic frequency declines as you move from the easternmost

Kissi tribe, through the Mende tribe, and continuing on to the Krahn, Kru, Grebo, and Webbo tribes on the western border.

Usually allelic frequencies can be correlated with some environmental factor, such as temperature, heavy-metal concentrations in the soil, pesticide use, or the prevalence of a disease or parasite.

The Wahlund Effect

Assume that two populations of equal size exist in Hardy–Weinberg equilibrium. If the two populations migrate and intermingle, they will not initially be in Hardy–Weinberg equilibrium. Observations will show that the combined population contains far fewer heterozygotes than expected. This unexpected low number of heterozygotes is known as the **Wahlund effect.** As soon as random mating begins within this combined population, however, the Hardy–Weinberg equilibrium becomes established.

Assume that two populations each contain the same two alleles, A_1 and A_2, at a specific locus. In population 1, the frequency of the A_1 allele is 0.1 (**table 23.12**). If this population is in Hardy–Weinberg equilibrium, then we would expect the frequency of the A_1A_1 genotype to be 0.01, the A_1A_2 genotype to be 0.18, and the A_2A_2 genotype to be 0.81.

In population 2, the frequency of the A_1 allele is 0.9 (see table 23.12). If this population is in Hardy–Weinberg equilibrium, then we would expect the frequency of the A_1A_1 genotype to be 0.81, the A_1A_2 genotype to be 0.18, and the A_2A_2 genotype to be 0.01.

Suppose the two populations are equal in size and combine. We would expect the A_1 allele frequency to now be 0.5 (the average of 0.1 and 0.9). Based on the Hardy–Weinberg equilibrium, we would expect the frequency of the A_1A_1 genotype to be 0.25, the A_1A_2 (heterozygote) genotype to be 0.50, and the A_2A_2 genotype to be 0.25 (see table 23.12). Instead, however, we would observe the frequency of the A_1A_1 genotype to be 0.42 (the average between 0.81 and 0.01), the A_1A_2 genotype to be 0.18, and the A_2A_2 genotype to be 0.042. Thus, there are far fewer heterozygotes than expected.

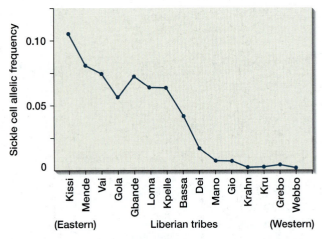

figure 23.13 The cline of the sickle cell allele in Liberia. The frequency of the sickle cell allele was determined in various tribes across Liberia. The X axis shows the linear arrangement of the tribes as they exist across the country with the Kissi being the easternmost tribe and the Webbo being the westermost tribe. There is a general trend of a declining allelic frequency as you move from east to west through the tribes across the country.

(Hedrick, *Genetics of Population*, 2005: Jones and Bartlett Publishers, Sudbury, MA. www.jbpub.com. Reprinted with permission.)

table 23.12 The Wahlund Effect

	Allelic frequency		Genotypic frequency		
	A_1	A_2	A_1A_1	A_1A_2	A_2A_2
Population 1	0.1	0.9	0.01	0.18	0.81
Population 2	0.9	0.1	0.81	0.18	0.01
Pooled populations (observed)	0.5	0.5	0.42	0.18	0.42
Pooled (expected)			0.25	0.50	0.25

The Wahlund effect is observed when a sample is selected from what is thought to be a single population, but is actually composed of separate gene pools. Most population geneticists who sample a population and find a deficiency of heterozygotes think of the Wahlund effect and population subdivision. An alternative possibility is inbreeding within the population.

Restating the Concepts

▶ Migration involves the movement of individuals between populations. If those individuals mate in the new population, gene flow occurs.

▶ The continent-island model of migration involves gene flow from a large population to a small one. In this model, the greater the number of immigrants to the small population or the greater the difference between the initial allelic frequencies of the two populations, then the larger will be the change in allelic frequency in the small population.

▶ Mutation and random genetic drift often have a small relative effect on gene frequencies. Selection on alleles that are introduced into the small population, by contrast, can have very large effects.

▶ A loss of heterozygosity is seen when subpopulations are combined, which is called the Wahlund effect. The Wahlund effect may exist for a limited number of loci.

23.7 Nonrandom Mating

The Hardy–Weinberg equilibrium assumes that random mating takes place in the population, but random mating does not often occur in a natural population. Among mammals, for example, males may try to select mates with a high fecundity, and females may tend to choose partners that are able to protect them and their offspring. We can classify nonrandom mating choice as based on either the level of physical resemblance or the level of genetic relatedness.

Inbreeding

Nonrandom mating may involve mate selection that is based on the level of genetic similarity between individuals. **Inbreeding** is the mating of individuals that are more closely related than would be expected if mates were chosen at random. The alternative is **outbreeding,** which is the mating of individuals that are genetically more distantly related than would be expected from a random mating. Inbreeding can occur because of geographic isolation or some physical barrier to gene flow.

Inbreeding in an infinitely large population with no genetic drift does not change allelic frequencies,

but inbreeding *does* change genotypic frequencies. One of the major effects of inbreeding is an increase in homozygosity (or an excess of homozygotes over that expected under random mating). This effect is similar to the decreased heterozygosity in the Wahlund effect. Inbreeding increases the average homozygosity of individuals (or populations) while making the differences among individuals (or populations) larger. In this respect inbreeding resembles random genetic drift and has consequences opposite to those of migration.

Inbreeding is measured by the **inbreeding coefficient (F)**—the probability that two alleles in an individual are identical because of their inheritance from a single allele in an ancestor. This single allele is inherited through both parental lines before inbreeding reunites the alleles in an individual.

An Example of the Inbreeding Coefficient

Figure 23.14 helps demonstrate what the inbreeding coefficient means. Imagine that we want to determine that the woman in generation IV is homozygous for the A_1 allele. The probability that her great-grandmother passed the A_1 allele to the woman's grandmother is 0.5; the probability that her grandmother passes the A_1 allele on to the woman's mother is 0.5; and the probability that the woman inherits the A_1 allele from her mother is 0.5. The probability that she will inherit the A_1 allele from her paternal side of the family is also 0.5 for every generation. Thus, the total probability that she inherits the A_1 allele from both her paternal and maternal ancestors is:

$$(0.5)(0.5)(0.5)(0.5)(0.5)(0.5) = 0.016 \text{ or } 1/64$$

However, the woman in generation IV could be homozygous for the A_2 allele, the A_3 allele, or the A_4

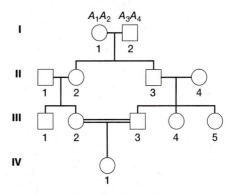

figure 23.14 Pedigree of a family that is used to calculate the inbreeding coefficient (*F*). The inbreeding coefficient is calculated for a locus containing four different alleles ($A_1, A_2, A_3,$ and A_4). The inbreeding coefficient is the probability of producing an individual in the fourth generation of a first cousin marriage (individual IV-1) who is homozygous for an allele that was present in an ancestor (either I-1 or I-2).

allele. Because she could be homozygous for four different alleles that would each consist of an identical ancestral allele, the inbreeding coefficient is:

$$F = \left(\frac{1}{64}\right)(4) = \frac{1}{16}$$

Inbreeding coefficients can also be estimated by the reduction in heterozygosity. In doing so, we use the following equation:

$$F = \frac{(H_e - H_o)}{H_e}$$

In this equation, H_e is the expected heterozygosity of a randomly mating population with the same allelic frequencies (based on the Hardy–Weinberg equilibrium). H_o is the observed heterozygosity in the population. In a completely randomly mating population, H_e would be equal to H_o and F would therefore equal 0. When $F = 0$, then no individual in the population is homozygous for an allele that is derived from a common ancestral allele. A value of $F = 1.0$ reveals that everyone in the population is homozygous for an allele that is derived from a common ancestral allele.

Effect of Inbreeding on Genotypic Frequency

Even though inbreeding in an infinitely large population with no random genetic drift does not change allelic frequencies, it does change genotypic frequencies. The genotypic frequencies expected under a given level of inbreeding can be calculated using p as the frequency of the A_1 allele, q as the frequency of the A_2 allele, and F as the inbreeding coefficient. Both homozygous genotypes (A_1A_1 and A_2A_2) increase by an amount that equals the product of the two allelic frequencies and the inbreeding coefficient (Fpq).

Let's look at a problem to see how inbreeding affects the genotypic frequencies (**table 23.13**). Assume that the frequency of the A_1 allele is 0.70 and the frequency of the A_2 allele is 0.30. If the inbreeding coefficient is 0.20, what are the expected genotypic frequencies, and how do these compare with the genotypic frequencies under Hardy–Weinberg equilibrium (see table 23.13)?

Under Hardy–Weinberg equilibrium, the expected homozygous genotypic frequencies are p^2 ($0.70^2 = 0.490$) and q^2 ($0.30^2 = 0.090$), and the heterozygous genotypic frequency is $2pq$ ($2 \times 0.70 \times 0.30 = 0.420$). Under inbreeding conditions, the expected homozygous genotypic frequencies are $p^2 + Fpq$ ($0.70^2 + [0.20 \times 0.70 \times 0.30] = 0.532$) and $q^2 + Fpq$ ($0.3^2 + [0.20 \times 0.70 \times 0.30] = 0.132$), and the heterozygous genotypic frequency is $2pq(1-F)$ ($2 \times 0.7 \times 0.3 \times [1 - 0.20] = 0.336$). As expected, inbreeding increased the homozygous genotypic frequencies, which also resulted in the decrease of the heterozygous genotypic frequency.

Because inbreeding increases the frequency of homozygotes, it can have a very pronounced effect on the expression of deleterious phenotypes. When inbreeding occurs, recessive alleles that are normally at a low frequency in the population because of their deleterious effects will be present in homozygous individuals at a much higher frequency. In a study of several recessive disorders in the Japanese and European populations, approximately 20–80% of the affected individuals were born to a first-cousin marriage (**fig. 23.15**). At the same time, only 5% of Japanese marriages and 2% of European marriages were between first cousins. A very high percentage of these recessive disorders are therefore found in a very small inbreeding portion of the population.

Relationship Between Inbreeding and Drift

You may have noted several similarities between random genetic drift and inbreeding—for instance, they both increase homozygosity. Indeed, the two are very difficult to separate in real populations, although conceptually they are quite different.

The predictions made from inbreeding versus drift for a particular locus within a population are not the same. Inbreeding does not change allelic frequencies, but random genetic drift does. This results in the fixing of a single homozygous genotype through random genetic drift, whereas inbreeding exhibits an increase in *both* homozygous genotypes (in a two-allele system). Random genetic drift also weakens the overall effect of selection; in contrast, inbreeding can increase the strength of selection against deleterious recessive alleles.

table 23.13   Genotypic Frequencies with Inbreeding

Genotypes	A_1A_1	A_1A_2	A_2A_2
Genotypic frequencies under inbreeding	$p^2 + Fpq$	$2pq(1-F)$	$q^2 + Fpq$
Expected genotypic frequencies under inbreeding	0.532	0.336	0.132
Genotypic frequencies in Hardy–Weinberg equilibrium	p^2	$2pq$	q^2
Expected genotypic frequencies in Hardy–Weinberg equilibrium	0.490	0.420	0.090

Inbreeding Depression

Inbreeding, as you have seen, increases the frequency of homozygous genotypes in a population. Rare, deleterious, recessive alleles exist within the population in heterozygotes. Because inbreeding increases the frequency of homozygotes, there will be a large increase in the frequency of homozygous recessive individuals (see fig. 23.15). This increased frequency of homozygous individuals expressing deleterious recessive alleles results in a lowered mean fitness for the inbred population. **Inbreeding depression** is the reduction in fitness due to inbreeding. An additional effect of inbreeding is the reduction of heterozygotes, in the cases when heterozygote advantage is operating.

Inbreeding alone does not change allelic frequencies; but it can influence the outcome of selection. Selection against homozygotes is usually different from that against heterozygotes. Increased numbers of homozygotes and the more efficient selection against the deleterious recessive allele in inbred populations has been labeled **purging.**

Assortative–Disassortative Mating

Two other forms of nonrandom mating are assortative mating and disassortative mating. **Assortative mating** occurs when mates are chosen, at least in part, because they share a common trait. Put more specifically, mates have a positive correlation for one or more phenotypic traits. In humans, mates are positively correlated for height, education level, and other factors.

Assortative mating has effects on genetic variation very similar to those of inbreeding in that it tends to increase homozygosity. One important distinction must be made between assortative mating and inbreeding, however. Whereas inbreeding tends to drive the entire genome toward homozygosity, assortative mating only increases homozygosity of the loci that are near the genes that encode the traits for which mates are being selected.

Disassortative mating occurs when individuals choose mates, at least in part, because they are different in some respect. That is, mates are negatively correlated for one or more phenotypic traits. One example of disassortative mating is the major histocompatability complex (MHC) in a number of animals, including humans. MHC genes are a component of the immune system in vertebrates. There is strong evidence that an amazing amount of allelic diversity at this complex is maintained through heterozygote advantage. But increasing evidence also shows that female mammals (including humans) choose mates disassortatively at these loci, increasing heterozygosity over that expected under random mating.

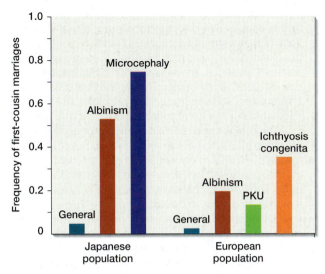

figure 23.15 Recessive disorders occur at a higher frequency in first-cousin marriages. Several different recessive disorders were examined in two different populations, Japanese and Europeans. For each recessive disorder, the frequency that the affected child is the result of a first-cousin marriage was calculated and plotted. As a reference, approximately 5% of the Japanese marriages are between first cousins (General, left) and 2% of the European marriages are between first cousins (General, right). Thus, all of these recessive disorders are much more frequent in first-cousin marriages than a purely random event.

(Hedrick, *Genetics of Population*, 2005: Jones and Bartlett Publishers, Sudbury, MA. www.jbpub.com. Reprinted with permission.)

 Restating the Concepts

► Nonrandom mating may take place in many ways. Inbreeding occurs when mates are chosen that are more closely related genetically than would be expected if mates were chosen at random. Outbreeding occurs when mates are chosen that are more distantly related than would be expected from random selection.

► Inbreeding does not alter the allelic frequency, but it does change the genotypic frequency by decreasing the number of heterozygotes. The loss of heterozygosity from inbreeding is seen throughout the genome.

► The increased homozygosity due to inbreeding can increase the expression of deleterious recessive alleles. This inbreeding depression can reduce the fitness of the population.

► Assortative mating is a positive correlation of mate choice with certain phenotypic traits. It may have effects similar to inbreeding, but does not lead to genome-wide homozygosity. Disassortative mating is a negative correlation of mate choice with phenotypic traits.

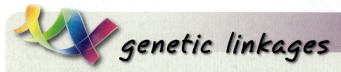

genetic linkages

In this chapter, we discussed the more global ramifications of genotype, phenotype, and genetic processes on populations. The basic principles and concepts presented earlier in this book, however, are still fundamental to the discussions in this chapter. The world is not a controlled environment, and some additional principles and tools are needed to study natural populations.

The basic principles of a monohybrid cross (chapter 2) are at the foundation of the allelic frequencies observed in the Hardy–Weinberg equilibrium. In fact, the random mating that is required for Hardy–Weinberg equilibrium allows us to apply the probabilities discussed in chapter 2 for a monohybrid cross to determine that $p^2 + 2pq + p^2 = 1.0$.

In chapter 5, we discussed deleterious and lethal mutations. In chapter 8, you learned about chromosomal mutations, how they are generated, and their genetic consequences in an individual. In chapter 18, we discussed various mechanisms that can lead to point mutations, and in chapter 20, methods of mutational analysis. It should not be surprising that selection attempts to remove mutations. Random genetic drift and the fixation of advantageous or deleterious alleles, however, may be new concepts.

This discussion on population genetics forms the basis for the next two chapters on quantitative genetics (chapter 24) and evolutionary genetics (chapter 25). Keep in mind that in natural populations, it is usually not possible to generate specific desired matings to test genetic principles. Our application of these principles must then take on a much more statistical analysis. The genotypes and phenotypes of natural populations, however, remain as specific and as real as those we discussed with the experiments on pea plants and *Drosophila*. This "real-world" application of genetics demonstrates the complexity, scope, and excitement of this field.

Chapter Summary Questions

1. For each term in column I, choose the best matching phrase in column II.

   Column I
   1. Additive allele effects
   2. Fitness
   3. Inbreeding depression
   4. Metapopulation
   5. Population genetics
   6. Purging
   7. Random mating
   8. Relative fitness

   Column II
   A. Spatially structured population consisting of two or more genetically dissimilar breeding groups linked by gene flow
   B. Each individual in the population has an equal chance of mating with any other individual in the population regardless of genotype or phenotype
   C. More efficient selection against deleterious recessive alleles with increases in the inbreeding coefficient
   D. Number of sexually mature offspring produced by an individual or genotype
   E. Number of sexually mature offspring produced by an individual or genotype relative to other individuals or genotypes
   F. Phenotypic value for the heterozygote is the mean value of the two homozygotes
   G. Reduction in fitness on inbreeding
   H. Study of patterns of genetic diversity within and among populations. Seeks to explain spatial and temporal patterns of genotypic and phenotypic change

2. What are the assumptions of Hardy–Weinberg equilibrium? Why are they necessary?

3. What do population/evolutionary geneticists mean when they speak of fitness?

4. What are the five major evolutionary forces influencing the amount and kind of genetic variation? How do they change genetic variation?

5. Discuss whether each of the following is necessary for evolution by natural selection. (a) Differential reproduction (b) Environmental change (c) Heritable variation in fitness (d) Sexual reproduction

6. What are the assumptions of selection models used in this chapter? Are the assumptions necessary for selection to work or are they made to keep the mathematics simple?

7. What determines the speed at which allelic frequencies change under selection?

8. Compare and contrast directional, disruptive, and stabilizing selection.

9. What kinds of selection increase genetic variation at the population level?

10. What are the major effects of random genetic drift? Is drift stronger in smaller or larger populations? Why?

11. Do random genetic drift and selection act independently of one another or do they interact?

12. What is the major influence of gene flow among populations? How does gene flow affect population fitness?

13. What are the effects of inbreeding on fitness? What are the effects of inbreeding on genetic diversity? How are these two effects connected?

14. How does one determine an inbreeding coefficient for an individual? How does one determine the inbreeding coefficient for a population?

15. Compare and contrast random genetic drift versus inbreeding.

16. "Population geneticists are interested in the forces that determine the amount and kind of genetic variation present in populations." What is meant by *kind* of genetic variation in this statement?

Solved Problems

PROBLEM 1. "Dominant phenotypes should always increase in frequency over recessive phenotypes, because in a mating between heterozygotes, 75% of the offspring show the dominant phenotype and the recessive phenotype only 25% of the time." Is this statement true or false? Why?

Answer: This statement is false. The frequency of the dominant phenotype relative to the recessive phenotype is determined by the frequency of the alleles and the mating patterns of the population. Whether the dominant phenotype is increasing or decreasing in frequency will depend on whether one phenotype has a fitness advantage over the other phenotype or in the absence of a selective advantage, or the whim of random genetic drift. The dominance of one allele over the other does not imply that it is superior (i.e., provides higher fitness on average).

PROBLEM 2. Describe how the genetic diversity generated by mutation differs from the standing genetic diversity found in natural populations.

Answer: The alleles found in natural populations are more recessive and of smaller effect than new mutations. This is because most new mutations are deleterious. Natural selection is more efficient at removing deleterious mutations that are dominant and of large effect, thus the genetic variation found standing in most natural populations is due to (partially) recessive deleterious alleles of small effect. Thus, this relates to the first problem. Dominant mutations are not inheritantly advantageous compared with recessive alleles, but most deleterious alleles found in wild populations are recessive.

PROBLEM 3. *Eichhornia paniculata* is an aquatic plant that usually exhibits polymorphism at a self-sterility allele, with populations having up to six different alleles. A study of 167 populations of this plant in northeastern Brazil, ranging in size from 6 individuals to 8000 individuals, demonstrated a strong correlation between population size and the number of alleles at this locus. In the largest populations, the alleles present in the population occurred in nearly equal frequencies and in the smaller population the ratios are more skewed. Populations with fewer that 25 individuals were often (42%) fixed for a single allele and without immigration (or mutation at this locus) are doomed to extinction. What best describes what is happening in these populations?

Answer: Self-sterility alleles evolved to decrease inbreeding, especially among potentially self-fertilizing plants. This represents a classic example of a locus that should be under frequency-dependent selection. We expect, in the absence of other forces, that selection will maintain the frequency of existing alleles equal to one another. If a new (unique) mutation occurs at this locus, it will have a selective advantage and will increase in frequency until all alleles are at their equilibrium (equal) frequencies again. The patterns seen in these plant populations are consistent with random genetic drift. Drift decreases genetic variation and is strongest in smaller populations. Among these populations, the number of individuals in the population and the number of alleles at this locus are positively correlated. Further, the equilibrium allelic frequencies expected under a selection only model can be disturbed by drift. Again, the smaller populations are more skewed in their allelic frequencies. Clearly, differences in the strength of random genetic drift compared with natural selection among the different populations are responsible for the differences in the number of alleles and their frequencies. The detrimental effects of drift can be seen by the loss of all self-sterility alleles but one in many of the very smallest populations, which means that all individuals in that population are unable to fertilize all other individuals in the population and in all likelihood such populations will soon go extinct.

Exercises and Problems

17. In a small town in Pennsylvania (population 3772), 500 people were sampled for their *MN* blood types. The estimated allelic frequencies in the town's population were: *M* = 0.59 and *N* = 0.41. Assuming Hardy–Weinberg equilibrium, how many people in the town do you expect to have each of the three possible genotypes (*MM*, *MN*, and *NN*)?

18. A particular human population has 500 *MM* individuals, 300 *MN* individuals, and 700 *NN* individuals. Calculate allelic frequencies and determine if the population is in Hardy–Weinberg equilibrium.

19. Naughty and nice are X-linked phenotypes determined by a single gene with two alleles, with nice (*N*) being completely dominant to naughty (*n*). On a certain island, avoided by sailors, 99% of the girls are nice. Alice decides to visit this island to see if she can finally meet a nice boy. What is the probability that Alice will be disappointed?

20. Dimples are a dominant phenotype, determined by one gene with two alleles (*D*, *d*). From data obtained from a sample of 400 American genetics students, 96 of the students had dimples. What are the frequencies of the two alleles in this population, assuming Hardy–Weinberg equilibrium?

21. Dimples are a dominant phenotype, determined by one gene with two alleles (*D*, *d*). A random sample of 2101 people in New Delhi (India), found 444 people with dimples and 1657 people without. What proportion of the individuals in this population is homozygous for the *D* allele (assume Hardy–Weinberg equilibrium)?

22. The ability to taste PTC is due to a dominant allele *T*. Among a sample of 215 individuals from a population in Vancouver, 150 could detect the taste of PTC and 65 could not. Calculate the allele frequencies of *T* and *t*. Is the population in Hardy–Weinberg equilibrium?

23. In a population of wolf spiders, 30 individuals are sampled at a particular microsatellite locus with five alleles (*A*, *B*, *C*, *D*, and *E*). The genotypes and their frequencies

follow. What are the allelic frequencies? Is the population in Hardy–Weinberg equilibrium?

$AA = 9, AB = 2, AC = 0, AD = 5, AE = 2, BB = 3, BC = 0, BD = 1,$
$BE = 1, CC = 2, CD = 0, CE = 1, DD = 1, DE = 1, EE = 2$

24. At a given locus in a population of wolf spiders, three different genotypes exist: W_1W_1, W_1W_2, and W_2W_2. The three genotypes produce the following numbers of offspring: 144.0, 133.2, and 102.9, respectively. Egg-to-adult survival rates for the three genotypes are: 1.6%, 2.5%, and 3.4%, respectively. What is the relative fitness of each genotype?

25. A population of saplings is genotyped at a single locus and found to be in Hardy–Weinberg equilibrium. Adult trees are surveyed again 15 years later. Is this population still in Hardy–Weinberg equilibrium? If not, what do you think caused the change?

A_1A_1 180
A_1A_2 180
A_2A_2 40

26. Imagine the following scenario. The A_1 allele is completely dominant to the A_2 allele with respect to fitness. The A_2 allele is lethal when homozygous. Start with equal frequencies for each allele (i.e., 0.5) and assume Hardy–Weinberg equilibrium. What is the frequency of the A_1 allele in the next generation? By how much is population fitness reduced due to the presence of the segregating A_2 allele in the initial population?

27. A scientist is studying a wild population of 1000 Japanese morning glories. It is easy to separate genotypes at a flower color locus by their phenotypes. Red individuals are homozygous for the R allele, yellow individuals are homozygous for the Y allele, and orange individuals are heterozygous. The original genotypic frequencies are in Hardy–Weinberg expectations ($RR = 0.49$, $RY = 0.42$, $YY = 0.09$). The genotypic frequencies in the next generation are presented in the following table. What could have caused such a change?

Genotype	Frequency
RR	531
RY	422
YY	47

28. A scientist is studying a different wild population of 1000 Japanese morning glories. In this population, the flower color phenotypes are controlled by the R and Y alleles as described in problem 27. The original genotypic frequencies are in Hardy–Weinberg expectations ($RR = 0.49$, $RY = 0.42$, $YY = 0.09$). The genotypic frequencies in the next generation are presented in the following table. What could have caused such a change?

Genotype	Frequency
RR	495
RY	410
YY	95

29. In a population of chimpanzees plagued by the poliovirus, the relative fitness and frequencies of the MHC genotypes are:

	Frequency	Relative Fitness
A_1A_1	0.50	0.80
A_1A_2	0.40	1.00
A_2A_2	0.10	0.40

What will the approximate frequencies of the A_1 and A_2 alleles be in the next generation? What will the equilibrium frequencies be?

30. In a population of rats, the relative fitness and frequencies of genotypes present are:

	Frequency	Relative Fitness
A_1A_1	0.4	1.0
A_1A_2	0.5	0.8
A_2A_2	0.1	0.6

What will the frequency of the A_1 allele be in the next generation?

31. A certain locus, M, affects the metabolic rates in lizards. Lizards that are homozygous for the M_1 allele run very fast and can avoid predators, but are at greater risk of starvation due to their high metabolism. Those that are homozygous for the M_2 allele are more susceptible to predators, but suffer a lower starvation risk. Heterozygotes have a weak, but significant reproductive advantage over either of the homozygotes. The relative fitnesses of the three genotypes are: M_1M_1 (0.85), M_1M_2 (1.0), M_2M_2 (0.85). What do you expect the equilibrium frequency of the M_1 allele to be, considering selection is the only force acting on this locus?

32. For many generations, the following genotypic frequencies were observed in a large population of dinosaurs: 4% AA, 32% Aa, and 64% aa. The climate changed abruptly and resulted in the death of all homozygous recessive dinosaurs ($s = 1$). In the next generation, the recessive homozygotes died shortly after birth. What percentage of newborn dinosaurs died during the generation following the environmental change? (Assume random mating.)

33. A gene, with two alternative alleles, is known to affect fitness in spiders. Genetic diversity at this locus is assayed in a large population (12,000 individuals) and a small population (40 individuals) of spiders. In the large population, the frequency of the D_1 allele is 1.0 (and the D_2 allele is 0.0), and in the small population the frequency of the D_1 allele is 0.8367 (and the D_2 allele is 0.1633). The relative fitness of the three genotypes is: $D_1D_1 = 1.0$, $D_1D_2 = 0.99$, and $D_2D_2 = 0.95$. If mating is at random, what is the relative fitness of the smaller population compared to the larger? What do you think is responsible for the differences in the allelic frequencies?

34. An allele is maintained at a frequency in accordance with mutation–selection balance. The allele is completely recessive, is deleterious with a selection coefficient (s) of 0.01, and has a mutation rate of 5.0×10^{-6}. What is the equilibrium frequency of the allele?

35. Fifteen mutineers from HMS *Bounty,* all of whom bear the NN blood group, arrive at Pitcairn Island. The mutineers are met by 15 native women, all bearing the MM blood group. In time, the mutineers become integrated into island society. Assuming random mating, no mutation, no selection (based on blood type or linked genes), and no genetic drift, what would you expect the blood group distribution to be among the 100 people existing on Pitcairn Island two generations later?

36. In which of these cases will immigration change the allele frequency of the A_1 allele in population II the fastest?
    a. 40 individuals from population I (frequency of the A_1 allele = 0.85) migrate to population II (60 individuals; frequency of the A_1 allele = 1.00)
    b. 8 individuals from population I (frequency of the A_1 allele = 0.70) migrate to population II (72 individuals; frequency of the A_1 allele = 0.50)
    c. 70 individuals from population I (frequency of the A_1 allele = 0.40) migrate to population II (630 individuals; frequency of the A_1 allele = 0.50)
    d. 60 individuals from population I (frequency of the A_1 allele = 1.00) migrate to population II (60 individuals; frequency of the A_1 allele = 0.05)
    e. 75 individuals from population I (frequency of the A_1 allele = 0.25) migrate to population II (675 individuals; frequency of the A_1 allele = 0.70)

37. In a randomly mating island population, N, the frequency of a recessive allele that causes an autoimmune disorder is 0.08. During a famine, individuals from the neighboring population M leave their island and travel to island N, where they mate at random with individuals of population N. The number of migrants is 30% of the total population of island N. The frequency of the deleterious recessive allele in the migrant population is 0.005. What is the ratio of affected individuals in the population now as compared with before migration?

38. Convert the pedigree below into a path diagram and determine the inbreeding coefficient of the inbred individual, assuming that the common ancestors are themselves not inbred.

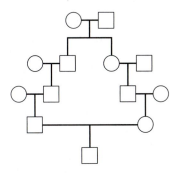

39. Two populations of wolf spiders are assayed for heterozygosity at 15 microsatellite loci. The populations are known to have been genetically isolated for 50 generations. In the larger population (12,000 spiders) heterozygosity averages 0.760 across the 15 loci, and in the smaller population (400 spiders) heterozygosity is 0.608.

What is the inbreeding coefficient of the smaller population relative to the larger?

40. Assume there is one locus, *A*, that determines fitness in the spider populations described in problem 39. Individuals homozygous for the A_1 allele have a relative fitness of 1.0. Individuals homozygous for the A_2 allele have a relative fitness of 0.6. The A_2 allele exists at a frequency of 0.1. What will be the population fitness under the following four scenarios?
    a. Additive allele effects and no inbreeding.
    b. Additive allele effects and an inbreeding coefficient of 0.2.
    c. The A_1 allele is completely dominant and there is no inbreeding.
    d. The A_1 allele is completely dominant and an inbreeding coefficient of 0.2.

    Compare the additive and dominance cases with and without inbreeding. What does this tell you about dominance and inbreeding depression?

41. A rare form of dwarfism is caused by a recessive mutation within a single gene. The mutation rate to the allele causing dwarfism is 9.13×10^{-6} (9.13 individuals in 1 million). It is estimated that 1 in 50,000 people have this form of dwarfism. What do you estimate the strength of selection against this allele to be?

42. Look at Figure 23.11. What two forces are interacting to cause the types of dynamics witnessed? What do you think would happen if the populations were made even smaller?

Chapter Integration Problem

A research group of 850 patients that were known to have been exposed repeatedly to disease X, were genotyped at their MHC class 1 type HLA locus (*B*, a specific gene within the major histocompatability complex). Three alleles were found at this locus, a wild-type allele (B_1) and two mutated alleles caused by two different point mutations in the same codon (B_2 and B_3). Among the genotyped individuals, variation in the likelihood of developing disease symptoms was strongly attributed to genetic variation at this locus. Individuals homozygous for the B_2 allele are 20% resistant to the disease, individuals homozygous for the B_3 allele are 50% resistant to the disease, and individuals heterozygous (B_2B_3) are 100% resistant to infection. Other allelic combinations provide no protection from infection.

Currently disease X is very rare. Both the B_2 and the B_3 alleles are mildly deleterious ($s = 0.03$ and 0.05, respectively) under normal circumstances (i.e., when not exposed to disease X), and both alleles are totally recessive to the wild-type allele. The two mutations are additive with respect to each other ($s = 0.04$ in heterozygotes).
    a. If the wild-type allele encodes a methionine at the mutated site, what are the possible codons that could be encoded by the B_2 and B_3 alleles?
    b. Given a forward mutation rate from the wild-type allele to each of the mutant alleles of 2.5×10^{-5}, and no backwards mutation, what do you expect the equilibrium frequency of both alleles to be under mutation–selection balance theory?

c. The B_2 allele has a frequency of 17.3% and the B_3 allele a frequency of 7.1% in a given population. Write out the frequency of each of the genotypes (assume Hardy–Weinberg equilibrium) and their relative fitness in the absence of disease X.

d. What are the new allelic frequencies after one generation of selection using the calculations from part **c**?

e. Disease X becomes pandemic, and everyone in a given community is exposed to the disease. The disease affects children younger than age 5 and is 100% lethal within 72 hours unless the individual has genetic resistance to the disease. Use the allelic frequencies given in part **c**, but calculate a new relative fitness for the cohort of children younger than 5 years of age exposed to the disease.

f. What are the new allelic frequencies after one generation of selection using the calculations from part **e**?

g. Disease X continues to be prevalent in the environment, so that the wild-type allele is lost from the population. What do you expect the long-term equilibrium frequency of the B_2 and B_3 alleles to be?

h. Disease X has now become stable and it is known that a random 40% of children will be exposed to the disease in any given generation. If current allele frequencies are $B_1 = 0.187$, $B_2 = 0.430$, and $B_3 = 0.383$, what will the allelic frequencies be after one generation of disease exposure at the current rate?

The following polyacrylamide gel shows the inheritance patterns for the PCR-amplified locus of interest for a set of parents and their children. The family was exposed to disease X as it spread into through the town. The father and the two youngest children (child #3 and child #4) survived and the rest of the family died.

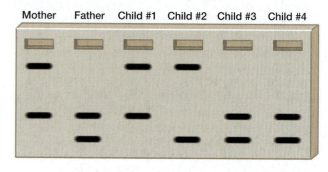

i. Given the information on which individuals live and which died, can you associate certain alleles with the bands on the gel?

j. For each of the children assign maternal or paternal origins to the alleles.

k. For this particular pair of parents, what proportion of their children would you expect be immune, partially immune, and susceptible to disease X?

Do you need additional review? Visit **www.mhhe.com/hyde** for practice tests, answers to end-of-chapter questions and problems, interactive exercises, and animation tutorials all designed to enhance your understanding of key genetic concepts.

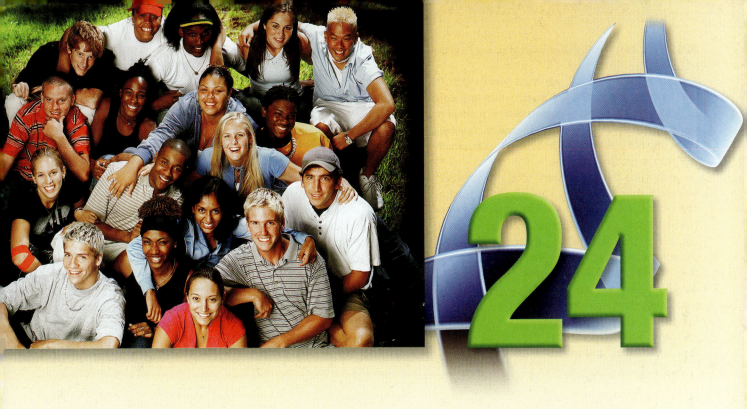

Quantitative Genetics

Essential Goals

1. Determine the proportion of variation in a trait that results from the environment and the genotype.

2. Understand how the mean phenotype of a given genotype is altered with changes in the environment.

3. Describe how and to what extent loci and alleles interact with each other to influence a phenotype.

4. Explain the differences observed due to natural and artificial selection.

5. Understand how the theory of quantitative genetics may affect the management of endangered species of plants and animals.

6. Estimate the number of loci involved in a trait and any limitations to this approach.

Chapter Outline

Photo: Many of the wide array of phenotypes that are observed in a population represent quantitative traits.

*I*n previous chapters, we dealt mostly with the inheritance of *discontinuous traits*, which are usually controlled by a single gene and exhibit a small number of phenotypes. Examples include green versus yellow pea seed color, straight versus curly *Drosophila* wings, and the human ABO blood group system.

Following the rediscovery of Mendel's work in 1900, biometricians studying continuously varying characters had argued that the variation in many important traits did not follow Mendelian rules. Height, intelligence, seed weight, and similar traits seemed to vary continuously and showed variable responses to the environment, rather than discrete phenotypes. These investigators thought that such traits could not be explained by any theory based on particulate inheritance that seemingly could lead only to discrete differences among individuals.

Most traits important to the evolution of organisms, food production, and medicine are influenced by multiple loci that interact with one another and with the environment. These *continuous traits* exhibit a spectrum of variation, rather than a few discrete phenotypes. Because the variation in phenotypes associated with continuous traits can be measured, they are termed *quantitative traits.*

Ronald Fisher, considered the father of both quantitative genetics and modern statistical analysis, proposed in 1918 the mathematical theory of how Mendelian inheritance could produce continuous variation. Quantitative genetic theory unified the observations of plant and animal breeders (and Darwin's thoughts on natural selection) with Mendel's observation of particulate inheritance and independent assortment.

Currently a new melding of molecular and quantitative genetics is underway. **Quantitative trait loci (QTLs),** the individual loci contributing to a quantitative trait, can now be mapped and a true unification of knowledge between the molecular function of a gene and its effects on the organism may be underway.

24.1 What Is Quantitative Genetics?

Quantitative genetics examines how genes interact with the environment to produce a continuous distribution of phenotypes for a given trait in a population. The methods of quantitative genetics attempt to predict the response of the trait to selection pressure, given data on the phenotype and the genetic relationships among individuals.

A quantitative trait can be the result of either a single locus or, more commonly, multiple loci that interact to produce the phenotype. The term **polygenic trait** is used for a character that is controlled by more than one locus. Because we rarely know the genes underlying these traits, and because the mode of inheritance can be very complicated, we must rely on statistical methods to analyze the traits of interest.

Applications of Quantitative Genetics

Knowledge of quantitative genetic theory is necessary for studying evolutionary biology, and it is central to understanding many issues in ecology and conservation biology. The traits on which natural selection acts most strongly, such as those that influence an organism's survival and adaptation to a changing environment, are almost entirely quantitative traits. Guidelines for conserving the world's biodiversity are based, in part, on the principles of quantitative genetics.

The National Institutes of Health (NIH) has also recognized the importance of quantitative genetics in medical research. The two leading causes of mortality in industrialized nations, heart disease and cancer, are both quantitative traits. Both of these diseases contain a complex genetic component with expression that depends on environmental influences. In less industrialized nations, 34% of all deaths are attributable to infectious and parasitic diseases. Susceptibilities to these diseases are also quantitative traits.

Advances in food production have long relied on quantitative genetics. Quantitative traits are very important to plant and animal breeders because traits such as grain yield, drought resistance, efficiency of food conversion, and meat quality are all quantitative traits.

Major Questions Addressed in Quantitative Genetics

Quantitative genetics can provide answers to the following questions:

- How many loci are involved in determining a trait?
- What percentage of a trait is attributable to each component locus?
- What would the predicted consequences of natural and artificial selection be for the trait? That is, how quickly would a trait be altered under different conditions?

Knowing the number of loci involved and the percentage of contribution allows predictions of how a trait can vary under specific conditions. Identifying the loci with greatest influence could enable modification of these loci, through molecular genetics techniques or standard hybridization. Finally, information about the potential effects of selection may suggest approaches for improving crops and maintaining endangered populations.

Restating the Concepts

▶ Quantitative genetics is the study of how multiple genetic loci interact with the environment to produce a specific trait (polygenic trait).

▶ Quantitative genetics is critical for the analysis of evolutionary biology, ecology, conservation biology, and many human diseases.

▶ Quantitative genetics attempts to identify the genetic loci involved in a polygenic trait and how they interact with the environment.

24.2 Polygenic Traits

Three different types of polygenic traits exist: *continuous, threshold,* and *meristic*. After a discussion of these three types of traits, we will consider how the range of phenotypes associated with a polygenic trait is produced, and how this explanation conforms to Mendelian inheritance.

Continuous Traits

Continuous traits (*metric characters*) are those that can take on an infinite number of values. Examples include height, growth rate, and metabolic rate.

The number of classes in a continuous trait is limited only by the precision of the measuring device. If precision is limited to centimeters, then for human height the classes would consist of centimeter values of 163, 164, 165, . . . 180, 181, 182, and so on. If precision is limited to millimeters, the height classes would now consist of millimeter values of 1630, 1631, 1632, . . . 1800, 1801, 1802, and so on. Continuous traits, or those that are approximately continuous, are the major focus of quantitative genetic studies.

Threshold Traits

Threshold traits are those for which an individual is classified as either having the trait or not—but the threshold is often arbitrary, and the underlying basis of the trait is multifactorial, controlled by both polygenic and environmental influences.

Heart disease is an example of a threshold trait. Unlike the case with a continuous trait, an individual either has heart disease or does not. But no single gene when mutated leads to heart disease. Rather, heart disease results from the interaction of a number of loci and environmental factors, such as diet and the amount of physical activity an individual gets. We know that some individuals have a high probability of developing heart disease and others have a low probability.

Placement of an individual along this spectrum of probability depends on both genetic factors, such as

a family history of heart disease, and environmental factors. Because the probability of an individual developing heart disease is continuous, this trait is treated as quantitative.

Meristic Traits

Meristic traits (*binary characteristics*) are those that vary by a whole number. Examples include the brood size of a dog, the number of flower petals produced by a plant, or the number of scales on a fish. Because these characteristics are under both polygenic and environmental control, they are considered quantitative traits; however, the variation within each of these traits is a whole number. A dog may have, for example, a brood size of 1, 2, 3, 4, 5, or 6 pups—but there is no such thing as a brood of 4.5 pups.

The number of phenotypic components in meristic traits is not limited by the precision of measurement. For meristic traits with a large or moderate number of possible values, the distribution of values for a trait is often approximately a normal distribution (described later on), and the trait can be treated as a continuous trait for analysis.

Generation of Continuous Variation

Because continuous traits are under genetic control, they are expected to follow Mendelian principles of inheritance. But it was unclear for a long time how discrete genetic units could produce continuous variation in a phenotype. The answer to this problem came from a series of experiments performed by Hermann Nilsson-Ehle in 1909. The results from these experiments were used by William Bateson and G. Udny Yule to propose the *multiple-gene hypothesis for quantitative traits*.

Nilsson-Ehle's Wheat Crosses

Nilsson-Ehle was studying the potential role of multiple genes in quantitative traits. He crossed wheat that contained red grain with wheat that had white grain (**fig. 24.1**). The F_1 progeny had an intermediate red grain, which was consistent with incomplete dominance (chapter 5). When he crossed two F_1 plants, however, he did not observe the 1:2:1 phenotypic ratio in the F_2 generation that is characteristic of incomplete dominance. Rather, he found 15/16 of the F_2 plants had some type of red grain and 1/16 had white grain. Closer examination of the red-grained plants revealed that there were four different shades of red, ranging from the parental red through the F_1 pink color to a lighter shade of pink.

Nilsson-Ehle reasoned that if 1/16 of the F_2 progeny were white, this phenotype was likely controlled by two genes, and that the white phenotype resulted from a plant's being homozygous recessive at both loci. Unlike the dihybrid crosses that we discussed in chapter 5, the wheat phenotypes did not fall in a 9:3:3:1 ratio. Nilsson-Ehle found that the phenotypes were present in a 1:4:6:4:1 ratio, which corresponded to the parental dark red: medium red: F_1 intermediate red: light red: white.

He predicted that the two genes controlling this phenotype each contained one allele that conferred the red phenotype and one allele that produced the white phenotype. The sum of the red alleles determined the shade of red that was produced (see fig. 24.1). This suggested that a plant that was heterozygous at both loci ($a^r a^w$; $b^r b^w$) would exhibit the same intermediate red phenotype as a plant that was homozygous for red at one locus and homozygous for white at the other locus ($a^r a^r$; $b^w b^w$ or $a^w a^w$; $b^r b^r$).

Applying the Multiple-Gene Hypothesis to Continuous Variation

Nilsson-Ehle's five phenotypes are still not the same as the continuous variation associated with a quantitative trait. Imagine that instead of two loci each containing two alleles, three loci with two alleles were contributing to the trait. In this case, we would expect seven phenotypic classes in a 1:6:15:20:15:6:1 ratio (**fig. 24.2**).

In the case of four loci that each possess two alleles, nine phenotypic classes would be produced in a 1:8:28:56:70:56:28:8:1 ratio. As the number of loci increases, the number of phenotypic classes increases accordingly.

Now consider the potential interaction between the environment and the genotype. When the environmental interaction with the phenotype is low, the phenotypic overlap between the different genotypes is minimal (**fig. 24.3**, bars). As the environmental interaction on the phenotype increases, the phenotypic overlap between the different genotypes also increases (fig. 24.3, lines flanking bars). For example, assume that under high nutrient and bright light conditions, the phenotype can be slightly shifted toward the deeper red at each genotype, and in low nutrient and low light conditions, the phenotype can be shifted slightly toward the lighter red for each genotype. This results in a broader phenotypic distribution for each genotype. Such interactions would increase overlapping phenotypes without an increase in the number of loci. In this way, the multiple-gene hypothesis can yield the continuous variation that is observed in many quantitative traits.

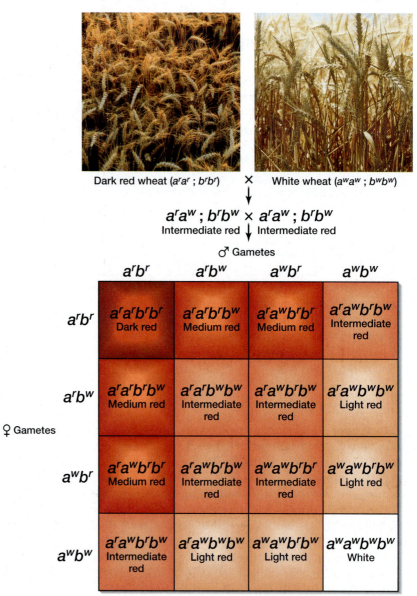

figure 24.1 Summary of Nilsson-Ehle's experiment using red-colored and white-colored wheat. The phenotypes of the two extreme pigmentation phenotypes, red-colored and white-colored, are shown. The pigmentation phenotype is controlled by two genes (*a* and *b*), each with two alleles (*r* and *w*). Crossing a dark red ($a^r a^r$; $b^r b^r$) with a white wheat plant ($a^w a^w$; $b^w b^w$) produces progeny with an intermediate red phenotype ($a^r a^w$; $b^r b^w$). These phenotypically intermediate F_1 progeny each produce 4 different gametes that are shown in the Punnett square. The phenotypic ratio of the F_2 progeny, 1:4:6:4:1, is determined by the total number of *r* alleles present in the plant, regardless of the genes (a^r and b^r are phenotypically equivalent).

Approximating the Number of Genes Controlling a Polygenic Trait

One aspect of quantitative genetics is determination of the number of genes that control a polygenic trait. The method used by Nilsson-Ehle to deduce that two genes were involved in the color of wheat is a fairly standard approach. If one of the two extreme phenotypes in the P generation (red or white grains in wheat) can be clearly identified in the F_2 generation, then a simple formula can be applied:

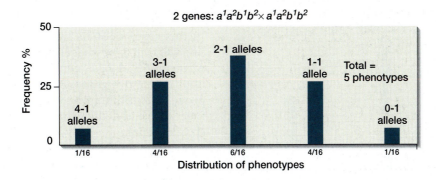

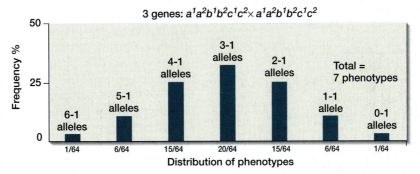

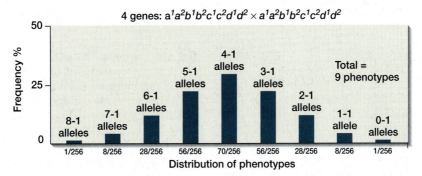

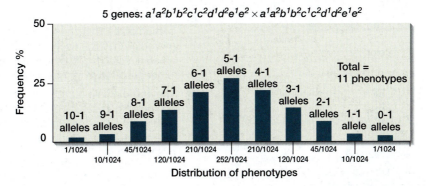

Figure 24.2   Increasing the number of genes contributing to a trait produces a normal distribution. In the examples shown, 2, 3, 4, and 5 genes, each gene pair possesses only two alleles (*1* and *2*) that affect the same phenotype. As the number of independently assorting genes contributing to a trait increases, the number of phenotypes also increases. If the two extreme phenotypes are the same in all the examples, then the increased number of phenotypes decreases the variation between adjacent phenotypes. The number of positive alleles (*1*) that are required to produce each phenotype is shown above each bar and the fraction of progeny that exhibit each phenotype is shown below each bar.

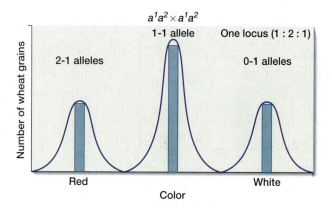

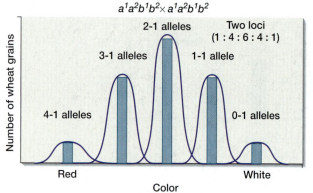

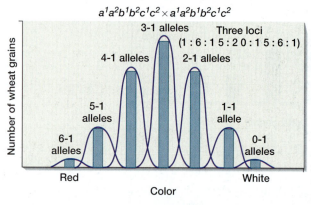

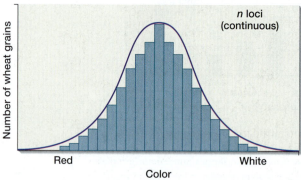

figure 24.3 The environment affects the phenotypic distribution. In the examples shown, increasing the number of interacting genes, each with two alleles, increases the number of phenotypes (blue bars). As the environment contributes to the phenotype, the distribution of each phenotype broadens (lines flanking each bar). As the number of phenotypes increases with more genes, the environmental effects further decrease the differences between adjacent phenotypes to produce a continuous distribution.

Fraction of F_2 progeny with one extreme phenotype $= 1/4^n$

In this case, n equals the number of genes controlling the trait. One-sixteenth (1/16) of the progeny had white grains in the F_2 generation, so $1/16 = 1/4^2$, which implies that two genes control the trait. If 1/256 of the progeny exhibited one of the extreme phenotypes, then we could deduce that four genes control the trait ($1/4^4 = 1/256$).

This simple approximation requires two major assumptions. First, every gene contributes equally to the trait. Second, the trait is not dramatically influenced by the environment. The ratios could be skewed significantly if the environmental conditions strongly affect the phenotype and these conditions are not consistent. As you saw in figure 23.3, the environmental influence can alter the phenotype to such an extent that two different genotypes could produce the same phenotype under different environmental conditions. Therefore it is important to realize that this approach is only an approximation.

Restating the Concepts

▶ Multiple genes can interact to control a single trait that can be classified as a continuous, threshold, or meristic trait. All these types of traits possess a quantitative genetic component underlying them.

▶ Polygenic traits involve the interaction between multiple genes to produce the phenotype. In this model, each gene has one allele that can function in an additive manner to produce the trait and one allele that does not contribute to the phenotype.

▶ The model assumes that each gene contributes a portion of the final phenotype. Although we discussed examples in which each locus contributes an equal amount, some loci in a polygenic trait may provide a larger percentage of the phenotype than other loci.

▶ The approximate number of loci that contribute to a polygenic trait can be deduced; however, unequal contributions by the different loci and environmental interactions may significantly affect the accuracy of this calculation.

24.3 Statistical Tools in Quantitative Genetics

The hallmark of quantitative genetics is the ability to apply a variety of statistical methods to analyze a trait. When determining the relative contributions of genetics and the environment, geneticists do not usually analyze the entire population. Instead, a subset of the population, called a **sample,** is evaluated.

The accuracy of the analysis depends on whether the sample is completely random and reflects the broader population. If the sample is nonrandom and is biased toward a particular subset of the population, then the results of the analysis will not reflect the true state of the population as a whole.

The Normal Distribution

A **normal distribution** is the frequency of values, such as classes of a trait, that occur randomly. The plot of the normal distribution is a bell-shaped, or Gaussian, curve (**fig. 24.4**). If a trait is quantitative and the random sample size is large enough, we expect to see a normal distribution. An example is seen in the distribution of heights of 83 genetics students at the University of Notre Dame in 2007 (**fig. 24.5**). In this example, the number of individuals in each column represents the number of individuals of a specific height. Notice that the general shape of the student height distribution is similar to the shape of the bell-shaped curve. Most quantitative traits in a population are normally distributed, with the largest number of individuals occurring in the middle of the curve.

Measures of Central Tendency

Two different sets of data can produce normal distribution curves that are clearly different (see fig. 24.4). One way to describe these two curves is to extract a single value from the data for comparison. The most common methods for defining a single value are to calculate the *mean, median,* and *mode.* These values represent different attempts to determine the **central tendency,** which are different midpoints in the data set.

The Mean

You are probably familiar with the arithmetic mean of a set of numbers. The **arithmetic mean** represents the average value of a set of numbers. The common formula for the arithmetic mean is:

$$\overline{X} = \frac{(x_1 + x_2 + \ldots + x_n)}{N}$$

where $\overline{X}$ is the arithmetic mean, x_1 through x_n are the individual phenotypic values for each member of the population or sample, and N is the number of individuals in the sample. In most cases, in which only a sample is analyzed and not the entire population, the resulting number is only an estimate of the true mean of the population.

For the heights of the 83 genetics students at the University of Notre Dame (see fig. 24.5), the arithmetic mean is calculated as 68.2 inches.

The Median

The **median,** another common measure of central tendency, is the value that divides the set of numbers into

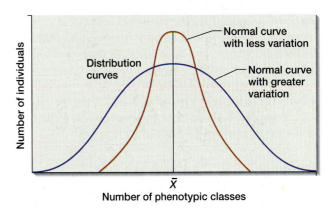

figure 24.4 Two different normal distribution curves with identical means (vertical line). The red curve, however, possesses less phenotypic variation than the black curve.

| Height (inches) | 60 | | 65 | | 70 | | 75 | |

| Number of students | 1 3 1 5 2 8 9 7 12 4 6 7 4 2 4 2 2 1 1 |

(a)

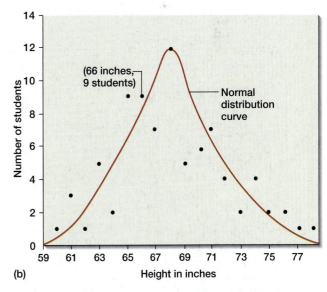

(b)

Figure 24.5 A normal distribution of human height. (a) Shown are 83 of the students in Dr. Hyde's genetics classes from 2005–2007 at the University of Notre Dame. The students are arranged by their height, with the height in inches and the number of students of each height shown below the photograph. (b) The normal distribution curve that plots the student heights (in inches along the X axis) against the number of students at each height (plotted along the Y axis).

two equal parts. In the case of an odd number of values, the median is the central value with an equal number of observations greater than and less than that value. In the case of an even number of values in the set, the median is the arithmetic mean of the two central values. In the set of student heights in figure 24.5, the median is 68 inches because 41 students are shorter than 68 inches, and 41 students are taller than 68 inches.

The Mode

The **mode** is the value that occurs most frequently in a set of numbers. In the example of college student heights, the mode is 68 inches because the greatest number of students (12) are 68 inches tall. The mode is a seldom-used measure of central tendency, although it does have significance in population descriptions.

Comparison of the Three Values

In an ideal normal distribution, the mean, median, and mode are all equal. As you can see in the bell-shaped curve in figure 24.4, a normal distribution is symmetrical about the mean—that is, an equal number of observations are above and below the mean value, which is the definition of median. The mean is the most common value—the definition of mode—with values increasingly farther from the mean occurring increasingly less frequently.

In the plot of student heights, the median and mode are the same, but the mean is slightly larger (**fig. 24.6**). This reveals that our quantitative data do not represent a perfectly normal distribution. This result

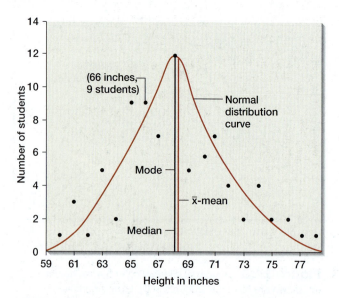

Figure 24.6   Central tendency of a normal distribution. The normal distribution curve plots 83 of Dr. Hyde's students from 2005–2007 at each height. The mean (red line) is 68.2 and the median (black line) is 68.0 for this curve. The mode, 68, is identical to the median. In an ideal normal distribution, all three values would be identical.

is not surprising based on the relatively small set of students that were analyzed. As the data set increases in size—for example, all the students attending the University of Notre Dame in 2007, the curve should approach a normal distribution of heights.

Measures of Dispersion

Figure 24.4 shows two different normal distribution curves with identical arithmetic means—and yet clearly these populations are different. How can we distinguish between these two curves, and what is the importance of the overall shape of the curve?

The dispersion of the data from the mean value—that is, how close or far the extreme values are—provides an indication of the width of the normal distribution curve. There are several ways to measure this dispersion. The *variance* measures how closely the values cluster about the mean. The *standard deviation,* derived from the variance, provides another benchmark for population evaluation.

Variance

Variance indicates the extent of the extremes of a population. A relatively small variance indicates that the values cluster tightly about the mean, and larger variances indicate a more dispersed distribution. The formula for the variance estimated from a sample is

$$s^2 = \sum \frac{(x_i - \overline{X})^2}{n-1}$$

Thus, the variance (s^2) equals the sum of the squared differences between each measured value and the arithmetic mean, which is then divided by one less than the entire sample size (n–1). In the example of student heights (fig. 24.5), the variance is 15.63 inches. **Table 24.1** shows how the variance was calculated.

Standard Deviation

The square root of the variance is called the **standard deviation,** symbolized as *s.* It can be thought of as estimating the mean distance of a randomly chosen value from the mean. In the student height example, the standard deviation is 3.95 inches. (The square root of 15.63, the variance, is equal to 3.95.)

A population can be described using only the mean and the standard deviation. This approach may seem to be a gross simplification; however, if the phenotypic values in the population are distributed in a bell-shaped curve—that is, the assumption of a normal distribution holds—then the expected proportion of individuals falling into any phenotypic class can be known. For example, 68.3% of the population should have phenotypic values that are between ±1.0 standard deviation of the mean, and 95.4% of the population should lie between ±2.0 standard deviations of the mean.

table 24.1   Calculating the Variance of the Heights of Students at the University of Notre Dame in 2007

Height (in inches)	Mean ($\bar{X}$)	$x_i - \bar{X}$	$(x_i - \bar{X})^2$	Number of Individuals of This Height	Sum of $(x_i - \bar{X})^2$ for Each Height ($[x_i - \bar{X}]^2 \times$ Number of Individuals)
60	68.2	8.2	67.24	1	67.24
61	68.2	7.2	51.84	3	155.52
62	68.2	6.2	38.44	1	38.44
63	68.2	5.2	27.04	5	135.20
64	68.2	4.2	17.64	2	35.28
65	68.2	3.2	10.24	9	92.16
66	68.2	2.2	4.84	9	43.56
67	68.2	1.2	1.44	7	10.08
68	68.2	0.2	0.04	12	0.48
69	68.2	0.8	0.64	5	3.20
70	68.2	1.8	3.24	6	19.44
71	68.2	2.8	7.84	7	54.88
72	68.2	3.8	14.44	4	57.76
73	68.2	4.8	23.04	2	46.08
74	68.2	5.8	33.64	4	134.56
75	68.2	6.8	46.24	2	92.48
76	68.2	7.8	60.84	2	121.68
77	68.2	8.8	77.44	1	77.84
78	68.2	9.8	96.04	1	96.04

$$\sum (x_i - \bar{X})^2 \quad 1281.52$$
$$n - 1 \quad 82$$
$$s^2 \quad 15.63$$

In the college student height example, 95.4% of the students would be expected to have heights between 60.3 and 76.1 inches. In reality, 80 of the 83 students (96.4%) are within this range.

Measures of Covariation

Very often geneticists want to see whether two or more traits, measured for a single individual, vary together. That is, if one individual has a larger than average value for trait 1, then is the same individual likely to have a larger value, or smaller value, than average for trait 2? **Covariation** is a measure of how two traits or variables change relative to each other. This analysis, however, does not try to suggest that one trait affects the second.

As an example, let's examine how the body weight of the adult female wolf spider relates to its fecundity (number of viable offspring). Our sample set consists of eight female wolf spiders. We can determine the weight of each spider and its fecundity (**table 24.2**). Thus, we have a pair of traits measured for each individual in the sample.

Visual examination of the eight pairs of values suggests that larger spiders have more offspring. This

table 24.2   Measurement of Body Mass (in g) and Fecundity (Number of Live Offspring) for Eight Adult Female Wolf Spiders (*Rabidosa rabida*)

Individual	Body Mass (g)	Fecundity
Spider 1	0.51	110
Spider 2	0.54	125
Spider 3	0.58	306
Spider 4	0.77	299
Spider 5	0.80	394
Spider 6	0.80	254
Spider 7	0.91	444
Spider 8	0.93	295

is the general meaning of a **correlation:** a relationship that exists between two things. The **covariance** is a statistical analysis of the amount of correlation between the two traits.

We can calculate the covariance of these data by summing the products of the difference between each measured value and the arithmetic mean for each trait

(x and y) and dividing by one less than the entire sample size (n–1).

$$\text{cov}_{XY} = \sum \frac{[(x_i - \bar{X})(y_i - \bar{Y})]}{n-1}$$

where $x_i - \bar{X}$ is the difference between each data point and the mean for one set of data, $y_i - \bar{Y}$ is the difference between each data point and the mean for the second set of data, and n is the number of individuals.

Using the data for the female wolf spiders shown in table 24.2, we can calculate the covariance (**table 24.3**). Notice that when the individual values for x and y are calculated, some are positive and others are negative. In a covariance calculation, if one data point is greater than the mean (positive value), it is important to determine whether the second data point is above (positive value) or below (negative value) the mean. From this analysis, the covariance works out to a value of 14.63.

More important than the covariance is the **correlation coefficient (r),** which signifies to what extent the variation in one quantitative trait is associated with the second trait. The correlation coefficient does not reveal what causes the variation, however, or whether the variation of one trait controls the variation observed for the second. The correlation coefficient is calculated by dividing the covariance by the standard deviation of both sets of quantitative measurements:

$$r = \frac{cov_{XY}}{s_X s_Y}$$

where s_X is the standard deviation for one data set (body mass) and s_Y is the standard deviation of the second data set (fecundity). Using these data, the correlation coefficient in the female wolf spider data is 0.76 (see table 24.3).

The value of the correlation coefficient ranges from –1 to +1. A correlation of 1.0 is a perfectly linear positive relationship, which means that an individual 1 standard deviation above the mean for one trait is also 1 standard deviation *above* the mean for the other trait. A correlation of –1.0 is a perfectly linear negative relationship, which means that an individual 1 standard deviation above the mean for one trait is also 1 standard deviation *below* the mean for the other trait. A correlation coefficient of 0 implies no relationship between the variables at all. The correlation coefficient is seldom if ever exactly 1 or exactly 0.

Correlation coefficient data may sometimes be misleading. First, the correlation coefficient is subject to sampling error, and a different sample of eight spiders is not likely to yield the same correlation coefficient. Second, the correlation coefficient does not demonstrate a cause-and-effect relationship, even if the coefficient is very large. In the spider example, the increased body mass may not directly cause the increased fecundity; instead, the larger body mass may attract more reproductively fit mates that produce the increased number of offspring.

It's Your Turn

The statistical analysis of quantitative data provides a convenient method to analyze a set of data and to compare different data sets. To test your knowledge of the various statistical tools, try problems 23, 24, and 30.

table 24.3   Calculating the Covariance (cov_{XY}) and Correlation Coefficient (r) of the Body Mass and Fecundity of the Female Wolf Spider

Body Mass (in g)	Mean	$x_i - \bar{X}$	Fecundity	Mean	$y_i - \bar{Y}$	$[(x_i - \bar{X})][(y_i - \bar{Y})]$
0.51	0.73	–0.22	110	278	–168	36.96
0.54	0.73	–0.19	125	278	–153	29.07
0.58	0.73	–0.15	306	278	28	–4.20
0.77	0.73	0.04	299	278	21	0.84
0.80	0.73	0.07	394	278	116	8.12
0.80	0.73	0.07	254	278	–24	–1.68
0.91	0.73	0.18	444	278	166	29.88
0.93	0.73	0.2	295	278	17	3.40
	s	0.165		s	116	
					$[(x_i - \bar{X})][(y_i - \bar{Y})]$	102.39
					cov_{XY}	14.63
					r	0.76

Analyzing Covariance and Predicting Values from the Data

You have likely observed in your friends that two tall parents often produce tall children, whereas two short parents usually have short children. In this example, the height of the parents likely correlates with the height of their children. In this case, tall and short are two phenotypes, and quantitative data can determine whether a relationship exists between them. You can examine this type of data in a sample of individuals.

As an example, let's examine wing length in *Drosophila*. You might think that the size of the fly's wings is a quantitative trait, and that large-winged flies are more likely to produce large-winged progeny. To determine whether these assumptions are correct, you must determine the average wing length of both parents (that is, the mid-parent length) and the length in the progeny (box table 24A).

From these data, you need to determine the sum of the X values and the sum of the Y values, which come out to 92.7 and 93.2, respectively. Using this information, you can determine the mean $\bar{X}$ value ($\Sigma x/n$) and mean $\bar{Y}$ value ($\Sigma y/n$) for the 37 data points. You find that the $\bar{X}$, 2.51 mm, is very similar to the $\bar{Y}$, 2.52 mm.

Your next step is to calculate the variance for both the X and Y values. The variance equation is:

$$\text{variance} = s^2 = \sum \frac{(x_i - \bar{X})^2}{n-1}$$

The variance of the midparent values (X) is 0.192, and the variance of the offspring values (Y) is 0.100. Calculating the square root of the variance yields the standard deviation (s_x or s_y), which is 0.438 for the midparent values and 0.316 for the offspring values.

Next, you need to find the covariance. To complete the covariance calculation, you use this equation:

$$\text{covariance} = \sum \frac{[(x_i - \bar{X})(y_i - \bar{Y})]}{n-1}.$$

Notice that you multiply the difference between each midparent value and the mean by the difference between each offspring value and the mean. The result is that if a midparent value is less than the average, a negative value is produced. If the corresponding offspring value is larger than its average, a positive value is produced. Multiplying two such values together produces a negative number. It is therefore possible to produce either positive or negative values for each data point.

If both values have the same sign, you obtain a positive covariance value, which means that both variables move in the same direction. If one value is positive and the other is negative, you obtain a negative covariance.

Using the data in the table, the covariance is calculated as 0.106. From this you are now able to determine the correlation coefficient ($r = cov_{XY}/s_X s_Y$), which requires knowing the covariance of the data set and the standard deviations for both sets of values. You obtained these values in the previous calculations. Using these data, you should find that the correlation coefficient is 0.766. Because this value is near 1.0, there appears to be some positive correlation between the midparent wing length and the offspring wing length.

table 24A Midparent and Offspring Wing Length in *Drosophila*

Midparent Wing Length (in mm, X)	Offspring Wing Length (in mm, Y)
1.5	2.0
1.7	2.0
1.9	2.2
2.0	2.0
2.0	2.2
2.0	2.2
2.1	1.9
2.1	2.2
2.1	2.5
2.2	2.3
2.3	2.2
2.3	2.6
2.4	2.0
2.4	2.3
2.4	2.4
2.4	2.6
2.4	2.6
2.4	2.6
2.4	2.7
2.4	2.7
2.6	2.7
2.6	2.7
2.6	2.8
2.6	2.9
2.8	2.7
2.8	2.7
2.9	2.5
2.9	2.7
2.9	2.7
2.9	3.0
3.0	2.8
3.0	2.8
3.0	2.9
3.1	3.0
3.2	2.4
3.2	2.8
3.2	2.9

Restating the Concepts

▶ If the sample set of quantitative data is sufficiently large, then the data exhibit a normal distribution.

▶ We can analyze the central tendency of a normal distribution by determining the mean, median, and mode. In an ideal normal distribution, all three of these values are the same.

▶ The dispersion of data from the mean is called the variance. The square root of the variance is the standard deviation. In a normal distribution, approximately 95% of the data will be within 2 standard deviations of the mean.

▶ In some cases, the relationship between two variables can be explained as correlation. Correlation does not suggest the change in one variable has a direct effect on the other variable. The degree of correlation can be quantified as the covariance, which varies from –1 to +1. The correlation coefficient is a handy measure of the type and strength of the correlation. A correlation coefficient of +1 suggests a linear positive relationship, 0 suggests that there is no relationship between the samples in the data set, and a correlation coefficient of –1 suggests that there is a linear negative relationship.

24.4 The Nature of Continuous Variation

A quantitative trait, by definition, is dependent on the interaction between multiple genes and the environment. For example, lean body mass is partly due to environmental factors (such as amount and type of exercise, level of nutrition, and caloric intake). But a genetic component of lean body mass also exists because some individuals gain lean body mass more quickly than others under similar environmental conditions.

Often, investigators studying quantitative traits are interested in determining the component of the trait that results from the genotype. For example, what is the relative genetic component in heart disease, relative to environmental factors such as diet and exercise? If we ignore the interaction between genetics and environment for now, the genotypic and environmental contributions to the phenotype can be summarized as follows:

$$P = G + E$$

where P is the phenotype of the individual, G is the genotypic contribution to the phenotype, and E is the environmental contribution to the individual's phenotype.

Moving from the individual level to the population level, we know that the mean phenotype of a popula-

tion is a function of the allelic frequencies within the population and their effects, including interactions among alleles. The mean phenotypic value for a given trait (M) that is determined by a single locus in a randomly mating population is:

$$M = p^2 (M_{11}) + 2pq (M_{12}) + q^2 (M_{22})$$

where p is equal to the proportion of the A_1 allele in the population, q is equal to the proportion of the A_2 allele. M_{11} is the phenotypic value for the A_1A_1 homozygote, M_{12} is the phenotypic value for the A_1A_2 heterozygote, and M_{22} is the phenotypic value for the A_2A_2 homozygote. When the effects of different loci combine *additively,* then the overall phenotypic mean is equal to the sum of the contributions of each locus. The preceding situation also assumes no genotypic–environmental interactions.

Genetic Basis of Quantitative Traits: Components of Phenotypic Variance

Just as the mean trait value can be broken down into subcomponents, the phenotypic variance for a quantitative trait can also be divided into different components. The components of variance consist of the same terms that make up the mean, but now we are speaking of the variance (differences) among individuals. V_P represents the phenotypic variance among individuals:

$$V_P = \sum \frac{(P_i - P_M)^2}{n-1}$$

where P_i is the phenotypic value for each individual in the population, P_M is the mean phenotypic value of the population, and n is the number of individuals in the population.

Because the phenotype of an individual is described by $P = G + E$, the variance in the population can be expressed as:

$$V_P = V_G + V_E$$

This equation again assumes that no interaction occurs between the genotype and phenotype, and that covariation between the genotype and the environment does not exist. The latter assumption is not generally true in *natural* populations. Offspring with a particular genetic propensity, for instance, may be provided more resources by their parents to further enhance that trait. As an example, an animal with a genetic propensity to be large and strong may also garner better territories and further increase its competitive advantage over rivals.

If these types of correlations between genotype and environment can be estimated, their effects can be handled with more advanced statistical treatment. If not, they will generally inflate the estimates of the genetic component, giving an inaccurate picture. In experimental situations, the environment is under the control of the researcher, and such problems can be avoided with carefully designed experiments.

These variance components can also be broken down further. For instance, the wolf spider's environment could include specific measures concerning total caloric consumption, crude protein consumption, humidity, and minimum overnight temperatures during winter. The most common breakdown of the equation, however, is the separation of the genetic contribution of variance into additive, dominance, and epistatic components. When we discuss heritability in a later section, it will become clear why this is often very important. The genetic component can be subdivided as follows:

$$V_{\mathrm{G}} = V_{\mathrm{A}} + V_{\mathrm{D}} + V_{\mathrm{I}}$$

Additive genetic variance (V_{A}) is the portion of the genetic variance that actually contributes to the resemblance between relatives and determines the response to selection. **Dominance variance (V_{D})** is the portion of the genetic variance that is due to dominant relationships among alleles at a locus. **Epistatic genetic variance (V_{I})** is that portion of the genetic variance that is due to interactions among alleles at different loci. What this actually means is explained later on when we discuss heritability.

Further Quantitative Genetic Phenomena

The **norm of reaction** is the pattern of phenotypes produced by a given genotype under a variety of environmental conditions. The phenotypic changes are often specific to a particular genotype.

Let's assume that we have seven different homozygous strains of *Drosophila*. If we count the average number of progeny (fecundity) of each strain at three different temperatures, we find that each homozygous genotype has a different fecundity at each temperature (**fig. 24.7**). Although some strains had the highest number of progeny at 14°C, other strains had the greatest number of progeny at 21°C. Therefore, not all genotypes interact in the same manner in a specific environment. The norm of reaction can be plotted for a genotype, allowing the phenotype to be predicted for the range of environmental conditions (**fig. 24.8**).

The components that contribute to the mean phenotype of an individual can now be expanded to

$$P = G + E + I$$

where I is the genotypic-environmental interaction term. Thus, in addition to the additive effects of the genotype and the environment, an interaction term is introduced that relates to the norm of reaction for a specific genotype (see fig. 24.7).

This genotypic–environmental interaction term is of great importance to a wide range of fields. In selective breeding programs for food production, it is important

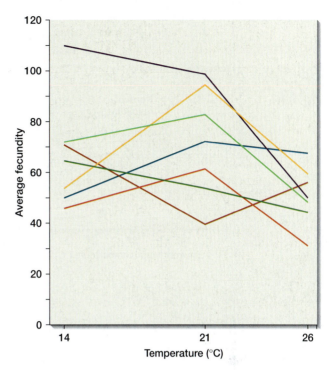

figure 24.7   Norm of reaction for fecundity in *Drosophila*. Seven different homozygous genotypes were raised at three different temperatures. The average fecundity of each fly was plotted against the temperature the flies were raised. Some genotypes had the most progeny at 14°C (purple line), while others had the most progeny at 21°C (gold line). No genotype, however, had the greatest number of progeny at 26°C. Each genotype is represented by a different colored line.

to carefully consider the range of temperatures, rainfall, diseases, soil conditions, and planting densities that a crop plant may face before deciding which genotype is most suited to those conditions. In medicine, the effectiveness of many drugs and the danger actually posed by risk factors for many disease processes (such as cancer and heart disease) are genotype specific. Current research is underway to target drug treatments and preventive medicine to specific genotypes.

Another phenomenon of interest to quantitative geneticists is the pattern of genetic correlation. Genetic correlations can arise through pleiotropy or linkage (chapters 5 and 6). There is growing evidence that most genes affect more than one trait.

Because of genetic correlations, selecting for an increase (or decrease) in one trait will usually produce a correlated response in one or more other traits. This correlated response can be either positive or negative, that is, the second trait may increase or decrease with selection for an increase in the first trait. Negative genetic correlations draw most of the attention, however. In artificial breeding programs, negative correlations can prevent a breeder from being able to select for two different desirable traits.

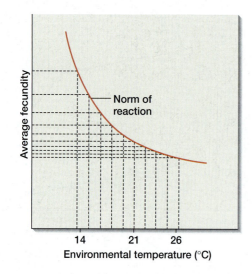

(a)

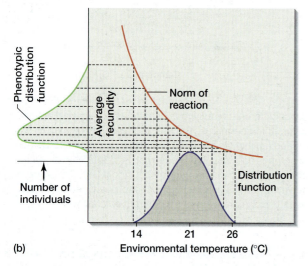

(b)

figure 24.8 Outcomes from the norm of reaction analysis. (a) For a given genotype, the norm of reaction curve (red) relates an environmental condition (temperature) with a phenotype (average fecundity). As seen in figure 24.7, norm of reactions do not have to be a straight line. By knowing the temperature the flies were reared at, you can read from the temperature up to the norm of reaction curve and then over to the Y axis to know how many bristles would be present. (b) In a population, not all the individuals will be exposed to the same environment. Thus, we generate a distribution function (purple curve) that reveals the number of individuals at each temperature. We can then generate a phenotypic distribution curve (green) that reveals the number of individuals with a specific fecundity (phenotype). Because the norm of reaction is curved (and not a straight line), the two distribution function curves will have different shapes.

 Restating the Concepts

▶ Variation is usually quantitative, not qualitative. Continuous variation arises from a large number of genes affecting a trait, a multitude of environmental factors, and the interactions among them.

▶ Models and mathematical equations have been developed that allow the components of the mean and variance of different traits to be analyzed.

▶ The norm of reaction describes the changes induced by the environment on a phenotype for a given genotype.

▶ Both pleiotropy and genetic linkage can affect the genetic correlation between two traits. These correlations can be either positive or negative.

24.5 Heritability

Heritability describes the amount of phenotypic variation in a population that is determined by the genotype. The estimated ratio of genetic variation to environmental variation does not reveal their relative importance or their actual contribution to a phenotype.

Developmental processes governed by genes lie at the base of every character. For example, the morphological structures that make *Homo sapiens* capable of speech depend on the development of structures such as the brain, vocal cords, and tongue. These structures are certainly under genetic control. However, *variation* in language is determined primarily by the environment, namely the languages to which someone is exposed during childhood.

It is important to understand what heritability tells us. For an individual, all phenotypic traits are the result of the interaction among genes and with the environment. Clearly someone's chance of developing heart disease, for example, results from the individual's genotype (did her parents suffer from heart disease), her environment (such as dietary factors), and an interaction between the two. Heritability does *not* correspond to the proportion of heart disease that is due to the genotype of an individual. Heritability does correspond to the proportion of the variation in the susceptibility to heart disease in a population.

It is very important to realize that estimates made on a population are good only for the population under study, and only for that particular environment. If the environment experienced by a population changes, the heritability may also change. Recall the variance equation described earlier

$$V_P = V_G + V_E$$

If the absolute magnitude of E is increased, then the $V_G:V_E$ ratio must also change. It follows that if the variation in the environment increases, the proportion of the phenotypic variation due to the environment should also increase, and as a result the heritability will decrease. If every individual in the population is a clone (genetically identical—as is the case with many crop plants), then any phenotypic variation observed must be due to environmental causes.

For example, **figure 24.9** shows the norm of reactions for 15 different genotypically identical populations in three different environments (low, medium, and high planting density). Because each population is genotypically identical, the phenotypic variation in each population must be due to the environment.

Note also that changes in the environment or in the genotype can also change the statistical mean of the phenotype. You can see this from the formulas. However, mean phenotype tells you nothing about whether the environment is a good environment or a poor one for that population. If the environment were to change, the mean phenotype could change, as could the heritability.

Heritability can be considered in two ways: as *broad-sense heritability*, which is the proportion of phenotypic variation due to genotypic variation; and as *narrow-sense heritability*, which is the proportion of phenotypic variation due to *additive* genetic variation. We now discuss both of these in turn.

Broad-Sense Heritability

Broad sense heritability, or H^2, is defined as the proportion of the genotypic variance (V_G) divided by the total phenotypic variation (V_P):

$$H^2 = \frac{V_G}{V_P}$$

The broad-sense heritability in a population ranges from 0.0 to 1.0. As the broad-sense heritability increases, the environmental variation diminishes.

Estimates of broad-sense heritability are generally performed by fixing either the genotype or the environment and then measuring phenotypic variation. This value is then compared when both components vary freely.

As an example, let's again consider a population of wolf spiders. Two "pure-breeding" lines of spiders are selected, one line with a mean weight of 0.40 g and another with a mean weight of 1.20 g. The selection of pure-breeding lines implies that the spiders for a specific phenotype were selected over several generations, and that one line is fixed for

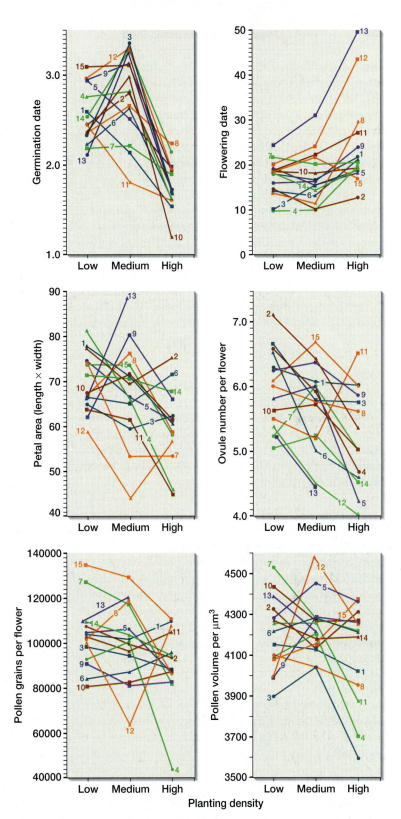

figure 24.9 Effect of population density on quantitative traits. Six different quantitative traits (germination date, flowering date, petal area, ovule number, number of pollen grains per flower, and pollen volume) were analyzed for 15 different paternal wild radish genotypes. Each genotype was planted at three different densities, low, medium, and high density, to produce these norms of reaction. For any specific genotype, the maximal phenotype was produced under different environmental conditions. For example, the maximal ovule number per flower for genotype 12 was observed at low density, but the maximal pollen volume was seen at medium planting density.

alleles for small size at all loci relating to body size. The other line is fixed for alleles for large size at all loci.

These lines are crossed to form a heterozygous F_1 generation. Keep in mind that every individual in the F_1 generation would be genetically identical with respect to the loci of concern. Therefore, all phenotypic differences can be attributed to differences in the environment experienced by individuals.

Imagine that V_P for body weight equals 0.010 g under these conditions (identical genotypes). The F_1 generation is then bred randomly among themselves for one or more generations, and when V_P for body mass is measured again it equals 0.050 g. This is the total phenotypic variance when both environmental and genetic components are allowed to vary. From this we can deduce that the genetic contribution alone must be 0.050 − 0.010, or 0.040 g. The broad-sense heritability can be calculated as

$$H^2 = \frac{V_G}{V_P} = \frac{0.040}{0.050} = 0.80$$

Therefore, 80% of the phenotypic variation is attributable to genetic variation.

Most quantitative traits in animals do not exhibit either very high or very low broad-sense heritability values, which suggests that both the genotype and environment contribute to these quantitative traits. However, broad-sense heritability is often very high in plants.

Narrow-Sense Heritability

The genotypic variance that is measured in broad-sense heritability encompasses the three types of genetic variances that we described earlier in this chapter: additive genetic variance, dominance variance, and epistatic variance. In contrast, the **narrow-sense heritability,** or h^2, is the proportion of phenotypic variance (V_P) that is due to only the additive genetic variance (V_A).

$$h^2 = \frac{V_A}{V_P}$$

Narrow-sense heritability is more important than broad-sense heritability in studying quantitative traits because it only reflects the variance resulting from the additive genetics component. At least for animals, narrow-sense heritability usually provides a more accurate prediction of the response to selection than broad-sense heritability. We will discuss more precisely how to work with additive genetic variation in the next section. The heritability of several traits from various organisms is summarized in **table 24.4**.

How Heritability Is Measured

One of the most common ways to measure heritability is by using the phenotypic resemblance among relatives and knowledge about their genetic relationships.

table 24.4   Heritabilities for Various Traits in Different Organisms

Organism	Trait	Narrow-Sense Heritability (h^2)
Fruit flies	Abdominal bristle number	0.53
	Body size	0.40
	Egg production	0.20
Dairy cattle	Milk yield	0.35
Humans	Fingerprint ridge count	0.92
	Schizophrenia	0.70
	Height	0.65
	Blood pressure (systolic)	0.60
	Score on IQ tests	0.50

HALLIBURTON, RICHARD, INTRODUCTION TO POPULATION GENETICS, 1st, © 2004. Reproduced by permission of Pearson Education, Inc., Upper Saddle River, New Jersey.

If variation among individuals is due in part to variation in genes, then genetic relatives should resemble one another more than individuals chosen from the population at random. This is only true if genetically related individuals are no more likely to share common environments than those that are not more genetically similar.

Remember, heritability is a measure of genetic differences among individuals, and not simply a measure of whether a trait is inherited. If all individuals in a population inherit the same alleles, there is no genetic variation and therefore no heritability.

Because heritability is a proportion, its values range between 0 and 1. This means that some of the differences among individuals are genetic and contribute to heritability, but other differences depend directly on the environment and do not contribute to heritability.

Cross-Fostering and Twin Studies

One way to calculate the narrow-sense heritability is to examine **cross-fostering,** which involves taking the progeny from one set of parents and having them reared by a different set of parents. For example, in a study of the heritability of beak size in song sparrows, young birds from nests at various stages of development (egg to hatchling) were collected and moved to randomly selected foster nests. The mean beak size of offspring and midparent values for both biological and foster parents were measured.

This study revealed a strong correlation between mean beak size of offspring and mean beak size of the biological parents, but no correlation between mean beak size in the offspring and beak size of the foster parents. Thus, virtually all the variation in beak size is due to genetic variation ($h^2 = 0.98$). This is a fairly typical manipulation that can be done in a laboratory or sometimes, as in this case, in wild populations.

For humans, twin studies are very important, particularly cases in which identical twins have been adopted by different parents as infants. Such cases provide an unintended experiment, because the twins are genetically identical and essentially cross-fostered in a random family environment. Such studies frequently reveal moderate to high heritabilities for a huge array of traits, including socially controversial ones such as IQ and sexual orientation.

These studies have sometimes been criticized because adoption agencies try to match the adoptive parents to the biological parents for a number of criteria, so that the environment experienced by the twins may not be uncorrelated in all aspects. There is no doubt that both genes and the environment play a role in determining human behaviors and other characteristics. The real controversy involves not how much of the variation is heritable, but whether traits can be improved through improving social conditions, which is actually a question about norms of reaction, not a question of heritability.

Another way to use twins is to compare monozygotic (identical) twins and dizygotic (fraternal) twins. Whereas monozygotic twins are genetically identical, dizygotic twins share approximately 50% of their genes—the same as any nontwin siblings would. Thus, the phenotypic variation between monozygotic twins will be equal to the environmental variance (V_E), whereas the phenotypic variance between dizygotic twins will be the result of the environmental variance and approximately half of the genotypic variance (V_G). The broad-sense heritability of a specific trait can then be determined by comparing the phenotypic variance of that trait between monozygotic and dizygotic twins.

Another way to utilize twins in studying heritability of traits is to determine the concordance of phenotypic expression. **Concordance** occurs when twins both express or both fail to express the same trait; **discordance** occurs when the twins express different phenotypes. It is then possible to compare the concordance between monozygotic and dizygotic twins. When monozygotic twins exhibit a significantly higher concordance value than dizygotic twins, there is likely a strong genetic contribution to the phenotype. It is important to realize that simply a high concordance for monozygotic twins does not reveal a strong genetic contribution, however.

The use of concordance is demonstrated for a number of human behavioral disorders that were measured in adults and children (**table 24.5**). In all cases, the monozygotic twins have a higher concordance than the dizygotic twins, which confirms a strong genetic component associated with all of these disorders.

Response to Selection: Realized Heritability

Genetic variation generated by mutation provides the raw material for evolution—the change in allele frequencies in a population over time. In contrast, environmental variation does not provide a basis for evolution. Differences in the phenotype caused by the environment are not inherited in the next generation, and this portion of the phenotype does not respond to selection. Thus, one straightforward way to estimate heritability is by applying selection to a trait. The narrow-sense heritability estimate generated this way is referred to as a *realized heritability*.

The formula for realized heritability is

$$h^2 = \frac{R}{S}$$

where R is the response to selection and S is the selection differential. The **response to selection** is the difference between the mean phenotype of the offspring

table 24.5 Twin Concordance for Adult and Childhood Behavioral Disorders

Disorder	Monozygotic Twins Concordance	Sample Size	Dizygotic Twins Concordance	Sample Size
Adult schizophrenia	.38	279	.11	461
Adult affective illness	.65	146	.14	278
Adult panic disorder	.24	67	.11	55
Adult bulimia nervosa	.23	35	.09	23
Childhood attention deficit	.58	69	.31	32
Childhood Tourette syndrome	.53	30	.08	13
Childhood autism	.64	45	.09	36

and the mean phenotype of the original population. The **selection differential** is the difference in the mean phenotype of the selected parents and the mean phenotype of the whole population. It is similar to, but not the same as, the selection coefficient for a single locus that you learned in chapter 23.

Even if $h^2 = 1.0$, the mean phenotype can still change with an alteration in the environment. A narrow-sense heritability of 1.0 implies that all the phenotypic variation within a population is due to additive genetic variation. This could be because the environment has no effect on the phenotype or because every individual experiences exactly the same environment.

For example, let's say you own 30 turkeys. The mean weight of your turkeys is 7.0 kg. You decide to use selective breeding to increase the weight of your turkeys, so you choose the 10 largest for breeding purposes. The mean weight of these 10 breeding turkeys is 9.0 kg. Let's say you randomly select five offspring from each turkey and weigh them once they reach adulthood. The mean weight of the new population of turkeys is 8.0 kg.

The response to selection (R) is 1.0 kg, which is the difference between the mean weight of the original population (7.0 kg) and the mean weight of the offspring (8.0 kg). The selection differential (S) is 2.0 kg, which is the difference between the mean weight of the selected breeding individuals (9.0 kg) and the mean weight of the population before selection (9.0 kg – 7.0 kg). The realized heritability of body weight in the turkeys can now be calculated as 1.0 kg/2.0 kg, or 0.50.

In other words, the realized heritability is the proportion of the selection differential applied that you actually get back as response. When the variation among individuals is entirely environmentally induced, selection should produce a response of 0. When the differences are entirely due to additive genetic causes, the net response would be 1; the weights of the offspring would be exactly the same as the selected parents.

The actual response to selection depends not only on the heritability, but also on the total amount of phenotypic variation. In other words, having a larger population to choose from usually results in larger gains during selection.

Restating the Concepts

▶ Heritability is a quantitative measure of the role that gene differences play in determining the phenotypic differences among individuals.

▶ Variation between phenotypes in a population arises from at least two sources: average differences between the genotypic and environmental variances. Estimates of genetic and environmental variance are specific to that population and to the environment in which the estimates are made.

▶ Broad-sense heritability is a measure of the proportion of a population's phenotypic variation that is due to genotypic variation.

▶ Narrow-sense heritability is the proportion of the population's phenotypic variation that is due to additive genetic variation, or that fraction of the population's phenotypic variation that is available for selection to act on and that determines the resemblance among genetic relatives.

24.6 Quantitative Genetics in Ecology

Understanding the variables that affect a population's ability to adapt and survive in a changing environment is a critical issue in evolutionary biology, conservation biology, and ecology. Evolution in a changing environment requires genetic variation for quantitative traits.

Forces Determining Genetic Variation in Populations

The maintenance of genetic variation for quantitative traits may be influenced by many factors, including the long-term effective population size, the strength and nature of the selection acting on the loci contributing to the trait of interest, and the rate at which mutations occur and the distribution of their effects.

As you saw for a single locus in chapter 23, the amount of genetic variation present for a quantitative trait is subject to random genetic drift. Thus, smaller populations tend to have less genetic variation than larger populations, and the traits are also often farther from their optimum due to random genetic drift. The implication is that smaller populations are generally less capable of adapting to changing environmental conditions.

Selection also affects levels of quantitative genetic variation for a trait. As we discussed in chapter 23, the three general forms of selection that can act on quantitative traits are (**fig. 24.10**)

- **Directional selection,** in which one extreme form of the trait has highest fitness.
- **Stabilizing selection,** in which the average form of the trait has higher fitness than does either extreme.
- **Disruptive selection,** in which both extreme forms of the trait have higher fitness than does the average.

Directional selection results in a change in the mean value of the trait toward the form that has highest fitness. In the absence of mutation, directional selection leads to a reduction of genetic variation; eventually, all individuals become fixed for the alleles

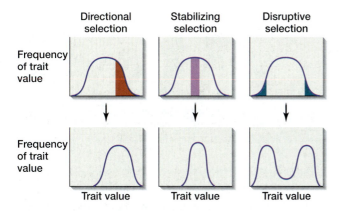

figure 24.10   Three different models of selection. The top row shows a normal distribution of a quantitative trait with the selected trait highlighted. In directional selection, one extreme trait has the highest fitness, which shifts the mean of the population towards the selected trait. Stabilizing selection, where the average trait has the highest fitness, results in the loss of both extreme traits without changing the mean trait value, which reduces the variation. Disruptive selection, where both extreme traits have higher fitness than the average, changes the normal distribution to two overlapping normal distributions, which increases the variation without changing the mean trait value.

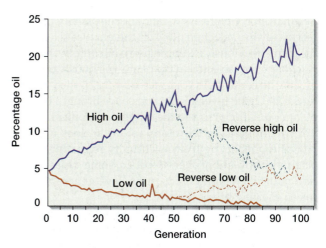

figure 24.11   Artificial selection of a quantitative trait, percentage of oil in maize. Two maize populations, with the same oil percentage, were selected for either high or low oil content for 100 generations (blue and red solid lines, respectively). The selection effectively increased the oil percentage in one population from 5% to over 20%, and decreased the oil in the second population to almost 0%. Reverse selection, such as selecting low oil in the high oil population, began in both populations in the 48th generation (blue and red dashed lines). This reverse selection returned the percentage of oil back to the original value. Thus, artificial selection can adjust quantitative traits in either direction. (Hedrick, *Genetics of Population*, 2005: Jones and Bartlett Publishers, Sudbury, MA. www.jbpub.com. Reprinted with permission.)

granting the highest fitness (heritability will become zero). Stabilizing selection acts against extreme trait values and may either decrease or help to maintain genetic variation, depending on the strength of selection, number of alleles, and their relative contributions. Disruptive selection results in an increase in both extremes, and a loss of intermediate forms.

Genetic variation of a quantitative trait may be maintained over time when there is directional selection, but the environment is variable. If different alleles are favored in different environments, and the environment changes too rapidly for one genotype to become fixed, genetic variation may be conserved.

A variable environment does contribute to the amount of genetic variation maintained for quantitative traits, but no agreement has been reached on whether it is an important contributor or fairly minor.

Most of the genetic variation for quantitative traits is believed to result from the same mechanisms as for a single locus: mutation–selection–drift balance. But the mutation rate in this case restores genetic variation much more quickly than it would for a single locus, because of the many loci contributing to a quantitative trait.

Long-Term Response to Selection

One of the major findings from countless selection experiments in a wide variety of organisms is that natural populations generally have large amounts of segregating genetic variation for most phenotypic traits. Virtually no trait that has been tested has failed to respond to selection in the laboratory. Even highly inbred populations usually retain substantial genetic variation.

There are some limits to variation, however. A number of artificial selection experiments have reached a plateau, usually because of exhausting all the additive genetic variation for the trait, negative correlations with fitness, or presence of linked alleles with lethal or deleterious effects.

Many long-term experiments never reach a plateau, however, if there is sufficient genetic variation, a large population size, and an effective selection process. A classic example is the Illinois corn selection experiments, started in 1986 (**fig. 24.11**). Researchers, starting with maize plants that contained slightly less than 5% oil, have selected plants for either high or low oil content for over 100 generations. During this time, the high-oil plants have increased their oil content over four-fold to greater than 20%, which is an increase of more than 22 standard deviation units compared with the original plants! This far exceeds the upper phenotypic limit in the original population, and the trait has yet to reach a plateau.

The plants with low oil content have reduced their oil content to less than 1%.

In the forty-eighth generation, some plants were chosen for reverse selection. Low oil content was selected in the plants previously exhibiting high oil content, and high oil content was selected in the plants previously possessing low oil content. As you can see by the dashed lines in figure 24.11, the reverse selections have brought the two plant populations back to the original oil content.

By implication, therefore, selection does not prevent the phenotype from eventually returning to the original population phenotype. In some cases, selection must be maintained every few generations to retain the new phenotype.

The power of natural selection acting on a range of phenotypic traits, over approximately a thousand generations, can be seen in the immense phenotypic variation among domestic dog breeds (**fig. 24.12**). This variation was accomplished even though all breeds of dogs are descended from wolves, little genetic change has occurred at the molecular level, and domestic dogs and wolves can readily hybridize. This history demonstrates the wide range of phenotypes that can be generated in selection experiments.

Conservation Genetics

Conservation genetics draws on principles from population, quantitative, and evolutionary genetics to accomplish two major goals:

1. Provide management guidelines for conserving endangered species of plants and animals, and
2. Provide guidance for the preservation and sustainable use of the earth's genetic resources to guarantee food security for present and future generations.

The ever-increasing human population has led to the decrease of wildlife populations and made them more vulnerable to extinction. As you have learned, reduced population size has many negative genetic repercussions for populations. Inbreeding depression reduces the fitness of populations and makes them more vulnerable to extinction. Random genetic drift leads to the loss of potentially adaptive genetic variation and causes maladaptive evolution to occur. A small population also has a reduced genetic variation and smaller input of advantageous mutations.

Conservation genetics can provide management guidelines for the preservation of a population—for example by estimating the minimum viable sizes for populations of endangered species. Organisms such as the California condor, peregrine falcon, Florida panther, and many other animal and plant species, have become endangered or are at high risk for extinction in the last 50 years (**fig. 24.13**). Careful monitoring of the population size and introducing the proper safeguards at the correct time have maintained or dramatically improved the size of some of these populations. One

figure 24.12   Production of different dog breeds by artificial selection. Selective breeding or artificial selection can affect multiple quantitative traits in an organism. By selecting for different quantitative traits, breeds preferentially exhibit different sizes, shapes, fur color, and behavior. This is how the various dog breeds were selected from wolves.

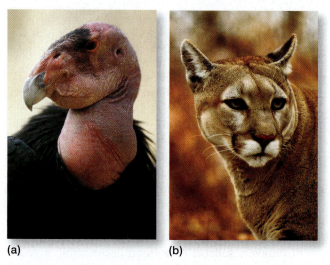

(a)   (b)

figure 24.13   Conservation genetics is the application of population genetic principles to preserve groups of organisms. Examples of endangered species that have benefited from conservation genetics include the California condor (a) and the Florida panther (b).

of the safeguards is breeding animals in captivity and nurturing the young to increase their chance of survival until they reach sexual maturity. This approach was used with the California condor, a program that, although controversial, has managed the successful release of condors back into the wild.

Another approach was used in the case of the Florida panther, a subspecies of the mountain lion restricted to southwestern Florida. In the 1990s, it was believed that only 50 to 70 Florida panthers remained in the wild. The population was dwindling in part due to reduced fitness of these animals. Nearly 70% of the Florida panther males exhibited undescended testicles (cryptorchidism), which was detrimental to the propagation of the species.

With the small population and the reduced fitness, it was clear that the Florida panther was quickly moving toward extinction. A decision was made to introduce to southwestern Florida eight females from the Texas panther (cougar) population, the nearest extant population source. It was anticipated that the Texas panthers would increase the genetic variation to the Florida panther population and might restore genetic fitness.

As expected, the F_1 and F_2 progeny lacked many of the Florida panther traits. The most striking was that of the five males that possessed Texas panther ancestry, all had properly descended testicles. In this way, the introduction of Texas panther alleles into the Florida panther population resulted in significant reduction in the expression of a deleterious trait.

Modern food production systems have led to the generation of a small number of highly inbred species of animals and crops (**fig. 24.14**). The number of species available for use is several orders of magnitude greater than the number actually being utilized. Many of these potential food sources could provide higher nutritional value at a lower cost, especially in the areas where they are indigenous. Furthermore, indigenous variants usually require less pesticide and fertilizer, which would decrease the environmental burden posed by pesticides and fertilizers. The selection for desirable traits and the inbreeding of the species has reduced the genetic variation in the population.

The lack of genetic variation within the species contributing to most of the human nutrition around the globe has made these varieties highly susceptible to diseases and insect pests. This situation raises the fear of possible future famines if we do not conserve wild and domestic sources of genetic resources that could

figure 24.14 Conservation genetics can also be applied to preserve plant species from a wide variety of environmental conditions. For example, conservation genetics has led to the production of crops that exhibit increased resistance to disease. A rice crop that is infected with wilt (left) is adjacent to one that is resistant (right).

prove valuable in the future. Conservation genetics is involved in regaining genetic variation in these species while also maintaining the beneficial nutritional and environmental traits that were previously selected.

Restating the Concepts

▶ Ecological genetics is the study of the ecological interactions of organisms with their biotic and abiotic environment.

▶ Selection changes a trait in a specific direction. To understand the potential for, and constraints on, the long-term population response to selection, the interaction between ecological and genetic issues must be considered.

▶ Conservation genetics seeks to maintain biological diversity and utilization of genetic resources for economic and environmental sustainability.

24.7 Quantitative Trait Loci Mapping

Because many quantitative traits are important to medicine and agriculture, identification of the genes that control these phenotypes is highly desirable. Yet the nature of multiple-loci interactions makes the identification of a locus that contributes only a fraction of the phenotype difficult.

The basic idea of **quantitative trait loci (QTL) mapping** is to use identifiable genetic markers to determine the linkage of the loci involved in determining the quantitative trait of interest. Often, this approach involves recombination mapping to identify the chromosome segments containing the desired QTLs. Statistics are then used to identify significant differences in phenotype associated with particular markers. If these differences exist, this implies the presence of a QTL linked to the associated markers. With a suitable set of markers spread throughout the entire genome, it is theoretically possible to map and characterize all the genes that affect a character.

QTL Mapping Theory

The theoretical basis for QTL mapping was laid out in the 1960s; however, its application awaited the discovery of DNA-based markers, such as single-nucleotide polymorphisms (SNPs), RFLPs, and microsatellites. Many of these markers were described in chapter 13. They are used to survey genetic variation at the DNA level and to allow the genotype at a given locus to be determined.

To perform a whole-genome QTL scan, a combination of many different kinds of markers is often required to provide sufficient genome coverage. To perform this mapping, markers are available for each chromosome from one end to the other and are spaced sufficiently close that recombination events only rarely occur between them. For practical purposes, this spacing is generally considered to be a map distance of less than 1.0 cM.

QTL Mapping Procedure

The standard method for mapping QTLs is to make a cross between two different genotypes, usually pure-breeding lines that have been selected for opposite and extreme values for the trait of interest. For example, one line exhibits a tall plant phenotype, and the other corresponds to short plants (**fig. 24.15**). Similar to the example given earlier for broad-sense heritability, these parental lines should be fixed for alleles with positive effects on the trait (blue chromosome line) and negative effects on the trait (red chromosome line). The F_1 offspring will be heterozygous at all loci that influence the trait and for which genetic variation exists.

The meiotic cells in the F_1 generation undergo recombination, and the resulting haploid gametes carry chromosomes that are a mosaic of the two parental genotypes.

These F_1 individuals are then mated with either siblings or the pure-breeding parental lines. The resulting F_2 individuals exhibit different extents of the quantitative trait, for example, varying heights. These F_2 individuals are then classified based on the level of the quantitative trait that they exhibit.

The genotypes of the F_2 individuals are then determined based on a large panel of molecular markers, which are scattered throughout the genome. Analysis of all of these markers produces a very large data set that requires computer analysis. For each individual, the quantitative trait and the molecular phenotype, namely the identity of the allele for all the molecular markers in the genome, are entered into a statistical program. The program searches for a molecular marker and a specific allele that appears with a particular phenotype; such a connection is termed **cosegregation.** The cosegregation of a phenotype and a molecular marker suggests that a quantitative trait locus is near the molecular marker.

Consider the F_2 offspring that are produced from a cross between F_1 siblings (see fig. 24.15). Assuming that the map of DNA markers is saturated, the genotype of each marker is then determined for every individual in the F_2 population. It is also necessary to measure the phenotype of each individual. If a chromosomal region is segregating for genetic variants that affect a trait, then the genotype of the markers in that region should correlate with the phenotype. The closer the QTL resides to the marker, the tighter that association will be.

For instance, the boxed region in the middle of the chromosome is associated with plant height (see fig. 24.15). Homozygous red individuals in this region are short, heterozygous individuals are intermediate, and homozygous blue individuals are tall. Regions of the chromosome containing no gene(s) affecting height do not show such an association. Computerized analysis is then employed to detect significant associations, estimate QTL positions, indicate relative strength of

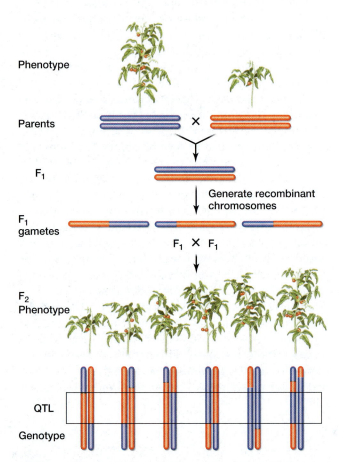

figure 24.15   Diagram of how quantitative trait loci (QTL) are mapped. Two individuals that exhibit extremes of the same trait (tall and short tomato plant height) are homozygous for all the alleles that affect plant height. Crossing a tall plant (blue genotype) with a short plant (red genotype) produces heterozygous F_1 progeny with an intermediate phenotype. Meioses in the F_1 individuals produce recombinant gametes that possess different amounts of the short height (red) and tall height (blue) genomic regions. Crossing two F_1 individuals produces F_2 progeny that exhibit a greater number of intermediate phenotypes (height). DNA markers are tested for cosegregation with either the tall or short height phenotype. This analysis revealed the presence of a genomic region (boxed) that cosegregates with the height phenotype (two blue regions corresponds to the tall phenotype, one blue region with an intermediate height, and no blue regions with the short phenotype).

their contribution, and identify dominance–epistatic interactions.

Other methods for QTL mapping look for marker–trait associations within natural populations. The methods and application of these studies are often quite different from the controlled crosses discussed here. Their great advantage is that they survey naturally occurring genetic variation. Although selection in laboratory lines may fix very rare mutations with large effects, it may be that these mutations have contributed little to genetic variation in the outbred natural population.

Experimental Results

One of the most basic questions that can be answered from QTL analysis is the approximate number of loci affecting a trait and the average effect of these loci. It is difficult to generalize from the large and rapidly growing number of QTL experiments performed because of differences in methodology and in the number of loci potentially at work. The number of loci, however, seems to be lower than what many investigators originally expected. Frequently 4 to 12 QTLs are found that account for 35–65% of the phenotypic variation for the trait (**table 24.6**). But the total number of loci affecting these traits may well number anywhere from 35 to 200.

When mapping QTLs in humans, such as the genetic components of breast cancer, investigators are usually unable to employ standard recombination mapping because the size of the family being studied is usually too small. Therefore, researchers often employ the lod (logarithm of odds) score that was described in chapter 13. Using DNA markers scattered throughout the human genome and several informative families, a geneticist may be able to determine the relative positions of the QTLs that have the largest contribution to the studied phenotype.

Returning to the example of human height, the genetic loci that contribute to this trait can be mapped by studying several families and noting the cosegregation of tall or short height with DNA markers known throughout the genome. Three of the DNA markers (**fig. 24.16**) exhibited very high lod scores or a low recombination frequency with three loci that had major contributions to the height phenotype. These three DNA markers (D6S2436, D9S301, and D12S375) serve as relative positions for the genes or QTLs on chromosomes 6, 9, and 12 that make a significant contribution to height. This does not suggest that these three genes provide all the genetic contribution to the height phenotype. With this mapping information, researchers can use DNA recombination techniques to clone the three loci that correspond to this quantitative trait.

We already discussed the generation of the human HapMap (see chapter 13). In 2007, a flurry of reports were published that used the HapMap to reveal the major genetic components that contribute to a variety of complex genetic diseases, such as heart disease, type 2 diabetes, schizophrenia, bipolar disorder, and autism. Because these complex genetic diseases are not caused by a single gene, analysis of specific pedigrees cannot reveal all, or most, of the affected genes. It is anticipated that correlating the haplotype and severity of the disease in the individuals of a population will reveal the location of the genes that contribute to these complex genetic diseases.

QTL mapping has verified what we already expected from "classic" quantitative genetics experiments and theory, namely that dominance, epistasis, and genetic correlations are extremely widespread. It is hoped, in the future, that QTL mapping will fulfill the promise of identifying the genetic components that underlie complex genetic diseases in humans and that control important traits in crops and domesticated animals.

table 24.6 Estimates of Number of QTLs Affecting Quantitative Traits in Various Organisms

Organism	Trait	Number of QTL Detected	Percent of Phenotypic Variation Explained by Detected QTLs
Tomato	Soluble liquids	4	44
	Fruit mass	6	58
	Fruit pH	5	48
Corn	Height	3–7	34–73
	Yield	3–13	10–87
Potato	Tuber shape	1	60
Aedes aegypti	*Plasmodium falciparum* resistance	2	67
Mouse	Epilepsy	2	50
Rat	Blood pressure	2	18–30

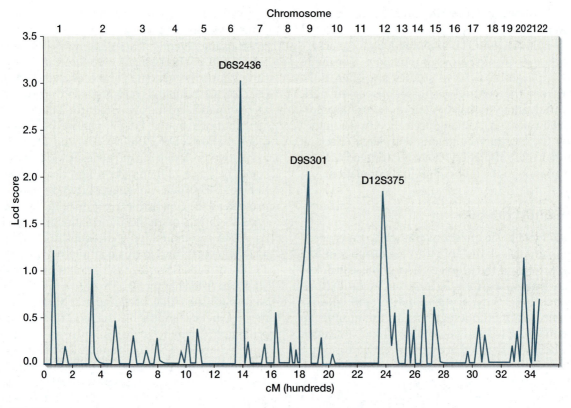

figure 24.16 Mapping QTLs controlling height in humans. DNA markers throughout the human genome were tested for cosegregation with height by plotting the relative position of the DNA markers against the Lod score of its cosegregation with the height phenotype. The distribution of the markers in the human genome in hundreds of centiMorgans (cM) is labeled along the X axis and the corresponding chromosome number is labeled along the top. The Lod score for the cosegregation of human height and DNA markers is labeled on the Y axis. Three DNA markers on chromosomes 6, 9, and 12 (D6S2436, D9S301, and D12S375) cosegregate at a high degree with QTLs that contribute most of the genetic component to height.

(Hedrick, *Genetics of Population,* 2005: Jones and Bartlett Publishers, Sudbury, MA. www.jbpub.com. Reprinted with permission.)

Restating the Concepts

▶ QTL mapping seeks to determine the number of loci affecting a quantitative trait, the location of those loci on the chromosome, their relative effects, and the interactions between alleles at a locus, between loci, and between the genotype and the environment.

▶ QTL mapping holds much revolutionary promise in fields such as evolution, plant and animal breeding, and the treatment of many human diseases.

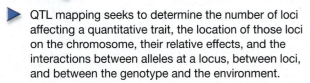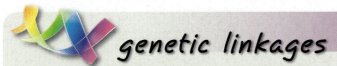

genetic linkages

In this chapter, we built on the discussion of population genetics in the preceding chapter. Whereas standard Mendelian genetics corresponds to one or more genes that interact to produce a particular phenotype (chapter 5), quantitative genetics involves two or more genes that each contributes a fraction of the final phenotype. Thus, the standard phenotypic ratios that were described for Mendelian genetics are not observed for quantitative traits. The approximate number of genes that are involved

in producing the quantitative trait can be determined, however. Once the number of genes is determined, it is possible to predict the phenotypic ratios of the offspring between two individuals who are heterozygous for each gene.

Quantitative data are often distributed in a Gaussian relationship as a normal curve, and these data can be analyzed by examining the central tendency of the data (mean, median, and mode) and the variation (or covariation) in the data. The chi square test described in chapter 2 can be used to statistically analyze data in the determination of a specific hypothesis, and using further statistical analysis, different components of heritability can be studied as well as the effects of selection on heritability.

Finally, we discussed the identification of quantitative trait loci (QTLs). This analysis is based on mating two homozygous individuals that exhibit opposite extremes of the desired phenotype. The progeny of this cross produce gametes that have undergone random recombination between the QTLs and DNA markers, which were described in chapters 12 and 13. When mapping QTLs in humans, where the number of progeny is small, the lod score analysis described in chapter 13 is employed. Using the lod score, multiple families may make up the sample used to calculate the relative position of the strongest components of the QTL. The generation of the human HapMap (chapter 13) will allow the analysis of QTLs within a population, rather than a pedigree. Analyzing a population, rather than pedigrees, potentially can reveal a greater number of the QTLs that underlie complex human genetic diseases.

In the final chapter of this text, we will discuss the application of genetics principles and technologies to the field of evolution and theories about how species arise under the pressures of natural selection.

Chapter Summary Questions

1. For each term in column I, choose the best matching phrase in column II.

Column I
1. Conservation genetics
2. Correlation coefficient
3. Cross-fostering
4. Heritability
5. Norm of reaction
6. Normal distribution
7. Quantitative genetics
8. Quantitative trait
9. QTL mapping
10. Realized heritability
11. Stabilizing selection
12. Standard deviation

Column II
A. Experimental design that randomizes the environment experienced by genetic relatives in order to remove genotype–environment covariances in heritability studies.
B. Frequency distribution that is symmetrical around the mean and the mean, median, and mode are all equivalent.
C. Heritability estimated from the response to selection.
D. Measure of dispersion around the mean.
E. Methodology for locating and describing the major loci contributing to a quantitative trait
F. Phenotypic changes in a single genotype across environments.
G. Phenotypic values closest to the mean have the highest average fitness.
H. Proportion of phenotypic variation in a population attributable to genotypic differences among individuals.
I. Quantitative measure of the strength of the relationship between two variables.
J. Study of how to best use genetic information to conserve endangered plants and animals and conserve genetic resources in order to provide food and medicines for people.
K. Study of phenotypic traits that are caused by multiple loci and environmental factors.
L. Trait for which the variation among individuals is determined by multiple loci and the environment.

2. How does "quantitative genetics" differ from the previous genetics topics dealt with in this book?

3. Define *broad-sense* and *narrow-sense* heritability. How do they differ and why are the differences important?

4. How was Mendel's theory of particulate inheritance made compatible with observations of continuous variation in phenotypes? How do you think this contributed to the acceptance of Darwin's theory of evolutionary change in small steps?

5. Why is the study of quantitative genetics important to ecology and evolution?

6. Why is the study of quantitative genetics important to medicine?

7. Why is the study of quantitative genetics important to food production?

8. Discuss why clinical depression and diabetes, though often treated as dichotomous variables, are in fact quantitative traits.

9. What forces determine genetic variation for quantitative traits? Why do you think there is typically so much genetic variation for quantitative traits, whereas the majority of loci are fixed for a single allele?

10. What usually prevents selection from going on indefinitely? What factors might prolong the response to selection?

11. What is conservation genetics? Why is it important to conserve genetic resources?

12. What is a minimum viable population size? What three genetic concerns are most important in conserving (avoiding extinction in) populations or species of endangered plants and animals?

13. Of the 10,000 known species that have been used for agriculture, only about 10 species account for 90% of global food production. Over the past 15 years, 1350 species of potential food sources have gone extinct in the wild and 2 more are being lost every week. What might be some of the advantages of conserving and fostering the use of these species?

14. What part does linkage (cosegregation) play in mapping QTLs?

15. How does a genotype–environment interaction differ from a norm of reaction? Why are genotype–environment interactions important?

16. Explain how stabilizing selection can help maintain genetic variation for quantitative traits.

17. Describe what has been discovered about the number of loci affecting a "typical" quantitative trait? What kind of generalizations can be made about the nature of their effects?

18. What are the two methods of QTL mapping discussed in the chapter? What are the advantages and disadvantages of both?

19. How do you expect the heritability and the mean phenotypic value for a trait to change under the following conditions?
    a. The environment becomes more variable.
    b. The quality of the environment decreases.
    c. Twenty five generations of positive artificial selection.

20. Explain what concordance and discordance for traits among mono- and dizygotic twins tells us about heritability.

Solved Problems

PROBLEM 1. At one inner-city school, children have IQ scores that are 10 points lower than the national average. A government official proposes to put more funding into the inner-city schools to raise their standard of education and close the IQ score gap. Another government official objects to this plan as a waste of money because studies have shown that IQ is 80% heritable. "Since IQ is determined primarily by genes, there is little that can be done to increase the academic performance of inner-city children." Is the second city official correct?

Answer: No. The second official does not understand that heritability describes the causes of variation among individuals, not the mean of groups. The fact that heritability for a trait (in this case IQ) is high or low has nothing to do with the mean trait for a group. Heritability also tells you nothing about how a population will respond to improvements in the environment. In fact, since heritability is specific to that population in a particular environment, there is no guarantee that heritability will remain the same if the environment is changed. Thus, it is impossible to predict from IQ estimates how heritability will respond to increases in educational funding for inner-city students. It is likely the extent that increased funding translates into better education and the improved education translates into higher IQ scores, that such a move should eliminate or

decrease the gap in IQ scores between the groups. Bottom line, it is not surprising that a population in a less desirable environment will have a lower mean than a genetically similar group in a more favorable environment. Heritability estimates cannot be used to predict changes in the mean phenotype with changes in the environment, because heritability attributes within population variation into genetic and other causes. The second official also confused broad-sense heritability with narrow-sense heritability.

PROBLEM 2. Two pure-breeding lines of Jersey cattle have mean weekly milk yields of 60 and 210 kg. The F_1 generation, which resulted from a cross between the two lines, had a mean of 135 kg and a variance of 16 pounds. The F_2 generation also had a mean of 135 pounds, but a variance of 80 pounds. What is the broad-sense heritability (H^2) for milk yield in this population?

Answer: All individuals in the F_1 generation, resulting from a cross of the two lines, would be heterozygous and genetically identical at all loci contributing to milk yield. Thus, $V_E = 16$. The F_2 generation experiences the same level of environmental variation but also is genetically variable, making $V_E + V_G = 80$. Therefore, $V_G = 80 - 16 = 64$. Broad-sense heritability is equal to $V_G / V_E + V_G$. In this case $H^2 = 64/80 = 0.8$.

Exercises and Problems

21. The following data are mean heterozygosity levels for 20 individual spiders, estimated from 15 microsatellite loci. Calculate the mean, median, and mode of the distribution. Do you think the sample was drawn from a normal distribution? Why or why not?

Mean Heterozygosity

0.526	0.568	0.614	0.730	0.643
0.638	0.603	0.569	0.620	0.456
0.637	0.629	0.576	0.647	0.750
0.476	0.431	0.524	0.804	0.487

22. Calculate the variance and standard deviation from the heterozygosity data in exercise 21.

23. Body weights of extinct mammals can be estimated by measuring the length and thickness of their bones and examining the size of the muscle attachments at joints. The following is the estimated weight (in kilograms) of 36 specimens of *Smilodon populator*, the largest of the saber-toothed cats. What is the mean, median, and mode of the distribution of body weights? What is

the estimated phenotypic variance (V_P) of body mass in these extinct carnivores?

	Weight in kg		Weight in kg		Weight in kg
Specimen 1	344	Specimen 13	219	Specimen 25	345
Specimen 2	492	Specimen 14	297	Specimen 26	306
Specimen 3	342	Specimen 15	336	Specimen 27	315
Specimen 4	209	Specimen 16	407	Specimen 28	306
Specimen 5	267	Specimen 17	343	Specimen 29	252
Specimen 6	465	Specimen 18	277	Specimen 30	316
Specimen 7	304	Specimen 19	323	Specimen 31	273
Specimen 8	303	Specimen 20	344	Specimen 32	395
Specimen 9	333	Specimen 21	333	Specimen 33	434
Specimen 10	255	Specimen 22	337	Specimen 34	302
Specimen 11	354	Specimen 23	341	Specimen 35	367
Specimen 12	334	Specimen 24	275	Specimen 36	350

24. A geneticist wished to know if variation in the number of egg follicles produced by chickens was inherited. As a first step in his experiment, he wanted to know if the number of eggs laid could be used to predict the number of follicles. If this were true, he could then avoid killing the chickens to obtain the data he needed. He obtained the following data from 14 chickens.

Chicken	Eggs	Follicles
1	39	37
2	29	34
3	46	52
4	28	26
5	31	32
6	25	25
7	49	55
8	57	65
9	51	44
10	21	25
11	42	45
12	38	26
13	34	29
14	47	30

Calculate a correlation coefficient and graph the data.

25. Imagine that in blue-breasted fairy wrens the intensity of blue in the breast feathers is determined by four different loci (A, B, C, and D), each with two alleles. The alleles act additively both within and between loci (i.e., no epistasis). At each locus, there exists an allele that produces blue pigment and a null allele that does not. Assuming no environmental influences, how many phenotypic classes will there be for breast color?

26. The following table contains data on human heights (in inches) for several different families. For each family, the height of the father, the height of the mother, and the mean height of the adult children is listed. The female heights

were adjusted upward to account for the mean height differences between the sexes. Calculate narrow-sense heritabilities using the father's and child's data, the mother's and child's data, and the midparent's and child's data. Which do you expect to have the highest correlation? Are there any differences between the heritability estimated from using the father alone versus the mother alone? Why might you expect there to be differences?

Father's Height (in inches)	Mother's adjusted height (in inches)	Mean adult child's height (females adjusted) (in inches)
67	69.5	69
70	71	70
69	74	72
73	72	73.5
73	73	74
72	72	71.5
70	71	72
70	69	68
65	66	67
72	68	68.5
72	70	71
67.5	68	69
72	70	69.5
73	74	73.5
70	68	67
69	70	71
73	74	72.5
68	72	72
72	66	67.5
72	65	69
66	69	70
73	71	69
74	76	75.5
71.5	71.5	72.5
69	73	70
68	72	69.5
74	72	72.5
68	67	74
68	65	67
74	72	73
72	70	70
69	70	70.5
69	67	71.5
75	69	71.5
76	75	74
66	68.5	68.5
77	70	74
73	76	72
70	68	69
71	69	70
67	72.5	70
63	69	70.5

27. In horses, the proportion of the face that is white can range from zero to one. In a particular population, the average proportion is 0.44. Parents are selected with a mean of 0.68. The offspring of these parents have a mean proportion of 0.62 of their face being white. What is the narrow-sense heritability of white facial coloring in this herd of horses?

28. Approximately 1250 swine were used in a study of the genetics of average backfat thickness (ABF), average daily weight gain (ADG) and loineye area (LEA). In addition to the information listed in the following table, genetic correlations were also estimated for these traits. The correlation coefficients are: ABF:ADG = 0.04, ABF: LEA –0.37, and ADG:LEA = –0.34.

	Phenotypic Mean	Phenotypic Variance	Additive Genetic Variance
Average daily growth	0.474 kg/day	0.013	0.040
Backfat thickness	1.00 cm	0.050	0.036
Loin eye area	31.0 cm	0.413	0.268

   (i)   Which of these traits would respond best to selection?
   (ii) A project is undertaken to increase average daily growth rate. Animals with growth rates of 0.65 kg of lean body mass are chosen to breed. What is the phenotypic mean for this trait in the next generation?
   (iii) Can you select for both increased loineye area and increased average daily growth?

29. Two genotypes of rice exist. The *Phil* genotype has a yield of 42 kg/acre during normal conditions and a yield of 37 kg/acre during drought conditions, phenotypic variances of 100 kg/acre and 144 kg/acre, respectively, and heritabilities of 0.75 and 0.50, respectively. The *Thai* genotype has a yield of 50 kg/acre during normal conditions and a yield of 30 kg/acre during drought conditions, phenotypic variances of 182 kg/acre and 506 kg/acre, respectively, and heritabilities of 0.67 and 0.44, respectively. Each genotype exists in sufficient numbers that selected individuals for breeding exceed the mean by two standard deviations in both normal and drought conditions. Which population will respond faster in one generation of selection in the normal environment? Which population will respond faster in one generation of selection in the stressful environment? Which genotype will have the highest yield in each environment after one generation of selection?

30. The data at right represent the number of offspring produced in a clutch and the mean weight (in milligrams) of the offspring produced by a species of wolf spider. What is the correlation coefficient of clutch size and mean offspring weight? Can you select for more offspring and larger offspring simultaneously?

Number of Offspring	Mean Weight of Offspring (mg)
177	1.41
132	1.29
194	1.44
199	1.16
138	1.38
132	1.14
92	1.63
120	1.33
126	1.51
197	1.27
100	1.60
121	1.65
207	1.16
102	1.27
13	1.54
155	1.29
134	1.27
8	1.25
159	1.26
61	1.48
158	1.33
89	1.35
76	1.45
107	1.21
152	1.18
123	1.38
99	1.52
83	1.33
85	1.41
207	1.26

31. Rice bran contains many phytochemicals with proposed health benefits. Research was conducted to quantify the levels of individual E-vitamers in an international germplasm collection, and to determine the effects of the environment on their levels. Three rice genotypes, *Indo, Phil,* and *Thai,* were planted in their country of origin and the other two countries. The following table provides data on E-vitamer content (in milligrams per gram) for each genotype in each environment. On a single figure, graph the norms or reaction for each genotype. Do you think there is an interaction between the genotype and environment for these strains of rice across the growing conditions in each country?

	Indonesia	Philippines	Thailand
Indo	0.323	0.277	0.180
Phil	0.260	0.355	0.250
Thai	0.180	0.311	0.443

32. The narrow-sense heritability of human height is 50% ($h^2 = 0.50$). Aliens choose individuals that exceed the mean height of the population by 3 inches for breeding purposes. Based on this information, would you expect the height of humans in the next generation:
   a. to increase about 1½ inches?
   b. to increase, but the increase depends on the amount of dominance?
   c. to remain unchanged, because the environmental component remains the same?
   d. to be unpredictable, because too many variables influence human height?
   e. to be unpredictable, because you did not tell us how tall the aliens are?

33. Tolerance of wheat (*Triticum aestivum*) to waterlogging is one of the limiting factors in wheat production. A study was conducted to estimate the narrow-sense heritability for grain yield under waterlogged conditions. Grain yield had a heritability (h^2) of 0.25 under these conditions and the mean yield was 136.5 kg/acre. The following were also estimated: $V_A = 140$, $V_D = 113$, $V_I = 176$.
   (i) What is the total phenotypic variance for grain yield under waterlogged conditions?
   (ii) What is the total environmental variance for grain yield under waterlogged conditions?
   (iii) What is the broad-sense heritability?
   (iv) If you select parent plants that are 1.5 standard deviations greater than the mean to begin selection, what will grain yield be after one generation of selection?

34. You determine the following variance components for leaf width in a particular species of plant:

Additive genetic variance (V_A)	4.0
Dominance genetic variance (V_D)	1.8
Epistatic variance (V_I)	0.5
Environmental variance (V_E)	2.5

Calculate the broad-sense and narrow-sense heritabilities.

35. If, in a population of swine, the narrow-sense heritability of maturation weight is 0.15, the phenotypic variance is 44 kg^2, the total genetic variance is 22 kg^2, and the epistatic variance is zero, calculate the dominance genetic variance and the environmental variance.

36. A group of 4-month-old hogs has an average weight of 76 kg. The average weight of selected breeders is 84 kg. If the narrow-sense heritability has been estimated at 40%, what is the expected average weight of the first generation of progeny after selection?

37. The following table shows concordance rates in humans for several traits in monozygotic (MZ) and dizygotic (DZ) twins.

	MZ	DZ
Adult schizophrenia	0.44	0.12
Breast cancer	0.19	0.10
Hypertension	0.41	0.13
Insulin-dependent diabetes	0.53	0.11
Nontraumatic epilepsy	0.58	0.09

Which of these diseases has the highest heritability? Which has the lowest?

38. Corn growing in an Indiana field had a mean lysine content of 2.0%, with a variance of 0.16. When grown in the greenhouse under controlled and uniform conditions, the mean lysine content was still 2.0%, but the variance was 0.09. What measure of heritability can you calculate? What is the heritability of lysine content?

39. Two pure-breeding lines of plants have mean heights of 26 and 42 inches. The F_1 generation, resulting from a cross of the two lines, had a mean of 34 inches and a variance of 20 inches. The F_2 generation also had a mean of 34 inches, but a variance of 60 inches. What is the broad-sense heritability (H^2) for plant height in this population?

40. The green lynx spider (*Peucetia viridans*) demonstrates maternal care by vigorously and diligently guarding her eggs. Imagine that artificial selection is carried out for increasing and decreasing the time that offspring are guarded, so that spiders might stop guarding before the eggs are hatched or continue to guard after the offspring hatch. The experiment started with 100 female lynx spiders. In each line, only females in the upper or lower (depending on selection line) twentieth percentile of the population distribution were allowed to breed in the next generation. The mean number of days a female spider will guard in the initial population was 12.1. Refer to the table at the top of the next page to answer the following questions.
   (i) Graph the heritability of guarding behavior in each generation for each line. What patterns do you see?
   (ii) Graph the selection differential in each generation for each line. What patterns do you see? Explain the patterns given what you know about heritability.

	Mean of Selected Parents	Population Mean
First generation of selection:		
High line	17.0	14.6
Low line	7.5	9.8
Second generation of selection:		
High line	19.2	16.6
Low line	5.6	8.5
Third generation of selection:		
High line	21.4	18.3
Low line	4.8	7.9
Fourth generation of selection		
High line	23.2	19.8
Low line	4.7	7.8

Chapter Integration Problem

A recent study published in the *Journal of Zoology* suggests that the American alligator, *Alligator mississippiensis,* has the most powerful bite force of any extant animal. Reading this, you decide to breed and sell attack alligators. However, your neighbor has the same idea, so you decide to perform selective breeding for even more dangerous alligators as a marketing ploy. You capture 300 adult alligators and measure their bite force. The mean bite force was found to be 1300 pounds per square inch (psi) with a total phenotypic variance of 28,561 psi. You select individuals with a mean trait value of 1600 psi to start your breeding program. The mean value for the offspring of these alligators is 1450 psi. In a separate experiment, you determine the broad-sense heritability of bite force to be 0.90.

    a. What is the nonadditive genetic variance (V_D and V_I) for bite force in this population of alligators?

    b. What is the environmental variance (V_E) for bite force in the described population?

    c. What is the narrow-sense heritability (h^2) of bite force in this population?

    d. Your neighbor used genetic engineering to produce alligators with a bite force of 1640 psi. A salesman visits your house and tells you that he can guarantee (money back if not satisfied) at least a 30% increase in the bite force of your alligators if you purchase his vitamin-enriched, high-protein, steroid-ridden alligator chow. Given the high broad-sense heritability of bit force in your alligators, should you purchase the alligator chow?

    e. If you make the diet and other environmental factors on your alligator farm more uniform for the individuals being raised there, will this affect heritability? How?

    f. If you raise the heritability by manipulating the environment, will you get an increase in the response to selection?

    g. As you compete with your neighbor to develop better attack alligators, you decide to do QTL mapping. You find that bite force is controlled by many loci of small affect. Does this bode better for his genetic engineering approach or your traditional artificial breeding approach? Why?

Do you need additional review? Visit **www.mhhe.com/hyde** for practice tests, answers to end-of-chapter questions and problems, interactive exercises, and animation tutorials all designed to enhance your understanding of key genetic concepts.

Evolutionary Genetics

Photo: The cactus ground-finch (*Geospiza scandens*) from Santa Cruz Island, Galápagos.

*T*he diversity of life on earth and the range of environments to which these life forms have adapted are amazing. From the frozen Antarctic to hot vents on the ocean's floor, the world teems with unique organisms that have an almost infinite variety of behaviors, metabolisms, and morphologies. These organisms range in size from the fungus *Armillaria ostoyae,* which can cover over 2000 acres, to viruses that can be seen only with an electron microscope.

In all cases observed, these diverse life forms have adapted to fit their environments. Adaptations are so prevalent that when Charles Darwin discovered an orchid in Madagascar with an 11-inch-long nectar-producing tube, he predicted that a moth with an 11-inch proboscis existed that feeds from the tube. Nearly 50 years later, Darwin's prediction was found to be true when the moth *Xanthopan morganii praedicta,* with a 12-inch proboscis was observed feeding from Darwin's orchid (*Angraecum sesquipedale*).

In addition to these wondrous adaptations, however, we find a number of things that do not make adaptive sense. For example, **vestigial features** are traits with no current function or that are potentially deleterious to the organism—for example, the human appendix. Modern whales, as another example, also retain vestigial features, such as pelvic and leg bones embedded within the musculature of their body walls and a number of small muscles devoted to nonexistent external ears.

These apparently nonadaptive traits make sense only if we assume that whales evolved from a land-living ancestor, and that evolution requires new features to be built from the foundation of earlier features. We now know from fossil and genetic evidence that whales evolved from a group of carnivorous hoofed mammals.

Populations change, or evolve, through natural selection and the other forces that perturb Hardy–Weinberg equilibrium. In chapters 23 and 24, we discussed the underlying principles that affect populations, the forces that act on populations, and the mechanisms that produce the quantitative traits that define the differences between individuals in a population. With this theoretical groundwork, we examine in this chapter the long-term process of evolution and speciation in natural populations.

25.1 Charles Darwin and the Principles of Natural Selection

The British naturalist Charles Darwin (**fig. 25.1**) was born in 1809. From 1831 to 1836, Darwin circled the world aboard the HMS *Beagle.* The primary purpose of the *Beagle's* voyage was to chart the coast of South America. During his travels on the *Beagle,* Darwin amassed observations (especially in South America and the Galápagos Islands) that led him to suggest a theory for the development of species.

Darwin's theory of evolution was published in 1859 in a book entitled *The Origin of Species by Means of Natural Selection, or the Preservation of Favoured Races in the Struggle for Life.* This book provided overwhelming support for evolution as well as a mechanism for the process. Darwin proposed that organisms become adapted to their environment by the process of natural selection. **Natural selection** allows individuals with desirable traits to possess an increased probability of survival in their environment. In outline, natural selection works according to the following principles:

1. *Variation is a characteristic of virtually every group of animals and plants.* Darwin

figure 25.1 Charles Darwin (1809–1882) was an English naturalist who first established the theory of evolution by natural selection.

saw phenotypic variation as an inherent property among individuals of all populations. Darwin proposed that some of these differences provided an advantage to those individuals in their environment. Look at the population of students in your class. You can observe a wide variety of heights, weights, athletic abilities, and numerous other characteristics.

2. *Phenotypic variation is heritable.* Even though Darwin preceded Mendel, he realized (as did most people) that offspring have characteristics that resemble their parents. Thus, the phenotypic variation that exists in a population will be present in the subsequent generation. We now know that this phenotypic variation is inherited as genotypic differences between the individuals.

3. *Every group of organisms overproduces offspring.* Most populations reproduce at an exponential rate, although resources in the environment are limited. The result is that a population is maintained at a relatively constant density over time. Thus, there is a competition among individuals to survive on the available resources.

4. *Individuals that do survive and reproduce pass on their genes in greater proportion.* This step is the cornerstone and the best known part of Darwin's theory. Among all the organisms competing for a limited array of resources, only the organisms best able to obtain and utilize these resources survive ("survival of the fittest"). If the favorable characteristics of these individuals can be inherited, these traits pass on to the next generation. In other words, these individuals have the greatest reproductive success.

Over time, therefore, as advantageous mutations arise, or if the environment changes, the characteristics of a population should change through the process of natural selection (for example, directional or disruptive selection, as described in chapter 23). A particularly well-adapted population in a stable environment may maintain its numbers through the forces of stabilizing selection. Nonrandom mating, genetic drift, and migration may also play a role in population differentiation.

Restating the Concepts

▶ Charles Darwin proposed in 1859 that organisms adapt to their environment by the process of natural selection. Darwin's best known studies described the diversity of life on the Galápagos Islands.

▶ One of the major points of natural selection is summarized as "survival of the fittest," which means that individuals possessing traits that increase their probability of surviving and reproducing will pass on their traits in a greater proportion of the population from one generation to the next.

25.2   What Is Evolutionary Genetics?

Evolutionary genetics takes the principles of population and quantitative genetics explained in chapters 23 and 24 and extends them over time to discover how genetic processes acted to create the diversity of life on earth. The framework of evolutionary genetics incorporates adaptation via natural selection, stochastic (random) processes, and biogeographical history.

Evolutionary genetics can be focused on several different levels, including

1. Molecular analysis of evolution. This field of emphasis examines the rate of nucleotide substitution for different mutations and the prevalence of recessive deleterious mutations.

2. Whole-genome analysis of evolution. Studies in this area include the presence, location, and frequency of transposable genetic elements, duplicated genes, and chromosomal rearrangements between different species.

3. The behavior and life history of a population or species. Topics may include studies of inbreeding avoidance and senescence (aging).

4. Evolutionary analysis of the similarities and differences among species and higher taxonomic groupings.

As with population and quantitative genetics, evolutionary genetics is directly relevant to vital issues in biology. Topics range from the generation of new strains of viruses (such as the avian flu), to the increased anti-

biotic resistance in pathogens, to the study and preservation of biodiversity throughout the world. Study of these topics involves discovery of the mechanisms by which these organisms function, including adaptations that increase the transmission of an individual's genetic information to the next generation. These adaptations can only be appreciated fully by considering them in an evolutionary genetics framework.

Has Evolution Occurred?

The fact that evolution has occurred is supported by overwhelming evidence from many independent research fields. A thorough review of all this evidence is beyond the scope of this chapter. All organisms, living and extinct, however, have clearly descended from one or a few original forms of life.

Darwin provided abundant evidence for evolution 150 years ago from biogeography, comparative anatomy, embryology, and palaeontology. Since that time, evidence has accumulated from many other fields, most spectacularly in molecular biology. Recent evidence includes, but is not limited to, the almost universal genetic code (see chapter 11) and the high amount of nucleotide sequences in many genes that play similar functional roles in very distantly related organisms (see histone genes in chapter 7). We also discussed in chapter 21 how a variety of genes important in development such as the homeobox genes, have related sequences and functions in organisms as diverse as the fruit fly and humans.

Inferences of common ancestry based on comparisons of the genome and genes of living species (and some extinct ones) are supported by direct fossil evidence of evolutionary transitions. The evolution of terrestrial amphibians from fishes, of reptiles from amphibians, of birds from dinosaurs, and of mammals from reptiles can all be traced in the fossil record.

Major Questions Addressed in Evolutionary Genetics

The answers to several major biological questions are sought through the study of evolutionary genetics. Among them are

1. Why and how does speciation occur?

2. What are the ancestor–descendant relationships (phylogeny) among species, both extant and extinct?

3. How rapidly, given the frequency and distribution of effects of mutations, do changes in genotype and phenotype normally occur?

4. How have processes such as mutation, natural selection, and random genetic drift given rise to the diverse behavioral, molecular, and morphological characteristics of different species?

The Role of Natural Selection in Evolution

Evolution has two major components: changes within a genetic lineage and the branching of lineages (speciation, described next). The major tenets of evolutionary genetics are as follows:

1. Populations have genetic variation that arises through mutation.
2. Populations evolve through changes in allelic frequencies via the interactions of natural selection, random genetic drift, and gene flow.
3. Most beneficial mutations have small phenotypic effects, and, therefore, most phenotypic changes are gradual.
4. Diversification arises via differential selection, mutation, and drift in genetically isolated lineages.
5. Over time, these processes can produce changes that lead to speciation or delineation at even higher taxonomic levels.

Natural selection is primarily responsible for the diversity of life on earth. A common consequence of natural selection is **adaptation,** an improvement in the average ability of the population's members to survive and reproduce in their environment. You learned in chapter 23 that natural selection acts as a mutational filter; it eliminates deleterious mutations and increases the frequency of beneficial mutations. Natural selection is therefore the ultimate cause of adaptation, but it cannot produce adaptation unless mutation and genetic variation occurs.

The conditions necessary for natural selection to occur are (1) reproductive excess accompanied by competition to survive and reproduce and (2) correlation of the differences in reproductive success with one or more heritable differences in phenotype.

Restating the Concepts

▶ Evolutionary genetics looks at the effects of the environment, gene flow, mutation, natural selection, and random genetic drift on the evolution of genes over time.

▶ Natural selection acts on genotypic (and resulting phenotypic) differences if a population exhibits reproductive excess and if difference in success of reproduction can be correlated with these differences.

25.3 Speciation

Natural selection results in a change in genotypic frequencies within a population of individuals, which usually produces a group of individuals that are better

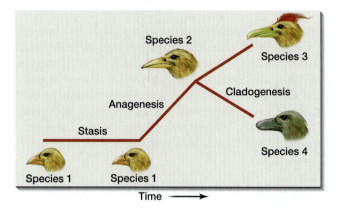

figure 25.2 Two types of speciation. In anagenesis, an organism changes over time until it is so different from its progenitor (species 1) that it is classified as a new species (species 2). In cladogenesis, speciation takes place as a branching process wherein one species (species 2) becomes two or more (species 3 and species 4) new species that differ from each other and from the progenitor species 2. Stasis is the time period when the species is undergoing minimal change.

adapted to the environment than their ancestors were. Over time, these changes in a population may result in new species being formed.

The generation of a new species can occur two ways (**fig. 25.2**). The first is **anagenesis,** which is a transformation of one species into a new species. In this case, only a single species is present at any given time. The second is **cladogenesis,** which occurs when one species splits into two new species. This second mechanism produces two species that are both present. The term **stasis** describes a time when a species is not actively undergoing either anagenesis or cladogenesis.

In order for species to maintain their separate identities and gene pools, pre- and post-reproductive barriers are often present between populations.

Classification of Species

The definition of a species is not simple. Rather, several different definitional methods have been used, and all these definitions have their own applications and strengths.

The Morphological Species Concept

Taxonomists and paleontologists, who often use preserved or fossilized specimens, typically use the **morphological species concept** as a working definition. Under this concept, two organisms are classified as belonging to the same species if they are morphologically similar. They are classified as belonging to two separate species if their differences are as great as those of two organisms belonging to two already recognized species.

The Biological Species Concept

The modern **biological species concept** defines a species as a group of individuals that can potentially interbreed to produce fertile offspring. This definition obviously cannot be used to classify the species of extinct organisms. Haploid and asexual species are also hard to classify using this method. A potential difficulty occurs when two organisms that normally do not interbreed in nature may do so in a laboratory setting. The interbreeding test carried out in a laboratory, which is done frequently, is not necessarily an adequate criterion for speciation.

Another problem arises in classifying groups that are geographically isolated from one another, such as populations on islands. Individuals in the populations are physically isolated, but in many cases they can interbreed freely when brought together with their mainland counterparts. So, although the concept of potentially interbreeding individuals comprising a species is useful in theory, more often than not biologists must also apply the morphological species concept to determine whether two populations belong to the same species.

Because speciation is a dynamic process, isolated subgroups of a population may be in various stages of becoming new species. The rate of successful interbreeding among individuals from these subgroups may range from 0 to 100%. Despite all of these potential limitations, the biological species concept is still widely accepted and very useful for most animals. For this reason, we will expand on the ways to generate reproductive isolation in the next section.

The Species Recognition Concept

If individuals with advantageous traits have the greatest chance of surviving and passing the traits on to their offspring, then individuals of the same species must recognize each other. The **species recognition concept** is based on this principle—that individuals of the same species have the ability to locate each other and mate successfully. Successful recognition can be based on visual or other sensory signals, behavioral cues, or even cellular recognition (between the egg and sperm). Species recognition would allow individuals of a single species, but physically isolated from each other, to recognize each other, even if their interaction fails to produce a successful mating and the generation of fertile offspring.

The Species Phylogeny Concept

DNA sequence analysis and molecular characterization of chromosomal organization has produced yet another definition of a species. The **species phylogeny concept** is based on species sharing one or more DNA sequences or molecular features, such that they may be classified as a single group of origin, or **clade.** This definition requires the ability to correctly identify the molecular relationships between the different species to properly assign single clades. With the significant increase in DNA sequences that has been generated from a wide variety of organisms, this definition has gained considerable attention.

In some cases, no decision can be made about the species status of a population. It is clear that a population has evolved, but it is not clear whether it has evolved enough to be called a new species. This is more of a problem for taxonomists and evolutionary biologists, however, than for the organisms themselves.

Thus, the definition of a species presents a true challenge for evolutionary biology, even though a single, precise definition remains unclear. The different definitions are based on the available data or on the discipline making the classification.

For the sake of the subsequent discussion, we define a species as a population that is reproductively isolated (the biological species concept or species recognition concept).

Mechanisms of Reproductive Isolation

How does one species become two? Basically, **reproductive isolating mechanisms** must evolve to prevent two subpopulations from interbreeding when they come into contact. These mechanisms are environmental, behavioral, mechanical, or physiological barriers that prevent individuals of two different species from producing viable offspring.

Prezygotic Mechanisms

Reproductive isolation can be achieved through a **prezygotic mechanism,** which prevents the fertilization of the egg and the production of a zygote. Prezygotic reproductive isolation can occur in four major ways:

1. **Residential isolation.** The populations live in the same region, but occupy different habitats.
2. **Seasonal or temporal isolation.** The populations exist in the same region, but are sexually mature at different times.
3. **Ethological isolation.** The populations are isolated by incompatible premating or courtship behavior. This form of isolation is observed only in animals.
4. **Mechanical isolation.** Cross-fertilization is prevented or restricted by incompatible differences in reproductive structures.

Postzygotic Mechanisms

Reproductive isolation can also be achieved through a **postzygotic mechanism,** which affects the zygotes after fertilization has taken place. Because these zygotes are produced from individuals of two different species, they are termed **hybrid zygotes.** Postzygotic isolation also can occur in four major ways:

1. **F_1 hybrid breakdown.** The F_1 hybrid zygotes are either unviable or exhibit a reduced viability.

2. **Developmental hybrid sterility.** The F_1 hybrids are viable but sterile, either because their gonads develop abnormally or because of failures during meiosis.

3. **Segregational hybrid sterility.** The F_1 hybrids are viable and sterile, but in this case, the sterility results from the abnormal distribution of whole chromosomes, chromosome segments, or combinations of genes into their gametes.

4. **F_2 breakdown.** In this case, the F_1 hybrids are normal, vigorous, and fertile, but the F_2 generation contains many compromised or sterile individuals.

Allopatric Speciation

Reproductive isolation acts as a barrier to **gene flow,** or the spread of alleles between populations. Genetic drift or selection can then act on these two populations differently.

Isolating mechanisms can evolve in two different ways, each of which defines a different mechanism of speciation. Usually, the mode of speciation is dictated by both the properties of the genetic systems of the organisms and stochastic, or accidental, events. For example, vertebrates tend to have different speciation modes than do insects that feed on plants.

The first mechanism that generates reproductive isolation is the appearance of a geographic barrier, such as a river or interstate highway, through the range of a species that physically separates and isolates the species into two populations (**fig. 25.3**). Physical isolation can also occur if migrants cross a particular barrier, such as a mountain range, and begin a new population (*founder effect*). Alternatively, some individuals may be transplanted, such as plants carried from Southern California to Texas, to create two separate populations of a single species. In all of these possibilities, the physically isolated populations may then evolve independently.

If reproductive isolating mechanisms evolve, then two distinct species are formed, and if these come together in the future, they remain distinct species. Speciation that occurs after physical separation of the populations, which results in the generation of reproductive isolation, is called **allopatric speciation.**

In many cases, it is difficult to prove allopatric speciation, because it occurs over very long periods of time. The proof of allopatric speciation usually requires both phylogenetic and biogeographic data. Here we briefly discuss two examples that appear to

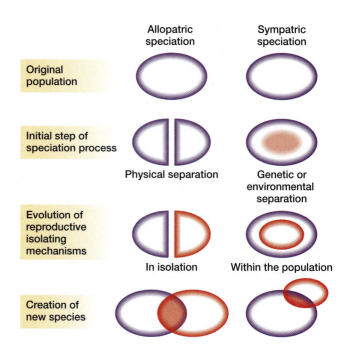

figure 25.3 Two mechanisms that create different reproductively isolated species from a single population. In allopatric speciation, physical separation of a species into two populations leads to reproductive isolation. This involves the creation of a physical barrier causing the separation. In sympatric speciation, reproductive isolation evolves while the new population is still in the vicinity of the parent population. Reproductive isolation may result from either genomic alterations (polyploidy or reciprocal translocations) or the new population entering a different environmental niche.

demonstrate the generation of a physical barrier prior to the creation of two species and the long timescale that is often involved.

The first example of allopatric speciation appears in the ground finches of the Galápagos Islands (**fig. 25.4**). These birds have been very well studied, not only because they present a striking case of speciation, but also because Darwin studied them and was strongly influenced by them in his views.

An original flock of finches somehow reached the Galápagos Islands from South America, 600 miles away, and spread through the various islands. Their movement may have been the result of a storm or some other event. Because the birds had limited ability to get from island to island, allopatric speciation took place. On each island, reproductive isolating mechanisms evolved in the finch population, while the finches evolved to fill certain niches not already filled on each island.

The woodpecker finch is an example of filling a niche. In South America, no finches existed that, like woodpeckers, fed on the insects under the bark of trees. This niche was unavailable for finches because many woodpecker species were already living in

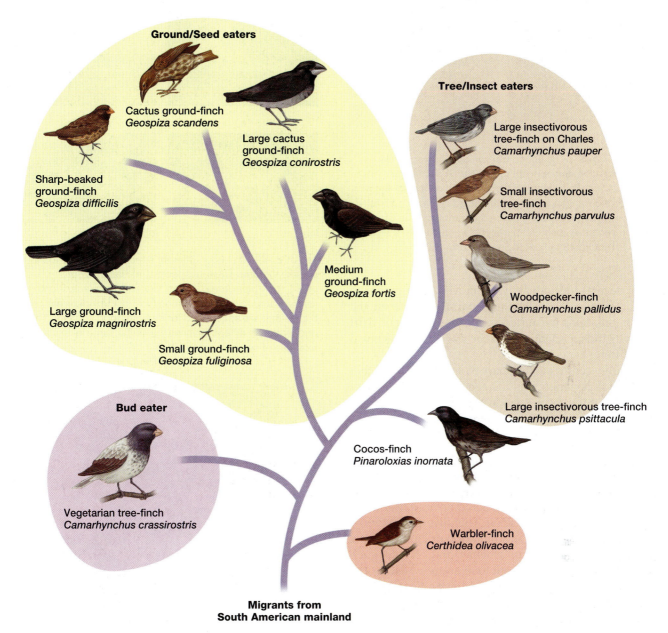

figure 25.4   Shown are 13 species of Darwin's finches. Apparently, a single group of finches migrated from the South American mainland to the Galápagos Islands. Depending on the island and the available niches, the 13 species evolved through allopatric speciation. They can be classified into five groups: vegetarian or bud eaters, ground or seed eaters, tree or insect eaters, the cocos-finch, and the warbler-finch.

South America. But the Galápagos Islands, being isolated from South America, have a **depauperate fauna,** a fauna lacking many types of animals and species found on the mainland. The islands lacked woodpeckers, so the insects beneath the bark of trees were available to birds as a food source. Finches that could make use of this resource would be at an advantage and would thus be favored by natural selection. On one island, a finch evolved to use this food resource. The woodpecker finch acts like a woodpecker by inserting cactus needles into holes in dead trees to extract insects.

Another case of allopatric speciation can be seen in antelope squirrels that populate the Grand Canyon rims (**fig. 25.5**). The Harris's antelope squirrel (*Ammospermophilus harrisi*) is found on the south rim of the canyon, and the white-tailed antelope squirrel (*Ammospermophilus leucurus*) is found on the north rim. Based on morphological criteria, these are two different species. Comparison of the genomic DNAs from the two different squirrels, however, reveals that they shared a common ancestor.

It is assumed that a single population of antelope squirrels existed on one rim of the canyon, and at some

856

A. harrisi   *A. leucurus*

figure 25.5 The Grand Canyon antelope squirrels underwent allopatric speciation with Harris's antelope squirrel (*Ammospermophilus harrisi*) found on the south rim of the Grand Canyon and the white-tailed antelope squirrel (*Ammospermophilus leucurus*) on the north rim. The Grand Canyon serves as a natural barrier between the two species, which is required for allopatric speciation.

point several individuals were moved or migrated to the other rim. The physical separation between the two populations permitted selection and genetic drift to occur independently until the populations evolved into different species.

Until recently, evolutionary biologists believed that allopatric speciation was the general rule. Many now believe that the second mode of speciation may occur frequently in certain groups of organisms.

Sympatric Speciation

The second mechanism to yield reproductive isolation, **sympatric speciation,** occurs when a population that occupies a large range diverges and enters a new niche in the same area (see fig. 25.3). This divergence can be a genetic change, such as a mutation or change in the chromosome number, or a behavioral change, such as exploitation of a different food source. Neither the original group nor the new group have a clear survival advantage; each simply represents an alternative.

Because no physical barrier separates the two groups, mechanisms must arise to direct the reproductive isolation. Typical mechanisms are a reduced fertility when individuals from the two groups mate, or a reduced viability of the hybrid offspring. In this case, assortative mating will reinforce mating between individuals from the same group. If the reinforcement is strong enough, the two groups become reproductively isolated and potentially lead to the generation of two species in a single area. This mode of speciation may be common in plants, insects, and parasites.

One example of sympatric speciation is found in the numerous species of cichlid fish in the Rift Valley lakes of East Africa. Lake Victoria alone has over 400 different species of cichlids (**fig. 25.6**). Obviously no physical barriers exist between these different species in a single lake. The different species exploit different feeding niches. Some feed on the algae on the lake surface, some feed on the bottom of the lake, some feed

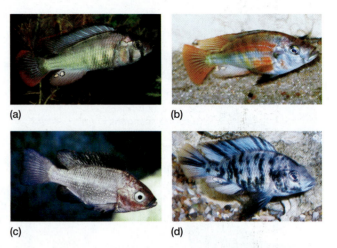

(a)   (b)   (c)   (d)

figure 25.6 Four different species of cichlids found in Lake Victoria. The four species, (a) *Haplochromis sp.83*, (b) *Mbipia lutea*, (c) *Oreochromis esculentus*, and (d) *Paralabidochromis chromogynos*, occupy different niches that are specialized for different food sources. Their occupation in different niches within the same lake is an example of sympatric speciation.

figure 25.7 The apple maggot fly, *Rhagoletis pomonella*. This species has expanded its host range since the nineteenth century from hawthorn to apple trees, cherry trees, and rose bushes. These changes to new host races are presumably the initial steps in sympatric speciation.

on insects, and others feed on other fish species. Thus, each species inhabits a niche based on the location of its food source within the single lake.

The different species within a single lake are more similar than species in different lakes. It is thought that an ancestral species existed in each lake and that this ancestral species underwent sympatric speciation based on the different food niches in each lake.

An example of incipient sympatric speciation has been seen recently in the host races of the apple maggot fly (*Rhagoletis pomonella*) in North America (**fig. 25.7**). This fly was found originally only on hawthorn plants.

The fly mates in the tree and lays its eggs on the hawthorn fruit. The development of the fly is tied to the development of the Hawthorn fruit. In the nineteenth century, however, it spread as a pest to newly introduced apple trees. Because apples develop in a timeframe slightly different from that of the Hawthorn fruit, flies that are attracted to one tree or the other become developmentally isolated from the sibling species.

Another form of sympatric speciation occurs when cytogenetic changes take place within some individuals in a population. These cytogenetic changes include polyploidy and translocations (see chapter 8). For example, if polyploid offspring cannot produce fertile hybrids with individuals from a parent population, then the polyploid is reproductively isolated and "instantaneous speciation" occurs. The generation of a new species through a change in ploidy is more frequently observed in plants than in animals.

Phyletic Gradualism Versus Punctuated Equilibrium

As mentioned earlier, the rates of speciation can vary enormously—from millions of years to a single generation, as in sympatric speciation through polyploidy. Two different models have been proposed to describe the process of evolutionary change. **Phyletic gradualism** describes evolution as a series of incremental steps that produces a new species over a long period of time (**fig. 25.8a**). In 1972, Niles Eldredge and Stephen J. Gould proposed a different model; the **punctuated equilibrium** model suggests that speciation and the accompanying morphological changes occur rapidly and these events are separated by long periods of time when little change occurs (fig. 25.8b).

Although these two models have very obvious differences, in practice they are difficult to tell apart. The fossil record appears to support the punctuated equilibrium model because fossils fail to exhibit the gradual changes that would be expected for phyletic gradualism. In the case of polyploid gametes, reproductive isolation is immediately generated, which can then lead to the rapid production of different species traits because of the absence of gene flow.

Regardless of which mechanism is explored, however, allopatric and sympatric speciation mechanisms apply to both.

Restating the Concepts

▶ Species can be distinguished based on morphological features, reproductive isolation, or the similarity of genome organization and DNA sequences. Ultimately, each species must be reproductively isolated from related species. Isolation

mechanisms can be classified as either prezygotic, which prevent fertilization, or postzygotic, which prevent survival or reproduction of offspring.

▶ Two models can describe different mechanisms of speciation. Allopatric speciation requires that a population is physically separated, and this barrier leads to reproductive isolation. In sympatric speciation, a physical barrier does not divide the population, but rather some members of the population acquire a new trait that allows them to enter a niche within the same environment. Because gene flow can still occur in sympatric speciation, any hybrids that are produced must be at a disadvantage for viability before reproductive isolation occurs.

▶ Evolution of a new species may occur through the gradual accumulation of small changes (phyletic gradualism) or through more rapid and sweeping changes followed by long periods with very little change (punctuated equilibrium).

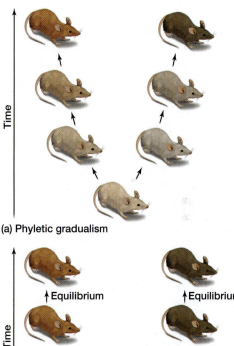

(a) Phyletic gradualism

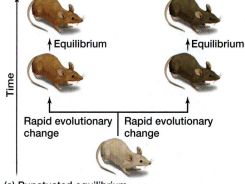

(a) Punctuated equilibrium

figure 25.8 Diagrammatic interpretation of two different models of cladogenesis. (a) Phyletic gradualism is depicted as a gradual divergence of two different populations from a common progenitor over time. This divergence usually is due to the accumulation of small genetic changes that ultimately result in different species. (b) Punctuated equilibrium is depicted as a rapid divergence of two groups into different species after long periods of no change.

25.4 Evolution at the Molecular Level

The ease of isolating DNA sequences from almost any organism, and the abundance of genomic DNA sequences that are now available on a variety of websites, make it convenient to define individuals in a species as sharing one or more DNA sequences. Although it is tempting to think that the definition of a species can be quickly solved through sequence comparisons, a number of questions must be asked. For example, how do we decide what sequences are important in assigning species? Are nucleotide and amino acid sequences an accurate reflection of evolutionary change and relationships? Should researchers look at the coding regions within genes, at the intron sequences, or at the sequences between genes? We consider these questions in the following sections.

Types of Molecular Data

The most obvious type of data that can be used for evolutionary studies is DNA sequence data. We can directly compare either the DNA sequence of a gene's open reading frame (ORF) or the sequence of the entire gene, which also includes both the promoter region and the introns.

Alternatively, we can compare the amino acid sequence of proteins, which are encoded by the genes. Two methods are used to generate this sequence: First, we can directly sequence the protein, which can be a laborious process, especially if it possesses posttranslational modifications. Or, we can determine the nucleotide sequence of the ORF and deduce the amino acid sequence. This latter method may not always produce the exact amino acid sequence of the protein, especially if posttranslational modifications have occurred. It is usually easier, however, to isolate and sequence a DNA fragment than to purify and sequence a protein.

We can compare the analogous sequences between two different species and observe the number of mutational events that have been produced since the two species diverged from each other during evolution. Assume that we are examining a region of a protein's amino acid sequence that is shown in **figure 25.9**. As you can see, two amino acid differences are present in this region of the protein from the two species. If we examine the genomic DNA sequence that encodes this protein sequence in the two species, however, we can identify five differences.

Because of the degeneracy in the genetic code, some mutations do not affect the encoded amino acid (chapter 11). Therefore, comparing the nucleotide sequences is a more accurate representation of the number of mutational events that have occurred between two species. Furthermore, because we do not know which sequence was the original (assuming that one is entirely the

Species 1: 5′ – TTA – CCG– TAC – ATG – GAC – TGG– 3′
– Leu – Pro – Tyr – Met – Asp – Trp

Species 2: 5′ – TTA – CCA – TAT – ATC – GCA– TGG– 3′
– Leu – Pro – Tyr – Ile – Ala – Trp

figure 25.9   Comparison of changes in the DNA nucleotide and amino acid sequences for the same gene from two different species. Notice that the nucleotide sequences (top) contain five differences (red), while the amino acid sequences (top) have only two differences (red) due to the degeneracy in the genetic code.

ancestral sequence), we assume that each species in this example diverged by an average of 2.5 nucleotides (total number of differences divided by 2).

Terminology of Gene Taxonomy

Two genes that are derived from a single ancestral gene are called homologous genes, or **homologs** (**fig. 25.10**). If these two genes are found in different species, they are called **orthologs,** whereas if they are present in the same species they are called **paralogs.** As an example, the human α- and β-globin genes are paralogs, but the human α-globin gene and the mouse α-globin gene are orthologs.

In figure 25.10, the ancestral α-gene underwent a gene duplication to produce two homologs, α and β. Each homolog can randomly accumulate mutations independently. If speciation occurs through cladogenesis, then two species are created that each contains an α- and β-gene. These genes continue to accumulate mutations independently of either their paralogous gene (present in the same species) or their orthologous gene (present in the other species).

Phylogenetic Analysis

We can examine the nucleotide sequences of orthologous genes and determine the number of nucleotide differences. From this information we can attempt to reconstruct the evolutionary history and degree of relatedness among members of a taxonomic group. The species can either be extant or extinct. The fundamental assumption behind **phylogenetic trees** (diagrams of relationships) is that groups that are very similar in nucleotide sequence share a more recent common ancestor than those that are less similar, and that the similarity can be attributed to inheritance from a common ancestor.

A similarity in the nucleotide sequence can be seen by comparing the human and mouse genomes. Using 1506 human–mouse orthologous gene pairs, investigators have found that 86% of the genes have the identical number of coding exons as their ortholog. Ninety-one percent of the orthologous coding exons are the

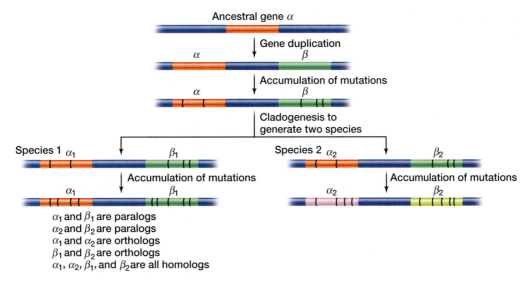

Ancestral gene α
↓ Gene duplication
α   β
↓ Accumulation of mutations
α   β
↓ Cladogenesis to generate two species

Species 1 α_1 β_1 Species 2 α_2 β_2

↓ Accumulation of mutations   ↓ Accumulation of mutations

α_1 β_1 α_2 β_2

α_1 and β_1 are paralogs
α_2 and β_2 are paralogs
α_1 and α_2 are orthologs
β_1 and β_2 are orthologs
α_1, α_2, β_1, and β_2 are all homologs

figure 25.10 Relationship of paralogs and orthologs. An ancestral gene is duplicated to produce two paralogs (α and β). Each paralog randomly accumulates mutations (black lines) and diverges in sequence. In cladogenesis, two species are generated and each contains a set of the duplicated genes. Random mutations continue to accumulate in all four genes to create further sequence divergence. The α_1 and α_2 genes are orthologs, the β_1 and β_2 genes are orthologs, and all four genes are homologs.

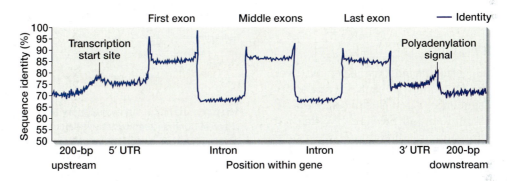

figure 25.11 Conservation of nucleotide sequences across orthologous genes. A total of 3,165 published human mRNA sequences were analyzed over 200 evenly spaced bases in the: 200 bp upstream of the transcription start site, the 5' untranslated region (UTR), exons, introns, 3' UTR, and 200 bp downstream of the gene. The percentage of nucleotide sequence identity in each of these regions was determined and then plotted to generate a "model" gene. As expected, sequence identity is higher in the exons relative to the introns. Very high nucleotide sequence identity at the junctions between each of the regions corresponds to conserved nucleotide sequences, such as near the transcription start site, the exon-intron splice junction, and the polyadenylation signal. (Reprinted with permission from Macmillan Publishers Ltd: *Nature,* vol. 420, p. 551, figure 25, 2002. Copyright © 2002.)

exact length, with another 8.5% differing by only three nucleotides (one codon). In contrast, only 1% of the orthologous introns have the same length. Therefore, the orthologs have a conserved structure.

We can consider a "typical" gene that contains 200 bp upstream of the transcriptional start site, a 5' and 3' untranslated region (UTR), and three exons separated by two introns. If the nucleotide sequence is compared in each of these regions with their ortholo-gous sequence, we find that the exons exhibit fairly high nucleotide sequence identity between the ortho-logs, followed by the 5' UTR (**fig. 25.11**). This finding reveals that some selection pressure acts to retain these sequences, and little selection pressure acts to preserve nucleotide sequences in the introns. For this reason,

investigators often use the coding regions of genes when performing phylogenetic analyses.

Gene Trees Versus Species Trees

Two different kinds of phylogenetic trees are in use. A **gene tree** shows the evolutionary relationships for a single homologous gene, both paralogs and homologs. A **species tree** shows the evolutionary relationships among species (or other taxonomic groupings) by examining multiple independent genes. Although these trees may appear to be equivalent, they are not necessarily the same. For example, if gene duplication occurred before speciation, then genes in different species may be more closely related than genes in a single species.

Let's look at a gene tree that is generated by the sequences of the small (16S or 18S) rRNA subunit gene from 64 different species (**fig. 25.12**). All organisms contain a conserved small rRNA subunit, which demonstrates the fundamental importance of the rRNA for life. This tree demonstrates that there are three major evolutionary groups: Bacteria, Archaea (archaeans), and Eukarya (eukaryotes). Even though the Archaea lack internal membranes, as do the Bacteria, this gene tree revealed that Archaea are as different from the Bacteria as they are from the Eukarya. The gene tree therefore revealed a distant evolutionary difference between these three groups that was not entirely obvious based on the cellular structure.

Additionally, this tree reveals that the eukaryotic organelles, mitochondria and chloroplasts, are more similar to prokaryotes than to the eukaryotic nuclear genomes (see chapter 19). Examination of other genes revealed both of these fundamental findings.

Neutral Theory

Neutral theory suggests that most of the genetic diversity we see at the molecular level (both nucleotide and amino acid sequence) is not due to selection, but rather is due to mutation and random genetic drift. Most mutations are therefore thought to have a small enough effect on fitness as to be essentially neutral (see chapter 23). These mutations may either be in the wobble position of the codon (see fig. 25.9) and not affect the encoded amino acid, or they may result in a different, but conserved, amino acid. This view says nothing about phenotypic evolution, and it does not imply that selection of beneficial alleles is unimportant. It only states that the extensive molecular genetic diversity we see is predominantly neutral or effectively neutral.

Neutral theory predicts that the greatest genetic variation should occur in the regions of the gene that are functionally less important. Inducing mutations in *E. coli* genes revealed that mutations falling in regions that differed little among various species impaired enzyme function, whereas mutations occurring in regions that varied widely among species had little effect on enzyme activity. Furthermore, functionally important genes, such as histones, are highly conserved among species. The H4 histone in the garden pea differs from the histone in domestic cattle by only 2 nucleotides, neither of which changes an amino acid.

As an extension of this prediction, neutral theory also predicts that molecular variation within species should be greater, and divergence among species more rapid, for genes with a smaller effect on the organism's fitness. Genes that encode unimportant proteins or do

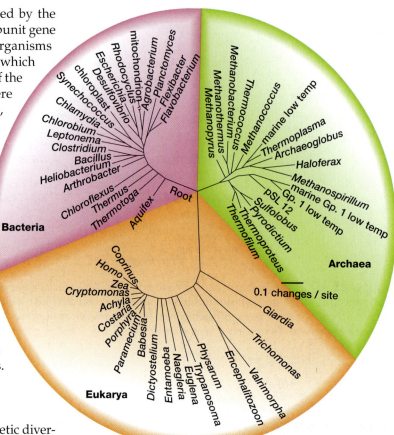

figure 25.12   An evolutionary tree based on the small rRNA subunit sequences. The nucleotide sequences of the small rRNA (16S or 18S) subunit from 64 different species were analyzed in pairs to identify the number of nucleotide differences, which was used to construct this phylogenetic tree. A relative scale of the mutation rate per distance between branches is shown.
(From Pace, N. R., Science, Figure 1, 276, p. 734–740, 1997. Reprinted with permission from AAAS.)

not code for functional proteins at all (*pseudogenes*), should display more nucleotide variation than genes that encode functionally important proteins. Studies of DNA variation have repeatedly confirmed this hypothesis.

The Molecular Clock

Using either the DNA or amino acid sequence for different genes, and from a number of organisms, allows us to make comparisons of substitution rates within orthologs. The **molecular clock** is based on the assumption that neutral substitutions occur at a constant rate over evolutionary time. However, this rate is specific to the gene or protein being studied. Consistent with the neutral theory, genes that encode proteins with highly conserved essential functions, such as histone H4 or cytochrome c, would have very

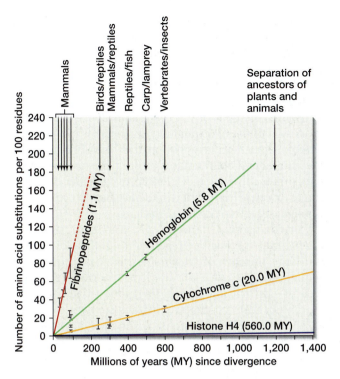

figure 25.13 The rate of the molecular clock varies with different proteins. The rate of neutral substitutions per 100 amino acids is shown for four proteins, fibrinopeptides, hemoglobin, cytochrome c, and histone H4. While the rate of amino acid substitutions is linear over time for each protein, it varies over 500-fold between proteins. The points in time when major evolutionary divergences occurred are shown at the top, with the scale of millions of years on the X axis.

table 25.1 Evolutionary Rates as Substitutions per Amino Acid Site per Year for Various Proteins

Protein	Evolutionary Rate
Fibrinopeptide	8.3
Pancreatic ribonuclease	2.1
Lysozyme	2.0
Hemoglobin α	1.2
Myoglobin	0.89
Insulin	0.44
Cytochrome c	0.3
Histone H4	0.01

low neutral substitution rates relative to genes and proteins that have highly variable activities in different species, such as the fibrinopeptides (**fig. 25.13**). In fact, this evolutionary rate can differ by nearly three orders of magnitude, depending on the gene or protein that is being studied (**table 25.1**). Based on this molecular clock, the time of genetic isolation between

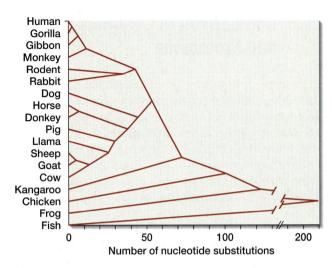

figure 25.14 This phylogenetic tree was constructed by comparing the number of nucleotide substitutions in hemoglobin, cytochrome c and fibrinopeptide A for each organism. The total number of nucleotide substitutions appears on the horizontal axis. This tree groups similar organisms and generally agrees with classical systematics.

two species can be calculated based on the number of neutral substitutions in an *orthologous* gene or protein. Using a variety of different genes, a composite phylogenetic tree is generated that reveals the evolutionary history of different species (**fig. 25.14**).

Restating the Concepts

▶ Phylogenetic analysis using gene trees examines the nucleotide or amino acid mutations in a single gene to reconstruct the evolutionary history of the different species. The more similar two species are, the fewer the number of neutral substitutions and the closer the evolutionary distance. In contrast, a species tree compares the multiple independent genes between different species to reconstruct the evolutionary history.

▶ The neutral theory predicts that essential functions, either regions within genes or the entire gene, will exhibit a lower rate of neutral substitutions from one species to another, relative either to regions in genes that do not affect enzymatic activity or to genes that are less important for the organism.

▶ Neutral theory is the basis for the molecular clock, which states that neutral substitution is constant over evolutionary time for a given homolog across species. Genes that encode proteins with highly conserved essential functions usually have very low neutral substitution rates relative to genes that encode proteins with highly variable activities in different species. These neutral substitution rates can vary by up to three orders of magnitude for different homologous genes.

25.5 Comparison of Genomes and Proteomes

In the preceding section, we discussed how changes at the nucleotide level can be used to examine the evolutionary distance between species. It is also possible to examine evolutionary changes in a more global way, which involves changes at the whole-genome level or in the complement of all the proteins expressed in an organism. These analyses can also reveal similarities and differences between species.

Comparison of Genomic Organization Between Species

As previously described, a change in the ploidy of an organism can result in "instantaneous" speciation; however, other types of changes at the genomic level can produce a difference between species. As you learned in chapter 8, reciprocal translocations cause the *trans*-heterozygote to be semisterile, which may also lead to sympatric speciation. If translocations can lead to speciation, then we would expect to find that related species would have this type of chromosomal rearrangement throughout the genome.

Examination of the banding patterns of the three largest human prophase chromosomes (chromosomes 1, 2, and 3) with the chromosomes from three different apes (chimpanzee, gorilla, and orangutan) reveals that they are quite similar (fig. 25.15). Several chromosomal rearrangements can be identified, however. For example, human chromosome 2 is present as two independent chromosomes in the apes, which accounts for the human genome having 23 pairs of chromosomes and the apes having 24 pairs.

You can also see that the banding pattern on human chromosome 3 is very similar to the banding pattern on the chimpanzee and gorilla chromosomes. However, the orangutan chromosome has a large pericentric inversion. This finding suggests that chimpanzees and gorillas, which both lack

the pericentric inversion, are more closely related to humans than are orangutans. This evolutionary relationship has been substantiated by a variety of other approaches.

Based on the phylogenetic tree shown in figure 25.16, we would expect that the chromosome inversion occurred at or after the split of the orangutan lineage, and that the fusion of the two smaller chromosomes into human chromosome 2 occurred at or after the split between the human and chimpanzee lineages. It is unlikely that the human chromosome split into two smaller chromosomes, because this event would have had to occur in all three lineages that produced the apes.

Chromosomal banding patterns can be informative, but the level of resolution is fairly low. With the large number of organisms that have now been sequenced and with the ability to perform fluorescent in situ hybridization (FISH) to localize genes on chromosomes (see chapter 14), researchers can now examine chromosome organization at a much higher level of resolution.

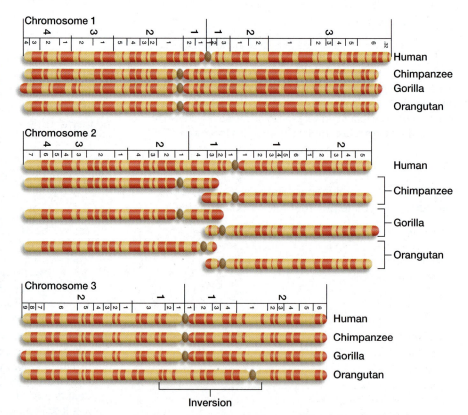

figure 25.15   Comparison of metaphase banding patterns of human chromosomes 1, 2, and 3 with the corresponding chromosomes in apes. Human chromosome 1 is similar to the three apes, except for the region flanking the centromere. Human chromosome 2 banding pattern is located on two different chromosomes in all three apes. Human chromosome 3 banding pattern is similar to the chimpanzee and gorilla, but the orangutan chromosome contains a pericentric inversion. The similarity between the human chromosome 3 and the chimpanzee and gorilla suggests that the orangutan diverged earlier from humans than either of the other two apes. The centromere is represented as the tan ring.

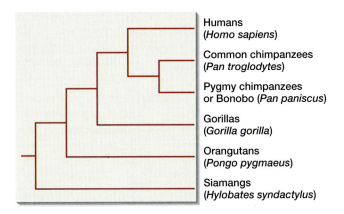

Humans
(*Homo sapiens*)

Common chimpanzees
(*Pan troglodytes*)

Pygmy chimpanzees
or Bonobo (*Pan paniscus*)

Gorillas
(*Gorilla gorilla*)

Orangutans
(*Pongo pygmaeus*)

Siamangs
(*Hylobates syndactylus*)

figure 25.16   This phylogenetic tree of humans and closely related hominoid species was generated by comparing the neutral nucleotide substitutions in the mitochondrial cytochrome oxidase subunit II gene. This tree is similar to the phylogenetic relationship we deduced from the chromosome banding patterns (figure 25.15). The pygmy chimpanzee has been renamed the bonobo.

If we compare the genes on human chromosome 22 with the orthologous genes in the mouse genome, we find that the genes are distributed on eight different mouse chromosomes. The striking result is that large blocks of genes remain together in both species, a phenomenon referred to as **synteny.** For example, the orthologous regions of human chromosome bands 22q13.3 and 22q13.2 are found on mouse chromosome 15. And, the orthologous region of human chromosome band 22q13.1 is found on mouse chromosomes 8, 10, and 15. This conservation of mouse chromosomal segments is observed throughout all the human autosomes to generate a synteny map (**fig. 25.17**). The rearrangement of chromosomal segments, rather than the random distribution of orthologous genes throughout the genome, is consistent with genomic shuffling contributing to speciation.

Comparison of the Proteome Between Species

The sequencing of genomes from a wide variety of species has allowed molecular geneticists to deduce most, if not all, of the proteins that are present in an organism. Proteomes of different species can then be compared. For example, proteins can be assigned into groups based on the known function of the protein or that of a related protein in another species.

When this analysis is performed with either the human or the mouse proteome, approximately 75% of the proteins can be assigned to functional groups. Furthermore, both the human and the mouse have similar percentages of proteins in their respective proteomes that fall into each functional group (**fig. 25.18**).

figure 25.17   A map of the 22 human autosomes and the X and Y chromosomes is color-coded to show the syntenic regions to each of the mouse chromosomes. Regions of the human chromosome that are not syntenic with the mouse are white. Notice that human chromosome 20 is entirely syntenic with a region of mouse chromosome 2. (Reprinted with permission from Macmillan Publishers Ltd: *Nature*, vol. 409, p. 910, 2001. Copyright © 2001.)

If we compare the human proteome with yeast, fruit fly, and worm, a clear increase in the complexity of the proteome is detected as we go from yeast to the invertebrates to humans. This complexity is not simply an increase in the number of proteins, but also includes an increase in the number of paralogs and an increase in the number of proteins with multiple functions. Therefore we can see that the increased complexity of life through evolution involves additional proteins, specialized proteins, and versatile protein functions.

Comparing the human proteome with all other known proteomes, only about 1% of the proteins are unique to humans. Similarly, less than 1% of mouse proteins in the proteome are present only in rodents. The uniqueness of a species results either from a very small percentage of its proteome or from how the various proteins function during development and adulthood.

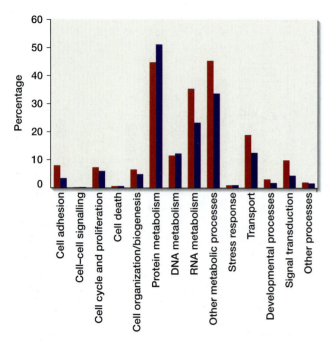

figure 25.18   Comparison of the cellular functions of the proteins in the human and mouse proteomes. The known human and mouse proteins were assigned cellular functions, also known as gene ontologies (GO), based on the presence of known functional domains within each protein. Mouse proteins are labeled in blue and human proteins are in red. Similar percentages of mouse and human proteins in their respective proteomes are assigned to each process.

(Reprinted with permission from Macmillan Publishers Ltd: *Nature*, vol. 420, p. 541, figure 18, 2002. Copyright © 2002.)

Restating the Concepts

▶ During evolution, large chromosomal rearrangements have occurred that might reproductively isolate new species through the semisterility of the hybrid offspring. These chromosomal rearrangements are visible through chromosome banding patterns and through genomic analysis of the clustering of orthologous genes between different species.

▶ Proteomic analysis has revealed that large numbers of related proteins have been conserved across all life forms. However, only a very small percentage of proteins are unique to a particular species. This suggests that species uniqueness results either from that very small percentage of proteins or from the specific ways in which proteins are expressed and function through development.

25.6 Applications of Evolutionary Genetics

In addition to its use in examining the evolutionary relationships between species, DNA sequence infor-

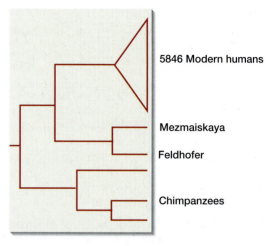

figure 25.19   A phylogenetic tree was constructed from the mitochondrial DNA sequences isolated from the skeletal remains of two Neanderthal populations (Mezmaiskaya and Feldhofer), 5,846 modern humans, and chimpanzees. This tree shows that chimpanzees diverged from modern humans before Neanderthals did over 600,000 years ago.

(Reprinted with permission from Macmillan Publishers Ltd: *Nature*, vol. 404, pp. 490–493, 2000. Copyright © 2000.)

mation can also provide a significant amount of information about changes that occur *within* a species.

Origin of Modern Humans

We can learn much about the origin of modern human beings (*Homo sapiens*) using genetic information. *Homo neanderthalensis* lived throughout Europe and western Asia from 300,000 to 30,000 years ago and overlapped by at least 30,000 years the presence of *Homo sapiens* in Europe. One evolutionary model suggests that *Homo sapiens* descended directly from *Homo neanderthalensis* and also additional populations throughout Africa and Asia. This is the "multiregional" hypothesis. An alternative model states that *Homo sapiens* originated from a single population of Neanderthals in Africa and migrated out to Europe and Asia. This is termed the "out of Africa" hypothesis.

To differentiate between these two possibilities, scientists analyzed mitochondrial DNA (mtDNA) sequences from different populations of modern humans and mtDNA isolated from skeletons of two different *Homo neanderthalensis* individuals (Mezmaiskaya and Feldhofer). Because the skeletal remains were found over 1000 miles from each other, they likely represent different Neanderthal populations. Because mitochondrial DNA accumulates mutations approximately 10-fold faster than nuclear DNA, its molecular clock is much faster. This increased rate of accumulating mutations allows scientists to examine evolution over a shorter period of time.

By comparing mtDNA sequences from different human populations, it was determined that the last common ancestor between *Homo sapiens* and *Homo neanderthalensis* lived almost 600,000 years ago (**fig. 25.19**). This analysis also revealed that *Homo neanderthalensis* did not interbreed with the direct ancestor of *Homo sapiens*.

Although this data revealed that the "multiregional" hypothesis was incorrect, it did confirm that *Homo neanderthalensis* was evolutionarily more closely related to *Homo sapiens* than to chimpanzees. But was there any support for the "out of Africa" model? African populations exhibit a greater number of mtDNA changes than other modern human populations, which is consistent with their mtDNA being the oldest and having the greatest amount of time to accumulate differences. Because mitochondria are inherited maternally in humans (chapter 19), this analysis reveals that the maternal human lineage, or the "mitochondrial Eve," originated in Africa.

In an analogous manner, the DNA sequence of the Y chromosome can be analyzed to trace the paternal origin of modern humans. When Y chromosome sequences in males from different populations throughout the world were analyzed, the data supported the model that the *Homo sapiens* male also originated in Africa. Thus, two independent lines of analysis, mtDNA and Y chromosome sequence analyses, generated similar predictions on the origins of *Homo sapiens*.

Once *Homo sapiens'* origin from a population in Africa was determined, questions arose as to how the human population spread throughout the world. By comparing mtDNA sequences, scientists determined that a group of early *Homo sapiens* migrated out of Africa, across the southern end of the Red Sea, and along the southern coast of Asia to Australia 60,000–75,000 years ago (**fig. 25.20**). A second wave of *Homo*

sapiens migrated up the Nile River to Eurasia and then fanned out across Europe and Asia approximately 45,000–50,000 years ago.

One example of the use of DNA analysis to clarify human origins is a study of the six indigenous tribal populations inhabiting the Andaman and Nicobar Islands in the Bay of Bengal. To examine when these tribes first inhabited these islands, scientists analyzed mtDNA from five Onge, five Great Andamanese, and five Nicobarese individuals (**fig. 25.21**). The mtDNA of all five Onge and all five Great Andamanese individuals were more similar to the Indian and East Asian populations than to the West Eurasian populations. In contrast, all five Nicobarese individuals possessed mtDNA sequences that were more similar to the West Eurasian mtDNA sequences. Thus, the Onge and Great Andamanese likely colonized the islands upon the first migration out of Africa, and the Nicobarese likely came to the islands after the second migration from Africa.

Analysis of Viral Origin

Many virus populations possess large amounts of genetic diversity, and this characteristic can frequently slow our ability to either control the spread of the virus or eradicate it. Genetics can be used to describe the origin of the virus, how the virus is propagated through

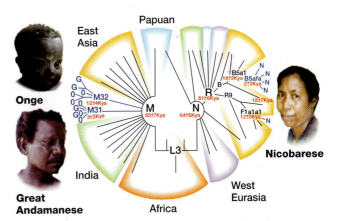

figure 25.21 To trace the inhabitation of the Indian islands, a phylogenetic tree was generated from mitochondrial DNA sequences from a variety of African, Eurasian, and Asian populations. This tree shows the relative positions of the Indian islanders, the Onge, Great Andamanese, and the Nicobarese. Because these island populations remained relatively isolated, it is possible to calculate when they separated from the mainland populations. This tree reveals that the Onge and Great Andamanese (labelled G and O) migrated to the islands from the original emigration from Africa through India, while the Nicobarese (labelled N) came to the islands much later, possibly from one of the European migrations to the east.
(From Thanaray, et al., *Science*, vol. 308, p. 996, 2005. Reprinted with permission from AAAS.)

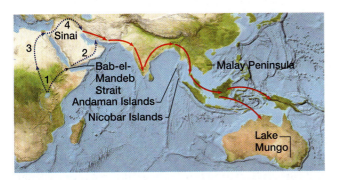

figure 25.20 *Homo sapiens* migrated out of Africa by either crossing the southern end of the Red Sea (1) and into Asia (2) or up the Nile River (3), crossing into Sinai (4) and then Asia. In both models, the migrants likely traveled along the Indian Ocean coast to reach Australia (approximately 46,000 years ago) and Borneo (approximately 45,000 years ago). Mitochondrial DNA analysis suggests that human settlements appeared in the Andaman and Nicobar islands approximately 60,000 years ago.
(From Forster and Matsumura, *Science*, vol. 308, p. 965, 2005. Reprinted with permission from AAAS.)

the host population, the amount of genetic variation present, and how these factors may influence the clinical symptoms of an infection. An evolutionary genetic analysis of a virus, such as human immunodeficiency virus (HIV), can have an effect at two different levels: on a global scale, and in individual patients.

Phylogenetic trees have shown that there are two major types of human immunodeficiency viruses throughout the world: HIV-1 and HIV-2. Both of these HIV types originated from the simian immunodeficiency virus (SIV). The HIV-1 strain is the more common and virulent HIV type, with HIV-2 primarily restricted to Africa.

Both HIV-1 and HIV-2 exhibit a large amount of genetic variation that can be organized into distinct "subtypes." These subtypes can be organized into phylogenetic trees based on their DNA sequence variation (**fig. 25.22**). The geographic distribution varies for each subtype, although most are found in Africa. These different subtypes may possess different biological properties. For example, subtype E, from Southeast Asia, appears to be more easily transmitted through sexual contact than other subtypes. This subtype may account for the dramatic spread of the virus in this part of the world, through prostitutes and heterosexual encounters.

If the correct subtypes can be identified through phylogenetic analysis, this identification would potentially allow the generation and administration of specific vaccines to combat viral subtypes.

An evolutionary genetics approach has also been used to examine the potential transmission of a virus between individuals. One case from the late 1980s involves a dentist in Florida who was infected with HIV. He continued to practice dentistry for approximately 2 years, until one of his patients tested positive for the HIV virus. On urging from the dentist, numerous additional patients were tested for HIV and several were also found to have the virus. Some of these infected patients had none of the risk factors associated with HIV infection. The immediate question was whether the dentist infected these patients during the dental procedures.

Viral samples were collected from the dentist, his HIV-positive patients, and several HIV-positive individuals in the area (these served as controls). DNA was

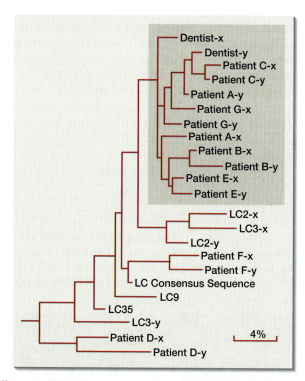

figure 25.23   To deduce the infection route of HIV in a population, a phylogenetic tree was generated from HIV strains isolated from an infected dentist, seven of his infected patients, and several local controls (unrelated infected individuals). Because an individual will possess many different viral variants, the two most divergent *env* gene sequences (x and y) from each individual were analyzed independently. Notice that five patients (in shaded box) have HIV strains that are similar to the dentist's strain, but diverged later than the dentist's strain. This suggests that these five patients were infected by the dentist and their virus underwent subsequent mutational events. In contrast, the other two infected patients (D and F) contained an HIV strain that diverged prior to the dentist's strain, which suggests they could not have been infected by the dentist.
(From Ou, Cy, et al., *Science*, vol. 256, p.1167, figure 1, 1992. Reprinted with permission from AAAS.)

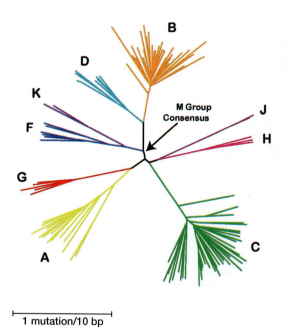

figure 25.22   This phylogenetic tree of 9 different HIV-1 subtypes is based on nucleotide differences. Each distinct viral subtype possesses multiple variants that are represented as individual lines of a common color.

extracted from these samples, and regions of the viral genome were PCR-amplified and sequenced. These DNA sequences were analyzed and arranged in a phylogenetic tree (**fig. 25.23**). The analysis revealed that the HIV samples from the dentist and five of his patients were more similar to each other than they were to the other two HIV-positive patients (patients F and D) and the area controls. Furthermore, the HIV sample from the dentist appeared to have an earlier placement in the phylogenetic tree than the HIV samples from the five patients. This result suggests that these five patients acquired their HIV infection from the dentist, and the other two HIV-positive patients were infected with HIV by a carrier or carriers other than the dentist.

There are two things to note. First, viral DNA, like mitochondrial DNA, acquires mutations very quickly, such that different viral sequences can be identified in different individuals who were all infected from the same source in approximately the same time period. This high mutation rate is partially due to the rapid replication and propagation of the virus in the host. Second, the transmission of HIV from a health-care worker to patients is extremely rare, and it should not be a matter of concern if normal precautions are followed.

Restating the Concepts

▶ Evolutionary genetics uses mitochondrial DNA and Y chromosome DNA sequences to trace the origin of modern humans and their migration out of Africa and across Asia. Based on these analyses, modern humans originated in Africa independently from *Homo neanderthalensis*.

▶ Evolutionary genetics has been used to identify a number of different "subtypes" of HIV-1 and HIV-2, with the subtypes sometimes possessing potentially unique biological properties. Such analyses can be important in finding treatments for diseases and limiting their transmission. The acquisition of mutations in the viral genome can also be used to trace the potential pathway of infection through individuals in a population.

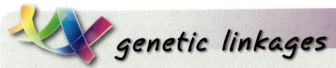

genetic linkages

In this chapter, we discussed the basic principles of evolution and speciation and the application of genetic techniques to this area of study. Natural selection, adaptation, and mutation, the driving forces of evolution, were previously discussed in chapter 23.

When genetic differences develop between two groups, speciation follows if reproductive isolation occurs. This isolation could result from a physical barrier (allopatric speciation) or a nonphysical barrier, such as a genetic barrier, without physical separation (sympatric speciation). In chapter 8, you learned about several mechanisms that can lead to reproductive isolation. "Instantaneous" speciation can occur when polyploid individuals are generated because they will be unable to reproduce successfully with their nonpolyploid ancestors. Chromosomal mutations, such as reciprocal translocations and large inversions, will lead to semisterility and to reduced fertility, respectively, and can also result in speciation.

You have seen how nuclear or mitochondrial DNA sequences can be used to generate phylogenetic trees.

Because mitochondria are inherited maternally in many animals (chapter 19), it is possible to trace the maternal lineage, much like it is possible to trace the paternal lineage through the inheritance of the Y chromosome in mammals and many other organisms (chapter 4). The DNA sequences used for a phylogenetic analysis are typically PCR-amplified and sequenced prior to analysis (chapter 12). The organization of genomic regions within a chromosome, using FISH analysis (chapter 14), can also provide a more global picture of genome evolution.

For many years, evolutionary studies concentrated on morphological and behavioral traits to classify organisms. With the development of recombinant DNA techniques and the ability to isolate small amounts of DNA from remnants of dead or extinct animals, phylogenetic analysis has become more quantitative. Phylogenetic analysis with DNA sequences has confirmed many of the evolutionary relationships that had previously been defined on the basis of gross traits.

Chapter Summary Questions

1. For each term in column I, choose the best matching phrase in column II.

Column I
1. Allopatric speciation
2. Anagenesis
3. Biological species concept
4. Cladogenesis
5. Gene flow
6. Homologs
7. Molecular clock
8. Morphological species concept
9. Natural selection
10. Phylogenetic tree
11. Postzygotic isolation
12. Prezygotic isolation
13. Sympatric speciation
14. Vestigial features

Column II
A. Speciation occurring due to niche specialization without the creation of a physical reproductive barrier
B. Neutral allele substitutions occurring at a constant rate between genetically isolated lineages over evolutionary time
C. Splitting of one species into two new species
D. Genes derived from a single ancestral gene
E. Spread of alleles among populations
F. Evolutionary relationships among genes, species, or higher taxonomic levels
G. Organisms classified as belonging to the same species if they are morphologically similar
H. Traits with no current function
I. Prevents production of a zygote
J. Speciation that occurs when physical barriers to gene flow cause reproductive isolation
K. Affects fitness of zygotes after fertilization
L. Transformation of one species to another
M. Mechanism of adaptive evolution; increase in the frequency of a beneficial allele through time
N. Defines a species as a group of individuals capable of reproducing fertile offspring

2. What is a species? What are the pros and cons of the various definitions of species provided in the chapter?

3. How do reproductive barriers and later, species, arise? Describe the process of allopatric and sympatric speciation.

4. In *Drosophila*, females in populations A and B produce an average of 250 offspring each. When the two populations are crossed, the AB females in the F_1 generation produce only 100 offspring each. Are populations A and B in the process of becoming different species?

5. A few plants of species Q ($2n = 14$) spontaneously double their chromosome number ($2n = 28$). The new genotype results in a plant that is named species R and is unable to successfully cross with individuals of the progenitor species Q. Why are QR hybrids sterile?

6. Define *evolution*. Why are *evolution* and *adaptation* not synonymous?

7. Explain the mechanisms of evolution.

8. Describe natural selection.

9. What limits the scope of phenotypic changes during evolution?

10. Discuss whether each of the following is necessary for evolution by natural selection.
   a. differential reproduction.
   b. environmental change
   c. heritable variation for fitness
   d. sexual reproduction

11. A person infected with *Neisseria gonorrhoeae* (the cause of gonorrhea) takes an antibiotic to stop the infection. At first the symptoms subside, but after several days the symptoms return and the infection seems resistant to any further treatment with the antibiotic. What is the cause of this phenomenon?

12. Explain the enigma of mutation. The majority of mutations are deleterious (harmful), yet mutations are said to be the building blocks of evolution. How is this possible?

13. A mutation has just occurred. What factors will affect whether the mutation is eventually lost, fixed, or maintained at an intermediate frequency in the population?

14. On the Web page of a genetics aficionado, the following proof that humans are evolving is offered: "Thirty percent of pregnancies end by spontaneous abortion of embryos and fetuses, 5% by stillbirths and infant deaths, and 3% by childhood deaths. Another 20% will survive to adulthood but never marry. Of those that do marry, 10% will have no children." Ignoring the fact that you don't have to be married in order to have children, is this proof that human populations are evolving? Why or why not?

15. What is meant by *constraint* in the molecular evolution of DNA and proteins?

16. How does the acceptance of neutral theory change our view of evolution?

17. In which codon position should the greatest abundance of variation occur? Why?

Solved Problems

PROBLEM 1. What are the roles of reproductive isolation in the process of evolution?

Answer: Reproductive isolation prevents individuals in two populations from mating with each other or producing viable offspring. These mechanisms can be prezygotic or postzygotic. They usually evolve while populations are isolated from one another, either physically (allopatric) or by entering a new niche (sympatric speciation).

PROBLEM 2. Compare and contrast the predictions from neutral theory versus selection theory.

Answer: Neutral theory predicts that genetic variation at the molecular level is due primarily to a balance between mutation and random genetic drift. This results in two major predictions: (i) deleterious recessive alleles should exist at low frequencies for most loci, especially those under weak selection, and (ii) since drift is a powerful force in this model, small populations should have less genetic diversity than larger populations. Selection theory predicts that selection is primarily responsible for the genetic variation seen at the molecular level and, if true, would result in very little genetic variation (since the fittest allele would be quickly and efficiently fixed in the population) and little difference would exist in the amount of genetic variation between large and small populations. The second prediction of neutral theory matches observation very closely. The first prediction does not fit either hypothesis closely. Thus, refinements to the neutral theory (the nearly neutral theory) allowing for weak selection and genetic hitchhiking have been developed to better fit the observed data amassed for populations.

Exercises and Problems

18. Dolphins and sharks share many morphological similarities, yet their DNA sequences suggest that they are not closely related evolutionarily. Explain how this can be so.

19. In the vertebrate eye, the nerves and blood vessels lie between the photosensitive cells and the source of the light needed to form an image. What could explain such a poor design?

20. Explain why evolutionary genetics is a predictive science with relevance to epidemiology and the origins of modern humans.

21. Two evolutionary mechanisms suggested for senescence are the *antagonistic pleiotropy* hypothesis and the *mutation accumulation* hypothesis. The first suggests that senescence is the primary result of alleles having beneficial effects early in life and deleterious late in life. The second hypothesis suggests that senescence results from the accumulation of deleterious alleles that act later in life, when natural selection is weaker. How would you test which of these hypotheses is correct?

22. Why does the nearly universal genetic code, the similarity of protein form and function in distantly related organisms, and the general concordance between the fossil record and phylogenetic analyses provide such strong evidence for evolution?

23. What are the necessary conditions for natural selection to occur? How does this lead to adaptation?

24. Two species of marine snail of the genus *Littorina* occupy rocky shores. The species differ in shell morphology, physiology, and mating behavior. Species *S* occupies the upper shores and species *T* the lower shores. Over a broad range of midshore zones the two species overlap in their distribution and often hybridize. The different shore types differ in the types of predators encountered, salinity levels, and a number of other factors. Monthly survival rates for species *S* in the upper shores is 44% as compared with only 5% for species *T*. Conversely, monthly survival for species *S* in the lower shores is only 8% compared with 53% for species *T*. Hybrids perform very poorly in upper and lower shores. However, in midshore zones the relative survival rates are: 25%, 30%, and 75%, for species *S*, species *T*, and hybrids, respectively. Hybrids are capable of producing viable offspring with each other or either parental species. Do you expect speciation to occur in this situation? Why or why not? What kind of speciation would it be?

25. What problems does the species recognition concept suffer from?

26. Why might you expect postzygotic reproductive isolating mechanisms to eventually lead to prezygotic reproductive isolating mechanisms?

27. Explain what you think is meant by the statement that allopatric speciation is passive and sympatric speciation active.

28. Why do islands and isolated lakes feature such rampant speciation, despite the fact that island species usually lack genetic variation compared with their mainland counterparts?

29. The species phylogeny concept describes species based on their genetic distances estimated from nucleotide differences often in neutral genes. What are some weaknesses in this concept?

30. The α and β families of the globin genes originated from a duplicated ancestral gene approximately 420 million years ago in an early vertebrate (fish) ancestor. Modern humans shared a common ancestor with Neanderthals about 600,000 years ago. Which is going to be more similar, the α- and β-*globin* genes from a Neanderthal or the α-*globin* gene in a modern human compared with a Neanderthal?

31. If the rate of amino acid substitutions per site per year is 2×10^{-9}, and the average number of amino acid

substitutions per site is 0.2, how long has it been since the two species diverged?

32. The globin gene family is composed of 14 homologous genes originally derived from a single ancestral globin gene. Sequence analysis suggests that the gene first duplicated 800 million years ago. Since that time, additional duplication events and chromosomal rearrangements have produced 14 genes clustered on three chromosomes in humans. All the globins are involved in oxygen binding, but the duplications have allowed specialization (e.g., myoglobin in muscles, hemoglobin in red blood cells).

    In fact, the number of duplicated genes in most organisms' genomes appears to be very large. For example, 40% of the approximately 13,600 coding sequences in the *D. melanogaster* genome appear to have arisen by gene duplication. Very similar numbers have been found in the human, mouse, *E. coli*, and *Arabadopis thaliana* genomes. A large proportion of these duplicate genes appear to have evolved new functions.

    What contributes to the success of duplicated genes across such an evolutionarily diverse group or organisms?

33. Looking at the following table, which listed organism is most closely related to humans evolutionarily? Which is least closely related? Does the information from the cytochrome *c* gene agree with your perception from similarities in morphology, physiology, and behavior?

Organism	Number of Amino Acid Differences Relative to the Human Cytochrome c Protein
Chimpanzee	0
Rhesus monkey	1
Rabbit	9
Cow	10
Pigeon	12
Bullfrog	20
Fruit fly	24
Wheat germ	37
Yeast	42

34. What is meant by an evolutionarily conserved gene? Can some parts of a gene be more conserved than other parts? Why?

35. What is the difference between a mutation rate and a substitution rate? Why is it important to differentiate between them?

36. There are three major genotypes of the influenza virus denoted A, B, and C. Type A influenza is the one responsible for major human pandemics and can be divided into numerous genotypes within the A strain of influenza. What practical information might come from a phylogenetic analysis of the influenza virus?

37. A plant is discovered with weak cancer-fighting properties. How might phylogenetic analysis help in the discovery of other plants with cancer-fighting properties?

Chapter Integration Problem

The following box shows the number of synonymous and nonsynonymous substitutions per site between *Drosophila melanogaster* and *Drosophila pseudoobscura* for five different genes.

Gene	Synonymous	Nonsynonymous
Adh	0.5734	0.0177
Gpdh	0.7617	0.0001
Tik	0.6842	0.0107
Ubx	0.5528	0.0082
Zen	0.7717	0.0434

a. Which is greater, the synonymous or nonsynonymous substitution rate? Does this agree with expectations? Why or why not?

b. Do the genes differ in their substitution rates? Do you expect the genes to have different or similar substitution rates? Why?

c. Comparing synonymous and nonsynonymous substitution rates, which is more variable among genes? Does this agree with theory?

d. Given the divergence in nucleotide sequences between these two species, do they represent true species?

e. If you had a difficult time answering part d, what additional information might you want to know before answering the question? What definition of species is best addressed with the data in the table?

Do you need additional review? Visit **www.mhhe.com/hyde** for practice tests, answers to end-of-chapter questions and problems, interactive exercises, and animation tutorials all designed to enhance your understanding of key genetic concepts.

glossary

a

acentric fragment A chromosomal piece without a centromere.

acrocentric chromosome A chromosome whose centromere lies very near one end.

activator A transcription factor (protein) that binds to a specific DNA sequence, often upstream of the transcription start site, that increases transcription of a gene.

active site The part of an enzyme where the actual enzymatic function is performed.

additive model A mechanism of quantitative inheritance in which alleles at different loci either add an amount to the phenotype or add nothing.

allele Alternative form of a gene.

allopatric speciation Speciation in which the evolution of reproductive isolating mechanisms occurs during the physical separation of the populations.

allopolyploidy Polyploidy produced by the hybridization of two species.

allosteric protein A protein whose shape changes when it binds a specific molecule. In the new shape, the protein's ability to interact with a second molecule is altered.

alternative splicing Various ways of removing introns, and some exons, from a single gene's pre-mRNA to generate different eukaryotic mRNAs that may encode different proteins.

amorph An allele that is equivalent to a nul allele that possesses no activity (e.g., encodes a completely nonfunctional protein); genetically the same as a deletion.

anagenesis The evolutionary process whereby one species evolves into another without any splitting of the phylogenetic tree. *See* cladogenesis.

anaphase The stage of mitosis and meiosis in which sister chromatids or homologous chromosomes are separated by spindle fibers.

anaphase-promoting complex (APC) Protein complex that breaks down cyclin B and promotes anaphase among its various roles in controlling the cell cycle. (Also called the cyclosome.)

aneuploidy The condition of a cell or of an organism that has additions or deletions of whole chromosomes.

anticodon The three-base sequence on the tRNA that is complementary to the codon on the mRNA.

antisense RNA RNA product that regulates another gene by base pairing with, and thus blocking, its mRNA.

antiterminator protein A protein that, when bound at its normal attachment site on specific RNAs, allows RNA polymerase to read through normal transcriptional terminator sequences (i.e., the N protein of phage λ).

antiterminator stem A configuration in the leader mRNA that does not terminate transcription in the attenuator-controlled amino acid operons.

AP endonucleases Endonucleases that initiate excision repair at apurinic and apyrimidinic sites on DNA.

apoptosis Programmed cell death.

aptamer domain A region of an RNA molecule that can fold and bind a molecule, which then affects either the trasncription or translation of the RNA.

archaea Highly specialized, bacteria-like organisms that make up the third kingdom of life on earth along with the bacteria and the eukaryotes. Identified by Carl Woese in 1977 based on rRNA sequences. Most are thermophilic, halophilic, or methanogenic.

architectural protein A protein that binds DNA and changes the conformation of the DNA (e.g., introduces a 90° bend in the DNA).

aster Configuration at the centrosome with microtubules radiating out in all directions.

attenuator region A control region following the promoter of repressible amino acid operons (i.e., the *trp* operon) that exerts transcriptional control of the polycistronic mRNA based on the translation of a small leader peptide gene.

autopolyploidy Polyploidy in which all the chromosomes come from the same species.

autoradiography A technique in which radioactive molecules make their locations known by exposing photographic film.

autosomes The nonsex chromosomes.

auxotroph Mutant bacterial organism that has specific nutritional requirements (i.e., cannot grow on minimal medium).

b

backcross The cross of an individual with one of its parents or with an individual with the same genotype as a parent.

back mutation A nucleotide change in a mutant gene that restores the original sequence and, hence, the wild-type phenotype (also known as reversion).

bacterial lawn A continuous growth of bacteria across the entire surface of a growth medium.

bacteriophages Bacterial viruses.

Balbiani rings The larger polytene chromosomal puffs. *See* chromosome puffs.

Barr body Heterochromatized X chromosome that is found in the nuclei of normal female mammals, but absent in the nuclei of normal males.

base excision repair The DNA excision-repair mechanism that replaces a nucleotide lacking its base (apurinic or apyrimidinic nucleotide) with a complete nucleotide.

base flipping A process whereby enzymes gain access to bases within the DNA double helix by first flipping the bases from the interior to the outside.

basic/helix-loop-helix/leucine zipper A DNA binding motif that consists of basic amino acids, followed by a helix-loop-helix domain and a leucine zipper, which holds two proteins together as a dimer.

bioinformatics The science of storing, retrieving, and analyzing genomic data.

bivalents Structures, formed during prophase of meiosis I, consisting of the synapsed homologous chromosomes. Equivalent to a tetrad of chromatids.

bottleneck A brief reduction in the size of a population, which usually leads to random genetic drift.

bouquet stage A stage during zygonema in which chromosome ends, attached to the nuclear membrane, come to lie near each other.

branch migration The process in which a crossover point between two duplexes slides along the duplexes.

C

cancer An informal term for a diverse class of diseases marked by abnormal cell proliferation.

cap A methylated guanosine added to the 5′ end of eukaryotic mRNA.

capsid The protein shell of a virus.

carcinogen A compund that increases the frequency of cancer in an organism. Many carcinogens are also mutagens, which increases the frequency of mutations in an organism.

carcinoma Tumor arising from epithelial tissue (e.g., glands, breasts, skin, linings of the urogenital and respiratory systems).

carrier An individual who is heterozygous for a recessive mutation.

catabolite activator protein (CAP) A protein that, when bound with cAMP, attaches to specific DNA sequences in sugar-metabolizing operons to activate transcription of these operons.

catabolite repression The suppression of transcription of specific sugar-metabolizing operons when glucose is present in the environment of the cell.

cDNA DNA synthesized from a mRNA template using the reverse transcriptase enzyme.

cell cycle The cycle of cell growth, replication of the genetic material, and nuclear and cytoplasmic division in eukaryotic cells.

centimorgan A chromosome-mapping unit, in which one centimorgan equals 1% recombinant offspring.

central dogma The original postulate that information can be transferred from DNA to RNA and then to protein, barring any transfer originating from the protein.

centric fragment A piece of chromosome containing a centromere.

centrioles Cylindrical organelles, found in eukaryotes (except in higher plants), that reside in the centrosome. Also called basal bodies when they organize flagella or cilia.

centromere Constrictions in eukaryotic chromosomes on which the kinetochore lies. Also, the DNA sequence within the constriction that is responsible for binding the kinetochore.

centrosome The spindle-microtubule organizing center in eukaryotes except for those, such as fungi, that use spindle pole bodies to organize the spindles.

charged tRNA A tRNA that is covalently attached to an amino acid; the form of the tRNA that binds to the mRNA in the ribosome during translation.

checkpoint Points in the cell cycle that can be stopped if certain conditions are not met.

chiasmata X-shaped configurations seen in tetrads during the later stages of prophase I of meiosis. They represent physical crossovers between nonsister chromatids (singular: *chiasma*).

chimeras Individuals made up of two or more genotypes. *See* mosaics.

chloroplast The organelle that carries out photosynthesis and starch grain formation.

chromatids The DNA-protein structure that corresponds to one double-stranded DNA molecule. At anaphase of meiosis II or mitosis, when the sister chromatids separate, each chromatid becomes a chromosome.

chromatin The DNA-protein complex that composes the eukaryotic chromosome.

chromatin remodeling The change in the structure or positioning of nucleosomes, usually to allow transcription.

chromatosome The core nucleosome plus the Hl protein, a unit that includes approximately 168 base pairs of DNA.

chromomeres Dark regions of chromatin condensation in eukaryotic chromosomes at meiosis, mitosis, or endomitosis.

chromosomal theory of inheritance The theory that genes are arranged on the chromosomes.

chromosome The form of the genetic material in viruses and cells. It is circular DNA molecule in most prokaryotes; either a DNA or an RNA molecule in viruses; a linear nucleoprotein complex in eukaryotes.

chromosome puffs Diffuse, uncoiled regions in polytene chromosomes where transcription is actively taking place.

chromosome walking A technique of isolating DNA clones that contain overlapping sequences.

cis Refers to a configuration where the mutations are located on the same chromosome.

cis-acting Mutations in a DNA sequence (e.g., an operator) that control the expression of genes on the same piece of DNA.

cis-trans complementation test A test to determine whether two mutations on opposite chromosomes are alleles; alleles fail to complement each other (do not produce a wild-type phenotype) in the *trans* configuration.

cistron Term coined by Seymour Benzer for the smallest non-complementing genetic unit using the *cis-trans* test; synonymous with *gene*.

cladogenesis The evolutionary process whereby one species splits into two or more species. *See* anagenesis.

clonal evolution theory The theory that cancer develops from sequential mutations (changes) in the genome of a single cell.

clone A group of cells arising from a single ancestor; also a term for an individual who is gentically identical to another individual.

codominance The phenomenon whereby a heterozygous individual expresses the phenotype corresponding to both alleles.

codon preference The situation where amino acids encoded by more than one codon, preferentially will use only a sub-set of the possible codons. The preferred codons usually correspond with abundant tRNAs.

codons The sequences of three RNA nucleotides that specify either an amino acid or termination of translation.

coefficient of coincidence The number of observed double crossovers divided by the number of expected double crossovers, based on the independent probablility of each of the single crossovers.

cointegrate An intermediate structure in transposition.

colony A large number of cells, which grow together on a solid medium, that originated from the same single cell.

complementarity The correspondence of DNA bases in the double helix so that adenine in one strand is opposite thymine in the other strand and cytosine in one strand is opposite guanine in the other. This relationship explains Chargaff's rule.

complementation The production of a wild-type phenotype by a cell or organism that contains two mutant genes. If complementation occurs, then the mutations are not allelic.

complementation group Another term for a cistron or gene.

complete medium A culture medium that is enriched to contain all of the growth requirements of a strain of organisms.

component of fitness A particular aspect in the life cycle of an organism upon which natural selection acts.

composite transposon A transposon constructed of two IS elements flanking a DNA region that occasionally contains antibiotic resistance genes.

concordance The amount of phenotypic similarity between individuals.

conditional-lethal mutant A mutant that is lethal under one condition but viable under another.

conditional mutant A mutant that exhibits the mutant phenotype under one environmental condition and the wild-type phenotype under a different environmental condition. See temperature-sensitive mutant.

conjugation A process whereby two bacterial cells come in contact and exchange genetic material in one direction.

consanguineous Meaning "between blood relatives"; usually refers to inbreeding or incestuous matings.

consensus sequence A sequence of the common nucleotides that are present in the homologous region of many different DNA or RNA samples (e.g., promoters).

conserved sequence A sequence found in many different DNA or RNA samples (e.g., promoters) that is invariant (identical) in the sample.

constitutive heterochromatin Highly condensed chromatin that surrounds the centromere and is thought to be transcriptionally inactive. *See* satellite DNA.

constitutive mutant   A mutant gene that is always transcribed, rather than being under some regulatory control.

contigs   Overlapping genomic DNA clones that cover a complete region of a chromosome.

continuous replication   Uninterrupted DNA replication in the 5′ to 3′ direction that is moving in the same direction as the replication fork.

continuous trait   A phenotype that is measured on a continuum and not defined in discrete units.

continuous variation   Variation measured on a continuum rather than in discrete units or categories (e.g., height in humans).

cooperativity   The ability of a protein to carry out a function and simultaneously aid another protein molecule to more easily perform the same task (e.g., transcription factor binding DNA).

corepressor   The metabolite that when bound to the repressor (of a repressible operon) forms a functional unit that can bind to an operator and block transcription.

correlation coefficient   A statistic that gives a measure of how close two variables are related.

cos **site**   The short single-stranded ends of λ DNA that are complementary and allow the linear λ molecule to circularize after infecting a bacterial cell.

cotransduction   The simultaneous transduction of two or more genes into a bacterial cell.

coupling   Allelic arrangement in which two mutations are on the same chromosome and wild-type alleles are on the homolog.

covariance   A statistical value measuring the simultaneous deviations of x and y variables from their means.

CpG islands   Stretches of CG repeats (in which CpG indicates sequential bases on the same strand of DNA, rather than a C-G basepair). These repeats are found in the promoters and beginning of the transcribed regions of many genes and can be methylated to repress transcription of the gene.

crossover suppression   The apparent lack of crossing over within an inversion loop in heterozygotes. This can be due to either real suppression of crossing over or an apparent suppression resulting from the death of zygotes carrying defective crossover chromosomes.

cyclin   Family of proteins involved in cell cycle control.

cyclin-dependent kinase (CDK)   Family of kinases (phosphorylating enzymes) that, when combined with a specific cyclin, are active in controlling checkpoints in the cell cycle.

cyclosome   Protein complex that breaks down cyclin B and promotes anaphase among its various roles in controlling the cell cycle. *See* anaphase-promoting complex (APC).

cytogenetics   The study of changes in the gross structure and number of chromosomes in cells.

cytokinesis   The division of the cytoplasm of a cell into two daughter cells. *See* karyokinesis.

d

degenerate code   A genetic code in which many different codons specify the same amino acid.

degrees of freedom   An estimate of the number of independent categories in a particular statistical test or experiment.

deleterious allele   A mutation that reduces the viability of a cell or organism.

denatured   The conversion of double-stranded DNA into single-stranded DNA by heating. Can also relate to the loss of any secondary structure in single-stranded RNA.

deterministic   Referring to events that have no random or probabilistic aspects, but proceed in a fixed, predictable fashion.

dicentric chromosome   A chromosome with two centromeres.

dideoxynucleotide   A 2′-deoxyribonucleotide that also lacks the 3′-hydroxyl, which prevents another nucleotide from being added to it during DNA sequencing.

dideoxy sequencing   A method of DNA sequencing that uses chain-terminating dideoxynucleotides.

dihybrid   An organism that is heterozygous at two different loci.

diploid   An organism having two copies of each chromosome per nucleus or cell.

directional selection   A type of selection that removes individuals from one end of a phenotypic distribution and thus causes a shift in the mean of the distribution.

disassortative mating   The mating of two individuals with dissimilar phenotypes.

discontinuous replication   DNA replication in short 5′ to 3′ segments that are moving away from the replication fork.

discontinuous trait   Phenotype that can be classified into discrete categories.

discontinuous variation   Variation that falls into discrete categories (e.g., the green and yellow color of garden peas).

disruptive selection   A type of selection that removes individuals from the center of a phenotypic distribution and thus causes the distribution to become bimodal.

DNA-DNA hybridization   The process of allowing two different single-stranded DNAs to base pair at complementary sequences (e.g., Southern hybridization). This technique is used to determine squence similarities and degrees of repetitiveness between a known sequence (probe) and target DNA sequences.

DNA fingerprint   A pattern of DNA restriction fragments that are separated by gel electrophoresis and identified using a probe for a variable sequence locus.

DNA-RNA hybridization   The process of allowing RNA to base pair with denatured DNA at complementary sequences (e.g., Northern hybridization). This technique is used to determine squence similarities and degrees of repetitiveness between a known sequence (probe) and target DNA sequences.

dominant   An allele that expresses itself even when heterozygous. Also, the trait controlled by that allele.

dominant-negative mutation   A dominant mutation in a gene that exhibits a phenotype that mimics the recessive loss-of-function phenotype.

dosage compensation   A mechanism by which species with sex chromosomes ensure that both the homogametic sex and heterogametic sex exhibit the same level of expression for genes on the sex chromosome.

double digest   The DNA fragments that are produced when two different restriction endonucleases act on the same segment of DNA.

double helix   The normal structure of DNA consisting of two helices rotating about the same axis.

downstream   A convention used when describing the position and direction of transcription by RNA polymerase on the DNA molecule. Downstream relates to the direction of transcription (3′), whereas upstream (5′) is either in the direction from which the polymerase came or farther away from the transcriptional start site before the start of the gene.

downstream promoter element (DPE)   A consensus sequence at about +128 to +134 of RNA polymerase II promoters that have initiator elements, but not TATA boxes.

dyad   Two sister chromatids attached to the same centromere.

e

electrophoresis   The separation of molecules by an electrical current.

electroporation   A technique to introduce DNA into cells by applying high-voltage electrical pulses.

elongation complex   The form of RNA polymerase II that actively carries out basal transcription.

elongation factors (EF-Ts, EF-Tu, EF-G)   Proteins that are necessary for the proper elongation and translocation processes at the ribosome during prokayotic translation.

embryonic stem (ES) cells   Cells that have the capacity to differentiate into any cell of an organism; a totipotent cell that is used in creating a mouse knockout mutant.

endogenote   Bacterial host chromosome.

endomitosis   Chromosomal replication without nuclear or cellular division that results in cells with many copies of the same chromosome, such as in the salivary glands of *Drosophila*.

endonucleases Enzymes that make nicks within the phosphodiester backbone of DNA.

endosymbiotic theory A theory that describes that eukaryotic organelles (e.g., mitochondria and chloroplasts) originated as a prokaryotic cell that was engulfed by a eukaryotic cell.

enhancer A eukaryotic DNA sequence that increases transcription of a gene by binding specific transcription factors.

epigenetic effect An environmentally induced change (e.g., DNA methylation) in the genetic material without altering the DNA sequence. Generally, a phenomenon of differential expression of alleles of a locus depending on the parent of origin.

epistasis The masking of the phenotype of one gene by allelic combinations at a second gene.

equational division The second meiotic division that does not reduce the number of chromosomes in the cell, but does reduce the amount of DNA by half.

euchromatin Eukaryotic chromosomal regions that are diffuse during interphase and presumably are the regions that are actively transcribed.

eukaryotes Organisms with true nuclei.

euploidy The condition of a cell or organism that has one or more complete sets of chromosomes.

evolution A change in phenotypic frequencies in a population.

evolutionary rates The rate of divergence between taxonomic groups, measured as the number of amino acid substitutions per million years.

excision repair A process whereby cells remove part of a damaged DNA strand and replace it through DNA synthesis, using the undamaged strand as a template.

exogenote The DNA that a bacterial cell has taken up through a sexual process (e.g., transduction).

exon In a eukaryotic gene that has intervening sequences (introns), a region that is present in the functional RNA molecule that is exported from the nucleus.

exonucleases Enzymes that hydrolyze the phosphodiester bond from either the 5' or 3' ends of polynucleotides to release nucleotides from a linear DNA or RNA molecule.

expressed sequence tag (EST) A nucleotide sequence that is generated from an end of a cDNA, which corresponds to a mRNA that was transcribed in a specific tissue.

expressivity The degree of expression of a genetically controlled trait.

f

feedback inhibition A posttranslational control mechanism in which the end product of a biochemical pathway inhibits the activity of the first enzyme of the same pathway.

fertility factor The plasmid (F plasmid) that allows a prokaryote to engage in conjugation with, and pass DNA into an F⁻ cell.

fitness The relative reproductive success (W) of a genotype as measured by survival, fecundity, or other life-history parameters.

5' untranslated region (5' UTR) The sequence on an mRNA that is located from the +1 nucleotide (first nucleotide transcribed) to the translation initiation codon.

fluorescent in situ hybridization (FISH) A technique in which a fluorescent dye is attached to a nucleotide probe that then binds to a specific site on a chromosome and makes itself visible by its fluorescence.

footprinting A techique to determine the length and sequence of a nucleic acid that is in contact with a protein. While the DNA is bound by the protein, it is digested with nucleases that degrade the unbound DNA. The protected DNA can be isolated and analyzed.

founder effect Genetic drift observed in a population founded by a small, nonrepresentative sample of a larger population.

F-pili Hairlike projections on an F⁺ or Hfr bacterium involved in anchorage to a F⁻ cell during conjugation.

frameshift A mutation in which there is an addition or deletion of nucleotides that causes the codon reading frame to shift.

frequency-dependent selection A selection whereby a genotype is at an advantage when rare and at a disadvantage when common.

g

gain-of-function allele A mutation that results in increased expression or activity in the encoded protein or the expression of the wild-type protein in a new time or location in the organism; often associated with a dominant phenotype.

gamete A germ cell having a haploid chromosomal complement. Gametes from parents of opposite sexes fuse to form zygotes.

gametic selection The forces acting to cause differential reproductive success of one allele over another in a heterozygote.

gametophyte The haploid stage of a plant life cycle that produces gametes (by mitosis). The gametophyte alternates with a diploid, sporophyte generation.

G-bands Eukaryotic chromosomal bands produced by treatment with Giemsa stain.

gene Inherited determinant of the phenotype.

gene amplification A process or processes by which the cell increases the number of repeats of a particular gene within the genome.

gene family A group of genes that has arose by duplication of an ancestral gene. The genes in the family may or may not have diverged.

gene flow The movement of genes from one population to another by interbreeding between individuals in the two populations.

gene pool All of the alleles available among the reproductive members of a population from which gametes can be drawn.

generalized transduction The transfer of any bacterial genomic region by particular bacteriophages.

genetic code The linear sequence of three-nucleotide long segments (codons) that specify the amino acids during the process of translation at the ribosome.

genetic load The relative decrease in the mean fitness of a population due to the presence of genotypes that have less than the highest fitness.

genetic polymorphism The occurrence together in the same population of more than one allele at the same locus, with the least frequent allele occurring more frequently than can be accounted for by mutation.

genic balance theory Bridges's theory that the sex of a fruit fly is determined by the ratio between the number of X chromosomes and number of sets of autosomal chromosomes.

genome The entire genetic complement of a prokaryote or virus or the haploid genetic complement of a eukaryote.

genomics The study of the mapping and sequencing of genomes of an organism.

genotype The genes that an organism possesses.

Giemsa stain A complex of stains specific for the phosphate groups of DNA.

group I introns Self-splicing introns that require an external guanine-containing nucleotide for splicing; the intron is released in a linear form.

group II introns Self-splicing introns that do not require an external nucleotide for splicing; the intron is removed in a lariat form.

G-tetraplex A structure of four guanines that can base pair to form a novel planar structure that may be involved with stabilizing the end of eukaryotic chromosomes in some organisms.

GTPase An enzymatic activity that hydrolyzes a high-energy bond in GTP and transfers the released energy to a separate reaction.

guide RNA (gRNA) RNA that base pairs with a different mRNA to idnetify where uridines are to be inserted into the mRNA (RNA editing) after transcription.

gynandromorphs Mosaic individuals having simultaneous aspects of both the male and the female phenotype.

h

haploid The state of having one copy of each chromosome per nucleus or cell.

haploinsufficiency In a diploid organism, the requirement of two copies of a specific chromosomal region or gene for the cell or organism to survive. A single wild-type copy of this region or gene results in death.

helix-turn-helix motif A DNA binding protein motif that consists of an alpha helix that binds the DNA and a second alpha helix that stabilizes the protein.

hemizygous The condition of loci present in only one copy in a diploid organism, such as loci on the X chromosome of the heterogametic sex of a diploid species.

heritability A measure of the degree to which the variance in the distribution of a phenotype is due to genetic causes. In the broad sense, it is measured by the total genetic variance divided by the total phenotypic variance. In the narrow sense, it is measured by the genetic variance due to additive genes divided by the total phenotypic variance.

heterochromatin Chromatin that remains tightly coiled (and darkly staining) throughout the cell cycle.

heteroduplex DNA Double-stranded DNA where each strand is derived from a different DNA molecule.

heterogametic sex The sex with two different sex chromosomes (e.g., males in mammals); during meiosis, the sex that produces two different types of gametes relative to the sex chromosomes.

heterogeneous nuclear mRNA (hnRNA) The original RNA transcripts found in eukaryotic nuclei before posttranscriptional modifications.

heterokaryon A cell that contains two or more nuclei from different origins.

heteroplasmy The existence of an individual or organism possessing genetic heterogeneity within the populations of mitochondria or chloroplasts.

heterozygote A diploid or polyploid with different alleles at a particular locus.

heterozygote advantage A selection model in which heterozygotes have the highest fitness.

Hfr High frequency of recombination. A strain of bacteria that has incorporated an F factor into its genome and then transfers the genomic DNA during conjugation.

histone acetyl transferases (HATs) Proteins that remodel chromatin by acetylating histones.

histones Arginine- and lysine-rich basic proteins making up a substantial portion of eukaryotic nucleoprotein.

Hogness box See TATA box.

holandric trait Trait controlled by a locus found only on the Y chromosome. Involves father-to-son transmission.

Holliday structure A junction between two cross-linked DNA double helices, usually at an intermediate stage in DNA recombination.

holoenzyme The complete enzyme, including all subunits.

homeobox A consensus sequence of about 180 base pairs discovered in *Drosophila* homeotic genes and later found in other developmentally important genes from yeast to human beings.

homeodomain The sixty amino acid peptide that is encoded in the homeobox.

homeotic gene Gene that controls the developmental fate of a cell type; mutations of the homeotic gene cause one cell type to follow the developmental pathway of another cell type.

homogametic sex The sex with two copies of the same sex chromosome (e.g., females in mammals); during meiosis, the sex that produces only one type of gamete relative to the sex chromosomes.

homologous chromosomes Members of a pair of essentially identical chromosomes that synapse during meiosis.

homologous recombination Breakage and reunion between homologous lengths of DNA mediated by RecA and RecBCD.

homoplasmy The existence of an individual containing plastids with only one genotype; usually referring to a single genetic identity in either the mitochondrial or chloroplast genomes.

homozygote A diploid or a polyploid with identical alleles at a locus.

hybrid Offspring of unlike parents.

hyperplasia Excessive cell growth that does not involve pathological changes to the cells.

hypervariable loci Loci with many alleles; especially those whose variation is due to variable numbers of tandem repeats.

hypomorph An allele that encodes a protein with reduced activity relative to the wild-type allele.

hypostatic gene A gene whose expression is masked by an epistatic gene.

i

idiogram A photograph or diagram of the chromosomes of a cell arranged in an orderly fashion. See karyotype.

imprinting The phenomenon in which there is differential expression of a gene depending on whether it was maternally or paternally inherited.

imprinting center (IC) A genomic DNA region that controls imprinting. The imprinting mark is almost certainly DNA methylation, which is able to turn off gene transcription.

inbreeding The mating of genetically related individuals.

inclusive fitness The expansion of the concept of the fitness of a genotype to include benefits accrued to relatives of an individual since relatives share parts of their genomes. An apparently altruistic act toward a relative may thus enhance the fitness of the individual performing the act.

incomplete dominance The situation in which both alleles of the heterozygote influence the phenotype, which is usually intermediate between the two homozygous phenotypes.

incomplete penetrance A phenomenon where a particular mutant genotype does not always exhibit the corresponding mutant phenotype.

inducer A small molecule that is bound by a DNA binding protein to activate transcription of a specific gene.

induction In regard to temperate phages, the process of causing a prophage to become virulent.

initiation codon The mRNA sequence AUG, which specifies methionine, the first amino acid used in the translation process.

initiation complex The complex formed for initiation of translation. In prokaryotes, it consists of the 30S ribosomal subunit, mRNA, N-formyl-methionine transfer RNA, and three initiation factors.

initiation factors (IFI, IF2, IF3) Proteins (prokaryotic with eukaryotic analogs) required for the proper initiation of translation.

initiator proteins Proteins that recognize the origin of replication on a replicon and take part in primosome construction.

initiator region (InR) A CT-rich area found in RNA polymerase II promoters that lack TATA boxes.

insertion sequences (IS) Small, simple transposons. *See* transposable genetic element.

intercalary heterochromatin Heterochromatin, other than centromeric heterochromatin, dispersed throughout eukaryotic chromosomes.

interference The effect of one crossing over effect has on a second crossing over event in the general region.

intergenic suppression A mutation at one locus that apparently restores the wild-type phenotype to a mutant at a different locus.

internal ribosome entry site (IRES) Sequence in eukaryotic mRNAs that allows ribosomes to initiate translation at a point other than the 5' cap.

interpolar microtubules Microtubules extending from one pole of the spindle and overlapping spindle fibers from the other pole, but not in contact with kinetochores.

intersex An organism with external sexual characteristics of both sexes.

intragenic suppression A second mutation within a mutant gene that results in the restoration of the original wild-type phenotype.

intron DNA sequences within a gene that are transcribed but removed during the posttranscriptional processing of the mRNA and before translation.

inversion The replacement of a section of a chromosome in the opposite orientation.

inverted repeat sequence A nucleotide sequence read in opposite orientations on the same double helix.

isochromosome A chromosome with two genetically and morphologically identical arms.

k

karyokinesis The process of nuclear division. *See* cytokinesis.

karyotype The chromosome complement of a cell.

kinetochore A complex of proteins that are associated with the centromere and attaches to the spindle fibers.

kinetochore microtubules Microtubules radiating from the centrosome and attached to kinetochores of chromosomes during mitosis and meiosis.

Klenow fragment Proteolytic fragment of *E. coli* DNA polymerase I that possesses both 5' to 3' polymerase activity and 3' to 5' exonuclease activity.

knockout mice Mice that were generated from ES cells that were manipulated to be homozygous for a DNA insertion into a specific gene.

l

lagging strand The DNA strand that is replicated discontinuously away from the replication fork.

lampbrush chromosomes Chromosomes of amphibian oocytes that have loops of actively transcribed genes that appear like an old-time lampbrush.

leader The length of mRNA from the 5' end to the initiation codon, AUG. *See* 5' UTR.

leader peptide gene A small gene within the attenuator control region of several repressible operons that are required for amino acid biosynthesis (e.g., tryptophan). Translation of the encoded peptide assesses the concentration of the amino acid in the cell and, in turn, acts to regulate the transcription of the operon.

leading strand Strand of DNA being replicated continuously towards the replication fork.

lethal alleles Mutations that cause the death of the cell or organism.

leucine zipper A DNA-binding protein motif in which leucines are located along one side of an alpha helix, which allows two proteins to interact by interdigitating the leucines on each protein.

level of significance The probability value in statistics used to reject the null hypothesis.

linkage groups Associations of loci on the same chromosome. In a species, there are as many linkage groups as there are homologous pairs of chromosomes.

linkage map The organization of loci along a chromosome, often with the relative distance between loci included.

linkage number The number of times one strand of a helix coils about the other.

linker A synthesized short DNA segment that contains a restriction site that can be ligated to blunt-ended DNA fragment to generate a restriction site at the end of the DNA fragment.

locus The position of a gene on a chromosome (plural: *loci*).

lod score The logarithm of odds that two loci are a specified recombination distance apart based on several independent pedigrees.

long interspersed elements (LINEs) Sequences of repetitive DNA, up to seven thousand base pairs in length, interspersed in eukaryotic chromosomes in many copies.

loss-of-function allele A mutation that exhibits reduced activity relative to wild type; usually associated with a recessive mutation.

Lyon hypothesis The hypothesis that the Barr body is an inactivated X chromosome.

lysate The contents released from a lysed cell.

lysis The breaking open of a cell by the destruction of its wall or membrane.

lysogen A bacterial cell that has an integrated phage (prophage) in its chromosome.

m

map unit The distance equal to 1% recombination between two loci.

maternal-effect gene A gene expressed in maternal tissue that influences the phenotype of the developing embryo.

maternal inheritance The transfer of the maternal parent's genotype to her offspring.

mating type In many species of microorganisms, individuals can be divided into two mating types. Mating can take place only between individuals of opposite mating types due to the interaction of cell surface components.

maturation-promoting factor (MPF) A protein complex of cyclin B and Cdc2p that initiates mitosis during the cell cycle. Also called the mitosis-promoting factor.

mean The arithmetic average; the sum of the data divided by the sample size.

mean fitness of the population The sum of the fitnesses of the genotypes of a population weighted by their proportions; hence, a weighted mean fitness.

median The central value in a set of numbers.

meiosis The nuclear process in eukaryotes that results in gametes or spores with half of the original number of chromosome sets per nucleus.

merodiploid A eukaryotic cell that is diploid for a limited amount of genomic DNA.

merozygote A bacterial cell that has a second copy of a particular chromosomal region, usually in the form of an exogenote on an F′ plasmid.

messenger RNA (mRNA) A RNA that is translated into a polypeptide at the ribosome.

metacentric chromosome A chromosome whose centromere is located in the middle.

metafemale A fruit fly with an X/A ratio greater than 1.0. Metafemales usually die prior to hatching as adults.

metamale A fruit fly with an X/A ratio below 0.5.

metaphase plate The plane of the equator of the spindle into which chromosomes are positioned during metaphase.

metastasis The migration of cancerous cells to other parts of the body.

micro RNA (miRNA) Short RNAs (21-23 nucleotides long that are associated with a protein complex (RISC) that can base pair with a complementary mRNA and block translation.

microsatellite DNA Repeats of very short sequences of DNA, such as CACACACA, dispersed throughout the eukaryotic genome. The loci can be studied by polymerase chain reaction amplification.

microtubule organizing center Active center from which microtubules are organized, such as the centrosome.

microtubules Hollow cylinders that are composed of the α and β subunits of tubulin, that produce, among other things, the spindle fibers.

mimicry A phenomenon in which an individual gains an advantage by looking like the individuals of a different species.

minisatellite Short DNA sequence of 10 to 100 nucleotides that is repeated in a tandem array at multiple loci in the eukaryotic genome.

minimal medium A culture medium for microorganisms that contains the minimal necessities for growth of wild-type individuals.

missense mutation Mutations that change a codon for one amino acid into a codon for a different amino acid.

mitochondrial DNA The genomic DNA, which is usually circular, that is present at multiple copies in every mitochondrion in a cell.

mitochondrion The eukaryotic cellular organelle in which the Krebs cycle and electron transport reactions take place to produce energy for the cell.

mitosis The nuclear division producing two daughter nuclei identical to the original nucleus.

mode The most frequent value in a set of numbers.

molecular chaperone A protein that aids in the folding of a second protein. The chaperone prevents proteins from forming structures that would be inactive.

molecular evolutionary clock A measurement of evolutionary time in nucleotide substitutions per year.

molecular mimicry The situation in which one type of molecule resembles another type in order to function. For example, the prokaryotic ribosomal release factors, RF1 and RF2, mimic the structure of a tRNA.

monohybrids Offspring of parents that differ in only one genetic characteristic. Usually implies heterozygosity at a single locus under study.

monosomic A diploid cell missing a single chromosome.

monovalent A single chromosome composed of two sister chromatids. Equivalent to a dyad.

morphogen A substance transported into or produced in a developing embryo that diffuses to form a gradient that helps determine cell differentiation.

morpholino An in vitro synthesized oligonucleotide that consists of a morpholine ring instead of deoxyribose (or ribose) that can base pair with a mRNA and block translation.

morphological species concept The idea that organisms are classified as the same species if they appear similar.

mosaics Individuals made up of two or more cell lines with different genotypes that originated in the same zygote.

multihybrid An organism that is heterozygous at numerous loci.

multiple-hit hypothesis The hypothesis that tumor development requires mutations in several different genes.

multiplicity of infection (moi) The ratio of number of bacteriophage to bacterial cells. A high moi increases the odds that a bacterial cell will be infected with multiple phages.

mutability The ability to change.

mutagen A substance that induces a mutation in DNA at a frequency that is greater than the generation of a spontaneous mutation.

mutant Alternative phenotype to the wild-type; the phenotype produced by alternative alleles.

mutation The process by which a gene or chromosome changes structurally; the end result of this process.

mutation rate The proportion of mutations per cell division in bacteria or single-celled organisms or the proportion of mutations per gamete in higher organisms.

n

natural selection The process in nature whereby one genotype leaves more offspring than another genotype because of superior life history attributes, such as survival of fecundity.

negative interference The phenomenon whereby a crossover event in a particular region facilitates the occurrence of another crossover in the same chromosomal region.

negative regulation The effect of a protein binding to DNA and repressing the transcription of a nearby gene.

N-end rule The life span of a protein is determined by its amino-terminal (N-terminal) amino acid.

neomorphic allele A mutation that encodes a protein with a function that is different than the wild type or expresses the wild-type protein in a new location or developmental time in the organism.

neoplasm New growth of abnormal tissue.

neutral gene hypothesis The hypothesis that most genetic variation in natural populations is not maintained by natural selection.

nondisjunction The failure of a pair of homologous chromosomes to separate properly during meiosis.

nonhistone proteins The proteins remaining in chromatin after the histones are removed.

nonparentals Another term for recombinants.

nonrecombinants In mapping studies, offspring that have the parental allele arrangements.

nonsense codon One of the three mRNA sequences (UAA, UAG, UGA) that signals the termination of translation.

nonsense mutations Mutations that change a codon for an amino acid into a nonsense codon.

nonsister chromatids Chromatids that represent different members of a homologous pair.

nontemplate strand DNA strand that does not serve as the template for transcription. The nontemplate strand will have the same sequence as the transcribed RNA, except for thymidines in place of uracils.

normal distribution A bell-shaped frequency curve with the relative position and shape defined by the mean and the standard deviation.

Northern blot A process where a membrane containing RNA, which was separated by gel electrophoresis, is then hybridized with a nucleic acid probe to identify specific RNAs on the membrane.

nuclease-hypersensitive site A region of a eukaryotic chromosome that is specifically vulnerable to nuclease attack because it is not wrapped as nucleosomes.

nucleolar organizer The chromosomal region that contains tandem repeats of the major rRNA gene, around which the the nucleolus forms.

nucleolus The globular, nuclear subdomain formed at the nucleolar organizer. Site of ribosome construction.

nucleoprotein The substance of eukaryotic chromosomes consisting of proteins and nucleic acids.

nucleoside A sugar-base compound that is a nucleotide precursor. Nucleotides are nucleoside phosphates.

nucleosome The basic structure of eukaryotic chromatin that is composed of approximately 165 base pairs of DNA wrapped around a histone structure that consists of two molecules each of H2a, H2b, H3, and H4.

nucleotide Basic structure of DNA and RNA that consists of a nitrogenous base, a sugar, and one or more phosphates.

nucleotide excision repair The DNA repair mechanism responsible for repairing thymine dimers and other lesions. Enzymes excise a short segment of one of the DNA strands and then repair and ligate the DNA.

null allele A mutant version of a gene that either does not express a protein or the protein entirely lacks any activity.

null hypothesis The statistical hypothesis that there are no differences between the observed data and the expected data based on the assumption of no experimental effect.

nullisomic A diploid cell missing both copies of the same chromosome.

o

Okazaki fragments Segments of newly replicated DNA produced during discontinuous DNA replication on the lagging strand.

oligonucleotide A short DNA fragment that is usually synthesized in vitro.

oncogene Genes capable of transforming a cell when expressed at a high level or constitutively. In the normal (nontransformed) cell, they are called proto-oncogenes.

oogenesis The process of ovum formation in female animals.

open reading frames (ORFs) Sequence of codons between the translation initiation and termination codons in a gene.

operator A DNA sequence that is recognized by DNA binding proteins to regulate the expression of a gene.

operon A sequence of adjacent genes that are transcribed on a polycistronic mRNA and are under a common mechanism of transcriptional regulation.

outbreeding The mating of genetically unrelated individuals.

ovum Egg. The one functional product of each meiosis in female animals.

p

palindrome A sequence of nucleotides that reads the same regardless of the direction from which one starts; the sites of recognition of type II restriction endonucleases.

paracentric inversion A chromosomal inversion that does not include the centromere in the inverted region.

parapatric speciation Speciation in which reproductive isolating mechanisms evolve when a population enters a new niche or habitat within the range of the parent species.

parentals In mapping studies, offspring that have the parental allele arrangements. *See* nonrecombinants.

partial digest A restriction digest that has not been allowed to go to completion, and thus contains pieces of DNA that have restriction endonuclease sites that have not been cleaved.

pedigree A representation of the ancestry of an individual or family; a family tree.

penetrance The production of a phenotype that corresponds to the organism's genotype.

pericentric inversion A chromosomal inversion that includes the centromere within the inverted region.

permissive temperature A temperature at which temperature-sensitive mutants exhibit the wild-type phenotype.

PEST hypothesis Rapid protein degradation that is signaled by a region that is enriched in proline (P), glutamic acid (E), serine (S), and threonine (T) amino acids.

phenocopy A phenotype that is controlled by the environment, but looks like a genetically controlled phenotype.

phenotype The observable traits of an organism.

phyletic gradualism The process of evolutionary change over a long period of time.

phylogenetic tree A diagram showing evolutionary lineages of organisms.

physical map Chromosomal map of DNA markers (e.g., microsatellites, expressed sequence tags) in which distances are in physical units of base pairs.

plaques Clear areas on a bacterial lawn caused by cell lysis due to phage attack.

plastid A chloroplast prior to the development of chlorophyll.

pleiotropy The phenomenon whereby a single mutation affects several apparently unrelated aspects of the phenotype.

pluripotent A cell that has the ability to differentiate into a wide, but not all, cell types in an organism.

point centromere The type of centromere, such as that found in Saccharomyces cerevisiae, that has defined sequences large enough to accommodate the binding of one spindle microtubule.

point mutations Small mutations that consist of a replacement, addition, or deletion of one or a few bases.

polar bodies The small cells that are the by-products of meiosis in female animals. One functional ovum and as many as three polar bodies result from meiosis of each primary oocyte.

pollen grain The male gametophyte in higher plants.

poly(A) tail A sequence of addenosine nucleotides that are added to the 3' end of eukaryotic mRNAs posttranscriptionally.

polycistronic mRNA Prokaryotic mRNAs that contain several genes within the same mRNA transcript.

polygenic inheritance The mechanism of genetic control of traits that exhibit continuous variation due to the effect of several different genes.

polymerase chain reaction (PCR) A method to amplify DNA segments rapidly in temperature-controlled cycles of DNA denaturation, oligonucleotide base pairing, and DNA synthesis.

polymerase cycling The process by which a DNA polymerase III enzyme completes an Okazaki fragment, releases it, and begins synthesis of the next Okazaki fragment.

polyploid Organism with greater than two chromosome sets.

population A group of organisms of the same species relatively isolated from other groups of the same species.

positive interference When the occurrence of one crossover event reduces the probability that a second crossover will occur in the same region.

positive regulation The binding of a protein to DNA activates transcription of a nearby gene.

posttranscriptional modifications Changes made to eukaryotic mRNA after transcription is completed (e.g. addition of the 5' cap, polyadenylation of 3' end, and removal of introns).

pre-initiation complex (PIC) The form of the RNA polymerase II enzyme with general transcription factors bound to the DNA. This is equivalent to the *E. coli* RNA polymerase enzyme in the closed promoter complex.

pre-mRNA The RNA molecule that is transcribed from eukaryotic DNA before any posttranscriptional modifications.

Pribnow box Consensus sequence of TATAAT in prokaryotic promoters centered at the position -10.

primary transcript The RNA molecule that is transcribed from eukaryotic DNA before any posttranscriptional modifications.

primase A RNA polymerase that creates the short RNA primer for initiation of Okazaki fragment synthesis during DNA replication.

primer In DNA replication, a length of DNA or RNA, which is base paired to a single-stranded DNA template, that provides a 3' end for the addition of another nucleotide.

primosome A complex of two proteins, a primase and helicase, that initiates RNA primers on the lagging DNA strand during DNA replication.

probability The expectation of the occurrence of a particular event.

probe In recombinant DNA work, a labeled (e.g., radioactive) nucleic acid that is used to identify a complementary DNA or RNA sequence by base pairing.

processivity The ability of an enzyme to repetitively continue its catalytic function without dissociating from its substrate.

product rule The rule that states that the probability that two independent events will both occur is the product of their separate probabilities.

prokaryotes Organisms that lack true nuclei.

promoter A DNA region that RNA polymerase binds to in order to initiate transcription.

prophage A temperate phage that has integrated into the bacterial host's genomic DNA.

propositus (proposita) The person through whom a pedigree was discovered.

proteasome A barrel-shaped cellular organelle that degrades proteins through the ubiquitin pathway.

proteome From *prote*ins of the gen*ome*; the complete set of proteins encoded in a particular genome. It is the protein analog to "genome."

proteomics The study of the complete set of proteins from a particular genome. It is the protein analog to "genomics."

proto-oncogene A cellular oncogene in an untransformed cell.

prototrophs Strains of organisms that can survive on minimal medium.

pseudoaautosomal gene A gene that occurs on both sex-determining heteromorphic chromosomes.

pseudodominance The phenomenon in which a recessive allele confers the phenotype when only one copy of the allele is present, as in hemizygous alleles or in deletion heterozygotes.

punctuated equilibrium The evolutionary process involving long periods without change (stasis) punctuated by short periods of rapid speciation.

purine A nitrogenous base, either adenine or gaunine, that is found in DNA and RNA.

pyrimidine Nitrogenous base of which thymine is found in DNA, uracil in RNA, and cytosine in both.

q

quantitative trait A phenotype that is exhibits a continum of values, rather than discrete values, that is often controlled genetically by several different genes.

quantitative trait loci (QTL) Chromosomal regions contributing to the inheritance of a quantitative trait. These regions may contain one or more polygenes that contribute to the phenotype.

quaternary structure The association of polypeptide subunits to form the final structure of a protein.

r

random genetic drift Changes in allelic frequency due to sampling error.

random mating The mating of individuals in a population such that the union of individuals with the trait under study occurs according to the product rule of probability.

realized heritability Heritability measured by a response to selection.

recessive An allele (or phenotype) that does not express itself in the heterozygous condition.

reciprocal cross A cross with the parental phenotypes of each sex reversed relative to the original cross; used to test the potential role of the sex chromosomes in generating the phenotype.

reciprocal translocation A chromosomal configuration in which the ends of two nonhomologous chromosomes are broken off and become attached to the nonhomologs.

recombinants In mapping studies, offspring with allelic arrangements made up of a combination of the original parental alleles.

recombination nodule Proteinaceous nodules found on bivalents during zygonema and pachynema associated with crossing over.

reductional division The first meiotic division that reduces the number of chromosomes and centromeres to half relative to the original cell.

regional centromere The type of centromere found in higher eukaryotes that can accommodate several spindle microtubules.

repetitive DNA DNA made up of copies of the same nucleotide sequence.

replicons A replicating genetic unit including a length of DNA and its site for the initiation of replication.

replisome The DNA-replicating structure at the Y-junction, consisting of two DNA polymerase III enzymes and a primosome (primase and DNA helicase).

repressor The protein product of a regulator gene that binds to specific sequences in the genomic DNA to inhibit the expression of inducible and repressible operons and genes.

reproductive isolating mechanisms Environmental, behavioral, mechanical, and physiological barriers that prevent two individuals of different populations from producing viable progeny.

reproductive success The relative production of offspring by a particular genotype.

repulsion Allelic arrangement in which the two mutations are located on homologous chromosomes.

restriction endonucleases Endonucleases that recognize specific DNA sequences, then cleave them. They protect cells from viral infection.

restriction fragment length polymorphism (RFLP) Variations (among individuals) in banding patterns of electrophoresed restriction digests due to the presence or absence of a restriction enzyme site.

restriction map A physical map of a segment of DNA showing the location and distance between restriction enzyme sites.

restriction site The sequence of DNA recognized by a restriction endonuclease.

restrictive temperature A temperature at which a temperature-sensitive mutant displays the mutant phenotype.

retrotransposons Transposable genetic elements found in eukaryotic DNA that move through the reverse transcription of an RNA intermediate.

reverse transcriptase An enzyme that can use RNA as a template to synthesize DNA.

reverse translation The prediction of the DNA sequence encoding a protein based on knowing the amino acid sequence of a protein and the genetic code.

reversion The return of a mutant to the wild-type phenotype by way of a second mutational event.

rho-dependent terminator A prokaryotic RNA sequence signaling the termination of transcription; termination requires the presence of the rho protein and a stem-loop structure in the RNA.

rho-independent terminator A stem-loop structure followed immediately by a string of uracils in a prokaryotic RNA that signal transcription termination in the absence of the rho protein.

rho protein A protein involved in the termination of some prokaryotic RNAs.

ribosomal RNA (rRNA) RNA components of the subunits of the ribosomes.

ribosomes Complexes at which translation takes place. They are made up of two subunits that are each composed of proteins and rRNAs.

ribozyme Catalytic or autocatalytic RNA.

RNA editing The insertion of uridines into mRNAs after transcription is completed; controlled by guide RNA. May also involve insertion of cytidines in some organisms or possible deletions of bases.

RNA polymerase The enzyme that polymerizes RNA by using DNA as a template.

Robertsonian fusion Fusion of two acrocentric chromosomes at their centromere.

rolling-circle replication A model of DNA replication that accounts for a circular DNA molecule (e.g., F plasmid) producing linear daughter double helices.

s

satellite DNA Highly repetitive eukaryotic DNA that is primarily located around centromeres.

scaffold The eukaryotic chromosomal structure that remains when DNA and histones have been removed.

scanning hypothesis Proposed mechanism by which the eukaryotic ribosome binds the capped 5' end of the mRNA and then moves along the mRNA to the first AUG initiation codon, where translation begins.

securin An inhibitory protein that prevents separin from acting on cohesin to separate sister chromatids.

segmentation genes Genes of developing embryos that determine the number and fate of segments.

selection coefficients (s, t) The sum of forces acting to lower the relative reproductive success of a genotype.

selection-mutation equilibrium An equilibrium allelic frequency resulting from the balance between selection against an allele and mutation re-creating this allele.

selective medium A culture medium enriched with a particular substance to allow the growth of particular strains of organisms.

selfed Fertilization in which the two gametes come from the same individual.

semiconservative replication The mode by which DNA replicates, where each strand acts as a template for a new double helix.

semisterility Nonviability of a proportion of gametes or zygotes.

separin An enzyme that breaks down cohesin and allows sister chromatids to separate at the start of anaphase of mitosis.

sequence-tagged sites (STSs) DNA sequences of 100-500 base pairs that are produced from randomly selected genomic DNA clones.

sex chromosomes Heteromorphic chromosomes whose distribution in a zygote determines the sex of the organism.

sex-determining region Y (SRY) The region on the Y chromosome that determines the male sex in mice. The equivalent gene in humans is the testis-determining factor gene (TDF).

sexduction A process whereby a bacterium gains access to and incorporates foreign DNA brought in by a F' vector during conjugation.

sex-influenced traits Traits that are controlled by autosomal genes, but exhibit different genetic properties in the two sexes (e.g., dominant in males and recessive in females).

sex-limited traits Traits that are expressed in only one sex, even though the corresponding gene is located on the autosome.

sex-linked The inheritance pattern of loci located on the sex chromosomes (usually the X chromosome in XY species); also refers to the loci themselves.

sex switch A gene in mammals, normally found on the Y chromosome, that directs the indeterminate gonads towards development as testes.

Shine-Dalgarno sequence A prokaryotic mRNA sequence that is upstream of the translation initiation codon that base pairs with the 3' end of the 16S rRNA in the small ribosomal subunit at the beginning of translation.

shotgun cloning The random cloning of pieces of genomic DNA from an organism without regard to the genes or sequences present in the cloned DNA.

shunting Process by which the first AUG codon in the mRNA is bypassed as the translational start codon for a AUG codon further downstream on the mRNA. This process is likely guided by the secondary structure of the mRNA.

siblings (sibs) Brothers and sisters.

sigma factor The protein, which is a subunit of RNA polymerase, that recognizes and binds the bacterial promoter.

single-nucleotide polymorphisms (SNPs) DNA differences between individuals that involve single base pairs that are located approximately every 1000 bases along the human genome. SNPs are useful for mapping disease genes.

sister chromatids Chromatids that are joined at the centromere, they represent the two DNA molecules that are the products of replicating a single DNA molecule.

site-specific recombination A crossover event, such as the integration of phage λ, that requires homology between the two DNA molecules and uses an enzyme specific for that recombination.

small nuclear ribonucleoproteins (snRNPs) Components of the spliceosome, the intron-removing apparatus in eukaryotic nuclei.

small nuclear RNAs (snRNAs) RNAs that are part of the spicesome, which are complementary to specific sequences on the pre-mRNA (e.g., the 5' and 3' splice junctions).

small nucleolar ribonucleoprotein particles (snoRNPs) Particles composed of RNA and protein found in the nucleolus that modify ribosomal RNAs, particularly by converting some uridines to pseudouridines and methylating some ribose sugars.

small nucleolar RNAs (snoRNAs) RNAs found in small nucleolar ribonucleoprotein particles (snoRNPs) that take part in modifying ribosomal RNA in the nucleolus.

somatic doubling A disruption of the mitotic process that produces a cell with twice the normal chromosome number.

Southern blotting A process where a membrane (e.g., nitrocellulose), which contains DNA fragments that were separated by agarose gel electrophoresis, is then hybridized to either a labeled DNA or labeled RNA probe to identify the complementary DNA fragment. The method is named after the developer of the technique, E. M. Southern.

specialized transduction Form of transduction based on faulty looping out by a temperate phage. Only neighboring loci to the attachment site can be transduced.

speciation A process whereby, over time, one species evolves into a different species (anagenesis) or one species diverges to become two or more species (cladogenesis).

species A group of organisms capable of interbreeding to produce fertile offspring.

spermatogenesis The process of sperm production.

sperm cells The gametes of males.

spindle fiber The microtubules that bind the kinetochore and drive the chromosomal movement during mitosis and meiosis.

spindle pole body Spindle microtubule organizing center found in fungi.

spliceosome Protein-RNA complex that removes introns in eukaryotic nuclear RNAs.

sporophyte The stage of a plant life cycle that produces spores by meiosis and alternates with the gametophyte stage.

stabilizing selection A type of selection that removes individuals from both ends of a phenotypic distribution, thus maintaining the same distributional mean.

standard deviation The square root of the variance.

standard error of the mean The standard deviation divided by the square root of the sample size. It is the standard deviation of a sample of means.

stem-loop structure A lollipop-shaped structure formed when a single-stranded nucleic acid molecule loops back on itself to form a complementary double helix (stem), topped by a loop.

stochastic A process with an indeterminate or random element as opposed to a deterministic process with no random element.

stringent response A translational control mechanism in prokaryotes that represses tRNA and rRNA synthesis during amino acid starvation.

submetacentric chromosome A chromosome whose centromere lies between its middle and its end, but closer to the middle.

subtelocentric chromosome A chromosome whose centromere lies between its middle and its end, but closer to the end.

sum rule The rule that states that the probability that two mutually exclusive events will occur is the sum of the probabilities of the individual events.

supergenes Several loci, which usually control related aspects of the phenotype, in close physical association.

suppression The phenomenon where a mutation (suppressor) restores the wild-type phenotype to a different mutation. The suppressor mutation may be in the same gene as the original mutation (intragenic suppression) or may be in a different gene (intergenic suppression).

surveillance mechanism Used to describe mechanisms that oversee the cell cycle checkpoints, where the cycle can be halted if certain conditions are not met.

survival of the fittest In evolutionary theory, survival of only those organisms that are best able to obtain and utilize resources (fittest). This is the cornerstone of Darwin's theory of evolution.

sympatric speciation Speciation in which reproductive isolating mechanisms evolve within the range and habitat of the parent species. This type of speciation may be common in parasites.

synapsis The point-by-point pairing of homologous chromosomes during zygonema or in certain dipteran tissues that undergo endomitosis.

synaptonemal complex A proteinaceous complex that is involved in mediating synapsis during the zygotene stage and then disintegrates.

syncitium A cell that has many nuclei not separated by cell membranes.

synteny Two loci located on the same chromosome that do not exhibit linkage by recombination (e.g., are very far apart).

 t

TATA-binding protein A protein, part of TFIID, that binds the TATA consensus sequence in eukaryotic promoters.

TATA box An invariant DNA sequence at about -25 in the promoter region of eukaryotic genes; analogous to the Pribnow box in prokaryotes (which is located around -10).

tautomeric shift Reversible shifts of proton positions in a molecule. The bases in nucleic acids shift between the keto and enol forms or between the amino and imino forms.

telocentric chromosome A chromosome whose centromere lies at one of its ends.

telomerase An enzyme that adds telomeric sequences to the ends of eukaryotic chromosomes.

telomere The ends of linear eukaryotic chromosomes.

temperate phage A phage that can enter into either the lytic phase or lysogeny with its host.

temperature-sensitive mutant An organism with an allele that exhibits the wild-type phenotype at the permissive temperature, but the mutant phenotype at the restrictive temperature.

template strand The DNA strand that serves as the DNA template for transcription, which will be complementary to the RNA sequence; also the DNA strand that is used by DNA polymerase in a DNA sequencing reaction.

terminator sequence A sequence in DNA that signals the termination of transcription to RNA polymerase.

terminator stem A configuration of the leader transcript that signals transcriptional termination in bacterial attenuator-controlled amino acid operons.

testcross The cross of an organism with a homozygous recessive organism.

testis-determining factor (*TDF*) General term for the gene determining maleness in human beings.

tetrads The meiotic configuration of four chromatids first seen in pachynema. There is one tetrad (bivalent) per homologous pair of chromosomes.

theta structure An intermediate structure formed during the replication of a circular DNA molecule.

three-point cross A cross involving three loci.

3' untranslated region (3' UTR) The length of mRNA from the nonsense codon to the 3' end or, in polycistronic messenger RNAs, from a nonsense codon to the next gene's leader.

titer The number of bacteriophage per unit volume.

t-loop A loop that forms at the end of mammalian telomeres by the interdigitation of the 3' free end into the DNA double helix.

topoisomerase An enzyme that can relieve (or create) supercoiling in DNA by creating transitory breaks in one (type I) or both (type II) strands of the helical backbone.

totipotent The state of a cell that can give rise to any and all adult cell types, as compared with a differentiated cell whose fate is determined.

trans Meaning "across" and referring usually to the geometric configuration of mutant alleles across from each other on a homologous pair of chromosomes.

***trans*-acting** Referring to mutations of, for example, a repressor gene, that acts through a diffusible protein product; the normal mode of action of most recessive mutations.

transcription The process whereby RNA is synthesized from a DNA template.

transcription factors Proteins that affect RNA polymerase in recognizing promoters and transcribing a gene. *See* activators and repressors.

transcriptomics The study of all the RNAs that are expressed in a cell or organism.

transducing particle A defective phage, carrying part of the host's genome.

transduction A process whereby a bacteriophage transfers genomic DNA from one bacterial cell to another bacterial cell; the recipient cell often recombines the foreign DNA into the recipient genome.

transfection The introduction of foreign DNA into eukaryotic cells.

transfer RNA (tRNA) Small RNA molecules that carry amino acids to the ribosome for polymerization. The tRNA contains a three base anticodon sequence that base pairs with the codon sequence in the mRNA in the ribosome.

transformation A process whereby prokaryotes take up DNA from the environment and incorporate it into their genomes, or the conversion of a eukaryotic cell into a cancerous one.

transgenic Eukaryotic organisms that have taken up foreign DNA.

transition mutation A mutation in which a purine-pyrimidine base pair replaces a base pair in the same purine-pyrimidine relationship.

translation The process of protein synthesis wherein the nucleotide sequence of the mRNA determines the amino acid sequence of the protein.

translocation A chromosomal configuration in which part of a chromosome becomes attached to a different chromosome. Also a part of the translation process in which the mRNA is shifted one codon in relation to the ribosome.

transposable element A region of the genome, flanked by inverted repeats, a copy of which can be inserted at another place; also called a transposon or a jumping gene.

transversion mutation A mutation in which a purine replaces a pyrimidine in the DNA, or vice versa.

trisomic A diploid cell with an extra chromosome.

tumor Abnormal growth of tissue.

tumor-suppressor genes Genes that normally prevent unlimited cellular growth. When both copies of the gene are mutated, cellular transformation follows. Examples are the *p53* gene and the genes for retinoblastoma and Wilm's tumor.

u

ubiquitin A peptide of twenty-six amino acids that are attached to proteins that are targeted to the proteosome for degradation.

unequal crossing over Nonreciprocal crossing over caused by the mismatching of homologous chromosomes. Usually occurs in regions of tandem repeats.

unique DNA A length of DNA with no repetitive nucleotide sequences.

upstream A convention on DNA related to the position and direction of transcription by RNA polymerase (5' to 3'). Downstream (or 3' to) is in the direction of transcription, whereas upstream (5' to) is in the direction from which the polymerase has come.

upstream element A sequence of about twenty AT-rich bases centered at 250 in promoters of prokaryotic genes that are expressed strongly.

v

variable-number-of-tandem-repeats (VNTR) loci Loci that are hypervariable because of tandem repeats. Presumably, variability is generated by unequal crossing over.

variance The average squared deviation about the mean of a set of data.

variegation Patchiness; a type of position effect that results when particular loci are contiguous with heterochromatin.

vector A DNA molecule that possesses an origin of replication that serves as a carrier for a cloned segment of foreign DNA.

virulent phage A bacteriophage that only enters the lytic cycle when it infects a bacterial cell.

w

western blotting A process where a membrane (e.g., nitrocellulose), which contains proteins that were separated by gel electrophoresis, is then incubated with a probe (usually and antibody) to identify a specific protein.

wild-type The phenotype of a particular organism when it is first seen in nature.

wobble Referring to the reduced constraint over the third base of an anticodon as compared with the other bases, thus allowing additional complementary base pairings.

x

X-inactivation center (*XIC*) Locus at which inactivation is initiated on the X chromosome in mammals.

X-linked Term used to describe either genes that are located on the X chromosome or traits that are controlled by genes that are mapped to the X chromosome. *See* sex-linked.

y

yeast artificial chromosome (YAC) Originating from a bacterial plasmid, a YAC contains additionally a yeast centromeric region (CEN) and a yeast origin of DNA replication (ARS). YACs are capable of possessing large pieces of foreign DNA after cloning.

Y-junction The point of active DNA replication where the double helix opens up so that each strand can serve as a template.

Y-linked Inheritance pattern of loci located on the Y chromosome only. Also refers to the loci themselves.

z

zinc finger Configuration of a DNA-binding protein that resembles a finger with a base, usually cysteines and histidines, binding a zinc ion. Discovered in a transcription factor in *Xenopus*.

credits

Photographs

About the Author
Photo of David Hyde with his dog, Tucker, was taken by Laura Schaefer Hyde.

Chapter 1
Opener: © Frank Lukasseck/Corbis; 1.1: The Granger Collection, New York; 1.2: © Kathy Talaro/Visuals Unlimited; 1.3: © Bettmann/Corbis; 1.4: Photo by A.F. Huettner/Courtesy Caltech Archives; 1.5: Roslin Institute; 1.10: © SPL/Photo Researchers, Inc.; 1.A: © SOVFOTO; 1.17: © K.G. Murti/Visuals Unlimited; 1.18: © Kalab/Custom Medical Stock Photo; 1.19: Nigel Cattlin/Photo Researchers, Inc.; 1.20: Zentrale für Unterrichtsmedien; 1.21a: Dan Tenaglia Photography; 1.21b, c: Elliot Meyerowitz; 1.22: altrendo nature/Getty Images; 1.23: © Inga Spence/Visuals Unlimited; 1.24: Baylor Coll. of Med./Peter Arnold, Inc.

Chapter 2
Opener: © Ariel Skelley/Corbis; 2.1: © SPL/Photo Researchers, Inc.; 2.2: Dr. Arne A. Anderberg/Naturhistoriska riksmuseet; 2.13: John Innes Archives, courtesy of the John Innes Foundation.

Chapter 3
Opener: Edward H. Hinchcliffe, Ph.D., Department of Biological Sciences, University of Notre Dame; 3.1: © Biophoto Associates; 3.2: Reproduced from *The Journal of Cell Biology*, 1968, Vol. 37, p. 381, by copyright permission of The Rockefeller University Press; 3.4b: © Biophoto Associates/Photo Researchers, Inc; 3.5 and 3.6: Reproduced courtesy of Dr. Thomas G. Brewster, Foundation of Blood Research, Scarborough, Maine; 3.12: Waheeb K. Heneen, "The centromeric region in the scanning electron microscope," *Hereditas*, 97 (1982): 311-14. Reproduced by permission; 3.15b: John M. Murray, Department of Anatomy, University of Pennsylvania. Cover of *BioTechniques*, volume 7, number 3, March 1989. Reproduced with permission; 3.19a-e: © Ed Reschke; 3.19f: © Ed Reschke/Peter Arnold Inc.; 3.19g: © Ed Reschke; 3.21a: Reproduced from D. von Wettstein 1971. *Proc. Natl. Acad. Sci.* USA. 68:851-855. Courtesy of D. von Wettstein, Washington State University; 3.23a: From M.I. Pigozzi and A.J. Solari, "Recombination Nodule Mapping and Chiasma Distribution in Spermatocytes of the Pigeon, Columba livia," in *Genome*, 42: 308-314, 1999. Reprinted by permission; 3.23b: Courtesy of Bernard John; 3.27: Courtesy of Dr. M.M. Rhoades, "Meiosis in maize," *Journal of Heredity*, 41: 59-67, 1950. Reproduced by permission.

Chapter 4
Opener: Thomas Cline/UC Berkeley photo; 4.2: from Stern C., Centerwall W.R., Sarkar S.S., "New data on the problem of Y-linkage of hairy pinnae" *Amer. Jour. of Human Genetics* 16:455-471, 1964; 4.3: © D. Robert & Lorri Franz/Corbis; 4.4a: © Bettmann/Corbis; 4.4b: From *Genetics*, 25 (1940): frontispiece. Courtesy of the Genetics Society of America; 4.5(both): © Phototake; 4.12: © Scott Camazine & Sue Trainor/Photo Researchers, Inc.; 4.16a-d: Bruce Baker; 4.17: © Michael Abbey/Photo Researchers, Inc.; 4.18b: © Photodisc Green/Getty Images RF; 4.21a: © SPL/Photo Researchers, Inc.; 4.21b: © Siebert/Custom Medical Stock Photo; 4.24: © Mary Evans Picture Library/Photo Researchers, Inc.

Chapter 5
Opener: © Custom Medical Stock Photo; 5.4: retrieved from the Mouse Genome Informatics Database, The Jackson Laboratory, Bar Harbor, Maine. World Wide Web (URL: http://www.informatics.jax.org/); 5.9a: © L.V. Bergman/The Bergman Collection; 5.12: © Jacques Jangoux/Photo Researchers, Inc.; 5.14 & 5.15: Annette Höggemeier; 5.17(white): Phillip Ruttenbur; (green): © Royalty-Free/Corbis; (yellow): U.S. Department of Agriculture.

Chapter 6
Opener: Courtesy of Bernard John; 6.1: American Philosophical Society; 6.11(both): © The McGraw-Hill Companies, Inc./Keith Maggert, photographer; 6.35: Dr. R.E. (Al) Rowland; 6.38: © Scott Camazine & Sue Trainor/Photo Researchers, Inc.

Chapter 7
Opener: © M. Freeman/Photolink/Getty Images; 7.3: Reproduced from the *Journal of Experimental Medicine*, 1944, 79: 137-158. © 1944 Rockefeller University Press; 7.6b: © Eye of Science/Photo Researchers, Inc.; 7.7a: © Biology Media/Photo Researchers, Inc.; 7.12a: Courtesy of Dr. Maurice H.F. Wilkins and Biophysics Department, King's College, London; 7.12b: Courtesy of Cold Spring Harbor Laboratory; 7.13: Reprinted by permission from Macmillan Publishers Ltd. From R.E. Franklin and R. Goslin, "Molecular configuration in sodium thymocucleate," *Nature* 171:740-41, 1953; 7.14: © A. Barrington Brown/Photo Researchers, Inc.; 7.15a: © John D. Baldeschwieler; 7.20: G. Marin and D.M. Prescott, "The frequency of sister chromatid exchanges following exposure to varying doses of 3H-thymidine or X-ray," *Journal of Cell Biology*, 21, (1964): 159-67, by copyright permission the Rockefeller University Press.; 7.21: D.E. Olins and A.L. Olins, "Nucleosomes: The structural quantum in chromosomes," *American Scientist*, 66: 704-11, November 1978. Reproduced by permission; 7.25a: Reprinted by permission from Macmillan Publishers Ltd. From Karolin Luger, et al., "Crystal structure of the nucleosome core particle at 2.8Å resolution" *Nature*, 389: 251-260, 1997; 7.26: Courtesy of Dr. Hans Ris; 7.29: J. Paulson and U. Laemmli, "The structure of histone-depleted metaphase chromosomes," *Cell*, 12:817-28, 1977. Micrograph courtesy of James R. Paulson; 7.30a: B.P. Kaufman, "Induced Chromosome Rearrangements in "Drosophila melanogaster," "*Journal of Heredity*," 30:178-90, 1939. Reproduced by permission of Oxford University Press; 7.30b: Jan-Erik Edstrom et al., *Developmental Biology* 91: 131-37, 1982, Figure 1B, Academic Press; 7.32: Reprinted by permission from B. Lambert, "Repeated DNA sequences in a Balbiani ring," *Journal of Molecular Biology*, 72:65-75, 1972. Copyright by Academic Press, Inc. (London) Ltd.; 7.33: H.D. Berendes, et al., "Experimental puffs in salivary gland chromosomes of Drosophila hydei," *Chromosome* (Berl.) 16:35-46, Fig 4a-b, 1965. © Springer-Verlag; 7.34: Joseph G. Gall, figure 2 in D.M. Prescott, ed., *Methods in Cell Physiology*, vol. 2 (New York: Academic Press, 1966), 39. Reproduced by permission; 7.36a: Dr. Gary Bauchan, USDA, Agricultural Research Service Beltsville, Maryland; 7.36b: Photograph Courtesy of Oncor, Inc. Gaithersburg, Maryland; 7.37: From B.R. Brinkley and J. Cartwright, Jr., *J. Cell Biology*, 50: 416-31, 1971; 7.41a: David Prescott, Boulder CO; 7.41b: Martin P. Horvath; 7.42c: From Jack D. Griffith, et al., "Mammalian telomeres end in a large duplex loop" in *Cell*, 97:503-14, May 14, 1999. © Cell Press.

Chapter 8
Opener: © Science VU/Visuals Unlimited; 8.5b & 8.7: © The McGraw-Hill Companies, Inc./Keith Maggert, photographer; 8.8a: Reproduced courtesy of Dr. Thomas G. Brewster, Foundation for Blood Research, Scarborough, Maine; 8.8b: Courtesy Lora Piepergerdes; 8.12: © The McGraw-Hill Companies, Inc./Keith Maggert, photographer; 8.15b: Steve Henikoff; 8.22a-c: © The McGraw-Hill Companies, Inc./Keith Maggert, photographer; 8.23: Reprinted by permission from Macmillan Publishers Ltd. From Ian Craig, "Methylation and the Fragile X," *Nature* 349:742, 1991; 8.27a: Reproduced courtesy of Dr. Thomas G. Brewster, Foundation for Blood Research, Scarborough, Maine; 8.27b: © Hattie Young/SPL/Photo Researchers, Inc.; 8.31: Reproduced courtesy of Dr. Jerome Lejeune, Institut de Progenese, Paris; 8.32: Reproduced courtesy of Dr. Thomas G. Brewster, Foundation for Blood Research, Scarborough, Maine; 8.33 & 8.34: UW Cytogenetic Services-WSLH.

Chapter 9

Opener: © Dr. Gopai Murti/Visuals Unlimited; 9.2a: Courtesy of Dr. Matthew Meselson. Photograph by Bud Gruce; 9.2b: Courtesy of Dr. Franklin W. Stahl; 9.6a: From J. Cairns, "The chromosome of E. coli," *Cold Spring Harbor Symposia on Quantitative Biology*, 28. © 1963 by Cold Spring Harbor Laboratory Press, Cold Spring Harbor, NY. Reprinted by permission; 9.9b: H. Kreigstein and D. Hogness, "Mechanism of DNA replication in Drosophilia chromosome: Structure of replication forks and evidence for bidirectionality," *Proceeding of the National Academy of Sciences USA*, 71 (1974): 135-39. Reproduced by permission; 9.25: © The McGraw-Hill Companies, Inc./Jeramia Ory, Ph.D. Figure adapted using coordinates from structure 1A35 in the Protein Data Bank (www.pdb.org).

Chapter 10

Opener: Jason D. Kahn, Univ. of Maryland; 10.15a, b: James A. Lake, *Journal of Molecular Biology* 105 (1976): 131-59. Reproduced by permission of Academic Press; 10.17c: Courtesy of Alexander Rich; 10.23: James A. Lake, *Journal of Molecular Biology* 105 (1976): 131-59. Reproduced by permission of Academic Press; 10.28b: Courtesy of O.L. Miller, Jr.; 10.30b: © O.L. Miller, B.R. Beatty, D.W. Fawcett/Visuals Unlimited; 10.34(left): Courtesy of Dr. Phillip A. Sharp; (right): Courtesy of Richard J. Roberts; 10.35a: Courtesy of Louise T. Chow and Thomas Broker.

Chapter 11

Opener: © Science VU - IBMRL/Visuals Unlimited; 11.2a: Photo by Richard Hartt, Pasadena/courtesy Caltech Archives; 11.2b: Courtesy of the Proceedings for the National Academy of Sciences; 11.24a,b: Courtesy of Rolf Hilgenfeld; 11.27b & 11.28b: Courtesy of Joachim Frank; 11.30: Reproduced courtesy of Dr. Barbara Hamkalo, *International Review of Cytology*, (1972) 33:7, fig 5. Copyright by Academic Press, Inc., Orlando, Florida; 11.31b: Courtesy of S.L. McKnight and O.L. Miller, Jr.; 11.35: Courtesy of Dr. R.W. Hendrix; 11.36a, b: Reprinted from Bernd Bakau and Arthur L. Horwich, "The Hsp70 and Hsp60 Chaperone Machines" in *Cell*, vol. 92, 351-366. Copyright 1988, with permission from Elsevier Science; 11.40: Courtesy of Marshall W. Nirenberg; 11.42: Courtesy of Dr. Philip Leder.

Chapter 12

Opener: © Gopal Murti/SPL/Photo Researchers, Inc.; 12.10b: Robert Duda, Ph.D.; 12.16(left): © David Hyde; 12.16(right): © Bruce Iverson; 12.21: Courtesy of Nick O. Bukanov; 12.29: Courtesy of Cellmark Diagnostics, Germantown, Maryland; 12.32: Courtesy of Richard J. Roberts; 12.35: From L. Johnston-Dow et al., *BioTechniques*, 5:754-765, 1987, Copyright © 1987, Eaton Publishing, Natick, MA. Reprinted with permission.

Chapter 13

Opener: R.L. Brinster and R.E. Hammer, School of Veterinary Medicine, University of Pennsylvania; 13.18a: Courtesy of Robert Turgeon and B. Gillian Turgeon, Cornell University; 13.26: Courtesy Maki Wakamiya; 13.28: © Dr. R.L. Brinster/Peter Arnold, Inc.; 13.32: © AP/Wide World Photo; 13.33: Anna Powers; 13.34: © AP/Wide World Photos; 13.35a, b: © Seoul National University/Handout/Reuters/Corbis; 13.36: Courtesy Monsanto; 13.37: Courtesy Ronald Carson, The Reproductive Science Center of Boston.

Chapter 14

14.2a: © Jim Varney/Photo Researchers, Inc.; 14.11: Jordan T. Shin, MD, PhD/Cardiovascular Research Center at MGH; 14.12: © Peter Menzel/ www.menzelphoto.com; 14.13-14.15: Vihtelic, T.S., Hyde, D.R., and O'Tousa, J.E. (1991). Isolation and characterization of the Drosophila retinal degeneration B (rdgB) gene. *Genetics* 127: 761-768; 14.16: Courtesy National Center for Biotechnology Information/National Library of Medicine/ National Institutes of Health; 14.25a, b: Engineering Services Inc. (ESI); 14.27: SMD (Stanford Microarray Database - http://smd.stanford.edu); 14.28b: Courtesy Dr. Breadan Kennedy, from Kennedy et al., *J. Biol. Chem.* 276: 14037-14043 fig. 6f; 14.30: Dr. Maode Lai; 14.31: Courtesy Hiroyuki Matsumoto, Ph.D. from Matsumoto & Pak 1984 *Science* 223: 184-18; 14.33a: PDB ID: 1d66 H.M. Berman, K. Henrick, H. Nakamura (2003): Announcing the worldwide Protein Data Bank. *Nature Structural Biology* 10 (12), p. 980 http://www.pdb.org/.

Chapter 15

Opener: © Oliver Meckes/MPI - Tubingen/Photo Researchers, Inc.; 15.1: Kindly supplied by I.D.J. Burdett and R.G.E. Murray; 15.2a: A.K. Kleinschmidt, et al., "Darstellung und Langen messungen des gesamten Deoxyribose-nucleinsaure-Inhaltes von T2-Basteriophagen," *Biochemica et Biophysica Acta*, 61: 5857-64, 1962. Reproduced by permission of Elsevier Science Publishers; 15.3: Photo by Robert Tamarin; 15.5: © Bruce Iverson; 15.6a: © Dr. E. Buttone/Peter Arnold, Inc.; 15.6b: © C. Case/Visuals Unlimited; 15.6c: © Michael G. Gabridge/Visuals Unlimited; 15.6d: © Cabisco/Visuals Unlimited; 15.8: Courtesy of Dr. Joshua Lederberg; 15.12: From *Molecular Biology of Bacterial Viruses* by Gunther S. Stent. © 1963, 1978 by W.H. Freeman and Company. Used with permission; 15.20: Courtesy of Wayne Rosenkrans and Dr. Sonia Guterman; 15.36: From *Molecular Biology of Bacterial Viruses* by Gunther S. Stent. © 1963, 1978 by W.H. Freeman and Company. Used with permission; 15.37: Courtesy of Dr. Seymour Benzer, 1970.

Chapter 16

Opener: © Biozentrum, Univ. of Basel/SPL/Photo Researchers, Inc.; 16.9: This image was made by Elizabeth Villa using VMD and is owned by the Theoretical and Computational Biophysics Group, an NIH Resource for Macromolecular Modeling and Bioinformatics, at the Beckman Institute, University of Illinois at Urbana-Champaign. Notations on the image were made by the book author; 16.16a: Thomas A. Steitz; 16.25a,b: Reprinted by permission from Macmillan Publishers Ltd. From Alfred A. Antson, et al., "Structure of the trp RNA-binding attenuation protein, TRAP, bound to RNA," *Nature*, 40: 2347-237, 1999

Chapter 17

Opener: D. Kosman et al. *Science* 2004 Aug. 6, 2004, vol. 305 p. 846; 17.4b: From Kennedy et al *J. Biol. Chem.* 276: 14037-14043, fig 6; 17.10: Fujioka, M, Emi-Sarker, Y, Yusibova, GL, Goto, T, and Jaynes, JB. Analysis of an even-skipped rescue transgene reveals both composite and discrete neuronal and early blastoderm enhancers, and multi-stripe positioning by gap gene repressor gradients. *Development* 1999; 126: 2527-2538; 17.14a,b: Courtesy of E.B. Lewis, California Institute of Technology; 17.14c: © Jeurgen Berger/ Photo Researchers, Inc.; 17.14d: © Science VU/Dr. F. R. Turner/Visuals Unlimited; 17.34: © Andrzej Bartke Ph.D./Southern Illinois University School of Medicine; 17.43: Baumeister et al. 1998 *Cell* 92: 357.

Chapter 18

Opener: © James King-Hilmes/SPL/Photo Researchers, Inc.; 18.1: Denis Stark M.B. B.S.; 18.22a,b: From S. Parikh, C. Mol, and J. Tainer, "Base excision repair enzyme family portrait," in *Structure*, 1997, 5:1543-1550, fig 1 a&b, p. 1544. Courtesy of J.A. Tainer, The Scripps Research Institute.; 18.24: © Yahia Albaili, DO, Dermatlas; http://www.dermatlas.org; 18.30a: © Science VU - IBMRL/Visuals Unlimited; 18.42a: Courtesy of the Barbara McClintock Papers, American Philosophical Society/National Library of Medicine; 18.42b: courtesy Dr. Neelima Sinha; Photo by author.

Chapter 19

Opener: © Professors P. Motta & T. Nagura/SPL/Photo Researchers, Inc.; 19.2: © Dr. Don W. Fawcett/Visuals Unlimited; 19.7a: © David M. Phillips/ Visuals Unlimited; 19.7b: © Dr. J. Burgess/SPL/Photo Researchers, Inc.; 19.8: From *Trends in Plant Science* vol. 7 (3) 2002 fig 2. Image courtesy Toshiyuki Mori; 19.9: W-L Chiu & BB Sears (1992) Electron microscopic localization of replication origins in Oenothera chloroplast DNA. *Mol Gen Genet* 232:33-39; 19.10a: Kuroiwa, T., Kawano, S. & Hizume, M. "Studies on mitochondrial structure and function in Physarum polycephalum. V. Behaviour of mitochondrial nucleoids throughout mitochondrial division cycle" *J. Cell Biol.* 1977 Mar: 72(3):687-694, Fig. C.; 19.10b: Gibbs, S.P., Mak, R., Ng, R. & Slankis, T. "The chloroplast nucleoid in Ochromonas danica. II. Evidence for an increase in plastid DNA during greening" *J. Cell Sci.* 1974 Dec: 16(3):579-91, Fig. 1. Reproduced with permission of the Company of Biologists; 19.17a: Wisconsin State Herbarium/Photographer: Clifford Orsted; 19.18a: © Gary Gaugler/Science Photo Library/Photo Researchers, Inc.; 19.24(both): Dr. Ronald A. Butow; 19.26: Barbara B. Sears.

Chapter 20

Opener: From van Eeden et al., 1996 *Development* vol. 123 fig. 4, p. 260. Courtesy Freek van Eeden. Reproduced with permission of the Company of Biologists; 20.2b: © P. Wegner/Peter Arnold Inc.; 20.12: Dr. Don Kane; 20.15(both): © David Hyde; 20.17: *Genes & Development*, Vol. 3, 1273-1287, © 1989 by Cold Spring Harbor Laboratory Press; 20.18a-d: Shiels et al., 2001, *Physiol. Genomics* 7: 179-186, 2001. Used with permission. The American Physiological Society; 20.20(left): © Herman Eisenbeiss/Photo Researchers, Inc.; (right): © Nigel Cattin/Photo Researchers, Inc.; (left): © Inga Spence/ Visuals Unlimited; 20.21(right): © Mark Smith/Photo Researchers, Inc.; 20.28a: Reprinted, with permission, from the *Annual Review of Neuroscience*, Volume 17 © 1994 by Annual Reviews www.annualreviews.org; 20.28b: Reinke R, Zipursky SL. 1988. "Cell-cell interaction in the Drosophila retina:

The bride of sevenless gene is required in photoreceptor cell R8 for R7 cell development." *Cell* 55: 321-30; 20.29a,b: U. Gaul et al., 1992, *Cell* vol. 68 pp.107-1019 from fig. 3, p. 1010; 20.32: Newsome et al. 2000 Development vol. 127 p. 851-860 fig. 1, p. 853; 20.38a-d: Frank-Kamenetsky 2002 *J. Biology*, vol. 1 article 10 fig. 3 p. 6

Chapter 21
Opener: Courtesy of Walter J. Gehring; 21.6b: F. Rudolf Turner; 21.7, 21.9a, b (nano): Courtesy Elizabeth R. Gavis; 21.9a (hunchback): Photograph courtesy of the Berkeley Drosophila Genome Project, Lawrence Berkeley National Laboratory; 21.9b (hunchback): Ernst A. Wimmer; 21.13: Jim Langeland, Steve Paddock and Sean Carroll/University of Wisconsin – Madison; 21.14a, b: Reprinted by permission from Macmillan Publishers Ltd. *Nature* 430: 368-371, 2004. Photo courtesy Johannes Jaeger; 21.15: Peter Lawrence see "The Making of a Fly" Blackwells, 1992; 21.20a, b: F. Rudolf Turner; 21.21c: Courtesy of E. B. Lewis, California Institute of Technology; 21.26: Cecilia Moens; 21.27a-c: Image courtesy Victoria Prince from *Development*, 2002, 129: 2339-2354. The Company of Biologists Ltd.; 21.30: Image courtesy Vincent Harley from Harley, Clarkson & Argentaro *Endocrine Reviews* Vol. 24, p. 468, fig. 2 (2003); 21.35a-e: Halder G. Callaerts P, Flister S, Walldorf U, Kloter U, Gehring WJ (1998) "Eyeless initiates the expression of both sine oculis and eyes absent during Drosophila compund eye development," *Development* 125: 2181-2191; 21.37a, b & 21.38a-c: Reinke and Zipursky 1988 "Cell-cell interaction in the Drosophila retina" *Cell* 55: 321-330; 21.40a-d: Karim et al. (1996) "A screen for genes that function downstream of Ras1 during Drosophila eye development" *Genetics* 143: 315-329 fig. 1 p. 318; 21.41a, b: Marco Vitale, Giorgio Zauli and Elisabetta Falcieri; 21.42(all): *Development*, 1999, 126: 5515-5522, © The Company of Biologists Ltd.

Chapter 22
Opener: Ruhong Li/University of California, Berkeley; 22.7b: The Centers for Disease Control; 22.B: From JOP. J Pancreas (Online) 2002; 3(5):144-151. Courtesy Department of Gastroenterology School of Medicine of Sao Paulo University.

Chapter 23
Opener: Lyle Buss, University of Florida, Entomology and Nematology Department; 23.2: Alejandro Calixto; 23.5a: © Perennou Nuridsany/ Photo Researchers, Inc.; 23.5b: © Michael Willmer Forbes Tweedie/Photo Researchers, Inc.; 23.6: Mark Smith; 23.10: © Robert Lindholm/Visuals Unlimited; 23.11: From Rich, Bell, and Wilson, "Genetic Drift in Small Populations of Tribolium," *Evolution*, 33:579-584. © Blackwell Publishers. Reprinted with permission.

Chapter 24
Opener: © Brownie Harris/Corbis; 24.1(both): © Photodisc Green/Getty Images; 24.5a: © The McGraw-Hill Companies, Inc./Photo by David Hyde and Wayne Falda; 24.12: © Carolyn A. McKeone/Photo Researchers, Inc.; 24.13a, b: © iStockphoto.com; 24.14: International Rice Research Institute http://www.ricephotos.org.

Chapter 25
Opener: © Frans Lanting/Photo Researchers, Inc.; 25.1: Painting by George Richmond, 1840. Downe House, Downe, Kent. © Archiv/Photo Researchers, Inc.; 25.5(left): © Paul & Joyce Berquist/Animals, Animals/ Earth Scenes; (right): © Gerald and Buff Corsi/Visuals Unlimited; 25.6a: © Heiko Bleher; 25.6b: Courtesy Greg Steeves; 25.6c: Ronald A. Anderson; 25.6d: Courtesy Greg Steeves; 25.7: Jeffrey L. Feder and Guy L. Bush, Zoology Department, Michigan State University; 25.21(Onge, Great Andamanese, Nicobarese): *Science*, 2005, vol. 308 p. 996 Thangaraj et al. Courtesy Dr. Thangaraj; 25.22: Bette Korber and John Mokili.

Line Art/Tables

Chapter 1
1.9: Reprinted from *Genomics* 22, by Van Soesten, et al., "Assignment of a Gene for Autosomal Recessive," pp. 499–504, © 1994, with permission from Elsevier.

Chapter 4
4.4: Reprinted from *Applied Animal Behavior Science*, 51, by Claudio Ciofi and Ian R. Swing, "Environmental Sex Determination in Reptiles," fig. 2 pp.251–265, © 1997, with permission from Elsevier.

Chapter 6
6.40: McKusick, Victor A., Mendelian Inheritance in Man, Twelfth Edition, volume 1: A Catalog of Human Genes and Genetic Disorders, pp. ccxviii–ccxix. © 1998 Johns Hopkins University Press. Reprinted with permission.

Chapter 7
Table 7.3: Reprinted from *The Nucleic Acids,* by E. Chargaff and J. Davidson, Academic Press, Copyright © 1955, with permission from Elsevier; 7.38: Data from L. Clarke and J. Carbon, "The Structure and Function of Yeast Centromeres," *Annual Review of Genetics,* 19:29–56, 1985.

Chapter 8
8.30: Reprinted from *The Lancet,* 2, by E. Hook, "Estimates of Maternal Age-Specific Risks of Down-Syndrome Birth in Women Age 34–41," pp. 33–34. © 1976, with permission from Elsevier.

Chapter 9
9.1: Nature by Watson and Crick. © 1953 Nature Publishing Group (permissions). Reproduced with permission of Nature Publishing Group (permissions) in the format Textbook via Copyright Clearance Center; 9.30: Data from Shippen-Lenz and Blackburn, *Science,* 247:550, 1990.

Chapter 10
Table 10.4: Data from R. G. Roeder, "The Role of General Initiation Factors in Transcription by RNA Polymerase II" in *Trends in Biochemical Sciences,* 21:327–35, 1996; 10.33: National Institutes of Health Research by David A. Konkel, et al., "The sequence of the chromosomal mouse β-globin major gene: Homolgies in capping, splicing and poly (A) Sites," *Cell,* 15:1125–32, 1978. Table 10.5: The Five Small Nuclear RNAs involved in Nuclear Messenger RNA Intron Removal, from *Annual Review of Genetics,* vol 28, © 1994 Annual Reviews. Reprinted with permission.

Chapter 11
Table 11.3: *Journal of Biological Chemistry* by Sasavage et al. © 1982 by American Society for Biochemistry & Molecular Biology. Reproduced with permission of American Society for Biochemistry and Molecular Biology in the format textbook via Copyright Clearance Center. 11.6: Data from S. Bonitz, et al., "Codon recognition rules in yeast mitochondria," *Proceedings of the National Academy of Sciences* 77:3167–70, 1980; 11.33: From C. Bernabeu and J.A. Lake, *Proceedings of the National Academy of Sciences*; 79:3111–15, 1982. Reprinted with permission.

Chapter 12
12.7: Reprinted with permission from Invitrogen.

Chapter 14
14.11: Courtesy: Srinivasan A and Fishman MC; 14.16: From NCBI. http://www.nlm.nih.gov. 14.32: Copyright © Sharon et al., Reprinted with permission.

Chapter 15
Table 15.1: Data from M. Rogosa, et al., *Journal of Bacteriology;* 54:13, 1947; Table 15.4: Reprinted from *Sexuality and the Genetics of Bacteria,* by E. Jacobs and E.L. Wollman, Academic Press, © 1961, with permission from Elsevier; 15.24: From F. Jacobs and E. L. Wollman, *Sexuality and the Genetics of Bacteria,* Academic Press, 1961; 15.44: Data from Symour Benzer, "The fine structure of the gene," *Scientific American,* 206: 70–84, January 1962.

Chapter 16
Table 16.3: Data from Bachmair, et al., *Science,* 234:179–86, 1986. 16.34: Reprinted from *Current Opinion in Structural Biology,* vol. 15 B.J. Tucker and R.R. Breaker, "Riboswitches as versatile gene control elements," pp. 342-348, figure 1, © 2005, with permission from Elsevier.

Chapter 18
Table 18.1: From E. Luria and M. Delbruck, *Genetics,* 28: 491. © 1943 Genetics Society of America. Reprinted with permission.

Chapter 24
Table 24.5: From "Genetic and Environmental Influences on Human Psychological Differences," by T. J. Bouchard and M. McGue, 2003, *Journal of Neurobiology,* 54: 4–45. © 2003 John Wiley & Sons. Reprinted with permission of Wiley-Liss, Inc. A subsidiary of John Wiley & Sons, Inc.

Chapter 25
Table 25.1: Data from M. Kimura, The Neutral Theory of Molecular Evolution, Cambridge University Press, 1983.

index